STUDENT SOLUTIONS MANUAL

TO ACCOMPANY

CALCULUS

WITH ANALYTIC GEOMETRY

FOURTH EDITION

ROBERT ELLIS
University of Maryland
at College Park

DENNY GULICK
University of Maryland
at College Park

HARCOURT BRACE JOVANOVICH, PUBLISHERS
and its subsidiary, Academic Press
San Diego New York Chicago Austin Washington, D.C.
London Sydney Tokyo Toronto

Preface

This Manual contains complete solutions to the odd-numbered exercises in *Calculus with Analytic Geometry*, Fourth Edition. Our aim has been to provide all the details necessary for full understanding of the solutions and of the various techniques involved. In each solution we have included all the important steps, so that the student can easily follow the line of reasoning. Where possible, solutions are patterned after the examples in the corresponding text section.

This Manual can be a valuable learning aid. We suggest that the student conscientiously attempt an exercise first, and only then refer to the manual, both for the answer and for the method of obtaining it.

The solutions to the exercises have been carefully checked. We would be grateful to have any residual errors in either the Solutions Manual or the textbook brought to our attention.

Robert Ellis
Denny Gulick

0-15-505691-3

Printed in the United States of America

Contents

1 FUNCTIONS

SECTION 1.1

1. Notice that $4 \cdot 16 = 64 > 63 = 9 \cdot 7$. Multiplying by the positive number $1/(9 \cdot 16)$, we have $(4 \cdot 16)/(9 \cdot 16) > (9 \cdot 7)/(9 \cdot 16)$ or $\frac{4}{9} > \frac{7}{16}$. Therefore, $a > b$.

3. Since $\pi^2 > (3.14)^2 = 9.8596$, we have $a > b$.

5. $(1.41)^2 = 1.9881 < 2$, so $\sqrt{2} > 1.41$.

7. Closed, bounded

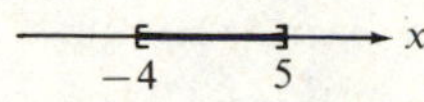

9. Open, unbounded

11. Closed, unbounded

13. Closed, unbounded

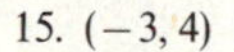

15. $(-3, 4)$

17. $(1, \infty)$

19. If $-6x - 2 > 5$, then $-6x > 7$, so $x < -\frac{7}{6}$. Thus the solution is $(-\infty, -\frac{7}{6})$.

21. If $-1 \le 2x - 3 < 4$, then $2 \le 2x < 7$, so $1 \le x < \frac{7}{2}$. Thus the solution is $[1, \frac{7}{2})$.

23. From the diagram we see that the solution is the union of $(-\infty, -\frac{1}{2}]$ and $[1, \infty)$.

25. From the diagram we see that the solution is the union of $(-\infty, -\frac{1}{3})$ and $(0, \frac{2}{3})$.

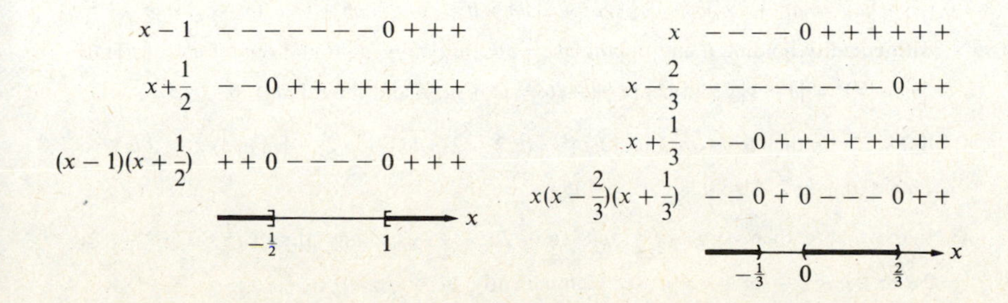

27. From the diagram we see that the solution is the union of $(-\infty, -3)$ and $(-1, \infty)$.

29. The given inequality is equivalent to $2x^2(2x - 3) \le 0$. From the diagram we see that the solution is $(-\infty, \frac{3}{2}]$.

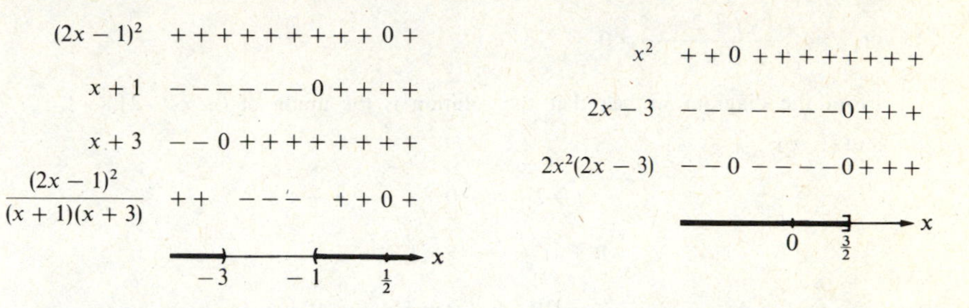

31. The given inequality is equivalent to $(8x^3 - 1)/x^2 > 0$. From the diagram we see that the solution is $(\frac{1}{2}, \infty)$.

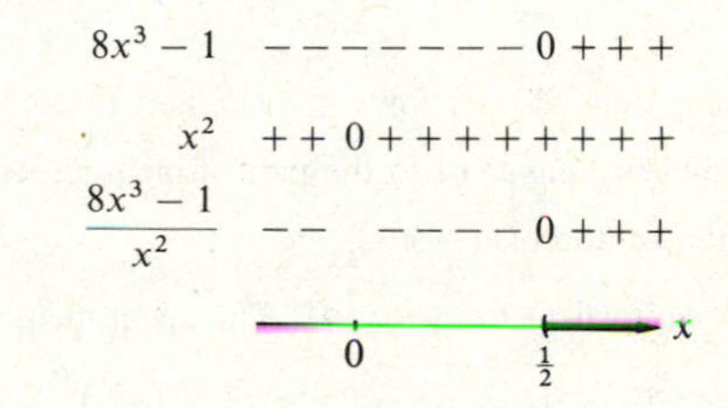

33. The given inequality is equivalent to

$$\frac{4x(x + \sqrt{6})(x - \sqrt{6})}{(x + 2)(x - 2)} < 0$$

From the diagram we see that the solution is the union of $(-\infty, -\sqrt{6})$, $(-2, 0)$, and $(2, \sqrt{6})$.

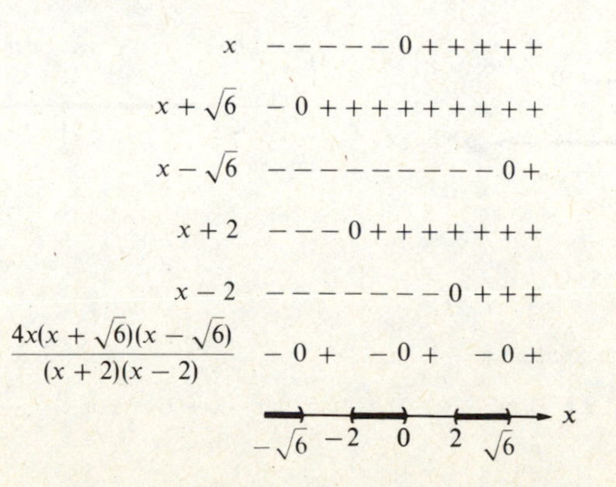

35. The given inequality is equivalent to

$$\frac{(t+2)(t-1)}{(t+1)^3(t-1)^3} \geq 0$$

or
$$\frac{t+2}{(t+1)^3(t-1)^2} \geq 0$$

From the diagram we see that the solution is the union of $(-\infty, -2]$, $(-1, 1)$, and $(1, \infty)$.

$t+2$ — — 0 + + + + + + + + +

$(t+1)^3$ — — — — 0 + + + + + + +

$(t-1)^2$ + + + + + + + + + 0 + +

$\dfrac{t+2}{(t+1)^3(t-1)^2}$ + + 0 — + + + + +

x: −2 −1 1

37. Observe that $\sqrt{9-6x}$ is defined only for $x \leq \frac{9}{6} = \frac{3}{2}$ and that $\sqrt{9-6x} > 0$ for $x < \frac{3}{2}$. Thus the given inequality is equivalent to the pair of inequalities $x < \frac{3}{2}$ and $2 - x > 0$ (that is, $x < 2$). Thus the solution is $(-\infty, \frac{3}{2})$.

39. The given inequality is equivalent to

$$\frac{1}{x+1} - \frac{3}{2} > 0, \quad \text{or} \quad \frac{-3(x+\frac{1}{3})}{2(x+1)} > 0$$

From the diagram we see that the solution is $(-1, -\frac{1}{3})$.

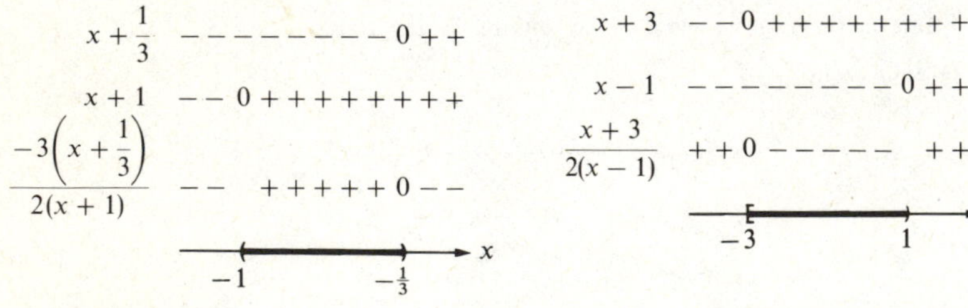

41. The given inequality is equivalent to

$$\frac{x+1}{x-1} - \frac{1}{2} \leq 0, \quad \text{or} \quad \frac{x+3}{2(x-1)} \leq 0$$

From the diagram we see that the solution is $[-3, 1)$.

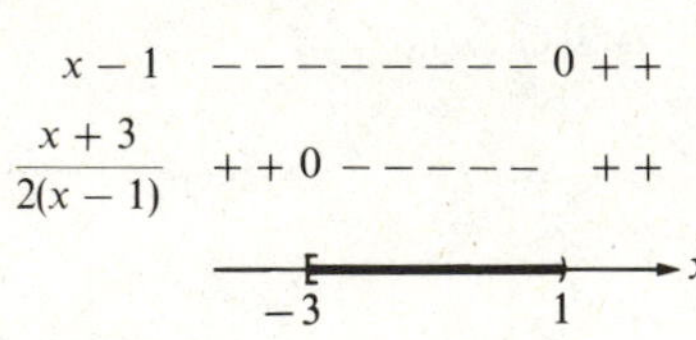

43. $-|-3| = -3$

45. $|-5| + |5| = 5 + 5 = 10$

47. $|x| = 1$ if $x = 1$ or $-x = 1$; the solution is $-1, 1$.

49. $|x-1| = 2$ if $x - 1 = 2$ (so that $x = 3$), or $-(x-1) = 2$ (so that $-x + 1 = 2$, or $x = -1$); the solution is $-1, 3$.

51. $|6x+5| = 0$ if $6x + 5 = 0$, or $x = -\frac{5}{6}$; the solution is $-\frac{5}{6}$.

53. If $|x| = |x|^2$, then either $|x| = 0$ or we may divide by $|x|$ to obtain $1 = |x|$ (so that $x = -1$ or $x = 1$). The solution is $-1, 0, 1$.

55. If $|x+1|^2 + 3|x+1| - 4 = 0$, then $(|x+1| + 4)(|x+1| - 1) = 0$. Since $|x+1| + 4 \neq 0$ it follows that $|x+1| - 1 = 0$, or $|x+1| = 1$. Thus either $x + 1 = 1$ (so that $x = 0$), or $-(x+1) = 1$ (so that $-x = 2$, or $x = -2$). The solution is $0, -2$.

57. If $|x+4| = |x-4|$, then either $x + 4 = x - 4$ (so that $4 = -4$, which is impossible), or $x + 4 = -(x-4)$ (so that $2x = 0$, or $x = 0$). The solution is 0.

59. If $|x-2| < 1$, then $-1 < x - 2 < 1$, or $1 < x < 3$. The solution is $(1, 3)$.

61. If $|x+1| < 0.01$, then $-0.01 < x + 1 < 0.01$, or $-1.01 < x < -0.99$. The solution is $(-1.01, -0.99)$.

63. If $|x+3| \geq 3$, then $x + 3 \geq 3$ (so that $x \geq 0$), or $-(x+3) \geq 3$ (so that $x \leq -6$). The solution is the union of $(-\infty, -6]$ and $[0, \infty)$.

65. If $|2x+1| \geq 1$, then either $2x + 1 \geq 1$ (so that $x \geq 0$), or $-(2x+1) \geq 1$, (so that $2x \leq -2$, or $x \leq -1$). The solution is the union of $(-\infty, -1]$ and $[0, \infty)$.

67. If $|2x - \frac{1}{3}| > \frac{2}{3}$, then either $2x - \frac{1}{3} > \frac{2}{3}$ (so that $2x > 1$, or $x > \frac{1}{2}$), or $-(2x - \frac{1}{3}) > \frac{2}{3}$ (so that $-2x > \frac{1}{3}$, or $x < -\frac{1}{6}$). The solution is the union of $(-\infty, -\frac{1}{6})$ and $(\frac{1}{2}, \infty)$.

69. Since $|4 - 2x| \geq 0 > -1$ for all x, the given inequality is equivalent to $|4 - 2x| < 1$, so that $-1 < 4 - 2x < 1$, or $-5 < -2x < -3$, or $\frac{5}{2} > x > \frac{3}{2}$. The solution is $(\frac{3}{2}, \frac{5}{2})$.

71. $[b, b]$ consists of all x satisfying $b \leq x \leq b$, or equivalently, $x = b$.

73. Yes, because if $x > 5$, then $x^2 > 25$.

75. No, because if $0 < x \leq 1$, then $1/x \geq 1 \geq x$.

77. a. If $a \geq 0$ and $b \geq 0$ (or $a \leq 0$ and $b \leq 0$), then $|ab| = ab = |a|\,|b|$. If $a \geq 0$ and $b \leq 0$ (or $a \leq 0$ and $b \geq 0$), then $|ab| = -(ab) = |a|\,|b|$. In either case, $|ab| = |a|\,|b|$.

b. If $b \geq 0$, then $b = |b| \geq -|b|$. If $b < 0$, then $-b = |b| > -|b|$, so $b = -|b| < |b|$. In either case, $-|b| \leq b \leq |b|$.

c. $|a - b| = |(-1)(b - a)| = |-1|\,|b - a| = |b - a|$.

79. $(|a| + |b|)^2 = |a|^2 + 2|a|\,|b| + |b|^2 = a^2 + 2|ab| + b^2 \geq a^2 + 2ab + b^2 = (a+b)^2 = |a+b|^2$, with equality holding if and only if $|ab| = ab$. But $|ab| = ab$ if and only if $ab \geq 0$. Thus $(|a| + |b|)^2 = |a+b|^2$, and hence $|a| + |b| = |a + b|$, if and only if $ab \geq 0$.

81. If $0 < a < b$, then $0 < (\sqrt{b/2} - \sqrt{a/2})^2 = (b/2 - 2\sqrt{ab/4} + a/2) = [(a+b)/2 - \sqrt{ab}]$, so $\sqrt{ab} < (a+b)/2$. Also $a = \sqrt{a^2} < \sqrt{ab}$.

83. If $0 < a < b$, then $a = \sqrt{aa} < \sqrt{ab}$, so $-2a > -2\sqrt{ab}$, and thus $(\sqrt{b} - \sqrt{a})^2 = b - 2\sqrt{ab} + a < b - 2a + a = b - a$. Consequently $\sqrt{b} - \sqrt{a} < \sqrt{b - a}$.

85. Assume $\sqrt{3} = p/q$, where p and q are integers such that at most one of them is divisible by 3. Then $3 = p^2/q^2$, or $p^2 = 3q^2$. Thus 3 divides p^2, so 3 divides p—say $p = 3a$. Then $3q^2 = p^2 = 9a^2$, so we have $q^2 = 3a^2$. Thus 3 divides q^2 and hence q. Therefore 3 divides both p and q, contradicting our assumption. Consequently $\sqrt{3}$ is irrational.

87. Let a and b be adjacent sides of the rectangle. Then $P = 2(a + b)$. By Exercise 81, $\sqrt{ab} \le (a + b)/2$, so $ab \le [(a + b)/2]^2 = |P/4|^2$. But ab is the area of the rectangle, whereas $|P/4|^2$ is the area of a square with perimeter P.

89. If A_R is the area of a rectangle of perimeter P, and A_S the area of a square of perimeter P, then by Exercise 87, $A_R \le A_S$. If A_C is the area of a circle of circumference (perimeter) P, then by Exercise 88, $A_S < A_C$. Thus $A_R < A_C$.

SECTION 1.2

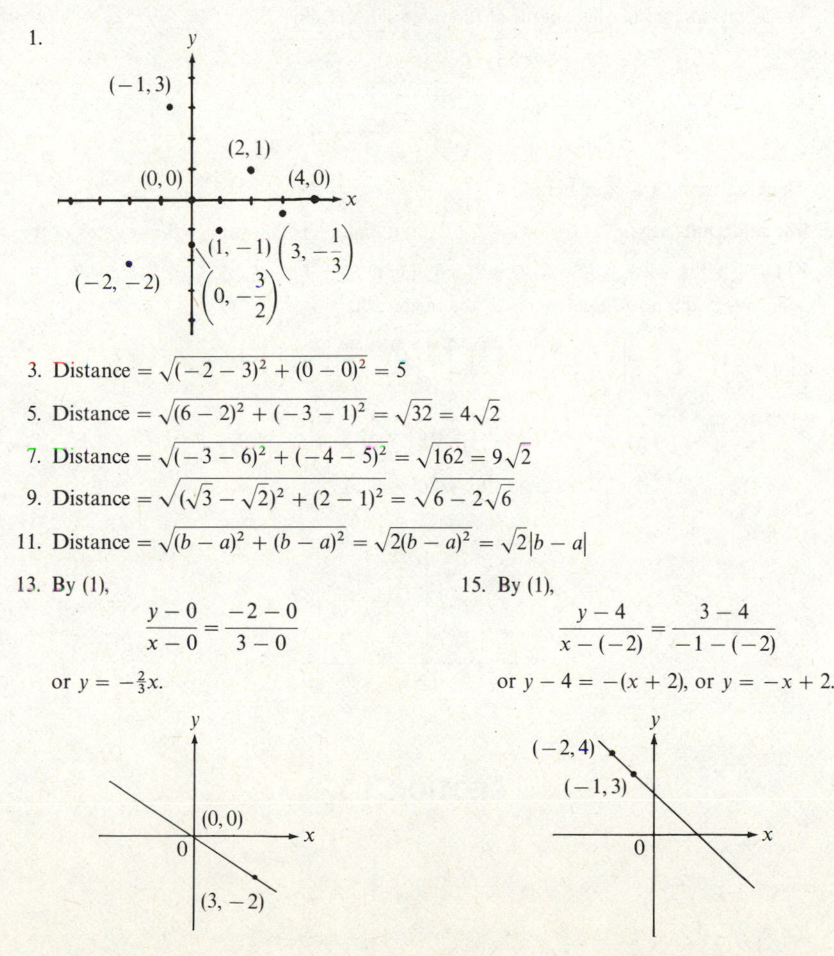

1.

3. Distance $= \sqrt{(-2-3)^2 + (0-0)^2} = 5$

5. Distance $= \sqrt{(6-2)^2 + (-3-1)^2} = \sqrt{32} = 4\sqrt{2}$

7. Distance $= \sqrt{(-3-6)^2 + (-4-5)^2} = \sqrt{162} = 9\sqrt{2}$

9. Distance $= \sqrt{(\sqrt{3}-\sqrt{2})^2 + (2-1)^2} = \sqrt{6 - 2\sqrt{6}}$

11. Distance $= \sqrt{(b-a)^2 + (b-a)^2} = \sqrt{2(b-a)^2} = \sqrt{2}|b-a|$

13. By (1),

$$\frac{y-0}{x-0} = \frac{-2-0}{3-0}$$

or $y = -\frac{2}{3}x$.

15. By (1),

$$\frac{y-4}{x-(-2)} = \frac{3-4}{-1-(-2)}$$

or $y - 4 = -(x + 2)$, or $y = -x + 2$.

17. By (3), $y - (-1) = 3(x - 2)$, or $y = 3x - 7$.

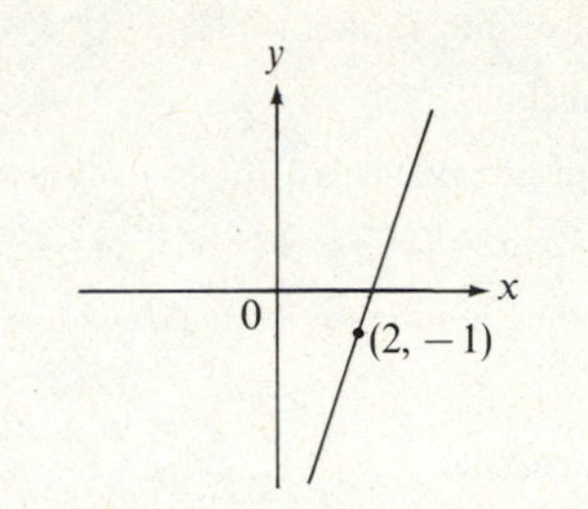

19. By (3), $y - \frac{1}{2} = -1(x - \frac{1}{2})$, or $y = -x + 1$.

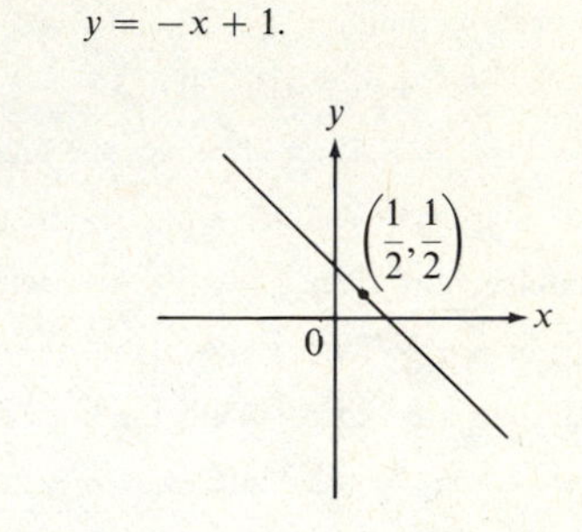

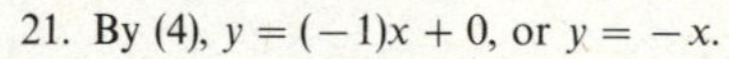

21. By (4), $y = (-1)x + 0$, or $y = -x$.

23. By (4), $y = 3x - 3$.

25. By (4), $m = -1$ and $b = 0$.

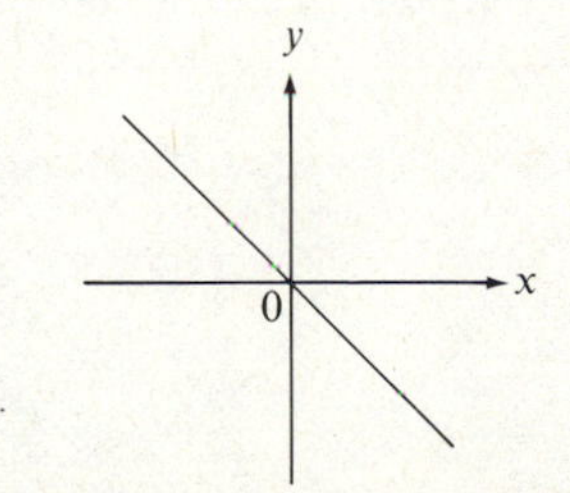

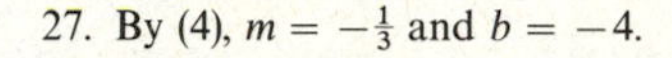

27. By (4), $m = -\frac{1}{3}$ and $b = -4$.

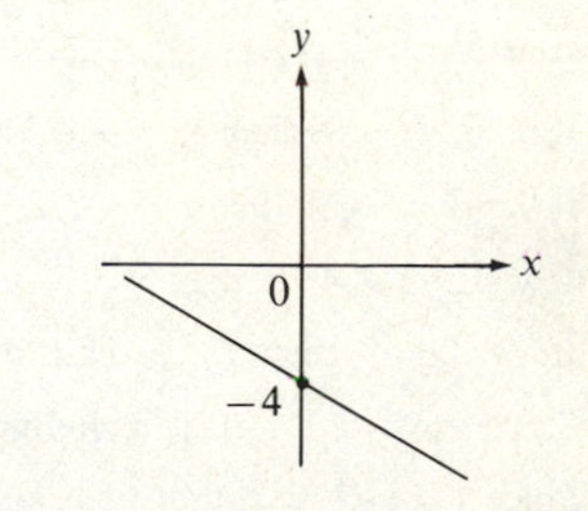

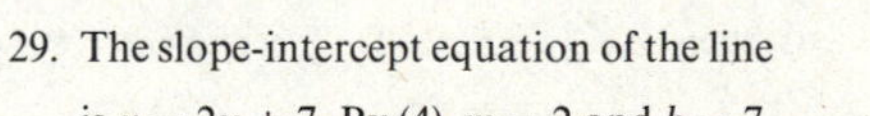

29. The slope-intercept equation of the line is $y = 2x + 7$. By (4), $m = 2$ and $b = 7$.

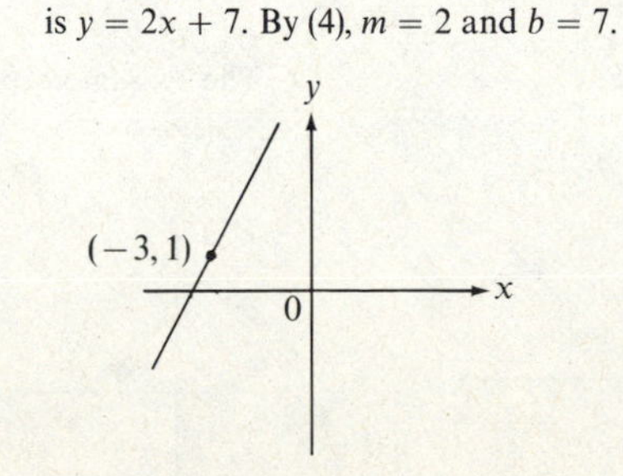

31. The slope-intercept equation of the line is $y = -2x + 4$. By (4), $m = -2$ and $b = 4$.

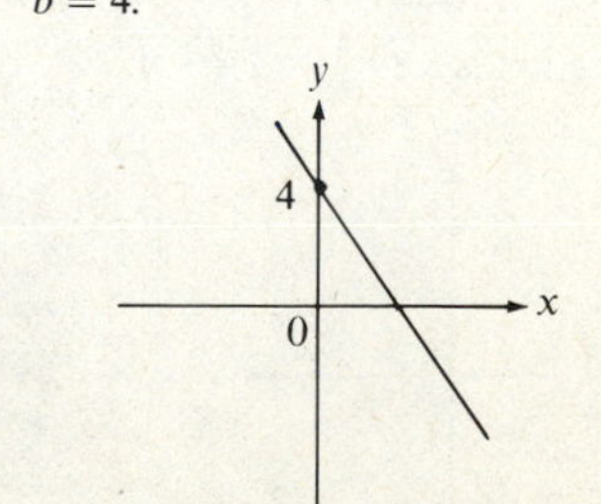

33. $m_1 = 2$; $m_2 = -\frac{1}{2}$. Since $m_1 m_2 = -1$, the lines are perpendicular. If (x, y) is the point of intersection, then $y = 2x + 3$ and $y = -\frac{1}{2}x - 1$, so that $2x + 3 = -\frac{1}{2}x - 1$, or $\frac{5}{2}x = -4$, or $x = -\frac{8}{5}$. Then $y = 2(-\frac{8}{5}) + 3 = -\frac{1}{5}$. The point of intersection is $(-\frac{8}{5}, -\frac{1}{5})$.

35. $m_1 = 1$; $m_2 = 1$. Since $m_1 = m_2$, the lines are parallel.

37. $m_1 = -\frac{2}{3}$; $m_2 = \frac{3}{2}$. Since $m_1 m_2 = -1$, the lines are perpendicular. If (x, y) is the point of intersection, then $y = -\frac{2}{3}x - \frac{1}{3}$ and $y = \frac{3}{2}x - \frac{5}{2}$, so $-\frac{2}{3}x - \frac{1}{3} = \frac{3}{2}x - \frac{5}{2}$, or $-\frac{13}{6}x = -\frac{13}{6}$, or $x = 1$. Then $y = -\frac{2}{3}(1) - \frac{1}{3} = -1$. The point of intersection is $(1, -1)$.

39. Both lines are vertical. Thus the lines are parallel.

41. $m_1 = -2$; $m_2 = -2$. Since $m_1 = m_2$, the lines are parallel.

43. The slope of ℓ is 3; the desired line has the same slope. From the point-slope equation, we get $y - (-1) = 3(x - 2)$, or $y = 3x - 7$.

45. The slope of ℓ is -1; the desired line has the same slope. From the point-slope equation, we get $y - 0 = -1(x - 0)$, or $y = -x$.

47. The slope of ℓ is $\frac{2}{3}$; the desired line has the same slope. From the point-slope equation, we get $y - 1 = \frac{2}{3}(x - 2)$, or $y = \frac{2}{3}x - \frac{1}{3}$.

49. The slope of ℓ is 2; from Theorem 1.3, the desired line has slope $-\frac{1}{2}$. By (3), an equation is $y - (-3) = -\frac{1}{2}(x - (-1))$, or $y = -\frac{1}{2}x - \frac{7}{2}$.

51. The slope of ℓ is $-\frac{2}{3}$; from Theorem 1.3, the desired line has slope $\frac{3}{2}$. By (3), an equation is $y - 3 = \frac{3}{2}(x - 2)$, or $y = \frac{3}{2}x$.

53. The slope of ℓ is 2; from Theorem 1.3, the desired line has slope $-\frac{1}{2}$. By (3), an equation is $y - (-5) = -\frac{1}{2}(x - 4)$, or $y = -\frac{1}{2}x - 3$.

55. a. If $0 = -x - 4$, then $x = -4$. Thus $a = -4$.
    b. If $0 = 2x - \sqrt{2}$, then $x = \sqrt{2}/2$. Thus $a = \sqrt{2}/2$.
    c. If $x = -3 - 4 \cdot 0$, then $x = -3$. Thus $a = -3$.
    d. If $x - 0 = 5$, then $x = 5$. Thus $a = 5$.
    e. Since $x = -2$ for all y, including $y = 0$, we have $a = -2$.
    f. Since $y = -7$, y cannot be 0, so there is no x intercept.

57. If $x = 0$, then $y = -2$, so $b = -2$. If $y = 0$, then $x = \frac{1}{2}$, so $a = \frac{1}{2}$. The two-intercept equation is $x/\frac{1}{2} + y/(-2) = 1$.

59. If $x = 0$, then $y = \frac{1}{3}$, so $b = \frac{1}{3}$. If $y = 0$, then $x = -\frac{1}{2}$, so $a = -\frac{1}{2}$. The two-intercept equation is $x/(-\frac{1}{2}) + y/\frac{1}{3} = 1$.

61. 63. 65.

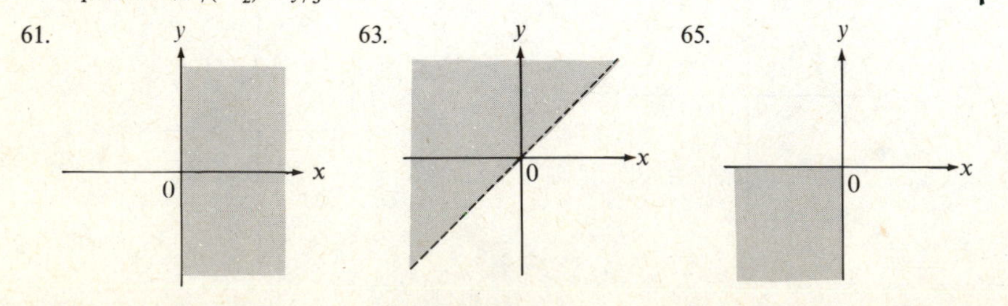

67. 69.

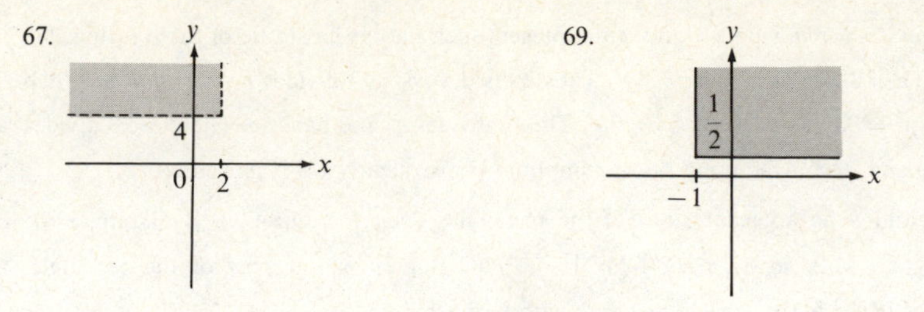

71. a. The requirement $(x, y) = (x, -y)$ is satisfied only when $y = -y$, or $y = 0$. Thus the region is the x axis.
    b. The requirement $(x, y) = (-x, y)$ is satisfied only when $x = -x$, or $x = 0$. Thus the region is the y axis.

73. Let d_1, d_2, and d_3 be the lengths of the three sides. Then

$$d_1 = \sqrt{(\sqrt{3} - 1 - (-1))^2 + (3 - 2)^2} = 2$$
$$d_2 = \sqrt{(-1 - (\sqrt{3} - 1))^2 + (4 - 3)^2} = 2$$
$$d_3 = \sqrt{(-1 - (-1))^2 + (4 - 2)^2} = 2$$

Thus the triangle is equilateral.

75. The midpoints are $(a/2, 0)$, $((a + b)/2, c/2)$, and $(b/2, c/2)$. The sum of the squares of the lengths of the sides is $a^2 + [(b - a)^2 + c^2] + (b^2 + c^2) = 2a^2 + 2b^2 + 2c^2 - 2ab$. The sum of the squares of the lengths of the medians is

$$\left[\left(b - \frac{a}{2}\right)^2 + c^2\right] + \left[\left(\frac{a + b}{2}\right)^2 + \left(\frac{c}{2}\right)^2\right] + \left[\left(\frac{b}{2} - a\right)^2 + \left(\frac{c}{2}\right)^2\right]$$
$$= b^2 - ab + \frac{a^2}{4} + c^2 + \frac{a^2}{4} + \frac{ab}{2} + \frac{b^2}{4} + \frac{c^2}{4} + \frac{b^2}{4} - ab + a^2 + \frac{c^2}{4}$$
$$= \tfrac{3}{4}(2a^2 + 2b^2 + 2c^2 - 2ab)$$

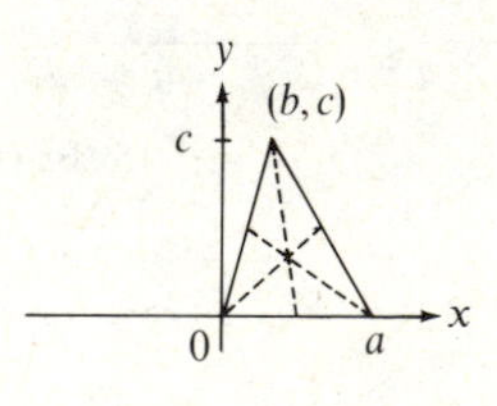

SECTION 1.3

1. $f(\sqrt{5}) = \sqrt{3}$; $f(\pi) = \sqrt{3}$

3. $f(0) = 1 - 0 + 0^3 = 1$; $f(-1) = 1 - (-1) + (-1)^3 = 1$

5. $g(\sqrt{2}) = \dfrac{1}{2(\sqrt{2})^2} = \dfrac{1}{4}$

7. $g(27) = \sqrt[3]{27} = 3$; $g(-\frac{1}{8}) = \sqrt[3]{-\frac{1}{8}} = -\frac{1}{2}$

9. $f(2) = \dfrac{2-1}{2^2+4} = \dfrac{1}{8}$

11. All real numbers

13. $[-2, 8]$

15. $[-2, \infty)$

17. All real numbers

19. Union of $(-\infty, -\sqrt{3}/3]$ and $[\sqrt{3}/3, \infty)$

21. All real numbers

23. All real numbers except 1

25. All real numbers except -4 and 4

27. All real numbers

29. Union of $[-4, -1]$ and $(0, 6)$

31. x is in the domain if $1 - \sqrt{9-x^2} \geq 0$ (so that $1 \geq \sqrt{9-x^2}$, or $1 \geq 9 - x^2$, or $x^2 \geq 8$) and $9 - x^2 \geq 0$ (so that $x^2 \leq 9$). Thus $8 \leq x^2 \leq 9$, so the domain is the union of $[-3, -2\sqrt{2}]$ and $[2\sqrt{2}, 3]$.

33. The set consisting of the number 1

35. If $x < 4$, then $f(x) = 3x - 2 < 3 \cdot 4 - 2 = 10$. Thus the range is $(-\infty, 10)$.

37. $f(x) \neq 0$. If y is any number except 0 and if $x = 1 + 1/y$, then

$$f(x) = \frac{1}{(1 + 1/y) - 1} = \frac{1}{1/y} = y$$

Thus the range consists of all real numbers except 0.

39. a. A function is described.
    b. A function is not described, because two values are assigned to the number -2.
    c. f is not a function because f assigns two values to every positive number.
    d. f is not a function because f assigns two values to every number.
    e. g is a function.
    f. g is a function.
    g. g is a function. (Note that $g(2) = 9$.)
    h. g is not a function because $g(1) = 2 - 3 = -1$ from the first line of the formula, whereas $g(1) = 3 - 3 = 0$ from the second line of the formula.
    i. f is a function.
    j. f is not a function, because it assigns two values to numbers such as $\sqrt{2}$: $(\sqrt{2})^2 = 2$, which is rational, so that $f(\sqrt{2}) = 2$; but $\sqrt{2}$ is irrational, so $f(\sqrt{2}) = \sqrt{2}$.

41. $f_1 = f_5$ and $f_2 = f_3$.

43. a. x is in the domain of f if and only if $x^2 - 1 \geq 0$, or $x^2 \geq 1$, so the domain of f is the union of $(-\infty, -1]$ and $[1, \infty)$. x is in the domain of g if and only if $x^2 - 1 \geq 0$ and $x + \sqrt{x^2-1} \neq 0$. The first condition requires x to be in $(-\infty, -1]$ or $[1, \infty)$. Since under this condition, $0 \leq x^2 - 1 < x^2$, we have $\sqrt{x^2-1} < |x|$. Therefore $\sqrt{x^2-1} \neq -x$, so $x + \sqrt{x^2-1} \neq x + (-x) = 0$. Thus the domain of f and the domain of g are equal to the union of $(-\infty, -1]$ and $[1, \infty)$.
    b. For all x in the domain of f,

$$f(x) = (x - \sqrt{x^2-1})\frac{x + \sqrt{x^2-1}}{x + \sqrt{x^2-1}} = \frac{x^2 - (x^2-1)}{x + \sqrt{x^2-1}} = \frac{1}{x + \sqrt{x^2-1}} = g(x)$$

Since f, g have the same domain and assign the same value to each x in that domain, $f = g$.

45. $f(x) = -\sqrt{x}$ for $x \geq 0$

47. $x = \dfrac{-6 \pm \sqrt{6^2 - 0}}{2} = \dfrac{-6 \pm 6}{2} = -6$ or 0

49. $x = \dfrac{-(-4) \pm \sqrt{16 - 4(1)(-5)}}{2} = \dfrac{4 \pm 6}{2} = -1$ or 5

51. There are no zeros, because $b^2 - 4ac = 4 - 4(3)(1) < 0$.

53. $t = \dfrac{-7 \pm \sqrt{49 - 4(2)(3)}}{4} = \dfrac{-7 \pm 5}{4} = -3$ or $-\dfrac{1}{2}$

55. If f has no zeros, then $b^2 - 4ac < 0$, so $ac > 0$. Thus $b^2 - 4a(-c) = b^2 + 4ac > 0$, so $\sqrt{b^2 + 4ac} > 0$. Hence we find that

$$\frac{-b + \sqrt{b^2 + 4ac}}{2a} \quad \text{and} \quad \frac{-b - \sqrt{b^2 + 4ac}}{2a}$$

are zeros of g.

57. $f(x) = \dfrac{1}{2}x^2\sqrt[3]{\dfrac{x}{5}}$ for $x \geq 0$

59. $A(x) = \dfrac{1}{2}x\left(\dfrac{\sqrt{3}}{2}x\right) = \dfrac{\sqrt{3}}{4}x^2$ for $x \geq 0$

61. If L is the length of the cylinder and r the radius of the hemispheres, then the volume V is given by $V = \pi r^2 L + \frac{4}{3}\pi r^3$. Since $V = 100$, we obtain $100 = \pi r^2 L + \frac{4}{3}\pi r^3$, so

$$L(r) = \frac{100 - \frac{4}{3}\pi r^3}{\pi r^2} = \frac{300 - 4\pi r^3}{3\pi r^2} \quad \text{for } r > 0$$

63. By (1) with $v_0 = 0$ and $h_0 = 784$, the height of the ball is given by $h(t) = -16t^2 + 784$. Therefore $h(3) = -16(3)^2 + 784 = 640$. Thus after 3 seconds the ball has fallen $784 - 640 = 144$ (feet). When the ball hits the ground, we have $h(t) = 0$, or $-16t^2 + 784 = 0$, or $t^2 = 49$. Thus $t = 7$, so the ball hits the ground after 7 seconds.

65. If t represents time in hours starting at noon and D distance in miles, then

$$D(t) = \begin{cases} 400t & \text{for } 0 \leq t < 2 \\ |400t - 800(t-2)| = |1600 - 400t| & \text{for } 2 \leq t \leq 5 \end{cases}$$

67. a. $R(0) = \dfrac{0}{c+0} = 0$ and $R(2) = \dfrac{2}{c+2d}$

$R(0) = 0$ indicates no response in the absence of the drug.
    b. Since $R(x)(c + dx) = x$, or equivalently, $cR(x) + dx\,R(x) = x$, we have $x - dx\,R(x) = cR(x)$, so that $x = cR(x)/[1 - dR(x)]$.

SECTION 1.4

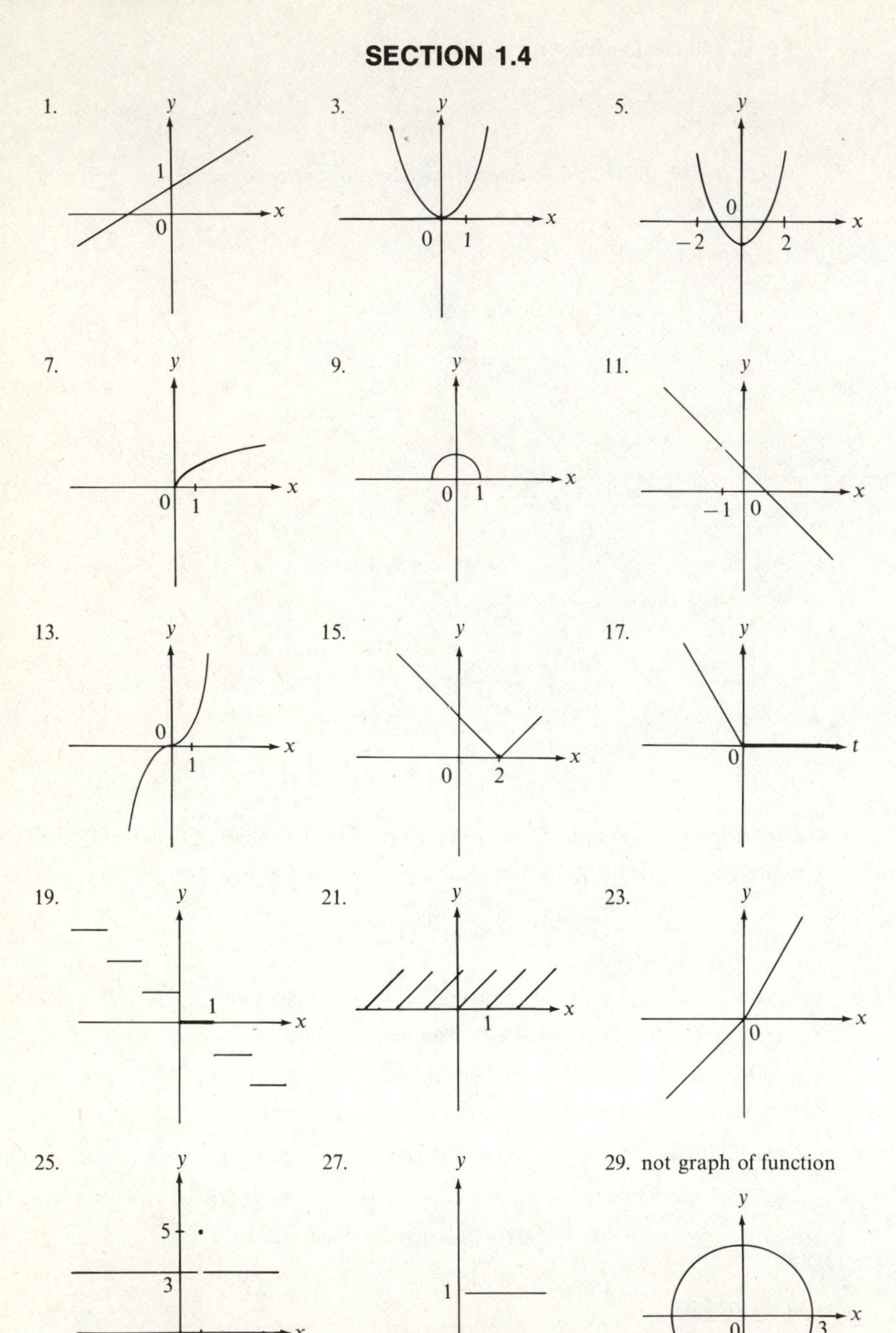

31. graph of function

33. not graph of function

35. not graph of function

37. graph consists of the two axes; not graph of function

39. not graph of function

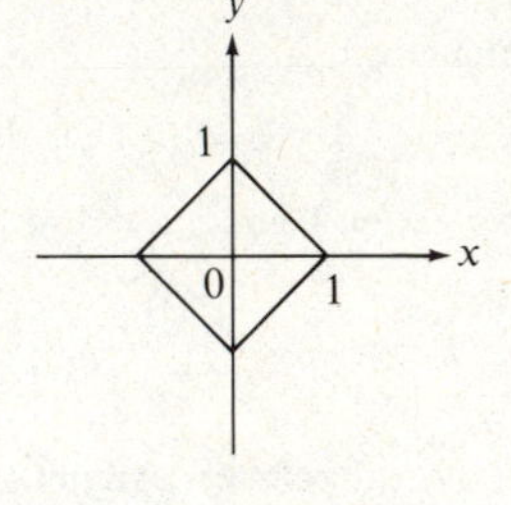

41. a, d, e, h

43. Notice that $P(x) = 25$ for $0 < x \leq 1$, $P(x) = 25 + 20 \cdot 1$ for $1 < x \leq 2$, $P(x) = 25 + 20 \cdot 2$ for $2 < x \leq 3$, etc. Notice also that $[-x] = -1$ for $0 < x \leq 1$, $[-x] = -2$ for $1 < x \leq 2$, $[-x] = -3$ for $2 < x \leq 3$, etc. Thus $[1 - x] = 0$ for $0 < x \leq 1$, $[1 - x] = -1$ for $1 < x \leq 2$, $[1 - x] = -2$ for $2 < x \leq 3$, etc., so $P(x) = 25 - 20[1 - x]$.

SECTION 1.5

			Symmetry		
	y intercepts	x intercepts	x axis	y axis	origin
1.	$-\sqrt{\frac{2}{3}}, \sqrt{\frac{2}{3}}$	-2	yes	no	no
3.	none	$-1, 1$	yes	yes	yes
5.	0	0	no	yes	no
7.	none	none	yes	yes	yes
9.	none	$-1, 1$	no	no	yes
11.	0	$[0, 1)$	no	no	no
13.	3	$-3, 3$	no	yes	no
15.	1	1	no	no	no

17. y intercept: 0 x intercept: 0

symmetry with respect to origin

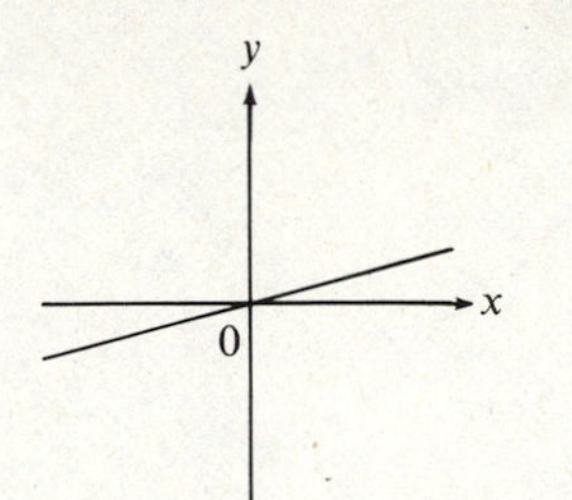

19. y intercept: -3 x intercepts: $-\sqrt{3}, \sqrt{3}$

symmetry with respect to y axis

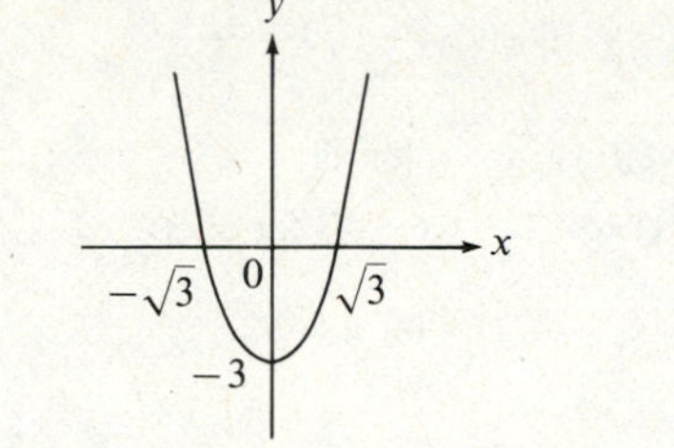

21. y intercepts: $-1, 1$

symmetry with respect to x axis, y axis, origin

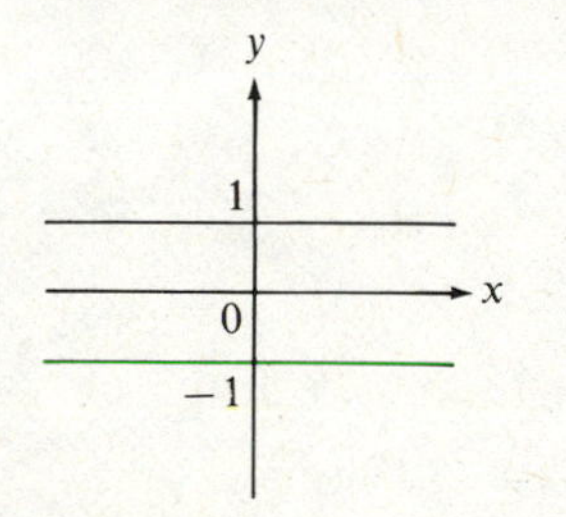

23. y intercepts: $-2, 2$ x intercept: 2

symmetry with respect to x axis

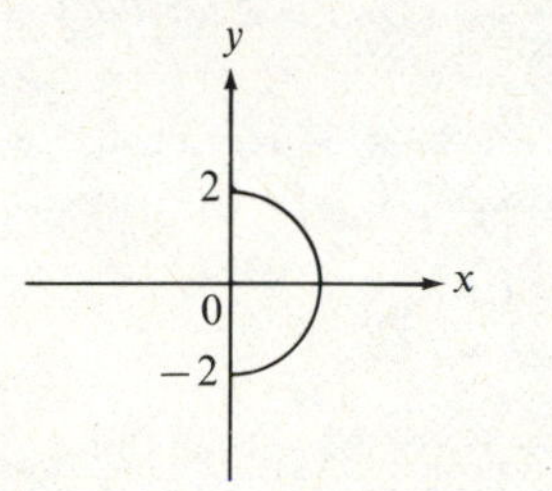

25. y intercept: 0 x intercept: 0

symmetry with respect to x axis, y axis, origin

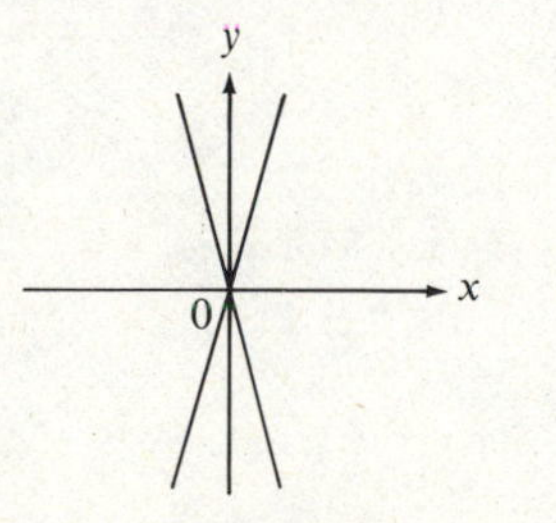

27. $(x-1)^2 + (y-3)^2 = 4$

Let $X = x - 1$, $Y = y - 3$.

Then $X^2 + Y^2 = 4$.

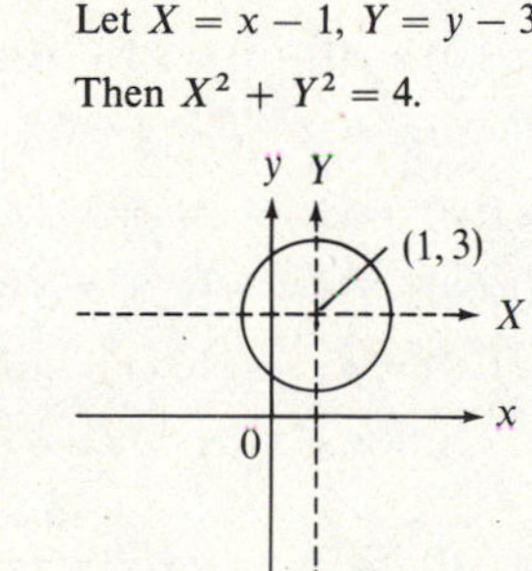

29. $x^2 - 2x + y^2 = 3$, so $(x-1)^2 + y^2 = 4$ Let $X = x - 1$, $Y = y$. Then $X^2 + Y^2 = 4$.

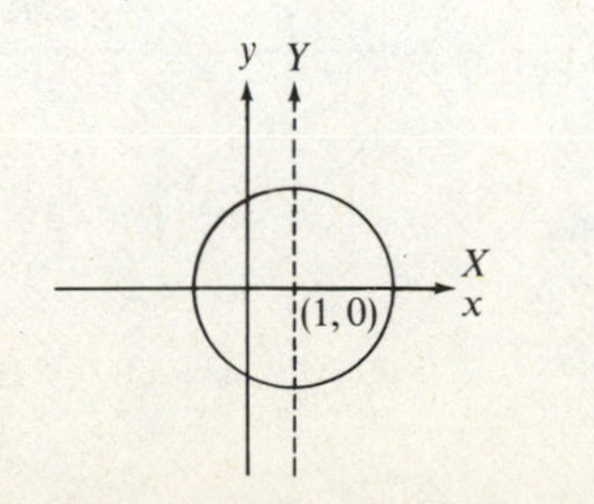

31. $x^2 + y - 3 = 0$, so $y - 3 = -x^2$

Let $X = x$, $Y = y - 3$. Then $Y = -X^2$.

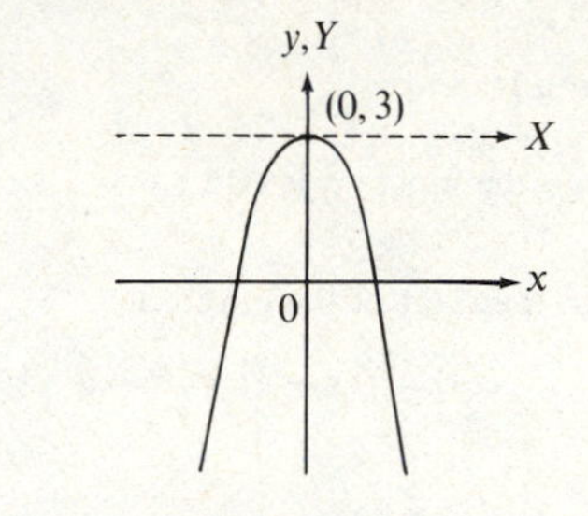

33. $x^2 + y^2 + 4x - 6y + 13 = 0$, so $(x+2)^2 + (y-3)^2 = 0$

Let $X = x + 2$, $Y = y - 3$. Then $X^2 + Y^2 = 0$.

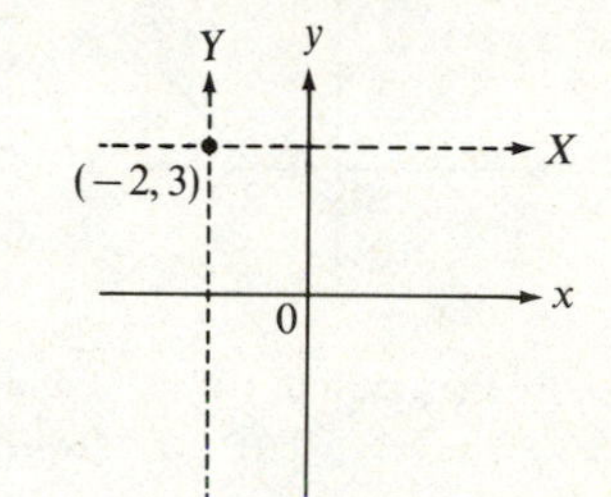

35. $y + 4 = \dfrac{1}{x+2}$

Let $X = x + 2$, $Y = y + 4$. Then $Y = 1/X$.

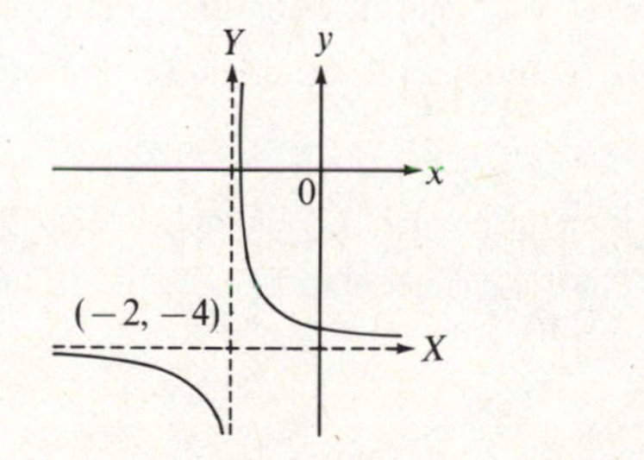

37. $x - 2 = |y - 2|$

Let $X = x - 2$, $Y = y - 2$. Then $X = |Y|$.

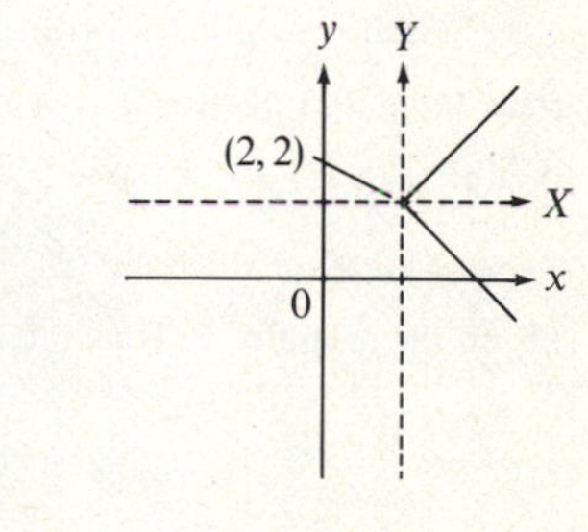

39. a. $f(-x) = -(-x) = x = -f(x)$, so f is odd.

b. $f(-x) = 5(-x)^2 - 3 = 5x^2 - 3 = f(x)$, so f is even.

c. $f(-x) = (-x)^3 + 1 = -x^3 + 1 \neq (x^3 + 1)$ or $-(x^3 + 1)$ if $x \neq 0$, so f is neither even nor odd.

d. $f(-x) = (-x-2)^2 = (-(x+2))^2 = (x+2)^2 \neq (x-2)^2$ or $-(x-2)^2$ if $x \neq 0$, so f is neither even nor odd.

e. $f(-x) = ((-x)^2 + 3)^3 = (x^2 + 3)^3 = f(x)$, so f is even.

f. $f(-x) = -x((-x)^2 + 1)^2 = -x(x^2 + 1)^2 = -f(x)$, so f is odd.

g. $\dfrac{-x}{(-x)^2 + 4} = -\dfrac{x}{x^2 + 4}$, so the function is odd.

h. $|-x| = |x|$, so the function is even.

i. $\dfrac{|-x|}{-x} = \dfrac{|x|}{-x} = -\dfrac{|x|}{x}$, so the function is odd.

41. The graph of g is 2 units to the right of the graph of f.

$$g(x) = f(x - 2) = |x - 2|$$

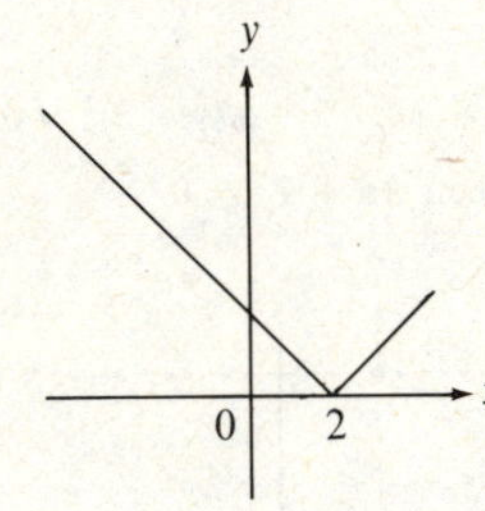

43. The graph of g is d units above the graph of f if $d \geq 0$, and is $-d$ units below the graph of f if $d < 0$.

45. Let (x, y) be on the graph. Since the graph is symmetric with respect to the x axis, $(x, -y)$ is on the graph. But then symmetry with respect to the y axis implies that $(-x, -y)$ is on the graph. Thus $(-x, -y)$ is on the graph whenever (x, y) is on the graph, so the graph is symmetric with respect to the origin. The converse is not true. For example, the graph of $xy = 1$ is symmetric with respect to the origin but not with respect to either axis.

47. Since $f(c - x) = f(c + x)$, the point $(c - x, y)$ is on the graph of f if and only if the point $(c + x, y)$ is on the graph of f. Thus the graph of f is symmetric with respect to the line $x = c$.

SECTION 1.6

1. $(f + g)(-1) = f(-1) + g(-1) = (-3) + (2) = -1$

3. $(fg)(\frac{1}{2}) = f(\frac{1}{2})g(\frac{1}{2}) = (-3)(\frac{11}{4}) = -\frac{33}{4}$

5. $f(g(1)) = f(2) = 6$

7. $\dfrac{f(x) - f(2)}{x - 2} = \dfrac{2x^2 + x - 10}{x - 2} = \dfrac{(x - 2)(2x + 5)}{x - 2} = 2x + 5$

9. $(f + g)(16) = f(16) + g(16) = \dfrac{15}{257} + 2 = \dfrac{529}{257}$

11. $(fg)(9) = f(9)g(9) = \dfrac{8}{82}\sqrt{3} = \dfrac{4\sqrt{3}}{41}$

13. $f(g(1)) = f(1) = 0$

15. $(f + g)(x) = x + 4$ for all x

$(fg)(x) = -2x^2 + 5x + 3$ for all x

$\left(\dfrac{f}{g}\right)(x) = \dfrac{2x + 1}{3 - x}$ for $x \neq 3$

17. $(f + g)(x) = \dfrac{2}{x - 1} + x - 1 = \dfrac{x^2 - 2x + 3}{x - 1}$ for $x \neq 1$

$(fg)(x) = \dfrac{2}{x - 1}(x - 1) = 2$ for $x \neq 1$ $\quad \left(\dfrac{f}{g}\right)(x) = \dfrac{2}{(x - 1)^2}$ for $x \neq 1$

19. $(f + g)(t) = t^{3/4} + t^2 + 3$ for $t \geq 0$

$(fg)(t) = t^{3/4}(t^2 + 3) = t^{11/4} + 3t^{3/4}$ for $t \geq 0$

$\left(\dfrac{f}{g}\right)(t) = \dfrac{t^{3/4}}{t^2 + 3}$ for $t \geq 0$

21. $(f + g)(t) = t^{2/3} + t^{3/5} - 1$ for all t

$(fg)(t) = t^{2/3}(t^{3/5} - 1) = t^{19/15} - t^{2/3}$ for all t

$\left(\dfrac{f}{g}\right)(t) = \dfrac{t^{2/3}}{t^{3/5} - 1}$ for $t \neq 1$

23. $(g \circ f)(x) = g(f(x)) = g(1 - x) = 2(1 - x) + 5 = -2x + 7$ for all x

$(f \circ g)(x) = f(g(x)) = f(2x + 5) = 1 - (2x + 5) = -2x - 4$ for all x

25. $(g \circ f)(x) = g(f(x)) = g(x^2) = \sqrt{x^2} = |x|$ for all x

$(f \circ g)(x) = f(g(x)) = f(\sqrt{x}) = (\sqrt{x})^2 = x$ for $x \geq 0$

27. $(g \circ f)(x) = g(f(x)) = g(\sqrt{x}) = (\sqrt{x})^2 - 5\sqrt{x} + 6 = x - 5\sqrt{x} + 6$ for $x \geq 0$

$(f \circ g)(x) = f(g(x)) = f(x^2 - 5x + 6) = \sqrt{x^2 - 5x + 6}$ for $x \leq 2$ or $x \geq 3$

29. $(g \circ f)(x) = g(f(x)) = g\left(\dfrac{1}{x - 1}\right) = \dfrac{1}{\dfrac{1}{x - 1} + 1} = \dfrac{x - 1}{x}$ for $x \neq 0, 1$

$(f \circ g)(x) = f(g(x)) = f\left(\dfrac{1}{x + 1}\right) = \dfrac{1}{\dfrac{1}{x + 1} - 1} = -\dfrac{x + 1}{x}$ for $x \neq -1, 0$

31. $(g \circ f)(x) = g(f(x)) = g(\sqrt{9 - x^2}) = [\sqrt{9 - x^2}]$ for $-3 \leq x \leq 3$

$(f \circ g)(x) = f(g(x)) = f([x]) = \sqrt{9 - [x]^2}$ for $-3 \leq x < 4$

33. Let $f(x) = x - 3$, $g(x) = \sqrt{x}$. Then $h(x) = g(x - 3) = g(f(x)) = (g \circ f)(x)$.

35. Let $f(x) = 3x^2 - 5\sqrt{x}$, $g(x) = x^{1/3}$. Then $h(x) = g(3x^2 - 5\sqrt{x}) = g(f(x)) = (g \circ f)(x)$.

37. Let $f(x) = x + 3$, $g(x) = 1/(x^2 + 1)$. Then $h(x) = g(x + 3) = g(f(x)) = (g \circ f)(x)$. Alternatively, let $f(x) = (x + 3)^2 + 1$ and $g(x) = 1/x$.

39. Let $f(x) = \sqrt{x} - 1$, $g(x) = \sqrt{x}$. Then $h(x) = g(\sqrt{x} - 1) = g(f(x)) = (g \circ f)(x)$. Alternatively, let $f(x) = \sqrt{x}$ and $g(x) = \sqrt{x - 1}$.

41. $g(x) = -|x - 2|$

43. If $0 \le x + 3 \le 4$, then $-3 \le x \le 1$; the domain of g is $[-3, 1]$.

45. All functions f such that $f(x) \ge 0$ for all x in the domain of f, that is, all functions with range contained in $[0, \infty)$.

47. All functions f such that $f(x) \ne 0$ for all x in the domain of f.

49. Since f is even and g is odd, $-x$ is in the domain of f and g and hence fg if x is. For such x, $(fg)(-x) = f(-x)g(-x) = f(x)(-g(x)) = -f(x)g(x) = -(fg)(x)$.

51. $f(f(x)) = f(a - x) = a - (a - x) = x$

53. $g(x) = f(x + p) - f(x) = [a(x + p) + b] - (ax + b) = ap$

55. $P(x) = R(x) - C(x) = 5x^2 - \frac{1}{10}x^4 - (4x^2 - 24x + 38) = -\frac{1}{10}x^4 + x^2 + 24x - 38$. Thus $P(1) = -\frac{1}{10} + 1 + 24 - 38 = -13.1$; $P(2) = -\frac{16}{10} + 4 + 48 - 38 = 12.4$. Since $P(1) = -13.1$, there is a loss when $x = 1$. Since $P(2) = 12.4$, there is a profit when $x = 2$.

57. a. $V(r(s)) = \dfrac{4}{3}\pi\left(\dfrac{1}{2}\sqrt{\dfrac{s}{\pi}}\right)^3 = \dfrac{1}{6\sqrt{\pi}}s^{3/2}$ for $s \ge 0$

b. $V(r(6)) = \dfrac{1}{6} \cdot 6\sqrt{\dfrac{6}{\pi}} = \sqrt{\dfrac{6}{\pi}}$

SECTION 1.7

1. a. $210° = \left(\dfrac{\pi}{180} \cdot 210\right)$ radians $= \dfrac{7\pi}{6}$ radians

b. $315° = \left(\dfrac{\pi}{180} \cdot 315\right)$ radians $= \dfrac{7\pi}{4}$ radians

c. $-405° = -\left(\dfrac{\pi}{180} \cdot 405\right)$ radians $= -\dfrac{9\pi}{4}$ radians

d. $1080° = \left(\dfrac{\pi}{180} \cdot 1080\right)$ radians $= 6\pi$ radians

e. $1° = \left(\dfrac{\pi}{180} \cdot 1\right)$ radian $= \dfrac{\pi}{180}$ radian

3. a. $\sin\dfrac{11\pi}{6} = \sin\left(-\dfrac{\pi}{6} + 2\pi\right) = \sin\left(-\dfrac{\pi}{6}\right) = -\sin\dfrac{\pi}{6} = -\dfrac{1}{2}$

b. $\sin\left(-\dfrac{2\pi}{3}\right) = -\sin\dfrac{2\pi}{3} = -\dfrac{\sqrt{3}}{2}$

c. $\cos\dfrac{5\pi}{4} = \cos\left(\pi + \dfrac{\pi}{4}\right) = -\cos\dfrac{\pi}{4} = -\dfrac{\sqrt{2}}{2}$

d. $\cos\left(-\dfrac{7\pi}{6}\right) = \cos\dfrac{7\pi}{6} = \cos\left(\pi + \dfrac{\pi}{6}\right) = -\cos\dfrac{\pi}{6} = -\dfrac{\sqrt{3}}{2}$

e. $\tan\dfrac{4\pi}{3} = \dfrac{\sin(4\pi/3)}{\cos(4\pi/3)} = \dfrac{\sin[\pi + (\pi/3)]}{\cos[\pi + (\pi/3)]} = \dfrac{-\sin(\pi/3)}{-\cos(\pi/3)} = \dfrac{-\sqrt{3}/2}{-\frac{1}{2}} = \sqrt{3}$

f. $\tan\left(-\dfrac{\pi}{4}\right) = \dfrac{\sin(-\pi/4)}{\cos(-\pi/4)} = \dfrac{-\sin(\pi/4)}{\cos(\pi/4)} = \dfrac{-\sqrt{2}/2}{\sqrt{2}/2} = -1$

g. $\cot\dfrac{\pi}{6} = \dfrac{\cos(\pi/6)}{\sin(\pi/6)} = \dfrac{\sqrt{3}/2}{\frac{1}{2}} = \sqrt{3}$

h. $\cot\left(-\dfrac{17\pi}{3}\right) = \dfrac{\cos[-(17\pi)/3]}{\sin[-(17\pi)/3]} = \dfrac{\cos[-6\pi + (\pi/3)]}{\sin(-6\pi + (\pi/3))} = \dfrac{\cos(\pi/3)}{\sin(\pi/3)} = \dfrac{\frac{1}{2}}{\sqrt{3}/2} = \dfrac{1}{\sqrt{3}} = \dfrac{\sqrt{3}}{3}$

i. $\sec 3\pi = \dfrac{1}{\cos 3\pi} = \dfrac{1}{\cos(2\pi + \pi)} = \dfrac{1}{\cos\pi} = \dfrac{1}{-1} = -1$

j. $\sec\left(-\dfrac{\pi}{3}\right) = \dfrac{1}{\cos(-\pi/3)} = \dfrac{1}{\cos(\pi/3)} = \dfrac{1}{\frac{1}{2}} = 2$

k. $\csc\dfrac{\pi}{2} = \dfrac{1}{\sin(\pi/2)} = \dfrac{1}{1} = 1$

l. $\csc[-(5\pi)/3] = \dfrac{1}{\sin[-(5\pi)/3]} = \dfrac{1}{\sin[-2\pi + (\pi/3)]} = \dfrac{1}{\sin(\pi/3)} = \dfrac{1}{\sqrt{3}/2} = \dfrac{2}{\sqrt{3}} = \dfrac{2\sqrt{3}}{3}$

5. $\tan x = \dfrac{\sin x}{\cos x} = -\dfrac{4}{3}$; $\cot x = \dfrac{1}{\tan x} = -\dfrac{3}{4}$; $\sec x = \dfrac{1}{\cos x} = -\dfrac{5}{3}$; $\csc x = \dfrac{1}{\sin x} = \dfrac{5}{4}$

7. $\cos x = \dfrac{1}{\sec x} = \dfrac{1}{\sqrt{5}} = \dfrac{\sqrt{5}}{5}$; $\sin x = \cos x \tan x = \dfrac{\sqrt{5}}{5} \cdot (-2) = -\dfrac{2\sqrt{5}}{5}$;

$\cot x = \dfrac{1}{\tan x} = -\dfrac{1}{2}$; $\csc x = \dfrac{1}{\sin x} = \dfrac{5}{-2\sqrt{5}} = -\dfrac{\sqrt{5}}{2}$

9. $7\pi/6$, $11\pi/6$

11. $5\pi/6$, $7\pi/6$

13. $\cos x = \cos 2x = 2\cos^2 x - 1$ implies that $2\cos^2 x - \cos x - 1 = 0$, or

$$(2\cos x + 1)(\cos x - 1) = 0$$

Now $2\cos x + 1 = 0$ if $\cos x = -\frac{1}{2}$, which happens for $x = 2\pi/3, 4\pi/3$; $\cos x - 1 = 0$ for $x = 0$. Solutions: $0, 2\pi/3, 4\pi/3$.

15. The union of $[0, 7\pi/6)$ and $(11\pi/6, 2\pi)$

17. The union of $[\pi/4, \pi/2)$ and $[5\pi/4, 3\pi/2)$

19. $\sec x < -1$ if $-1 < \cos x < 0$, which happens for x in $(\pi/2, 3\pi/2)$

21. $|\cos x| < \frac{1}{2}\sqrt{3}$ if $-\frac{1}{2}\sqrt{3} < \cos x < \frac{1}{2}\sqrt{3}$. Now $\cos x < \frac{1}{2}\sqrt{3}$ for x in $(\pi/6, 11\pi/6)$, whereas $-\frac{1}{2}\sqrt{3} < \cos x$ for x in $[0, 5\pi/6)$ and $(7\pi/6, 2\pi)$. Solution: The union of $(\pi/6, 5\pi/6)$ and $(7\pi/6, 11\pi/6)$.

23. The union of $(0, \pi/4]$, $(\pi/2, 3\pi/4]$, $(\pi, 5\pi/4]$, and $(3\pi/2, 7\pi/4]$

25. y intercept: -1 x intercepts: $\pi/2 + n\pi$ for any integer n symmetric with respect to y axis; an even function.
(Note: $\sin[x - (\pi/2)] = \sin x \cos(\pi/2) - \cos x \sin(\pi/2) = -\cos x$)

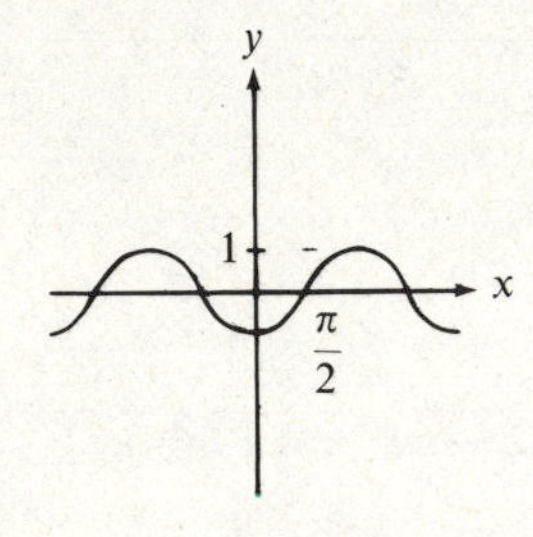

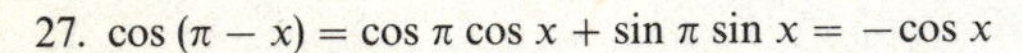
27. $\cos(\pi - x) = \cos \pi \cos x + \sin \pi \sin x = -\cos x$

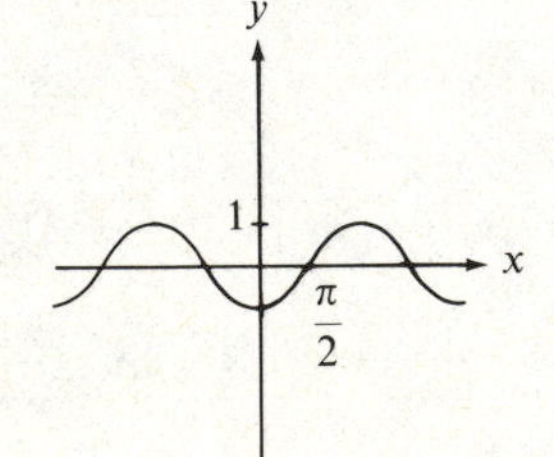

29. y intercepts: none x intercepts: $(\pi/2) + n\pi$ for any integer n symmetric with respect to origin; an odd function
(Let $X = x + (\pi/2)$, $Y = y$; equation becomes $Y = \tan X$.)

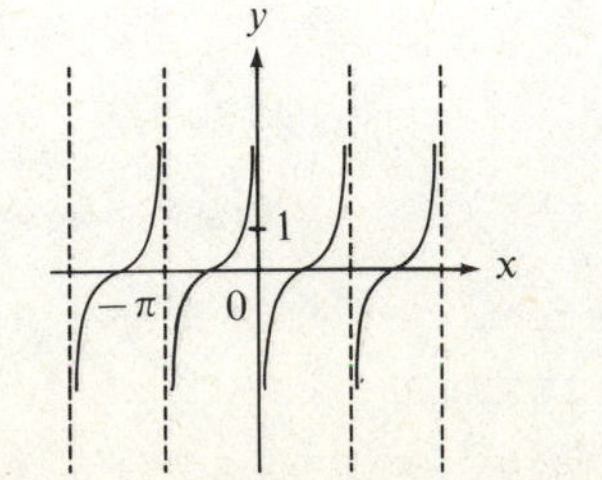

31. y intercept: 1 x intercepts: none symmetric with respect to y axis; an even function
$\sec(2\pi - x) = \sec(-x) = \sec x$; see Figure 1.50(c)

33. y intercept: 0 x intercepts: $n\pi/2$ for any integer n symmetric about origin; an odd function

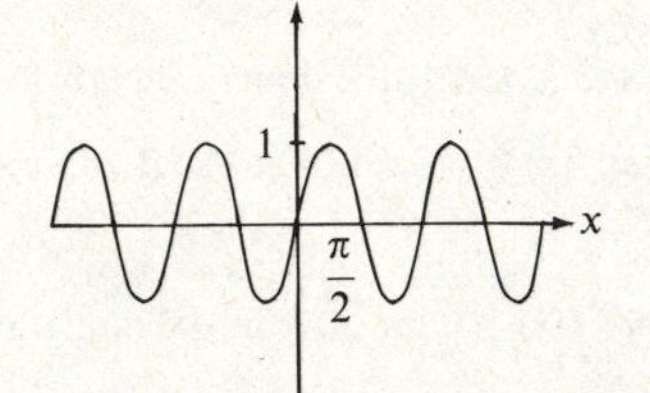

35. $\sin \frac{7\pi}{12} = \sin\left(\frac{\pi}{3} + \frac{\pi}{4}\right) = \sin \frac{\pi}{3} \cos \frac{\pi}{4} + \cos \frac{\pi}{3} \sin \frac{\pi}{4} = \frac{\sqrt{3}}{2} \cdot \frac{\sqrt{2}}{2} + \frac{1}{2} \cdot \frac{\sqrt{2}}{2} = \frac{\sqrt{2}}{4}(\sqrt{3} + 1)$

37. $\sin \frac{11\pi}{12} = \sin\left(\frac{3\pi}{4} + \frac{\pi}{6}\right) = \sin \frac{3\pi}{4} \cos \frac{\pi}{6} + \cos \frac{3\pi}{4} \sin \frac{\pi}{6} = \frac{\sqrt{2}}{2} \cdot \frac{\sqrt{3}}{2} + \left(-\frac{\sqrt{2}}{2}\right)\frac{1}{2}$
$= \frac{\sqrt{2}}{4}(\sqrt{3} - 1)$

39. $\tan \frac{5\pi}{12} = \frac{\sin(5\pi/12)}{\cos(5\pi/12)}$

First,

$$\sin \frac{5\pi}{12} = \sin\left(\frac{\pi}{4} + \frac{\pi}{6}\right) = \sin \frac{\pi}{4} \cos \frac{\pi}{6} + \cos \frac{\pi}{4} \sin \frac{\pi}{6} = \frac{\sqrt{2}}{2} \cdot \frac{\sqrt{3}}{2} + \frac{\sqrt{2}}{2} \cdot \frac{1}{2} = \frac{\sqrt{2}}{4}(\sqrt{3} + 1)$$

Next,

$$\cos \frac{5\pi}{12} = \cos\left(\frac{\pi}{4} + \frac{\pi}{6}\right) = \cos \frac{\pi}{4} \cos \frac{\pi}{6} - \sin \frac{\pi}{4} \sin \frac{\pi}{6} = \frac{\sqrt{2}}{2} \cdot \frac{\sqrt{3}}{2} - \frac{\sqrt{2}}{2} \cdot \frac{1}{2} = \frac{\sqrt{2}}{4}(\sqrt{3} - 1)$$

It follows that

$$\tan \frac{5\pi}{12} = \frac{(\sqrt{2}/4)(\sqrt{3} + 1)}{(\sqrt{2}/4)(\sqrt{3} - 1)} = \frac{\sqrt{3} + 1}{\sqrt{3} - 1} = \frac{\sqrt{3} + 1}{\sqrt{3} - 1} \frac{\sqrt{3} + 1}{\sqrt{3} + 1} = 2 + \sqrt{3}$$

41. $2\sin^2 x + \sin x - 1 = 0$ if and only if $(2\sin x - 1)(\sin x + 1) = 0$, so the solution consists of all x for which $\sin x = \frac{1}{2}$ or $\sin x = -1$. Solution: $(\pi/6) + 2n\pi$, $(5\pi/6) + 2n\pi$, $(3\pi/2) + 2n\pi$ for any integer n.

43. If $\cos x \neq 0$, then

$$\frac{\sin^2 x + \cos^2 x}{\cos^2 x} = \frac{1}{\cos^2 x} \quad \text{or} \quad \left(\frac{\sin x}{\cos x}\right)^2 + 1 = \left(\frac{1}{\cos x}\right)^2$$

Thus $1 + \tan^2 x = \sec^2 x$ whenever $\tan x$ and $\sec x$ are defined.

45. a. By (9) and (11), $\sin(\pi - x) = \sin\pi\cos(-x) + \cos\pi\sin(-x) = -\sin(-x) = \sin x$.

b. By (9) and (12),

$$\sin[(3\pi/2) - x] = \sin(3\pi/2)\cos(-x) + \cos(3\pi/2)\sin(-x) = -\cos(-x) = -\cos x$$

c. By (10) and (12),

$$\cos(\pi - x) = \cos\pi\cos(-x) - \sin\pi\sin(-x) = -\cos(-x) = -\cos x$$

d. By (10) and (11),

$$[\cos(3\pi/2) - x] = \cos(3\pi/2)\cos(-x) - \sin(3\pi/2)\sin(-x) = \sin(-x) = -\sin x$$

47. $m_1 = 4,\ m_2 = \dfrac{-2}{3}$; $\tan\theta = \dfrac{m_2 - m_1}{1 + m_1 m_2} = \dfrac{(-2/3) - 4}{1 + (-2/3)(4)} = \dfrac{14}{5}$.

49. a. π b. π c. 2π d. 2π e. $2\pi/3$ f. π g. π h. π

51. Let d be the distance between the beacon and the illuminated point. Then $\sec\theta = d/2$, so $d = 2\sec\theta$.

53. If one side of the angle is a diameter and the chord has length x, then $\sin\theta = x/1 = x$. Since inscribed angles subtending equal arcs are equal, if θ is any other angle that subtends the same arc, it equals an angle one of whose sides is a diameter.

55. Let the distance between the motorist and the intersection be $60 + x$. Then

$$x = 40\cot\frac{\pi}{6} = 40\sqrt{3}$$

and thus the distance is $60 + 40\sqrt{3} \approx 129.3$ feet.

57. Using the notation in the diagram, we have $y = 2\cos\theta$ and $x = 2\sin\theta$, so the cross-sectional area equals $2y + 2(\frac{1}{2}xy) = 2y + xy = 4(\cos\theta)(1 + \sin\theta)$. Thus the volume equals $40(\cos\theta)(1 + \sin\theta)$.

59. a. Let θ be as in the figure, so that $\theta = \frac{1}{2}(2\pi/n) = \pi/n$. Next let x be half the length of one side of the polygon, as in the figure. Then $x = r\sin(\pi/n)$. Thus the perimeter of the n-sided polygon is given by $p_n(r) = 2nr\sin(\pi/n)$.

b. From part (a), $2x = 2r\sin(\pi/n)$ or $r = 2x/[2\sin(\pi/n)]$. For the Pentagon, $2x = 921$ and $n = 5$, so $r = 921/[2\sin(\pi/5)] \approx 783.4$ (feet).

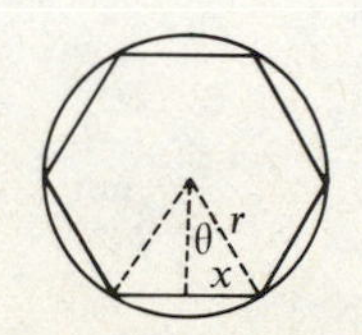

CHAPTER 1 REVIEW

1. From the diagram we see that the solution is the union of $(-\infty, -\frac{3}{2})$ and $(4, \infty)$.

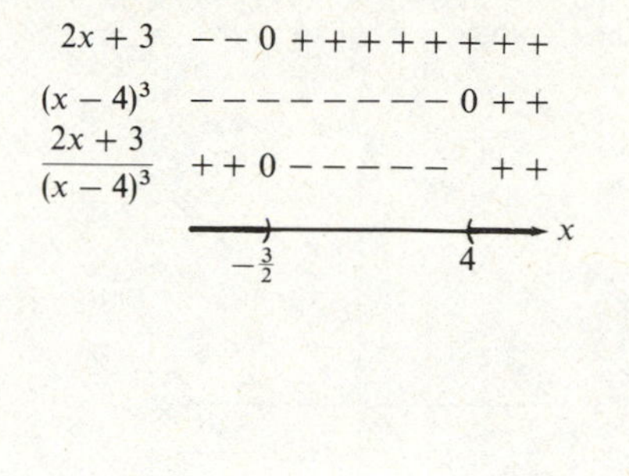

3. The given inequality is equivalent to

$$\frac{2}{3 - x} - 4 \geq 0, \text{ or } \frac{4(x - \frac{5}{2})}{3 - x} \geq 0$$

From the diagram we see that the solution is $[\frac{5}{2}, 3)$.

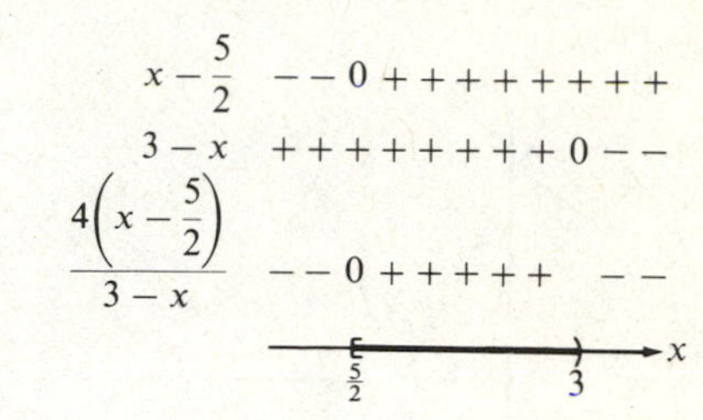

5. If $|4 - 6x| < \frac{1}{2}$, then $-\frac{1}{2} < 4 - 6x < \frac{1}{2}$, or $-\frac{9}{2} < -6x < -\frac{7}{2}$, or $\frac{7}{12} < x < \frac{3}{4}$. The solution is $(\frac{7}{12}, \frac{3}{4})$.

7. a. For $a \geq 0$ we have $|a| = a$, so $\dfrac{(|a| + a)}{2} = \dfrac{(a + a)}{2} = a$.

b. For $a < 0$, we have $|a| = -a$, so $\dfrac{(|a| + a)}{2} = \dfrac{(-a + a)}{2} = 0$.

9. $\dfrac{y - 1}{x - 0} = \dfrac{-3 - 1}{1 - 0} = -4$, or

$y = -4x + 1$

11. $y = -\frac{1}{3}x + 6$

13. $m_1 = 2$; $m_2 = -\frac{1}{2}$. Since $m_1 m_2 = -1$, the lines are perpendicular.

15. The slope of ℓ is $-\frac{1}{2}$.

a. The desired line has slope $-\frac{1}{2}$ also, so an equation is $y - (-2) = -\frac{1}{2}(x - 1)$, or $y = -\frac{1}{2}x - \frac{3}{2}$.

b. The desired line has slope 2, so an equation is $y - (-2) = 2(x - 1)$, or $y = 2x - 4$.

17. The first and fourth lines are parallel, with slope 1; the second and third lines are parallel, with slope -1. Thus the lines form a rectangle. The points of intersection are $(1, 1)$, $(5, 5)$, $(-3, 5)$, and $(1, 9)$. The lengths of the sides of the figure are

$$\sqrt{(5-1)^2+(5-1)^2} = \sqrt{(5-1)^2+(5-9)^2} = \sqrt{(1-(-3))^2+(9-5)^2}$$
$$= \sqrt{(-3-1)^2+(5-1)^2} = 4\sqrt{2}$$

Since the sides all have the same length, the rectangle is a square.

19. All numbers except $-3, -1, 0$.

21.

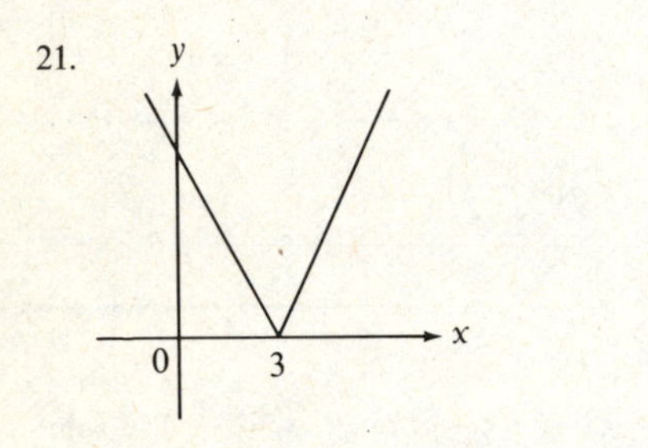

23.

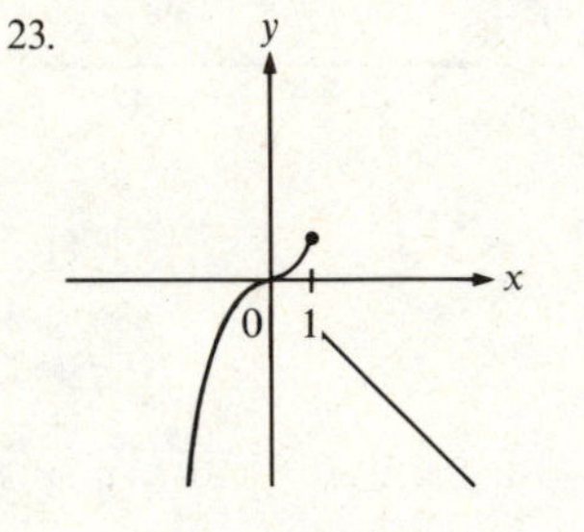

25. y intercept: 1 $\quad x$ intercept: 1

not symmetric with respect to either axis or origin

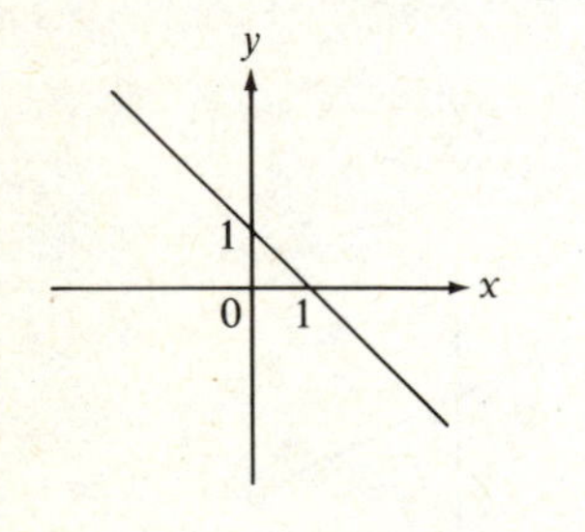

27. y intercepts: $-\sqrt{3}, \sqrt{3}$

x intercepts: $-\sqrt{3}, \sqrt{3}$

symmetric with respect to the x axis, y axis and origin

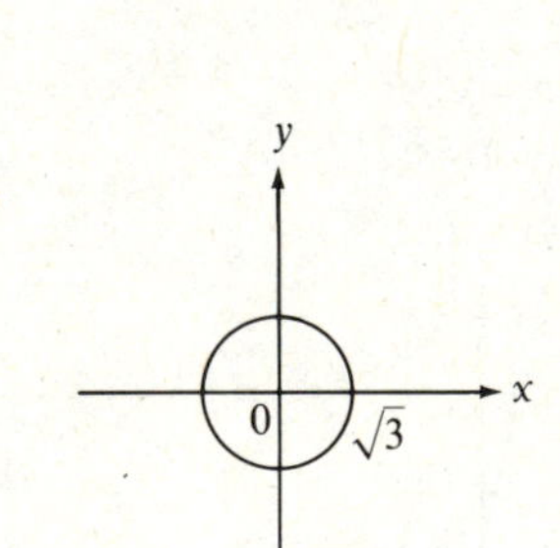

29. $x^2 + 6x + y + 4 = 0$

$(x^2 + 6x + 9) + y + 4 = 9$

$(x + 3)^2 + (y - 5) = 0$

Let $X = x + 3$, $Y = y - 5$.

The equation becomes $X^2 + Y = 0$, or $Y = -X^2$.

31. $x - 2 = 1/(y + 3)$

Let $X = x - 2$, $Y = y + 3$.

The equation becomes $X = 1/Y$.

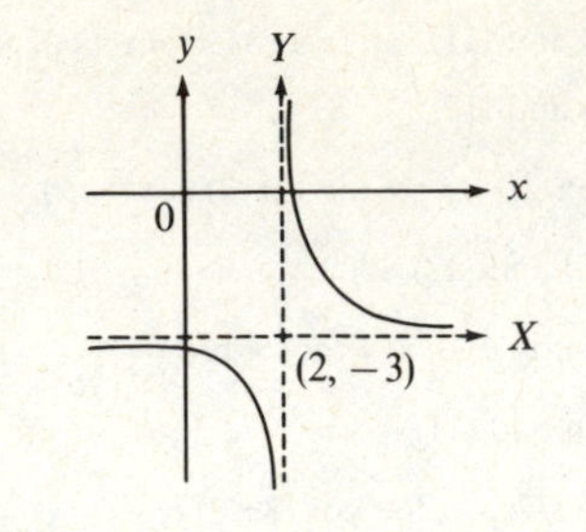

33. $$(f - g)(x) = \frac{x+2}{x^2-4x+3} - \frac{x+1}{x^2-2x-3}$$
$$= \frac{x+2}{(x-1)(x-3)} - \frac{x+1}{(x-3)(x+1)}$$
$$= \frac{x+2}{(x-1)(x-3)} - \frac{1}{x-3} = \frac{(x+2)-(x-1)}{(x-1)(x-3)}$$
$$= \frac{3}{(x-1)(x-3)} \quad \text{for } x \neq -1, 1, \text{ and } 3$$

$$\left(\frac{f}{g}\right)(x) = \frac{(x+2)/(x^2-4x+3)}{(x+1)/(x^2-2x-3)} = \frac{(x+2)(x-3)(x+1)}{(x-1)(x-3)(x+1)} = \frac{x+2}{x-1} \quad \text{for } x \neq -1, 1, \text{ and } 3$$

35. a. Domain of f: $[-1, \infty)$; domain of g: $[-2, \infty)$; domain of h: the union of $(-\infty, -2]$ and $[-1, \infty)$.

b. Domain of fg: $[-1, \infty)$. The domain of fg is smaller than the domain of h (although the rules of fg and h are the same).

37. If $x \neq 1$, then $x/(x-1) \neq 1$, so x is in the domain of $f \circ f$.

a. $$f(f(x)) = \frac{x/(x-1)}{[x/(x-1)]-1} = \frac{x}{x-(x-1)} = x$$

b. By part (a), $f(f(f(f(x)))) = f(f(x)) = x$ for $x \neq 1$.

39. The domain of f is $(-\infty, \infty)$ if $c = 0$ and $a \neq 0$, and in this case, $f(x) = -x - (b/a)$, so $f(f(x)) = -[-x - (b/a)] - b/a = x$. Next, the domain consists of all $x \neq a/c$ if $c \neq 0$. In this case, if $f(x) = a/c$, then $(ax + b)/(cx - a) = a/c$, so that $acx + bc = acx - a^2$, and hence $a^2 + bc = 0$. By hypothesis $a^2 + bc \neq 0$, so $f(x) \neq a/c$ for all x in the domain of f. Consequently the domain of $f \circ f$ is the domain of f. Finally,

$$f(f(x)) = f\left(\frac{ax-b}{cx-a}\right) = \frac{a[(ax+b)/(cx-a)]+b}{c[(ax+b)/(cx-a)]-a} = \frac{a^2x+ab+bcx-ab}{acx+bc-acx+a^2}$$
$$= \frac{(a^2+bc)x}{a^2+bc} = x$$

41. $$\cos x = \frac{\sin x}{\tan x} = \frac{-2/3}{-2\sqrt{5}/5} = \frac{\sqrt{5}}{3}; \quad \cot x = \frac{1}{\tan x} = -\frac{5}{2\sqrt{5}} = -\frac{\sqrt{5}}{2};$$

$$\sec x = \frac{1}{\cos x} = \frac{3}{\sqrt{5}} = \frac{3\sqrt{5}}{5}; \quad \csc x = \frac{1}{\sin x} = -\frac{3}{2}$$

43. $\cos x < -\frac{1}{2}$ for all numbers in the interval $(2\pi/3, 4\pi/3)$.

45. Notice that if $X = x$ and $Y = y - 1$, then $y = 1 - \sin x$ becomes
$Y = -\sin X$ y intercept: 1
x intercepts: $\dfrac{\pi}{2} + 2n\pi$ for any integer n
not symmetric with respect to either axis or origin

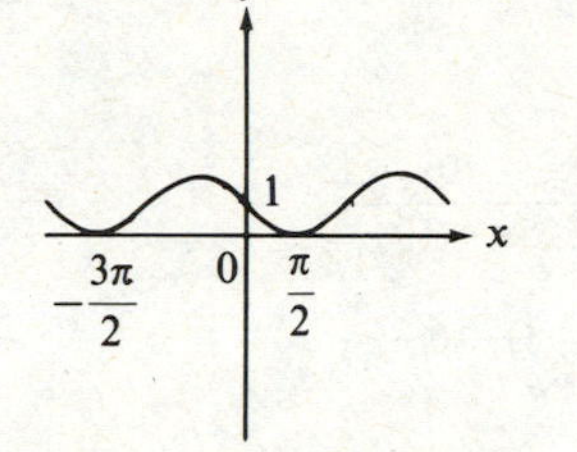

47. Notice that if $X = x - (\pi/4)$ and $Y = y$, then $y = \tan [x - (\pi/4)]$ becomes
$Y = \tan X$ y intercept: -1
x intercepts: $(\pi/4) + n\pi$ for any integer n
not symmetric with respect to either axis or to origin.

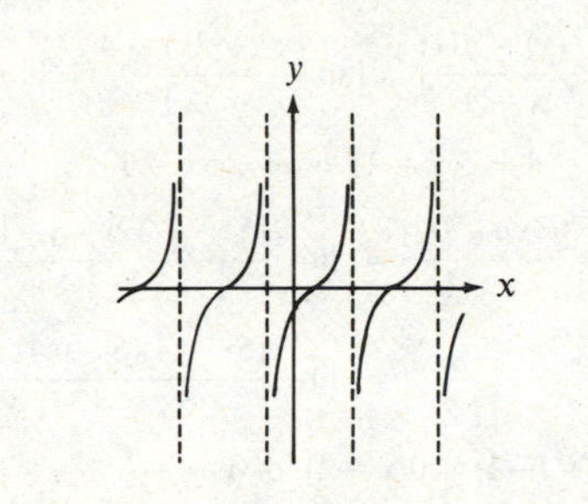

49. $f(-x) = (-x)^2 - \cos(-x) = x^2 - \cos x = f(x)$, so f is an even function.

51. a. $a \sin(x + b) = a \sin x \cos b + a \cos x \sin b = \sin x + \sqrt{3} \cos x$ if $a \cos b = 1$ and $a \sin b = \sqrt{3}$. Thus $\tan b = (a \sin b)/(a \cos b) = \sqrt{3}/1 = \sqrt{3}$, so $b = (\pi/3) + n\pi$ for any integer n. If $b = (\pi/3) + 2n\pi$, then $a = 1/(\cos b) = 1/[\cos(\pi/3)] = 1/\frac{1}{2} = 2$. If $b = (4\pi/3) + 2n\pi$, then $a = 1/(\cos b) = 1/[\cos(4\pi/3)] = 1/(-\frac{1}{2}) = -2$. Thus the solutions are $a = 2$, $b = (\pi/3) + 2n\pi$ for any integer n, and $a = -2$, $b = (4\pi/3) + 2n\pi$ for any integer n.

b. $a \cos(x + b) = a \cos x \cos b - a \sin x \sin b = \sin x + \sqrt{3} \cos x$ if $a \cos b = \sqrt{3}$ and $-a \sin b = 1$. Thus $\tan b = (a \sin b)/(a \cos b) = -1/\sqrt{3}$, so $b = -\pi/6 + n\pi$ for any integer n. If $b = -(\pi/6) + 2n\pi$, then

$$a = \frac{\sqrt{3}}{\cos b} = \frac{\sqrt{3}}{\cos(-\pi/6)} = \frac{\sqrt{3}}{\sqrt{3}/2} = 2$$

If $b = (5\pi/6) + 2n\pi$, then $a = \sqrt{3}/(\cos b) = \sqrt{3}/\cos(5\pi/6) = \sqrt{3}/(-\sqrt{3}/2) = -2$. Thus the solutions are $a = 2$, $b = (-\pi/6) + 2n\pi$ for any integer n, and $a = -2$, $b = (5\pi/6) + 2n\pi$ for any integer n.

53. Let P be the perimeter of the square and the equilateral triangle. Then each side of the square has length $\frac{1}{4}P$, so the area of the square is $\frac{1}{16}P^2$. Each side of the triangle has length $\frac{1}{3}P$, so by Exercise 59 of Section 1.3, the area of the triangle is $\frac{1}{4}\sqrt{3}\,(\frac{1}{3}P)^2 = (\sqrt{3}/36)P^2$. Since $\sqrt{3}/36 < \frac{1}{16}$, the square has the larger area. However, one can make the area of a rectangle with perimeter P as small as one wishes by making the length of one side of the rectangle small enough. Thus an equilateral triangle can have a larger area or smaller area than a rectangle with the same perimeter.

55. The fourth vertex could be $(1, 3)$ (opposite $(2, 0)$), $(5, 1)$ (opposite $(0, 1)$), or $(-1, -1)$ (opposite $(3, 2)$).

2

LIMITS AND CONTINUITY

SECTION 2.1

1. 3   3. 0   5. $-\frac{1}{11}$ 7. $\frac{2}{5}$

9. $\lim_{x\to-2}\frac{x^2-4}{x+2}=\lim_{x\to-2}\frac{(x+2)(x-2)}{x+2}=\lim_{x\to-2}(x-2)=-4$

11. $\lim_{x\to-5}\frac{x^2+4x-5}{x+5}=\lim_{x\to-5}\frac{(x-1)(x+5)}{x+5}=\lim_{x\to-5}(x-1)=-6$

13. $\lim_{x\to1}\frac{x^3-1}{x-1}=\lim_{x\to1}\frac{(x-1)(x^2+x+1)}{x-1}=\lim_{x\to1}(x^2+x+1)=3$

15. $\lim_{x\to0}3(\sin^2x+\cos^2x)=\lim_{x\to0}3=3$

17. $\lim_{x\to-1}-\frac{|x|}{x}=\lim_{x\to-1}-\frac{-x}{x}=\lim_{x\to-1}1=1$

19. 1.0033467, 1.0000333, 1.0000003, 1.0033467, 1.0000333, 1.0000003; $\lim_{x\to0}f(x)=1$

21. 0.04995835, 0.00499996, 0.0005, −0.04995835, −0.00499996, −0.0005; $\lim_{x\to0}f(x)=0$

23. $\frac{f(x)-f(3)}{x-3}=\frac{x^2-3^2}{x-3}=\frac{(x-3)(x+3)}{x-3}=x+3$; $\lim_{x\to3}(x+3)=6$, so the slope is 6.

25. $\frac{f(x)-f(4)}{x-4}=\frac{(2x^2+1)-33}{x-4}=\frac{2x^2-32}{x-4}=\frac{2(x-4)(x+4)}{x-4}=2(x+4)$;
$\lim_{x\to4}2(x+4)=16$, so the slope is 16.

27. b or d   29. a. −6   b. $-\frac{2}{3}$

31. average velocity $=\frac{f(t)-f(2)}{t-2}=\frac{(2t^2+1)-9}{t-2}=\frac{2(t^2-4)}{t-2}=\frac{2(t-2)(t+2)}{t-2}=2(t+2)$;

$v(t)=\lim_{t\to2}2(t+2)=8$ (miles per minute)

SECTION 2.2

1. −5   3. $\frac{1}{2}$ 5. $2(-1)+5=3$ 7. $\left|\frac{3}{2}\right|=\frac{3}{2}$

9. $\lim_{x\to0}f(x)=\lim_{x\to0}2x=0$

11. $\lim_{x\to3}\frac{f(x)-f(3)}{x-3}=\lim_{x\to3}\frac{\pi-\pi}{x-3}=0$; ℓ: $y-\pi=0(x-3)$, or $y=\pi$

13. $\lim_{x\to0}\frac{f(x)-f(0)}{x-0}=\lim_{x\to0}\frac{(-x^2+2)-2}{x-0}=\lim_{x\to0}\frac{-x^2}{x}=\lim_{x\to0}(-x)=0$;
ℓ: $y-2=0(x-0)$, or $y=2$

15. $\lim_{x\to1/2}\frac{f(x)-f(\frac12)}{x-\frac12}=\lim_{x\to1/2}\frac{(4x^2+\frac12)-\frac32}{x-\frac12}=\lim_{x\to1/2}\frac{4(x^2-\frac14)}{x-\frac12}=\lim_{x\to1/2}\frac{4(x-\frac12)(x+\frac12)}{x-\frac12}$
$=\lim_{x\to1/2}4(x+\frac12)=4$; ℓ: $y-\frac32=4(x-\frac12)$, or $y=4x-\frac12$

17. $\lim_{x\to1}\frac{f(x)-f(1)}{x-1}=\lim_{x\to1}\frac{(x^2+3x)-4}{x-1}=\lim_{x\to1}\frac{(x-1)(x+4)}{x-1}=\lim_{x\to1}(x+4)=5$;
ℓ: $y-4=5(x-1)$, or $y=5x-1$

19. $\lim_{x\to1/5}\frac{f(x)-f(\frac15)}{x-\frac15}=\lim_{x\to1/5}\frac{(5x^2-2x)-(-\frac15)}{x-\frac15}=\lim_{x\to1/5}\frac{25x^2-10x+1}{5x-1}$
$=\lim_{x\to1/5}\frac{(5x-1)(5x+1)}{5x-1}=\lim_{x\to1/5}(5x-1)=0$;
ℓ: $y-(-\frac15)=0(x-\frac15)$, or $y=-\frac15$

21. $v(2)=\lim_{t\to2}\frac{f(t)-f(2)}{t-2}=\lim_{t\to2}\frac{(2t+1)-5}{t-2}=\lim_{t\to2}\frac{2(t-2)}{t-2}=\lim_{t\to2}2=2$

23. $v(4)=\lim_{t\to4}\frac{f(t)-f(4)}{t-4}=\lim_{t\to4}\frac{-16t^2-(-16)4^2}{t-4}=\lim_{t\to4}\frac{-16(t^2-16)}{t-4}$
$=\lim_{t\to4}\frac{-16(t-4)(t+4)}{t-4}=\lim_{t\to4}-16(t+4)=-128$

25. $v(2)=\lim_{t\to2}\frac{f(t)-f(2)}{t-2}=\lim_{t\to2}\frac{(-16t^2+8t+54)-6}{t-2}=\lim_{t\to2}\frac{-16t^2+8t+48}{t-2}$
$=\lim_{t\to2}\frac{-8(2t^2-t-6)}{t-2}=\lim_{t\to2}\frac{-8(t-2)(2t+3)}{t-2}=\lim_{t\to2}-8(2t+3)=-56$

27. The graphs in (b) and (c)

29. If $b=0$, the result to be proved is $\lim_{x\to a}c=c$, which was proved in Example 1. If $b\neq0$, then for any $\varepsilon>0$ let $\delta=\varepsilon/|b|$. If $0<|x-a|<\delta$, then $|(bx+c)-(ba+c)|=|bx-ba|=|b|\,|x-a|<|b|\delta=\varepsilon$. Therefore $\lim_{x\to a}(bx+c)=ba+c$.

31. Let $\varepsilon=1$ and choose $\delta>0$ such that if $0<|x-5|<\delta$, then $|f(x)-4|<1$. For $0<|x-5|<\delta$ we have $|f(x)-2|=|2-(4-f(x))|\geq2-|4-f(x)|>2-1=1$. Thus any open interval containing 5 also contains a value of x for which $|f(x)-2|\geq1=\varepsilon$. Hence $\lim_{x\to5}f(x)\neq2$.

33. Let $L=\lim_{x\to a}g(x)$ and let $\varepsilon>0$. Let $\delta_1>0$ be such that if $0<|x-a|<\delta_1$, then $f(x)=g(x)$. Let $\delta_2>0$ be such that if $0<|x-a|<\delta_2$, then $|g(x)-L|<\varepsilon$. Choose $\delta>0$ to be less than both δ_1 and δ_2. If $0<|x-a|<\delta$, then $f(x)=g(x)$ and $|g(x)-L|<\varepsilon$, so $|f(x)-L|<\varepsilon$. Thus $\lim_{x\to a}f(x)=L=\lim_{x\to a}g(x)$.

35. $v(1) = \lim_{t\to 1} \frac{h(t) - h(1)}{t - 1} = \lim_{t\to 1} \frac{(-16t^2 + 144) - (-16 \cdot 1^2 + 144)}{t - 1} = \lim_{t\to 1} \frac{-16(t^2 - 1^2)}{t - 1}$

$= \lim_{t\to 1} -16(t + 1) = \lim_{t\to 1} (-16t - 16) = -16 - 16 = -32$ (feet per second)

37. By (10), $h(t) = -16t^2 - 16t + 144$.

a. $v(2) = \lim_{t\to 2} \frac{h(t) - h(2)}{t - 2} = \lim_{t\to 2} \frac{(-16t^2 - 16t + 144) - (-16 \cdot 2^2 - 16 \cdot 2 + 144)}{t - 2}$

$= \lim_{t\to 2} \frac{-16(t^2 - 2^2) - 16(t - 2)}{t - 2} = \lim_{t\to 2} \frac{-16(t - 2)(t + 2) - 16(t - 2)}{t - 2}$

$= \lim_{t\to 2} [-16(t + 2) - 16] = \lim_{t\to 2} (-16t - 48)$

$= -16 \cdot 2 - 48 = -80$ (feet per second).

b. The speed after 2 second equals $|v(2)| = |-80| = 80$ (feet per second).

39. By (10) with $v_0 = -16$ and $h_0 = 0$, $h(t) = -16t^2 - 16t$. Therefore

$v(3) = \lim_{t\to 3} \frac{h(t) - h(3)}{t - 3} = \lim_{t\to 3} \frac{(-16t^2 - 16t) - (-16 \cdot 3^2 - 16 \cdot 3)}{t - 3}$

$= \lim_{t\to 3} \frac{-16(t^2 - 3^2) - 16(t - 3)}{t - 3} = \lim_{t\to 3} \frac{-16(t - 3)(t + 3) - 16(t - 3)}{t - 3}$

$= \lim_{t\to 3} [-16(t + 3) - 16] = \lim_{t\to 3} (-16t - 64)$

$= -16 \cdot 3 - 64 = -112$ (feet per second).

SECTION 2.3

1. By the Constant Multiple Rule and (8), $\lim_{x\to 16} -\frac{1}{2}\sqrt{x} = -\frac{1}{2} \lim_{x\to 16} \sqrt{x} = -\frac{1}{2}\sqrt{16} = -2$.

3. By (4), $\lim_{x\to -1} (-5x^2 + 2x - \frac{1}{2}) = -5(-1)^2 + 2(-1) - \frac{1}{2} = -\frac{15}{2}$.

5. By the Sum and Constant Multiple Rules, along with (8) of this section and (7) of Section 2.2, $\lim_{x\to 4} (3x^2 - 5\sqrt{x} - 6|x|) = 3(4)^2 - 5\sqrt{4} - 6|4| = 14$.

7. By the Product Rule and (4),

$\lim_{x\to\sqrt{2}} (x^2 + 5)(\sqrt{2}x + 1) = \lim_{x\to\sqrt{2}} (x^2 + 5) \lim_{x\to\sqrt{2}} (\sqrt{2}x + 1) = [(\sqrt{2})^2 + 5][\sqrt{2} \cdot \sqrt{2} + 1]$

$= 21$

9. By the Product, Sum, Constant Multiple and Difference Rules and by (6),

$\lim_{y\to 0} (2y + \frac{1}{2})(3y^{2/3} - 9) = \lim_{y\to 0} (2y + \frac{1}{2}) \lim_{y\to 0} (3y^{2/3} - 9) = [2(0) + \frac{1}{2}][3(0)^{2/3} - 9] = -\frac{9}{2}$

11. By the Product and Sum Rules and (7),

$\lim_{y\to 64} (\sqrt[3]{y} + \sqrt{y})^2 = \left[\lim_{y\to 64} (\sqrt[3]{y} + \sqrt{y})\right]^2 = (\sqrt[3]{64} + \sqrt{64})^2 = 144$

13. By (5), $\lim_{x\to 2} -\frac{1}{6x} = -\frac{1}{6(2)} = -\frac{1}{12}$.

15. By (5), $\lim_{y\to -2} \frac{4y - 1}{5y + 4} = \frac{4(-2) - 1}{5(-2) + 4} = \frac{3}{2}$.

17. By the Quotient and Sum Rules and (6),

$$\lim_{t\to 4} \frac{t^{-3/2} + 1}{t^{1/2} - 4} = \frac{\lim_{t\to 4} (t^{-3/2} + 1)}{\lim_{t\to 4} (t^{1/2} - 4)} = \frac{4^{-3/2} + 1}{4^{1/2} - 4} = -\frac{9}{16}$$

19. $\lim_{x\to -1} \frac{x^2 - 1}{x + 1} = \lim_{x\to -1} \frac{(x - 1)(x + 1)}{x + 1} = \lim_{x\to -1} (x - 1) = -1 - 1 = -2$

21. $\lim_{x\to 1} \frac{x^3 - 1}{x - 1} = \lim_{x\to 1} \frac{(x - 1)(x^2 + x + 1)}{x - 1} = \lim_{x\to 1} (x^2 + x + 1) = 1 + 1 + 1 = 3$

23. $\lim_{x\to -2} \frac{x^2 - 16}{4 - x^2} = -\lim_{x\to -2} \frac{x^4 - 16}{x^2 - 4} = -\lim_{x\to -2} \frac{(x^2 - 4)(x^2 + 4)}{x^2 - 4} = -\lim_{x\to -2} (x^2 + 4)$

$= -((-2)^2 + 4) = -8$

25. $\lim_{x\to 9} \frac{x - 9}{\sqrt{x} - 3} = \lim_{x\to 9} \frac{(\sqrt{x} - 3)(\sqrt{x} + 3)}{\sqrt{x} - 3} = \lim_{x\to 9} (\sqrt{x} + 3) = \sqrt{9} + 3 = 6$

27. $\lim_{y\to 1/27} \frac{y^{2/3} - \frac{1}{9}}{y^{1/3} - \frac{1}{3}} = \lim_{y\to 1/27} \frac{(y^{1/3} - \frac{1}{3})(y^{1/3} + \frac{1}{3})}{y^{1/3} - \frac{1}{3}} = \lim_{y\to 1/27} (y^{1/3} + \frac{1}{3}) = (\frac{1}{27})^{1/3} + \frac{1}{3} = \frac{2}{3}$

29. $\lim_{y\to 1/2} \frac{6y - 3}{y(1 - 2y)} = -3 \lim_{y\to 1/2} \frac{1 - 2y}{y(1 - 2y)} = -3 \lim_{y\to 1/2} \frac{1}{y} = -3\left(\frac{1}{1/2}\right) = -6$

31. $\lim_{x\to 0} \left(1 - \frac{1}{x}\right) = \lim_{x\to 0} (x - 1) = 0 - 1 = -1$

33. $\lim_{x\to 0} \frac{1 + 1/x}{2 + 1/x} = \lim_{x\to 0} \frac{x(1 + 1/x)}{x(2 + 1/x)} = \lim_{x\to 0} \frac{x + 1}{2x + 1} = 1$

35. $\lim_{x\to -1} \frac{f(x) - f(-1)}{x - (-1)} = \lim_{x\to -1} \frac{(x^2 + 4x + 1) - (-2)}{x + 1} = \lim_{x\to -1} \frac{x^2 + 4x + 3}{x + 1}$

$= \lim_{x\to -1} \frac{(x + 1)(x + 3)}{x + 1} = \lim_{x\to -1} (x + 3) = 2$

$\ell: y - (-2) = 2(x - (-1))$, or $y + 2 = 2(x + 1)$

37. $\lim_{x\to 2} \frac{f(x) - f(2)}{x - 2} = \lim_{x\to 2} \frac{1/x - 1/2}{x - 2} = \lim_{x\to 2} \frac{(2 - x)/(2x)}{x - 2} = \lim_{x\to 2} -\frac{1}{2x} = -\frac{1}{4}$

$\ell: y - \frac{1}{2} = -\frac{1}{4}(x - 2)$

39. $\lim_{x\to -1} \frac{f(x) - f(-1)}{x - (-1)} = \lim_{x\to -1} \frac{1/(x + 3) - 1/2}{x + 1} = \lim_{x\to -1} \frac{\frac{-1 - x}{2(x + 3)}}{x + 1} = \lim_{x\to -1} \frac{-1}{2(x + 3)} = -\frac{1}{4}$

$\ell: y - \frac{1}{2} = -\frac{1}{4}(x + 1)$

41. $\lim_{x\to 16}\dfrac{f(x)-f(16)}{x-16}=\lim_{x\to 16}\dfrac{\sqrt{x}-4}{x-16}=\lim_{x\to 16}\dfrac{(\sqrt{x}-4)(\sqrt{x}+4)}{(x-16)(\sqrt{x}+4)}=\lim_{x\to 16}\dfrac{x-16}{(x-16)(\sqrt{x}+4)}$

$=\lim_{x\to 16}\dfrac{1}{\sqrt{x}+4}=\dfrac{1}{8}$; $\ell: y-4=\frac{1}{8}(x-16)$

43. a. $\lim_{x\to 0}\dfrac{f(x)-f(0)}{x-0}=\lim_{x\to 0}\dfrac{x^2-0}{x-0}=\lim_{x\to 0}x=0$; $\lim_{x\to 0}\dfrac{g(x)-g(0)}{x-0}=\lim_{x\to 0}\dfrac{x^3-0}{x-0}=\lim_{x\to 0}x^2=0$.

Since the tangent lines through (0, 0) have the same slope, they are the same.

b. $\lim_{x\to 0}\dfrac{f(x)-f(0)}{x-0}=\lim_{x\to 0}\dfrac{(x^2+1)-1}{x-0}=\lim_{x\to 0}x=0$

$\lim_{x\to 0}\dfrac{g(x)-g(0)}{x-0}=\lim_{x\to 0}\dfrac{(-x^2+1)-1}{x-0}=\lim_{x\to 0}(-x)=0$

Since the tangent lines through (0, 1) have the same slope, they are the same.

45. a. $\lim_{x\to a}\dfrac{f(x)-f(a)}{x-a}=\lim_{x\to a}\dfrac{(1/x)-(1/a)}{x-a}=\lim_{x\to a}\dfrac{a-x}{ax(x-a)}=\lim_{x\to a}\dfrac{-1}{ax}=\dfrac{-1}{a^2}$

Thus an equation of the tangent line is $y-1/a=(-1/a^2)(x-a)$, or $y=-(1/a^2)x+2/a$.

b. The x intercept of the tangent line is $2a$, and the y intercept is $2/a$. Thus the area A of the triangle is $\frac{1}{2}(2a)(2/a)=2$, so the area is independent of a.

47. Since $g(x)=(fg)(x)/f(x)$ (if $f(x)\neq 0$), and $\lim_{x\to\sqrt{2}}(fg)(x)$ and $\lim_{x\to\sqrt{2}}f(x)$ both exist, the Quotient Rule tells us that $\lim_{x\to\sqrt{2}}g(x)$ exists. Moreover,

$$\lim_{x\to\sqrt{2}}g(x)=\frac{\lim_{x\to\sqrt{2}}(fg)(x)}{\lim_{x\to\sqrt{2}}f(x)}=-\frac{\sqrt{2}}{3}$$

49. Let $f(x)=1/(x-1)$ and $g(x)=-1/(x-1)$. Then $\lim_{x\to 1}f(x)$ and $\lim_{x\to 1}g(x)$ do not exist. However $f(x)+g(x)=1/(x-1)+-1/(x-1)=0$ for $x\neq 1$. Thus $\lim_{x\to 1}(f(x)+g(x))=\lim_{x\to 1}0=0$.

51. Let $f(x)=\begin{cases}0 & \text{for } x<2\\ 1 & \text{for } x\geq 2\end{cases}$ and $g(x)=\begin{cases}1 & \text{for } x<2\\ 0 & \text{for } x\geq 2\end{cases}$

Then $\lim_{x\to 2}f(x)$ and $\lim_{x\to 2}g(x)$ do not exist. However, $f(x)g(x)=0$ for all x. Thus $\lim_{x\to 2}f(x)g(x)=\lim_{x\to 2}0=0$.

53. By (4) with x replaced by t and a replaced by t_0, $\lim_{t\to t_0}h(t)=\lim_{t\to t_0}(-16t^2+v_0t+h_0)=-16t_0^2+v_0t_0+h_0$.

SECTION 2.4

1. By (5), $\lim_{x\to\pi/3}(\sqrt{3}\sin x-2x)=\sqrt{3}\sin\dfrac{\pi}{3}-2\left(\dfrac{\pi}{3}\right)=\dfrac{3}{2}-\dfrac{2\pi}{3}$.

3. By (5), $\lim_{x\to-\pi/3}3x^2\cos x=3\left(-\dfrac{\pi}{3}\right)^2\cos\left(-\dfrac{\pi}{3}\right)=\dfrac{\pi^2}{6}$.

5. By (5), $\lim_{y\to 2\pi/3}\dfrac{\pi\sin y\cos y}{y}=\dfrac{\pi\sin 2\pi/3\cos 2\pi/3}{2\pi/3}=-\dfrac{3\sqrt{3}}{8}$.

7. Let $y=2x-16$. Then $\lim_{x\to 7}y=\lim_{x\to 7}(2x-16)=-2$. By the Substitution Rule, $\lim_{x\to 7}(2x-16)^4=\lim_{y\to -2}y^4=(-2)^4=16$.

9. Let $y=9-x^2$. Then $\lim_{x\to\sqrt{5}}(9-x^2)=4$. By the Substitution Rule, $\lim_{x\to\sqrt{5}}(9-x^2)^{-5/2}=\lim_{y\to 4}y^{-5/2}=4^{-5/2}=\frac{1}{32}$.

11. Let $y=2x$. Then $\lim_{x\to-\pi/6}y=\lim_{x\to-\pi/6}2x=-\pi/3$. By the Substitution Rule and (5), $\lim_{x\to-\pi/6}\cos 2x=\lim_{y\to-\pi/3}\cos y=\cos(-\pi/3)=\frac{1}{2}$.

13. Let $y=\sin t$. By (2), $\lim_{t\to 0}y=\lim_{t\to 0}\sin t=\sin 0=0$. By the Substitution Rule, $\lim_{t\to 0}\sin^{4/3}t=\lim_{y\to 0}y^{4/3}=0$.

15. By the Substitution Rule, (5), and the comment following Example 1, $\lim_{t\to\pi/6}\sin^3 t=\lim_{y\to 1/2}y^3=\frac{1}{8}$ and $\lim_{t\to\pi/6}\sec^4 t=\lim_{y\to 2/\sqrt{3}}y^4=\frac{16}{9}$. By the Product Rule,

$$\lim_{t\to\pi/6}\sin^3 t\sec^4 t=\tfrac{1}{8}\cdot\tfrac{16}{9}=\tfrac{2}{9}$$

17. Let $y=\csc w$. By the comment following Example 1, $\lim_{w\to\pi/2}y=\lim_{w\to\pi/2}\csc w=\csc\pi/2=1$. By the Substitution Rule, $\lim_{w\to\pi/2}\cos^2(\csc w)=\lim_{y\to 1}\cos^2 y$. Let $z=\cos y$. By (5), $\lim_{y\to 1}z=\lim_{y\to 1}\cos y=\cos 1$. By the Substitution Rule, $\lim_{y\to 1}\cos^2 y=\lim_{z\to\cos 1}z^2=(\cos 1)^2=\cos^2 1$.

19. By Example 2 and the Substitution Rule with $y=3x$,

$$\lim_{x\to 0}\frac{\sin 3x}{5x}=\lim_{y\to 0}\frac{\sin y}{\frac{5}{3}y}=\frac{3}{5}\lim_{y\to 0}\frac{\sin y}{y}=\frac{3}{5}$$

21. By Example 2 and the Substitution Rule with $y=x^{1/3}$,

$$\lim_{x\to 0}\frac{\sin x^{1/3}}{x^{1/3}}=\lim_{y\to 0}\frac{\sin y}{y}=1$$

23. Using Example 3 and (5), we have

$$\lim_{t\to 0}\frac{\cos^2 t-1}{t}=\lim_{t\to 0}\frac{(\cos t-1)(\cos t+1)}{t}=\lim_{t\to 0}\frac{\cos t-1}{t}\cdot\lim_{t\to 0}(\cos t+1)=0(1+1)=0$$

25. Using Example 2 and (5), we have

$$\lim_{y\to 0}\frac{\tan y}{y}=\lim_{y\to 0}\left(\frac{\sin y}{y}\cdot\frac{1}{\cos y}\right)=\lim_{y\to 0}\frac{\sin y}{y}\cdot\lim_{y\to 0}\frac{1}{\cos y}=1\cdot\frac{1}{\cos 0}=1$$

27. Using Example 2 and (5), we have

$$\lim_{y\to 0}\frac{\sin y-\tan y}{y}=\lim_{y\to 0}\frac{\sin y}{y}-\lim_{y\to 0}\left(\frac{\sin y}{y}\cdot\frac{1}{\cos y}\right)=1-\lim_{y\to 0}\frac{\sin y}{y}\cdot\lim_{y\to 0}\frac{1}{\cos y}$$

$$=1-1\cdot 1=0$$

29. Using Example 2 and the Substitution Rule with $y = 2x$, we have

$$\lim_{x\to 0} \frac{\sin^2 2x}{8x^2} = \lim_{y\to 0} \frac{\sin^2 y}{2y^2} = \frac{1}{2} \lim_{y\to 0}\left(\frac{\sin y}{y}\cdot\frac{\sin y}{y}\right) = \frac{1}{2}\left(\lim_{y\to 0}\frac{\sin y}{y}\right)^2 = \frac{1}{2}\cdot 1^2 = \frac{1}{2}$$

31. Using the comment following Example 2, we have

$$\lim_{x\to\pi} \frac{\tan^2 x}{1+\sec x} = \lim_{x\to\pi}\frac{\sec^2 x - 1}{1+\sec x} = \lim_{x\to\pi}\frac{(\sec x + 1)(\sec x - 1)}{1+\sec x}$$
$$= \lim_{x\to\pi}(\sec x - 1) = \sec \pi - 1 = -2$$

33. Using (5), we have

$$\lim_{x\to 0}\frac{\cos(x+\pi/2)}{x} = \lim_{x\to 0}\frac{\cos x \cos(\pi/2) - \sin x \sin(\pi/2)}{x} = \lim_{x\to 0}\frac{-\sin x}{x} = -1$$

35. Using (5), we have $\lim_{x\to a}\sec x = \lim_{x\to a}\dfrac{1}{\cos x} = \dfrac{1}{\lim_{x\to a}\cos x} = \dfrac{1}{\cos a} = \sec a.$

37. By Example 3, $\lim_{x\to 0}\dfrac{f(x)-f(0)}{x-0} = \lim_{x\to 0}\dfrac{\cos x - 1}{x} = 0$; ℓ: $y - 1 = 0(x-0)$, or $y = 1$.

39. Using the Substitution Rule with $y = 4 - x^2$, we have

$$\lim_{x\to 1}\frac{f(x)-f(1)}{x-1} = \lim_{x\to 1}\frac{\sqrt{4-x^2}-\sqrt{3}}{x-1} = \lim_{x\to 1}\frac{(\sqrt{4-x^2}-\sqrt{3})(\sqrt{4-x^2}+\sqrt{3})}{(x-1)(\sqrt{4-x^2}+\sqrt{3})}$$
$$= \lim_{x\to 1}\frac{4-x^2-3}{(x-1)(\sqrt{4-x^2}+\sqrt{3})} = \lim_{x\to 1}\frac{1-x^2}{(x-1)(\sqrt{4-x^2}+\sqrt{3})}$$
$$= \lim_{x\to 1}\frac{(1-x)(1+x)}{(x-1)(\sqrt{4-x^2}+\sqrt{3})} = \lim_{x\to 1} -\frac{1+x}{\sqrt{4-x^2}+\sqrt{3}}$$
$$= -\frac{\lim_{x\to 1}(1+x)}{\lim_{y\to 3}(\sqrt{y}+\sqrt{3})}$$
$$= \frac{-2}{2\sqrt{3}} = -\frac{\sqrt{3}}{3};\quad \ell: y - \sqrt{3} = -\frac{\sqrt{3}}{3}(x-1)$$

41. By hypothesis, $-M|x-a| \le f(x) \le M|x-a|$ for $x \ne a$. Also $\lim_{x\to a} M|x-a| = 0$. By the Squeezing Theorem, $\lim_{x\to a} f(x) = 0$.

43. a. Let $y = f(x)$. Then $\lim_{x\to a} y = \lim_{x\to a} f(x) = L$. By the Substitution Rule and Exercise 28 in Section 2.2, $\lim_{x\to a}|f(x)| = \lim_{y\to L}|y| = |L|$.

b. Suppose $\lim_{x\to a}|f(x)| = 0$, and let ε be any positive number. By Definition 2.1, there is a number $\delta > 0$ such that

$$\text{if } 0 < |x-a| < \delta, \text{ then } ||f(x)| - 0| < \varepsilon$$

Since $||f(x)| - 0| = |f(x)| = |f(x) - 0|$, it follows that

$$\text{if } 0 < |x-a| < \delta, \text{ then } |f(x) - 0| < \varepsilon$$

This implies that $\lim_{x\to a} f(x) = 0$.

c. Let $f(x) = -1$ for $x \le 0$ and $f(x) = 1$ for $x > 0$. Then $\lim_{x\to 0} f(x)$ does not exist. But $|f(x)| = 1$ for all x, so that $\lim_{x\to 0}|f(x)| = \lim_{x\to 0} 1 = 1$.

45. a. $\lim_{x\to 0} x = 0$ and $|\cos(1/x)| \le 1$ for $x \ne 0$, so by Exercise 44, $\lim_{x\to 0} x\cos(1/x) = 0$.

b. $\lim_{x\to 0} x = 0$ and $|\sin(1/x)| \le 1$ for $x \ne 0$, so by Exercise 44, $\lim_{z\to 0} x\sin(1/x) = 0$.

47. Let $y = x^n$. Since x^n approaches 0 as x approaches 0, it follows that y approaches 0, so by the Substitution Rule, $\lim_{x\to 0} f(x^n) = \lim_{y\to 0} f(y) = L$.

49. a. $m_r = \dfrac{\sqrt{1-a^2}-0}{a-0} = \dfrac{\sqrt{1-a^2}}{a}$ b. $m_t = \dfrac{-1}{m_r} = \dfrac{-a}{\sqrt{1-a^2}}$

c. $m_x = \dfrac{\sqrt{1-x^2}-\sqrt{1-a^2}}{x-a}$

d. $$\lim_{x\to a} m_x = \lim_{x\to a}\frac{\sqrt{1-x^2}-\sqrt{1-a^2}}{x-a}$$
$$= \lim_{x\to a}\frac{\sqrt{1-x^2}-\sqrt{1-a^2}}{x-a}\,\frac{\sqrt{1-x^2}+\sqrt{1-a^2}}{\sqrt{1-x^2}+\sqrt{1-a^2}}$$
$$= \lim_{x\to a}\frac{(1-x^2)-(1-a^2)}{(x-a)(\sqrt{1-x^2}+\sqrt{1-a^2})}$$
$$= \lim_{x\to a}\frac{-(x^2-a^2)}{(x-a)(\sqrt{1-x^2}+\sqrt{1-a^2})}$$
$$= \lim_{x\to a}\frac{-(x+a)}{\sqrt{1-x^2}+\sqrt{1-a^2}} = \frac{-a}{\sqrt{1-a^2}}$$

SECTION 2.5

1. $\lim_{x\to -2^+}(x^3+3x-5) = \lim_{x\to -2}(x^3+3x-5) = (-2)^3 + 3(-2) - 5 = -19$

3. $\lim_{x\to 2^-}\dfrac{x^2-4}{x-2} = \lim_{x\to 2^-}\dfrac{(x-2)(x+2)}{x-2} = \lim_{x\to 2^-}(x+2) = \lim_{x\to 2}(x+2) = 4$

5. $\lim_{x\to 1^+}\dfrac{x^2+3x-4}{x^2-1} = \lim_{x\to 1^+}\dfrac{(x-1)(x+4)}{(x-1)(x+1)} = \lim_{x\to 1^+}\dfrac{x+4}{x+1} = \lim_{x\to 1}\dfrac{x+4}{x+1} = \dfrac{5}{2}$

7. Since $t - 5 > 0$ for $t > 5$, $\lim_{t\to 5^+}(|t-5|)/(5-t) = \lim_{t\to 5^+}(t-5)/(5-t) = \lim_{t\to 5^+}(-1) = -1$.

9. $\lim_{t\to 2^-}(t - \sqrt{4-2t}) = \lim_{t\to 2^-} t - \lim_{t\to 2^-}\sqrt{4-2t} = 2 - \lim_{t\to 2^-}\sqrt{4-2t}$. Let $y = 4 - 2t$. As t approaches 2 from the left, $-2t$ approaches -4 from the right, so $4 - 2t$ and thus y approaches 0 from the right. By the version of the Substitution Rule for one-sided limits, $\lim_{t\to 2^-}\sqrt{4-2t} = \lim_{y\to 0^+}\sqrt{y} = 0$. Therefore $\lim_{t\to 2^-}(t - \sqrt{4-2t}) = 2 - 0 = 2$.

11. If $x < 0$, then $x^3 < 0$. Thus $4/x^3 < 0$, so $\lim_{x\to 0^-}(4/x^3) = -\infty$.

13. For all x except $x = 0$, we have $-1/x^2 < 0$. Thus $\lim_{x\to 0^+}(-1/x^2) = \lim_{x\to 0^-}(-1/x^2) = -\infty$. Therefore $\lim_{x\to 0}(-1/x^2) = -\infty$.

15. If $y < 2$, then $y - 2 < 0$, so $3/(y-2) < 0$. Therefore $\lim_{y\to 2^-} [3/(y-2)] = -\infty$.

17. If $y < -1$, then $y + 1 < 0$, so $\pi/(y+1) < 0$. Therefore $\lim_{y\to -1^-} [\pi/(y+1)] = -\infty$.

19. For $0 < z < \pi/2$, $\tan z > 0$. Since $\tan z = (\sin z)/(\cos z)$, $\lim_{z\to\pi/2} \cos z = 0$, and $\lim_{z\to\pi/2} \sin z = 1$, we have $\lim_{z\to\pi/2^-} \tan z = \infty$.

21. If $x > \frac{3}{2}$, then $3 - 2x < 0$, so $\sqrt{3-2x}$ is not defined. Thus $\lim_{x\to 3/2^+} \sqrt{3-2x}$ does not exist.

23. If $x > 5$ then $x - 5 > 0$, so $\sqrt{x-5}$ is defined. By the version of the Substitution Rule for one-sided limits, $\lim_{x\to 5^+} \sqrt{x-5} = \lim_{y\to 0^+} \sqrt{y} = 0$. Thus $\lim_{x\to 5^+} 1/(x\sqrt{x-5}) = \infty$.

25. By the versions of the Product and Substitution Rules for one-sided limits,

$$\lim_{x\to 0^+} \sqrt{x\cos 2x} = \lim_{x\to 0^+} \sqrt{x} \lim_{x\to 0^+} \sqrt{\cos 2x} = \lim_{x\to 0^+} \sqrt{x} \lim_{y\to 0^+} \sqrt{\cos y}$$
$$= \lim_{x\to 0^+} \sqrt{x} \lim_{z\to 1^-} \sqrt{z} = 0 \cdot 1 = 0$$

27. $\displaystyle\lim_{x\to 0} \sqrt{\frac{1}{x^2}} = \lim_{x\to 0} \frac{1}{\sqrt{x^2}} = \lim_{x\to 0} \frac{1}{|x|} = \infty.$

29. If $x < -\frac{1}{2}$, then $x + \frac{1}{2} < 0$ and $4x - 7 < 0$. Thus $(4x-7)/(x+\frac{1}{2}) > 0$, so

$$\lim_{x\to -1/2^-} \frac{4x-7}{x+\frac{1}{2}} = \lim_{x\to -1/2^-} (4x-7)\cdot\frac{1}{x+\frac{1}{2}} = \infty$$

31. If $0 < x < 2$, then $x + 1 > 0$ and $x^2 - 4 < 0$, so $(x+1)/[2(x^2-4)] < 0$, and thus

$$\lim_{x\to 2^-} \frac{x+1}{2(x^2-4)} = \lim_{x\to 2^-} \left[\frac{x+1}{2}\cdot\frac{1}{x^2-4}\right] = -\infty$$

33. If $4 < x < 6$ and $x \neq 5$, then $x^2 - 3 > 0$ and $(x-5)^2 > 0$, so $(x^2-3)/(x-5)^2 > 0$, and thus

$$\lim_{x\to 5} \frac{x^2-3}{(x-5)^2} = \lim_{x\to 5} \left[(x^2-3)\cdot\frac{1}{(x-5)^2}\right] = \infty$$

35. If $y < 3$, then $3 - y > 0$ and $\lim_{y\to 3^-} \sqrt{3-y} = 0$. Thus $\lim_{y\to 3^-} (-1/\sqrt{3-y}) = -\infty$.

37. If $-1 < y < 1$, then $1 + y > 0$ and $1 - y > 0$, so

$$\frac{\sqrt{1-y^2}}{y-1} = \frac{\sqrt{1+y}\sqrt{1-y}}{-(1-y)} = \frac{-\sqrt{1+y}}{\sqrt{1-y}} \le 0$$

Thus $$\lim_{y\to 1^-} \frac{\sqrt{1-y^2}}{y-1} = \lim_{y\to 1^-} \left[(-\sqrt{1+y})\cdot\frac{1}{\sqrt{1-y}}\right] = -\infty$$

39. For $y < 0$, $\sqrt{y}$ is undefined. Therefore $\lim_{y\to 0^-} [(5+\sqrt{1+y^2})/\sqrt{y}]$ does not exist.

41. If $-\pi/2 < t < 0$, then $\sin 2t < 0$ and $\lim_{t\to 0^-} \sin 2t = 0$. Thus

$$\lim_{t\to 0^-} \csc 2t = \lim_{t\to 0^-} \left(\frac{1}{\sin 2t}\right) = -\infty$$

43. $$\lim_{t\to 0} \frac{1-\cos t}{t^2} = \lim_{t\to 0} \left[\frac{1-\cos t}{t^2}\cdot\frac{1+\cos t}{1+\cos t}\right] = \lim_{t\to 0} \left[\frac{\sin^2 t}{t^2}\cdot\frac{1}{1+\cos t}\right]$$
$$= \lim_{t\to 0} \frac{\sin t}{t} \lim_{t\to 0} \frac{\sin t}{t} \lim_{t\to 0} \frac{1}{1+\cos t} = 1\cdot 1\cdot\tfrac{1}{2} = \tfrac{1}{2}$$

45. For $x \neq 3, -3$, we have

$$\frac{1}{x-3} - \frac{6}{x^2-9} = \frac{x+3}{(x-3)(x+3)} - \frac{6}{(x-3)(x+3)} = \frac{x-3}{(x-3)(x+3)} = \frac{1}{x+3}$$

Thus $$\lim_{x\to 3^-} \left(\frac{1}{x-3} - \frac{6}{x^2-9}\right) = \lim_{x\to 3^-} \frac{1}{x+3} = \frac{1}{6}$$

47. Since $\lim_{h\to 0^+} (1-\sqrt{h}) = 1$, we have $\lim_{h\to 0^+} (1/h - 1/\sqrt{h}) = \lim_{h\to 0^+} (1/h)(1-\sqrt{h}) = \infty$.

49. $\lim_{x\to -2^-} f(x) = \lim_{x\to -2^-} (-1+4x) = -9$; $\lim_{x\to -2^+} f(x) = \lim_{x\to -2^+} (-9) = -9$.
Thus $\lim_{x\to -2} f(x) = -9$.

51. $$\lim_{x\to 3^+} f(x) = \lim_{x\to 3^+} \left(3 + \frac{1}{x-3}\right) = \lim_{x\to 3^+} \frac{3x-8}{x-3} = \lim_{x\to 3^+} \left[(3x-8)\cdot\frac{1}{x-3}\right] = \infty.$$

53. $\lim_{x\to -4^+} [1/(x+4)] = \infty$, so $x = -4$ is a vertical asymptote.

55. $\lim_{x\to\sqrt{2}^+} (2x-9)/(x+\sqrt{2}) = -\infty$, so $x = -\sqrt{2}$ is a vertical asymptote.

57. $$\lim_{x\to 2^+} \frac{x^2-1}{x^2-4} = \infty = \lim_{x\to 2^-} \frac{x^2-1}{x^2-4}$$
so $x = 2$ and $x = -2$ are vertical asymptotes.

59. $\lim_{x\to 4^-} f(x) = \infty = \lim_{x\to -4^+} f(x)$, so $x = 4$ and $x = -4$ are vertical asymptotes. Since $x^2 + 1 > 0$ for all x, there are no other vertical asymptotes.

61. If $x \neq -2, 3$, then

$$\frac{x^2-4x-12}{x^2-x-6} = \frac{(x+2)(x-6)}{(x+2)(x-3)} = \frac{x-6}{x-3}$$

Thus $\lim_{x\to 3^-} f(x) = \lim_{x\to 3^-} (x-6)/(x-3) = \infty$, so $x = 3$ is a vertical asymptote.

63. If $x \neq -5, -2, 0$, then

$$\frac{x^2+2x-15}{x^3+7x^2+10x} = \frac{(x-3)(x+5)}{x(x+2)(x+5)} = \frac{x-3}{x(x+2)}$$

Thus $\lim_{x\to 0^-} f(x) = \infty = \lim_{x\to -2^+} f(x)$, so $x = -2$ and $x = 0$ are vertical asymptotes.

65. $\displaystyle\frac{x+1/x}{x^4+1} = \frac{x^2+1}{x(x^4+1)}$, so $\displaystyle\lim_{x\to 0^+} f(x) = \infty$, and thus $x = 0$ is a vertical asymptote.

67. $\lim_{x\to 0} (\sin x)/x = 1$, so there are no vertical asymptotes.

69. $\lim_{x\to(\pi/2+n\pi)^-} \tan x = \infty$ for any integer n, so $x = \pi/2 + n\pi$ is a vertical asymptote for any integer n.

71. $\lim_{x\to 0}\frac{f(x)-f(0)}{x-0}=\lim_{x\to 0}\frac{x^{1/5}-0}{x-0}=\lim_{x\to 0}\frac{1}{x^{4/5}}=\infty$

Thus there is a vertical tangent line ℓ at (0, 0); ℓ: $x = 0$.

73. $\lim_{x\to 0}\frac{f(x)-f(0)}{x-0}=\lim_{x\to 0}\frac{(1-5x^{3/5})-1}{x-0}=\lim_{x\to 0}\frac{-5}{x^{2/5}}=-\infty$

Thus there is a vertical tangent line ℓ at (0, 1); ℓ: $x = 0$.

75. Two-sided limit: (a); right-hand limit: (a), (e), (f); left-hand limit: (a), (b), (f); none: (c), (d)

77. For example, if $f(x) = 1/(x-a)^2$ and $g(x) = f(x) - 2$, then $\lim_{x\to a} f(x) = \infty = \lim_{x\to a} g(x)$. But $f(x) - g(x) = 2$, so $\lim_{x\to a}[f(x) - g(x)] = 2$.

79. Since $a > 0$ and $b > 0$, we have

$$\lim_{x\to 1}\frac{x}{a+bx}=\frac{1}{a+b\cdot 1}=\frac{1}{a+b}$$

Therefore
$$\lim_{x\to 1^-}\frac{x}{a+bx}=\lim_{x\to 1}\frac{x}{a+bx}=\frac{1}{a+b}$$

81. By (10) in Section 2.2, until the rock hits the bottom of the well, its position is given by $f(t) = -16t^2 - 48t$. When it hits the bottom, we have $-64 = f(t) = -16t^2 - 48t$, so that $-16(t^2 + 3t - 4) = 0$, or $(t + 4)(t - 1) = 0$. Thus the rock hits the bottom after 1 second. Since

$$\lim_{t\to 1^-}\frac{f(t)-f(1)}{t-1}=\lim_{t\to 1^-}\frac{-16t^2-48t-(-64)}{t-1}=\lim_{t\to 1^-}\frac{-16(t^2+3t-4)}{t-1}$$
$$=\lim_{t\to 1^-}\frac{-16(t-1)(t+4)}{t-1}=\lim_{t\to 1^-}-16(t+4)=-80$$

the velocity of the rock will be -80 feet per second when it hits the bottom.

SECTION 2.6

1. Continuous at 2

3. Continuous at 0

5. Not continuous from either the left or the right at -1

7. Not continuous from either the left or the right at 0

9. Continuous at $\frac{1}{2}$; continuous from the left at 1

11. Continuous at 0; continuous from the left at 1

13. Continuous from the right at $\pi/2$

15. Continuous at $\pi/2$

17. f is a polynomial, so f is continuous on $(-\infty, \infty)$.

19. Since $x + 3$ is continuous on $(-\infty, \infty)$, the square root function is continuous on $[0, \infty)$, and $x + 3 \geq 0$ for x in $[-3, \infty)$, Theorem 2.10 implies that f is continuous on $[-3, \infty)$.

21. Since $\sin x$ and x are both continuous on $(-\infty, \infty)$, and since $x \neq 0$ on $(-\infty, 0)$ and $(0, \infty)$, Theorem 2.9 implies that f is continuous on $(-\infty, 0)$ and $(0, \infty)$. Since $\lim_{x\to 0} f(x) = \lim_{x\to 0} (\sin x)/x = 1 = f(0)$, f is also continuous at 0. Therefore f is continuous on $(-\infty, \infty)$.

23. Since $x^2 - x^4$ is continuous on $(-\infty, \infty)$, the square root function is continuous on $[0, \infty)$, and $x^2 - x^4 > 0$ for x in (0, 1), Theorem 2.10 implies then that $\sqrt{x^2 - x^4}$ is continuous on (0, 1). Finally, the constant function 1 is continuous on $(-\infty, \infty)$. Therefore Theorem 2.9 implies that f is continuous on (0, 1).

25. Observe that $1 - x \geq 0$ and $3x - 2 > 0$ for $\frac{2}{3} < x \leq 1$, so that $(1 - x)/(3x - 2) \geq 0$ for $\frac{2}{3} < x \leq 1$. Since $(1 - x)/(3x - 2)$, being a rational function, is continuous on $(\frac{2}{3}, 1]$, and since the square root function is continuous on $[0, \infty)$, it follows from Theorem 2.10 that f is continuous on $(\frac{2}{3}, 1]$.

27. This follows from the final remark of the second paragraph following Definition 2.8. Alternatively, one could apply Theorem 2.9 to the quotient of the continuous functions $\sin x$ and $\cos x$, noting that $\cos x \neq 0$ for x in $(-\pi/2, \pi/2)$.

29. Since $1/x$ is continuous on $(0, \infty)$ and $\cot x = (\cos x)/(\sin x)$ is continuous on $(0, \pi)$ and $0 \leq 1/x \leq \pi$ for x in $(1/\pi, \infty)$, Theorem 2.10 implies that f is continuous on $(1/\pi, \infty)$.

31. a. Since $\lim_{x\to 3} f(x) = \lim_{x\to 3} (x^2 - 9)/(x - 3) = \lim_{x\to 3} (x + 3) = 6$, let $f(3) = 6$. Then f is continuous at 3.

    b. Since $\lim_{x\to -2} f(x) = \lim_{x\to -2} (x^2 + 5x + 6)/(x + 2) = \lim_{x\to -2} (x + 3) = 1$, let $f(-2) = 1$. Then f is continuous at -2.

    c. Since $\lim_{x\to 1^+} f(x) = \infty$, we cannot define $f(1)$ so f will be continuous at 1.

    d. Since $\lim_{x\to 0^-} f(x) = \lim_{x\to 0^-} (-1) = -1$ and $\lim_{x\to 0^+} f(x) = \lim_{x\to 0^+} 1 = 1$, we cannot define $f(0)$ so f will be continuous at 0.

    e. Since $\lim_{x\to 0} f(x) = 0$, let $f(0) = 0$. Then f is continuous at 0.

    f. Since $\lim_{x\to \pi/2^-} f(x) = \infty$, we cannot define $f(\pi/2)$ so f will be continuous at $\pi/2$.

33. b and f

35. Let $f(x) = x + 2$. Since f is continuous at -3 and g is continuous at $f(-3) = -1$, Theorem 2.10 implies that the function $h = g \circ f$ is continuous at -3.

37. Since f and g are continuous at a, we have $\lim_{x\to a} f(x) = f(a)$ and $\lim_{x\to a} g(x) = g(a)$. From the Difference Rule, $\lim_{x\to a}[f(x) - g(x)] = \lim_{x\to a} f(x) - \lim_{x\to a} g(x) = f(a) - g(a)$. Thus $f - g$ is continuous at a.

39. Let a be any real number and let $0 < \varepsilon < 1$. If δ is any positive number, then in the interval $(a - \delta, a + \delta)$ there is an irrational number x and a rational number y. If a is rational, then $|f(x) - f(a)| = |1 - 0| > \varepsilon$, whereas if a is irrational, then $|f(y) - f(a)| = |0 - 1| > \varepsilon$. Thus f is not continuous at a.

41. Let $\varepsilon > 0$, and let $\delta = \varepsilon/M$. If $|x - a| < \delta$, then $|f(x) - f(a)| \le M|x - a| < M\delta = M(\varepsilon/M) = \varepsilon$. Thus $\lim_{x \to a} f(x) = f(a)$, so that f is continuous at a.

43. Since the function $273.15 + t$ is a polynomial, it is continuous on $[-273.15, \infty)$. Thus S is continuous on $[-273.15, \infty)$ by Theorem 2.9.

SECTION 2.7

1. Let $f(x) = x^4 - x - 1$. Then f is continuous on $[-1, 1]$. Since $f(-1) = 1$ and $f(1) = -1$, the Intermediate Value Theorem implies that there is a c in $[-1, 1]$ such that $f(c) = 0$, that is, $c^4 - c - 1 = 0$.

3. Let $f(x) = x^2 + 1/x$. Then f is continuous on $[-2, -\frac{1}{2}]$. Since $f(-2) = \frac{7}{2}$ and $f(-\frac{1}{2}) = -\frac{7}{4}$, the Intermediate Value Theorem implies that there is a c in $[-2, -\frac{1}{2}]$ such that $f(c) = 1$, that is, $c^2 + (1/c) = 1$.

5. Let $f(x) = x^3 + x^2 + x - 2$. Then f is continuous on $[-1, 1]$. Since $f(-1) = -3$ and $f(1) = 1$, the Intermediate Value Theorem implies that there is a c in $[-1, 1]$ such that $f(c) = 0$, that is, $c^3 + c^2 + c - 2 = 0$.

7. Let $f(x) = \cos x - x$. Then f is continuous on $[0, \pi/2]$. Since $f(0) = 1$ and $f(\pi/2) = -\pi/2$, the Intermediate Value Theorem implies that there is a c in $[0, \pi/2]$ such that $f(c) = 0$, that is, such that $\cos c = c$.

9. Let $f(x) = (2 - x)^2(40 - 8x)$. Then $f(x) = 8(2 - x)^2(5 - x) = 0$ if $x = 2$ or 5.

Interval	c	$f(c)$	Sign of $f(x)$ on interval
$(-\infty, 2)$	0	160	+
$(2, 5)$	3	16	+
$(5, \infty)$	6	-128	$-$

Therefore $(2 - x)^2(40 - 8x) > 0$ on the union of $(-\infty, 2)$ and $(2, 5)$.

11. Let $f(x) = (x^4 + x)(x + 3)$. Then $f(x) = x(x^3 + 1)(x + 3) = 0$ if $x = -3, -1$, or 0.

Interval	c	$f(c)$	Sign of $f(x)$ on interval
$(-\infty, -3)$	-4	-252	$-$
$(-3, -1)$	-2	14	+
$(-1, 0)$	$-\frac{1}{2}$	$-\frac{35}{32}$	$-$
$(0, \infty)$	1	8	+

Therefore $(x^4 + x)(x + 3) \le 0$ on the union of $(-\infty, -3]$ and $[-1, 0]$.

13. Let $f(x) = (x - 1)(x - 3)/[(2x + 1)(2x - 1)]$. Then $(2x + 1)(2x - 1) = 0$ if $x = -\frac{1}{2}$ or $\frac{1}{2}$, so the domain of f is the union of $(-\infty, -\frac{1}{2})$, $(-\frac{1}{2}, \frac{1}{2})$ and $(\frac{1}{2}, \infty)$. Next, $f(x) = 0$ if $x = 1$ or 3.

Interval	c	$f(c)$	Sign of $f(x)$ on interval
$(-\infty, -\frac{1}{2})$	-1	$\frac{8}{3}$	+
$(-\frac{1}{2}, \frac{1}{2})$	0	-3	$-$
$(\frac{1}{2}, 1)$	$\frac{3}{4}$	$\frac{9}{20}$	+
$(1, 3)$	2	$-\frac{1}{15}$	$-$
$(3, \infty)$	4	$\frac{1}{21}$	+

Therefore $(x - 1)(x - 3)/[(2x + 1)(2x - 1)] \ge 0$ on the union of $(-\infty, -\frac{1}{2})$, $(\frac{1}{2}, 1]$ and $[3, \infty)$.

15. Let $f(x) = -2(x - 1)(x^2 + 2x + 4)/(27 - x^3)$. Then $27 - x^3 = 0$ if $x = 3$, so the domain of f is the union of $(-\infty, 3)$ and $(3, \infty)$. Next, since $x^2 + 2x + 4 = (x + 1)^2 + 3 > 0$ for all x, $f(x) = 0$ if $x = 1$.

Interval	c	$f(c)$	Sign of $f(x)$ on interval
$(-\infty, 1)$	0	$\frac{8}{27}$	+
$(1, 3)$	2	$-\frac{24}{19}$	$-$
$(3, \infty)$	4	$\frac{168}{37}$	+

Therefore $\dfrac{-2(x - 1)(x^2 + 2x + 4)}{27 - x^3} < 0$ on $(1, 3)$.

17. Since $f(1) < 0$ and $f(2) > 0$, we let $a = 1$ and $b = 2$, and assemble the following table:

Interval	Length	Midpoint c	$f(c)$
$[1, 2]$	1	$\frac{3}{2}$	$\frac{1}{4}$
$[1, \frac{3}{2}]$	$\frac{1}{2}$	$\frac{5}{4}$	$-\frac{7}{16}$
$[\frac{5}{4}, \frac{3}{2}]$	$\frac{1}{4}$	$\frac{11}{8}$	$-\frac{7}{64}$
$[\frac{11}{8}, \frac{3}{2}]$	$\frac{1}{8}$		

Since the length of $[\frac{11}{8}, \frac{3}{2}]$ is $\frac{1}{8}$, and neither $\frac{11}{8}$ nor $\frac{3}{2}$ is a zero of f, the midpoint $\frac{23}{16}$ of $[\frac{11}{8}, \frac{3}{2}]$ is less than $\frac{1}{16}$ from a zero of f.

19. Since $f(1) < 0$ and $f(2) > 0$, we let $a = 1$ and $b = 2$, and assemble the following table:

Interval	Length	Midpoint c	$f(c)$
$[1, 2]$	1	$\frac{3}{2}$	$\frac{3}{8}$
$[1, \frac{3}{2}]$	$\frac{1}{2}$	$\frac{5}{4}$	$-\frac{67}{64}$
$[\frac{5}{4}, \frac{3}{2}]$	$\frac{1}{4}$	$\frac{11}{8}$	$-\frac{205}{512}$
$[\frac{11}{8}, \frac{3}{2}]$	$\frac{1}{8}$		

Since the length of $[\frac{11}{8}, \frac{3}{2}]$ is $\frac{1}{8}$, and neither $\frac{11}{8}$ nor $\frac{3}{2}$ is a zero of f, the midpoint $\frac{23}{16}$ of $[\frac{11}{8}, \frac{3}{2}]$ is less than $\frac{1}{16}$ from a zero of f.

21. Since $f(0) = 1$ and $f(1) = (\cos 1) - 1 < 0$, we let $a = 0$ and $b = 1$, and assemble the following table:

Interval	Length	Midpoint c	$f(c)$
$[0, 1]$	1	$\frac{1}{2}$	$(\cos \frac{1}{2}) - \frac{1}{2} \approx .378$
$[\frac{1}{2}, 1]$	$\frac{1}{2}$	$\frac{3}{4}$	$(\cos \frac{3}{4}) - \frac{3}{4} \approx -.018$
$[\frac{1}{2}, \frac{3}{4}]$	$\frac{1}{4}$	$\frac{5}{8}$	$(\cos \frac{5}{8}) - \frac{5}{8} \approx .186$
$[\frac{5}{8}, \frac{3}{4}]$	$\frac{1}{8}$		

Since the length of $[\frac{5}{8}, \frac{3}{4}]$ is $\frac{1}{8}$, and neither $\frac{5}{8}$ nor $\frac{3}{4}$ is a zero of f, the midpoint $\frac{11}{16}$ of $[\frac{5}{8}, \frac{3}{4}]$ is less than $\frac{1}{16}$ from a zero of f.

23. Let $f(x) = x^2 - 5$, so that $\sqrt{5}$ is the zero of f. Since $f(2) < 0$ and $f(3) > 0$, the zero lies in $(2, 3)$. We let $a = 2$ and $b = 3$, and assemble the following table:

Interval	Length	Midpoint c	$f(c)$
$[2, 3]$	1	$\frac{5}{2}$	$\frac{5}{4}$
$[2, \frac{5}{2}]$	$\frac{1}{2}$	$\frac{9}{4}$	$\frac{1}{16}$
$[2, \frac{9}{4}]$	$\frac{1}{4}$	$\frac{17}{8}$	$-\frac{31}{64}$
$[\frac{17}{8}, \frac{9}{4}]$	$\frac{1}{8}$		

Since the length of $[\frac{17}{8}, \frac{9}{4}]$ is $\frac{1}{8}$, and neither $\frac{17}{8}$ nor $\frac{9}{4}$ is a zero of f, the midpoint $\frac{35}{16}$ of $[\frac{17}{8}, \frac{9}{4}]$ is less than $\frac{1}{16}$ from $\sqrt{5}$.

25. Let $f(x) = x^2 - 0.7$, so that $\sqrt{0.7}$ is the zero of f. Since $f(0) < 0$ and $f(1) > 0$, the zero lies in $(0, 1)$. We let $a = 0$ and $b = 1$, and assemble the following table:

Interval	Length	Midpoint c	$f(c)$
$[0, 1]$	1	$\frac{1}{2}$	$-\frac{9}{20}$
$[\frac{1}{2}, 1]$	$\frac{1}{2}$	$\frac{3}{4}$	$-\frac{11}{80}$
$[\frac{3}{4}, 1]$	$\frac{1}{4}$	$\frac{7}{8}$	$\frac{21}{320}$
$[\frac{3}{4}, \frac{7}{8}]$	$\frac{1}{8}$		

Since the length of $[\frac{3}{4}, \frac{7}{8}]$ is $\frac{1}{8}$, and neither $\frac{3}{4}$ nor $\frac{7}{8}$ is a zero of f, the midpoint $\frac{13}{16}$ of $[\frac{3}{4}, \frac{7}{8}]$ is less than $\frac{1}{16}$ from $\sqrt{0.7}$.

27. Let p be any number, and let $f(x) = x^3$. If $p = 0$ and $c = 0$, then $c^3 = p$. If $p > 0$, then $f(0) = 0$, whereas $f(p + 1) = (p + 1)^3 \geq p + 1 > p > 0$. The Intermediate Value Theorem implies that there exists a number c in $[0, p + 1]$ such that $c^3 = f(c) = p$. If $p < 0$, then $f(0) = 0$ and $f(p - 1) = (p - 1)^3 < p - 1 < p < 0$. The Intermediate Value Theorem implies that there is a number c in $[p - 1, 0]$ such that $c^3 = f(c) = p$. In any case, c is the desired cube root of p.

29. Let $f(x) = \pi x^2$. Notice that f is continuous, with $f(0) = 0$ and $f(10) = 100\pi > 300$. By the Intermediate Value Theorem there is a c in $[0, 10]$ such that $f(c) = 200$.

CHAPTER 2 REVIEW

1. By (5) of Section 2.3, $\lim_{x \to 2} \dfrac{-4x + 3}{x^2 - 1} = \dfrac{-4(2) + 3}{2^2 - 1} = -\dfrac{5}{3}$.

3. By the Product Rule, $\lim_{x \to 1} x\sqrt{x + 3} = (\lim_{x \to 1} x)(\lim_{x \to 1} \sqrt{x + 3}) = \lim_{x \to 1} \sqrt{x + 3}$. Let $y = x + 3$, so that y approaches 4 as x approaches 1. By the Substitution Rule, $\lim_{x \to 1} \sqrt{x + 3} = \lim_{y \to 4} \sqrt{y} = \sqrt{4} = 2$. Thus $\lim_{x \to 1} x\sqrt{x + 3} = \lim_{x \to 1} \sqrt{x + 3} = 2$.

5. Let $y = x + 3\pi/4$. Since y approaches π as x approaches $\pi/4$, the Substitution Rule implies that $\lim_{x \to \pi/4} \sec^3 (x + 3\pi/4) = \lim_{y \to \pi} \sec^3 y = \sec^3 \pi = (-1)^3 = -1$.

7. Let $y = 6v$. Since y approaches 0 as v approaches 0, it follows from the Substitution Rule, the Constant Multiple Rule, and Example 2 of Section 2.4 that

$$\lim_{v \to 0} \frac{5 \sin 6v}{4v} = \lim_{y \to 0} \frac{5 \sin y}{4(y/6)} = \frac{15}{2} \lim_{y \to 0} \frac{\sin y}{y} = \frac{15}{2}$$

9. If $x \neq -10$, then $(x + 10)^2 > 0$ and $\lim_{x \to -10} (x + 10)^2 = 0$. Thus

$$\lim_{x \to -10} \frac{4x}{(x + 10)^2} = -\infty$$

11. If $x < 0$, then $-x > 0$, so $\sqrt{-x} > 0$ and thus $9 + \sqrt{-x} > 9$, so $\sqrt{9 + \sqrt{-x}} > 0$. Using the Substitution Rule twice, we find that

$$\lim_{x \to 0^-} \sqrt{9 + \sqrt{-x}} = \lim_{y \to 0^+} \sqrt{9 + y} = \lim_{z \to 9^+} \sqrt{z} = \sqrt{9} = 3$$

13. First notice that

$$\frac{\sqrt{2 + 3w} - \sqrt{2 - 3w}}{w} = \frac{\sqrt{2 + 3w} - \sqrt{2 - 3w}}{w} \frac{\sqrt{2 + 3w} + \sqrt{2 - 3w}}{\sqrt{2 + 3w} + \sqrt{2 - 3w}}$$

$$= \frac{(2 + 3w) - (2 - 3w)}{w(\sqrt{2 + 3w} + \sqrt{2 - 3w})} = \frac{6}{\sqrt{2 + 3w} + \sqrt{2 - 3w}}$$

By the Substitution Rule, $\lim_{w \to 0} \sqrt{2 + 3w} = \lim_{y \to 2} \sqrt{y} = \sqrt{2}$ and $\lim_{w \to 0} \sqrt{2 - 3w} = \lim_{y \to 2} \sqrt{y} = \sqrt{2}$, so it follows that

$$\lim_{w \to 0} \frac{\sqrt{2 + 3w} - \sqrt{2 - 3w}}{w} = \lim_{w \to 0} \frac{6}{\sqrt{2 + 3w} + \sqrt{2 - 3w}} = \frac{6}{\sqrt{2} + \sqrt{2}} = \frac{3}{2}\sqrt{2}$$

15. Let $f(x) = \cos(1/w)$. If $w = 1/2n\pi$ for any positive integer n, then $0 < w < 1/n$ and $f(w) = 1$. If $w = 1/(2n\pi + \pi)$ for any positive integer n, then $0 < w < 1/n$ and $f(w) = -1$. Thus $\lim_{w \to 0^+} f(w)$ does not exist.

17. Since $\lim_{x\to 4^+} (3-2|x|)/(x-4) = -\infty$, $x = 4$ is a vertical asymptote of f.

19. We have
$$\frac{x^2+3x-4}{x^2-5x-14} = \frac{(x+4)(x-1)}{(x-7)(x+2)}$$

Since $\lim_{x\to 7^+} f(x) = \infty = \lim_{x\to -2^+} f(x)$, $x = -2$ and $x = 7$ are vertical asymptotes of f.

21. Since
$$\lim_{x\to -2^+} f(x) = \lim_{x\to -2^+} \frac{(x-2)^2}{(x-2)(x+2)} = \lim_{x\to -2^+} \frac{x-2}{x+2} = -\infty$$

$x = -2$ is a vertical asymptote of f. (Since $\lim_{x\to 2} f(x) = \lim_{x\to 2} (x-2)/(x+2) = 0$, $x = 2$ is *not* a vertical asymptote of f.)

23. $f(x)$ is defined for all x in $(-\infty, -\sqrt{13}]$ or in $[\sqrt{13}, \infty)$. Thus f is continuous from the left at $-\sqrt{13}$.

25. Since $\lim_{x\to 2^+} f(x) = \lim_{x\to 2^+} (3x+4) = 10 = f(2)$ but $\lim_{x\to 2^-} f(x) = \lim_{x\to 2^-} (3x-4) = 2 \neq f(2)$, f is continuous from the right at 2.

27. Since f is a rational function whose denominator is not zero for x in the interval $(2, \infty)$, f is continuous on $(2, \infty)$.

29. If x is in $[\sqrt{2}/3, 4)$, then $3x - \sqrt{2} \geq 0$ and $4 - x > 0$. Thus for such x, $(3x - \sqrt{2})/(4-x)$ is nonnegative, so on $[\sqrt{2}/3, 4)$, the rational function $(3x-\sqrt{2})/(4-x)$ is continuous. Since the square root function is continuous for x in $[0, \infty)$, Theorem 2.15 implies that f is continuous on $[\sqrt{2}/3, 4)$.

31. Since $\lim_{x\to 2} f(x) = \lim_{x\to 2} [(x+8)(x-2)]/(x-2) = \lim_{x\to 2} (x+8) = 10$, we can define $f(2) = 10$ to make f continuous at 2.

33. a. Number b. Does not exist c. Does not exist d. Does not exist
e. Number f. ∞

35. Let $f(x) = (x+4)/[(x^2-4)(1+x^2)]$. Then $(x^2-4)(1+x^2) = 0$ if $x = -2$ or 2, so the domain of f is the union of $(\infty, -2)$, $(-2, 2)$ and $(2, \infty)$. Next, $x + 4 = 0$ if $x = -4$.

Interval	c	$f(c)$	Sign of $f(x)$ on interval
$(-\infty, -4)$	-5	$-\frac{1}{546}$	$-$
$(-4, -2)$	-3	$\frac{1}{50}$	$+$
$(-2, 2)$	0	-1	$-$
$(2, \infty)$	3	$\frac{7}{50}$	$+$

Therefore $\dfrac{x+4}{(x^2-4)(1+x^2)} \geq 0$ on the union of $[-4, -2)$ and $(2, \infty)$.

37. Let $f(x) = \sqrt{x+1} - x^2$. Since f is continuous on $[1, 2]$, $f(1) > 0$ and $f(2) < 0$, the Intermediate Value Theorem implies that there is a c in $[1, 2]$ such that $f(c) = 0$, that is, a solution of the given equation. For the bisection method we let $a = 1$ and $b = 2$, and assemble the following table:

Interval	Length	Midpoint c	$f(c)$
$[1, 2]$	1	$\frac{3}{2}$	$\sqrt{\frac{3}{2}+1} - \frac{9}{4} \approx -.0253$
$[1, \frac{3}{2}]$	$\frac{1}{2}$	$\frac{5}{4}$	$\sqrt{\frac{5}{4}+1} - \frac{25}{16} \approx .556$
$[\frac{5}{4}, \frac{3}{2}]$	$\frac{1}{4}$	$\frac{11}{8}$	$\sqrt{\frac{11}{8}+1} - \frac{121}{64} \approx .282$
$[\frac{11}{8}, \frac{3}{2}]$	$\frac{1}{8}$		

Since the length of $[\frac{11}{8}, \frac{3}{2}]$ is $\frac{1}{8}$, and neither $\frac{11}{8}$ nor $\frac{3}{2}$ is a zero of f, the midpoint $\frac{23}{16}$ of $[\frac{11}{8}, \frac{3}{2}]$ is less than $\frac{1}{16}$ from a zero of f, and hence from a solution of the equation $\sqrt{x+1} = x^2$.

39. Let $f(x) = \sin^2 x - \cos x$. Since f is continuous on $[0, \pi/2]$, $f(0) < 0$ and $f(\pi/2) > 0$, the Intermediate Value Theorem implies that there is a c in $[0, \pi/2]$ such that $f(c) = 0$, that is, a solution of the given equation. For the bisection method we let $a = 0$ and $b = \pi/2$, and assemble the following table:

Interval	Length	Midpoint c	$f(c)$
$[0, \pi/2]$	$\pi/2$	$\pi/4$	$\sin^2 \pi/4 - \cos \pi/4 \approx -.207$
$[\pi/4, \pi/2]$	$\pi/4$	$3\pi/8$	$\sin^2 3\pi/8 - \cos 3\pi/8 \approx .471$
$[\pi/4, 3\pi/8]$	$\pi/8$	$5\pi/16$	$\sin^2 5\pi/16 - \cos 5\pi/16 \approx .136$
$[\pi/4, 5\pi/16]$	$\pi/16$	$9\pi/32$	$\sin^2 9\pi/32 - \cos 9\pi/32 \approx -.0368$
$[9\pi/32, 5\pi/16]$	$\pi/32$		

Since the length of $[9\pi/32, 5\pi/16]$ is $\pi/32$, and neither $9\pi/32$ nor $5\pi/16$ is a zero of f, the midpoint $19\pi/64$ of $[9\pi/32, 5\pi/16]$ is less than $\frac{1}{16}$ from a zero of f, and hence from a solution of the equation $\sin^2 x = \cos x$.

41. It does not include the phrase "provided the limits of the two functions both exist."

43. a. Since the absolute value function is continuous, the result follows from Theorem 2.10 since $|f|$ is a composite of f and the absolute value function.

b. $|f(x)| = 1$ for all x, so $|f|$ is continuous at 2. Since $\lim_{x\to 2^+} f(x) = 1 \neq f(2)$, f is not continuous at 2.

45. a. $m_1 = \lim_{x\to 0} \dfrac{f(x)-f(0)}{x-0} = \lim_{x\to 0} \dfrac{2\sin x - 0}{x-0} = 2$

$m_2 = \lim_{x\to 1} \dfrac{g(x)-g(0)}{x-1} = \lim_{x\to 1} \dfrac{(-1/x^2)-(-1)}{x-1} = \lim_{x\to 1} \dfrac{x^2-1}{x^2(x-1)} = \lim_{x\to 1} \dfrac{x+1}{x^2} = 2$

Since $m_1 = m_2$, the tangent lines are parallel.

b. $m_1 = \lim_{x \to -1} \frac{f(x) - f(-1)}{x - (-1)} = \lim_{x \to -1} \frac{x^3 - (-1)}{x + 1} = \lim_{x \to -1} (x^2 - x + 1) = 3$

$m_2 = \lim_{x \to 4} \frac{g(x) - g(4)}{x - 4} = \lim_{x \to 4} \frac{(3/8)x^2 - 6}{x - 4} = \lim_{x \to 4} (3/8)(x + 4) = 3$

Since $m_1 = m_2$, the tangent lines are parallel.

c. $m_1 = \lim_{x \to -4} \frac{f(x) - f(-4)}{x - (-4)} = \lim_{x \to -4} \frac{(x^2 + 5x + 1) - (-3)}{x + 4} = \lim_{x \to -4} \frac{x^2 + 5x + 4}{x + 4}$

$= \lim_{x \to -4} (x + 1) = -3$

$m_2 = \lim_{x \to 3} \frac{g(x) - g(3)}{x - 3} = \lim_{x \to 3} \frac{(-3x - 7) - (-16)}{x - 3} = \lim_{x \to 3} \frac{-3x + 9}{x - 3} = -3$

Since $m_1 = m_2$, the tangent lines are parallel.

47. First we find the point of intersection of the line and the graph of f:

$$\sqrt{x} = \frac{x + 4}{4}$$

$$4\sqrt{x} = x + 4$$

$$16x = x^2 + 8x + 16$$

$$x^2 - 8x + 16 = 0$$

$$(x - 4)^2 = 0, \text{ or } x = 4$$

Since $\lim_{x \to 4} \frac{f(x) - f(4)}{x - 4} = \lim_{x \to 4} \frac{\sqrt{x} - 2}{x - 4} = \lim_{x \to 4} \frac{1}{\sqrt{x} + 2} = \frac{1}{4}$

the tangent line at (4, 2) is $y - 2 = \frac{1}{4}(x - 4)$, or $4y = x + 4$. The point of tangency is (4, 2).

49. Let the origin be at the top surface of the river. Then by (10) of Section 2.2 with $v_0 = 0$, the position of the stone after t seconds is given by $h(t) = -16t^2 + h_0$, where h_0 is the distance between the river and the top of the bridge.

a. Since the stone enters the water after 6 seconds, we have $h(6) = 0$. Thus $0 = -16 \cdot 6^2 + h_0 = -576 + h_0$, so that $h_0 = 576$ (feet).

b. $v(6) = \lim_{t \to 6} \frac{h(t) - h(6)}{t - 6} = \lim_{t \to 6} \frac{(-16t^2 + 576) - 0}{t - 6} = \lim_{t \to 6} \frac{-16(t^2 - 36)}{t - 6}$

$= \lim_{t \to 6} -16(t + 6) = -192$

Thus the stone hits the water with a speed of 192 feet per second.

3 DERIVATIVES

SECTION 3.1

1. $f'(4) = \lim_{x \to 4} \frac{f(x) - f(4)}{x - 4} = \lim_{x \to 4} \frac{5 - 5}{x - 4} = \lim_{x \to 4} 0 = 0$

3. $f'(1) = \lim_{x \to 1} \frac{f(x) - f(1)}{x - 1} = \lim_{x \to 1} \frac{(2x + 3) - 5}{x - 1} = \lim_{x \to 1} \frac{2x - 2}{x - 1} = \lim_{x \to 1} 2 = 2$

5. $f'(-1) = \lim_{x \to -1} \frac{f(x) - f(-1)}{x - (-1)} = \lim_{x \to -1} \frac{(x^2 - 2) - (-1)}{x + 1}$

$= \lim_{x \to 1} \frac{(x - 1)(x + 1)}{x + 1} = \lim_{x \to -1} (x - 1) = -2$

7. $f'(-2) = \lim_{x \to -2} \frac{f(x) - f(-2)}{x - (-2)} = \lim_{x \to -2} \frac{(1/x) - (-1/2)}{x + 2} = \lim_{x \to -2} \frac{x + 2}{2x(x + 2)} = \lim_{x \to -2} \frac{1}{2x} = -\frac{1}{4}$

9. For x near $\sqrt{2}$, $f(x) = x$. Thus

$$f'(\sqrt{2}) = \lim_{x \to \sqrt{2}} \frac{f(x) - f(\sqrt{2})}{x - \sqrt{2}} = \lim_{x \to \sqrt{2}} \frac{x - \sqrt{2}}{x - \sqrt{2}} = \lim_{x \to \sqrt{2}} 1 = 1$$

11. $\lim_{x \to 2^-} \frac{f(x) - f(2)}{x - 2} = \lim_{x \to 2^-} \frac{x^2 - 4}{x - 2} = \lim_{x \to 2^-} \frac{(x - 2)(x + 2)}{x - 2} = \lim_{x \to 2^-} (x + 2) = 4;$

$\lim_{x \to 2^+} \frac{f(x) - f(2)}{x - 2} = \lim_{x \to 2^+} \frac{(4x - 4) - 4}{x - 2} = \lim_{x \to 2^+} 4 = 4.$ Thus $f'(2) = \lim_{x \to 2} \frac{f(x) - f(2)}{x - 2} = 4.$

13. $f'(x) = \lim_{t \to x} \frac{f(t) - f(x)}{t - x} = \lim_{t \to x} \frac{-\pi - (-\pi)}{t - x} = \lim_{t \to x} 0 = 0$

15. $f'(x) = \lim_{t \to x} \frac{f(t) - f(x)}{t - x} = \lim_{t \to x} \frac{-5t^2 - (-5x^2)}{t - x} = \lim_{t \to x} \frac{-5(t - x)(t + x)}{t - x} = \lim_{t \to x} -5(t + x)$

$= -5(2x) = -10x$

17. $g'(x) = \lim_{t \to x} \frac{g(t) - g(x)}{t - x} = \lim_{t \to x} \frac{t^3 - x^3}{t - x} = \lim_{t \to x} \frac{(t - x)(t^2 + tx + t^2)}{t - x}$

$= \lim_{t \to x} (t^2 + tx + x^2) = (x^2 + x^2 + x^2) = 3x^2$

19. $k'(x) = \lim_{t\to x} \dfrac{k(t) - k(x)}{t - x} = \lim_{t\to x} \dfrac{[(1/t^2) - \sqrt{7}] - [(1/x^2) - \sqrt{7}]}{t - x} = \lim_{t\to x} \dfrac{x^2 - t^2}{t^2x^2(t - x)}$

$= \lim_{t\to x} \dfrac{(x - t)(x + t)}{t^2x^2(t - x)} = \lim_{t\to x} \dfrac{-(t + x)}{t^2x^2} = \dfrac{-2x}{x^4} = -\dfrac{2}{x^3}$

21. $\dfrac{dy}{dx} = \lim_{t\to x} \dfrac{\frac{7}{3} - \frac{7}{3}}{t - x} = \lim_{t\to x} 0 = 0$

23. $\dfrac{dy}{dx} = \lim_{t\to x} \dfrac{(3t^2 + 1) - (3x^2 + 1)}{t - x} = \lim_{t\to x} \dfrac{3(t - x)(t + x)}{t - x} = \lim_{t\to x} 3(t + x) = 3(2x) = 6x$

25. $\left.\dfrac{dy}{dx}\right|_{x=2} = \lim_{x\to 2} \dfrac{0.25 - 0.25}{x - 2} = \lim_{x\to 2} 0 = 0$

27. $\left.\dfrac{dy}{dx}\right|_{x=2} = \lim_{x\to 2} \dfrac{(x^2 - 3) - 1}{x - 2} = \lim_{x\to 2} \dfrac{(x - 2)(x + 2)}{x - 2} = \lim_{x\to 2} (x + 2) = 4$

29. $\lim_{x\to 0} \dfrac{f(x) - f(0)}{x - 0} = \lim_{x\to 0} \dfrac{x^{1/3} - 0}{x - 0} = \lim_{x\to 0} \dfrac{1}{x^{2/3}} = \infty$; no derivative at 0.

31. $\lim_{x\to 0^-} \dfrac{f(x) - f(0)}{x - 0} = \lim_{x\to 0^-} \dfrac{(-x - x) - 0}{x - 0} = \lim_{x\to 0^-} (-2) = -2;$

$\lim_{x\to 0^+} \dfrac{f(x) - f(0)}{x - 0} = \lim_{x\to 0^+} \dfrac{(x - x) - 0}{x - 0} = \lim_{x\to 0^+} 0 = 0$; no derivative at 0.

33. $\lim_{x\to 3} \dfrac{g(x) - g(3)}{x - 3} = \lim_{x\to 3} \dfrac{(x + 3) - 6}{x - 3} = \lim_{x\to 3} 1 = 1$; $g'(3) = 1$.

35. $\lim_{x\to 0^-} \dfrac{k(x) - k(0)}{x - 0} = \lim_{x\to 0^-} \dfrac{(-x^2 + 4x) - (-1)}{x - 0} = \lim_{x\to 0^-} \left(-x + 4 + \dfrac{1}{x}\right) = -\infty$; no derivative at 0.

37. $f'(-2) = 2(-2) = -4$ from Example 3. Thus ℓ: $y - 4 = -4(x - (-2))$, or $y = -4x - 4$.

39. $f'(4) = \frac{1}{2} \cdot 4^{-1/2} = \frac{1}{4}$ from Example 4. Thus ℓ: $y - 2 = \frac{1}{4}(x - 4)$, or $y = \frac{1}{4}x + 1$.

41. a. $f(x) = x^4$, $a = 2$;

$$f'(2) = \lim_{x\to 2} \frac{x^4 - 16}{x - 2} = \lim_{x\to 2} \frac{(x - 2)(x + 2)(x^2 + 4)}{x - 2} = \lim_{x\to 2} (x + 2)(x^2 + 4) = 32$$

b. $x^3 + 2x^2 + 4x + 8 = \dfrac{(x^3 + 2x^2 + 4x + 8)(x - 2)}{(x - 2)} = \dfrac{x^4 - 16}{x - 2}$;

$f(x) = x^4$, $a = 2$; $f'(2) = 32$ by part (b).

43. By the Substitution Rule, with $x = a - h$, we find that

$$\lim_{h\to 0} \frac{f(a - h) - f(a)}{h} = -\lim_{h\to 0} \frac{f(a - h) - f(a)}{-h}$$
$$= -\lim_{x\to a} \frac{f(x) - f(a)}{x - a} = -f'(a)$$

45. a. Since $f(-a) = f(a)$, we use the solution of Exercise 43 to deduce that

$$f'(-a) = \lim_{h\to 0} \frac{f(-a + h) - f(-a)}{h} = \lim_{h\to 0} \frac{f(a - h) - f(a)}{h} = -f'(a) = -2$$

b. Since $f(-a) = -f(a)$, we use the solution of Exercise 43 to deduce that

$$f'(-a) = \lim_{h\to 0} \frac{f(-a + h) - f(-a)}{h} = \lim_{h\to 0} \frac{-f(a - h) + f(a)}{h}$$
$$= -\lim_{h\to 0} \frac{f(a - h) - f(a)}{h} = f'(a) = 2$$

47. No. Let $f(x) = 1$ and $g(x) = x$ for all x. Then $f(1) = 1 = g(1)$, but $f'(1) = 0$ and $g'(1) = 1$.

49. $\lim_{x\to a} g(x) = \lim_{x\to a} \dfrac{f(x) - f(a)}{x - a} = f'(a) = g(a)$. Thus g is continuous at a.

51. Because $v(t) = 96 - 32t$ for $0 < t < 8$ by Exercise 52, $v(t) = 0$ if $96 - 32t = 0$, that is, $t = 3$.

53. $m_C(50) = C'(50) = \lim_{x\to 50} \dfrac{C(x) - C(50)}{x - 50} = \lim_{x\to 50} \dfrac{(10{,}000 + 3/x) - (10{,}000 + \frac{3}{50})}{x - 50}$

$= \lim_{x\to 50} \dfrac{3(50 - x)}{50x(x - 50)} = \lim_{x\to 50} \dfrac{-3}{50x} = -\dfrac{3}{2500}$

55. $m_R(16) = R'(16) = \lim_{x\to 16} \dfrac{R(x) - R(16)}{x - 16} = \lim_{x\to 16} \dfrac{450x^{1/2} - 450 \cdot 16^{1/2}}{x - 16}$

$= \lim_{x\to 16} \left[\dfrac{450(x^{1/2} - 16^{1/2})}{x - 16} \cdot \dfrac{x^{1/2} + 16^{1/2}}{x^{1/2} + 16^{1/2}}\right] = \lim_{x\to 16} \dfrac{450(x - 16)}{(x - 16)(x^{1/2} + 16^{1/2})}$

$= \lim_{x\to 16} \dfrac{450}{x^{1/2} + 16^{1/2}} = \dfrac{450}{16^{1/2} + 16^{1/2}} = \dfrac{225}{4}$

57. Since the cost of each thousand gallons has been increased by $\frac{1}{2}$ thousand dollars, the cost is given by

$$C(x) = \tfrac{1}{2}x + (3 + 12x - 2x^2) = 3 + \frac{25x}{2} - 2x^2 \quad \text{for } 0 \le x \le 3$$

Thus $m_C(1) = \lim_{x\to 1} \dfrac{C(x) - C(1)}{x - 1} = \lim_{x\to 1} \dfrac{(3 + (25x/2) - 2x^2) - (3 + (25/2) - 2)}{x - 1}$

$= \lim_{x\to 1} \dfrac{\frac{25}{2}(x - 1) - 2(x^2 - 1)}{x - 1} = \lim_{x\to 1} [\frac{25}{2} - 2(x + 1)] = \frac{17}{2}$

Thus the marginal cost is increased by $\frac{1}{2}$ thousand dollars per thousand gallons.

59. We set up a coordinate system so that the port is at the origin, the positive x and y axes point east and north, respectively, and the units represent nautical miles. By assumption, after t hours the northbound ship is at $(0, 15t)$ and the other ship is at $(-20t, 0)$. If $D(t)$ represents the distance between the ships t hours after they leave port, then

$D(t)=\sqrt{(-20t-0)^2+(0-15t)^2}=\sqrt{400t^2+225t^2}=25t$. Thus

$$D'(t_0)=\lim_{t\to t_0}\frac{D(t)-D(t_0)}{t-t_0}=\lim_{t\to t_0}\frac{25t-25t_0}{t-t_0}=25$$

so the distance is increasing at the constant rate of 25 knots.

61. We set up a coordinate system with the starting point at the origin, one boat traveling in the positive direction along the x axis, and the other in the first quadrant along the line $y=x$.

a. By hypothesis after t minutes one boat is at $(2t, 0)$. The other boat is $2t$ units from the origin on the line $y=x$, so $2t=\sqrt{x^2+y^2}=\sqrt{x^2+x^2}=x\sqrt{2}$; thus $x=\sqrt{2}t$, so $y=\sqrt{2}t$, and the second boat is at $(\sqrt{2}t, \sqrt{2}t)$. The distance $D(t)$ between the boats t minutes after starting is given by $D(t)=\sqrt{(\sqrt{2}t-2t)^2+(\sqrt{2}t-0)^2}=\sqrt{(8-4\sqrt{2})t^2}=t\sqrt{8-4\sqrt{2}}$. Then

$$D'(t_0)=\lim_{t\to t_0}\frac{D(t)-D(t_0)}{t-t_0}=\lim_{t\to t_0}\frac{t\sqrt{8-4\sqrt{2}}-t_0\sqrt{8-4\sqrt{2}}}{t-t_0}=\sqrt{8-4\sqrt{2}}$$

so the rate of the increase is $\sqrt{8-4\sqrt{2}}$ meters per minute.

b. This time t minutes after starting the boats are at $(at, 0)$ and $((a/\sqrt{2})t, (a/\sqrt{2})t)$, where a is to be determined and is the speed of each boat. The distance $D(t)$ between the boats t minutes after starting is given by

$$D(t)=\sqrt{\left(\frac{a}{\sqrt{2}}t-at\right)^2+\left(\frac{a}{\sqrt{2}}t\right)^2}=\sqrt{(2-\sqrt{2})a^2t^2}\doteq at\sqrt{2-\sqrt{2}}$$

Now

$$D'(t_0)=\lim_{t\to t_0}\frac{D(t)-D(t_0)}{t-t_0}=\lim_{t\to t_0}\frac{at\sqrt{2-\sqrt{2}}-at_0\sqrt{2-\sqrt{2}}}{t-t_0}=a\sqrt{2-\sqrt{2}}$$

By hypothesis $D'(t_0)=3$, so $a\sqrt{2-\sqrt{2}}=3$, and thus $a=3/\sqrt{2-\sqrt{2}}$. Thus the speed of the boats is $3/\sqrt{2-\sqrt{2}}$ meters per minute.

SECTION 3.2

1. By Example 1, $f'(x)=0$ for all x, so $f'(1)=0$.

3. By (1), $f'(x)=2x$ for all x, so $f'(3/2)=3$ and $f'(0)=0$.

5. By (1), $f'(x)=4x^3$, so $f'(\sqrt[3]{2})=4(\sqrt[3]{2})^3=4\cdot 2=8$.

7. By (1), $f'(x)=10x^9$, so $f'(1)=10$.

9. By (4), $f'(t)=-\sin t$, so $f'(0)=0$ and $f'(-\pi/3)=\sqrt{3}/2$.

11. $$f'(x)=\lim_{t\to x}\frac{f(t)-f(x)}{t-x}=\lim_{t\to x}\frac{(-2t-1)-(-2x-1)}{t-x}=\lim_{t\to x}(-2)=-2$$

13. $$f'(x)=\lim_{t\to x}\frac{f(t)-f(x)}{t-x}=\lim_{t\to x}\frac{t^5-x^5}{t-x}=\lim_{t\to x}\frac{(t-x)(t^4+t^3x+t^2x^2+tx^3+x^4)}{t-x}=5x^4$$

15. $$f'(x)=\lim_{t\to x}\frac{f(t)-f(x)}{t-x}=\lim_{t\to x}\frac{(t^2+t)-(x^2+x)}{t-x}$$
$$=\lim_{t\to x}\frac{(t^2-x^2)+(t-x)}{t-x}=\lim_{t\to x}[(t+x)+1]=2x+1$$

17. $$f'(x)=\lim_{t\to x}\frac{f(t)-f(x)}{t-x}=\lim_{t\to x}\frac{[(t^2-1)/(t^2+1)]-[(x^2-1)/(x^2+1)]}{t-x}$$
$$=\lim_{t\to x}\frac{(t^2-1)(x^2+1)-(x^2-1)(t^2+1)}{(t^2+1)(x^2+1)(t-x)}=\lim_{t\to x}\frac{2(t-x)(t+x)}{(t^2+1)(x^2+1)(t-x)}$$
$$=\lim_{t\to x}\frac{2(t+x)}{(t^2+1)(x^2+1)}=\frac{4x}{(x^2+1)^2}$$

19. $$\frac{dy}{dx}=\lim_{h\to 0}\frac{-3\cos(x+h)-(-3\cos x)}{h}=-3\lim_{h\to 0}\frac{\cos(x+h)-\cos x}{h}$$
$=(-3)(-\sin x)=3\sin x$ by the discussion preceding (4).

21. If $x\neq 0$, then
$$\frac{dy}{dx}=\lim_{t\to x}\frac{t^{2/3}-x^{2/3}}{t-x}=\lim_{t\to x}\frac{(t^{1/3}-x^{1/3})(t^{1/3}+x^{1/3})}{(t^{1/3}-x^{1/3})(t^{2/3}+t^{1/3}x^{1/3}+x^{2/3})}$$
$$=\lim_{t\to x}\frac{t^{1/3}+x^{1/3}}{t^{2/3}+t^{1/3}x^{1/3}+x^{2/3}}=\frac{2x^{1/3}}{3x^{2/3}}=\frac{2}{3x^{1/3}}=\frac{2}{3}x^{-1/3}$$

23. If $x>1$, then
$$\frac{dy}{dx}=\lim_{t\to x}\frac{\sqrt{t-1}-\sqrt{x-1}}{t-x}=\lim_{t\to x}\frac{\sqrt{t-1}-\sqrt{x-1}}{t-x}\cdot\frac{\sqrt{t-1}+\sqrt{x-1}}{\sqrt{t-1}+\sqrt{x-1}}$$
$$=\lim_{t\to x}\frac{(t-1)-(x-1)}{(t-x)(\sqrt{t-1}+\sqrt{x-1})}=\lim_{t\to x}\frac{1}{\sqrt{t-1}+\sqrt{x-1}}=\frac{1}{2\sqrt{x-1}}$$

25. For all x,
$$\lim_{t\to x}\frac{f(t)-f(x)}{t-x}=\lim_{t\to x}\frac{(t^2+t)-(x^2+x)}{t-x}=\lim_{t\to x}\frac{(t^2-x^2)+(t-x)}{t-x}$$
$$=\lim_{t\to x}\frac{(t-x)(t+x)+(t-x)}{t-x}=\lim_{t\to x}(t+x+1)=2x+1$$

Thus f is differentiable on $(-\infty, \infty)$.

27. For $x>4$,
$$\lim_{t\to x}\frac{f(t)-f(x)}{t-x}=\lim_{t\to x}\frac{[1/(4-t)]-[1/(4-x)]}{t-x}=\lim_{t\to x}\frac{(4-x)-(4-t)}{(4-t)(4-x)(t-x)}$$
$$=\lim_{t\to x}\frac{1}{(4-t)(4-x)}=\frac{1}{(4-x)^2}$$

Therefore f is differentiable on $(4, \infty)$.

29. For $x \ge 1$, $f(x) = x - 1$, and for $x > 1$,

$$\lim_{t\to x} \frac{f(t) - f(x)}{t - x} = \lim_{t\to x} \frac{(t-1) - (x-1)}{t - x} = \lim_{t\to x} 1 = 1$$

Therefore f is differentiable on $(1, \infty)$. Also,

$$\lim_{t\to 1^+} \frac{f(t) - f(1)}{t - 1} = \lim_{t\to 1^+} \frac{t - 1 - 0}{t - 1} = 1$$

Thus f is differentiable on $[1, \infty)$.

31. a. $f'(x) = \lim_{t\to x} \dfrac{-2t^2 + 2x^2}{t - x} = \lim_{t\to x} [-2(t + x)] = -4x.$

Since $f'(a) = 12$, we have $-4a = 12$, so $a = -3$.

b. $f'(x) = \lim_{t\to x} \dfrac{(3t + t^2) - (3x + x^2)}{t - x} = \lim_{t\to x} [3 + (t + x)] = 3 + 2x.$

Since $f'(a) = 13$, we have $3 + 2a = 13$, so $a = 5$.

c. $f'(x) = \lim_{t\to x} \dfrac{(1/t) - (1/x)}{t - x} = \lim_{t\to x} \dfrac{-(t - x)}{tx(t - x)} = \lim_{t\to x} \dfrac{-1}{tx} = -\dfrac{1}{x^2}$

Since $f'(a) = -\frac{1}{9}$, we have $-1/a^2 = -\frac{1}{9}$. Thus $a = 3$ or $a = -3$.

d. $f'(x) = \cos x$ by (3). Since $f'(a) = \sqrt{3}/2$, we have $\cos a = \sqrt{3}/2$. Thus $a = -\pi/6 + 2n\pi$ or $a = \pi/6 + 2n\pi$, where n is any integer.

33. $v(t) = f'(t) = \lim_{h\to 0} \dfrac{f(t + h) - f(t)}{h} = \lim_{h\to 0} \dfrac{-3\sin(t + h) - (-3\sin t)}{h}$

$$= -3 \lim_{h\to 0} \frac{\sin(t + h) - \sin t}{h} = -3\cos t$$

by the discussion preceding (3). In particular, $v(\pi/6) = -3\cos(\pi/6) = -\frac{3}{2}\sqrt{3}$.

35. If $0 < x < 6$, then $R'(x) = \lim_{t\to x} (1432t - 1432x)/(t - x) = 1432$, so R is differentiable on $(0, 6)$. If $x > 6$, then $R'(x) = \lim_{t\to x} (8592 - 8592)/(t - x) = 0$, so R is differentiable on $(6, \infty)$. Since

$$\lim_{x\to 6^-} \frac{R(x) - R(6)}{x - 6} = \lim_{x\to 6^-} \frac{1432x - 8592}{x - 6} = \lim_{x\to 6^-} 1432 = 1432$$

and

$$\lim_{x\to 6^+} \frac{R(x) - R(6)}{x - 6} = \lim_{x\to 6^+} \frac{8592 - 8592}{x - 6} = 0$$

R is not differentiable at 6.

37. The area of a circle of radius r is given by $A(r) = \pi r^2$ for $r > 0$. Thus

$$A'(r) = \lim_{t\to r} \frac{A(t) - A(r)}{t - r} = \lim_{t\to r} \frac{\pi t^2 - \pi r^2}{t - r} = \lim_{t\to r} \pi(t + r) = 2\pi r$$

(which is the circumference of the circle).

39. a. The altitude of the triangle with side of length x is $(\sqrt{3}/2)x$, so the area of the triangle is given by $A(x) = \frac{1}{2}[(\sqrt{3}/2)x](x) = (\sqrt{3}/4)x^2$. Then

$$A'(x) = \lim_{t\to x} \frac{(\sqrt{3}/4)t^2 - (\sqrt{3}/4)x^2}{t - x} = \lim_{t\to x} \frac{\sqrt{3}}{4}(t + x) = \frac{\sqrt{3}}{2}x \quad \text{for } x > 0$$

b. If $A'(x) = A(x)$ and $x > 0$, then $(\sqrt{3}/2)x = (\sqrt{3}/4)x^2$, or $2x = x^2$, so that $x = 2$.

41. If $r > 0$, then

$$F'(r) = \lim_{t\to r} \frac{(k/t^2) - (k/r^2)}{t - r} = \lim_{t\to r} \frac{k(r^2 - t^2)}{t^2r^2(t - r)} = \lim_{t\to r} \frac{-k(r + t)}{t^2r^2} = \frac{-2k}{r^3} < 0$$

43. a. By (1), $f'(x) = 2x$, so that $f'(-1) = -2$. Thus the slope of the tangent line at $(-1, 1)$ is -2, so that the slope of the normal line at $(-1, 1)$ is $\frac{1}{2}$. An equation of the normal line is $y - 1 = \frac{1}{2}(x + 1)$.

b. $f'(-1) = \lim_{x\to -1} \dfrac{f(x) - f(-1)}{x - (-1)} = \lim_{x\to -1} \dfrac{(2x - 3) - (-5)}{x + 1} = \lim_{x\to -1} \dfrac{2(x + 1)}{x + 1}$

$$= \lim_{x\to -1} 2 = 2.$$

Thus the slope of the tangent line at $(-1, -5)$ is 2, so that the slope of the normal line is $-\frac{1}{2}$. An equation of the normal line is $y - (-5) = -\frac{1}{2}(x - (-1))$, or $y + 5 = -\frac{1}{2}(x + 1)$.

c. By (3), $f'(x) = \cos x$, so that $f'(0) = 1$. Thus the slope of the tangent line at $(0, 0)$ is 1, so that the slope the normal line is -1. An equation of the normal line is $y - 0 = -1(x - 0)$, or $y = -x$.

SECTION 3.3

1. $f'(x) = -12x^2$

3. $f'(x) = 6x - 4$

5. $f'(x) = 16x^3 + 9x^2 + 4x + 1$

7. $f'(t) = \dfrac{36}{t^{10}}$

9. $g'(x) = 2(x + 5) + (2x - 3) \cdot 1 = 4x + 7$

11. $g'(x) = \left(-\dfrac{1}{x^2}\right)\left(2 - \dfrac{1}{x}\right) + \left(1 + \dfrac{1}{x}\right)\left(\dfrac{1}{x^2}\right) = -\dfrac{1}{x^2} + \dfrac{2}{x^3}$

13. $g'(x) = 12x^{-4} - 2\sin x$

15. $f'(z) = -6z^2 + 4\sec z\tan z$

17. $f'(z) = 2z\sin z + z^2\cos z$

19. $f(x) = \sin x\sin x$, so $f'(x) = \cos x\sin x + \sin x\cos x = 2\sin x\cos x = \sin 2x$

21. $f'(x) = -4\tan x\sec x + (-4x)\sec^2 x\sec x + (-4x)\tan x\sec x\tan x$

$$= -4\tan x\sec x - 4x\sec^3 x - 4x\tan^2 x\sec x$$

23. $f'(x) = \dfrac{2(4x-1)-(2x+3)4}{(4x-1)^2} = -\dfrac{14}{(4x-1)^2}$

25. $f'(t) = \dfrac{1(t^2+4t+4)-(t+2)(2t+4)}{(t^2+4t+4)^2} = -\dfrac{1}{t^2+4t+4}$

27. $f'(t) = \dfrac{(2t+5)(t^2+t-20)-(t^2+5t+4)(2t+1)}{(t^2+t-20)^2} = \dfrac{-4(t^2+12t+26)}{(t^2+t-20)^2}$

29. $f'(x) = \dfrac{(-\sin x)\sin x - \cos x \cos x}{(\sin x)^2} = \dfrac{-1}{\sin^2 x} = -\csc^2 x$

31. $f'(y) = \dfrac{1}{2\sqrt{y}} \sec y + \sqrt{y} \sec y \tan y$

33. $\dfrac{dy}{dx} = 3(2x^2-5x)+(3x+1)(4x-5) = 18x^2-26x-5$

35. $\dfrac{dy}{dx} = 2x - \dfrac{2}{x^3}$

37. $\dfrac{dy}{dx} = \dfrac{2(5-3x)-(2x-1)(-3)}{(5-3x)^2} = \dfrac{7}{(5-3x)^2}$

39. $\dfrac{dy}{dx} = \dfrac{3x^2(x^4+1)-(x^3-1)(4x^3)}{(x^4+1)^2} = \dfrac{-x^6+4x^3+3x^2}{(x^4+1)^2}$

41. $\dfrac{dy}{dx} = (-\csc x \cot x)\sec x + \csc x(\sec x \tan x) = -\csc^2 x + \sec^2 x$

43. $\dfrac{dy}{dx} = \dfrac{(\sin x + x\cos x)(x^2+1)-(x\sin x)(2x)}{(x^2+1)^2}$

$= \dfrac{(x^3+x)\cos x + (1-x^2)\sin x}{(x^2+1)^2}$

45. $f'(x) = 63/x^{10}$; $f'(1) = 63$.

47. $f'(x) = (1/\pi)(6x-4)$; $f'(-2) = -16/\pi$

49. $f'(x) = \dfrac{(\cos x)\sqrt{x} - (\sin x)(1/2\sqrt{x})}{x} = \dfrac{2x\cos x - \sin x}{2x^{3/2}}$

$f'\left(\dfrac{\pi}{2}\right) = \dfrac{-1}{2(\pi/2)^{3/2}} = -\sqrt{2}\pi^{-3/2}$

51. $f'(x) = 2x-3$, so $f'(2) = 4-3 = 1$. Thus ℓ: $y-(-6) = 1(x-2)$, or $y = x-8$.

53. $f'(x) = \cos x - (-\sin x) = \cos x + \sin x$, so $f'(\pi/2) = 1$. Thus ℓ: $y - 1 = 1(x-(\pi/2))$, or $y = x + 1 - (\pi/2)$.

55. $(f+g)(x) = |x| - |x| = 0$, so $(f+g)'(x) = 0$ for all x. By Example 4 in Section 3.2, $f'(x)$ exists only for $x \neq 0$, and the Constant Multiple Rule therefore implies that $g'(x)$ also exists only for $x \neq 0$.

57. Since $f(x) = [(fg)(x)]/g(x)$ for all x in the domain of fg for which $g(x) \neq 0$, and since $(fg)'(a)$ and $g'(a)$ exist with $g(a) \neq 0$, it follows from Theorem 3.7 that $f'(a)$ exists.

59. a. By Theorem 3.6, $h'(a) = (fg)'(a) = f'(a)g(a) + f(a)g'(a)$. Dividing by $h(a)$, which equals $f(a)g(a)$, we find that

$$\frac{h'(a)}{h(a)} = \frac{f'(a)g(a)}{f(a)g(a)} + \frac{f(a)g'(a)}{f(a)g(a)} = \frac{f'(a)}{f(a)} + \frac{g'(a)}{g(a)}$$

b. By Theorem 3.7, $k'(a) = \left(\dfrac{f}{g}\right)'(a) = \dfrac{f'(a)g(a)-f(a)g'(a)}{[g(a)]^2}$. Since $k(a) = \dfrac{f(a)}{g(a)}$, we have $\dfrac{1}{k(a)} = \dfrac{g(a)}{f(a)}$, so that

$$\frac{k'(a)}{k(a)} = \frac{f'(a)g(a)-f(a)g'(a)}{[g(a)]^2} \cdot \frac{g(a)}{f(a)} = \frac{f'(a)g(a)-f(a)g'(a)}{g(a)f(a)} = \frac{f'(a)}{f(a)} - \frac{g'(a)}{g(a)}$$

61. a. By the solution of Example 5, we have $v(t) = -32t + 24$ for $0 < t < 4$, so $v(1) = -8$ and $v(2) = -40$ (feet per second).

b. From (2) we find that the height $h(t)$ of the ball at time t is given by $h(t) = -16t^2 + 48t + 160$, so that $v(t) = h'(t) = -32t + 48$. Thus $v(1) = 16$ (feet per second).

63. a. From (2), the height of the ball at any time t before the ball hits the ground is given by $h(t) = -16t^2 + 128t$. Since $h(t) = 0$ for $t = 0$ and $t = 8$, the ball hits the ground after 8 seconds. Thus $h(t) = -16t^2 + 128t$ for $0 \leq t \leq 8$, so that $v(t) = -32t + 128$ for $0 < t < 8$. The velocity of the ball will be 0 when the height of the ball is maximum. Since $v(t) = 0$ only for $t = 4$, the maximum height is $h(4) = 256$ (feet).

b. $v(8) = \lim\limits_{t\to 8^-} \dfrac{h(t)-h(8)}{t-8} = \lim\limits_{t\to 8^-} \dfrac{(-16t^2+128t)-0}{t-8}$

$= \lim\limits_{t\to 8^-} (-16t) = -128$ (feet per second)

65. From (2) we find that the height of the ball at any time t before the ball hits the ground is given by $h(t) = -16t^2 - 16t + 96 = -16(t+3)(t-2)$. Since $h(2) = 0$, the ball hits the ground after 2 seconds. Thus $h(t) = -16t^2 - 16t + 96$ for $0 \leq t \leq 2$, so that $v(t) = h'(t) = -32t - 16$ for $0 < t < 2$.

67. a. By (2) the height of the lowest point of the chandelier is given by $h(t) = -16t^2 + 30$ until the chandelier hits something. It would hit your head when $h(t) = 6$, that is, $-16t^2 + 30 = 6$, or $16t^2 = 24$, or $t = \sqrt{1.5}$. Thus you have $\sqrt{1.5} \approx 1.225$ seconds to get out of the way.

b. Since $v(t) = h'(t) = -32t$, it follows from part (a) that the chandelier would hit your head with velocity $v(\sqrt{1.5}) = -32\sqrt{1.5}$. Thus the chandelier will be moving at the rate of $32\sqrt{1.5} \approx 39.19$ feet per second when it hits your head.

c. If you were only 5 feet tall, the chandelier would hit your head when $h(t) = 5$, that is, $-16t^2 + 30 = 5$, or $16t^2 = 25$, or $t = \frac{5}{4} = 1.25$. Since $v(1.25) = -32(1.25) = -40$, the speed would be $40 - 32\sqrt{1.5} \approx 40 - 39.19 = 0.81$ feet per second greater.

69. Since $\mu > 0$, we have $\mu \sin x + \cos x > 0$ for $0 \le x \le \pi/2$. Since the sine and cosine functions are differentiable on $(-\infty, \infty)$, so is the function $\mu \sin x + \cos x$. By the Quotient Rule, the function $50\mu/(\mu \sin x + \cos x)$ is differentiable at every point in $[0, \pi/2]$. Thus F is differentiable on $[0, \pi/2]$; $F'(x) = [-50\mu(\mu \cos x - \sin x)]/(\mu \sin x + \cos x)^2$.

71. $f'(x) = \dfrac{100nkx^{n-1}(1 + kx^n) - (100kx^n)(nkx^{n-1})}{(1 + kx^n)^2} = \dfrac{100nkx^{n-1}}{(1 + kx^n)^2}$ for $x > 0$.

73. $P(x) = R(x) - C(x) = \sqrt{x} - \dfrac{x + 3}{\sqrt{x} + 1} = \dfrac{x + \sqrt{x} - (x + 3)}{\sqrt{x} + 1} = \dfrac{\sqrt{x} - 3}{\sqrt{x} + 1}$

for $1 \le x \le 15$. Thus $P(x) = 0$ if $\sqrt{x} - 3 = 0$, that is, if $x = 9$. Since the function $(\sqrt{x} - 3)/(\sqrt{x} + 1)$ is differentiable on $(0, \infty)$, it follows that P is differentiable on $[1, 15]$, and

$$m_P(x) = P'(x) = \frac{[1/(2\sqrt{x})](\sqrt{x} + 1) - (\sqrt{x} - 3)[1/(2\sqrt{x})]}{(\sqrt{x} + 1)^2}$$

$$= \frac{(\sqrt{x} + 1) - (\sqrt{x} - 3)}{2\sqrt{x}(\sqrt{x} + 1)^2} = \frac{2}{\sqrt{x}(\sqrt{x} + 1)^2} \ne 0 \quad \text{for } 1 \le x \le 15$$

Therefore for no value of x is the marginal profit equal to 0.

SECTION 3.4

1. $f'(x) = \frac{9}{4}x^{5/4}$

3. $f'(x) = \frac{2}{3}x^{-1/3} - 7(-\frac{1}{3})x^{-4/3} = \frac{2}{3}x^{-1/3} + \frac{7}{3}x^{-4/3}$

5. $f'(x) = 400(4 - 3x^2)^{399}(-6x) = -2400x(4 - 3x^2)^{399}$

7. $f'(x) = 3\left(\dfrac{x - 1}{x + 1}\right)^2\left(\dfrac{(x + 1) - (x - 1)}{(x + 1)^2}\right) = \dfrac{6(x - 1)^2}{(x + 1)^4}$

9. $f'(x) = \sqrt{2 - 7x^2} + x\dfrac{1}{2\sqrt{2 - 7x^2}}(-14x) = \dfrac{2 - 14x^2}{\sqrt{2 - 7x^2}}$

11. $f'(t) = (\cos 5t)(5) = 5 \cos 5t$

13. $f'(t) = 4 \sin^3 t \cos t + 4 \cos^3 t(-\sin t) = 4 \sin^3 t \cos t - 4 \cos^3 t \sin t$

15. $g'(x) = 6 \tan^5 x \sec^2 x$

17. $g'(x) = \frac{1}{3}(1 - \sin x)^{-2/3}(-\cos x) = (-\frac{1}{3} \cos x)(1 - \sin x)^{-2/3}$

19. $f'(x) = [-\sin(\sin x)] \cos x$

21. $f'(x) = \dfrac{1}{2}\left(x^2 + \dfrac{1}{x^2}\right)^{-1/2}\left(2x - \dfrac{2}{x^3}\right) = \left(x^2 + \dfrac{1}{x^2}\right)^{-1/2}\left(x - \dfrac{1}{x^3}\right)$

23. $f'(x) = \dfrac{-1}{(x\sqrt{5 - 2x})^2}\left[\sqrt{5 - 2x} + x\dfrac{-2}{2\sqrt{5 - 2x}}\right] = -\dfrac{1}{x^2(5 - 2x)}\left[\dfrac{(5 - 2x) - x}{\sqrt{5 - 2x}}\right]$
$= \dfrac{3x - 5}{x^2(5 - 2x)^{3/2}}$

25. $f'(x) = \cos\dfrac{1}{x} + x\left(-\sin\dfrac{1}{x}\right)\left(-\dfrac{1}{x^2}\right) = \cos\dfrac{1}{x} + \dfrac{1}{x}\sin\dfrac{1}{x}$

27. $g'(z) = \frac{1}{2}[2z - (2z)^{1/3}]^{-1/2}[2 - \frac{1}{3}(2z)^{-2/3}(2)] = [2z - (2z)^{1/3}]^{-1/2}[1 - \frac{1}{3}(2z)^{-2/3}]$

29. $g'(z) = [2 \cos(3z^6)][-\sin(3z^6)][18z^5] = -36z^5 \cos(3z^6) \sin(3z^6)$

31. $f'(x) = \frac{1}{4}(\sec(\tan x))^{-3/4}(\sec(\tan x) \tan(\tan x)) \sec^2 x = \frac{1}{4}\sqrt[4]{\sec(\tan x)} \tan(\tan x) \sec^2 x$

33. $f'(x) = [2 \cot(2\sqrt{3x + 1})][-\csc^2(2\sqrt{3x + 1})]\dfrac{2}{2\sqrt{3x + 1}}(3)$
$= \dfrac{-6 \cot(2\sqrt{3x + 1}) \csc^2(2\sqrt{3x + 1})}{\sqrt{3x + 1}}$

35. $\dfrac{dy}{dx} = 3(-\frac{2}{3})x^{-5/3} = -2x^{-5/3}$

37. $\dfrac{dy}{dx} = -\sqrt{1 + 3x^2} - x \cdot \dfrac{1}{2}(1 + 3x^2)^{-1/2}(6x) = -\dfrac{1 + 6x^2}{\sqrt{1 + 3x^2}}$

39. $\dfrac{dy}{dx} = \dfrac{2}{3}\left(\dfrac{1}{x \sin x}\right)^{-1/3}\left[\dfrac{-1}{(x \sin x)^2}\right][\sin x + x \cos x] = -\dfrac{2}{3}\dfrac{\sin x + x \cos x}{(x \sin x)^{5/3}}$

41. $\dfrac{dy}{dx} = [3 \tan^2(\frac{1}{2}x)][\sec^2(\frac{1}{2}x)](\frac{1}{2}) = \frac{3}{2} \tan^2(\frac{1}{2}x) \sec^2(\frac{1}{2}x)$

43. $\dfrac{d}{dx}(y^5) = 5y^4\dfrac{dy}{dx}$

45. $\dfrac{d}{dx}\left(\dfrac{2}{y}\right) = \dfrac{-2}{y^2}\dfrac{dy}{dx}$

47. $\dfrac{d}{dx}(\sin\sqrt{y}) = (\cos\sqrt{y})\dfrac{1}{2\sqrt{y}}\dfrac{dy}{dx} = \left(\dfrac{1}{2\sqrt{y}}\cos\sqrt{y}\right)\dfrac{dy}{dx}$

49. $\dfrac{d}{dx}(x^3y^2) = 3x^2y^2 + x^3\left(2y\dfrac{dy}{dx}\right) = 3x^2y^2 + 2x^3y\dfrac{dy}{dx}$

51. $\dfrac{d}{dx}(\sqrt{x^2 + y^2}) = \dfrac{1}{2\sqrt{x^2 + y^2}}\left[2x + 2y\dfrac{dy}{dx}\right] = \dfrac{x + y(dy/dx)}{\sqrt{x^2 + y^2}}$

53. $f'(x) = -2(x + 1)^{-3}(1)$, so $f'(0) = -2$. Thus ℓ: $y - 1 = -2(x - 0)$, or $y = -2x + 1$.

55. $f'(x) = (-2)(-\sin 3x)(3) = 6 \sin 3x$, so $f'(\pi/3) = 6 \sin \pi = 0$. Thus ℓ: $y = 2$.

57. $(g \circ f)'(-3) = g'(f(-3))\, f'(-3) = g'(2)f'(-3) = (\sqrt{2})4 = 4\sqrt{2}$
$(f \circ g)'(0) = f'(g(0))g'(0) = f'(-3)g'(0) = 4(-7) = -28$
$(g \circ f)'(0) = g'(f(0))f'(0) = g'(1)f'(0) = 13 \cdot 2 = 26$
$(f \circ g)'(2) = f'(g(2))g'(2) = f'(0)g'(2) = 2(\sqrt{2}) = 2\sqrt{2}$
$(g \circ f)'(2) = g'(f(2))f'(2) = g'(-3)f'(2) = 11 \cdot 2 = 22$

59. Let $f(x) = x^{1/n}$ and $g(x) = x$, so that $g(x) = (x^{1/n})^n = (f(x))^n$. Then $g'(x) = n(f(x))^{n-1}f'(x)$. Since $g(x) = x$, we know that $g'(x) = 1$. Thus $1 = n(f(x))^{n-1}f'(x)$, so

$$f'(x) = \frac{1}{n}(f(x))^{1-n} = \frac{1}{n}(x^{1/n})^{1-n} = \frac{1}{n}x^{(1/n)-1}$$

61. Since the cosine function and the polynomial $\pi t/24$ are differentiable on $(-\infty, \infty)$, $\cos(\pi t/24)$ and hence F are differentiable on $[0, 24]$. Also

$$F'(t) = \frac{336{,}000}{\pi}\left(\sin\frac{\pi t}{24}\right)\left(\frac{\pi}{24}\right) = 14{,}000 \sin\frac{\pi t}{24}$$

63. a. $\dfrac{dE}{dv} = 100(.4)\left(1 - \dfrac{v}{V}\right)^{-0.6}\left(\dfrac{-1}{V}\right) = \dfrac{-40}{V}\left(1 - \dfrac{v}{V}\right)^{-0.6}$

b. $\dfrac{dE}{dV} = 100(.4)\left(1 - \dfrac{v}{V}\right)^{-0.6}\left(\dfrac{v}{V^2}\right) = \dfrac{40v}{V^2}\left(1 - \dfrac{v}{V}\right)^{-0.6}$

65. a. $v'(r) = \dfrac{1}{2}\left(\dfrac{192{,}000}{r} + v_0^2 - 48\right)^{-1/2}\dfrac{-192{,}000}{r^2}$

$$= -\frac{96{,}000}{r^2}\left(\frac{192{,}000}{r} + v_0^2 - 48\right)^{-1/2}$$

b. If $v_0 = 8$, then

$$v'(24{,}000) = -\frac{96{,}000}{(24{,}000)^2}\left(\frac{192{,}000}{24{,}000} + 16\right)^{-1/2}$$

$$= -\frac{\sqrt{6}}{72{,}000} \text{ (miles per second per mile)}$$

67. $\dfrac{dV}{dt} = \dfrac{dV}{dr}\dfrac{dr}{dt} = [\tfrac{4}{3}\pi(3r^2)](10) = 40\pi r^2$

69. $\dfrac{dA}{dh} = \dfrac{dA}{dx}\dfrac{dx}{dh} = \left(\dfrac{\sqrt{3}}{2}x\right)\left(\dfrac{2\sqrt{3}}{3}\right) = x = \tfrac{2}{3}\sqrt{3}\,h$, so that $\left.\dfrac{dA}{dh}\right|_{h=\sqrt{3}} = \dfrac{2\sqrt{3}}{3}(\sqrt{3}) = 2$

71. Since $D'(x) = \tfrac{1}{2}(3 - 2x)^{-1/2}(-2) = -(3 - 2x)^{-1/2}$, we have $D'(x) < 0$ for $0 < x < \tfrac{3}{2}$.

SECTION 3.5

1. $f'(x) = 5$; $f''(x) = 0$

3. $f'(x) = -60x^4 + 2x^3 - \tfrac{1}{2}(1 - x)^{-1/2}(-1) = -60x^4 + 2x^3 + \tfrac{1}{2}(1 - x)^{-1/2}$

$$f''(x) = -240x^3 + 6x^2 + \frac{1}{2}\left(\frac{-1}{2}\right)(1 - x)^{-3/2}(-1) = -240x^3 + 6x^2 + \tfrac{1}{4}(1 - x)^{-3/2}$$

5. $f'(x) = 2(-2)(1 - 4x)^{-3}(-4) = 16(1 - 4x)^{-3}$

$f''(x) = 16(-3)(1 - 4x)^{-4}(-4) = 192(1 - 4x)^{-4}$

7. $f'(x) = a(-n)x^{-n-1}$; $f''(x) = (-an)(-n - 1)x^{-n-2} = an(n + 1)x^{-n-2}$

9. $f'(x) = \dfrac{-3x^2}{(x^3 - 1)^2}$

$$f''(x) = \frac{-6x(x^3 - 1)^2 + 3x^2(2)(x^3 - 1)(3x^2)}{(x^3 - 1)^4}$$

$$= \frac{6x(x^3 - 1)(-x^3 + 1 + 3x^3)}{(x^3 - 1)^4} = \frac{6x(2x^3 + 1)}{(x^3 - 1)^3}$$

11. $f'(x) = \dfrac{5}{2}\pi x^{3/2} + \dfrac{(-\sin x)x - \cos x}{x^2} = \dfrac{5}{2}\pi x^{3/2} - \dfrac{\sin x}{x} - \dfrac{\cos x}{x^2}$

$$f''(x) = \frac{5}{2}\pi\left(\frac{3}{2}\right)x^{1/2} - \frac{x\cos x - \sin x}{x^2} - \frac{-x^2\sin x - 2x\cos x}{x^4}$$

$$= \frac{15}{4}\pi x^{1/2} - \frac{\cos x}{x} + \frac{2\sin x}{x^2} + \frac{2\cos x}{x^3}$$

13. $f'(x) = \sec x \tan x$

$f''(x) = (\sec x \tan x)\tan x + \sec x(\sec^2 x)$

$= \sec x \tan^2 x + \sec^3 x$

15. $f'(x) = \cot(-4x) + x(-\csc^2(-4x))(-4) = \cot(-4x) + 4x\csc^2(-4x)$

$f''(x) = -\csc^2(-4x)(-4) + 4\csc^2(-4x)$

$+ 4x[2\csc(-4x)(-\csc(-4x))\cot(-4x)(-4)]$

$= 8\csc^2(-4x) + 32x\csc^2(-4x)\cot(-4x)$

17. $f'(x) = \tan^3 2x + x[3(\tan^2 2x)(\sec^2 2x)(2)] = \tan^3 2x + 6x\tan^2 2x\sec^2 2x$

$f''(x) = 3[\tan^2 2x\sec^2 2x](2) + 6\tan^2(2x)\sec^2 2x + 6x[(2\tan 2x\sec^4 2x)(2)$

$+ (\tan^2 2x)2(\sec^2 2x\tan 2x)(2)]$

$= 12\tan^2 2x\sec^2 2x + 24x(\tan 2x\sec^4 2x + \tan^3 2x\sec^2 2x)$

19. $\dfrac{dy}{dx} = \dfrac{3}{2}x^{1/2}$; $\dfrac{d^2y}{dx^2} = \dfrac{3}{4}x^{-1/2}$

21. $\dfrac{dy}{dx} = 3(x^4 - \tan x)^2(4x^3 - \sec^2 x)$

$$\frac{d^2y}{dx^2} = 3[2(x^4 - \tan x)(4x^3 - \sec^2 x)^2 + (x^4 - \tan x)^2\cdot(12x^2 - 2\sec^2 x\tan x)]$$

23. $\dfrac{dy}{dx} = 2ax + b$; $\dfrac{d^2y}{dx^2} = 2a$

25. $\dfrac{dy}{dx} = \dfrac{-1}{(3 - x)^2}(-1) = \dfrac{1}{(3 - x)^2}$; $\dfrac{d^2y}{dx^2} = (-2)\dfrac{1}{(3 - x)^3}(-1) = \dfrac{2}{(3 - x)^3}$

27. $\dfrac{dy}{dx} = -\csc x\cot x$

$$\frac{d^2y}{dx^2} = -(-\csc x\cot x)\cot x - \csc x(-\csc^2 x)$$

$$= \csc x\cot^2 x + \csc^3 x$$

29. $\dfrac{dy}{dx} = \cos x - \sin x$; $\dfrac{d^2y}{dx^2} = -\sin x - \cos x$

31. $f'(x) = -8x$; $f''(x) = -8$; $f^{(3)}(x) = 0$

33. $f'(x) = 2x\cos x^2$; $f''(x) = 2\cos x^2 - 4x^2 \sin x^2$

$$f^{(3)}(x) = -4x\sin x^2 - 8x\sin x^2 - 8x^3\cos x^2 = -12x\sin x^2 - 8x^3\cos x^2$$

35. $f'(x) = \dfrac{-1}{x^2}$; $f''(x) = \dfrac{2}{x^3}$; $f^{(3)}(x) = \dfrac{-6}{x^4}$

37. $f'(x) = \dfrac{3(4x+5) - (3x)(4)}{(4x+5)^2} = \dfrac{15}{(4x+5)^2}$

$$f''(x) = \frac{15(-2)(4)}{(4x+5)^3} = \frac{-120}{(4x+5)^3}$$

$$f^{(3)}(x) = \frac{(-120)(-3)(4)}{(4x+5)^4} = \frac{1440}{(4x+5)^4}$$

39. $\dfrac{dy}{dx} = 6x$; $\dfrac{d^2y}{dx^2} = 6$; $\dfrac{d^3y}{dx^3} = 0$

41. $\dfrac{dy}{dx} = -\dfrac{3}{70}x^{-5/2}$; $\dfrac{d^2y}{dx^2} = \dfrac{3}{28}x^{-7/2}$; $\dfrac{d^3y}{dx^3} = -\dfrac{3}{8}x^{-9/2}$

43. $\dfrac{dy}{dx} = 2x\sin\dfrac{1}{x} + x^2\left(\cos\dfrac{1}{x}\right)\left(\dfrac{-1}{x^2}\right) = 2x\sin\dfrac{1}{x} - \cos\dfrac{1}{x}$

$$\frac{d^2y}{dx^2} = 2\sin\frac{1}{x} + 2x\left(\cos\frac{1}{x}\right)\left(\frac{-1}{x^2}\right) + \left(\sin\frac{1}{x}\right)\left(\frac{-1}{x^2}\right) = 2\sin\frac{1}{x} - \frac{2}{x}\cos\frac{1}{x} - \frac{1}{x^2}\sin\frac{1}{x}$$

$$\frac{d^3y}{dx^3} = 2\left(\cos\frac{1}{x}\right)\left(\frac{-1}{x^2}\right) + \frac{2}{x^2}\cos\frac{1}{x} + \frac{2}{x}\left(\sin\frac{1}{x}\right)\left(\frac{-1}{x^2}\right) + \frac{2}{x^3}\sin\frac{1}{x} - \frac{1}{x^2}\left(\cos\frac{1}{x}\right)\left(\frac{-1}{x^2}\right) = \frac{1}{x^4}\cos\frac{1}{x}$$

45. $\dfrac{dy}{dx} = 3ax^2 + 2bx + c$; $\dfrac{d^2y}{dx^2} = 6ax + 2b$; $\dfrac{d^3y}{dx^3} = 6a$

47. $f'(x) = 24x^7 + \frac{9}{2}x^5 - 3x^{-1/4} - 2x^{-2}$

$$f''(x) = 168x^6 + \tfrac{45}{2}x^4 + \tfrac{3}{4}x^{-5/4} + 4x^{-3}$$

$$f^{(3)}(x) = 1008x^5 + 90x^3 - \tfrac{15}{16}x^{-9/4} - 12x^{-4}$$

$$f^{(4)}(x) = 5040x^4 + 270x^2 + \tfrac{135}{64}x^{-13/4} + 48x^{-5}$$

49. $f'(x) = \pi\cos\pi x$; $f''(x) = -\pi^2\sin\pi x$; $f^{(3)}(x) = -\pi^3\cos\pi x$

$$f^{(4)}(x) = \pi^4\sin\pi x$$

51. $v(t) = f'(t) = -32t + 3$; $a(t) = v'(t) = f''(t) = -32$

53. $v(t) = f'(t) = 2\cos t + 3\sin t$;

$$a(t) = v'(t) = f''(t) = -2\sin t + 3\cos t$$

55. Let $f(x) = c_n x^n + c_{n-1}x^{n-1} + \cdots + c_1 x + c_0$, where $c_n \neq 0$. Then $f'(x) = nc_n x^{n-1} + (n-1)c_{n-1}x^{n-2} + \cdots + c_1$. Thus f' is a polynomial of degree $n-1$. In the same way we find that f'' is a polynomial of degree $n-2$. Continuing, we see that $f^{(n)}$ is a polynomial of degree 0, that is, $f^{(n)}$ is a constant function. Thus $f^{(n+1)} = 0$, and consequently $f^{(n+2)} = 0$ also.

57. a. $f^{(3)}(x) = (f'')'(x) = f'(x)$

b. $f^{(35)}(x) = (f^{(4)})^{(31)}(x) = f^{(31)}(x)$

59. Let $g(x) = 1/x$ and $h(x) = 1/(x+1)$. Then

$$g'(x) = -\frac{1}{x^2}, \quad g''(x) = \frac{(-2)(-1)}{x^3}, \quad g^{(3)}(x) = \frac{(-3)(-2)(-1)}{x^4}$$

and in general,

$$g^{(n)}(x) = \frac{(-1)^n n!}{x^{n+1}}$$

Similarly,

$$h^{(n)}(x) = \frac{(-1)^n n!}{(x+1)^{n+1}}$$

Since $f(x) = g(x) - h(x)$, it follows that

$$f^{(n)}(x) = g^{(n)}(x) - h^{(n)}(x) = \frac{(-1)^n n!}{x^{n+1}} - \frac{(-1)^n n!}{(x+1)^{n+1}} = (-1)^n n!\left[\frac{1}{x^{n+1}} - \frac{1}{(x+1)^{n+1}}\right]$$

61. $h'(x) = g'(f(x))f'(x)$; $h''(x) = g''(f(x))(f'(x))^2 + g'(f(x))f''(x)$

63. $a = \dfrac{dv}{dt} = \dfrac{dv}{dx}\dfrac{dx}{dt} = \dfrac{dv}{dx}v = \left(\dfrac{x}{50} - \dfrac{11}{10}\right)\left(\dfrac{x^2}{100} - \dfrac{11x}{10} + 25\right)$

If $x = -1$, then

$$a = \left(\frac{-1}{50} - \frac{11}{10}\right)\left(\frac{1}{100} + \frac{11}{10} + 25\right) = -29.2432 \text{ (feet per second per second)}$$

SECTION 3.6

1. $6y\dfrac{dy}{dx} = -8x^3$; $\dfrac{dy}{dx} = -\dfrac{4x^3}{3y}$

3. $2y\dfrac{dy}{dx} + \dfrac{dy}{dx} = \dfrac{(1-x) - (1+x)(-1)}{(1-x)^2} = \dfrac{2}{(1-x)^2}$; $\dfrac{dy}{dx} = \dfrac{2}{(2y+1)(1-x)^2}$

5. $\sec y \tan y\dfrac{dy}{dx} - \sec^2 x = 0$; $\dfrac{dy}{dx} = \dfrac{\sec^2 x}{\sec y \tan y}$

7. $\dfrac{(\cos y)(dy/dx)(y^2+1)-(\sin y)2y(dy/dx)}{(y^2+1)^2}=3;\ \dfrac{dy}{dx}=\dfrac{3(y^2+1)^2}{(y^2+1)\cos y-2y\sin y}$

9. $2x+\left(2xy^2+2x^2y\dfrac{dy}{dx}\right)+3y^2\dfrac{dy}{dx}=0;\ \dfrac{dy}{dx}=\dfrac{-2x-2xy^2}{2x^2y+3y^2}$

11. $2x+2y\dfrac{dy}{dx}=\dfrac{(2y(dy/dx))x^2-y^2(2x)}{x^4}=\dfrac{2y}{x^2}\dfrac{dy}{dx}-\dfrac{2y^2}{x^3}$

$\dfrac{dy}{dx}=\dfrac{x+(y^2/x^3)}{(y/x^2)-y}=\dfrac{x^4+y^2}{xy-x^3y}$

13. $\dfrac{1}{2}(xy)^{-1/2}\left(y+x\dfrac{dy}{dx}\right)+\dfrac{1}{2}(x+2y)^{-1/2}\left(1+2\dfrac{dy}{dx}\right)=0$

$\dfrac{dy}{dx}=\dfrac{-y(xy)^{-1/2}-(x+2y)^{-1/2}}{x(xy)^{-1/2}+2(x+2y)^{-1/2}}$

15. $\dfrac{1}{2}(x^2+y^2)^{-1/2}\left(2x+2y\dfrac{dy}{dx}\right)=\dfrac{2(x^2+y^2)^{1/2}-2x\left(\frac{1}{2}\right)(x^2+y^2)^{-1/2}\left(2x+2y\frac{dy}{dx}\right)}{x^2+y^2},$

or $(x^2+y^2)\left(x+y\dfrac{dy}{dx}\right)=2(x^2+y^2)-2x\left(x+y\dfrac{dy}{dx}\right)$

$\dfrac{dy}{dx}=\dfrac{2y^2-x^3-xy^2}{y^3+x^2y+2xy}$

17. $2x+2y\dfrac{dy}{dx}=\dfrac{dy}{dx};\ \dfrac{dy}{dx}=\dfrac{2x}{1-2y}$. At $(0,1)$, $\dfrac{dy}{dx}=\dfrac{0}{1-2}=0.$

19. $y+x\dfrac{dy}{dx}=0;\ \dfrac{dy}{dx}=-\dfrac{y}{x}$. At $(-2,-1)$, $\dfrac{dy}{dx}=\dfrac{-(-1)}{-2}=-\dfrac{1}{2}.$

21. $3x^2+2\left(y+x\dfrac{dy}{dx}\right)=0;\ \dfrac{dy}{dx}=-\dfrac{3x^2+2y}{2x}$. At $(1,2)$, $\dfrac{dy}{dx}=\dfrac{3+2\cdot 2}{-2}=-\dfrac{7}{2}.$

23. $2x+\dfrac{y-x(dy/dx)}{y^2}=0;\ \dfrac{dy}{dx}=\dfrac{2x+(1/y)}{x/y^2}=\dfrac{2xy^2+y}{x}.$

At $(1,-\frac{1}{3})$, $\dfrac{dy}{dx}=\dfrac{2(1)(-\frac{1}{3})^2+(-\frac{1}{3})}{1}=-\dfrac{1}{9}.$

25. $\dfrac{1}{2\sqrt{x}}(\sqrt{y}+2)+(\sqrt{x}+1)\left(\dfrac{1}{2\sqrt{y}}\dfrac{dy}{dx}\right)=0;\ \dfrac{dy}{dx}=\dfrac{\sqrt{y}+2}{2\sqrt{x}}\cdot\dfrac{-2\sqrt{y}}{\sqrt{x}+1}=-\dfrac{y+2\sqrt{y}}{x+\sqrt{x}}.$

At $(1,4)$, $\dfrac{dy}{dx}=-\dfrac{4+2\cdot 2}{1+1}=-4$

27. $\cos x=-(\sin y)\dfrac{dy}{dx};\ \dfrac{dy}{dx}=-\dfrac{\cos x}{\sin y}.$

At $\left(\dfrac{\pi}{6},\dfrac{\pi}{3}\right)$, $\dfrac{dy}{dx}=-\dfrac{\cos(\pi/6)}{\sin(\pi/3)}=-\dfrac{\sqrt{3}/2}{\sqrt{3}/2}=-1$

29. $y^2+2xy\dfrac{dy}{dx}=0;\ \dfrac{dy}{dx}=-\dfrac{y}{2x}$. At $(2,-3)$, $\dfrac{dy}{dx}=-\dfrac{(-3)}{2(2)}=\dfrac{3}{4}.$

Thus ℓ: $y-(-3)=\frac{3}{4}(x-2)$, or $y=\frac{3}{4}x-\frac{9}{2}$.

31. $\cos(x+y)\left(1+\dfrac{dy}{dx}\right)=2;\ \dfrac{dy}{dx}=\dfrac{2}{\cos(x+y)}-1.$

At $(0,\pi)$, $\dfrac{dy}{dx}=\dfrac{2}{\cos(0+\pi)}-1=-2-1=-3.$

Thus ℓ: $y-\pi=-3(x-0)$, or $y=-3x+\pi$.

33. $2x-4y^3\dfrac{dy}{dx}=0;\ \dfrac{dy}{dx}=\dfrac{x}{2y^3}$

$\dfrac{d^2y}{dx^2}=\dfrac{2y^3-x(6y^2(dy/dx))}{(2y^3)^2}=\dfrac{2y^3-6xy^2[x/(2y^3)]}{4y^6}=\dfrac{2y^3-(3x^2/y)}{4y^6}=\dfrac{2y^4-3x^2}{4y^7}$

35. $2x\sin 2y+2x^2(\cos 2y)\dfrac{dy}{dx}=0;\ \dfrac{dy}{dx}=-\dfrac{2x\sin 2y}{2x^2\cos 2y}=-\dfrac{\tan 2y}{x}$

$\dfrac{d^2y}{dx^2}=-\dfrac{(\sec^2 2y)(2(dy/dx))x-\tan 2y}{x^2}=-\dfrac{(\sec^2 2y)2\left(-\frac{\tan 2y}{x}\right)x-\tan 2y}{x^2}$

$=\dfrac{\tan 2y\,(2\sec^2 2y+1)}{x^2}$

37. $2y\dfrac{dy}{dt}-2x\dfrac{dx}{dt}=0$, so $\dfrac{dy}{dt}=\dfrac{2x(dx/dt)}{2y}=\dfrac{x}{y}\dfrac{dx}{dt}$

39. $\dfrac{dx}{dt}\sin y+x\cos y\dfrac{dy}{dt}=0$, so $\dfrac{dy}{dt}=-\dfrac{\sin y}{x\cos y}\dfrac{dx}{dt}=-\dfrac{\tan y}{x}\dfrac{dx}{dt}$

41. $\dfrac{dy}{dt}=-\sin(xy^2)\left[\dfrac{dx}{dt}y^2+2xy\dfrac{dy}{dt}\right]$, so $\dfrac{dy}{dt}=-\dfrac{y^2\sin(xy^2)(dx/dt)}{1+2xy\sin(xy^2)}$

43. a. $5y^4\dfrac{dy}{dx}+\dfrac{dy}{dx}+1=0;\ \dfrac{dy}{dx}=\dfrac{-1}{5y^4+1}$

b. $5y^4\dfrac{dy}{dx}+3y^2\dfrac{dy}{dx}+3x^2+\dfrac{dy}{dx}=0;\ \dfrac{dy}{dx}=\dfrac{-3x^2}{5y^4+3y^2+1}$

c. $5=3y^2\dfrac{dy}{dx}+\cos y\dfrac{dy}{dx}+\dfrac{dy}{dx};\ \dfrac{dy}{dx}=\dfrac{5}{3y^2+\cos y+1}$

45. $3x^2+3y^2\dfrac{dy}{dx}=3y+3x\dfrac{dy}{dx};\ \dfrac{dy}{dx}=\dfrac{x^2-y}{x-y^2}$, and at $(\frac{3}{2},\frac{3}{2})$, $\dfrac{dy}{dx}=\dfrac{\frac{9}{4}-\frac{3}{2}}{\frac{3}{2}-\frac{9}{4}}=-1.$

Tangent line: $y-\frac{3}{2}=-1(x-\frac{3}{2})$, or $y=-x+3$. Normal line: $y-\frac{3}{2}=x-\frac{3}{2}$, or $y=x$. The normal line passes through the origin.

47. The curves intersect at (x, y) such that $x^2 + y^2 = (x-1)^2 + y^2$, or $x^2 = (x-1)^2 = x^2 - 2x + 1$, or $x = \frac{1}{2}$. Thus $y^2 = 1 - x^2 = 1 - \frac{1}{4} = \frac{3}{4}$, so the points of intersection are $(\frac{1}{2}, \sqrt{3}/2)$ and $(\frac{1}{2}, -\sqrt{3}/2)$. To obtain dy/dx, we proceed as follows:

$$x^2 + y^2 = 1 \qquad (x-1)^2 + y^2 = 1$$

$$2x + 2y\frac{dy}{dx} = 0 \qquad 2(x-1) + 2y\frac{dy}{dx} = 0$$

$$\frac{dy}{dx} = \frac{-x}{y} \qquad \frac{dy}{dx} = \frac{1-x}{y}$$

At $(\frac{1}{2}, \sqrt{3}/2)$ the slopes are $m_1 = -\frac{1}{2}/(\sqrt{3}/2) = -1/\sqrt{3}$ and $m_2 = \frac{1}{2}/(\sqrt{3}/2) = 1/\sqrt{3}$. Thus

$$\tan\theta = \frac{(1/\sqrt{3}) - (-1/\sqrt{3})}{1 + (-1/\sqrt{3})(1/\sqrt{3})} = \sqrt{3}$$

so $\theta = \pi/3$. At $(\frac{1}{2}, -\sqrt{3}/2)$ the slopes are $m_1 = -\frac{1}{2}/(-\sqrt{3}/2) = 1/\sqrt{3}$ and $m_2 = \frac{1}{2}/(-\sqrt{3}/2) = -1/\sqrt{3}$. Thus

$$\tan\theta = \frac{-1/\sqrt{3} - 1/\sqrt{3}}{1 + (1/\sqrt{3})(-1/\sqrt{3})} = -\sqrt{3}$$

so $\theta = \frac{2}{3}\pi$.

49. Differentiating implicitly, we obtain $2(x^2 + y^2)(2x + 2y(dy/dx)) = 2x - 2y(dy/dx)$. If $dy/dx = 0$, then $(x^2 + y^2)(2x) = x$, that is, $(2x^2 + 2y^2 - 1)x = 0$. Since $x \neq 0$ by hypothesis, it follows that $2x^2 + 2y^2 - 1 = 0$, or $y^2 = \frac{1}{2} - x^2$. Substituting for y^2 in the original equation, we obtain $(x^2 + \frac{1}{2} - x^2)^2 = x^2 - (\frac{1}{2} - x^2)$, so that $\frac{1}{4} = 2x^2 - \frac{1}{2}$, or $x = \pm\sqrt{\frac{3}{8}} = \pm\sqrt{6}/4$. Then $y^2 = \frac{1}{2} - x^2 = \frac{1}{2} - \frac{3}{8} = \frac{1}{8}$. Thus the points at which the tangent lines are horizontal are $(\sqrt{6}/4, \sqrt{2}/4)$, $(\sqrt{6}/4, -\sqrt{2}/4)$, $(-\sqrt{6}/4, \sqrt{2}/4)$, and $(-\sqrt{6}/4, -\sqrt{2}/4)$.

SECTION 3.7

1. As in (2), $dV/dt = 4\pi r^2(dr/dt)$, and we are to determine dV/dt at the instant t_0 when $r = 4$. Since $dr/dt = -\frac{1}{2}$ by hypothesis, we have

$$\left.\frac{dV}{dt}\right|_{t=t_0} = 4\pi r^2\left.\frac{dr}{dt}\right|_{t=t_0} = 4\pi(4^2)(-\tfrac{1}{2}) = -32\pi$$

Thus the volume decreases at the rate of 32π cubic inches per minute when $t = t_0$.

3. Here

$$4\pi r^2\frac{dr}{dt} = \frac{dV}{dt} = -\frac{2}{V} = -\frac{2}{4\pi r^3/3} = -\frac{3}{2\pi r^3}$$

and we are to determine dr/dt at the instant t_0 when $r = \frac{1}{2}$. We have

$$\left.\frac{dr}{dt}\right|_{t=t_0} = -\frac{3}{(2\pi r^3)4\pi r^2} = -\frac{3}{8\pi^2 r^5} = -\frac{3}{8\pi^2(1/32)} = -\frac{12}{\pi^2} \text{ (inches per hour)}$$

Thus the radius decreases at the rate of $12/\pi^2$ inches per hour.

5. Here $4\pi r^2(dr/dt) = dV/dt = \frac{d}{dt}(4\sqrt{t}) = 2/\sqrt{t}$ for $t > 0$, and we are to determine dr/dt when $t = 64$. Since $V = 4\sqrt{64} = 32$ for $t = 64$, we have $32 = V = \frac{4}{3}\pi r^3$ at that instant, so $r^3 = 96/(4\pi) = 24/\pi$ and thus $r = (24/\pi)^{1/3}$ when $t = 64$. Then

$$\left.\frac{dr}{dt}\right|_{t=64} = \frac{2}{\sqrt{t}}\frac{1}{4\pi r^2} = \frac{2}{\sqrt{64}}\frac{1}{4\pi(24/\pi)^{2/3}} = \frac{1}{64\pi^{1/3}3^{2/3}} \text{ (inches per second)}$$

Thus after 64 seconds the radius increases at the rate of $1/(64\pi^{1/3}3^{2/3})$ inches per second.

7. Let A denote the area and r the radius of the circular pool. Then $A = \pi r^2$ and $dA/dt = 2\pi r(dr/dt)$, and we are to determine dr/dt at the instant t_0 when $r = 10$. Since $dA/dt = 3$ by hypothesis, we have

$$3 = \left.\frac{dA}{dt}\right|_{t=t_0} = 2\pi r\left.\frac{dr}{dt}\right|_{t=t_0} = 2\pi(10)\left.\frac{dr}{dt}\right|_{t=t_0}, \quad \text{so} \quad \left.\frac{dr}{dt}\right|_{t=t_0} = \frac{3}{2\pi(10)} = \frac{3}{20\pi}$$

Thus the radius increases at the rate of $3/(20\pi)$ inches per minute when $t = t_0$.

9. Let x be the distance from the bottom of the ladder to the wall and y the distance from the ground to the top of the ladder. Then $x^2 + y^2 = 13^2$, so $2x(dx/dt) + 2y(dy/dt) = 0$. We are to find dx/dt at the instant t_0 when $x = 3$. Because $x^2 + y^2 = 13^2$, it follows that if $x = 3$, then $y^2 = 13^2 - x^2 = 13^2 - 3^2 = 160$, so $y = \sqrt{160} = 4\sqrt{10}$. Since $dy/dt = 1$ by hypothesis, for $t = t_0$ the equation $2x(dx/dt) + 2y(dy/dt) = 0$ become

$$2(3)\left.\frac{dx}{dt}\right|_{t=t_0} + 2(4\sqrt{10}) = 0, \quad \text{so that} \quad \left.\frac{dx}{dt}\right|_{t=t_0} = -\frac{8\sqrt{10}}{6} = -\frac{4}{3}\sqrt{10}$$

Thus the base of the ladder approaches the wall at the rate of $\frac{4}{3}\sqrt{10}$ feet per second when $t = t_0$.

11. Let x be the distance from the bottom of the board to the wall and y the distance from the ground to the top of the board, so $x^2 + y^2 = 5^2$, and thus $2x(dx/dt) + 2y(dy/dt) = 0$.

a. We are to find dx/dt at the instant t_0 when $x = 4$ and $dy/dt = -2$. If $x = 4$, then $y^2 = 5^2 - x^2 = 5^2 - 4^2 = 9$, so $y = 3$. We obtain

$$2(4)\left.\frac{dx}{dt}\right|_{t=t_0} + 2(3)(-2) = 0, \quad \text{so} \quad \left.\frac{dx}{dt}\right|_{t=t_0} = \frac{-2(3)(-2)}{2(4)} = \frac{3}{2}$$

Thus the bottom end slides at the rate of 3/2 feet per second when $t = t_0$.

b. Let A denote the area of the region, so $A = \frac{1}{2}xy$, and $dA/dt = \frac{1}{2}(dx/dt)y + \frac{1}{2}x(dy/dt)$. We are to find dA/dt at the instant t_0 when $x = 4$, $y = 3$, $dx/dt = \frac{3}{2}$, and $dy/dt = -2$. We obtain

$$\left.\frac{dA}{dt}\right|_{t=t_0} = \frac{1}{2}\left(\frac{3}{2}\right)3 + \frac{1}{2}(4)(-2) = -\frac{7}{4}$$

Thus the area shrinks at the rate of $\frac{7}{4}$ square feet per second when $t = t_0$.

13. From (8) we have $dh/dt = [9/(\pi h^2)](dV/dt)$, and we are to find dV/dt at the instant t_0 when $h = 2$. Since $dh/dt = \frac{1}{2}$ by hypothesis, we obtain

$$\frac{1}{2} = \frac{9}{\pi(2)^2}\left.\frac{dV}{dt}\right|_{t=t_0}, \quad \text{so that} \quad \left.\frac{dV}{dt}\right|_{t=t_0} = \frac{1}{2}\frac{\pi(2)^2}{9} = \frac{2\pi}{9}$$

Thus the volume increases at the rate of $2\pi/9$ cubic inches per second when $t = t_0$.

15. Since $x = \tan\theta$ and since $d\theta/dt = 10\pi$, the Chain Rule implies that

$$\frac{dx}{dt} = (\sec^2\theta)\left(\frac{d\theta}{dt}\right) = 10\pi\sec^2\theta$$

At any time t_0 at which the spotlight shines on the sand 2 miles from the lighthouse, $\sec\theta = 2$, so that $(dx/dt)|_{t=t_0} = 40\pi$ (miles per minute).

17. Let x be the height of the bottom end of the box, and y the distance the bottom end has traveled at any time (see the figure). Then by similar triangles, $x/5 = y/20$, so that $x = \frac{1}{4}y$. We are to find dx/dt. Since $dy/dt = 3$ by hypothesis, we obtain $dx/dt = \frac{1}{4}(dy/dt) = \frac{1}{4}(3) = \frac{3}{4}$. Thus the bottom of the box rises at the rate of $\frac{3}{4}$ foot per second.

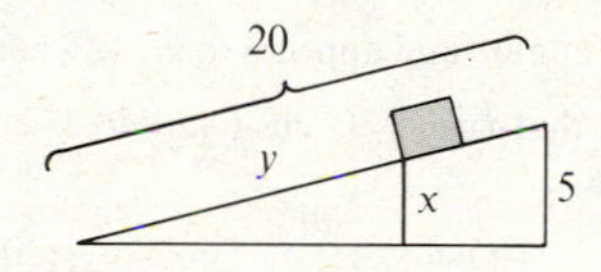

19. Let x be the distance along the pulley and y the distance from the bow to the dock at any given time. Then from the figure, $x^2 = y^2 + 5^2$, so that $2x(dx/dt) = 2y(dy/dt)$. We are to find dx/dt at the instant t_0 when $y = 12$. If $y = 12$, then $x^2 = y^2 + 5^2 = 12^2 + 5^2 = 169$, so $x = 13$. Since $dy/dt = -2$ by hypothesis, we obtain

$$2(13)\left.\frac{dx}{dt}\right|_{t=t_0} = 2(12)(-2), \quad \text{so} \quad \left.\frac{dx}{dt}\right|_{t=t_0} = -\frac{24}{13}$$

Thus the rope is pulled in at the rate of 24/13 feet per second when $t = t_0$.

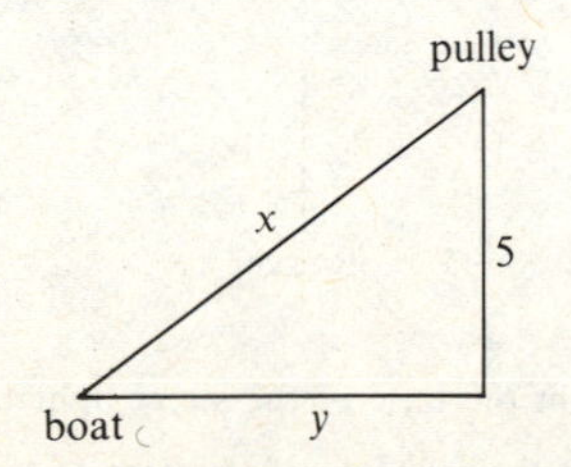

21. Assume that $x > 0$, as in the figure. By hypothesis, $y = x^2 + 1$, so that $dy/dt = 2x(dx/dt)$. We are to find dx/dt at the instant t_0 when $y = 2501$. If $y = 2501$, then $x^2 = y - 1 = 2501 - 1 = 2500$, so $x = 50$. Since $dy/dt = -100$ by hypothesis, we obtain

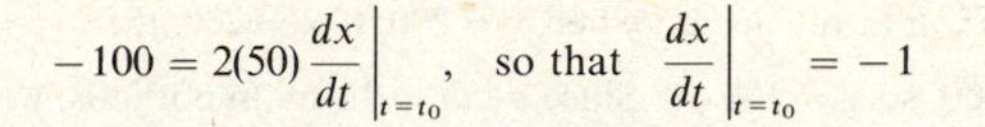

$$-100 = 2(50)\left.\frac{dx}{dt}\right|_{t=t_0}, \quad \text{so that} \quad \left.\frac{dx}{dt}\right|_{t=t_0} = -1$$

Thus the shadow moves at the rate of 1 foot per second when $t = t_0$.

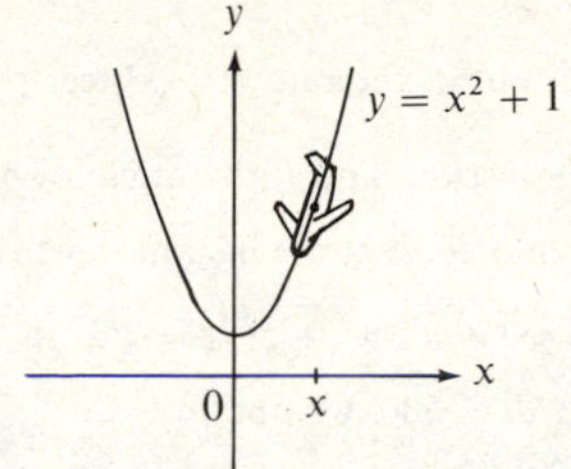

23. Let x be the distance between the runner and third base. Then $\tan\theta = x/90$, so that $(\sec^2\theta)(d\theta/dt) = \frac{1}{90}(dx/dt)$. We are to find $d\theta/dt$ at the instant t_0 when $x = 30$. Since the runner is approaching third base at the rate of 24 feet per second, we have $dx/dt = -24$. Moreover, if $x = 30$, then $\tan\theta = \frac{30}{90} = \frac{1}{3}$, and thus $\sec^2\theta = 1 + \tan^2\theta = 1 + (\frac{1}{3})^2 = \frac{10}{9}$. Therefore

$$\left.\frac{d\theta}{dt}\right|_{t=t_0} = \frac{1}{90\sec^2\theta}\left.\frac{dx}{dt}\right|_{t=t_0} = \frac{1}{90(\frac{10}{9})}(-24) = -\frac{6}{25}$$

Consequently the angle θ is changing at the rate of $\frac{6}{25}$ radians per second when the runner is 30 feet from third base.

25. Let x be the distance between the sports car and the intersection, and y the distance between the police car and sports car (see the figure). Then $y^2 = x^2 + (\frac{1}{4})^2$, so $2y(dy/dt) = 2x(dx/dt)$. We are to find x under the condition that $dx/dt = -40$ and $dy/dt = -25$. This yields $2y(-25) = 2x(-40)$, so that $y = \frac{8}{5}x$. Since $x^2 = y^2 - (\frac{1}{4})^2$, we substitute $y = \frac{8}{5}x$ to obtain $x^2 = (\frac{8}{5}x)^2 - (\frac{1}{4})^2$, and thus $x^2 = \frac{64}{25}x^2 - \frac{1}{16}$, so $\frac{39}{25}x^2 = \frac{1}{16}$, or $x^2 = 25/(39 \cdot 16)$. Therefore $x = \frac{5}{156}\sqrt{39} \approx \frac{1}{5}$. Thus the sports car would be $5\sqrt{39}/156$ ($\approx\frac{1}{5}$) mile from the intersection.

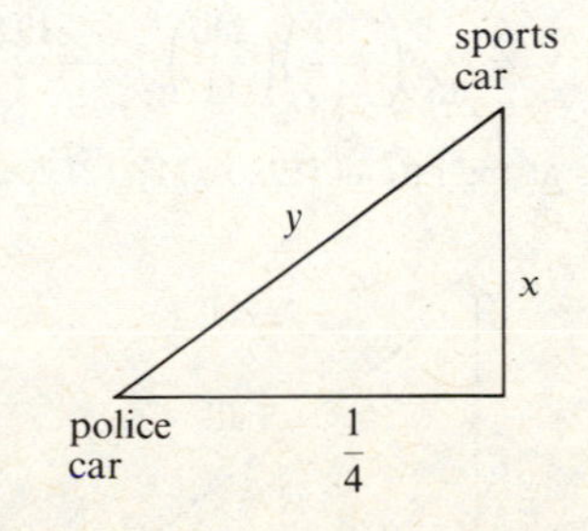

27. Let x be the length of string let out, and y the ground distance from the holder to a position directly below the kite. Then $x^2 = y^2 + 100^2$, so that $2x(dx/dt) = 2y(dy/dt)$. We

are to find dx/dt at the instant when $x = 200$. If $x = 200$, then $y^2 = x^2 - 100^2 = 200^2 - 100^2 = 30{,}000$, so $y = 100\sqrt{3}$. Since $dy/dt = 10$ by hypothesis, we have

$$2(200)\left.\frac{dx}{dt}\right|_{t=t_0} = 2(100\sqrt{3})(10), \quad \text{so that} \quad \left.\frac{dx}{dt}\right|_{t=t_0} = 5\sqrt{3}$$

Thus the string must be let out at the rate of $5\sqrt{3}$ feet per second when $t = t_0$.

29. Let x be the altitude of the rocket, and θ the angle of elevation (see the figure). Then $\tan\theta = x/5$, and we are to find $d\theta/dt$ at the instant t_0 when $x = 2$. Now $(\sec^2\theta)(d\theta/dt) = \frac{1}{5}(dx/dt)$, and if $x = 2$, then $\sec\theta = \sqrt{x^2 + 5^2}/5 = \sqrt{2^2 + 5^2}/5 = \sqrt{29}/5$, so that $\sec^2\theta = \frac{29}{25}$. Since at that instant $dx/dt = 300$, we obtain

$$\left.\frac{29}{25}\frac{d\theta}{dt}\right|_{t=t_0} = \frac{1}{5}(300), \quad \text{so that} \quad \left.\frac{d\theta}{dt}\right|_{t=t_0} = \frac{1500}{29}$$

Thus the angle of elevation increases at the rate of $\frac{1500}{29}$ radians per hour when $t = t_0$.

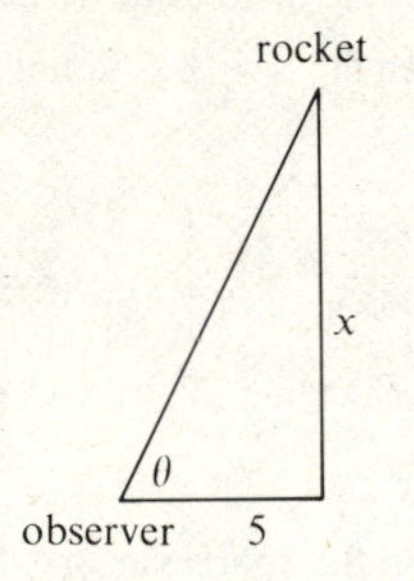

31. Let x be the distance from the shadow to the point directly beneath the light, and s the distance from the ball to the ground. We are to find dx/dt at the instant t_0 when $s = 5$. Notice that $s/(x - 15) = 16/x$, or $s = 16(1 - (15/x))$, so that $ds/dt = (240/x^2)/(dx/dt)$. But $s = 16 - 16t^2$, so that $ds/dt = -32t$. Thus $-32t = ds/dt = (240/x^2)/(dx/dt)$, or $dx/dt = -2tx^2/15$. If $s = 5$, then $5 = 16(1 - (15/x))$, so that $x = \frac{240}{11}$. Also if $s = 5$, then $5 = 16 - 16t^2$, so that $t = \sqrt{11}/4$. Thus if $s = 5$, then

$$\left.\frac{dx}{dt}\right|_{t=t_0} = \frac{-2}{15}\left(\frac{\sqrt{11}}{4}\right)\left(\frac{240}{11}\right)^2 = \frac{-1920\sqrt{11}}{121}$$

so that the shadow moves at the rate of $(1920\sqrt{11})/121$ feet per second when $t = t_0$.

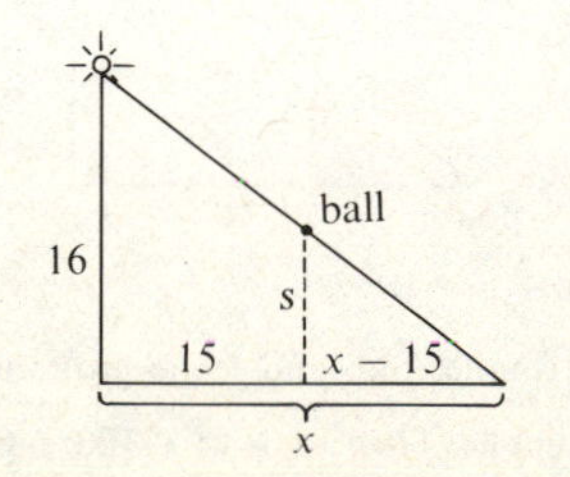

33. Using the notation in the figure, we have $x = 3000\tan\theta$, so that

$$\frac{dx}{dt} = 3000(\sec^2\theta)(d\theta/dt)$$

When the distance between the helicopter and the searchlight is 5000 feet, we have $\sec\theta = \frac{5000}{3000} = \frac{5}{3}$. Since $dx/dt = 100$, it follows that $100 = 3000(\frac{5}{3})^2(d\theta/dt)$, so that $d\theta/dt = \frac{3}{250}$ (radians per second) when the distance between the helicopter and the searchlight is 5000 feet.

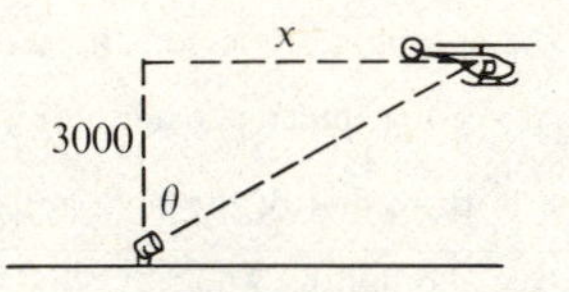

35. Let x be one half the width of the sign seen by the observer, and θ the angle shown in the figure. We are to find dx/dt at the instant t_0 when $x = 3$ and $dx/dt > 0$. Since the sign makes 10 revolutions per minute and appears to grow when $t = t_0$, we have $d\theta/dt = 20\pi$. Notice that $x = 5\sin\theta$, so that $dx/dt = (5\cos\theta)(d\theta/dt)$. If $x = 3$, then $\cos\theta = \frac{4}{5}$. Therefore

$$\left.\frac{dx}{dt}\right|_{t=t_0} = (5\cos\theta)\left.\frac{d\theta}{dt}\right|_{t=t_0} = 5\left(\frac{4}{5}\right)(20\pi) = 80\pi$$

Thus half the width changes at the rate of 80π feet per minute, and the total width changes at the rate of 160π feet per minute when $t = t_0$.

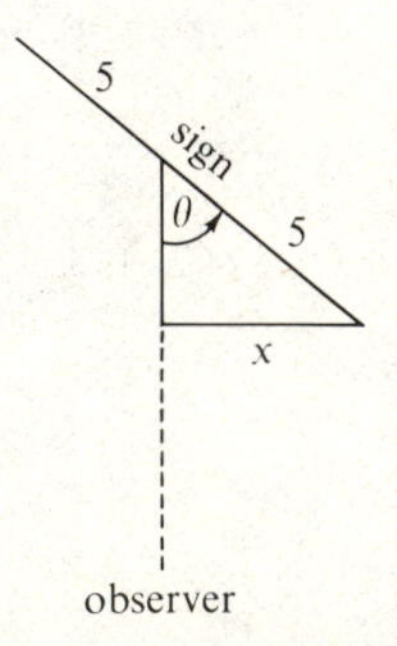

37. Let x be the distance from the base of the street light to the forelegs of the deer and let y be the distance from the rear legs of the deer to the tip of the shadow. We want y as a function of x.

Notice that if the deer is close to the base of the light, the tip is formed by a ray just passing over the deer's rump. Since the rump is 4 feet high and the deer is 5 feet long, by using similar triangles we have

$$\frac{y}{4} = \frac{x+y+5}{20} \quad \text{or} \quad y = \frac{x+5}{4}$$

Thus for "small" x, $dy/dt = \frac{1}{4}(dx/dt) = -\frac{3}{4}$ (feet per second).

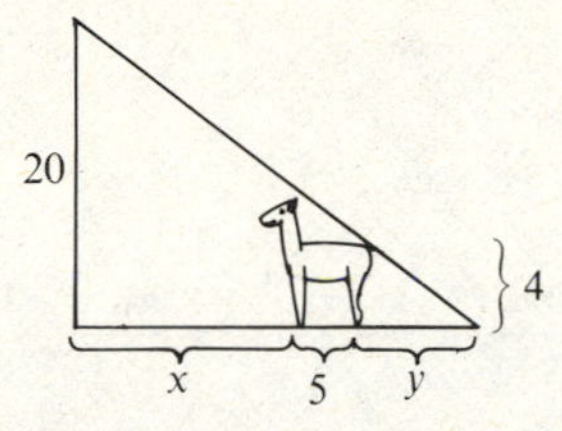

Now if the deer is far from the base of the light, the tip of the shadow is formed by light just passing over the deer's head. Since the head is 6 feet high and the deer is 5 feet long, by using similar triangles we have

$$\frac{y+5}{6} = \frac{x+y+5}{20} \quad \text{or} \quad y = \frac{3}{7}x - 5$$

Thus for "large" x, $dy/dt = \frac{3}{7}(dx/dt) = -9/7$ (feet per second).

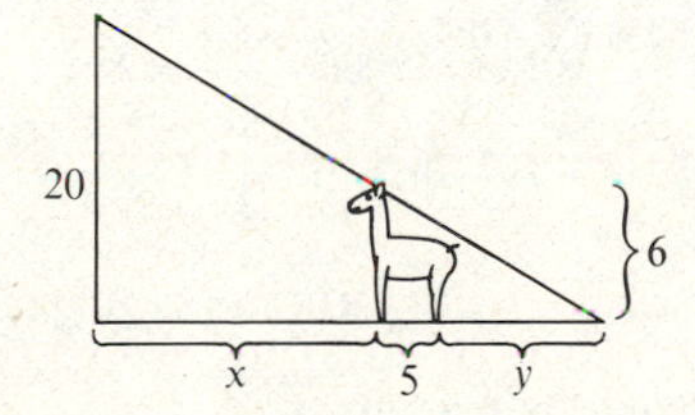

We now determine the boundary between "x small" and "x large". That boundary occurs when the same ray of light just passes over the head and just passes over the rump. At that point $(x+5)/4 = y = \frac{3}{7}x - 5$. Solving for x, we find the boundary at $x = 35$ (feet).

For (a) we conclude that when the deer is 48 feet from the street light, x is "large", so the shadow is decreasing at $\frac{9}{7}$ feet per second. For (b) we conclude that when the deer is 24 feet from the street light, x is "small", so the shadow is decreasing at $\frac{3}{4}$ feet per second.

SECTION 3.8

Exercises 1–12 can be solved with either (2) or (7). We will use (2).

1. Let $f(x) = \sqrt{x}$, $a = 100$, and $h = 1$. Then $f'(x) = \frac{1}{2}x^{-1/2}$, $f'(100) = \frac{1}{20}$, and $\sqrt{101} \approx f(100) + f'(100)h = 10 + \frac{1}{20}(1) = 10.05$.

3. Let $f(x) = \sqrt[3]{x}$, $a = 27$, and $h = 2$. Then $f'(x) = \frac{1}{3}x^{-2/3}$, $f'(27) = \frac{1}{27}$, and $\sqrt[3]{29} \approx f(27) + f'(27)h = 3 + \frac{1}{27}(2) \approx 3.07407$.

5. Let $f(x) = \sqrt{x}$, $a = 36$, and $h = 0.36$. Then $f'(x) = \frac{1}{2}x^{-1/2}$, $f'(36) = \frac{1}{12}$, and $\sqrt{36.36} \approx f(36) + f'(36)h = 6 + \frac{1}{12}(0.36) = 6.03$.

7. Let $f(x) = x^{4/3}$, $a = 27$, and $h = 1$. Then $f'(x) = \frac{4}{3}x^{1/3}$, $f'(27) = 4$, and $(28)^{4/3} \approx f(27) + f'(27)h = 81 + 4(1) = 85$.

9. Let $f(x) = \cos x$, $a = \pi/6$, and $h = -\pi/78$. Then

$$f'(x) = -\sin x, \qquad f'\left(\frac{\pi}{6}\right) = \frac{-1}{2}$$

and

$$\cos\left(\frac{2\pi}{13}\right) \approx f\left(\frac{\pi}{6}\right) + f'\left(\frac{\pi}{6}\right)h = \frac{\sqrt{3}}{2} + \left(\frac{-1}{2}\right)\left(\frac{-\pi}{78}\right) = \frac{\sqrt{3}}{2} + \frac{\pi}{156} \approx .886164$$

11. Let $f(x) = \sec x$, $a = \pi/4$, and $h = -\pi/68$. Then $f'(x) = \sec x \tan x$, $f'(\pi/4) = \sqrt{2}$, and $\sec(4\pi/17) \approx f(\pi/4) + f'(\pi/4)h = \sqrt{2} + \sqrt{2}(-\pi/68) \approx 1.34888$.

13. $df = f'(4)h = \left(\frac{1}{2\sqrt{4}}\right)(.2) = .05$

15. $df = f'(2)h = \frac{1}{2}(1 + 2^3)^{-1/2}(12)(.01) = .02$

17. $df = 15x^2\,dx$

19. $df = -\sin x \cos(\cos x)\,dx$

21. $du = 2x^3(1 + x^4)^{-1/2}\,dx$

23. a. By (8) and Theorem 3.4,

$$d(u+v) = \frac{d}{dx}(u+v)\,dx = \left(\frac{du}{dx} + \frac{dv}{dx}\right)dx = \left(\frac{du}{dx}\right)dx + \left(\frac{dv}{dx}\right)dx = du + dv$$

b. By (8) and Theorem 3.5, $d(cu) = \frac{d}{dx}(cu)\,dx = c\left(\frac{du}{dx}\right)dx = c\,du$.

c. By (8) and Theorem 3.6,

$$d(uv) = \frac{d}{dx}(uv)\,dx = \left(\frac{du}{dx}v + u\frac{dv}{dx}\right)dx = v\left(\frac{du}{dx}\right)dx + u\left(\frac{dv}{dx}\right)dx = v\,du + u\,dv$$

d. By (8) and Theorem 3.7,

$$d\left(\frac{u}{v}\right) = \left(\frac{d}{dx}\left(\frac{u}{v}\right)\right)dx = \left[\frac{v\left(\frac{du}{dx}\right) - u\left(\frac{dv}{dx}\right)}{v^2}\right]dx$$

$$= \frac{v\left(\frac{du}{dx}\right)dx - u\left(\frac{dv}{dx}\right)dx}{v^2} = \frac{v\,du - u\,dv}{v^2}$$

25. We let $f(x) = x^3 - 3x - 1$, so that $f'(x) = 3x^2 - 3$. Letting the initial value of c be 2 and using a computer, we obtain 1.879385 for the desired approximate solution.

27. We let $f(x) = x^3 - 2x - 5$, so that $f'(x) = 3x^2 - 2$. Letting the initial value of c be 2 and using a computer, we obtain 2.094552 for the desired approximate solution.

29. We let $f(x) = 2x^3 - 5x - 3$, so that $f'(x) = 6x^2 - 5$. Letting the initial value of c be -0.5 and using a computer, we obtain -0.822876 for the desired approximate solution.

31. We let $f(x) = \tan x - x$, so that $f'(x) = \sec^2 x - 1$. Letting the initial value of c be 4.5 and using a computer, we obtain 4.493410 for the desired approximate solution.

33. We let $f(x) = x^2 - 15$, so that $f'(x) = 2x$. Using the Newton-Raphson method with initial value 4 for c, and using a computer, we find that $\sqrt{15} \approx 3.872983$.

35. We let $f(x) = x^3 - 9$, so that $f'(x) = 3x^2$. Using the Newton-Raphson method with initial value 2 for c, and using a computer, we find that $\sqrt[3]{9} \approx 2.080084$.

37. Notice that $f'(x) = 4x^3 + 4x - 1$. Letting the initial value of c be first -0.5 and then 0.5 and using a computer, we obtain -0.18 and 0.60, respectively, for the approximate zeros of f.

39. By (10), $c_{n+1} = c_n - 0 = c_n$. Thus $f(c_{n+1}) = f(c_n) = 0$ and $f'(c_{n+1}) = f'(c_n) \neq 0$, so by using (10) again with n replaced by $n + 1$, we find that $c_{n+2} = c_{n+1} = c_n$. In general, $c_m = c_n$ for $m \geq n$.

41. Notice that $f'(x) = 4x^3 + 2x + 8$. Letting the initial value of c be -1 and using a computer, we obtain 0.123078 for the approximate zero of f. However, 0.123078 lies outside the interval $[-2, 0]$, and hence is not the desired zero.

43. Notice that $f'(x) = (1 - x)/(x + 1)^3$. Letting the initial value of c be 2 and using a computer, we find that the successive values of c become larger. For example, $c_{10} > 2604$.

45. Using the Difference and Quotient Rules for differentiation, we find that

$$N'_f(x) = 1 - \frac{f'(x)f'(x) - f(x)f''(x)}{[f'(x)]^2} = \frac{[f'(x)]^2 - [f'(x)]^2 + f(x)f''(x)}{[f'(x)]^2} = \frac{f(x)f''(x)}{[f'(x)]^2}$$

47. The volume of a ball of radius r is given by $V(r) = \frac{4}{3}\pi r^3$. Since $V'(r) = 4\pi r^2$, we use (2) to determine that the volume of the material in the ball is, $V(5.137) - V(5) \approx V'(5)(.137) = 4\pi 5^2(.137) \approx 43.0398$ (cubic inches).

CHAPTER 3 REVIEW

1. $f'(x) = -12x^2 - \dfrac{4}{x^3}$

3. $f'(x) = \dfrac{2(5x+7) - (2x-9)5}{(5x+7)^2} = \dfrac{59}{(5x+7)^2}$

5. $g'(x) = \dfrac{(2x-1)^2 - x[2(2x-1)(2)]}{(2x-1)^4} = -\dfrac{2x+1}{(2x-1)^3}$

7. $f'(t) = (-\sin t)\sin 2t + (\cos t)(2\cos 2t) = -\sin t \sin 2t + 2\cos t \cos 2t$

9. $f'(t) = 5\tan t + 5t\sec^2 t + 9\sec 3t \tan 3t$

11. $\dfrac{dy}{dx} = 12x^2 - \sqrt{3} - \dfrac{2}{5x^2}$

13. $\dfrac{dy}{dx} = [-\csc^2 x^3](3x^2) = -3x^2\csc^2 x^3$

15. $\dfrac{dy}{dx} = \dfrac{\cos x\,(1 - \sec x) - \sin x\,(-\sec x \tan x)}{(1 - \sec x)^2} = \dfrac{\cos x - 1 + \tan^2 x}{(1 - \sec x)^2}$

17. $\dfrac{dy}{dx} = 3x^2\sqrt{x^2 - 4} + x^3[\frac{1}{2}(x^2 - 4)^{-1/2}(2x)] = 3x^2\sqrt{x^2 - 4} + x^4/\sqrt{x^2 - 4}$

19. $f'(x) = 9x^2 - 4x$, so $f'(1) = 5$. Thus ℓ: $y - 5 = 5(x - 1)$, or $y = 5x$.

21. $f'(x) = \cos\sqrt{2}x - x(\sqrt{2}\sin\sqrt{2}x)$, so $f'(0) = 1$. Thus ℓ: $y - 0 = 1(x - 0)$, or $y = x$.

23. $f'(x) = \sqrt{x - 1} + x\dfrac{1}{2\sqrt{x - 1}}$, so $f'(5) = 2 + \dfrac{5}{4} = \dfrac{13}{4}$.

Thus ℓ: $y - 10 = \frac{13}{4}(x - 5)$, or $y = \frac{13}{4}x - \frac{25}{4}$.

25. $\displaystyle\lim_{x\to 0^-}\frac{f(x) - f(0)}{x - 0} = \lim_{x\to 0^-}\frac{2\sin x - 0}{x} = 2\lim_{x\to 0^-}\frac{\sin x}{x} = 2;$

$\displaystyle\lim_{x\to 0^+}\frac{f(x) - f(0)}{x - 0} = \lim_{x\to 0^+}\frac{3x^2 + 2x - 0}{x} = \lim_{x\to 0^+}(3x + 2) = 2.$

Therefore $f'(0) = 2$. Thus ℓ: $y - 0 = 2(x - 0)$, or $y = 2x$.

27. $f'(x) = 3x^{11} - 36x^5$; $f''(x) = 33x^{10} - 180x^4$

29. $f'(t) = 3t(t^2 + 9)^{1/2}$; $f''(t) = 3(t^2 + 9)^{1/2} + 3t^2(t^2 + 9)^{-1/2}$

31. $f'(x) = 2\tan x \sec^2 x$; $f''(x) = 2\sec^4 x + 4\tan^2 x \sec^2 x$

33. $9y^2\dfrac{dy}{dx} - \left(8xy + 4x^2\dfrac{dy}{dx}\right) + \left(y + x\dfrac{dy}{dx}\right) = 0$; $\dfrac{dy}{dx} = \dfrac{8xy - y}{9y^2 - 4x^2 + x}$

35. $\dfrac{dy}{dx}(\sqrt{x} + 1) + y\left(\dfrac{1}{2\sqrt{x}}\right) = 1$; $\dfrac{dy}{dx} = \dfrac{2\sqrt{x} - y}{2\sqrt{x}(\sqrt{x} + 1)}$

37. $3y^2\dfrac{dy}{dx} + \cos(xy^2)\left[y^2 + 2xy\dfrac{dy}{dx}\right] = 0$; $\dfrac{dy}{dx} = \dfrac{-y^2\cos xy^2}{3y^2 + 2xy\cos xy^2} = -\dfrac{y\cos xy^2}{3y + 2x\cos xy^2}$

39. $6x^2 - (4\cos 4y)\dfrac{dy}{dx} = 2xy + x^2\dfrac{dy}{dx}$; $\dfrac{dy}{dx} = \dfrac{6x^2 - 2xy}{x^2 + 4\cos 4y}$. At $(1, 0)$, $\dfrac{dy}{dx} = \dfrac{6}{1 + 4} = \dfrac{6}{5}$

41. $\dfrac{dx}{dt}y + x\dfrac{dy}{dt} = 0$, so $\dfrac{dy}{dt} = -\dfrac{y}{x}\dfrac{dx}{dt}$

43. $df = (2x \cos x - x^2 \sin x)\,dx$

45. $df = 7(x - \sin x)^6(1 - \cos x)\,dx$

47. Let $f(x) = 1 + \sqrt{x}$. Then $df = f'(x)\,dx = (1/2\sqrt{x})\,dx$, so if $x = 9$ and $dx = 1$, then $f(9) = 4$ and $df = \frac{1}{6}$, so $1 + \sqrt{10} = f(10) \approx f(9) + df = 4 + \frac{1}{6} = \frac{25}{6}$.

49. Let $f(x) = \sec x$. Then $df = f'(x)\,dx = \sec x \tan x\,dx$, so if $x = \pi/4$ and $dx = \frac{1}{100}\pi$, then

$$f\left(\frac{\pi}{4}\right) = \sec\frac{\pi}{4} = \sqrt{2} \quad\text{and}\quad df = \left(\sec\frac{\pi}{4}\tan\frac{\pi}{4}\right)\left(\frac{1}{100}\pi\right) = \frac{\sqrt{2}}{100}\pi$$

so $\sec 0.26\pi = f(0.26\pi) \approx f(\pi/4) + df = \sqrt{2} + (\sqrt{2}/100)\pi = \sqrt{2}\,(1 + \pi/100)$.

51. We let $f(x) = 2\sin x - x$, so that $f'(x) = 2\cos x - 1$. Letting the initial value of c be 2 and using a computer, we obtain 1.895 for the approximate zero of f, that is, the approximate solution of the given equation.

53. To find the point of intersection, we solve the equation

$$x^2 + 1 = x^2 - \cos\left(\frac{\pi}{x^2+1}\right), \quad\text{or}\quad \cos\left(\frac{\pi}{x^2+1}\right) = -1$$

so that $\pi/(x^2 + 1) = \pi + 2n\pi$ for some integer n. Since $0 < \pi/(x^2 + 1) \le \pi$, the only solution is $x = 0$. Since

$$f'(x) = 2x \quad\text{and}\quad g'(x) = 2x - \frac{2\pi x}{(x^2+1)^2}\sin\left(\frac{\pi}{x^2+1}\right)$$

we have $f'(0) = 0 = g'(0)$. Thus the tangent lines at the point of intersection are identical.

55. $$\lim_{x\to a}\frac{f(x) - f(a)}{x^{1/2} - a^{1/2}} = \lim_{x\to a}\left[\frac{f(x) - f(a)}{x^{1/2} - a^{1/2}}\,\frac{x^{1/2} + a^{1/2}}{x^{1/2} + a^{1/2}}\right]$$
$$= \lim_{x\to a}\frac{f(x) - f(a)}{x - a}\cdot\lim_{x\to a}(x^{1/2} + a^{1/2}) = 2a^{1/2}f'(a)$$

57. $\dfrac{dy}{dx} = a\cos x - b\sin x$; $\dfrac{d^2y}{dx^2} = -a\sin x - b\cos x = -y$. Thus $\dfrac{d^2y}{dx^2} + y = 0$.

59. a. $\dfrac{dV}{dr} = \dfrac{2\pi}{3}rh$ b. $\dfrac{dV}{dh} = \dfrac{\pi}{3}r^2$

c. Differentiating the equation $V = \pi r^2 h/3$ implicitly with respect to V, we obtain $1 = (\pi/3)r^2(dh/dV)$, so that $dh/dV = 3/\pi r^2$.

61. Let x be the length of the shadow and y the distance from the person to the tower. By the figure, $(x + y)/40 = x/6$, so that $6x + 6y = 40x$ and thus $x = \frac{3}{17}y$. We are to show that dx/dt is constant and negative. Now $dx/dt = \frac{3}{17}(dy/dt)$, and since $dy/dt = -150$ by hypothesis, we obtain $dx/dt = \frac{3}{17}(-150) = -\frac{450}{17}$, a negative constant.

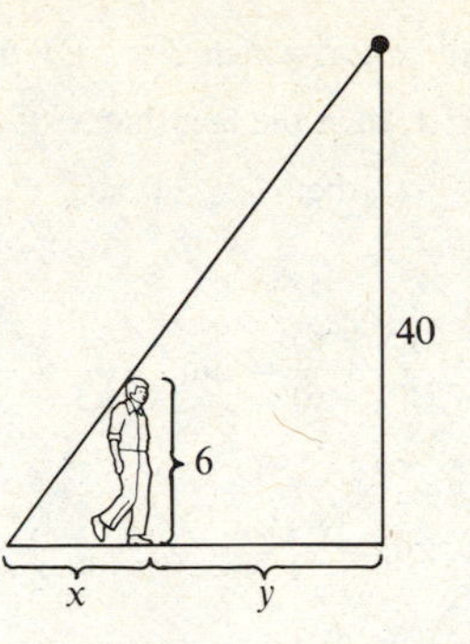

63. $m_C(x) = 12 - 3x^2$ and $m_R(x) = \frac{45}{4}$, so we need to determine the value of a for which $0 < a < 2$ and $12 - 3a^2 = \frac{45}{4}$, or $3a^2 = \frac{3}{4}$. Thus $a = \frac{1}{2}$.

CUMULATIVE REVIEW (CHAPTERS 1–2)

1. The given inequality is equivalent to $\dfrac{x(x + \sqrt{3})(x - \sqrt{3})}{(1 - x)^3} > 0$. From the diagram we see that the solution is the union of $(-\sqrt{3}, 0)$ and $(1, \sqrt{3})$.

x	− − − − − − 0 + + + + + +
$x + \sqrt{3}$	− 0 + + + + + + + + + + +
$x - \sqrt{3}$	− − − − − − − − − − − 0 +
$(1 - x)^3$	+ + + + + + + + 0 − − − −
$\dfrac{x(x + \sqrt{3})(x - \sqrt{3})}{(1 - x)^3}$	− 0 + + + + 0 − + + 0 −

Number line (x): $-\sqrt{3}$, 0, 1, $\sqrt{3}$

3. The given inequality is equivalent to $3 - 2 < 1/|x| < 3 + 2$, or $\frac{1}{5} < |x| < 1$. Thus the solution is the union of $(-1, -\frac{1}{5})$ and $(\frac{1}{5}, 1)$.

5. Since

$$f(x) = 6x^2 - x - 2 = (2x + 1)(3x - 2)$$

the inequality $f(x) > 0$ is equivalent to $(2x + 1)(3x - 2) > 0$. From the diagram we see that the solution is the union of $(-\infty, -\frac{1}{2})$ and $(\frac{2}{3}, \infty)$.

$2x + 1$	− − − 0 + + + + + + + +
$3x - 2$	− − − − − − − − 0 + + +
$(2x + 1)(3x - 2)$	+ + + 0 − − − − 0 + + +

Number line (x): $-\frac{1}{2}$, $\frac{2}{3}$

7. The domain consists of all x for which $x^2 - 1 \geq 0$ (hence $x \leq -1$ or $x \geq 1$) and $\sqrt{x^2 - 1} - x \geq 0$. But if $x \geq 1$, then the fact that $x^2 - 1 \leq x^2$ implies that $\sqrt{x^2 - 1} < x$ (so $\sqrt{x^2 - 1} - x < 0$); if $x \leq -1$, then $-x \geq 0$, so $\sqrt{x^2 - 1} - x \geq 0$. Thus the domain is $(-\infty, -1]$.

9. $\lim_{x\to 2} \frac{x^2 - 3x + 2}{x^2 - 5x + 6} = \lim_{x\to 2} \frac{(x-1)(x-2)}{(x-2)(x-3)} = \lim_{x\to 2} \frac{x-1}{x-3} = -1$

11. Since $|x^2| = x^2$, we have

$$\lim_{x\to 0} \frac{|x^3| - x^2}{x^3 + x^2} = \lim_{x\to 0} \frac{|x|x^2 - x^2}{x^3 + x^2} = \lim_{x\to 0} \frac{|x| - 1}{x + 1} = \frac{0 - 1}{0 + 1} = -1$$

13. $$\lim_{x\to 2} \frac{f(x) - f(2)}{x - 2} = \lim_{x\to 2} \frac{\sqrt{2x^2 - 4} - 2}{x - 2} = \lim_{x\to 2} \frac{\sqrt{2x^2 - 4} - 2}{x - 2} \frac{\sqrt{2x^2 - 4} + 2}{\sqrt{2x^2 - 4} + 2}$$
$$= \lim_{x\to 2} \frac{2x^2 - 4 - 4}{(x-2)(\sqrt{2x^2 - 4} + 2)} = \lim_{x\to 2} \frac{2(x-2)(x+2)}{(x-2)\sqrt{2x^2 - 4} + 2}$$
$$= \lim_{x\to 2} \frac{2(x+2)}{\sqrt{2x^2 - 4} + 2} = 2$$

15.

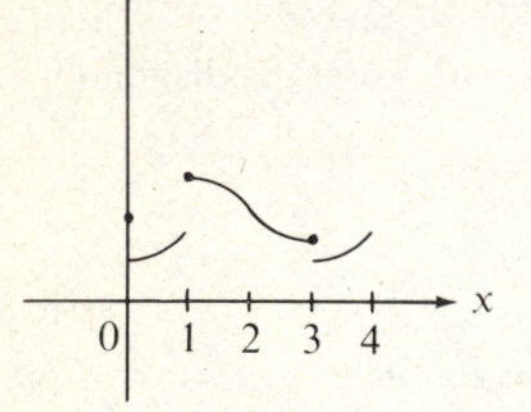

4

APPLICATIONS OF THE DERIVATIVE

SECTION 4.1

1. $f'(x) = 2x + 4 = 2(x + 2)$, so $f'(x) = 0$ for $x = -2$. Critical point: -2.

3. $f'(x) = 12x^3 + 12x^2 - 24x = 12x(x + 2)(x - 1)$, so $f'(x) = 0$ for $x = -2, 0$, or 1. Critical points: $-2, 0, 1$.

5. $g'(x) = 1 - (1/x^2) = (x^2 - 1)/x^2$, so $g'(x) = 0$ for $x = \pm 1$. Critical points: $-1, 1$.

7. $k'(t) = -2t/(t^2 + 4)^2$, so $k'(t) = 0$ for $t = 0$. Critical point: 0.

9. $k'(t) = \frac{6}{5}t^{-3/5}$, so $k'(t)$ is never zero, but k' does not exist at 0. Critical point: 0.

11. $f'(x) = \cos x$, so $f'(x) = 0$ for $x = (\pi/2) + n\pi$ for any integer n. Critical points: $(\pi/2) + n\pi$ for any integer n.

13. $f'(x) = 1 + \cos x$, so $f'(x) = 0$ for $x = \pi + 2n\pi = (2n + 1)\pi$ for any integer n. Critical points: $(2n + 1)\pi$ for any integer n.

15. $f'(z) = 1$ for $z > 2$, $f'(z) = -1$ for $z < 2$, and $f'(2)$ does not exist. Critical point: 2.

17. Since $f'(x) = 2x - 1$, the only critical point in $(0, 2)$ is $\frac{1}{2}$. Thus the extreme values of f on $[0, 2]$ can occur only at $0, \frac{1}{2}$, or 2. Since $f(0) = 0$, $f(\frac{1}{2}) = -\frac{1}{4}$ and $f(2) = 2$, the minimum value of f on $[0, 2]$ is $f(\frac{1}{2})$ which equals $-\frac{1}{4}$, and the maximum value is $f(2)$, which equals 2.

19. Since $g'(x) = 4x^3 - 4$, the only critical point of g in $(-4, 4)$ is 1. Thus the extreme values of g on $[-4, 4]$ can occur only at -4, 1, or 4. Since $g(-4) = 272$, $g(1) = -3$, and $g(4) = 240$, the minimum value of g on $[-4, 4]$ is $g(1)$, which equals -3, and the maximum value is $g(-4)$, which equals 272.

21. Since $f'(t) = 1/(2t^2)$ is never zero and is defined for every positive number t, there are no critical points. Because f can have an extreme value on $(0, \infty)$ only at a critical point or at an end point in the interval, and because neither exists, f has neither maximum nor minimum values.

23. Since $k'(z) = z/\sqrt{1+z^2}$, the only critical point of k in $(-2, 3)$ is 0. Thus the extreme values of k on $[-2, 3]$ can occur only at -2, 0 or 3. Since $k(-2) = \sqrt{5}$, $k(0) = 1$ and $k(3) = \sqrt{10}$, the minimum value of k on $[-2, 3]$ is $k(0)$, which equals 1, and the maximum value is $k(3)$, which equals $\sqrt{10}$.

25. Since $f'(x) = 1/(2\sqrt{x})$ for $x > 0$ and $f'(x) = -1/(2\sqrt{-x})$ for $x < 0$ and $f'(0)$ does not exist, the only critical point of f in $(-1, 2)$ is 0. Thus the only possible extreme value of f on $(-1, 2)$ can occur at 0. Observe that for $-1 < x < 2$ we have $f(x) = \sqrt{|x|} \geq 0 = f(0)$, so that the minimum value of f on $(-1, 2)$ is $f(0)$, which equals 0, and there is no maximum value.

27. Since $f'(x) = -\frac{1}{3}x^{-2/3}\cos\sqrt[3]{x}$ for $x \neq 0$ and $f'(0)$ does not exist, the only critical point of f in $(-\pi^3/27, \pi^3/8)$ is 0. Thus the extreme values of f on $[-\pi^3/27, \pi^3/8]$ can occur at $-\pi^3/27$, 0, or $\pi^3/8$. Since $f(-\pi^3/27) = -\sin(-\pi/3) = \frac{1}{2}\sqrt{3}$, $f(0) = 0$, and $f(\pi^3/8) = -\sin(\pi/2) = -1$, the minimum value of f on $[-\pi^3/27, \pi^3/8]$ is $f(\pi^3/8)$, which equals -1, and the maximum value is $f(-\pi^3/27)$, which equals $\frac{1}{2}\sqrt{3}$.

29. Since $f'(x) = \frac{1}{2}\sec^2(x/2) > 0$ for x in $(-\pi/2, \pi/6)$, f has no critical points in $(-\pi/2, \pi/6)$. Along with the fact that $(-\pi/2, \pi/6)$ has no endpoints, this means that f has no extreme value on $(-\pi/2, \pi/6)$.

31. Since $f'(x) = 3x^2 - 6x = 3x(x-2)$, f has critical points 0 and 2. On $[-1, 1]$, -1, 0, and 1 are the only candidates at which extreme values may occur. Since $f(-1) = -4$, $f(0) = 0$, and $f(1) = -2$, it follows that 0 is the maximum value of f on that interval. Therefore $f(0)$, which equals 0, is a relative maximum value. On $[1, 3]$, 1, 2, and 3 are the only candidates at which extreme values may occur. Since $f(1) = -2$, $f(2) = -4$, and $f(3) = 0$, it follows that -4 is the minimum value of f on that interval. Therefore $f(2)$, which equals -4, is a relative minimum value.

33. Since $f'(x) = 3x^2 + 6x - 9 = 3(x+3)(x-1)$, f has critical points at -3 and 1. On $[-4, -2]$, -4, -3, and -2 are the only candidates at which extreme values may occur. Since $f(-4) = 18$, $f(-3) = 25$, and $f(-2) = 20$, it follows that 25 is the maximum value of f on that interval. Therefore $f(-3)$, which equals 25, is a relative maximum value. On $[0, 2]$, 0, 1, and 2 are the only candidates at which extreme values may occur. Since $f(0) = -2$, $f(1) = -7$, and $f(2) = 0$, -7 is the minimum value of f on that interval. Therefore $f(1)$, which equals -7, is a relative minimum value.

35. Since $g'(x) = 5x^4 - 20 = 5(x^2+2)(x+\sqrt{2})(x-\sqrt{2})$, f has critical points at $-\sqrt{2}$ and $\sqrt{2}$. On $[-2\sqrt{2}, 0]$, $-2\sqrt{2}$, $-\sqrt{2}$, and 0 are the only candidates at which extreme values may occur. Since $g(-2\sqrt{2}) = -88\sqrt{2}$, $g(-\sqrt{2}) = 16\sqrt{2}$, and $g(0) = 0$, it follows that $16\sqrt{2}$ is the maximum value of g on that interval. Therefore $g(-\sqrt{2})$, which equals $16\sqrt{2}$, is a relative maximum value. Since g is an odd function, it follows that $g(\sqrt{2})$, which equals $-16\sqrt{2}$, is a relative minimum value.

37. Since $f'(x) = -\cos x$, f has critical points at $\pi/2 + n\pi$ for every integer n. Since $f(x) = -3 - \sin x \geq -4 = -3 - \sin(\pi/2 + n\pi)$ for any even integer n, $f(\pi/2 + n\pi)$, which equals -4, is a relative minimum value for any even integer n. Since $f(x) = -3 - \sin x \leq -2 = -3 - \sin(\pi/2 + n\pi)$ for any odd integer n, $f(\pi/2 + n\pi)$, which equals -2, is a relative maximum value for any odd integer n.

39. Since $f'(x) = \pi\sec\pi x\tan\pi x$, f has critical points at n for any integer n. Since $f(x) = \sec\pi x \geq 1 = f(n)$ for $n - \frac{1}{2} < x < n + \frac{1}{2}$ and n even, $f(n)$, which equals 1, is a relative minimum value for any even integer n. Analogously, $f(x) = \sec\pi x \leq -1 = f(n)$ for $n - \frac{1}{2} < x < n + \frac{1}{2}$ and n odd, so that $f(n)$, which equals -1, is a relative maximum value for any odd integer n.

41. Since $f'(x) = 4x^3 + 4x - 1$, we see that f' is continuous and $f'(0) < 0$ and $f'(1) > 0$. Thus from the Intermediate Value Theorem it follows that there is a critical point of f in $[0, 1]$. To approximate such a critical point we use the bisection method on f', with $a = 0$ and $b = 1$:

Interval	Length	Midpoint c	$f'(c)$
$[0, 1]$	1	$\frac{1}{2}$	$\frac{3}{2}$
$[0, \frac{1}{2}]$	$\frac{1}{2}$	$\frac{1}{4}$	$\frac{1}{16}$
$[0, \frac{1}{4}]$	$\frac{1}{4}$	$\frac{1}{8}$	$-\frac{63}{128}$
$[\frac{1}{8}, \frac{1}{4}]$	$\frac{1}{8}$		

Since the length of $[\frac{1}{8}, \frac{1}{4}]$ is $\frac{1}{8}$, and neither $\frac{1}{8}$ nor $\frac{1}{4}$ is a zero of f', the midpoint $\frac{3}{16}$ of $[\frac{1}{8}, \frac{1}{4}]$ is less than $\frac{1}{16}$ from a zero of f', and hence from a critical point of f.

43. Let $g(x) = f'(x) = x^3 + 3x^2 + 2$. We apply the Newton-Raphson method to g to find an approximate zero of g in $[-4, -3]$. Notice that $g'(x) = 3x^2 + 6x$. Letting the initial value of c be -3.5 and using a computer, we obtain -3.195823 for the desired approximate critical point of f.

45. Let $g(x) = f'(x) = x^3 + 3x^2 - 1$. We apply the Newton-Raphson method to g to approximate the zeros of g. Notice that $g'(x) = 3x^2 + 6x$. Letting the initial values of c be -3, -0.6 and 0.5 successively and using a computer, we obtain -2.879385, -0.652704, and 0.532089, respectively, for the desired approximate critical points of f.

47. $f(x_0)$ is the maximum value of f on $[a, b]$ if and only if $f(x) \leq f(x_0)$ for x in $[a, b]$ if and only if $-f(x) \geq -f(x_0)$ for x in $[a, b]$ if and only if $g(x) \geq g(x_0)$ for x in $[a, b]$ if and only if $g(x_0)$ is the minimum value of g on $[a, b]$.

49. Assume that $f'(c) = \lim_{x \to c} [f(x) - f(c)]/(x - c) < 0$. Then by (10) of Section 2.3, for x in some open interval I about c the inequality $[f(x) - f(c)]/(x - c) < 0$ holds. If x is in I and $x > c$, then $x - c > 0$, so that

$$f(x) - f(c) = (x - c)\left(\frac{f(x) - f(c)}{x - c}\right) < 0$$

Therefore $f(x) < f(c)$, so $f(c)$ is not a minimum value. If x is in I and $x < c$, then $x - c < 0$, so that

$$f(x) - f(c) = (x - c)\left(\frac{f(x) - f(c)}{x - c}\right) > 0$$

Therefore $f(x) > f(c)$, so $f(c)$ is not a maximum value. Thus if $f'(c) < 0$, then f does not have an extreme value at c.

51. Let $g(x) = f(x) - mx$ for $a \le x \le b$. Since $g'(x) = f'(x) - m$ and $f'(a) < m < f'(b)$, we have $g'(a) = f'(a) - m < 0$ and $g'(b) = f'(b) - m > 0$. Since $g'(a) < 0$, we must have $[g(x) - g(a)]/(x - a) < 0$ for x in an interval $(a, a + \delta)$. But then

$$(x - a)\left(\frac{g(x) - g(a)}{x - a}\right) < 0$$

so $g(x) < g(a)$ for x in $(a, a + \delta)$. Similarly, since $g'(b) > 0$, we must have

$$\frac{[g(x) - g(b)]}{(x - b)} > 0$$

for x in an interval $(b - \delta, b)$. But then

$$g(x) - g(b) = (x - b)\left(\frac{g(x) - g(b)}{x - b}\right) < 0 \quad \text{for } b - \delta < x < b$$

and hence $g(x) < g(b)$ for x in $(b - \delta, b)$. Thus g does not assume its minimum value at a or b. But since g is continuous on $[a, b]$, the Maximum-Minimum Theorem says that g has a minimum value on $[a, b]$, which must occur on (a, b). Since g is differentiable on (a, b), Theorem 4.3 says that there is a number c in (a, b) such that $g'(c) = 0$. But $g'(c) = f'(c) - m$. Thus $f'(c) = m$.

53. Let x and y be nonnegative numbers with $x + y = 1$. Then $0 \le x \le 1$ and $xy = x(1 - x)$. Let $f(x) = x(1 - x)$ for $0 \le x \le 1$. We are to find the largest value of f on $[0, 1]$. Now $f'(x) = 1 - 2x$, so that $f'(x) = 0$ only for $x = \frac{1}{2}$. Thus the maximum value of f can occur only at $0, \frac{1}{2}$, or 1. Since $f(0) = f(1) = 0$ and $f(\frac{1}{2}) = \frac{1}{4}$, the maximum value of f, and hence the largest product, is $\frac{1}{4}$. The smallest product is 0, which occurs for $x = 0$ and $y = 1$, or $x = 1$ and $y = 0$.

55. Since $P(x_0)$ is the maximum of P on $[0, \infty)$, it is also the maximum value of P on $[0, 2x_0]$. By Theorem 4.3, $P'(x_0) = 0$, and since $P'(x_0) = R'(x_0) - C'(x_0)$, it follows that $R'(x_0) = C'(x_0)$.

SECTION 4.2

1. $\dfrac{f(4) - f(0)}{4 - 0} = \dfrac{-8 - 0}{4} = -2$; $f'(x) = 2x - 6$

   We must find c in $(0, 4)$ such that $f'(c) = -2$, which means that $2c - 6 = -2$, or $c = 2$.

3. $\dfrac{f(0) - f(-2)}{0 - (-2)} = \dfrac{0 - 4}{2} = -2$; $f'(x) = 3x^2 - 6$

   We must find c in $(-2, 0)$ such that $f'(c) = -2$, which means that $3c^2 - 6 = -2$, or $c^2 = \frac{4}{3}$. Since $-2 < c < 0$, we have $c = -\frac{2}{3}\sqrt{3}$.

5. $\dfrac{f(1) - f(-2)}{1 - (-2)} = \dfrac{5 - (-4)}{3} = 3$; $f'(x) = 3x^2$

   We must find c $(-2, 1)$ such that $f'(c) = 3$, which means that $3c^2 = 3$, or $c^2 = 1$. Since $-2 < c < 1$, we have $c = -1$.

7. $\dfrac{f(2) - f(-2)}{2 - (-2)} = \dfrac{3 - (-25)}{4} = 7$; $f'(x) = 3x^2 - 6x + 3$

   We must find c in $(-2, 2)$ such that $f'(c) = 7$, which means that $3c^2 - 6c + 3 = 7$, or $3c^2 - 6c - 4 = 0$. The roots are $1 + \sqrt{21}/3$ and $1 - \sqrt{21}/3$. Since $-2 < c < 2$, we have $c = 1 - \sqrt{21}/3$.

9. $\dfrac{f(8) - f(1)}{8 - 1} = \dfrac{3 - 2}{7} = \dfrac{1}{7}$; $f'(x) = \frac{1}{3}x^{-2/3}$

   We must find c in $(1, 8)$ such that $f'(c) = \frac{1}{7}$, which means that $\frac{1}{3}c^{-2/3} = \frac{1}{7}$, so that $c = (\frac{7}{3})^{3/2}$.

11. Since $f'(x) = 2Ax + B$, c must satisfy the equation

$$2Ac + B = \frac{(Ab^2 + Bb + C) - (Aa^2 + Ba + C)}{b - a}$$

$$= \frac{A(b^2 - a^2) + B(b - a)}{b - a} = A(b + a) + B$$

   Therefore $c = (b + a)/2$, the midpoint of $[a, b]$.

13. By the Mean Value Theorem, there exists a number c in (a, b) such that

$$\frac{f(b) - f(a)}{b - a} = f'(c), \quad \text{so that} \quad \left|\frac{f(b) - f(a)}{b - a}\right| = |f'(c)| \le M$$

   Thus $|f(b) - f(a)| \le M|b - a|$. Equivalently, $f(a) - M(b - a) \le f(b) \le f(a) + M(b - a)$.

15. If $f(x) = x^{2/3}$, then $f'(x) = \frac{2}{3}x^{-1/3}$, and for $27 \le x \le 28$, we have $|f'(x)| \le \frac{2}{9}$. From Exercise 13, $|28^{2/3} - 27^{2/3}| \le \frac{2}{9}(28 - 27) = \frac{2}{9}$. Thus $9 - \frac{2}{9} \le 28^{2/3} \le 9 + \frac{2}{9} = \frac{83}{9}$. But $28^{2/3} > 9$, so $9 < 28^{2/3} \le \frac{83}{9}$.

17. Let $f(x) = \sqrt{x}$. If x is in $[2.89, 3]$, then

$$|f'(x)| = \left|\frac{1}{2\sqrt{x}}\right| \leq \frac{1}{2\sqrt{2.89}} = \frac{1}{3.4} < .295$$

With $a = 2.89$ and $b = 3$, Exercise 13 implies that $|\sqrt{3} - 1.7| = |f(3) - f(2.89)| < (0.295)(3 - 2.89) = 0.03245$.

19. Let $f(t)$ be the distance in feet that the car skidded in t seconds after the skid began. Then $f(0) = 0$ and $f(9) = 400$. Assuming that f is continuous on $[0, 9]$ and differentiable on $(0, 9)$, we conclude from the Mean Value Theorem that for some t_0 in $(0, 9)$, $f'(t_0) = [f(9) - f(0)]/(9 - 0) = \frac{400}{9} > 44$. That is, at some instant *after* the brakes were applied, the velocity was greater than 44 feet per second, or 30 miles per hour. Therefore the car must have been speeding before it skidded.

21. Let $f(x) = (x - 1)\sin x$. Then f is continuous on $[0, 1]$ and differentiable on $(0, 1)$. Since $f(0) = f(1) = 0$ and $f'(x) = \sin x + (x - 1)\cos x$, Rolle's Theorem implies the existence of a number c in $(0, 1)$ such that $\sin c + (c - 1)\cos c = 0$, or $\tan c = 1 - c$.

23. f satisfies the hypothesis of Rolle's Theorem. Thus there is c in $(0, 1)$ such that $f'(c) = 0$. Since $f'(x) = mx^{m-1}(x - 1)^n + nx^m(x - 1)^{n-1} = x^{m-1}(x - 1)^{n-1}[m(x - 1) + nx]$, the only value of c in $(0, 1)$ for which $f'(c) = 0$ satisfies $m(c - 1) + nc = 0$, that is, $c = m/(m + n)$. The point $m/(m + n)$ divides the interval $[0, 1]$ into the intervals $[0, m/(m + n)]$ and $[m/(m + n), 1]$. The lengths of these intervals are $m/(m + n)$ and $1 - m/(m + n) = n/(m + n)$. Thus the ratio of the lengths is m/n.

25. Assume that $a < x < y < z < b$ and $g(x) = g(y) = g(z) = 0$. By Rolle's Theorem there are numbers c in (x, y) and d in (y, z) such that $g'(c) = 0$ and $g'(d) = 0$.

27. a. Suppose b and d are fixed points, with $b < d$. Then $f(b) = b$ and $f(d) = d$. Now apply the Mean Value Theorem to f on the interval $[b, d]$ to find that there is a number c such that

$$f'(c) = \frac{f(d) - f(b)}{d - b} = \frac{d - b}{d - b} = 1$$

which contradicts the assumption that $f'(x) < 1$ for every number x. Thus f has at most one fixed point.

b. Since $f'(x) = \frac{1}{2}\cos 2x \leq \frac{1}{2} < 1$, we conclude from part (a) that f has at most one fixed point. Since $f(0) = \sin 0 = 0$, 0 is the only fixed point.

SECTION 4.3

1. C 3. $\frac{1}{2}x^2 + C$ 5. $-\frac{1}{3}x^3 + C$

7. $-3x^2 + 5x + C$ 9. $-\cos x + C$

11. Let $g(x) = -2x$. Then $g'(x) = -2 = f'(x)$, so $f(x) = g(x) + C = -2x + C$ for the appropriate constant C. Since $f(0) = 0$, we have $0 = f(0) = -2(0) + C = C$. Thus $f(x) = -2x$.

13. Let $g(x) = -\frac{1}{2}x^2$. Then $g'(x) = -x = f'(x)$, so $f(x) = g(x) + C = -\frac{1}{2}x^2 + C$ for the appropriate constant C. Since $f(-1) = \sqrt{2}$, we have $\sqrt{2} = f(-1) = -\frac{1}{2}(-1)^2 + C$, so that $C = \frac{1}{2} + \sqrt{2}$. Thus $f(x) = -\frac{1}{2}x^2 + \frac{1}{2} + \sqrt{2}$.

15. Let $g(x) = \frac{1}{3}x^3$. Then $g'(x) = x^2 = f'(x)$, so $f(x) = g(x) + C = \frac{1}{3}x^3 + C$ for the appropriate constant C. Since $f(0) = -5$, we have $-5 = f(0) = \frac{1}{3}(0)^3 + C = C$, so that $C = -5$. Thus $f(x) = \frac{1}{3}x^3 - 5$.

17. Let $g(x) = \sin x$. Then $g'(x) = \cos x = f'(x)$, so $f(x) = g(x) + C = \sin x + C$ for the appropriate constant C. Since $f(\pi/3) = 1$, we have $1 = f(\pi/3) = \sin(\pi/3) + C = (\sqrt{3}/2) + C$, so $C = 1 - (\sqrt{3}/2)$. Thus $f(x) = \sin x + 1 - (\sqrt{3}/2)$.

19. Let $g(x) = 0$. Then $g'(x) = 0 = f''(x) = (f')'(x)$, so by Theorem 4.7, $f'(x) = g(x) + C_1 = C_1$ for some constant C_1. Now let $h(x) = C_1 x$. Then $h'(x) = C_1 = f'(x)$, so by Theorem 4.7, $f(x) = h(x) + C_2 = C_1 x + C_2$ for some constant C_2.

21. By Exercise 19, $f(x) = C_1 x + C_2$, where C_1 and C_2 are constants. Thus $f'(x) = C_1$. Since $f'(0) = -1$ and $f(0) = 2$ by hypothesis, it follows that $C_1 = -1$ and $2 = f(0) = C_1 \cdot 0 + C_2 = C_2$. Thus $f(x) = -x + 2$.

23. Let $g(x) = -\cos x$. Then $g'(x) = \sin x = f''(x) = (f')'(x)$, so by Theorem 4.7, $f'(x) = g(x) + C_1 = -\cos x + C_1$ for some constant C_1. Now let $h(x) = -\sin x + C_1 x$. Then $h'(x) = -\cos x + C_1 = f'(x)$, so by Theorem 4.7, $f(x) = h(x) + C_2 = -\sin x + C_1 x + C_2$ for some constant C_2. Since $f'(\pi) = -2$, we have $-2 = f'(\pi) = -\cos\pi + C_1 = 1 + C_1$, so $C_1 = -3$. Since $f(0) = 4$, we have $4 = f(0) = -\sin 0 + C_1 \cdot 0 + C_2$, so $C_2 = 4$. Thus $f(x) = -\sin x - 3x + 4$.

25. Since $f^{(4)}(x) = (f')^{(3)}(x)$, it follows from the solution of Exercise 24 that $f'(x) = D_1 x^2 + D_2 x + D_3$, where D_1, D_2, and D_3 are constants. Now let $g(x) = \frac{1}{3}D_1 x^3 + \frac{1}{2}D_2 x^2 + D_3 x$. Then $g'(x) = D_1 x^2 + D_2 x + D_3 = f'(x)$, so by Theorem 4.7, $f(x) = g(x) + D_4 = \frac{1}{3}D_1 x^3 + \frac{1}{2}D_2 x^2 + D_3 x + D_4$, where D_4 is some constant. Letting $C_1 = \frac{1}{3}D_1$, $C_2 = \frac{1}{2}D_2$, $C_3 = D_3$, and $C_4 = D_4$, we have $f(x) = C_1 x^3 + C_2 x^2 + C_3 x + C_4$.

27. $f'(x) = 2x + 1$, so $f'(x) < 0$ for $x < -\frac{1}{2}$ and $f'(x) > 0$ for $x > -\frac{1}{2}$. Moreover, $f'(x) = 0$ only for $x = -\frac{1}{2}$. By Theorem 4.8, f is strictly decreasing on $(-\infty, -\frac{1}{2}]$ and strictly increasing on $[-\frac{1}{2}, \infty)$.

29. $f'(x) = 3x^2 - 2x + 1 = 3(x - \frac{1}{3})^2 + \frac{2}{3} > 0$ for all x. By Theorem 4.8, f is strictly increasing on $(-\infty, \infty)$.

31. $f'(x) = 4x^3 - 6x^2 = 2x^2(2x - 3)$, so $f'(x) < 0$ for $x < \frac{3}{2}$, and $f'(x) > 0$ for $x > \frac{3}{2}$. Moreover, $f'(x) = 0$ only for $x = 0$ or $\frac{3}{2}$. By Theorem 4.8, f is strictly decreasing on $(-\infty, \frac{3}{2}]$ and is strictly increasing on $[\frac{3}{2}, \infty)$.

33. $f'(x) = 5x^4 + 3x^2 - 2 = (x^2 + 1)(5x^2 - 2)$, so $f'(x) < 0$ for $-\sqrt{2/5} < x < \sqrt{2/5}$, and $f'(x) > 0$ for $x < -\sqrt{2/5}$ and for $x > \sqrt{2/5}$. Moreover, $f'(x) = 0$ only for $x = -\sqrt{2/5}$ or $\sqrt{2/5}$. By Theorem 4.8, f is strictly increasing on $(-\infty, -\sqrt{2/5}]$ and on $[\sqrt{2/5}, \infty)$, and is strictly decreasing on $[-\sqrt{2/5}, \sqrt{2/5}]$.

35. Notice that the domain of g is $[-4, 4]$. Also $g'(x) = -x/\sqrt{16 - x^2}$, so $g'(x) > 0$ for $-4 < x < 0$ and $g'(x) < 0$ for $0 < x < 4$. Moreover, $g'(x) = 0$ only for $x = 0$. By Theorem 4.8, g is strictly increasing on $[-4, 0]$ and is strictly decreasing on $[0, 4]$.

37. Notice that -3 is not in the domain of g. Also $g'(x) = -1/(x + 3)^2 < 0$ for $x \neq -3$. We conclude from Theorem 4.8 that g is strictly decreasing on $(-\infty, -3)$ and on $(-3, \infty)$.

39. $k'(x) = -2x/(x^2 + 1)^2$, so $k'(x) > 0$ for $x < 0$ and $k'(x) < 0$ for $x > 0$. Moreover, $k'(x) = 0$ only for $x = 0$. By Theorem 4.8, k is strictly increasing on $(-\infty, 0]$ and is strictly decreasing on $[0, \infty)$.

41. Notice that $f(t)$ is not defined for $t = \pi/2 + n\pi$ for any integer n. Also $f'(t) = \sec^2 t > 0$ for all t in the domain of f. By Theorem 4.8, f is strictly increasing on each interval of the form $(\pi/2 + n\pi, \pi/2 + (n + 1)\pi)$, where n is any integer.

43. $f'(t) = -2 \sin t - 1$, so $f'(t) < 0$ if $\sin t > -\frac{1}{2}$ and $f'(t) > 0$ if $\sin t < -\frac{1}{2}$. Thus $f'(t) < 0$ for $-\pi/6 + 2n\pi < t < 7\pi/6 + 2n\pi$ for any integer n, and $f'(t) > 0$ for $7\pi/6 + 2n\pi < t < 11\pi/6 + 2n\pi$ for any integer n. By Theorem 4.8, f is strictly decreasing on $[-\pi/6 + 2n\pi, 7\pi/6 + 2n\pi]$ for any integer n and is strictly increasing on $[7\pi/6 + 2n\pi, 11\pi/6 + 2n\pi]$ for any integer n.

45. a. Since $F' = f$ and $G' = g$, we have $(F + G)' = F' + G' = f + g$.
    b. Since $F' = f$ and $G' = g$, we have $(F - G)' = F' - G' = f - g$.
    c. Since $F' = f$ we have $(cF)' = cF' = cf$.
    d. No. If $f(x) = g(x) = x$, then $(fg)(x) = x^2$. If $F(x) = G(x) = \frac{1}{2}x^2$, then $F' = f$ and $G' = g$, $(FG)(x) = \frac{1}{4}x^4$, and $(FG)'(x) = x^3$. Thus $(FG)' \neq fg$.
    e. No. If $f(x) = g(x) = x$, then $(f/g)(x) = 1$ for $x \neq 0$. If $F(x) = G(x) = \frac{1}{2}x^2$, then $F' = f$ and $G' = g$, $(F/G)(x) = 1$ and $(F/G)'(x) = 0$ for $x \neq 0$. Thus $(F/G)' \neq f/g$.

47. Let $f(x) = x^4 - 4x$. Then $f'(x) = 4x^3 - 4 = 4(x - 1)(x^2 + x + 1) \geq 0$ for $x > 1$. By Exercise 46(a), $x^4 - 4x = f(x) \geq f(1) = -3$ for $x \geq 1$.

49. Let $f(x) = \frac{1}{4}x + 1/x$. Then

$$f'(x) = \frac{1}{4} - \frac{1}{x^2} = \frac{(x + 2)(x - 2)}{4x^2} \geq 0 \quad \text{for } x > 2$$

By Exercise 46(a), $\frac{1}{4}x + 1/x = f(x) \geq f(2) = 1$ for $x \geq 2$.

51. a. Let $u < v$. Then $f'(x) = 1 - \cos x \geq 0$ for all x in $[u, v]$ and is zero for only finitely many values (any of the form $2n\pi$ for some integer n) in $[u, v]$. By Theorem 4.8(a), f is strictly increasing on $[u, v]$. Therefore $f(u) < f(v)$. Since u, v were arbitrary numbers, f is strictly increasing on $(-\infty, \infty)$.
    b. Since f is strictly increasing by part (a), if $x < 0$ then $x - \sin x = f(x) < f(0) = 0$. Therefore $x < \sin x$ on $(-\infty, 0)$. For $x > 0$, $x - \sin x = f(x) > f(0) = 0$ by part (a). Thus $x > \sin x$ on $(0, \infty)$.

53. Let $f(x) = \tan x - x$. Then $f'(x) = \sec^2 x - 1 \geq 0$ for x in $[0, \pi/2)$, with $f'(x) \neq 0$ for x in $(0, \pi/2)$. By Theorem 4.8(a), f is strictly increasing on $[0, \pi/2)$. Since $f(0) = 0$, it follows that $f(x) > f(0) = 0$ for x in $(0, \pi/2)$. Thus $\tan x - x > 0$, or equivalently $\tan x > x$, for x in $(0, \pi/2)$.

55. Let $f(x) = \cos x$ and $g(x) = 1 - x^2/2$. Then $f(0) = 1 = g(0)$, and by Exercise 51(b), $f'(x) = -\sin x > -x = g'(x)$ for $x > 0$. It follows from Exercise 52(b) that $\cos x = f(x) > g(x) = 1 - x^2/2$ for $x > 0$.

57. a. Let $f(x) = \cos x - 2x$. Since $f(0) = 1$, $f(\pi/2) = -\pi$, and f is continuous on $[0, \pi/2]$, there is an x in $(0, \pi/2)$ such that $f(x) = 0$ by the Intermediate Value Theorem. If there were two numbers, u and v, with $f(u) = f(v) = 0$, then by Rolle's Theorem there would be a number c between them such that $f'(c) = 0$. But $f'(x) = -\sin x - 2 \leq -1 < 0$ for all x, so there can be no such c. Therefore there is at most one x such that $f(x) = 0$. Combining the two parts, we conclude that there is exactly one x such that $f(x) = 0$, which is equivalent to $\cos x - 2x = 0$, or $\cos x = 2x$.
    b. Let $f(x) = \cos x - 2x$, so that $f'(x) = -\sin x - 2$. Letting the initial value of c be 1 and using a computer, we obtain 0.450184 as the desired approximate zero of f, and hence approximate solution of $\cos x = 2x$.

59. Taking $x = y = 0$ in the equation $f(x + y) = f(x) + f(y)$, we have $f(0) = f(0) + f(0)$, so that $f(0) = 0$. Taking $y = h$ in the equation $f(x + y) = f(x) + f(y)$, we find that $f(x + h) = f(x) + f(h)$, so that $f(x + h) - f(x) = f(h)$. It now follows from (2) in Section 3.2 that for any x,

$$f'(x) = \lim_{h \to 0} \frac{f(x + h) - f(x)}{h} = \lim_{h \to 0} \frac{f(h)}{h} = \lim_{h \to 0} \frac{f(h) - f(0)}{h} = f'(0)$$

Letting $f'(0) = c$, we have $f'(x) = c$, so that by Theorem 4.7, $f(x) = cx + C$ for some constant C. Some $f(0) = 0$, it follows that $C = 0$, and thus that $f(x) = cx$ for all x.

61. Let the time period be from t_1 to t_2, and let $f(t)$ be the position of the particle at any time t. Then by hypothesis, $f'(t) = v(t) = 0$ for $t_1 < t < t_2$. It follows from Theorem 4.7(a) that f is constant on $[t_1, t_2]$; that is, for t between t_1 and t_2, we have $f(t) = f(t_1)$, which means that the particle stands still during that period.

63. a. $\frac{dT}{dW} = \frac{1}{3\sqrt{1 - S^2/3L^2}} > 0$ for all W

so T is an increasing function of W.

b. $\frac{dT}{dS} = \left(\frac{-1}{2}\right)\frac{W}{3}\left(1 - \frac{S^2}{3L^2}\right)^{-3/2}\left(\frac{-2S}{3L^2}\right) = \frac{WS}{9L^2}\left(1 - \frac{S^2}{3L^2}\right)^{-3/2} > 0$

so T is an increasing function of S.

c. $\frac{dT}{dL} = \left(\frac{-1}{2}\right)\left(\frac{W}{3}\right)\left(1 - \frac{S^2}{3L^2}\right)^{-3/2}\left(\frac{2S^2}{3L^3}\right) = \frac{-WS^2}{9L^3}\left(1 - \frac{S^2}{3L^2}\right)^{-3/2} < 0$

so T is a decreasing function of L.

SECTION 4.4

1. $f'(x) = 2x + 6 = 2(x + 3)$, so f' changes from negative to positive at -3.

3. $f'(x) = 8x^3 - 8x = 8x(x - 1)(x + 1)$, so f' changes from negative to positive at -1 and 1, and from positive to negative at 0.

5. $f'(t) = \frac{(2t - 1)(t^2 + t + 1) - (t^2 - t + 1)(2t + 1)}{(t^2 + t + 1)^2} = \frac{2(t + 1)(t - 1)}{(t^2 + t + 1)^2}$

so f' changes from positive to negative at -1 and from negative to positive at 1.

7. $f'(t) = \cos t + \frac{1}{2}$, so $f'(t) > 0$ if $\cos t > -\frac{1}{2}$, and $f'(t) < 0$ if $\cos t < -\frac{1}{2}$. Thus f' changes from positive to negative at $2\pi/3 + 2n\pi$ for any integer n, and changes from negative to positive at $4\pi/3 + 2n\pi$ for any integer n.

9. $f'(x) = -6x + 3 = 3(1 - 2x)$, so f' changes from positive to negative at $\frac{1}{2}$. By the First Derivative Test, $f(\frac{1}{2}) = \frac{31}{4}$ is a relative maximum value of f.

11. $f'(x) = 3x^2 + 6x = 3x(x + 2)$, so f' changes from positive to negative at -2 and from negative to positive at 0. By the First Derivative Test, $f(-2) = 8$ is a relative maximum value and $f(0) = 4$ is a relative minimum value of f.

13. $g'(x) = 8x + \frac{1}{x^2} = \frac{8x^3 + 1}{x^2} = \frac{(2x + 1)(4x^2 - 2x + 1)}{x^2}$

Since $4x^2 - 2x + 1 > 0$ for all x, g' changes from negative to positive at $-\frac{1}{2}$. By the First Derivative Test, $g(-\frac{1}{2}) = 3$ is a relative minimum value of g.

15. $f'(x) = \frac{(16 + x^3) - x(3x^2)}{(16 + x^3)^2} = \frac{16 - 2x^3}{(16 + x^3)^2} = \frac{-2(x - 2)(x^2 + 2x + 4)}{(16 + x^3)^2}$

Since $x^2 + 2x + 4 > 0$ for all x, f' changes from positive to negative at 2. By the First Derivative Test, $f(2) = 1/12$ is a relative maximum value of f.

17. $f'(x) = \sqrt{1 - x^2} - \frac{x^2}{\sqrt{1 - x^2}} = \frac{1 - 2x^2}{\sqrt{1 - x^2}} = \frac{(1 - \sqrt{2}x)(1 + \sqrt{2}x)}{\sqrt{1 - x^2}}$

so f' changes from negative to positive at $-\frac{1}{2}\sqrt{2}$ and from positive to negative at $\frac{1}{2}\sqrt{2}$. By the First Derivative Test, $f(-\frac{1}{2}\sqrt{2}) = -\frac{1}{2}$ is a relative minimum value and $f(\frac{1}{2}\sqrt{2}) = \frac{1}{2}$ is a relative maximum value of f.

19. $k'(x) = -\sin x + \frac{1}{2}$, so $k'(x) > 0$ if $\sin x < \frac{1}{2}$, and $k'(x) < 0$ if $\sin x > \frac{1}{2}$. Thus k' changes from positive to negative at $\pi/6 + 2n\pi$ for any integer n, and from negative to positive at $5\pi/6 + 2n\pi$ for any integer n. By the First Derivative Test, $k(\pi/6 + 2n\pi) = \frac{1}{2}\sqrt{3} + \pi/12 + n\pi$ is a relative maximum value of k for any integer n, and $k(5\pi/6 + 2n\pi) = -\frac{1}{2}\sqrt{3} + 5\pi/12 + n\pi$ is a relative minimum value of k for any integer n.

21. $k'(x) = \cos\left(\frac{x^2}{1 + x^2}\right)\frac{2x(1 + x^2) - x^2(2x)}{(1 + x^2)^2} = \frac{2x}{(1 + x^2)^2}\cos\left(\frac{x^2}{1 + x^2}\right)$

Since $0 \le x^2/(1 + x^2) \le 1 < \pi/2$, we have $\cos(x^2/(1 + x^2)) > 0$ for all x. Thus k' changes from negative to positive at 0. By the First Derivative Test, $k(0) = 0$ is a relative minimum value of k.

23. $f'(x) = -8x + 3$, so $f'(x) = 0$ if $x = \frac{3}{8}$. Since $f''(x) = -8 < 0$ for all x, the Second Derivative Test implies that $f(3/8) = -\frac{7}{16}$ is a relative maximum value of f.

25. $f'(x) = 3x^2 - 6x - 24 = 3(x - 4)(x + 2)$, so $f'(x) = 0$ if $x = 4$ or $x = -2$. Since $f''(x) = 6x - 6$, so that $f''(4) = 18 > 0$ and $f''(-2) = -18 < 0$, the Second Derivative Test implies that $f(4) = -79$ is a relative minimum value of f, and $f(-2) = 29$ is a relative maximum value of f.

27. $f'(x) = 12x^3 - 12x^2 - 9x = 3x(2x + 1)(2x - 3)$, so $f'(x) = 0$ if $x = 0$, $x = -\frac{1}{2}$, or $x = \frac{3}{2}$. Since $f''(x) = 36x^2 - 24x - 9$, so that $f''(0) = -9 < 0$, $f''(-\frac{1}{2}) = 12 > 0$ and $f''(\frac{3}{2}) = 36 > 0$, the Second Derivative Test implies that $f(0) = \frac{1}{2}$ is a relative minimum value of f, $f(-\frac{1}{2}) = \frac{1}{16}$ is a relative minimum value of f, and $f(\frac{3}{2}) = -\frac{127}{16}$ is a relative minimum value of f.

29. $f'(t) = 2t - 1/t^2$, so $f'(t) = 0$ if $2t = 1/t^2$, or $t^3 = \frac{1}{2}$, or $t = 1/\sqrt[3]{2}$. Since $f''(t) = 2 + 2/t^3$, so that $f''(1/\sqrt[3]{2}) = 2 + 2/\frac{1}{2} = 6 > 0$, the Second Derivative Test implies that $f(1/\sqrt[3]{2}) = 1/(2^{2/3}) + \sqrt[3]{2} + 1$ is a relative minimum value of f.

31. $f'(t) = \cos t - \sin t$, so $f'(t) = 0$ if $\sin t = \cos t$, that is, if $x = \pi/4 + n\pi$ for any integer n. Next, $f''(t) = -\sin t - \cos t$. If n is even, then $f''(\pi/4 + n\pi) = f''(\pi/4) = -\sqrt{2} < 0$, so the Second Derivative Test implies that $f(\pi/4 + n\pi) = \sqrt{2}$ is a relative maximum value of f. Analogously, if n is odd, then $f''(\pi/4 + n\pi) = f''(\pi/4 + \pi) = f''(5\pi/4) = \sqrt{2} > 0$, so the Second Derivative Test implies that $f(\pi/4 + n\pi) = -\sqrt{2}$ is a relative minimum value of f!

33. $f'(x) = 2x + 8 = 2(x + 4)$, so that $f'(x) = 0$ for $x = -4$. Also $f''(x) = 2 > 0$ for all x. By the Second Derivative Test, $f(-4) = -4$ is a relative minimum value of f.

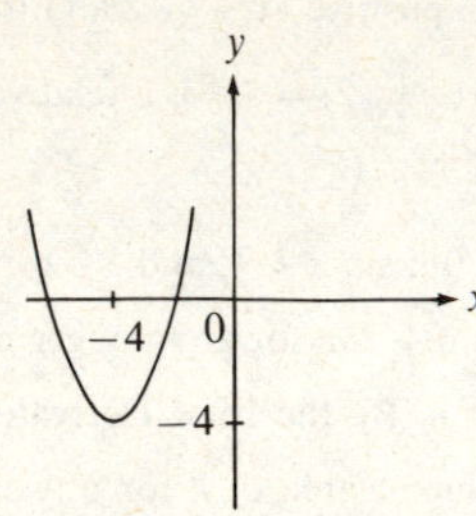

35. $f'(x) = 3x^2 + 3 = 3(x^2 + 1) > 0$ for all x, so that f has no critical values, and therefore no relative extreme values.

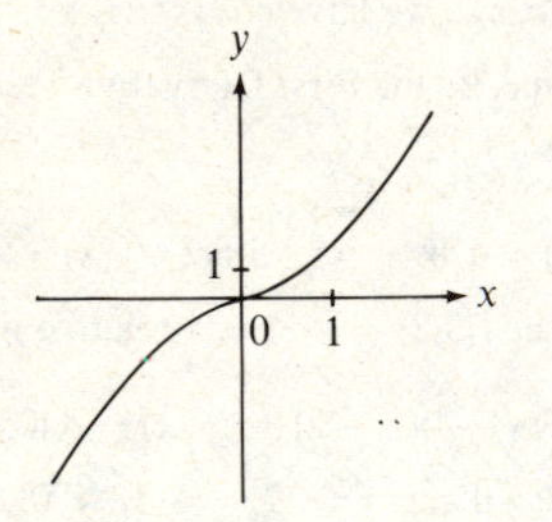

37. $f'(x) = 4x^3 + 4 = 4(x + 1)(x^2 - x + 1)$, so that since $x^2 - x + 1 > 0$ for all x, $f'(x) = 0$ for $x = -1$. Next $f''(x) = 12x^2$, so that $f''(-1) = 12 > 0$. By the Second Derivative Test, $f(-1) = -3$ is a relative minimum value of f.

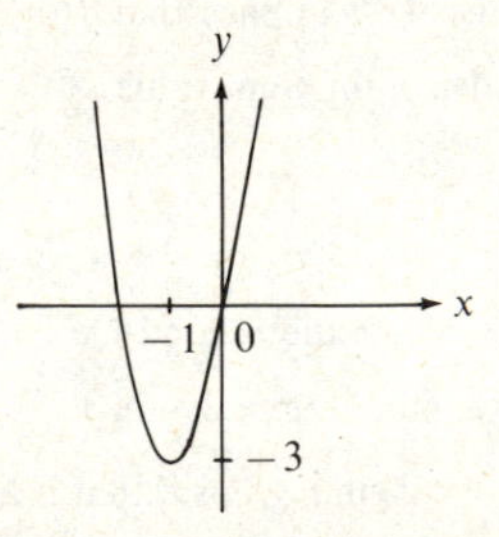

39. $f'(x) = 2(x^2 - 1)(2x) = 4x(x + 1)(x - 1)$, so that $f'(x) = 0$ for $x = 0$, $x = 1$, or $x = -1$. Next, $f''(x) = 12x^2 - 4$, so that $f''(0) = -4 < 0$, $f''(1) = 8 > 0$, and $f''(-1) = 8 > 0$. By the Second Derivative Test, $f(0) = 1$ is a relative maximum value of f, and $f(1) = f(-1) = 0$ is a relative minimum value of f.

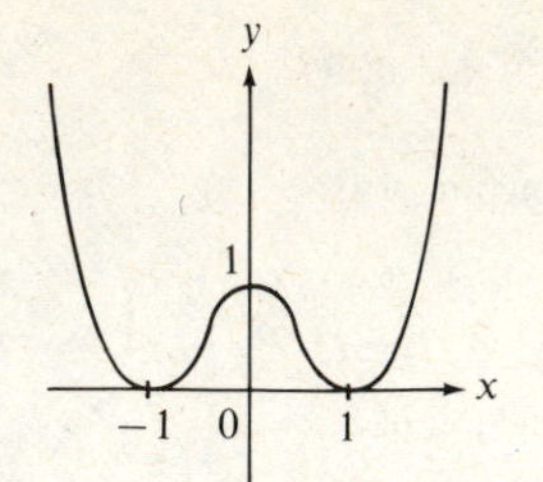

41. $f'(x) = 2(x - 2)(x + 1)^2 + (x - 2)^2 2(x + 1) = 2(x - 2)(x + 1)(2x - 1)$, so that $f'(x) = 0$ for $x = 2$, $x = -1$, or $x = \frac{1}{2}$. Since f' changes from negative to positive at -1 and 2, and positive to negative at $\frac{1}{2}$, the First Derivative Test implies that $f(-1) = f(2) = 0$ is a relative minimum value of f, and $f(\frac{1}{2}) = \frac{81}{16}$ is a relative maximum value of f. (The graph in Exercise 41 is the same as that in Exercise 40 shifted right 2 units.)

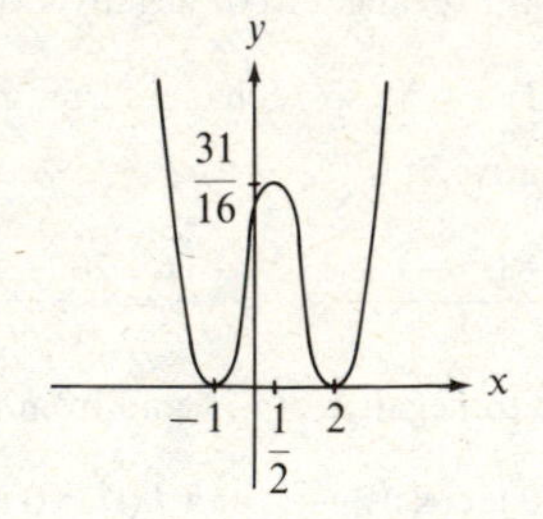

43. $f'(x) = 5x^4$ and $f''(x) = 20x^3$, so that $f'(0) = f''(0) = 0$. But $f'(x) > 0$ if $x \neq 0$, so f is strictly increasing, and thus has no relative extreme values.

45. $f'(x) = 5x^4 - 3x^2 = x^2(5x^2 - 3)$ and $f''(x) = 20x^3 - 6x = 2x(10x^2 - 3)$, so that $f'(0) = f''(0) = 0$. But $f'(x) < 0$ for x in $(-\sqrt{\frac{3}{5}}, \sqrt{\frac{3}{5}})$ and $x \neq 0$, so that f is strictly decreasing on $[-\sqrt{\frac{3}{5}}, \sqrt{\frac{3}{5}}]$, and thus f has no relative extreme value at 0.

47. $f'(x) = 3(x - 2)^2(x + 1) + (x - 2)^3 = (x - 2)^2(4x + 1)$ and $f''(x) = 2(x - 2)(4x + 1) + 4(x - 2)^2$, so that $f'(2) = f''(2) = 0$. But $f'(x) > 0$ if $x > -\frac{1}{4}$ and $x \neq 2$, so that f is strictly increasing on $[-\frac{1}{4}, \infty)$, and thus has no relative extreme value at 2.

49. Assume that $f''(c) > 0$. Since $f'(c) = 0$, (9) in Section 2.3 implies that $f'(x)/(x - c) = [f'(x) - f'(c)]/(x - c) > 0$ for all x in some interval $(c - \delta, c + \delta)$. Therefore if $c - \delta < x < c$, then $f'(x) < 0$ since $x - c < 0$, while if $c < x < c + \delta$, then $f'(x) > 0$ since $x - c > 0$. This means that f' changes from negative to positive at c, and hence by the First Derivative Test, f has a relative minimum value at c.

SECTION 4.5

1. Let x and y be the positive numbers, so $0 < x < 18$ and $0 < y < 18$. If P denotes their product, we seek to maximize P. Since $x + y = 18$, we have $P = xy = x(18 - x) = 18x - x^2$. Now $P'(x) = 18 - 2x$, so $P'(x) = 0$ for $x = 9$. Since $P''(x) = -2 < 0$ for all x, by Theorem 4.12 we know that the maximum value of P occurs for $x = 9$. The corresponding value of y is $y = 18 - 9 = 9$. The numbers are 9 and 9.

3. Let x be the length of a side of the base, h the height, V the volume, and S the surface area. We must minimize S. By hypothesis $S = 4xh + x^2$ and $4 = V = x^2h$, so $h = 4/x^2$. Thus $S = 4x(4/x^2) + x^2 = 16/x + x^2$. Next, $S'(x) = -16/x^2 + 2x$, so $S'(x) = 0$ if $-16/x^2 + 2x = 0$, that is, $2x^3 = 16$, or $x = 2$. Since $S''(x) = 32/x^3 + 2 > 0$ for all $x > 0$, it follows from Theorem 4.12 that the surface area is minimum for $x = 2$. Then $h = 4/2^2 = 1$, so the dimensions are 2 meters on a side of the base, and 1 meter in height.

5. Let A denote the area of the region enclosed by the track, with other notation as in Example 3. Then $A = 2rx + \pi r^2$, and we are to maximize A. Since $x = 220 - \pi r$, we have $A = 2r(220 - \pi r) + \pi r^2 = 440r - \pi r^2$ for $0 < r \leq 220/\pi$. Now $A'(r) = 440 - 2\pi r$, so $A'(r) = 0$ if $440 - 2\pi r = 0$, that is, $r = 440/(2\pi) = 220/\pi$. Since $A''(r) = -2\pi < 0$ for all r, it follows from Theorem 4.12 that the area A is maximum if $r = 220/\pi$. But then $x = 220 - \pi(220/\pi) = 0$, so the track is circular.

7. Let x and h be as in the figure, and let P be the perimeter and A the area. We are to maximize A. By hypothesis the triangle is equilateral, so its altitude is $\frac{1}{2}\sqrt{3}x$. Thus $A = xh + \frac{1}{2}x(\frac{1}{2}\sqrt{3}x) = xh + \frac{1}{4}\sqrt{3}x^2$ and $12 = P = 2h + 3x$. Then $h = -\frac{3}{2}x$, so $A = x(6 - \frac{3}{2}x) + \frac{1}{4}\sqrt{3}x^2 = 6x + (\frac{1}{4}\sqrt{3} - \frac{3}{2})x^2$ for $0 < x \leq 4$. Now $A'(x) = 6 + (\frac{1}{2}\sqrt{3} - 3)x$, so $A'(x) = 0$ if $x = -6/(\sqrt{3}/2 - 3) = 12/(6 - \sqrt{3})$. Since $A''(x) = \frac{1}{2}\sqrt{3} - 3 < 0$ for all x, it follows from Theorem 4.12 that A is maximum if $x = 12/(6 - \sqrt{3})$. Since

$$h = 6 - \frac{3}{2}\frac{12}{6 - \sqrt{3}} = \frac{18 - 6\sqrt{3}}{6 - \sqrt{3}}$$

the maximum amount of light will enter if

$$x = \frac{12}{6 - \sqrt{3}} \approx 2.8 \text{ (feet)} \quad \text{and} \quad h = \frac{18 - 6\sqrt{3}}{6 - \sqrt{3}} \approx 1.8 \text{ (feet)}$$

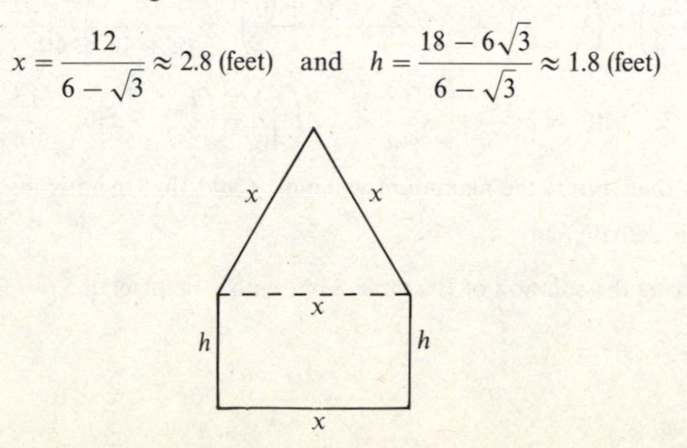

9. Let $t = 0$ correspond to 3 p.m. and set up a coordinate system as in the diagram, with the origin at the location of the oil tanker at time 0. At any time t the position of the oil tanker on the x axis is $-15t$ and the position of the luxury liner on the y axis is $25 + 25t$. Thus the distance D between them is given by $D = \sqrt{(0 + 15t)^2 + (25 + 25t - 0)^2}$. We will minimize $E = D^2 = 225t^2 + 625(1 + t)^2$. Now $E'(t) = 450t + 1250(1 + t) = 1700t + 1250$, so $E'(t) = 0$ if $1700t = -1250$, or $t = -\frac{1250}{1700} = -\frac{25}{34}$. Since $E''(t) = 1700 > 0$ for

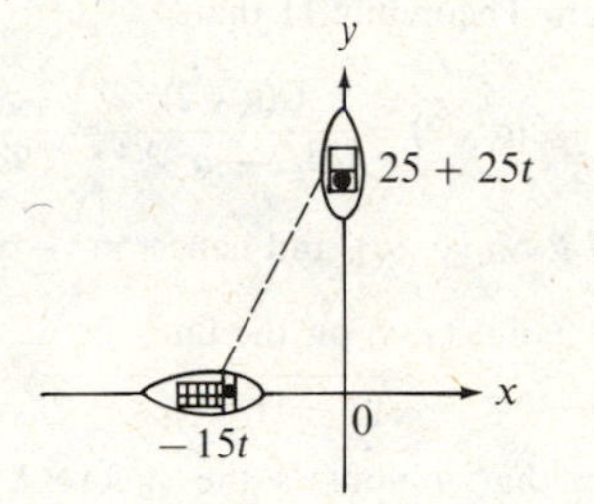

all t, it follows from Theorem 4.12 that E and hence D is minimum if $t = -\frac{25}{34}$, which corresponds to approximately 44 minutes before 3 p.m., that is, 2:16 p.m.

11. Let $r(x) = dx/dt = kx(a - x)$. Then $r'(x) = ka - 2kx$, so $r'(x) = 0$ if $ka - 2kx = 0$, that is, $x = a/2$. Since $r''(x) = -2k < 0$, Theorem 4.12 implies that $r(x)$ is maximized for $x = a/2$. Thus the value of x for which dx/dt is maximum is $a/2$.

13. $$f'(\theta) = q \csc^2 \theta - \frac{r \cos \theta}{\sin^2 \theta} = \frac{q - r \cos \theta}{\sin^2 \theta}$$

so that $f'(\theta) = 0$ if $\cos \theta = q/r$. Also

$$f''(\theta) = \frac{r \sin \theta (\sin^2 \theta) - (q - r \cos \theta)2 \sin \theta \cos \theta}{\sin^4 \theta} = \frac{r(\sin^2 \theta + 2 \cos^2 \theta) - 2q \cos \theta}{\sin^3 \theta}$$
$$= \frac{r(1 + \cos^2 \theta) - 2q \cos \theta}{\sin^3 \theta} > \frac{r(1 + \cos^2 \theta - 2 \cos \theta)}{\sin^3 \theta}$$
$$= \frac{r(1 - \cos \theta)^2}{\sin^3 \theta} > 0 \quad \text{for } \theta \text{ in } \left(0, \frac{\pi}{2}\right)$$

By Theorem 4.12, $f(\theta)$ is minimum for the value of θ in $(0, \pi/2)$ satisfying $\cos \theta = q/r$.

15. If we let $p = 6ab$, $q = 3b^2/2$, and $r = 3\sqrt{3}b^2/2$ in Exercise 13, then $q < r$, so Exercise 13 implies that the minimum value of S is attained for θ_0 such that $\cos \theta_0 = q/r = 1/\sqrt{3}$.

17. We must determine the maximum value of P on $[0, 4]$. Now

$$P'(t) = -6(20 - t)^2t + 2(20 - t)^3 = (20 - t)^2(-6t + 40 - 2t) = (20 - t)^2(40 - 8t)$$

so $P'(t) > 0$ for $0 \leq t \leq 4$. Thus the net profit increases the longer the stocks are kept, so the company should retain them the full four-year period.

19. Since $E(x) < 0$ for $x < 0$ and $E(x) > 0$ for $x > 0$, it suffices to find the maximum value of E on $[0, \infty)$.

$$E'(x) = \frac{Q(x^2 + a^2)^{3/2} - Qx(x^2 + a^2)^{1/2}(3x)}{(x^2 + a^2)^3}$$

$$= \frac{Q(x^2 + a^2) - 3Qx^2}{(x^2 + a^2)^{5/2}} = \frac{-2Qx^2 + Qa^2}{(x^2 + a^2)^{5/2}}$$

Thus $E'(x) = 0$ if $Qa^2 = 2Qx^2$, or $x = a/\sqrt{2}$. Since $E'(x) > 0$ if $0 < x < a/\sqrt{2}$ and $E'(x) < 0$ if $x > a/\sqrt{2}$, it follows from Theorem 4.11 that

$$E(a/\sqrt{2}) = \frac{Q(a/\sqrt{2})}{(a^2/2 + a^2)^{3/2}} = \frac{2\sqrt{3}Q}{9a^2}$$

is the maximum value of E on $[0, \infty)$, and hence on $(-\infty, \infty)$.

21. Since $y = 2x - 4$ for any point (x, y) on the line, the distance between (x, y) and $(1, 3)$ is $\sqrt{(x-1)^2 + [(2x-4) - 3]^2} = \sqrt{(x-1)^2 + (2x-7)^2}$. This distance is minimized for the same value of x that minimizes the square E of the distance, so let $E = (x-1)^2 + (2x-7)^2$. Then $E'(x) = 2(x-1) + 4(2x-7) = 10x - 30 = 10(x-3)$. Thus $E'(x) = 0$ for $x = 3$. Since $E'(x) < 0$ for $x < 3$ and $E'(x) > 0$ for $x > 3$, it follows from Theorem 4.11 that E has its minimum value at 3. If $x = 3$, then $y = 2(3) - 4 = 2$. Thus $(3, 2)$ is the point on the line $y = 2x - 4$ that is closest to the point $(1, 3)$.

23. Let x and y be as in the figure. Since triangles ABC and OBD are similar, we have $y/x = 1/(x-1)$, so $y = x/(x-1)$. Thus the area A of triangle OBD is given by

$A = \frac{1}{2}xy = x^2/[2(x-1)]$ for $x > 1$.

Then

$$A'(x) = \frac{(2x)2(x-1) - 2x^2}{4(x-1)^2} = \frac{2x^2 - 4x}{4(x-1)^2} = \frac{x(x-2)}{2(x-1)^2} \quad \text{for } x > 1$$

Thus $A'(x) = 0$ for $x = 2$. Since $A'(x) < 0$ for $1 < x < 2$ and $A'(x) > 0$ for $x > 2$, it follows from Theorem 4.11 that A has its minimum value at 2. If $x = 2$, then $y = 2/(2-1) = 2$, and the length of the hypotenuse BD is $\sqrt{2^2 + 2^2} = 2\sqrt{2}$. Thus the lengths of the sides of the triangle with smallest area are 2, 2, and $2\sqrt{2}$.

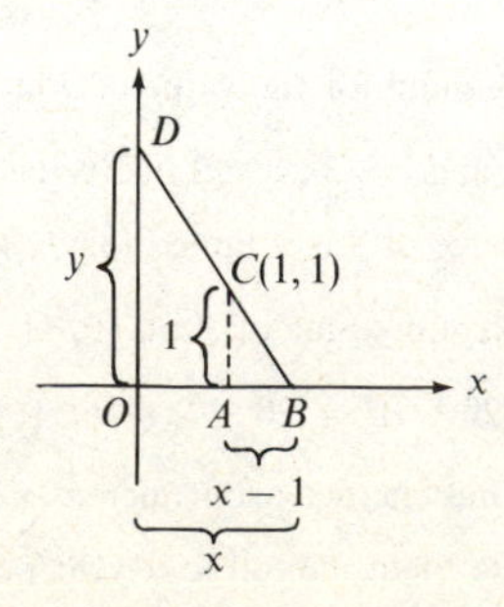

25. Let x be the length of the base, y the common length of the other two sides, and A the area of the triangle. We must maximize A. By hypothesis $x + 2y = 3$, so $y = \frac{3}{2} - \frac{1}{2}x$, so by the Pythagorean Theorem, the height h of the triangle is given by $h = \sqrt{y^2 - \frac{1}{4}x^2} = \sqrt{(\frac{3}{2} - \frac{1}{2}x)^2 - \frac{1}{4}x^2} = \frac{1}{2}\sqrt{9 - 6x}$. Thus $A = \frac{1}{2}xh = \frac{1}{2}x(\frac{1}{2}\sqrt{9-6x}) = \frac{1}{4}x\sqrt{9-6x}$ for $0 \le x \le \frac{3}{2}$. Now

$$A'(x) = \frac{1}{4}\sqrt{9-6x} - \frac{3x}{4\sqrt{9-6x}} = \frac{9-9x}{4\sqrt{9-6x}}$$

so $A'(x) = 0$ if $9 - 9x = 0$, that is, $x = 1$. Since $A'(x) > 0$ if $x < 1$ and $A'(x) < 0$ if $x > 1$, it follows from Theorem 4.11 that the area A is maximum if $x = 1$, so $y = \frac{3}{2} - \frac{1}{2} \cdot 1 = 1$, making the triangle equilateral.

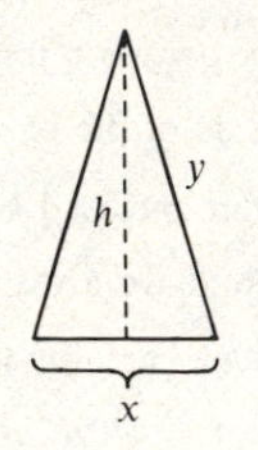

27. Let x be the length of the sides of the square, r the radius of the circle, and A the sum of the areas of the square and circle. Thus $A = x^2 + \pi r^2$, and we are to maximize and minimize A. Notice that $L = 4x + 2\pi r$, so $r = (L - 4x)/(2\pi)$. Thus

$$A = x^2 + \pi\left(\frac{L-4x}{2\pi}\right)^2 \quad \text{for } 0 \le x \le \frac{L}{4}$$

Then
$$A'(x) = 2x + 2\pi\left(\frac{L-4x}{2\pi}\right)\left(-\frac{4}{2\pi}\right) = 2x - \frac{2L}{\pi} + \frac{8x}{\pi}$$

so $A'(x) = 0$ if $x - L/\pi + 4x/\pi = 0$, that is, $x = L/\pi \cdot 1/(1 + 4/\pi) = L/(\pi + 4)$. Since $A''(x) = 2 + 8/\pi > 0$ for $0 < x < L/4$, it follows from Theorem 4.12 that the minimum value of A is

$$A\left(\frac{L}{\pi+4}\right) = \left(\frac{L}{\pi+4}\right)^2 + \pi\left(\frac{L - 4L/(\pi+4)}{2\pi}\right)^2 = \frac{L^2}{(\pi+4)^2} + \frac{\pi L^2}{4(\pi+4)^2} = \frac{L^2}{4(\pi+4)}$$

Since
$$A(0) = \pi\left(\frac{L}{2\pi}\right)^2 = \frac{L^2}{4\pi} \text{ and } A\left(\frac{L}{4}\right) = \left(\frac{L}{4}\right)^2 + \pi(0) = \frac{L^2}{16}$$

it follows that $A(0)$ is the maximum value of A and that there is no maximum value if the wire is actually cut.

29. a. Following the solution of Example 5 but with 10 replacing 5, we find that T is given by

$$T = \frac{\sqrt{x^2+4}}{3} + \frac{10-x}{4} \quad \text{for } 0 < x < 10$$

Thus $T'(x)$ is the same as in Example 5, so $T'(x) = 0$ for $x = \frac{6}{7}\sqrt{7}$. As in Example 5, it follows from Theorem 4.12 that T is minimized for $x = \frac{6}{7}\sqrt{7}$. Thus the ranger should walk toward the point $\frac{6}{7}\sqrt{7}$ miles down the road.

b. This time T is given by

$$T = \frac{\sqrt{x^2+4}}{3} + \frac{\frac{1}{2} - x}{4} \quad \text{for } 0 \le x \le \tfrac{1}{2}$$

Thus $T'(x)$ is the same as in Example 5. However, since $\frac{6}{7}\sqrt{7} > \frac{1}{2}$, there are no critical points of T in $(0, \frac{1}{2})$, so the minimum value of T must occur at 0 or $\frac{1}{2}$. Since

$$T(0) = \frac{\sqrt{0^2+4}}{3} + \frac{\frac{1}{2} - 0}{4} = \frac{19}{24} \quad \text{and} \quad T(\tfrac{1}{2}) = \frac{\sqrt{(\frac{1}{2})^2+4}}{3} + 0 = \frac{\sqrt{17}}{6}$$

it follows that T is minimized for $x = \frac{1}{2}$. Thus the ranger should walk directly toward the car.

c. T is given by

$$T = \frac{\sqrt{x^2+4}}{3} + \frac{c - x}{4} \quad \text{for } 0 \le x \le c$$

so $T'(x)$ is as given in Example 5. If $c > \frac{6}{7}\sqrt{7}$, it follows as in Example 5 that the ranger should walk toward the point $\frac{6}{7}\sqrt{7}$ miles down the road. However, if $c \le \frac{6}{7}\sqrt{7}$, it follows as in part (b) that the ranger should walk directly toward the car.

31. Let x and y denote, respectively, the width and the length of the printing on the page. We are to minimize the area A of the page. By hypothesis $xy = 35$, so $y = 35/x$, and thus $A = (x+2)(y+4) = (x+2)(35/x+4) = 43 + 70/x + 4x$ for $x > 0$. Now $A'(x) = -70/x^2 + 4$, so $A'(x) = 0$ if $x^2 = \frac{70}{4}$, or $x = \frac{1}{2}\sqrt{70}$. Since $A''(x) = 140/x^3 > 0$ for $x > 0$, it follows from Theorem 4.12 that the minimum value of A is

$$A(\tfrac{1}{2}\sqrt{70}) = 43 + 70/(\tfrac{1}{2}\sqrt{70}) + 4(\tfrac{1}{2}\sqrt{70}) = 43 + 4\sqrt{70}$$

Thus the minimum area is $43 + 4\sqrt{70}$ square inches.

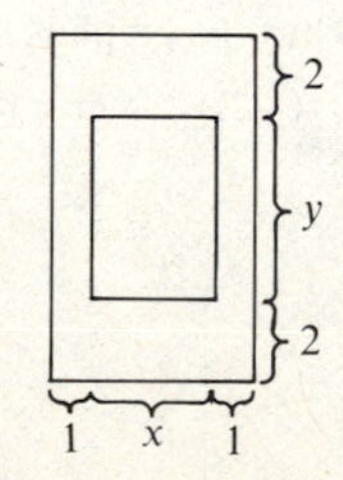

33. Let x be $\frac{1}{2}$ the length of the base of the rectangle, y the height, and A the area. We are to maximize A. Since triangles BCD and EFD are similar, it follows that $\frac{12}{3} = y/(3-x)$, so that $y = 4(3-x) = 12 - 4x$. Thus $A = 2xy = 2x(12-4x) = 24x - 8x^2$. Now $A'(x) = 24 - 16x$, so $A'(x) = 0$ if $x = \frac{3}{2}$. Since $A''(x) = -16$ for all x, it follows from Theorem 4.12 that $A(\frac{3}{2}) = 24(\frac{3}{2}) - 8(\frac{3}{2})^2 = 18$ is the maximum area of an inscribed rectangle.

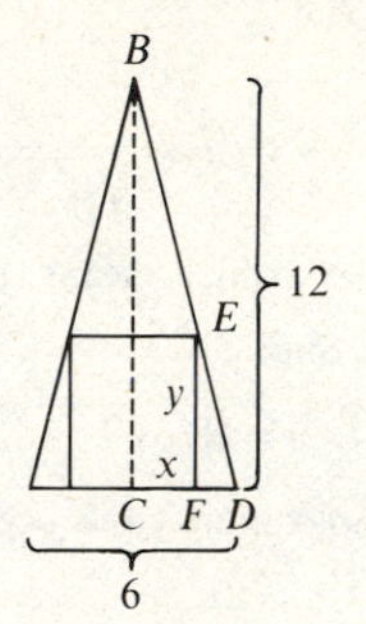

35. Let r be the radius of the can, h the height, V the volume, and S the surface area. We must minimize S. By hypothesis, $V = \pi r^2 h$, so $h = V/\pi r^2$, and thus

$$S = (2\pi r)h + 2(\pi r^2) = 2\pi r V/(\pi r^2) + 2\pi r^2 = 2V/r + 2\pi r^2$$

Now $S'(r) = -2V/r^2 + 4\pi r$, so $S'(r) = 0$ if $-2V/r^2 + 4\pi r = 0$, that is, $2V/r^2 = 4\pi r$, or $r^3 = V/(2\pi)$. Therefore $r = \sqrt[3]{V/(2\pi)}$. Since $S''(r) = 4V/r^3 + 4\pi > 0$ for $r > 0$, it follows from Theorem 4.12 that the smallest surface area is obtained for $r = \sqrt[3]{V/(2\pi)}$.

37. Let V be the volume of the cylinder, h its height, and r its radius. We are to maximize V. From similar triangles ABC and ADE, we deduce that $H/R = h/(R-r)$, so that $h = (H/R)(R-r)$ and thus $V = \pi r^2 h = \pi r^2[(H/R)(R-r)] = \pi H r^2 - (\pi H/R)r^3$ for $0 \le r \le R$. Now $V'(r) = 2\pi H r - (3\pi H/R)r^2$, so $V'(r) = 0$ if $2\pi H r - (3\pi H/R)r^2 = 0$, that is, $r = 0$ or $r = \frac{2}{3}R$. Since $V(0) = 0 = V(R)$ and $V(\frac{2}{3}R) = \pi H(\frac{2}{3}R)^2 - (\pi H/R)(\frac{2}{3}R)^3 = \frac{4}{27}\pi R^2 H$, it follows that the maximum possible volume is $\frac{4}{27}\pi R^2 H$.

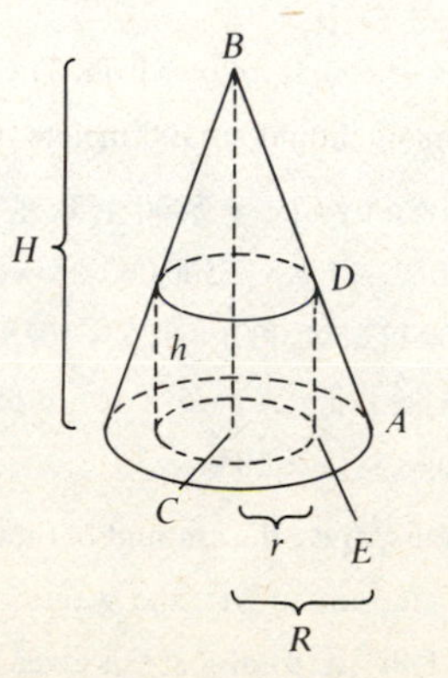

39. Let x and h be as in the figure. We are to maximize the area A, which is given by

$$A = \tfrac{1}{2}h[L + (L + 2x)] = h(L + x)$$

By the Pythagorean Theorem, $h = \sqrt{L^2 - x^2}$, so that $A = (L + x)\sqrt{L^2 - x^2}$ for $0 \le x \le L$. Now

$$A'(x) = \sqrt{L^2 - x^2} + (L + x)\frac{-x}{\sqrt{L^2 - x^2}} = \frac{L^2 - Lx - 2x^2}{\sqrt{L^2 - x^2}}$$

so $A'(x) = 0$ if $L^2 - Lx - 2x^2 = 0$, which means $(L + x)(L - 2x) = 0$. Since $x \ge 0$ and $L > 0$, this means that $x = L/2$. Since

$$A(0) = L^2,\ A(L/2) = (\tfrac{3}{2}L)\sqrt{L^2 - \tfrac{1}{4}L^2} = [(3\sqrt{3})/4]L^2$$

and $A(L) = 0$, it follows that maximum area occurs for $x = L/2$, in which case the length of the fourth side is $2x + L = 2L$.

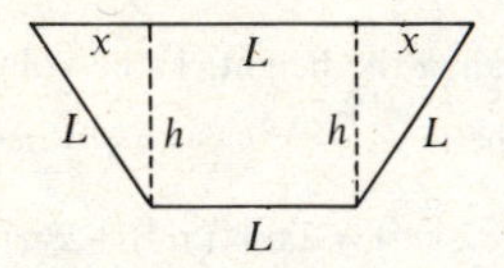

41. Let x be the distance from the person to the quieter highway, so $300 - x$ is the distance to the noisier highway. The total intensity of noise is given for $0 < x < 300$ by

$$f(x) = k\left(\frac{1}{x^2}\right) + k\left(\frac{8}{(300 - x)^2}\right)$$

where k is a positive constant. Then

$$f'(x) = \frac{-2k}{x^3} + \frac{16k}{(300 - x)^3} = 2k\frac{8x^3 - (300 - x)^3}{x^3(300 - x)^3}$$

Thus $f'(x) = 0$ if $8x^3 - (300 - x)^3 = 0$, or $2x = 300 - x$, so that $x = 100$. Since

$$f''(x) = \frac{6k}{x^4} + \frac{48k}{(300 - x)^4}$$

we have $f''(x) > 0$ for $0 < x < 300$. It follows from Theorem 4.12 that $f(100)$ is the minimum value of f, so the person should sit 100 meters from the quieter highway.

43. The total cost per day is given by $C(x) = 5000 + 3x + x^2/2{,}500{,}000$ so the total cost per unit is given by $U(x) = 5000/x + 3 + x/2{,}500{,}000$. Now $U'(x) = -5000/x^2 + 1/2{,}500{,}000$, so that $U'(x) = 0$ if $x^2 = (5000)(2{,}500{,}000) = 125 \times 10^8$, or $x = 5\sqrt{5} \times 10^4$. Since $U'(x) < 0$ if $x < 5\sqrt{5} \times 10^4$ and $U'(x) > 0$ if $x > 5\sqrt{5} \times 10^4$, it follows from Theorem 4.11 that U has a minimum value at $x = 5\sqrt{5} \times 10^4$.

45. Let x be the number of pickers, and t the amount of time (in hours) needed for harvesting. The wages of the pickers amount to $6xt$, the wages of the supervisor amount to $10t$, and the union collects $10x$. Thus the total cost C is given for $x > 0$ by $C = 6xt + 10t + 10x$. Since 62,500 tomatoes are to be picked, $625xt = 62{,}500$, so that $t = 100/x$. Thus $C(x) = 600 + 1000/x + 10x$. Then $C'(x) = -1000/x^2 + 10$, and $C'(x) = 0$ if $x = 10$. Since $C''(x) = 2000/x^3$, it follows that $C''(x) > 0$ for $x > 0$. Theorem 4.12 then implies that the minimum cost occurs when the farmer hires 10 pickers. The minimum cost is $C(10) = \$800$.

SECTION 4.6

1. $f'(x) = -3x + 1$; $f''(x) = -3$. Thus the graph is concave downward on $(-\infty, \infty)$. Also $f(\frac{1}{3}) = \frac{1}{6}$ is the maximum value of f.

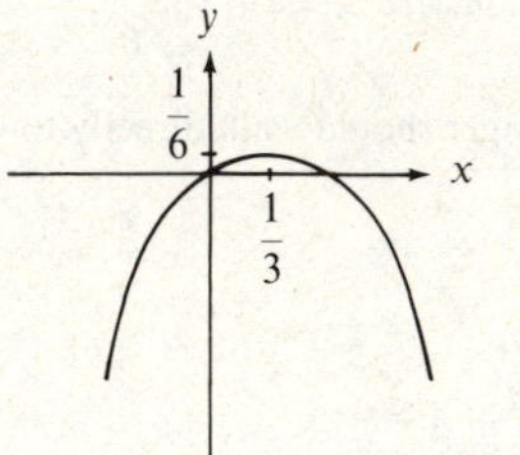

3. $f'(x) = 3x^2$; $f''(x) = 6x$. Thus the graph is concave upward on $(0, \infty)$ and concave downward on $(-\infty, 0)$.

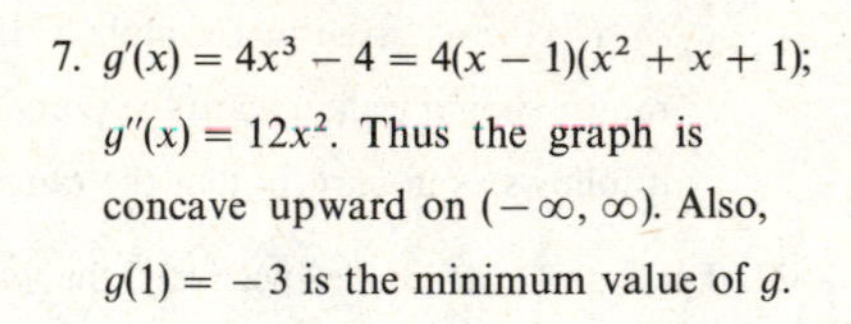

5. $g'(x) = 3x^2 - 12x + 12 = 3(x - 2)^2$; $g''(x) = 6x - 12 = 6(x - 2)$. Thus the graph is concave upward on $(2, \infty)$ and concave downward on $(-\infty, 2)$.

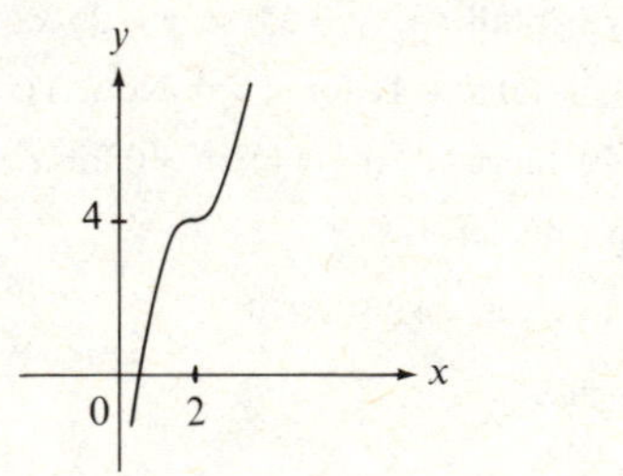

7. $g'(x) = 4x^3 - 4 = 4(x - 1)(x^2 + x + 1)$; $g''(x) = 12x^2$. Thus the graph is concave upward on $(-\infty, \infty)$. Also, $g(1) = -3$ is the minimum value of g.

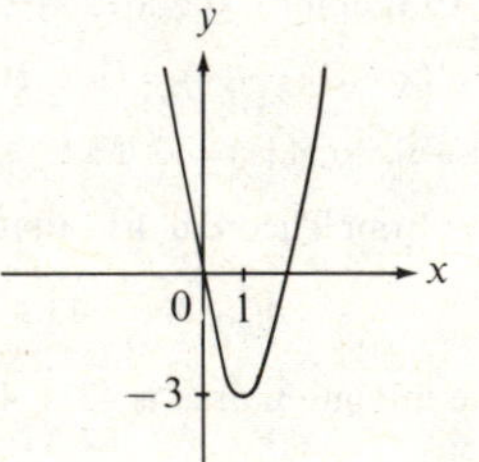

9. $f'(x) = 1 - (1/x)^2$; $f''(x) = 2/x^3$. Thus the graph is concave upward on $(0, \infty)$ and concave downward on $(-\infty, 0)$. Also, $f(-1) = -2$ is a relative maximum value, and $f(1) = 2$ is a relative minimum value.

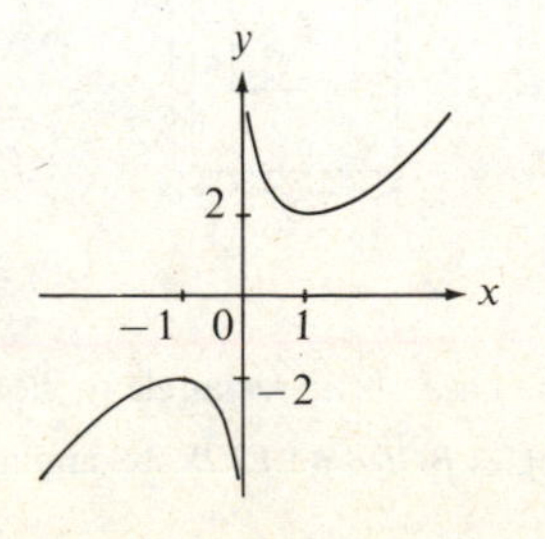

11. Note: The domain of f consists of all numbers x such that $x \geq 1$.

$$f'(x) = \sqrt{x-1} + \frac{x}{2\sqrt{x-1}} = \frac{3x-2}{2\sqrt{x-1}}$$

$$f''(x) = \frac{6\sqrt{x-1} - (3x-2)/\sqrt{x-1}}{4(x-1)} = \frac{3x-4}{4(x-1)^{3/2}}$$

Thus the graph is concave upward on $(\frac{4}{3}, \infty)$ and concave downward on $(1, \frac{4}{3})$.

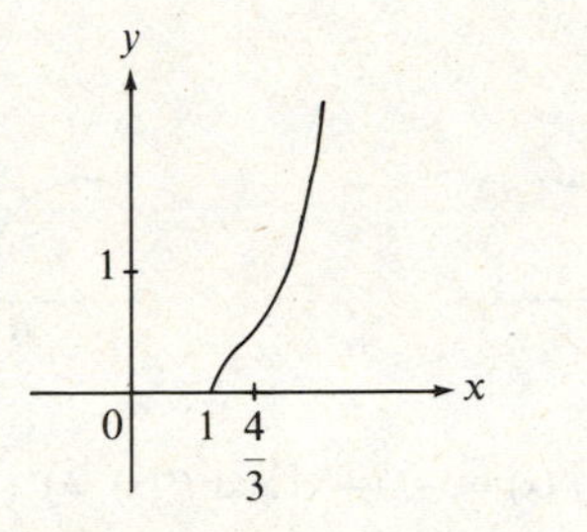

13. Refer to Example 4 and the paragraph following it. We have $f'(x) = 2\cos 2x$ and $f''(x) = -4\sin 2x = -4f(x)$. Thus the graph is concave upward on $(n\pi + \pi/2, (n+1)\pi)$ for any integer n and concave downward on $(n\pi, n\pi + \pi/2)$ for any integer n. Furthermore, for any integer n, $f(n\pi + \pi/4) = 1$ is the maximum value, and $f(n\pi - \pi/4) = -1$ the minimum value of f.

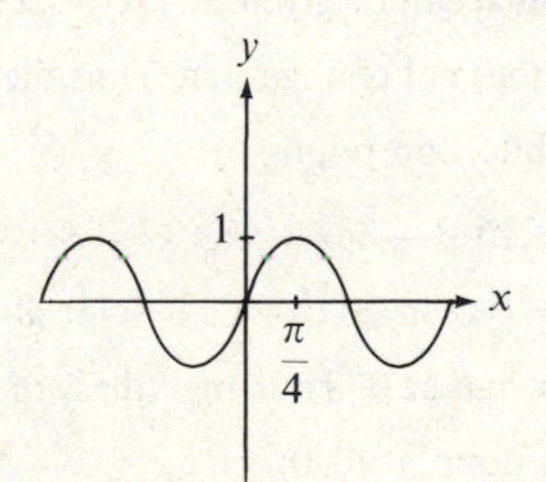

15. $f'(x) = \sec x \tan x$; $f''(x) = \sec x \tan^2 x + \sec^3 x = \sec x\,((\sec^2 x - 1) + \sec^2 x) = \sec x\,(2\sec^2 x - 1)$. Since $|\sec x| \geq 1$ for all x in the domain, $2\sec^2 x - 1 > 0$ for all x in the domain, so f and f'' have the same sign. Thus the graph of f is concave upward on $(-\pi/2 + 2n\pi, \pi/2 + 2n\pi)$ and concave downward on $(\pi/2 + 2n\pi, 3\pi/2 + 2n\pi)$ for any integer n. Also $f(2n\pi) = 1$ is a relative minimum value and $f((2n+1)\pi) = -1$ is a relative maximum value of f.

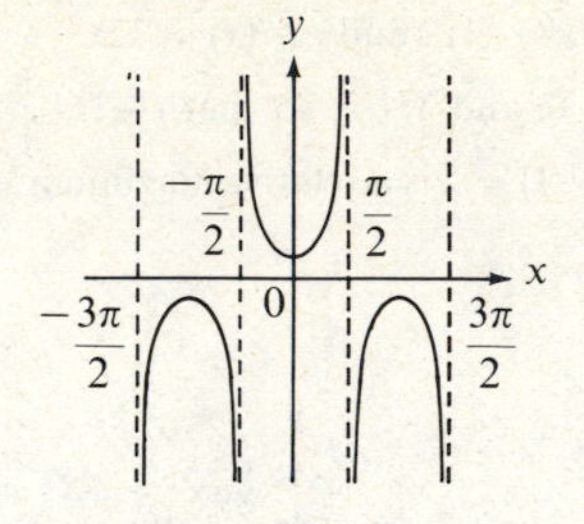

17. $f'(x) = 3(x+2)^2$ and $f''(x) = 6(x+2)$. Thus f'' changes sign at -2, so $(-2, 0)$ is an inflection point.

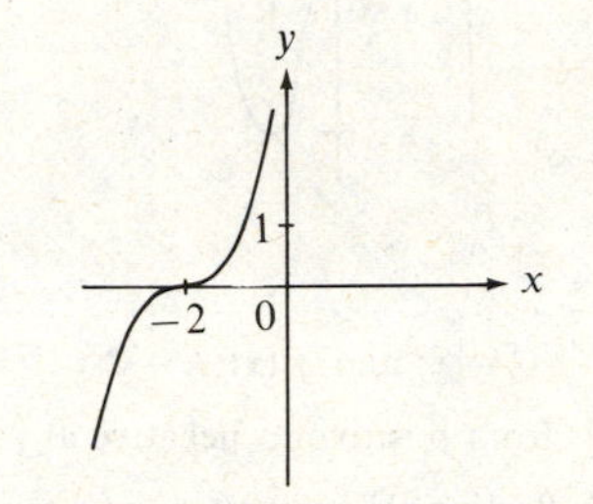

19. $f'(x) = 3x^2 + 6x - 9 = 3(x+3)(x-1)$ and $f''(x) = 6x + 6 = 6(x+1)$. Thus f'' changes sign at -1, so $(-1, 9)$ is an inflection point. Also $f(-3) = 25$ is a relative maximum value and $f(1) = -7$ is a relative minimum value of f.

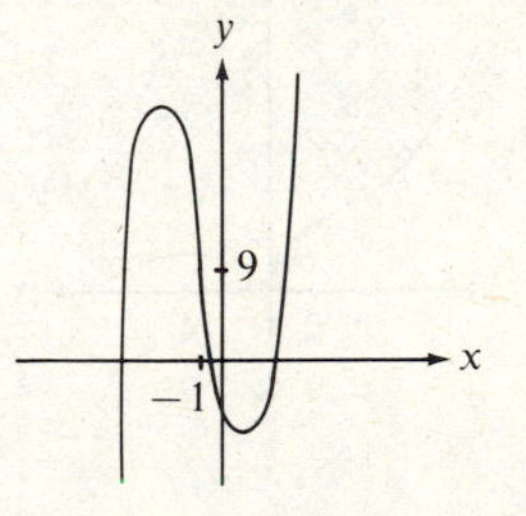

21. $g'(x) = 12x^3 + 12x^2 = 12x^2(x+1)$ and $g''(x) = 36x^2 + 24x = 12x(3x+2)$. Thus f'' changes sign at 0 and $-\frac{2}{3}$, so $(0, 0)$ and $(-\frac{2}{3}, -\frac{16}{27})$ are inflection points. Also $g(-1) = -1$ is the minimum value of g.

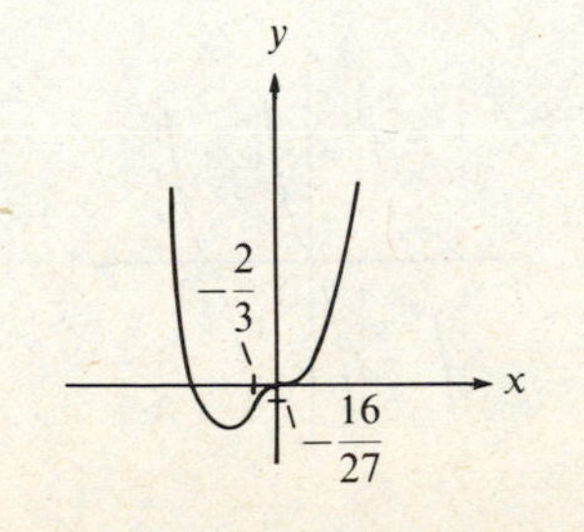

23. $g'(x) = 9x^8 - 9x^2 = 9x^2(x^6 - 1)$ and $g''(x) = 72x^7 - 18x = 18x(4x^6 - 1)$. Thus g'' changes sign at $-1/\sqrt[3]{2}$, 0, and $1/\sqrt[3]{2}$, so that $(-1/\sqrt[3]{2}, \frac{11}{8})$, $(0, 0)$, and $(1/\sqrt[3]{2}, -\frac{11}{8})$ are inflection points. Also $g(-1) = 2$ is a relative maximum value and $g(1) = -2$ is a relative minimum value of g.

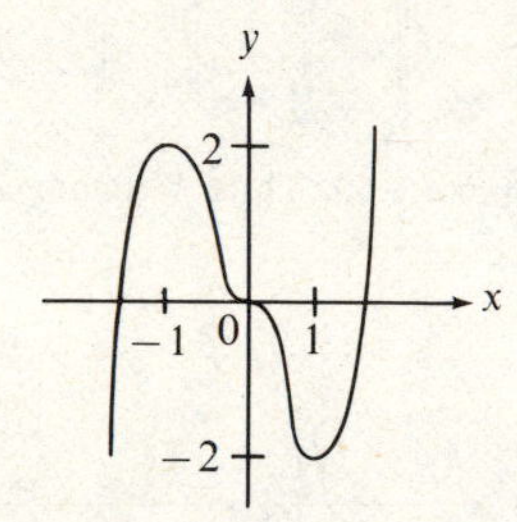

25. $g'(x) = \frac{4}{9}x^{-1/3} - x^{2/3} = x^{-1/3}(\frac{4}{9} - x)$ and $g''(x) = -\frac{4}{27}x^{-4/3} - \frac{2}{3}x^{-1/3} = -\frac{2}{3}x^{-4/3}(\frac{2}{9} + x)$ for $x \neq 0$. Since g'' changes from positive to negative at $-\frac{2}{9}$, the point $(-\frac{2}{9}, g(-\frac{2}{9})) = (-\frac{2}{9}, \frac{4}{5}(\frac{2}{9})^{2/3})$ is a point of inflection. Finally $g(\frac{4}{9}) = \frac{3}{5}(\frac{4}{9})^{2/3}$ is a relative maximum value of g.

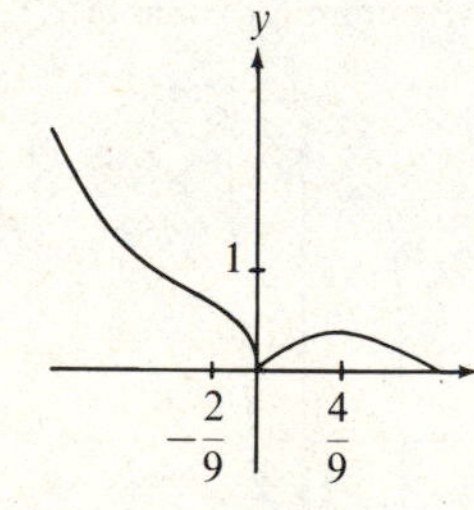

27. $f(t) = \sec^2 t$ and $f''(t) = 2\sec^2 t \tan t$. Thus f'' changes sign at $n\pi$, for any integer n, so $(n\pi, 0)$ is an inflection point for any integer n.

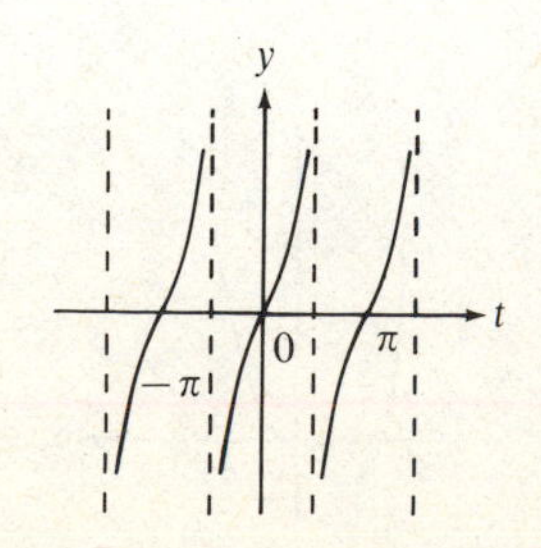

29.

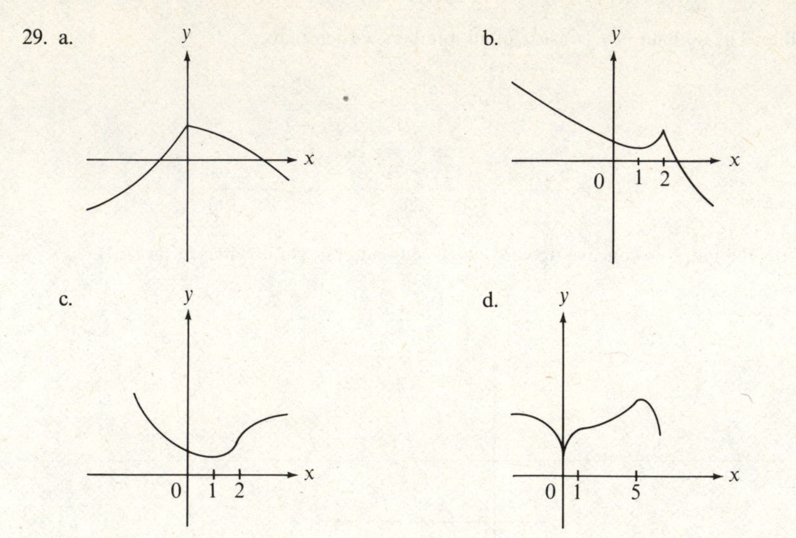

31. a. Since $f(x) = f(-x)$, $f'(x) = -f'(-x)$ and $f''(x) = f''(-x)$. Thus the graph if concave upward on $(-\infty, 0)$.

    b. Since $f(x) = -f(-x)$, $f'(x) = f'(-x)$ and $f''(x) = -f''(-x)$. Thus the graph is concave downward on $(-\infty, 0)$.

33. Let $f(x) = g(x) = x^2 - 1$, for $-1 < x < 1$. Then $f''(x) = g''(x) = 2$, so the graphs of f and g are concave upward on $(-1, 1)$. If $k(x) = f(x)g(x) = x^4 - 2x^2 + 1$, then $k'(x) = 4x^3 - 4x$ and $k''(x) = 12x^2 - 4$, so that $k''(x) < 0$ for $-1/\sqrt{3} < x < 1/\sqrt{3}$. Thus the graph of k is concave downward on $(-1/\sqrt{3}, 1/\sqrt{3})$, and in particular at 0.

35. Any second degree polynomial can be given by $f(x) = ax^2 + bx + c$. Then $f'(x) = 2ax + b$ and $f''(x) = 2a$, so that f'' does not change sign. Thus the graph of a second degree polynomial does not have an inflection point.

37. $f'(x) = 5x^4 - 3cx^2$; $f''(x) = 20x^3 - 6cx = x(20x^2 - 6c)$. If $c = 0$ then $f''(x) = 20x^3$, and f'' changes sign at 0. If $c \neq 0$ then $20x^2 - 6c \neq 0$ for all x such that $x^2 < \frac{3}{10}|c|$, that is, $|x| < \sqrt{\frac{3}{10}|c|}$, so f'' changes sign at 0. Thus regardless of the value of c, f'' changes sign at 0, so f has an inflection point at $(0, 0)$.

39. If f is a polynomial of degree n, $n \geq 2$, then f'' is a polynomial of degree $n - 2$. Since f'' is defined for all x and has at most $n - 2$ real zeros, the graph of f can have at most $n - 2$ inflection points.

41. Suppose that $f''(c) > 0$. Since f'' is continuous on an open interval containing c, by (9) of Section 2.3 there is an open interval I centered at c such that $f'' > 0$ on I. Thus f'' does not change sign at c, and hence f does not have an inflection point at $(c, f(c))$, as assumed. Similarly, the assumption that $f''(c) < 0$ leads to a contradiction. Therefore $f''(c) = 0$.

SECTION 4.7

1. $\lim_{x\to\infty} \dfrac{2}{x-3} = \lim_{x\to\infty} \dfrac{2/x}{1-3/x} = \dfrac{0}{1-0} = 0$

3. $\lim_{x\to\infty} \dfrac{x}{3x+2} = \lim_{x\to\infty} \dfrac{1}{3+2/x} = \dfrac{1}{3+0} = \dfrac{1}{3}$

5. $\lim_{x\to\infty} \dfrac{2x^2+x-1}{x^2-x+4} = \lim_{x\to\infty} \dfrac{2+1/x-1/x^2}{1-1/x+4/x^2} = \dfrac{2+0-0}{1-0+0} = 2$

7. $\lim_{t\to\infty} \dfrac{t}{t^{1/2}+2t^{-1/2}} = \lim_{t\to\infty}\left(\dfrac{t}{t+2}\cdot t^{1/2}\right) = \lim_{t\to\infty}\left(\dfrac{1}{1+2/t}\cdot t^{1/2}\right) = \infty$

9. Since

$$-\frac{1}{\sqrt{x^2-1}} \le \frac{\cos x}{\sqrt{x^2-1}} \le \frac{1}{\sqrt{x^2-1}} \quad \text{for } x < -1$$

and since $\quad \lim_{x\to-\infty} -\dfrac{1}{\sqrt{x^2-1}} = \lim_{x\to-\infty} \dfrac{1}{\sqrt{x^2-1}} = 0$

the Squeezing Theorem for limits at $-\infty$ implies that $\lim_{x\to-\infty} (\cos x)/\sqrt{x^2-1} = 0$.

11. $\lim_{x\to-\infty} \dfrac{x^2}{4x^3-9} = \lim_{x\to-\infty} \dfrac{1/x}{4-9/x^3} = \dfrac{0}{4-0} = 0$

13. For $x > \frac{1}{2}$,

$$x - \sqrt{4x^2-1} = (x-\sqrt{4x^2-1})\frac{x+\sqrt{4x^2-1}}{x+\sqrt{4x^2-1}} = \frac{1-3x^2}{x+\sqrt{4x^2-1}} = \frac{1/x-3x}{1+\sqrt{4-1/x^2}}$$

so $\quad \lim_{x\to\infty} (x-\sqrt{4x^2-1}) = \lim_{x\to\infty} \dfrac{1/x-3x}{1+\sqrt{4-1/x^2}} = -\infty$

The exercise can also be solved as follows:

$$\lim_{x\to\infty} (x-\sqrt{4x^2-1}) = \lim_{x\to\infty}\left(x - x\sqrt{4-\frac{1}{x^2}}\right) = \lim_{x\to\infty} x\left(1-\sqrt{4-\frac{1}{x^2}}\right) = -\infty$$

15. Since $\lim_{x\to\infty} 1/x = 0$ and $\lim_{y\to0^+} \tan y = 0$, the Substitution Rule for limits at ∞ (with $y = 1/x$) implies that $\lim_{x\to\infty} \tan(1/x) = \lim_{y\to0^+} \tan y = 0$.

17. $\lim_{x\to\infty} \dfrac{1}{x-2} = \lim_{x\to-\infty} \dfrac{1}{x-2} = 0$

so $y = 0$ is a horizontal asymptote; $f'(x) = -1/(x-2)^2 < 0$ for $x \ne 2$, so f is decreasing on $(-\infty, 2)$ and $(2, \infty)$: $f''(x) = 2/(x-2)^3$, so the graph of f is concave upward on $(2, \infty)$ and concave downward on $(-\infty, 2)$;

$$\lim_{x\to2^+} \frac{1}{x-2} = \infty \quad \text{and} \quad \lim_{x\to2^-} \frac{1}{x-2} = -\infty$$

so $x = 2$ is a vertical asymptote.

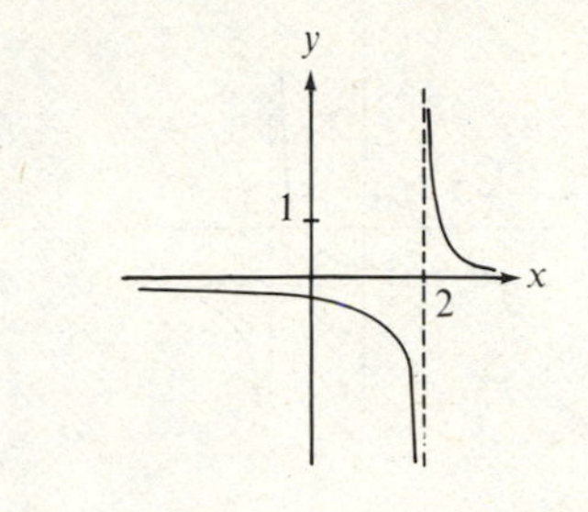

19. $\lim_{x\to\infty} \dfrac{3x}{2x-4} = \lim_{x\to-\infty} \dfrac{3x}{2x-4} = \dfrac{3}{2}$

so $y = \frac{3}{2}$ is a horizontal asymptote;

$$f'(x) = \frac{3(2x-4)-3x(2)}{(2x-4)^2} = \frac{-12}{(2x-4)^2} = \frac{-3}{(x-2)^2} \quad \text{for } x \ne 2$$

so f is decreasing on $(-\infty, 2)$ and $(2, \infty)$; $f''(x) = 6/(x-2)^3$, so the graph of f is concave upward on $(2, \infty)$ and concave downward on $(-\infty, 2)$;

$$\lim_{x\to2^+} \frac{3x}{2x-4} = \infty \quad \text{and} \quad \lim_{x\to2^-} \frac{3x}{2x-4} = -\infty$$

so $x = 2$ is a vertical asymptote.

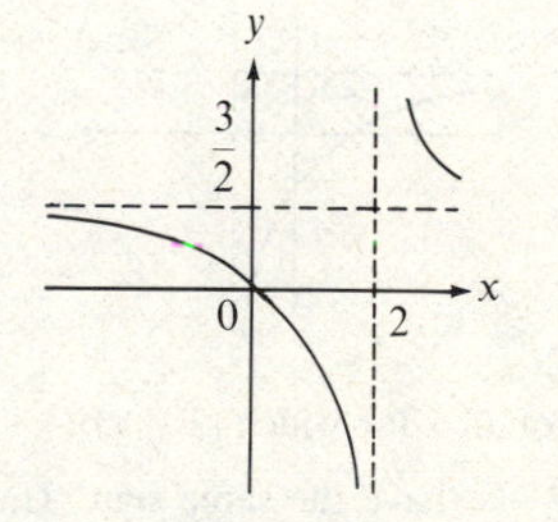

21. For $x \ne -3$,

$$f(x) = \frac{2x(x+3)}{(3-x)(3+x)} = \frac{2x}{3-x}, \quad \text{so} \quad \lim_{x\to\infty} f(x) = \lim_{x\to-\infty} f(x) = -2$$

and thus $y = -2$ is a horizontal asymptote;

$$f'(x) = \frac{2(3-x)-2x(-1)}{(3-x)^2} = \frac{6}{(3-x)^2} > 0 \quad \text{for } x \ne -3 \text{ and } 3$$

so f is increasing on $(-\infty, -3)$, $(-3, 3)$, and $(3, \infty)$; $f''(x) = 12/(3-x)^3$ for $x \neq -3$ and 3, so the graph of f is concave upward on $(-\infty, -3)$ and $(-3, 3)$ and concave downward on $(3, \infty)$; $\lim_{x\to 3^+} 2x/(3-x) = -\infty$ and $\lim_{x\to 3^-} 2x/(3-x) = \infty$, so $x = 3$ is a vertical asymptote.

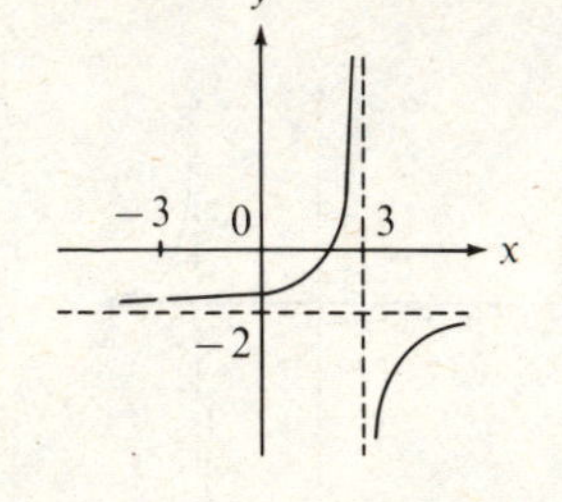

23. $$\lim_{x\to\infty} \frac{x+2}{x-1} = \lim_{x\to-\infty} \frac{x+2}{x-1} = 1$$

so $y = 1$ is a horizontal asymptote;

$$f'(x) = \frac{1(x-1)-(x+2)(1)}{(x-1)^2} = \frac{-3}{(x-1)^2} < 0 \text{ for } x \neq 1$$

so f is decreasing on $(-\infty, 1)$ and $(1, \infty)$; $f''(x) = 6/(x-1)^3$, so the graph of f is concave upward on $(1, \infty)$ and concave downward on $(-\infty, 1)$;

$$\lim_{x\to 1^+} \frac{x+2}{x-1} = \infty \quad\text{and}\quad \lim_{x\to 1^-} \frac{x+2}{x-1} = -\infty$$

so $x = 1$ is a vertical asymptote.

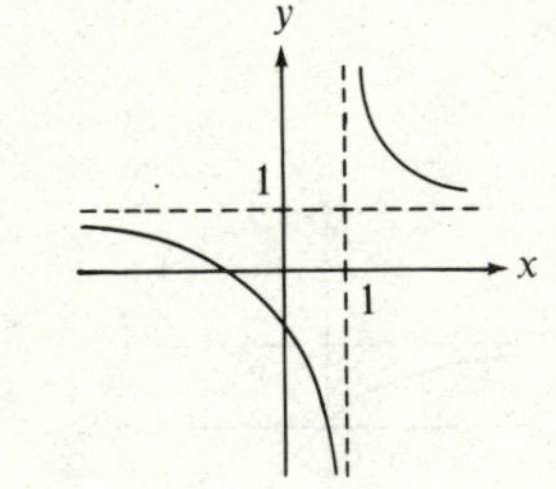

25. The domain of f consists of all x for which $(3-x)/(4-x) \geq 0$, that is, all x for which $3 - x = 0$, or $3 - x$ and $4 - x$ have the same sign. Thus the domain of f consists of $(-\infty, 3]$ and $(4, \infty)$.

$$\lim_{x\to\infty} \sqrt{\frac{3-x}{4-x}} = \lim_{x\to-\infty} \sqrt{\frac{3-x}{4-x}} = 1$$

so $y = 1$ is a horizontal asymptote;

$$f'(x) = \frac{1}{2}\left(\frac{3-x}{4-x}\right)^{-1/2}\left[\frac{-(4-x)-(3-x)(-1)}{(4-x)^2}\right] = \frac{-1}{2(4-x)^2}\left(\frac{3-x}{4-x}\right)^{-1/2}$$

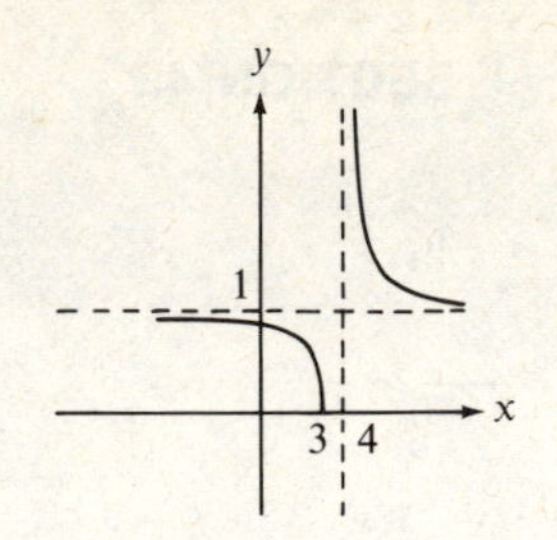

so f is decreasing on $(-\infty, 3]$ and $(4, \infty)$;

$$f''(x) = \frac{-1}{(4-x)^3}\left(\frac{3-x}{4-x}\right)^{-1/2} + \frac{1}{4(4-x)^2}\left(\frac{3-x}{4-x}\right)^{-3/2}\left[\frac{-(4-x)-(3-x)(-1)}{(4-x)^2}\right]$$
$$= \frac{-1}{(4-x)^3}\left(\frac{3-x}{4-x}\right)^{-1/2}\left[1 + \frac{1}{4(4-x)}\left(\frac{3-x}{4-x}\right)^{-1}\right]$$
$$= \frac{-1}{(4-x)^3}\left(\frac{3-x}{4-x}\right)^{-1/2}\left[1 + \frac{1}{4(3-x)}\right] = \frac{4x-13}{4(3-x)(4-x)^3}\left(\frac{3-x}{4-x}\right)^{-1/2}$$

so the graph of f is concave upward on $(4, \infty)$ and concave downward on $(-\infty, 3)$; $\lim_{x\to 4^+} \sqrt{(3-x)/(4-x)} = \infty$, so $x = 4$ is a vertical asymptote.

27. $$\lim_{x\to\infty} \frac{1}{x^2-4} = \lim_{x\to-\infty} \frac{1}{x^2-4} = 0$$

so $y = 0$ is a horizontal asymptote; $f'(x) = -2x/(x^2-4)^2$, so f is increasing on $(-\infty, -2)$ and $(-2, 0]$, and decreasing on $[0, 2)$ and $(2, \infty)$;

$$f''(x) = \frac{-2(x^2-4)^2 - (-2x)2(x^2-4)(2x)}{(x^2-4)^4} = \frac{2(3x^2+4)}{(x^2-4)^3}$$

so the graph of f is concave upward on $(-\infty, -2)$ and $(2, \infty)$ and concave downward on $(-2, 2)$;

$$\lim_{x\to 2^+} \frac{1}{x^2-4} = \lim_{x\to -2^-} \frac{1}{x^2-4} = \infty \quad\text{and}\quad \lim_{x\to 2^-} \frac{1}{x^2-4} = \lim_{x\to -2^+} \frac{1}{x^2-4} = -\infty$$

so $x = 2$ and $x = -2$ are vertical asymptotes.

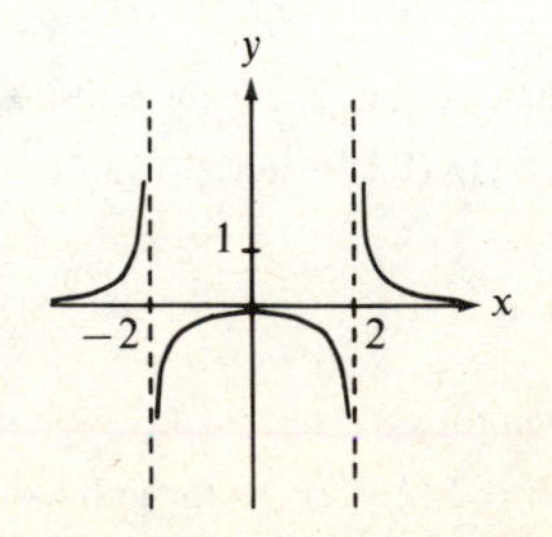

29. If $x > 0$, then

$$f(x) = \frac{x^2}{x^2+1} = \frac{1}{1+(1/x^2)}, \quad \text{so} \quad \lim_{x\to\infty} f(x) = 1$$

Thus $y = 1$ is a horizontal asymptote. If $x < 0$, then $f(x) = -x^2/(x^2+1)$, so $\lim_{x\to-\infty} f(x) = -1$. Thus $y = -1$ is a horizontal asymptote.

31. Let $N > 0$ and choose $M < 0$ such that if $x < M$, then $f(x)/g(x) > \frac{1}{2}$ and $g(x) > 2N$. It follows that if $x < M$, then $f(x) = [f(x)/g(x)]g(x) > \frac{1}{2} \cdot 2N = N$. Thus $\lim_{x\to-\infty} f(x) = \infty$.

33. $\lim\limits_{t\to\infty} S(t) = \lim\limits_{t\to\infty}\left(\frac{a}{t} + b\right) = b$

35. Assuming that the end of the universe is "at infinity," we compute the work required as

$$\lim_{x\to\infty} W(x) = \lim_{x\to\infty} GMm\left(\frac{1}{3960} - \frac{1}{x}\right) = \frac{GMm}{3960}$$

SECTION 4.8

1. $f'(x) = 3(x^2+1)$; $f''(x) = 6x$; increasing on $(-\infty, \infty)$; concave upward on $(0, \infty)$ and concave downward on $(-\infty, 0)$; inflection point is $(0, 2)$.

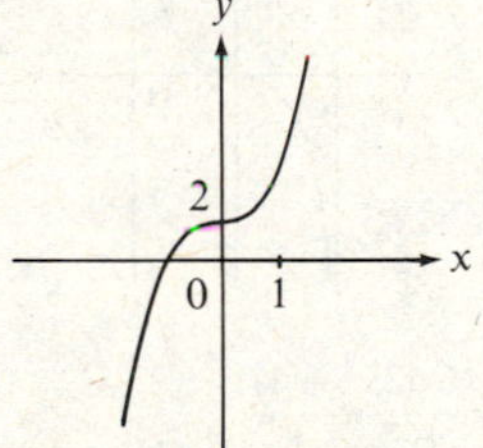

3. $f'(x) = 4x^3 + 24x^2 + 72x$; $f''(x) = 12x^2 + 48x + 72$; relative minimum value is $f(0) = -3$; increasing on $[0, \infty)$ and decreasing on $(-\infty, 0]$; concave upward on $(-\infty, \infty)$.

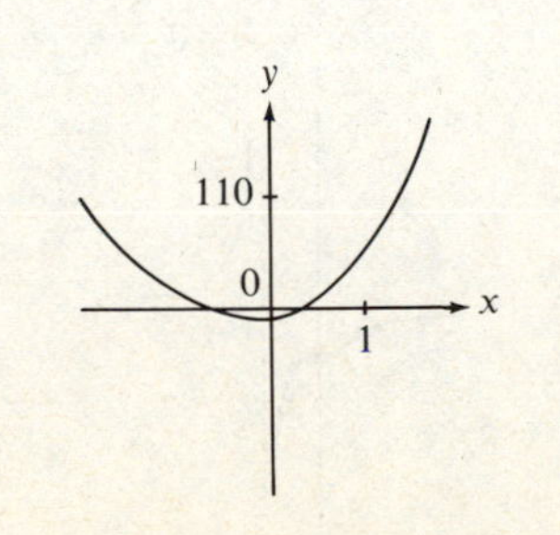

5. $g'(x) = 1 - 4/x^2$; $g''(x) = 8/x^3$; relative maximum value is $g(-2) = -4$; relative minimum value is $g(2) = 4$; increasing on $(-\infty, -2]$ and $[2, \infty)$, and decreasing on $[-2, 0)$ and $(0, 2]$; concave upward on $(0, \infty)$ and concave downward on $(-\infty, 0)$; vertical asymptote is $x = 0$; symmetry with respect to the origin.

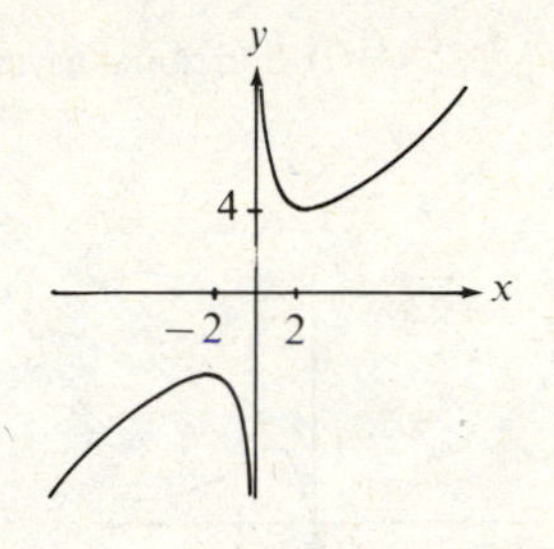

7. $g'(x) = \dfrac{2-x}{x^3}$; $\quad g''(x) = \dfrac{2(x-3)}{x^4}$

relative maximum value is $g(2) = \frac{1}{4}$; increasing on $(0, 2]$, and decreasing on $(-\infty, 0)$ and on $[2, \infty)$; concave upward on $(3, \infty)$ and concave downward on $(-\infty, 0)$ and $(0, 3)$; inflection point is $(3, \frac{2}{9})$; vertical asymptote is $x = 0$; horizontal asymptote is $y = 0$.

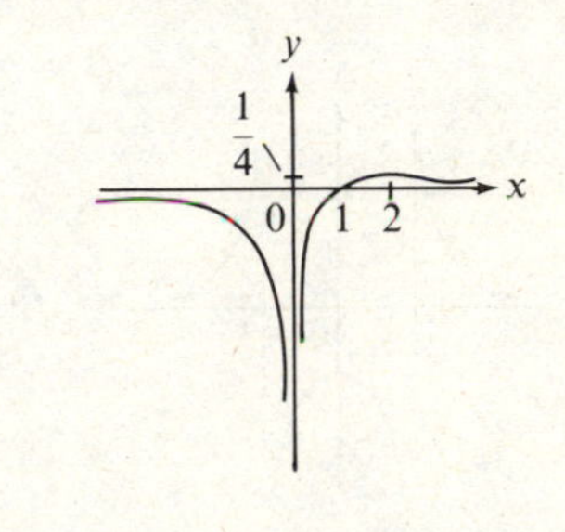

9. $k'(x) = \dfrac{2}{(1-x)^2}$; $\quad k''(x) = \dfrac{4}{(1-x)^3}$

increasing on $(-\infty, 1)$ and $(1, \infty)$; concave upward on $(-\infty, 1)$ and concave downward on $(1, \infty)$; vertical asymptote is $x = 1$; horizontal asymptote is $y = -1$.

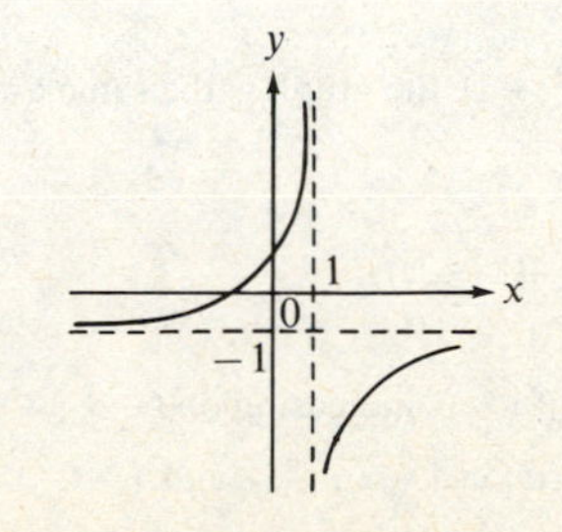

11. $k'(x) = \dfrac{4(1-4x^2)}{(1+4x^2)^2}$; $k''(x) = \dfrac{32x(-3+4x^2)}{(1+4x^2)^3}$

relative maximum value is $k(\frac{1}{2}) = 1$; relative minimum value is $k(-\frac{1}{2}) = -1$; increasing on $[-\frac{1}{2}, \frac{1}{2}]$, and decreasing on $(-\infty, -\frac{1}{2}]$ and $[\frac{1}{2}, \infty)$; concave upward on $(-\sqrt{3}/2, 0)$ and $(\sqrt{3}/2, \infty)$, and concave downward on $(-\infty, -\sqrt{3}/2)$ and $(0, \sqrt{3}/2)$; inflection points are $(-\sqrt{3}/2, -\sqrt{3}/2)$, $(0, 0)$, and $(\sqrt{3}/2, \sqrt{3}/2)$; horizontal asymptote is $y = 0$; symmetry with respect to the origin.

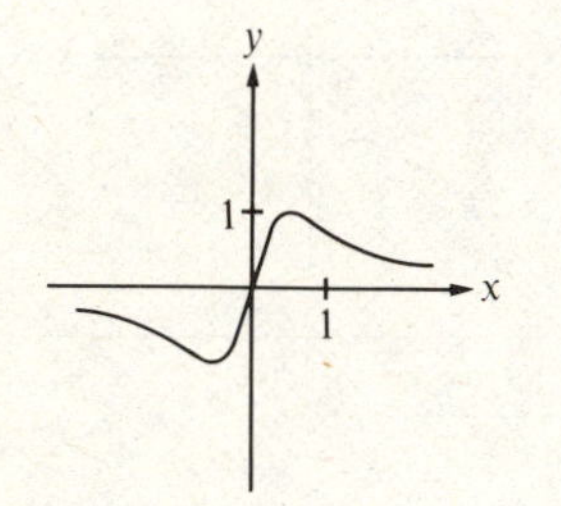

13. $f'(t) = \dfrac{-2t}{(t^2-1)^2}$; $f''(t) = \dfrac{6t^2+2}{(t^2-1)^3}$

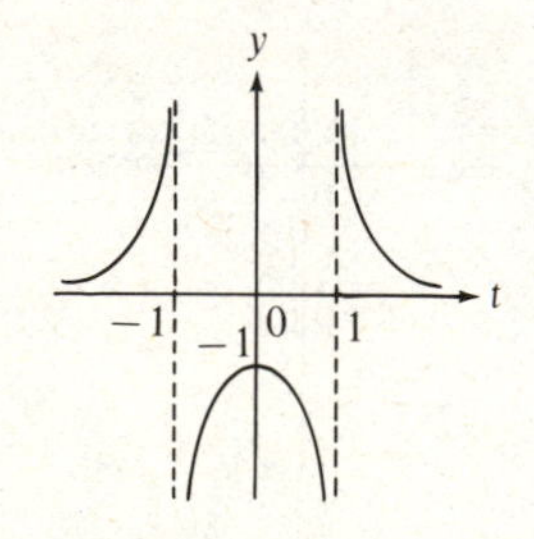

relative maximum value is $f(0) = -1$; increasing on $(-\infty, -1)$ and $(-1, 0]$, and decreasing on $[0, 1)$ and $(1, \infty)$; concave upward on $(-\infty, -1)$ and $(1, \infty)$, and concave downward on $(-1, 1)$; vertical asymptotes are $t = -1$ and $t = 1$; horizontal asymptote is $y = 0$; symmetry with respect to the y axis.

15. Since $t^2/(t^2-1) = 1 + 1/(t^2-1)$, the graph is the same as the graph in Exercise 13, but translated upward one unit.

17. $f'(z) = \dfrac{4z}{(z+1)^3}$; $f''(z) = \dfrac{4(1-2z)}{(z+1)^4}$

relative minimum value is $f(0) = 0$; increasing on $(-\infty, -1)$ and $[0, \infty)$, and decreasing on $(-1, 0]$; concave upward on $(-\infty, -1)$ and $(-1, \frac{1}{2})$, and concave downward on $(\frac{1}{2}, \infty)$; inflection point is $(\frac{1}{2}, \frac{2}{9})$; vertical asymptote is $z = -1$; horizontal asymptote is $y = 2$.

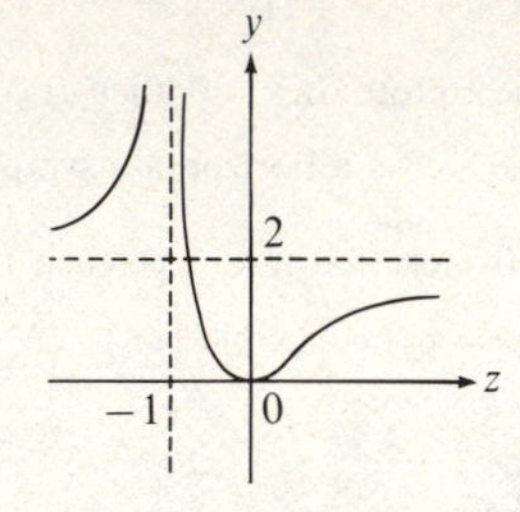

19. $f'(x) = \dfrac{x^2(3-x^2)}{(1-x^2)^2}$; $f''(x) = \dfrac{2x(x^2+3)}{(1-x^2)^3}$

relative maximum value is $f(\sqrt{3}) = -\frac{3}{2}\sqrt{3}$; relative minimum value is $f(-\sqrt{3}) = \frac{3}{2}\sqrt{3}$; increasing on $[-\sqrt{3}, -1)$, $(-1, 1)$, and $(1, \sqrt{3}]$, and decreasing on $(-\infty, -\sqrt{3}]$ and on $(\sqrt{3}, \infty)$; concave upward on $(-\infty, -1)$ and on $(0, 1)$, and concave downward on $(-1, 0)$ and $(1, \infty)$; inflection point is $(0, 0)$; vertical asymptotes are $x = -1$ and $x = 1$; symmetry with respect to the origin.

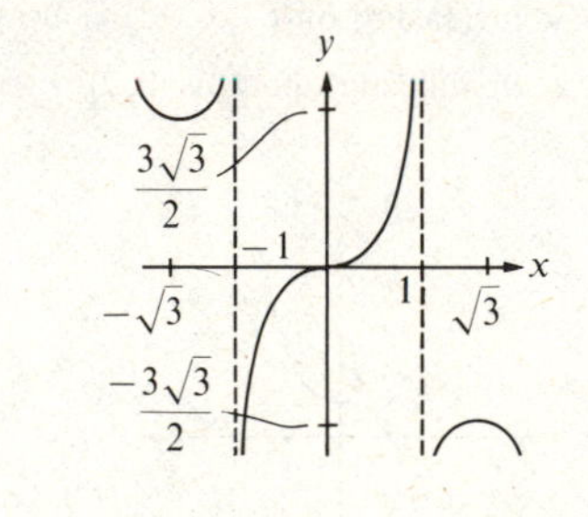

21. $f'(x) = \dfrac{-3x^2}{(x^3-1)^2}$; $f''(x) = \dfrac{6x(2x^3+1)}{(x^3-1)^3}$

decreasing on $(-\infty, 1)$ and $(1, \infty)$; concave upward on $(-\sqrt[3]{\frac{1}{2}}, 0)$ and $(1, \infty)$, and concave downward on $(-\infty, -\sqrt[3]{\frac{1}{2}})$ and $(0, 1)$; inflection points are $(-\sqrt[3]{\frac{1}{2}}, -\frac{2}{3})$ and $(0, -1)$; vertical asymptote is $x = 1$; horizontal asymptote is $y = 0$.

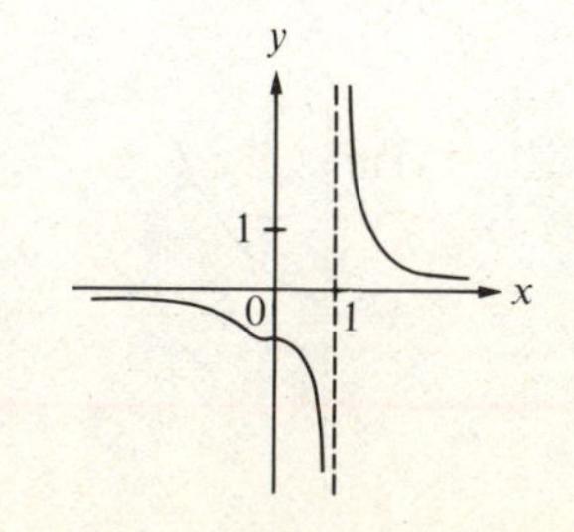

23. $f'(x) = \frac{1}{2}(x-1)^{-1/2} + \frac{1}{2}(x+1)^{-1/2}$; $f''(x) = (-1/4)(x-1)^{-3/2} - \frac{1}{4}(x+1)^{-3/2}$; increasing on $[1, \infty)$; concave downward on $(1, \infty)$.

25. $f'(x) = \dfrac{x(2-x^2)}{(1-x^2)^{3/2}}$; $f''(x) = \dfrac{2+x^2}{(1-x^2)^{5/2}}$

relative minimum value is $f(0) = 0$; increasing on $[0, 1)$ and decreasing on $(-1, 0)$; concave upward on $(-1, 1)$; vertical asymptotes are $x = -1$ and $x = 1$; symmetry with respect to the y axis.

27. $f'(x) = \dfrac{1+2x^2}{(1+x^2)^{1/2}}$; $f''(x) = \dfrac{x(3+2x^2)}{(1+x^2)^{3/2}}$

increasing on $(-\infty, \infty)$; concave upward on $(0, \infty)$, and concave downward on $(-\infty, 0)$; inflection point is $(0, 0)$; symmetry with respect to the origin.

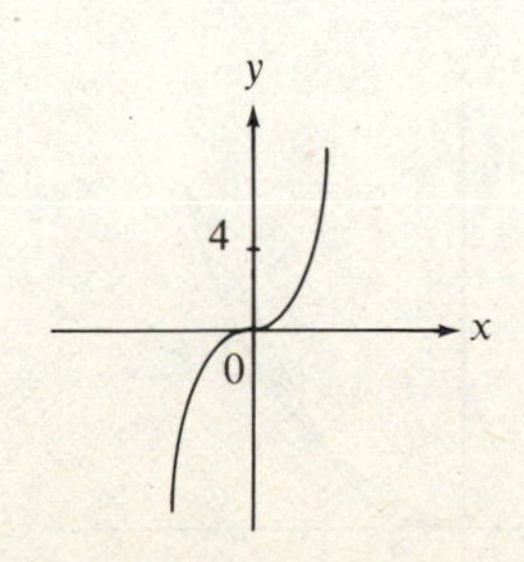

29. $f'(x) = \dfrac{x^{-1/2}}{2(1+x^{1/2})^2}$; $f''(x) = \dfrac{-(x^{-3/2}+3x^{-1})}{4(1+x^{1/2})^3}$

increasing on $[0, \infty)$; concave downward on $(0, \infty)$; horizontal asymptote is $y = 1$.

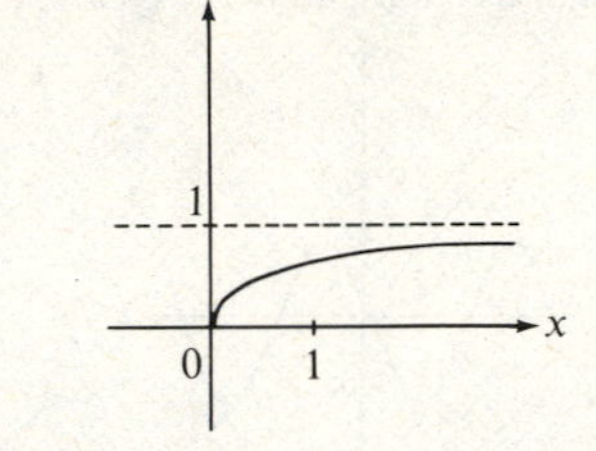

31. $f'(x) = 4(1 - x^{1/3})$; $f''(x) = -\frac{4}{3}x^{-2/3}$

relative maximum value is $f(1) = 1$; increasing on $(-\infty, 1]$ and decreasing on $[1, \infty)$; concave downward on $(-\infty, \infty)$.

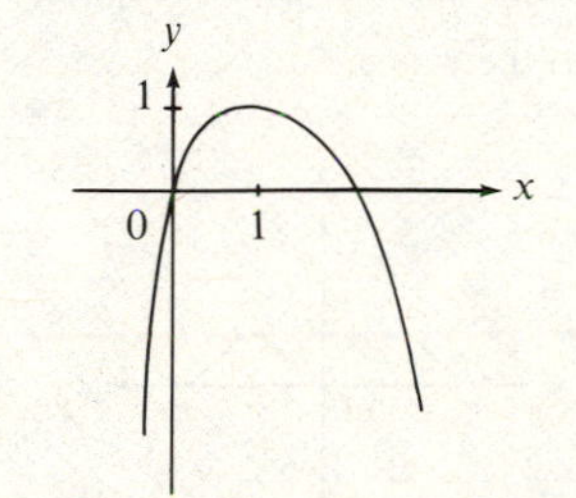

33. $g'(x) = \begin{cases} \cos x & \text{for } 2n\pi < x < (2n+1)\pi \\ -\cos x & \text{for } (2n+1)\pi < x < (2n+2)\pi \end{cases}$, for any integer n

$g''(x) = -|\sin x|$ for $x \neq n\pi$, for any integer n

relative maximum value is $g(\pi/2 + n\pi) = 1$; relative minimum value is $g(n\pi) = 0$; increasing on $[n\pi, \pi/2 + n\pi]$ and decreasing on $[\pi/2 + n\pi, (n+1)\pi]$; concave downward on $(n\pi, (n+1)\pi)$; symmetry with respect to the y axis.

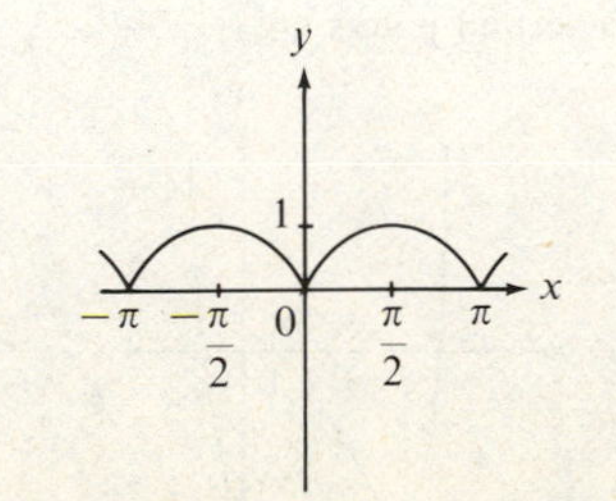

35. $g'(x) = \sqrt{3}\cos x - \sin x$; $g''(x) = -\sqrt{3}\sin x - \cos x = -g(x)$; relative maximum value is $g(\pi/3 + 2n\pi) = 2$ for any integer n; relative minimum value is $g(-2\pi/3 + 2n\pi) = -2$ for any integer n; increasing on $[-2\pi/3 + 2n\pi, \pi/3 + 2n\pi]$ and decreasing on $[\pi/3 + 2n\pi, 4\pi/3 + 2n\pi]$; concave upward on $(-7\pi/6 + 2n\pi, -\pi/6 + 2n\pi)$ and concave downward on $(-\pi/6 + 2n\pi, 5\pi/6 + 2n\pi)$; inflection points are $(-\pi/6 + n\pi, 0)$.

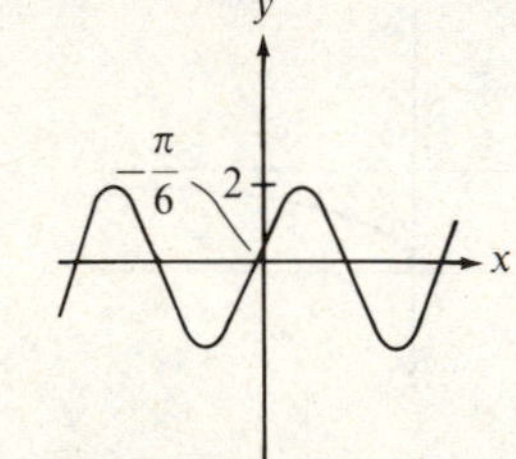

37. $g'(t) = 2\sin t\cos t = \sin 2t$; $g''(t) = 2\cos 2t$; relative maximum value is $g(n\pi + \pi/2) = 1$ for any integer n; relative minimum value is $g(n\pi) = 0$ for any integer n; increasing on $[n\pi, n\pi + \pi/2]$ and decreasing on $[n\pi + \pi/2, (n+1)\pi]$; concave upward on $(n\pi - \pi/4, n\pi + \pi/4)$ concave downward on $(n\pi + \pi/4, n\pi + 3\pi/4)$; inflection points are $(\pi/4 + n\pi/2, \frac{1}{2})$: symmetry with respect to the y axis

39. $f'(x) = 4x^3 + 12x^2 + 8x = 4x(x+1)(x+2)$

$f''(x) = 12x^2 + 24x + 8 = 12(x + 1 - \frac{1}{3}\sqrt{3})(x + 1 + \frac{1}{3}\sqrt{3})$

by the Newton-Raphson method, the zeros of f are approximately -2.554 and 0.554; relative maximum value is $f(-1) = -1$; relative minimum value is $f(-2) = f(0) = -2$; increasing on $[-2, -1]$ and $[0, \infty)$, and decreasing on $(-\infty, -2]$ and $[-1, 0]$; concave upward on $(-\infty, -1 - \frac{1}{3}\sqrt{3})$ and $(-1 + \frac{1}{3}\sqrt{3}, \infty)$, and concave downward on $(-1 - \frac{1}{3}\sqrt{3}, -1 + \frac{1}{3}\sqrt{3})$; inflection points are $(-1 - \frac{1}{3}\sqrt{3}, -\frac{14}{9})$ and $(-1 + \frac{1}{3}\sqrt{3}, -\frac{14}{9})$.

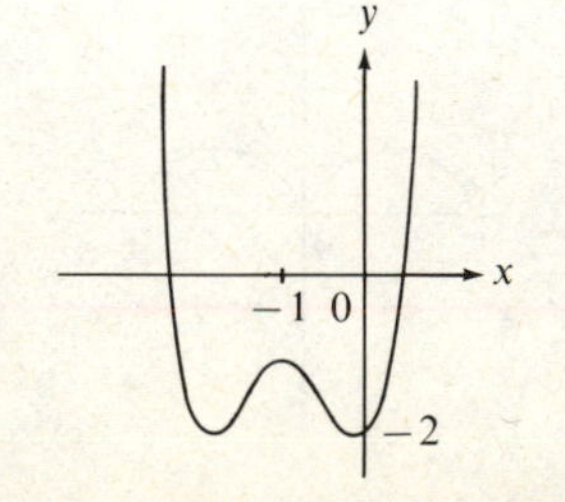

41. $f'(x) = \dfrac{-x^2 + 2x + 1}{x^2(x+1)^2}$; $\quad f''(x) = \dfrac{2x^3 - 6x^2 - 6x - 2}{x^3(x+1)^3}$

relative minimum value is $f(1 - \sqrt{2}) = \sqrt{2}/(3\sqrt{2} - 4)$; relative maximum value is $f(1 + \sqrt{2}) = \sqrt{2}/(4 + 3\sqrt{2})$; increasing on $[1 - \sqrt{2}, 0)$ and $(0, 1 + \sqrt{2})$, and decreasing on $(-\infty, -1)$, $(-1, 1 - \sqrt{2}]$ and $(1 + \sqrt{2}, \infty)$; by the Newton-Raphson method the zero c of f'' is approximately 3.847; concave upward on $(-1, 0)$ and (c, ∞), and concave downward on $(-\infty, -1)$ and $(0, c)$; inflection point is $(c, f(c))$, which is approximately $(3.847, 0.153)$; vertical asymptotes are $x = -1$ and $x = 0$; horizontal asymptote is $y = 0$.

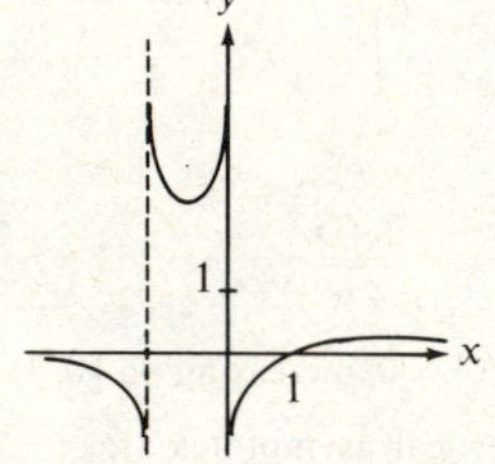

43. The graphs intersect at (x, y) such that $x^2 = y = x$, or $x^2 - x = 0$, or $x = 0$ and $x = 1$. For $0 \le x \le 1$, $f(x) \le g(x)$. Since $g'(x) = 1$ and $f'(x) = 2x$, both functions are increasing on $[0, 1]$. The graph of f is concave upward on $(0, 1)$.

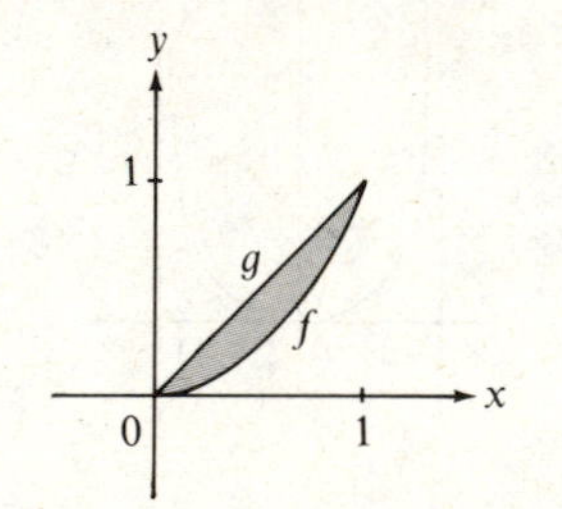

45. The graphs intersect at (x, y) such that $x^3 + x = y = 3x^2 - x$, or $x^3 - 3x^2 + 2x = 0$, or $x = 0$, 1, and 2. For $0 < x < 1$, $f(x) \ge g(x)$, and for $1 < x < 2$, $f(x) \le g(x)$. Since $f'(x) = 3x^2 + 1$ and $f''(x) = 6x$, f is increasing on $[0, 2]$ and its graph is concave upward on $(0, 2)$. Since $g'(x) = 6x - 1$ and $g''(x) = 6$, g is increasing on $[\frac{1}{6}, 2]$ and decreasing on $[0, \frac{1}{6}]$; its graph is concave upward on $(0, 2)$.

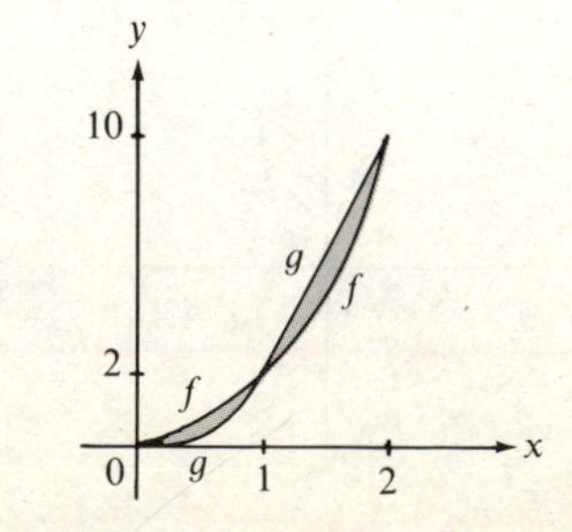

47. The graphs intersect at (x, y) such that

$$\frac{2x}{\sqrt{1+x^2}} = y = \frac{x}{\sqrt{1-x^2}}, \quad \text{or} \quad \frac{4x^2}{1+x^2} = \frac{x^2}{1-x^2}, \quad \text{or} \quad 3x^2 = 5x^4$$

so that $x = 0$ and $x = \pm\sqrt{3/5}$. Since

$$g'(x) = \frac{2}{(1+x^2)^{3/2}} \quad \text{and} \quad g''(x) = \frac{-6x}{(1+x^2)^{5/2}}$$

g is increasing on $[-\sqrt{\frac{3}{5}}, \sqrt{\frac{3}{5}}]$, and the graph of g is concave upward on $[-\sqrt{\frac{3}{5}}, 0)$ and concave downward on $(0, \sqrt{\frac{3}{5}})$. Since

$$k'(x) = \frac{1}{(1-x^2)^{3/2}} \quad \text{and} \quad k''(x) = \frac{3x}{(1-x^2)^{5/2}}$$

k is increasing on $[-\sqrt{\frac{3}{5}}, \sqrt{\frac{3}{5}}]$, and the graph of k is concave downward on $(-\sqrt{\frac{3}{5}}, 0)$ and concave upward on $(0, \sqrt{\frac{3}{5}})$.

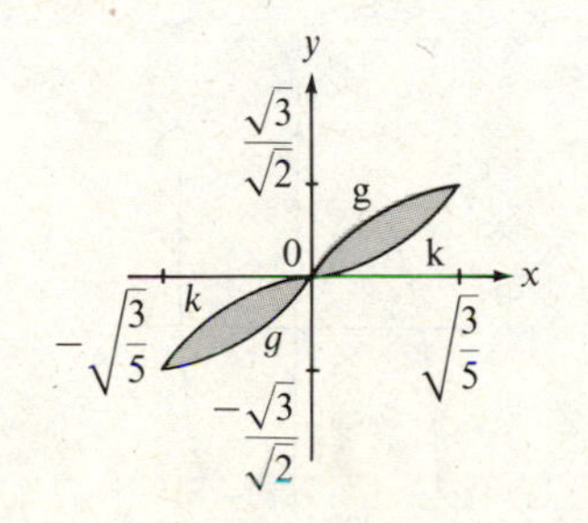

49.

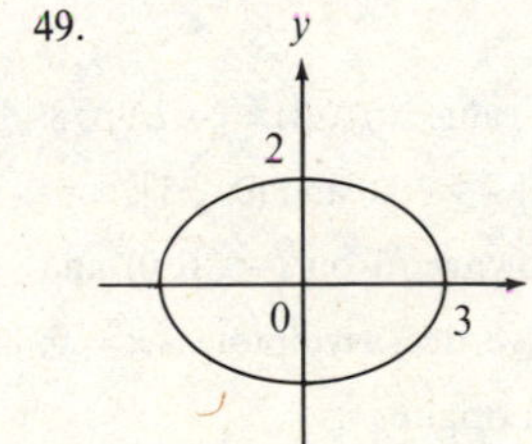

51.

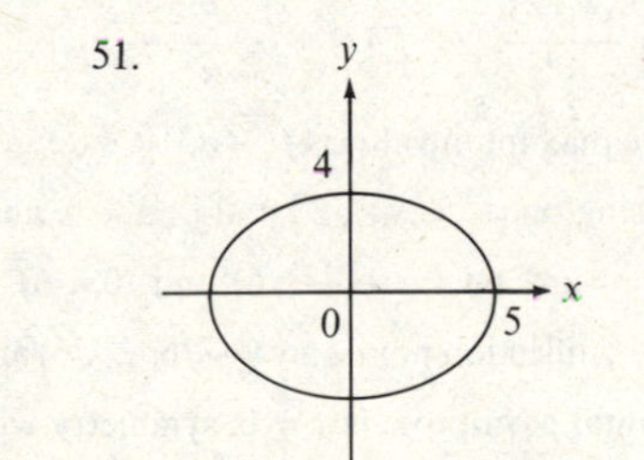

53.

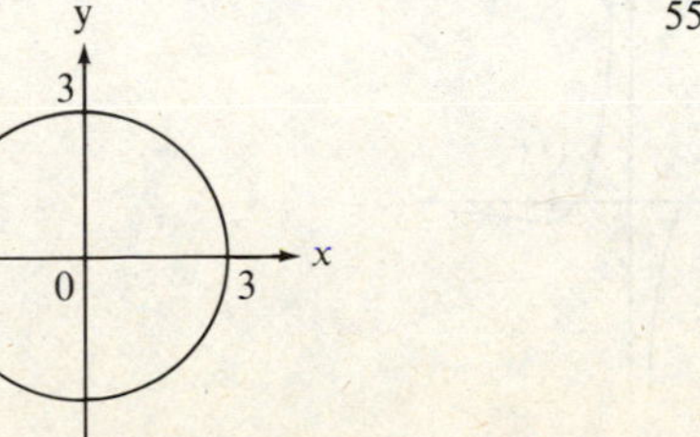

55.

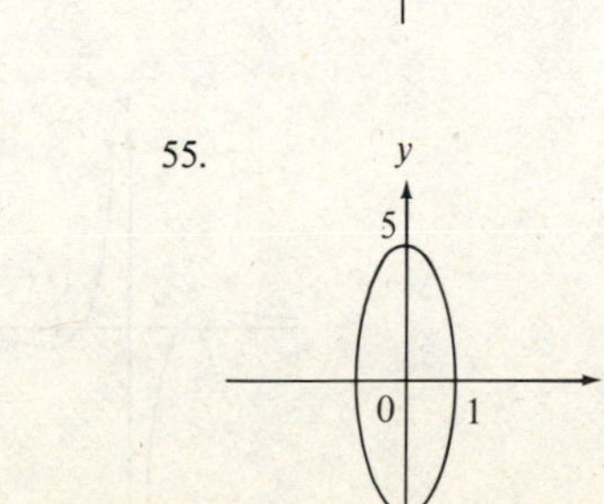

57.

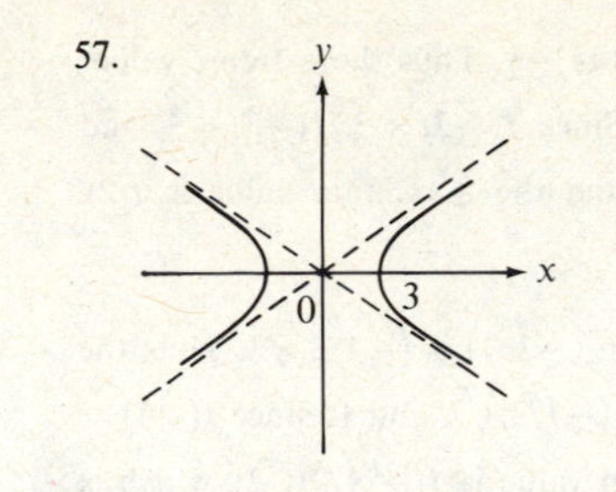

59.

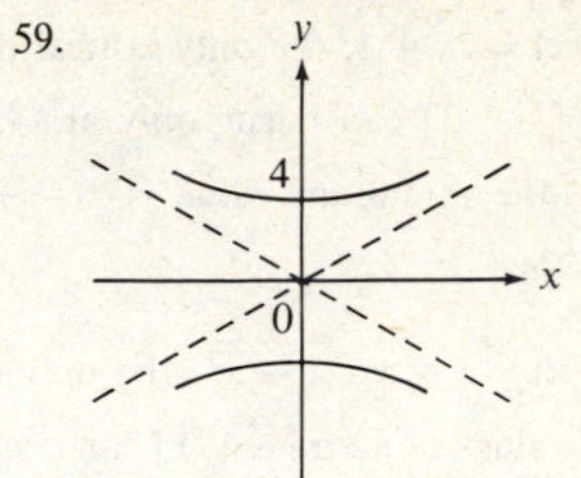

61.

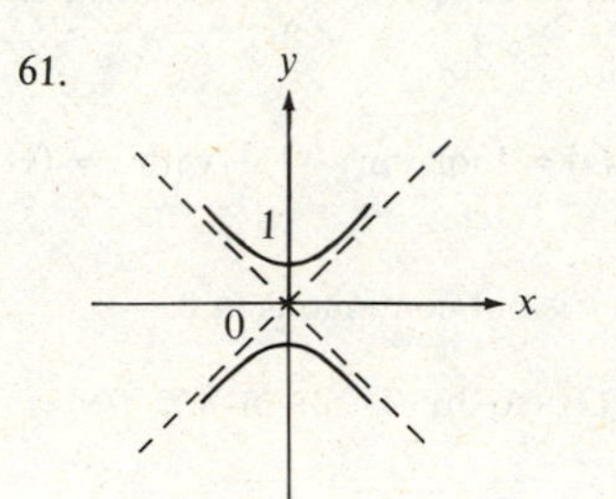

63. $x^2 - 2x + 4y^2 - 3 = 0$

$(x^2 - 2x + 1) + 4y^2 = 4$

$(x-1)^2 + 4y^2 = 4$

Let $X = x - 1$ and $Y = y$. Then $X^2 + 4Y^2 = 4$.

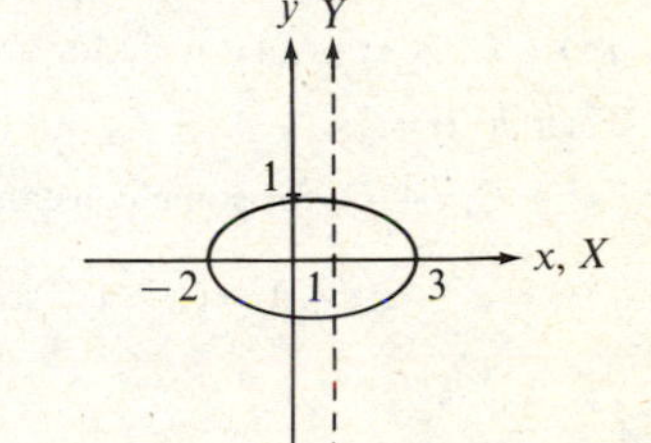

65. $8x^2 - 8x + 4y^2 + 4y = 33$

$(8x^2 - 8x + 2) + (4y^2 + 4y + 1) = 36$

$8(x - \frac{1}{2})^2 + 4(y + \frac{1}{2})^2 = 36$

Let $X = x - \frac{1}{2}$ and $Y = y + \frac{1}{2}$. Then $8X^2 + 4Y^2 = 36$.

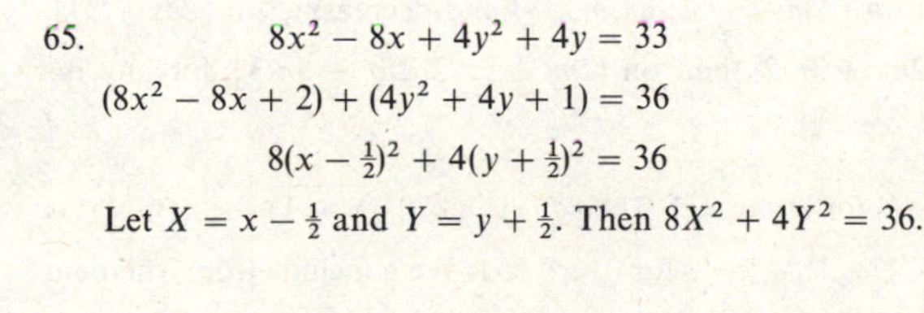

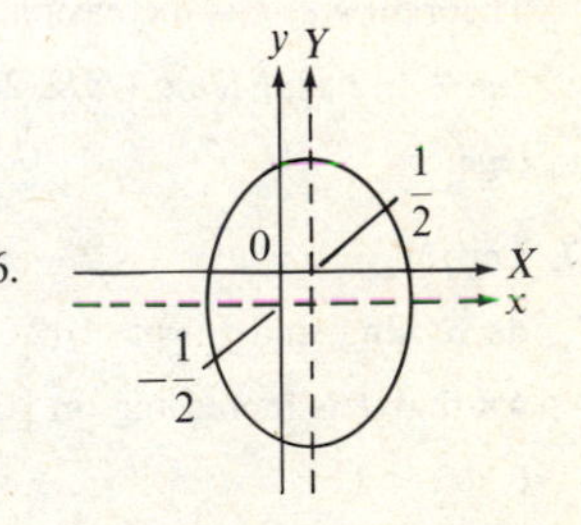

CHAPTER 4 REVIEW

1. The domain of f is $(-\infty, 2]$;

$$f'(x) = 2x\sqrt{2-x} - \frac{x^2}{2\sqrt{2-x}} = \frac{x(8-5x)}{2\sqrt{2-x}}$$

so $f'(x) = 0$ for $x = 0$ and $\frac{8}{5}$, and $f'(x)$ is undefined for $x = 2$. Critical points: $0, \frac{8}{5}, 2$.

3. Since $f'(x) = 2x + 1$, the only critical point in $(-2, 2)$ is $-\frac{1}{2}$. Thus the extreme values of f on $[-2, 2]$ can occur only at -2, $-\frac{1}{2}$ and 2. Since $f(-2) = 3$, $f(-\frac{1}{2}) = \frac{3}{4}$, and $f(2) = 7$, the minimum value is $f(-\frac{1}{2})$, which is $\frac{3}{4}$, and the maximum value is $f(2)$, which is 7.

5. Since $f'(x) = 1 + x/\sqrt{1 - x^2}$, the only critical point in $(-1, 1)$ is $(-1/2)\sqrt{2}$. Thus the extreme values of f on $[-1, 1]$ can occur only at $-1, (-1/2)\sqrt{2}$, and 1. Since $f(-1) = -1$, $f((-1/2)\sqrt{2}) = -\sqrt{2}$, and $f(1) = 1$, the minimum value is $f((-1/2)\sqrt{2})$, which is $-\sqrt{2}$, and the maximum value is $f(1)$, which is 1.

7. a. $f(-1) = (-1) + 1 = 0$ and $f(1) = 1 - 1 = 0$, and $f'(x) = 1$ for x in $(-1, 1)$ with $x \neq 0$. Also $f'(0)$ does not exist.

   b. This does not contradict Rolle's Theorem because f is not continuous at 0.

9. Let $g(x) = \frac{1}{3}x^3 + \cos x$. Then $g'(x) = x^2 - \sin x = f'(x)$, so by Theorem 4.7, $f(x) = g(x) + C = \frac{1}{3}x^3 + \cos x + C$.

11. Let $g(x) = \frac{1}{3}x^3 - 4x$. Then $g'(x) = x^2 - 4 = f''(x) = (f')'(x)$ so by Theorem 4.7, $f'(x) = g(x) + C_1 = \frac{1}{3}x^3 - 4x + C_1$ for some constant C_1. Now let $h(x) = \frac{1}{12}x^4 - 2x^2 + C_1x$. Then $h'(x) = \frac{1}{3}x^3 - 4x + C_1 = f'(x)$, so by Theorem 4.7, $f(x) = h(x) + C_2 = \frac{1}{12}x^4 - 2x^2 + C_1x + C_2$ for some constant C_2.

13. $f'(x) = x^2 - 2x + 1 = (x - 1)^2$, so $f'(x) \geq 0$ for all x. By Theorem 4.8, f is increasing on $(-\infty, \infty)$.

15. $f'(x) = \cos x - \frac{1}{8}\sec^2 x$, so $f'(x) > 0$ if $\cos x > \frac{1}{8}\sec^2 x$, or $\cos^3 x > \frac{1}{8}$, or $\cos x > \frac{1}{2}$. By Theorem 4.8, f is increasing on $[2n\pi - \pi/3, 2n\pi + \pi/3]$ and decreasing on $[2n\pi + \pi/3, 2n\pi + \pi/2)$, on $(2n\pi + \pi/2, 2n\pi + 3\pi/2)$, and on $(2n\pi + 3\pi/2, 2n\pi + 5\pi/3]$, for any integer n.

17. Let $f(x) = \sqrt{x + 3} - \sqrt{3} - x/4$ for $0 \leq x \leq 1$. Then $f'(x) = 1/(2\sqrt{x + 3}) - \frac{1}{4}$. Thus f' is decreasing, and $f'(x) > 1/(2\sqrt{1 + 3}) - \frac{1}{4} = 0$ for $0 < x < 1$. We conclude from Theorem 4.8 that f is increasing on $[0, 1]$. Since $f(0) = 0$, it follows that $\sqrt{x + 3} \geq \sqrt{3} + x/4$ for $0 \leq x \leq 1$.

19. $f'(x) = 12x^3 - 30x^2 + 12x = 6x(2x - 1)(x - 2)$; $f''(x) = 36x^2 - 60x + 12$. The relative extreme values must occur at the critical points 0, $\frac{1}{2}$, and 2. Since $f''(0) = 12 > 0$, $f''(\frac{1}{2}) = -9 < 0$, and $f''(2) = 36 > 0$, the Second Derivative Test implies that $f(0) = 3$ is a relative minimum value, $f(\frac{1}{2}) = \frac{55}{16}$ is a relative maximum value, and $f(2) = -5$ is a relative minimum value.

21. $f'(x) = 2(x + 1)(x - 2)^4 + 4(x + 1)^2(x - 2)^3 = 6x(x + 1)(x - 2)^3$. The relative extreme values must occur at the critical points -1, 0, and 2. Since $f'(x) < 0$ for x in $(-\infty, -1)$ or $(0, 2)$, and since $f'(x) > 0$ for x in $(-1, 0)$ or $(2, \infty)$, the First Derivative Test implies that $f(-1) = f(2) = 0$ is a relative minimum value, and $f(0) = 16$ is a relative maximum value.

23. $f'(x) = 2x^3 + 3x^2 - 12x$; $f''(x) = 6x^2 + 6x - 12 = 6(x + 2)(x - 1)$. Thus the graph is concave upward on $(-\infty, -2)$ and $(1, \infty)$, and concave downward on $(-2, 1)$.

25. $f'(x) = \dfrac{-4x^3}{(1 + x^4)^2}$ and $f''(x) = \dfrac{4x^2(5x^4 - 3)}{(1 + x^4)^3}$

    Thus the graph is concave upward on $(-\infty, -\sqrt[4]{\frac{3}{5}})$ and on $(\sqrt[4]{\frac{3}{5}}, \infty)$, and concave downward on $(-\sqrt[4]{\frac{3}{5}}, \sqrt[4]{\frac{3}{5}})$.

27. $f'(x) = 3x^2 - 6$; $f''(x) = 6x$; relative maximum value is $f(-\sqrt{2}) = 4\sqrt{2} - 1$; relative minimum value is $f(\sqrt{2}) = -4\sqrt{2} - 1$; increasing on $(-\infty, -\sqrt{2}]$ and $[\sqrt{2}, \infty)$, and decreasing on $[-\sqrt{2}, \sqrt{2}]$; concave upward on $(0, \infty)$ and concave downward on $(-\infty, 0)$; inflection point is $(0, -1)$.

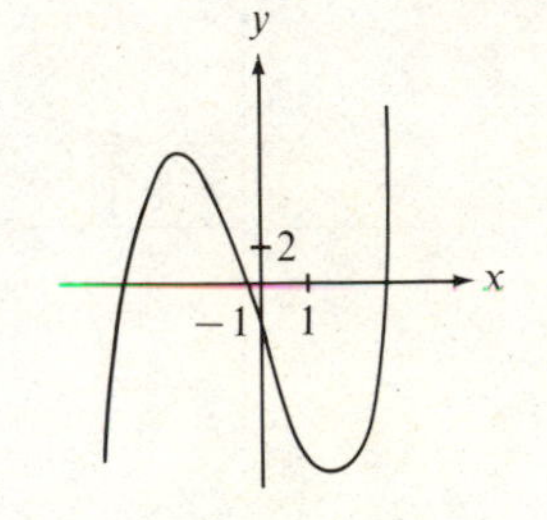

29. $f'(x) = \dfrac{x^2 - 3}{x^4}$; $f''(x) = \dfrac{12 - 2x^2}{x^5}$

    relative maximum value is $f(-\sqrt{3}) = \frac{2}{9}\sqrt{3}$; relative minimum value is $f(\sqrt{3}) = (-2/9)\sqrt{3}$; increasing on $(-\infty, -\sqrt{3}]$ and $[\sqrt{3}, \infty)$, and decreasing on $[-\sqrt{3}, 0)$ and $(0, \sqrt{3}]$; concave upward on $(-\infty, -\sqrt{6})$ and $(0, \sqrt{6})$, and concave downward on $(-\sqrt{6}, 0)$ and $(\sqrt{6}, \infty)$; inflection points are $(-\sqrt{6}, \frac{5}{36}\sqrt{6})$ and $(\sqrt{6}, -\frac{5}{36}\sqrt{6})$; vertical asymptote is $x = 0$; horizontal asymptote is $y = 0$; symmetry with respect to the origin.

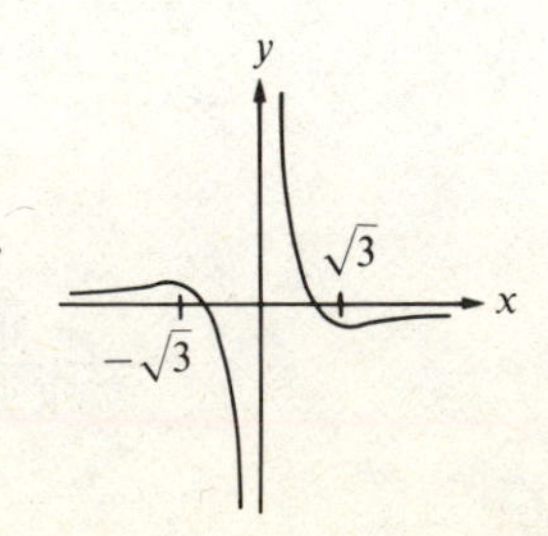

31. $k'(x) = \dfrac{-10x}{(x^2-4)^2}$; $\quad k''(x) = \dfrac{10(3x^2+4)}{(x^2-4)^3}$

relative maximum value is $k(0) = -1/4$; increasing on $(-\infty, -2)$ and $(-2, 0]$ and decreasing on $[0, 2)$ and $(2, \infty)$; concave upward on $(-\infty, -2)$ and $(2, \infty)$, and concave downward on $(-2, 2)$; vertical asymptotes are $x = -2$ and $x = 2$; horizontal asymptote is $y = 1$; symmetry with respect to the y axis.

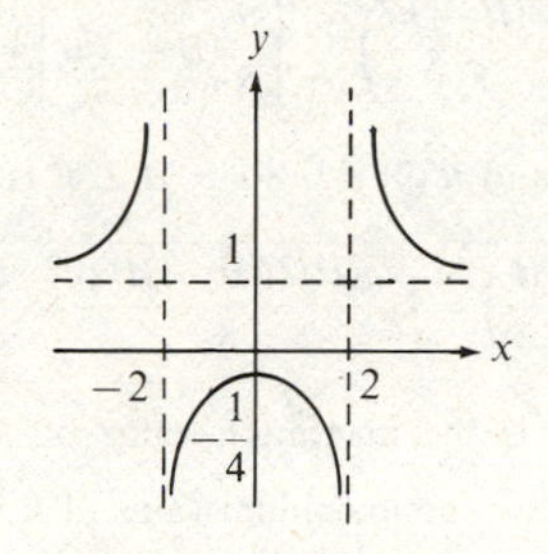

33. The graphs intersect at (x, y) such that $x^2 + 4x = y = x - 2$, or $x^2 + 3x + 2 = 0$, or $x = -2$ or $x = -1$. For $-2 \le x \le -1$, $f(x) \le g(x)$. Since $f'(x) = 2x + 4$ and $f''(x) = 2$, f has a relative minimum value at -2, and the graph of f is concave upward on $(-2, -1)$; g is strictly increasing on $[-2, -1]$.

35. 37.

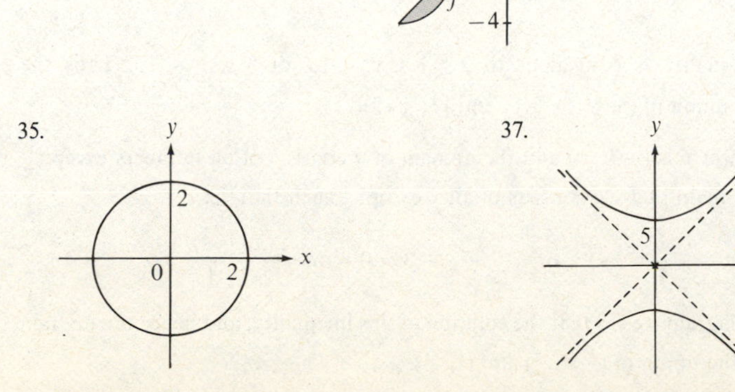

39.

$$9x^2 - 18x + y^2 + 4y = 3$$
$$(9x^2 - 18x + 9) + (y^2 + 4y + 4) = 16$$
$$9(x-1)^2 + (y+2)^2 = 16$$

Let $X = x - 1$ and $Y = y + 2$. Then $9X^2 + Y^2 = 16$.

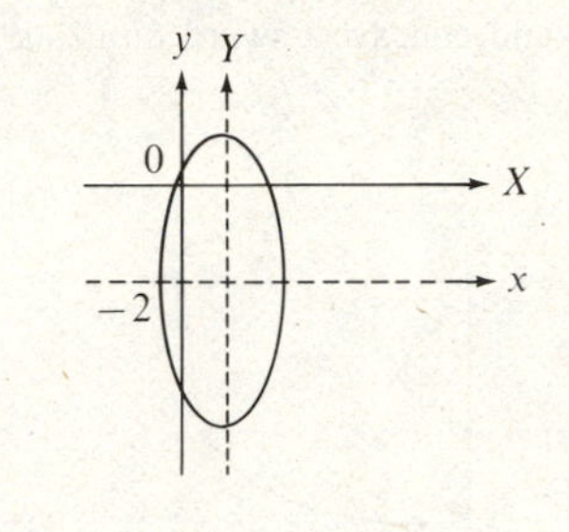

41. $D'(v) = \dfrac{\sqrt{3}}{48}\left(120v^{1/2} - \dfrac{5}{2}v^{3/2}\right) = \dfrac{5\sqrt{3}}{96}v^{1/2}(48 - v)$

so $D'(v) = 0$ if $v = 48$. Since $D'(v) > 0$ for $0 < v < 48$ and $D'(v) < 0$ for $48 < v < 75$, Theorem 4.11 implies that the maximum value of D occurs for $v = 48$ (miles per hour).

43. The square of the distance between the point $(4, 0)$ and the point $(x, x^{1/2})$ on the curve $y = x^{1/2}$ is given by $f(x) = (x-4)^2 + (x^{1/2} - 0)^2 = x^2 - 7x + 16$. We will minimize f in order to find the closest point on the graph. Now $f'(x) = 2x - 7$, so that $f'(x) = 0$ if $x = \frac{7}{2}$. Since $f''(x) = 2$ for $x > 0$, Theorem 4.12 implies that the minimum square of the distance, and hence the minimum distance, occurs at $(\frac{7}{2}, \sqrt{\frac{7}{2}})$.

45. Let x be the length in feet of one side of the base, y the height in feet of the toolshed, and C the cost of the material. We are to minimize C. Since the volume is 800 cubic feet, we have $x^2y = 800$, so that $y = 800/x^2$. Since the floor costs $6x^2$ dollars, the roof costs $2x^2$ dollars, and each side $5xy$ dollars, C is given by $C(x) = 6x^2 + 2x^2 + 4(5xy) = 8x^2 + 20x(800/x^2) = 8(x^2 + 2000/x)$ for $x > 0$. Thus $C'(x) = 8(2x - 2000/x^2)$ so that $C'(x) = 0$ if $2x = 2000/x^2$, that is, for $x^3 = 1000$, or $x = 10$. Since $C''(x) = 16 + 32{,}000/x^3 > 0$ for $x > 0$, Theorem 4.12 implies that the minimum value of C is $C(10)$. For $x = 10$ we have $y = 800/10^2 = 8$, so the base should be 10 feet on a side, and the shed should be 8 feet tall.

47. The revenue the company receives from x passengers is given by $R(x) = x(1000 - 2x) = 1000x - 2x^2$. We are to maximize R. Since $R'(x) = 1000 - 4x$, we find that $R'(x) = 0$ for $x = 250$. Since $R''(x) = -4 < 0$ for $x > 0$, Theorem 4.12 implies that the maximum revenue is $R(250) = 125{,}000$ dollars.

49. $f'(v) = \dfrac{v}{c^2}\left(1 - \dfrac{v^2}{c^2}\right)^{-3/2}$

and

$$f''(v) = \frac{1}{c^2}\left(1 - \frac{v^2}{c^2}\right)^{-3/2} + \frac{3v^2}{c^4}\left(1 - \frac{v^2}{c^2}\right)^{-5/2} = \frac{1}{c^2}\left(1 + \frac{2v^2}{c^2}\right)\left(1 - \frac{v^2}{c^2}\right)^{-5/2}$$

The function is increasing and concave upward on $(0, c)$. The line $v = c$ is a vertical asymptote.

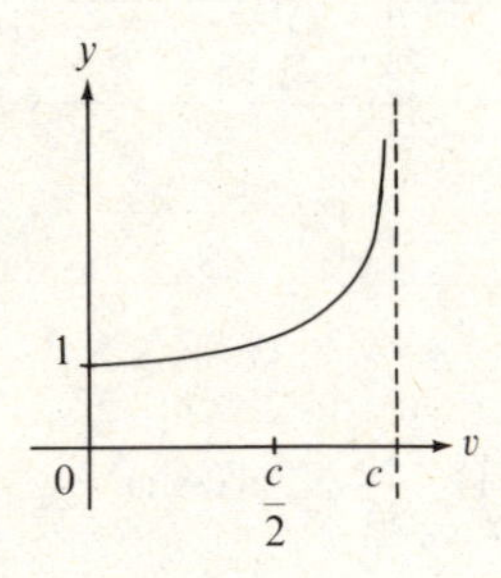

51. Let x be the length of the fence parallel to the river, y the length of each fence perpendicular to the river, and P the total length of fence. We are to minimize P. Then $xy = 6400$, so that $y = 6400/x$, and thus $P(x) = x + 4y = x + 4(6400/x) = x + 25{,}600/x$. Since $P'(x) = 1 - 25{,}600/x^2$, we have $P'(x) = 0$ if $x = 160$. Since $P''(x) = 51{,}200/x^3$, we have $P''(x) > 0$ for all $x > 0$. Thus Theorem 4.12 implies that the total length of fence is minimized if $x = 160$ feet.

53. We are to minimize C. Now $C'(x) = a - b/x^2$, so that $C'(x) = 0$ if $x = \sqrt{b/a}$. Since $C''(x) = 2b/x^3$, and $C''(x) > 0$ for all $x > 0$, Theorem 4.12 implies that the total cost is minimized if the cross-sectional area of wire is $\sqrt{b/a}$. The purchase price of wire of cross-sectional area $\sqrt{b/a}$ is $a\sqrt{b/a} = \sqrt{ab}$, and the cost due to power loss is $b/\sqrt{b/a} = \sqrt{ab}$, so the purchase price and cost due to power loss are the same if the cross-sectional area is $\sqrt{b/a}$, and hence when the total cost is minimized.

55. Let P denote the profit; we are to maximize P. If the firm borrows money at x percent, then it can borrow kx^2 dollars, where k is a constant. Its income from lending kx^2 dollars is $\frac{18}{100}(kx^2)$ dollars, whereas the cost of borrowing kx^2 dollars is $(x/100)(kx^2)$ dollars. Therefore the profit in dollars is given by $P(x) = \frac{18}{100}kx^2 - (k/100)x^3 = (k/100)(18x^2 - x^3)$. Thus $P'(x) = (k/100)(36x - 3x^2) = (3kx/100)(12 - x)$, so that $P'(x) = 0$ for $x = 0$ or $x = 12$. Since $P'(x) > 0$ for $0 < x < 12$ and $P'(x) < 0$ for $x > 12$, Theorem 4.11 implies that the maximum profit occurs for $x = 12$, so the firm should borrow at 12% in order to maximize its annual profit.

57. Since the initial velocity of a droplet is 0, (1) of Section 1.3 (with $v_0 = 0$ and $h_0 = h$) implies that the height at time t_0 is 0 if $-16t_0^2 + H - h = 0$, that is, if $t_0 = \sqrt{(H - h)/16}$. Therefore by the hint,

$$R = \sqrt{2gh}\sqrt{\frac{H - h}{16}} = \sqrt{\frac{gh(H - h)}{8}}$$

We wish to find the maximum value of R as a function of h. Now

$$R'(h) = \frac{1}{2}\left(\frac{gh(H - h)}{8}\right)^{-1/2}\left[\frac{g}{8}(H - 2h)\right] = 0 \quad \text{if} \quad h = \frac{H}{2}$$

Since $R'(h) > 0$ if $h < H/2$ and $R'(h) < 0$ if $h > H/2$, R is maximum for $h = H/2$. Since

$$R\left(\frac{H}{2}\right) = \sqrt{\frac{g(H/2)(H - H/2)}{8}} = \frac{H}{4}\sqrt{\frac{g}{2}}$$

it follows that $(H/4)\sqrt{g/2}$ is the maximum value of R. Taking $g = 32$, we obtain $R(H/2) = (H/4)\sqrt{\frac{32}{2}} = H$. Thus the maximum value of R is H.

CUMULATIVE REVIEW (CHAPTERS 1–3)

1. Since $4x^6 - 1 = 0$ for $x = -1/\sqrt[3]{2}$ and $x = 1/\sqrt[3]{2}$, it follows that $4x^6 - 1 > 0$ for $x < -1/\sqrt[3]{2}$ and for $x > 1/\sqrt[3]{2}$, and $4x^6 - 1 < 0$ for $-1/\sqrt[3]{2} < x < 1/\sqrt[3]{2}$. From the diagram we see that the solution of the given inequality is the union of $(-1/\sqrt[3]{2}, 0)$ and $(1/\sqrt[3]{2}, \infty)$.

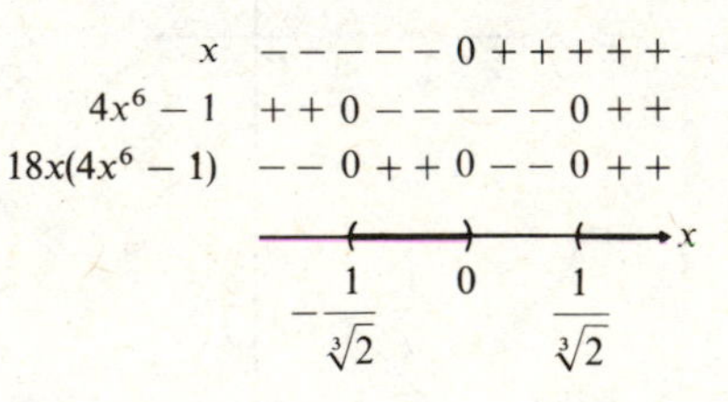

3. The given inequality is equivalent to $2 \le 1 + x^4 < 16$, or $1 \le x^4 < 15$. Thus the solution is the union of $(-\sqrt[4]{15}, -1]$ and $[1, \sqrt[4]{15})$.

5. a. The domain of f is $(-3, \infty)$ and the domain of g consists of all numbers except $\frac{1}{2}$. Thus the domain of $f \circ g$ consists of all x except $\frac{1}{2}$ such that

$$\frac{1}{2x - 1} > -3, \quad \text{or} \quad \frac{1}{2x - 1} + 3 > 0, \quad \text{or} \quad \frac{2(3x - 1)}{2x - 1} > 0$$

From the diagram we see that the solution of this inequality, and hence the domain of $f \circ g$, is the union of $(-\infty, \frac{1}{3})$ and $(\frac{1}{2}, \infty)$.

$3x-1$	$- - 0 + + + + + + + + +$
$2x-1$	$- - - - - - - - - 0 + +$
$\frac{2(3x-1)}{2x-1}$	$+ + 0 - - - - - \quad + +$

x: $\frac{1}{3}$, $\frac{1}{2}$

b. $(f \circ g)(x) = f(g(x)) = f\left(\frac{1}{2x-1}\right) = \frac{1}{\sqrt{1/(2x-1)+3}} = \sqrt{\frac{2x-1}{6x-2}}$

7. $\lim_{x\to 4^+} \frac{x(x+4)}{16-x^2} = \lim_{x\to 4^+} \frac{x(x+4)}{(4-x)(4+x)} = \lim_{x\to 4^+} \frac{x}{4-x} = -\infty$

9. $\lim_{x\to 0^+} \frac{\sin 2x - 2\sqrt{x}\sin x + 4x^2}{x} = \lim_{x\to 0^+}\left(\frac{\sin 2x}{x} - 2\sqrt{x}\frac{\sin x}{x} + 4x\right)$

$= \lim_{x\to 0^+} 2\frac{\sin 2x}{2x} - 2\lim_{x\to 0^+}\sqrt{x}\lim_{x\to 0^+}\frac{\sin x}{x} + 4\lim_{x\to 0^+} x = 2(1) - 2(0)(1) + 4(0) = 2$

11. a. $\lim_{x\to 1^-} f(x) = \lim_{x\to 1^-}(x-1)^2 = 0;\quad \lim_{x\to 1^+} f(x) = \lim_{x\to 1^+}(x^3-x^2) = 0.$

Thus $\lim_{x\to 1} f(x) = 0 = f(1)$, so f is continuous at 1.

b. $\lim_{x\to 1^-}\frac{f(x)-f(1)}{x-1} = \lim_{x\to 1^-}\frac{(x-1)^2-0}{x-1} = \lim_{x\to 1^-}(x-1) = 0$

$\lim_{x\to 1^+}\frac{f(x)-f(1)}{x-1} = \lim_{x\to 1^+}\frac{x^3-x^2}{x-1} = \lim_{x\to 1^+}\frac{x^2(x-1)}{x-1} = \lim_{x\to 1^+} x^2 = 1$

since the two one-sided limits differ, it follows that $\lim_{x\to 1}[f(x)-f(1)]/(x-1)$ fails to exist, so f is not differentiable at 1.

13. $f'(x) = \frac{3\sqrt{x^2+2x+2} - (3x+1)[1/(2\sqrt{x^2+2x+2})](2x+2)}{(\sqrt{x^2+2x+2})^2} = \frac{2x+5}{(x^2+2x+2)^{3/2}}$

15. $\frac{d}{dx}[3(2x+1)^{5/2} - 5(2x+1)^{3/2}] = 3\left(\frac{5}{2}\right)(2x+1)^{3/2}(2) - 5\left(\frac{3}{2}\right)(2x+1)^{1/2}(2)$

$= 15(2x+1)^{1/2}[(2x+1)-1] = 30x(2x+1)^{1/2}$

17. Differentiating the given equation implicitly, we have

$$y^2 + 2xy\frac{dy}{dx} + 3\frac{dy}{dx} - 4 = 0, \quad \text{so} \quad \frac{dy}{dx} = \frac{4-y^2}{2xy+3}$$

19. The area A is given by $A = \pi r^2$. Differentiating implicitly, we have $dA/dt = 2\pi r(dr/dt)$. Since $dA/dt = 2\pi\sqrt{r}$ by hypothesis, it follows that $2\pi\sqrt{r} = dA/dt = 2\pi r(dr/dt)$, so $dr/dt = 1/\sqrt{r}$. When the radius is increasing at the rate of 2 feet per minute, we have $2 = dr/dt = 1/\sqrt{r}$, so $r = (\frac{1}{2})^2 = \frac{1}{4}$. At that time, $A = \pi(\frac{1}{4})^2 = \frac{1}{16}\pi$ (square feet).

5

THE INTEGRAL

SECTION 5.1

1. $L_f(\mathscr{P}) = 7\left(\frac{-5}{2} - (-3)\right) + 7\left(\frac{-3}{2} - \left(\frac{-5}{2}\right)\right) + 7\left(0 - \left(\frac{-3}{2}\right)\right) = 21$

$U_f(\mathscr{P}) = 7\left(\frac{-5}{2} - (-3)\right) + 7\left(\frac{-3}{2} - \left(\frac{-5}{2}\right)\right) + 7\left(0 - \left(\frac{-3}{2}\right)\right) = 21$

3. $L_f(\mathscr{P}) = 1\left(\frac{-1}{2} - (-1)\right) + \frac{3}{2}\left(0 - \left(\frac{-1}{2}\right)\right) + 2\left(\frac{1}{2} - 0\right) + \frac{5}{2}\left(1 - \frac{1}{2}\right) + 3\left(\frac{3}{2} - 1\right)$

$+ \frac{7}{2}\left(2 - \frac{3}{2}\right) = \frac{27}{4}$

$U_f(\mathscr{P}) = \frac{3}{2}\left(\frac{-1}{2} - (-1)\right) + 2\left(0 - \left(\frac{-1}{2}\right)\right) + \frac{5}{2}\left(\frac{1}{2} - 0\right) + 3\left(1 - \frac{1}{2}\right) + \frac{7}{2}\left(\frac{3}{2} - 1\right)$

$+ 4\left(2 - \frac{3}{2}\right) = \frac{33}{4}$

5. $L_f(\mathscr{P}) = \frac{1}{16}\left(\frac{-1}{2} - (-1)\right) + 0\left(0 - \left(\frac{-1}{2}\right)\right) + 0\left(\frac{1}{2} - 0\right) + \frac{1}{16}\left(1 - \frac{1}{2}\right) + 1\left(\frac{3}{2} - 1\right)$

$+ \frac{81}{16}\left(2 - \frac{3}{2}\right) = \frac{99}{32}$

$U_f(\mathscr{P}) = 1\left(\frac{-1}{2} - (-1)\right) + \frac{1}{16}\left(0 - \left(\frac{-1}{2}\right)\right) + \frac{1}{16}\left(\frac{1}{2} - 0\right) + 1\left(1 - \frac{1}{2}\right) + \frac{81}{16}\left(\frac{3}{2} - 1\right)$

$+ 16\left(2 - \frac{3}{2}\right) = \frac{371}{32}$

7. $L_f(\mathscr{P}) = 0\left(\frac{\pi}{4} - 0\right) + \frac{\sqrt{2}}{2}\left(\frac{\pi}{2} - \frac{\pi}{4}\right) = \frac{\pi\sqrt{2}}{8}$

$U_f(\mathscr{P}) = \frac{\sqrt{2}}{2}\left(\frac{\pi}{4} - 0\right) + 1\left(\frac{\pi}{2} - \frac{\pi}{4}\right) = \frac{\pi}{4}\left(\frac{\sqrt{2}}{2} + 1\right)$

9. $L_f(\mathscr{P}) = \frac{1}{2}\left(\frac{-\pi}{6} - \left(\frac{-\pi}{3}\right)\right) + \frac{\sqrt{3}}{2}\left(0 - \left(\frac{-\pi}{6}\right)\right) + \frac{\sqrt{3}}{2}\left(\frac{\pi}{6} - 0\right) + \frac{1}{2}\left(\frac{\pi}{3} - \frac{\pi}{6}\right)$

$= \frac{\pi}{6}(1 + \sqrt{3})$

$$U_f(\mathscr{P}) = \frac{\sqrt{3}}{2}\left(\frac{-\pi}{6} - \left(\frac{-\pi}{3}\right)\right) + 1\left(0 - \left(\frac{-\pi}{6}\right)\right) + 1\left(\frac{\pi}{6} - 0\right) + \frac{\sqrt{3}}{2}\left(\frac{\pi}{3} - \frac{\pi}{6}\right)$$
$$= \frac{\pi}{6}(\sqrt{3} + 2)$$

11. $L_f(\mathscr{P}) = 3(1 - 0) + 7(2 - 1) = 10$
    $U_f(\mathscr{P}) = 7(1 - 0) + 17(2 - 1) = 24$

13. Since $\Delta x_k = \frac{1}{9}$ for $1 \le k \le 9$, we obtain

$$L_f(\mathscr{P}) = 0\left(\frac{1}{9}\right) + \frac{1}{3}\left(\frac{1}{9}\right) + \frac{\sqrt{2}}{3}\left(\frac{1}{9}\right) + \frac{\sqrt{3}}{3}\left(\frac{1}{9}\right) + \frac{\sqrt{4}}{3}\left(\frac{1}{9}\right) + \frac{\sqrt{5}}{3}\left(\frac{1}{9}\right) + \frac{\sqrt{6}}{3}\left(\frac{1}{9}\right)$$
$$+ \frac{\sqrt{7}}{3}\left(\frac{1}{9}\right) + \frac{\sqrt{8}}{3}\left(\frac{1}{9}\right)$$
$$\approx 0.603926$$
$$U_f(\mathscr{P}) = \frac{1}{3}\left(\frac{1}{9}\right) + \frac{\sqrt{2}}{3}\left(\frac{1}{9}\right) + \frac{\sqrt{3}}{3}\left(\frac{1}{9}\right) + \frac{\sqrt{4}}{3}\left(\frac{1}{9}\right) + \frac{\sqrt{5}}{3}\left(\frac{1}{9}\right) + \frac{\sqrt{6}}{3}\left(\frac{1}{9}\right) + \frac{\sqrt{7}}{3}\left(\frac{1}{9}\right)$$
$$+ \frac{\sqrt{8}}{3}\left(\frac{1}{9}\right) + \frac{\sqrt{9}}{3}\left(\frac{1}{9}\right)$$
$$\approx 0.715037$$

15. $L_f(\mathscr{P}) = \sqrt{\frac{97}{81}}(\frac{1}{3}) + \sqrt{\frac{17}{16}}(\frac{1}{6}) + 1(\frac{1}{2}) + 1(\frac{1}{4}) + \sqrt{\frac{257}{256}}(\frac{1}{4}) + \sqrt{\frac{17}{16}}(\frac{1}{6}) + \sqrt{\frac{97}{81}}(\frac{1}{3})$
    ≈ 2.07362
    $U_f(\mathscr{P}) = \sqrt{2}(\frac{1}{3}) + \sqrt{\frac{97}{81}}(\frac{1}{6}) + \sqrt{\frac{17}{16}}(\frac{1}{2}) + \sqrt{\frac{257}{256}}(\frac{1}{4}) + \sqrt{\frac{17}{16}}(\frac{1}{4}) + \sqrt{\frac{97}{81}}(\frac{1}{6}) + \sqrt{2}(\frac{1}{3})$
    ≈ 2.33115

17. $L_f(\mathscr{P}) = 0(0 - (-1)) + 1(2 - 0) = 2$
    $U_f(\mathscr{P}) = 1(0 - (-1)) + 3(2 - 0) = 7$
    $L_f(\mathscr{P}') = 0(0 - (-1)) + 1(1 - 0) + 2(2 - 1) = 3$
    $U_f(\mathscr{P}') = 1(0 - (-1)) + 2(1 - 0) + 3(2 - 1) = 6$

19. $L_f(\mathscr{P}) = 0\left(\frac{\pi}{2} - 0\right) + 0\left(\pi - \frac{\pi}{2}\right) = 0$

$$U_f(\mathscr{P}) = 1\left(\frac{\pi}{2} - 0\right) + 1\left(\pi - \frac{\pi}{2}\right) = \pi$$
$$L_f(\mathscr{P}') = 0\left(\frac{\pi}{4} - 0\right) + \frac{\sqrt{2}}{2}\left(\frac{\pi}{2} - \frac{\pi}{4}\right) + \frac{\sqrt{2}}{2}\left(\frac{3\pi}{4} - \frac{\pi}{2}\right) + 0\left(\pi - \frac{3\pi}{4}\right) = \frac{\sqrt{2}\pi}{4}$$
$$U_f(\mathscr{P}') = \frac{\sqrt{2}}{2}\left(\frac{\pi}{4} - 0\right) + 1\left(\frac{\pi}{2} - \frac{\pi}{4}\right) + 1\left(\frac{3\pi}{4} - \frac{\pi}{2}\right) + \frac{\sqrt{2}}{2}\left(\pi - \frac{3\pi}{4}\right) = \frac{\pi}{4}(\sqrt{2} + 2)$$

21. $L_f(\mathscr{P}) = 0\left(\frac{\pi}{2} - 0\right) + \left(\frac{\pi}{2} + 1\right)\left(\pi - \frac{\pi}{2}\right) = \left(\frac{\pi}{2} + 1\right)\frac{\pi}{2}$

$$U_f(\mathscr{P}) = \left(\frac{\pi}{2} + 1\right)\left(\frac{\pi}{2} - 0\right) + \pi\left(\pi - \frac{\pi}{2}\right) = \left(\frac{3\pi}{2} + 1\right)\frac{\pi}{2}$$
$$L_f(\mathscr{P}') = 0\left(\frac{\pi}{4} - 0\right) + \left(\frac{\pi}{4} + \frac{\sqrt{2}}{2}\right)\left(\frac{\pi}{2} - \frac{\pi}{4}\right) + \left(\frac{\pi}{2} + 1\right)\left(\frac{3\pi}{4} - \frac{\pi}{2}\right)$$
$$+ \left(\frac{3\pi}{4} + \frac{\sqrt{2}}{2}\right)\left(\pi - \frac{3\pi}{4}\right) = \frac{\pi}{4}\left(\frac{3\pi}{2} + \sqrt{2} + 1\right)$$
$$U_f(\mathscr{P}') = \left(\frac{\pi}{4} + \frac{\sqrt{2}}{2}\right)\left(\frac{\pi}{4} - 0\right) + \left(\frac{\pi}{2} + 1\right)\left(\frac{\pi}{2} - \frac{\pi}{4}\right) + \left(\frac{3\pi}{4} + \frac{\sqrt{2}}{2}\right)\left(\frac{3\pi}{4} - \frac{\pi}{2}\right)$$
$$+ \pi\left(\pi - \frac{3\pi}{4}\right) = \frac{\pi}{4}\left(\frac{5\pi}{2} + \sqrt{2} + 1\right)$$

23. $L_f(\mathscr{P}) = \frac{4}{9}(\frac{1}{2}) + \frac{1}{4}(\frac{1}{2}) + \frac{4}{25}(\frac{1}{2}) + \frac{1}{9}(\frac{1}{2}) = \frac{869}{1800}$

25. a. $L_f(\mathscr{P}) = 0(1) + 1(5) + 6(1) = 11$; $L_f(\mathscr{P}') = 0(4) + 4(3) = 12$.
    Thus $L_f(\mathscr{P}) \ne L_f(\mathscr{P}')$.

    b. $L_f(\mathscr{P}'') = 0\left(\frac{7 - \sqrt{5}}{2}\right) + \left(\frac{7 - \sqrt{5}}{2}\right)\left(\frac{7 + \sqrt{5}}{2}\right) = \frac{49 - 5}{4} = 11.$
    Thus $L_f(\mathscr{P}) = L_f(\mathscr{P}'')$.

27. Let $\mathscr{P} = \{x_0, x_1, \ldots, x_n\}$. In order to find $U_f(\mathscr{P})$ we must be able to find the maximum value of f on $[x_0, x_1] = [0, x_1]$, and this is not possible since $\lim_{x \to 0^+} 1/x = \infty$. We can find the minimum value of f on each subinterval determined by $\mathscr{P}$. Thus we can find $L_f(\mathscr{P})$.

29. Let $\mathscr{P}$ be a partition such that $L_f(\mathscr{P}) = U_f(\mathscr{P})$. Then $m_k = M_k$ for each k, so that f is constant on each subinterval $[x_{k-1}, x_k]$. Moreover, $f(a) = f(x_1) = f(x_2) = \cdots = f(x_n)$. Thus f is constant on $[a, b]$.

SECTION 5.2

1. $L_f(\mathscr{P}) = -2(1) + 0(1) + 2(1) + 4(1) = 4$
   $U_f(\mathscr{P}) = 0(1) + 2(1) + 4(1) + 6(1) = 12$

3. $L_f(\mathscr{P}) = \frac{1}{2}(\frac{1}{2}) + 0(\frac{1}{2}) + 0(\frac{1}{2}) + \frac{1}{2}(\frac{1}{2}) + 1(\frac{1}{2}) + \frac{3}{2}(\frac{1}{2}) + 2(\frac{1}{2}) + \frac{5}{2}(\frac{1}{2}) = 4$
   $U_f(\mathscr{P}) = 1(\frac{1}{2}) + \frac{1}{2}(\frac{1}{2}) + \frac{1}{2}(\frac{1}{2}) + 1(\frac{1}{2}) + \frac{3}{2}(\frac{1}{2}) + 2(\frac{1}{2}) + \frac{5}{2}(\frac{1}{2}) + 3(\frac{1}{2}) = 6$

5. $L_f(\mathscr{P}) = \left(\frac{-3\sqrt{2}}{2}\right)\left(\frac{\pi}{4}\right) + 0\left(\frac{\pi}{4}\right) = \frac{-3\sqrt{2}\pi}{8}$; $U_f(\mathscr{P}) = 0\left(\frac{\pi}{4}\right) + \frac{3\sqrt{2}}{2}\left(\frac{\pi}{4}\right) = \frac{3\sqrt{2}\pi}{8}$

7. By Example 1, $\int_{-2}^{3} 4\,dx = 4[3 - (-2)] = 20$

9. By Example 1, $\int_{-1}^{1} -\frac{1}{3}\,dx = \frac{-1}{3}[1 - (-1)] = -\frac{2}{3}$

11. By Definition 5.4 and Example 1, $\int_{2}^{0} \pi\,dx = -\int_{0}^{2} \pi\,dx = -\pi(2 - 0) = -2\pi$

13. By Example 2, $\int_{-3}^{3} x\,dx = \frac{1}{2}[3^2 - (-3)^2] = 0$

15. By Definition 5.4 and Example 2, $\int_{1/3}^{-1/3} x\,dx = -\int_{-1/3}^{1/3} x\,dx = -\frac{1}{2}[(\frac{1}{3})^2 - (-\frac{1}{3})^2] = 0$

17. By Definition 5.4 and (1),
$$\int_{-1/2}^{-7/2} -x\,dx = -\int_{-7/2}^{-1/2} -x\,dx = -[-\tfrac{1}{2}((-\tfrac{1}{2})^2 - (-\tfrac{7}{2})^2)] = -6$$

19. By Definition 5.4, $\int_{-5}^{-5} x^2\,dx = 0$

21. By Definition 5.4 and (2), $\int_{2}^{-3} x^2\,dx = -\int_{-3}^{2} x^2\,dx = -\frac{1}{3}[2^3 - (-3)^3] = -\frac{35}{3}$

23. $A = \int_{-2}^{3} \frac{5}{2}\,dx = \frac{5}{2}(3 - (-2)) = \frac{25}{2}$

25. $\int_{1}^{4} x\,dx = \frac{1}{2}(4^2 - 1^2) = \frac{15}{2}$

27. $A = \int_{-3}^{-1} x^2\,dx = \frac{1}{3}[(-1)^3 - (-3)^3] = \frac{26}{3}$

29. Let $f(x) = x^2 - x$ for $1 \le x \le 3$.

left sum: $f(1)(1) + f(2)(1) = 0(1) + 2(1) = 2$

right sum: $f(2)(1) + f(3)(1) = 2(1) + 6(1) = 8$

midpoint sum: $f(\frac{3}{2})(1) + f(\frac{5}{2})(1) = \frac{3}{4} + \frac{15}{4} = \frac{9}{2}$

31. Let $f(x) = \sin \pi x$ for $0 \le x \le 2$.

left sum:
$$f(0)\left(\frac{1}{2}\right) + f\left(\frac{1}{2}\right)\left(\frac{1}{2}\right) + f(1)(1) = (\sin 0)\left(\frac{1}{2}\right) + \left(\sin \frac{\pi}{2}\right)\left(\frac{1}{2}\right) + (\sin \pi)(1)$$
$$= 0\left(\frac{1}{2}\right) + 1\left(\frac{1}{2}\right) + 0(1) = \frac{1}{2}$$

right sum:
$$f\left(\frac{1}{2}\right)\left(\frac{1}{2}\right) + f(1)\left(\frac{1}{2}\right) + f(2)(1) = \left(\sin \frac{\pi}{2}\right)\left(\frac{1}{2}\right) + (\sin \pi)\left(\frac{1}{2}\right) + (\sin 2\pi)(1)$$
$$= 1\left(\frac{1}{2}\right) + 0\left(\frac{1}{2}\right) + 0(1) = \frac{1}{2}$$

midpoint sum:
$$f\left(\frac{1}{4}\right)\left(\frac{1}{2}\right) + f\left(\frac{3}{4}\right)\left(\frac{1}{2}\right) + f\left(\frac{3}{2}\right)(1) = \left(\sin \frac{\pi}{4}\right)\left(\frac{1}{2}\right) + \left(\sin \frac{3\pi}{4}\right)\left(\frac{1}{2}\right) + \left(\sin \frac{3\pi}{2}\right)(1)$$
$$= \frac{\sqrt{2}}{2}\left(\frac{1}{2}\right) + \frac{\sqrt{2}}{2}\left(\frac{1}{2}\right) + (-1)(1)$$
$$= \frac{\sqrt{2}}{2} - 1 \approx -.292893$$

33. Let $f(x) = 1/x$ for $1 \le x \le 5$.

left sum: $f(1)(1) + f(2)(1) + f(3)(1) + f(4)(1) = 1(1) + \frac{1}{2}(1) + \frac{1}{3}(1) + \frac{1}{4}(1) = \frac{25}{12} \approx 2.08333$

right sum: $f(2)(1) + f(3)(1) + f(4)(1) + f(5)(1) = \frac{1}{2}(1) + \frac{1}{3}(1) + \frac{1}{4}(1) + \frac{1}{5}(1)$
$= \frac{77}{60} \approx 1.28333$

midpoint sum: $f(\frac{3}{2})(1) + f(\frac{5}{2})(1) + f(\frac{7}{2})(1) + f(\frac{9}{2})(1) = \frac{2}{3}(1) + \frac{2}{5}(1) + \frac{2}{7}(1) + \frac{2}{9}(1)$
$= \frac{496}{315} \approx 1.57460$

35. a. $\int_{\pi/4}^{3\pi/4} \frac{\sin x}{x}\,dx \approx \frac{2\sqrt{2}}{\pi}\left(\frac{\pi}{4}\right) + \frac{2}{\pi}\left(\frac{\pi}{4}\right) \approx 1.20711$

b. $\int_{\pi/4}^{3\pi/4} \frac{\sin x}{x}\,dx \approx \frac{2}{\pi}\left(\frac{\pi}{4}\right) + \frac{2\sqrt{2}}{3\pi}\left(\frac{\pi}{4}\right) \approx .735702$

37. a. $\int_0^1 \sin \pi x^2\,dx \approx 0\left(\frac{1}{2}\right) + \frac{\sqrt{2}}{2}\left(\frac{\sqrt{2}}{2} - \frac{1}{2}\right) + 1\left(\frac{\sqrt{3}}{2} - \frac{\sqrt{2}}{2}\right) + \frac{\sqrt{2}}{2}\left(1 - \frac{\sqrt{3}}{2}\right)$
$\approx .400100$

b. $\int_0^1 \sin \pi x^2\,dx \approx \frac{\sqrt{2}}{2}\left(\frac{1}{2}\right) + 1\left(\frac{\sqrt{2}}{2} - \frac{1}{2}\right) + \frac{\sqrt{2}}{2}\left(\frac{\sqrt{3}}{2} - \frac{\sqrt{2}}{2}\right) + 0\left(1 - \frac{\sqrt{3}}{2}\right)$
$\approx .673033$

39. $A \approx f(0)(\frac{1}{4}) + f(\frac{1}{4})(\frac{1}{4}) + f(\frac{1}{2})(\frac{1}{4}) + f(\frac{3}{4})(\frac{1}{4}) = 0(\frac{1}{4}) + \frac{7}{8}(\frac{1}{4}) + 2(\frac{1}{4}) + \frac{27}{8}(\frac{1}{4}) = \frac{25}{16}$

41. $A \approx f(0)(\frac{1}{2}) + f(\frac{1}{2})(\frac{1}{2}) + f(1)(1) = 0(\frac{1}{2}) + \frac{1}{3}(\frac{1}{2}) + \frac{1}{2}(1) = \frac{2}{3}$

43. $A \approx f(1)(\frac{1}{50}) + f(\frac{51}{50})(\frac{1}{50}) + f(\frac{52}{50})(\frac{1}{50}) + \cdots + f(\frac{149}{50})(\frac{1}{50})$
$= 1(\frac{1}{50}) + \frac{(51)^2}{(50)^2}(\frac{1}{50}) + \frac{(52)^2}{(50)^2}(\frac{1}{50}) + \cdots + \frac{(149)^2}{(50)^2}(\frac{1}{50}) \approx 8.5868$

45. For $0 \le x \le 1$, we have $x^6 \le x$. Thus $\int_0^1 x^6\,dx \le \int_0^1 x\,dx$.

47. For $x \ge 1$ we have $x^6 \ge x$ and hence $1/x^6 \le 1/x$. Thus $\int_1^2 (1/x^6)\,dx \le \int_1^2 (1/x)\,dx$.

49. For $0 \le x \le \pi/4$ we have $\sin x \le \sqrt{2}/2 \le \cos x$. Thus $\int_0^{\pi/4} \sin x\,dx \le \int_0^{\pi/4} \cos x\,dx$.

51. a. Let $\mathscr{P} = \{x_0, x_1, \ldots, x_n\}$ be any partition of $[a, b]$, and for any k between 1 and n let $\Delta x_k = x_k - x_{k-1}$. From the first set of inequalities in Example 2,
$$x_0\,\Delta x_1 + x_1\,\Delta x_2 + \cdots + x_{n-1}\,\Delta x_n < \tfrac{1}{2}(b^2 - a^2)$$
whereas from the second set of inequalities in Example 2,
$$x_1\,\Delta x_1 + x_2\,\Delta x_2 + \cdots + x_n\,\Delta x_n > \tfrac{1}{2}(b^2 - a^2)$$
Suppose $c > 0$. For any k between 1 and n, $m_k = cx_{k-1}$ and $M_k = cx_k$. Thus $L_f(\mathscr{P}) = m_1\Delta x_1 + m_2\,\Delta x_2 + \cdots + m_n\,\Delta x_n = cx_0\,\Delta x_1 + cx_1\,\Delta x_2 + \cdots + cx_{n-1}\,\Delta x_n = c[x_0\,\Delta x_1 + x_1\,\Delta x_2 + \cdots + x_{n-1}\,\Delta x_n] < c[\frac{1}{2}(b^2 - a^2)]$, whereas $U_f(\mathscr{P}) = M_1\,\Delta x_1 + M_2\,\Delta x_2 + \cdots + M_n\,\Delta x_n = cx_1\,\Delta x_1 + cx_2\,\Delta x_2 + \cdots + cx_n\,\Delta x_n = c[x_1\,\Delta x_1 + x_2\,\Delta x_2 + \cdots + x_n\,\Delta x_n] > c[\frac{1}{2}(b^2 - a^2)]$. Therefore for any partition $\mathscr{P}$ of $[a, b]$,
$$L_f(\mathscr{P}) \le \tfrac{c}{2}(b^2 - a^2) \le U_f(\mathscr{P})$$
and we conclude from Definition 5.2 that $\int_a^b cx\,dx = (c/2)(b^2 - a^2)$ if $c > 0$.

Suppose $c < 0$. For any k between 1 and n, $m_k = cx_k$ and $M_k = cx_{k-1}$. Thus $L_f(\mathscr{P}) = m_1\,\Delta x_1 + m_2\,\Delta x_2 + \cdots + m_n\,\Delta x_n = cx_1\,\Delta x_1 + cx_2\,\Delta x_2 + \cdots + cx_n\,\Delta x_n = c[x_1\,\Delta x_1 + x_2\,\Delta x_2 + \cdots + x_n\,\Delta x_n] < c[\frac{1}{2}(b^2 - a^2)]$, whereas $U_f(\mathscr{P}) = M_1\,\Delta x_1 + M_2\,\Delta x_2 + \cdots + M_n\,\Delta x_n = cx_0\,\Delta x_1 + cx_1\,\Delta x_2 + \cdots + cx_{n-1}\,\Delta x_n = c[x_0\,\Delta x_1 + x_1\,\Delta x_2 + \cdots + x_{n-1}\,\Delta x_n] > c[\frac{1}{2}(b^2 - a^2)]$. Therefore, for any partition $\mathscr{P}$ of $[a, b]$,

$$L_f(\mathscr{P}) \le \frac{c}{2}(b^2 - a^2) \le U_f(\mathscr{P})$$

and we conclude from Definition 5.2 that $\int_a^b cx\,dx = (c/2)(b^2 - a^2)$ if $c < 0$.

If $c = 0$, then $\int_a^b cx\,dx = \int_a^b 0\,dx = 0(b - a) = 0 = (c/2)(b^2 - a^2)$ by Example 1.

b. If $a > b$, then $\int_a^b cx\,dx = (c/2)(b^2 - a^2)$ from (a). If $a < b$, then by Definition 5.4 and part (a),

$$\int_a^b cx\,dx = -\int_b^a cx\,dx = -\frac{c}{2}(a^2 - b^2) = \frac{c}{2}(b^2 - a^2)$$

If $a = b$, then by Definition 5.4, $\int_a^b cx\,dx = 0 = (c/2)(b^2 - a^2)$.

53. a. Let $\mathscr{P} = \{x_0, x_1, \ldots, x_n\}$ be any partition of $[a, b]$. Then for any k between 1 and n we have

$$m_k = x_{k-1}^2, \qquad M_k = x_k^2 \quad \text{and} \quad x_{k-1}^2 < \frac{x_k^2 + x_k x_{k-1} + x_{k-1}^2}{3} < x_k^2$$

Thus

$$\begin{aligned} L_f(\mathscr{P}) &= m_1\,\Delta x_1 + m_2\,\Delta x_2 + \cdots + m_n\,\Delta x_n \\ &= x_0^2(x_1 - x_0) + x_1^2(x_2 - x_1) + \cdots + x_{n-1}^2(x_n - x_{n-1}) \\ &< \frac{x_1^2 + x_1x_0 + x_0^2}{3}(x_1 - x_0) + \frac{x_2^2 + x_2x_1 + x_1^2}{3}(x_2 - x_1) + \cdots \\ &\quad + \frac{x_n^2 + x_nx_{n-1} + x_{n-1}^2}{3}(x_n - x_{n-1}) \\ &= \tfrac{1}{3}[(x_1^3 - x_0^3) + (x_2^3 - x_1^3) + \cdots + (x_n^3 - x_{n-1}^3)] = \tfrac{1}{3}(b^3 - a^3) \end{aligned}$$

Similarly

$$\begin{aligned} U_f(\mathscr{P}) &= M_1\,\Delta x_1 + M_2\,\Delta x_2 + \cdots + M_n\,\Delta x_n \\ &= x_1^2(x_1 - x_0) + x_2^2(x_2 - x_1) + \cdots + x_n^2(x_n - x_{n-1}) \\ &> \frac{x_1^2 + x_1x_0 + x_0^2}{3}(x_1 - x_0) + \frac{x_2^2 + x_2x_1 + x_1^2}{3}(x_2 - x_1) + \cdots \\ &\quad + \frac{x_n^2 + x_nx_{n-1} + x_{n-1}^2}{3}(x_n - x_{n-1}) \\ &= \tfrac{1}{3}(b^3 - a^3) \end{aligned}$$

Since for any partition $\mathscr{P}$ of $[a, b]$, $L_f(\mathscr{P}) \le \frac{1}{3}(b^3 - a^3) \le U_f(\mathscr{P})$, we conclude from Definition 5.2 that $\int_a^b x^3\,dx = \frac{1}{3}(b^3 - a^3)$.

b. If $a < b$, then by part (a), $\int_a^b x^2\,dx = \frac{1}{3}(b^3 - a^3)$. If $a > b$, then by Definition 5.4 and part (a), $\int_a^b x^2\,dx = -\int_b^a x^2\,dx = -\frac{1}{3}(a^3 - b^3) = \frac{1}{3}(b^3 - a^3)$. If $a = b$, then by Definition 5.4, $\int_a^b x^2\,dx = 0 = \frac{1}{3}(b^3 - a^3)$.

55. $\int_a^b x^n\,dx = \dfrac{1}{n+1}(b^{n+1} - a^{n+1})$

57. Let t_k be the midpoint of $[x_{k-1}, x_k]$. Then $t_k = -t_{n-k+1}$ for $1 \le k \le n$. Since $f(-x) = -f(x)$, we have $f(t_k) = f(-t_{n-k+1}) = -f(t_{n+k-1})$, so that

$$\sum_{k=1}^{n} f(t_k)\,\Delta x_k = \sum_{k=1}^{n} f(t_k)\frac{2a}{n} = 0 = \int_{-a}^{a} f(x)\,dx$$

SECTION 5.3

1. $\int_3^5 7\,dx = 7(5 - 3) = 14$

3. $\int_{17}^{100} 1\,dr = 1(100 - 17) = 83$

5. $\int_2^{-1} -10\,du = -10(-1 - 2) = 30$

7. $\int_0^1 x\,dx = \frac{1}{2}(1^2 - 0^2) = \frac{1}{2}$, $\int_1^2 x\,dx = \frac{1}{2}(2^2 - 1^2) = \frac{3}{2}$, so $\int_0^1 x\,dx + \int_1^2 x\,dx = \frac{1}{2} + \frac{3}{2} = 2$; $\int_0^2 x\,dx = \frac{1}{2}(2^2 - 0^2) = 2$

9. $\int_1^0 y^2\,dy = -\int_0^1 y^2\,dy = -\frac{1}{3}(1^3 - 0^3) = -\frac{1}{3}$, $\int_0^2 y^2\,dy = \frac{1}{3}(2^3 - 0) = \frac{8}{3}$, so $\int_1^0 y^2\,dy + \int_0^2 y^2\,dy = -\frac{1}{3} + \frac{8}{3} = \frac{7}{3}$; $\int_1^2 y^2\,dy = \frac{1}{3}(2^3 - 1^3) = \frac{7}{3}$

11. $\int_0^2 f(x)\,dx + \int_3^0 f(x)\,dx = \int_3^0 f(x)\,dx + \int_0^2 f(x)\,dx = \int_3^2 f(x)\,dx$, so $a = 3$ and $b = 2$.

13. $\int_a^b f(t)\,dt = \int_5^3 f(t)\,dt + \int_3^1 f(t)\,dt = \int_5^1 f(t)\,dt$, so $a = 5$ and $b = 1$.

15. $m = M = 13$; $39 = 13[0 - (-3)] \le \int_{-3}^0 13\,dx \le 13[0 - (-3)] = 39$

17. $m = 0$, $M = 9$; $0 = 0[3 - (-1)] \le \int_{-1}^3 x^2\,dx \le 9[3 - (-1)] = 36$

19. $m = \frac{1}{3}$, $M = \frac{1}{2}$; $\frac{1}{3} = \frac{1}{3}(3 - 2) \le \int_2^3 (1/x)\,dx \le \frac{1}{2}(3 - 2) = \frac{1}{2}$

21. $m = \dfrac{1}{2}$, $M = \dfrac{\sqrt{2}}{2}$; $\dfrac{\pi}{24} = \dfrac{1}{2}\left(\dfrac{\pi}{3} - \dfrac{\pi}{4}\right) \le \int_{\pi/4}^{\pi/3} \cos x\,dx \le \dfrac{\sqrt{2}}{2}\left(\dfrac{\pi}{3} - \dfrac{\pi}{4}\right) = \dfrac{\sqrt{2}\pi}{24}$

23. $m = 0$, $M = \sqrt{3}$; $0 = 0\left(\dfrac{\pi}{3} - 0\right) \le \int_0^{\pi/3} \tan t\,dt \le \sqrt{3}\left(\dfrac{\pi}{3} - 0\right) = \dfrac{\sqrt{3}\pi}{3}$

25. $A = \int_{-1}^1 f(x)\,dx = \int_{-1}^0 -x\,dx + \int_0^1 x^2\,dx = -\frac{1}{2}[0^2 - (-1)^2] + \frac{1}{3}[1^3 - 0^3] = \frac{5}{6}$

27. The average value of f on $[a, b]$ is

$$\frac{1}{b - a}\int_a^b f(x)\,dx = \frac{1}{b - a}\int_a^b c\,dx = \frac{1}{b - a}[c(b - a)] = c$$

29. a. By definition, the average value of f on $[a, b]$ is

$$\frac{1}{b-a}\int_a^b f(x)\,dx = \frac{1}{b-a}\int_a^b x^2\,dx = \frac{1}{b-a}\left[\frac{1}{3}(b^3-a^3)\right]$$
$$= \frac{1}{3}\frac{b^3-a^3}{b-a} = \frac{1}{3}(b^2+ba+a^2) = \frac{1}{3}(a^2+ab+b^2)$$

b. By the Mean Value Theorem for Integrals there is a number c in $[a, b]$ such that $\int_a^b f(x)\,dx = f(c)(b-a)$, so

$$c^2 = f(c) = \frac{1}{b-a}\int_a^b f(x)\,dx = \frac{1}{3}(a^2+ab+b^2)$$

31. $\int_0^{\pi/2} \sin x\,dx = \int_0^{\pi/6} \sin x\,dx + \int_{\pi/6}^{\pi/2} \sin x\,dx$; $\int_0^{\pi/6} \sin x\,dx \geq \int_0^{\pi/6} 0\,dx = 0$, and since $\sin x \geq \frac{1}{2}$ for $\pi/6 \leq x \leq \pi/2$, we have $\int_{\pi/6}^{\pi/2} \sin x\,dx \geq \int_{\pi/6}^{\pi/2} \frac{1}{2}\,dx = \frac{1}{2}(\pi/2 - \pi/6) = \pi/6 > 0$. Therefore $\int_0^{\pi/2} \sin x\,dx \geq 0 + \pi/6 = \pi/6 > 0$.

33. If $c < a < b$, then by the Addition Property we have

$$\int_c^b f(x)\,dx = \int_c^a f(x)\,dx + \int_a^b f(x)\,dx$$

so that $$\int_a^b f(x)\,dx = -\int_c^a f(x)\,dx + \int_c^b f(x)\,dx = \int_a^c f(x)\,dx + \int_c^b f(x)\,dx$$

If $b < a < c$, then by the Addition Property we have

$$\int_b^c f(x)\,dx = \int_b^a f(x)\,dx + \int_a^c f(x)\,dx$$

so that

$$\int_a^b f(x)\,dx = -\int_b^a f(x)\,dx = \int_a^c f(x)\,dx - \int_b^c f(x)\,dx$$
$$= \int_a^c f(x)\,dx + \int_c^b f(x)\,dx$$

SECTION 5.4

1. $F'(x) = x(1+x^3)^{29}$

3. $F(y) = \int_y^2 \frac{1}{t^3}\,dt = -\int_2^y \frac{1}{t^3}\,dt$, so $F'(y) = -\frac{1}{y^3}$

5. Let $G(x) = \int_0^x t\sin t\,dt$, so $F(x) = G(x^2)$. Since $G'(x) = x\sin x$, the Chain Rule implies that $F'(x) = [G'(x^2)](2x) = (x^2\sin x^2)(2x) = 2x^3\sin x^2$.

7. Notice that

$$G(y) = \int_y^{y^2} (1+t^2)^{1/2}\,dt = \int_y^0 (1+t^2)^{1/2}\,dt + \int_0^{y^2} (1+t^2)^{1/2}\,dt$$
$$= -\int_0^y (1+t^2)^{1/2}\,dt + \int_0^{y^2} (1+t^2)^{1/2}\,dt.$$

Let $H(y) = \int_0^y (1+t^2)^{1/2}\,dt$ and $K(y) = \int_0^{y^2} (1+t^2)^{1/2}\,dt$. Then $K(y) = H(y^2)$ and $G(y) = -H(y) + K(y)$, so that $G'(y) = -H'(y) + K'(y)$. Now $H'(y) = (1+y^2)^{1/2}$ and by the Chain Rule, $K'(y) = [H'(y^2)](2y) = 2y(1+y^4)^{1/2}$. Therefore

$$G'(y) = -(1+y^2)^{1/2} + 2y(1+y^4)^{1/2}.$$

9. Let $G(x) = \int_0^x (1+t^2)^{4/5}\,dt$. Then $G'(x) = (1+x^2)^{4/5}$ and

$$F(x) = \frac{d}{dx}G(4x) = [G'(4x)](4) = 4(1+16x^2)^{4/5}$$

so that $F'(x) = \frac{16}{5}(1+16x^2)^{-1/5}(32x) = (512x/5)(1+16x^2)^{-1/5}$.

11. $\int_0^1 4\,dx = 4x\Big|_0^1 = 4$

13. $\int_1^3 -y\,dy = -\frac{1}{2}y^2\Big|_1^3 = -9/2 - (-1/2) = -4$

15. $\int_1^{-3} 3u\,du = \frac{3}{2}u^2\Big|_1^{-3} = \frac{27}{2} - \frac{3}{2} = 12$

17. $\int_0^1 x^{100}\,dx = \frac{1}{101}x^{101}\Big|_0^1 = \frac{1}{101}$

19. $\int_{-1}^1 u^{1/3}\,du = \frac{3}{4}u^{4/3}\Big|_{-1}^1 = \frac{3}{4} - \frac{3}{4}(-1)^{4/3} = 0$

21. $\int_1^4 x^{-7/9}\,dx = \frac{9}{2}x^{2/9}\Big|_1^4 = \frac{9}{2}(4^{2/9} - 1)$

23. $\int_{-1.5}^{2\pi} (5-x)\,dx = (5x - \frac{1}{2}x^2)\Big|_{-1.5}^{2\pi} = (10\pi - 2\pi^2) - (-7.5 - 2.25/2)$
$= 10\pi - 2\pi^2 + 8.625$

25. $\int_{-4}^{-1} (5x+14)\,dx = (\frac{5}{2}x^2 + 14x)\Big|_{-4}^{-1} = (\frac{5}{2} - 14) - (40 - 56) = \frac{9}{2}$

27. $\int_{-\pi}^{\pi/3} \cos x\,dx = \sin x\Big|_{-\pi}^{\pi/3} = \sqrt{3}/2 - 0 = \sqrt{3}/2$

29. $\int_{\pi/3}^{-\pi/4} \sin t\,dt = -\cos t\Big|_{\pi/3}^{-\pi/4} = -\sqrt{2}/2 - (-1/2) = \frac{1}{2} - \sqrt{2}/2$

31. $\int_1^2 \frac{1}{y^4}\,dy = \frac{-1}{3y^3}\Big|_1^2 = \frac{-1}{24} - \left(\frac{-1}{3}\right) = \frac{7}{24}$

33. $\int_{\pi/6}^{\pi/2} \csc^2 t\,dt = -\cot t\Big|_{\pi/6}^{\pi/2} = 0 - (-\sqrt{3}) = \sqrt{3}$

35. $\int_0^{\pi/2} \left(\frac{d}{dx}\sin^5 x\right)dx = \sin^5 x\Big|_0^{\pi/2} = 1 - 0 = 1$

37. $A = \int_{-1}^1 x^4\,dx = \frac{1}{5}x^5\Big|_{-1}^1 = \frac{1}{5} - (-1/5) = \frac{2}{5}$

39. $A = \int_0^{2\pi/3} \sin x\,dx = -\cos x\Big|_0^{2\pi/3} = \frac{1}{2} - (-1) = \frac{3}{2}$

41. $A = \int_1^4 x^{1/2}\,dx = \frac{2}{3}x^{3/2}\Big|_1^4 = \frac{16}{3} - \frac{2}{3} = \frac{14}{3}$

43. $A = \int_0^{\pi/4} \sec^2 x\,dx = \tan x\Big|_0^{\pi/4} = 1$

45. $A = \int_0^{\pi/2} \cos^2 x\,dx = \left(\frac{x}{2} + \frac{\sin 2x}{4}\right)\Big|_0^{\pi/2} = \frac{\pi}{4} - 0 = \frac{\pi}{4}$

47. a. $\int_0^x f(t)\,dt = \int_0^x t\,dt = \frac{1}{2}x^2$; $\frac{d}{dx}\int_0^x f(t)\,dt = \frac{d}{dx}\left(\frac{1}{2}x^2\right) = x = f(x)$

b. $\int_0^x f(t)\,dt = \int_0^x -2t^2\,dt = \frac{-2}{3}t^3\Big|_0^x = \frac{-2}{3}x^3$;

$$\frac{d}{dx}\int_0^x f(t)\,dt = \frac{d}{dx}\left(\frac{-2}{3}x^3\right) = -2x^2 = f(x)$$

c. $\int_0^x f(t)\,dt = \int_0^x -\sin t\,dt = \cos t\Big|_0^x = \cos x - 1$;

$$\frac{d}{dx}\int_0^x f(t)\,dt = \frac{d}{dx}(\cos x - 1) = -\sin x = f(x)$$

d. $\int_0^x f(t)\,dt = \int_0^x 10t^4\,dt = 2t^5\Big|_0^x = 2x^5$; $\frac{d}{dx}\int_0^x f(t)\,dt = \frac{d}{dx}(2x^5) = 10x^4 = f(x)$

49. Since $x^2 + 4x$ is an increasing function on $[1, 2]$, the first sum is the lower sum of $\int_1^2 (x^2 + 4x)\,dx$ for the partition $\mathscr{P}$, and the last sum is the upper sum of $\int_1^2 (x^2 + 4x)\,dx$ for $\mathscr{P}$. Since $L_f(\mathscr{P}) \le I \le U_f(\mathscr{P})$ for every partition $\mathscr{P}$, it follows that

$$I = \int_1^2 (x^2 + 4x)\,dx = (\tfrac{1}{3}x^3 + 2x^2)\Big|_1^2 = (\tfrac{8}{3} + 8) - (\tfrac{1}{3} + 2) = \tfrac{25}{3}$$

51. The velocity in miles per second is $v/3600$. The distance D traveled during the first 5 seconds is given (approximately) by

$$D \approx \frac{v(1)}{3600}(1) + \frac{v(2)}{3600}(1) + \frac{v(3)}{3600} + \frac{v(4)}{3600}(1) + \frac{v(5)}{3600}(1)$$

$$= \frac{5}{3600} + \frac{15}{3600} + \frac{50}{3600} + \frac{200}{3600} + \frac{500}{3600} = \frac{770}{3600} \approx .213889 \text{ (miles)}$$

53. $C(x) - C(2) = \int_2^x m_C(t)\,dt = \int_2^x (3 - 0.1t)\,dt = (3t - 0.05t^2)\Big|_2^x = (3x - 0.05x^2) - (6 - 0.2) = 3x - 0.05x^2 - 5.8$. Since $C(2) = 10.98$, it follows that $C(x) = 10.98 + 3x - 0.05x^2 - 5.8$, so that $C(30) = 10.98 + 3(30) - 0.05(30)^2 - 5.8 = 50.18$ (dollars).

55. a. For $0 \le t \le 10$, we have $f(t) - f(0) = \int_0^t v(s)\,ds = \int_0^t (10s - s^2)\,ds = (5s^2 - \frac{1}{3}s^3)\Big|_0^t = 5t^2 - \frac{1}{3}t^3$.

b. $a(t) = v'(t) = 10 - 2t$, so that $a(t) = 0$ if $t = 5$. Since $f(0) = 0$, it follows from (a) that $f(5) = f(5) - f(0) = 5(5^2) - \frac{1}{3}(5^3) = \frac{250}{3}$.

57. The amount of flow per day is $\int_0^{24} F'(t)\,dt$, and by (5) and the hint,

$$\int_0^{24} F'(t)\,dt = -\frac{336{,}600}{\pi}\cos\frac{\pi t}{24}\Big|_0^{24} = -\frac{336{,}000}{\pi}(\cos \pi - \cos 0)$$

$$= \frac{672{,}000}{\pi} \text{ (tons)}$$

59. We need to coordinate our units of measure—let time be in hours. Then $v(0) = 60$, $v(\frac{1}{30}) = 0$, and the acceleration is

$$\frac{v(\frac{1}{30}) - v(0)}{\frac{1}{30} - 0} = \frac{0 - 60}{\frac{1}{30}} = -1800$$

Therefore $v(t) = v(0) + \int_0^t a(s)\,ds = v(0) + \int_0^t -1800\,ds = 60 - 1800t$, so that if we let $f(0) = 0$, then

$$f(t) = f(0) + \int_0^t v(s)\,ds = f(0) + \int_0^t (60 - 1800s)\,ds = 0 + (60s - 900s^2)\Big|_0^t = 60t - 900t^2$$

After 2 minutes the position of the train is $f(\frac{1}{30})$, and $f(\frac{1}{30}) = 60(\frac{1}{30}) - 900(\frac{1}{30})^2 = 1$. Thus the train was 1 mile from the cow when the brakes were applied.

61. $\pi\int_{-r}^r (r^2 - x^2)\,dx = \pi\left(r^2x - \frac{x^3}{3}\right)\Big|_{-r}^r = \pi\left[\left(r^3 - \frac{r^3}{3}\right) - \left(-r^3 + \frac{r^3}{3}\right)\right] = \frac{4}{3}\pi r^3$

63. a. By the Fundamental Theorem (or by (7)), $C(b) - C(a) = \int_a^b m_C(x)\,dx$. Then

$$\frac{\int_a^b m_C(x)\,dx}{b - a} = \frac{C(b) - C(a)}{b - a} = \text{average cost between the } a\text{th and } b\text{th units produced}$$

as defined in Section 3.1.

b. $\frac{\int_1^4 m_C(x)\,dx}{4 - 1} = \frac{\int_1^4 (1/x^{1/2})\,dx}{4 - 1} = \frac{2x^{1/2}\Big|_1^4}{3} = \frac{4 - 2}{3}$

$= \frac{2}{3}$ (thousand dollars per thousand umbrellas)

65. $W = (2500\pi)(62.5)\int_0^{100} (100 - y)\,dy = 156{,}250\pi\left(100y - \frac{y^2}{2}\right)\Big|_0^{100}$

$= 156{,}250\pi(10{,}000 - 5{,}000) = 781{,}250{,}000\pi$ (foot-pounds)

67. Area of shaded region $= \int_0^\pi \sin\theta\,d\theta = -\cos\theta\Big|_0^\pi = -(-1 - 1) = 2$. Area of rectangle is 2π, so the required proportion is $2/(2\pi) = 1/\pi$.

69. If the needle is 2 inches long, then we must have $y \le 2\sin\theta$. The area of the shaded region $= \int_0^\pi 2\sin\theta\,d\theta = -2\cos\theta\Big|_0^\pi = -2(-1 - 1) = 4$. The area of the rectangle is 5π, so the proportion is $4/(2\pi) = 2/\pi$.

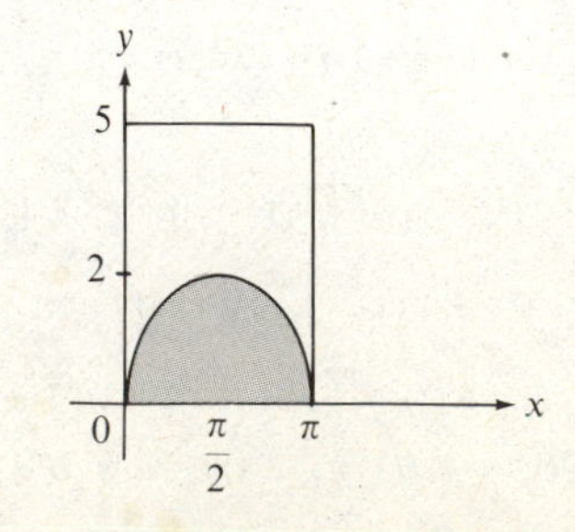

SECTION 5.5

1. $\int (2x - 7)\,dx = x^2 - 7x + C$

3. $\int (2x^{1/3} - 3x^{3/4} + x^{2/5})\,dx = \frac{3}{2}x^{4/3} - \frac{12}{7}x^{7/4} + \frac{5}{7}x^{7/5} + C$

5. $\int \left(t^5 - \frac{1}{t^4}\right) dt = \frac{t^6}{6} + \frac{1}{3t^3} + C$

7. $\int (2\cos x - 5x)\,dx = 2\sin x - \frac{5}{2}x^2 + C$

9. $\int (3\csc^2 x - x)\,dx = -3\cot x - \frac{1}{2}x^2 + C$

11. $\int (2t+1)^2\,dt = \int (4t^2 + 4t + 1)\,dt = \frac{4}{3}t^3 + 2t^2 + t + C$

13. $\int_{-1}^{2} (3x - 4)\,dx = (\frac{3}{2}x^2 - 4x)\Big|_{-1}^{2} = -2 - \frac{11}{2} = -\frac{15}{2}$

15. $\int_{\pi/4}^{\pi/2} (-7\sin x + 3\cos x)\,dx = (7\cos x + 3\sin x)\Big|_{\pi/4}^{\pi/2} = 3 - 5\sqrt{2}$

17. $\int_{-\pi/4}^{-\pi/2} \left(3x - \frac{1}{x^2} + \sin x\right) dx = \left(\frac{3}{2}x^2 + \frac{1}{x} - \cos x\right)\Big|_{-\pi/4}^{-\pi/2}$
$$= \left(\frac{3\pi^2}{8} - \frac{2}{\pi}\right) - \left(\frac{3\pi^2}{32} - \frac{4}{\pi} - \frac{\sqrt{2}}{2}\right) = \frac{9\pi^2}{32} + \frac{2}{\pi} + \frac{\sqrt{2}}{2}$$

19. $\int_{\pi/3}^{\pi/4} (3\sec^2\theta + 4\csc^2\theta)\,d\theta = (3\tan\theta - 4\cot\theta)\Big|_{\pi/3}^{\pi/4}$
$$= -1 - \left(3\sqrt{3} - \frac{4\sqrt{3}}{3}\right) = -1 - \frac{5}{3}\sqrt{3}$$

21. $\int_{1}^{1/3} (3t+2)^3\,dt = \int_{1}^{1/3} (27t^3 + 54t^2 + 36t + 8)\,dt$
$$= \left(\frac{27}{4}t^4 + 18t^3 + 18t^2 + 8t\right)\Big|_{1}^{1/3} = \frac{-136}{3}$$

23. $\int_{\pi/2}^{\pi} \left(\pi\sin x - 2x + \frac{5}{x^2} + 2\pi\right) dx = \left(-\pi\cos x - x^2 - \frac{5}{x} + 2\pi x\right)\Big|_{\pi/2}^{\pi} = \pi + \frac{5}{\pi} + \frac{\pi^2}{4}$

25. $\int_{-1}^{1} (2x+5)(2x-5)\,dx = \int_{-1}^{1} (4x^2 - 25)\,dx = (\frac{4}{3}x^3 - 25x)\Big|_{-1}^{1} = -\frac{142}{3}$

27. $\int_4^7 |x - 5|\,dx = \int_4^5 (5 - x)\,dx + \int_5^7 (x - 5)\,dx$
$$= (5x - \tfrac{1}{2}x^2)\Big|_4^5 + (\tfrac{1}{2}x^2 - 5x)\Big|_5^7 = \tfrac{1}{2} + 2 = \tfrac{5}{2}$$

29. $\int_{-3}^{4} |-5x + 2|\,dx = \int_{-3}^{2/5} (-5x + 2)\,dx + \int_{2/5}^{4} (5x - 2)\,dx$
$$= (-\tfrac{5}{2}x^2 + 2x)\Big|_{-3}^{2/5} + (\tfrac{5}{2}x^2 - 2x)\Big|_{2/5}^{4} = \tfrac{613}{10}$$

31. $\int_0^{\pi/2} f(x)\,dx = \int_0^{\pi/4} \sec^2 x\,dx + \int_{\pi/4}^{\pi/2} \csc^2 x\,dx = \tan x\Big|_0^{\pi/4} - \cot x\Big|_{\pi/4}^{\pi/2} = 1 + 1 = 2$

33. Since $F'(x) = 20x(1 + x^2)^9$, we have $\int 20x(1+x^2)^9\,dx = (1 + x^2)^{10} + C$.

35. Since $F'(x) = x\cos x + 2\sin x$, we have $\int (x\cos x + 2\sin x)\,dx = x\sin x - \cos x + C$.

37. Since $F'(x) = 21\sin^6 x\cos x$, we have $\int 21\sin^6 x\cos x\,dx = 3\sin^7 x + C$.

39. $A = \int_{-1}^{1} (3x^2 + 4)\,dx = (x^3 + 4x)\Big|_{-1}^{1} = 5 - (-5) = 10$

41. $A = \int_1^4 \left(3\sqrt{x} - \frac{1}{\sqrt{x}}\right) dx = (2x^{3/2} - 2x^{1/2})\Big|_1^4 = (16 - 4) - (2 - 2) = 12$

43. $A = \int_{\pi/4}^{\pi/2} (2\sin x + 3\cos x)\,dx = (-2\cos x + 3\sin x)\Big|_{\pi/4}^{\pi/2} =$
$$(0 + 3) - \left[-2\left(\frac{\sqrt{2}}{2}\right) + 3\left(\frac{\sqrt{2}}{2}\right)\right] = 3 - \frac{\sqrt{2}}{2}$$

45. a. Since $0 \le \sin x \le x$ for $0 \le x \le 1$ we have $0 \le \sin x^2 \le x^2$ for $0 \le x \le 1$. Then by Corollary 5.20, $0 \le \int_0^1 \sin(x^2)\,dx \le \int_0^1 x^2\,dx = \frac{1}{3}x^3\Big|_0^1 = \frac{1}{3}$.

b. Since $0 \le \sin x \le x$ for $0 \le x \le 1$, we have $0 \le \sin^{3/2} x \le x^{3/2}$ for $0 \le x \le 1$. Then by Corollary 5.20, $0 \le \int_0^{\pi/6} \sin^{3/2} x\,dx \le \int_0^{\pi/6} x^{3/2}\,dx = \frac{2}{5}x^{5/2}\Big|_0^{\pi/6} = \frac{2}{5}(\pi/6)^{5/2}$.

47. $\int_0^2 f(x)\,dx = \int_0^2 (1 - x)\,dx = (x - \frac{1}{2}x^2)\Big|_0^2 = (2 - 2) - 0 = 0$, so
$$\left|\int_0^2 f(x)\,dx\right| = 0$$
However,
$$\int_0^2 |f(x)|\,dx = \int_0^1 (1 - x)\,dx + \int_1^2 -(1 - x)\,dx = (x - \tfrac{1}{2}x^2)\Big|_0^1 - (x - \tfrac{1}{2}x^2)\Big|_1^2$$
$$= (\tfrac{1}{2} - 0) - (0 - \tfrac{1}{2}) = 1$$
Thus
$$\left|\int_0^2 f(x)\,dx\right| < \int_0^2 |f(x)|\,dx$$

49. Since $\tan x$ is an increasing function on $(-\pi/2, \pi/2)$, it follows that $-\sqrt{3} = \tan(-\pi/3) \le \tan x \le \tan(-\pi/4) = -1$ for x in $[-\pi/3, -\pi/4]$. Thus for in $[-\pi/3, -\pi/4]$, we have $|\tan x| \le \sqrt{3}$. By Exercise 48,
$$\left|\int_{-\pi/3}^{-\pi/4} \tan x\,dx\right| \le \sqrt{3}\left(-\frac{\pi}{4} + \frac{\pi}{3}\right) = \frac{1}{12}\sqrt{3}\pi$$

51. a. Fix any x in $(a, b]$. By the Mean Value Theorem there is a number c in (a, x) such that $[f(x) - f(a)]/(x - a) = f'(c)$. Since $|f'(c)| \le M$ and $f(a) = 0$ by hypothesis, we have
$$\frac{|f(x)|}{|x - a|} = \frac{|f(x) - f(a)|}{|x - a|} = |f'(c)| \le M, \quad \text{or} \quad |f(x)| \le M|x - a| = M(x - a)$$

b. By (a) and (5) we have
$$\left|\int_a^b f(x)\,dx\right| \le \int_a^b |f(x)|\,dx \le \int_a^b M(x - a)\,dx = M\int_a^b (x - a)\,dx = M\,\frac{(x-a)^2}{2}\Big|_a^b$$
$$= M\,\frac{(b-a)^2}{2}$$

53. If $g(x) \le f(x)$ for $b \le x \le a$, then $-f(x) \le -g(x)$, so that Corollary 5.20 implies that
$$\int_a^b g(x)\,dx = \int_b^a -g(x)\,dx \ge \int_b^a -f(x)\,dx = \int_a^b f(x)\,dx$$

55. We will use induction. The result is valid for $n = 2$ by Theorem 5.17. Assume the result is valid for $n - 1$ functions, and let $f_1, f_2, \ldots, f_n$ be continuous on an interval I. Then

$$\int [f_1(x) + f_2(x) + \cdots + f_{n-1}(x) + f_n(x)]\,dx$$
$$= \int [f_1(x) + f_2(x) + \cdots + f_{n-1}(x)]\,dx + \int f_n(x)\,dx$$
$$= \left[\int f_1(x)\,dx + \int f_2(x)\,dx + \cdots + \int f_{n-1}(x)\,dx\right] + \int f_n(x)\,dx$$

which is equivalent to the desired equation.

57. a. Since $(x/2) + 2y(dy/dx) = 0$, so that $dy/dx = -x/4y$, the slope of the tangent line at $(\sqrt{2}, 1/\sqrt{2})$ is

$$\left.\frac{dy}{dx}\right|_{x=\sqrt{2}} = \frac{-\sqrt{2}}{4/\sqrt{2}} = \frac{-1}{2}$$

The equation of the tangent line is $y - 1/\sqrt{2} = (-1/2)(x - \sqrt{2})$, or $y = \sqrt{2} - \frac{1}{2}x$. Since $\sqrt{1 - (x^2/4)} \le \sqrt{2} - \frac{1}{2}x$ for $0 \le x \le 2$, we have

$$\int_0^2 \sqrt{1 - \frac{x^2}{4}}\,dx \le \int_0^2 \left(\sqrt{2} - \frac{1}{2}x\right)dx = \left.\left(\sqrt{2}x - \frac{1}{4}x^2\right)\right|_0^2 = 2\sqrt{2} - 1$$

b. The triangle inscribed in the quarter ellipse has vertices (0, 0), (2, 0), (0, 1), so the area of the triangle is 1.

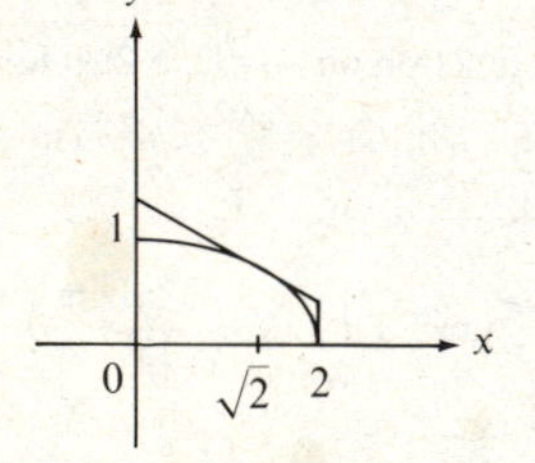

59. Let $p(r) = \int_a^b [f(x) + rg(x)]^2\,dx$. Then by the Addition Property and (2),

$$p(r) = \int_a^b [(f(x))^2 + 2rf(x)g(x) + r^2(g(x))^2]\,dx$$
$$= \int_a^b (f(x))^2\,dx + 2r\int_a^b f(x)g(x)\,dx + r^2\int_a^b (g(x))^2\,dx$$
$$= Ar^2 + Br + C$$

where $A = \int_a^b (g(x))^2\,dx$, $B = 2\int_a^b f(x)g(x)\,dx$ and $C = \int_a^b (f(x))^2\,dx$. For any given real number r we have $[f(x) + rg(x)]^2 \ge 0$ for all x in $[a, b]$, so by Corollary 5.10, $p(r) \ge 0$ for any r. But then p has at most one zero, so by the quadratic formula the discriminant of p, which is $B^2 - 4AC$, is nonpositive. Thus $B^2 \le 4AC$. This is equivalent to the desired inequality.

SECTION 5.6

1. Let $u = 4x - 5$, so that $du = 4\,dx$. Then

$$\int \sqrt{4x-5}\,dx = \int \sqrt{u}\,\tfrac{1}{4}\,du = \tfrac{1}{4}\int \sqrt{u}\,du = \tfrac{1}{4}(\tfrac{2}{3}u^{3/2}) + C = \tfrac{1}{6}(4x-5)^{3/2} + C$$

3. Let $u = \pi x$, so that $du = \pi\,dx$. Then

$$\int \cos \pi x\,dx = \int (\cos u)\frac{1}{\pi}\,du = \frac{1}{\pi}\int \cos u\,du = \frac{1}{\pi}\sin u + C = \frac{1}{\pi}\sin \pi x + C$$

5. Let $u = x^2$, so that $du = 2x\,dx$. Then

$$\int x\cos x^2\,dx = \int (\cos u)\,\tfrac{1}{2}\,du = \tfrac{1}{2}\int \cos u\,du = \tfrac{1}{2}\sin u + C = \tfrac{1}{2}\sin x^2 + C.$$

7. Let $u = \cos t$, so that $du = -\sin t\,dt$. Then

$$\int \cos^{-4} t \sin t\,dt = \int u^{-4}(-1)\,du = \tfrac{1}{3}u^{-3} + C = \tfrac{1}{3}\cos^{-3} t + C$$

9. Let $u = t^2 - 3t + 1$, so that $du = (2t - 3)\,dt$. Then

$$\int \frac{2t-3}{(t^2-3t+1)^{7/2}}\,dt = \int \frac{1}{u^{7/2}}\,du = \int u^{-7/2}\,du$$
$$= -\frac{2}{5}u^{-5/2} + C = -\frac{2}{5}\frac{1}{(t^2-3t+1)^{5/2}} + C$$

11. Let $v = x + 1$, so that $dv = dx$ and $x - 1 = v - 2$. Then

$$\int (x-1)\sqrt{x+1}\,dx = \int (v-2)\sqrt{v}\,dv = \int (v^{3/2} - 2v^{1/2})\,dv = \tfrac{2}{5}v^{5/2} - -2(\tfrac{2}{3}v^{3/2}) + C$$
$$= \tfrac{2}{5}(x+1)^{5/2} - \tfrac{4}{3}(x+1)^{3/2} + C$$

13. Let $u = 3 + \sec x$, so that $du = \sec x \tan x\,dx$. Then

$$\int \sec x \tan x\sqrt{3 + \sec x}\,dx = \int \sqrt{u}\,du = \tfrac{2}{3}u^{3/2} + C = \tfrac{2}{3}(3 + \sec x)^{3/2} + C$$

15. Let $u = x^3 + 1$, so that $du = 3x^2\,dx$. Then

$$\int 3x^2(x^3+1)^{12}\,dx = \int u^{12}\,du = \tfrac{1}{13}u^{13} + C = \tfrac{1}{13}(x^3+1)^{13} + C$$

17. Let $u = x^2 + 3x + 4$, so that $du = (2x + 3)\,dx$. Then

$$\int (2x+3)(x^2+3x+4)^5\,dx = \int u^5\,du = \tfrac{1}{6}u^6 + C = \tfrac{1}{6}(x^2+3x+4)^6 + C$$

19. Let $u = 3x + 7$, so that $du = 3\,dx$. Then

$$\int \sqrt{3x+7}\,dx = \int u^{1/2}\,\tfrac{1}{3}\,du = \tfrac{1}{3}\int u^{1/2}\,du = \tfrac{1}{3}(\tfrac{2}{3}u^{3/2}) + C = \tfrac{2}{9}(3x+7)^{3/2} + C$$

21. Let $u = 1 + 2x + 4x^2$, so that $du = (2 + 8x)\,dx$. Then

$$\int (1+4x)\sqrt{1+2x+4x^2}\,dx = \int (\sqrt{u})\tfrac{1}{2}\,du = \tfrac{1}{2}\int u^{1/2}\,du = \tfrac{1}{2}(\tfrac{2}{3}u^{3/2}) + C$$
$$= \tfrac{1}{3}(1+2x+4x^2)^{3/2} + C$$

23. Let $u = \pi x$, so that $du = \pi\,dx$. If $x = -1$, then $u = -\pi$; if $x = 3$, then $u = 3\pi$. Thus

$$\int_{-1}^{3} \sin \pi x\,dx = \int_{-\pi}^{3\pi} (\sin u)\frac{1}{\pi}\,du = \frac{1}{\pi}\int_{-\pi}^{3\pi} \sin u\,du = \frac{1}{\pi}(-\cos u)\Big|_{-\pi}^{3\pi}$$
$$= \frac{1}{\pi}[-(-1) + (-1)] = 0$$

25. Let $u = \sin t$, so that $du = \cos t\,dt$. Then

$$\int \sin^6 t \cos t\,dt = \int u^6\,du = \tfrac{1}{7}u^7 + C = \tfrac{1}{7}\sin^7 t + C$$

27. Let $u = \sin 2z$, so that $du = 2\cos 2z\,dz$. Then

$$\int \sqrt{\sin 2z}\,\cos 2z\,dz = \int \sqrt{u}\,\tfrac{1}{2}\,du = \tfrac{1}{2}\int \sqrt{u}\,du = \tfrac{1}{2}(\tfrac{2}{3}u^{3/2}) + C = \tfrac{1}{3}(\sin 2z)^{3/2} + C$$

29. Let $u = \cos z$, so that $du = -\sin z\,dz$. If $z = 0$, then $u = 1$; if $z = \pi/4$, then $u = \sqrt{2}/2$. Thus

$$\int_0^{\pi/4} \frac{\sin z}{\cos^2 z}\,dz = \int_1^{\sqrt{2}/2} \frac{1}{u^2}(-1)\,du = \frac{1}{u}\Big|_1^{\sqrt{2}/2} = \frac{1}{\sqrt{2}/2} - 1 = \sqrt{2} - 1$$

31. Let $u = \sqrt{z}$, so that $du = \tfrac{1}{2}(1/\sqrt{z})\,dz$. Then

$$\int \left(\frac{1}{\sqrt{z}}\right)\sec^2 \sqrt{z}\,dz = \int (\sec^2 u)(2)\,du = 2\int \sec^2 u\,du = 2\tan u + C = 2\tan\sqrt{z} + C$$

33. Let $u = w^2 + 1$, so that $du = 2w\,dw$. Then

$$\int w\left(\sqrt{w^2+1} + \frac{1}{\sqrt{w^2+1}}\right)dw = \int \left(\sqrt{u} + \frac{1}{\sqrt{u}}\right)\frac{1}{2}\,du = \frac{1}{2}\int \left(\sqrt{u} + \frac{1}{\sqrt{u}}\right)du$$
$$= \tfrac{1}{2}(\tfrac{2}{3}u^{3/2} + 2u^{1/2}) + C = \tfrac{1}{3}(w^2+1)^{3/2} + (w^2+1)^{1/2} + C$$

35. Let $u = 1 + 4x^{1/3}$, so that $du = \tfrac{4}{3}x^{-2/3}\,dx$. If $x = 1$, then $u = 5$; if $x = 8$, then $u = 9$. Thus

$$\int_1^8 x^{-2/3}\sqrt{1 + 4x^{1/3}}\,dx = \int_5^9 \sqrt{u}(\tfrac{3}{4})\,du = \tfrac{3}{4}\int_5^9 \sqrt{u}\,du = \tfrac{3}{4}(\tfrac{2}{3}u^{3/2})\Big|_5^9 = \tfrac{1}{2}(27 - 5\sqrt{5}).$$

37. Let $u = x + 2$, so that $du = dx$ and $x = u - 2$. Then

$$\int x\sqrt{x+2}\,dx = \int (u-2)\sqrt{u}\,du = \int (u^{3/2} - 2u^{1/2})\,du$$
$$= \tfrac{2}{5}u^{5/2} - \tfrac{4}{3}u^{3/2} + C = \tfrac{2}{5}(x+2)^{5/2} - \tfrac{4}{3}(x+2)^{3/2} + C$$

39. Let $u = 6 - 2x$, so that $du = -2\,dx$ and $4x = 2(6 - u)$. If $x = 1$, then $u = 4$; if $x = 3$, then $u = 0$. Thus

$$\int_1^3 4x\sqrt{6-2x}\,dx = \int_4^0 2(6-u)\sqrt{u}(-\tfrac{1}{2})\,du = -\int_4^0 (6u^{1/2} - u^{3/2})\,du$$
$$= \int_0^4 (6u^{1/2} - u^{3/2})\,du = (4u^{3/2} - \tfrac{2}{5}u^{5/2})\Big|_0^4$$
$$= [4(8) - \tfrac{2}{5}(32)] - 0 = \tfrac{96}{5}$$

41. Let $u = 1 - 8t$, so that $du = -8dt$ and $t^2 = \tfrac{1}{64}(1-u)^2$. Then

$$\int t^2\sqrt{1-8t}\,dt = \int \tfrac{1}{64}(1-u)^2\sqrt{u}(-\tfrac{1}{8})\,du = -\tfrac{1}{512}\int (u^{1/2} - 2u^{3/2} + u^{5/2})\,du$$
$$= -\tfrac{1}{512}(\tfrac{2}{3}u^{3/2} - \tfrac{4}{5}u^{5/2} + \tfrac{2}{7}u^{7/2}) + C$$
$$= -\tfrac{1}{256}[\tfrac{1}{3}(1-8t)^{3/2} - \tfrac{2}{5}(1-8t)^{5/2} + \tfrac{1}{7}(1-8t)^{7/2}] + C$$

43. Let $u = t + 2$, so that $du = dt$ and $t^2 = (u-2)^2$. If $t = -1$, then $u = 1$; if $t = 2$, then $u = 4$. Thus

$$\int_{-1}^{2} \frac{t^2}{\sqrt{t+2}}\,dt = \int_1^4 \frac{(u-2)^2}{\sqrt{u}}\,du = \int_1^4 (u^{3/2} - 4u^{1/2} + 4u^{-1/2})\,du$$
$$= (\tfrac{2}{5}u^{5/2} - \tfrac{8}{3}u^{3/2} + 8u^{1/2})\Big|_1^4$$
$$= [\tfrac{2}{5}(32) - \tfrac{8}{3}(8) + 8(2)] - [\tfrac{2}{5} - \tfrac{8}{3} + 8] = \tfrac{26}{15}$$

45. $A = \int_0^3 \sqrt{x+1}\,dx$. Let $u = x + 1$, so that $du = dx$. If $x = 0$, then $u = 1$; if $x = 3$, then $u = 4$. Thus $A = \int_0^3 \sqrt{x+1}\,dx = \int_1^4 \sqrt{u}\,du = \tfrac{2}{3}(4^{3/2} - 1^{3/2}) = \tfrac{14}{3}$.

47. $A = \int_1^2 [x/(x^2+1)^2]\,dx$. Let $u = x^2 + 1$, so that $du = 2x\,dx$. If $x = 1$, then $u = 2$; if $x = 2$, then $u = 5$. Thus

$$A = \int_1^2 \frac{x}{(x^2+1)^2}\,dx = \int_2^5 \frac{1}{u^2}\left(\frac{1}{2}\right)du = \frac{1}{2}\int_2^5 \frac{1}{u^2}\,du = -\frac{1}{2}\frac{1}{u}\Big|_2^5 = -\frac{1}{2}\left(\frac{1}{5} - \frac{1}{2}\right) = \frac{3}{20}$$

49. $A = \displaystyle\int_{1/8}^{1/3} \frac{1}{x^2}\left(1 + \frac{1}{x}\right)^{1/2} dx$

Let $u = 1 + 1/x$, so that $du = -(1/x^2)\,dx$. If $x = \tfrac{1}{8}$, then $u = 9$; if $x = \tfrac{1}{3}$, then $u = 4$. Thus

$$A = \int_{1/8}^{1/3} \frac{1}{x^2}\left(1 + \frac{1}{x}\right)^{1/2} dx = \int_9^4 u^{1/2}(-1)\,du = -\int_9^4 u^{1/2}\,du = -\frac{2}{3}u^{3/2}\Big|_9^4 = \frac{38}{3}$$

51. a. We have $\int_a^{a+k\pi} \sin^2 x\,dx = (\tfrac{1}{2}x - \tfrac{1}{4}\sin 2x)\Big|_a^{a+k\pi} = [\tfrac{1}{2}(a + k\pi) - \tfrac{1}{4}\sin 2(a + k\pi)] - [\tfrac{1}{2}a - \tfrac{1}{4}\sin 2a] = \tfrac{1}{2}k\pi$.

b. We have $\int_a^{a+k\pi} \cos^2 x\,dx = (\tfrac{1}{2}x + \tfrac{1}{4}\sin 2x)\Big|_a^{a+k\pi} = [\tfrac{1}{2}(a + k\pi) + \tfrac{1}{4}\sin 2(a + k\pi)] - [\tfrac{1}{2}a + \tfrac{1}{4}\sin 2a] = \tfrac{1}{2}k\pi$.

53. a. Let $u = ax + b$, so that $du = a\,dx$. Then

$$\int f(ax+b)\,dx = \int f(u)\frac{1}{a}\,du = \frac{1}{a}\int f(u)\,du = \frac{1}{a}F(u) + C = \frac{1}{a}F(ax+b) + C$$

b. If $f(x) = \sin x$, then since $\int \sin x\,dx = -\cos x + C$, part (a) implies that

$$\int \sin(ax+b)\,dx = -\frac{1}{a}\cos(ax+b) + C$$

c. If $f(x) = x^n$ with $n \neq 0, -1$, then since $\int x^n\,dx = [1/(n+1)]x^{n+1} + C$, part (a) implies that

$$\int (ax+b)^n\,dx = \frac{1}{a}\left[\frac{1}{n+1}(ax+b)^{n+1}\right] + C = \frac{1}{a(n+1)}(ax+b)^{n+1} + C$$

55. Let $u = -x$, so that $du = -dx$. If $x = -a$, then $u = a$; if $x = 0$, then $u = 0$. Thus

$$\int_{-a}^{0} f(x)\,dx = \int_a^0 [f(-u)](-1)\,du = \int_0^a f(-u)\,du = \int_0^a f(-x)\,dx$$

where we have replaced u by x in the last equation.

57. a. Since $\sin x$ is odd, it follows from Exercise 56(b) that $\int_{-\pi/3}^{\pi/3} \sin x\, dx = 0$.

b. Since $\cos t$ is even, it follows from Exercise 56(c) that

$$\int_{-\pi/4}^{\pi/4} \cos t\, dt = 2\int_0^{\pi/4} \cos t\, dt = 2(\sin t)\Big|_0^{\pi/4} = 2\left(\frac{\sqrt{2}}{2} - 0\right) = \sqrt{2}$$

59. Let $u = x - a$, so that $du = dx$. If $x = x_1$, then $u = x_1 - a$; if $x = x_2$, then $u = x_2 - a$. Thus

$$W = \int_{x_1}^{x_2} \frac{Gm_1m_2}{(x-a)^2}\, dx = \int_{x_1-a}^{x_2-a} \frac{Gm_1m_2}{u^2}\, du = Gm_1m_2 \int_{x_1-a}^{x_2-a} \frac{1}{u^2}\, du = Gm_1m_2\left(-\frac{1}{u}\right)\Big|_{x_1-a}^{x_2-a}$$

$$= Gm_1m_2\left(-\frac{1}{x_2-a} + \frac{1}{x_1-a}\right) = \frac{Gm_1m_2(x_2-x_1)}{(x_1-a)(x_2-a)}$$

SECTION 5.7

1. $\int_1^3 \frac{1}{x}\, dx = \ln x\Big|_1^3 = \ln 3 - \ln 1 = \ln 3$

3. $\int_{1/9}^{1/4} \frac{-1}{3x}\, dx = -\frac{1}{3}\int_{1/9}^{1/4} \frac{1}{x}\, dx = -\frac{1}{3}\ln x\Big|_{1/9}^{1/4} = -\frac{1}{3}\left(\ln\frac{1}{4} - \ln\frac{1}{9}\right)$

$$= -\frac{1}{3}(-\ln 4 + \ln 9) = \frac{1}{3}\ln\frac{4}{9}$$

5. $\int_{-4}^{-12} \frac{2}{t}\, dt = 2\int_{-4}^{-12} \frac{1}{t}\, dt = 2\ln|t|\Big|_{-4}^{-12} = 2(\ln 12 - \ln 4) = 2\ln 3$

7. The domain is $(-1, \infty)$; $f'(x) = 1/(x+1)$.

9. The domain consists of all x such that $x^2 - 1 > 0$, that is, $(-\infty, -1)$ and $(1, \infty)$;

$$g'(x) = \ln(x^2-1) + x\left[\frac{1}{x^2-1}(2x)\right] = \ln(x^2-1) + \frac{2x^2}{x^2-1}$$

11. The domain consists of all x such that $(x-3)/(x-2) > 0$, that is, $(-\infty, 2)$ and $(3, \infty)$; since $f(x) = \frac{1}{2}\ln(x-3)/(x-2)$, it follows that

$$f'(x) = \frac{1}{2}\frac{1}{(x-3)/(x-2)}\frac{1}{(x-2)^2} = \frac{1}{2(x-3)(x-2)}$$

13. The domain consists of $(0, 1)$ and $(1, \infty)$;

$$f'(x) = \frac{(1/x)(x-1) - \ln x}{(x-1)^2} = \frac{x - 1 - x\ln x}{x(x-1)^2}$$

15. The domain is $(0, \infty)$; $f'(t) = [\cos(\ln t)](1/t)$.

17. The domain consists of all x such that $\ln x > 0$, that is, $(1, \infty)$; $f'(x) = 1/(\ln x)\cdot 1/x = 1/(x\ln x)$.

19. By implicit differentiation,

$$\ln(y^2+x) + x\left[\frac{1}{y^2+x}\left(2y\frac{dy}{dx} + 1\right)\right] = 5\frac{dy}{dx}$$

so that

$$\left(5 - \frac{2xy}{y^2+x}\right)\frac{dy}{dx} = \ln(y^2+x) + \frac{x}{y^2+x}, \quad \text{and thus} \quad \frac{dy}{dx} = \frac{(y^2+x)\ln(y^2+x) + x}{5(y^2+x) - 2xy}$$

21. The domain consists of $(-\infty, 0)$ and $(0, \infty)$.

$$f(x) = \begin{cases} \ln x & \text{for } x > 0 \\ \ln(-x) & \text{for } x < 0 \end{cases}$$

$$f'(x) = \frac{1}{x}; \qquad f''(x) = \frac{-1}{x^2}$$

No critical points or inflection points; concave downward on $(-\infty, 0)$ and on $(0, \infty)$; vertical asymptote is $x = 0$; symmetric with respect to the y axis

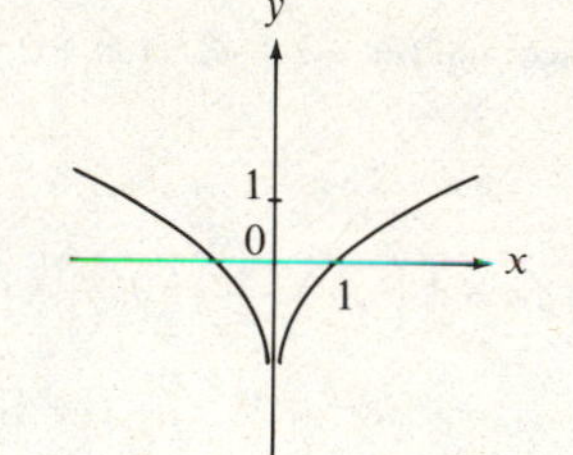

23. The domain is $(-\infty, \infty)$.

$$g'(x) = \frac{\cos x}{2 + \sin x}; \qquad g''(x) = \frac{-(1 + 2\sin x)}{(2 + \sin x)^2}$$

Relative maximum value is $g(\pi/2 + 2n\pi) = \ln 3$ for any integer n; relative minimum value is $g(3\pi/2 + 2n\pi) = 0$ for any integer n; inflection points are $(-\pi/6 + 2n\pi, \ln\frac{3}{2})$ and $(7\pi/6 + 2n\pi, \ln\frac{3}{2})$; concave downward on $(-\pi/6 + 2n\pi, 7\pi/6 + 2n\pi)$ and concave upward on $(7\pi/6 + 2n\pi, 11\pi/6 + 2n\pi)$

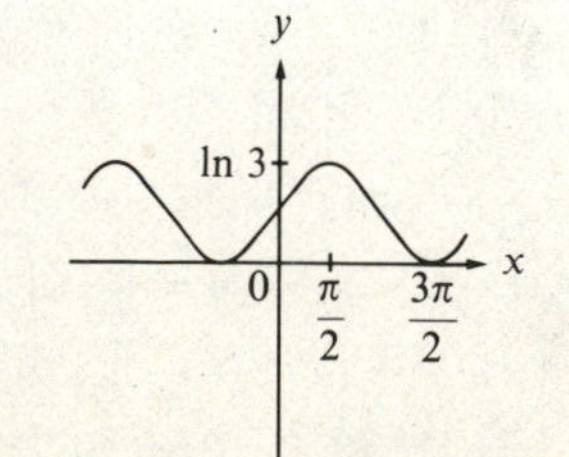

25. The domain is $(0, \infty)$.

$$f'(x) = 2\frac{\ln x}{x}; \qquad f''(x) = \frac{2(1/x)x - 2\ln x}{x^2} = \frac{2(1-\ln x)}{x^2}$$

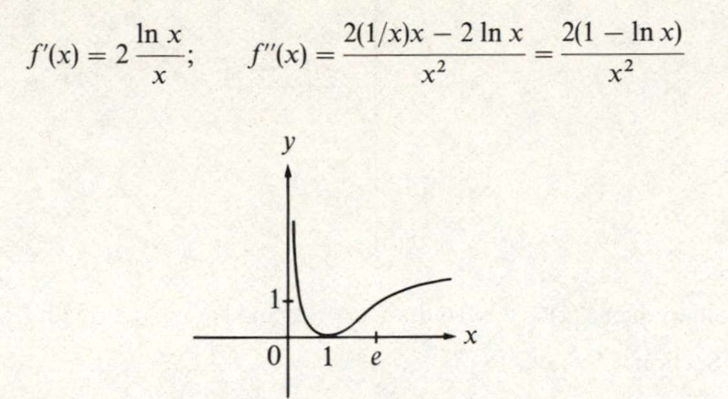

Relative minimum value is $f(1) = 0$; inflection point is $(e, 1)$; concave upward on $(0, e)$ and concave downward on (e, ∞); vertical asymptote is $x = 0$

27. $f'(x) = -2/x^3 - 1/x$. Letting the initial value of c be 1 and using a computer, we obtain 1.531584 for the approximate zero of f.

29. Let $u = x - 1$, so that $du = dx$. Then

$$\int \frac{1}{x-1}\,dx = \int \frac{1}{u}\,du = \ln|u| + C = \ln|x-1| + C$$

31. Let $u = x^2 + 4$, so that $du = 2x\,dx$. Then

$$\int \frac{x}{x^2+4}\,dx = \int \frac{1}{u}\frac{1}{2}\,du = \frac{1}{2}\int \frac{1}{u}\,du = \frac{1}{2}\ln|u| + C = \frac{1}{2}\ln(x^2+4) + C$$

33. Let $u = x^4 - 4$, so that $du = 4x^3dx$. Then

$$\int \frac{x^3}{x^4-4}\,dx = \int \frac{1}{u}\frac{1}{4}\,du = \frac{1}{4}\int \frac{1}{u}\,du = \frac{1}{4}\ln|u| + C = \frac{1}{4}\ln|x^4-4| + C$$

35. Let $u = 1 - 3\cos x$, so that $du = 3\sin x\,dx$. If $x = 0$, then $u = -2$; if $x = \pi/3$, then $u = -\frac{1}{2}$. Thus

$$\int_0^{\pi/3} \frac{\sin x}{1 - 3\cos x}\,dx = \int_{-2}^{-1/2} \frac{1}{u}\left(\frac{1}{3}\right)du = \frac{1}{3}\int_{-2}^{-1/2}\frac{1}{u}\,du = \frac{1}{3}\ln|u|\Big|_{-2}^{-1/2}$$
$$= \frac{1}{3}\left(\ln\frac{1}{2} - \ln 2\right) = -\frac{2}{3}\ln 2$$

37. Let $u = 1 + \sqrt{x}$, so that $du = (1/2\sqrt{x})\,dx$. If $x = 1$, then $u = 2$; if $x = 4$, then $u = 3$. Thus

$$\int_1^4 \frac{1}{\sqrt{x}(1+\sqrt{x})}\,dx = \int_2^3 \frac{1}{u}(2)\,du = 2\int_2^3 \frac{1}{u}\,du = 2\ln|u|\Big|_2^3 = 2(\ln 3 - \ln 2) = 2\ln\frac{3}{2}$$

39. Let $u = \ln z$, so that $du = (1/z)\,dz$. Then

$$\int \frac{\ln z}{z}\,dz = \int u\,du = \frac{1}{2}u^2 + C = \frac{1}{2}(\ln z)^2 + C$$

41. Let $u = \ln(\ln t)$, so that

$$du = \frac{1}{\ln t}\cdot\frac{1}{t}\,dt = \frac{1}{t\ln t}\,dt$$

Then
$$\int \frac{\ln(\ln t)}{t\ln t}\,dt = \int u\,du = \frac{1}{2}u^2 + C = \frac{1}{2}(\ln(\ln t))^2 + C$$

43. Note that $\int \cot t\,dt = \int (\cos t/\sin t)\,dt$. Let $u = \sin t$, so that $du = \cos t\,dt$. Then

$$\int \cot t\,dt = \int \frac{\cos t}{\sin t}\,dt = \int \frac{1}{u}\,du = \ln|u| + C = \ln|\sin t| + C$$

45. Note that

$$\int \frac{x}{1 + x\tan x}\,dx = \int \frac{x}{1 + x\dfrac{\sin x}{\cos x}}\,dx = \int \frac{x\cos x}{\cos x + x\sin x}\,dx$$

Let $u = \cos x + x\sin x$, so that $du = (-\sin x + \sin x + x\cos x)\,dx = x\cos x\,dx$. Then

$$\int \frac{x}{1 + x\tan x}\,dx = \int \frac{x\cos x}{\cos x + x\sin x}\,dx = \int \frac{1}{u}\,du$$
$$= \ln|u| + C = \ln|\cos x + x\sin x| + C$$

47. $A = \int_e^{e^2} \frac{1}{x}\,dx = \ln|x|\Big|_e^{e^2} = \ln e^2 - \ln e = 2\ln e - \ln e = 1$

49. $A = \int_{-2}^{-\sqrt{3}} [x/(2 - x^2)]\,dx$. Let $u = 2 - x^2$, so that $du = -2x\,dx$. If $x = -2$, then $u = -2$; if $x = -\sqrt{3}$, then $u = -1$. Thus

$$A = \int_{-2}^{-\sqrt{3}} \frac{x}{2 - x^2}\,dx = \int_{-2}^{-1}\frac{1}{u}\left(-\frac{1}{2}\right)du = -\frac{1}{2}\int_{-2}^{-1}\frac{1}{u}\,du = -\frac{1}{2}\ln|u|\Big|_{-2}^{-1}$$
$$= -\tfrac{1}{2}(\ln 1 - \ln 2) = \tfrac{1}{2}\ln 2$$

51. $A = \int_{\pi/4}^{\pi/3} \sec^3 x\tan x\,dx = \int_{\pi/4}^{\pi/3} \sec^2 x(\sec x\tan x)\,dx$. Let $u = \sec x$, so that $du = \sec x\tan x\,dx$. If $x = \pi/4$, then $u = \sqrt{2}$; if $x = \pi/3$, then $u = 2$. Thus

$$A = \int_{\pi/4}^{\pi/3} \sec^3 x\tan x\,dx = \int_{\sqrt{2}}^{2} u^2\,du = \tfrac{1}{3}u^3\Big|_{\sqrt{2}}^{2} = \tfrac{8}{3} - \tfrac{2}{3}\sqrt{2}$$

53. $\ln(a^3b^2) = \ln a^3 + \ln b^2 = 3\ln a + 2\ln b = 3(2.3) + 2(4.7) = 16.3$.

55. a. By repeated use of the Addition Property, we have

$$\int_1^3 \frac{1}{t}\,dt = \int_1^{5/4}\frac{1}{t}\,dt + \int_{5/4}^{6/4}\frac{1}{t}\,dt + \cdots + \int_{11/4}^{3}\frac{1}{t}\,dt$$

b. By using (a) and the Comparison Property, we have

$$\ln 3 = \int_1^3 \frac{1}{t}\,dt = \int_1^{5/4}\frac{1}{t}\,dt + \int_{5/4}^{6/4}\frac{1}{t}\,dt + \cdots + \int_{11/4}^{3}\frac{1}{t}\,dt > \frac{1}{5/4}\cdot\frac{1}{4} + \frac{1}{6/4}\cdot\frac{1}{4} + \cdots$$
$$+ \frac{1}{3}\cdot\frac{1}{4} = \frac{1}{4}\left(\frac{4}{5} + \frac{4}{6} + \cdots + \frac{4}{12}\right) = \frac{1}{5} + \frac{1}{6} + \cdots + \frac{1}{12} > 1$$

c. Since $\ln e = 1$ and $\ln x$ is a strictly increasing function, it follows that $3 > e$.

57. Let $f(x) = \ln x^r$ and $g(x) = r \ln x$. Then $f'(x) = (1/x^r)rx^{r-1} = r/x$ and $g'(x) = r/x$, so $f'(x) = g'(x)$. Moreover $f(1) = \ln 1^r = \ln 1 = 0$ and $g(1) = r \ln 1 = 0$. By Theorem 4.7(b), $f = g$, so $\ln b^r = r \ln b$.

59. a. By the Law of Logarithms, $\ln (b/c) = \ln b(1/c) = \ln b + \ln (1/c)$, and by (5), $\ln (1/c) = -\ln c$. Therefore $\ln (b/c) = \ln b - \ln c$.

b. Let $f(x) = \ln (x/c)$ and $g(x) = \ln x - \ln c$. Then

$$f'(x) = \frac{1}{x/c}\left(\frac{1}{c}\right) = \frac{1}{x} \quad \text{and} \quad g'(x) = \frac{1}{x}$$

so $f'(x) = g'(x)$. Moreover, $f(c) = \ln (c/c) = \ln 1 = 0$ and $g(c) = \ln c - \ln c = 0$. By Theorem 4.7(b), $f = g$, so $\ln (b/c) = \ln b - \ln c$.

61. Since $dy/dt = -1/t^2$, an equation of the tangent line at (1, 1) is $y - 1 = -1(x - 1)$, or $y = -x + 2$. For $0 < h < 1$, the area of the region below the line $y = -x + 2$ on $[1, 1 + h]$ equals $\int_1^{1+h} (-x + 2)\, dx = (-\frac{1}{2}x^2 + 2x)\Big|_1^{1+h} = h - \frac{1}{2}h^2$. For $-1 < h < 0$, the area of the region below the line $y = -x + 2$ on $[1 + h, 1]$ equals $\int_{1+h}^{1} (-x + 2)\, dx = -\int_1^{1+h} (-x + 2)\, dx = -(h - \frac{1}{2}h^2)$.

63. a. Since $1/t \le 1/\sqrt{t}$ for $t \ge 1$, we have

$$\ln x = \int_1^x \frac{1}{t}\, dt \le \int_1^x \frac{1}{\sqrt{t}}\, dt = 2\sqrt{t}\Big|_1^x = 2(\sqrt{x} - 1) \quad \text{for } x \ge 1$$

b. Since $0 \le \ln x \le 2(\sqrt{x} - 1)$ for $x \ge 1$ by (a), we have $0 \le (\ln x)/x \le (2\sqrt{x} - 2)/x$. But $\lim_{x\to\infty} (2\sqrt{x} - 2)/x = \lim_{x\to\infty} (2/\sqrt{x} - 2/x) = 0$. Thus by the Squeezing Theorem, $\lim_{x\to\infty} (\ln x)/x = 0$.

c. By (b) and (5),

$$\lim_{x\to 0^+} x \ln x = \lim_{y\to\infty} \frac{1}{y} \ln \frac{1}{y} = \lim_{y\to\infty} \frac{-\ln y}{y} = 0$$

65. a. $f'(x) = \dfrac{(1/x)x - \ln x}{x^2} = \dfrac{1 - \ln x}{x^2}$

Since $\ln e = 1$ and $\ln x$ is strictly increasing, we have $f'(x) > 0$ for $0 < x < e$ and $f'(x) < 0$ for $x > e$. Thus f is strictly increasing on $(0, e]$ and strictly decreasing on $[e, \infty)$.

b. By (a), $f(e) = 1/e$ is the maximum value of f.

c. $f''(x) = \dfrac{-(1/x)x^2 - 2x(1 - \ln x)}{x^4} = \dfrac{-3 + 2 \ln x}{x^3}$

If $\ln x_0 = \frac{3}{2}$, then $x_0 \approx 4.5$, and the graph is concave downward on $(0, x_0)$ and concave upward on (x_0, ∞). By Exercise 63(b), $y = 0$ is a horizontal asymptote of f. Since $\lim_{x\to 0^+} \ln x = -\infty$ and $\lim_{x\to 0^+} 1/x = \infty$, we have $\lim_{x\to 0^+} (\ln x)/x = -\infty$, so that the line $x = 0$ is a vertical asymptote.

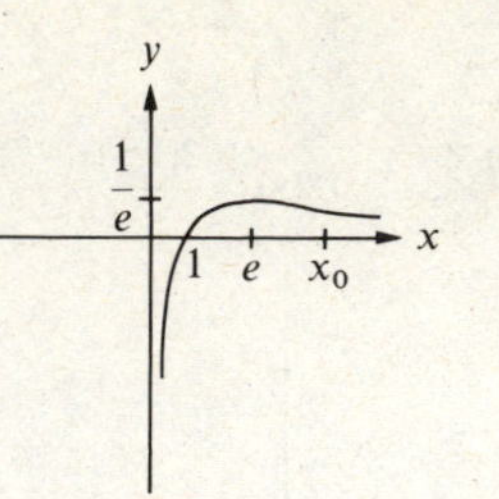

67. If $f(x) = \ln x$, then $f'(x) = 1/x$, so that $f(x) \ge 0$ and $|f'(x)| \le 1$ on $[1, 2]$. Thus by Exercise 51(b) of Section 5.5,

$$0 \le \int_1^2 \ln x\, dx = \left|\int_1^2 \ln x\, dx\right| \le 1(\tfrac{1}{2}) = \tfrac{1}{2}$$

69. a. We use induction. If $n = 2$, then by Theorem 5.23,

$$\ln |f(x)| = \ln |f_1(x)f_2(x)| = \ln |f_1(x)||f_2(x)| = \ln |f_1(x)| + \ln |f_2(x)|$$

If we assume the formula for $n - 1$ and let $f = f_1 f_2 \cdots f_n$, then

$$\begin{aligned}\ln |f(x)| &= \ln |f_1(x)f_2(x)\cdots f_n(x)| = \ln |f_1(x)f_2(x)\cdots f_{n-1}(x)||f_n(x)| \\ &= \ln |f_1(x)f_2(x)\cdots f_{n-1}(x)| + \ln |f_n(x)| = \ln |f_1(x)| + \ln |f_2(x)| + \cdots \\ &\quad + \ln |f_{n-1}(x)| + \ln |f_n(x)|\end{aligned}$$

b. By (a) and Exercise 68,

$$\begin{aligned}\frac{f'(x)}{f(x)} &= \frac{d}{dx} \ln |f(x)| = \frac{d}{dx} \ln |f_1(x)f_2 \cdots f_n(x)| \\ &= \frac{d}{dx} [\ln |f_1(x)| + \ln |f_2(x)| + \cdots + \ln |f_n(x)|] \\ &= \frac{f_1'(x)}{f_1(x)} + \frac{f_2'(x)}{f_2(x)} + \cdots + \frac{f_n'(x)}{f_n(x)}\end{aligned}$$

Thus $$f'(x) = f(x)\left[\frac{f_1'(x)}{f_1(x)} + \frac{f_2'(x)}{f_2(x)} + \cdots + \frac{f_n'(x)}{f_n(x)}\right]$$

71. a. $f(1) = f(1 \cdot 1) = f(1) + f(1) = 2f(1)$. Therefore $f(1) = 0$.

b. $f(y) = f(x \cdot y/x) = f(x) + f(y/x)$. Thus $f(y/x) = f(y) - f(x)$.

c. If $0 < |h| < x$, then $x + h > 0$, so that $x + h$ is in the domain of f. By part (b), $f((x + h)/x) = f(x + h) - f(x)$. Thus

$$\frac{f(x + h) - f(x)}{h} = \frac{f(x + h/x)}{h} = \frac{f(1 + h/x)}{h} = \frac{1}{x}\frac{f(1 + h/x) - f(1)}{h/x}$$

d. Fix $x > 0$. Since $\lim_{h\to 0} (h/x) = 0$, it follows from (c) and the Substitution Rule that

$$\begin{aligned}\lim_{h\to 0} \frac{f(x + h) - f(x)}{h} &= \lim_{h\to 0} \frac{1}{x}\frac{f(1 + h/x) - f(1)}{h/x} \\ &= \lim_{y\to 0} \frac{1}{x}\frac{f(1 + y) - f(1)}{y} = \frac{1}{x} f'(1)\end{aligned}$$

73. a. $\ln w = 4.4974 + 3.135 \ln 1 = 4.4974$, so $w \approx 89.7834$ (kilograms)

b. $\ln w = 4.4974 + 3.135 \ln \frac{1}{2}$, so $w \approx 10.2204$ (kilograms)

SECTION 5.8

1. $x^2 + 2x \le 0$ on $[-1, 0]$, whereas $x^2 + 2x \ge 0$ on $[0, 3]$. Thus

$$A = \int_{-1}^{0} -(x^2 + 2x)\,dx + \int_0^3 (x^2 + 2x)\,dx = -(\tfrac{1}{3}x^3 + x^2)\Big|_{-1}^{0} + (\tfrac{1}{3}x^3 + x^2)\Big|_0^3$$
$$= \tfrac{2}{3} + 18 = \tfrac{56}{3}$$

3. $\cos x - \sin x \ge 0$ on $[0, \pi/4]$, whereas $\cos x - \sin x \le 0$ on $[\pi/4, \pi/3]$. Thus

$$A = \int_0^{\pi/4} (\cos x - \sin x)\,dx + \int_{\pi/4}^{\pi/3} (\sin x - \cos x)\,dx = (\sin x + \cos x)\Big|_0^{\pi/4}$$
$$+ (-\cos x - \sin x)\Big|_{\pi/4}^{\pi/3} = (\sqrt{2} - 1) + \left(\frac{-1}{2} - \frac{\sqrt{3}}{2} + \sqrt{2}\right) = 2\sqrt{2} - \frac{3}{2} - \frac{\sqrt{3}}{2}$$

5. $x\sqrt{1 - x^2} \le 0$ on $[-1, 0]$, whereas $x\sqrt{1 - x^2} \ge 0$ on $[0, 1]$. Thus

$$A = \int_{-1}^{0} -x\sqrt{1 - x^2}\,dx + \int_0^1 x\sqrt{1 - x^2}\,dx \overset{u = 1 - x^2}{=} \int_0^1 \sqrt{u}\,\tfrac{1}{2}\,du + \int_1^0 \sqrt{u}\,(-\tfrac{1}{2})\,du$$
$$= \tfrac{1}{2}\int_0^1 \sqrt{u}\,du + \tfrac{1}{2}\int_0^1 \sqrt{u}\,du = \int_0^1 \sqrt{u}\,du = \tfrac{2}{3}u^{3/2}\Big|_0^1 = \tfrac{2}{3}$$

7. $x/(x^2 - 1) \ge 0$ on $[-\frac{1}{2}, 0]$, whereas $x/(x^2 - 1) \le 0$ on $[0, \frac{1}{3}]$. Thus

$$A = \int_{-1/2}^{0} \frac{x}{x^2 - 1}\,dx + \int_0^{1/3} \frac{-x}{x^2 - 1}\,dx \overset{u = x^2 - 1}{=} \int_{-3/4}^{-1} \frac{1}{2u}\,du + \int_{-1}^{-8/9} -\frac{1}{2u}\,du$$
$$= \tfrac{1}{2}\ln|u|\Big|_{-3/4}^{-1} + (-\tfrac{1}{2}\ln|u|)\Big|_{-1}^{-8/9} = -\tfrac{1}{2}\ln\tfrac{3}{4} - \tfrac{1}{2}\ln\tfrac{8}{9} = \tfrac{1}{2}\ln\tfrac{3}{2}$$

9. $f(x) \ge g(x)$ for $-2 \le x \le 1$, so

$$A = \int_{-2}^{1} (x^2 - x^3)\,dx = (\tfrac{1}{3}x^3 - \tfrac{1}{4}x^4)\Big|_{-2}^{1} = \tfrac{1}{12} + \tfrac{20}{3} = \tfrac{27}{4}$$

11. $g(x) - k(x) = x^2 + 3x + 2 = (x + 1)(x + 2)$, so $g(x) - k(x) \ge 0$ for $-3 \le x \le -2$ and for $-1 \le x \le 0$, whereas $g(x) - k(x) \le 0$ for $-2 \le x \le -1$. Thus

$$A = \int_{-3}^{-2} (x^2 + 3x + 2)\,dx + \int_{-2}^{-1} (-x^2 - 3x - 2)\,dx + \int_{-1}^{0} (x^2 + 3x + 2)\,dx$$
$$= \left(\frac{1}{3}x^3 + \frac{3}{2}x^2 + 2x\right)\Big|_{-3}^{-2} + \left(\frac{-1}{3}x^3 - \frac{3}{2}x^2 - 2x\right)\Big|_{-2}^{-1} + \left(\frac{1}{3}x^3 + \frac{3}{2}x^2 + 2x\right)\Big|_{-1}^{0}$$
$$= \tfrac{5}{6} + \tfrac{1}{6} + \tfrac{5}{6} = \tfrac{11}{6}$$

13. Since $\sec x \ge \tan x$ and $\sec x \ge 0$ for $-\pi/3 \le x \le \pi/6$, we have $f(x) \ge g(x)$ for $-\pi/3 \le x \le \pi/6$. Thus

$$A = \int_{-\pi/3}^{\pi/6} (\sec^2 x - \sec x \tan x)\,dx = (\tan x - \sec x)\Big|_{-\pi/3}^{\pi/6}$$
$$= (\tfrac{1}{3}\sqrt{3} - \tfrac{2}{3}\sqrt{3}) - (-\sqrt{3} - 2) = \tfrac{2}{3}\sqrt{3} + 2$$

15. $g(x) \ge k(x)$ for $-\pi/4 \le x \le 0$ and $g(x) \le k(x)$ for $0 \le x \le \pi/4$. Thus

$$A = \int_{-\pi/4}^{0} \left(\frac{1}{2} - \frac{1}{2}\cos 2x - \tan x\right)dx + \int_0^{\pi/4} -\left(\frac{1}{2} - \frac{1}{2}\cos 2x - \tan x\right)dx$$
$$= \left(\frac{x}{2} - \frac{\sin 2x}{4} + \ln|\cos x|\right)\Big|_{-\pi/4}^{0} + \left(-\frac{x}{2} + \frac{\sin 2x}{4} - \ln|\cos x|\right)\Big|_0^{\pi/4}$$
$$= \left(\frac{\pi}{8} - \frac{1}{4} - \ln\frac{\sqrt{2}}{2}\right) + \left(-\frac{\pi}{8} + \frac{1}{4} - \ln\frac{\sqrt{2}}{2}\right) = -2\ln\frac{\sqrt{2}}{2} = \ln 2$$

17. $g(x) \ge f(x)$ for $-1 \le x \le 0$ and $f(x) \ge g(x)$ for $0 \le x \le 3$. Thus

$$A = \int_{-1}^{0} (x^2 - x\sqrt{2x + 3})\,dx + \int_0^3 -(x^2 - x\sqrt{2x + 3})\,dx$$
$$= \frac{1}{3}x^3\Big|_{-1}^{0} - \int_{-1}^{0} x\sqrt{2x + 3}\,dx - \frac{1}{3}x^3\Big|_0^3 + \int_0^3 x\sqrt{2x + 3}\,dx$$
$$\overset{u = 2x + 3}{=} \frac{1}{3} - \int_1^3 \frac{1}{2}(u - 3)\sqrt{u}\,\frac{1}{2}\,du - 9 + \int_3^9 \frac{1}{2}(u - 3)\sqrt{u}\,\frac{1}{2}\,du$$
$$= \frac{1}{3} - \frac{1}{4}\left(\frac{2}{5}u^{5/2} - 2u^{3/2}\right)\Big|_1^3 - 9 + \frac{1}{4}\left(\frac{2}{5}u^{5/2} - 2u^{3/2}\right)\Big|_3^9$$
$$= \frac{1}{3} - \frac{1}{4}\left[\left(\frac{18\sqrt{3}}{5} - 6\sqrt{3}\right) - \left(\frac{2}{5} - 2\right)\right] - 9 + \frac{1}{4}\left[\left(\frac{2}{5}\cdot 243 - 54\right) - \left(\frac{18\sqrt{3}}{3} - 6\sqrt{3}\right)\right]$$
$$= \frac{6}{5}\sqrt{3} + \frac{26}{15}$$

19. The graphs intersect at (x, y) if $x^3 = y = x^{1/3}$, or $x^3 - x^{1/3} = 0$, or $x^{1/3}(x^{8/3} - 1) = 0$, or $x = -1$, 0, or 1. Also $f(x) \ge g(x)$ on $[-1, 0]$ and $g(x) \ge f(x)$ on $[0, 1]$. Thus

$$A = \int_{-1}^{0} (x^3 - x^{1/3})\,dx + \int_0^1 (x^{1/3} - x^3)\,dx = (\tfrac{1}{4}x^4 - \tfrac{3}{4}x^{4/3})\Big|_{-1}^{0} + (\tfrac{3}{4}x^{4/3} - \tfrac{1}{4}x^4)\Big|_0^1$$
$$= \tfrac{1}{2} + \tfrac{1}{2} = 1$$

21. The graphs intersect at (x, y) if $x^2 + 1 = y = 2x + 9$, or $x^2 - 2x - 8 = 0$, or $x = -2$ or $x = 4$. Also $g(x) \ge f(x)$ on $[-2, 4]$. Thus

$$A = \int_{-2}^{4} [(2x + 9) - (x^2 + 1)]\,dx = \int_{-2}^{4} (2x - x^2 + 8)\,dx = (x^2 - \tfrac{1}{3}x^3 + 8x)\Big|_{-2}^{4}$$
$$= (16 - \tfrac{64}{3} + 32) - (4 + \tfrac{8}{3} - 16) = 36$$

23. The graphs intersect at (x, y) if $x^3 + 1 = y = (x + 1)^2$, or $x^3 - x^2 - 2x = 0$, or $x(x + 1)(x - 2) = 0$, or $x = -1$, 0, or 2. Also $f(x) \ge g(x)$ on $[-1, 0]$ and $g(x) \ge f(x)$ on $[0, 2]$. Thus

$$A = \int_{-1}^{0} [(x^3 + 1) - (x + 1)^2]\,dx + \int_0^2 [(x + 1)^2 - (x^3 + 1)]\,dx$$
$$= \int_{-1}^{0} (x^3 - x^2 - 2x)\,dx + \int_0^2 (-x^3 + x^2 + 2x)\,dx$$
$$= (\tfrac{1}{4}x^4 - \tfrac{1}{3}x^3 - x^2)\Big|_{-1}^{0} + (-\tfrac{1}{4}x^4 + \tfrac{1}{3}x^3 + x^2)\Big|_0^2$$
$$= -(\tfrac{1}{4} + \tfrac{1}{3} - 1) + (-4 + \tfrac{8}{3} + 4) = \tfrac{37}{12}$$

25. The graphs intersect at (x, y) if $6x = y^2 = (x^2/6)^2$, or $x^4 = 216x$, or $x = 0$ or 6. Thus the graphs intersect at $(0, 0)$ and $(6, 6)$, so

$$A = \int_0^6 \left(\sqrt{6x} - \frac{1}{6}x^2\right) dx = \left(\frac{2\sqrt{6}}{3}x^{3/2} - \frac{1}{18}x^3\right)\Bigg|_0^6 = \left(\frac{2\sqrt{6}}{3}\right)(6\sqrt{6}) - 12 = 12$$

27. The graphs intersect at (x, y) if $\frac{1}{2}(y^2 + 5) = x = y + 4$, or $y^2 - 2y - 3 = 0$, or $y = -1$ or 3. Thus the graphs intersect at $(3, -1)$ and $(7, 3)$. Thus

$$A = \int_{-1}^{3} [(y + 4) - \tfrac{1}{2}(y^2 + 5)]\, dy = \int_{-1}^{3} (y - \tfrac{1}{2}y^2 + \tfrac{3}{2})\, dy$$
$$= (\tfrac{1}{2}y^2 - \tfrac{1}{6}y^3 + \tfrac{3}{2}y)\Big|_{-1}^{3} = (\tfrac{9}{2} - \tfrac{9}{2} + \tfrac{9}{2}) - (\tfrac{1}{2} + \tfrac{1}{6} - \tfrac{3}{2}) = \tfrac{16}{3}$$

29. The graphs of $y = x + 2$ and $y = \frac{1}{3}(2 - x)$ intersect at $(-1, 1)$; the graphs of $y = x + 2$ and $y = -3x + 6$ intersect at $(1, 3)$; and the graphs of $y = \frac{1}{3}(2 - x)$ and $y = -3x + 6$ intersect at $(2, 0)$. We obtain

$$A = \int_{-1}^{1} [(x + 2) - \tfrac{1}{3}(2 - x)]\, dx + \int_{1}^{2} [(-3x + 6) - \tfrac{1}{3}(2 - x)]\, dx$$
$$= \int_{-1}^{1} (\tfrac{4}{3}x + \tfrac{4}{3})\, dx + \int_{1}^{2} (\tfrac{16}{3} - \tfrac{8}{3}x)\, dx$$
$$= \tfrac{1}{3}(2x^2 + 4x)\Big|_{-1}^{1} + \tfrac{1}{3}(16x - 4x^2)\Big|_{1}^{2} = \tfrac{8}{3} + \tfrac{4}{3} = 4$$

31. The graphs intersect at (x, y) if $y^2 - y = x = y - y^2$, or $2(y^2 - y) = 0$, or $y = 0$ or 1. Thus

$$A = \int_0^1 [(y - y^2) - (y^2 - y)]\, dy = \int_0^1 2(y - y^2)\, dy = (y^2 - \tfrac{2}{3}y^3)\Big|_0^1 = \tfrac{1}{3}$$

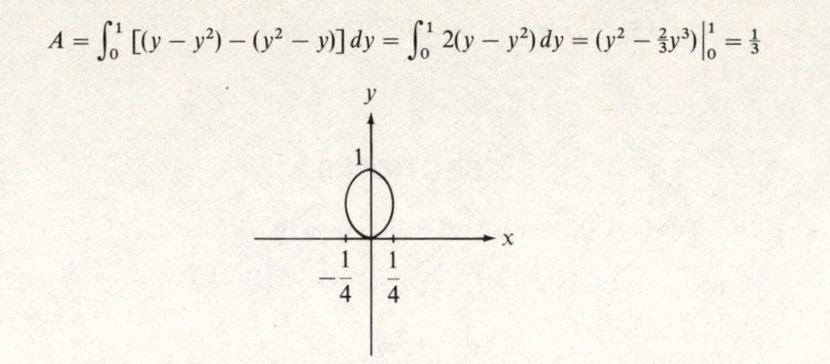

33. The graphs intersect at (x, y) if $y^2 = x = 6 - y - y^2$, or $2y^2 + y - 6 = 0$, or $(2y - 3)(y + 2) = 0$, or $y = -2$ or $\frac{3}{2}$. Thus

$$A = \int_{-2}^{3/2} [(6 - y - y^2) - y^2]\, dy = \int_{-2}^{3/2} (6 - y - 2y^2)\, dy = (6y - \tfrac{1}{2}y^2 - \tfrac{2}{3}y^3)\Big|_{-2}^{3/2}$$
$$= (9 - \tfrac{9}{8} - \tfrac{9}{4}) - (-12 - 2 + \tfrac{16}{3}) = \tfrac{343}{24}$$

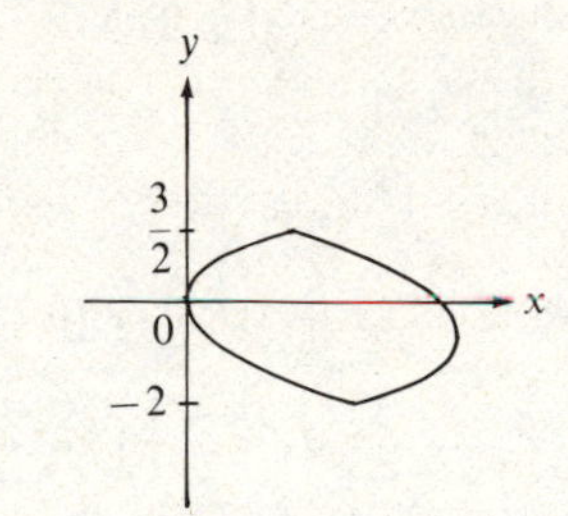

CHAPTER 5 REVIEW

1. $$L_f(\mathscr{P}) = f\left(\frac{3}{2}\right)\left(\frac{1}{2}\right) + f(2)\left(\frac{1}{2}\right) + f\left(\frac{5}{2}\right)\left(\frac{1}{2}\right) + f(3)\left(\frac{1}{2}\right)$$
$$= \frac{1}{3/2}\left(\frac{1}{2}\right) + \frac{1}{2}\left(\frac{1}{2}\right) + \frac{1}{5/2}\left(\frac{1}{2}\right) + \frac{1}{3}\left(\frac{1}{2}\right) = \frac{19}{20}$$
$$U_f(\mathscr{P}) = f(1)\left(\frac{1}{2}\right) + f\left(\frac{3}{2}\right)\left(\frac{1}{2}\right) + f(2)\left(\frac{1}{2}\right) + f\left(\frac{5}{2}\right)\left(\frac{1}{2}\right)$$
$$= \frac{1}{1}\left(\frac{1}{2}\right) + \frac{1}{3/2}\left(\frac{1}{2}\right) + \frac{1}{2}\left(\frac{1}{2}\right) + \frac{1}{5/2}\left(\frac{1}{2}\right) = \frac{77}{60}$$

3. $\int (x^{3/5} - 8x^{5/3})\, dx = \frac{5}{8}x^{8/5} - 8(\frac{3}{8}x^{8/3}) + C = \frac{5}{8}x^{8/5} - 3x^{8/3} + C$

5. $\int (x^3 - 3x + 2 - 2/x)\, dx = \frac{1}{4}x^4 - \frac{3}{2}x^2 + 2x - 2\ln|x| + C$

7. Let $u = 1 + \sqrt{x + 1}$, so that $du = 1/(2\sqrt{x + 1})\, dx$. Then

$$\int \frac{1 + \sqrt{x + 1}}{\sqrt{x + 1}}\, dx = \int u(2)\, du = 2\int u\, du = 2(\tfrac{1}{2}u^2) + C = (1 + \sqrt{x + 1})^2 + C$$

9. Let $u = \cos 3t$, so that $du = -3 \sin 3t\,dt$. Then

$$\int \cos^3 3t \sin 3t\,dt = \int u^3(-\tfrac{1}{3})\,du = -\tfrac{1}{3}\int u^3\,du = -\tfrac{1}{3}(\tfrac{1}{4}u^4) + C = -\tfrac{1}{12}\cos^4 3t + C$$

11. Let $u = 1 + \sqrt{x}$, so that $du = 1/(2\sqrt{x})\,dx$ and $2\sqrt{x} = 2(u-1)$. Then

$$\int \sqrt{1+\sqrt{x}}\,dx = \int \sqrt{1+\sqrt{x}}\,2\sqrt{x}\left(\frac{1}{2\sqrt{x}}\right)dx = \int \sqrt{u}[2(u-1)]\,du = \int 2(u^{3/2} - u^{1/2})\,du$$
$$= 2(\tfrac{2}{5}u^{5/2} - \tfrac{2}{3}u^{3/2}) + C = \tfrac{4}{5}(1+\sqrt{x})^{5/2} - \tfrac{4}{3}(1+\sqrt{x})^{3/2} + C$$

13. $$\int_{-1}^{-2}\left(x^{2/3} - \frac{5}{x^3}\right)dx = \left(\frac{3}{5}x^{5/3} + \frac{5}{2x^2}\right)\bigg|_{-1}^{-2}$$
$$= \frac{3}{5}[(-2)^{5/3} - (-1)^{5/3}] + \frac{5}{2}\left[\frac{1}{(-2)^2} - \frac{1}{(-1)^2}\right]$$
$$= \frac{3}{5}[(-2)2^{2/3} + 1] + \frac{5}{2}\left(\frac{1}{4} - 1\right) = -\frac{51}{40} - \frac{6}{5}(2^{2/3})$$

15. $$\int_{-5\pi/3}^{\pi/4} 5\cos x\,dx = 5\sin x\bigg|_{-5\pi/3}^{\pi/4} = 5\sin\frac{\pi}{4} - 5\sin\left(-\frac{5\pi}{3}\right) = \frac{5}{2}(\sqrt{2} - \sqrt{3})$$

17. $$\int_{-8}^{-2}\frac{-1}{5u}\,du = -\frac{1}{5}\ln|u|\bigg|_{-8}^{-2} = -\frac{1}{5}\ln 2 + \frac{1}{5}\ln 8 = -\frac{1}{5}\ln 2 + \frac{1}{5}\ln 2^3$$
$$= -\frac{1}{5}\ln 2 + \frac{3}{5}\ln 2 = \frac{2}{5}\ln 2$$

19. Let $u = x^3 + 9x + 1$, so that $du = (3x^2 + 9)\,dx = 3(x^2 + 3)\,dx$. If $x = 0$, then $u = 1$; if $x = 2$, then $u = 27$. Thus

$$\int_0^2 (x^2+3)(x^3+9x+1)^{1/3}\,dx = \int_1^{27} u^{1/3}\tfrac{1}{3}\,du = \tfrac{1}{3}\int_1^{27} u^{1/3}\,du$$
$$= \tfrac{1}{3}(\tfrac{3}{4}u^{4/3})\Big|_1^{27} = \tfrac{1}{4}(81 - 1) = 20$$

21. Let $u = 1 + \sin t$, so that $du = \cos t\,dt$. If $t = -\pi/4$ then $u = 1 - \sqrt{2}/2$, and if $t = \pi/2$ then $u = 2$. Thus

$$\int_{-\pi/4}^{\pi/2}\frac{\cos t}{1+\sin t}\,dt = \int_{1-\sqrt{2}/2}^{2}\frac{1}{u}\,du = \ln u\Big|_{1-\sqrt{2}/2}^{2}$$
$$= \ln 2 - \ln\left(\frac{2-\sqrt{2}}{2}\right) = 2\ln 2 - \ln(2 - \sqrt{2})$$

23. Let $u = x/(x+1)$, so that

$$du = \frac{(x+1) - x}{(x+1)^2}\,dx = \frac{1}{(x+1)^2}\,dx$$

If $x = \frac{1}{26}$ then $u = \frac{1}{27}$, and if $x = \frac{1}{7}$ then $u = \frac{1}{8}$. Thus

$$\int_{1/26}^{1/7}\frac{1}{x^2}\left(\frac{x+1}{x}\right)^{1/3}dx = \int_{1/26}^{1/7}\left(\frac{x+1}{x}\right)^2\left(\frac{x+1}{x}\right)^{1/3}\frac{1}{(x+1)^2}\,dx = \int_{1/27}^{1/8} u^{-7/3}\,du$$
$$= -\frac{3}{4}u^{-4/3}\bigg|_{1/27}^{1/8} = -\frac{3}{4}(16 - 81) = \frac{195}{4}$$

25. $$A = \int_2^4\left(\frac{7}{4}x^2\sqrt{x} + \frac{1}{\sqrt{x}}\right)dx = \int_2^4\left(\frac{7}{4}x^{5/2} + x^{-1/2}\right)dx = \left(\frac{1}{2}x^{7/2} + 2x^{1/2}\right)\bigg|_2^4$$
$$= (64 + 4) - (4\sqrt{2} + 2\sqrt{2}) = 68 - 6\sqrt{2}$$

27. $x^2 - 4x + 3 = (x-1)(x-3) \ge 0$ on $[0, 1]$ and $[3, 4]$, and $x^2 - 4x + 3 \le 0$ on $[1, 3]$. Thus

$$A = \int_0^1 (x^2 - 4x + 3)\,dx - \int_1^3 (x^2 - 4x + 3)\,dx + \int_3^4 (x^2 - 4x + 3)\,dx$$
$$= (\tfrac{1}{3}x^3 - 2x^2 + 3x)\Big|_0^1 - (\tfrac{1}{3}x^3 - 2x^2 + 3x)\Big|_1^3 + (\tfrac{1}{3}x^3 - 2x^2 + 3x)\Big|_3^4$$
$$= (\tfrac{1}{3} - 2 + 3) - [(9 - 18 + 9) - (\tfrac{1}{3} - 2 + 3)] + [(\tfrac{64}{3} - 32 + 12) - (9 - 18 + 9)] = 4$$

29. The graphs intersect at (x, y) if $2x^5 + 5x^4 = y = 2x^5 + 20x^2$, or $5x^4 - 20x^2 = 0$, or $x = -2$, 0 or 2. Also $g(x) \ge f(x)$ on $[-2, 2]$. Thus

$$A = \int_{-2}^{2}[(2x^5 + 20x^2) - (2x^5 + 5x^4)]\,dx = \int_{-2}^{2}(20x^2 - 5x^4)\,dx = (\tfrac{20}{3}x^3 - x^5)\Big|_{-2}^{2}$$
$$= (\tfrac{160}{3} - 32) - (-\tfrac{160}{3} + 32) = \tfrac{128}{3}$$

31. The graphs intersect at (x, y) if $2y^3 + y^2 + 5y - 7 = x = y^3 + 4y^2 + 3y - 7$, or $y^3 - 3y^2 + 2y = 0$, or $y = 0$, 1, or 2. Also $2y^3 + y^2 + 5y - 7 \ge y^3 + 4y^2 + 3y - 7$ on $[0, 1]$, and $y^3 + 4y^2 + 3y - 7 \ge 2y^3 + y^2 + 5y - 7$ on $[1, 2]$. Thus

$$A = \int_0^1 [(2y^3 + y^2 + 5y - 7) - (y^3 + 4y^2 + 3y - 7)]\,dy + \int_1^2 [(y^3 + 4y^2 + 3y - 7)$$
$$- (2y^3 + y^2 + 5y - 7)]\,dy = \int_0^1 (y^3 - 3y^2 + 2y)\,dy + \int_1^2 (-y^3 + 3y^2 - 2y)\,dy$$
$$= (\tfrac{1}{4}y^4 - y^3 + y^2)\Big|_0^1 + (-\tfrac{1}{4}y^4 + y^3 - y^2)\Big|_1^2$$
$$= (\tfrac{1}{4} - 1 + 1) + [(-4 + 8 - 4) - (-\tfrac{1}{4} + 1 + 1)] = \tfrac{1}{2}$$

33. The domain is the union of $(-\infty, -2)$ and $(2, \infty)$.

$$f'(x) = \frac{2x}{x^2 - 4}; \qquad f''(x) = \frac{-8 - 2x^2}{(x^2 - 4)^2}$$

decreasing on $(-\infty, -2)$ and increasing on $(2, \infty)$; concave downward on $(-\infty, -2)$ and $(2, \infty)$; vertical asymptotes are $x = -2$ and $x = 2$; symmetric with respect to the y axis

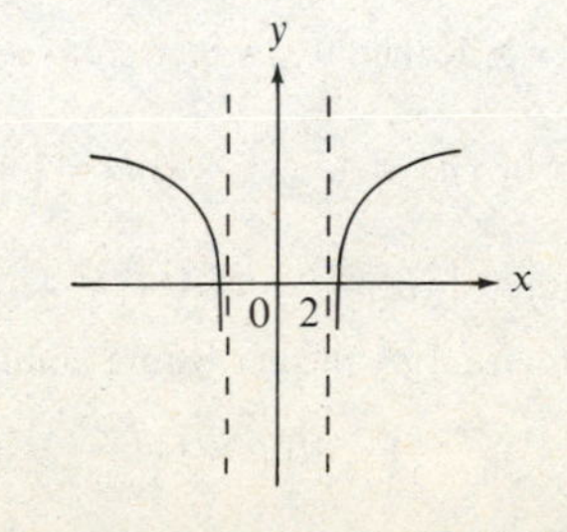

35. $F'(x) = x\sqrt{1 + x^5}$

37. Let $G(x) = \int_1^x (1/t)\,dt$, so that $F(x) = G(\ln x)$. Since $G'(x) = 1/x$, the Chain Rule implies that

$$F'(x) = [G'(\ln x)]\frac{1}{x} = \frac{1}{\ln x}\cdot\frac{1}{x} = \frac{1}{x \ln x}$$

39. $f'(x) = \left\{\sin(\ln x) + x[\cos(\ln x)]\frac{1}{x}\right\} - \left\{\cos(\ln x) + x[-\sin(\ln x)]\frac{1}{x}\right\} = 2\sin(\ln x)$

41. $f'(x) = \dfrac{1}{\ln\cos x + 2\ln\sec x}\left(\dfrac{-\sin x}{\cos x} + \dfrac{2\sec x\tan x}{\sec x}\right) = \dfrac{\tan x}{\ln\cos x + 2\ln\sec x}$.

Alternatively, $f(x) = \ln(-\ln\cos x)$, so

$$f'(x) = \frac{1}{-\ln\cos x}\cdot\frac{\sin x}{\cos x} = \frac{\tan x}{-\ln\cos x}$$

43. a. The integrals in (i) and (iii) cannot be easily evaluated by such a substitution. For the integral in (ii) let $u = x^2 + 6$, so that $du = 2x\,dx$. Then

$$\int x\sqrt{x^2 + 6}\,dx = \int \sqrt{u}\tfrac{1}{2}\,du = \tfrac{1}{3}u^{3/2} + C = \tfrac{1}{3}(x^2 + 6)^{3/2} + C$$

b. The integrals in (i) and (ii) cannot be easily evaluated by such a substitution. For the integral in (iii) let $u = \sqrt{x}$, so that $du = 1/(2\sqrt{x})\,dx$. Then

$$\int (1/\sqrt{x})\sin\sqrt{x}\,dx = \int (\sin u)2\,du = -2\cos u + C = -2\cos\sqrt{x} + C$$

c. The integrals in (i) and (ii) cannot be easily evaluated by such a substitution. For the integral in (iii) let $u = \ln(x + 1)$, so that $du = [1/(x + 1)]\,dx$. Then

$$\int \frac{\ln(x + 1)}{x + 1}\,dx = \int u\,du = \tfrac{1}{2}u^2 + C = \tfrac{1}{2}[\ln(x + 1)]^2 + C$$

45. By the General Comparison Property,

$$\frac{26}{3} = \frac{1}{3}x^3\Big|_1^3 = \int_1^3 \sqrt{x^4}\,dx \le \int_1^3 \sqrt{1 + x^4}\,dx \le \int_1^3 \sqrt{2x^4}\,dx = \frac{\sqrt{2}}{3}x^3\Big|_1^3 = \frac{26}{3}\sqrt{2}$$

47. a. If $1 \le t \le x$, then $1/t \le 1$, so the General Comparison Property implies that

$$\int_1^x (1/t)\,dt \le \int_1^x 1\,dt$$

b. Since $\int_1^x (1/t)\,dt = \ln x$ and $\ln x \ge 0$ for $x \ge 1$, and since $\int_1^x 1\,dt = x - 1$, part (a) implies that $0 \le \ln x \le x - 1$.

c. From (b) we find that if $x \ge 1$, then $0 \le x\ln x \le x(x - 1) = x^2 - x$. Thus

$$0 \le \int_1^2 x\ln x\,dx \le \int_1^2 (x^2 - x)\,dx = \left(\frac{x^3}{3} - \frac{x^2}{2}\right)\Big|_1^2 = \frac{5}{6}$$

49. $a(t) = -4$, so that $v(t) = v(0) + \int_0^t a(s)\,ds = 44 + \int_0^t -4\,ds = 44 - 4t$, so that $v(t) = 0$ if $t = 11$. Thus the distance traveled by the car before coming to a stop is

$$x(11) - x(0) = \int_0^{11} v(t)\,dt = \int_0^{11} (44 - 4t)\,dt = (44t - 2t^2)\Big|_0^{11} = 484 - 242 = 242 \text{ (feet)}$$

51. $f'(x) = \dfrac{-5(3\ln x - 2)}{(3x\ln x - 5x + 10)^2}$

so $f'(x_0) = 0$ if $\ln x_0 = \frac{2}{3}$, or $x_0 \approx 1.95$ (years). Since $f'(x) > 0$ for $x < x_0$ and $f'(x) < 0$ for $x > x_0$, it follows that $f(x_0)$ is the maximum value of f. Thus a child learns best at approximately 1.95 years of age.

53. Let H denote the average rate of heat production, so

$$H = \frac{\int_0^{1/60} (110\sin 120\pi t)^2 R\,dt}{1/60} = 60(110)^2 R\int_0^{1/60} \sin^2 120\pi t\,dt$$

Let $u = 120\pi t$, so that $du = 120\pi\,dt$. Then

$$H = \frac{60(110)^2 R}{120\pi}\int_0^{2\pi} \sin^2 u\,du = \frac{6050}{\pi}R\left(\frac{1}{2}u - \frac{1}{4}\sin 2u\right)\Big|_0^{2\pi} = 6050R$$

55. a. Since $dv/dm = -u_0/m$, integration yields $v = -u_0\ln m + C$. By hypothesis, at $t = 0$ we have $v = v_0$ and $m = m_0$. Therefore $v_0 = -u_0\ln m_0 + C$, so $C = v_0 + u_0\ln m_0$. Consequently $v = -u_0\ln m + (v_0 + u_0\ln m_0) = v_0 + u_0\ln(m_0/m)$.

b. By hypothesis, $u_0 = 3 \times 10^3$ and $v - v_0 = 3 \times 10^7$, so the equation

$$v = v_0 + u_0\ln(m_0/m)$$

becomes $3 \times 10^7 = (3 \times 10^3)\ln(m_0/m)$, or $10^4 = \ln(m_0/m)$. Thus $m_0/m = e^{(10^4)}$, so that $m/m_0 = e^{-(10^4)} \approx 0$.

CUMULATIVE REVIEW (CHAPTERS 1–4)

1. Since $2 + \sin x > 0$ and $(x - 3)^2 \ge 0$ for all x, the given inequality is equivalent to

$$-\frac{1}{(x - 3)^2}\left(\frac{2 - x}{4 - x}\right)^{-1} < 0, \quad\text{or}\quad \frac{x - 4}{(x - 3)^2(2 - x)} < 0$$

From the diagram we see that the solution is the union of $(-\infty, 2)$ and $(4, \infty)$.

$x - 4$	− − − − − − − − 0 + +
$2 - x$	+ + 0 − − − − − − − −
$(x - 3)^2$	+ + + + + 0 + + + + +
$\dfrac{x - 4}{(x - 3)^2(2 - x)}$	− − + + + + 0 − −
x	2 3 4

3. $$\lim_{x\to\pi/2^-} \frac{\cos x}{\sin x - 1} = \lim_{x\to\pi/2^-}\left(\frac{\cos x}{\sin x - 1}\,\frac{\sin x + 1}{\sin x + 1}\right) = \lim_{x\to\pi/2^-} \frac{\cos x(\sin x + 1)}{\sin^2 x - 1}$$

$$= \lim_{x\to\pi/2^-} \frac{\cos x(\sin x + 1)}{-\cos^2 x} = \lim_{x\to\pi/2^-} \frac{\sin x + 1}{-\cos x} = -\infty$$

5. Let $f(x) = \tan x$, so that $f'(x) = \sec^2 x$. Then

$$\lim_{x\to\pi/4} \frac{\tan x - 1}{x - \pi/4} = \lim_{x\to\pi/4} \frac{f(x) - f(\pi/4)}{x - \pi/4} = f'\left(\frac{\pi}{4}\right) = \sec^2\left(\frac{\pi}{4}\right) = 2$$

7. $f'(x) = -\left[\csc^2\left(\frac{1}{x^2+1}\right)\right]\frac{-2x}{(x^2+1)^2} = \frac{2x}{(x^2+1)^2}\csc^2\left(\frac{1}{x^2+1}\right)$

9. $f'(x) = 6x^2 - 18x + 14 = 6(x - \frac{3}{2})^2 + \frac{1}{2} > 0$ for all x, so f is strictly increasing.

11. Since $f'(x) = c + 3/x^4$, the equations $f(x) = 0$ and $f'(x) = 0$ become $cx - 1/x^3 - \frac{1}{2} = 0$ and $c + 3/x^4 = 0$. From the second of these we conclude that $cx = -3/x^3$, so the first becomes $-3/x^3 - 1/x^3 - \frac{1}{2} = 0$, or $x^3 = -8$, or $x = -2$. Since $cx = -3/x^3$, this implies that $c = -3/(-2)^4 = -\frac{3}{16}$. Thus if $c = -\frac{3}{16}$, then the equations $f(x) = 0$ and $f'(x) = 0$ have the same root, namely -2.

13.

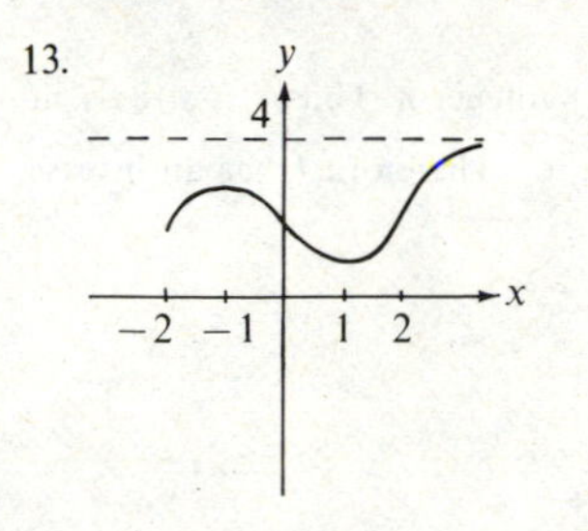

15. $f'(x) = \frac{1}{x^2} - \frac{2}{x^3}$; $\quad f''(x) = -\frac{2}{x^3} + \frac{6}{x^4}$

relative minimum value is $f(2) = \frac{3}{4}$; increasing on $(-\infty, 0)$ and $[2, \infty)$, and decreasing on $(0, 2]$; concave upward on $(-\infty, 0)$ and $(0, 3)$, and concave downward on $(3, \infty)$; inflection point is $(3, \frac{7}{9})$; vertical asymptote is $x = 0$; horizontal asymptote is $y = 1$

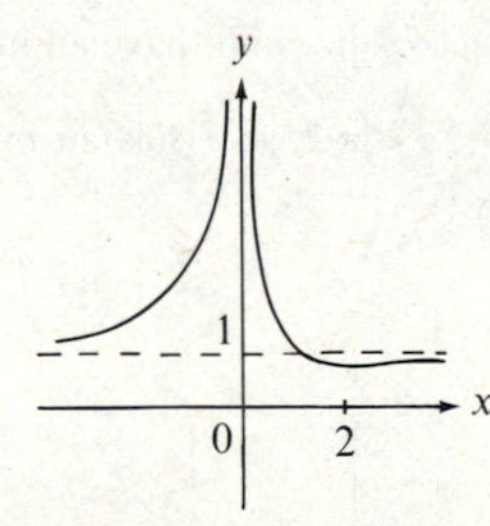

17. a. By (1) of Section 1.3, the height $h(t)$ of the flower pot t seconds after it is thrown is given by $h(t) = -16t^2 - 32t + 128$, so $v(t) = h'(t) = -32t - 32$. Thus $v(1) = -64$ (feet per second).

b. The pot hits the ground at the time $t > 0$ such that $h(t) = 0$, or $-16(t^2 + 2t - 8) = 0$, or $-16(t + 4)(t - 2) = 0$, or $t = 2$. Thus it takes 2 seconds for the pot to hit the ground.

19. From the given information it follows that if the toll is set at $3 + x$ dollars, then the number of cars using the toll road per day would be $24{,}000 - (20x)300 = 24{,}000 - 6000x$. Thus the total revenue R would be given by $R = (24{,}000 - 6000x)(x + 3) = -6000x^2 + 6000x + 72{,}000$. Then $R'(x) = -12{,}000x + 6000 = 0$ for $x = \frac{1}{2}$. Since $R''(x) = -12{,}000 < 0$, the revenue is maximized for $x = \frac{1}{2}$, which means that the toll would be \$3.50.

6
INVERSE FUNCTIONS

SECTION 6.1

1. $f'(x) = 1 > 0$, so f has an inverse. Domain of f^{-1}: $(-\infty, \infty)$; range of f^{-1}: $(-\infty, \infty)$.

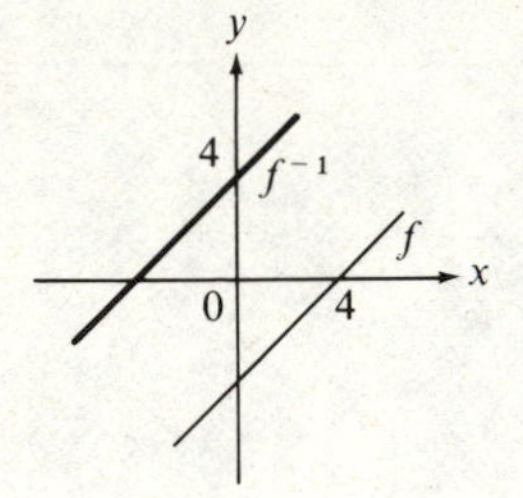

3. $f'(x) = 5x^4 \geq 0$, and $f'(x) > 0$ for $x \neq 0$, so f has an inverse. Domain of f^{-1}: $(-\infty, \infty)$; range of f^{-1}: $(-\infty, \infty)$.

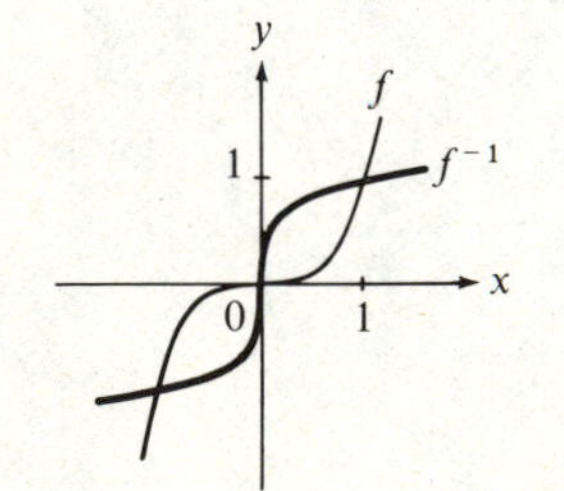

5. $f(-x) = f(x)$ for all x, so f does not have an inverse.

7. $g'(x) = \frac{1}{6}x^{-5/6} > 0$ for $x > 0$, so g has an inverse. Domain of g^{-1}: $[0, \infty)$; range of g^{-1}: $[0, \infty)$.

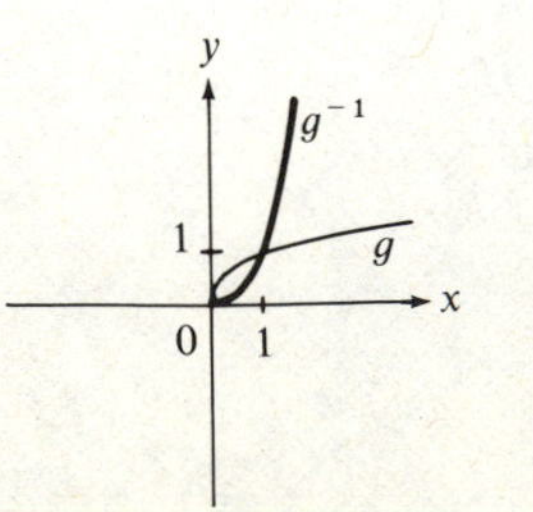

9. $f'(t) = -\frac{1}{2}(4 - t)^{-1/2} < 0$ for $t < 4$, so f has an inverse. Domain of f^{-1}: $[0, \infty)$; range of f^{-1}: $(-\infty, 4]$.

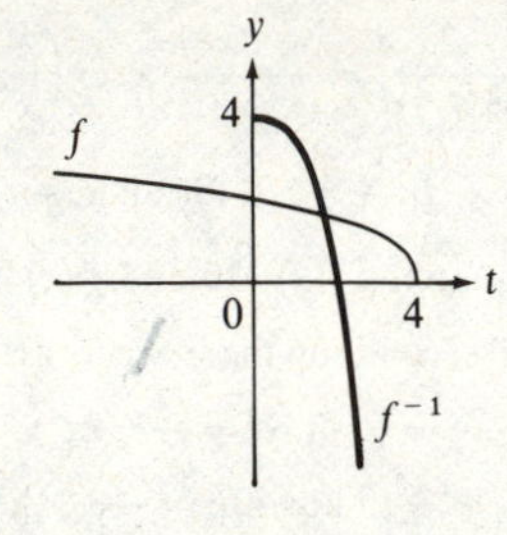

11. $f(x) = 0$ for $x \leq 0$, so f does not have an inverse.

13. $f'(x) = 1 - \cos x \geq 0$, and $f'(x) > 0$ for $x \neq 2n\pi$ for any integer n. Thus f is strictly increasing on any bounded interval and hence on $(-\infty, \infty)$. Therefore f has an inverse. Domain of f^{-1}: $(-\infty, \infty)$; range of f^{-1}: $(-\infty, \infty)$.

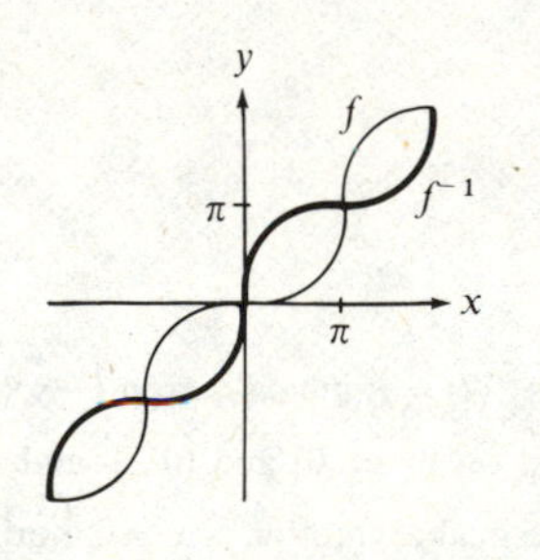

15. $f(n\pi) = 0$ for every integer n, so f does not have an inverse.

17. $f'(z) = \sec z \tan z > 0$ for $0 < z < \pi/2$, so f has an inverse. Domain of f^{-1}: $(1, \infty)$; range of f^{-1}: $(0, \pi/2)$.

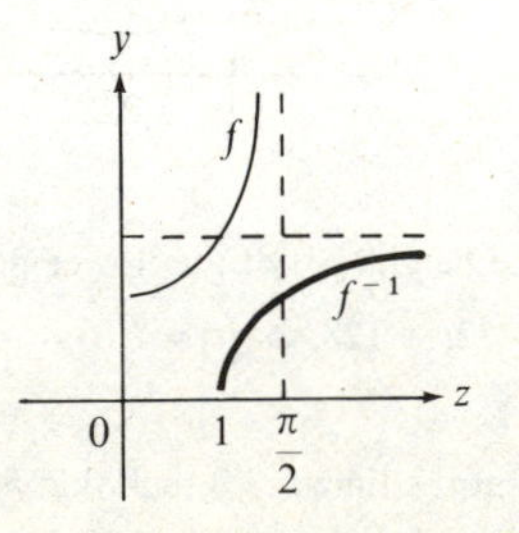

19. $k(-x) = k(x)$ for $x \neq 0$, so k does not have an inverse.

21. $f(2\pi + x) = f(x)$ for all x, so f does not have an inverse by Exercise 20.

23. $f(2\pi + x) = f(x)$ for all x in the domain of f, so f does not have an inverse by Exercise 20.

25. $y = -4x^3 - 1$; $x^3 = \dfrac{y+1}{-4}$; $x = -\sqrt[3]{\dfrac{y+1}{4}}$, so $f^{-1}(x) = -\sqrt[3]{\dfrac{x+1}{4}}$.

27. $y = \sqrt{1+x}$; $y^2 = 1 + x$; $x = y^2 - 1$, so $g^{-1}(x) = x^2 - 1$ for $x \geq 0$.

29. $y = \dfrac{1-2x}{5x}$; $5xy = 1 - 2x$; $x(5y+2) = 1$; $x = \dfrac{1}{5y+2}$, so $f^{-1}(x) = \dfrac{1}{5x+2}$.

31. $y = \dfrac{t-1}{t+1}$; $y(t+1) = t - 1$; $t(y-1) = -1 - y$; $t = \dfrac{y+1}{1-y}$, so $k^{-1}(t) = \dfrac{t+1}{1-t}$.

33. $f'(x) = 2x$. Since $f'(x) > 0$ for $x > 0$, f has an inverse on $[0, \infty)$. Since $f'(x) < 0$ for $x < 0$, f has an inverse on $(-\infty, 0]$.

35. $f'(t) = 4t^3 + 4t = 4t(t^2 + 1)$. Since $f'(t) > 0$ for $t > 0$, f has an inverse on $[0, \infty)$. Since $f'(t) < 0$ for $t < 0$, f has an inverse on $(-\infty, 0]$.

37. $f'(x) = 3x^2 - 5$. Since $f'(x) > 0$ for $x > \sqrt{\frac{5}{3}}$, f has an inverse on $[\sqrt{\frac{5}{3}}, \infty)$. Since $f'(x) > 0$ for $x < -\sqrt{\frac{5}{3}}$, f has an inverse on $(-\infty, -\sqrt{\frac{5}{3}}]$. Since $f'(x) < 0$ for $-\sqrt{\frac{5}{3}} < x < \sqrt{\frac{5}{3}}$, f has an inverse on $[-\sqrt{\frac{5}{3}}, \sqrt{\frac{5}{3}}]$.

39. $f'(u) = -2u/(1+u^2)^2$. Since $f'(u) > 0$ for $u < 0$, f has an inverse on $(-\infty, 0]$. Since $f'(u) < 0$ for $u > 0$, f has an inverse on $[0, \infty)$.

41. $f'(x) = \sec^2 x$ for $x \neq \pi/2 + n\pi$ for any integer n. Since $f'(x) > 0$ for $\pi/2 + n\pi < x < \pi/2 + (n+1)\pi$, f has an inverse on any interval of the form $(\pi/2 + n\pi, \pi/2 + (n+1)\pi)$, where n is an integer.

43. $f'(x) = -\csc x \cot x = \dfrac{-\cos x}{\sin^2 x}$. If n is any integer, then $f'(x) < 0$ for $2n\pi - \pi/2 < x < 2n\pi$ and for $2n\pi < x < 2n\pi + \pi/2$, and $f'(x) > 0$ for $2n\pi + \pi/2 < x < (2n+1)\pi$ and for $(2n+1)\pi < x < (2n+1)\pi + \pi/2$. Thus f has an inverse on any interval of the form $[n\pi - \pi/2, n\pi)$ or of the form $(n\pi, n\pi + \pi/2]$, where n is an integer.

45. Since $f'(x) = \sqrt{1+x^4} > 0$ for all x, f has an inverse.

47. By Definition 6.2, the range of f is the same as the domain of f^{-1}. If $f = f^{-1}$, then the domain of f^{-1} is the same as the domain of f. Therefore if $f = f^{-1}$, then the range of f is the same as the domain of f.

49. a. If f is a polynomial of even degree, then $f(-x) = f(x)$ for all x, so f cannot have an inverse.

    b. If f is a polynomial of odd degree, then f can have an inverse. An example is given by $f(x) = x^3 + 5x$.

51. Let $y = (g \circ f)(x)$. Then

$$(f^{-1} \circ g^{-1})(y) = (f^{-1} \circ g^{-1})((g \circ f)(x)) = f^{-1}(g^{-1}(g(f(x)))) = f^{-1}(f(x)) = x$$

Let $x = (f^{-1} \circ g^{-1})(y)$. Then

$$(g \circ f)(x) = (g \circ f)(f^{-1} \circ g^{-1})(y) = g(f(f^{-1}(g^{-1}(y)))) = g(g^{-1}(y)) = y$$

Thus $g \circ f$ has an inverse, and $(g \circ f)^{-1} = f^{-1} \circ g^{-1}$ by Definitions 6.1 and 6.2.

53. Since f^{-1} exists, Exercise 51 says that $g \circ f$ has an inverse, and since $f^{-1}(x) = -x$, we find that

$$(g \circ f)^{-1}(x) = (f^{-1} \circ g^{-1})(x) = f^{-1}(g^{-1}(x)) = -g^{-1}(x)$$

55. Let $k(x) = (1/a)f^{-1}(x)$. Then $g(k(x)) = f(a((1/a)f^{-1}(x))) = f(f^{-1}(x)) = x$, and $k(g(x)) = (1/a)f^{-1}(f(ax)) = (1/a)(ax) = x$. Thus g has an inverse, and $g^{-1}(x) = k(x) = (1/a)f^{-1}(x)$.

SECTION 6.2

1. $f(-1) = 6$. Since $f'(x) = 3x^2$, we have $f'(-1) = 3$. Thus $(f^{-1})'(6) = 1/[f'(-1)] = \frac{1}{3}$.

3. $f(0) = 0$. Since $f'(x) = 1 + \cos x$, we have $f'(0) = 2$. Thus $(f^{-1})'(0) = 1/[f'(0)] = \frac{1}{2}$.

5. $f(1) = 0$. Since $f(x) = 4/x$, we have $f'(1) = 4$. Thus $(f^{-1})'(0) = 1/[f'(1)] = \frac{1}{4}$.

7. $f(-1) = -2$. Since $f'(t) = 3 + 3/t^4$, we have $f'(-1) = 6$. Thus $(f^{-1})'(-2) = 1/[f'(-1)] = \frac{1}{6}$.

9. $\dfrac{dx}{dy} = \dfrac{1}{dy/dx} = \dfrac{1}{9x^8 + 7}$

11. $\dfrac{dx}{dy} = \dfrac{1}{dy/dx} = \dfrac{1}{[1/(x^3+1)]3x^2} = \dfrac{x^3+1}{3x^2}$

13. $\dfrac{dx}{dy} = \dfrac{1}{dy/dx} = \dfrac{1}{\cos x}$ for $-\dfrac{\pi}{2} < x < \dfrac{\pi}{2}$

15. Since $f'(x) = \sqrt{1+x^4}$, we have $f'(1) = \sqrt{2}$. Thus $(f^{-1})'(c) = 1/[f'(1)] = 1/\sqrt{2} = \sqrt{2}/2$.

17. $$\left.\frac{dy}{dx}\right|_{x=2} = 12 \quad \text{and} \quad \left.\frac{dx}{dy}\right|_{y=8} = \frac{1}{12} = \frac{1}{(dy/dx)|_{x=2}}$$

but

$$\left.\frac{dx}{dy}\right|_{y=2} = \frac{1}{3(2)^{2/3}} \neq \frac{1}{12} = \frac{1}{(dy/dx)|_{x=2}}$$

19. a. $f(\pi) = \pi$, $f'(x) = 1 + \cos x$, and $f'(\pi) = 1 - 1 = 0$. By Exercise 18, $(f^{-1})'(\pi)$ does not exist.

    b. $f(0) = -4$, $f'(x) = 5x^4 + 3x^2$, and $f'(0) = 5 \cdot 0 + 3 \cdot 0 = 0$. By Exercise 18, $(f^{-1})'(-4)$ does not exist.

21. Let $f(x) = [(x-1)^{1/3}+1]^{1/2}$. If $y = f(x)$, then $y^2 = (x-1)^{1/3}+1$, so $x = (y^2-1)^3+1$. Thus f^{-1} exists, and $f^{-1}(y) = (y^2-1)^3+1$. Now we use Exercise 20(b), with $a = 0$ and $b = 1$. Since $f(0) = 0$ and $f(1) = 1$, we obtain

$$\begin{aligned}\int_0^1 [(x-1)^{1/3}+1]^{1/2}\,dx &= \int_0^1 f(x)\,dx = 1f(1) - 0f(0) - \int_0^1 f^{-1}(y)\,dy \\ &= 1 - \int_0^1 [(y^2-1)^3+1]\,dy = 1 - \int_0^1 (y^6 - 3y^4 + 3y^2)\,dy \\ &= 1 - \left(\tfrac{1}{7}y^7 - \tfrac{3}{5}y^5 + y^3\right)\Big|_0^1 = \tfrac{16}{35}\end{aligned}$$

23. From (1) with $a = f^{-1}(x)$ and $c = x$, we have

$$(f^{-1})'(x) = \frac{1}{f'(f^{-1}(x))}$$

Differentiating both sides of this equation, we obtain

$$\begin{aligned}(f^{-1})''(x) &= \frac{-f''(f^{-1}(x))\cdot(f^{-1})'(x)}{[f'(f^{-1}(x))]^2} \\ &= \frac{-f''(f^{-1}(x))}{[f'(f^{-1}(x))]^2}\cdot\frac{1}{f'(f^{-1}(x))} = \frac{-f''(f^{-1}(x))}{[f'(f^{-1}(x))]^3}\end{aligned}$$

SECTION 6.3

1. By (9), $\ln\sqrt{e} = \ln(e^{1/2}) = \frac{1}{2}$.

3. By (9), $\ln e^e = e$.

5. By (8), $e^{\ln 3x} = 3x$.

7. By (14), $f'(x) = 4e^{4x}$.

9. By (14), $f'(x) = e^{(x^5)}(5x^4) = 5x^4e^{(x^5)}$.

11. $f'(x) = \dfrac{e^x(e^x-1)-(e^x+1)(e^x)}{(e^x-1)^2} = \dfrac{-2e^x}{(e^x-1)^2}$

13. By (14), $\dfrac{dy}{dz} = (\sec^2 e^{3z})(3e^{3z}) = 3e^{3z}\sec^2 e^{3z}$.

15. By (14), $\dfrac{dy}{dz} = (-e^{-z})\sin az + e^{-z}(a\cos az) = e^{-z}(-\sin az + a\cos az)$.

17. $2xe^y + x^2e^y\dfrac{dy}{dx} = \dfrac{1}{xy}\left(y + x\dfrac{dy}{dx}\right)$, so $\left(x^2e^y - \dfrac{1}{y}\right)\dfrac{dy}{dx} = \dfrac{1}{x} - 2xe^y$ and thus

$$\frac{dy}{dx} = \frac{1/x - 2xe^y}{x^2e^y - 1/y} = \frac{y - 2x^2ye^y}{x^3ye^y - x}$$

19. $f'(x) = 4e^{4x}$, $f''(x) = 16e^{4x}, \ldots, f^{(n)}(x) = 4^ne^{4x}$

21. $\dfrac{dy}{dx} = e^{2x} + 2xe^{2x} = (1+2x)e^{2x}$, so $x\dfrac{dy}{dx} = (xe^{2x})(1+2x) = y(1+2x)$.

23. $f'(x) = -e^{-x}$; $f''(x) = e^{-x}$; concave upward on $(-\infty, \infty)$; $\lim_{x\to\infty} e^{-x} = 0$, so that $y = 0$ is a horizontal asymptote.

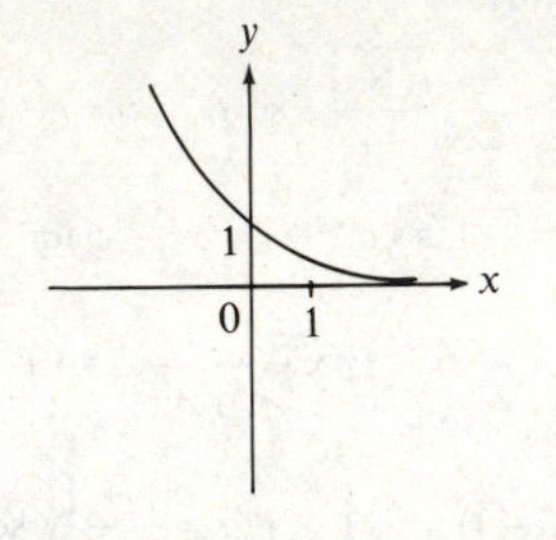

25. $f'(x) = e^x\cos\sqrt{3}x - \sqrt{3}e^x\sin\sqrt{3}x$; $f''(x) = -2e^x\cos\sqrt{3}x - 2\sqrt{3}e^x\sin\sqrt{3}x$; relative maximum values are

$$f\left(\frac{\pi(1+12n)}{6\sqrt{3}}\right) = \frac{\sqrt{3}}{2}e^{\pi(1+12n)/6\sqrt{3}} \quad \text{for any integer } n$$

relative minimum values are

$$f\left(\frac{\pi(7+12n)}{6\sqrt{3}}\right) = -\frac{\sqrt{3}}{2}e^{\pi(7+12n)/6\sqrt{3}} \quad \text{for any integer } n$$

increasing on $[\pi(7+12n)/(6\sqrt{3}), \pi(13+12n)/(6\sqrt{3})]$ and decreasing on $[\pi(1+12n)/(6\sqrt{3}), \pi(7+12n)/(6\sqrt{3})]$; concave upward on $(\pi(5+12n)/(6\sqrt{3}), \pi(11+12n)/(6\sqrt{3}))$ and concave downward on $(\pi(-1+12n)/(6\sqrt{3}), \pi(5+12n)/(6\sqrt{3}))$; inflection points are

$$\left(\frac{\pi(-1+12n)}{6\sqrt{3}}, \frac{\sqrt{3}}{2}e^{\pi(-1+12n)/6\sqrt{3}}\right) \quad\text{and}\quad \left(\frac{\pi(5+12n)}{6\sqrt{3}}, -\frac{\sqrt{3}}{2}e^{\pi(5+12n)/6\sqrt{3}}\right)$$

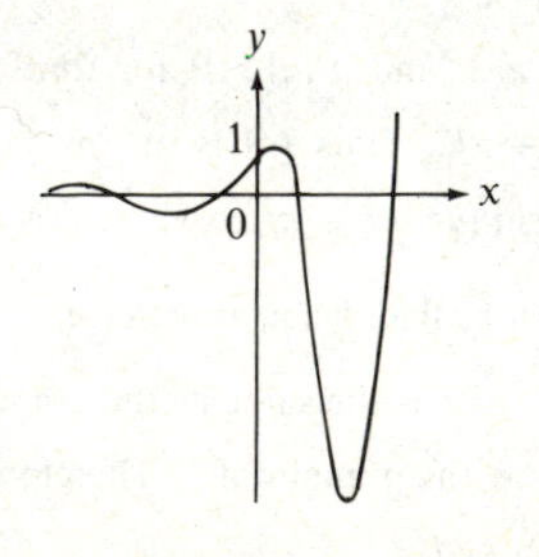

27. $f'(x) = \dfrac{e^x - e^{-x}}{e^x + e^{-x}}$; $f''(x) = \dfrac{4}{(e^x + e^{-x})^2}$;

$f'(x) = 0$ if $e^x = e^{-x}$, or $e^{2x} = 1$, or $x = 0$; $f''(0) = 1 > 0$; relative minimum value is $f(0) = \ln 2$; concave upward on $(-\infty, \infty)$; symmetric with respect to the y axis.

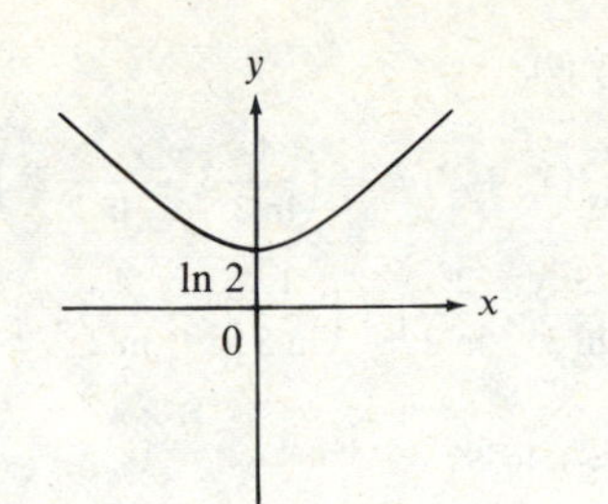

29. $\int e^{-4x}\,dx \overset{u=-4x}{=} \int -\frac{1}{4}e^u\,du = -\frac{1}{4}e^u + C = -\frac{1}{4}e^{-4x} + C$

31. $\int_0^{\sqrt{2}} ye^{(y^2)}\,dy \overset{u=y^2}{=} \int_0^2 \frac{1}{2}e^u\,du = \frac{1}{2}e^u\Big|_0^2 = \frac{1}{2}(e^2 - 1)$

33. $\int_0^{\pi/2} (\cos y)e^{\sin y}\,dy \overset{u=\sin y}{=} \int_0^1 e^u\,du = e^u\Big|_0^1 = e - 1$

35. $\int \frac{e^t}{e^t + 1}\,dt \overset{u=e^t+1}{=} \int \frac{1}{u}\,du = \ln|u| + C = \ln(e^t + 1) + C$

37. $\int \frac{e^{2t}}{\sqrt{e^{2t} - 4}}\,dt \overset{u=e^{2t}-y}{=} \int \frac{1}{2\sqrt{u}}\,du = \sqrt{u} + C = \sqrt{e^{2t} - 4} + C$

39. $\int \frac{e^{-t}\ln(1 + e^{-t})}{1 + e^{-t}}\,dt \overset{u=\ln(1+e^{-t})}{=} \int -u\,du = -\frac{1}{2}u^2 + C = -\frac{1}{2}[\ln(1 + e^{-t})]^2 + C$

41. $\int \frac{1}{e^{-x/2}}\,dx = \int e^{x/2}\,dx \overset{u=x/2}{=} \int 2e^u\,du = 2e^u + C = 2e^{x/2} + C$

43. $\int \frac{1}{1 + e^x}\,dx = \int \frac{e^{-x}}{e^{-x} + 1}\,dx \overset{u=e^{-x}+1}{=} \int -\frac{1}{u}\,du = -\ln|u| + C = -\ln(e^{-x} + 1) + C$

45. $\int e^{(x - e^x)}\,dx = \int e^x e^{-(e^x)}\,dx \overset{u=e^x}{=} \int e^{-u}\,du = -e^{-u} + C = -e^{-(e^x)} + C$

47. Since $f'(x) = -6e^{-3x}$, we have $f'(0) = -6e^{-3(0)} = -6$. Thus an equation of the line tangent to the graph of f at $(0, 2)$ is $y - 2 = -6(x - 0)$, or $y = -6x + 2$.

49. If the graphs intersect at (x, y), then $e^x = e^{1-x}$, so that $x = 1 - x$, and thus $x = \frac{1}{2}$. If $x = \frac{1}{2}$, then $f(x) = e^{1/2} = g(x)$. Thus the graphs intersect at $(\frac{1}{2}, e^{1/2})$.

51. If the graphs intersect at (x, y), then $e^{3x} = 2e^{-x}$, or $e^{4x} = 2$, so that $4x = \ln 2$, and thus $x = \frac{1}{4}\ln 2$. If $x = \frac{1}{4}\ln 2$, then $f(x) = e^{(3/4)\ln 2} = 2^{3/4}$ and $g(x) = 2e^{-(1/4)\ln 2} = 2(2^{-1/4}) = 2^{3/4}$. Thus the graphs intersect at $(\frac{1}{4}\ln 2, 2^{3/4})$.

53. Since $e^x > 0$ for all x and $\ln x \le 0$ for $0 < x \le 1$, any point of intersection must lie to the right of the line $x = 1$. However, $f(1) = e > 0 = g(1)$, and $f'(x) = e^x > 1 \ge 1/x = g'(x)$ for $x > 1$. Thus $f(x) > g(x)$ for $x > 1$, so that the graphs do not intersect.

55. $y = \frac{e^x - 1}{e^x + 1}$; $(e^x + 1)y = e^x - 1$; $e^x = \frac{1 + y}{1 - y}$; $x = \ln\left(\frac{1 + y}{1 - y}\right)$, so $f^{-1}(x) = \ln\left(\frac{1 + x}{1 - x}\right)$

57. Let $f(x) = e^{-x} - x$, so that $f'(x) = -e^{-x} - 1$. Using the Newton-Raphson method with initial value 0 for c, and using a computer, we find that 0.567143 is an approximate zero of f, and hence the desired approximate solution of $e^{-x} = x$.

59. $A = \int_0^1 e^{-x}\,dx = -e^{-x}\Big|_0^1 = -e^{-1} + 1 = 1 - e^{-1}$

61. The graphs intersect at (x, y) if $e^{2x} = y = e^{-2x}$, or $e^{4x} = 1$, or $x = 0$. Since $e^{2x} \ge e^{-2x}$ on $[0, \frac{1}{2}]$, we have

$$A = \int_0^{1/2} (e^{2x} - e^{-2x})\,dx = \left(\frac{1}{2}e^{2x} + \frac{1}{2}e^{-2x}\right)\Bigg|_0^{1/2} = \frac{1}{2}(e + e^{-1}) - 1$$

63. Let $f(x) = \ln x$, so that $f^{-1}(y) = e^y$. Also let $a = 1$ and $b = e$, so that $f(a) = \ln 1 = 0$ and $f(b) = \ln e = 1$. Therefore by the formula,

$$\int_1^e \ln x\,dx = e\ln e - 1\ln 1 - \int_0^1 f^{-1}(y)\,dy = e - \int_0^1 e^y\,dy = e - e^y\Big|_0^1 = e - (e - 1) = 1$$

65. $\frac{f(10)}{f(0)} = \frac{ce^{-14}}{c} = e^{-14} \approx 8.31529 \times 10^{-7}$

67. a. Comparing the formula for $f(t)$ with the formula for $f(x)$ in Exercise 66, we see that $a = 449$.

b. Comparing the formula for $f(t)$ with the formula for $f(x)$ in Exercise 66, we see that $b = e^{5.4094}$ and $ka = 1.0235$. The time t at which the population grows fastest occurs when $f''(t) = 0$. By the formula given in Exercise 66(b), this happens if $-1 + be^{-kat} = 0$, or $1 = be^{-kat}$, or $e^{kat} = b$, or $kat = \ln b$. Thus

$$t = \frac{\ln b}{ka} = \frac{\ln e^{5.4094}}{1.0235} = \frac{5.4094}{1.0235} \approx 5.28520 \text{ (days)}$$

69. a. $dy/dt = [c/(a - b)](-be^{-bt} + ae^{-at})$, so that $dy/dt = 0$ if $-be^{-bt} + ae^{-at} = 0$, or $be^{-bt} = ae^{-at}$, or $e^{t(a-b)} = a/b$, or $t = [1/(a - b)]\ln a/b$. Since $dy/dt > 0$ for $t < [1/(a - b)]\ln a/b$ and $dy/dt < 0$ for $t > [1/(a - b)]\ln a/b$, y has its maximum value at $[1/(a - b)]\ln a/b$ by Theorem 4.11. The maximum value is

$$\frac{c}{a - b}\left(e^{-b/(a-b)\ln a/b} - e^{-a/(a-b)\ln a/b}\right)$$

b. $\lim_{t\to\infty} y = \lim_{t\to\infty} \frac{c}{a - b}(e^{-bt} - e^{-at}) = \frac{c}{a - b}(0 - 0) = 0$

Thus the values of y approach 0 as t becomes very large.

SECTION 6.4

1. By (12), $\log_9 3 = \log_9 (9^{1/2}) = \frac{1}{2}$.

3. By (12), $\log_4 (4^x) = x$.

5. By (11), $7^{\log_7 2x} = 2x$.

7. By (5), $f'(x) = (\ln 5)5^x$.

9. By (5), $f'(x) = (\ln 3)3^{5x-7}(5) = (5\ln 3)3^{5x-7}$.

11. By (5), $g'(x) = (\ln a)a^x \sin bx + ba^x \cos bx$.

13. By (13), $g(x) = x \log_2 x = \dfrac{x \ln x}{\ln 2}$, so

$$g'(x) = \frac{1}{\ln 2}\left(\ln x + x \cdot \frac{1}{x}\right) = \frac{\ln x}{\ln 2} + \frac{1}{\ln 2} = \log_2 x + \frac{1}{\ln 2}.$$

15. By (8), $y = t^t = e^{t \ln t}$, so $dy/dt = e^{t \ln t}(\ln t + t \cdot 1/t) = t^t(\ln t + 1)$.

17. By (8), $y = t^{2/t} = e^{(2 \ln t)/t}$, so

$$\frac{dy}{dt} = e^{(2 \ln t)/t}\left(\frac{(2/t)t - 2 \ln t}{t^2}\right) = t^{2/t}\left(\frac{2 - 2 \ln t}{t^2}\right)$$

19. By (8), $f(x) = (\cos x)^{\cos x} = e^{(\cos x) \ln (\cos x)}$, so

$$f'(x) = e^{(\cos x) \ln (\cos x)}\left[(-\sin x) \ln (\cos x) + (\cos x)\left(\frac{-\sin x}{\cos x}\right)\right]$$

$$= -\sin x (\cos x)^{\cos x}[\ln (\cos x) + 1]$$

21. Using (7) and the Chain Rule, we have $f'(x) = \sqrt{2}(2x)^{\sqrt{2}-1}(2) = 2\sqrt{2}(2x)^{\sqrt{2}-1}$.

23. $f'(x) = \ln 2(2^x + 2^{-x})$; $f''(x) = (\ln 2)^2(2^x - 2^{-x})$; $f''(x) = 0$ if $2^x = 2^{-x}$, or $2^{2x} = 1$, or $x = 0$; $f''(x) < 0$ if $x < 0$ and $f''(x) > 0$ if $x > 0$; concave downward on $(-\infty, 0)$ and concave upward on $(0, \infty)$; inflection point: $(0, 0)$; symmetric with respect to the origin.

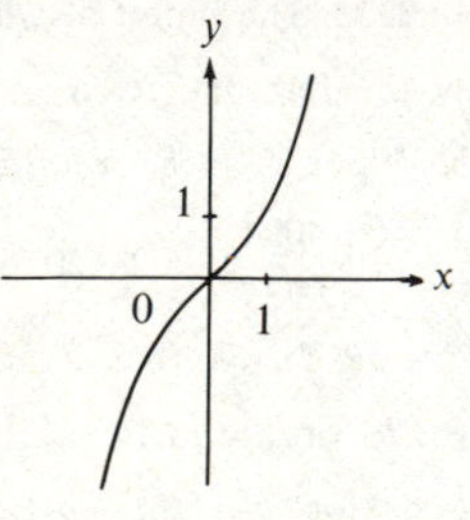

25. By (13), $\log_3 x = \log_2 x$ if and only if $(\ln x)/(\ln 3) = (\ln x)/(\ln 2)$, which holds if and only if $\ln x = 0$, so that $x = 1$. Thus the graphs intersect at $(1, 0)$.

27. By (9), $\int 2^x\,dx = \dfrac{1}{\ln 2} 2^x + C$.

29. By (9), $\int_{-2}^{0} 3^{-x}\,dx \overset{u=-x}{=} \int_{2}^{0} 3^u(-1)\,du = \dfrac{-1}{\ln 3} 3^u\Big|_2^0 = -\dfrac{1}{\ln 3}(1 - 3^2) = \dfrac{8}{\ln 3}$.

31. By (9), $\int x \cdot 5^{-x^2}\,dx \overset{u=-x^2}{=} \int -\dfrac{1}{2} 5^u\,du = -\dfrac{1}{2} \cdot \dfrac{1}{\ln 5} 5^u + C = \dfrac{-1}{2 \ln 5} 5^{-x^2} + C$.

33. By (10), $\int x^{2\pi}\,dx = \dfrac{1}{2\pi + 1} x^{2\pi+1} + C$.

35. By (9), $A = \int_1^2 x \cdot 2^{(x^2)}\,dx \overset{u=x^2}{=} \int_1^4 2^u(\tfrac{1}{2})\,du = \dfrac{1}{2 \ln 2} 2^u\Big|_1^4 = \dfrac{1}{2 \ln 2}(2^4 - 2^1) = \dfrac{7}{\ln 2}$.

37. $3^x \geq 2^x$ on $[0, 2]$, so by (9),

$$A = \int_0^2 (3^x - 2^x)\,dx = \left(\frac{1}{\ln 3} 3^x - \frac{1}{\ln 2} 2^x\right)\Big|_0^2$$

$$= \left(\frac{3^2}{\ln 3} - \frac{2^2}{\ln 2}\right) - \left(\frac{1}{\ln 3} 3^0 - \frac{1}{\ln 2} 2^0\right) = \frac{8}{\ln 3} - \frac{3}{\ln 2}$$

39. By (13), $\log_3 5 = \dfrac{\ln 5}{\ln 3} \approx 1.46497$.

41. By (13), $\log_\pi e = \dfrac{\ln e}{\ln \pi} = \dfrac{1}{\ln \pi} \approx 0.873569$.

43. Suppose $0 < b < a \leq e$. Since $(\ln x)/x$ is strictly increasing on $(0, e]$, we have $(\ln a)/a > (\ln b)/b$, so $b \ln a > a \ln b$ and thus $\ln a^b > \ln b^a$. Since $\ln x$ is strictly increasing, it follows that $a^b > b^a$. Now suppose $e \leq a < b$. Since $(\ln x)/x$ is strictly decreasing on $[e, \infty)$, it follows that $(\ln a)/a > (\ln b)/b$, so that $b \ln a > a \ln b$ and thus $\ln a^b > \ln b^a$. Because $\ln x$ is strictly increasing, we conclude that $a^b > b$.

45. By (13), $f(x) = (x \ln x)/\ln a$, so $f'(x) = (1/\ln a)(\ln x + x \cdot 1/x) = (1/\ln a)(\ln x + 1)$ and $f''(x) = 1/(x \ln a)$. Thus the graph of f is concave upward on $(0, \infty)$ if $\ln a > 0$, that is, $a > 1$, and the graph of f is concave downward on $(0, \infty)$ if $\ln a < 0$, that is, $0 < a < 1$.

47. $a^{b+c+d} = a^{(b+c)+d} = a^{b+c}a^d = (a^b a^c)a^d = a^b a^c a^d$

49. a. In Exercise 48(a), let $x = b$. Then $\log_b a \log_a b = \log_b b = 1$. Thus $\log_b a = 1/\log_a b$.
    b. Taking $b = 1/a$ in Exercise 48(a), we have $\log_{1/a} x = (\log_{1/a} a) \log_a x$. Since $a = (1/a)^{-1}$, it follows that $\log_{1/a} a = \log_{1/a} (1/a)^{-1} = (-1) \log_{1/a} (1/a) = -1$. Therefore

$$\log_{1/a} x = -\log_a x$$

51. Let x be the amplitude of the earthquake's largest wave 100 kilometers from the epicenter, and a the corresponding amplitude of a zero-level earthquake. By (14), $\log (x/a) = 2$, so $x/a = 10^2 = 100$. But $a = 0.001$. Therefore $x = 100\,a = 0.1$ (millimeters).

53. Let x_1 be the maximal amplitude of an earthquake of amplitude 8.5, and x_2 the maximal amplitude of an earthquake of amplitude 8.4. By (14), $8.5 = \log (x_1/a)$ and $8.4 = \log (x_2/a)$. Thus

$$\log \frac{x_2}{x_1} = \log \frac{x_2/a}{x_1/a} = \log \frac{x_2}{a} - \log \frac{x_1}{a} = 8.4 - 8.5 = -0.1$$

Therefore $x_2/x_1 = 10^{-0.1} \approx 0.794328$.

55. Let x_1 be the intensity of a whisper and x_2 the intensity of conversation. Then $20 = 10 \log (x_1/I_0)$ and $65 = 10 \log (x_2/I_0)$, so

$$\log \frac{x_1}{x_2} = \log \frac{x_1/I_0}{x_2/I_0} = \log \frac{x_1}{I_0} - \log \frac{x_2}{I_0} = \frac{20}{10} - \frac{65}{10} = -4.5$$

Therefore $x_2/x_1 \approx 10^{-4.5} \approx 3.16228 \times 10^{-5}$.

57. If $y = 1000x$, then

$$L(y) = L(x) = 10 \log \frac{1000x}{I_0} - 10 \log \frac{x}{I_0} = 10 \log \frac{1000x/I_0}{x/I_0} = 10 \log 1000 = 10(3) = 30$$

Thus the difference in the noise levels is 30 decibels.

59. If $L(y) = L(x) - 0.6$, then $10 \log (y/I_0) = 10 \log (x/I_0) - 0.6$, so $\log (x/I_0) - \log (y/I_0) = 0.06$. Thus

$$\log \frac{x}{y} = \log \frac{x/I_0}{y/I_0} = \log \frac{x}{I_0} - \log \frac{y}{I_0} = 0.06$$

so that $x/y = 10^{0.06}$. Therefore the ratio of the intensities is $10^{0.06} \approx 1.15$.

61. a. Taking the derivatives of both sides of (15) with respect to t, we obtain $N'(10^7 - P(t))(-P'(t)) = 10^7$. Since $P'(t) = 10^7 - P(t)$, this gives us $N'(10^7 - P(t))$. $(P(t) - 10^7) = 10^7$ for $t > 0$.

b. Let $x = 10^7 - P(t)$. Since $0 \le P(t) < 10^7$, we have $0 < x \le 10^7$. From part (a) we obtain $N'(x)(-x) = 10^7$, or $N'(x) = -10^7/x$ for $0 < x < 10^7$. Since $(d/dx)(-10^7 \ln x) = -10^7/x$, it follows from Theorem 4.7 that $N(x) = -10^7 \ln x + C$ for $0 < x < 10^7$, where C is a constant.

c. Since $N(10^7) = 0$, we have $0 = N(10^7) = -10^7 \ln 10^7 + C$, so that $C = 10^7 \ln 10^7$. Thus $N(x) = -10^7 \ln x + 10^7 \ln 10^7$ for $0 < x < 10^7$.

d. Let $b = e^{-1/10^7}$. By Exercise 49, we have

$$\log_b e = \frac{1}{\log_e b} = \frac{1}{-1/10^7} = -10^7$$

Thus by Exercise 48(a), $-10^7 \ln x = \log_b e \ln x = \log_b e \log_e x = \log_b x$.

e. Combining the results of (c) and (d), we have $N(x) = -10^7 \ln x + 10^7 \ln 10^7 = \log_b x + C$ for $0 < x < 10^7$, where $C = 10^7 \ln 10^7$.

SECTION 6.5

1. a. By (4) we have $f(t) = 4f(0)$ if $f(0)e^{kt} = 4f(0)$, or $e^{kt} = 4$. Since $k = \frac{1}{2} \ln 2$ by the solution of Example 1, this means that $e^{(t/2) \ln 2} = 4$, or $2^{t/2} = 4$. Thus $t/2 = 2$, or $t = 4$. Therefore it takes 4 days for the algae to quadruple in number.

b. As in (a), if $f(t) = 3f(0)$, then $2^{t/2} = 3$, so $(t/2) \ln 2 = \ln 3$, or $t = (2 \ln 3)/\ln 2 \approx 3.16993$. Therefore it takes approximately 3.16993 days for the algae to triple in number.

3. Let $f(t)$ be the number of beetles t days after there are 1200 beetles. Then $f(0) = 1200$, and we seek the value of t for which $f(t) = 1500$. Since the doubling time is 6 days and 20 hours, which is the same as $\frac{41}{6}$ days, we have $f(\frac{41}{6}) = 2f(0)$, so by (4), $2f(0) = f(\frac{41}{6}) = f(0)e^{41k/6}$. Thus $e^{41k/6} = 2$, so that $k = (6 \ln 2)/41$. If $f(t) = 1500$, then by (4), $1500 = f(0)e^{kt} = 1200e^{kt}$, or $e^{kt} = \frac{5}{4}$, so that $t = (1/k) \ln 1.25 = (41 \ln 1.25)/(6 \ln 2) \approx 2.19984$. Thus there were 1200 beetles approximately 2.19984 days ago.

5. Let $f(t)$ be the population in millions of the country with a doubling time of 20 years t years after its population is 50,000,000, and let $g(t)$ be the population of the country with a doubling time of 10 years. From the hypotheses we notice that $f(0) = 50$ and $g(0) = 20$. We seek the value of t for which $f(t) = g(t)$. By (4), there are constants k_1 and k_2 such that $f(t) = f(0)e^{k_1 t} = 50e^{k_1 t}$ and $g(t) = 20e^{k_2 t}$. Since $f(20) = 2f(0)$, we have $2f(0) = f(20) = f(0)e^{20k_1}$, or $e^{20k_1} = 2$, so that $k_1 = \frac{1}{20} \ln 2$. Similarly, since $g(10) = 2g(0)$, we have $k_2 = \frac{1}{10} \ln 2$. Now if $f(t) = g(t)$, then $50e^{k_1 t} = 20e^{k_2 t}$, or $e^{(k_2 - k_1)t} = 2.5$, so that

$$t = \frac{1}{k_2 - k_1} = \frac{1}{\frac{1}{10} \ln 2 - \frac{1}{20} \ln 2} \ln 2.5 = \frac{20 \ln 2.5}{\ln 2} \approx 26.4386$$

Thus it will be approximately 26.4386 years until the two countries have the same population.

7. a. The population halves in any time interval of duration $-(\ln 2)/k$ if for any time t, $f(t + (-\ln 2)/k) = \frac{1}{2} f(t)$. By hypothesis, $f(t)$ decays exponentially, so that $f(t - (\ln 2)/k) = f(0)e^{k[t - (\ln 2)/k]} = f(0)e^{kt - \ln 2} = f(0)e^{kt}e^{-\ln 2} = f(t)\frac{1}{2} = \frac{1}{2} f(t)$.

b. Since $h = -(\ln 2)/k$, we have $k = -(\ln 2)/h$, so that $f(t) = f(0)e^{kt} = f(0)e^{-(\ln 2)t/h} = f(0)(e^{-\ln 2})^{t/h} = f(0)(\frac{1}{2})^{t/h}$.

9. Let $f(t)$ be the amount of radium remaining after t years. Then $f(t) \approx f(0)e^{kt}$, where $1590 \approx (-\ln 2)/k$, or $k \approx (-\ln 2)/1590$, by Exercise 7. We seek t such that $f(t)/f(0) = \frac{9}{10}$. We have $\frac{9}{10} = f(t)/f(0) \approx e^{(-\ln 2/1590)t}$, so that $x \approx -1590/(\ln 2) \ln \frac{9}{10} \approx 241.685$ (years).

11. Let $f(t)$ be the amount of C^{14} present t years after 13,000 B.C. and let $g(t)$ be the amount of C^{14} present t years after 12,300 B.C. Then $f(t) = f(0)e^{kt}$ and $g(t) = g(0)e^{kt}$ for $t \ge 0$, where $k = -1.24 \times 10^{-4}$ (by Example 2) and $f(0) = g(0)$. The amounts present in 2000 A.D. are $f(15{,}000)$ and $g(14{,}300)$. Furthermore,

$$\frac{g(14{,}300)}{g(0)} - \frac{f(15{,}000)}{f(0)} = e^{14{,}300k} - e^{15{,}000k} \approx 0.169789 - 0.155673 = 0.014116$$

Thus the difference is approximately 1.4116%.

13. Let $f(t)$ be the amount (in milligrams) of iodine 131 t days after delivery. We are to determine $f(0)$, and are given that $f(2) = 100$ and $f(8.14) = \frac{1}{2} f(0)$. By (4), $\frac{1}{2} f(0) = f(8.14) = f(0)e^{8.14k}$, or $e^{8.14k} = \frac{1}{2}$, so that $k = (1/8.14) \ln \frac{1}{2}$. Also by (4) and the hypothesis, $100 = f(2) = f(0)e^{2k}$, so that $f(0) = 100e^{-2k} = 100e^{-(2/8.14) \ln (1/2)} = 100e^{(\ln 2)/4.07} \approx 118.567$. Thus approximately 118.567 milligrams of iodine 131 should be purchased.

15. By (8), $f(t) \approx e^{(-1.25 \times 10^{-4})t}$, so that $f(1600) \approx e^{(-1.25 \times 10^{-4})1600} = e^{-.2} \approx 0.818731$.

17. a. $p(0) \approx 29.92$ (inches of mercury)

b. $p(5) \approx (29.92)e^{(-.2)5} \approx 11.0070$ (inches of mercury)

c. $p(10) \approx (29.92)e^{(-.2)10} \approx 4.04923$ (inches of mercury)

19. Let $f(t)$ be the amount of sodium pentobarbitol in the blood stream after t hours. Then $f(t) = f(0)e^{kt}$ for $t \geq 0$. By hypothesis, $f(5) = \frac{1}{2}f(0)$, so that $k = (-\ln 2)/5$. To anesthetize a dog weighing 10 kilograms for one-half an hour we need $f(\frac{1}{2}) = 20(10) = 200$ milligrams, so that

$$200 = f(\tfrac{1}{2}) = f(0)e^{-[\ln 2)/5](1/2)} = f(0)e^{-(\ln 2)/10}$$

and thus $f(0) = 200e^{(\ln 2)/10} \approx 214.355$ (milligrams).

21. For $A(10) = 2S$, we need p to satisfy $2 = A(10)/S = e^{10p/100} = e^{p/10}$, and thus $p = 10 \ln 2 \approx 6.93147$ (%).

23. Let $f(t)$ be the amount of sugar after t minutes. Then $f(t) = f(0)e^{kt} = e^{kt}$ for $t \geq 0$. By hypothesis $f(1) = \frac{3}{4}$, so that $\frac{3}{4} = e^{1k} = e^k$, and thus $k = \ln \frac{3}{4}$. Consequently $f(t) = \frac{1}{2} = e^{(\ln 3/4)t}$ if $t = (\ln \frac{1}{2})/(\ln \frac{3}{4}) \approx 2.40942$ (minutes).

25. a. Since $f'(t) = kf(t)$ for $t > 0$, we have $k = f'(t)/f(t)$. Thus if $a > 0$, then $k(t-a) = \int_a^t k\,ds = \int_a^t [f'(s)/f(s)]\,ds = \ln f(s)\Big|_a^t = \ln f(t) - \ln f(a) = \ln [f(t)/f(a)]$ for $t > 0$.

b. From (a) it follows that $e^{k(t-a)} = e^{\ln (f(t)/f(a))} = [f(t)/f(a)]$, so that $f(t) = f(a)e^{k(t-a)}$ for $t > 0$. Holding t fixed and letting a tend to 0, we find that $f(t) = \lim_{a \to 0^+} f(a)e^{k(t-a)} = f(0)e^{k(t-0)} = f(0)e^{kt}$ for $t > 0$. Since $f(0) = f(0)e^{k0}$, we have $f(t) = f(0)e^{kt}$ for $t \geq 0$.

27. Since $D(t) = P(0) - P(t)$ and $P(0) = e^{\lambda t}P(t)$, we have $D(t) = P(t)(e^{\lambda t} - 1)$, or $D(t)/P(t) = e^{\lambda t} - 1$. Thus $\lambda t = \ln \left(\dfrac{D(t)}{P(t)} + 1\right)$, or $t = \dfrac{1}{\lambda} \ln \left(\dfrac{D(t)}{P(t)} + 1\right)$.

29. a. $t = (1.885)10^9 \ln \left[9.068\left(\dfrac{1.95 \times 10^{-12}}{2.885 \times 10^{-8}}\right) + 1\right] \approx 1.15499 \times 10^6$ (years)

b. $\dfrac{D(t)}{P(t)} = \dfrac{1}{9.068}(e^{4.19/1.885} - 1) \approx 0.907946$

SECTION 6.6

1. Since $\sin \dfrac{\pi}{3} = \dfrac{\sqrt{3}}{2}$ and $-\dfrac{\pi}{2} \leq \dfrac{\pi}{3} \leq \dfrac{\pi}{2}$, it follows that $\arcsin \dfrac{\sqrt{3}}{2} = \dfrac{\pi}{3}$.

3. Since $\cos \dfrac{\pi}{4} = \dfrac{\sqrt{2}}{2}$ and $0 \leq \dfrac{\pi}{4} \leq \pi$, it follows that $\arccos \dfrac{\sqrt{2}}{2} = \dfrac{\pi}{4}$.

5. Since $\tan \left(-\dfrac{\pi}{6}\right) = -\dfrac{1}{\sqrt{3}}$ and $-\dfrac{\pi}{2} < -\dfrac{\pi}{6} < \dfrac{\pi}{2}$, it follows that $\arctan \left(-\dfrac{1}{\sqrt{3}}\right) = -\dfrac{\pi}{6}$.

7. Since $\cot \dfrac{\pi}{6} = \sqrt{3}$ and $0 < \dfrac{\pi}{6} < \pi$, it follows that $\operatorname{arccot} \sqrt{3} = \dfrac{\pi}{6}$.

9. Since $\sec \dfrac{5\pi}{4} = -\sqrt{2}$ and $\pi \leq \dfrac{5\pi}{4} < \dfrac{3\pi}{2}$, it follows that $\operatorname{arcsec}(-\sqrt{2}) = \dfrac{5\pi}{4}$.

11. Since $\csc \dfrac{\pi}{3} = \dfrac{2\sqrt{3}}{3}$ and $0 < \dfrac{\pi}{3} < \dfrac{\pi}{2}$, it follows that $\operatorname{arccsc} \dfrac{2\sqrt{3}}{3} = \dfrac{\pi}{3}$.

13. Since $\cos \dfrac{5\pi}{6} = -\dfrac{\sqrt{3}}{2}$ and $0 \leq \dfrac{5\pi}{6} \leq \pi$, it follows that $\cos^{-1}\left(\dfrac{-\sqrt{3}}{2}\right) = \dfrac{5\pi}{6}$.

15. $\sin \left(\arcsin \left(-\dfrac{1}{2}\right)\right) = \sin \left(-\dfrac{\pi}{6}\right) = -\dfrac{1}{2}$

17. $\cos (\arctan (-1)) = \cos \left(-\dfrac{\pi}{4}\right) = \dfrac{\sqrt{2}}{2}$

19. $\tan (\operatorname{arcsec} \sqrt{2}) = \tan \dfrac{\pi}{4} = 1$

21. $\csc (\operatorname{arccot} (-\sqrt{3})) = \csc \dfrac{5\pi}{6} = 2$

23. $\arcsin \left(\cos \dfrac{\pi}{6}\right) = \arcsin \dfrac{1}{2}\sqrt{3} = \dfrac{\pi}{3}$

25. We will evaluate $\cos (\arcsin x)$ by evaluating $\cos y$ for the value of y in $[-\pi/2, \pi/2]$ such that $\arcsin x = y$, that is, $\sin y = x$. From the diagram we have $\cos (\arcsin x) = \cos y = \sqrt{1 - x^2}$.

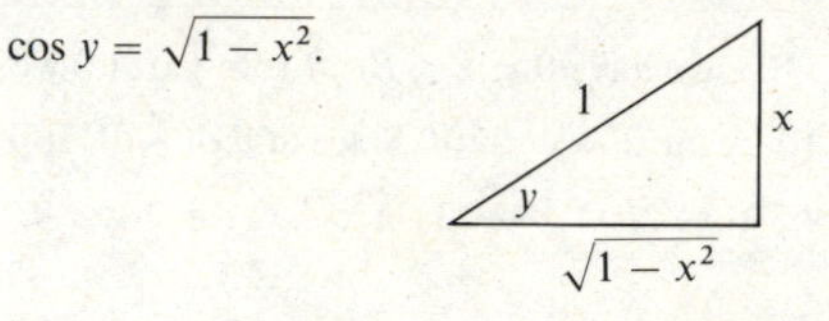

27. We will evaluate $\sec (\arctan x)$ by evaluating $\sec y$ for the value of y in $(-\pi/2, \pi/2)$ such that $\arctan x = y$, that is, $\tan y = x$. From the diagram we have $\sec (\arctan x) = \sec y = \sqrt{x^2 + 1}$.

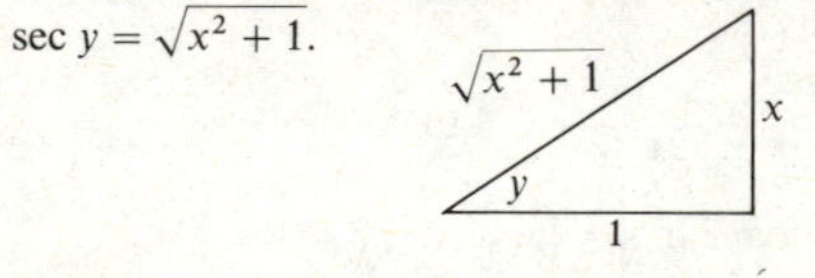

29. We will evaluate $\cos (\operatorname{arccot} x^2)$ by evaluating $\cos y$ for the value of y in $(0, \pi)$ such that $\operatorname{arccot} x^2 = y$, that is, $\cot y = x^2$. Since $x^2 \geq 0$, we have $0 < y \leq \pi/2$. We find from the diagram that $\cos (\operatorname{arccot} x^2) = \cos y = x^2/\sqrt{x^4 + 1}$.

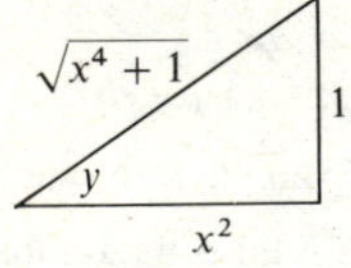

31. Using a trigonometric identity for $\cos 2a$ and the fact that $\cos (\arcsin x) = \sqrt{1 - x^2}$ (see Exercise 25), we find that $\cos (2 \arcsin x) = \cos^2 (\arcsin x) - \sin^2 (\arcsin x) = (\sqrt{1 - x^2})^2 - x^2 = (1 - x^2) - x^2 = 1 - 2x^2$.

33. Using (7), we find that $\dfrac{d}{dx} \arccos x = \dfrac{-1}{\sqrt{1 - x^2}}$, so that by the Chain Rule,

$$f'(x) = \frac{-1}{\sqrt{1 - (-3x)^2}}(-3) = \frac{3}{\sqrt{1 - 9x^2}}$$

35. Using (4), we find that $f'(t) = \dfrac{1}{(\sqrt{t})^2+1}\left(\dfrac{1}{2\sqrt{t}}\right) = \dfrac{1}{2(t+1)\sqrt{t}}$.

37. By (8), $\dfrac{d}{dx}\operatorname{arccot} x = \dfrac{-1}{x^2+1}$, so that if $f(x) = \operatorname{arccot}\sqrt{1-x^2}$, then

$$f'(x) = \frac{-1}{(\sqrt{1-x^2})^2+1}\left(\frac{-x}{\sqrt{1-x^2}}\right) = \frac{x}{(2-x^2)\sqrt{1-x^2}}$$

39. By (9), $\dfrac{d}{dx}\operatorname{arcsec} x = \dfrac{1}{x\sqrt{x^2-1}}$, so that

$$f'(x) = \frac{1}{\ln x\sqrt{(\ln x)^2-1}}\left(\frac{1}{x}\right) = \frac{1}{x\ln x\sqrt{(\ln x)^2-1}}$$

41. $\displaystyle\int \frac{1}{x^2+16}\,dx = \frac{1}{4}\arctan\frac{x}{4} + C$

43. $\displaystyle\int \frac{1}{9x^2+16}\,dx = \frac{1}{9}\int\frac{1}{x^2+\frac{16}{9}}\,dx = \frac{1}{9}\cdot\frac{3}{4}\arctan\frac{3x}{4} + C = \frac{1}{12}\arctan\frac{3x}{4} + C$

45. $\displaystyle\int \frac{1}{2x^2+4x+6}\,dx = \frac{1}{2}\int\frac{1}{(x+1)^2+2}\,dx \overset{u=x+1}{=} \frac{1}{2}\int\frac{1}{u^2+2}\,du$

$$= \frac{1}{2}\left(\frac{1}{\sqrt{2}}\arctan\frac{u}{\sqrt{2}}\right) + C = \frac{\sqrt{2}}{4}\arctan\frac{x+1}{\sqrt{2}} + C$$

47. $\displaystyle\int \frac{1}{\sqrt{9-4x^2}}\,dx = \frac{1}{2}\int\frac{1}{\sqrt{\frac{9}{4}-x^2}}\,dx = \frac{1}{2}\arcsin\frac{2x}{3} + C$

49. $\displaystyle\int \frac{1}{x\sqrt{x^2-25}}\,dx = \frac{1}{5}\operatorname{arcsec}\frac{x}{5} + C$

51. $\displaystyle\int \frac{x^3}{\sqrt{1-x^8}}\,dx \overset{u=x^4}{=} \frac{1}{4}\int\frac{1}{\sqrt{1-u^2}}\,du = \frac{1}{4}\arcsin u + C = \frac{1}{4}\arcsin x^4 + C$

53. $\displaystyle\int \frac{e^{-x}}{1+e^{-2x}}\,dx \overset{u=e^{-x}}{=} \int\frac{-1}{1+u^2}\,du = -\arctan u + C = -\arctan e^{-x} + C$

55. $\displaystyle\int \frac{\arctan 2x}{1+4x^2}\,dx \overset{u=\arctan 2x}{=} \frac{1}{2}\int u\,du = \frac{1}{4}u^2 + C = \frac{1}{4}(\arctan 2x)^2 + C$

57. $\displaystyle\int \frac{\cos t}{9+\sin^2 t}\,dt \overset{u=\sin t}{=} \int\frac{1}{9+u^2}\,du = \frac{1}{3}\arctan\frac{u}{3} + C = \frac{1}{3}\arctan\left(\frac{1}{3}\sin t\right) + C$

59. $\displaystyle\int \frac{\cos 4x}{(\sin 4x)\sqrt{16\sin^2 4x-4}}\,dx \overset{u=u\sin 4x}{=} \int\frac{1}{\frac{1}{4}u\sqrt{u^2-4}}\left(\frac{1}{16}\right)du = \frac{1}{4}\int\frac{1}{u\sqrt{u^2-4}}\,du$

$$= \frac{1}{8}\operatorname{arcsec}\frac{u}{2} + C = \frac{1}{8}\operatorname{arcsec}(2\sin 4x) + C$$

61. $\displaystyle\int_0^2 \frac{1}{\sqrt{16-x^2}}\,dx = \arcsin\frac{x}{4}\Big|_0^2 = \arcsin\tfrac{1}{2} - \arcsin 0 = \tfrac{1}{6}\pi$

63. $\displaystyle\int_{-2}^{2\sqrt{3}-2} \frac{1}{u^2+4u+8}\,du = \int_{-2}^{2\sqrt{3}-2}\frac{1}{(u+2)^2+4}\,du = \frac{1}{2}\arctan\left(\frac{u+2}{2}\right)\Big|_{-2}^{2\sqrt{3}-2}$

$$= \frac{1}{2}(\arctan\sqrt{3} - \arctan 0) = \tfrac{1}{6}\pi$$

65. The graphs intersect at (x, y) if $1/(x^2-2x+4) = y = \frac{1}{3}$, or $3 = x^2-2x+4$, or $x^2-2x+1 = 0$, or $x = 1$. Since $\frac{1}{3} \ge 1/(x^2-2x+4)$ on $[0, 1]$, we have

$$A = \int_0^1\left(\frac{1}{3} - \frac{1}{x^2-2x+4}\right)dx = \int_0^1\left(\frac{1}{3} - \frac{1}{(x-1)^2+3}\right)dx$$

$$= \left(\frac{x}{3} - \frac{1}{\sqrt{3}}\arctan\frac{x-1}{\sqrt{3}}\right)\Big|_0^1$$

$$= \left(\frac{1}{3}-0\right) - \left(0 - \frac{1}{\sqrt{3}}\arctan\frac{-1}{\sqrt{3}}\right) = \frac{1}{3} - \frac{1}{18}\pi\sqrt{3}$$

67. Let $f(x) = \arcsin x$, so that $f^{-1}(y) = \sin y$. Also let $a = 0$ and $b = 1$, so that $f(a) = \arcsin 0 = 0$ and $f(b) = \arcsin 1 = \pi/2$. Therefore by the formula,

$$\int_0^1 \arcsin x\,dx = 1\arcsin 1 - 0\arcsin 0 - \int_0^{\pi/2}\sin y\,dy = \frac{\pi}{2} + \cos y\Big|_0^{\pi/2} = \frac{\pi}{2} - 1$$

69. Let $f(x) = \arctan x$. Then $f'(x) = 1/(1+x^2)$ and $f''(x) = -2x/(1+x^2)^2$. Since $f'(0)$ exists, $f''(x) > 0$ for $x < 0$, and $f''(x) < 0$ for $x < 0$, there is an inflection point at $x = 0$.

71. $\displaystyle\frac{d}{dx}\left(\arctan\frac{x}{\sqrt{1-x^2}}\right) = \frac{1}{1+[x/\sqrt{1-x^2}]^2}\;\frac{\sqrt{1-x^2}+(x^2/\sqrt{1-x^2})}{1-x^2}$

$$= (1+x^2)\cdot\frac{1}{(1-x^2)^{3/2}} = \frac{1}{\sqrt{1-x^2}} = \frac{d}{dx}(\arcsin x)$$

Thus by Theorem 4.7 there is a constant C such that $\arctan(x/\sqrt{1-x^2}) = \arcsin x + C$. For $x = 0$ we obtain $0 = \arctan 0 = \arcsin 0 + C = C$, so that $C = 0$ and thus

$$\arctan\frac{x}{\sqrt{1-x^2}} = \arcsin x$$

73. a. $\displaystyle\arctan\frac{1}{2} + \arctan\frac{1}{3} = \arctan\frac{\frac{1}{2}+\frac{1}{3}}{1-(\frac{1}{2})(\frac{1}{3})} = \arctan 1 = \frac{\pi}{4}$

b. $\displaystyle 2\arctan\frac{1}{3} + \arctan\frac{1}{7} = \arctan\frac{\frac{1}{3}+\frac{1}{3}}{1-(\frac{1}{3})(\frac{1}{3})} + \arctan\frac{1}{7} = \arctan\frac{3}{4} + \arctan\frac{1}{7}$

$$= \arctan\frac{\frac{3}{4}+\frac{1}{7}}{1-(\frac{3}{4})(\frac{1}{7})} = \arctan 1 = \frac{\pi}{4}$$

c. $\displaystyle\arctan\frac{120}{119} - \arctan\frac{1}{239} = \arctan\frac{\frac{120}{119}-\frac{1}{239}}{1+(\frac{120}{119})(\frac{1}{239})} = \arctan 1 = \frac{\pi}{4}$

d. Using Exercise 72 twice, we find that

$$4\arctan\frac{1}{5} = 2\arctan\frac{\frac{1}{5}+\frac{1}{5}}{1-(\frac{1}{5})(\frac{1}{5})} = 2\arctan\frac{5}{12}$$

$$= \arctan\frac{\frac{5}{12}+\frac{5}{12}}{1-(\frac{5}{12})(\frac{5}{12})} = \arctan\frac{120}{119}$$

Now by (c) we obtain $4\arctan\frac{1}{5} - \arctan\frac{1}{239} = \arctan\frac{120}{119} - \arctan\frac{1}{239} = \pi/4$.

75. We will use trigonometric identities for $\cos 2a$ and $\sin(\pi/2 - a)$, along with the fact that $\cos(\arcsin x) = \sqrt{1-x^2}$ (see Exercise 25).

a. Notice that

$$\sin\left(\frac{\pi}{2} - 2\arcsin\sqrt{1-\frac{x}{6}}\right) = \cos\left(2\arcsin\sqrt{1-\frac{x}{6}}\right)$$

$$= \cos^2\left(\arcsin\sqrt{1-\frac{x}{6}}\right) - \sin^2\left(\arcsin\sqrt{1-\frac{x}{6}}\right)$$

$$= \left[\sqrt{1-\left(\sqrt{1-\frac{x}{6}}\right)^2}\right]^2 - \left(\sqrt{1-\frac{x}{6}}\right)^2 = \left[1-\left(1-\frac{x}{6}\right)\right] - \left(1-\frac{x}{6}\right) = \frac{x}{3} - 1$$

Since $-\pi/2 \le \pi/2 - 2\arcsin\sqrt{1-x/6} \le \pi/2$ it follows that $\arcsin(x/3 - 1) = \pi/2 - 2\arcsin\sqrt{1-x/6}$.

b. Notice that

$$\sin\left(2\arcsin\frac{\sqrt{x}}{\sqrt{6}} - \frac{\pi}{2}\right) = -\sin\left(\frac{\pi}{2} - 2\arcsin\frac{\sqrt{x}}{\sqrt{6}}\right)$$

$$= -\cos\left(2\arcsin\frac{\sqrt{x}}{\sqrt{6}}\right) = \sin^2\left(\arcsin\frac{\sqrt{x}}{\sqrt{6}}\right) - \cos^2\left(\arcsin\frac{\sqrt{x}}{\sqrt{6}}\right)$$

$$= \left(\frac{\sqrt{x}}{\sqrt{6}}\right)^2 - \left[\sqrt{1-\left(\frac{\sqrt{x}}{\sqrt{6}}\right)^2}\right]^2 = \frac{x}{6} - \left(1-\frac{x}{6}\right) = \frac{x}{3} - 1$$

Since $-\pi/2 \le 2\arcsin(\sqrt{x}/\sqrt{6}) - \pi/2 \le \pi/2$, it follows that $\arcsin(x/3 - 1) = 2\arcsin(\sqrt{x}/\sqrt{6}) - \pi/2$.

77. $$f'(x) = \frac{1}{1+x^2} + \frac{1}{1+1/x^2}\left(-\frac{1}{x^2}\right) = \frac{1}{1+x^2} - \frac{1}{1+x^2} = 0$$

on $(-\infty, 0)$ and on $(0, \infty)$. Thus on $(-\infty, 0)$ there is a constant c_1 such that $f(x) = c_1$; since $f(-1) = \arctan(-1) + \arctan(-1) = -\pi/4 - \pi/4 = -\pi/2$, it follows that on $(-\infty, 0)$, $f(x) = -\pi/2$. Similarly, on $(0, \infty)$ there is a constant c_2 such that $f(x) = c_2$; since $f(1) = \arctan 1 + \arctan 1 = \pi/4 + \pi/4 = \pi/2$, it follows that on $(0, \infty)$, $f(x) = \pi/2$.

79. Let x denote the distance in feet from the floor to the bottom of the painting, and let V denote the viewing angle. We are to maximize V. If the bottom of the painting is below eye level, as in the first figure, then

$$V = \theta + \phi = \arctan\frac{x}{10} + \arctan\frac{5-x}{10} = \arctan\frac{x}{10} - \arctan\frac{x-5}{10} \quad \text{for } x > 5$$

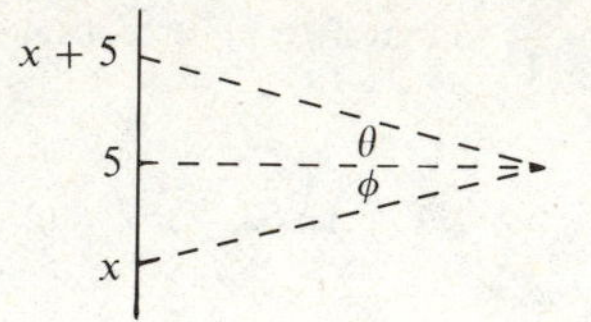

If the bottom of the painting is above eye level, as in the second figure, then

$$V = \theta = (\theta + \phi) - \phi = \arctan\frac{x}{10} - \arctan\frac{x-5}{10} \quad \text{for } 0 \le x \le 5$$

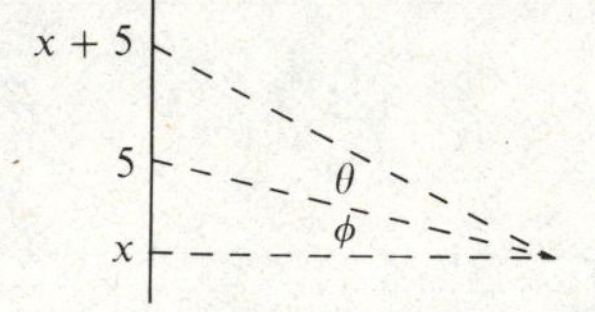

Therefore we are to maximize V with respect to x, where

$$V(x) = \arctan\frac{x}{10} - \arctan\frac{x-5}{10} \quad \text{with } 0 \le x$$

Now $$V'(x) = \frac{1}{1+(x/10)^2}\cdot\frac{1}{10} - \frac{1}{1+[(x-5)/10]^2}\cdot\frac{1}{10}$$

so that $V'(x) = 0$ if

$$1 + \frac{x^2}{10^2} = 1 + \frac{(x-5)^2}{10^2}$$

or $x^2 = (x-5)^2$, or $x^2 = x^2 - 10x + 25$, or $x = 2.5$. Since $V'(x) > 0$ for $0 \le x < 2.5$ and $V'(x) < 0$ for $x > 2.5$, the First Derivative Test implies that V is maximum for $x = 2.5$. Consequently the bottom of the painting should be 2.5 feet above the floor. If 10 is replaced by 15 (or any positive value), a similar analysis would lead again to the equation $x^2 = (x-5)^2$, so the answer would be the same.

SECTION 6.7

1. $$\sinh 0 = \frac{e^0 - e^{-0}}{2} = 0$$

3. $$\tanh 0 = \frac{\sinh 0}{\cosh 0} = \frac{e^0 - e^{-0}}{e^0 + e^{-0}} = \frac{0}{1} = 0$$

5. $$\coth(-1) = \frac{\cosh(-1)}{\sinh(-1)} = \frac{e^{-1} + e^1}{e^{-1} - e^1} = \frac{1 + e^2}{1 - e^2}$$

7. $\sinh(\ln 3) = \dfrac{e^{\ln 3} - e^{-\ln 3}}{2} = \dfrac{3 - \frac{1}{3}}{2} = \dfrac{4}{3}$

9. $\coth(\ln 4) = \dfrac{\cosh(\ln 4)}{\sinh(\ln 4)} = \dfrac{e^{\ln 4} + e^{-\ln 4}}{e^{\ln 4} - e^{-\ln 4}} = \dfrac{4 + \frac{1}{4}}{4 - \frac{1}{4}} = \dfrac{17}{15}$

11. $\operatorname{sech}(\ln\sqrt{2}) = \dfrac{1}{\cosh(\ln\sqrt{2})} = \dfrac{2}{e^{\ln\sqrt{2}} + e^{-\ln\sqrt{2}}} + \dfrac{2}{\sqrt{2} + 1/\sqrt{2}} = \dfrac{2\sqrt{2}}{3}$

13. $\sinh(\ln x) = \dfrac{e^{\ln x} - e^{-\ln x}}{2} = \dfrac{x - 1/x}{2} = \dfrac{x^2 - 1}{2x}$

15. $\tanh(\ln x) = \dfrac{\sinh(\ln x)}{\cosh(\ln x)} = \left(\dfrac{x^2 - 1}{2x}\right)\Big/\left(\dfrac{x^2 + 1}{2x}\right) = \dfrac{x^2 - 1}{x^2 + 1}$ from Exercises 13 and 14

17. $\dfrac{d}{dx}\tanh x = \dfrac{d}{dx}\dfrac{\sinh x}{\cosh x} = \dfrac{\cosh^2 x - \sinh^2 x}{\cosh^2 x} = \dfrac{1}{\cosh^2 x} = \operatorname{sech}^2 x$

19. $\dfrac{d}{dx}\operatorname{sech} x = \dfrac{d}{dx}\dfrac{1}{\cosh x} = \dfrac{-\sinh x}{\cosh^2 x} = -\operatorname{sech} x \tanh x$

21. $f'(x) = (-\operatorname{sech}\sqrt{x}\tanh\sqrt{x})\left(\dfrac{1}{2\sqrt{x}}\right) = \dfrac{-1}{2\sqrt{x}}\operatorname{sech}\sqrt{x}\tanh\sqrt{x}$

23. $f'(x) = (2\sinh\sqrt{1 - x^2}\cosh\sqrt{1 - x^2})\left(\dfrac{-x}{\sqrt{1 - x^2}}\right)$

$= \dfrac{-2x}{\sqrt{1 - x^2}}\sinh\sqrt{1 - x^2}\cosh\sqrt{1 - x^2}$

25. $f'(x) = \sinh(\arctan e^{2x})\left(\dfrac{1}{1 + e^{4x}}\right)(2e^{2x}) = \left(\dfrac{2e^{2x}}{1 + e^{4x}}\right)\sinh(\arctan e^{2x})$

27. Since $\dfrac{dy}{dx} = \dfrac{\dfrac{1}{\sqrt{1 + x^2}}\sqrt{1 + x^2} - (\sinh^{-1} x)\left(\dfrac{x}{\sqrt{1 + x^2}}\right)}{1 + x^2} = \dfrac{1}{1 + x^2} - \dfrac{xy}{1 + x^2}$,

it follows that $(1 + x^2)(dy/dx) = 1 - xy$, or $(1 + x^2)(dy/dx) + xy = 1$

29. $\int \operatorname{sech}^2 x\,dx = \tanh x + C$

31. $\int \operatorname{sech} x\,dx = \int \dfrac{\operatorname{sech} x(\operatorname{sech} x + \tanh x)}{\operatorname{sech} x + \tan x}\,dx = \int \dfrac{\operatorname{sech}^2 x + \tanh x \operatorname{sech} x}{\operatorname{sech} x + \tanh x}\,dx$

$= \ln|\operatorname{sech} x + \tanh x| + C$

33. $\int_5^{10} \dfrac{1}{\sqrt{x^2 + 1}}\,dx = \sinh^{-1} 10 - \sinh^{-1} 5 = \ln(10 + \sqrt{101}) - \ln(5 + \sqrt{26})$

$= \ln\left(\dfrac{10 + \sqrt{101}}{5 + \sqrt{26}}\right)$

35. $A = \int_{-4}^{4} 4\cosh\dfrac{x}{4}\,dx \overset{u = x/4}{=} 4\int_{-1}^{1}(\cosh u)\cdot 4\,du$

$= 16\sinh u\Big|_{-1}^{1} = 16(\sinh 1 - \sinh(-1)) = 32\sinh 1 = 16(e - e^{-1})$

37. If $x \geq 0$, then $\cosh x = \frac{1}{2}(e^x + e^{-x}) > \frac{1}{2}e^x = \frac{1}{2}e^{|x|}$.

If $x < 0$, then $\cosh x = \frac{1}{2}(e^x + e^{-x}) > \frac{1}{2}e^{-x} = \frac{1}{2}e^{|x|}$.

Thus $\cosh x > \frac{1}{2}e^{|x|}$ for all x.

39. a. From Exercise 38(a) we have $(\cosh x + \sinh x)^n = (e^x)^n = e^{nx} = \cosh nx + \sinh nx$ for all x.

b. From Exercise 38(b) we have $(\cosh x - \sinh x)^n = (e^{-x})^n = e^{-(nx)} = \cosh nx - \sinh nx$.

41. $\cosh^2 x + \sinh^2 x = \left(\dfrac{e^x + e^{-x}}{2}\right)^2 + \left(\dfrac{e^x - e^{-x}}{2}\right)^2 = \dfrac{e^{2x} + 2 + e^{-2x}}{4} + \dfrac{e^{2x} - 2 + e^{-2x}}{4}$

$= \dfrac{2e^{2x} + 2e^{-2x}}{4} = \dfrac{e^{2x} + e^{-2x}}{2} = \cosh 2x$

$2\sinh^2 x + 1 = 2\left(\dfrac{e^x - e^{-x}}{2}\right)^2 + 1 = \dfrac{e^{2x} - 2 + e^{-2x}}{2} + 1$

$= \dfrac{e^{2x} + e^{-2x}}{2} = \cosh 2x$

Thus $\cosh 2x = \cosh^2 x + \sinh^2 x = 2\sinh^2 x + 1$.

43. Let the shaded region have area A. Since the area of the triangle with vertices $(0, 0)$, $(\cosh t, 0)$, and $(\cosh t, \sinh t)$ is $\frac{1}{2}(\sinh t)(\cosh t)$, it follows that

$$A = \tfrac{1}{2}(\sinh t)(\cosh t) - \int_1^{\cosh t}\sqrt{x^2 - 1}\,dx$$

From Exercise 40 we know that $2\sinh t\cosh t = \sinh 2t$, and from Exercise 42, we have

$$\int_1^{\cosh t}\sqrt{x^2 - 1}\,dx = \frac{\sinh 2t}{4} - \frac{t}{2}$$

Thus $\quad A = \dfrac{1}{4}\sinh 2t - \left(\dfrac{\sinh 2t}{4} - \dfrac{t}{2}\right) = \dfrac{t}{2}$

45. a. $\sinh^{-1}\sqrt{x^2 - 1} = \ln\left[\sqrt{x^2 - 1} + \sqrt{(\sqrt{x^2 - 1})^2 + 1}\right]$

$= \ln(\sqrt{x^2 - 1} + \sqrt{x^2}) = \ln(\sqrt{x^2 - 1} + x)$

$= \cosh^{-1} x$ for $x \geq 1$

b. $\cosh^{-1}\sqrt{x^2 + 1} = \ln\left[\sqrt{x^2 + 1} + \sqrt{(\sqrt{x^2 + 1})^2 - 1}\right]$

$= \ln(\sqrt{x^2 + 1} + \sqrt{x^2}) = \ln(\sqrt{x^2 + 1} + x)$

$= \sinh^{-1} x$ for $x \geq 0$

SECTION 6.8

1. $\lim_{x\to a}(x^{16}-a^{16})=0=\lim_{x\to a}(x-a)$; $\lim_{x\to a}\dfrac{x^{16}-a^{16}}{x-a}=\lim_{x\to a}\dfrac{16x^{15}}{1}=16a^{15}$

3. $\lim_{x\to 0}(\cos x-1)=0=\lim_{x\to 0}x$; $\lim_{x\to 0}\dfrac{\cos x-1}{x}=\lim_{x\to 0}\dfrac{-\sin x}{1}=0$

5. $\lim_{x\to 0^+}\sin 8\sqrt{x}=0\ \lim_{x\to 0^+}\sin 5\sqrt{x}$; $\lim_{x\to 0^+}\dfrac{\sin 8\sqrt{x}}{\sin 5\sqrt{x}}=\lim_{x\to 0^+}\dfrac{(\cos 8\sqrt{x})(4/\sqrt{x})}{(\cos 5\sqrt{x})[5/(2\sqrt{x})]}$

$$=\lim_{x\to 0^+}\frac{8\cos 8\sqrt{x}}{5\cos 5\sqrt{x}}=\frac{8}{5}$$

7. $\lim_{x\to\infty}(e^{1/x}-1)=0=\lim_{x\to\infty}\dfrac{1}{x}$; $\lim_{x\to\infty}\dfrac{e^{1/x}-1}{1/x}=\lim_{x\to\infty}\dfrac{e^{1/x}(-1/x^2)}{-1/x^2}=\lim_{x\to\infty}e^{1/x}=1$

9. $\lim_{x\to\pi/2^-}\tan x=\infty=\lim_{x\to\pi/2^-}(\sec x+1)$;

$$\lim_{x\to\pi/2^-}\frac{\tan x}{\sec x+1}=\lim_{x\to\pi/2^-}\frac{\sec^2 x}{\sec x\tan x}=\lim_{x\to\pi/2^-}\frac{\sec x}{\tan x}=\lim_{x\to\pi/2^-}\frac{1}{\sin x}=1$$

11. $\lim_{x\to\infty}x=\infty=\lim_{x\to\infty}\ln x$; $\lim_{x\to\infty}\dfrac{x}{\ln x}=\lim_{x\to\infty}\dfrac{1}{1/x}=\lim_{x\to\infty}x=\infty$

13. $\lim_{x\to\infty}\ln(1+x)=\infty=\lim_{x\to\infty}\ln x$;

$$\lim_{x\to\infty}\frac{\ln(1+x)}{\ln x}=\lim_{x\to\infty}\frac{1/(1+x)}{1/x}=\lim_{x\to\infty}\frac{x}{1+x}=\lim_{x\to\infty}\frac{1}{(1/x)+1}=1$$

15. $\lim_{x\to 0}(1-\cos x)=0=\lim_{x\to 0}\sin x$; $\lim_{x\to 0}\dfrac{1-\cos x}{\sin x}=\lim_{x\to 0}\dfrac{\sin x}{\cos x}=0$

17. $\lim_{x\to\pi/2^-}\tan 4x=0=\lim_{x\to\pi/2^-}\tan 2x$; $\lim_{x\to\pi/2^-}\dfrac{\tan 4x}{\tan 2x}=\lim_{x\to\pi/2^-}\dfrac{4\sec^2 4x}{2\sec^2 2x}=2$

19. The conditions for applying l'Hôpital's Rule twice are met;

$$\lim_{x\to 0}\frac{1-\cos 2x}{1-\cos 3x}=\lim_{x\to 0}\frac{2\sin 2x}{3\sin 3x}=\lim_{x\to 0}\frac{4\cos 2x}{9\cos 3x}=\frac{4}{9}$$

21. $\lim_{x\to 0}(\sqrt{1+x}-\sqrt{1-x})=0=\lim_{x\to 0}x$;

$$\lim_{x\to 0}\frac{\sqrt{1+x}-\sqrt{1-x}}{x}=\lim_{x\to 0}\frac{1/(2\sqrt{1+x})+1/(2\sqrt{1-x})}{1}=1$$

23. $\lim_{x\to 0^+}\sin x=0=\lim_{x\to 0^+}(e^{\sqrt{x}}-1)$;

$$\lim_{x\to 0^+}\frac{\sin x}{e^{\sqrt{x}}-1}=\lim_{x\to 0^+}\frac{\cos x}{(1/2\sqrt{x})e^{\sqrt{x}}}=\lim_{x\to 0^+}\frac{2\sqrt{x}\cos x}{e^{\sqrt{x}}}=0$$

25. $\lim_{x\to 0}(5^x-3^x)=1-1=0=\lim_{x\to 0}x$; $\lim_{x\to 0}\dfrac{5^x-3^x}{x}=\lim_{x\to 0}\dfrac{(\ln 5)5^x-(\ln 3)3^x}{1}=\ln 5-\ln 3$

27. $\lim_{x\to 0^+}\ln|\ln x|=\infty=\lim_{x\to 0^+}\dfrac{1}{x}$;

$$\lim_{x\to 0^+}x\ln|\ln x|=\lim_{x\to 0^+}\frac{\ln|\ln x|}{1/x}=\lim_{x\to 0^+}\frac{1/(x\ln x)}{-1/x^2}=\lim_{x\to 0^+}\frac{-x}{\ln x}=0$$

29. $\lim_{x\to 1^-}\left(\dfrac{\pi}{2}-\arcsin x\right)=0=\lim_{x\to 1^-}\sqrt{1-x^2}$;

$$\lim_{x\to 1^-}\frac{\pi/2-\arcsin x}{\sqrt{1-x^2}}=\lim_{x\to 1^-}\frac{-1/\sqrt{1-x^2}}{-x/\sqrt{1-x^2}}=\lim_{x\to 1^-}\frac{1}{x}=1$$

31. $\lim_{x\to 1}(\ln x-x+1)=0=\lim_{x\to 1}(x^3-3x+2)$;

$$\lim_{x\to 1}\frac{\ln x-x+1}{x^3-3x+2}=\lim_{x\to 1}\frac{1/x-1}{3x^2-3}=\lim_{x\to 1}\frac{1-x}{3x(x-1)(x+1)}=\lim_{x\to 1}\frac{-1}{3x(x+1)}=\frac{-1}{6}$$

33. $\lim_{x\to 0}(\csc x-\cot x)=\lim_{x\to 0}\dfrac{1-\cos x}{\sin x}=0$ by Exercise 15

35. $\lim_{x\to 0^+}\ln(\sin x)=-\infty$; $\lim_{x\to 0^+}\csc x=\infty$

$$\lim_{x\to 0^+}\sin x\ln(\sin x)=\lim_{x\to 0^+}\frac{\ln(\sin x)}{\csc x}=\lim_{x\to 0^+}\frac{\cos x/\sin x}{-\csc x\cot x}=\lim_{x\to 0^+}(-\sin x)=0$$

37. $\lim_{x\to 0^+}\left(\ln\dfrac{1}{x}\right)^x=\lim_{x\to 0^+}e^{x\ln(\ln(1/x))}=e^{\lim_{x\to 0^+}x\ln(\ln(1/x))}=e^{\lim_{x\to 0^+}[\ln(\ln(1/x))/(1/x)]}$;

$$\lim_{x\to 0^+}\ln\left(\ln\frac{1}{x}\right)=\infty=\lim_{x\to 0^+}\frac{1}{x};$$

$$\lim_{x\to 0^+}\frac{\ln(\ln(1/x))}{1/x}=\lim_{x\to 0^+}\frac{\frac{1}{\ln(1/x)}\frac{1}{1/x}\left(\frac{-1}{x^2}\right)}{-1/x^2}=\lim_{x\to 0^+}\frac{x}{\ln(1/x)}=0$$

Thus $\lim_{x\to 0^+}(\ln(1/x))^x=e^0=1$.

39. $\lim_{x\to\pi/2^-}\ln(\cos x)=-\infty$, $\lim_{x\to\pi/2^-}\tan x=\infty$;

$$\lim_{x\to\pi/2^-}\frac{\ln\cos x}{\tan x}=\lim_{x\to\pi/2^-}\frac{(-\sin x)/(\cos x)}{\sec^2 x}=\lim_{x\to\pi/2^-}(-\sin x\cos x)=0$$

41. $\lim_{x\to\infty}x\sin\dfrac{1}{x}=\lim_{x\to\infty}\dfrac{\sin(1/x)}{1/x}$; $\lim_{x\to\infty}\sin\dfrac{1}{x}=0=\lim_{x\to\infty}\dfrac{1}{x}$;

$$\lim_{x\to\infty}x\sin\frac{1}{x}=\lim_{x\to\infty}\frac{\sin(1/x)}{1/x}=\lim_{x\to\infty}\frac{(-1/x^2)\cos(1/x)}{-1/x^2}=\lim_{x\to\infty}\cos\frac{1}{x}=1$$

43. $\lim_{x\to\infty}\dfrac{\ln(x^2+1)}{\ln x}=\lim_{x\to\infty}\dfrac{2x/(x^2+1)}{1/x}=\lim_{x\to\infty}\dfrac{2x^2}{x^2+1}=\lim_{x\to\infty}\dfrac{2}{1+1/x^2}=2$

45. $\lim_{x\to\infty}e^{(e^x)}=\infty=\lim_{x\to\infty}e^x$; $\lim_{x\to\infty}\dfrac{e^{(e^x)}}{e^x}=\lim_{x\to\infty}\dfrac{e^xe^{(e^x)}}{e^x}=\lim_{x\to\infty}e^{(e^x)}=\infty$

47. $\lim_{x\to\infty}\left(1+\frac{1}{x^2}\right)^x = \lim_{x\to\infty} e^{x\ln(1+1/x^2)} = e^{\lim_{x\to\infty} x\ln(1+1/x^2)} = \lim_{e^{x\to\infty}} \frac{\ln(1+1/x^2)}{1/x}$;

$\lim_{x\to\infty}\frac{\ln(1+1/x^2)}{1/x} = \lim_{x\to\infty}\frac{[x^2/(x^2+1)](-2/x^3)}{-1/x^2} = \lim_{x\to\infty}\frac{2x}{x^2+1} = \lim_{x\to\infty}\frac{2/x}{1+1/x^2} = 0$

Thus $\lim_{x\to\infty}(1+1/x^2)^x = e^0 = 1$.

49. $\lim_{x\to\infty}\ln(\ln x) = \infty = \lim_{x\to\infty}\sqrt{x}$;

$\lim_{x\to\infty} x^{-1/2}\ln(\ln x) = \lim_{x\to\infty}\frac{\ln(\ln x)}{\sqrt{x}} = \lim_{x\to\infty}\frac{[1/(\ln x)](1/x)}{1/(2\sqrt{x})} = \lim_{x\to\infty}\frac{2}{\sqrt{x}\ln x} = 0$

51. $\lim_{x\to a}(a^2-ax) = 0 = \lim_{x\to a}(a-\sqrt{ax})$; $\lim_{x\to a}\frac{a^2-ax}{a-\sqrt{ax}} = \lim_{x\to a}\frac{-a}{-a/(2\sqrt{ax})} = \lim_{x\to a} 2\sqrt{ax} = 2a$

53. $f'(x) = e^{-x} - xe^{-x}$; $f''(x) = -2e^{-x} + xe^{-x}$; relative maximum value is $f(1) = e^{-1}$; concave downward on $(-\infty, 2)$; concave upward on $(2, \infty)$; $\lim_{x\to\infty} xe^{-x} = \lim_{x\to\infty}(x/e^x) = \lim_{x\to\infty}(1/e^x) = 0$, so $y = 0$ is a horizontal asymptote. Also $\lim_{x\to-\infty} xe^{-x} = -\infty$.

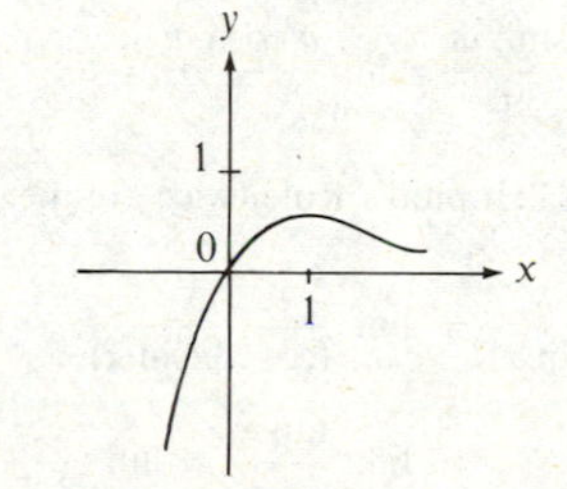

55. The conditions for applying l'Hôpital's Rule twice are met;

$$\lim_{x\to0^+} f(x) = \lim_{x\to0^+} x(\ln x)^2 = \lim_{x\to0^+}\frac{(\ln x)^2}{1/x} = \lim_{x\to0^+}\frac{(2\ln x)(1/x)}{-1/x^2}$$
$$= \lim_{x\to0^+}\frac{2\ln x}{-1/x} = \lim_{x\to0^+}\frac{2/x}{1/x^2} = \lim_{x\to0^+} 2x = 0 = f(0)$$

57. a. $\lim_{x\to\pi/2}\sin x \neq 0$ and $\lim_{x\to\pi/2} x \neq 0$. Thus l'Hôpital's Rule does not apply.

b. By the Quotient Rule for Limits, $\lim_{x\to\pi/2}\frac{\sin x}{x} = \frac{\lim_{x\to\pi/2}\sin x}{\lim_{x\to\pi/2} x} = \frac{\sin\pi/2}{\pi/2} = \frac{2}{\pi}$.

59. Applying l'Hôpital's Rule several times, we find that

$$\lim_{x\to0}\frac{e^{-1/x^2}}{x} = \lim_{x\to0}\frac{e^{-1/x^2}(2/x^3)}{1} = \lim_{x\to0}\frac{2e^{-1/x^2}}{x^3} = \lim_{x\to0}\frac{2e^{-1/x^2}(2/x^3)}{3x^2} = \lim_{x\to0}\frac{4e^{-1/x^2}}{3x^5}$$
$$= \lim_{x\to0}\frac{4e^{-1/x^2}(2/x^3)}{x5x^4} = \lim_{x\to0}\frac{8e^{-1/x^2}}{15x^7}$$

Thus each time we apply l'Hôpital's Rule we obtain a fraction with a higher power of x in the denominator, so that we are unable to evaluate the limit in this manner.

61. $\lim_{x\to0^+}(x^x)^x = \lim_{x\to0^+} e^{x\ln(x^x)} = \lim_{x\to0^+} e^{x^2\ln x} = e^{\lim_{x\to0^+} x^2\ln x}$

Since $\lim_{x\to0^+} x^2\ln x = \lim_{x\to0^+}(\ln x)/(1/x^2)$, and since $\lim_{x\to0^+}\ln x = -\infty$ and $\lim_{x\to0^+} 1/x_2 = \infty$. l'Hôpital's Rule implies that

$$\lim_{x\to0^+}\frac{\ln x}{1/x^2} = \lim_{x\to0^+}\frac{1/x}{-2/x^3} = \lim_{x\to0^+}\left(-\frac{x^2}{2}\right) = 0$$

Thus $\lim_{x\to0^+}(x^x)^x = e^0 = 1$.

63. Suppose $\lim_{x\to a} g(x) = 0$ and $\lim_{x\to a}\frac{f(x)}{g(x)}$ exists, with $\lim_{x\to a}\frac{f(x)}{g(x)} = L$.

It follows from the Product Rule for Limits that $\lim_{x\to a}(g(x))(f(x)/g(x))$ exists, and that

$$\lim_{x\to a} f(x) = \lim_{x\to a}\left(g(x)\frac{f(x)}{g(x)}\right) = 0\cdot L = 0$$

65. Since $f(1) - f(-1) = 1 - 1 = 0$ and $g(1) - g(-1) = 1 + 1 = 2$, it follows that $[f(1) - f(-1)]/[g(1) - g(-1)] = 0$. But $f'(c)/g'(c)$ could only be 0 if $f'(c) = 0$ and $g'(c) \neq 0$. However $f'(c) = 0$ only if $c = 0$, and in that case $g'(c) = g'(0) = 0$ also. Thus there is no c such that $f'(c)/g'(c) = 0$. The Generalized Mean Value Theorem is not violated since it is not true that $g'(x) \neq 0$ for $-1 < x < 1$. (In fact the conditions of Exercise 64 are not even met.)

67. If the vertex of R on the graph of e^{-x} is (z, e^{-z}), then the base of R has length z and the area A of R is given by ze^{-z}. Then $dA/dz = e^{-z} - ze^{-z} = (1-z)e^{-z}$. Since $dA/dz > 0$ if $0 < z < 1$ and $dA/dz < 0$ if $z > 1$, it follows that A is maximum if $z = 1$. Thus $A \leq 1\cdot e^{-1} = e^{-1}$.

CHAPTER 6 REVIEW

1. $f'(x) = 9x^2 + 25x^4 \geq 0$ for all x, and $f'(x) = 0$ only for $x = 0$, so f has an inverse.

3. $f'(x) = 1/x^2 > 0$ for x in $(-\infty, 0)$ and $(0, \infty)$, and $f(x) > 1$ for $x < 0$, whereas $f(x) < 1$ for $x > 0$. Thus f has an inverse.

5. $g(x + 2\pi) = g(x)$ for all x, so g does not have an inverse.

7. $y = \frac{3x-2}{-x+1}$; $(-x+1)y = 3x - 2$; $x(y+3) = y + 2$; $x = \frac{y+2}{y+3}$; $f^{-1}(x) = \frac{x+2}{x+3}$

9. Since $f(-x) = f(x)$ for every x in $(-a, a)$, f cannot have an inverse by (5) of Section 6.1.

11. $\frac{dy}{dx} = \frac{1}{1+2^x}(\ln 2)2^x = \frac{(\ln 2)2^x}{1+2^x}$

13. $\frac{dy}{dx} = \frac{1}{1+\sinh^2 x}\cosh x = \frac{\cosh x}{\cosh^2 x} = \frac{1}{\cosh x}$

15. $\dfrac{dy}{dx} = \dfrac{1}{\sqrt{1-(1-x^2)^{2/3}}}\,\dfrac{1}{3}(1-x^2)^{-2/3}(-2x) = \dfrac{-2x}{3(1-x^2)^{2/3}\sqrt{1-(1-x^2)^{2/3}}}$

17. Since $y = \dfrac{\ln(\arctan x^2)}{\ln 4}$, it follows that

$$\frac{dy}{dx} = \frac{1}{(\ln 4)(\arctan x^2)}\cdot\frac{2x}{1+x^4} = \frac{2x}{(\ln 4)(1+x^4)\arctan x^2}$$

19. $\displaystyle\int \frac{e^x}{\sqrt{1+e^x}}\,dx \overset{u=1+e^x}{=} \int \frac{1}{\sqrt{u}}\,du = 2\sqrt{u} + C = 2\sqrt{1+e^x} + C$

21. $\displaystyle\int \frac{e^x}{\sqrt{1+e^{2x}}}\,dx \overset{u=e^x}{=} \int \frac{1}{\sqrt{1+u^2}}\,du = \sinh^{-1} u + C = \sinh^{-1}(e^x) + C$

23. $\displaystyle\int \frac{e^x}{e^x+e^{-x}}\,dx \overset{u=e^x}{=} \int \frac{1}{u+1/u}\,du = \int \frac{u}{u^2+1}\,du \overset{v=u^2}{=} \int \frac{1}{v+1}\,\frac{1}{2}\,dv = \frac{1}{2}\int \frac{1}{v+1}\,dv$

$$= \tfrac{1}{2}\ln|v+1| + C = \tfrac{1}{2}\ln(u^2+1) + C = \tfrac{1}{2}\ln(e^{2x}+1) + C$$

25. $\displaystyle\int_0^1 x^2\, 5^{-x^3}\,dx \overset{u=x^3}{=} \int_0^1 5^{-u}\,\frac{1}{3}\,du = \left.\frac{-5^{-u}}{3\ln 5}\right|_0^1 = \frac{1}{3\ln 5}\left(1-\frac{1}{5}\right) = \frac{4}{15\ln 5}$

27. $\displaystyle\int \frac{3}{1+4t^2}\,dt \overset{u=2t}{=} \int \frac{3}{1+u^2}\,\frac{1}{2}\,du = \tfrac{3}{2}\arctan u + C = \tfrac{3}{2}\arctan 2t + C$

29. $\displaystyle\int_{-5/4}^{5/4} \frac{1}{\sqrt{25-4t^2}}\,dt = \frac{1}{2}\int_{-5/4}^{5/4} \frac{1}{\sqrt{25/4-t^2}}\,dt = \left.\frac{1}{2}\arcsin\frac{2t}{5}\right|_{-5/4}^{5/4}$

$$= \frac{1}{2}\left[\arcsin\frac{1}{2} - \arcsin\left(\frac{-1}{2}\right)\right] = \frac{\pi}{6}$$

31. $\displaystyle\int 2^x \sinh 2^x\,dx \overset{u=2^x}{=} \int \sinh u\cdot\frac{1}{\ln 2}\,du = \frac{\cosh u}{\ln 2} + C = \frac{\cosh 2^x}{\ln 2} + C$

33. $\displaystyle\int \frac{x}{x^4+4x^2+10}\,dx = \int \frac{x}{(x^2+2)^2+6}\,dx \overset{u=x^2+2}{=} \int \frac{1}{u^2+6}\,\frac{1}{2}\,du$

$$= \frac{1}{2}\,\frac{1}{\sqrt{6}}\arctan\frac{u}{\sqrt{6}} + C = \frac{\sqrt{6}}{12}\arctan\left(\frac{x^2+2}{\sqrt{6}}\right) + C$$

35. Let $y = \sinh^{-1} x$, so $x = \sinh y$. Note that $\cosh y = \sqrt{1+\sinh^2 y} = \sqrt{1+x^2}$. Then

$$\frac{d}{dx}\sinh^{-1} x = \frac{1}{\dfrac{d}{dy}\sinh y} = \frac{1}{\cosh y} = \frac{1}{\sqrt{1+x^2}}$$

37. a. Let $f(x) = e^x - x - 1$. Then $f'(x) = e^x - 1$ and $f''(x) = e^x$. Since $f'(x) = 0$ for $x = 0$ and $f''(x) > 0$ for all x, $f(0) = 0$ is the absolute minimum value of f. Thus $f(x) \ge f(0) = 0$ for all x, so $e^x \ge 1 + x$ for all x.

b. Since $e^x \ge 1 + x$ for all x, we must have $e^{-x} \ge 1 - x$ for all x, or $e^x \le 1/(1-x)$ for $0 \le x < 1$.

c. Since $e^x \ge 1 + x$ for all x and $e^x \le 1/(1-x)$ for $0 \le x < 1$, we must have $(x+1)/x = 1 + 1/x \le e^{1/x} \le 1/(1-1/x) = x/(x-1)$ for $x > 1$. The logarithm function is strictly increasing, and thus

$$\ln\left(\frac{x+1}{x}\right) \le \ln(e^{1/x}) = \frac{1}{x} \le \ln\left(\frac{x}{x-1}\right) \quad \text{for } x > 1$$

d. $\ln 1.05 = \ln\frac{21}{20} \le \frac{1}{20}$ (using (c) with $x = 20$), and $\ln 1.05 = \ln\frac{21}{20} \ge \frac{1}{21}$ (using (c) with $x = 21$). Thus $\frac{1}{21} \le \ln 1.05 \le \frac{1}{20}$.

39. a. For $0 < a < b$ the logarithmic function is continuous on $[a, b]$ and differentiable on (a, b). Thus there is a number c in (a, b) such that $(\ln b - \ln a)/(b-a) = 1/c$. Since $1/b < 1/c < 1/a$, we find that $1/b \le (\ln b - \ln a)/(b-a) \le 1/a$. Thus $(b-a)/b \le \ln b - \ln a \le (b-a)/a$. These inequalities also hold if $0 < a = b$, because then the three expressions are 0.

b. If we take $a = 1$ in part (a), then for $b \ge 1$ we have $(b-1)/b \le \ln b \le b - 1$.

c. If we take $b = 1.05$ in part (b), we obtain $\frac{1}{21} = (1.05-1)/1.05 \le \ln 1.05 \le 1.05 - 1 = \frac{1}{20}$.

41. $\displaystyle\lim_{t\to 0}\sinh at = 0 = \lim_{t\to 0} t$; $\displaystyle\lim_{t\to 0}\frac{\sinh at}{t} = \lim_{t\to 0}\frac{a\cosh at}{1} = a$

43. The conditions for applying l'Hôpital's Rule twice are met;

$$\lim_{x\to 0^+}\frac{\ln x}{\ln(\sin x)} = \lim_{x\to 0^+}\frac{1/x}{(\cos x)/(\sin x)} = \lim_{x\to 0^+}\frac{\tan x}{x} = \lim_{x\to 0^+}\frac{\sec^2 x}{1} = 1$$

45. $\displaystyle\lim_{x\to 0^+}\left(\frac{1}{x} - \frac{1}{\arctan x}\right) = \lim_{x\to 0^+}\frac{\arctan x - x}{x\arctan x}$;

the conditions for applying l'Hôpital's Rule twice are met;

$$\lim_{x\to 0^+}\left(\frac{1}{x} - \frac{1}{\arctan x}\right) = \lim_{x\to 0^+}\frac{\arctan x - x}{x\arctan x} = \lim_{x\to 0^+}\frac{\dfrac{1}{x^2+1} - 1}{\arctan x + \dfrac{x}{x^2+1}}$$

$$= \lim_{x\to 0^+}\frac{\dfrac{-2x}{(x^2+1)^2}}{\dfrac{1}{x^2+1} + \dfrac{1-x^2}{(x^2+1)^2}} = \lim_{x\to 0^+}(-x) = 0$$

47. $\displaystyle\lim_{x\to\infty}\left(\frac{x+1}{x-1}\right)^x = \lim_{x\to\infty} e^{x\ln[(x+1)/(x-1)]} = e^{\lim_{x\to\infty} x\ln[(x+1)/(x-1)]}$;

$\displaystyle\lim_{x\to\infty} x\ln\frac{x+1}{x-1} = \lim_{x\to\infty}\frac{\ln[(x+1)/(x-1)]}{1/x}$ and $\displaystyle\lim_{x\to\infty}\ln\frac{x+1}{x-1} = 0 = \lim_{x\to\infty}\frac{1}{x}$;

Thus

$$\lim_{x\to\infty} x \ln \frac{x+1}{x-1} = \lim_{x\to\infty} \frac{\ln \frac{x+1}{x-1}}{1/x} = \lim_{x\to\infty} \frac{\frac{1}{x+1} - \frac{1}{x-1}}{-1/x^2} = \lim_{x\to\infty} \frac{2x^2}{x^2-1} = 2$$

so that $\lim_{x\to\infty} \left(\frac{x+1}{x-1}\right)^x = e^2$.

49. $\lim_{x\to\infty} \frac{f(x)}{g(x)} = \lim_{x\to\infty} \frac{x+\sin x}{x} = \lim_{x\to\infty} \left(1 + \frac{\sin x}{x}\right) = 1 + 0 = 1$; $\frac{f'(x)}{g'(x)} = \frac{1+\cos x}{1} = 1 + \cos x$, but since $\lim_{x\to\infty} 1 = 1$ and $\lim_{x\to\infty} \cos x$ does not exist, it follows that $\lim_{x\to\infty} [f'(x)/g'(x)]$ does not exist.

51. $f'(x) = (2x)^x(\ln 2x + 1)$, $f''(x) = (2x)^x[(\ln 2x + 1)^2 + 1/x]$; relative minimum value is $f(\frac{1}{2}e^{-1}) = 1/e^{1/(2e)} \approx 0.831986$; concave upward on $(0, \infty)$. Finally $\lim_{x\to 0^+} f(x) = \lim_{x\to 0^+} (2x)^x = \lim_{x\to 0^+} 2^x \lim_{x\to 0^+} x^x$. Since $\lim_{x\to 0^+} 2^x = 2^0 = 1$, and since $\lim_{x\to 0^+} x^x = 1$ by Example 8 of Section 6.8, it follows that $\lim_{x\to 0^+} (2x)^x = 1 \cdot 1 = 1$. Also $\lim_{x\to\infty} (2x)^x = \infty$.

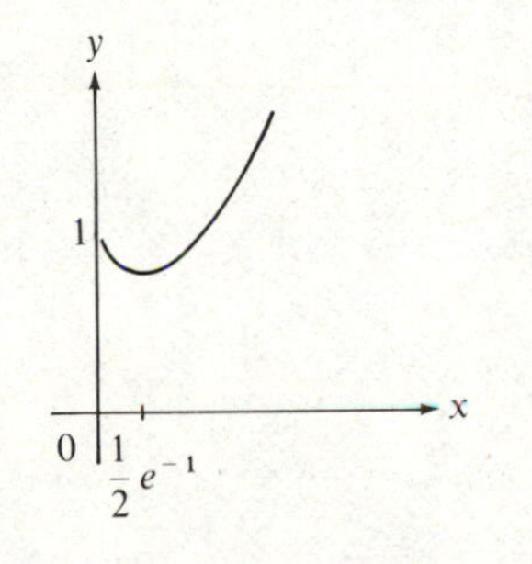

53. $f'(x) = -2e^{-2x} + 3e^{-3x}$, $f''(x) = 4e^{-2x} - 9e^{-3x}$; $f'(x) = 0$ if $2e^{-2x} = 3e^{-3x}$, so $x = \ln \frac{3}{2}$; $f''(\ln \frac{3}{2}) = 4(\frac{3}{2})^{-2} - 9(\frac{3}{2})^{-3} = -8/9 < 0$, so $f(\ln \frac{3}{2}) = \frac{4}{27}$ is a relative maximum value. Next, $f''(x) = 0$ if $4e^{-2x} = 9e^{-3x}$, so $x = \ln \frac{9}{4}$; graph is concave downward on $(0, \ln \frac{9}{4})$ and concave upward on $(\ln \frac{9}{4}, \infty)$, with inflection point $(\ln \frac{9}{4}, \frac{80}{729})$. Since $\lim_{x\to\infty} (e^{-2x} - e^{-3x}) = 0$, $y = 0$ is a horizontal asymptote.

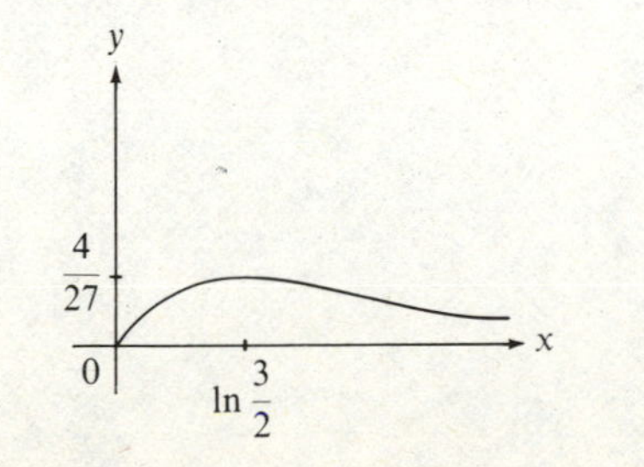

55. Since $x^2 \geq x^2 - 1$ on $[\sqrt{2}, 2]$, it follows that $f(x) \geq g(x)$ on $[\sqrt{2}, 2]$. Thus

$$A = \int_{\sqrt{2}}^{2} \left(\frac{x}{\sqrt{x^2-1}} - \frac{\sqrt{x^2-1}}{x}\right) dx = \int_{\sqrt{2}}^{2} \frac{x^2 - (x^2-1)}{x\sqrt{x^2-1}}\, dx = \int_{\sqrt{2}}^{2} \frac{1}{x\sqrt{x^2-1}}\, dx$$

$$= \operatorname{arcsec} x \Big|_{\sqrt{2}}^{2} = \operatorname{arcsec} 2 - \operatorname{arcsec} \sqrt{2} = \frac{\pi}{3} - \frac{\pi}{4} = \frac{\pi}{12}$$

57. Let x_1 be the intensity of an earthquake that measures 6.1 on the Richter scale, and M the magnitude of an earthquake that is 100 times as intense. By (14) of Section 6.4, we have

$$M - 6.1 = \log \frac{100\, x}{a} - \log \frac{x}{a} = \log \frac{100\, x/a}{x/a} = \log 100 = 2$$

Therefore $M = 2 + 6.1 = 8.1$.

59. a. Using the fact that the half-life of U^{238} is 4.5×10^9 years and the half-life of U^{235} is 7.1×10^8 years, we have $f(a) = f(0)(\frac{1}{2})^{a/(4.5\times 10^9)}$ and $g(a) = g(0)(\frac{1}{2})^{a/(7.1\times 10^8)}$. Thus $f(0)(\frac{1}{2})^{a/(4.5\times 10^9)} = (99.27/0.72)g(0)(\frac{1}{2})^{a/(7.1\times 10^8)}$, so that $(\frac{1}{2})^{a[1/(4.5\times 10^9) - 1/(7.1\times 10^8)]} = 99.27/0.72$, since $f(0) = g(0)$. Therefore

$$a\left(\frac{1}{7.1\times 10^8} - \frac{1}{4.5\times 10^9}\right) = \log_2 \frac{99.27}{0.72}$$

or

$$a = \frac{(4.5)(7.1)(10^9)}{37.9} \log_2 \frac{99.27}{0.72} \approx 5.99144 \times 10^9 \text{ (years)}$$

b. We have $f(a) = f(0)(\frac{1}{2})^{a/(4.5\times 10^9)}$ and $g(a) = g(0)(\frac{1}{2})^{a/(7.1\times 10^8)}$. If $a = 4.5 \times 10^9$, then

$$\frac{f(0)}{g(0)} = \frac{2f(a)}{2^{45/7.1}g(a)} = 2^{(-37.9/7.1)}\left(\frac{99.27}{0.72}\right)$$

$$= e^{(-37.9/7.1)\ln 2}\left(\frac{99.27}{0.72}\right) \approx 3.40862$$

CUMULATIVE REVIEW (CHAPTERS 1–5)

1. $$\lim_{x\to 3^+} \frac{\sqrt{x^2-9}}{x-3} = \lim_{x\to 3^+} \frac{\sqrt{x-3}\sqrt{x+3}}{(\sqrt{x-3})^2} = \lim_{x\to 3^+} \frac{\sqrt{x+3}}{\sqrt{x-3}} = \infty$$

3. Since

$$-|\sin x| \leq \sin x \sin \frac{1}{x} \leq |\sin x| \quad \text{and} \quad \lim_{x\to 0} |\sin x| = 0 = \lim_{x\to 0} (-|\sin x|)$$

the Squeezing Theorem implies that $\lim_{x\to 0} \sin x \sin (1/x) = 0$.

5. $$f'(x) = \sqrt{\ln x} + x \frac{1}{2\sqrt{\ln x}} \cdot \frac{1}{x} = \sqrt{\ln x} + \frac{1}{2\sqrt{\ln x}}$$

$$f'(e^4) = \sqrt{\ln (e^4)} + \frac{1}{2\sqrt{\ln (e^4)}} = \sqrt{4} + \frac{1}{2\sqrt{4}} = \frac{9}{4}$$

7. $f'(x) = \cos x - (x-1)\sin x$

$f''(x) = -\sin x - \sin x - (x-1)\cos x = -2\sin x - (x-1)\cos x$

$f^{(3)}(x) = -2\cos x - \cos x + (x-1)\sin x = -3\cos x + (x-1)\sin x$

$f^{(4)}(x) = 3\sin x + \sin x + (x-1)\cos x = 4\sin x + (x-1)\cos x$

In general,

$$f^{(2n)}(x) = (-1)^n[2n\sin x + (x-1)\cos x]$$

Thus $$f^{(24)}(x) = 24\sin x + (x-1)\cos x$$

9. If h denotes the height of the equilateral triangle, then the area A is given by $A = (\frac{1}{2}h)(2h/\sqrt{3}) = h^2/\sqrt{3}$. Since the area is by hypothesis growing at the rate of 9 square inches per minute, $9 = dA/dt = (2h/\sqrt{3})(dh/dt)$, so $dh/dt = (9\sqrt{3})/2h$. When $A = \sqrt{3}$, we have $\sqrt{3} = h^2/\sqrt{3}$, so $h = \sqrt{3}$, and thus $dh/dt = (9\sqrt{3})/(2\sqrt{3}) = \frac{9}{2}$ (inches per minute).

11. $f'(x) = 5x^4 - 10x^2 + 5 = 5(x^2-1)^2$; $f''(x) = 20x^3 - 20x = 20x(x^2-1)$; increasing on $(-\infty, \infty)$; concave upward on $(-1, 0)$ and $(1, \infty)$, and concave downward on $(-\infty, -1)$ and $(0, 1)$; inflection points are $(-1, -\frac{8}{3})$, $(0, 0)$, and $(1, \frac{8}{3})$; symmetry with respect to the origin.

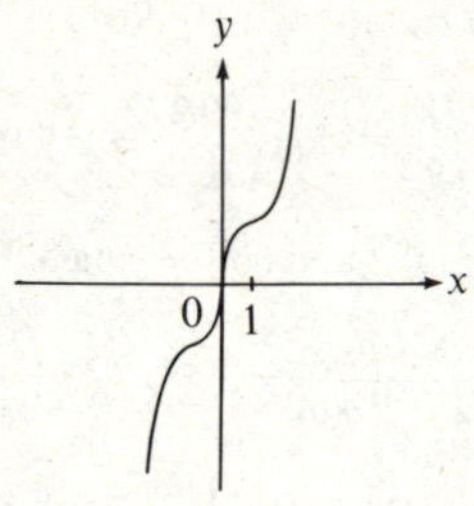

13. $a(t) = \frac{1}{2}t + \cos t$, and we are to find the maximum value of a on $[0, \pi/2]$. Now $a'(t) = \frac{1}{2} - \sin t$, so that $a'(t) = 0$ if $\sin t = \frac{1}{2}$. Since t must lie in $[0, \pi/2]$, it follows that $t = \pi/6$. Since $a''(t) = -\cos t < 0$ for t in $(0, \pi/2)$, we conclude that $a(\pi/6)$ is the maximum value, so that the acceleration is maximum in $[0, \pi/2]$ for $t = \pi/6$.

15. It suffices to find the points on the graph of $y = 2/(1+x^2)$ for which the square S of the distance to the origin is minimized. For (x, y) on the graph, S is given by

$$S = (x-0)^2 + (y-0)^2 = x^2 + \left(\frac{2}{1+x^2}\right)^2 = x^2 + \frac{4}{(1+x^2)^2}$$

Thus $$S'(x) = 2x - \frac{16x}{(1+x^2)^3} = 2x\left(1 - \frac{8}{(1+x^2)^3}\right)$$

so that $S'(x) = 0$ if $x = -1$, $x = 0$, or $x = 1$. Since S' changes from negative to positive at -1 and 1, and from positive to negative at 0, it follows that S assumes its minimum value at -1 and 1 (at which points S takes on the same value, namely 2). Thus the points $(-1, 1)$ and $(1, 1)$ are the points on the graph that are closest to the origin.

17. Since $g(x) = \int_x^{x+\pi} \sin^{2/3} t\,dt = \int_0^{x+\pi} \sin^{2/3} t\,dt - \int_0^x \sin^{2/3} t\,dt$, we have

$$g'(x) = \sin^{2/3}(x+\pi) - \sin^{2/3} x = (-\sin x)^{2/3} - \sin^{2/3} x = 0$$

Thus g is a constant function, by Theorem 4.7(a).

19. Let $u = 2 + \cos x$, so that $du = -\sin x\,dx$; if $x = 0$, then $u = 3$, and if $x = \pi$, then $u = 1$;

$$\int_0^\pi \frac{\sin x}{2+\cos x}\,dx = \int_3^1 \frac{1}{u}(-1)\,du = -\ln u\Big|_6^1 = \ln 3$$

21. Let $u = \ln t$, so that $du = (1/t)\,dt$;

$$\int \frac{1}{t}(\ln t)^{5/3}\,dt = \int u^{5/3}\,du = \frac{3}{8}u^{8/3} + C = \frac{3}{8}(\ln t)^{8/3} + C$$

7 TECHNIQUES OF INTEGRATION

SECTION 7.1

1. $u = x$, $dv = \sin x\,dx$; $du = dx$, $v = -\cos x$; $\int x \sin x\,dx = x(-\cos x) - \int (-\cos x)\,dx = -x\cos x + \sin x + C$.

3. $u = \ln x$, $dv = x\,dx$; $du = (1/x)\,dx$, $v = \frac{1}{2}x^2$; $\int x \ln x\,dx = \frac{1}{2}x^2 \ln x - \int (\frac{1}{2}x^2)(1/x)\,dx = \frac{1}{2}x^2 \ln x - \frac{1}{2}\int x\,dx = \frac{1}{2}x^2 \ln x - \frac{1}{4}x^2 + C$.

5. $u = (\ln x)^2$, $dv = dx$; $du = [(2\ln x)/x]\,dx$, $v = x$;

$$\int (\ln x)^2\,dx = (\ln x)^2 x - \int x[(2\ln x)/x]\,dx = x(\ln x)^2 - 2\int \ln x\,dx$$

By Example 3, $-2\int \ln x\,dx = -2x\ln x + 2x + C$. Thus $\int (\ln x)^2\,dx = x(\ln x)^2 - 2x\ln x + 2x + C$.

7. $u = \ln x$, $dv = x^3\,dx$; $du = (1/x)\,dx$, $v = \frac{1}{4}x^4$;

$$\int x^3 \ln x\,dx = (\ln x)(\tfrac{1}{4}x^4) - \int (\tfrac{1}{4}x^4)(1/x)\,dx = \tfrac{1}{4}x^4 \ln x - \tfrac{1}{4}\int x^3\,dx = \tfrac{1}{4}x^4 \ln x - \tfrac{1}{16}x^4 + C$$

9. $u = x^2$, $dv = e^{4x}\,dx$; $du = 2x\,dx$, $v = \frac{1}{4}e^{4x}$; $\int x^2 e^{4x}\,dx = \frac{1}{4}x^2 e^{4x} - \int \frac{1}{2}xe^{4x}\,dx = \frac{1}{4}x^2e^{4x} - \frac{1}{2}\int xe^{4x}\,dx$. For $\int xe^{4x}\,dx$, let $u = x$, $dv = e^{4x}\,dx$; $du = dx$, $v = \frac{1}{4}e^{4x}$; $\int xe^{4x}\,dx = \frac{1}{4}xe^{4x} - \int \frac{1}{4}e^{4x}\,dx = \frac{1}{4}xe^{4x} - \frac{1}{16}e^{4x} + C_1$. Thus $\int x^2e^{4x}\,dx = \frac{1}{4}x^2e^{4x} - \frac{1}{8}xe^{4x} + \frac{1}{32}e^{4x} + C$.

11. $u = x^3$, $dv = \cos x\,dx$; $du = 3x^2\,dx$, $v = \sin x$; $\int x^3 \cos x\,dx = x^3 \sin x - \int 3x^2 \sin x\,dx = x^3 \sin x - 3\int x^2 \sin x\,dx$. Since $\int x^2 \sin x\,dx = -x^2\cos x + 2x\sin x + 2\cos x + C_1$ by the solution of Exercise 10, it follows that $\int x^3\cos x\,dx = x^3\sin x + 3x^2\cos x - 6x\sin x - 6\cos x + C$.

13. $u = \cos 3x$, $dv = e^{3x}\,dx$; $du = -3\sin 3x\,dx$, $v = \frac{1}{3}e^{3x}$; $\int e^{3x}\cos 3x\,dx = \frac{1}{3}e^{3x}\cos 3x + \int e^{3x}\sin 3x\,dx$. For $\int e^{3x}\sin 3x\,dx$, let $u = \sin 3x$, $dv = e^{3x}\,dx$; $du = 3\cos 3x\,dx$, $v = \frac{1}{3}e^{3x}$; $\int e^{3x}\sin 3x\,dx = \frac{1}{3}e^{3x}\sin 3x - \int e^{3x}\cos 3x\,dx$. Thus $\int e^{3x}\cos 3x\,dx = \frac{1}{3}e^{3x}\cos 3x + \frac{1}{3}e^{3x}\sin 3x - \int e^{3x}\cos 3x\,dx$, so $2\int e^{3x}\cos 3x\,dx = \frac{1}{3}e^{3x}\cos 3x + \frac{1}{3}e^{3x}\sin 3x + C_1$ and therefore $\int e^{3x}\cos 3x\,dx = \frac{1}{6}e^{3x}\cos 3x + \frac{1}{6}e^{3x}\sin 3x + C$.

15. $u = t$, $dv = 2^t\,dt$; $du = dt$, $v = (1/\ln 2)2^t$;

$$\int t\cdot 2^t\,dt = \frac{t}{\ln 2}2^t - \int \frac{1}{\ln 2}2^t\,dt = \frac{t}{\ln 2}2^t - \frac{1}{(\ln 2)^2}2^t + C$$

17. $u = t^2$, $dv = 4^t\,dt$; $du = 2t\,dt$, $v = (1/\ln 4)4^t$;

$$\int t^2 4^t\,dt = \frac{t^2}{\ln 4}4^t - \int \frac{2t}{\ln 4}4^t\,dt = \frac{t^2}{\ln 4}4^t - \frac{2}{\ln 4}\int t\cdot 4^t\,dt$$

For $\int t\cdot 4^t\,dt$, let $u = t$, $dv = 4^t\,dt$; $du = dt$, $v = \dfrac{1}{\ln 4}4^t$;

$$\int t\cdot 4^t\,dt = \frac{t}{\ln 4}4^t - \int \frac{1}{\ln 4}4^t\,dt = \frac{t}{\ln 4}4^t - \frac{1}{(\ln 4)^2}4^t + C_1$$

Thus

$$\int t^2 4^t\,dt = \left(\frac{t^2}{\ln 4} - \frac{2t}{(\ln 4)^2} + \frac{2}{(\ln 4)^3}\right)4^t + C$$

19. $u = t$, $dv = \sinh t\,dt$; $du = dt$, $v = \cosh t$;

$$\int t\sinh t\,dt = t\cosh t - \int \cosh t\,dt = t\cosh t - \sinh t + C$$

21. $u = \arctan x$, $dv = dx$; $du = 1/(1 + x^2)\,dx$, $v = x$;

$$\int \arctan x\,dx = x\arctan x - \int \frac{x}{1 + x^2}\,dx$$

For $\int [x/(1 + x^2)]\,dx$ substitute $u = 1 + x^2$, so that $du = 2x\,dx$. Then

$$\int \frac{x}{1 + x^2}\,dx = \int \frac{1}{u}\frac{1}{2}\,du = \frac{1}{2}\int \frac{1}{u}\,du = \frac{1}{2}\ln|u| + C = \frac{1}{2}\ln(1 + x^2) + C$$

Then $\int \arctan x\,dx = x\arctan x - \frac{1}{2}\ln(1 + x^2) + C$.

23. $u = \arccos(-7x)$, $dv = dx$; $du = -(-7/\sqrt{1 - (-7x)^2})\,dx = (7/\sqrt{1 - 49x^2})\,dx$, $v = x$;

$$\int \arccos(-7x)\,dx = x\arccos(-7x) - \int \frac{7x}{\sqrt{1 - 49x^2}}\,dx$$

$$= x\arccos(-7x) - 7\int \frac{x}{\sqrt{1 - 49x^2}}\,dx$$

For $\int (x/\sqrt{1 - 49x^2})\,dx$ substitute $u = 1 - 49x^2$, so that $du = -98\,x\,dx$. Then

$$\int \frac{x}{\sqrt{1 - 49x^2}}\,dx = \int \frac{1}{\sqrt{u}}\left(-\frac{1}{98}\right)du = -\frac{1}{49}\sqrt{u} + C_1 = -\frac{1}{49}\sqrt{1 - 49x^2} + C_1$$

Thus $\int \arccos(-7x)\,dx = x\arccos(-7x) + \frac{1}{7}\sqrt{1 - 49x^2} + C$.

25. By Exercise 24,

$$\int x^n \ln x\,dx = \frac{1}{n + 1}x^{n+1}\ln x - \frac{1}{(n + 1)^2}x^{n+1} + C$$

$$\int x^n \ln x^m\,dx = m\int x^n \ln x\,dx = \frac{m}{n + 1}x^{n+1}\ln x - \frac{m}{(n + 1)^2}x^{n+1} + C$$

27. By the solution of Exercise 26, we have $\int \cos(\ln x)\,dx = x\cos(\ln x) + \int \sin(\ln x)\,dx$ and $\int \sin(\ln x)\,dx = x\sin(\ln x) - \int \cos(\ln x)\,dx$, so that $\int \cos(\ln x)\,dx = \frac{1}{2}[x\cos(\ln x) + x\sin(\ln x)] + C$.

29. $u = x,\ dv = e^{5x}\,dx;\ du = dx,\ v = \frac{1}{5}e^{5x};$

$$\int_0^1 xe^{5x}\,dx = \tfrac{1}{5}xe^{5x}\Big|_0^1 - \int_0^1 \tfrac{1}{5}e^{5x}\,dx = \tfrac{1}{5}e^5 - \tfrac{1}{5}\int_0^1 e^{5x}\,dx$$
$$= \tfrac{1}{5}e^5 - \left(\tfrac{1}{25}e^{5x}\right)\Big|_0^1 = \tfrac{1}{5}e^5 - \tfrac{1}{25}e^5 + \tfrac{1}{25}e^0 = \tfrac{4}{25}e^5 + \tfrac{1}{25}$$

31. $u = t^2,\ dv = \cos t\,dt;\ du = 2t\,dt,\ v = \sin t;\ \int_0^\pi t^2\cos t\,dt = t^2\sin t\Big|_0^\pi - \int_0^\pi 2t\sin t\,dt = -2\int_0^\pi t\sin t\,dt$. By the solution of Exercise 1, $-2\int_0^\pi t\sin t\,dt = -2(-t\cot t + \sin t)\Big|_0^\pi = -2(-\pi\cos\pi + 0) + 2\cdot 0 = 2\pi$. Thus $\int_0^\pi t^2\cos t\,dt = -2\pi$.

33. $u = x,\ dv = \sec^2 x\,dx;\ du = dx,\ v = \tan x;$

$$\int_{-\pi/3}^{\pi/4} x\sec^2 x\,dx = x\tan x\Big|_{-\pi/3}^{\pi/4} - \int_{\pi/3}^{\pi/4}\tan x\,dx = \frac{\pi}{4} - \frac{\pi\sqrt{3}}{3} + (\ln|\cos x|)\Big|_{-\pi/3}^{\pi/4}$$
$$= \frac{\pi}{3} - \frac{\pi\sqrt{3}}{3} + \ln\frac{1}{\sqrt{2}} - \ln\frac{1}{2} = \frac{\pi}{4} - \frac{\pi\sqrt{3}}{3} + \frac{1}{2}\ln 2$$

35. Substitute $u = x + 1$, so that $du = dx$. If $x = 0$ then $u = 1$, and if $x = 1$ then $u = 2$. Thus by the solution of Example 3,

$$\int_0^1 \ln(x+1)\,dx = \int_1^2 \ln u\,du = (u\ln u - u)\Big|_1^2 = (2\ln 2 - 2) + 1 = 2\ln 2 - 1$$

37. Substitute $u = ax$, so that $du = a\,dx$ and $x = u/a$. Then

$$\int x\sin ax\,dx = \int \frac{u}{a}(\sin u)\frac{1}{a}\,du = \frac{1}{a^2}\int u\sin u\,du$$

By the solution of Exercise 1, $\int u\sin u\,du = -u\cos u + \sin u + C_1$. Thus

$$\int x\sin ax\,dx = \frac{1}{a^2}(-ax\cos ax + \sin ax + C_1) = -\frac{x}{a}\cos ax + \frac{1}{a^2}\sin ax + C$$

39. Substitute $u = \cos x$, so that $du = -\sin x\,dx$. Then

$$\int \sin x\arctan(\cos x)\,dx = -\int \arctan u\,du$$

By the solution of Exercise 21,

$$\int \arctan u\,du = u\arctan u - \tfrac{1}{2}\ln(1 + u^2) + C_1$$

Thus

$$\int \sin x\arctan(\cos x)\,dx = -\cos x\arctan(\cos x) + \tfrac{1}{2}\ln(1 + \cos^2 x) + C$$

41. Substitute $s = \sqrt{t}$, so that $ds = [1/2\sqrt{t}]\,dt$. Then $\int \cos\sqrt{t}\,dt = \int(\cos s)(2s)\,ds = 2\int s\cos s\,ds$. By the solution of Example 1, $\int s\cos s\,ds = s\sin s + \cos s + C_1$. Thus $\int \cos\sqrt{t}\,dt = 2\int s\cos s\,ds = 2s\sin s + 2\cos s + C = 2\sqrt{t}\sin\sqrt{t} + 2\cos\sqrt{t} + C$.

43. Substitute $u = x/2$, so that $du = \frac{1}{2}dx$. If $x = 0$ then $u = 0$, and if $x = \pi/2$ then $u = \pi/4$.

Thus $$\int_0^{\pi/2}\cos^3\frac{x}{2}\,dx = 2\int_0^{\pi/4}\cos^3 u\,du$$

By (11) with $n = 3$ we have

$$\int_0^{\pi/4}\cos^3 u\,du = \frac{1}{3}\cos^2 u\sin u\Big|_0^{\pi/4} + \frac{2}{3}\int_0^{\pi/4}\cos u\,du$$
$$= \left(\frac{1}{3}\left(\frac{\sqrt{2}}{2}\right)^3 - 0\right) + \left(\frac{2}{3}\sin u\right)\Big|_0^{\pi/4}$$
$$= \frac{\sqrt{2}}{12} + \frac{\sqrt{2}}{3} = \frac{5}{12}\sqrt{2}$$

Therefore $$\int_0^{\pi/2}\cos^3\frac{x}{2}\,dx = 2\int_0^{\pi/4}\cos^3 u\,du = 2\left(\frac{5}{12}\sqrt{2}\right) = \frac{5\sqrt{2}}{6}$$

45. Using the formula for $\int \sin^n x\,dx$ in (10) and then the one for $\int \sin^2 x\,dx$, we have

$$\int \sin^4 x\,dx = \frac{-1}{4}\sin^3 x\cos x + \frac{3}{4}\int \sin^2 x\,dx = \frac{-1}{4}\sin^3 x\cos x - \frac{3}{16}\sin 2x + \frac{3}{8}x + C$$
$$= \frac{-1}{4}\sin^3 x\cos x - \frac{3}{8}\sin x\cos x + \frac{3}{8}x + C$$

47. $u = \cos^{n-1} x,\ dv = \cos x\,dx;\ du = -(n-1)\cos^{n-2} x\sin x\,dx,\ v = \sin x;$

$$\int \cos^n x\,dx = \int \cos^{n-1} x\cos x\,dx$$
$$= (\cos^{n-1} x)(\sin x) - \int [\sin x][-(n-1)\cos^{n-2} x\sin x]\,dx$$
$$= \cos^{n-1} x\sin x + (n-1)\int \cos^{n-2} x\sin^2 x\,dx$$
$$= \cos^{n-1} x\sin x + (n-1)\int (\cos^{n-2} x - \cos^n x)\,dx$$

Thus $n\int \cos^n x\,dx = \cos^{n-1} x\sin x + (n-1)\int \cos^{n-2} x\,dx$, so that

$$\int \cos^n x\,dx = \frac{1}{n}\cos^{n-1} x\sin x + \frac{n-1}{n}\int \cos^{n-2} x\,dx$$

49. $u = \dfrac{1}{(x^2 + a^2)^n},\ dv = dx;\ du = \dfrac{-2nx}{(x^2 + a^2)^{n+1}}\,dx,\ v = x;$

$$\int \frac{1}{(x^2 + a^2)^n}\,dx = \frac{x}{(x^2 + a^2)^n} + \int \frac{2nx^2}{(x^2 + a^2)^{n+1}}\,dx = \frac{x}{(x^2 + a^2)^n} + \int \frac{2n(x^2 + a^2 - a^2)}{(x^2 + a^2)^{n+1}}\,dx$$
$$= \frac{x}{(x^2 + a^2)^n} + \int \frac{2n}{(x^2 + a^2)^n}\,dx - \int \frac{2na^2}{(x^2 + a^2)^{n+1}}\,dx$$

Thus $$(1 - 2n)\int \frac{1}{(x^2 + a^2)^n}\,dx = \frac{x}{(x^2 + a^2)^n} - \int \frac{2na^2}{(x^2 + a^2)^{n+1}}\,dx$$

so $$\int \frac{1}{(x^2 + a^2)^{n+1}}\,dx = \frac{x}{2na^2(x^2 + a^2)^n} + \frac{2n - 1}{2na^2}\int \frac{1}{(x^2 + a^2)^n}\,dx$$

51. By Exercise 48 with $n = 3$, $\int (\ln x)^3\,dx = x(\ln x)^3 - 3\int (\ln x)^2\,dx$.
By Exercise 48 with $n = 2$, $\int (\ln x)^2\,dx = x(\ln x)^2 - 2\int \ln x\,dx$.
By Example 3, $\int \ln x\,dx = x\ln x - x + C_1$. Thus

$$\int (\ln x)^3\,dx = x(\ln x)^3 - 3x(\ln x)^2 + 6x\ln x - 6x + C$$

53. a. $u = \sin bx$, $dv = e^{ax}\,dx$; $du = b\cos bx\,dx$, $v = (1/a)e^{ax}$;

$$\int e^{ax}\sin bx\,dx = \frac{1}{a}e^{ax}\sin bx - \int \frac{b}{a}e^{ax}\cos bx\,dx = \frac{1}{a}e^{ax}\sin bx - \frac{b}{a}\int e^{ax}\cos bx\,dx$$

For $\int e^{ax}\cos bx\,dx$, let $u = \cos bx$, $dv = e^{ax}\,dx$; $du = -b\sin bx\,dx$, $v = (1/a)e^{ax}$; $\int e^{ax}\cos bx\,dx = (1/a)e^{ax}\cos bx + \int (b/a)\,e^{ax}\sin bx\,dx$. Thus

$$\int e^{ax}\sin bx\,dx = \frac{1}{a}e^{ax}\sin bx - \frac{b}{a^2}e^{ax}\cos bx - \frac{b^2}{a^2}\int e^{ax}\sin bx\,dx$$

so that $(1 + b^2/a^2)\int e^{ax}\sin bx\,dx = (1/a)e^{ax}\sin bx - (b/a^2)e^{ax}\cos bx + C_1$. Since $1 + b^2/a^2 = (a^2 + b^2)/a^2$, we find that

$$\int e^{ax}\sin bx\,dx = \frac{a^2}{a^2+b^2}\left(\frac{1}{a}e^{ax}\sin bx - \frac{b}{a^2}e^{ax}\cos bx\right) + C$$
$$= \frac{e^{ax}}{a^2+b^2}(a\sin bx - b\cos bx) + C$$

b. $u = \cos bx$, $dv = e^{ax}\,dx$; $du = -b\sin bx\,dx$, $v = (1/a)\,e^{ax}$;

$$\int e^{ax}\cos bx\,dx = \frac{1}{a}e^{ax}\cos bx - \int -\frac{b}{a}e^{ax}\sin bx\,dx = \frac{1}{a}e^{ax}\cos bx + \frac{b}{a}\int e^{ax}\sin bx\,dx$$

Therefore by part (a),

$$\int e^{ax}\cos bx\,dx = \frac{1}{a}e^{ax}\cos bx + \frac{b}{a}\frac{e^{ax}}{a^2+b^2}(a\sin bx - b\cos bx) + C$$
$$= \frac{e^{ax}}{a^2+b^2}\left[\frac{a^2+b^2}{a}\cos bx + b\sin bx - \frac{b^2}{a}\cos bx\right] + C$$
$$= \frac{e^{ax}}{a^2+b^2}[a\cos bx + b\sin bx] + C$$

55. a. By Example 1, the moment M is given by

$$M = \int_0^{\pi/2} x\cos x\,dx = (x\sin x + \cos x)\Big|_0^{\pi/2} = \frac{\pi}{2} - 1$$

b. By Example 1, the moment M is given by

$$M = \int_{-\pi/2}^{\pi/2} x\cos x\,dx = (x\sin x + \cos x)\Big|_{-\pi/2}^{\pi/2} = 0$$

c. By Exercise 3, the moment M is given by

$$M = \int_1^2 x\ln x\,dx = \left(\frac{1}{2}x^2\ln x - \frac{x^2}{4}\right)\Big|_1^2 = 2\ln 2 - \frac{3}{4}$$

57. For $\frac{1}{e} \le x \le 1$, $f(x) \le 0$; for $1 \le x \le e$, $f(x) \ge 0$. Using the solution of Exercise 3, we find that the area A is given by

$$A = -\int_{1/e}^1 x\ln x\,dx + \int_1^e x\ln x\,dx = -\left(\frac{1}{2}x^2\ln x - \frac{1}{4}x^2\right)\Big|_{1/e}^1 + \left(\frac{1}{2}x^2\ln x - \frac{1}{4}x^2\right)\Big|_1^e$$
$$= -\left[-\frac{1}{4} - \left(-\frac{1}{2e^2} - \frac{1}{4e^2}\right)\right] + \left[\frac{1}{2}e^2 - \frac{1}{4}e^2 + \frac{1}{4}\right] = \frac{1}{2} + \frac{1}{4}e^2 - \frac{3}{4e^2}$$

59. Using the solution of Exercise 21, we find that the area A is given by

$$A = \int_0^1 \arctan x\,dx = \left(x\arctan x - \frac{1}{2}\ln(1+x^2)\right)\Big|_0^1$$
$$= \left(\arctan 1 - \frac{1}{2}\ln 2\right) - 0 = \frac{\pi}{4} - \frac{1}{2}\ln 2$$

61. By (12), $c = \int_0^{\pi/2} \sin^4 x\,dx = \left(-\tfrac{1}{4}\sin^3 x\cos x - \tfrac{3}{8}\sin x\cos x + \tfrac{3}{8}x\right)\Big|_0^{\pi/2} = \tfrac{3}{16}\pi$.

SECTION 7.2

1. $u = \cos x$, $du = -\sin x\,dx$;

$$\int \sin^3 x\cos^2 x\,dx = -\int (-\sin x)(1 - \cos^2 x)\cos^2 x\,dx$$
$$= -\int (1 - u^2)u^2\,du = \int (-u^2 + u^4)\,du = \frac{-1}{3}u^3 + \frac{1}{5}u^5 + C$$
$$= \frac{-1}{3}\cos^3 x + \frac{1}{5}\cos^5 x + C$$

3. $u = \sin 3x$, $du = 3\cos 3x\,dx$;

$$\int \sin^3 3x\cos 3x\,dx = \tfrac{1}{3}\int u^3\,du = \tfrac{1}{12}u^4 + C = \tfrac{1}{12}\sin^4 3x + C$$

5. First let $u = 1/x$, so that $du = -(1/x^2)\,dx$. Then

$$\int \frac{1}{x^2}\sin^5\frac{1}{x}\cos^2\frac{1}{x}\,dx = \int -\sin^5 u\cos^2 u\,du = -\int \sin^5 u\cos^2 u\,du$$

For $\int \sin^5 u\cos^2 u\,du$, let $v = \cos u$, so that $dv = -\sin u\,du$. Then $\int \sin^5 u\cos^2 u\,du = \int (1 - \cos^2 u)^2(\sin u)\cos^2 u\,du = \int -(1 - v^2)^2 v^2\,dv = -\int (v^2 - 2v^4 + v^6)\,dv = -(\tfrac{1}{3}v^3 - \tfrac{2}{5}v^5 + \tfrac{1}{7}v^7) + C_1 = -\tfrac{1}{3}\cos^3 u + \tfrac{2}{5}\cos^5 u - \tfrac{1}{7}\cos^7 u + C_1$. Thus

$$\int \frac{1}{x^2}\sin^5\frac{1}{x}\cos^2\frac{1}{x}\,dx = \frac{1}{3}\cos^3\frac{1}{x} - \frac{2}{5}\cos^5\frac{1}{x} + \frac{1}{7}\cos^7\frac{1}{x} + C$$

7. By (1), $\int \sin^2 y\cos^2 y\,dy = \int (\sin y\cos y)^2\,dy = \int (\tfrac{1}{2}\sin 2y)^2\,dy = \tfrac{1}{4}\int \sin^2 2y\,dy$. Next, let $u = 2y$, so that $du = 2y$. Then $\int \sin^2 2y\,dy = \int (\sin^2 u)\tfrac{1}{2}\,du = \tfrac{1}{2}\int \sin^2 u\,du$. By (4), $\int \sin^2 u\,du = \tfrac{1}{2}u - \tfrac{1}{4}\sin 2u + C_1$. Thus $\int \sin^2 y\cos^2 y\,dy = \tfrac{1}{4}\int \sin^2 2y\,dy = \tfrac{1}{8}\int \sin^2 u\,du = \tfrac{1}{8}(\tfrac{1}{2}u - \tfrac{1}{4}\sin 2u) + C = \tfrac{1}{8}y - \tfrac{1}{32}\sin 4y + C$.

9. By (1), $\int \sin^4 x \cos^4 x\,dx = \int \frac{1}{16}(2\sin x\cos x)^4\,dx = \frac{1}{16}\int \sin^4 2x\,dx$. Next, let $u = 2x$, so that $du = 2\,dx$. Then $\frac{1}{16}\int \sin^4 2x\,dx = \frac{1}{16}\int (\sin^4 u)\frac{1}{2}\,du = \frac{1}{32}\int \sin^4 u\,du$. By (12) of Section 7.1,

$$\frac{1}{32}\int \sin^4 u\,du = \frac{1}{32}\left(-\frac{1}{4}\sin^3 u\cos u - \frac{3}{8}\sin u\cos u + \frac{3}{8}u\right) + C$$
$$= -\frac{1}{128}\sin^3 2x\cos 2x - \frac{3}{256}\sin 2x\cos 2x + \frac{3}{128}x + C.$$

11. $u = \sin x$, $du = \cos x\,dx$; $\int \sin^{-10} x\cos^3 x\,dx = \int \sin^{-10} x(1 - \sin^2 x)\cos x\,dx = \int (u^{-10} - u^{-8})\,du = (-1/9)u^{-9} + \frac{1}{7}u^{-7} + C = (-1/9)\sin^{-9} x + \frac{1}{7}\sin^{-7} x + C.$

13. $$\int (1 + \sin^2 x)(1 + \cos^2 x)\,dx = \int \left(1 + \frac{1 - \cos 2x}{2}\right)\left(1 + \frac{1 + \cos 2x}{2}\right)dx$$
$$= \int \left(\frac{3 - \cos 2x}{2}\right)\left(\frac{3 + \cos 2x}{2}\right)dx = \frac{1}{4}\int (9 - \cos^2 2x)\,dx$$

Using (5) we find that $\frac{1}{4}\int (9 - \cos^2 2x)\,dx = \frac{9}{4}x - \frac{1}{8}x - \frac{1}{32}\sin 4x + C = \frac{17}{8}x - \frac{1}{32}\sin 4x + C.$

15. $u = \cos x$, $du = -\sin x\,dx$; if $x = 0$ then $u = 1$, and if $x = \pi/4$ then $u = \sqrt{2}/2$;

$$\int_0^{\pi/4} \frac{\sin^3 x}{\cos^2 x}\,dx = \int_0^{\pi/4} \frac{(1 - \cos^2 x)}{\cos^2 x}\sin x\,dx = \int_1^{\sqrt{2}/2} \frac{-(1 - u^2)}{u^2}\,du = \int_1^{\sqrt{2}/2}\left(1 - \frac{1}{u^2}\right)du$$
$$= \left(u + \frac{1}{u}\right)\Bigg|_1^{\sqrt{2}/2} = \left(\frac{\sqrt{2}}{2} + \sqrt{2}\right) - (1 + 1) = \frac{3}{2}\sqrt{2} - 2$$

17. $u = \tan x$, $du = \sec^2 x\,dx$; $\int \tan^5 x\sec^2 x\,dx = \int u^5\,du = \frac{1}{6}u^6 + C = \frac{1}{6}\tan^6 x + C.$

19. $u = \tan t$, $du = \sec^2 t\,dt$; $\int_0^{\pi/4} \tan^5 t\sec^4 t\,dt = \int_0^{\pi/4} \tan^5 t(\tan^2 t + 1)\sec^2 t\,dt = \int_0^1 u^5(u^2 + 1)\,du = \int_0^1 (u^7 + u^5)\,du = \left(\frac{1}{8}u^8 + \frac{1}{6}u^6\right)\Big|_0^1 = \frac{7}{24}.$

21. $u = \sec x$, $du = \sec x\tan x\,dx$; $\int_{5\pi/4}^{4\pi/3} \tan^3 x\sec x\,dx = \int_{5\pi/4}^{4\pi/3} (\sec^2 x - 1)(\sec x\tan x)\,dx = \int_{-\sqrt{2}}^{-2} (u^2 - 1)\,du = \left(\frac{1}{3}u^3 - u\right)\Big|_{-\sqrt{2}}^{-2} = -\frac{2}{3} - \frac{1}{3}\sqrt{2}.$

23. $u = \sec\sqrt{x}$; $du = [1/(2\sqrt{x})]\sec\sqrt{x}\tan\sqrt{x}\,dx$;

$$\int \frac{1}{\sqrt{x}}\tan^3\sqrt{x}\sec^3\sqrt{x}\,dx = \int (\tan^2\sqrt{x})(\sec^2\sqrt{x})\left(\frac{1}{\sqrt{x}}\sec\sqrt{x}\tan\sqrt{x}\right)dx$$
$$= \int [(\sec^2\sqrt{x} - 1)\sec^2\sqrt{x}]\frac{1}{\sqrt{x}}\sec\sqrt{x}\tan\sqrt{x}\,dx$$
$$= \int (u^2 - 1)u^2(2)\,du = 2\int (u^4 - u^2)\,du = \frac{2}{5}u^5 - \frac{2}{3}u^3 + C$$
$$= \frac{2}{5}\sec^5\sqrt{x} - \frac{2}{3}\sec^3\sqrt{x} + C$$

25. $u = \tan x$, $du = \sec^2 x\,dx$; $\int \tan^3 x\sec^4 x\,dx = \int \tan^3 x(\tan^2 x + 1)\sec^2 x\,dx = \int u^3(u^2 + 1)\,du = \int (u^5 + u^3)\,du = \frac{1}{6}u^6 + \frac{1}{4}u^4 + C = \frac{1}{6}\tan^6 x + \frac{1}{4}\tan^4 x + C$

27. $u = \sec x$, $du = \sec x\tan x\,dx$; $\int \tan x\sec^5 x\,dx = \int u^4\,du = \frac{1}{5}u^5 + C = \frac{1}{5}\sec^5 x + C$

29. $u = \cot x$, $du = -\csc^2 x\,dx$;
$\int \cot^3 x\csc^2 x\,dx = -\int u^3\,du = (-1/4)u^4 + C = (-1/4)\cot^4 x + C.$

31. $u = \csc x$, $du = -\csc x\cot x\,dx$;

$$\int_{\pi/4}^{\pi/2} \cot^3 x\csc^3 x\,dx = \int_{\pi/4}^{\pi/2} (\csc^2 x - 1)\csc^2 x(\cot x\csc x)\,dx$$
$$= -\int_{\sqrt{2}}^1 (u^4 - u^2)\,du = \left(\frac{-1}{5}u^5 + \frac{1}{3}u^3\right)\Bigg|_{\sqrt{2}}^1 = \frac{2}{15}(\sqrt{2} + 1)$$

33. $\int \cot x\csc^{-2} x\,dx = \int \frac{\cos x}{\sin x}\sin^2 x\,dx = \int \cos x\sin x\,dx = \frac{1}{2}\sin^2 x + C$

35. $u = \cos x$, $du = -\sin x\,dx$;

$$\int \frac{\tan x}{\cos^3 x}\,dx = \int \frac{\sin x}{\cos^4 x}\,dx = \int \frac{1}{u^4}(-1)\,du$$
$$= \frac{1}{3u^3} + C = \frac{1}{3\cos^3 x} + C$$

37. $$\int \frac{\tan^2 x}{\sec^5 x}\,dx = \int \frac{\sin^2 x}{\cos^2 x}\cos^5 x\,dx = \int \sin^2 x\cos^3 x\,dx = \int \sin^2 x(1 - \sin^2 x)\cos x\,dx$$
$$= \int (\sin^2 x - \sin^4 x)\cos x\,dx$$

If $u = \sin x$ then $du = \cos x\,dx$, so

$$\int \frac{\tan^2 x}{\sec^5 x}\,dx = \int (\sin^2 x - \sin^4 x)\cos x\,dx = \int (u^2 - u^4)\,du$$
$$= \frac{1}{3}u^3 - \frac{1}{5}u^5 + C = \frac{1}{3}\sin^3 x - \frac{1}{5}\sin^5 x + C$$

39. $\int \frac{\tan x}{\sec^2 x}\,dx = \int \frac{\sin x}{\cos x}\cos^2 x\,dx = \int \sin x\cos x\,dx = \frac{1}{2}\sin^2 x + C$

41. $\int \tan^2 x\,dx = \int (\sec^2 x - 1)\,dx = \tan x - x + C$

43. $\int \tan^4 x\,dx = \int \tan^2 x(\sec^2 x - 1)\,dx = \int \tan^2 x\sec^2 x\,dx - \int \tan^2 x\,dx$. For $\int \tan^2 x\sec^2 x\,dx$, let $u = \tan x$, so that $du = \sec^2 x\,dx$. Then $\int \tan^2 x\sec^2 x\,dx = \int u^2\,du = \frac{1}{3}u^3 + C_1 = \frac{1}{3}\tan^3 x + C_1$. By Exercise 41, $\int \tan^2 x\,dx = \tan x - x + C_2$. Thus $\int \tan^4 x\,dx = \frac{1}{3}\tan^3 x - \tan x + x + C.$

45. By (8) with $a = 2$ and $b = 3$, $\int \sin 2x\cos 3x\,dx = \frac{1}{2}\int (\sin(-x) + \sin 5x)\,dx = \frac{1}{2}\cos(-x) - \frac{1}{10}\cos 5x + C = \frac{1}{2}\cos x - \frac{1}{10}\cos 5x + C.$

47. By (8) with $a = -4$ and $b = -2$, $\int \sin(-4x)\cos(-2x)\,dx = \frac{1}{2}\int (\sin(-2x) + \sin(-6x))\,dx = \frac{1}{4}\cos(-2x) + \frac{1}{12}\cos(-6x) + C = \frac{1}{4}\cos 2x + \frac{1}{12}\cos 6x + C.$

49. By (8) with $a = \frac{1}{2}$ and $b = \frac{2}{3}$, $\int \sin\frac{1}{2}x\cos\frac{2}{3}x\,dx = \frac{1}{2}\int (\sin(-\frac{1}{6}x) + \sin\frac{7}{6}x)\,dx = 3\cos(-\frac{1}{6}x) - \frac{3}{7}\cos\frac{7}{6}x + C = 3\cos\frac{1}{6}x - \frac{3}{7}\cos\frac{7}{6}x + C.$

51. $\int \sin 2x \sin 3x\,dx = \int [-\frac{1}{2}\cos(2+3)x + \frac{1}{2}\cos(2-3)x]\,dx$

$$= \frac{-1}{10}\sin 5x + \frac{1}{-2}\sin(-x) + C = \frac{-1}{10}\sin 5x + \frac{1}{2}\sin x + C$$

53. $\int \cos 5x \cos(-3x)\,dx = \int [\frac{1}{2}\cos(5-3)x + \frac{1}{2}\cos(5+3)x]\,dx = \frac{1}{4}\sin 2x + \frac{1}{16}\sin 8x + C$

55. $\int_{\pi/4}^{\pi/2} \frac{1}{1+\cos x}\,dx = \int_{\pi/4}^{\pi/2} \frac{1}{1+\cos x}\frac{1-\cos x}{1-\cos x}\,dx = \int_{\pi/4}^{\pi/2} \frac{1-\cos x}{\sin^2 x}\,dx$

$$= \int_{\pi/4}^{\pi/2} (\csc^2 x - \csc x \cot x)\,dx = (-\cot x + \csc x)\Big|_{\pi/4}^{\pi/2} = 2 - \sqrt{2}$$

57. $\int \frac{1+\cos x}{\sin x}\,dx = \int \frac{1+\cos x}{\sin x}\frac{1-\cos x}{1-\cos x}\,dx = \int \frac{\sin x}{1-\cos x}\,dx = \ln|1-\cos x| + C$

59. a. If $m \neq n$, then by (8) with $a = m$ and $b = n$,

$$\int_{-\pi}^{\pi} \sin mx \cos nx\,dx = \int_{-\pi}^{\pi} \left\{\frac{1}{2}\sin[(m-n)x] + \frac{1}{2}\sin[(m+n)x]\right\}dx$$

$$= \left\{-\frac{1}{2}\frac{1}{m-n}\cos[(m-n)x] - \frac{1}{2}\frac{1}{m+n}\cos[(m+n)x]\right\}\Bigg|_{-\pi}^{\pi} = 0$$

since $\cos[(m-n)\pi] = \cos[(m-n)(-\pi)]$ and $\cos[(m+n)\pi] = \cos[(m+n)(-\pi)]$ for all integers m and n. If $m = n$, then $\int_{-\pi}^{\pi} \sin mx \cos nx\,dx = \int_{-\pi}^{\pi} \sin mx \cos mx\,dx = \int_{-\pi}^{\pi} \frac{1}{2}\sin 2mx\,dx = [-1/(4m)]\cos 2mx\Big|_{-\pi}^{\pi} = 0$, since $\cos 2m\pi = \cos[2m(-\pi)]$ for all integers m.

b. If $m \neq n$, then by the solution to Exercise 52,

$$\int_{-\pi}^{\pi} \cos mx \cos nx\,dx = \left[\frac{1}{2(m+n)}\sin(m+n)x + \frac{1}{2(m-n)}\sin(m-n)x\right]\Bigg|_{-\pi}^{\pi} = 0$$

since $\sin k\pi = 0$ for all integers k. If $m = n$, then by (3),

$$\int_{-\pi}^{\pi} \cos mx \cos nx\,dx = \int_{-\pi}^{\pi} \cos^2 mx\,dx = \int_{-\pi}^{\pi} \left(\frac{1}{2} + \frac{1}{2}\cos 2mx\right)dx$$

$$= \left(\frac{1}{2}x + \frac{1}{4m}\sin 2mx\right)\Bigg|_{-\pi}^{\pi} = \left(\frac{1}{2}\pi + 0\right) - \left(-\frac{1}{2}\pi + 0\right) = \pi$$

since $\sin k\pi = 0$ for all integers k.

c. If $m \neq n$, then by the solution to Exercise 50,

$$\int_{-\pi}^{\pi} \sin mx \sin nx\,dx = \left[\frac{-1}{2(m+n)}\sin(m+n)x + \frac{1}{2(m-n)}\sin(m-n)x\right]\Bigg|_{-\pi}^{\pi} = 0$$

since $\sin k\pi = 0$ for all integers k. If $m = n$, then by (2),

$$\int_{-\pi}^{\pi} \sin mx \sin nx\,dx = \int_{-\pi}^{\pi} \sin^2 mx\,dx = \int_{-\pi}^{\pi} \left(\frac{1}{2} - \frac{1}{2}\cos 2mx\right)dx$$

$$= \left(\frac{1}{2}x - \frac{1}{4m}\sin 2mx\right)\Bigg|_{-\pi}^{\pi} = \left(\frac{1}{2}\pi - 0\right) - \left(-\frac{1}{2}\pi - 0\right) = \pi$$

since $\sin k\pi = 0$ for all integers k.

61. The area is given by $A = \int_0^{\pi/2} \sin^2 x \cos^3 x\,dx$, so by the solution of Example 1, $A = (\frac{1}{3}\sin^3 x - \frac{1}{5}\sin^5 x)\Big|_0^{\pi/2} = \frac{1}{3} - \frac{1}{5} = \frac{2}{15}$.

63. The area is given by $A = \int_0^{\pi/6} \sin 2x \cos 3x\,dx$, so by the solution of Exercise 45,

$$A = (\tfrac{1}{2}\cos x - \tfrac{1}{10}\cos 5x)\Big|_0^{\pi/6} = [\tfrac{1}{2}(\tfrac{1}{2}\sqrt{3}) - \tfrac{1}{10}(-\tfrac{1}{2}\sqrt{3})] - (\tfrac{1}{2} - \tfrac{1}{10}) = \tfrac{3}{10}\sqrt{3} - \tfrac{2}{5}$$

65. Since $\frac{1}{4}\tan x \sec^4 x - \tan^3 x = \tan x\,(\frac{1}{4}\sec^4 x - (\sec^2 x - 1)) =$ $\frac{1}{4}\tan x\,(\sec^4 x - 4\sec^2 x + 4) = \frac{1}{4}\tan x\,(\sec^2 x - 2)^2 \geq 0$ on $[0, \pi/3]$, we have $A = \int_0^{\pi/3} (\frac{1}{4}\tan x \sec^4 x - \tan^3 x)\,dx = \int_0^{\pi/3} [\frac{1}{4}\tan x\,(\tan^2 x + 1)\sec^2 x -$ $\tan x\,(\sec^2 x - 1)]\,dx = \int_0^{\pi/3} (\frac{1}{4}\tan^3 x \sec^2 x - \frac{3}{4}\tan x \sec^2 x + \tan x)\,dx =$ $(\frac{1}{16}\tan^4 x - \frac{3}{8}\tan^2 x - \ln|\cos x|)\Big|_0^{\pi/3} = (\frac{9}{16} - \frac{9}{8} - \ln\frac{1}{2}) - \ln 1 = \ln 2 - \frac{9}{16}$

SECTION 7.3

1. $x = \frac{1}{2}\sin u$, $dx = \frac{1}{2}\cos u\,du$; if $x = 0$ then $u = 0$, and if $x = \frac{1}{2}$ then $u = \pi/2$;

$$\int_0^{1/2} \sqrt{1-4x^2}\,dx = \int_0^{\pi/2} \sqrt{1-\sin^2 u}\,(\tfrac{1}{2}\cos u)\,du = \tfrac{1}{2}\int_0^{\pi/2} \cos^2 u\,du$$

$$= \frac{1}{4}\int_0^{\pi/2} (1+\cos 2u)\,du = \frac{1}{4}\left(u + \frac{1}{2}\sin 2u\right)\Bigg|_0^{\pi/2} = \frac{\pi}{8}$$

3. $x = 2\sin u$, $dx = 2\cos u\,du$; if $x = -2$ then $u = -\pi/2$, and if $x = 2$ then $u = \pi/2$;

$$\int_{-2}^{2} \sqrt{1 - \frac{x^2}{4}}\,dx = \int_{-\pi/2}^{\pi/2} \sqrt{1-\sin^2 u}\,(2\cos u)\,du = 2\int_{-\pi/2}^{\pi/2} \cos^2 u\,du$$

$$= \int_{-\pi/2}^{\pi/2} (1+\cos 2u)\,du = (u + \tfrac{1}{2}\sin 2u)\Big|_{-\pi/2}^{\pi/2} = \pi$$

5. $t = 3\tan u$, $dt = 3\sec^2 u\,du$;

$$\int \frac{1}{(9+t^2)^2}\,dt = \int \frac{1}{(9+9\tan^2 u)^2}\,3\sec^2 u\,du = \frac{1}{27}\int \frac{\sec^2 u}{\sec^4 u}\,du = \frac{1}{27}\int \cos^2 u\,du$$

$$= \tfrac{1}{54}\int (1+\cos 2u)\,du = \tfrac{1}{54}(u + \tfrac{1}{2}\sin 2u) + C$$

Now $t = 3\tan u$ implies that $u = \arctan t/3$, and from the figure, $\sin u = t/\sqrt{9+t^2}$ and $\cos u = 3/\sqrt{9+t^2}$, so that $\sin 2u = 2\sin u \cos u = 6t/(9+t^2)$. Then

$$\int \frac{1}{(9+t^2)^2}\,dt = \frac{1}{54}u + \frac{1}{108}\sin 2u + C = \frac{1}{54}\arctan\frac{t}{3} + \frac{t}{18(9+t^2)} + C$$

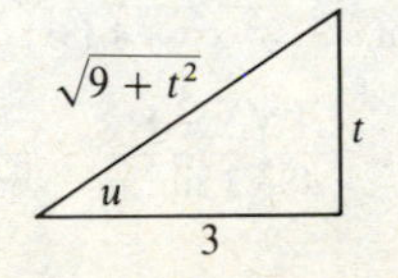

7. $x = \tan u$, $dx = \sec^2 u\,du$; $\sin u = x/\sqrt{x^2+1}$;

$$\int \frac{1}{(x^2+1)^{3/2}}\,dx = \int \frac{1}{(\tan^2 u + 1)^{3/2}} \sec^2 u\,du = \int \frac{\sec^2 u}{\sec^3 u}\,du = \int \cos u\,du$$

$$= \sin u + C = \frac{x}{\sqrt{x^2+1}} + C$$

9. $x = \sqrt{\frac{2}{3}}\tan u$, $dx = \sqrt{\frac{2}{3}}\sec^2 u\,du$; $\sin u = (\sqrt{3}x)/\sqrt{3x^2+2}$;

$$\int \frac{1}{(3x^2+2)^{5/2}}\,dx = \int \frac{1}{(2\tan^2 u + 2)^{5/2}}\left(\sqrt{\frac{2}{3}}\sec^2 u\right)du = \frac{1}{4\sqrt{3}}\int \frac{1}{\sec^5 u}\sec^2 u\,du$$

$$= \frac{1}{4\sqrt{3}}\int \cos^3 u\,du = \frac{1}{4\sqrt{3}}\int (1-\sin^2 u)\cos u\,du$$

$$= \frac{1}{4\sqrt{3}}\left(\sin u - \frac{1}{3}\sin^3 u\right) + C = \frac{x}{4\sqrt{3x^2+2}} - \frac{x^3}{4(3x^2+2)^{3/2}} + C$$

Thus $\int_0^1 \frac{1}{(3x^2+2)^{5/2}}\,dx = \left(\frac{x}{4\sqrt{3x^2+2}} - \frac{x^3}{4(3x^2+2)^{3/2}}\right)\Bigg|_0^1 = \frac{\sqrt{5}}{25}.$

11. $x = \frac{2}{3}\sec u$, $dx = \frac{2}{3}\sec u \tan u\,du$;

$$\int \frac{1}{(9x^2-4)^{5/2}}\,dx = \int \frac{1}{(4\sec^2 u - 4)^{5/2}}\,\frac{2}{3}\sec u \tan u\,du = \int \frac{\sec u \tan u}{48\tan^5 u}\,du$$

$$= \frac{1}{48}\int \frac{\sec u}{\tan^4 u}\,du = \frac{1}{48}\int \frac{\cos^3 u}{\sin^4 u}\,du$$

$$= \frac{1}{48}\int \frac{1-\sin^2 u}{\sin^4 u}\cos u\,du$$

$$= \frac{1}{48}\int\left(\frac{1}{\sin^4 u} - \frac{1}{\sin^2 u}\right)\cos u\,du = \frac{1}{48}\left(\frac{-1}{3\sin^3 u} + \frac{1}{\sin u}\right) + C$$

Now $x = \frac{2}{3}\sec u$ implies that $\sin u = \sqrt{9x^2-4}/(3x)$. Thus

$$\int \frac{1}{(9x^2-4)^{5/2}}\,dx = \frac{1}{48}\left(\frac{-1}{3\sin^3 u} + \frac{1}{\sin u}\right) + C$$

$$= \frac{-3x^3}{16(9x^2-4)^{3/2}} + \frac{x}{16(9x^2-4)^{1/2}} + C$$

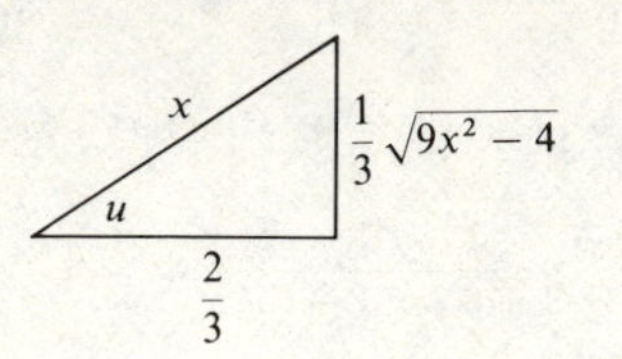

13. $x = \sqrt{3}\sin u$, $dx = \sqrt{3}\cos u\,du$;

$$\int \frac{1}{(3-x^2)^{3/2}}\,dx = \int \frac{1}{(3-3\sin^2 u)^{3/2}}(\sqrt{3}\cos u)\,du = \frac{1}{3}\int \sec^2 u\,du$$

$$= \frac{1}{3}\tan u + C = \frac{1}{3}\frac{x}{\sqrt{3-x^2}} + C$$

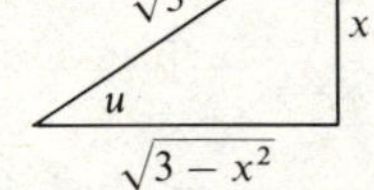

15. $x = \frac{5}{2}\sin u$, $dx = \frac{5}{2}\cos u\,du$; if $x = 0$ then $u = 0$, and if $x = \frac{5}{4}$ then $u = \pi/6$;

$$\int_0^{5/4} \frac{1}{\sqrt{25-4x^2}}\,dx = \int_0^{\pi/6} \frac{1}{\sqrt{25-25\sin^2 u}}\left(\frac{5}{2}\cos u\right)du$$

$$= \tfrac{1}{2}\int_0^{\pi/6} 1\,du = \tfrac{1}{2}u\Big|_0^{\pi/6} = \tfrac{1}{12}\pi$$

17. $$\int \frac{1}{\sqrt{4x^2+4x+2}}\,dx = \int \frac{1}{\sqrt{(4x^2+4x+1)+1}}\,dx = \int \frac{1}{\sqrt{(2x+1)^2+1}}\,dx$$

Let $2x + 1 = \tan u$, so that $2\,dx = \sec^2 u\,du$. Then

$$\int \frac{1}{\sqrt{(2x+1)^2+1}}\,dx = \int \frac{1}{\sqrt{\tan^2 u + 1}}\,\frac{1}{2}\sec^2 u\,du$$

$$= \tfrac{1}{2}\int \sec u\,du = \tfrac{1}{2}\ln|\sec u + \tan u| + C$$

Since $\sec u = \sqrt{\tan^2 u + 1} = \sqrt{(2x+1)^2+1} = \sqrt{4x^2+4x+2}$, we have

$$\int \frac{1}{\sqrt{4x^2+4x+2}}\,dx = \frac{1}{2}\ln\left|\sqrt{4x^2+4x+2} + (2x+1)\right| + C$$

19. $w = (1/\sqrt{2})\sin u$, $dw = (1/\sqrt{2})\cos u\,du$;

$$\int \frac{1}{(1-2w^2)^{5/2}}\,dw = \int \frac{1}{(1-\sin^2 u)^{5/2}}\left(\frac{1}{\sqrt{2}}\cos u\right)du = \frac{1}{\sqrt{2}}\int \sec^4 u\,du$$

$$= \frac{1}{\sqrt{2}}\int (\tan^2 u + 1)\sec^2 u\,du = \frac{1}{\sqrt{2}}\left(\frac{1}{3}\tan^3 u + \tan u\right) + C$$

$$= \frac{1}{\sqrt{2}}\left(\frac{1}{3} - \frac{2\sqrt{2}w^3}{(1-2w^2)^{3/2}} + \frac{\sqrt{2}w}{(1-2w^2)^{1/2}}\right) + C$$

$$= \frac{2w^3}{3(1-2w^2)^{3/2}} + \frac{w}{(1-2w^2)^{1/2}} + C$$

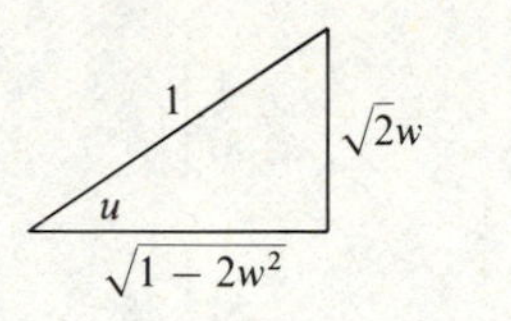

21. $\int_0^1 \frac{1}{2x^2-2x+1}\,dx = \int_0^1 \frac{1}{2(x-\frac{1}{2})^2+\frac{1}{2}}\,dx$

Let $x - \frac{1}{2} = \frac{1}{2}\tan u$, so that $dx = \frac{1}{2}\sec^2 u\,du$; if $x = 0$ then $u = -\pi/4$, and if $x = 1$ then $u = \pi/4$. Then

$$\int_0^1 \frac{1}{2x^2-2x+1}\,dx = \int_0^1 \frac{1}{2(x-\frac{1}{2})^2+\frac{1}{2}}\,dx = \int_{-\pi/4}^{\pi/4} \frac{1}{\frac{1}{2}\tan^2 u + \frac{1}{2}}\left(\frac{1}{2}\sec^2 u\right)du$$

$$= \int_{-\pi/4}^{\pi/4} 1\,du = \frac{\pi}{2}$$

23. $\int \sqrt{x-x^2} = \int \sqrt{\frac{1}{4}-(x-\frac{1}{2})^2}\,dx$. Let $x - \frac{1}{2} = \frac{1}{2}\sin u$, so that $dx = \frac{1}{2}\cos u\,du$. Then

$$\int \sqrt{\tfrac{1}{4}-(x-\tfrac{1}{2})^2}\,dx = \int \sqrt{\tfrac{1}{4}-\tfrac{1}{4}\sin^2 u}\,(\tfrac{1}{2}\cos u)\,du = \tfrac{1}{4}\int \cos^2 u\,du$$

$$= \tfrac{1}{4}\int (\tfrac{1}{2}+\tfrac{1}{2}\cos 2u)\,du = \tfrac{1}{4}(\tfrac{1}{2}u + \tfrac{1}{4}\sin 2u) + C$$

Now $x - \frac{1}{2} = \frac{1}{2}\sin u$ implies that $2x - 1 = \sin u$, so that $u = \arcsin(2x-1)$. From the figure, we see that $\cos u = (\sqrt{x-x^2})/\frac{1}{2} = 2\sqrt{x-x^2}$, so that $\sin 2u = 2\sin u\cos u = [2(2x-1)][2\sqrt{x-x^2}] = 4(2x-1)\sqrt{x-x^2}$. Thus

$$\int \sqrt{x-x^2}\,dx = \int \sqrt{\tfrac{1}{4}-(x-\tfrac{1}{2})^2}\,dx = \tfrac{1}{4}\{\tfrac{1}{2}\arcsin(2x-1) + \tfrac{1}{4}[4(2x-1)(\sqrt{x-x^2})]\} + C$$

$$= \tfrac{1}{8}\arcsin(2x-1) + \tfrac{1}{4}(2x-1)\sqrt{x-x^2} + C$$

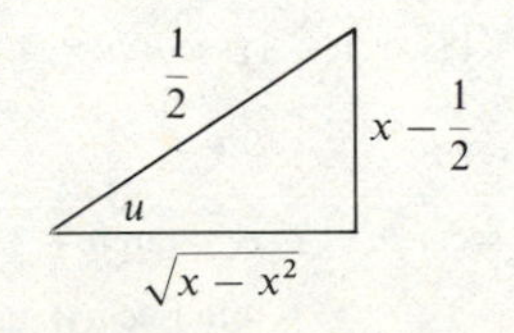

25. $x = \frac{1}{3}\sec u$, $dx = \frac{1}{3}\sec u\tan u\,du$;

$$\int \frac{x^2}{\sqrt{9x^2-1}}\,dx = \int \frac{\frac{1}{9}\sec^2 u}{\sqrt{\sec^2 u - 1}}\,\frac{1}{3}\sec u\tan u\,du = \int \frac{1}{27}\sec^3 u\,du$$

By (7) of Section 7.2, $\int \frac{1}{27}\sec^3 u\,du = \frac{1}{54}\sec u\tan u + \frac{1}{54}\ln|\sec u + \tan u| + C$. Now $x = \frac{1}{3}\sec u$ implies that $\sec u = 3x$ and $\tan u = \sqrt{9x^2-1}$. Thus

$$\int \frac{x^2}{\sqrt{9x^2-1}}\,dx = \frac{1}{54}\sec u\tan u + \frac{1}{54}\ln|\sec u + \tan u| + C$$

$$= \tfrac{1}{18}x\sqrt{9x^2-1} + \tfrac{1}{54}\ln\left|3x + \sqrt{9x^2-1}\right| + C$$

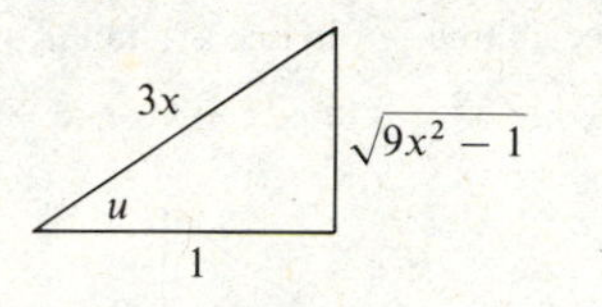

27. $x = 2\tan u$, $dx = 2\sec^2 u\,du$, $\csc u = \sqrt{x^2+4}/x$, $\cot u = 2/x$;

$$\int \frac{1}{x\sqrt{x^2+4}}\,dx = \int \frac{1}{2\tan u\sqrt{4\tan^2 u + 4}}(2\sec^2 u)\,du = \frac{1}{2}\int \frac{\sec u}{\tan u}\,du = \frac{1}{2}\int \csc u\,du$$

$$= \frac{-1}{2}\ln|\csc u + \cot u| + C = \frac{-1}{2}\ln\left|\frac{\sqrt{x^2+4}}{x} + \frac{2}{x}\right| + C$$

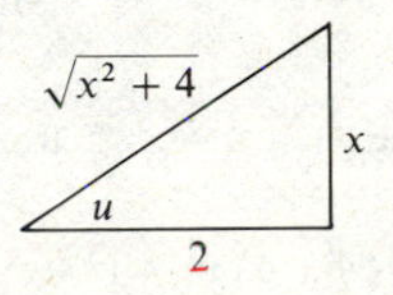

29. $x = \frac{3}{2}\sec u$, $dx = \frac{3}{2}\sec u\tan u\,du$;

$$\int \frac{1}{x^2\sqrt{4x^2-9}}\,dx = \int \frac{1}{\frac{9}{4}\sec^2 u\sqrt{9\sec^2 u - 9}}\,\frac{3}{2}\sec u\tan u\,du = \int \frac{2}{9\sec u}\,du$$

$$= \tfrac{2}{9}\int \cos u\,du = \tfrac{2}{9}\sin u + C$$

Now $x = \frac{3}{2}\sec u$ implies that $\sin u = \sqrt{4x^2-9}/(2x)$. Thus

$$\int \frac{1}{x^2\sqrt{4x^2-9}}\,dx = \frac{2}{9}\sin u + C = \frac{\sqrt{4x^2-9}}{9x} + C$$

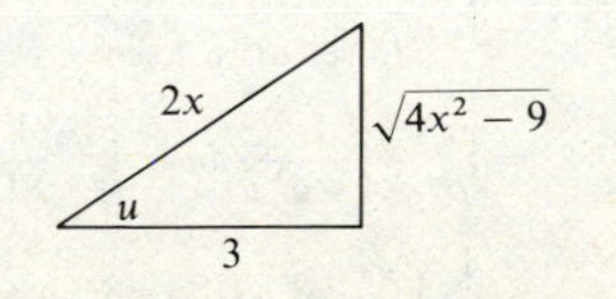

31. $x = \tan u$, $dx = \sec^2 u\,du$, $\sec u = \sqrt{1+x^2}$; from (7) in Section 7.2 we find that

$$\int \frac{x^2}{\sqrt{1+x^2}}\,dx = \int \frac{\tan^2 u}{\sqrt{1+\tan^2 u}}\sec^2 u\,du = \int \tan^2 u \sec u\,du$$
$$= \int (\sec^2 u - 1)\sec u\,du = \int (\sec^3 u - \sec u)\,du$$
$$= (\tfrac{1}{2}\sec u \tan u + \tfrac{1}{2}\ln|\sec u + \tan u|) - \ln|\sec u + \tan u| + C$$
$$= \tfrac{1}{2}\sec u \tan u - \tfrac{1}{2}\ln|\sec u + \tan u| + C$$
$$= \tfrac{1}{2}x\sqrt{1+x^2} - \tfrac{1}{2}\ln\left|\sqrt{1+x^2} + x\right| + C$$

$\sqrt{1+x^2}$, x, u, 1

33. $x = 2\sec u$, $dx = 2\sec u \tan u\,du$; if $x = 2$ then $u = 0$, and if $x = 2\sqrt{2}$ then $u = \pi/4$;

$$\int_2^{2\sqrt{2}} \frac{\sqrt{x^2-4}}{x}\,dx = \int_0^{\pi/4} \frac{\sqrt{4\sec^2 u - 4}}{2\sec u}(2\sec u \tan u)\,du = 2\int_0^{\pi/4} \tan^2 u\,du$$
$$= 2\int_0^{\pi/4} (\sec^2 u - 1)\,du = 2(\tan u - u)\Big|_0^{\pi/4} = 2 - \frac{\pi}{2}$$

35. $x = 2\tan u$, $dx = 2\sec^2 u\,du$; $\sec u = \sqrt{4+x^2}/2$; from (7) in Section 7.2 we have

$$\int \sqrt{4+x^2}\,dx = \int \sqrt{4+4\tan^2 u}\,(2\sec^2 u)\,du = 4\int \sec^3 u\,du$$
$$= 2\sec u \tan u + 2\ln|\sec u + \tan u| + C$$
$$= \frac{x\sqrt{4+x^2}}{2} + 2\ln\left|\frac{\sqrt{4+x^2}}{2} + \frac{x}{2}\right| + C$$

$\sqrt{4+x^2}$, x, u, 2

37. $x = 3\sec u$, $dx = 3\sec u \tan u\,du$; if $x = 3\sqrt{2}$ then $u = \pi/4$, and if $x = 6$ then $u = \pi/3$. Thus

$$\int_{3\sqrt{2}}^{6} \frac{1}{x^4\sqrt{x^2-9}}\,dx = \int_{\pi/4}^{\pi/3} \frac{1}{(3\sec u)^4\sqrt{9\sec^2 u - 9}}\,3\sec u \tan u\,du$$
$$= \int_{\pi/4}^{\pi/3} \frac{1}{81\sec^3 u}\,du = \int_{\pi/4}^{\pi/3} \frac{1}{81}\cos^3 u\,du$$
$$= \frac{1}{81}\int_{\pi/4}^{\pi/3} \cos u\,(1 - \sin^2 u)\,du$$

Let $v = \sin u$, so that $dv = \cos u\,du$. If $u = \pi/4$ then $v = \sqrt{2}/2$, and if $u = \pi/3$ then $v = \sqrt{3}/2$. Thus

$$\frac{1}{81}\int_{\pi/4}^{\pi/3} \cos u\,(1 - \sin^2 u)\,du = \frac{1}{81}\int_{\sqrt{2}/2}^{\sqrt{3}/2} (1 - v^2)\,dv = \frac{1}{81}\left(v - \frac{1}{3}v^3\right)\Big|_{\sqrt{2}/2}^{\sqrt{3}/2}$$
$$= \frac{1}{81}\left[\left(\frac{\sqrt{3}}{2} - \frac{1}{3}\frac{3\sqrt{3}}{8}\right) - \left(\frac{\sqrt{2}}{2} - \frac{1}{3}\frac{\sqrt{2}}{4}\right)\right]$$
$$= \frac{1}{81}\left(\frac{3\sqrt{3}}{8} - \frac{5\sqrt{2}}{12}\right)$$

39. $$\int \frac{x}{\sqrt{2x^2+12x+19}}\,dx = \int \frac{x}{\sqrt{2(x^2+6x+9)+1}}\,dx \int \frac{x}{\sqrt{2(x+3)^2+1}}\,dx$$

Let $\sqrt{2}(x+3) = \tan u$, so that $\sqrt{2}\,dx = \sec^2 u\,du$ and $x = (1/\sqrt{2})\tan u - 3$. Then

$$\int \frac{x}{\sqrt{2(x+3)^2+1}}\,dx = \int \frac{(1/\sqrt{2})\tan u - 3}{\sqrt{\tan^2 u + 1}}\frac{1}{\sqrt{2}}\sec^2 u\,du = \frac{1}{2}\int (\tan u - 3\sqrt{2})\sec u\,du$$
$$= \tfrac{1}{2}\int \tan u \sec u\,du - \tfrac{3}{2}\sqrt{2}\int \sec u\,du$$
$$= \tfrac{1}{2}\sec u - \tfrac{3}{2}\sqrt{2}\ln|\sec u + \tan u| + C$$

Since $\sqrt{2}(x+3) = \tan u$, we have $\sec u = \sqrt{[\sqrt{2}(x+3)]^2 + 1} = \sqrt{2x^2+12x+19}$, and thus

$$\int \frac{x}{\sqrt{2x^2+12x+19}}\,dx = \frac{1}{2}\sqrt{2x^2+12x+19}$$
$$- \tfrac{3}{2}\sqrt{2}\ln\left|\sqrt{2x^2+12x+19} + \sqrt{2}(x+3)\right| + C$$

41. $$\int \sqrt{x^2+6x+5}\,dx = \int \sqrt{(x+3)^2 - 4}\,dx$$

Let $x + 3 = 2\sec u$, so that $dx = 2\sec u \tan u\,du$;

$$\int \sqrt{(x+3)^2 - 4}\,dx = \int \sqrt{4\sec^2 u - 4}\,(2\sec u \tan u)\,du = \int 4\tan^2 u \sec u\,du$$
$$= 4\int (\sec^2 u - 1)\sec u\,du = 4\int \sec^3 u\,du - 4\int \sec u\,du$$

By (6) and (7) of Section 7.2,

$$4\int \sec^3 u\,du - 4\int \sec u\,du = 2\sec u \tan u + 2\ln|\sec u + \tan u|$$
$$- 4\ln|\sec u + \tan u| + C$$
$$= 2\sec u \tan u - 2\ln|\sec u + \tan u| + C$$

Now $x + 3 = 2\sec u$ implies that $\sec u = (x+3)/2$ and $\tan u = \sqrt{x^2+6x+5}/2$. Thus

$$\int \sqrt{x^2+6x+5}\,dx = 2\sec u \tan u - 2\ln|\sec u + \tan u| + C$$
$$= \frac{x+3}{2}\sqrt{x^2+6x+5} - 2\ln\left|\frac{x+3}{2} + \frac{1}{2}\sqrt{x^2+6x+5}\right| + C$$

$x+3$ $\quad \sqrt{(x+3)^2-4}$ $\quad u$ $\quad 2$

43. Let $v = e^w$, so that $dv = e^w\,dw$. Then $\int e^w\sqrt{1+e^{2w}}\,dw = \int \sqrt{1+v^2}\,dv$. Let $v = \tan u$, so that $dv = \sec^2 u\,du$. Then $\int \sqrt{1+v^2}\,dv = \int \sqrt{1+\tan^2 u}\,(\sec^2 u)\,du = \int \sec^3 u\,du$, and by (7) of Section 7.2, $\int \sec^3 u\,du = \frac{1}{2}\sec u \tan u + \frac{1}{2}\ln|\sec u + \tan u| + C$. By the figure, $\sec u = \sqrt{1+v^2}$, so that

$$\int e^w\sqrt{1+e^{2w}}\,dw = \int\sqrt{1+v^2}\,dv = (v/2)\sqrt{1+v^2} + \tfrac{1}{2}\ln\left|\sqrt{1+v^2}+v\right| + C$$
$$= \tfrac{1}{2}e^w\sqrt{1+e^{2w}} + \tfrac{1}{2}\ln\left|\sqrt{1+e^{2w}}+e^w\right| + C$$

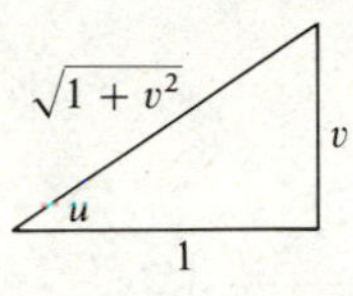

45. For integration by parts, let $u = \arcsin x$, $dv = x\,dx$, so that $du = (1/\sqrt{1-x^2})\,dx$, $v = \frac{1}{2}x^2$. Then

$$\int x \arcsin x\,dx = \frac{1}{2}x^2 \arcsin x - \int \frac{x^2/2}{\sqrt{1-x^2}}\,dx = \frac{1}{2}x^2\arcsin x - \frac{1}{2}\int\frac{x^2}{\sqrt{1-x^2}}\,dx$$

For $\int (x^2/\sqrt{1-x^2})\,dx$ let $x = \sin w$, so that $dx = \cos w\,dw$. Then

$$\int \frac{x^2}{\sqrt{1-x^2}}\,dx = \int \frac{\sin^2 w}{\sqrt{1-\sin^2 w}}\cos w\,dw = \int \sin^2 w\,dw = \int\left(\frac{1}{2}-\frac{1}{2}\cos 2w\right)dw$$
$$= \tfrac{1}{2}w - \tfrac{1}{4}\sin 2w + C_1 = \tfrac{1}{2}w - \tfrac{1}{2}\sin w\cos w + C_1$$

Now $x = \sin w$, so from the figure, $\cos w = \sqrt{1-x^2}$, and thus $\int (x^2/\sqrt{1-x^2})\,dx = \frac{1}{2}w - \frac{1}{2}\sin w \cos w + C_1 = \frac{1}{2}\arcsin x - \frac{1}{2}x\sqrt{1-x^2} + C_1$. Consequently $\int x\arcsin x\,dx = \frac{1}{2}x^2\arcsin x - \frac{1}{4}\arcsin x + \frac{1}{4}x\sqrt{1-x^2} + C$.

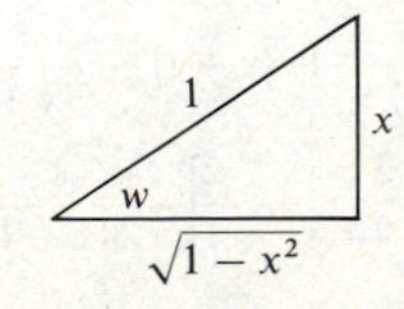

47. The area A is given by $A = \int_0^1 \sqrt{1-x^2}\,dx$. Let $x = \sin u$, so that $dx = \cos u\,du$. If $x = 0$ then $u = 0$, and if $x = 1$ then $u = \pi/2$. Thus

$$A = \int_0^1 \sqrt{1-x^2}\,dx = \int_0^{\pi/2}\sqrt{1-\sin^2 u}\cos u\,du = \int_0^{\pi/2}\cos^2 u\,du$$
$$= \int_0^{\pi/2}\left(\tfrac{1}{2}+\tfrac{1}{2}\cos 2u\right)du = \left(\tfrac{1}{2}u + \tfrac{1}{4}\sin 2u\right)\Big|_0^{\pi/2} = \pi/4$$

49. The area A is given by $A = \int_0^3 \sqrt{9+x^2}\,dx$. Let $x = 3\tan u$, so that $dx = 3\sec^2 u\,du$. If $x = 0$ then $u = 0$, and if $x = 3$ then $u = \pi/4$. By (7) of Section 7.2,

$$A = \int_0^3 \sqrt{9+x^2}\,dx = \int_0^{\pi/4}\sqrt{9+9\tan^2 u}\,(3\sec^2 u)\,du = 9\int_0^{\pi/4}\sqrt{1+\tan^2 u}\,\sec^2 u\,du$$
$$= 9\int_0^{\pi/4}\sec^3 u\,du = \left(\tfrac{9}{2}\sec u\tan u + \tfrac{9}{2}\ln|\sec u + \tan u|\right)\Big|_0^{\pi/4} = \tfrac{9}{2}\sqrt{2} + \tfrac{9}{2}\ln(\sqrt{2}+1)$$

51. By Exercise 50, the amount would be $\pi(750)(640) = 480{,}000\pi$ (square feet).

53. a. Let $x = R + r\sin\theta$, $dx = r\cos\theta\,d\theta$. Then

$$\int_{R-r}^{R+r} 4\pi x\sqrt{r^2-(x-R)^2}\,dx = 4\pi\int_{-\pi/2}^{\pi/2}(R + r\sin\theta)\sqrt{r^2 - r^2\sin^2\theta}\,(r\cos\theta)\,d\theta$$
$$= 4\pi\int_{-\pi/2}^{\pi/2} Rr^2\cos^2\theta\,d\theta + 4\pi\int_{-\pi/2}^{\pi/2} r^3\sin\theta\cos^2\theta\,d\theta$$
$$= 4\pi Rr^2\int_{-\pi/2}^{\pi/2}\left(\frac{1}{2}+\frac{1}{2}\cos 2\theta\right)d\theta - \frac{4\pi r^3}{3}\cos^3\theta\Big|_{-\pi/2}^{\pi/2}$$
$$= 4\pi Rr^2\left(\frac{\theta}{2}+\frac{1}{4}\sin 2\theta\right)\Big|_{-\pi/2}^{\pi/2} - 0 = 2\pi^2 Rr^2$$

b. Since $2\pi^2(4)(2)^2 > 2\pi^2(6)(1)^2$, the doughnut having $R = 4$ and $r = 2$ should cost more.

SECTION 7.4

1. $$\int \frac{x}{x+1}\,dx = \int\left(1 - \frac{1}{x+1}\right)dx = x - \ln|x+1| + C$$

3. $$\int \frac{x^2}{x^2-1}\,dx = \int\left(1 + \frac{1}{(x+1)(x-1)}\right)dx;\quad \frac{1}{(x+1)(x-1)} = \frac{A}{x+1} + \frac{B}{x-1};$$

$A(x-1) + B(x+1) = 1$; $A + B = 0$, $-A + B = 1$; $A = -1/2$, $B = \frac{1}{2}$;

$$\int \frac{x^2}{x^2-1}\,dx = \int\left(1 - \frac{1}{2(x+1)} + \frac{1}{2(x-1)}\right)dx$$
$$= x - \frac{1}{2}\ln|x+1| + \frac{1}{2}\ln|x-1| + C$$
$$= x + \frac{1}{2}\ln\left|\frac{x-1}{x+1}\right| + C$$

5. $$\frac{x^2+4}{x(x-1)^2} = \frac{A}{x} + \frac{B}{x-1} + \frac{C}{(x-1)^2};\quad A(x-1)^2 + Bx(x-1) + Cx = x^2 + 4;$$

$A + B = 1$, $-2A - B + C = 0$, $A = 4$; $A = 4$, $B = -3$, $C = 5$;

$$\int \frac{x^2+4}{x(x-1)^2}\,dx = \int\left(\frac{4}{x} - \frac{3}{x-1} + \frac{5}{(x-1)^2}\right)dx$$
$$= 4\ln|x| - 3\ln|x-1| - \frac{5}{x-1} + C_1$$
$$= \ln\left|\frac{x^4}{(x-1)^3}\right| - \frac{5}{x-1} + C_1$$

7. $\dfrac{5}{(x-2)(x+3)} = \dfrac{A}{x-2} + \dfrac{B}{x+3}$; $A(x+3) + B(x-2) = 5$; $A + B = 0$, $3A - 2B = 5$;

$A = 1$, $B = -1$;

$$\int_3^4 \frac{5}{(x-2)(x+3)}\,dx = \int_3^4 \left(\frac{1}{x-2} - \frac{1}{x+3}\right) dx = (\ln|x-2| - \ln|x+3|)\Big|_3^4$$
$$= \ln 2 - \ln 7 + \ln 6 = \ln \tfrac{12}{7}$$

9. $\dfrac{3t}{t^2 - 8t + 15} = \dfrac{A}{t-5} + \dfrac{B}{t-3}$; $A(t-3) + B(t-5) = 3t$; $A + B = 3$, $-3A - 5B = 0$;

$A = \frac{15}{2}$, $B = -9/2$;

$$\int \frac{3t}{t^2 - 8t + 15}\,dt = \int \left(\frac{15}{2(t-5)} - \frac{9}{2(t-3)}\right) dt = \frac{15}{2}\ln|t-5| - \frac{9}{2}\ln|t-3| + C$$

11. $$\int_{-1}^0 \frac{x^2+x+1}{x^2+1}\,dx = \int_{-1}^0 \left(1 + \frac{x}{x^2+1}\right) dx = \left[x + \frac{1}{2}\ln(x^2+1)\right]\Bigg|_{-1}^0$$
$$= -[(-1) + \tfrac{1}{2}\ln 2] = 1 - \tfrac{1}{2}\ln 2$$

13. $\displaystyle\int \frac{x^2+x+1}{x^2-1}\,dx = \int \left(1 + \frac{x+2}{x^2-1}\right) dx$; $\dfrac{x+2}{x^2-1} = \dfrac{A}{x+1} + \dfrac{B}{x-1}$; $A(x-1) + B(x+1) =$

$x + 2$; $A + B = 1$, $-A + B = 2$; $A = -1/2$, $B = \frac{3}{2}$;

$$\int \frac{x^2+x+1}{x^2-1}\,dx = \int \left(1 - \frac{1}{2(x+1)} + \frac{3}{2(x-1)}\right) dx$$
$$= x - \tfrac{1}{2}\ln|x+1| + \tfrac{3}{2}\ln|x-1| + C$$

15. $$\int_0^1 \frac{u-1}{u^2+u+1}\,du = \frac{1}{2}\int_0^1 \frac{2u+1}{u^2+u+1}\,du - \frac{3}{2}\int_0^1 \frac{1}{(u+\frac{1}{2})^2 + \frac{3}{4}}\,du;$$

$$\frac{1}{2}\int_0^1 \frac{2u+1}{u^2+u+1}\,du = \frac{1}{2}\ln(u^2+u+1)\Big|_0^1 = \frac{1}{2}\ln 3;$$

$$-\frac{3}{2}\int_0^1 \frac{1}{(u+\frac{1}{2})^2+\frac{3}{4}}\,du \overset{v=u+1/2}{=} -\frac{3}{2}\int_{1/2}^{3/2} \frac{1}{v^2+\frac{3}{4}}\,dv = \frac{-3}{2}\left(\frac{2}{\sqrt{3}}\right)\arctan\frac{2v}{\sqrt{3}}\Bigg|_{1/2}^{3/2} = \frac{-\sqrt{3}\pi}{6}$$

Thus $\displaystyle\int_0^1 \frac{u-1}{u^2+u+1}\,du = \frac{1}{2}\ln 3 - \frac{\sqrt{3}\pi}{6}$

17. $\dfrac{3x}{(x-2)^2} = \dfrac{A}{x-2} + \dfrac{B}{(x-2)^2}$; $A(x-2) + B = 3x$; $A = 3$, $-2A + B = 0$; $A = 3$, $B = 6$;

$$\int \frac{3x}{(x-2)^2}\,dx = \int \left(\frac{3}{x-2} + \frac{6}{(x-2)^2}\right) dx = 3\ln|x-2| - \frac{6}{x-2} + C$$

19. $\dfrac{-x}{x(x-1)(x-2)} = \dfrac{-1}{(x-1)(x-2)} = \dfrac{A}{x-1} + \dfrac{B}{x-2}$; $A(x-2) + B(x-1) = -1$;

$A + B = 0$, $-2A - B = -1$; $A = 1$, $B = -1$;

$$\int \frac{-x}{x(x-1)(x-2)}\,dx = \int \left(\frac{1}{x-1} - \frac{1}{x-2}\right) dx$$
$$= \ln|x-1| - \ln|x-2| + C$$
$$= \ln\left|\frac{x-1}{x-2}\right| + C$$

21. $$\int \frac{u^3}{(u+1)^2}\,du = \int \left(u - 2 + \frac{3u+2}{(u+1)^2}\right) du$$

$\dfrac{3u+2}{(u+1)^2} = \dfrac{A}{u+1} + \dfrac{B}{(u+1)^2}$; $A(u+1) + B = 3u + 2$; $A = 3$, $A + B = 2$; $A = 3$, $B = -1$;

$$\int \frac{u^3}{(u+1)^2}\,du = \int \left(u - 2 + \frac{3}{u+1} - \frac{1}{(u+1)^2}\right) du$$
$$= \frac{1}{2}u^2 - 2u + 3\ln|u+1| + \frac{1}{u+1} + C$$

23. $$\int \frac{1}{(1-x^2)^2}\,dx = \int \left(\frac{1}{2(x+1)} - \frac{1}{2(x-1)}\right)^2 dx$$
$$= \int \left(\frac{1}{4(x+1)^2} - \frac{1}{2(x+1)(x-1)} + \frac{1}{4(x-1)^2}\right) dx$$
$$= \int \left[\frac{1}{4(x+1)^2} + \frac{1}{2}\left(\frac{1}{2(x+1)} - \frac{1}{2(x-1)}\right) + \frac{1}{4(x-1)^2}\right] dx$$
$$= \frac{-1}{4(x+1)} + \frac{1}{4}\ln\left|\frac{x+1}{x-1}\right| - \frac{1}{4(x-1)} + C$$

25. $\dfrac{x}{(x+1)^2(x-2)} = \dfrac{A}{x+1} + \dfrac{B}{(x+1)^2} + \dfrac{C}{x-2}$;

$A(x+1)(x-2) + B(x-2) + C(x+1)^2 = x$; $A + C = 0$, $-A + B + 2C = 1$,

$-2A - 2B + C = 0$; $A = -2/9$, $B = \frac{1}{3}$, $C = \frac{2}{9}$;

$$\int \frac{x}{(x+1)^2(x-2)}\,dx = \int \left(\frac{-2}{9(x+1)} + \frac{1}{3(x+1)^2} + \frac{2}{9(x-2)}\right) dx$$
$$= \frac{-2}{9}\ln|x+1| - \frac{1}{3(x+1)} + \frac{2}{9}\ln|x-2| + C_1$$
$$= \frac{2}{9}\ln\left|\frac{x-2}{x+1}\right| - \frac{1}{3(x+1)} + C_1$$

27. $\dfrac{-x^3+x^2+x+3}{(x+1)(x^2+1)^2} = \dfrac{A}{x+1} + \dfrac{Bx+C}{x^2+1} + \dfrac{Dx+E}{(x^2+1)^2}$;

$A(x^2+1)^2 + (Bx+C)(x+1)(x^2+1) + (Dx+E)(x+1) = -x^3 + x^2 + x + 3$;

$A + B = 0$, $B + C = -1$, $2A + B + C + D = 1$, $B + C + D + E = 1$, $A + C + E = 3$;

$A = 1$, $B = -1$, $C = 0$, $D = 0$, $E = 2$.

With the help of the evaluation of $\int [1/(x^2+1)^2]\,dx$ in Exercise 26, we find that

$$\int \frac{-x^3+x^2+x+3}{(x+1)(x^2+1)^2}\,dx = \int \left(\frac{1}{x+1} - \frac{x}{x^2+1} + \frac{2}{(x^2+1)^2}\right) dx$$

$$= \ln|x+1| - \frac{1}{2}\ln(x^2+1) + 2\left(\frac{1}{2}\arctan x + \frac{1}{2}\frac{x}{x^2+1}\right) + C_1$$

$$= \ln|x+1| - \frac{1}{2}\ln(x^2+1) + \arctan x + \frac{x}{x^2+1} + C_1$$

29. $\dfrac{x^2-1}{x^3+3x+4} = \dfrac{(x+1)(x-1)}{(x+1)(x^2-x+4)} = \dfrac{x-1}{x^2-x+4}$;

$$\int \frac{x^2-1}{x^3+3x+4}\,dx = \int \left(\frac{2x-1}{2(x^2-x+4)} - \frac{1}{2(x^2-x+4)}\right) dx$$

$$= \frac{1}{2}\ln|x^2-x+4| - \frac{1}{2}\int \frac{1}{(x-\frac{1}{2})^2+\frac{15}{4}}\,dx$$

$$= \frac{1}{2}\ln|x^2-x+4| - \frac{1}{2}\left(\frac{2}{\sqrt{15}}\arctan\frac{x-\frac{1}{2}}{\sqrt{15}/2}\right) + C$$

$$= \frac{1}{2}\ln|x^2-x+4| - \frac{1}{\sqrt{15}}\arctan\frac{\sqrt{15}(2x-1)}{15} + C$$

31. Let $u = \sqrt{x+1}$, so $du = [1/(2\sqrt{x+1})]\,dx$ and $dx = 2u\,du$;

$$\int \frac{1}{x\sqrt{x+1}}\,dx = \int \frac{2u}{(u^2-1)u}\,du = \int \frac{2}{(u+1)(u-1)}\,du = \int \left(\frac{-1}{u+1} + \frac{1}{u-1}\right) du$$

$$= -\ln|u+1| + \ln|u-1| + C = \ln\left|\frac{\sqrt{x+1}-1}{\sqrt{x+1}+1}\right| + C$$

33. Let $u = \sqrt[6]{x}$, so $du = \frac{1}{6}x^{-5/6}\,dx$ and $dx = 6u^5\,du$;

$$\int \frac{\sqrt{x}}{1+\sqrt[3]{x}}\,dx = \int \frac{u^3}{1+u^2}(6u^5)\,du = 6\int \frac{u^8}{1+u^2}\,du$$

$$= 6\int \left(u^6 - u^4 + u^2 - 1 + \frac{1}{u^2+1}\right) du$$

$$= 6(\tfrac{1}{7}u^7 - \tfrac{1}{5}u^5 + \tfrac{1}{3}u^3 - u + \arctan u) + C$$

$$= \tfrac{6}{7}x^{7/6} - \tfrac{6}{5}x^{5/6} + 2x^{1/2} - 6x^{1/6} + 6\arctan x^{1/6} + C$$

35. Let $u = \sqrt{\dfrac{x+1}{x-1}}$, so $dx = \dfrac{-4u}{(u^2-1)^2}\,du$ and $x = \dfrac{u^2+1}{u^2-1}$;

$$\int_{-5/3}^{-1} \sqrt{\frac{x+1}{x-1}}\,dx = \int_{1/2}^{0} \frac{-4u^2}{(u^2-1)^2}\,du = -4\int_{1/2}^{0}\left(\frac{u}{u^2-1}\right)^2 du$$

$$= -4\int_{1/2}^{0}\left(\frac{1}{2(u+1)} + \frac{1}{2(u-1)}\right)^2 du$$

$$= -\int_{1/2}^{0}\left(\frac{1}{(u+1)^2} + \frac{2}{(u+1)(u-1)} + \frac{1}{(u-1)^2}\right) du$$

$$= -\int_{1/2}^{0}\left(\frac{1}{(u+1)^2} - \frac{1}{u+1} + \frac{1}{u-1} + \frac{1}{(u-1)^2}\right) du$$

$$= \left(\frac{1}{u+1} + \ln|u+1| - \ln|u-1| + \frac{1}{u-1}\right)\Big|_{1/2}^{0} = \frac{4}{3} - \ln 3$$

37. Let $u = \sin x$, so that $du = \cos x\,dx$. Then

$$\int \frac{\sin^2 x\cos x}{\sin^2 x + 1}\,dx = \int \frac{u^2}{u^2+1}\,du = \int \left(1 - \frac{1}{u^2+1}\right) du = u - \arctan u + C$$

$$= \sin x - \arctan(\sin x) + C$$

39. Let $u = e^x$, so that $du = e^x\,dx$;

$$\int \frac{e^x}{1-e^{3x}}\,dx = \int \frac{1}{1-u^3}\,du = \int \frac{1}{(1-u)(1+u+u^2)}\,du = \int \left(\frac{A}{1-u} + \frac{Bu+C}{1+u+u^2}\right) du;$$

$A(1+u+u^2) + (Bu+C)(1-u) = 1$; $A - B = 0$, $A + B - C = 0$, $A + C = 1$, so $A = \frac{1}{3}$, $B = \frac{1}{3}$, $C = \frac{2}{3}$. Therefore

$$\int \frac{1}{(1-u)(1+u+u^2)}\,du = \int \left[\frac{1}{3(1-u)} + \frac{u+2}{3(1+u+u^2)}\right] du$$

$$= \frac{1}{3}\int \frac{1}{1-u}\,du + \frac{1}{6}\int \frac{2u+1}{1+u+u^2}\,du + \frac{1}{2}\int \frac{1}{(u+\frac{1}{2})^2+\frac{3}{4}}\,du$$

$$= -\frac{1}{3}\ln|1-u| + \frac{1}{6}\ln(1+u+u^2) + \frac{1}{2}\frac{2}{\sqrt{3}}\arctan\frac{u+\frac{1}{2}}{\sqrt{3}/2} + C$$

$$= -\frac{1}{3}\ln|1-e^x| + \frac{1}{6}\ln(1+e^x+e^{2x})$$

$$+ \frac{1}{\sqrt{3}}\arctan\frac{2e^x+1}{\sqrt{3}} + C$$

41. To integrate by parts, let $u = \arctan x$, $dv = x\,dx$, so that $du = [1/(1+x^2)]\,dx$, $v = \frac{1}{2}x^2$. Then

$$\int x\arctan x\,dx = \frac{1}{2}x^2\arctan x - \int \frac{1}{2}x^2\frac{1}{1+x^2}\,dx = \frac{1}{2}x^2\arctan x - \frac{1}{2}\int \frac{x^2}{1+x^2}\,dx$$

Now
$$\int \frac{x^2}{1+x^2}\,dx = \int \left(1 - \frac{1}{1+x^2}\right) dx.$$

Thus

$$\int x\arctan x\,dx = \frac{1}{2}x^2\arctan x - \frac{1}{2}\int \left(1 - \frac{1}{1+x^2}\right) dx$$

$$= \tfrac{1}{2}x^2\arctan x - \tfrac{1}{2}x + \tfrac{1}{2}\arctan x + C$$

43. To integrate by parts, let $u = \ln(x^2+1)$, $dv = dx$, so that $du = [2x/(x^2+1)]\,dx$, $v = x$. Then

$$\int \ln(x^2+1)\,dx = x\ln(x^2+1) - \int x\frac{2x}{x^2+1}\,dx = x\ln(x^2+1) - 2\int \frac{x^2}{x^2+1}\,dx$$

Now $\int \frac{x^2}{x^2+1}\,dx = \int \left(1 - \frac{1}{x^2+1}\right) dx = x - \arctan x + C_1$

Thus $\int \ln(x^2+1)\,dx = x\ln(x^2+1) - 2x + 2\arctan x + C$

45. $$\int \frac{1}{2+\sin x}\,dx = \int \frac{1}{2 + 2u/(1+u^2)}\left(\frac{2}{1+u^2}\right) du = \int \frac{1}{1+u+u^2}\,du$$
$$= \int \frac{1}{(u+\frac{1}{2})^2 + \frac{3}{4}}\,du = \frac{2}{\sqrt{3}}\arctan\frac{2(u+\frac{1}{2})}{\sqrt{3}} + C$$
$$= \frac{2}{\sqrt{3}}\arctan\frac{1}{\sqrt{3}}\left(2\tan\frac{x}{2} + 1\right) + C$$

47. $$\int \frac{1}{2\cos x + \sin x}\,dx = \int \frac{1}{\dfrac{2(1-u^2)}{1+u^2} + \dfrac{2u}{1+u^2}}\left(\frac{2}{1+u^2}\right) du$$
$$= \int \frac{1}{1+u-u^2}\,du = \int \frac{1}{\frac{5}{4} - (u-\frac{1}{2})^2}\,du$$
$$\overset{v=u-1/2}{=} \int \frac{1}{\frac{5}{4} - v^2}\,dv = \int \frac{1}{(\sqrt{5}/2 - v)(\sqrt{5}/2 + v)}\,dv$$
$$= \frac{1}{\sqrt{5}}\int\left(\frac{1}{\sqrt{5}/2 - v} + \frac{1}{\sqrt{5}/2 + v}\right) dv$$
$$= \frac{-1}{\sqrt{5}}\ln\left|\frac{\sqrt{5}}{2} - v\right| + \frac{1}{\sqrt{5}}\ln\left|\frac{\sqrt{5}}{2} + v\right| + C$$
$$= \frac{\sqrt{5}}{5}\ln\left|\frac{\sqrt{5} - 1 + 2\tan(x/2)}{\sqrt{5} + 1 - 2\tan(x/2)}\right| + C$$

49. $A = \int_0^3 \frac{x^3}{x^2+1}\,dx = \int_0^3 \left(x - \frac{x}{x^2+1}\right) dx = \left[\frac{1}{2}x^2 - \frac{1}{2}\ln(x^2+1)\right]\Big|_0^3$
$= (\frac{9}{2} - \frac{1}{2}\ln 10) + \frac{1}{2}\ln 1 = \frac{9}{2} - \frac{1}{2}\ln 10$

51. The graphs intersect at (x, y) if $x^2/[(x-2)(x^2+1)] = y = 1/(x-3)$, or $x^3 - 3x^2 = x^3 - 2x^2 + x - 2$, or $x^2 + x - 2 = 0$, or $x = 1$ or -2. Since $x^2/[(x-2)(x^2+1)] \geq 1/(x-3)$ on $[-2, 1]$, the area is given by

$$A = \int_{-2}^{1}\left(\frac{x^2}{(x-2)(x^2+1)} - \frac{1}{x-3}\right) dx$$

Now $$\frac{x^2}{(x-2)(x^2+1)} = \frac{B}{x-2} + \frac{Cx+D}{x^2+1}$$

$B(x^2+1) + (Cx+D)(x-2) = x^2$; $B + C = 1$, $(-2C + D) = 0$, $B - 2D = 0$; $B = \frac{4}{5}$, $C = \frac{1}{5}$, $D = \frac{2}{5}$; thus

$$A = \int_{-2}^{1}\left(\frac{4}{5(x-2)} + \frac{x+2}{5(x^2+1)} - \frac{1}{x-3}\right) dx$$
$$= \int_{-2}^{1}\left(\frac{4}{5}\frac{1}{x-2} + \frac{1}{10}\cdot\frac{2x}{x^2+1} + \frac{2}{5}\cdot\frac{1}{x^2+1} - \frac{1}{x-3}\right) dx$$
$$= \left(\frac{4}{5}\ln|x-2| + \frac{1}{10}\ln(x^2+1) + \frac{2}{5}\arctan x - \ln|x-3|\right)\Big|_{-2}^{1}$$
$$= \left(\frac{1}{10}\ln 2 + \frac{\pi}{10} - \ln 2\right) - \left(\frac{4}{5}\ln 4 + \frac{1}{10}\ln 5 + \frac{2}{5}\arctan(-2) - \ln 5\right)$$
$$= -\frac{5}{2}\ln 2 + \frac{9}{10}\ln 5 + \frac{\pi}{10} + \frac{2}{5}\arctan 2$$

SECTION 7.5

1. By Formula 77 in the Table, with $a = 3$, we have $\int \sqrt{x^2+9}\,dx = (x/2)\sqrt{x^2+9} + \frac{9}{2}\ln|x + \sqrt{x^2+9}| + C$.

3. By Formula 49 in the Table, with $a = 5$ and $b = \frac{1}{2}$, we have
$$\int_0^1 e^{5x}\sin\frac{1}{2}x\,dx = \frac{e^{5x}}{25 + \frac{1}{4}}\left(5\sin\frac{1}{2}x - \frac{1}{2}\cos\frac{1}{2}x\right)\Big|_0^1$$
$$= \tfrac{4}{101}e^5[(5\sin\tfrac{1}{2} - \tfrac{1}{2}\cos\tfrac{1}{2}) + \tfrac{1}{2}]$$

5. To use Formula 87 in the Table, we make the substitution $u = 2x$, and note that $a = 3$ in Formula 87:
$$\int \frac{1}{4x^2 - 9}\,dx \overset{u=2x}{=} \frac{1}{2}\int \frac{1}{u^2 - 9}\,du = \frac{1}{12}\ln\left|\frac{u-3}{u+3}\right| + C = \frac{1}{12}\ln\left|\frac{2x-3}{2x+3}\right| + C$$

7. To use Formula 115 in the Table, we make the substitution $u = \frac{1}{2}x$, and note that $a = 10$ in Formula 115:
$$\int \frac{\sqrt{10x - \frac{1}{4}x^2}}{x}\,dx \overset{u=1/2x}{=} 2\int \frac{\sqrt{20u - u^2}}{2u}\,du = \int \frac{\sqrt{2\cdot 10u - u^2}}{u}\,du$$
$$= \sqrt{2\cdot 10u - u^2} + 10\arccos\left(1 - \frac{u}{10}\right) + C$$
$$= \sqrt{10x - \frac{1}{4}x^2} + 10\arccos\left(1 - \frac{x}{20}\right) + C$$

9. To use Formula 63 in the Table, we make the substitution $u = \sqrt{x}$, and note that $a = 1$ and $b = 2$ in Formula 63:
$$\int \frac{e^{\sqrt{x}}}{\sqrt{x}}\sinh 2\sqrt{x}\,dx \overset{u=\sqrt{x}}{=} 2\int e^u \sinh 2u\,du = \frac{2e^u}{1^2 - 2^2}(\sinh 2u - 2\cosh 2u) + C$$
$$= \tfrac{2}{3}e^{\sqrt{x}}(2\cosh 2\sqrt{x} - \sinh 2\sqrt{x}) + C$$

11. To use Formula 27 in the Table, we make the substitution $u = e^x$:
$$\int_0^1 e^{2x}\cos e^x\,dx = \int_0^1 (e^x\cos e^x)e^x\,dx \overset{u=e^x}{=} \int_1^e u\cos u\,du = (u\sin u + \cos u)\Big|_1^e$$
$$= (e\sin e + \cos e) - (\sin 1 + \cos 1) = e\sin e + \cos e - \sin 1 - \cos 1$$

13. To use Formula 49 in the Table, we make the substitution $u = \ln x$, and note that $a = 1$ and $b = 1$ in Formula 49:

$$\int \sin(\ln x)\,dx = \int x[\sin(\ln x)]\frac{1}{x}\,dx \overset{u=\ln x}{=} \int e^u \sin u\,du = \frac{e^u}{2}(\sin u - \cos u) + C$$

$$= \frac{e^{\ln x}}{2}[\sin(\ln x) - \cos(\ln x)] + C = \tfrac{1}{2}x[\sin(\ln x) - \cos(\ln x)] + C$$

15. To use Formula 114 in the Table, we make the substitution $u = \sqrt{x}$, and note that $a = 1$ in Formula 114:

$$\int \sqrt{2\sqrt{x} - x}\,dx \overset{u=\sqrt{x}}{=} 2\int u\sqrt{2u - u^2}\,du = \frac{2u^2 - u - 3}{3}\sqrt{2u - u^2} + \arccos(1 - u) + C$$

$$= \frac{2x - \sqrt{x} - 3}{3}\sqrt{2\sqrt{x} - x} + \arccos(1 - \sqrt{x}) + C$$

17. By Formula 122 in the Table, with $a = 2$, we have

$$\int x\sqrt{\frac{2 + x}{2 - x}}\,dx = -\frac{4 + x}{2}\sqrt{4 - x^2} + 2\arcsin\frac{x}{2} + C$$

SECTION 7.6

Let T and S be the approximations by the Trapezoidal Rule and Simpson's Rule, respectively.

1. $T = \frac{2}{4}(1 + 2(\frac{1}{2}) + \frac{1}{3}) = \frac{7}{6}$; $S = \frac{2}{6}(1 + 4(\frac{1}{2}) + \frac{1}{3}) = \frac{10}{9}$

3. $T = \frac{6}{12}(1 + 2(\frac{1}{2}) + 2(\frac{1}{3}) + 2(\frac{1}{4}) + 2(\frac{1}{5}) + 2(\frac{1}{6}) + \frac{1}{7}) = \frac{283}{140} \approx 2.02143$

   $S = \frac{6}{18}(1 + 4(\frac{1}{2}) + 2(\frac{1}{3}) + 4(\frac{1}{4}) + 2(\frac{1}{5}) + 4(\frac{1}{6}) + \frac{1}{7}) = \frac{617}{315} \approx 1.95873$

5. $T = \frac{2}{4}(1 + 2\sqrt{2} + 3) = 2 + \sqrt{2} \approx 3.41421$; $S = \frac{2}{6}(1 + 4\sqrt{2} + 3) = \dfrac{4 + 4\sqrt{2}}{3} \approx 3.21895$

7. $T = \frac{4}{8}[-1 + 2(2\ln 2 - 2) + 2(3\ln 3 - 3) + 2(4\ln 4 - 4) + (5\ln 5 - 5)] \approx 2.25090$

   $S = \frac{4}{12}[-1 + 4(2\ln 2 - 2) + 2(3\ln 3 - 3) + 4(4\ln 4 - 4) + (5\ln 5 - 5)] \approx 2.12158$

9. $T = \dfrac{2}{8}\left[0 + 2\cdot\dfrac{\sqrt{2}}{2} + 2\cdot 0 + 2\cdot\dfrac{\sqrt{2}}{2} + 0\right] = \dfrac{1}{2}\sqrt{2}$

   $S = \dfrac{2}{12}\left[0 + 4\cdot\dfrac{\sqrt{2}}{2} + 2\cdot 0 + 4\cdot\dfrac{\sqrt{2}}{2} + 0\right] = \dfrac{2}{3}\sqrt{2}$

11. $T = \dfrac{\pi}{8}\left[1 + 2\dfrac{1}{1 + \sqrt{2}/2} + 2\left(\dfrac{1}{2}\right) + 2\dfrac{1}{1 + \sqrt{2}/2} + 1\right] \approx 2.09825$

    $S = \dfrac{\pi}{12}\left[1 + 4\dfrac{1}{1 + \sqrt{2}/2} + 2\left(\dfrac{1}{2}\right) + 4\dfrac{1}{1 + \sqrt{2}/2} + 1\right] \approx 2.01227$

13. $S = \frac{1}{18}(1 + 4\sqrt{1 + (\frac{1}{6})^6} + 2\sqrt{1 + (\frac{1}{3})^6} + 4\sqrt{1 + (\frac{1}{2})^6} + 2\sqrt{1 + (\frac{2}{3})^6} + 4\sqrt{1 + (\frac{5}{6})^6} + \sqrt{2}) \approx 1.06412$

15. $S = \frac{2}{30}[\ln(1 + 1) + 4\ln((\frac{6}{5})^2 + 1) + 2\ln((\frac{7}{5})^2 + 1) + 4\ln((\frac{8}{5})^2 + 1) + 2\ln((\frac{9}{5})^2 + 1) + 4\ln(4 + 1) + 2\ln((\frac{11}{5})^2 + 1) + 4\ln((\frac{12}{5})^2 + 1) + 2\ln((\frac{13}{5})^2 + 1) + 4\ln((\frac{14}{5})^2 + 1) + \ln(9 + 1)] \approx 3.14191$

17. Let $f(x) = \ln x$. To use (3) we calculate that $f'(x) = 1/x$ and $f''(x) = -1/x^2$, so that $|f''(x)| \le 1$ for x in $[1, 2]$. Thus we take $M = 1$. Therefore by (3),

$$E_n^T \le \frac{(2 - 1)^3 1}{12n^2} = \frac{1}{12n^2}$$

so that $E_n^T \le 0.01$ if $1/(12n^2) \le 0.01$, that is, if $n \ge \sqrt{100/12}$. Since $2 < \sqrt{100/12} < 3$, we take $n = 3$.

19. Let $f(x) = x\ln x - x$. To use (8) we calculate that $f'(x) = \ln x + 1 - 1 = \ln x$, $f''(x) = 1/x$, $f^{(3)}(x) = -1/x^2$, $f^{(4)}(x) = 2/x^3$, so that $|f^{(4)}(x)| \le 2$ for x in $[1, 5]$. Thus we take $M = 2$. Therefore by (8),

$$E_n^S \le \frac{(5 - 1)^5 2}{180n^4} = \frac{2(4^5)}{180n^4} = \frac{512}{45n^4}$$

so that $E_n^S \le 10^{-4}$ if $512/45n^4 \le 10^{-4}$, that is, if $n \ge 10\sqrt[4]{512/45}$. Since $18 < 10\sqrt[4]{512/45} < 19$, and since n must be even, we take $n = 20$.

21. Let $f(x) = 1/x$. To use (3) we calculate that $f'(x) = -1/x^2$ and $f''(x) = 2/x^3$, so that $|f''(x)| \le 2$ for x in $[1, 3]$. Thus we take $M = 2$. Therefore by (3),

$$E_n^T \le \frac{(3 - 1)^3(2)}{12n^2} = \frac{4}{3n^2}$$

so that $E_4^T \le 4/[3(4^2)] = \frac{1}{12}$. The estimate by the Trapezoidal Rule is $T = \frac{2}{8}[1 + 2(\frac{2}{3}) + 2(\frac{1}{2}) + 2(\frac{2}{5}) + \frac{1}{3}] = \frac{67}{60}$. Thus

$$\frac{67}{60} - \frac{1}{12} \le \int_1^3 \frac{1}{x}\,dx \le \frac{67}{60} + \frac{1}{12}, \quad \text{or more simply,} \quad \frac{31}{30} \le \int_1^3 \frac{1}{x}\,dx \le \frac{6}{5}$$

23. Let $f(x) = 1/x$. To use (8) we calculate that $f'(x) = -1/x^2$, $f''(x) = 2/x^3$, $f^{(3)}(x) = -6/x^4$, $f^{(4)}(x) = 24/x^5$, so that $|f^{(4)}(x)| \le 24$ for x in $[1, 3]$. Thus we take $M = 24$. Therefore by (8),

$$E_n^S \le \frac{(3 - 1)^5 24}{180n^4} = \frac{64}{15n^4}$$

so that $E_4^S \le 64/[15(4^4)] = \frac{1}{60}$. The estimate by Simpson's Rule is $S = \frac{2}{12}[1 + 4(\frac{2}{3}) + 2(\frac{1}{2}) + 4(\frac{2}{5}) + \frac{1}{3}] = \frac{396}{360} = \frac{11}{10}$. Thus

$$\frac{11}{10} - \frac{1}{60} \le \int_1^3 \frac{1}{x}\,dx \le \frac{11}{10} + \frac{1}{60}, \quad \text{or more simply,} \quad \frac{13}{12} \le \int_1^3 \frac{1}{x}\,dx \le \frac{67}{60}.$$

25. $$A = \int_{-\pi/3}^{\pi/3} \frac{1}{1 + \cos x}\,dx \approx \frac{2\pi}{36}\left[\frac{1}{1 + \frac{1}{2}} + 4\frac{1}{1 + \sqrt{3}/2} + 2\left(\frac{1}{2}\right) + 4\frac{1}{1 + \sqrt{3}/2} + \frac{1}{1 + \frac{1}{2}}\right]$$

$$\approx 1.15550$$

27. a. $T = \frac{1}{12}(4 + 2(\frac{144}{37}) + 2(\frac{36}{10}) + 2(\frac{16}{5}) + 2(\frac{36}{13}) + 2(\frac{144}{61}) + 2) = 3.13696$
$S = \frac{1}{12}(4 + 4(\frac{64}{17}) + 2(\frac{16}{5}) + 4(\frac{64}{25}) + 2) = 3.14157$

b. $E_6^T \approx |3.14159 - 3.13696| = 0.00463$; $E_4^S \approx |3.14159 - 3.14157| = .00002$

29. a. If $f(x) = 1/x$, then $f'(x) = -1/x^2$, $f''(x) = 2/x^3$, $f^{(3)}(x) = -6/x^4$, $f^{(4)}(x) = 24/x^5$, so $M = \max_{1 \le x \le 8} |24/x^5| = 24$. Thus $E_n^S \le [(8-1)^5 24]/(180n^4)$, so $E_n^S \le 10^{-4}$ if $7^5(24)/(180n^4) \le 10^{-4}$, that is, $n^4 \ge 7^5(24)10^4/180$, or $n \ge 68.8$. Therefore, since n must be even, we take $n = 70$.

b. Since $M = \max_{1 \le x \le 2} 24/x^5 = 24$, it follows that $E_n^s \le [(2-1)^5 24]/(180n^4) = 2/(15n^4)$ and thus $3E_n^S \le 10^{-4}$ if $6/(15n^4) \le 10^{-4}$, that is, $n^4 \ge (6)10^4/15$, or $n \ge 7.96$. Therefore we take $n = 8$.

31. a. From the figure we see that the trapezoid above $[x_{k-1}, x_k]$ on the x axis contains the region bounded by the graph of f. Adding the corresponding areas, we obtain $T_n \ge \int_a^b f(x)\,dx$.

b. This time the trapezoids are contained in the regions determined by the graph of f, so $T_n \le \int_a^b f(x)\,dx$.

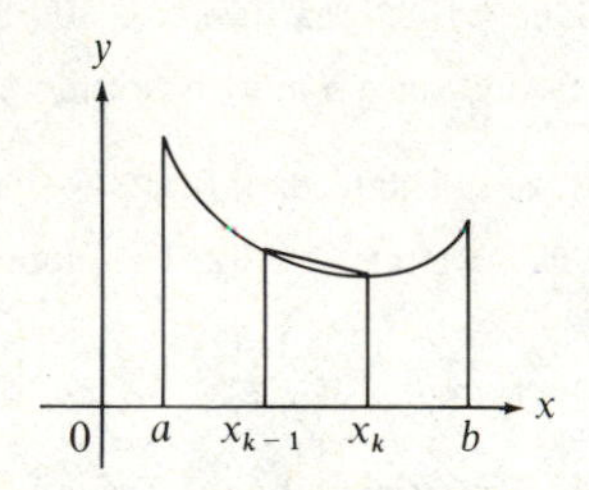

33. If $f(x) = c_3x^3 + c_2x^2 + c_1x + c_0$, then $f^{(4)}(x) = 0$ for $a \le x \le b$, so that by (8) the error E_n^S by using Simpson's Rule satisfies

$$0 \le E_n^S \le \frac{(b-a)^5 \cdot 0}{180n^4} = 0$$

35. a. If k is odd, then

$$\int_{-a}^{a} x^k\,dx = \frac{1}{k+1}x^{k+1}\Big|_{-a}^{a} = 0$$

The estimate by Simpson's Rule with $n = 2$ is

$$S = \frac{2a}{6}((-a)^k + 4(0) + a^k) = 0$$

Thus $$S = \int_{-a}^{a} x^k\,dx \quad \text{if } k \text{ is odd}$$

b. If $k = 2$, then $\int_{-a}^{a} x^k\,dx = \int_{-a}^{a} x^2\,dx = \frac{1}{3}x^3\Big|_{-a}^{a} = \frac{2}{3}a^3$. The estimate by Simpson's Rule with $n = 2$ is

$$S = \frac{2a}{6}((-a)^2 + 4(0) + a^2) = \frac{2}{3}a^3$$

Thus $$S = \int_{-a}^{a} x^2\,dx \quad \text{if } k = 2$$

c. In general, if k is even, then

$$\int_{-a}^{a} x^k\,dx = \frac{1}{k+1}x^{k+1}\Big|_{-a}^{a} = \frac{2}{k+1}a^{k+1} \quad \text{and} \quad S = \frac{2a}{6}((-a)^k + 4(0) + a^k) = \frac{2}{3}a^{k+1}$$

so that $S = \int_{-a}^{a} x^k\,dx$ only if $k = 2$.

37. We have $x_1 - x_0 = x_2 - x_1 = h$ and $x_2 - x_0 = 2h$. Thus

$$\int_{x_0}^{x_2} f(x_0)\,dx = f(x_0)(x_2 - x_0) = 2hf(x_0)$$

$$\int_{x_0}^{x_2} \frac{f(x_1) - f(x_0)}{h}(x - x_0)\,dx = \frac{f(x_1) - f(x_0)}{h}\cdot\frac{1}{2}(x - x_0)^2\Big|_{x_0}^{x_2} = 2h[f(x_1) - f(x_0)]$$

by integration by parts with $u = (x - x_0)$, $dv = (x - x_1)\,dx$, $du = dx$, and $v = \frac{1}{2}(x - x_1)^2$, we have

$$\begin{aligned}
&\int_{x_0}^{x_2} \frac{f(x_0) - 2f(x_1) + f(x_2)}{2h^2}(x - x_0)(x - x_1)\,dx \\
&= \frac{f(x_0) - 2f(x_1) + f(x_2)}{2h^2}\left[(x - x_0)\cdot\frac{1}{2}(x - x_1)^2\Big|_{x_0}^{x_2} - \int_{x_0}^{x_2}\frac{1}{2}(x - x_1)^2\,dx\right] \\
&= \frac{f(x_0) - 2f(x_1) + f(x_2)}{2h^2}\left[h^3 - \frac{1}{6}(x - x_1)^3\Big|_{x_0}^{x_2}\right] \\
&= \frac{h}{3}[f(x_0) - 2f(x_1) + f(x_2)]
\end{aligned}$$

Adding these integrals, we find that

$$\begin{aligned}
\int_{x_0}^{x_2} p(x)\,dx &= 2h\,f(x_0) + 2h\,[f(x_1) - f(x_0)] + \frac{h}{3}[f(x_0) - 2f(x_1) + f(x_2)] \\
&= \frac{h}{3}[f(x_0) + 4f(x_1) + f(x_2)]
\end{aligned}$$

SECTION 7.7

1. converges; $\int_0^1 \frac{1}{x^{0.9}}\,dx = \lim_{c \to 0^+} \int_c^1 \frac{1}{x^{0.9}}\,dx = \lim_{c \to 0^+} 10x^{0.1}\Big|_c^1 = \lim_{c \to 0^+} 10(1 - c^{0.1}) = 10$

3. diverges; $\int_3^4 \frac{1}{(t-4)^2}\,dt = \lim_{c \to 4^-} \int_3^c \frac{1}{(t-4)^2}\,dt = \lim_{c \to 4^-} \frac{-1}{t-4}\Big|_3^c = \lim_{c \to 4^-}\left(\frac{-1}{c-4} - 1\right) = \infty$

5. converges; $\int_3^4 \frac{1}{\sqrt[3]{x-3}}\,dx = \lim_{c \to 3^+} \int_c^4 \frac{1}{\sqrt[3]{x-3}}\,dx = \lim_{c \to 3^+} \frac{3}{2}(x-3)^{2/3}\Big|_c^4$
$= \lim_{c \to 3^+} (\frac{3}{2} - \frac{3}{2}(c-3)^{2/3}) = \frac{3}{2}$

7. diverges; $\int_0^{\pi/2} \sec^2\theta\,d\theta = \lim_{c \to \pi/2^-} \int_0^c \sec^2\theta\,d\theta = \lim_{c \to \pi/2^-} \tan\theta\Big|_0^c = \lim_{c \to \pi/2^-} \tan c = \infty$

9. converges; let $u = 1 + \cos x$, so that $du = -\sin x\,dx$; then

$$\int_0^\pi \frac{\sin x}{\sqrt{1+\cos x}}\,dx = \lim_{c\to\pi^-}\int_0^c \frac{\sin x}{\sqrt{1+\cos x}}\,dx = \lim_{c\to\pi^-}\int_2^{1+\cos c}\frac{-1}{u^{1/2}}\,du$$
$$= \lim_{c\to\pi^-}(-2u^{1/2})\Big|_2^{1+\cos c} = \lim_{c\to\pi^-}[-2(1+\cos c)^{1/2} + 2\sqrt{2}] = 2\sqrt{2}$$

11. diverges; $\int_1^2 \frac{1}{w\ln w}\,dw = \lim_{c\to1^+}\int_c^2 \frac{1}{w\ln w}\,dw = \lim_{c\to1^+}\ln(\ln w)\Big|_c^2$

$$= \lim_{c\to1^+}[\ln(\ln 2) - \ln(\ln c)] = \infty$$

13. converges. Let $d = \frac{1}{2}$. Then

$$\int_0^1 \frac{3x^2-1}{\sqrt[3]{x^3-x}}\,dx = \lim_{c\to0^+}\int_c^{1/2}\frac{3x^2-1}{\sqrt[3]{x^3-x}}\,dx + \lim_{p\to1^-}\int_{1/2}^p \frac{3x^2-1}{\sqrt[3]{x^3-x}}\,dx$$
$$= \lim_{c\to0^+}\frac{3}{2}(x^3-x)^{2/3}\Big|_c^{1/2} + \lim_{p\to1^-}\frac{3}{2}(x^3-x)^{2/3}\Big|_{1/2}^p$$
$$= \lim_{c\to0^+}\left[\frac{3}{2}\left(\frac{-3}{8}\right)^{2/3} - \frac{3}{2}(c^3-c)^{2/3}\right] + \lim_{p\to1^-}\left[\frac{3}{2}(p^3-p)^{2/3} - \frac{3}{2}\left(\frac{-3}{8}\right)^{2/3}\right]$$
$$= \tfrac{3}{2}(\tfrac{3}{8})^{2/3} - \tfrac{3}{2}(\tfrac{3}{8})^{2/3} = 0$$

15. diverges. Let $d = \pi/2$. Then

$$\int_0^\pi \csc^2 x\,dx = \lim_{c\to0^+}\int_c^{\pi/2}\csc^2 x\,dx + \lim_{p\to\pi^-}\int_{\pi/2}^p \csc^2 x\,dx$$
$$= \lim_{c\to0^+}(-\cot x)\Big|_c^{\pi/2} + \lim_{p\to\pi^-}(-\cot x)\Big|_{\pi/2}^p$$
$$= \lim_{c\to0^+}(0+\cot c) + \lim_{p\to\pi^-}(-\cot p + 0)$$

Neither one-sided limit exists, so the integral diverges.

17. diverges; let $u = e^t$, so that $du = e^t\,dt$ and $(1/u)\,du = dt$; then

$$\int_0^1 \frac{1}{e^t - e^{-t}}\,dt = \lim_{c\to0^+}\int_c^1 \frac{1}{e^t-e^{-t}}\,dt = \lim_{c\to0^+}\int_{e^c}^e \frac{1}{u-1/u}\,\frac{1}{u}\,du$$
$$= \lim_{c\to0^+}\int_{e^c}^e \frac{1}{u^2-1}\,du = \lim_{c\to0^+}\int_{e^c}^e\left(\frac{1}{2}\frac{1}{u-1} - \frac{1}{2}\frac{1}{u+1}\right)du$$
$$= \lim_{c\to0^+}\left(\frac{1}{2}\ln|u-1| - \frac{1}{2}\ln|u+1|\right)\Big|_{e^c}^e = \lim_{c\to0^+}\left(\frac{1}{2}\ln\left|\frac{u-1}{u+1}\right|\right)\Big|_{e^c}^e$$
$$= \lim_{c\to0^+}\left(\frac{1}{2}\ln\left|\frac{e-1}{e+1}\right| - \frac{1}{2}\ln\left|\frac{e^c-1}{e^c+1}\right|\right)$$

Since $\lim_{c\to0^+}(e^c-1)/(e^c+1) = 0$, it follows that

$$\lim_{c\to0^+}\left(\frac{1}{2}\ln\left|\frac{e-1}{e+1}\right| - \frac{1}{2}\ln\left|\frac{e^c-1}{e^c+1}\right|\right) = \infty$$

so the integral diverges.

19. diverges;

$$\int_{-1}^2\left(\frac{1}{x}+\frac{1}{x^2}\right)dx = \lim_{c\to0^-}\int_{-1}^c\left(\frac{1}{x}+\frac{1}{x^2}\right)dx + \lim_{p\to0^+}\int_p^2\left(\frac{1}{x}+\frac{1}{x^2}\right)dx$$
$$= \lim_{c\to0^-}\left(\ln|x| - \frac{1}{x}\right)\Big|_{-1}^c + \lim_{p\to0^+}\left(\ln|x| - \frac{1}{x}\right)\Big|_p^2$$
$$= \lim_{c\to0^-}\left(\ln|c| - \frac{1}{c} - 1\right) + \lim_{p\to0^+}\left(\ln 2 - \frac{1}{2} - \ln p + \frac{1}{p}\right)$$

Since $\lim_{p\to0^+}(\ln 2 - \frac{1}{2} - \ln p + 1/p) = \infty$, the integral diverges.

21. converges;

$$\int_0^2 \frac{1}{(x-1)^{1/3}}\,dx = \lim_{c\to1^-}\int_0^c \frac{1}{(x-1)^{1/3}}\,dx + \lim_{p\to1^+}\int_p^2 \frac{1}{(x-1)^{1/3}}\,dx$$
$$= \lim_{c\to1^-}\tfrac{3}{2}(x-1)^{2/3}\Big|_0^c + \lim_{p\to1^+}\tfrac{3}{2}(x-1)^{2/3}\Big|_p^2$$
$$= \lim_{c\to1^-}(\tfrac{3}{2}(c-1)^{2/3} - \tfrac{3}{2}) + \lim_{p\to1^+}(\tfrac{3}{2} - \tfrac{3}{2}(p-1)^{2/3}) = 0$$

23. converges; $\int_0^1 \frac{1}{\sqrt{1-t^2}}\,dt = \lim_{c\to1^-}\int_0^c \frac{1}{\sqrt{1-t^2}}\,dt = \lim_{c\to1^-}\arcsin t\Big|_0^c$

$$= \lim_{c\to1^-}(\arcsin c - \arcsin 0) = \arcsin 1 = \frac{\pi}{2}$$

25. diverges;

$$\int_0^\infty \frac{1}{x}\,dx = \lim_{a\to0^+}\int_a^1 \frac{1}{x}\,dx + \lim_{b\to\infty}\int_1^b \frac{1}{x}\,dx = \lim_{a\to0^+}\ln|x|\Big|_a^1 + \lim_{b\to\infty}\ln|x|\Big|_1^b$$
$$= \lim_{a\to0^+}(0 - \ln a) + \lim_{b\to\infty}(\ln b - 0)$$

Since both limits are infinite, the integral diverges.

27. converges;

$$\int_0^\infty \frac{1}{(2+x)^\pi}\,dx = \lim_{b\to\infty}\int_0^b (2+x)^{-\pi}\,dx = \lim_{b\to\infty}\frac{1}{-\pi+1}(2+x)^{-\pi+1}\Big|_0^b$$
$$= \lim_{b\to\infty}\frac{1}{-\pi+1}((2+b)^{-\pi+1} - 2^{-\pi+1}) = \frac{1}{\pi-1}2^{-\pi+1}$$

29. diverges; $\int_0^\infty \sin y\,dy = \lim_{b\to\infty}\int_0^b \sin y\,dy = \lim_{b\to\infty}(-\cos y)\Big|_0^b = \lim_{b\to\infty}(1-\cos b)$

The limit does not exist, so the integral diverges.

31. converges; $\int_0^\infty \frac{1}{(1+x)^3}\,dx = \lim_{b\to\infty}\int_0^b \frac{1}{(1+x)^3}\,dx = \lim_{b\to\infty}\left(\frac{-1}{2}\right)\frac{1}{(1+x)^2}\Big|_0^b$

$$= \lim_{b\to\infty}\left(\frac{-1}{2}\frac{1}{(1+b)^2} + \frac{1}{2}\right) = \frac{1}{2}$$

33. diverges; $\int_0^\infty \frac{x}{1+x^2}\,dx = \lim_{b\to\infty}\int_0^b \frac{x}{1+x^2}\,dx = \lim_{b\to\infty}\frac{1}{2}\ln(1+x^2)\Big|_0^b$

$$= \lim_{b\to\infty}\tfrac{1}{2}\ln(1+b^2) = \infty$$

35. diverges;

$$\int_3^\infty \ln x\,dx = \lim_{b\to\infty}\int_3^b \ln x\,dx = \lim_{b\to\infty}(x\ln x - x)\Big|_3^b = \lim_{b\to\infty}(b\ln b - b - 3\ln 3 + 3)$$

Since $\lim_{b\to\infty}(b\ln b - b) = \lim_{b\to\infty} b(\ln b - 1) = \infty$, the integral diverges.

37. converges; let $u = \ln x$, so that $du = (1/x)\,dx$; then

$$\int_2^\infty \frac{1}{x(\ln x)^3}\,dx = \lim_{b\to\infty}\int_2^b \frac{1}{x(\ln x)^3}\,dx$$

$$= \lim_{b\to\infty}\int_{\ln 2}^{\ln b}\frac{1}{u^3}\,du = \lim_{b\to\infty}\frac{-1}{2u^2}\Big|_{\ln 2}^{\ln b}$$

$$= \lim_{b\to\infty}\left[-\frac{1}{2(\ln b)^2} + \frac{1}{2(\ln 2)^2}\right] = \frac{1}{2(\ln 2)^2}$$

39. diverges; let $d = -4$; then

$$\int_{-\infty}^0 \frac{1}{(x+3)^2}\,dx = \lim_{a\to-\infty}\int_a^{-4}\frac{1}{(x+3)^2}\,dx + \lim_{c\to-3^-}\int_{-4}^c \frac{1}{(x+3)^2}\,dx$$

$$+ \lim_{p\to-3^+}\int_p^0 \frac{1}{(x+3)^2}\,dx$$

$$= \lim_{a\to-\infty}\frac{-1}{x+3}\Big|_a^{-4} + \lim_{c\to-3^-}\frac{-1}{x+3}\Big|_{-4}^c + \lim_{p\to-3^+}\frac{-1}{x+3}\Big|_p^0$$

$$= \lim_{a\to-\infty}\left(1 + \frac{1}{a+3}\right) + \lim_{c\to-3^-}\left(\frac{-1}{c+3} - 1\right) + \lim_{p\to-3^+}\left(-\frac{1}{3} + \frac{1}{p+3}\right)$$

The latter two one-sided limits do not exist, so the integral diverges.

41. diverges; let $u = \sqrt{x} + 1$, so that $du = 1/(2\sqrt{x})\,dx$; then

$$\int_1^\infty \frac{1}{\sqrt{x}(\sqrt{x}+1)}\,dx = \lim_{b\to\infty}\int_1^b \frac{1}{\sqrt{x}(\sqrt{x}+1)}\,dx$$

$$= \lim_{b\to\infty} 2\int_2^{\sqrt{b}+1}\frac{1}{u}\,du$$

$$= \lim_{b\to\infty} 2\ln|u|\Big|_2^{\sqrt{b}+1}$$

$$= \lim_{b\to\infty}(2\ln(\sqrt{b}+1) - 2\ln 2)$$

$$= \infty$$

43. diverges; $\int_0^\infty e^{4x}\,dx = \lim_{b\to\infty}\int_0^b e^{4x}\,dx = \lim_{b\to\infty}\tfrac{1}{4}e^{4x}\Big|_0^b = \lim_{b\to\infty}(\tfrac{1}{4}e^{bx} - \tfrac{1}{4}) = \infty$

45. converges; let $u = x$, $dv = e^{-x}\,dx$, so $du = dx$, $v = -e^{-x}$;

$$\int_0^\infty xe^{-x}\,dx = \lim_{b\to\infty}\int_0^b xe^{-x}\,dx = \lim_{b\to\infty}\left(-xe^{-x}\Big|_0^b + \int_0^b e^{-x}\,dx\right)$$

$$= \lim_{b\to\infty}\left(-xe^{-x}\Big|_0^b - e^{-x}\Big|_0^b\right) = \lim_{b\to\infty}[(-be^{-b}+0) - (e^{-b}-1)]$$

$$= \left(-\lim_{b\to\infty} be^{-b}\right) - \lim_{b\to\infty} e^{-b} + 1 = \left(-\lim_{b\to\infty} be^{-b}\right) + 1$$

Since $\lim_{b\to\infty} b = \infty = \lim_{b\to\infty} e^b$, l'Hôpital's Rule implies that

$$\lim_{b\to\infty} be^{-b} = \lim_{b\to\infty}\frac{b}{e^b} = \lim_{b\to\infty}\frac{1}{e^b} = 0$$

Thus $$\int_0^\infty xe^{-x}\,dx = 1$$

47. diverges; for $\int(1/\sqrt{x^2-1})\,dx$ let $x = \sec u$, so that $dx = \sec u\tan u\,du$; thus

$$\int\frac{1}{\sqrt{x^2-1}}\,dx = \int\frac{1}{\sqrt{\sec^2 u - 1}}\sec u\tan u\,du = \int\sec u\,du = \ln|\sec u + \tan u| + C$$

Now $x = \sec u$ implies that $\tan u = \sqrt{x^2-1}$. Thus

$$\int\frac{1}{\sqrt{x^2-1}}\,dx = \ln|\sec u + \tan u| + C = \ln|x + \sqrt{x^2-1}| + C$$

Since $$\int_1^\infty \frac{1}{\sqrt{x^2-1}}\,dx = \lim_{a\to1^+}\int_a^2 \frac{1}{\sqrt{x^2-1}}\,dx + \lim_{b\to\infty}\int_2^b \frac{1}{\sqrt{x^2-1}}\,dx$$

and since

$$\lim_{b\to\infty}\int_2^b \frac{1}{\sqrt{x^2-1}}\,dx = \lim_{b\to\infty}\ln|x + \sqrt{x^2-1}|\Big|_2^b$$

$$= \lim_{b\to\infty}[\ln|b + \sqrt{b^2-1}| - \ln(2+\sqrt{3})] = \infty$$

it follows that $\int_1^\infty (1/\sqrt{x^2-1})\,dx$ diverges.

49. converges;

$$\int_1^\infty \frac{1}{t\sqrt{t^2-1}}\,dt = \lim_{a\to1^+}\int_a^2 \frac{1}{t\sqrt{t^2-1}}\,dt + \lim_{b\to\infty}\int_2^b \frac{1}{t\sqrt{t^2-1}}\,dt$$

$$= \lim_{a\to1^+}\operatorname{arcsec} t\Big|_a^2 + \lim_{b\to\infty}\operatorname{arcsec} t\Big|_2^b$$

$$= \lim_{a\to1^+}(\operatorname{arcsec} 2 - \operatorname{arcsec} a) + \lim_{b\to\infty}(\operatorname{arcsec} b - \operatorname{arcsec} 2)$$

$$= \left(\frac{\pi}{3} - 0\right) + \left(\frac{\pi}{2} - \frac{\pi}{3}\right) = \frac{1}{2}\pi$$

51. converges; since $t^2+4t+8=(t+2)^2+4$, let $u=t+2$, so that $du=dt$;

$$\int_{-2}^{\infty}\frac{1}{t^2+4t+8}\,dt=\lim_{b\to\infty}\int_{-2}^{b}\frac{1}{(t+2)^2+4}\,dt=\lim_{b\to\infty}\int_0^{b+2}\frac{1}{u^2+4}\,du$$
$$=\lim_{b\to\infty}\frac{1}{2}\arctan\frac{u}{2}\Big|_0^{b+2}=\lim_{b\to\infty}\frac{1}{2}\arctan\frac{b+2}{2}=\frac{1}{4}\pi$$

53. diverges; $\int_{-\infty}^{\infty}x\,dx=\lim_{a\to-\infty}\int_a^0 x\,dx+\lim_{b\to\infty}\int_0^b x\,dx=\lim_{a\to-\infty}\frac{x^2}{2}\Big|_a^0+\lim_{b\to\infty}\frac{x^2}{2}\Big|_0^b$

$$=\lim_{a\to-\infty}\frac{-1}{2}a^2+\lim_{b\to\infty}\frac{1}{2}b^2$$

Neither limit exists, so the integral diverges.

55. diverges; let $u=x$, $dv=\sin x\,dx$, so $du=dx$, $v=-\cos x$; then

$$\int_{-\infty}^{\infty}x\sin x\,dx=\lim_{a\to-\infty}\int_a^0 x\sin x\,dx+\lim_{b\to\infty}\int_0^b x\sin x\,dx$$
$$=\lim_{a\to-\infty}\left[-x\cos x\Big|_a^0+\int_a^0\cos x\,dx\right]$$
$$+\lim_{b\to\infty}\left[-x\cos x\Big|_0^b+\int_0^b\cos x\,dx\right]$$
$$=\lim_{a\to-\infty}\left[(0+a\cos a)+(\sin x)\Big|_a^0\right]$$
$$+\lim_{b\to\infty}\left[-b\cos b+0)+(\sin x)\Big|_0^b\right]$$
$$=\lim_{a\to-\infty}[a\cos a+0-\sin a]+\lim_{b\to\infty}[-b\cos b+\sin b-0]$$

Neither limit exists, so the integral diverges.

57. converges; let $u=x^4+1$, so that $du=4x^3\,dx$; then

$$\int_{-\infty}^{\infty}\frac{x^3}{(x^4+1)^2}\,dx=\lim_{a\to-\infty}\int_a^0\frac{x^3}{(x^4+1)^2}\,dx+\lim_{b\to\infty}\int_0^b\frac{x^3}{(x^4+1)^2}\,dx$$
$$=\lim_{a\to-\infty}\frac{1}{4}\int_{a^4+1}^{1}\frac{1}{u^2}\,du+\lim_{b\to\infty}\frac{1}{4}\int_1^{b^4+1}\frac{1}{u^2}\,du$$
$$=\lim_{a\to-\infty}\frac{-1}{4u}\Big|_{a^4+1}^{1}+\lim_{b\to\infty}\frac{-1}{4u}\Big|_1^{b^4+1}$$
$$=\lim_{a\to-\infty}\left[-\frac{1}{4}+\frac{1}{4(a^4+1)}\right]+\lim_{b\to\infty}\left[-\frac{1}{4(b^4+1)}+\frac{1}{4}\right]=0$$

59. converges; since $x^2-6x+10=(x-3)^2+1$, let $u=x-3$, so that $du=dx$; then

$$\int_{-\infty}^{\infty}\frac{1}{x^2-6x+10}\,dx=\lim_{a\to-\infty}\int_a^3\frac{1}{(x-3)^2+1}\,dx+\lim_{b\to\infty}\int_3^b\frac{1}{(x-3)^2+1}\,dx$$
$$=\lim_{a\to-\infty}\int_{a-3}^{0}\frac{1}{u^2+1}\,du+\lim_{b\to\infty}\int_0^{b-3}\frac{1}{u^2+1}\,du$$
$$=\lim_{a\to-\infty}\arctan u\Big|_{a-3}^{0}+\lim_{b\to\infty}\arctan u\Big|_0^{b-3}$$
$$=\lim_{a\to-\infty}-(\arctan(a-3))+\lim_{b\to\infty}\arctan(b-3)$$
$$=-\left(-\frac{\pi}{2}\right)+\frac{\pi}{2}=\pi$$

61. a. $\dfrac{1}{x(x+1)}=\dfrac{A}{x}+\dfrac{B}{x+1}=\dfrac{A(x+1)+Bx}{x(x+1)}$; $A+B=0$, $A=1$;

thus $B=-1$, so that $\dfrac{1}{x(x+1)}=\dfrac{1}{x}-\dfrac{1}{x+1}$.

b. No, since both $\int_1^{\infty}(1/x)\,dx$ and $\int_1^{\infty}[1/(x+1)]\,dx$ diverge, whereas

$$\int_1^{\infty}\frac{1}{x(x+1)}\,dx=\lim_{b\to\infty}\int_1^b\frac{1}{x(x+1)}\,dx=\lim_{b\to\infty}\int_1^b\left(\frac{1}{x}-\frac{1}{x+1}\right)dx$$
$$=\lim_{b\to\infty}(\ln|x|-\ln|x+1|)\Big|_1^b=\lim_{b\to\infty}[\ln b-\ln(b+1)+\ln 2]$$
$$=\lim_{b\to\infty}\left(\ln\frac{b}{b+1}+\ln 2\right)=\ln 2$$

63. $A=\int_{-\infty}^{0}\frac{1}{(x-3)^2}\,dx=\lim_{a\to-\infty}\int_a^0\frac{1}{(x-3)^2}\,dx=\lim_{a\to-\infty}\frac{-1}{x-3}\Big|_a^0$

$$=\lim_{a\to-\infty}\left(\frac{1}{3}+\frac{1}{a-3}\right)=\frac{1}{3}$$

so the region has finite area.

65. $A=\int_2^{\infty}\frac{\ln x}{x}\,dx=\lim_{b\to\infty}\int_2^b\frac{\ln x}{x}\,dx=\lim_{b\to\infty}\frac{1}{2}(\ln x)^2\Big|_2^b=\lim_{b\to\infty}\left[\frac{1}{2}(\ln b)^2-\frac{1}{2}(\ln 2)^2\right]=\infty$,

so the region has infinite area.

67. $A=\int_2^{\infty}\frac{1}{\sqrt{x+1}}\,dx=\lim_{b\to\infty}\int_2^b\frac{1}{\sqrt{x+1}}\,dx=\lim_{b\to\infty}2\sqrt{x+1}\Big|_2^b$

$$=\lim_{b\to\infty}(2\sqrt{b+1}-2\sqrt{3})=\infty$$

so the region has infinite area.

69. a. $A_1=\int_{-2}^{2}\frac{1}{\sqrt{2\pi}}e^{-x^2/2}\,dx$

$$\approx\frac{4}{24}\frac{1}{\sqrt{2\pi}}(e^{-2}+4e^{-9/8}+2e^{-1/2}+4e^{-1/8}+2e^{0}$$
$$+4e^{-1/8}+2e^{-1/2}+4e^{-9/8}+e^{-2})$$
$$\approx 0.954402$$

b. $A_2 = \int_{-5}^{5} \frac{1}{\sqrt{2\pi}} e^{-x^2/2}\,dx \approx \frac{10}{150}\frac{1}{\sqrt{2\pi}}(e^{-(5^2)/2} + 4e^{-(4.8)^2/2} + 2e^{-(4.6)^2/2} + \cdots + e^{5^2/2})$

≈ 0.978946

c. 1

71. By the General Comparison Property, $\int_a^b g(x)\,dx \le \int_a^b f(x)\,dx$ for any $b > a$. For any given M, choose N such that if $b > N$, then $\int_a^b g(x)\,dx > M$. Then $\int_a^b f(x)\,dx > M$, so $\lim_{b\to\infty} \int_a^b f(x)\,dx = \infty$. Consequently $\int_a^\infty f(x)\,dx$ diverges.

73. Let $n \ge 1$. For integration by parts, let $u = x^n$, $dv = e^{-x}\,dx$, so that $du = nx^{n-1}\,dx$, $v = -e^{-x}$. Then

$$I_n = \int_0^\infty x^n e^{-x}\,dx = \lim_{b\to\infty} \int_0^b x^n e^{-x}\,dx = \lim_{b\to\infty}\left[-x^n e^{-x}\Big|_0^b + \int_0^b nx^{n-1}e^{-x}\,dx\right]$$

$$= \lim_{b\to\infty}\left(-b^n e^{-b} + n\int_0^b x^{n-1}e^{-x}\,dx\right)$$

Since $\lim_{x\to\infty}(e^x/e^n) = \infty$ by the comment following Example 6 of Section 6.8, it follows that $\lim_{b\to\infty}(-b^n e^{-b}) = -\lim_{b\to\infty}(b^n/e^b) = 0$ for every positive integer n, and thus

$$I_n = \lim_{b\to\infty} n\int_0^b x^{n-1}e^{-x}\,dx = n\int_0^\infty x^{n-1}e^{-x}\,dx = nI_{n-1}$$

Inductively we have $I_n = nI_{n-1} = n(n-1)I_{n-2} = \cdots = n(n-1)(n-2)\cdots 2I_1$. Now by the above calculation with $n = 1$ we have

$$I_1 = \lim_{b\to\infty}\left(-be^{-b} + \int_0^b e^{-x}\,dx\right) = \lim_{b\to\infty}\left(-be^{-b} - e^{-x}\Big|_0^b\right) = \lim_{b\to\infty}(-be^{-b} - e^{-b} + 1) = 1$$

since we have shown above that $\lim_{b\to\infty}(-b/e^{-b}) = 0$. Therefore

$$I_n = n(n-1)(n-2)\cdots 2\cdot 1$$

75. $M = -\frac{1}{A}\int_0^\infty tkf(t)\,dt = -\frac{1}{A}\int_0^\infty tkAe^{kt}\,dt = -\int_0^\infty tke^{kt}\,dt$

For integration by parts, let $u = t$, $dv = ke^{kt}\,dt$, so that $du = dt$, $v = e^{kt}$. Then

$$M = \lim_{b\to\infty} -\int_0^b tke^{kt}\,dt = \lim_{b\to\infty} -\left(te^{kt}\Big|_0^b - \int_0^b e^{kt}\,dt\right) = \lim_{b\to\infty} -\left(be^{kb} - \frac{1}{k}e^{kt}\Big|_0^b\right) = -\frac{1}{k}$$

a. If $k = -1.24 \times 10^{-4}$, then $M = 1/1.24 \times 10^4 \approx 8064.52$ (years).

b. If $k = -4.36 \times 10^{-4}$, then $M = 1/4.36 \times 10^4 \approx 2293.58$ (years).

77. By Exercise 74, $\frac{d}{dx}\int_x^\infty g(t)\,dt = -g(x)$, so that $g(x) = -\frac{d}{dx}(cx^{-1.5}) = 1.5cx^{-2.5}$ for $x \ge s$.

CHAPTER 7 REVIEW

1. $u = \ln(x^2+9)$, $dv = dx$; $du = [2x/(x^2+9)]\,dx$, $v = x$;

$$\int \ln(x^2+9)\,dx = x\ln(x^2+9) - \int \frac{2x^2}{x^2+9}\,dx = x\ln(x^2+9) - 2\int\left(1 - \frac{9}{x^2+9}\right)dx$$

$$= x\ln(x^2+9) - 2x + 18\left(\frac{1}{3}\right)\arctan\frac{x}{3} + C$$

$$= x\ln(x^2+9) - 2x + 6\arctan\frac{x}{3} + C$$

3. $u = x$, $dv = \csc^2 x\,dx$; $du = dx$, $v = -\cot x$;

$$\int x\csc^2 x\,dx = -x\cot x + \int \cot x\,dx = -x\cot x + \ln|\sin x| + C$$

5. $u = x$, $dv = \cosh x\,dx$; $du = dx$, $v = \sinh x$;

$$\int x\cosh x\,dx = x\sinh x - \int \sinh x\,dx = x\sinh x - \cosh x + C$$

7. $\int x\cos^2 x\,dx = \int x(\frac{1}{2} + \frac{1}{2}\cos 2x)\,dx = \frac{1}{2}\int x\,dx + \frac{1}{2}\int x\cos 2x\,dx = \frac{1}{4}x^2 + \frac{1}{2}\int x\cos 2x\,dx$.

For $\int x\cos 2x\,dx$, let $u = x$, $dv = \cos 2x\,dx$; $du = dx$, $v = \frac{1}{2}\sin 2x$;

$$\int x\cos 2x\,dx = \frac{1}{2}x\sin 2x - \frac{1}{2}\int \sin 2x\,dx = \frac{1}{2}x\sin 2x + \frac{1}{4}\cos 2x + C_1$$

Thus $\int x\cos^2 x\,dx = \frac{1}{4}x^2 + \frac{1}{4}x\sin 2x + \frac{1}{8}\cos 2x + C$.

9. $u = \sin x^3$, $du = 3x^2\cos x^3\,dx$;

$$\int x^2\sin x^3\cos x^3\,dx = \frac{1}{3}\int u\,du = \frac{1}{6}u^2 + C = \frac{1}{6}\sin^2 x^3 + C$$

11. $\int \tan^5 x\,dx = \int \tan^3 x(\sec^2 x - 1)\,dx = \int \tan^3 x\sec^2 x\,dx - \int \tan^3 x\,dx$

$$= \int \tan^3 x\sec^2 x\,dx - \int \tan x(\sec^2 x - 1)\,dx$$

$$= \int \tan^3 x\sec^2 x\,dx - \int \tan x\sec^2 x\,dx + \int \tan x\,dx$$

For the first two integrals let $u = \tan x$, so that $du = \sec^2 x\,dx$. Then

$$\int \tan^3 x\sec^2 x\,dx - \int \tan x\sec^2 x\,dx = \int u^3\,du - \int u\,du = \frac{1}{4}u^4 - \frac{1}{2}u^2 + C_1$$

$$= \frac{1}{4}\tan^4 x - \frac{1}{2}\tan^2 x + C_1$$

Thus $\int \tan^5 x\,dx = \frac{1}{4}\tan^4 x - \frac{1}{2}\tan^2 x - \ln|\cos x| + C$.

13. $u = x^2$, $dv = x\cos x^2\,dx$; $du = 2x\,dx$, $v = \frac{1}{2}\sin x^2$;

$$\int x^3\cos x^2\,dx = \frac{1}{2}x^2\sin x^2 - \int \frac{1}{2}(2x)\sin x^2\,dx = \frac{1}{2}x^2\sin x^2 + \frac{1}{2}\cos x^2 + C$$

15. $u = 1 - 3t$, $du = -3\,dt$, so $t = \frac{1}{3}(1-u)$;

$$\int t^2\sqrt{1-3t}\,dt = \int [\tfrac{1}{3}(1-u)]^2\sqrt{u}(-\tfrac{1}{3})\,du = -\frac{1}{27}\int (u^{1/2} - 2u^{3/2} + u^{5/2})\,du$$

$$= -\frac{1}{27}(\tfrac{2}{3}u^{3/2} - \tfrac{4}{5}u^{5/2} + \tfrac{2}{7}u^{7/2}) + C$$

$$= -\frac{2}{81}(1-3t)^{3/2} + \frac{4}{135}(1-3t)^{5/2} - \frac{2}{189}(1-3t)^{7/2} + C$$

17. $\int \frac{\cos x}{1+\cos x}\,dx = \int \frac{\cos x}{1+\cos x}\frac{1-\cos x}{1-\cos x}\,dx = \int \frac{\cos x-(1-\sin^2 x)}{\sin^2 x}\,dx$

$= \int \left(\frac{\cos x}{\sin^2 x} - \csc^2 x + 1\right)dx = \frac{-1}{\sin x} + \cot x + x + C$

$= -\csc x + \cot x + x + C$

19. $x+1 = 3\tan u$, $dx = 3\sec^2 u\,du$;

$$\int \frac{x^2}{(x^2+2x+10)^{5/2}}\,dx = \int \frac{x^2}{((x+1)^2+9)^{5/2}}\,dx = \int \frac{(3\tan u - 1)^2}{(9\tan^2 u + 9)^{5/2}}\,3\sec^2 u\,du$$

$$= \frac{1}{81}\int \frac{9\tan^2 u - 6\tan u + 1}{\sec^5 u}\sec^2 u\,du$$

$$= \frac{1}{9}\int \sin^2 u\cos u\,du - \frac{2}{27}\int \sin u\cos^2 u\,du$$

$$+ \frac{1}{81}\int (1-\sin^2 u)\cos u\,du$$

$$= \frac{1}{27}\sin^3 u + \frac{2}{81}\cos^3 u + \frac{1}{81}\sin u - \frac{1}{243}\sin^3 u + C$$

$$= \frac{8}{243}\frac{(x+1)^3}{(x^2+2x+10)^{3/2}} + \frac{2}{3}\frac{1}{(x^2+2x+10)^{3/2}}$$

$$+ \frac{1}{81}\frac{x+1}{(x^2+2x+10)^{1/2}} + C$$

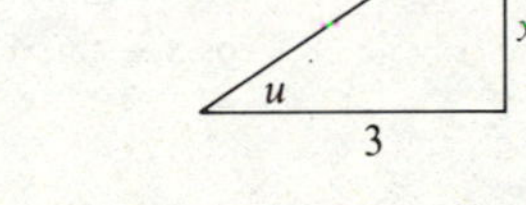

21. $x = \tan u$, $dx = \sec^2 u\,du$;

$$\int \frac{x^4}{(x^2+1)^2}\,dx = \int \frac{\tan^4 u}{(\tan^2 u+1)^2}\sec^2 u\,du = \int \frac{\tan^4 u}{\sec^2 u}\,du = \int \frac{\sin^4 u}{\cos^4 u}\cos^2 u\,du$$

$$= \int \frac{\sin^4 u}{\cos^2 u}\,du = \int \frac{(1-\cos^2 u)^2}{\cos^2 u}\,du = \int \frac{1 - 2\cos^2 u + \cos^4 u}{\cos^2 u}\,du$$

$$= \int (\sec^2 u - 2 + \cos^2 u)\,du = \tan u - 2u + \int \left(\tfrac{1}{2} + \tfrac{1}{2}\cos 2u\right)du$$

$$= \tan u - 2u + \tfrac{1}{2}u + \tfrac{1}{4}\sin 2u + C = \tan u - \tfrac{3}{2}u + \tfrac{1}{2}\sin u\cos u + C$$

Now $u = \arctan x$, and by the figure, $\sin u = x/\sqrt{x^2+1}$ and $\cos u = 1/\sqrt{x^2+1}$, so that

$$\int \frac{x^4}{(x^2+1)^2}\,dx = x - \frac{3}{2}\arctan x + \frac{1}{2}\frac{x}{x^2+1} + C$$

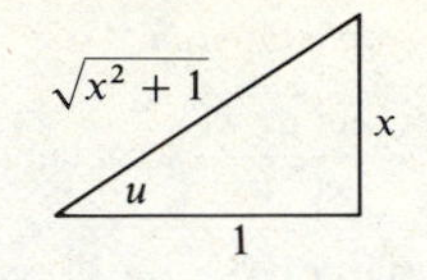

23. $\int \frac{x}{x^2+3x-18}\,dx = \int \frac{x}{(x+6)(x-3)}\,dx$; $\frac{x}{(x+6)(x-3)} = \frac{A}{x+6} + \frac{B}{x-3}$;

$A(x-3) + B(x+6) = x$; $A + B = 1$, $-3A + 6B = 0$; $A = \frac{2}{3}$, $B = \frac{1}{3}$;

$$\int \frac{x}{x^2+3x-18}\,dx = \int \left(\frac{2}{3}\frac{1}{x+6} + \frac{1}{3}\frac{1}{x-3}\right)dx = \frac{2}{3}\int \frac{1}{x+6}\,dx + \frac{1}{3}\int \frac{1}{x-3}\,dx$$

$$= \tfrac{2}{3}\ln|x+6| + \tfrac{1}{3}\ln|x-3| + C$$

25. Improper integral;

$$\int_{-2}^{1} \frac{1}{3x+4}\,dx = \lim_{c\to -4/3^-}\int_{-2}^{c}\frac{1}{3x+4}\,dx + \lim_{p\to -4/3^+}\int_{p}^{1}\frac{1}{3x+4}\,dx$$

$$= \lim_{c\to -4/3^-}\tfrac{1}{3}\ln|3x+4|\Big|_{-2}^{c} + \lim_{p\to -4/3^+}\tfrac{1}{3}\ln|3x+4|\Big|_{p}^{1}$$

$$= \lim_{c\to -4/3^-}\left(\tfrac{1}{3}\ln|3c+4| - \tfrac{1}{3}\ln 2\right) + \lim_{p\to -4/3^+}\left(\tfrac{1}{3}\ln 7 - \tfrac{1}{3}\ln|3p+4|\right)$$

Neither limit exists, so the integral diverges.

27. Proper integral; $u = x$, $dv = \sec x\tan x\,dx$, $du = dx$, $v = \sec x$;

$$\int_0^{\pi/4} x\sec x\tan x\,dx = x\sec x\Big|_0^{\pi/4} - \int_0^{\pi/4}\sec x\,dx$$

$$= x\sec x\Big|_0^{\pi/4} - \ln|\sec x + \tan x|\Big|_0^{\pi/4} = \frac{\pi\sqrt{2}}{4} - \ln(\sqrt{2}+1)$$

29. Proper integral; $u = 1 + \sqrt{x}$, $du = \frac{1}{2}x^{-1/2}\,dx$, so $dx = 2(u-1)\,du$;

$$\int_1^4 \frac{1}{1+\sqrt{x}}\,dx = \int_2^3 \frac{2(u-1)}{u}\,du = \int_2^3 \left(2 - \frac{2}{u}\right)du$$

$$= (2u - 2\ln|u|)\Big|_2^3 = 6 - 2\ln 3 - 4 + 2\ln 2 = 2 + \ln\tfrac{4}{9}$$

31. Proper integral; $u = \sin x$, $du = \cos x\,dx$;

$$\int_0^{\pi/4}\sin^4 x\cos^3 x\,dx = \int_0^{\pi/4}\sin^4 x(1-\sin^2 x)\cos x\,dx = \int_0^{\sqrt{2}/2}(u^4 - u^6)\,du$$

$$= \left(\frac{1}{5}u^5 - \frac{1}{7}u^7\right)\Big|_0^{\sqrt{2}/2} = \frac{\sqrt{2}}{40} - \frac{\sqrt{2}}{112} = \frac{9\sqrt{2}}{560}$$

33. Proper integral; $u = \tan x$, $du = \sec^2 x\,dx$;

$$\int_0^{\pi/4}(\tan^3 x + \tan^5 x)\,dx = \int_0^{\pi/4}\tan^3 x(1+\tan^2 x)\,dx = \int_0^{\pi/4}\tan^3 x\sec^2 x\,dx$$

$$= \int_0^1 u^3\,du = \frac{u^4}{4}\Big|_0^1 = \frac{1}{4}$$

35. Proper integral; $x = \sec u$, $dx = \sec u \tan u\, du$;

$$\int_1^{\sqrt{2}} \frac{\sqrt{x^2-1}}{x^2}\,dx = \int_0^{\pi/4} \frac{\sqrt{\sec^2 u - 1}}{\sec^2 u} \sec u \tan u\, du$$
$$= \int_0^{\pi/4} \frac{\sin^2 u}{\cos u}\,du = \int_0^{\pi/4} \frac{1-\cos^2 u}{\cos u}\,du$$
$$= \int_0^{\pi/4} (\sec u - \cos u)\,du = (\ln|\sec u + \tan u| - \sin u)\Big|_0^{\pi/4}$$
$$= \ln(\sqrt{2}+1) - \frac{\sqrt{2}}{2}$$

37. Proper integral; $u = x^{3/2}$, $du = \frac{3}{2}x^{1/2}\,dx$; then $u = \sin t$, $du = \cos t\, dt$;

$$\int_0^{3\sqrt{2}/2} \frac{\sqrt{x}}{\sqrt{1-x^3}}\,dx = \frac{2}{3}\int_0^{1/2} \frac{1}{\sqrt{1-u^2}}\,du = \frac{2}{3}\int_0^{\pi/6} \frac{1}{\sqrt{1-\sin^2 t}}\cos t\, dt$$
$$= \frac{2}{3}\int_0^{\pi/6} 1\, dt = \frac{2}{3}t\Big|_0^{\pi/6} = \frac{1}{9}\pi$$

39. Proper integral; $x = \tan u$, $dx = \sec^2 u\, du$;

$$\int_0^{\sqrt{3}} \sqrt{x^2+1}\,dx = \int_0^{\pi/3} \sqrt{\tan^2 u + 1}\sec^2 u\, du = \int_0^{\pi/3} \sec^3 u\, du$$
$$\underset{(7)\text{ of Section 7.2}}{=} \left(\tfrac{1}{2}\sec u \tan u + \tfrac{1}{2}\ln|\sec u + \tan u|\right)\Big|_0^{\pi/3}$$
$$= \tfrac{1}{2}(2)(\sqrt{3}) + \tfrac{1}{2}\ln|2+\sqrt{3}| = \sqrt{3} + \tfrac{1}{2}\ln(2+\sqrt{3})$$

41. $\displaystyle\int_{-5}^0 \frac{x}{x^2+4x-5}\,dx = \int_{-5}^0 \frac{x}{(x+5)(x-1)}\,dx$, so the integral is improper. Next,

$$\frac{x}{(x+5)(x-1)} = \frac{A}{x+5} + \frac{B}{x-1} = \frac{A(x-1)+B(x+5)}{(x+5)(x-1)}$$

$A(x-1) + B(x+5) = x$; $A + B = 1$, $-A + 5B = 0$; $A = \frac{5}{6}$, $B = \frac{1}{6}$. Thus

$$\int_{-5}^0 \frac{x}{x^2+4x-5}\,dx = \lim_{c\to -5^+} \int_c^0 \left(\frac{5}{6}\frac{1}{x+5} + \frac{1}{6}\frac{1}{x-1}\right)dx$$
$$= \lim_{c\to -5^+} \left(\frac{5}{6}\int_c^0 \frac{1}{x+5}\,dx + \frac{1}{6}\int_c^0 \frac{1}{x-1}\,dx\right)$$
$$= \lim_{c\to -5^+} \left(\frac{5}{6}\ln|x+5|\Big|_c^0 + \frac{1}{6}\ln|x-1|\Big|_c^0\right)$$
$$= \lim_{c\to -5^+} \left(\tfrac{5}{6}\ln 5 - \tfrac{5}{6}\ln(c+5) - \tfrac{1}{6}\ln|c-1|\right)$$

Since $\lim_{c\to -5^+} \ln(c+5) = -\infty$, the integral diverges.

43. Improper integral;

$$\int_0^{\pi/2} \frac{1}{1-\sin x}\,dx = \lim_{c\to\pi/2^-} \int_0^c \frac{1}{1-\sin x}\,dx = \lim_{c\to\pi/2^-} \int_0^c \frac{1}{1-\sin x}\frac{1+\sin x}{1+\sin x}\,dx$$
$$= \lim_{c\to\pi/2^-} \int_0^c \frac{1+\sin x}{\cos^2 x}\,dx = \lim_{c\to\pi/2^-} \int_0^c \left(\sec^2 x + \frac{\sin x}{\cos^2 x}\right)dx$$
$$= \lim_{2\to\pi/2^-} \left(\tan x\Big|_0^c + \int_0^c \frac{\sin x}{\cos^2 x}\,dx\right)$$

For $\int_0^c (\sin x)/(\cos^2 x)\,dx$ let $u = \cos x$, so $du = -\sin x\, dx$. Then

$$\int_0^c \frac{\sin x}{\cos^2 x}\,dx = \int_1^{\cos c} -\frac{1}{u^2}\,du = \frac{1}{u}\Big|_1^{\cos c} = \frac{1}{\cos c} - 1 = \sec c - 1$$

Thus
$$\int_0^{\pi/2} \frac{1}{1-\sin x}\,dx = \lim_{c\to\pi/2^-} (\tan c + \sec c - 1)$$

Since $\lim_{c\to\pi/2^-} \tan c = \infty = \lim_{c\to\pi/2^-} \sec c$, the integral diverges.

45. Improper integral; $u = \ln x$, $dv = x\,dx$, $du = (1/x)\,dx$, $v = \frac{1}{2}x^2$; $\int x\ln x\,dx = \frac{1}{2}x^2\ln x - \int \frac{1}{2}x\,dx = \frac{1}{2}x^2 \ln x - \frac{1}{4}x^2 + C$. Thus

$$\int_0^1 x\ln x\,dx = \lim_{c\to 0^+}\int_c^1 x\ln x\,dx = \lim_{c\to 0^+}\left(\tfrac{1}{2}x^2\ln x - \tfrac{1}{4}x^2\right)\Big|_c^1$$
$$= \lim_{c\to 0^+}\left(-\tfrac{1}{4} - \tfrac{1}{2}c^2\ln c + \tfrac{1}{4}c^2\right) = -\tfrac{1}{4}$$

since $\lim_{x\to 0^+} c\ln c = 0$ by Example 7 of Section 6.8.

47. Improper integral;

$$\int_1^\infty \frac{1}{x(\ln x)^2}\,dx = \lim_{c\to 1^+}\int_c^2 \frac{1}{x(\ln x)^2}\,dx + \lim_{b\to\infty}\int_2^b \frac{1}{x(\ln x)^2}\,dx$$
$$= \lim_{c\to 1^+} \frac{-1}{\ln x}\Big|_c^2 + \lim_{b\to\infty}\frac{-1}{\ln x}\Big|_2^b$$
$$= \lim_{c\to 1^+}\left(\frac{1}{\ln c} - \frac{1}{\ln 2}\right) + \lim_{b\to\infty}\left(\frac{1}{\ln 2} - \frac{1}{\ln b}\right)$$

Since $\lim_{c\to 1^+} (1/(\ln c)) - 1/(\ln 2)) = \infty$, the integral diverges.

49. Improper integral; by Exercise 53(b) of Section 7.1, with $a = -1$ and $b = 1$,

$$\int e^{-x}\cos x\,dx = \tfrac{1}{2}e^{-x}(-\cos x + \sin x) + C$$

Thus

$$\int_0^\infty e^{-x}\cos x\,dx = \lim_{b\to\infty}\int_0^b e^{-x}\cos x\,dx = \lim_{b\to\infty}\tfrac{1}{2}e^{-x}(\sin x - \cos x)\Big|_0^b$$
$$= \lim_{b\to\infty}\left[\tfrac{1}{2}e^{-b}(\sin b - \cos b) + \tfrac{1}{2}\right] = \tfrac{1}{2}$$

51. Improper integral;

$$\int_1^\infty \frac{1}{x(x^2+1)}\,dx = \lim_{b\to\infty}\int_1^b \frac{1}{x(x^2+1)}\,dx \overset{x=\tan u}{=} \lim_{b\to\infty}\int_{\pi/4}^{\arctan b}\frac{1}{\tan u(\sec^2 u)}\sec^2 u\,du$$
$$= \lim_{b\to\infty}\int_{\pi/4}^{\arctan b}\cot u\,du = \lim_{b\to\infty}\ln|\sin u|\Big|_{\pi/4}^{\arctan b}$$

By the figure, if $u = \arctan b$, then $\sin u = b/\sqrt{b^2+1}$, so that

$$\lim_{b\to\infty}\ln|\sin u|\Big|_{\pi/4}^{\arctan b} = \lim_{b\to\infty}\left(\ln\frac{b}{\sqrt{b^2+1}} - \ln\left(\sin\frac{\pi}{4}\right)\right) = \lim_{b\to\infty}\left(\ln 1 - \ln\frac{\sqrt{2}}{2}\right)$$
$$= \ln\sqrt{2} = \tfrac{1}{2}\ln 2$$

Therefore the integral converges, and $\int_1^{\infty} 1/[x(x^2+1)]\,dx = \frac{1}{2}\ln 2$.

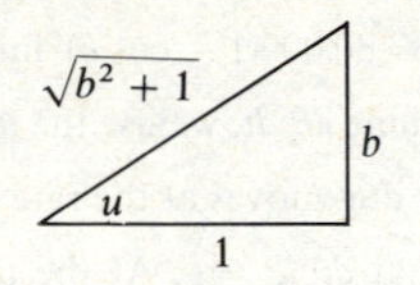

53. $u = \ln x$, $dv = (1/x)\,dx$, $du = (1/x)\,dx$, $v = \ln x$;

$$\int \frac{\ln x}{x}\,dx = (\ln x)^2 - \int \frac{\ln x}{x}\,dx, \quad \text{so that} \quad 2\int \frac{\ln x}{x}\,dx = (\ln x)^2 + C_1$$

and thus $\int (\ln x)/x\,dx = \frac{1}{2}(\ln x)^2 + C$.

55. a. $u = 1 - x$, $du = -dx$

$$\int_0^1 x^n(1-x)^m\,dx = \int_1^0 -(1-u)^n u^m\,du = \int_0^1 u^m(1-u)^n\,du = \int_0^1 x^m(1-x)^n\,dx$$

b. By part (a) with $m = 10$ and $n = 2$, we obtain

$$\int_0^1 x^2(1-x)^{10}\,dx = \int_0^1 x^{10}(1-x)^2\,dx = \int_0^1 (x^{10} - 2x^{11} + x^{12})\,dx$$
$$= \left(\tfrac{1}{11}x^{11} - \tfrac{1}{6}x^{12} + \tfrac{1}{13}x^{13}\right)\Big|_0^1 = \tfrac{1}{858}$$

57. $x = 1 - 3\cos^2 u$, $dx = 6\sin u\cos u\,du$; then $(2+x)/(1-x) = \tan^2 u$;

$$\int_{-2}^{-1/2} \left(\frac{2+x}{1-x}\right)^{1/2} dx = \int_0^{\pi/4} \tan u\,(6\sin u\cos u)\,du = 6\int_0^{\pi/4} \sin^2 u\,du$$
$$= 3\left(u - \frac{1}{2}\sin 2u\right)\Big|_0^{\pi/4} = \frac{3\pi}{4} - \frac{3}{2}$$

59. Using the given identities, we have

$$\int \frac{\tan(\pi/4 + x/2)}{\sec^2(x/2)}\,dx = \int \frac{\sin(\pi/2 + x)}{1 + \cos(\pi/2 + x)}\,\frac{1 + \cos x}{2}\,dx$$
$$= \int \left(\frac{\cos x}{1 - \sin x}\right)\left(\frac{1 + \cos x}{2}\right) dx$$
$$= \frac{1}{2}\int \frac{(\cos x + \cos^2 x)(1 + \sin x)}{(1 - \sin x)(1 + \sin x)}\,dx$$
$$= \frac{1}{2}\int \frac{(\cos x + \cos^2 x)(1 + \sin x)}{\cos^2 x}\,dx$$
$$= \frac{1}{2}\int (\sec x + 1 + \tan x + \sin x)\,dx$$
$$= \frac{1}{2}(\ln|\sec x + \tan x| + \ln|\sec x| - \cos x + x) + C$$
$$= \frac{1}{2}\ln|\sec^2 x + \sec x\tan x| - \frac{1}{2}\cos x + \frac{x}{2} + C$$

61. a. $\int_0^1 \sqrt{1-x^2}\,dx \approx \frac{1}{20}\left(1 + 2\sqrt{\frac{99}{100}} + 2\sqrt{\frac{24}{25}} + 2\sqrt{\frac{91}{100}} + 2\sqrt{\frac{21}{25}} + 2\sqrt{\frac{3}{4}} + 2\sqrt{\frac{16}{25}}\right.$
$\left. + 2\sqrt{\frac{51}{100}} + 2\sqrt{\frac{9}{25}} + 2\sqrt{\frac{19}{100}} + 0\right)$
≈ 0.776130

b. $\int_0^1 \sqrt{1-x^2}\,dx \approx \frac{1}{30}\left(1 + 4\sqrt{\frac{99}{100}} + 2\sqrt{\frac{24}{25}} + 4\sqrt{\frac{91}{100}} + 2\sqrt{\frac{21}{25}} + 4\sqrt{\frac{3}{4}} + 2\sqrt{\frac{16}{25}}\right.$
$\left. + 4\sqrt{\frac{51}{100}} + 2\sqrt{\frac{9}{25}} + 4\sqrt{\frac{19}{100}} + 0\right)$
≈ 0.781752

63. a. By the solution of Exercise 62 it suffices to use the Trapezoidal Rule with $n = 5$. We obtain

$$\int_2^{2.5} \sqrt{x^2-1}\,dx \approx \tfrac{1}{20}(\sqrt{3} + 2\sqrt{3.41} + 2\sqrt{3.84} + 2\sqrt{4.29} + 2\sqrt{4.76} + \sqrt{5.25})$$
$$\approx 1.00709$$

b. By the solution of Exercise 62 it suffices to use Simpson's Rule with $n = 2$. We obtain

$$\int_2^{2.5} \sqrt{x^2-1}\,dx \approx \tfrac{1}{12}(\sqrt{3} + 4\sqrt{4.0625} + \sqrt{5.25}) \approx 1.00713$$

65. $A = \int_{-3}^0 \sqrt{9-x^2}\,dx$. Let $x = 3\sin u$, so that $dx = 3\cos u\,du$. Then

$$\int_{-3}^0 \sqrt{9-x^2}\,dx = \int_{-\pi/2}^0 \sqrt{9 - 9\sin^2 u}\;3\cos u\,du = 9\int_{-\pi/2}^0 \cos^2 u\,du$$
$$= 9\left(\frac{1}{2}u + \frac{1}{4}\sin 2u\right)\Big|_{-\pi/2}^0 = 0 - 9\left(-\frac{\pi}{4}\right) = \frac{9}{4}\pi$$

67. $A = \displaystyle\int_\pi^{3\pi/2} \left|\frac{\cos x}{1 + \sin x}\right| dx = \int_\pi^{3\pi/2} \frac{-\cos x}{1 + \sin x}\,dx = -\lim_{c\to 3\pi/2^-} \int_\pi^c \frac{\cos x}{1 + \sin x}\,dx$

$$= -\lim_{c\to 3\pi/2^-} \ln(1 + \sin x)\Big|_\pi^c = -\lim_{c\to 3\pi/2^-} \ln(1 + \sin c) = \infty$$

so the region has infinite area.

69. $A = \displaystyle\int_{-\infty}^{\infty} \left|\frac{x^3}{2 + x^4}\right| dx = \lim_{a\to\infty} \int_a^0 \frac{-x^3}{2 + x^4}\,dx + \lim_{b\to\infty} \int_0^b \frac{x^3}{2 + x^4}\,dx$

$$= \lim_{a\to-\infty} \left(\frac{-1}{4}\ln(2 + x^4)\right)\Big|_a^0 + \lim_{b\to\infty} \frac{1}{4}\ln(2 + x^4)\Big|_0^b$$
$$= \lim_{a\to-\infty} \left[\tfrac{1}{4}\ln(2 + a^4) - \tfrac{1}{4}\ln 2\right] + \lim_{b\to\infty} \left[\tfrac{1}{4}\ln(2 + b^4) - \tfrac{1}{4}\ln 2\right]$$

Neither limit exists, so the area is infinite.

71. $A = \displaystyle\int_{-\infty}^{\infty} \left|\frac{x^3}{(2 + x^4)^2}\right| dx = \lim_{a\to-\infty} \int_a^0 \frac{-x^3}{(2 + x^4)^2}\,dx + \lim_{b\to\infty} \int_0^b \frac{x^3}{(2 + x^4)^2}\,dx$

$$= \lim_{a\to-\infty} \frac{1}{4(2 + x^4)}\Big|_a^0 + \lim_{b\to\infty} \frac{-1}{4(2 + x^4)}\Big|_0^b$$
$$= \lim_{a\to-\infty} \left[\frac{1}{8} - \frac{1}{4(2 + a^4)}\right] + \lim_{b\to\infty} \left[\frac{1}{8} - \frac{1}{4(2 + b^4)}\right] = \frac{1}{4}$$

CUMULATIVE REVIEW (CHAPTERS 1–6)

1. $\lim_{x\to\sqrt{2}} \dfrac{2\sqrt{2}-2x}{8-4x^2} = \lim_{x\to\sqrt{2}} \dfrac{2(\sqrt{2}-x)}{4(\sqrt{2}-x)(\sqrt{2}+x)} = \lim_{x\to\sqrt{2}} \dfrac{1}{2(\sqrt{2}+x)} = \dfrac{1}{4\sqrt{2}} = \dfrac{1}{8}\sqrt{2}$

3. $f'(x) = \dfrac{-e^{-x}(1+e^x) - (1+e^{-x})e^x}{(1+e^x)^2} = \dfrac{-e^{-x} - 2 - e^x}{(1+e^x)^2}$

5. $f'(x) = 1/x - \frac{1}{4}x$, so

$$\sqrt{1+(f'(x))^2} = \sqrt{1+\left(\frac{1}{x}-\frac{1}{4}x\right)^2} = \sqrt{1+\frac{1}{x^2}-\frac{1}{2}+\frac{1}{16}x^2}$$
$$= \sqrt{\frac{1}{x^2}+\frac{1}{2}+\frac{1}{16}x^2} = \sqrt{\left(\frac{1}{x}+\frac{1}{4}x\right)^2}$$
$$= \frac{1}{x}+\frac{1}{4}x$$

7. a. The inequality $(x-4)/(2x+6) \ge \frac{3}{20}$ is equivalent to $(x-4)/(2x+6) - \frac{3}{20} \ge 0$, or $[14(x-7)]/[40(x+3)] \ge 0$. From the diagram we see that the solution is the union of $(-\infty, -3)$ and $[7, \infty)$. Thus $f(x) \ge \frac{3}{20}$ for $x \ge 7$.

$x-7$	− − − − − − − − − 0 + +
$x+3$	− − 0 + + + + + + + + +
$\dfrac{14(x-7)}{40(x+3)}$	+ + − − − − − − 0 + +

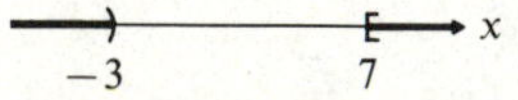

b. $f'(x) = \dfrac{2x+6-(x-4)(2)}{(2x+6)^2} = \dfrac{14}{(2x+6)^2} > 0$ for $x \ge 7$.

Thus f is increasing on $[7, \infty)$. Since $f(7) = \frac{3}{20}$, it follows that $f(x) \ge \frac{3}{20}$ for $x \ge 7$.

9. $f'(x) = 4x^3 - 16x = 4x(x^2-4)$; $f''(x) = 12x^2 - 16 = 4(3x^2-4)$; relative maximum value is $f(0) = 0$; relative minimum value is $f(-2) = f(2) = -16$; increasing on $[-2, 0]$ and $[2, \infty)$; decreasing on $(-\infty, -2]$ and $[0, 2]$; concave upward on $(-\infty, -\frac{2}{3}\sqrt{3})$ and $(\frac{2}{3}\sqrt{3}, \infty)$; concave downward on $(-\frac{2}{3}\sqrt{3}, \frac{2}{3}\sqrt{3})$; inflection points are $(-\frac{2}{3}\sqrt{3}, -\frac{80}{9})$ and $(\frac{2}{3}\sqrt{3}, -\frac{80}{9})$; symmetry with respect to the y axis.

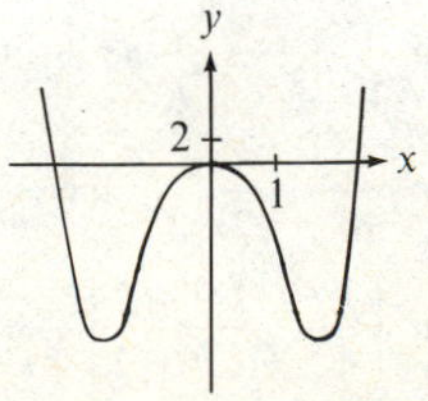

11. Using the notation in the figure and the Law of Cosines, we have $D^2 = (200)^2 + (200)^2 - 2(200)(200)\cos\theta = 80{,}000(1-\cos\theta)$. We are to find dD/dt at the moment that $D = 200$. Differentiating $D^2 = 80{,}000(1-\cos\theta)$ implicitly, we find that $2D(dD/dt) = 80{,}000(\sin\theta)(d\theta/dt)$. To determine $d\theta/dt$, we use the fact that if s is the arc corresponding to D, then $s = 200\theta$. Since the dog moves at the rate of 50 feet per second, it follows that $50 = ds/dt = 200(d\theta/dt)$, so that $d\theta/dt = \frac{1}{4}$. At the instant $D = 200$, the triangle in the figure is equilateral, so $\theta = \pi/3$. Therefore at that instant

$$\frac{dD}{dt} = \frac{80{,}000(\sin(\pi/3))\frac{1}{4}}{2(200)} = 25\sqrt{3} \text{ (feet per second)}$$

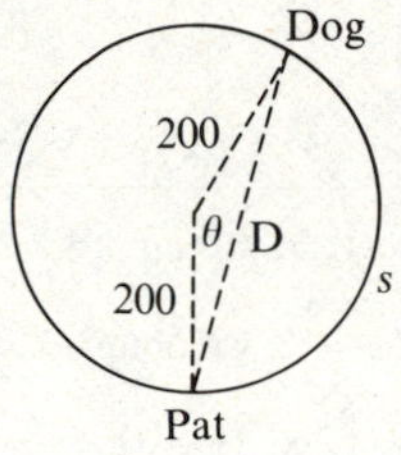

13. Let $u = 4 + x^2$, so that $du = 2x\,dx$;

$$\int \frac{x}{4+x^2}\,dx = \int \frac{1}{u}\left(\frac{1}{2}\right) du = \frac{1}{2}\ln|u| + C = \frac{1}{2}\ln(4+x^2) + C$$

15. $\int e^{-3x}(e^{5x}+1)\,dx = \int (e^{-3x}e^{5x} + e^{-3x})\,dx = \int (e^{2x} + e^{-3x})\,dx$

$$= \int e^{2x}\,dx + \int e^{-3x}\,dx = \tfrac{1}{2}e^{2x} - \tfrac{1}{3}e^{-3x} + C$$

17. Let $f(t)$ be the position of the car at time t. Then

$$f(2) - f(0) = \int_0^2 f'(t)\,dt = \int_0^2 v(t)\,dt = \int_0^2 \left(40 + \frac{40}{4+t^2}\right) dt = \left(40t + \frac{40}{2}\arctan\frac{t}{2}\right)\Big|_0^2$$
$$= (80 + 20\arctan 1) - (0 + 20\arctan 0) = 80 + 20\left(\frac{\pi}{4}\right) = 80 + 5\pi$$

Thus the car travels $80 + 5\pi \approx 95.7080$ (miles) in the two hours.

19. For $x > 0$ and $y \ge 0$, we have $y^2 - x^7 = 8x^4$, so $y = \sqrt{8x^4 + x^7} = x^2\sqrt{1+x^3}$. Thus $A = \int_1^2 x^2\sqrt{8+x^3}\,dx$. Let $u = 8 + x^3$, so that $du = 3x^2\,dx$. If $x = 1$, then $u = 9$, and if $x = 2$, then $u = 16$. Thus

$$A = \int_1^2 x^2\sqrt{8+x^3}\,dx = \int_9^{16} \sqrt{u}\left(\tfrac{1}{3}\right) du = \tfrac{2}{9}u^{3/2}\Big|_9^{16} = \tfrac{2}{9}(64-27) = \tfrac{74}{9}$$

8 APPLICATIONS OF THE INTEGRAL

SECTION 8.1

1. $V=\int_0^1 \pi(x^2)^2\,dx = \frac{\pi}{5}x^5\Big|_0^1 = \frac{\pi}{5}$

3. $V=\int_0^{\sqrt{3}}\pi(\sqrt{3-x^2})^2\,dx = \pi\int_0^{\sqrt{3}}(3-x^2)\,dx = \pi(3x-\frac{1}{3}x^3)\Big|_0^{\sqrt{3}} = 2\sqrt{3}\pi$

5. $V=\int_{-\pi/4}^0 \pi\sec^2 x\,dx = \pi\tan x\Big|_{-\pi/4}^0=\pi$

7. $V=\int_0^1\pi(\sqrt{x}e^x)^2\,dx = \pi\int_0^1 xe^{2x}\,dx \overset{\text{parts}}{=} \pi(\frac{1}{2}xe^{2x}-\frac{1}{4}e^{2x})\Big|_0^1$

$= \pi[(\frac{1}{2}e^2-\frac{1}{4}e^2)-(0-\frac{1}{4})] = \frac{\pi}{4}(e^2+1)$

9. $V=\int_1^4\pi(x^{3/2})^2\,dx=\pi\int_1^4x^3\,dx=\frac{\pi}{4}x^4\Big|_1^4=\frac{\pi}{4}(256)-\frac{\pi}{4}=\frac{255}{4}\pi$

11. $V=\int_1^2\pi[x(x^3+1)^{1/4}]^2\,dx=\pi\int_1^2x^2(x^3+1)^{1/2}\,dx$

$\overset{u=x^3+1}{=}\pi\int_2^9 u^{1/2}\cdot\frac{1}{3}\,du=\frac{\pi}{3}\left(\frac{2}{3}u^{3/2}\right)\Big|_2^9=6\pi-\frac{4\sqrt{2}}{9}\pi$

13. $V=\int_1^2\pi(e^y)^2\,dy=\pi\int_1^2e^{2y}\,dy=\frac{\pi}{2}e^{2y}\Big|_1^2=\frac{\pi}{2}e^4-\frac{\pi}{2}e^2=\frac{\pi}{2}(e^4-e^2)$

15. $V=\int_1^2\pi(\sqrt{1+y^3})^2\,dy=\pi\int_1^2(1+y^3)\,dy=\pi(y+\frac{1}{4}y^4)\Big|_1^2=\pi[(2+4)-(1+\frac{1}{4})]=\frac{19}{4}\pi$

17. $V=\int_1^3\pi[(\sqrt{x+1})^2-(\sqrt{x-1})^2]\,dx=\pi\int_1^3 2\,dx=2\pi x\Big|_1^3=4\pi$

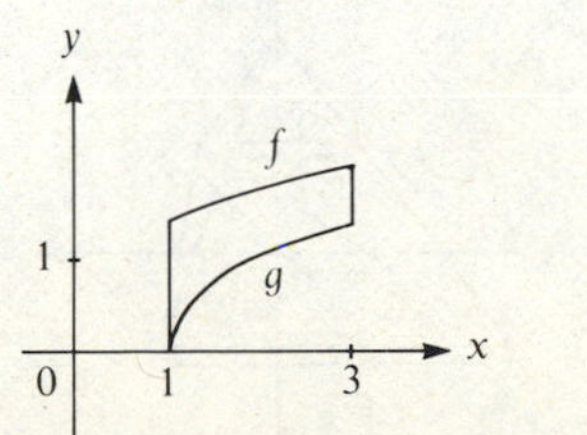

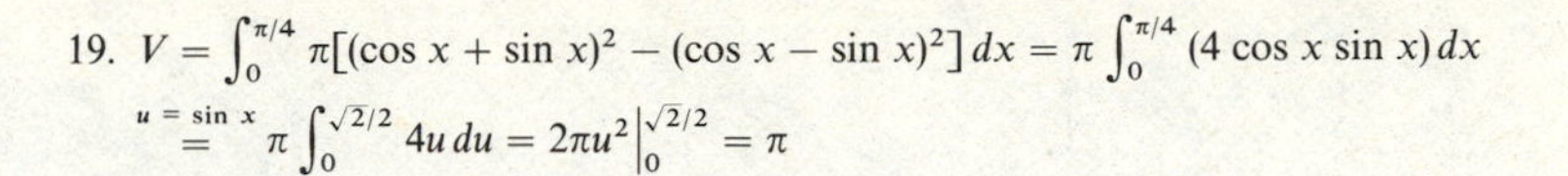

19. $V=\int_0^{\pi/4}\pi[(\cos x+\sin x)^2-(\cos x-\sin x)^2]\,dx=\pi\int_0^{\pi/4}(4\cos x\sin x)\,dx$

$\overset{u=\sin x}{=}\pi\int_0^{\sqrt{2}/2}4u\,du=2\pi u^2\Big|_0^{\sqrt{2}/2}=\pi$

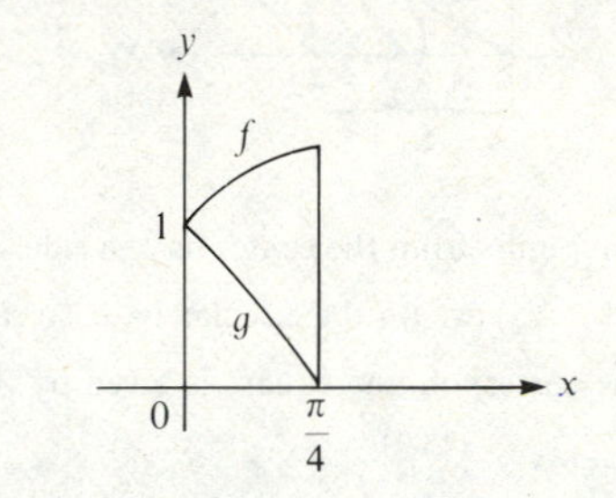

21. The graphs of $y=x^2/2+3$ and $y=12-x^2/2$ intersect for (x, y) such that $x^2/2+3=y=12-x^2/2$, or $x^2=9$, so that $x=-3$ or $x=3$. Since $x^2/2+3\le 12-x^2/2$ for $-3\le x\le 3$, it follows that

$$V=\int_{-3}^3\pi\left[\left(12-\frac{x^2}{2}\right)^2-\left(\frac{x^2}{2}+3\right)^2\right]dx$$
$$=\pi\int_{-3}^3(135-15x^2)\,dx=\pi(135x-5x^3)\Big|_{-3}^3=540\pi$$

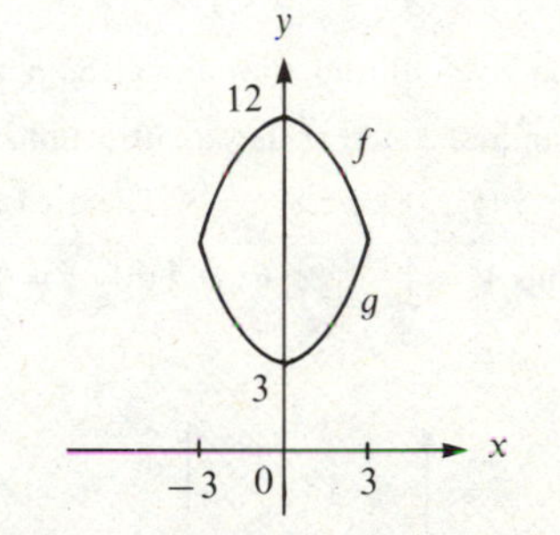

23. The graphs of $y=5x$ and $y=x^2+2x+2$ intersect for (x, y) such that

$$5x=y=x^2+2x+2,\quad\text{or}\quad x^2-3x+2=0$$

so that $x=1$ or $x=2$. Since $5x\ge x^2+2x+2$ for $1\le x\le 2$, it follows that

$$V=\int_1^2\pi[(5x)^2-(x^2+2x+2)^2]\,dx=\pi\int_1^2(-x^4-4x^3+17x^2-8x-4)\,dx$$
$$=\pi(-\tfrac{1}{5}x^5-x^4+\tfrac{17}{3}x^3-4x^2-4x)\Big|_1^2$$
$$=\pi[(-\tfrac{32}{5}-16+\tfrac{136}{3}-16-8)-(-\tfrac{1}{5}-1+\tfrac{17}{3}-4-4)]=\tfrac{37}{15}\pi$$

25. Place the base so that L_1 lies along the positive x axis with the vertex opposite L_2 at the origin. Then the diameter of the semicircle x units from the origin is x and the cross-sectional area is given by $A(x)=\frac{1}{2}\pi(x/2)^2=\frac{1}{8}\pi x^2$, so $V=\int_0^4\frac{1}{8}\pi x^2\,dx=\frac{1}{24}\pi x^3\Big|_0^4=\frac{8}{3}\pi$.

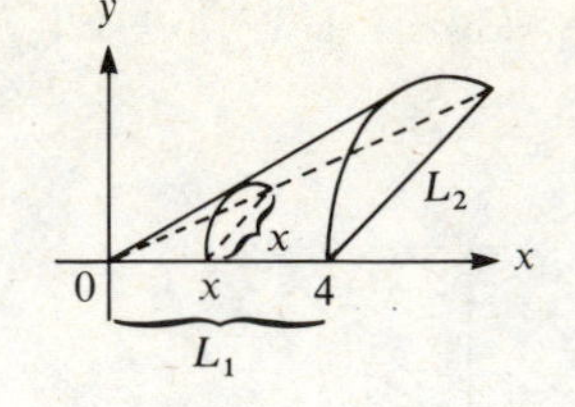

27. Let a square cross-section x units from the center have a side $s(x)$ units long. The points $(x, -\sqrt{1-x^2})$ and $(x, \sqrt{1-x^2})$ are on the circular base, so $s(x) = 2\sqrt{1-x^2}$. Thus the cross-sectional area of the corresponding square is given by $A(x) = (s(x))^2 = 4(1-x^2)$, so $V = \int_{-1}^{1} 4(1-x^2)\,dx = (4x - \frac{4}{3}x^3)\Big|_{-1}^{1} = \frac{16}{3}$.

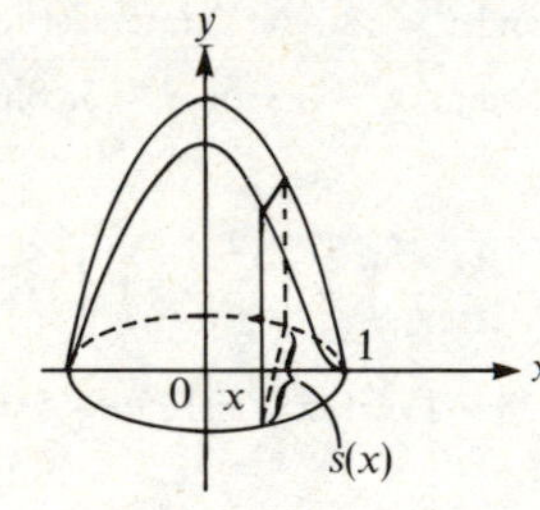

29. Place the base so that the given altitude lies along the positive x axis with the vertex at the origin. Then the length of a side of the square x units from the origin is $2z$, where by similar triangles $z/x = 5/(5\sqrt{3})$, or $z = x/\sqrt{3}$. The area $A(x)$ of that square is given by $A(x) = (2z)^2 = \frac{4}{3}x^2$. Thus $V = \int_0^{5\sqrt{3}} \frac{4}{3}x^2\,dx = \frac{4}{9}x^3\Big|_0^{5\sqrt{3}} = \frac{4}{9}(125)(3\sqrt{3}) = \frac{500}{3}\sqrt{3}$.

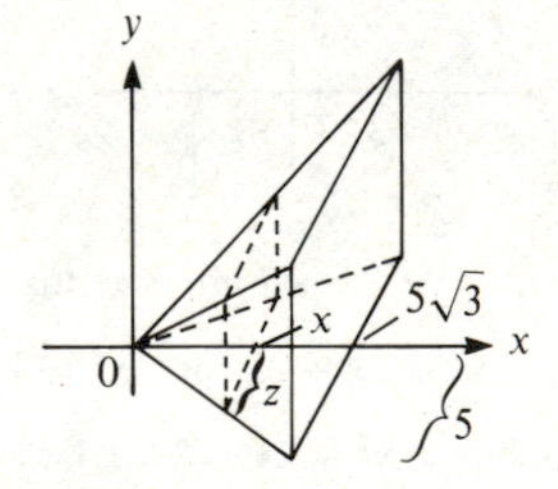

31. $V = \int_a^b \pi[f(x) - c]^2\,dx$

33. $V = \int_0^1 \pi(1 - e^{-2x})^2\,dx = \pi \int_0^1 (1 - 2e^{-2x} + e^{-4x})\,dx = \pi(x + e^{-2x} - \frac{1}{4}e^{-4x})\Big|_0^1$
$= \pi[(1 - e^{-2} - \frac{1}{4}e^{-4}) - (1 - \frac{1}{4})] = \pi(\frac{1}{4} - e^{-2} - \frac{1}{4}e^{-4})$

35. The graphs of $y = x^2 - x + 1$ and $y = 2x^2 - 4x + 3$ intersect for (x, y) such that

$$x^2 - x + 1 = y = 2x^2 - 4x + 3, \quad \text{or} \quad x^2 - 3x + 2 = 0$$

so that $x = 1$ or $x = 2$. Since $x^2 - x + 1 \geq 2x^2 - 4x + 3$ for $1 \leq x \leq 2$, it follows from Exercise 34 that

$$V = \int_1^2 \pi[(x^2 - x + 1) - 1]^2\,dx - \int_1^2 \pi[(2x^2 - 4x + 3) - 1]^2\,dx$$
$$= \pi \int_1^2 (x^4 - 2x^3 + x^2)\,dx - \pi \int_1^2 4(x-1)^4\,dx$$
$$= \pi(\tfrac{1}{5}x^5 - \tfrac{1}{2}x^4 + \tfrac{1}{3}x^3)\Big|_1^2 - \tfrac{4}{5}\pi(x-1)^5\Big|_1^2$$
$$= \pi[(\tfrac{32}{5} - 8 + \tfrac{8}{3}) - (\tfrac{1}{5} - \tfrac{1}{2} + \tfrac{1}{3})] - \tfrac{4}{5}\pi = \tfrac{7}{30}\pi$$

37. We follow the solution of Example 2, with h replacing 4 and a replacing 3. Then the length $s(y)$ of a side of the cross-section at y satisfies, by similar triangles,

$$\frac{s(y)}{a} = \frac{h-y}{h}, \quad \text{so} \quad s(y) = \frac{a}{h}(h-y)$$

so the cross-sectional area is given by $A(y) = [s(y)]^2 = (a^2/h^2)(h-y)^2$. Then

$$V = \int_0^h \frac{a^2}{h^2}(h-y)^2\,dy = -\frac{a^2}{3h^2}(h-y)^3\Big|_0^h = \frac{a^2h^3}{3h^2} = \frac{1}{3}a^2h$$

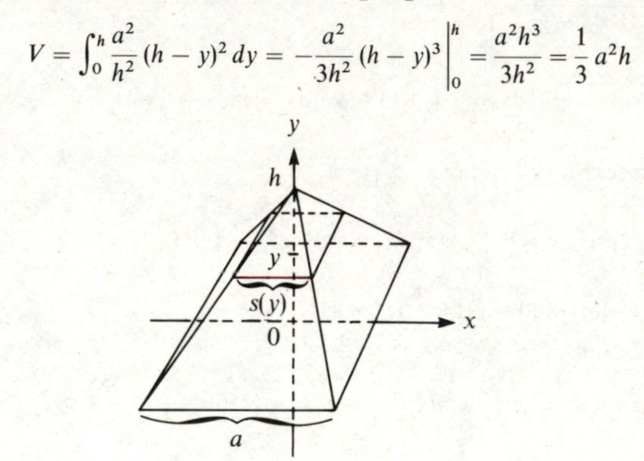

39. Let a triangular cross-section y feet above the base have a side $s(y)$ feet long. By similar triangles, we have $s(y)/(h-y) = a/h$, so $s(y) = (a/h)(h-y)$. Thus the cross-sectional area is given by

$$A(y) = \tfrac{1}{2}s(y)\left(\frac{\sqrt{3}}{2}s(y)\right) = \frac{\sqrt{3}}{4}\frac{a^2}{h^2}(h-y)^2,$$

so
$$V = \int_0^h A(y)\,dy = \int_0^h \frac{\sqrt{3}}{4}\frac{a^2}{h^2}(h-y)^2\,dy = -\frac{\sqrt{3}}{12}\frac{a^2}{h^2}(h-y)^3\Big|_0^h = \frac{\sqrt{3}}{12}a^2h$$

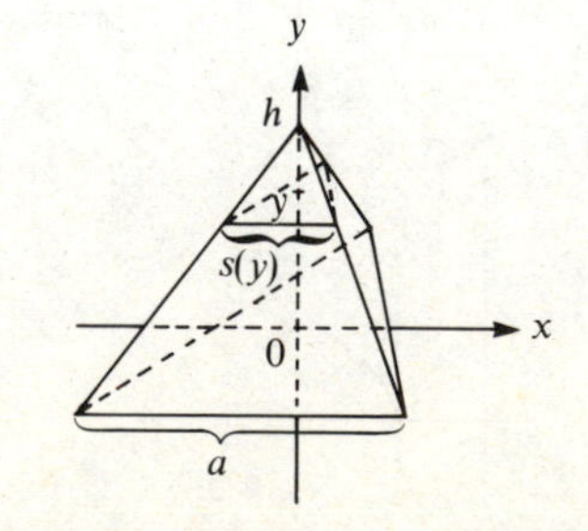

41. Let a square cross-section x feet from the center have a side $s(x)$ feet long. Then $s(x) = 20\sqrt{1 - x^2/400} = \sqrt{400 - x^2}$. Thus the cross-sectional area is given by $A(x) = (s(x))^2 = 400 - x^2$, so that

$$V = \int_{-20}^{20} (400 - x^2)\,dx = (400x - \tfrac{1}{3}x^3)\Big|_{-20}^{20} = \frac{32{,}000}{3} \text{ (cubic feet)}$$

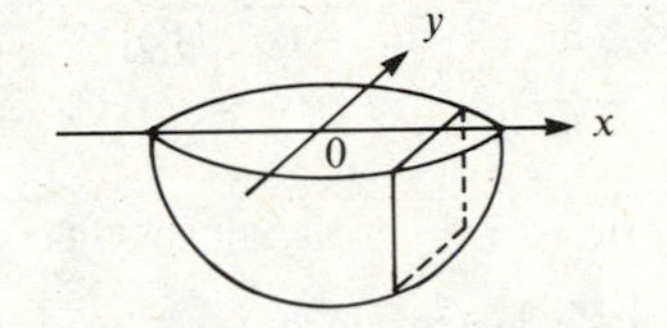

43. We can use the result of Exercise 42 with $a = \frac{1}{2}$ and $b = 1$. Then $V = \frac{4}{3}\pi(\frac{1}{2})(1)^2 = 2\pi/3$ (cubic centimeters).

45. The volume is given by

$$V = \int_{-\sqrt{3}}^{\sqrt{3}} \pi[(4 - x^2)^2 - 1^2]\,dx = \pi \int_{-\sqrt{3}}^{\sqrt{3}} (15 - 8x^2 + x^4)\,dx$$
$$= \pi\left(15x - \frac{8}{3}x^3 + \frac{x^5}{5}\right)\Big|_{-\sqrt{3}}^{\sqrt{3}} = \frac{88\sqrt{3}}{5}\pi$$

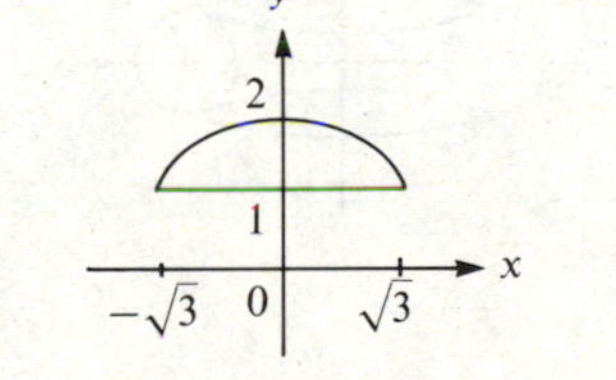

47. a. $\lim_{b\to\infty} \int_1^b \pi\left(\frac{1}{x}\right)^2 dx = \lim_{b\to\infty} \pi \int_1^b \frac{1}{x^2}\,dx = \lim_{b\to\infty} \pi\left(-\frac{1}{x}\right)\Big|_1^b = \lim_{b\to\infty} \pi\left(-\frac{1}{b} + 1\right) = \pi$

so the volume is π.

b. $\lim_{b\to\infty} \int_1^b \pi\left(\frac{1}{x^2}\right)^2 dx = \lim_{b\to\infty} \pi \int_1^b \frac{1}{x^4}\,dx = \lim_{b\to\infty} \frac{-\pi}{3x^3}\Big|_1^b = \lim_{b\to\infty}\left(-\frac{\pi}{3b^3} + \frac{\pi}{3}\right) = \frac{\pi}{3}$

so the volume is $\frac{\pi}{3}$.

c. $\lim_{b\to\infty} \int_1^b \pi\left(\frac{1}{x^{1/2}}\right)^2 dx = \lim_{b\to\infty} \pi \int_0^b \frac{1}{x}\,dx = \lim_{b\to\infty} \ln x\Big|_1^b = \lim_{b\to\infty} \ln b = \infty$

so the volume is infinite.

49. Let $A_1(x)$ be the cross-sectional area at x for the first solid and $A_2(x)$ the cross-sectional area at x for the second solid. Then $A_1(x) = A_2(x)$ for each x in $[a, b]$. By (1) the volume of the first solid is given by $V_1 = \int_a^b A_1(x)\,dx$, and the volume of the second is given by $V_2 = \int_a^b A_2(x)\,dx$. Since $A_1(x) = A_2(x)$ for all x, we must have $V_1 = V_2$.

SECTION 8.2

1. $V = \int_0^{\sqrt{3}} 2\pi x\sqrt{x^2 + 1}\,dx \overset{u = x^2+1}{=} \pi \int_1^4 u^{1/2}\,du = \frac{2\pi}{3}u^{3/2}\Big|_1^4 = \frac{14\pi}{3}$

3. $V = \int_0^1 2\pi x e^{2x+1}\,dx \overset{\text{parts}}{=} 2\pi\left(\frac{1}{2}xe^{2x+1}\Big|_0^1 - \int_0^1 \frac{1}{2}e^{2x+1}\,dx\right) = 2\pi\left(\frac{1}{2}e^3 - \frac{1}{4}e^{2x+1}\Big|_0^1\right)$

$= 2\pi\left(\frac{1}{4}e^3 + \frac{1}{4}e\right) = \frac{1}{2}\pi e(e^2 + 1)$

5. $V = \int_1^2 2\pi x\sqrt{x - 1}\,dx \overset{u = x-1}{=} 2\pi \int_0^1 (u + 1)u^{1/2}\,du = 2\pi \int_0^1 (u^{3/2} + u^{1/2})\,du$

$= 2\pi\left(\frac{2}{5}u^{5/2} + \frac{2}{3}u^{3/2}\right)\Big|_0^1 = \frac{32\pi}{15}$

7. $V = \int_1^3 2\pi x \ln x\,dx \overset{\text{parts}}{=} 2\pi\left(\frac{1}{2}x^2 \ln x\right)\Big|_1^3 - 2\pi \int_1^3 \frac{1}{2}x\,dx = 9\pi \ln 3 - \frac{\pi}{2}x^2\Big|_1^3$

$= 9\pi \ln 3 - 4\pi$

9. $V = \int_0^1 2\pi y(y^2\sqrt{1 + y^4})\,dy \overset{u = 1+y^4}{=} 2\pi \int_1^2 \sqrt{u}\cdot\frac{1}{4}du = 2\pi\cdot\frac{1}{4}\cdot\frac{2}{3}u^{3/2}\Big|_1^2 = \frac{1}{3}\pi(2\sqrt{2} - 1)$

11. $V = \int_0^{\sqrt{2}/2} 2\pi \frac{y}{\sqrt{1 - y^4}}\,dy \overset{u = y^2}{=} 2\pi \int_0^{1/2} \frac{1}{\sqrt{1 - u^2}}\cdot\frac{1}{2}\,du = \pi \arcsin u\Big|_0^{1/2} = \frac{\pi^2}{6}$

13. Since $f(x) = 1 \geq x - 2 = g(x)$ for $1 \leq x \leq 3$,

$$V = \int_1^3 2\pi x[1 - (x - 2)]\,dx = 2\pi \int_1^3 (3x - x^2)\,dx = 2\pi(\tfrac{3}{2}x^2 - \tfrac{1}{3}x^3)\Big|_1^3$$
$$= 2\pi[(\tfrac{27}{2} - 9) - (\tfrac{3}{2} - \tfrac{1}{3})] = \tfrac{20}{3}\pi$$

15. Since $f(x) = \cos x \geq \sin x = g(x)$ for $0 \leq x \leq \pi/4$,

$$V = \int_0^{\pi/4} 2\pi x(\cos x - \sin x)\,dx \overset{\text{parts}}{=} 2\pi x(\sin x + \cos x)\Big|_0^{\pi/4} - 2\pi \int_0^{\pi/4} (\sin x + \cos x)\,dx$$
$$= \frac{\pi^2\sqrt{2}}{2} + 2\pi(\cos x - \sin x)\Big|_0^{\pi/4} = \tfrac{1}{2}\sqrt{2}\pi^2 - 2\pi$$

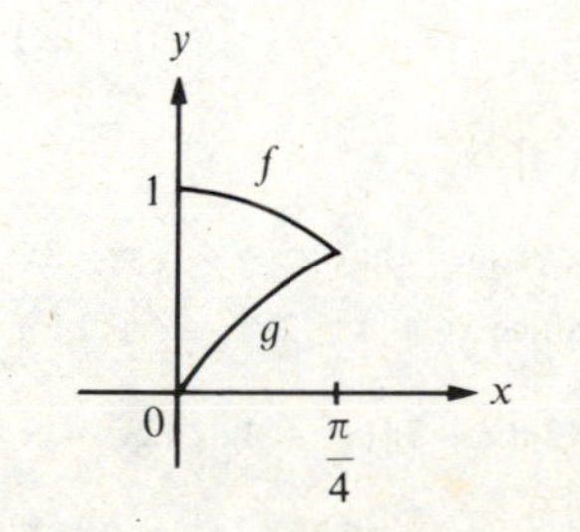

17. Let $0 \leq y \leq 1$. Notice that $y^2 + 1 \geq y\sqrt{1 + y^3}$ if $y^4 + 2y^2 + 1 = (y^2 + 1)^2 \geq y^2(1 + y^3) = y^2 + y^5$ or, equivalently, $y^4 - y^5 + y^2 + 1 \geq 0$, which is valid for $0 \leq y \leq 1$. Thus

$f(y) \ge g(y)$ for $0 \le y \le 1$, so that

$$V = \int_0^1 2\pi y[(y^2+1) - y\sqrt{1+y^3}]\,dy$$
$$= 2\pi\left[\left(\frac{1}{4}y^4 + \frac{1}{2}y^2\right)\Big|_0^1 - \int_0^1 y^2\sqrt{1+y^3}\,dy\right] \overset{u=1+y^3}{=} 2\pi\left(\frac{3}{4} - \int_1^2 \sqrt{u}\cdot\frac{1}{3}\,du\right)$$
$$= 2\pi\left(\frac{3}{4} - \frac{1}{3}\cdot\frac{2}{3}u^{3/2}\Big|_1^2\right) = 2\pi\left(\frac{3}{4} - \frac{4}{9}\sqrt{2} + \frac{2}{9}\right) = \frac{\pi}{18}(35 - 16\sqrt{2})$$

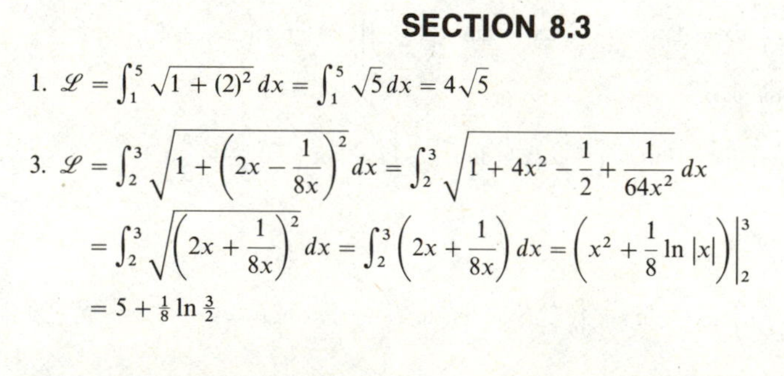

19. The graphs intersect for (x, y) such that $2x = y = x^2$, so that $x = 0$ or $x = 2$. Since $2x \ge x^2$ for $0 \le x \le 2$, we have

$$V = \int_0^2 2\pi x(2x - x^2)\,dx = 2\pi\int_0^2 (2x^2 - x^3)\,dx = 2\pi(\tfrac{2}{3}x^3 - \tfrac{1}{4}x^4)\Big|_0^2 = 2\pi(\tfrac{16}{3} - 4) = \tfrac{8}{3}\pi$$

21. The graphs intersect for (x, y) such that $|x - 2| = y = \frac{1}{2}(x-2)^2 + \frac{1}{2}$. If $x > 2$, then the equations reduce to $x - 2 = \frac{1}{2}(x-2)^2 + \frac{1}{2}$, so that $(x-2)^2 - 2(x-2) + 1 = 0$, or $((x-2)-1)^2 = 0$, so that $x = 3$. If $x < 2$, then the first equations reduce to $-(x-2) = \frac{1}{2}(x-2)^2 + \frac{1}{2}$, so that $(x-2)^2 + 2(x-2) + 1 = 0$, or $((x-2)+1)^2 = 0$, so that $x = 1$. Since $\frac{1}{2}(x-2)^2 + \frac{1}{2} \ge |x-2|$ for $1 \le x \le 3$, we have

$$V = \int_1^2 2\pi x[\tfrac{1}{2}(x-2)^2 + \tfrac{1}{2} + (x-2)]\,dx + \int_2^3 2\pi x[\tfrac{1}{2}(x-2)^2 + \tfrac{1}{2} - (x-2)]\,dx$$
$$= 2\pi\int_1^2 (\tfrac{1}{2}x^3 - x^2 + \tfrac{1}{2}x)\,dx + 2\pi\int_2^3 (\tfrac{1}{2}x^3 - 3x^2 + \tfrac{9}{2}x)\,dx$$
$$= 2\pi(\tfrac{1}{8}x^4 - \tfrac{1}{3}x^3 + \tfrac{1}{4}x^2)\Big|_1^2 + 2\pi(\tfrac{1}{8}x^4 - x^3 + \tfrac{9}{4}x^2)\Big|_2^3$$
$$= 2\pi[(2 - \tfrac{8}{3} + 1) - (\tfrac{1}{8} - \tfrac{1}{3} + \tfrac{1}{4})] + 2\pi[(\tfrac{81}{8} - 27 + \tfrac{81}{4}) - (2 - 8 + 9)] = \tfrac{4}{3}\pi$$

23. $V = \int_0^1 2\pi(x+1)x^4\,dx = 2\pi\int_0^1 (x^5 + x^4)\,dx = 2\pi(\tfrac{1}{6}x^6 + \tfrac{1}{5}x^5)\Big|_0^1 = \tfrac{11}{15}\pi$

25. $V = \int_a^b 2\pi(x - c)[f(x) - g(x)]\,dx$

27. The graphs intersect for (x, y) such that $x^2 + 4 = y = 2x^2 + x + 2$, or $x^2 + x - 2 = 0$, so that $x = -2$ or $x = 1$. Since $x^2 + 4 \ge 2x^2 + x + 2$ for $-2 \le x \le 1$, we have

$$V = \int_{-2}^1 2\pi(x+5)[(x^2+4) - (2x^2 + x + 2)]\,dx$$
$$= 2\pi\int_{-2}^1 (-x^3 - 6x^2 - 3x + 10)\,dx$$
$$= 2\pi(-\tfrac{1}{4}x^4 - 2x^3 - \tfrac{3}{2}x^2 + 10x)\Big|_{-2}^1$$
$$= 2\pi[(-\tfrac{1}{4} - 2 - \tfrac{3}{2} + 10) - (-4 + 16 - 6 - 20)] = \tfrac{81}{2}\pi$$

29. Let $f(x) = h - (h/a)x$ for $0 \le x \le a$. A cone of radius a and height h is obtained by revolving the region between the graph of f and the x axis in $[0, a]$ about the y axis. By (3) the volume is given by

$$V = \int_0^a 2\pi x\left(h - \frac{h}{a}x\right)dx = 2\pi\left(\frac{hx^2}{2} - \frac{hx^3}{3a}\right)\Big|_0^a = \frac{1}{3}\pi a^2 h$$

31. $$V = \int_{b-r}^{b+r} 2\pi x[\sqrt{r^2 - (x-b)^2} - (-\sqrt{r^2 - (x-b)^2})]\,dx$$
$$= 4\pi\int_{b-r}^{b+r} x\sqrt{r^2 - (x-b)^2}\,dx$$
$$\overset{x=b+r\sin u}{=} 4\pi\int_{-\pi/2}^{\pi/2} (b + r\sin u)\cdot\sqrt{r^2 - r^2\sin^2 u}(r\cos u)\,du$$
$$= 4\pi r^2\int_{-\pi/2}^{\pi/2} (b\cos^2 u + r\sin u\cos^2 u)\,du$$
$$= 4\pi r^2\int_{-\pi/2}^{\pi/2}\left(\frac{b}{2} + \frac{b}{2}\cos 2u + r\sin u\cos^2 u\right)du$$
$$= 4\pi r^2\left(\frac{b}{2}u + \frac{b}{4}\sin 2u - \frac{r}{3}\cos^3 u\right)\Big|_{-\pi/2}^{\pi/2} = 2\pi^2 br^2$$

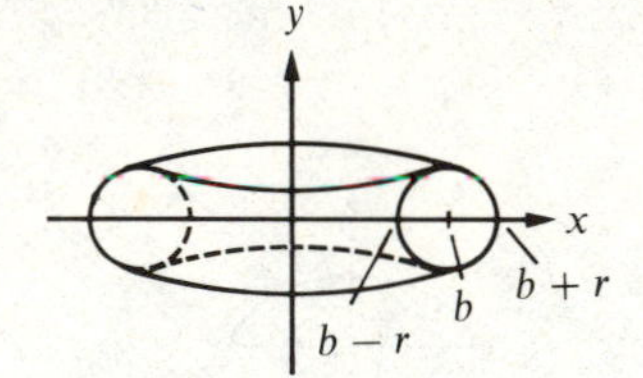

SECTION 8.3

1. $\mathscr{L} = \int_1^5 \sqrt{1 + (2)^2}\,dx = \int_1^5 \sqrt{5}\,dx = 4\sqrt{5}$

3. $$\mathscr{L} = \int_2^3 \sqrt{1 + \left(2x - \frac{1}{8x}\right)^2}\,dx = \int_2^3 \sqrt{1 + 4x^2 - \frac{1}{2} + \frac{1}{64x^2}}\,dx$$
$$= \int_2^3 \sqrt{\left(2x + \frac{1}{8x}\right)^2}\,dx = \int_2^3\left(2x + \frac{1}{8x}\right)dx = \left(x^2 + \frac{1}{8}\ln|x|\right)\Big|_2^3$$
$$= 5 + \tfrac{1}{8}\ln\tfrac{3}{2}$$

5. $$\mathscr{L} = \int_1^2 \sqrt{1 + \left(4x^3 - \frac{1}{16x^3}\right)^2}\,dx = \int_1^2 \sqrt{1 + 16x^6 - \frac{1}{2} + \frac{1}{256x^6}}\,dx$$
$$= \int_1^2 \sqrt{\left(4x^3 + \frac{1}{16x^3}\right)^2}\,dx = \int_1^2\left(4x^3 + \frac{1}{16x^3}\right)dx = \left(x^4 - \frac{1}{32x^2}\right)\Big|_1^2$$
$$= (16 - \tfrac{1}{128}) - (1 - \tfrac{1}{32}) = \tfrac{1923}{128}$$

7. $\mathscr{L} = \int_1^2 \sqrt{1 + \left[\frac{2x}{1+x^2} - \frac{1}{8}\left(x + \frac{1}{x}\right)\right]^2}\,dx = \int_1^2 \sqrt{1 + \frac{4x^2}{(1+x^2)^2} - \frac{1}{2} + \frac{1}{64}\frac{(1+x^2)^2}{x^2}}\,dx$

$= \int_1^2 \sqrt{\left[\frac{2x}{1+x^2} + \frac{1}{8}\left(\frac{1+x^2}{x}\right)\right]^2}\,dx = \int_1^2 \left[\frac{2x}{1+x^2} + \frac{1}{8}\left(\frac{1}{x} + x\right)\right] dx$

$= \left[\ln(1+x^2) + \frac{1}{8}(\ln|x| + \frac{1}{2}x^2)\right]\Big|_1^2$

$= (\ln 5 + \frac{1}{8}\ln 2 + \frac{1}{4}) - (\ln 2 + \frac{1}{16}) = \ln 5 - \frac{7}{8}\ln 2 + \frac{3}{16}$

9. $\mathscr{L} = \int_{\pi/4}^{\pi/3} \sqrt{1 + \left(\frac{-1}{4}\cos x + \sec x\right)^2}\,dx = \int_{\pi/4}^{\pi/3} \sqrt{1 + \frac{1}{16}\cos^2 x - \frac{1}{2} + \sec^2 x}\,dx$

$= \int_{\pi/4}^{\pi/3} \sqrt{(\frac{1}{4}\cos x + \sec x)^2}\,dx = \int_{\pi/4}^{\pi/3} (\frac{1}{4}\cos x + \sec x)\,dx$

$= (\frac{1}{4}\sin x + \ln|\sec x + \tan x|)\Big|_{\pi/4}^{\pi/3} = [\frac{1}{8}\sqrt{3} + \ln(2+\sqrt{3})] - [\frac{1}{8}\sqrt{2} + \ln(\sqrt{2}+1)]$

$= \frac{1}{8}(\sqrt{3} - \sqrt{2}) + \ln\frac{2+\sqrt{3}}{\sqrt{2}+1}$

11. $\mathscr{L} = \int_0^1 \sqrt{1 + \left[x^2 + 1 - \frac{1}{4(1+x^2)}\right]^2}\,dx = \int_0^1 \sqrt{1 + (x^2+1)^2 - \frac{1}{2} + \left[\frac{1}{4(1+x^2)}\right]^2}\,dx$

$= \int_0^1 \sqrt{(x^2+1)^2 + \frac{1}{2} + \left[\frac{1}{4(1+x^2)}\right]^2}\,dx = \int_0^1 \sqrt{\left[x^2 + 1 + \frac{1}{4(1+x^2)}\right]^2}\,dx$

$= \int_0^1 \left[x^2 + 1 + \frac{1}{4(1+x^2)}\right] dx = \left(\frac{1}{3}x^3 + x + \frac{1}{4}\arctan x\right)\Big|_0^1 = \frac{4}{3} + \frac{\pi}{16}$

13. $\mathscr{L} = \int_{\sqrt{3}}^{\sqrt{8}} \sqrt{1 + \left(\frac{1}{x}\right)^2}\,dx = \int_{\sqrt{3}}^{\sqrt{8}} \sqrt{1 + \frac{1}{x^2}}\,dx$

$= \int_{\sqrt{3}}^{\sqrt{8}} \frac{1}{x}\sqrt{x^2+1}\,dx \overset{u=\sqrt{x^2+1}}{=} \int_2^3 \frac{u^2}{u^2-1}\,du = \int_2^3 \left(1 + \frac{1}{2}\frac{1}{u-1} - \frac{1}{2}\frac{1}{u+1}\right) du$

$= [u + \frac{1}{2}\ln(u-1) - \frac{1}{2}\ln(u+1)]\Big|_2^3 = (3 + \frac{1}{2}\ln 2 - \frac{1}{2}\ln 4) - (2 - \frac{1}{2}\ln 3)$

$= 1 - \frac{1}{2}\ln 2 + \frac{1}{2}\ln 3$

15. $\mathscr{L} = \int_0^1 \sqrt{1 + (2x)^2}\,dx = \int_0^1 \sqrt{1 + 4x^2}\,dx$. Let $x = \frac{1}{2}\tan u$. Then, by (7) of Section 7.2,

$$\int \sqrt{1+4x^2}\,dx = \int \sqrt{1+\tan^2 u}(\tfrac{1}{2}\sec^2 u)\,du = \tfrac{1}{2}\int \sec^3 u\,du$$
$$= \tfrac{1}{4}[\sec u \tan u + \ln|\sec u + \tan u|] + C$$
$$= \tfrac{1}{4}[2x\sqrt{1+4x^2} + \ln(\sqrt{1+4x^2} + 2x) + C$$

Thus

$$\mathscr{L} = \int_0^1 \sqrt{1+(2x)^2}\,dx = \tfrac{1}{4}[2x\sqrt{1+4x^2} + \ln(\sqrt{1+4x^2} + 2x)]\Big|_0^1$$
$$= \tfrac{1}{2}\sqrt{5} + \tfrac{1}{4}\ln(\sqrt{5} + 2)$$

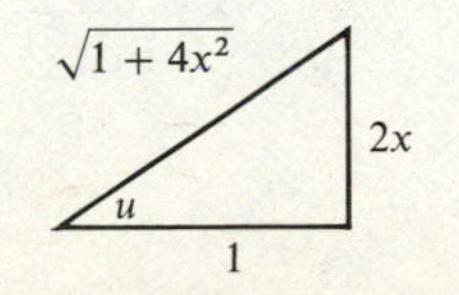

17. $\mathscr{L} = \int_2^3 \sqrt{1 + (\sqrt{x^2-1})^2}\,dx = \int_2^3 \sqrt{x^2}\,dx = \int_2^3 x\,dx = \frac{1}{2}x^2\Big|_2^3 = \frac{5}{2}$

19. $\mathscr{L} = \int_{2\pi/3}^{3\pi/4} \sqrt{1 + (\sqrt{\tan^2 x - 1})^2}\,dx = \int_{2\pi/3}^{3\pi/4} |\tan x|\,dx$

$= \int_{2\pi/3}^{3\pi/4} -\tan x\,dx = \ln|\cos x|\Big|_{2\pi/3}^{3\pi/4} = \ln\frac{\sqrt{2}}{2} - \ln\frac{1}{2} = \frac{1}{2}\ln 2$

21. $\mathscr{L} = \int_{25}^{100} \sqrt{1 + (\sqrt{\sqrt{x} - 1})^2}\,dx = \int_{25}^{100} x^{1/4}\,dx = \frac{4}{5}x^{5/4}\Big|_{25}^{100} = \frac{4}{5}(10^{5/2} - 5^{5/2})$

23. $\mathscr{L} = \int_{-1}^1 \sqrt{1 + \left(\frac{-3x}{2\sqrt{4-x^2}}\right)^2}\,dx = \int_{-1}^1 \sqrt{1 + \frac{9x^2}{4(4-x^2)}}\,dx$

$= \int_{-1}^1 \frac{1}{2}\sqrt{\frac{16+5x^2}{4-x^2}}\,dx \approx \frac{1}{6}(\sqrt{7} + 4\sqrt{4} + \sqrt{7}) = \frac{1}{3}(\sqrt{7} + 4) \approx 2.21525$

25. The length would be given by

$$\mathscr{L} = \int_0^{\ln 2} \sqrt{1 + \cosh^2 x}\,dx = \int_0^{\ln 2} \sqrt{1 + \tfrac{1}{4}(e^{2x} + 2 + e^{-2x})}\,dx$$

Neither integrand can be simplified as in Exercise 10, so the integration appears to be impossible.

27. $\mathscr{L} = \int_0^{26} \sqrt{1 + \left(\frac{-4x}{169} + \frac{4}{13}\right)^2}\,dx$. To evaluate the integral we first let $\tan u = \frac{4x}{169} - \frac{4}{13}$.

Then $\sec u = \sqrt{1 + \left(\frac{4x}{169} - \frac{4}{13}\right)^2}$, and by (7) of Section 7.2 we obtain

$$\int \sqrt{1 + \left(\frac{-4x}{169} + \frac{4}{13}\right)^2}\,dx = \int \sqrt{1+\tan^2 u}\left(\frac{169}{4}\right)\sec^2 u\,du = \frac{169}{4}\int \sec^3 u\,du$$
$$= \frac{169}{8}(\sec u \tan u + \ln|\sec u + \tan u|) + C$$
$$= \frac{169}{8}\left[\sqrt{1 + \left(\frac{4x}{169} - \frac{4}{13}\right)^2}\left(\frac{4x}{169} - \frac{4}{13}\right) + \ln\left|\sqrt{1 + \left(\frac{4x}{169} - \frac{4}{13}\right)^2} + \left(\frac{4x}{169} - \frac{4}{13}\right)\right|\right] + C$$

Thus

$$\mathscr{L} = \left[\sqrt{1 + \left(\frac{4x}{169} - \frac{4}{13}\right)^2}\left(\frac{x}{2} - \frac{13}{2}\right) + \frac{169}{8}\ln\left|\sqrt{1 + \left(\frac{4x}{169} - \frac{4}{13}\right)^2} + \left(\frac{4x}{169} - \frac{4}{13}\right)\right|\right]\Bigg|_0^{26}$$
$$= 13\sqrt{1 + \frac{16}{169}} + \frac{169}{8}\ln\left|\frac{4}{13} + \sqrt{1 + \frac{16}{169}}\right| - \frac{169}{8}\ln\left|-\frac{4}{13} + \sqrt{1 + \frac{16}{169}}\right|$$
$$= \sqrt{185} + \frac{169}{8}\left(\ln\frac{4+\sqrt{185}}{13} - \ln\frac{-4+\sqrt{185}}{13}\right) \approx 26.4046 \text{ (feet)}$$

29. $\mathscr{L} = \int_0^2 \sqrt{1 + \left(\frac{4\pi}{16}\cos 4\pi x\right)^2}\,dx = \int_0^2 \sqrt{1 + \frac{\pi^2}{16}\cos^2 4\pi x}\,dx$

$\approx \frac{2}{24}\left(\sqrt{1 + \frac{\pi^2}{16}} + 4\sqrt{1 + \frac{\pi^2}{16}\cos^2 \pi} + 2\sqrt{1 + \frac{\pi^2}{16}\cos^2 2\pi} + 4\sqrt{1 + \frac{\pi^2}{16}\cos^2 3\pi}\right.$

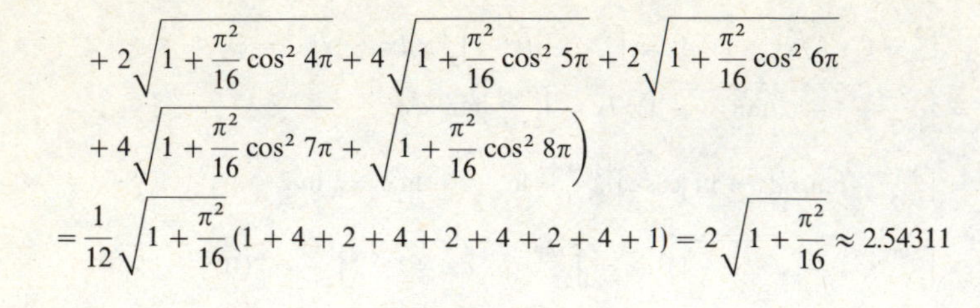

$$+2\sqrt{1+\frac{\pi^2}{16}\cos^2 4\pi}+4\sqrt{1+\frac{\pi^2}{16}\cos^2 5\pi}+2\sqrt{1+\frac{\pi^2}{16}\cos^2 6\pi}$$
$$+4\sqrt{1+\frac{\pi^2}{16}\cos^2 7\pi}+\sqrt{1+\frac{\pi^2}{16}\cos^2 8\pi}\Bigg)$$
$$=\frac{1}{12}\sqrt{1+\frac{\pi^2}{16}}\,(1+4+2+4+2+4+2+4+1)=2\sqrt{1+\frac{\pi^2}{16}}\approx 2.54311$$

SECTION 8.4

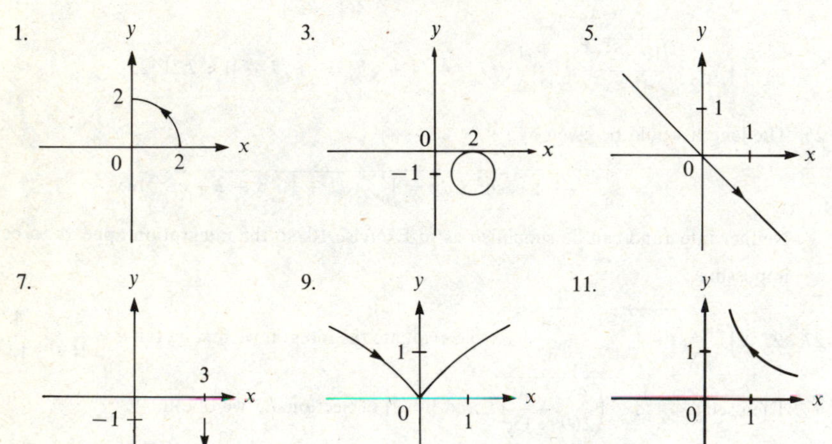

13. $x^2+y^2=\dfrac{4t^2}{(1+t^2)^2}+\dfrac{1-2t^2+t^4}{(1+t^2)^2}=\dfrac{1+2t^2+t^4}{(1+t^2)^2}=\dfrac{(1+t^2)^2}{(1+t^2)^2}=1$

15. $\mathcal{L}=\int_0^{\sqrt{3}}\sqrt{(2t)^2+(2)^2}\,dt=\int_0^{\sqrt{3}}2\sqrt{t^2+1}\,dt$
$\overset{t=\tan u}{=}\int_0^{\pi/3}2\sqrt{\tan^2 u+1}\sec^2 u\,du=\int_0^{\pi/3}2\sec^3 u\,du,$

so by (7) of Section 7.2,

17. $\mathcal{L}=\int_0^{\pi}\sqrt{(e^t\sin t+e^t\cos t)^2+(e^t\cos t-e^t\sin t)^2}\,dt$
$=\int_0^{\pi}\sqrt{e^{2t}(\sin^2 t+2\sin t\cos t+2\cos^2 t-2\cos t\sin t+\sin^2 t)}\,dt$
$=\int_0^{\pi}e^t\sqrt{2}\,dt=\sqrt{2}e^t\Big|_0^{\pi}=\sqrt{2}(e^{\pi}-1)$

19. $\mathcal{L}=\int_0^{1/2}\sqrt{\left(\frac{1}{\sqrt{1-t^2}}\right)^2+\left(\frac{-t}{1-t^2}\right)^2}\,dt=\int_0^{1/2}\sqrt{\frac{1}{1-t^2}+\frac{t^2}{(1-t^2)^2}}\,dt$
$=\int_0^{1/2}\sqrt{\frac{1-t^2+t^2}{(1-t^2)^2}}\,dt=\int_0^{1/2}\frac{1}{1-t^2}\,dt=\int_0^{1/2}\frac{1}{2}\left(\frac{1}{1-t}+\frac{1}{1+t}\right)dt$
$=\left(-\frac{1}{2}\ln|1-t|+\frac{1}{2}\ln|1+t|\right)\Big|_0^{1/2}=\frac{1}{2}\ln\left|\frac{1+t}{1-t}\right|\Big|_0^{1/2}=\frac{1}{2}\ln 3$

21. a. $\mathcal{L}=\int_0^{2\pi}\sqrt{(-3r\cos^2 t\sin t)^2+(3r\sin^2 t\cos t)^2}\,dt$
$=\int_0^{2\pi}\sqrt{9r^2\cos^2 t\sin^2 t(\cos^2 t+\sin^2 t)}\,dt$
$=\int_0^{2\pi}3r\sqrt{\cos^2 t\sin^2 t}\,dt=\int_0^{2\pi}3r|\cos t\sin t|\,dt$
$=\int_0^{\pi/2}3r\cos t\sin t\,dt+\int_{\pi/2}^{\pi}-3r\cos t\sin t\,dt$
$+\int_{\pi}^{3\pi/2}3r\cos t\sin t\,dt+\int_{3\pi/2}^{2\pi}-3r\cos t\sin t\,dt=\frac{3}{2}r\sin^2 t\Big|_0^{\pi/2}$
$-\frac{3}{2}r\sin^2 t\Big|_{\pi/2}^{\pi}+\frac{3}{2}r\sin^2 t\Big|_{\pi}^{3\pi/2}-\frac{3}{2}r\sin^2 t\Big|_{3\pi/2}^{2\pi}=6r$

b. From the parametric equations $x=r\cos^3 t$ and $y=r\sin^3 t$, we find that $x^{2/3}+y^{2/3}=r^{2/3}\cos^2 t+r^{2/3}\sin^2 t=r^{2/3}$.

c. From (6) we obtain $y^{2/3}=r^{2/3}-x^{2/3}$, so that if $y\geq 0$, then $y=(r^{2/3}-x^{2/3})^{3/2}$. Now if $f(x)=(r^{2/3}-x^{2/3})^{3/2}$ then $f'(x)=\frac{3}{2}(r^{2/3}-x^{2/3})^{1/2}(-\frac{2}{3}x^{-1/3})=-x^{-1/3}(r^{2/3}-x^{2/3})^{1/2}$ for $x>0$, so that

$$\tfrac{1}{4}\mathcal{L}=\lim_{a\to0^+}\int_a^r\sqrt{1+[f'(x)]^2}\,dx=\lim_{a\to0^+}\int_a^r\sqrt{1+[-x^{-1/3}(r^{2/3}-x^{2/3})^{1/2}]^2}\,dx$$
$$=\lim_{a\to0^+}\int_a^r\sqrt{1+x^{-2/3}(r^{2/3}-x^{2/3})}\,dx=\lim_{a\to0^+}\int_a^r\sqrt{x^{-2/3}r^{2/3}}\,dx$$
$$=\lim_{a\to0^+}\int_a^r x^{-1/3}r^{1/3}\,dx=\lim_{a\to0^+}\tfrac{3}{2}x^{2/3}r^{1/3}\Big|_a^r=\lim_{a\to0^+}(\tfrac{3}{2}r-\tfrac{3}{2}a^{2/3}r^{1/3})=\tfrac{3}{2}r$$

This is one-fourth the value obtained in part (a).

SECTION 8.5

1. $S=\int_{-1/2}^{3/2}2\pi\sqrt{4-x^2}\sqrt{1+\left(\frac{-x}{\sqrt{4-x^2}}\right)^2}\,dx$
$=2\pi\int_{-1/2}^{3/2}\sqrt{4-x^2}\sqrt{1+\frac{x^2}{4-x^2}}\,dx=2\pi\int_{-1/2}^{3/2}2\,dx=4\pi x\Big|_{-1/2}^{3/2}=8\pi$

3. $S=\int_2^6 2\pi\sqrt{x}\sqrt{1+\left(\frac{1}{2\sqrt{x}}\right)^2}\,dx=2\pi\int_2^6\sqrt{x}\sqrt{1+\frac{1}{4x}}\,dx=\pi\int_2^6\sqrt{4x+1}\,dx$
$=\frac{\pi}{6}(4x+1)^{3/2}\Big|_2^6=\frac{\pi}{6}(125-27)=\frac{49}{3}\pi$

5. $S = \int_1^2 2\pi(x^2 - \frac{1}{8}\ln x)\sqrt{1 + \left(2x - \dfrac{1}{8x}\right)^2}\,dx$

$= 2\pi \int_1^2 (x^2 - \frac{1}{8}\ln x)\sqrt{1 + \left(4x^2 - \dfrac{1}{2} + \dfrac{1}{64x^2}\right)}\,dx$

$= 2\pi \int_1^2 (x^2 - \frac{1}{8}\ln x)\sqrt{\left(2x + \dfrac{1}{8x}\right)^2}\,dx = 2\pi \int_1^2 (x^2 - \frac{1}{8}\ln x)\left(2x + \dfrac{1}{8x}\right)dx$

$= 2\pi \int_1^2 (2x^3 + \frac{1}{8}x)\,dx - 2\pi \int_1^2 \frac{1}{4}x \ln x\,dx - 2\pi \int_1^2 \dfrac{1}{64}\dfrac{\ln x}{x}\,dx$

$= 2\pi\left(\dfrac{1}{2}x^4 + \dfrac{1}{16}x^2\right)\Big|_1^2 - \dfrac{\pi}{2}\left(\dfrac{1}{2}x^2 \ln x - \dfrac{1}{4}x^2\right)\Big|_1^2 - \dfrac{\pi}{64}(\ln x)^2\Big|_1^2$

$= 2\pi\left(\dfrac{33}{4} - \dfrac{9}{16}\right) - \dfrac{\pi}{2}\left[(2\ln 2 - 1) + \dfrac{1}{4}\right] - \dfrac{\pi}{64}(\ln 2)^2 = \pi\left[\dfrac{63}{4} - \ln 2 - \dfrac{1}{64}(\ln 2)^2\right]$

7. $S = \int_0^\pi 2\pi \sin x\sqrt{1 + \cos^2 x}\,dx \overset{u=\cos x}{=} 2\pi \int_1^{-1} \sqrt{1 + u^2}(-1)\,du$

$\overset{u=\tan v}{=} -2\pi \int_{\pi/4}^{-\pi/4} \sec^3 v\,dv,$

so by (7) of Section 7.2,

$$S = -2\pi(\tfrac{1}{2}\sec v \tan v + \tfrac{1}{2}\ln|\sec v + \tan v|)\Big|_{\pi/4}^{-\pi/4}$$
$$= \pi[2\sqrt{2} + \ln(\sqrt{2} + 1) - \ln(\sqrt{2} - 1)] = \pi[2\sqrt{2} + \ln(3 + 2\sqrt{2})]$$

9. $S = \int_{\sqrt{3}}^{2\sqrt{2}} (2\pi)t\sqrt{t^2 + 1}\,dt = \dfrac{2\pi}{3}(t^2 + 1)^{3/2}\Big|_{\sqrt{3}}^{2\sqrt{2}} = \dfrac{2\pi}{3}(27 - 8) = \dfrac{38}{3}\pi$

11. $S = \int_0^{\pi/2} 2\pi \sin t \cos t\sqrt{(2\sin t \cos t)^2 + (\cos^2 t - \sin^2 t)^2}\,dt$

$= 2\pi \int_0^{\pi/2} \sin t \cos t\sqrt{\sin^2 2t + \cos^2 2t}\,dt = 2\pi \int_0^{\pi/2} \sin t \cos t\,dt = \pi \sin^2 t\Big|_0^{\pi/2} = \pi$

13. $S = \int_{-a}^a 2\pi\sqrt{r^2 - x^2}\sqrt{1 + \left(\dfrac{-x}{\sqrt{r^2 - x^2}}\right)^2}\,dx = 2\pi \int_{-a}^a \sqrt{r^2 - x^2}\sqrt{1 + \dfrac{x^2}{r^2 - x^2}}\,dx$

$= 2\pi \int_{-a}^a r\,dx = 4\pi ra$

15. $S = \int_0^{\pi/2} 2\pi \sin^n t\sqrt{(-n\cos^{n-1} t \sin t)^2 + (n \sin^{n-1} t \cos t)^2}\,dt$

$= 2n\pi \int_0^{\pi/2} \sin^n t \sin t \cos t\sqrt{\cos^{2n-4} t + \sin^{2n-4} t}\,dt$

We can carry out the integration if $2n - 4 = 0$ or 2, which means that $n = 2$ or 3. (Exercise 14 uses $n = 3$.)

17. The area of the lateral surface of the barrel is given by

$S = \int_{-\arcsin 1/\sqrt{3}}^{\arcsin 1/\sqrt{3}} 2\pi \cos x\sqrt{1 + (-\sin x)^2}\,dx = 2\pi \int_{-\arcsin 1/\sqrt{3}}^{\arcsin 1/\sqrt{3}} \cos x\sqrt{1 + \sin^2 x}\,dx$

$\overset{u=\sin x}{=} 2\pi \int_{-1/\sqrt{3}}^{1/\sqrt{3}} \sqrt{1 + u^2}\,du \overset{u=\tan v}{=} 2\pi \int_{-\pi/6}^{\pi/6} \sec^3 v\,dv$

$= 2\pi[\frac{1}{2}\sec v \tan v + \frac{1}{2}\ln|\sec v + \tan v|]\Big|_{-\pi/6}^{\pi/6}$

$= \pi\left[\dfrac{2}{\sqrt{3}} \cdot \dfrac{1}{\sqrt{3}} + \ln\left(\dfrac{2}{\sqrt{3}} + \dfrac{1}{\sqrt{3}}\right) - \dfrac{2}{\sqrt{3}}\left(\dfrac{-1}{\sqrt{3}}\right) - \ln\left(\dfrac{2}{\sqrt{3}} - \dfrac{1}{\sqrt{3}}\right)\right] = \pi(\frac{4}{3} + \ln 3)$

Each end of the barrel is a disk with radius $\cos\left(\arcsin \dfrac{1}{\sqrt{3}}\right) = \dfrac{\sqrt{2}}{\sqrt{3}}$, so the area of each end is $\pi\left(\dfrac{\sqrt{2}}{\sqrt{3}}\right)^2 = \dfrac{2}{3}\pi$. Therefore the total surface area of the barrel is $\pi(\frac{4}{3} + \ln 3) + 2(\frac{2}{3}\pi) = \pi(\frac{8}{3} + \ln 3)$.

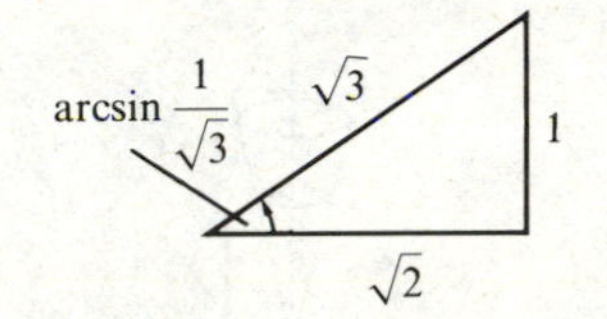

19. $S = \int_a^b 2\pi\sqrt{1 - x^2}\sqrt{1 + \left(\dfrac{-x}{\sqrt{1 - x^2}}\right)^2}\,dx = 2\pi \int_a^b \sqrt{1 - x^2}\sqrt{1 + \dfrac{x^2}{1 - x^2}}\,dx$

$= 2\pi \int_a^b 1\,dx = 2\pi(b - a)$

SECTION 8.6

1. $W = \int_0^{60} 60\left(1 - \dfrac{x^2}{20{,}000}\right)dx = 60\left(x - \dfrac{x^3}{60{,}000}\right)\Big|_0^{60} = 3384$ (foot-pounds)

3. $f(x) = 10$; $W = \int_0^8 10\,dx = 10\,x\Big|_0^8 = 80$ (foot-pounds)

5. $W = \int_0^\pi 10^4 \sin x\,dx = -10^4 \cos x\Big|_0^\pi = 2 \times 10^4$ (mile-pounds)

7. $W = \int_0^{1/6} 5\left(\dfrac{1}{6} - x\right)dx = 5\left(\dfrac{x}{6} - \dfrac{x^2}{2}\right)\Big|_0^{1/6} = \dfrac{5}{72}$ (foot-pounds)

9. $W = \int_3^6 30x\,dx = 15x^2\Big|_3^6 = 405$ (foot-pounds)

11. By hypothesis, $\int_0^{-1} kx\,dx = 6$. But $\int_0^{-1} kx\,dx = (kx^2/2)\Big|_0^{-1} = k/2$, so that $k/2 = 6$, or $k = 12$. Thus $W = \int_0^2 12x\,dx = 6x^2\Big|_0^2 = 24$ (foot-pounds).

13. As in Example 3, we place the origin at the bottom of the pool. Since the water is to be pumped up to 3 feet above the top of the tank, we have $\ell = 11$, so a particle of water x feet from the bottom is to be raised $11 - x$ feet, and the water to be pumped extends

from 0 to 5 on the x axis. As in Example 3, we have $A(x) = 100\pi$, so by (4),

$$W = \int_0^5 62.5(11 - x)(100\pi)\,dx = 6250\pi(11x - \tfrac{1}{2}x^2)\Big|_0^5$$
$$= 265{,}625\pi \text{ (foot-pounds)}$$

15. We position the x axis as in the figure, with the origin at the vertex of the cone. Then $\ell = 12$ and a particle of water x feet from the vertex is to be raised $12 - x$ feet. Moreover, the water to be pumped extends from 0 to 12 on the x axis. By similar triangles, we have $r(x)/x = \frac{3}{12} = \frac{1}{4}$, so that $r(x) = x/4$ and hence $A(x) = \pi[r(x)]^2 = \pi x^2/16$. Thus by (4),

$$W = \int_0^{12} 62.5(12 - x)\frac{\pi}{12}x^2\,dx = \frac{62.5\pi}{16}\int_0^{12}(12x^2 - x^3)\,dx = \frac{62.5\pi}{16}\left(4x^3 - \frac{1}{4}x^4\right)\Bigg|_0^{12}$$
$$= 6750\pi \text{ (foot-pounds)}$$

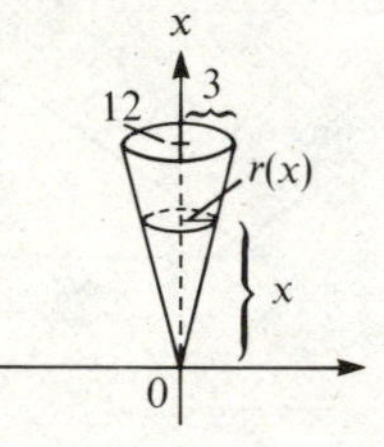

17. a. If the tank is positioned with respect to the x axis as in the figure, then $\ell = 0$ and the gasoline to be pumped extends from -4 to 0 on the x axis. Moreover, the width $w(x)$ of the cross-section at x is given by $w(x) = 2\sqrt{16 - x^2}$, so $A(x) = 10w(x) = 20\sqrt{16 - x^2}$. Thus by (4) with 62.5 replaced by 42, we have

$$W = \int_{-4}^0 42(0 - x)20\sqrt{16 - x^2}\,dx \overset{u = 16 - x^2}{=} 840\int_0^{16}\sqrt{u}(\tfrac{1}{2})\,du = 420(\tfrac{2}{3}u^{3/2})\Big|_0^{16}$$
$$= 17{,}920 \text{ (foot-pounds)}$$

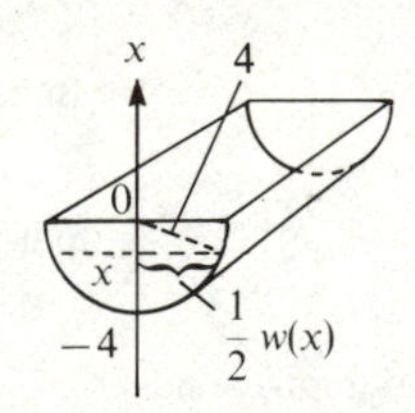

b. If the tank is positioned with respect to the x axis as in the figure, then $\ell = 4$ and the gasoline to be pumped extends from 0 to 4 on the x axis. As in the solution of part (a), we have $A(x) = 20\sqrt{16 - x^2}$. Thus

$$W = \int_0^4 42(4 - x)20\sqrt{16 - x^2}\,dx = 3360\int_0^4\sqrt{16 - x^2}\,dx - 840\int_0^4 x\sqrt{16 - x^2}\,dx$$

Now

$$3360\int_0^4\sqrt{16 - x^2}\,dx \overset{x = 4\sin u}{=} 3360\int_0^{\pi/2}\sqrt{16 - 16\sin^2 u}(4\cos u)\,du$$
$$= (3360)(16)\int_0^{\pi/2}\cos^2 u\,du$$
$$= (3360)(16)\int_0^{\pi/2}(\tfrac{1}{2} + \tfrac{1}{2}\cos 2u)\,du$$
$$= (3360)(16)\left(\frac{u}{2} + \frac{1}{4}\sin 2u\right)\Bigg|_0^{\pi/2} = 13{,}440\pi$$

and

$$840\int_0^4 x\sqrt{16 - x^2}\,dx = -\tfrac{840}{3}(16 - x^2)^{3/2}\Big|_0^4 = 17{,}920$$

Thus

$$W = 13{,}440\pi - 17{,}920 \text{ (foot-pounds)}$$

19. a. $h'(t) = 8 - 32t$, so $h'(t) = 0$ for $t = \frac{1}{4}$. The maximum height is $h(\frac{1}{4}) = 7$ (feet). Since $f(h) = -0.2$, we have

$$W = \int_7^0 -0.2\,dh = (-0.2)h\Big|_7^0 = 1.4 \text{ (foot-pounds)}$$

b. We change to 6 the upper limit on the integral in the solution of part (a). Thus

$$W = \int_7^6 -0.2\,dh = (-0.2)h\Big|_7^6 = 0.2 \text{ (foot-pound)}$$

21. a. $f(x) = 200$; $W = \int_0^{80} 200\,dx = 200(80) = 16{,}000$ (foot-pounds).

b. When the bucket is x feet above the ground, the chain is $80 - x$ feet long, so $f(x) = 200 + (80 - x) = 280 - x$. Thus $W = \int_0^{80}(280 - x)\,dx = (280x - \frac{1}{2}x^2)\Big|_0^{80} = 19{,}200$ (foot-pounds).

23. When the bucket has been raised x feet, $x/2$ seconds have elapsed, so that $\frac{1}{2}(x/2)$ pounds of water have leaked out. Therefore the total weight of water and bucket is given by $f(x) = 21 - x/4$. Since the bucket is empty when $x/4 = 20$, or $x = 80$, we have

$$W = \int_0^{80}\left(21 - \frac{x}{4}\right)dx = \left(21x - \frac{x^2}{8}\right)\Bigg|_0^{80} = 880 \text{ (foot-pounds)}$$

25. Let $s(x)$ denote the length of the side of a cross-section x feet above ground. Then $s(x)/754 = (482 - x)/482$, so $s(x) = \frac{754}{482}(482 - x)$. The cross-sectional area is given by $A(x) = (s(x))^2 = (\frac{754}{482})^2(482 - x)^2$. Let $\mathscr{P} = \{x_0, x_1, \ldots, x_n\}$ be any partition of $[0, 482]$,

and for $1 \le k \le n$ let t_k be any point in $[x_{k-1}, x_k]$. Then the amount of work ΔW_k required to lift the portion of the pyramid that will reside between the heights x_{k-1} and x_k is approximately $(150A(t_k)\Delta x_k)(t_k)$ (weight × distance). Thus the total work W, which is the sum of $\Delta W_1, \Delta W_2, \ldots, \Delta W_n$, is approximately $\sum_{k=1}^{n} 150 t_k A(t_k)\Delta x_k$, so that

$$W = \int_0^{482} 150xA(x)\,dx = 150\int_0^{482} x(\tfrac{754}{482})^2(482-x)^2\,dx$$
$$= \frac{150(754)^2}{(482)^2}\int_0^{482} ((482)^2x - 964x^2 + x^3)\,dx$$
$$= \frac{150(754)^2}{(482)^2}\left[\frac{(482)^2}{2}x^2 - \frac{964}{3}x^3 + \frac{1}{4}x^4\right]\Bigg|_0^{482} = \frac{150(754)^2(482)^2}{12}$$
$$\approx 1.65100 \times 10^{12} \text{ (foot-pounds)}$$

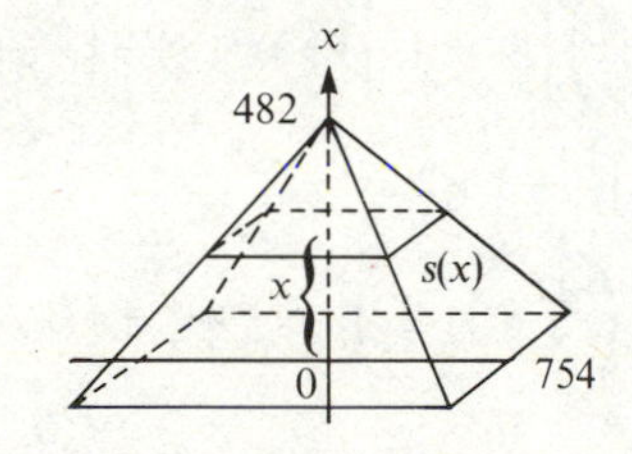

27. Set up a coordinate system as in the figure, and let $\mathscr{P} = \{x_0, x_1, \ldots, x_n\}$ be a partition of $[0, 4]$. The portion of the triangle in the interval $[x_{k-1}, x_k]$ has approximate weight $2y_k\Delta x_k$ units, and must rise approximately x_k feet. By similar triangles, $y_k/(4 - x_k) = \frac{3}{4}$, so that $y_k = \frac{3}{4}(4 - x_k)$. Thus the work ΔW_k required to raise the portion in the interval $[x_{k-1}, x_k]$ is approximately $2[\frac{3}{4}(4 - x_k)\Delta x_k](x_k)$. Then the total work W, which is the sum of $\Delta W_1, \Delta W_2, \ldots, \Delta W_n$, is approximately $\sum_{k=1}^{n} \frac{3}{2}(4 - x_k)x_k\Delta x_k$. Therefore

$$W = \int_0^4 \tfrac{3}{2}(4-x)x\,dx = \tfrac{3}{2}\int_0^4 (4x - x^2)\,dx = \tfrac{3}{2}(2x^2 - \tfrac{1}{3}x^3)\Big|_0^4 = 16 \text{ (foot-pounds)}$$

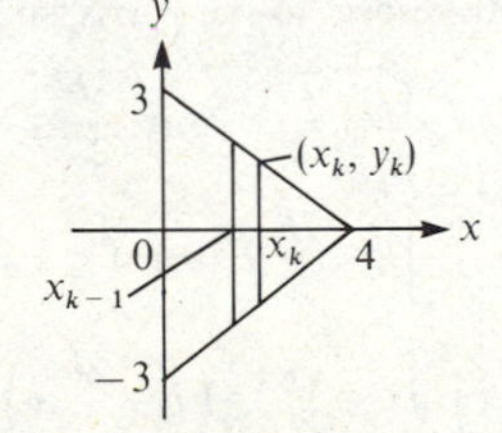

SECTION 8.7

1. Take the origin of the x axis at the axis of revolution of the seesaw, with the positive x axis to the right. Then the moment of the two children is $30(-1.5) + 20(2) = -5$. Thus the right side will rise.

3. Take the origin of the x axis at the axis of revolution of the seesaw, with the positive x axis pointing toward the 10-kilogram child. If the 15-kilogram child sits at x and the 20-kilogram child sits at $-x$, then the three children will be in equilibrium if $10(1) + 15(x) + 20(-x) = 0$, or $10 - 5x = 0$, or $x = 2$. Thus the 15-kilogram child should sit 2 meters from the axis of revolution and on the same side as the 10-kilogram child, while the 20-kilogram child should sit on the opposite side, 2 meters from the axis of revolution.

5. $$\mathscr{M}_x = \int_0^2 \tfrac{1}{2}[x^2 - (-2)^2]\,dx = \tfrac{1}{2}(\tfrac{1}{3}x^3 - 4x)\Big|_0^2 = \frac{-8}{3}$$
$$\mathscr{M}_y = \int_0^2 x[x - (-2)]\,dx = \int_0^2 (x^2 + 2x)\,dx = (\tfrac{1}{3}x^3 + x^2)\Big|_0^2 = \tfrac{20}{3}$$
$$A = \int_0^2 (x - (-2))\,dx = (\tfrac{1}{2}x^2 + 2x)\Big|_0^2 = 6$$
$$\bar{x} = \frac{\mathscr{M}_y}{A} = \frac{20/3}{6} = \frac{10}{9};\quad \bar{y} = \frac{\mathscr{M}_x}{A} = \frac{-8/3}{6} = \frac{-4}{9};\quad (\bar{x}, \bar{y}) = \left(\frac{10}{9}, \frac{-4}{9}\right)$$

7. $$\mathscr{M}_x = \int_0^2 \tfrac{1}{2}[(2-x)^2 - (-(2-x))^2]\,dx = 0$$
$$\mathscr{M}_y = \int_0^2 x[(2-x) - (-(2-x))]\,dx = \int_0^2 (4x - 2x^2)\,dx = (2x^2 - \tfrac{2}{3}x^3)\Big|_0^2 = \tfrac{8}{3}$$
$$A = \int_0^2 [(2-x) - (-(2-x))]\,dx = \int_0^2 (4 - 2x)\,dx = (4x - x^2)\Big|_0^2 = 4$$
$$\bar{x} = \frac{\mathscr{M}_y}{A} = \frac{8/3}{4} = \frac{2}{3};\quad \bar{y} = \frac{\mathscr{M}_x}{A} = \frac{0}{4} = 0;\quad (\bar{x}, \bar{y}) = (\tfrac{2}{3}, 0)$$

9. $$\mathscr{M}_x = \int_1^2 \tfrac{1}{2}[(x+1)^4 - (x-1)^4]\,dx = \tfrac{1}{2}[\tfrac{1}{5}(x+1)^5 - \tfrac{1}{5}(x-1)^5]\Big|_1^2 = 21$$
$$\mathscr{M}_y = \int_1^2 x[(x+1)^2 - (x-1)^2]\,dx = \int_1^2 4x^2\,dx = \tfrac{4}{3}x^3\Big|_1^2 = \tfrac{28}{3}$$
$$A = \int_1^2 [(x+1)^2 - (x-1)^2]\,dx = \int_1^2 4x\,dx = 2x^2\Big|_1^2 = 6$$
$$\bar{x} = \frac{\mathscr{M}_y}{A} = \frac{28/3}{6} = \frac{14}{9};\quad \bar{y} = \frac{\mathscr{M}_x}{A} = \frac{21}{6} = \frac{7}{2};\quad (\bar{x}, \bar{y}) = (\tfrac{14}{9}, \tfrac{7}{2})$$

11. $$\mathscr{M}_x = \int_0^1 \frac{1}{2}[(\sqrt{1-x^2})^2 - (-(1+x))^2]\,dx = \frac{1}{2}\int_0^1 (-2x - 2x^2)\,dx$$
$$= \frac{1}{2}\left(-x^2 - \frac{2}{3}x^3\right)\Bigg|_0^1 = \frac{-5}{6}$$
$$\mathscr{M}_y = \int_0^1 x[\sqrt{1-x^2} - (-(1+x))]\,dx = \int_0^1 (x\sqrt{1-x^2} + x + x^2)\,dx$$
$$= \left(\frac{-1}{3}(1-x^2)^{3/2} + \frac{1}{2}x^2 + \frac{1}{3}x^3\right)\Bigg|_0^1 = \frac{7}{6}$$
$$A = \int_0^1 [\sqrt{1-x^2} - (-(1+x))]\,dx = \int_0^1 \sqrt{1-x^2}\,dx + \int_0^1 (1+x)\,dx$$
$$\overset{x=\sin u}{=} \int_0^{\pi/2} \sqrt{1-\sin^2 u}\cos u\,du + \left(x + \frac{1}{2}x^2\right)\Bigg|_0^1 = \int_0^{\pi/2} \cos^2 u\,du + \frac{3}{2}$$
$$= \left(\frac{u}{2} + \frac{1}{4}\sin 2u\right)\Bigg|_0^{\pi/2} + \frac{3}{2} = \frac{\pi}{4} + \frac{3}{2}$$

$$\bar{x} = \frac{\mathcal{M}_y}{A} = \frac{7/6}{\pi/4 + \frac{3}{2}} = \frac{14}{3\pi + 18};\ \bar{y} = \frac{\mathcal{M}_x}{A} = \frac{-5/6}{\pi/4 + \frac{3}{2}} = \frac{-10}{3\pi + 18};$$

$$(\bar{x}, \bar{y}) = \left(\frac{14}{3\pi + 18}, \frac{-10}{3\pi + 18}\right)$$

13. $$\mathcal{M}_x = \int_0^{\pi/2} \tfrac{1}{2}[(\sin x + \cos x)^2 - (\sin x - \cos x)^2]\,dx = \tfrac{1}{2}\int_0^{\pi/2} 4 \sin x \cos x\,dx$$

$$= \int_0^{\pi/2} \sin 2x\,dx = \frac{-1}{2}\cos 2x\Big|_0^{\pi/2} = 1$$

$$\mathcal{M}_y = \int_0^{\pi/2} x[(\sin x + \cos x) - (\sin x - \cos x)]\,dx = \int_0^{\pi/2} 2x \cos x\,dx$$

$$\overset{\text{parts}}{=} 2x \sin x\Big|_0^{\pi/2} - 2\int_0^{\pi/2} \sin x\,dx = \pi + 2\cos x\Big|_0^{\pi/2} = \pi - 2$$

$$A = \int_0^{\pi/2} [(\sin x + \cos x) - (\sin x - \cos x)]\,dx = \int_0^{\pi/2} 2\cos x\,dx = 2 \sin x\Big|_0^{\pi/2} = 2$$

$$\bar{x} = \frac{\mathcal{M}_y}{A} = \frac{\pi - 2}{2} = \frac{\pi}{2} - 1;\ \bar{y} = \frac{\mathcal{M}_x}{A} = \frac{1}{2};\ (\bar{x}, \bar{y}) = \left(\frac{\pi}{2} - 1, \frac{1}{2}\right)$$

15. $$\mathcal{M}_x = \int_1^2 \tfrac{1}{2}[(1 + \ln x)^2 - (1 - \ln x)^2]\,dx = \tfrac{1}{2}\int_1^2 4 \ln x\,dx \overset{\text{parts}}{=} 2(x \ln x - x)\Big|_1^2 = 4 \ln 2 - 2$$

$$\mathcal{M}_y = \int_1^2 x[(1 + \ln x) - (1 - \ln x)]\,dx = \int_1^2 2x \ln x\,dx \overset{\text{parts}}{=} x^2 \ln x\Big|_1^2 - \int_1^2 x\,dx$$

$$= 4 \ln 2 - \tfrac{1}{2}x^2\Big|_1^2 = 4 \ln 2 - \tfrac{3}{2}$$

$$A = \int_1^2 [(1 + \ln x) - (1 - \ln x)]\,dx = \int_1^2 2 \ln x\,dx = (2x \ln x - 2x)\Big|_1^2 = 4 \ln 2 - 2$$

$$\bar{x} = \frac{\mathcal{M}_y}{A} = \frac{4 \ln 2 - \frac{3}{2}}{4 \ln 2 - 2} = \frac{8 \ln 2 - 3}{8 \ln 2 - 4};\ \bar{y} = \frac{\mathcal{M}_x}{A} = \frac{4 \ln 2 - 2}{4 \ln 2 - 2} = 1;\ (\bar{x}, \bar{y}) = \left(\frac{8 \ln 2 - 3}{8 \ln 2 - 4}, 1\right)$$

17. The graphs of f and g intersect for (x, y) such that $2 - x^2 = y = |x|$, which means $2 - x^2 = x$ for $x \geq 0$ and $2 - x^2 = -x$ for $x < 0$. Thus $x = -1$ or $x = 1$.

$$\mathcal{M}_x = \int_{-1}^1 \tfrac{1}{2}[(2 - x^2)^2 - |x|^2]\,dx = \tfrac{1}{2}\int_{-1}^1 (4 - 5x^2 + x^4)\,dx = \tfrac{1}{2}(4x - \tfrac{5}{3}x^3 + \tfrac{1}{5}x^5)\Big|_{-1}^1 = \tfrac{38}{15}$$

$$\mathcal{M}_y = \int_{-1}^1 x[(2 - x^2) - |x|]\,dx = \int_{-1}^0 (2x + x^2 - x^3)\,dx + \int_0^1 (2x - x^2 - x^3)\,dx$$

$$= \left(x^2 + \frac{x^3}{3} - \frac{x^4}{4}\right)\Big|_{-1}^0 + \left(x^2 - \frac{x^3}{3} - \frac{x^4}{4}\right)\Big|_0^1 = \frac{-5}{12} + \frac{5}{12} = 0$$

$$A = \int_{-1}^1 (2 - x^2 - |x|)\,dx = \int_{-1}^0 (2 + x - x^2)\,dx + \int_0^1 (2 - x - x^2)\,dx$$

$$= \left(2x + \frac{x^2}{2} - \frac{x^3}{3}\right)\Big|_{-1}^0 + \left(2x - \frac{x^2}{2} - \frac{x^3}{3}\right)\Big|_0^1 = \frac{7}{6} + \frac{7}{6} = \frac{7}{3}$$

$$\bar{x} = \frac{\mathcal{M}_y}{A} = 0;\ \bar{y} = \frac{\mathcal{M}_x}{A} = \frac{38/15}{7/3} = \frac{38}{35};\ (\bar{x}, \bar{y}) = (0, \tfrac{38}{35})$$

19. The graphs of the lines $y = x + 2$, $y = -3x + 6$, and $y = (2 - x)/3$ intersect in pairs for (x, y) such that $x + 2 = -3x + 6$, $x + 2 = (2 - x)/3$ or $-3x + 6 = (2 - x)/3$. From these equations we obtain $x = 1$, $x = -1$, and $x = 2$, respectively. To use Definition 8.4 we let

$$f(x) = \begin{cases} x + 2 & \text{for } -1 \leq x \leq 1 \\ -3x + 6 & \text{for } \quad 1 \leq x \leq 2 \end{cases}$$

and $g(x) = (2 - x)/3$ for $-1 \leq x \leq 2$.

$$\mathcal{M}_x = \int_{-1}^1 \frac{1}{2}\left[(x + 2)^2 - \left(\frac{2 - x}{3}\right)^2\right]dx + \frac{1}{2}\int_1^2 \left[(-3x + 6)^2 - \left(\frac{2 - x}{3}\right)^2\right]dx$$

$$= \tfrac{4}{9}\int_{-1}^1 (x^2 + 5x + 4)\,dx + \tfrac{49}{9}\int_1^2 (x^2 - 4x + 4)\,dx$$

$$= \tfrac{4}{9}(\tfrac{1}{3}x^3 + \tfrac{5}{2}x^2 + 4x)\Big|_{-1}^1 + \tfrac{40}{9}(\tfrac{1}{3}x^3 - 2x^2 + 4x)\Big|_1^2 = \tfrac{104}{27} + \tfrac{40}{27} = \tfrac{16}{3}$$

$$\mathcal{M}_y = \int_{-1}^1 x\left[(x + 2) - \left(\frac{2 - x}{3}\right)\right]dx + \int_1^2 x\left[(-3x + 6) - \left(\frac{2 - x}{3}\right)\right]dx$$

$$= \frac{4}{3}\int_{-1}^1 (x^2 + x)\,dx + \frac{8}{3}\int_1^2 (-x^2 + 2x)\,dx = \frac{4}{3}\left(\frac{1}{3}x^3 + \frac{1}{2}x^2\right)\Big|_{-1}^1 + \frac{8}{3}\left(\frac{-1}{3}x^3 + x^2\right)\Big|_1^2$$

$$= \tfrac{8}{9} + \tfrac{16}{9} = \tfrac{8}{3}$$

$$A = \int_{-1}^1 \left[(x + 2) - \left(\frac{2 - x}{3}\right)\right]dx + \int_1^2 \left[(-3x + 6) - \left(\frac{2 - x}{3}\right)\right]dx$$

$$= \frac{4}{3}\int_{-1}^1 (x + 1)\,dx + \frac{8}{3}\int_1^2 (-x + 2)\,dx$$

$$= \frac{4}{3}\left(\frac{1}{2}x^2 + x\right)\Big|_{-1}^1 + \frac{8}{3}\left(\frac{-1}{2}x^2 + 2x\right)\Big|_1^2 = \tfrac{8}{3} + \tfrac{4}{3} = 4$$

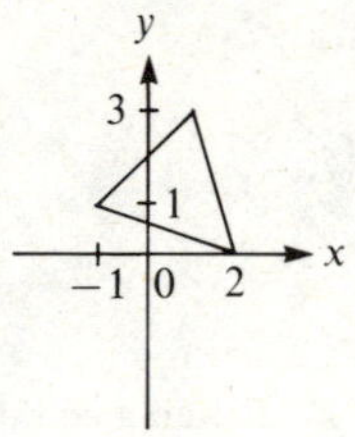

$$\bar{x} = \frac{\mathcal{M}_y}{A} = \frac{8/3}{4} = \frac{2}{3};\ \bar{y} = \frac{\mathcal{M}_x}{A} = \frac{16/3}{4} = \frac{4}{3};\ (\bar{x}, \bar{y}) = (\tfrac{2}{3}, \tfrac{4}{3})$$

21. $(\bar{x}, \bar{y}) = (0, 3)$

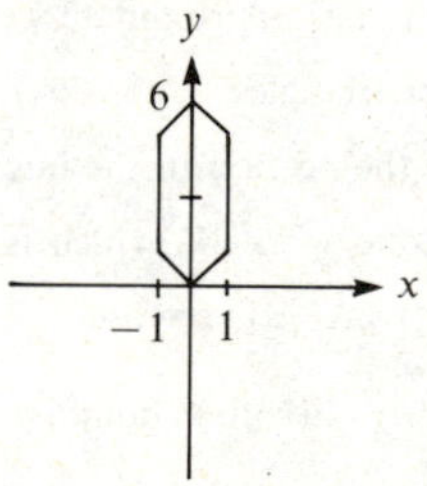

23. $(\bar{x}, \bar{y}) = (0, 0)$

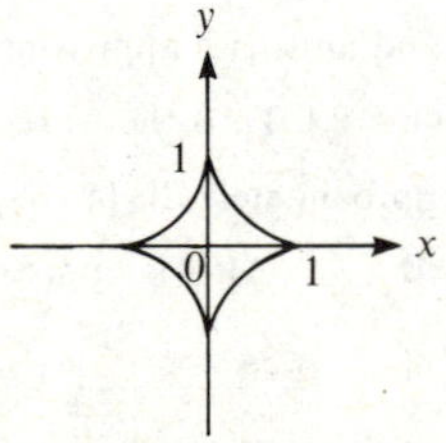

25. The triangle is symmetric with respect to the y axis, so $\bar{x} = 0$. Let

$$f(x) = \begin{cases} h + \dfrac{2h}{b}x & \text{for } \dfrac{-b}{2} \leq x \leq 0 \\[2ex] h - \dfrac{2h}{b}x & \text{for } 0 \leq x \leq \dfrac{b}{2} \end{cases}$$

Then $$\mathcal{M}_x = \int_{-b/2}^0 \frac{1}{2}\left(h + \frac{2h}{b}x\right)^2 dx + \int_0^{b/2} \frac{1}{2}\left(h - \frac{2h}{b}x\right)^2 dx$$

$$= \frac{1}{2}\left(\frac{b}{2h}\right)\left(\frac{1}{3}\right)\left(h + \frac{2h}{b}x\right)^3\Big|_{-b/2}^0 + \frac{1}{2}\left(\frac{-b}{2h}\right)\left(\frac{1}{3}\right)\left(h - \frac{2h}{b}x\right)^3\Big|_0^{b/2} = \frac{bh^2}{6}$$

$$A = \frac{bh}{2};\ \bar{y} = \frac{bh^2/6}{bh/2} = \frac{h}{3}$$

$(\bar{x}, \bar{y}) = (0, h/3)$, which is the "centroid" of the triangle.

27. Let $f(x) = (h/a^2)x^2$ and $g(x) = 0$.

$$\mathcal{M}_x = \int_0^a \frac{1}{2}\left(\frac{h}{a^2}x^2\right)^2 dx = \frac{h^2}{2a^4}\left(\frac{1}{5}x^5\right)\Big|_0^a = \frac{ah^2}{10}$$

$$\mathcal{M}_y = \int_0^a x\left(\frac{h}{a^2}x^2\right)dx = \frac{hx^4}{4a^2}\Big|_0^a = \frac{a^2h}{4}$$

$$A = \int_0^a \frac{h}{a^2}x^2 dx = \frac{hx^3}{3a^2}\Big|_0^a = \frac{ah}{3}$$

$$\bar{x} = \frac{a^2h/4}{ah/3} = \frac{3a}{4};\ \bar{y} = \frac{ah^2/10}{ah/3} = \frac{3h}{10};\ \text{thus } (\bar{x}, \bar{y}) = \left(\frac{3a}{4}, \frac{3h}{10}\right)$$

29. The area A of the semicircular region and the volume V of the sphere generated by revolving the semicircular region R about its diameter are given, respectively, by $A = (\pi/2)r^2$ and $V = \frac{4}{3}\pi r^3$. The Theorem of Pappus and Guldin then implies that $\frac{4}{3}\pi r^3 = 2\pi\bar{x}(\frac{1}{2}\pi r^2)$, so that $\bar{x} = 4r/3\pi$. Since $\bar{y} = 0$ by symmetry, the center of gravity of the semicircular region is $(4r/3\pi, 0)$.

31. Set up a coordinate system as in the figure, so that the center of gravity of the square is $(0, \sqrt{2})$ and the square is revolved about the x axis. Since $\bar{y} = \sqrt{2}$ and since the area A of the square is 4, the Theorem of Pappus and Guldin says that the volume V of the solid is given by $V = 2\pi\bar{y}A = 2\pi\sqrt{2}(4) = 8\pi\sqrt{2}$.

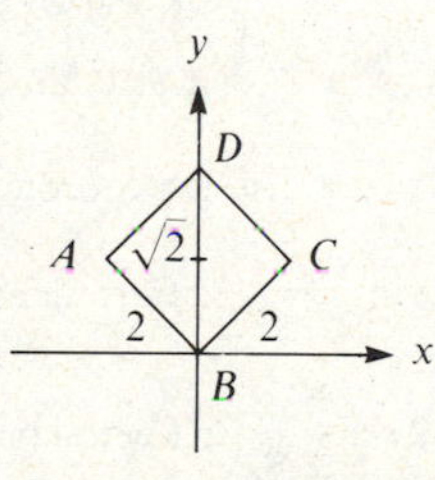

33. Substituting $u = -x$, we obtain

$$\int_{-a}^0 xf(x)\,dx = -\int_a^0 (-u)f(-u)\,du = -\int_0^a uf(u)\,du = -\int_0^a xf(x)\,dx$$

Thus $$\mathcal{M}_y = \int_{-a}^a xf(x)\,dx = \int_{-a}^0 xf(x)\,dx + \int_0^a xf(x)\,dx = 0$$

35. If ℓ is any line tangent to the circle which has radius r, and if the area of R is A, then the Pappus-Guldin Theorem says that the volume V of the solid of revolution is given by $V = (2\pi r)A$.

37. $\mathcal{L} = \pi r/2$ since the wire is a quarter circle of radius r. Let $f(t) = r\cos t$ and $g(t) = r\sin t$, so that $f'(t) = -r\sin t$ and $g'(t) = r\cos t$. Then

$$\mathcal{M}_x = \int_0^{\pi/2} r\sin t\sqrt{(-r\sin t)^2 + (r\cos t)^2}\,dt = r^2\int_0^{\pi/2}\sin t\,dt = -r^2\cos t\Big|_0^{\pi/2} = r^2$$

Likewise

$$\mathcal{M}_y = \int_0^{\pi/2} r\cos t\sqrt{(-r\sin t)^2 + (r\cos t)^2}\,dt = r^2\int_0^{\pi/2}\cos t\,dt = r^2\sin t\Big|_0^{\pi/2} = r^2$$

Therefore $$(\bar{x}, \bar{y}) = \left(\frac{r^2}{\pi r/2}, \frac{r^2}{\pi r/2}\right) = \left(\frac{2r}{\pi}, \frac{2r}{\pi}\right)$$

Since $(2r/\pi)^2 + (2r/\pi)^2 = (8/\pi^2)r^2 \neq r^2$, the centroid does not lie on the quarter circle.

SECTION 8.8

1. Following the solution of Example 1, but with $c = 3$ instead of 6, we find that

$$F = \int_0^3 62.5\,(3 - x)(\tfrac{1}{2}x)\,dx = 31.25\int_0^3 (3x - x^2)\,dx = 31.25\,(\tfrac{3}{2}x^2 - \tfrac{1}{3}x^3)\Big|_0^3$$
$$= (31.25)\tfrac{9}{2} = 140.625 \text{ (pounds)}$$

3. We take the origin at water level. Then $c = 0$ and $w(x)/2 = [x - (-1)]/\sqrt{3}$, so that $w(x) = [(2\sqrt{3})/3](x + 1)$;

$$F = \int_{-1}^0 62.5(0 - x)\frac{2\sqrt{3}}{3}(x + 1)\,dx = \frac{125\sqrt{3}}{3}\left(-\frac{1}{3}x^3 - \frac{1}{2}x^2\right)\Big|_{-1}^0 = \frac{125\sqrt{3}}{18}$$

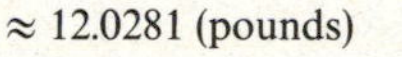

≈ 12.0281 (pounds)

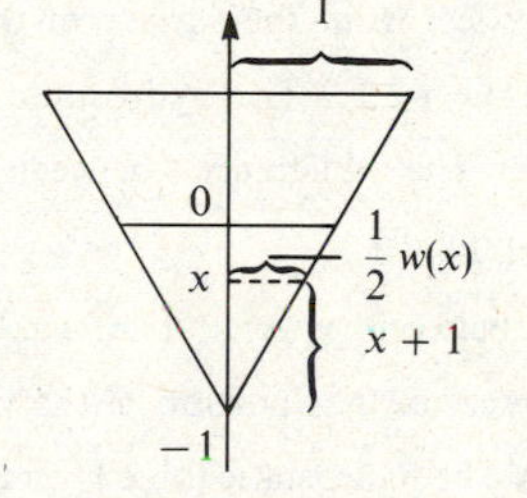

5. For the triangles pointing downward, $c = 0$ and $w(x)/2 = (x + 3)/\sqrt{3}$, so that $w(x) = [(2\sqrt{3})/3](x + 3)$;

$$F = \int_{-3}^{-3+\sqrt{3}} 62.5(0 - x)\frac{2\sqrt{3}}{3}(x + 3)\,dx = \frac{125\sqrt{3}}{3}\left(-\frac{1}{3}x^3 - \frac{3}{2}x^2\right)\Big|_{-3}^{-3+\sqrt{3}}$$
$$= \frac{125\sqrt{3}}{3}\left(\frac{9}{2} - \sqrt{3}\right) \approx 199.760 \text{ (pounds)}$$

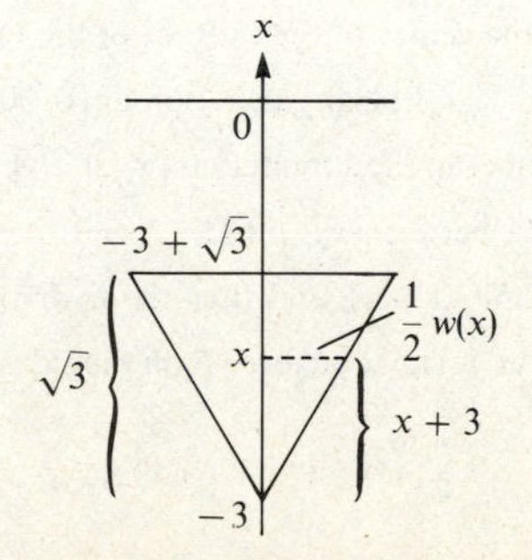

For the triangles pointing upward, $c = 0$ and

$$\frac{w(x)}{2} = \frac{-3+\sqrt{3}-x}{\sqrt{3}}, \text{ so that } w(x) = \frac{2\sqrt{3}}{3}(-3+\sqrt{3}-x)$$

$$F = \int_{-3}^{-3+\sqrt{3}} 62.5(0-x)\frac{2\sqrt{3}}{3}(-3+\sqrt{3}-x)\,dx$$
$$= \frac{125\sqrt{3}}{3}\left[\frac{(3-\sqrt{3})}{2}x^2 + \frac{1}{3}x^3\right]\Bigg|_{-3}^{-3+\sqrt{3}}$$
$$= \frac{125\sqrt{3}}{3}\left(\frac{9}{2} - \frac{\sqrt{3}}{2}\right) \approx 262.260 \text{ (pounds)}$$

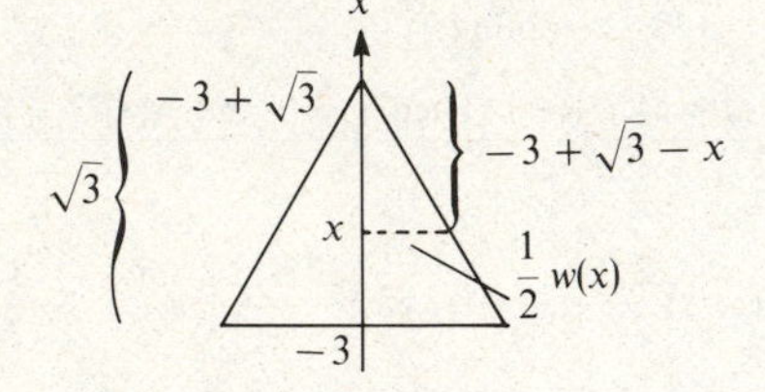

7. If the top of the block is x feet from the surface of the water, the hydrostatic force on the top is $F_t = 62.5(x)(1) = 62.5x$. The hydrostatic force on the bottom is $F_b = 62.5(x+1)(1) = 62.5(x+1)$. The difference between these forces is $F_b - F_t = 62.5(x+1) - 62.5x = 62.5$ (pounds).

9. The portion of the plate between x_{k-1} and x_k has area approximately equal to $w(t_k)(\Delta x_k \sec\theta)$. Thus the force on that portion of the plate is approximately equal to $(62.5)(c - t_k)(w(t_k)\sec\theta\,\Delta x_k)$. The hydrostatic force F on the plate is approximately equal to

$$\sum_{k=1}^{n} (62.5)(c - t_k)(\sec\theta\, w(t_k)\,\Delta x_k)$$

which is a Riemann sum for $62.5(c - x)(\sec\theta)w$. Thus the hydrostatic force is given by

$$F = 62.5\sec\theta \int_a^b (c - x)w(x)\,dx$$

11. Let us place the origin at the center of the bottom of the tank, with the x axis pointing upward. Let $\mathscr{P} = \{x_0, x_1, \ldots, x_n\}$ be any partition of $[0, 30]$. For each k between 1 and n, let t_k be an arbitrary number in the subinterval $[x_{k-1}, x_k]$. Then the area of the portion S_k of the surface of the tank between x_{k-1} and x_k is $40\pi\,\Delta x_k$. The pressure at any point on S_k is approximately $(62.5)(30 - t_k)$, and thus the hydrostatic force on S_k is approximately $(62.5)(30 - t_k)(40\pi\,\Delta x_k)$. The total force F on the sides of the tank is approximately

$$\sum_{k=1}^{n} (62.5)(30 - t_k)(40\pi\,\Delta x_k)$$

which is a Riemann sum for $(62.5)(30 - x)(40\pi)$. Thus

$$F = 62.5\int_0^{30} 40\pi(30 - x)\,dx \overset{u=30-x}{=} 62.5\int_{30}^{0} 40\pi u(-1)\,du$$
$$= 62.5\int_0^{30} 40\pi x\,dx$$

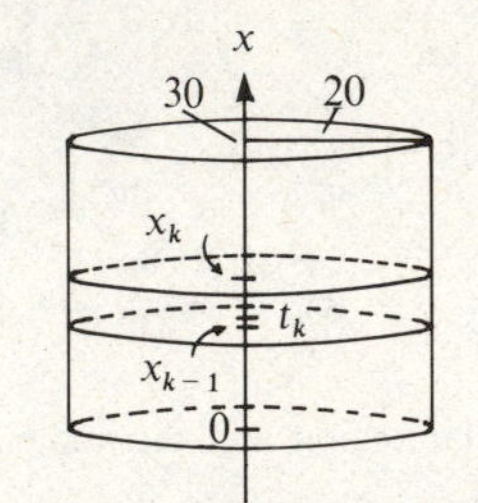

SECTION 8.9

1. a. $x = 3\cos\frac{\pi}{4} = \frac{3\sqrt{2}}{2}$; $y = 3\sin\frac{\pi}{4} = \frac{3\sqrt{2}}{2}$; Cartesian coordinates: $\left(\frac{3\sqrt{2}}{2}, \frac{3\sqrt{2}}{2}\right)$

   b. $x = -2\cos\left(\frac{-\pi}{6}\right) = -\sqrt{3}$; $y = -2\sin\left(\frac{-\pi}{6}\right) = 1$;

   Cartesian coordinates: $(-\sqrt{3}, 1)$

   c. $x = 3\cos\frac{7\pi}{3} = \frac{3}{2}$; $y = 3\sin\frac{7\pi}{3} = \frac{3\sqrt{3}}{2}$; Cartesian coordinates: $\left(\frac{3}{2}, \frac{3\sqrt{3}}{2}\right)$

   d. $x = 5\cos 0 = 5$; $y = 5\sin 0 = 0$; Cartesian coordinates: $(5, 0)$

   e. $x = -2\cos\frac{\pi}{2} = 0$; $y = -2\sin\frac{\pi}{2} = -2$; Cartesian coordinates: $(0, -2)$

   f. $x = -2\cos\frac{3\pi}{2} = 0$; $y = -2\sin\frac{3\pi}{2} = 2$; Cartesian coordinates: $(0, 2)$

   g. $x = 4\cos\frac{3\pi}{4} = -2\sqrt{2}$; $y = 4\sin\frac{3\pi}{4} = 2\sqrt{2}$; Cartesian coordinates: $(-2\sqrt{2}, 2\sqrt{2})$

   h. $x = 0\cos\frac{6\pi}{7} = 0$; $y = 0\sin\frac{6\pi}{7} = 0$; Cartesian coordinates: $(0, 0)$

   i. $x = -1\cos\frac{23\pi}{3} = \frac{-1}{2}$; $y = -1\sin\frac{23\pi}{3} = \frac{\sqrt{3}}{2}$; Cartesian coordinates: $\left(\frac{-1}{2}, \frac{\sqrt{3}}{2}\right)$

   j. $x = -1\cos\left(\frac{-23\pi}{3}\right) = \frac{-1}{2}$; $y = -1\sin\left(\frac{-23\pi}{3}\right) = \frac{-\sqrt{3}}{2}$; Cartesian coordinates: $\left(\frac{-1}{2}, \frac{-\sqrt{3}}{2}\right)$

   k. $x = 1\cos\frac{3\pi}{2} = 0$; $y = 1\sin\frac{3\pi}{2} = -1$; Cartesian coordinates: $(0, -1)$

1. $x = 3\cos\left(\dfrac{-5\pi}{6}\right) = \dfrac{-3\sqrt{3}}{2}$; $y = 3\sin\left(\dfrac{-5\pi}{6}\right) = \dfrac{-3}{2}$; Cartesian coordinates: $\left(\dfrac{-3\sqrt{3}}{2}, \dfrac{-3}{2}\right)$

3. If $2x + 3y = 4$, then $2(r\cos\theta) + 3(r\sin\theta) = 4$, so $r = 4/(2\cos\theta + 3\sin\theta)$.

5. If $x^2 + 9y^2 = 1$, then $r^2\cos^2\theta + 9(r^2\sin^2\theta) = 1$, so

$$r^2 = \frac{1}{\cos^2\theta + 9\sin^2\theta} = \frac{1}{8\sin^2\theta + 1}, \text{ or } r = \frac{1}{\sqrt{8\sin^2\theta + 1}}$$

7. If $(x^2 + y^2)^2 = x^2 - y^2$, then $(r^2)^2 = r^2\cos^2\theta - r^2\sin^2\theta$, so $r^4 = r^2\cos 2\theta$, or $r^2 = \cos 2\theta$.

9. If $x^2 + y^2 = x(x^2 - 3y^2)$, then it follows that $r^2 = r\cos\theta\,(r^2\cos^2\theta - 3r^2\sin^2\theta)$, and thus

$$r^2 = \frac{r}{\cos\theta\,(\cos^2\theta - 3\sin^2\theta)}$$

11. If $y^2 = x^2(3 - x)/(1 + x)$, then

$$r^2\sin^2\theta = \frac{r^2\cos^2\theta\,(3 - r\cos\theta)}{1 + r\cos\theta}$$

so $\sin^2\theta\,(1 + r\cos\theta) = \cos^2\theta\,(3 - r\cos\theta)$, or

$$r = \frac{3\cos^2\theta - \sin^2\theta}{\sin^2\theta\cos\theta + \cos^3\theta} = \frac{3\cos^2\theta - \sin^2\theta}{\cos\theta}$$

13. If $r = 3\cos\theta$, then $r^2 = 3r\cos\theta$, or $x^2 + y^2 = 3x$.

15. If $\cot\theta = 3$, then $\cos\theta = 3\sin\theta$, so $r\cos\theta = 3r\sin\theta$, or $x = 3y$.

17. If $r = \sin 2\theta = 2\sin\theta\cos\theta$, then $r^3 = 2(r\sin\theta)(r\cos\theta)$, so $(x^2 + y^2)^{3/2} = 2xy$, or $(x^2 + y^2)^3 = 4x^2y^2$.

19. If $r \neq 0$, the equation $r = 1 + \cos\theta$ is equivalent to the equation $r^2 = r + r\cos\theta$ since division by r is permissible. If $r = 0$ (which corresponds to the origin), then $r^2 = r + r\cos\theta$ is satisfied by (r, θ) for any θ, whereas $r = 1 + \cos\theta$ is satisfied by (r, π). Thus the origin is on the polar graph of both equations, so the polar graphs are the same.

21. $r = 5$ 23. $r = 0$ 25. $\theta = \dfrac{-7\pi}{6}$

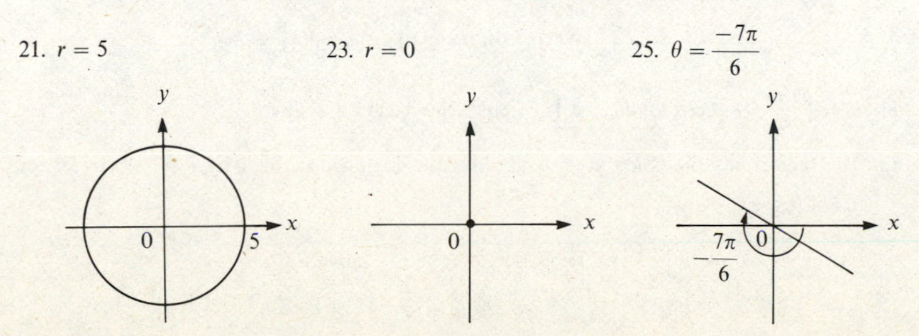

27. $r\sin\theta = 5$ 29. $r = \dfrac{-3}{2}\cos\theta$

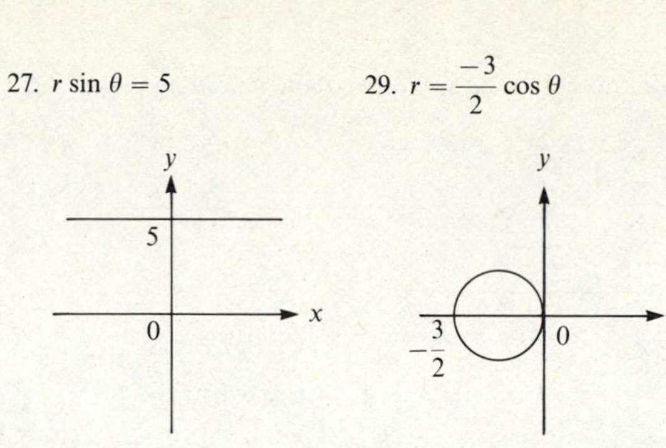

31. $r\cos\left(\theta - \dfrac{\pi}{3}\right) = 2$

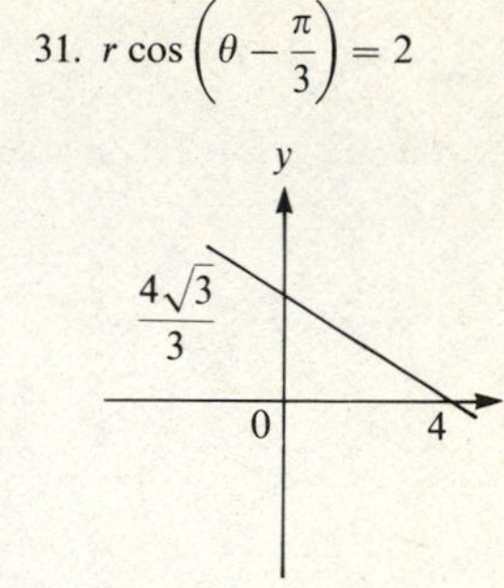

33. $r = 2\cot\theta\csc\theta$ 35. $r(\sin\theta + \cos\theta) = 1$ 37. $r^2(\cos^2\theta - 9\sin^2\theta) = 9$

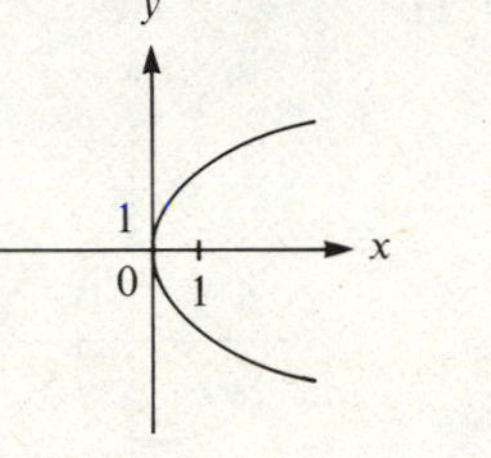

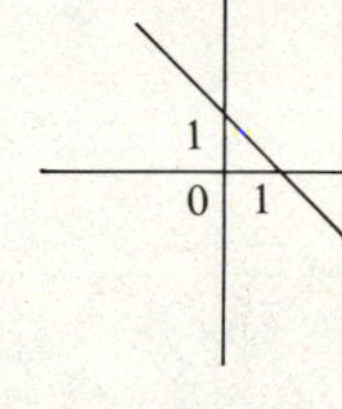

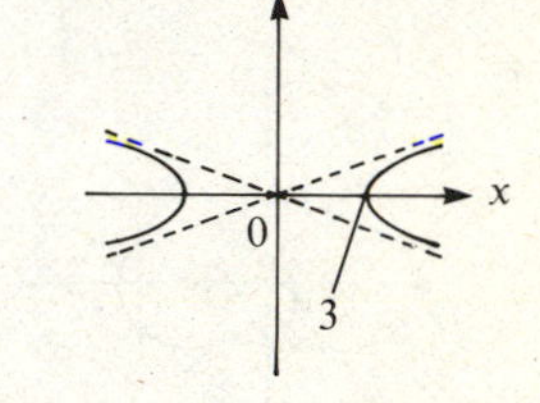

39. $r = \dfrac{-1}{\cos\theta + 4\sin\theta}$ 41. $r = 2\cos 2\theta$; symmetry with respect to both axes and origin. 43. $r = -4\sin 3\theta$; symmetry with respect to the y axis.

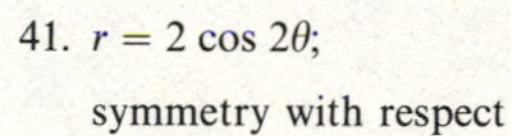

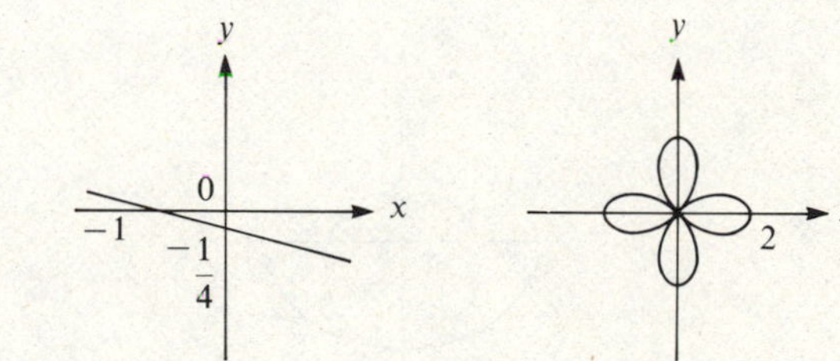

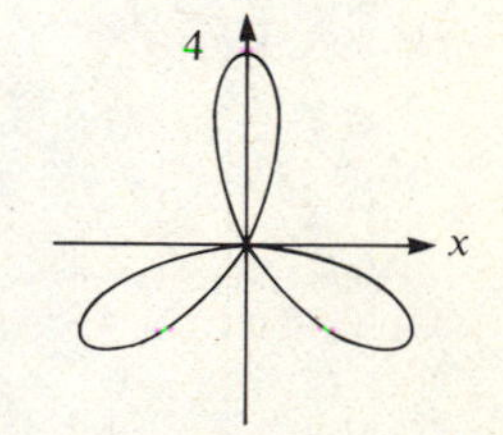

45. $r = 2\cos 6\theta$; symmetry with respect to both axes and origin.

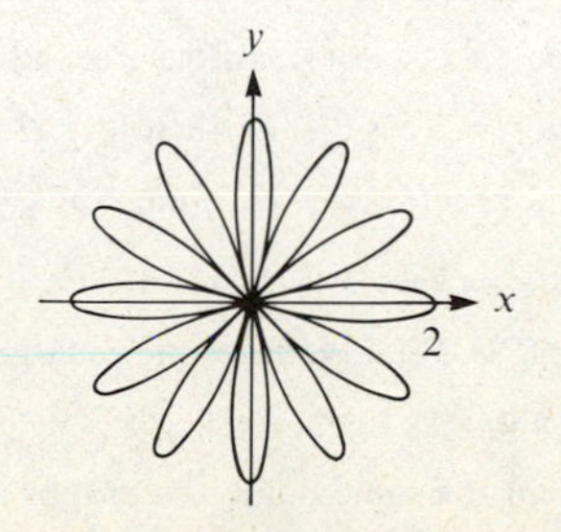

47. $r = \sin\frac{\theta}{2}$; symmetry with respect to both axes and origin (note if (r, θ) on graph then $(-r, \theta + 2\pi)$ on graph)

49. $r^2 = 25\cos\theta$; symmetry with respect to both axes and origin.

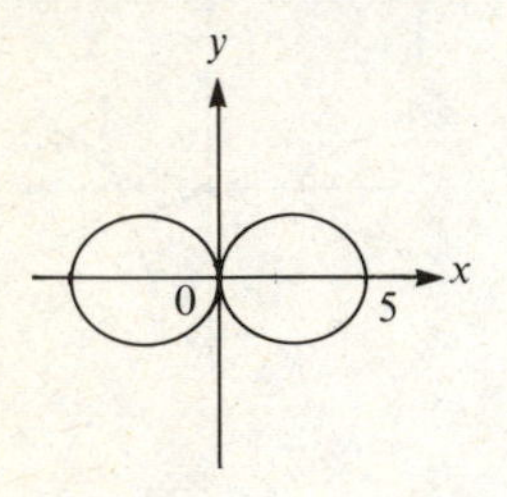

51. $r^2 = 4\cos 2\theta$; symmetry with respect to both axes and origin.

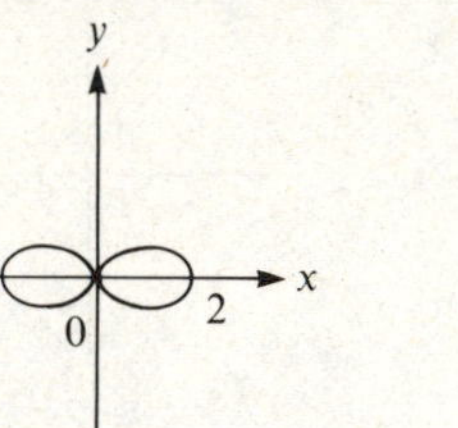

53. $r = 2 - \cos\theta$; symmetry with respect to the x axis.

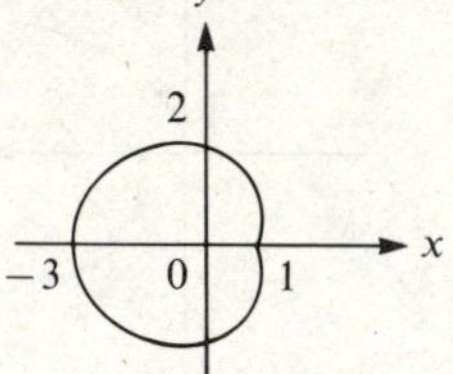

55. $r = 3\tan\theta$; symmetry with respect to both axes and origin.

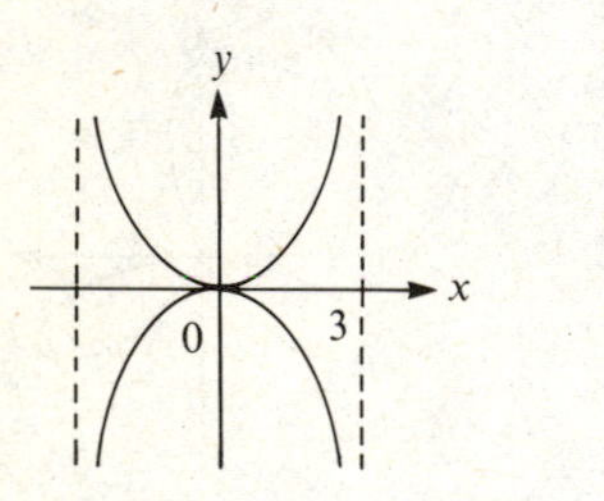

57. $r = 2\theta$; symmetry with respect to the y axis.

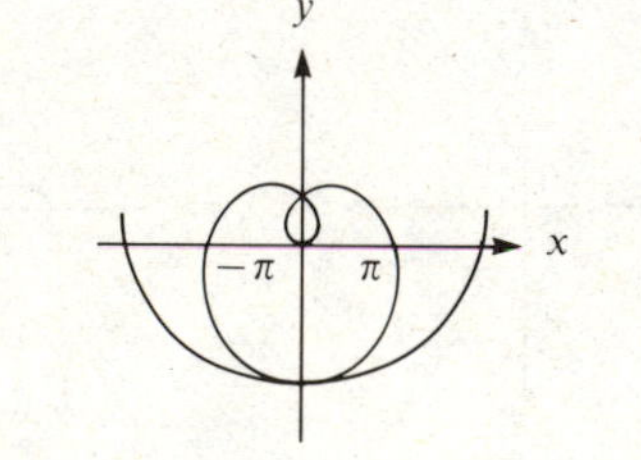

59. a. Suppose (r, θ) satisfies $r = 3(\cos\theta + 1)$. Then $3(\cos(\theta + \pi) - 1) = 3(-\cos\theta - 1) = -3(\cos\theta + 1) = -r$, so $(-r, \theta + \pi)$ satisfies the equation $r = 3(\cos\theta - 1)$. Conversely, if (r, θ) satisfies $r = 3(\cos\theta - 1)$, then $(-r, \theta + \pi)$ satisfies $r = 3(\cos\theta + 1)$. Since (r, θ) and $(-r, \theta + \pi)$ represent the same point, the graphs are the same.

b. Suppose (r, θ) satisfies $r = 2(\sin\theta + 1)$. Then $2(\sin(\theta + \pi) - 1) = 2(-\sin\theta - 1) = -2(\sin\theta + 1) = -r$, so $(-r, \theta + \pi)$ satisfies the equation $r = 2(\sin\theta - 1)$. Conversely, if (r, θ) satisfies $r = 2(\sin\theta - 1)$, then $(-r, \theta + \pi)$ satisfies $r = 2(\sin\theta + 1)$. Since (r, θ) and $(-r, \theta + \pi)$ represent the same point, the graphs are the same.

c. Since (r, θ) and $(r, \theta - 2\pi)$ represent the same point, the graphs are the same.

61. Let (r, θ) be a point on the graph. Then

$$1 = \sqrt{(r\cos\theta - 1)^2 + (r\sin\theta)^2}\sqrt{(r\cos\theta + 1)^2 + (r\sin\theta)^2}$$

Squaring both sides, we obtain

$$1 = (r^2 - 2r\cos\theta + 1)\cdot(r^2 + 2r\cos\theta + 1) = r^4 - 4r^2\cos^2\theta + 2r^2 + 1$$

so that $r^2[r^2 + 2(1 - 2\cos^2\theta)] = 0$. Thus either $r = 0$ or $r^2 = -2(1 - 2\cos^2\theta) = 2\cos 2\theta$. Since $(0, \pi/4)$ satisfies the equation $r^2 = 2\cos 2\theta$, we may simply write $r^2 = 2\cos 2\theta$, even when $r = 0$.

SECTION 8.10

1. By (2),

$$\mathscr{L} = \int_{-\pi/2}^{\pi/2} \sqrt{(-2\sin\theta)^2 + (2\cos\theta)^2}\,d\theta = \int_{-\pi/2}^{\pi/2} \sqrt{4(\sin^2\theta + \cos^2\theta)}\,d\theta$$
$$= \int_{-\pi/2}^{\pi/2} 2\,d\theta = 2\pi$$

3. By (2),

$$\mathscr{L} = \int_0^{2\pi} \sqrt{(2\sin\theta)^2 + (2 - 2\cos\theta)^2}\,d\theta = \int_0^{2\pi} \sqrt{4\sin^2\theta + 4 - 8\cos\theta + 4\cos^2\theta}\,d\theta$$
$$= \int_0^{2\pi} \sqrt{8 - 8\cos\theta}\,d\theta = 2\int_0^{\pi} 4\sqrt{\frac{1 - \cos\theta}{2}}\,d\theta = 8\int_0^{\pi} \sin\frac{\theta}{2}\,d\theta = -16\cos\frac{\theta}{2}\bigg|_0^{\pi} = 16$$

5. By (2),

$$\mathscr{L} = \int_{-\pi/2}^{3\pi/2} \sqrt{\left(\sin\frac{\theta}{2}\cos\frac{\theta}{2}\right)^2 + \left(\sin^2\frac{\theta}{2}\right)^2}\,d\theta = \int_{-\pi/2}^{3\pi/2} \sqrt{\sin^2\frac{\theta}{2}\left(\cos^2\frac{\theta}{2} + \sin^2\frac{\theta}{2}\right)}\,d\theta$$
$$= \int_{-\pi/2}^{3\pi/2} \sqrt{\sin^2\frac{\theta}{2}}\,d\theta = \int_0^{2\pi} \sqrt{\sin^2\frac{\theta}{2}}\,d\theta = \int_0^{2\pi} \sin\frac{\theta}{2}\,d\theta = -2\cos\frac{\theta}{2}\bigg|_0^{2\pi} = 4$$

7. $A = \int_0^{2\pi} (\frac{1}{2})4^2\,d\theta = 8\theta\Big|_0^{2\pi} = 16\pi$

9. The entire region is obtained from $r = 3\sin\theta$ for $0 \le \theta \le \pi$;

$$A = \int_0^{\pi} (\tfrac{1}{2})9\sin^2\theta\,d\theta = \tfrac{9}{2}(\tfrac{1}{2}\theta - \tfrac{1}{4}\sin 2\theta)\Big|_0^{\pi} = \tfrac{9}{4}\pi$$

11. $A = \int_{\pi/2}^{3\pi/2} \frac{1}{2}(-2\cos\theta)^2\,d\theta = 2\int_{\pi/2}^{3\pi/2} \cos^2\theta\,d\theta = 2(\frac{1}{2}\theta + \frac{1}{4}\sin 2\theta)\Big|_{\pi/2}^{3\pi/2} = \pi$

13. The region has the same area A as does the region described by $r = -9\cos 2\theta$ for $\pi/4 \le \theta \le \pi/2$. Now

$$A = \int_{\pi/4}^{\pi/2} \tfrac{1}{2}(-9\cos 2\theta)^2\,d\theta = \tfrac{81}{2}\int_{\pi/4}^{\pi/2} \cos^2 2\theta\,d\theta$$
$$= \tfrac{81}{2}\int_{\pi/4}^{\pi/2} (\tfrac{1}{2} - \tfrac{1}{2}\cos 4\theta)\,d\theta = \tfrac{81}{2}(\tfrac{1}{2}\theta - \tfrac{1}{8}\sin 4\theta)\Big|_{\pi/4}^{\pi/2} = \tfrac{81}{16}\pi$$

15. The three leaves have the same area. The area of one of them is given by

$$A_1 = \int_{-\pi/6}^{\pi/6} \tfrac{1}{2}(\tfrac{1}{2}\cos 3\theta)^2\,d\theta = \tfrac{1}{8}\int_{-\pi/6}^{\pi/6} \cos^2 3\theta\,d\theta$$
$$= \frac{1}{8}\int_{-\pi/6}^{\pi/6} (\tfrac{1}{2} + \tfrac{1}{2}\cos 6\theta)\,d\theta = \tfrac{1}{8}(\tfrac{1}{2}\theta + \tfrac{1}{12}\sin 6\theta)\Big|_{-\pi/6}^{\pi/6} = \tfrac{1}{48}\pi$$

Thus the total area $A = 3A_1 = \pi/16$.

17. $A = \int_0^{2\pi} \tfrac{1}{2}(4)(1 - \sin\theta)^2\,d\theta = 2\int_0^{2\pi} (1 - 2\sin\theta + \sin^2\theta)\,d\theta$

$= 2\int_0^{2\pi} (\tfrac{3}{2} - 2\sin\theta - \tfrac{1}{2}\cos 2\theta)\,d\theta = 2(\tfrac{3}{2}\theta + 2\cos\theta - \tfrac{1}{4}\sin 2\theta)\Big|_0^{2\pi} = 6\pi$

19. $A = \int_0^{2\pi} \tfrac{1}{2}(4 + 3\cos\theta)^2\,d\theta = \tfrac{1}{2}\int_0^{2\pi} (16 + 24\cos\theta + 9\cos^2\theta)\,d\theta$

$= \tfrac{1}{2}\int_0^{2\pi} (\tfrac{41}{2} + 24\cos\theta + \tfrac{9}{2}\cos 2\theta)\,d\theta$

$= \tfrac{1}{2}(\tfrac{41}{2}\theta + 24\sin\theta + \tfrac{9}{4}\sin 2\theta)\Big|_0^{2\pi} = \tfrac{41}{2}\pi$

21. The two leaves of the region have the same area. The area of one of them is given by

$$A_1 = \int_{-\pi/2}^{\pi/2} \tfrac{1}{2}(\sqrt{25\cos\theta})^2\,d\theta = \tfrac{25}{2}\int_{-\pi/2}^{\pi/2} \cos\theta\,d\theta = \tfrac{25}{2}\sin\theta\Big|_{-\pi/2}^{\pi/2} = 25$$

Thus the total area $A = 2A_1 = 50$.

23. $f(\theta) = 5,\ g(\theta) = 1;\ A = \int_0^{2\pi} \tfrac{1}{2}(5^2 - 1^2)\,d\theta = 12\theta\Big|_0^{2\pi} = 24\pi$

25. $f(\theta) = 1,\ g(\theta) = \begin{cases} \sin\theta & \text{for } 0 \le \theta \le \pi \\ 0 & \text{for } \pi \le \theta \le 2\pi \end{cases};$

$A = \int_0^{\pi} \tfrac{1}{2}(1^2 - \sin^2\theta)\,d\theta + \int_{\pi}^{2\pi} \tfrac{1}{2}(1^2)\,d\theta = \tfrac{1}{2}(\tfrac{1}{2}\theta + \tfrac{1}{4}\sin 2\theta)\Big|_0^{\pi} + \tfrac{1}{2}\theta\Big|_{\pi}^{2\pi}$

$= \dfrac{\pi}{4} + \dfrac{\pi}{2} = \dfrac{3}{4}\pi$

27. $f(\theta) = 1,\ g(\theta) = \begin{cases} \sqrt{\cos 2\theta} & \text{for } \dfrac{-\pi}{4} \le \theta \le \dfrac{\pi}{4} \text{ and } \dfrac{3\pi}{4} \le \theta \le \dfrac{5\pi}{4} \\ 0 & \text{for } \dfrac{\pi}{4} \le \theta \le \dfrac{3\pi}{4} \text{ and } \dfrac{5\pi}{4} \le \theta \le \dfrac{7\pi}{4} \end{cases};$

$A = \int_{-\pi/4}^{\pi/4} \tfrac{1}{2}[1^2 - (\sqrt{\cos 2\theta})^2]\,d\theta$

$+ \int_{\pi/4}^{3\pi/4} \tfrac{1}{2}(1^2)\,d\theta + \int_{3\pi/4}^{5\pi/4} \tfrac{1}{2}[1^2 - (\sqrt{\cos 2\theta})^2]\,d\theta + \int_{5\pi/4}^{7\pi/4} \tfrac{1}{2}(1^2)\,d\theta$

$= \dfrac{1}{2}\int_{-\pi/4}^{\pi/4} (1 - \cos 2\theta)\,d\theta + \dfrac{\pi}{4} + \dfrac{1}{2}\int_{3\pi/4}^{5\pi/4} (1 - \cos 2\theta)\,d\theta + \dfrac{\pi}{4}$

$= \dfrac{1}{2}\left(\theta - \dfrac{1}{2}\sin 2\theta\right)\Big|_{-\pi/4}^{\pi/4} + \dfrac{1}{2}\left(\theta - \dfrac{1}{2}\sin 2\theta\right)\Big|_{3\pi/4}^{5\pi/4} + \dfrac{\pi}{2}$

$= \left(\dfrac{\pi}{4} - \dfrac{1}{2}\right) + \left(\dfrac{\pi}{4} - \dfrac{1}{2}\right) + \dfrac{\pi}{2} = \pi - 1$

29. $f(\theta) = 2 + \cos\theta,\ g(\theta) = \begin{cases} 0 & \text{for } \dfrac{-\pi}{2} \le \theta \le \dfrac{\pi}{2} \\ -\cos\theta & \text{for } \dfrac{\pi}{2} \le \theta \le \dfrac{3\pi}{2} \end{cases};$

$A = \int_{-\pi/2}^{\pi/2} \tfrac{1}{2}(2 + \cos\theta)^2\,d\theta + \int_{\pi/2}^{3\pi/2} \tfrac{1}{2}[(2 + \cos\theta)^2 - (-\cos\theta)^2]\,d\theta$

$= \tfrac{1}{2}\int_{-\pi/2}^{\pi/2} (4 + 4\cos\theta + \cos^2\theta)\,d\theta + \tfrac{1}{2}\int_{\pi/2}^{3\pi/2} (4 + 4\cos\theta)\,d\theta$

$= \tfrac{1}{2}(\tfrac{9}{2}\theta + 4\sin\theta + \tfrac{1}{4}\sin 2\theta)\Big|_{-\pi/2}^{\pi/2} + \tfrac{1}{2}(4\theta + 4\sin\theta)\Big|_{\pi/2}^{3\pi/2}$

$= \left(\dfrac{9\pi}{4} + 4\right) + (2\pi - 4) = \dfrac{17}{4}\pi$

31. The circle and the cardioid intersect for (r, θ) such that $r = 0$ or $\cos\theta = r = 1 - \cos\theta$, so that $\theta = -\pi/3$ or $\theta = \pi/3$. By symmetry we obtain

$$A = 2\left[\int_0^{\pi/3} \frac{1}{2}(1 - \cos\theta)^2\,d\theta + \int_{\pi/3}^{\pi/2} \frac{1}{2}\cos^2\theta\,d\theta\right]$$
$$= \int_0^{\pi/3} (1 - 2\cos\theta + \cos^2\theta)\,d\theta + \int_{\pi/3}^{\pi/2} \cos^2\theta\,d\theta$$
$$= (\tfrac{3}{2}\theta - 2\sin\theta + \tfrac{1}{4}\sin 2\theta)\Big|_0^{\pi/3} + (\tfrac{1}{2}\theta + \tfrac{1}{4}\sin 2\theta)\Big|_{\pi/3}^{\pi/2}$$
$$= \left(\frac{\pi}{2} - \frac{7}{8}\sqrt{3}\right) + \left(\frac{\pi}{12} - \frac{\sqrt{3}}{8}\right) = \frac{7}{12}\pi - \sqrt{3}$$

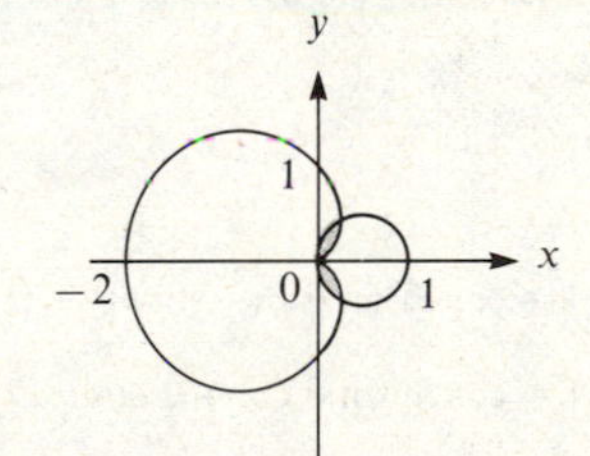

33. The two cardioids intersect for (r, θ) such that $r = 0$ or $1 + \cos\theta = r = 1 + \sin\theta$, so that $\cos\theta = \sin\theta$, and thus $\theta = \pi/4$ or $\theta = 5\pi/4$. Thus

$A = \int_{\pi/4}^{5\pi/4} \tfrac{1}{2}[(1 + \sin\theta)^2 - (1 + \cos\theta)^2]\,d\theta$

$= \tfrac{1}{2}\int_{\pi/4}^{5\pi/4} (2\sin\theta + \sin^2\theta - 2\cos\theta - \cos^2\theta)\,d\theta$

$= \tfrac{1}{2}\int_{\pi/4}^{5\pi/4} (2\sin\theta - 2\cos\theta - \cos 2\theta)\,d\theta$

$= \tfrac{1}{2}(-2\cos\theta - 2\sin\theta - \tfrac{1}{2}\sin 2\theta)\Big|_{\pi/4}^{5\pi/4} = 2\sqrt{2}$

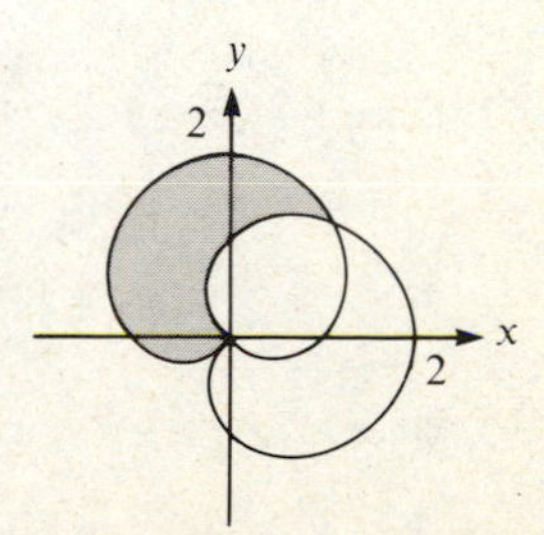

35. a. If
$$y^2 = \frac{x^2(1+x)}{1-x}$$
then
$$r^2\sin^2\theta = \frac{r^2\cos^2\theta\,(1+r\cos\theta)}{1-r\cos\theta}$$
and for $r \neq 0$ this means that
$$\sin^2\theta\,(1-r\cos\theta) = \cos^2\theta\,(1+r\cos\theta)$$
so that
$$r\cos\theta\,(\cos^2\theta+\sin^2\theta) = \sin^2\theta - \cos^2\theta$$
or
$$r\cos\theta = 1-2\cos^2\theta, \quad\text{or}\quad r = \sec\theta - 2\cos\theta$$
Now observe that if $x \neq 0$, then $(1+x)/(1-x) = y^2/x^2 \geq 0$, so that x can assume only the values in $[-1, 1)$, whereas y can assume any value. Since $x = r\cos\theta = 1-2\cos^2\theta$ and $y = x\tan\theta$, it follows that all the required values of x and y are assumed if θ assumes the values in $(-\pi/2, \pi/2)$.

b. The loop is the graph of $r = \sec\theta - 2\cos\theta$ for $-\pi/4 \leq \theta \leq \pi/4$, which has the same area as the region bounded by the graph of $r = 2\cos\theta - \sec\theta$ for $-\pi/4 \leq \theta \leq \pi/4$. Its area is given by
$$\begin{aligned} A &= \int_{-\pi/4}^{\pi/4} \tfrac{1}{2}(2\cos\theta - \sec\theta)^2\,d\theta \\ &= \tfrac{1}{2}\int_{-\pi/4}^{\pi/4}(4\cos^2\theta - 4 + \sec^2\theta)\,d\theta \\ &= [2(\tfrac{1}{2}\theta + \tfrac{1}{4}\sin 2\theta) - 2\theta + \tfrac{1}{2}\tan\theta]\Big|_{-\pi/4}^{\pi/4} \\ &= 2 - \tfrac{1}{2}\pi \end{aligned}$$

37. By (6),
$$\begin{aligned} S &= \int_0^\pi 2\pi(1+\cos\theta)\sin\theta\sqrt{(-\sin\theta)^2 + (1+\cos\theta)^2}\,d\theta \\ &= \int_0^\pi 2\pi(1+\cos\theta)\sin\theta\sqrt{2+2\cos\theta}\,d\theta \\ &= 2\sqrt{2}\pi\int_0^\pi (1+\cos\theta)^{3/2}\sin\theta\,d\theta \\ &= -\tfrac{4}{5}\sqrt{2}\pi\,(1+\cos\theta)^{5/2}\Big|_0^\pi = \tfrac{32}{5}\pi \end{aligned}$$

CHAPTER 8 REVIEW

1. Since $1+\sqrt{x} \geq 1 \geq e^{-x}$ for $0 \leq x \leq 1$,
$$\begin{aligned} V &= \int_0^1 \pi[(1+\sqrt{x})^2 - (e^{-x})^2]\,dx = \pi\int_0^1 (1+2\sqrt{x}+x-e^{-2x})\,dx \\ &= \pi(x + \tfrac{4}{3}x^{3/2} + \tfrac{1}{2}x^2 + \tfrac{1}{2}e^{-2x})\Big|_0^1 = \pi(\tfrac{7}{3} + \tfrac{1}{2}e^{-2}) \end{aligned}$$

3. Since $e^{x^2} \geq 1 \geq e^{-x^2}$ for $0 \leq x \leq 1$,
$$V = \int_0^1 2\pi x(e^{x^2} - e^{-x^2})\,dx \overset{u=x^2}{=} 2\pi\int_0^1 (e^u - e^{-u})\tfrac{1}{2}\,du = \pi(e^u + e^{-u})\Big|_0^1 = \pi(e + e^{-1} - 2)$$

5. a. $V = \int_0^2 2\pi x(x^3)\,dx = \frac{2}{5}\pi x^5\Big|_0^2 = \frac{64}{5}\pi$

b. $\mathscr{M}_x = \frac{1}{2}\int_0^2 (x^3)^2\,dx = \frac{1}{14}x^7\Big|_0^2 = \frac{64}{7}$; $\mathscr{M}_y = \int_0^2 x(x^3)\,dx = \frac{1}{5}x^5\Big|_0^2 = \frac{32}{5}$

$$A = \int_0^2 x^3\,dx = \tfrac{1}{4}x^4\Big|_0^2 = 4$$
$$\bar{x} = \frac{32/5}{4} = \frac{8}{5};\ \bar{y} = \frac{64/7}{4} = \frac{16}{7};\ (\bar{x}, \bar{y}) = (\tfrac{8}{5}, \tfrac{16}{7})$$

c. $A = 4$, $b = \frac{8}{5}$; $V = (2\pi)(\frac{8}{5})(4) = \dfrac{64\pi}{5}$

7. a. $V_1 = \int_1^3 2\pi x\left[\left(x + \frac{c}{x}\right) - x\right]dx = \int_1^3 2\pi c\,dx = 4\pi c$

b. $$\begin{aligned} V_2 &= \int_1^3 \pi\left[\left(x+\frac{c}{x}\right)^2 - x^2\right]dx = \pi\int_1^3\left(2c + \frac{c^2}{x^2}\right)dx = \pi\left(2cx - \frac{c^2}{x}\right)\Big|_1^3 \\ &= \pi\left[\left(6c - \frac{c^2}{3}\right) - (2c - c^2)\right] = 4\pi c + \tfrac{2}{3}\pi c^2 \end{aligned}$$

c. $4\pi c + \frac{2}{3}\pi c^2 = V_2 = V_1 = 4\pi c$ only if $c = 0$.

9. Let a hexagonal cross-section x feet from the base have a side $s(x)$ feet long. Then $s(x)/2 = (10-x)/10$, so that $s(x) = \frac{1}{5}(10-x)$. The area is $A(x) = 6(\sqrt{3}/4)(s(x))^2 = [(3\sqrt{3})/50](10-x)^2$. Thus the volume is given by
$$V = \int_0^{10}\frac{3\sqrt{3}}{50}(10-x)^2\,dx = \frac{3\sqrt{3}}{50}\left(\frac{-1}{3}\right)(10-x)^3\Big|_0^{10} = 20\sqrt{3}\text{ (cubic feet)}$$

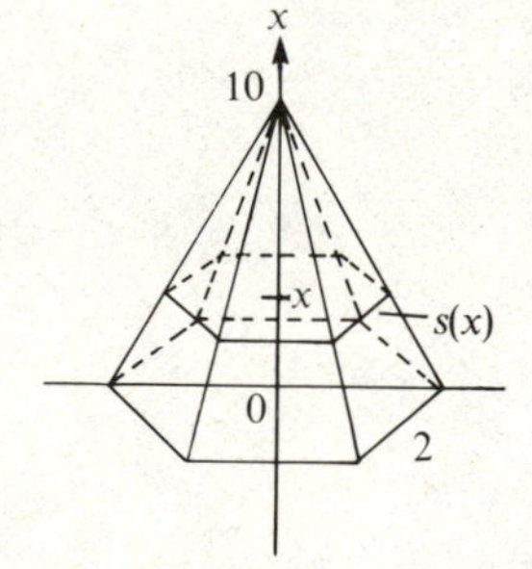

11. $$\begin{aligned} \mathscr{L} &= \int_{\pi/6}^{5\pi/6}\sqrt{1 + \left(\frac{\cos x}{\sin x}\right)^2}\,dx = \int_{\pi/6}^{5\pi/6}\sqrt{1+\cot^2 x}\,dx \\ &= \int_{\pi/6}^{5\pi/6}\csc x\,dx = -\ln|\csc x + \cot x|\Big|_{\pi/6}^{5\pi/6} \\ &= \ln(2+\sqrt{3}) - \ln(2-\sqrt{3}) \\ &= 2\ln(2+\sqrt{3}) = \ln(7+4\sqrt{3}) \end{aligned}$$

13.

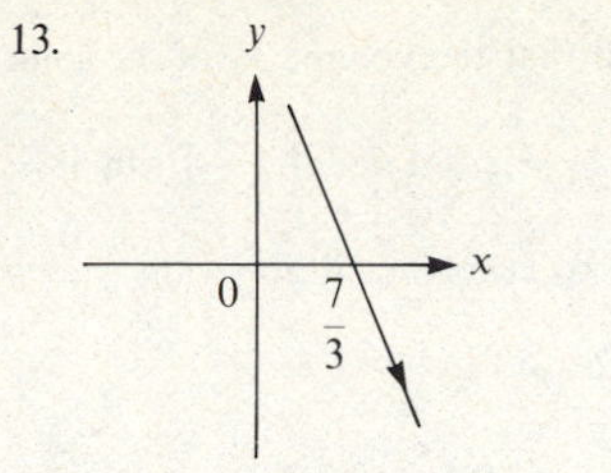

15. $\mathscr{L} = \int_0^1 \sqrt{(\cosh 2t - 1)^2 + (2 \sinh t)^2}\, dt = \int_0^1 \sqrt{(2 \sinh^2 t)^2 + (2 \sinh t)^2}\, dt$

$= \int_0^1 \sqrt{4 \sinh^4 t + 4 \sinh^2 t}\, dt = \int_0^1 2 \sinh t \sqrt{\sinh^2 t + 1}\, dt = \int_0^1 2 \sinh t \cosh t\, dt$

$= \sinh^2 t \Big|_0^1 = \sinh 1 = \frac{1}{2}(e - e^{-1})$

17. $\mathscr{L} = \int_{-\pi/3}^{\pi/3} \sqrt{1 + (\sec^2 x)^2}\, dx = \int_{-\pi/3}^{\pi/3} \sqrt{1 + \sec^4 x}\, dx$

$\approx \dfrac{2\pi/3}{12}(\sqrt{17} + 4\sqrt{\frac{25}{9}} + 2\sqrt{2} + 4\sqrt{\frac{25}{9}} + \sqrt{17}) \approx 4.25999$

19. $S = \int_0^1 2\pi(e^x + \frac{1}{4}e^{-x})\sqrt{1 + (e^x - \frac{1}{4}e^{-x})^2}\, dx$

$= 2\pi \int_0^1 (e^x + \frac{1}{4}e^{-x})\sqrt{1 + (e^{2x} - \frac{1}{2} + \frac{1}{16}e^{-2x})}\, dx$

$= 2\pi \int_0^1 (e^x + \frac{1}{4}e^{-x})\sqrt{e^{2x} + \frac{1}{2} + \frac{1}{16}e^{-2x}}\, dx = 2\pi \int_0^1 (e^x + \frac{1}{4}e^{-x})(e^x + \frac{1}{4}e^{-x})\, dx$

$= 2\pi \int_0^1 (e^{2x} + \frac{1}{2} + \frac{1}{16}e^{-2x})\, dx = 2\pi(\frac{1}{2}e^{2x} + \frac{1}{2}x - \frac{1}{32}e^{-2x})\Big|_0^1 = \pi(e^2 - \frac{1}{16}e^{-2} + \frac{1}{16})$

21. $S = \int_1^{\sqrt{3}} (2\pi)\frac{1}{2}t^2\sqrt{(t^2)^2 + (t)^2}\, dt = 2\pi \int_1^{\sqrt{3}} \frac{1}{2}t^2\sqrt{t^4 + t^2}\, dt$

$= 2\pi \int_1^{\sqrt{3}} \frac{1}{2}t^3\sqrt{t^2 + 1}\, dt \overset{u = t^2+1}{=} 2\pi \int_2^4 \frac{1}{4}(u - 1)\sqrt{u}\, du$

$= \frac{1}{2}\pi \int_2^4 (u^{3/2} - u^{1/2})\, du = \frac{1}{2}\pi(\frac{2}{5}u^{5/2} - \frac{2}{3}u^{3/2})\Big|_2^4$

$= \frac{1}{2}\pi(\frac{64}{5} - \frac{16}{3}) - \frac{1}{2}\pi(\frac{8}{5}\sqrt{2} - \frac{4}{3}\sqrt{2})$

$= \dfrac{\pi}{15}(56 - 2\sqrt{2})$

23. Let $f(x) = -15$, $g(x) = -\sqrt{400 - x^2}$ for $0 \le x \le 5\sqrt{7}$. Then

$V = \int_0^{5\sqrt{7}} 2\pi x[-15 - (-\sqrt{400 - x^2})]\, dx$

$= 2\pi\left[\dfrac{-15}{2}x^2 - \dfrac{1}{3}(400 - x^2)^{3/2}\right]\Bigg|_0^{5\sqrt{7}}$

$= \dfrac{1375\pi}{3}$ (cubic feet)

25. $\ell = 0$, $A(x) = 9\pi$; $W = \int_0^{20} (150)9\pi x\, dx = 675\pi x^2\Big|_0^{20} = 270{,}000\pi$ (foot-pounds)

27. a. If the tank is positioned as in the figure, with the origin in the center of the tank, then $\ell = 11$, and at a distance x from the center of the tank, the width $w(x)$ of a rectangular cross-section is given by $w(x) = 2\sqrt{25 - x^2}$. Thus the cross-sectional area is given by $A(x) = 40\sqrt{25 - x^2}$. Therefore

$W = \int_{-5}^{5} 42(11 - x)(40\sqrt{25 - x^2})\, dx$

$= 18{,}480 \int_{-5}^{5} \sqrt{25 - x^2}\, dx - 1680 \int_{-5}^{5} x\sqrt{25 - x^2}\, dx$

$\overset{x = 5\sin u}{=} 462{,}000 \int_{-\pi/2}^{\pi/2} \cos^2 u\, du + 1680[-\frac{1}{3}(25 - x^2)^{3/2}]\Big|_{-5}^{5}$

$= 462{,}000 \int_{-\pi/2}^{\pi/2} (\frac{1}{2} + \frac{1}{2}\cos 2u)\, du + 0 = 462{,}000(\frac{1}{2}u + \frac{1}{4}\sin 2u)\Big|_{-\pi/2}^{\pi/2}$

$= 231{,}000\pi$ (foot-pounds)

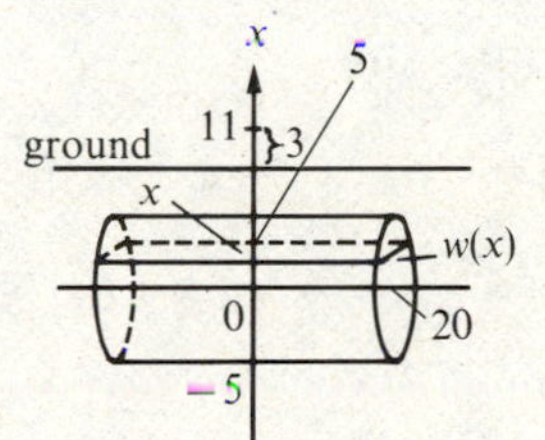

b. The figure is the same, except that there are hemispheres at each end of the tank. At a distance x from the center of the tank, the radius $r(x)$ of each hemisphere is given by $r(x) = \sqrt{25 - x^2}$, so the total cross-sectional area is given by $A(x) = \pi(25 - x^2) + 40\sqrt{25 - x^2}$. Using the calculations of part (a), we find that

$W = \int_{-5}^{5} 42(11 - x)[\pi(25 - x^2) + 40\sqrt{25 - x^2}]\, dx$

$= 42\pi \int_{-5}^{5} (275 - 25x - 11x^2 + x^3)\, dx + 42 \int_{-5}^{5} (11 - x)(40\sqrt{25 - x^2})\, dx$

$= 42\pi(275x - \frac{25}{2}x^2 - \frac{11}{3}x^3 + \frac{1}{4}x^4)\Big|_{-5}^{5} + 231{,}000\pi$

$= 77{,}000\pi + 231{,}000\pi = 308{,}000\pi$ (foot-pounds)

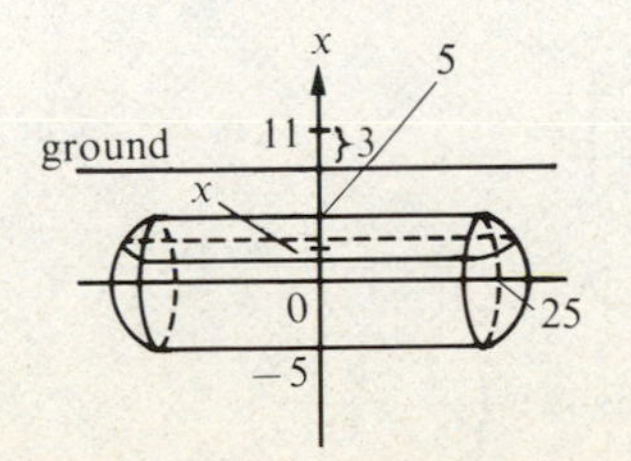

29. $\mathcal{M}_x = \int_0^2 \frac{1}{2}[(1 + 2x - x^2)^2 - (x^2 - 2x + 1)^2]\,dx = \int_0^2 (4x - 2x^2)\,dx = (2x^2 - \frac{2}{3}x^3)\Big|_0^2 = \frac{8}{3}$

$\mathcal{M}_y = \int_0^2 x[(1 + 2x - x^2) - (x^2 - 2x + 1)]\,dx = \int_0^2 (4x^2 - 2x^3)\,dx$

$= (\frac{4}{3}x^3 - \frac{1}{2}x^4)\Big|_0^2 = \frac{32}{3} - 8 = \frac{8}{3}$

$A = \int_0^2 [(1 + 2x - x^2) - (x^2 - 2x + 1)]\,dx = \int_0^2 (4x - 2x^2)\,dx = (2x^2 - \frac{2}{3}x^3)\Big|_0^2 = \frac{8}{3}$

$\bar{x} = \frac{\mathcal{M}_y}{A} = \frac{8/3}{8/3} = 1;\ \bar{y} = \frac{\mathcal{M}_x}{A} = \frac{8/3}{8/3} = 1;\ (\bar{x}, \bar{y}) = (1, 1)$

31. By the hint, the graphs of $y^2 = 3x$ and $y = x^2 - 2x$ intersect for (x, y) such that $x = 0$ or $x = 3$. To use Definition 6.4 we let $f(x) = \sqrt{3x}$ and $g(x) = x^2 - 2x$ for $0 \le x \le 3$. Then

$\mathcal{M}_x = \int_0^3 \frac{1}{2}[(\sqrt{3x})^2 - (x^2 - 2x)^2]\,dx = \frac{1}{2}\int_0^3 (3x - x^4 + 4x^3 - 4x^2)\,dx$

$= \frac{1}{2}(\frac{3}{2}x^2 - \frac{1}{5}x^5 + x^4 - \frac{4}{3}x^3)\Big|_0^3 = \frac{99}{20}$

$\mathcal{M}_y = \int_0^3 x[\sqrt{3x} - (x^2 - 2x)]\,dx = \int_0^3 (\sqrt{3}x^{3/2} - x^3 + 2x^2)\,dx$

$= (\frac{2}{5}\sqrt{3}x^{5/2} - \frac{1}{4}x^4 + \frac{2}{3}x^3)\Big|_0^3 = \frac{171}{20}$

$A = \int_0^3 [\sqrt{3x} - (x^2 - 2x)]\,dx = (\frac{2}{3}\sqrt{3}x^{3/2} - \frac{1}{3}x^3 + x^2)\Big|_0^3 = 6$

$\bar{x} = \frac{171/20}{6} = \frac{171}{120} = \frac{57}{40};\ \bar{y} = \frac{99/20}{6} = \frac{33}{40};\ (\bar{x}, \bar{y}) = (\frac{57}{40}, \frac{33}{40})$

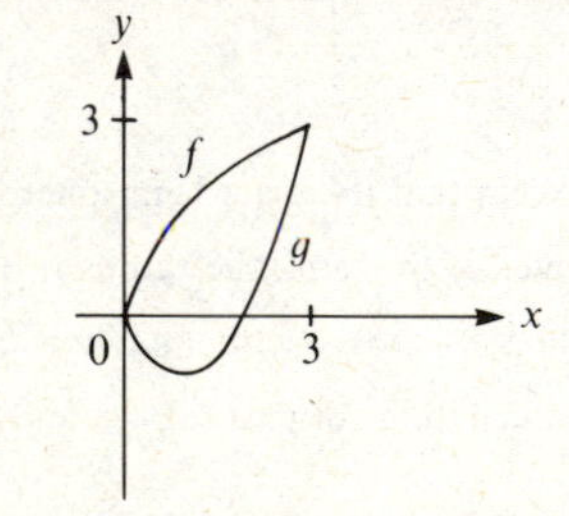

33. a. $\mathcal{M}_x = \int_1^3 \frac{1}{2}\left[\left(x + \frac{c}{x}\right)^2 - x^2\right]dx = \frac{1}{2}\int_1^3 \left(2c + \frac{c^2}{x^2}\right)dx$

$= \frac{1}{2}\left(2cx - \frac{c^2}{x}\right)\Big|_1^3 = \frac{1}{2}\left[\left(6c - \frac{c^2}{3}\right) - (2c - c^2)\right] = 2c + \frac{1}{3}c^2$

$\mathcal{M}_y = \int_1^3 x\left[\left(x + \frac{c}{x}\right) - x\right]dx = \int_1^3 c\,dx = 2c$

$A = \int_1^3 \left[\left(x + \frac{c}{x}\right) - x\right]dx = \int_1^3 \frac{c}{x}\,dx = c \ln x\Big|_1^3 = c \ln 3$

$\bar{x} = \frac{\mathcal{M}_y}{A} = \frac{2c}{c \ln 3} = \frac{2}{\ln 3};\ \bar{y} = \frac{\mathcal{M}_x}{A} = \frac{2c + c^2/3}{c \ln 3} = \frac{6 + c}{3 \ln 3}$

Thus the center of gravity is $\left(\frac{2}{\ln 3}, \frac{6 + c}{3 \ln 3}\right)$.

b. Using the Theorem of Pappus and Guldin, we find that the volume V_1 of the solid obtained by revolving R about the y axis is given by $V_1 = 2\pi\bar{x}A = 2\pi\left(\frac{2}{\ln 3}\right)(c \ln 3) = 4\pi c$. Similarly, the volume V_2 of the solid obtained by revolving R about the x axis is given by $V_2 = 2\pi\bar{y}A = 2\pi\left(\frac{6 + c}{3 \ln 3}\right)(c \ln 3) = 4\pi c + \frac{2}{3}\pi c^2$.

35. We take the origin at the bottom of the dam.

a. $c = 50$, and from the figure, we see that $[\frac{1}{2}w(x) - 50]/50 = x/100$, so $w(x) = x + 100$;

$F = \int_0^{50} 62.5(50 - x)(x + 100)\,dx = 62.5 \int_0^{50} (5000 - 50x - x^2)\,dx$

$= 62.5(5000x - 25x^2 - \frac{1}{3}x^3)\Big|_0^{50} = \frac{27{,}343{,}750}{3}$ (pounds)

b. $c = 100$, and as in part (a), $w(x) = x + 100$;

$F = \int_0^{100} 62.5(100 - x)(x + 100)\,dx = 62.5 \int_0^{100} (10{,}000 - x^2)\,dx$

$= 62.5(10{,}000x - \frac{1}{3}x^3)\Big|_0^{100} = \frac{125{,}000{,}000}{3}$ (pounds)

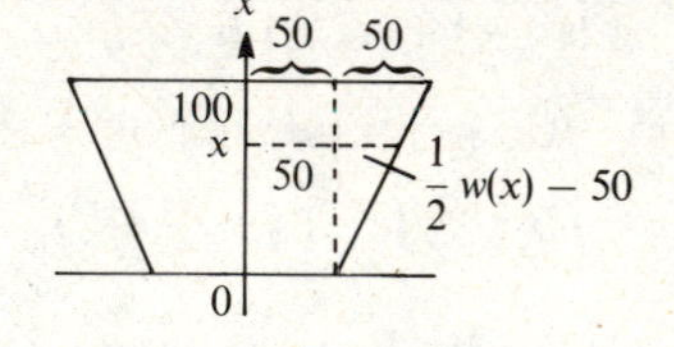

37. $r = 2 \cos \theta - 2$; symmetry with respect to x axis.

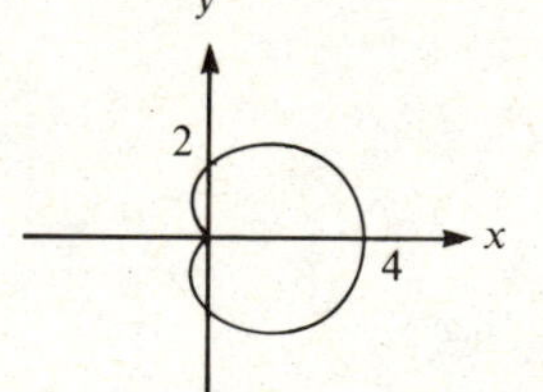

39. $r^2 = \frac{1}{4} \cos 2\theta$; symmetry with respect to both axes and origin

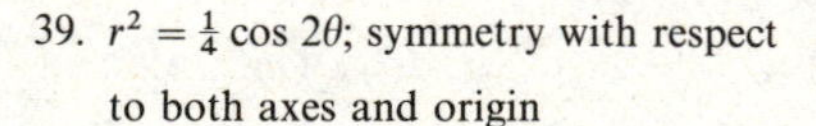

41. $\mathcal{L} = \int_0^\pi \sqrt{\cos^2 \theta + (1 + \sin \theta)^2}\,d\theta = \int_0^\pi \sqrt{2 + 2 \sin \theta}\,d\theta$

$= 2\int_0^\pi \sqrt{\frac{1 + \cos(\pi/2 - \theta)}{2}}\,d\theta = 2\int_0^\pi \cos\left(\frac{\pi/2 - \theta}{2}\right)d\theta$

$= 2\int_0^\pi \cos\left(\frac{\pi}{4} - \frac{\theta}{2}\right)d\theta = -2 \sin\left(\frac{\pi}{4} - \frac{\theta}{2}\right)\Big|_0^\pi$

$= -2 \sin\left(-\frac{\pi}{4}\right) + 2 \sin \frac{\pi}{4} = 2\sqrt{2}$

43. The graphs intersect for (r, θ) such that $1 = r^2 = 2\sin 2\theta$, which means that $2\theta = \pi/6$ or $2\theta = 5\pi/6$, and thus $\theta = \pi/12$ or $5\pi/12$. The area of the portion of the region in the first quadrant is given by

$$A_1 = \int_{\pi/12}^{5\pi/12} \tfrac{1}{2}[(\sqrt{2\sin 2\theta})^2 - 1^2]\,d\theta = \tfrac{1}{2}\int_{\pi/12}^{5\pi/12} (2\sin 2\theta - 1)\,d\theta$$
$$= \frac{1}{2}(-\cos 2\theta - \theta)\Big|_{\pi/12}^{5\pi/12} = \frac{1}{2}\sqrt{3} - \frac{\pi}{6}$$

Thus the total area $A = 2A_1 = \sqrt{3} - \pi/3$.

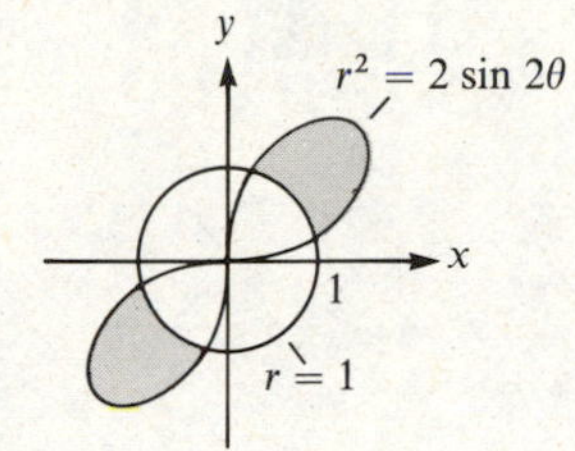

CUMULATIVE REVIEW (CHAPTERS 1–7)

1. The conditions for applying l'Hôpital's Rule three times are met;

$$\lim_{x\to 0} \frac{\sin x - \sin(\sin x)}{x^3} = \lim_{x\to 0} \frac{\cos x - \cos(\sin x)\cos x}{3x^2} = \lim_{x\to 0} \cos x \lim_{x\to 0} \frac{1 - \cos(\sin x)}{3x^2}$$
$$= \lim_{x\to 0} \frac{1 - \cos(\sin x)}{3x^2} = \lim_{x\to 0} \frac{\sin(\sin x)\cos x}{6x}$$
$$= \lim_{x\to 0} \frac{\sin(\sin x)}{6x} \lim_{x\to 0} \cos x = \lim_{x\to 0} \frac{\sin(\sin x)}{6x}$$
$$= \lim_{x\to 0} \frac{\cos(\sin x)\cos x}{6}$$

Let $y = \sin x$, so that y approaches 0 as x approaches 0. Then

$$\lim_{x\to 0} \frac{\cos(\sin x)\cos x}{6} = \lim_{x\to 0} \cos(\sin x) \lim_{x\to 0} \frac{\cos x}{6} = \lim_{y\to 0} \cos y \lim_{x\to 0} \frac{\cos x}{6} = 1 \cdot \frac{1}{6} = \frac{1}{6}$$

Thus the given limit equals $\frac{1}{6}$.

3. $\sqrt{cx+1} - \sqrt{x} = (\sqrt{cx+1} - \sqrt{x})\dfrac{(\sqrt{cx+1} + \sqrt{x})}{\sqrt{cx+1} + \sqrt{x}} = \dfrac{(c-1)x+1}{\sqrt{cx+1} + \sqrt{x}}$

so that if $c = 1$, then

$$\lim_{x\to\infty} (\sqrt{cx+1} - \sqrt{x}) = \lim_{x\to\infty} \frac{1}{\sqrt{x+1} + \sqrt{x}} = 0$$

If $0 \le c < 1$, then

$$\lim_{x\to\infty} (\sqrt{cx+1} - \sqrt{x}) = \lim_{x\to\infty} \frac{(c-1)x+1}{\sqrt{cx+1} + \sqrt{x}} = -\infty$$

and if $c > 1$, then

$$\lim_{x\to\infty} (\sqrt{cx+1} - \sqrt{x}) = \lim_{x\to\infty} \frac{(c-1)x+1}{\sqrt{cx+1} + \sqrt{x}} = \infty$$

Finally, if $c < 0$, then $\sqrt{cx+1}$ is not defined for $x > 1/|c|$, so $\lim_{x\to\infty} (\sqrt{cx+1} - \sqrt{x})$ is meaningless. Thus $\lim_{x\to\infty} (\sqrt{cx+1} - \sqrt{x})$ exists only for $c = 1$.

5. $f'(x) = \dfrac{2e^{2x}(e^x+1) - e^{2x}e^x}{(e^x+1)^2} = \dfrac{e^{3x} + 2e^{2x}}{(e^x+1)^2}$

7. Differentiating the equation $a^2 - a = 2v^2 - 6v$ implicitly, we have

$$2a\frac{da}{dt} - \frac{da}{dt} = 4v\frac{dv}{dt} - 6\frac{dv}{dt}, \quad \text{so that} \quad (2a-1)\frac{da}{dt} = (4v-6)\frac{dv}{dt}$$

At the instant at which $da/dt = 6(dv/dt)$, this becomes

$$(2a-1)6\frac{dv}{dt} = (4v-6)\frac{dv}{dt}$$

Since $dv/dt = a > 0$ by hypothesis, it follows that $(2a-1)6 = 4v - 6$, so that $v = 3a$.

9. $A(x) = \int_0^x (e^t - 1)\,dt = (e^t - t)\big|_0^x = e^x - x - 1$. Thus finding the value of $x > 0$ for which $A(x) = 1$ is equivalent to solving the equation $e^x - x - 1 = 1$, or $e^x - x - 2 = 0$. To that end we let $f(x) = e^x - x - 2$, so that $f'(x) = e^x - 1$, and apply the Newton-Raphson method with initial value of c equal to 1. By computer we obtain 1.15 as the desired approximate zero of f, and hence the approximate value of x for which $A(x) = 1$.

11. $f'(x) = \dfrac{(1)(x-3)^2 - 2x(x-3)}{(x-3)^4} = -\dfrac{x+3}{(x-3)^3}$;

$f''(x) = -\dfrac{(1)(x-3)^3 - 3(x+3)(x-3)^2}{(x-3)^6} = \dfrac{2(x+6)}{(x-3)^4}$;

relative minimum value is $f(-3) = -\frac{1}{12}$; increasing on $[-3, 3)$, and decreasing on $(-\infty, -3]$ and $(3, \infty)$; concave upward on $(-6, 3)$ and $(3, \infty)$, and concave downward on $(-\infty, -6)$; inflection point is $(-6, -\frac{2}{27})$; vertical asymptote is $x = 3$; horizontal asymptote is $y = 0$

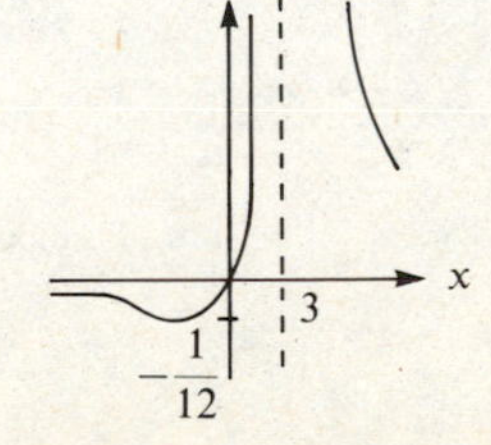

13. Using the notation in the figure, we find that the area A and circumference C are given by $A = \frac{1}{2}xy = \frac{1}{2}\sin\theta\cos\theta$ and $C = 1 + x + y = 1 + \sin\theta + \cos\theta$. Thus the ratio R is given by

$$R = \frac{C}{A} = \frac{1 + \sin\theta + \cos\theta}{\frac{1}{2}\sin\theta\cos\theta} = 2(\csc\theta\sec\theta + \sec\theta + \csc\theta) \quad \text{for } 0 < \theta < \frac{\pi}{2}$$

We are to minimize R. Notice that

$$\begin{aligned} R'(\theta) &= 2(-\csc\theta\cot\theta\sec\theta + \csc\theta\sec\theta\tan\theta + \sec\theta\tan\theta - \csc\theta\cot\theta) \\ &= 2\left(\frac{-1}{\sin^2\theta} + \frac{1}{\cos^2\theta} + \frac{\sin\theta}{\cos^2\theta} - \frac{\cos\theta}{\sin^2\theta}\right) \\ &= \frac{2(-\cos^2\theta + \sin^2\theta + \sin^3\theta - \cos^2\theta)}{\sin^2\theta\cos^2\theta} \\ &= \frac{2(\sin^2\theta + \sin^3\theta - \cos^2\theta - \cos^3\theta)}{\sin^2\theta\cos^2\theta} \end{aligned}$$

If $\theta = \pi/4$, then $R'(\theta) = 0$ since $\sin(\pi/4) = \cos(\pi/4)$. Moreover, since $\sin\theta < \cos\theta$ on $(0, \pi/4)$ and $\sin\theta > \cos\theta$ on $(\pi/4, \pi/2)$, it follows that $R'(\theta) < 0$ for $0 < \theta < \pi/4$ and $R'(\theta) > 0$ for $\pi/4 < \theta < \pi/2$. Consequently the ratio R is minimum for $\theta = \pi/4$, that is, when the right triangle is isosceles.

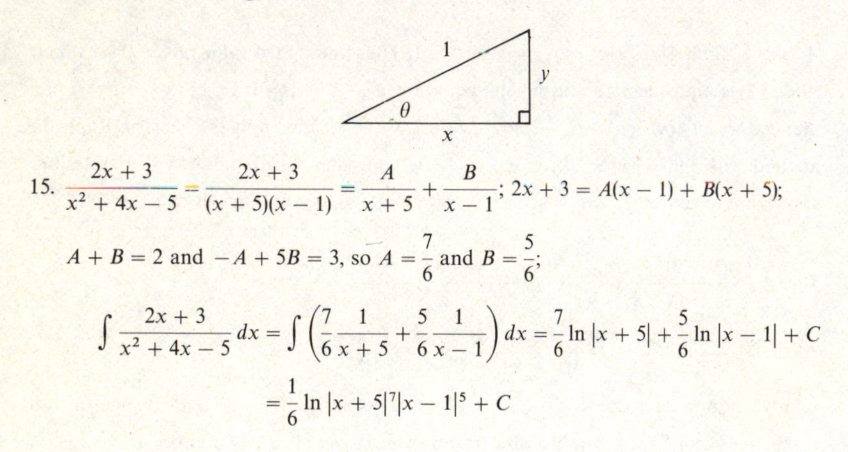

15. $\dfrac{2x+3}{x^2+4x-5} = \dfrac{2x+3}{(x+5)(x-1)} = \dfrac{A}{x+5} + \dfrac{B}{x-1}$; $2x + 3 = A(x-1) + B(x+5)$;

$A + B = 2$ and $-A + 5B = 3$, so $A = \dfrac{7}{6}$ and $B = \dfrac{5}{6}$;

$$\begin{aligned} \int \frac{2x+3}{x^2+4x-5}\,dx &= \int \left(\frac{7}{6}\frac{1}{x+5} + \frac{5}{6}\frac{1}{x-1}\right) dx = \frac{7}{6}\ln|x+5| + \frac{5}{6}\ln|x-1| + C \\ &= \frac{1}{6}\ln|x+5|^7|x-1|^5 + C \end{aligned}$$

17. $\displaystyle\int \frac{1}{\sqrt{-x^2+6x-8}}\,dx = \int \frac{1}{\sqrt{1-(x-3)^2}}\,dx = \arcsin(x-3) + C$

19. $$\begin{aligned} \int_0^{\pi/4} \tan 2x\,dx &= \lim_{b\to\pi/4^-} \int_0^b \tan 2x\,dx \overset{u=2x}{=} \lim_{b\to\pi/4^-} \int_0^{2b} \tan u\,du \\ &= \lim_{b\to\pi/4^-} \left(-\ln|\cos u|\Big|_0^{2b}\right) = \lim_{b\to\pi/4^-} (-\ln|\cos 2b|) = \infty, \end{aligned}$$

so the given integral diverges.

21. For $0 \le x \le \pi$, we have $e^x \ge e^{-x}$ and $\sin x \ge 0$, so that $e^x \sin x \ge e^{-x}\sin x$. For $-\dfrac{\pi}{2} \le x < 0$, we have $e^x < e^{-x}$ and $\sin x < 0$, so that $e^x \sin x \ge e^{-x}\sin x$. Thus $e^x \sin x \ge e^{-x}\sin x$ for $-\dfrac{\pi}{2} \le x \le \pi$, so that $A = \int_{-\pi/2}^{\pi} (e^x \sin x - e^{-x}\sin x)\,dx$. By Exercise 53 of Section 7.1, $A = \left[\dfrac{e^x}{2}(\sin x - \cos x) - \dfrac{e^{-x}}{2}(-\sin x - \cos x)\right]\Big|_{-\pi/2}^{\pi} = \dfrac{1}{2}(e^\pi - e^{-\pi}) - \dfrac{1}{2}(-e^{-\pi/2} - e^{\pi/2}) = \dfrac{1}{2}(e^\pi - e^{-\pi} + e^{-\pi/2} + e^{\pi/2})$.

9 SEQUENCES AND SERIES

SECTION 9.1

1. Since $f(x) = \sin x$, we have $f'(x) = \cos x$, $f''(x) = -\sin x$, and $f^{(3)}(x) = -\cos x$, so that $f(0) = 0$, $f'(0) = 1$, $f''(0) = 0$, and $f^{(3)}(0) = -1$. Therefore $p_0(x) = 0$, $p_1(x) = x$, $p_2(x) = x$ and $p_3(x) = x - x^3/3! = x - x^3/6$.

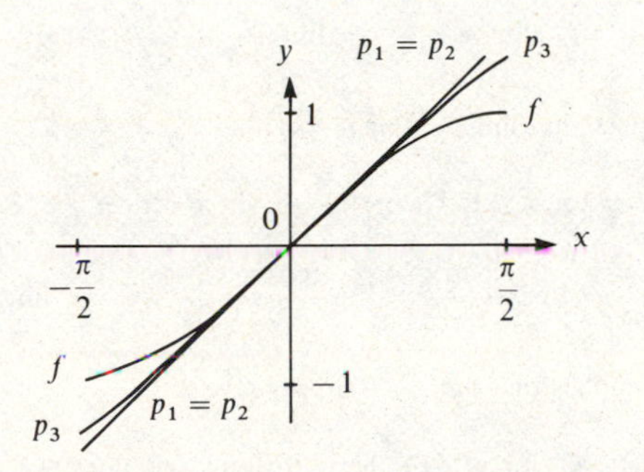

3. $f(x) = x^2 - x - 2$, $f'(x) = 2x - 1$, $f''(x) = 2$, $f^{(k)}(x) = 0$ for $k \geq 3$; $p_0(x) = -2$, $p_1(x) = -2 - x$, $p_n(x) = -2 - x + 2x^2/2 = -2 - x + x^2$ for $n \geq 2$.

5. $f(x) = \dfrac{1}{1+x}$, $f'(x) = \dfrac{-1}{(1+x)^2}$, $f''(x) = \dfrac{2}{(1+x)^3}$, $f^{(3)}(x) = \dfrac{-6}{(1+x)^4}$,

$f^{(k)}(x) = \dfrac{(-1)^k k!}{(1+x)^{k+1}}$; $f^{(k)}(0) = (-1)^k k!$;

$$p_n(x) = 1 - x + \frac{2!}{2!}x^2 - \frac{3!}{3!}x^3 + \cdots + \frac{(-1)^n n!}{n!}x^n = 1 - x + x^2 - x^3 + \cdots + (-1)^n x^n$$

7. $f(x) = e^{-x}$, $f'(x) = -e^{-x}$, $f''(x) = e^{-x}$, $f^{(3)}(x) = -e^{-x}$, $f^{(k)}(x) = (-1)^k e^{-x}$;

$f^{(k)}(0) = (-1)^k$; $p_n(x) = 1 - x + \dfrac{x^2}{2!} - \dfrac{x^3}{3!} + \cdots + \dfrac{(-1)^n}{n!}x^n$

9. $f(x) = \cosh x$, $f'(x) = \sinh x$, $f''(x) = \cosh x$; $f^{(2k)}(x) = \cosh x$, $f^{(2k+1)}(x) = \sinh x$;

$f^{(2k)}(0) = 1$, $f^{(2k+1)}(0) = 0$; $p_{2n+1}(x) = p_{2n}(x) = 1 + \dfrac{x^2}{2!} + \dfrac{x^4}{4!} + \cdots + \dfrac{x^{2n}}{(2n)!}$

11. $f(x) = \sin x$, $f'(x) = \cos x$, $f''(x) = -\sin x$, $f^{(3)}(x) = -\cos x$, $f^{(4)}(x) = \sin x$,

$f^{(2k)}(x) = (-1)^k \sin x$, $f^{(2k+1)}(x) = (-1)^k \cos x$; $f^{(2k)}(0) = 0$, $f^{(2k+1)}(0) = (-1)^k$;

$$p_{2n+2}(x) = p_{2n+1}(x) = x - \frac{x^3}{3!} + \frac{x^5}{5!} - \frac{x^7}{7!} + \cdots + \frac{(-1)^n}{(2n+1)!}x^{2n+1}$$

13. Let $f(x) = e^x$. Then $f^{(n+1)}(x) = e^x$. We want $|r_n(\frac{1}{2})| < 0.001$. Since $0 < t_{1/2} < \frac{1}{2}$, we find by (6) that

$$\left|r_n\left(\frac{1}{2}\right)\right| = \frac{e^{t_{1/2}}}{(n+1)!}\left(\frac{1}{2}\right)^{n+1} < \frac{e^{1/2}}{(n+1)!}\left(\frac{1}{2}\right)^{n+1} < \frac{4^{1/2}}{(n+1)!}\left(\frac{1}{2}\right)^{n+1} = \frac{1}{2^n(n+1)!}$$

Thus $|r_n(\frac{1}{2})| < 0.001$ if $n \geq 4$. From Example 1 we have $p_4(\frac{1}{2}) = 1 + \frac{1}{2} + \frac{1}{2}(\frac{1}{2})^2 + \frac{1}{6}(\frac{1}{2})^3 + \frac{1}{24}(\frac{1}{2})^4 \approx 1.64844$.

15. Let $f(x) = \ln(1+x)$. Then $f^{(n+1)}(x) = (-1)^n n!/(1+x)^{n+1}$. We want $|r_n(.1)| < 0.00001$. Since $0 < t_{.1} < 1$. we find by (6) that

$$|r_n(.1)| = \left|\frac{(-1)^n n!}{(1+t_{.1})^{n+1}(n+1)!}(-1)^{n+1}\right| < \frac{(.1)^{n+1}}{n+1}$$

Thus $|(r_n(.1)| < 0.00001$ if $n \geq 4$. From Example 2 we have $p_4(.1) = (.1) - (.1)^2/2 + (.1)^3/3 - (.1)^4/4 \approx 0.0953083$.

17. Let $f(x) = \cos x$. Then $f^{(2n)}(x) = (-1)^n \cos x$ and $f^{(2n+1)}(x) = (-1)^{n+1} \sin x$. We want $|r_n(-\pi/3)| < 0.001$. Since $|f^{(n)}(t)| \leq 1$ for all t and all n, we find by (6) that

$$\left|r_n\left(\frac{-\pi}{3}\right)\right| = \left|\frac{f^{(n+1)}(t_{-\pi/3})}{(n+1)!}\left(\frac{-\pi}{3}\right)^{n+1}\right| \leq \frac{1}{(n+1)!}\left(\frac{\pi}{3}\right)^{n+1}$$

Thus $|r_n(-\pi/3)| < 0.001$ if $n \geq 6$. By the solution of Exercise 12, we have $p_6(x) = 1 - x^2/2! + x^4/4! - x^6/6!$, so

$$p_6\left(\frac{-\pi}{3}\right) = 1 - \frac{1}{2}\left(\frac{-\pi}{3}\right)^2 + \frac{1}{24}\left(\frac{-\pi}{3}\right)^4 - \frac{1}{720}\left(\frac{-\pi}{3}\right)^6 \approx 0.499965$$

19. Let $f(x) = \ln[(1+x)/(1-x)]$. Then $f(\frac{1}{5}) = \ln\frac{3}{2}$ and

$$f^{(n+1)}(x) = \frac{(-1)^n n!}{(1+x)^{n+1}} + \frac{n!}{(1-x)^{n+1}}$$

We want $|r_n(\frac{1}{5})| < 0.01$. Since $0 < t_{1/5} < \frac{1}{5}$, we find by (6) that

$$\left|r_n\left(\frac{1}{5}\right)\right| = \left|\frac{f^{(n+1)}(t_{1/5})}{(n+1)!}\left(\frac{1}{5}\right)^{n+1}\right| = \left|\frac{(-1)^n n!}{(1+t_{1/5})^{n+1}} + \frac{n!}{(1-t_{1/5})^{n+1}}\right|\frac{1}{(n+1)!5^{n+1}}$$

$$\leq \left(n! + \frac{n!}{(4/5)^{n+1}}\right)\frac{1}{(n+1)!5^{n+1}} = \left(1 + \left(\frac{5}{4}\right)^{n+1}\right)\frac{1}{(n+1)5^{n+1}}$$

Thus $|r_n(\frac{1}{5})| < 0.01$ if $n \geq 2$. By the solution of Exercise 10, we have $p_2(x) = p_1(x) = 2x$, so $p_2(\frac{1}{5}) = p_1(\frac{1}{5}) = 2(\frac{1}{5}) = 0.4$.

21. $f(x) = e^{-(x^2)}$, $f'(x) = -2xe^{-(x^2)}$, $f''(x) = -2e^{-(x^2)} + 4x^2e^{-(x^2)}$,

$f^{(3)}(x) = 12xe^{-(x^2)} - 8x^3e^{-(x^2)}$; $f(0) = 0$, $f'(0) = 0$, $f''(0) = -2$, $f^{(3)}(0) = 0$;

$p_3(x) = 1 - (-2/2!)x^2 = 1 - x^2$

23. $f(x) = \ln(\cos x)$, $f'(x) = (-\sin x)/(\cos x) = -\tan x$, $f''(x) = -\sec^2 x$; $f(0) = 0$, $f'(0) = 0$, $f''(0) = -1$; $p_2(x) = -\frac{1}{2}x^2$

25. $f(x) = \sec x$, $f'(x) = \sec x \tan x$, $f''(x) = \sec x \tan^2 x + \sec^3 x$, $f^{(3)}(x) = \sec x \tan^3 x + 5\sec^3 x \tan x$; $f(0) = 1$, $f'(0) = 0$, $f''(0) = 1$, $f^{(3)}(0) = 0$; $p_3(x) = p_2(x) = 1 + \frac{1}{2}x^2$

27. $f(x) = e^{-1/x^2}$ for $x \neq 0$, $f(0) = 0$; $f'(x) = (2/x^3)e^{-1/x^2}$ for $x \neq 0$; by l'Hôpital's Rule,

$$f'(0) = \lim_{x\to 0} \frac{f(x) - f(0)}{x - 0} = \lim_{x\to 0} \frac{e^{-1/x^2} - 0}{x - 0} = \lim_{x\to 0} \frac{1/x}{e^{1/x^2}}$$

$$= \lim_{x\to 0} \frac{-1/x^2}{(-2/x^3)e^{1/x^2}} = \lim_{x\to 0} \frac{x}{2} e^{-1/x^2} = 0$$

by l'Hôpital's Rule,

$$f''(0) = \lim_{x\to 0} \frac{f'(x) - f'(0)}{x - 0} \lim_{x\to 0} \frac{(2/x^3)e^{-1/x^2} - 0}{x - 0} = \lim_{x\to 0} \frac{2/x^4}{e^{1/x^2}} = \lim_{x\to 0} \frac{-8/x^5}{(-2/x^3)e^{1/x^2}}$$

$$= \lim_{x\to 0} \frac{4/x^2}{e^{1/x^2}} \lim_{x\to 0} \frac{-8/x^3}{(-2/x^3)e^{1/x^2}} = \lim_{x\to 0} 4e^{-1/x^2} = 0$$

$p_2(x) = 0$.

29. a. $f(x) = \sqrt{1 + x}$, $f'(x) = \dfrac{1}{2(1 + x)^{1/2}}$, $f''(x) = -\dfrac{1}{4(1 + x)^{3/2}}$;

$f(0) = 1$, $f'(0) = \frac{1}{2}$, $f''(0) = -\frac{1}{4}$; $p_2(x) = 1 + \frac{1}{2}x - \frac{1}{8}x^2$

b. $f(1) = \sqrt{2}$, and $p_2(1) = 1 + \frac{1}{2} - \frac{1}{8} = 1.375$ is the desired approximation to $\sqrt{2}$, whose value is 1.41421 (accurate to 6 digits).

c. $f(0.1) = \sqrt{1.1}$, and $p_2(0.1) = 1 + \frac{1}{2}(0.1) - \frac{1}{8}(0.1)^2 = 1.04875$ is the desired approximation to $\sqrt{1.1}$, whose value is 1.04881 (accurate to 6 digits).

SECTION 9.2

1. $\frac{1}{3}, \frac{1}{4}, \frac{1}{5}, \frac{1}{6}$

3. $0, \frac{1}{3}, \frac{1}{2}, \frac{3}{5}$

5. Let $\varepsilon > 0$, and let N be any integer. If $n \geq N$, then $|(-2) - (-2)| = 0 < \varepsilon$. Thus $\lim_{n\to\infty} (-2) = -2$.

7. Let $\varepsilon > 0$, and let N be any integer greater than $1/\varepsilon$. If $n \geq N$, then $1/n \leq 1/N$, so that

$$\left|\frac{3n + 1}{n} - 3\right| = \left|3 + \frac{1}{n} - 3\right| = \frac{1}{n} \leq \frac{1}{N} < \varepsilon$$

Thus $\lim_{n\to\infty} (3n + 1)/n = 3$.

9. Let M be any number, and let N be any positive integer such that $N > M^2$. If $n \geq N$, then $\sqrt{n} \geq \sqrt{N} > \sqrt{M^2} = |M| \geq M$. Thus $\lim_{n\to\infty} \sqrt{n} = \infty$.

11. Let M be any number, and let N be any positive integer such that $N > \ln|M|$. If $k \geq N$, then $e^k \geq e^N > e^{\ln|M|} = |M| \geq M$, so that $\lim_{k\to\infty} e^k = \infty$.

13. Let $f(x) = \pi + 1/x$ for $x \geq 1$. Then $f(n) = \pi + 1/n$ for $n \geq 1$. Since $\lim_{x\to\infty} (\pi + 1/x) = \pi$, Theorem 9.5 implies that $\lim_{n\to\infty} (\pi + 1/n) = \pi$.

15. Since $0 < 0.8 < 1$, it follows from (6) with $r = 0.8$ that $\lim_{j\to\infty} (0.8)^j = 0$.

17. Since $e^{-n} = (1/e)^n$ and $0 < 1/e < 1$, it follows from (6) with $r = 1/e$ that $\lim_{n\to\infty} e^{-n} = 0$.

19. Let $f(x) = (x + 3)/(x^2 - 2)$ for $x \geq 2$. Then $f(n) = (n + 3)/(n^2 - 2)$ for $n \geq 2$. Since

$$\lim_{x\to\infty} \frac{x + 3}{x^2 - 2} = \lim_{x\to\infty} \frac{1/x + 3/x^2}{1 - 2/x^2} = \frac{0 + 0}{1 - 0} = 0$$

Theorem 9.5 implies that $\lim_{n\to\infty} (n + 3)/(n^2 - 2) = 0$.

21. Let $f(x) = (2x^2 - 4)/(-x - 5)$ for $x \geq 1$. Then $f(n) = (2n^2 - 4)/(-n - 5)$ for $n \geq 1$. Since

$$\lim_{x\to\infty} \frac{2x^2 - 4}{-x - 5} = \lim_{x\to\infty} \frac{2x - 4/x}{-1 - 5/x} = -\infty$$

Theorem 9.5 implies that $\lim_{n\to\infty} (2n^2 - 4)/(-n - 5) = -\infty$.

23. Let $f(x) = x \sin \pi/x$ for $x \geq 1$. Then $f(n) = n \sin \pi/n$ for $n \geq 1$. By l'Hôpital's Rule,

$$\lim_{x\to\infty} x \sin\frac{\pi}{x} = \lim_{x\to\infty} \frac{\sin \pi/x}{1/x} = \lim_{x\to\infty} \frac{(\cos \pi/x)(-\pi/x^2)}{-1/x^2} = \lim_{x\to\infty} \left(\pi \cos\frac{\pi}{x}\right) = \pi$$

so Theorem 9.5 implies that $\lim_{n\to\infty} n \sin \pi/n = \pi$.

25. Let $f(x) = (1 + 0.05/x)^x$ for $x \geq 1$. Then $f(n) = (1 + 0.05/n)^n$ for $n \geq 1$. Since $\ln f(x) = x \ln(1 + 0.05/x)$ for $x \geq 1$ and $\lim_{x\to\infty} \ln(1 + 0.05/x) = 0 = \lim_{x\to\infty} 1/x$, l'Hôpital's Rule implies that

$$\lim_{x\to\infty} \ln f(x) = \lim_{x\to\infty} x \ln\left(1 + \frac{0.05}{x}\right) = \lim_{x\to\infty} \frac{\ln(1 + 0.05/x)}{1/x} = \lim_{x\to\infty} \frac{\dfrac{1}{1 + 0.05/x}(-0.05/x^2)}{-1/x^2}$$

$$= \lim_{x\to\infty} \frac{0.05}{1 + 0.05/x} = 0.05$$

Therefore $\lim_{x\to\infty} f(x) = e^{0.05}$, so that by Theorem 9.5, $\lim_{n\to\infty} (1 + 0.05/n)^n = e^{0.05}$.

27. Let $f(x) = (1 + x)^{1/(2x)}$ for $x \geq 1$. Then $f(k) = (1 + k)^{1/(2k)}$ for $k \geq 1$. Since $\ln f(x) = [1/(2x)] \ln(1 + x)$ for $x \geq 1$, and $\lim_{x\to\infty} \ln(1 + x) = \infty = \lim_{x\to\infty} 2x$, l'Hôpital's Rule implies that

$$\lim_{x\to\infty} \ln f(x) = \lim_{x\to\infty} \frac{\ln(1 + x)}{2x} = \lim_{x\to\infty} \frac{1/(1 + x)}{2} = \lim_{x\to\infty} \frac{1}{2(1 + x)} = 0$$

Therefore $\lim_{x\to\infty} f(x) = e^0 = 1$, so that by Theorem 9.5, $\lim_{k\to\infty} (1 + k)^{1/(2k)} = 1$.

29. Let $f(x) = (1/\sqrt{2}) \cos 1/x$ for $x = 1$. Then $f(k) = (1/\sqrt{2}) \cos 1/k$ for $k \geq 1$. Since the cosine function is continuous at 0, we have $\lim_{x\to\infty} (1/\sqrt{2}) \cos 1/x = (1/\sqrt{2}) \cos 0 = 1/\sqrt{2}$. Next, let $g(x) = \arcsin x$. Since the arcsine function is continuous on $[-1, 1]$, it follows from the Substitution Theorem with $y = (1/\sqrt{2}) \cos 1/x$ that $\lim_{x\to\infty} g(f(x)) = \lim_{x\to\infty} \arcsin [(1/\sqrt{2}) \cos 1/x] = \lim_{y\to 1/\sqrt{2}} \arcsin y = \arcsin 1/\sqrt{2} = \pi/4$. Therefore by Theorem 9.5, $\lim_{k\to\infty} \arcsin [(1/\sqrt{2}) \cos 1/k] = \pi/4$.

31. Observe that $\int_{-1/n}^{1/n} e^x\,dx = e^x \Big|_{-1/n}^{1/n} = e^{1/n} - e^{-1/n}$. Since $\lim_{n\to\infty} e^{1/n} = 1$ by Exercise 18, we find that

$$\lim_{n\to\infty} \int_{-1/n}^{1/n} e^x\,dx = \lim_{n\to\infty} (e^{1/n} - e^{-1/n}) = \lim_{n\to\infty} \left(e^{1/n} - \frac{1}{e^{1/n}}\right) = 1 - \frac{1}{1} = 0$$

33. Since $\lim_{n\to\infty} (-4n) = -\infty$, the sequence diverges.

35. Since $\lim_{n\to\infty} 1/(n^2 - 1) = 0$, the sequence converges to 0.

37. Since $\lim_{n\to\infty} (-\frac{1}{3})^n = 0$ by Example 6, the sequence converges to 0.

39. For $n \geq 1$ let $\mathscr{P}_n = \{0, 1/n, 2/n, \ldots, (n-1)/n, 1\}$ and for $1 \leq k \leq n$ let $x_k = k/n$. Then $\mathscr{P}_n$ is a partition of $[0, 1]$, $\Delta x_k = 1/n$, $\|\mathscr{P}_n\| = 1/n$, and x_k is in the kth subinterval $[(k-1)/n, k/n]$. The corresponding Riemann sum for $\int_0^1 x\,dx$ is

$$\frac{1}{n}\Delta x_1 + \frac{2}{n}\Delta x_2 + \frac{3}{n}\Delta x_3 + \cdots + \frac{n-1}{n}\Delta x_{n-1} + \frac{n}{n}\Delta x_n = \frac{1}{n^2} + \frac{2}{n^2} + \frac{3}{n^2} + \cdots + \frac{n}{n^2} = a_n$$

Thus $\lim_{n\to\infty} a_n = \int_0^1 x\,dx = \frac{1}{2}x^2\Big|_0^1 = \frac{1}{2}$.

41. Suppose $\lim_{n\to\infty} a_n = L_1$ and $\lim_{n\to\infty} a_n = L_2$. Then for any $\varepsilon > 0$ there are integers N_1 and N_2 such that if $n \geq N_1$, then $|a_n - L_1| < \varepsilon/2$, and if $n \geq N_2$, then $|a_n - L_2| < \varepsilon/2$. Let n be an integer greater than both N_1 and N_2. Then $|L_2 - L_1| = |(a_n - L_1) - (a_n - L_2)| \leq |a_n - L_1| + |a_n - L_2| < \varepsilon/2 + \varepsilon/2 = \varepsilon$. Since $|L_2 - L_1| < \varepsilon$ for any $\varepsilon > 0$, we have $L_2 = L_1$. Thus the limit of a convergent sequence is unique.

43. By hypothesis, $M > 0$ and $r > 0$. Suppose that $p_n > M$. Then $M - p_n < 0$ and hence $\dfrac{r(M - p_n)}{M} p_n < 0$. Therefore (10) implies that $p_{n+1} = p_n + \dfrac{r(M - p_n)}{M} p_n < p_n$.

45. Let p_0, p_1, and p_2 denote the population in 1900, 1910, and 1920, respectively. Using (8) and the values of p_0 and p_1 given in Example 11, we have

$$r = \frac{p_1}{p_0} \approx \frac{92.2 \times 10^6}{76.2 \times 10^6} \approx 1.21$$

Using (8) again, we find that $p_2 = rp_1 \approx (1.21)(92.2 \times 10^6) \approx 112 \times 10^6$, which means that the population in 1920 would have been approximately 112 million. This is not as close to the actual 1920 census figure of 106,021,537 as the number obtained in the solution of Example 11. Thus the prediction is better when we use a Verhulst sequence.

47. In general, if the amount of money in the account at the beginning of a year is P, then the interest earned during the year is $(.01)rP$, so that the amount of money in the account at the end of the year is $P + (.01)rP = P(1 + (.01)r)$. Let a_n be the amount of money in the account at the end of the nth year. Then $a_1 = 1000(1 + (.01)r)$ and $a_n = a_{n-1}(1 + (.01)r)$ for $n > 1$. By induction it follows that $a_n = 1000(1 + (.01)r)^n$.

49. a. If the interest is compounded n times a year, then each interest period lasts $1/n$ years and the interest rate for each interest period is r/n percent. Thus the amount of money in the account at the end of $1/n$ years is $P(1 + (.01)r/n)$, the amount after $2/n$ years is $P(1 + (.01)r/n)(1 + (.01)r/n) = P(1 + (0.1)r/n)^2$, the amount after $3/n$ years is $P(1 + (.01)r/n)^2(1 + (.01)r/n) = P(1 + (.01)r/n)^3$, and by induction the amount at the end of one year (after n interest periods) is $P(1 + (.01)r/n)^n$.

b. By Theorem 9.5,

$$\lim_{n\to\infty} P\left(1 + \frac{(.01)r}{n}\right)^n = \lim_{x\to\infty} P\left(1 + \frac{(.01)r}{x}\right)^x = P \lim_{x\to\infty} \left(1 + \frac{(.01)r}{x}\right)^x$$

Let $f(x) = (1 + (.01)r/x)^x$. Then $\ln f(x) = x \ln (1 + (.01)r/x)$. By l'Hôpital's Rule,

$$\lim_{x\to\infty} \ln f(x) = \lim_{x\to\infty} \frac{\ln (1 + (.01)r/x)}{1/x} = \lim_{x\to\infty} \frac{\left(\dfrac{1}{1 + (.01)r/x}\right)\left(\dfrac{-(.01)r}{x^2}\right)}{-1/x^2}$$

$$= \lim_{x\to\infty} \frac{(.01)r}{1 + (.01)r/x} = (.01)r$$

Thus $\lim_{x\to\infty} f(x) = e^{(.01)r}$, so that $\lim_{n\to\infty} P(1 + (.01)r/n)^n = Pe^{(.01)r}$.

c. If interest is compounded continuously, then by part (b) the amount after 1 year is $1000e^{.05} \approx 1051.27$. If interest is compounded quarterly, then by part (a) the amount after 1 year is $1000(1 + (.01)5/4)^4 = 1000(1.0125)^4 \approx 1050.95$. Thus the difference is approximately 32 cents.

SECTION 9.3

1. Since $1 \leq 1 + 2/n \leq 3$ for $n \geq 1$, the sequence is bounded.

3. Since $n + 1/(2n) \geq n$ for $n \geq 1$, the sequence is unbounded.

5. Since $\cosh k = \frac{1}{2}(e^k + e^{-k}) \geq \frac{1}{2}e^k$ for $k \geq 10$, and since $\lim_{x\to\infty} \frac{1}{2}e^x = \infty$, the sequence is unbounded.

7. By (1), $\lim_{n\to\infty} (2 + 1/n) = \lim_{n\to\infty} 2 + \lim_{n\to\infty} 1/n = 2 + 0 = 2$.

9. By (1) and (2),

$$\lim_{n\to\infty} \left(\frac{1}{n} - \frac{1}{n+1}\right) = \lim_{n\to\infty} \frac{1}{n} + \lim_{n\to\infty} \frac{-1}{n+1} = \lim_{n\to\infty} \frac{1}{n} - \lim_{n\to\infty} \frac{1}{n+1} = 0 - 0 = 0$$

11. By (2) and Example 6 of Section 9.2,

$$\lim_{n\to\infty}\frac{2^{n+1}}{5^{n+2}}=\lim_{n\to\infty}\frac{2\cdot 2^n}{5^2 5^n}=\lim_{n\to\infty}\frac{2}{25}\left(\frac{2}{5}\right)^n=\frac{2}{25}\lim_{n\to\infty}\left(\frac{2}{5}\right)^n=\frac{2}{25}\cdot 0=0$$

13. By (1), $\lim_{k\to\infty}(k+1)/k=\lim_{k\to\infty}(1+1/k)=\lim_{k\to\infty}1+\lim_{k\to\infty}1/k=1+0=1$.

15. By (1) – (4), $$\lim_{n\to\infty}\frac{4n^3-5}{6n^3+3}=\lim_{n\to\infty}\frac{4-5/n^3}{6+3/n^3}=\frac{4-5(0)^3}{6+3(0)^3}=\frac{2}{3}$$

17. By Examples 8 and 9 of Section 9.2, $\lim_{n\to\infty}\sqrt[n]{3}=1$ and $\lim_{n\to\infty}\sqrt[n]{n}=1$. Thus by (3) we have $\lim_{n\to\infty}\sqrt[n]{3n}=\lim_{n\to\infty}\sqrt[n]{3}\lim_{n\to\infty}\sqrt[n]{n}\,1\cdot 1=1$

19. By (4), $$\lim_{n\to\infty}\frac{\sqrt{n}+1}{\sqrt{n}-1}=\lim_{n\to\infty}\frac{1+1/\sqrt{n}}{1-1/\sqrt{n}}=\frac{1+0}{1-0}=1$$

21. By (1) and (4),

$$\lim_{n\to\infty}\sqrt{n}(\sqrt{n+1}-\sqrt{n})=\lim_{n\to\infty}\sqrt{n}(\sqrt{n+1}-\sqrt{n})\frac{\sqrt{n+1}+\sqrt{n}}{\sqrt{n+1}+\sqrt{n}}=\lim_{n\to\infty}\frac{\sqrt{n}[(n+1)-n]}{\sqrt{n+1}+\sqrt{n}}$$

$$=\lim_{n\to\infty}\frac{\sqrt{n}}{\sqrt{n+1}+\sqrt{n}}=\lim_{n\to\infty}\frac{1}{\sqrt{1+1/n}+1}=\frac{1}{1+1}=\frac{1}{2}$$

23. Observe that $-1/n\le(-1)^n/n\le 1/n$ for $n\ge 1$. Since $\lim_{n\to\infty}-1/n=0=\lim_{n\to\infty}1/n$, it follows from (5) that $\lim_{n\to\infty}(-1)^n/n=0$.

25. Since $0\le 1/ne^n\le 1/n$ for $n\ge 1$, and since $\lim_{n\to\infty}0=0=\lim_{n\to\infty}1/n$, it follows from (5) that $\lim_{n\to\infty}1/ne^n=0$.

27. Since $0\le[\ln(1+1/j)]/j\le(\ln 2)/j$ for $j\ge 1$, and since $\lim_{j\to\infty}(\ln 2)/j=0=\lim_{j\to\infty}0$, it follows from (5) that $\lim_{j\to\infty}[\ln(1+1/j)]/j=0$.

29. Since $0<e^{-2n}<e^{-n}$ for $n\ge 1$, and since $\lim_{n\to\infty}e^{-n}=0=\lim_{n\to\infty}0$ by Exercise 17 of Section 9.2, it follows from (5) that $\lim_{n\to\infty}e^{-2n}=0$. It follows from (1) and (4) that

$$\lim_{n\to\infty}\tanh n=\lim_{n\to\infty}\frac{1-e^{-2n}}{1+e^{-2n}}=\frac{1-0}{1+0}=1$$

31. By Figure 9.9, the iterates of 0 for $f_{-1/4}$ converge. By Example 6 and the comment following it, the iterates must converge to one of the two fixed points of $f_{-1/4}$, which are $\frac{1}{2}(1+\sqrt{1-4(-1/4)})=\frac{1}{2}(1+\sqrt{2})$ and $\frac{1}{2}(1-\sqrt{1-4(-1/4)})=\frac{1}{2}(1-\sqrt{2})$. The iterates converge to $\frac{1}{2}(1-\sqrt{2})$, as the accompanying figure suggests.

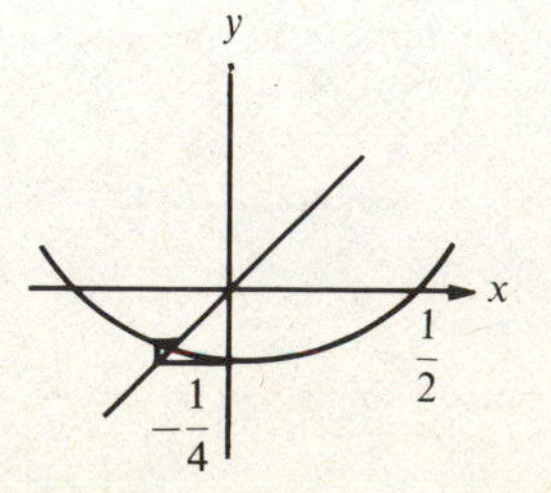

33. By Figure 9.9, the iterates of 0 for $f_{1/2}$ are unbounded, and hence do not converge. The accompanying figure supports this claim.

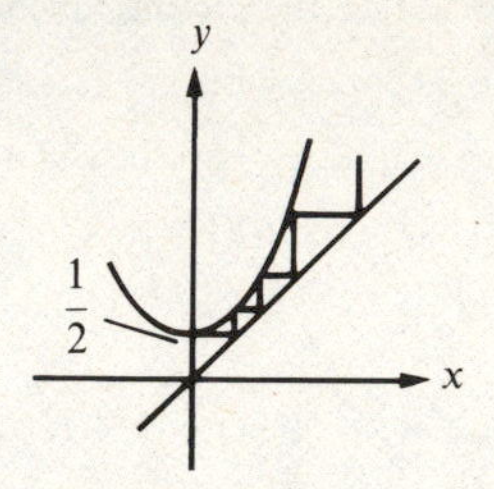

35. The sequence begins −0.5, −0.25, −0.4375, −0.308594, −0.40477, −0.336161, −0.386996, −0.350234, and converges to the number −0.366025 . . . . The numbers in the sequence are alternately greater than and less than the limit.

37. a. b.

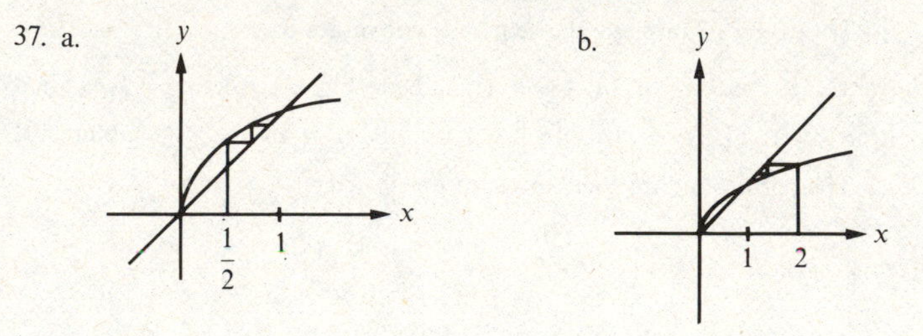

39. a. 29

b. Let $f(x)=\cos x-x$, so that $f'(x)=-\sin x-1$. Using the Newton–Raphson method with initial value 0 and using a calculator, we find that the fixed point is approximately 0.7390851.

41. We will prove the assertion by induction. In the first place, $a_1=0^2+c=c>0=a_0$. If $a_{n+1}\ge a_n$ for a given positive integer n, then $a_{n+2}=a_{n+1}^2+c\ge a_n^2+c=a_{n+1}$. It follows by induction that $a_{n+1}\ge a_n$ for all $n\ge 0$. Therefore if $c>0$, then $\{a_n\}_{n=0}^{\infty}$ is increasing.

43. If $\lim_{n\to\infty}(a_n+b_n)$ existed, then since $b_n=(a_n+b_n)-a_n$, (1) and (2) would imply that $\lim_{n\to\infty}b_n$ would exist, which contradicts the hypothesis. Thus $\{a_n+b_n\}_{n=1}^{\infty}$ diverges.

45. a. First, $a_1=\sqrt{2}<2$. Next, if $a_n<2$, then $a_{n+1}=(\sqrt{2})^{a_n}<(\sqrt{2})^2=2$, so by induction, $a_n<2$ for $n\ge 1$. Therefore $\{a_n\}_{n=1}^{\infty}$ is bounded. Next, we will show that $\{a_n\}_{n=1}^{\infty}$ is increasing. Since $a_n<2$, it follows from Exercise 65(a) of Section 5.7 that $(\ln a_n)/a_n<(\ln 2)/2=\ln\sqrt{2}$, so that $\ln a_n<a_n\ln\sqrt{2}=\ln(\sqrt{2}^{a_n})=\ln a_{n+1}$. Since $\ln x$ is strictly increasing, we conclude that $a_n<a_{n+1}$. By Theorem 9.7 the bounded, increasing sequence $\{a_n\}_{n=1}^{\infty}$ converges to a number L, and $L\le 2$ since $a_n<2$ for $n\ge 1$.

b. By part (a), $\lim_{n\to\infty} a_n = L$, so that $\lim_{n\to\infty} a_{n+1} = L$. Thus $\ln L = \lim_{n\to\infty} \ln a_{n+1} = \lim_{n\to\infty} \ln(\sqrt{2})^{a_n} = \lim_{n\to\infty} (a_n \ln\sqrt{2}) = (\ln\sqrt{2}) \lim_{n\to\infty} a_n = (\ln\sqrt{2})L$. Therefore $(\ln L)/L = \ln\sqrt{2} = (\ln 2)/2$. Since $0 < L < e$ and $(\ln x)/x$ is strictly increasing on $(0, e)$ by Exercise 65(a) in Section 5.7, it follows that $L = 2$.

47. Let $b_n = a_{n+1}/a_n$ and $b = \lim_{n\to\infty} b_n = \lim_{n\to\infty} b_{n-1}$. Since $a_{n+1} = a_n + a_{n-1}$ for $n \geq 2$, we have $a_{n+1}/a_n = 1 + a_{n-1}/a_n$, so that $b_n = 1 + 1/b_{n-1}$. Thus $b = \lim_{n\to\infty} b_n = \lim_{n\to\infty} (1 + 1/b_{n-1}) = 1 + 1/b$, or $b = 1 + 1/b$, or $b^2 - b - 1 = 0$. Since $b \geq 0$, it follows that $b = (1 + \sqrt{1+4})/2 = (1+\sqrt{5})/2$.

SECTION 9.4

1. $s_4 = 1 + 1 + 1 + 1 = 4$

3. $s_4 = 1 + \frac{1}{3} + \frac{1}{9} + \frac{1}{27} = \frac{40}{27}$

5. $s_4 = \frac{1}{2} - \frac{1}{3} + \frac{1}{4} - \frac{1}{5} = \frac{13}{60}$

7. $\lim\limits_{n\to\infty} \left(1 + \frac{1}{n}\right) = 1$; series diverges.

9. $\lim\limits_{n\to\infty} (-1)^n \frac{1}{n^2} = 0$; series could converge.

11. $\lim\limits_{n\to\infty} \sin\left(\frac{\pi}{2} - \frac{1}{n}\right) = \sin\frac{\pi}{2} = 1$; series diverges.

13. $\lim\limits_{n\to\infty} n \sin\frac{1}{n} = \lim\limits_{n\to\infty} \frac{\sin 1/n}{1/n} = 1$; series diverges.

15. $s_j = j$; $\lim\limits_{j\to\infty} s_j = \lim\limits_{j\to\infty} j = \infty$; series diverges.

17. $s_{2j} = 0$ and $s_{2j+1} = 1$ for $j \geq 1$; $\lim_{j\to\infty} s_j$ does not exist; series diverges.

19. $s_j = \left(\frac{1}{2} - \frac{1}{3}\right) + \left(\frac{1}{3} - \frac{1}{4}\right) + \cdots + \left(\frac{1}{j+1} - \frac{1}{j+2}\right) = \frac{1}{2} - \frac{1}{j+2}$;

$\lim\limits_{j\to\infty} s_j = \lim\limits_{j\to\infty} \left(\frac{1}{2} - \frac{1}{j+2}\right) = \frac{1}{2}$; $\sum\limits_{n=1}^{\infty} \left(\frac{1}{n+1} - \frac{1}{n+2}\right) = \frac{1}{2}$

21. $s_j = (1 - 8) + (8 - 27) + \cdots + (j^3 - (j+1)^3) = 1 - (j+1)^3$;

$\lim\limits_{j\to\infty} s_j = \lim\limits_{j\to\infty} (1 - (j+1)^3) = -\infty$; series diverges.

23. $s_n = (a_1 - a_2) + (a_2 - a_3) + \cdots + (a_{n-1} - a_n) + (a_n - a_{n+1}) = a_1 - a_{n+1}$. Thus $\lim_{n\to\infty} s_n$ exists if and only if $\lim_{n\to\infty} a_{n+1}$, or equivalently, $\lim_{n\to\infty} a_n$, exists. If $\lim_{n\to\infty} a_n$ exists, then

$$\lim_{n\to\infty} s_n = \lim_{n\to\infty} (a_1 - a_{n+1}) = a_1 - \lim_{n\to\infty} a_{n+1} = a_1 - \lim_{n\to\infty} a_n$$

25. Since the series is a geometric series with ratio $\frac{4}{7} < 1$, it converges;

$$\sum_{n=1}^{\infty} 5\left(\frac{4}{7}\right)^n = \frac{5(\frac{4}{7})}{1 - \frac{4}{7}} = \frac{20}{3}$$

27. Since the series is a geometric series with ratio $-.3$ and $|-.3| < 1$, it converges;

$$\sum_{n=0}^{\infty} (-1)^n(.3) = \frac{1}{1 - (-.3)} = \frac{10}{13}$$

29. Since the series is a geometric series with ratio $\frac{1}{2} < 1$, it converges;

$$\sum_{n=1}^{\infty} 5\left(\frac{1}{2}\right)^{n+1} = \sum_{n=1}^{\infty} \frac{5}{2}\left(\frac{1}{2}\right)^n = \frac{\frac{5}{2}(\frac{1}{2})}{1 - \frac{1}{2}} = \frac{5}{2}$$

31. Since the series is a geometric series with ratio $\frac{3}{5} < 1$, it converges;

$$\sum_{n=1}^{\infty} \frac{3^{n+3}}{5^{n-1}} = \sum_{n=1}^{\infty} (27)(5)\left(\frac{3}{5}\right)^n = \frac{135(\frac{3}{5})}{1 - \frac{3}{5}} = \frac{405}{2}$$

33. Since the series is a geometric series with ratio $-3/2$ and $|-3/2| > 1$, it diverges.

35. $\frac{1}{7} + \frac{1}{7^2} + \frac{1}{7^3} + \cdots = \frac{1}{7} + \frac{1}{7^2} + \frac{1}{7^3} + \sum\limits_{n=4}^{\infty} \left(\frac{1}{7}\right)^n = \frac{57}{343} + \sum\limits_{n=4}^{\infty} \left(\frac{1}{7}\right)^n$

37. $\sum\limits_{n=1}^{\infty} \frac{1}{n^2+1} = \frac{1}{2} + \frac{1}{5} + \frac{1}{10} + \sum\limits_{n=4}^{\infty} \frac{1}{n^2+1} = \frac{4}{5} + \sum\limits_{n=4}^{\infty} \frac{1}{n^2+1}$

39. $\sum\limits_{n=1}^{\infty} \frac{1}{n^2} = 1 + \frac{1}{4} + \frac{1}{9} + \sum\limits_{n=4}^{\infty} \frac{1}{n^2} = \frac{49}{36} + \sum\limits_{n=4}^{\infty} \frac{1}{n^2}$

41. $0.72727272\ldots = 72\left(\frac{1}{100}\right) + 72\left(\frac{1}{100}\right)^2 + 72\left(\frac{1}{100}\right)^3 + \cdots = \sum\limits_{n=1}^{\infty} 72\left(\frac{1}{100}\right)^n$

$$= \frac{72(\frac{1}{100})}{1 - \frac{1}{100}} = \frac{8}{11}$$

43. $0.232232232\ldots = 232\left(\frac{1}{1000}\right) + 232\left(\frac{1}{1000}\right)^2 + 232\left(\frac{1}{1000}\right)^3 + \cdots$

$$= \sum_{n=1}^{\infty} 232\left(\frac{1}{1000}\right)^n = \frac{232(\frac{1}{1000})}{1 - \frac{1}{1000}} = \frac{232}{999}$$

45. $27.56123123123\ldots = 27.56 + 123\left(\frac{1}{10}\right)^5 + 123\left(\frac{1}{10}\right)^8 + 123\left(\frac{1}{10}\right)^{11} + \cdots$

$$= \frac{2756}{100} + \sum_{n=1}^{\infty} \frac{123}{100}\left(\frac{1}{1000}\right)^n = \frac{2756}{100} + \frac{\frac{123}{100}(\frac{1}{1000})}{1 - \frac{1}{1000}}$$

$$= \frac{2756}{100} + \frac{123}{99900} = \frac{917789}{33300}$$

47. $0.86400000 = \frac{864}{1000} = \frac{108}{125}$

49. $\sum\limits_{n=4}^{\infty} (-1)^{n+1} \frac{1}{n} = \sum\limits_{n=1}^{\infty} (-1)^{n+1} \frac{1}{n} - \left(1 - \frac{1}{2} + \frac{1}{3}\right) = \ln 2 - \frac{5}{6}$

51. If $-1 < r < 1$, then $0 \leq r^2 < 1$, so that by Exercise 50 (with r replaced by r^2), $\sum_{n=0}^{\infty} (-1)^n r^{2n} = 1/(1 + r^2)$.

53. If $r > 0$, then either $r > 1$ or $1/r \geq 1$, so that either $\sum_{n=0}^{\infty} (1/r)^n$ or $\sum_{n=1}^{\infty} r^n$ diverges.

55. a. If $q \geq 2$, then

$$0 < q! \sum_{n=q+1}^{\infty} \frac{1}{n!} = q!\left[\frac{1}{(q+1)!} + \frac{1}{(q+1)!(q+2)} + \frac{1}{(q+1)!(q+2)(q+3)} + \cdots\right]$$

$$= \frac{q!}{(q+1)!}\left[1 + \frac{1}{q+2} + \frac{1}{(q+2)(q+3)} + \cdots\right]$$

$$= \frac{1}{q+1}\left[1 + \frac{1}{q+2} + \frac{1}{(q+2)(q+3)} + \cdots\right]$$

$$< \frac{1}{3}\left(1 + \frac{1}{4} + \frac{1}{4^2} + \frac{1}{4^3} + \cdots\right)$$

$$= \frac{1}{3}\frac{1}{1-\frac{1}{4}} = \frac{4}{9} < 1$$

b. Suppose e is rational. Then $e = p/q$ for some positive integers p and q with $q \geq 2$. Then

$$\frac{p}{q} = \sum_{n=0}^{\infty} \frac{1}{n!} = \sum_{n=0}^{q} \frac{1}{n!} + \sum_{n=q+1}^{\infty} \frac{1}{n!}, \quad \text{or} \quad p = q\sum_{n=0}^{q} \frac{1}{n!} + q\sum_{n=q+1}^{\infty} \frac{1}{n!}$$

Then $p(q-1)! = q!\sum_{n=0}^{q} 1/n! + q!\sum_{n=q+1}^{\infty} 1/n!$. By hypothesis $p(q-1)!$ is an integer, and $q!/n!$ is an integer for every n satisfying $0 \leq n \leq q$, so that $q!\sum_{n=0}^{q} 1/n!$ is an integer. This implies that $q!\sum_{n=q+1}^{\infty} 1/n!$ must be an integer, contradicting the result of part (a). Consequently e is irrational.

57. a. 11 b. 31

59. During the first month the person earns 1500 and spends $p1500$, so that $w_1 = 1500 - p1500 = 1500(1-p)$. If $n \geq 2$, then during the nth month the person earns 1500 and spends $p(1500 + w_{n-1})$. Thus

$$w_n = 1500 + w_{n-1} - p(1500 + w_{n-1}) = (1500 + w_{n-1})(1-p)$$

If

$$w_{n-1} = 1500(1-p) + 1500(1-p)^2 + \cdots + 1500(1-p)^{n-1},$$

then

$$w_n = (1500 + w_{n-1})(1-p) = 1500(1-p) + 1500(1-p)^2 + \cdots + 1500(1-p)^n$$

which is the nth partial sum of the geometric series $\sum_{n=1}^{\infty} 1500(1-p)^n$. Since $0 < p < 1$ by hypothesis, it follows that $0 < 1 - p < 1$, so that the series converges, and

$$\lim_{n\to\infty} w_n = \sum_{n=1}^{\infty} 1500(1-p)^n = \frac{1500(1-p)}{1-(1-p)} = \frac{1500(1-p)}{p}$$

61. a. Initially there is one equilateral triangle whose sides have length 1. At each successive step in generating the Koch snowflake, each line segment of the nth step is replaced by 4 line segments, each having 1/3 the length of the line segment being replaced. Consequently after n steps there are $3 \cdot 4^n$ sides of length $1/3^n$. It follows that for the $(n+1)$st step there are $3 \cdot 4^n$ new equilateral triangles. The side of such a triangle has length $1/3^{n+1}$, and thus the area of the triangle is given by

$$\frac{\sqrt{3}}{4}\left(\frac{1}{3^{n+1}}\right)^2 = \frac{\sqrt{3}}{4}\cdot\frac{1}{3^{2n+2}} = \frac{\sqrt{3}}{36}\cdot\frac{1}{3^{2n}}$$

Therefore in the $(n+1)$st step the total amount of area of the $3 \cdot 4^n$ new triangles is

$$3 \cdot 4^n\left(\frac{\sqrt{3}}{36}\cdot\frac{1}{3^{2n}}\right) = \frac{\sqrt{3}}{12}\cdot\frac{4^n}{9^n} = \frac{\sqrt{3}}{12}\left(\frac{4}{9}\right)^n$$

Consequently the area A of the Koch snowflake is given by

$$A = \frac{\sqrt{3}}{4} + \sum_{n=0}^{\infty} \frac{\sqrt{3}}{12}\left(\frac{4}{9}\right)^n = \frac{\sqrt{3}}{4} + \frac{\sqrt{3}}{12}\cdot\frac{1}{1-4/9} = \frac{\sqrt{3}}{4} + \frac{\sqrt{3}}{12}\cdot\frac{9}{5} = \frac{2}{5}\sqrt{3}$$

b. As we noted in the solution of part (a), there are $3 \cdot 4^n$ sides of length $1/3^n$ after n steps. Thus the length of the boundary after n steps is $(3 \cdot 4^n)(1/3^n) = 3(\frac{4}{3})^n$. Since $\lim_{n\to\infty} 3(\frac{4}{3})^n = \infty$, the boundary of the snowflake has infinite length.

63. The distance run by Achilles is

$$100 + 10 + 1 + \frac{1}{10} + \cdots = \sum_{n=0}^{\infty} 100\left(\frac{1}{10}\right)^n = \frac{100}{1-\frac{1}{10}} = \frac{1000}{9} \text{ (yards)}$$

The distance run by the tortoise during the same time is

$$10 + 1 + \frac{1}{10} + \frac{1}{100} + \cdots = \sum_{n=0}^{\infty} 10\left(\frac{1}{10}\right)^n = \frac{10}{1-\frac{1}{10}} = \frac{100}{9} \text{ (yards)}$$

Since $\frac{1000}{9} - \frac{100}{9} = \frac{900}{9} = 100$, Achilles catches up with the tortoise after running $\frac{1000}{9}$ (yards).

65. a. By the end of the 1st minute the ant has covered 1 foot, which is $\frac{1}{3}$ of the band. The band is then stretched so it is 6 feet long. By the end of the 2nd minute the ant has covered another foot, which is $\frac{1}{6}$ of the stretched band. Thus at the end of 2 minutes the ant has covered altogether $\frac{1}{3} + \frac{1}{6}$ of the band. The band then is stretched so it is 9 feet long. In general, during the jth minute the band is $3j$ feet long, and until the ant reaches the end of the band it covers an additional foot during that minute, which means covering $1/(3j)$ of the stretched band. Thus at the end of the jth minute the ant has covered altogether $\frac{1}{3} + \frac{1}{6} + \frac{1}{9} + \cdots + 1/(3j)$ of the band. So we need to find the smallest positive integer j so that $\frac{1}{3} + \frac{1}{6} + \frac{1}{9} + \cdots + 1/(3j) \geq 1$, that is, $\frac{1}{3}(1 + \frac{1}{2} + \frac{1}{3} + \cdots + 1/j) \geq 1$, or $\sum_{n=1}^{j} 1/n \geq 3$. But by Exercise 57(a), this happens if $j = 11$. Thus the ant reaches the other end of the band, and it takes between 10 and 11 minutes to do so.

b. For $n \geq 1$ let a_n be the distance in feet from the ant to A at the end of n minutes (just after the band has been stretched). During the first minute the ant crawls 1 foot and hence is $\frac{1}{20}$ of the way between the two ends. Thus after the band is stretched an additional 20 feet, the ant is still $\frac{1}{20}$ of the way between the two ends, so that

$a_1 = \frac{1}{20}(40) = 2$. By an argument similar to that used in part (a), we have $a_n = [(a_{n-1} + 1)/(20n)][20(n+1)]$, so that $a_n/[20(n+1)] = a_{n-1}/(20n) + 1/(20n)$. By induction it follows that $a_n/[20(n+1)] = \frac{1}{20}(1 + \frac{1}{2} + \frac{1}{3} + \cdots + 1/n)$. The ant will reach the other end of the band during the $(n+1)$st minute if n satisfies $a_n + 1 \geq 20(n+1)$, or $a_n/[20(n+1)] + 1/[20(n+1)] \geq 1$, or $\frac{1}{20}(1 + \frac{1}{2} + \frac{1}{3} + \cdots + 1/(n+1)) \geq 1$. By the comments following the solution of Example 4 the ant reaches the other end, and it takes between 272,400,599 and 272,400,600 minutes.

SECTION 9.5

1. Since $1/(n+1)^2 \leq 1/n^2$ for $n > 1$, and since $\sum_{n=1}^{\infty} 1/n^2$ is a convergent p series (with $p = 2$), $\sum_{n=1}^{\infty} 1/(n+1)^2$ converges by the Comparison Test.

3. Since $1/\sqrt{n^3+1} \leq 1/n^{3/2}$ for $n \geq 1$, and since $\sum_{n=1}^{\infty} 1/n^{3/2}$ is a convergent p series (with $p = \frac{3}{2}$), $\sum_{n=1}^{\infty} 1/\sqrt{n^3+1}$ converges by the Comparison Test.

5. Since $1/\sqrt{n^2+1} \geq 1/\sqrt{n^2+2n+1} = 1/(n+1)$ for $n \geq 1$, and since $\sum_{n=1}^{\infty} 1/(n+1) = \sum_{n=2}^{\infty} 1/n$ diverges, so does $\sum_{n=1}^{\infty} 1/\sqrt{n^2+1}$ by the Comparison Test.

7. Since $1/e^{(n^2)} < 1/e^n = (1/e)^n$ for $n \geq 1$, and since $\sum_{n=1}^{\infty} (1/e)^n$ is a convergent geometric series, $\sum_{n=1}^{\infty} 1/e^{(n^2)}$ converges by the Comparison Test.

9. Since $1/(n-1)(n-2) \leq 1/(n-2)^2$ for $n \geq 3$, and since $\sum_{n=3}^{\infty} 1/(n-2)^2 = \sum_{n=1}^{\infty} 1/n^2$ converges, so does $\sum_{n=3}^{\infty} 1/(n-1)(n-2)$ by the Comparison Test.

11. Since
$$\lim_{n\to\infty} \frac{(n^2-1)/(n^3-n-1)}{1/n} = \lim_{n\to\infty} \frac{n^3-n}{n^3-n-1} = \lim_{n\to\infty} \frac{1-1/n^2}{1-1/n^2-1/n^3} = 1$$
and since $\sum_{n=2}^{\infty} 1/n$ diverges, the Limit Comparison Test implies that $\sum_{n=2}^{\infty} (n^2-1)/(n^3-n-1)$ diverges.

13. Since $n/7^n < n/6^n = (n/2^n)(1/3^n) < (\frac{1}{3})^n$ for $n \geq 1$, and since $\sum_{n=1}^{\infty} (\frac{1}{3})^n$ is a convergent geometric series, the Comparison Test implies that $\sum_{n=1}^{\infty} n/7^n$ converges.

15. Since
$$\lim_{n\to\infty} \frac{\sqrt{n}/(n^2-3)}{1/n^{3/2}} = \lim_{n\to\infty} \frac{n^2}{n^2-3} = \lim_{n\to\infty} \frac{1}{1-3/n^2} = 1$$
and since $\sum_{n=4}^{\infty} 1/n^{3/2}$ is a convergent p series (with $p = \frac{3}{2}$), the Limit Comparison Test implies that $\sum_{n=4}^{\infty} \sqrt{n}/(n^2-3)$ converges.

17. Since
$$\lim_{n\to\infty} \frac{n/(n^3+1)^{3/7}}{1/n^{2/7}} = \lim_{n\to\infty} \frac{n^{9/7}}{(n^3+1)^{3/7}} = \lim_{n\to\infty} \frac{1}{(1+1/n^3)^{3/7}} = 1$$
and since $\sum_{n=1}^{\infty} 1/n^{2/7}$ is a divergent p series (with $p = \frac{2}{7}$), the Limit Comparison Test implies that $\sum_{n=1}^{\infty} n/(n^3+1)^{3/7}$ diverges.

19. Since
$$\lim_{n\to\infty} \frac{1/(n\sqrt{n^2-1})}{1/n^2} = \lim_{n\to\infty} \frac{n}{\sqrt{n^2-1}} = \lim_{n\to\infty} \frac{1}{\sqrt{1-1/n^2}} = 1$$
and since $\sum_{n=2}^{\infty} 1/n^2$ is a convergent p series (with $p = 2$), the Limit Comparison Test implies that $\sum_{n=2}^{\infty} 1/(n\sqrt{n^2-1})$ converges.

21. Since $0 \leq (\arctan n)/(n^2+1) \leq \pi/2n^2$ for $n \geq 1$, and since $\sum_{n=1}^{\infty} \pi/2n^2$ converges by Example 1, the Comparison Test implies that $\sum_{n=1}^{\infty} (\arctan n)/(n^2+1)$ converges.

23. Since $(\ln n)/n^2 \leq 2\sqrt{n}/n^2 = 2/n^{3/2}$ for $n \geq 1$, and since $\sum_{n=1}^{\infty} 2/n^{3/2}$ converges by Example 1, the Comparison Test implies that $\sum_{n=1}^{\infty} (\ln n)/n^2$ converges.

25. Since $0 < 1/(\ln n)^n < 1/(\ln 3)^n$ for $n \geq 3$, since $1/\ln 3 < 1$, and since $\sum_{n=3}^{\infty} 1/(\ln 3)^n$ is a convergent geometric series, the Comparison Test implies that $\sum_{n=3}^{\infty} 1/(\ln n)^n$ converges. Thus $\sum_{n=2}^{\infty} 1/(\ln n)^n$ also converges.

27. Since
$$\lim_{n\to\infty} \frac{\sin 1/n}{1/n} = \lim_{x\to 0^+} \frac{\sin x}{x} = 1$$
and since $\sum_{n=1}^{\infty} 1/n$ diverges, the Limit Comparison Test implies that $\sum_{n=1}^{\infty} \sin 1/n$ diverges.

29. If $p \leq 0$, then $1/[n(\ln n)^p] \geq 1/n$ for $n \geq 3$. But $\sum_{n=3}^{\infty} 1/n$ diverges, and thus $\sum_{n=3}^{\infty} 1/[n(\ln n)^p]$ diverges by the Comparison Test, so that $\sum_{n=2}^{\infty} 1/[n(\ln n)^p]$ also diverges. Let $p > 0$, and let $f(x) = 1/[x(\ln x)^p]$. Then f is positive, continuous, and decreasing on $[3, \infty)$. If $p = 1$, the series diverges by Example 2. If $p \neq 1$, then
$$\int_3^{\infty} \frac{1}{x(\ln x)^p}\,dx = \lim_{b\to\infty} \int_3^b \frac{1}{x(\ln x)^p}\,dx = \lim_{b\to\infty} \left(\frac{1}{-p+1}(\ln x)^{-p+1}\right)\Big|_3^b$$
$$= \lim_{b\to\infty} \frac{1}{-p+1}[(\ln b)^{-p+1} - (\ln 3)^{-p+1}]$$
If $0 < p < 1$, this limit is ∞. If $p > 1$, this limit is $[1/(p-1)](\ln 3)^{1-p}$. Thus by the Integral Test the series converges only for $p > 1$.

31. Since $\sum_{n=1}^{\infty} b_n$ converges, we have $\lim_{n\to\infty} b_n = 0$. Since $b_n \geq 0$ this means there is a positive integer N such that $0 = b_n \leq 1$ for $n \geq N$. Then $0 \leq a_n b_n \leq a_n$ for $n \geq N$. Since $\sum_{n=1}^{\infty} a_n$ converges by hypothesis, the Comparison Test implies that $\sum_{n=N}^{\infty} a_n b_n$ converges. Thus $\sum_{n=1}^{\infty} a_n b_n$ converges.

33. The proportion of fuel used during the first j hundred miles is the jth partial sum of the series $\sum_{n=1}^{\infty} 1/(n+1)^2 = \sum_{n=2}^{\infty} 1/n^2$. Since $\sum_{n=1}^{\infty} 1/n^2 = \pi^2/6$, we have $\sum_{n=2}^{\infty} 1/n^2 = \pi^2/6 - 1 < 1$. Thus the rocket never uses up all its fuel.

SECTION 9.6

1. $\lim_{n\to\infty} \frac{(n+1)!/2^{n+1}}{n!/2^n} = \lim_{n\to\infty} \frac{n+1}{2} = \infty$; the series diverges.

3. $\lim_{n\to\infty} \frac{(n+1)!3^{n+1}/10^{n+1}}{n!3^n/10^n} = \lim_{n\to\infty} \frac{3(n+1)}{10} = \infty$; the series diverges.

5. $\lim_{n\to\infty} \sqrt[n]{\left(\frac{n}{2n+5}\right)^n} = \lim_{n\to\infty} \frac{n}{2n+5} = \frac{1}{2} < 1$; the series converges.

7. $\lim_{n\to\infty} \frac{(2n+2)!/[(n+1)!]^2}{(2n)!/(n!)^2} = \lim_{n\to\infty} \frac{(2n+2)(2n+1)}{(n+1)^2} = 4 > 1$; the series diverges.

9. $\lim_{n\to\infty} \sqrt[n]{n^{100}e^{-n}} = \lim_{n\to\infty} (\sqrt[n]{n})^{100}e^{-1} = e^{-1} < 1$; the series converges.

11. $\lim_{n\to\infty} \frac{(n+1)^{1.7}/(1.7)^{n+1}}{n^{1.7}/(1.7)^n} = \lim_{n\to\infty} \frac{(n+1)^{1.7}}{(1.7)n^{1.7}} = \lim_{n\to\infty} \frac{1}{(1.7)}\left(1+\frac{1}{n}\right)^{1.7} = \frac{1}{1.7} < 1;$

the series converges.

13. $\lim_{n\to\infty} \frac{(n+1)!/e^{n+1}}{n!/e^n} = \lim_{n\to\infty} \frac{n+1}{e} = \infty$; the series diverges.

15. $\lim_{n\to\infty} \sqrt[n]{\frac{1}{(\ln n)^n}} = \lim_{n\to\infty} \frac{1}{\ln n} = 0$; the series converges.

17. Since
$$\frac{1\cdot 3\cdot 5\cdots(2n-1)}{2\cdot 4\cdot 6\cdots(2n)} = \frac{3}{2}\cdot\frac{5}{4}\cdots\frac{2n-1}{2n-2}\cdot\frac{1}{2n} > \frac{1}{2n}$$
and since $\sum_{n=1}^{\infty} 1/2n$ diverges, the given series diverges by the Comparison Test.

19.
$$\lim_{n\to\infty} \frac{\dfrac{(2n+2)!}{(n+1)!(2n+2)^{n+1}}}{\dfrac{(2n)!}{n!(2n)^n}} = \lim_{n\to\infty} \frac{(2n+2)(2n+1)(2n)^n}{(n+1)(2n+2)^{n+1}} = \lim_{n\to\infty} \left(\frac{2n+1}{n+1}\right)\left(\frac{n}{n+1}\right)^n$$
$$= \lim_{n\to\infty} \frac{2+1/n}{(1+1/n)(1+1/n)^n} = \frac{2}{e} < 1$$
the series converges.

21. Since $\sin 1/n! \le 1/n!$ and $\cos 1/n! \ge \cos 1$ for $n \ge 1$, we have
$$\frac{\sin 1/n!}{\cos 1/n!} \le \frac{1}{n!\cos 1}$$
It follows from Example 1 of Section 9.4 that $\sum_{n=1}^{\infty} 1/(n!\cos 1)$ converges. By the Comparison Test $\sum_{n=1}^{\infty} (\sin 1/n!)/(\cos 1/n!)$ converges.

23. Let $b_n = (n/(2n+1))^n$ for $n \ge 1$. Then
$$\lim_{n\to\infty} \sqrt[n]{b_n} = \lim_{n\to\infty} \frac{n}{2n+1} = \lim_{n\to\infty} \frac{1}{2+1/n} = \frac{1}{2} < 1$$
Thus $\sum_{n=1}^{\infty} b_n$ converges by the Root Test. Since $0 \le a_n \le b_n$ for $n \ge 1$, it follows from the Comparison Test that $\sum_{n=1}^{\infty} a_n$ converges.

25. a. $\sqrt[n]{a_n} = \begin{cases} 0 & \text{for } n \text{ even} \\ \dfrac{n}{2n+1} & \text{for } n \text{ odd} \end{cases}$

Since $n/(2n+1) \ge \frac{1}{3}$ for $n \ge 1$, it follows that $\lim_{n\to\infty} \sqrt[n]{a_n}$ does not exist. The solution of Exercise 23 shows that $\sum_{n=1}^{\infty} a_n$ converges.

b. Let $a_n = n^n$ for $n \ge 1$. Then $\sqrt[n]{a_n} = n$, so that $\lim_{n\to\infty} \sqrt[n]{a_n}$ does not exist. Since $\lim_{n\to\infty} a_n = \lim_{n\to\infty} n^n = \infty$, $\sum_{n=1}^{\infty} n^n$ does not converge.

SECTION 9.7

1. Since $\{1/(2n+1)\}_{n=1}^{\infty}$ is a decreasing, nonnegative sequence with $\lim_{n\to\infty} 1/(2n+1) = 0$, the series converges.

3. Since $\lim_{n\to\infty} a_n = \lim_{n\to\infty} (2n+1)/(5n+1) = \frac{2}{5} \ne 0$, the series diverges.

5. Let $f(x) = (x+2)/(x^2+3x+5)$. Then $f'(x) = (-x^2-4x-1)/(x^2+3x+5)^2 < 0$ for $x > 0$, so f is decreasing on $[1, \infty)$. Thus $\{(n+2)/(n^2+3n+5)\}_{n=1}^{\infty}$ is a nonnegative, decreasing sequence, and $\lim_{n\to\infty} (n+2)/(n^2+3n+5) = 0$. Therefore the series converges.

7. Let $f(x) = (\ln x)/x$. Since $f'(x) = (1-\ln x)/x^2 < 0$ for $x > e$, the function f is decreasing one $[e, \infty)$. Thus $\{(\ln n)/n\}_{n=3}^{\infty}$ is a nonnegative, decreasing sequence, and $\lim_{n\to\infty} (\ln n)/n = 0$. Therefore $\sum_{n=3}^{\infty} (-1)^n[(\ln n)/n]$ converges, and hence $\sum_{n=1}^{\infty} (-1)^n[(\ln n)/n]$ converges.

9. Let $f(x) = (\ln x)^p/x$. Then
$$f'(x) = \frac{p(\ln x)^{p-1} - (\ln x)^p}{x^2} = \frac{(\ln x)^{p-1}(p - \ln x)}{x^2} < 0 \quad \text{for } x > e^p$$
Moreover, by l'Hôpital's Rule,
$$\lim_{x\to\infty} \frac{(\ln x)^p}{x} = \lim_{x\to\infty} \frac{p(\ln x)^{p-1}}{x} = \cdots = \lim_{x\to\infty} \frac{p!}{x} = 0$$
Thus $\{(\ln n)^p/n\}_{n=3^p}^{\infty}$ is a nonnegative, decreasing sequence with $\lim_{n\to\infty} (\ln n)^p/n = 0$. Thus $\sum_{n=3^p}^{\infty} (-1)^n[(\ln n)^p/n]$ converges, and hence $\sum_{n=1}^{\infty} (-1)^n[(\ln n)^p/n]$ converges.

11. Since
$$\frac{a_{n+1}}{a_n} = \frac{(n+1)!/100^{n+1}}{n!/100^n} = \frac{n+1}{100} > 1 \quad \text{for } n \ge 100$$
we have $a_{n+1} > a_n$ for $n \ge 100$. The series diverges.

13. Since $\lim_{n\to\infty} n^2/(2n+1) = \infty$, the series diverges.

15. For any $p > 0$, $\{1/n^p\}_{n=1}^{\infty}$ is a nonnegative, decreasing sequence with $\lim_{n\to\infty} 1/n^p = 0$. Thus $\sum_{n=1}^{\infty} (-1)^n(1/n^p)$ converges for any $p > 0$.

17. $\left|\sum_{n=1}^{\infty} (-1)^n \frac{1}{n^2} - s_4\right| \le a_5 = \frac{1}{25}$

19. $\left|\sum_{n=1}^{\infty} (-1)^n \frac{1}{\sqrt{n+4}} - s_4\right| \le a_5 = \frac{1}{3}$

21. Since $1/(1 + n + 6n^2) < 0.01$ for $n \ge 4$, it follows that $s_2 = -1/27 + \frac{1}{58} \approx -0.0197957$ approximates the sum with an error less than 0.01.

23. $\{1/(3n+4)\}_{n=1}^{\infty}$ is a nonnegative, decreasing sequence and $\lim_{n\to\infty} 1/(3n+4) = 0$. Thus $\sum_{n=1}^{\infty} (-1)^{n+1} [1/(3n+4)]$ converges. But notice that $\sum_{n=1}^{\infty} 1/(3n+4)$ diverges because $1/(3n+1) \ge 1/7n$ for $n \ge 1$ and $\sum_{n=1}^{\infty} 1/7n$ diverges, so $\sum_{n=1}^{\infty} (-1)^{n+1} [1/(3n+4)]$ converges conditionally.

25. $\lim_{n\to\infty} \frac{(n+1)^{n+1}/(n+1)!}{n^n/n!} = \lim_{n\to\infty} \left(\frac{n+1}{n}\right)^n = \lim_{n\to\infty} \left(1 + \frac{1}{n}\right)^n = e > 1$; the series diverges.

27. Since $\lim_{n\to\infty} 1/n^{1/n} = 1$, the series diverges.

29. Since

$$\int_2^{\infty} \frac{1}{x(\ln x)^2}\,dx = \lim_{b\to\infty} \int_2^b \frac{1}{x(\ln x)^2}\,dx = \lim_{b\to\infty} \left(\frac{-1}{\ln x}\right)\Bigg|_2^b = \lim_{b\to\infty} \left(\frac{-1}{\ln b} + \frac{1}{\ln 2}\right) = \frac{1}{\ln 2}$$

the Integral Test implies that $\sum_{n=2}^{\infty} 1/[n(\ln n)^2]$ converges. Thus $\sum_{n=2}^{\infty} (-1)^{n+1}\{1/[n(\ln n)^2]\}$ converges absolutely.

31. $\lim_{n\to\infty} \sqrt[n]{\left(\frac{1}{\ln n}\right)^n} = \lim_{n\to\infty} \frac{1}{\ln n} = 0$ so $\sum_{n=2}^{\infty} (-1)^{n+1}[1/(\ln n)^n]$ converges absolutely.

33. $$\lim_{n\to\infty} \frac{\dfrac{1\cdot 3\cdot 5\cdots(2n+3)}{2\cdot 5\cdot 8\cdots(3n+5)}}{\dfrac{1\cdot 3\cdot 5\cdots(2n+1)}{2\cdot 5\cdot 8\cdots(3n+2)}} = \lim_{n\to\infty} \frac{2n+3}{3n+5} = \lim_{n\to\infty} \frac{2 + \frac{3}{n}}{3 + \frac{5}{n}} = \frac{2}{3}$$

The series converges absolutely.

35. $\lim_{n\to\infty} \frac{(n+2)^2/(n+1)!}{(n+1)^2/n!} = \lim_{n\to\infty} \frac{(n+2)^2}{(n+1)^3} = 0$, so $\lim_{n\to\infty} \frac{(n+1)^2}{n!} = 0$.

37. $\lim_{n\to\infty} \sqrt[n]{\frac{x^{2n}}{n}} = \lim_{n\to\infty} \frac{x^2}{\sqrt[n]{n}} = x^2$

If $|x| < 1$, then $x^2 < 1$, so $\lim_{n\to\infty} x^{2n}/n = 0$. If $|x| = 1$, then $\lim_{n\to\infty} |x^{2n}/n| = \lim_{n\to\infty} 1/n = 0$, so $\lim_{n\to\infty} x^{2n}/n = 0$.

39. $\lim_{n\to\infty} \frac{(n+1)!|x|^{n+1}/(n+1)^{n+1}}{n!|x|^n/n^n} = \lim_{n\to\infty} |x|\left(\frac{n}{n+1}\right)^n = \lim_{n\to\infty} |x| \frac{1}{(1+1/n)^n} = \frac{|x|}{e}$

so that if $|x| < e$, then $\lim_{n\to\infty} n!x^n/n^n = 0$.

41. The series $\sum_{n=1}^{\infty} (-1)^n(1/\sqrt{n})$ converges (conditionally), but $((-1)^n(1/\sqrt{n}))^2 = 1/n$ and $\sum_{n=1}^{\infty} 1/n$ diverges.

43. Since $\lim_{n\to\infty} a_n = a \ne 0$, there is an integer N such that if $n \ge N$, then $|a_n - a| < |a|/2$, so that $|a|/2 < |a_n| < 3|a|/2$, and thus $2/(3|a|) < 1/|a_n| < 2/|a|$. Notice that $|1/a_{n+1} - 1/a_n| = |(a_{n+1} - a_n)/a_{n+1}a_n|$, and that for $n \ge N$, we have

$$\frac{4}{9a^2}|a_{n+1} - a_n| < \left|\frac{a_{n+1} - a_n}{a_{n+1}a_n}\right| < \frac{4}{a^2}|a_{n+1} - a_n|$$

Since $\sum_{n=N}^{\infty} (4/(9a^2))|a_{n+1} - a_n|$, $\sum_{n=N}^{\infty} (4/a^2)|a_{n+1} - a_n|$, and $\sum_{n=N}^{\infty} |a_{n+1} - a_n|$ all converge or all diverge, the same is true of $\sum_{n=1}^{\infty} |a_{n+1} - a_n|$ and $\sum_{n=1}^{\infty} |1/a_{n+1} - 1/a_n|$, and thus both converge or both diverge.

SECTION 9.8

1. $\lim_{n\to\infty} \sqrt[n]{\frac{|x|^n}{n^2}} = \lim_{n\to\infty} \frac{|x|}{(\sqrt[n]{n})^2} = |x|$

the series converges for $|x| < 1$ and diverges for $|x| > 1$. Since $\sum_{n=1}^{\infty} 1/n^2$ converges and hence $\sum_{n=1}^{\infty} (-1)^n/n^2$ converges, the interval of convergence is $[-1, 1]$.

3. $\lim_{n\to\infty} \sqrt[n]{\frac{1}{\sqrt{n}3^n}|x|^n} = \lim_{n\to\infty} \frac{|x|}{3(\sqrt[n]{n})^{1/2}} = \frac{|x|}{3}$

the series converges for $|x| < 3$ and diverges for $|x| > 3$. Since $\sum_{n=1}^{\infty} 1/\sqrt{n}$ diverges and and $\sum_{n=1}^{\infty} (-1)^n/\sqrt{n}$ converges, the interval of convergence is $[-3, 3)$.

5. $\lim_{n\to\infty} \frac{|x|^{2n+2}/(n+2)}{|x|^{2n}/(n+1)} = \lim_{n\to\infty} \frac{n+1}{n+2}x^2 = x^2$

the series converges for $|x| < 1$ and diverges for $|x| > 1$. Since $\sum_{n=0}^{\infty} (-1)^n/(n+1)$ converges, the interval of convergence is $[-1, 1]$.

7. $\lim_{n\to\infty} \frac{|x|^{n+2}/(2n+1)}{|x|^{n+1}/(2n-1)} = \lim_{n\to\infty} \frac{2n-1}{2n+1}|x| = |x|$

the series converges for $|x| < 1$ and diverges for $|x| > 1$. The alternating series

$$\sum_{n=1}^{\infty} \frac{(-1)^n}{2n-1} = \sum_{n=1}^{\infty} \frac{(-1)^n 1^{n+1}}{2n-1}$$

converges, and $$\sum_{n=1}^{\infty} \frac{-1}{2n-1} = \sum_{n=1}^{\infty} \frac{(-1)^n(-1)^{n+1}}{2n-1}$$

diverges, so the interval of convergence is $(-1, 1]$.

9. $\lim_{n\to\infty} \sqrt[n]{\frac{|x|^n}{n^n}} = \lim_{n\to\infty} \frac{|x|}{n} = 0$; the interval of convergence is $(-\infty, \infty)$.

11. $\lim_{n\to\infty} \dfrac{(n+1)!|x|^{n+1}/(2n+2)!}{n!|x|^n/(2n)!} = \lim_{n\to\infty} \dfrac{n+1}{(2n+1)(2n+2)}|x| = 0$

the interval of convergence is $(-\infty, \infty)$.

13. $\lim_{n\to\infty} \dfrac{2^{n+1}|x|^{n+2}/(n+1)3^{n+3}}{2^n|x|^{n+1}/n3^{n+2}} = \lim_{n\to\infty} \dfrac{2n}{3(n+1)}|x| = \dfrac{2}{3}|x|$

the series converges for $|x| < \frac{3}{2}$ and diverges for $|x| > \frac{3}{2}$. Since $\sum_{n=1}^{\infty} 1/(6n)$ diverges and $\sum_{n=1}^{\infty} (-1)^{n+1}/(6n)$ converges, the interval of convergence is $[-\frac{3}{2}, \frac{3}{2})$.

15. By l'Hôpital's Rule,

$$\lim_{x\to\infty} \frac{\ln(x+1)}{\ln x} = \lim_{x\to\infty} \frac{1/(x+1)}{1/x} = \lim_{x\to\infty} \frac{x}{x+1} = 1$$

so we have $$\lim_{n\to\infty} \frac{|x|^{n+1}\ln(n+1)}{|x|^n \ln n} = \lim_{n\to\infty} \frac{\ln(n+1)}{\ln n}|x| = |x|$$

the series converges for $|x| < 1$ and diverges for $|x| > 1$. Since $\sum_{n=2}^{\infty} \ln n$ and $\sum_{n=2}^{\infty} (-1)^n \ln n$ both diverge, the interval of convergence is $(-1, 1)$.

17. By l'Hôpital's Rule, $\lim_{x\to\infty} \ln(x+1)/\ln x = 1$, so

$$\lim_{n\to\infty} \frac{|x|^{n+1}\ln(n+1)/(n+1)^2}{|x|^n(\ln n)/n^2} = \lim_{n\to\infty} \left(\frac{n}{n+1}\right)^2 \frac{\ln(n+1)}{\ln n}|x| = |x|$$

the series converges for $|x| < 1$ and diverges for $|x| > 1$. By the solution of Exercise 23 of Section 9.5, $\sum_{n=2}^{\infty} (\ln n)/n^2$ converges and hence $\sum_{n=2}^{\infty} (-1)^n(\ln n)/n^2$ converges. The interval of convergence is $[-1, 1]$.

19. $\lim_{n\to\infty} \dfrac{|x|^{(n+1)^2}}{|x|^{n^2}} = \lim_{n\to\infty} |x|^{2n+1}$, and $\lim_{n\to\infty} |x|^{2n+1} = 0$ if $|x| < 1$

whereas $\lim_{n\to\infty} |x|^{2n+1} = \infty$ if $|x| > 1$. Thus the series converges for $|x| < 1$ and diverges for $|x| > 1$. Since $\sum_{n=0}^{\infty} 1^{(n^2)}$ and $\sum_{n=0}^{\infty} (-1)^{(n^2)}$ both diverge, the interval of convergence is $(-1, 1)$.

21. $\lim_{n\to\infty} \dfrac{|x|^{n+1}(n+1)!/(n+1)^{n+1}}{|x|^n n!/n^n} = \lim_{n\to\infty} \dfrac{1}{(1+1/n)^n}|x| = \dfrac{|x|}{e}$, so $R = e$.

23. $\lim_{n\to\infty} \dfrac{1^2\cdot 3^2\cdot 5^2\cdots(2n+1)^2|x|^{2n+2}/[2^2\cdot 4^2\cdot 6^2\cdots(2n)^2(2n+2)^2]}{1^2\cdot 3^2\cdot 5^2\cdots(2n-1)^2|x|^{2n}/[2^2\cdot 4^2\cdot 6^2\cdots(2n)^2]} = \lim_{n\to\infty} \dfrac{(2n+1)^2}{(2n+2)^2}x^2 = x^2$

so $R = 1$.

25. $\lim_{n\to\infty} \dfrac{1\cdot 3\cdot 5\cdots(2n-1)(2n+1)|x|^{n+1}/2^{n+1}[1\cdot 4\cdot 7\cdots(3n-2)(3n+1)]}{1\cdot 3\cdot 5\cdots(2n-1)|x|^n/2^n[1\cdot 4\cdot 7\cdots(3n-2)]} =$

$\lim_{n\to\infty} \dfrac{2n+1}{2(3n+1)}|x| = \dfrac{|x|}{3}$, so $R = 3$.

27. $f(x) = \sum_{n=1}^{\infty} (n+1)x^n$ for x in $(-1, 1)$; $f'(x) = \sum_{n=1}^{\infty} n(n+1)x^{n-1}$ for x in $(-1, 1)$;

$\int_0^x f(t)\,dt = \sum_{n=1}^{\infty} x^{n+1}$ for x in $(-1, 1)$.

29. $f(x) = \sum_{n=1}^{\infty} \dfrac{5}{n} x^{(n^2)}$ for x in $[-1, 1)$; $f'(x) = \sum_{n=1}^{\infty} 5nx^{n^2-1}$ for x in $(-1, 1)$;

$\int_0^x f(t)\,dt = \sum_{n=1}^{\infty} \dfrac{5}{n(n^2+1)} x^{n^2+1}$ for x in $(-1, 1)$.

31. Since $\sin x = \sum_{n=0}^{\infty} (-1)^n x^{2n+1}/(2n+1)!$, it follows that

$$\sin x^2 = \sum_{n=0}^{\infty} \frac{(-1)^n(x^2)^{2n+1}}{(2n+1)!} = \sum_{n=0}^{\infty} \frac{(-1)^n x^{4n+2}}{(2n+1)!}$$

By the Integration Theorem,

$$\int_0^2 \sin x^2\,dx = \int_0^2 \sum_{n=0}^{\infty} \frac{(-1)^n x^{4n+2}}{(2n+1)!}\,dx = \sum_{n=0}^{\infty} \int_0^2 \frac{(-1)^n x^{4n+2}}{(2n+1)!}\,dx$$
$$= \sum_{n=0}^{\infty} \left(\frac{(-1)^n x^{4n+3}}{(4n+3)(2n+1)!}\bigg|_0^2\right) = \sum_{n=0}^{\infty} \frac{(-1)^n 2^{4n+3}}{(4n+3)(2n+1)!}$$

By the Alternating Series Test,

$$\left|\sum_{n=m+1}^{\infty} \frac{(-1)^n 2^{4n+3}}{(4n+3)(2n+1)!}\right| < 10^{-2} \quad \text{if} \quad \frac{2^{4(m+1)+3}}{[4(m+1)+3][2(m+1)+1)]!} < 10^{-2}$$

which occurs for $m = 4$. Thus

$$\sum_{n=0}^{4} \frac{(-1)^n 2^{4n+3}}{(4n+3)(2n+1)!} = \frac{2^3}{3(1)} - \frac{2^7}{7(3!)} + \frac{2^{11}}{11(5!)} - \frac{2^{15}}{15(7!)} + \frac{2^{19}}{19(9!)} \approx 0.813166$$

is the desired approximation.

33. Since $e^x = \sum_{n=0}^{\infty} x^n/n!$ by (3), it follows that

$$\frac{1-e^{-x}}{x} = \frac{1 - \sum_{n=0}^{\infty} (-1)^n x^n/n!}{x} = \sum_{n=1}^{\infty} \frac{(-1)^{n+1}x^{n-1}}{n!} \quad \text{for } x \neq 0$$

By the Integration Theorem,

$$\int_0^1 \frac{1-e^{-x}}{x}\,dx = \lim_{c\to 0^+} \int_c^1 \frac{1-e^{-x}}{x}\,dx = \lim_{c\to 0^+} \int_c^1 \sum_{n=1}^{\infty} \frac{(-1)^{n+1}x^{n-1}}{n!}\,dx$$
$$= \lim_{c\to 0^+} \sum_{n=1}^{\infty} \int_c^1 \frac{(-1)^{n+1}x^{n-1}}{n!}\,dx = \lim_{c\to 0^+} \sum_{n=1}^{\infty} \left(\frac{(-1)^{n+1}x^n}{n(n!)}\bigg|_c^1\right)$$
$$= \lim_{c\to 0^+} \left(\sum_{n=1}^{\infty} \frac{(-1)^{n+1}}{n(n!)} - \sum_{n=1}^{\infty} \frac{(-1)^{n+1}c^n}{n(n!)}\right) = \sum_{n=1}^{\infty} \frac{(-1)^{n+1}}{n(n!)}$$

By the Alternating Series Test,

$$\left|\sum_{n=m+1}^{\infty} \frac{(-1)^{n+1}}{n(n!)}\right| < 10^{-3} \quad \text{if} \quad \frac{1}{(m+1)[(m+1)!]} < 10^{-3}$$

which occurs for $m = 5$. Thus

$$\sum_{n=1}^{5} \frac{(-1)^{n+1}}{n(n!)} = 1 - \frac{1}{2(2!)} + \frac{1}{3(3!)} - \frac{1}{4(4!)} + \frac{1}{5(5!)} \approx 0.796806$$

is the desired approximation.

35. From (1) we have $1/(1-x)=\sum_{n=0}^{\infty} x^n$ for $|x|<1$, so that

$$\frac{x^2}{1+x}=\sum_{n=0}^{\infty} x^2(-x)^n=\sum_{n=0}^{\infty}(-1)^n x^{n+2} \quad \text{for } |x|<1$$

By the Integration Theorem,

$$\int_0^{1/2}\frac{x^2}{1+x}\,dx=\int_0^{1/2}\sum_{n=0}^{\infty}(-1)^n x^{n+2}\,dx=\sum_{n=0}^{\infty}\int_0^{1/2}(-1)^n x^{n+2}\,dx$$
$$=\sum_{n=0}^{\infty}\left(\frac{(-1)^n x^{n+3}}{n+3}\right)\Bigg|_0^{1/2}=\sum_{n=0}^{\infty}\frac{(-1)^n}{2^{n+3}(n+3)}$$

By the Alternating Series Test,

$$\left|\sum_{n=m+1}^{\infty}\frac{(-1)^n}{2^{n+3}(n+3)}\right|<10^{-3} \quad \text{if} \quad \frac{1}{2^{(m+1)+3}[(m+1)+3]}<10^{-3}$$

which occurs for $m=4$. Thus

$$\sum_{n=0}^{4}\frac{(-1)^n}{2^{n+3}(n+3)}=\frac{1}{2^3(3)}-\frac{1}{2^4(4)}+\frac{1}{2^5(5)}-\frac{1}{2^6(6)}+\frac{1}{2^7(7)}\approx 0.0308036$$

is the desired approximation.

37. From (3) we have $e^x=\sum_{n=0}^{\infty} x^n/n!$, so that $e^{(x^2)}=\sum_{n=0}^{\infty}(x^2)^n/n!=\sum_{n=0}^{\infty} x^{2n}/n!$. By the Integration Theorem,

$$\int_{-1}^{0} e^{(x^2)}\,dx=\int_{-1}^{0}\sum_{n=0}^{\infty}\frac{x^{2n}}{n!}\,dx=\sum_{n=0}^{\infty}\int_{-1}^{0}\frac{x^{2n}}{n!}\,dx=\sum_{n=0}^{\infty}\left(\frac{x^{2n+1}}{(2n+1)n!}\Bigg|_{-1}^{0}\right)$$
$$=\sum_{n=0}^{\infty}\frac{-(-1)^{2n+1}}{(2n+1)n!}=\sum_{n=0}^{\infty}\frac{1}{(2n+1)n!}$$

If $a_n=1/(2n+1)n!$ for $n\geq 0$, then $a_{n+1}/a_n\leq\frac{1}{3}$, so $\sum_{n=m+1}^{\infty} a_n\leq\sum_{n=m+1}^{\infty} 1/3^n$. But

$$\sum_{n=m+1}^{\infty}\frac{1}{3^n}=\frac{(\frac{1}{3})^{m+1}}{1-\frac{1}{3}}=\frac{1}{3^m\cdot 2}$$

by the Geometric Series Theorem, and $1/(3^m\cdot 2)<10^{-3}$, if $m=6$. Thus

$$\sum_{n=0}^{6}\frac{1}{(2n+1)n!}=1+\frac{1}{3(1)}+\frac{1}{5(2!)}+\frac{1}{7(3!)}+\frac{1}{9(4!)}+\frac{1}{11(5!)}+\frac{1}{13(6!)}\approx 1.46264$$

is the desired approximation.

39. Since $\sum_{n=0}^{\infty} x^n=1/(1-x)$ for $|x|<1$, the Differentiation Theorem implies that $\sum_{n=1}^{\infty} nx^{n-1}=1/(1-x)^2$ for $|x|<1$. Then $\sum_{n=1}^{\infty} nx^n=x(\sum_{n=1}^{\infty} nx^{n-1})=x/(1-x)^2$ for $|x|<1$.

41. a. By the Differentiation Theorem,

$$f'(x)=\sum_{n=1}^{\infty}\frac{(-1)^n 2n}{(2n)!}x^{2n-1}=\sum_{n=1}^{\infty}\frac{(-1)^n}{(2n-1)!}x^{2n-1}=-\sum_{n=0}^{\infty}\frac{(-1)^n}{(2n+1)!}x^{2n+1}=-g(x)$$

whereas
$$g'(x)=\sum_{n=0}^{\infty}\frac{(-1)^n(2n+1)}{(2n+1)!}x^{2n}=\sum_{n=0}^{\infty}\frac{(-1)^n}{(2n)!}x^{2n}=f(x)$$

b. Since $f'(x)=-g(x)$ and $g'(x)=f(x)$, it follows that $f''(x)=-g'(x)=-f(x)$ and $g''(x)=f'(x)=-g(x)$.

c. The cosine and sine functions have the properties of (a) and (b), respectively.

43. Since $e^x=\sum_{n=0}^{\infty} x^n/n!$, we have

$$e^x-1-x=\sum_{n=2}^{\infty}\frac{x^n}{n!} \quad \text{and} \quad \frac{e^x-1-x}{x^2}=\sum_{n=2}^{\infty}\frac{x^{n-2}}{n!}=\sum_{n=0}^{\infty}\frac{x^n}{(n+2)!}$$

The latter series converges for all x and defines a continuous function of x, with domain $(-\infty,\infty)$. Thus

$$\lim_{x\to 0}\frac{e^x-1-x}{x^2}=\lim_{x\to 0}\left(\sum_{n=0}^{\infty}\frac{x^n}{(n+2)!}\right)=\frac{1}{2}+\sum_{n=1}^{\infty}\frac{0^n}{(n+2)!}=\frac{1}{2}$$

45. From (8) we have $\ln(1+x)=\sum_{n=1}^{\infty}((-1)^{n+1}/n)x^n$ for $|x|<1$. Thus $\ln(1+x^2)=\sum_{n=1}^{\infty}((-1)^{n+1}/n)x^{2n}$ for $|x|<1$. Since

$$\lim_{n\to\infty}\frac{|x|^{2(n+1)}/(n+1)}{|x|^{2n}/n}=\lim_{n\to\infty}\frac{n}{n+1}|x|^2=|x|^2$$

the radius of convergence is 1.

47. a. $$\ln\frac{1}{1-x}=-\ln(1-x)=-\sum_{n=1}^{\infty}\frac{(-1)^{n+1}}{n}(-x)^n=\sum_{n=1}^{\infty}\frac{x^n}{n} \quad \text{for } |x|<1$$

b. $$\ln 2=\ln\frac{1}{1-\frac{1}{2}}=\sum_{n=1}^{\infty}\frac{1}{n}\left(\frac{1}{2}\right)^n=\sum_{n=1}^{\infty}\frac{1}{n2^n}$$

c. Since $$\left|\ln 2-\sum_{n=1}^{N-1}\frac{1}{n2^n}\right|=\sum_{n=N}^{\infty}\frac{1}{n2^n}\leq\sum_{n=N}^{\infty}\frac{1}{N2^n}=\frac{1}{N}\frac{1/2^N}{1-\frac{1}{2}}=\frac{1}{N2^{N-1}}$$

we need to find N such that $1/(N2^{N-1})<0.01$. But if $N=6$, then $1/N2^{N-1}=1/(6\cdot 2^5)<0.01$, so $\sum_{n=1}^{5} 1/(n2^n)=\frac{1}{2}+\frac{1}{8}+\frac{1}{24}+\frac{1}{64}+\frac{1}{160}\approx .688542$ is the desired estimate of ln 2.

49. $$\left|4\frac{(-1)^4(\frac{1}{5})^9}{9}\right|=\tfrac{4}{9}(\tfrac{1}{5})^9<2.276\times 10^{-7} \quad \text{and} \quad \left|\frac{(-1)^1(\frac{1}{239})^3}{3}\right|=\frac{1}{3}\left(\frac{1}{239}\right)^3<.245\times 10^{-7}$$

Thus the error introduced in approximating $\pi/4$ is less than $2.276\times 10^{-7}+.245\times 10^{-7}=2.521\times 10^{-7}$.

51. a. From (1), $$\frac{1}{1+t^4}=\frac{1}{1-(-t^4)}=\sum_{n=0}^{\infty}(-t^4)^n=\sum_{n=0}^{\infty}(-1)^n t^{4n}$$

Then $t^2/(1+t^4)=\sum_{n=0}^{\infty}(-1)^n t^{4n+2}$ for $|t|<1$.

b. $$\int_0^{1/2}\frac{t^2}{1+t^4}\,dt=\int_0^{1/2}\left(\sum_{n=0}^{\infty}(-1)^n t^{4n+2}\right)dt=\sum_{n=0}^{\infty}(-1)^n\left(\int_0^{1/2}t^{4n+2}\,dt\right)$$
$$=\sum_{n=0}^{\infty}(-1)^n\left(\frac{t^{4n+3}}{4n+3}\Bigg|_0^{1/2}\right)=\sum_{n=0}^{\infty}\frac{(-1)^n}{4n+3}\left(\frac{1}{2}\right)^{4n+3}$$

53. a. By (3), $e^x = \sum_{n=0}^{\infty} \frac{1}{n!} x^n$, so $xe^x = \sum_{n=0}^{\infty} \frac{1}{n!} x^{n+1}$.

b. On the one hand, by integration by parts with $u = x$, $dv = e^x\,dx$, we obtain

$$\int_0^1 xe^x\,dx = xe^x\Big|_0^1 - \int_0^1 e^x\,dx = e - e^x\Big|_0^1 = e - e + 1 = 1$$

On the other hand, using the power series expansion of (a), we obtain

$$\int_0^1 xe^x\,dx = \int_0^1 \left(\sum_{n=0}^{\infty} \frac{1}{n!} x^{n+1}\right) dx = \sum_{n=0}^{\infty} \frac{1}{n!}\left(\int_0^1 x^{n+1}\,dx\right)$$
$$= \sum_{n=0}^{\infty} \frac{1}{n!}\left(\frac{x^{n+2}}{n+2}\Big|_0^1\right) = \sum_{n=0}^{\infty} \frac{1}{n!(n+2)}$$

Thus $\sum_{n=0}^{\infty} 1/n!(n+2) = 1$.

55. $$\lim_{n\to\infty} \left|\frac{c_{n+1}x^{n+1}}{c_n x^n}\right| = \lim_{n\to\infty} \left|\frac{c_{n+1}}{c_n}\right| |x| = |x| \lim_{n\to\infty} \left|\frac{c_{n+1}}{c_n}\right|$$

so that $\sum_{n=0}^{\infty} c_n x^n$ converges for $|x| < \dfrac{1}{\lim_{n\to\infty} |c_{n+1}/c_n|}$ and diverges for $|x| > \dfrac{1}{\lim_{n\to\infty} |c_{n+1}/c_n|}$. Therefore $R = \dfrac{1}{\lim_{n\to\infty} |c_{n+1}/c_n|}$, and thus $\lim_{n\to\infty} |c_{n+1}/c_n| = 1/R$.

57. Since $[(2j-1)/(2j)]^2 < 1$ for $j = 1, 2, \ldots, n$, we have

$$\frac{1^2 \cdot 3^2 \cdot 5^2 \cdots (2n-1)^2}{2^2 \cdot 4^2 \cdot 6^2 \cdots (2n)^2} \leq \frac{1^2}{2^2} = \frac{1}{4}$$

Thus if $x^2 < 0.018$, then

$$\sum_{n=1}^{\infty} \frac{1^2 \cdot 3^2 \cdot 5^2 \cdots (2n-1)^2}{2^2 \cdot 4^2 \cdot 6^2 \cdots (2n)^2} x^{2n} \leq \sum_{n=1}^{\infty} \frac{1}{4} x^{2n} \leq \sum_{n=1}^{\infty} \frac{1}{4}(0.018)^n = \frac{1}{4}\left(\frac{0.018}{1-0.018}\right) < 0.005$$

SECTION 9.9

1. $f(x) = x^4 - x + 2$, $f'(x) = 4x^3 - 1$, $f''(x) = 12x^2$, $f^{(3)}(x) = 24x$; $f(-1) = 4$, $f'(-1) = -5$, $f''(-1) = 12$, $f^{(3)}(-1) = -24$; $p_3(x) = 4 - 5(x+1) + (12/2!)(x+1)^2 - (24/3!)(x+1)^3 = 4 - 5(x+1) + 6(x+1)^2 - 4(x+1)^3$.

3. $f(x) = \sqrt{x}$, $f'(x) = 1/(2\sqrt{x})$, $f''(x) = -1/(4x^{3/2})$, $f^{(3)}(x) = 3/(8x^{5/2})$; $f(4) = 2$, $f'(4) = \frac{1}{4}$, $f''(4) = -1/32$, $f^{(3)}(4) = \frac{3}{256}$; $p_3(x) = 2 + \frac{1}{4}(x-4) - \frac{1}{32} \cdot (1/2!)(x-4)^2 + \frac{3}{256} \cdot (1/3!)(x-4)^3 = 2 + \frac{1}{4}(x-4) - \frac{1}{64}(x-4)^2 + \frac{1}{512}(x-4)^3$.

5. $g(x) = \cos x$, $g'(x) = -\sin x$, $g''(x) = -\cos x$, $g^{(3)}(x) = \sin x$;

$$g\left(\frac{\pi}{4}\right) = \frac{\sqrt{2}}{2},\ g'\left(\frac{\pi}{4}\right) = \frac{-\sqrt{2}}{2},\ g''\left(\frac{\pi}{4}\right) = \frac{-\sqrt{2}}{2},\ g^{(3)}\left(\frac{\pi}{4}\right) = \frac{\sqrt{2}}{2};$$

$$p_3(x) = \sqrt{2}/2 - (\sqrt{2}/2)(x - \pi/4) - \frac{\sqrt{2}}{2} \cdot \frac{1}{2!}\left(x - \frac{\pi}{4}\right)^2 + \frac{\sqrt{2}}{2} \cdot \frac{1}{3!}\left(x - \frac{\pi}{4}\right)^3$$
$$= \frac{\sqrt{2}}{2} - \frac{\sqrt{2}}{2}\left(x - \frac{\pi}{4}\right) - \frac{\sqrt{2}}{4}\left(x - \frac{\pi}{4}\right)^2 + \frac{\sqrt{2}}{12}\left(x - \frac{\pi}{4}\right)^3$$

7. $k(x) = \csc x$, $k'(x) = -\csc x \cot x$, $k''(x) = \csc x \cot^2 x + \csc^3 x$, $k^{(3)}(x) = -\csc x \cot^3 x - 5\csc^3 x \cot x$; $k(\pi/2) = 1$, $k'(\pi/2) = 0$, $k''(\pi/2) = 1$; $k^{(3)}(\pi/2) = 0$; $p_3(x) = 1 + (1/2!)(x - \pi/2)^2 = 1 + \frac{1}{2}(x - \pi/2)^2$.

9. $f(x) = 4x^2 - 2x + 1$, $f'(x) = 8x - 2$, $f''(x) = 8$, $f^{(n)}(x) = 0$ for $n \geq 3$; $f(0) = 1$, $f'(0) = -2$, $f''(0) = 8$, $f^{(n)}(0) = 0$ for $n \geq 3$, so the Taylor series about 0 is $f(x) = 1 - 2x + (8/2!)x^2 = 1 - 2x + 4x^2$, and $r_n(x) = 0$ for $n \geq 3$. $f(-3) = 43$, $f'(-3) = -26$, $f''(-3) = 8$, $f^{(n)}(-3) = 0$ for $n \geq 3$, so the Taylor series about -3 is given by

$$43 - 26(x+3) + 8 \cdot \left(\frac{1}{2!}\right)(x+3)^2 = 43 - 26(x+3) + 4(x+3)^2$$

and $r_n(x) = 0$ for $n \geq 3$ and all x.

11. $f(x) = \cos x$, $f'(x) = -\sin x$, $f''(x) = -\cos x$, $f^{(3)}(x) = \sin x$, $f^{(4)}(x) = \cos x$; $f^{(2k)}(x) = (-1)^k \cos x$ and $f^{(2k+1)}(x) = (-1)^{k+1} \sin x$, so that $f^{(2k)}(0) = (-1)^k$ and $f^{(2k+1)}(0) = 0$ for any nonnegative integer k; the Taylor series about 0 is

$$1 - \frac{1}{2!}x^2 + \frac{1}{4!}x^4 - \frac{1}{6!}x^6 + \cdots + (-1)^n \frac{1}{(2n)!}x^2 + \cdots = \sum_{n=0}^{\infty} \frac{(-1)^n}{(2n)!} x^{2n}$$
$$|r_n(x)| = \left|\frac{f^{(n+1)}(t_x)}{(n+1)!} x^{n+1}\right| \leq \frac{|x|^{n+1}}{(n+1)!}$$

and for any x, $\lim_{n\to\infty} |x|^{n+1}/(n+1)! = 0$ by Example 7 in Section 9.7. Thus $\lim_{n\to\infty} |r_n(x)| = 0$ for all x, so that $\lim_{n\to\infty} r_n(x) = 0$ for all x.

13. $f(x) = \sin x$, $f'(x) = \cos x$, $f''(x) = -\sin x$, $f^{(3)}(x) = -\cos x$, $f^{(4)}(x) = \sin x$; $f^{(2k)}(x) = (-1)^k \sin x$ and $f^{(2k+1)}(x) = (-1)^k \cos x$, so that we have $f^{(2k)}(\pi/3) = (-1)^k(\sqrt{3}/2)$ and $f^{(2k+1)}(\pi/3) = (-1)^k\frac{1}{2}$ for any nonnegative integer k; the Taylor series about $\pi/3$ is

$$\frac{\sqrt{3}}{2} + \frac{1}{2}\left(x - \frac{\pi}{3}\right) - \frac{\sqrt{3}}{2} \cdot \frac{1}{2!}\left(x - \frac{\pi}{3}\right)^2 - \frac{1}{2} \cdot \frac{1}{3!}\left(x - \frac{\pi}{3}\right)^3 + \cdots$$
$$= \sum_{n=0}^{\infty} \frac{(-1)^n}{2}\left[\frac{\sqrt{3}}{(2n)!}\left(x - \frac{\pi}{3}\right)^{2n} + \frac{1}{(2n+1)!}\left(x - \frac{\pi}{3}\right)^{2n+1}\right]$$

$$|r_n(x)| = \left|\frac{f^{(n+1)}(t_x)}{(n+1)!}\left(x - \frac{\pi}{3}\right)^{n+1}\right| \leq \frac{|x - \pi/3|^{n+1}}{(n+1)!}, \quad \text{and} \quad \lim_{n\to\infty} \frac{|x - \pi/3|^{n+1}}{(n+1)!} = 0$$

so $\lim_{n\to\infty} |r_n(x)| = 0$, and thus $\lim_{n\to\infty} r_n(x) = 0$ for all x.

15. $f^{(k)}(x) = e^x$ and $f^{(k)}(-2) = e^{-2}$ for any nonnegative integer k; the Taylor series about -2 is

$$e^{-2} + e^{-2}(x+2) + e^{-2} \cdot \frac{1}{2!}(x+2)^2 + e^{-2} \cdot \frac{1}{3!}(x+2)^3 + \cdots = \sum_{n=0}^{\infty} \frac{e^{-2}}{n!}(x+2)^n$$

$$|r_n(x)| = \left| \frac{f^{(n+1)}(t_x)}{(n+1)!}(x+2)^{n+1} \right| \le \frac{1+e^x}{(n+1)!}|x+2|^{n+1}, \quad \text{and} \quad \lim_{n\to\infty} \frac{1+e^x}{(n+1)!}|x+2|^{n+1} = 0$$

so $\lim_{n\to\infty} |r_n(x)| = 0$, and thus $\lim_{n\to\infty} r_n(x) = 0$ for all x.

17. $f(x) = \dfrac{1}{x}$, $f^{(k)}(x) = \dfrac{(-1)^k k!}{x^{k+1}}$, $f^{(k)}(3) = \dfrac{(-1)^k k!}{3^{k+1}}$ for any nonnegative integer k;

the Taylor series about 3 is

$$\frac{1}{3} - \frac{1}{3^2}(x-3) + \frac{2!}{3^3} \cdot \frac{1}{2!}(x-3)^2 - \frac{3!}{3^4} \cdot \frac{1}{3!}(x-3)^3 + \cdots = \sum_{n=0}^{\infty} \frac{(-1)^n}{3^{n+1}}(x-3)^n$$

19. $f(x) = \ln x$, $f^{(k)}(x) = \dfrac{(-1)^{k+1}(k-1)!}{x^k}$, $f^{(k)}(2) = \dfrac{(-1)^{k+1}(k-1)!}{2^k}$ for any positive integer k; the Taylor series about 2 is

$$\ln 2 + \frac{1}{2}(x-2) - \frac{1}{2^2} \cdot \frac{1}{2!}(x-2)^2 + \frac{2!}{2^3} \cdot \frac{1}{3!}(x-2)^3 - \frac{3!}{2^4} \cdot \frac{1}{4!}(x-2)^4 + \cdots$$

$$= \ln 2 + \sum_{n=1}^{\infty} \frac{(-1)^{n+1}}{n2^n}(x-2)^n$$

21. Let $m = 1$ and let $g(x) = xf(x)$. Then $g^{(k)}(x) = kf^{(k-1)}(x) + xf^{(k)}(x)$, and $g^{(k)}(0) = kf^{(k-1)}(0)$ for any nonnegative integer k. Thus

$$\sum_{n=0}^{\infty} \frac{g^{(n)}(0)}{n!}x^n = \sum_{n=1}^{\infty} \frac{nf^{(n-1)}(0)}{n!}x^n = \sum_{n=1}^{\infty} \frac{f^{(n-1)}(0)}{(n-1)!}x^n = \sum_{n=0}^{\infty} \frac{f^{(n)}(0)}{n!}x^{n+1}$$

By hypothesis, $c_n = f^{(n)}(0)/n!$, so

$$\sum_{n=0}^{\infty} \frac{g^{(n)}(0)}{n!}x^n = \sum_{n=0}^{\infty} c_n x^{n+1}$$

so that $\sum_{n=0}^{\infty} c_n x^{n+1}$ is the Taylor series of g and thus of xf. By induction, the Taylor series of $x^m f$ is $\sum_{n=0}^{\infty} c_n x^{n+m}$ for any integer $m \ge 0$.

23. $\cos x^2 = \displaystyle\sum_{n=0}^{\infty} \frac{(-1)^n}{(2n)!}(x^2)^{2n} = \sum_{n=0}^{\infty} \frac{(-1)^n}{(2n)!}x^{4n}$

25. $\ln(1-2x) = \displaystyle\sum_{n=0}^{\infty} \frac{(-1)^n}{n+1}(-2x)^{n+1} = \sum_{n=0}^{\infty} -\frac{2^{n+1}}{n+1}x^{n+1}$

27. Since $\ln(1+x) = \sum_{n=0}^{\infty} [(-1)^n/(n+1)]x^{n+1}$, we have

$$x \ln(1+x^2) = x \sum_{n=0}^{\infty} \frac{(-1)^n}{n+1}(x^2)^{n+1} = \sum_{n=0}^{\infty} \frac{(-1)^n}{n+1}x^{2n+3}$$

29. $2^x = e^{x \ln 2} = \displaystyle\sum_{n=0}^{\infty} \frac{1}{n!}(x \ln 2)^n = \sum_{n=0}^{\infty} \frac{(\ln 2)^n}{n!}x^n$

31. $\dfrac{x-1}{x+1} = 1 - \dfrac{2}{x+1} = 1 - \dfrac{2}{2+(x-1)} = 1 - \dfrac{1}{1+\frac{1}{2}(x-1)} = 1 - \displaystyle\sum_{n=0}^{\infty} \frac{(-1)^n}{2^n}(x-1)^n$

$$= \sum_{n=1}^{\infty} \frac{(-1)^{n+1}}{2^n}(x-1)^n = \sum_{n=0}^{\infty} \frac{(-1)^n}{2^{n+1}}(x-1)^{n+1}$$

33. $\cos^2 x = \dfrac{1}{2}(1 + \cos 2x) = \dfrac{1}{2}\left(1 + \displaystyle\sum_{n=0}^{\infty} \frac{(-1)^n}{2n!}(2x)^{2n}\right) = 1 + \sum_{n=1}^{\infty} \frac{(-1)^n 2^{2n-1}}{(2n)!}x^{2n}$

35. For $x \ne 0$,

$$\frac{\sin x}{x} = \frac{1}{x}\sum_{n=0}^{\infty} \frac{(-1)^n}{(2n+1)!}x^{2n+1} = \sum_{n=0}^{\infty} \frac{(-1)^n}{(2n+1)!}x^{2n}, \quad \text{so} \quad f(x) = \sum_{n=0}^{\infty} \frac{(-1)^n}{(2n+1)!}x^{2n}$$

37. $\dfrac{-3x+2}{2x^2-3x+1} = \dfrac{-3x+2}{(2x-1)(x-1)} = \dfrac{A}{2x-1} + \dfrac{B}{x-1} = \dfrac{A(x-1)+B(2x-1)}{(2x-1)(x-1)}$

$A + 2B = -3$ and $A + B = -2$ imply that $B = -1$, so $A = -2 - B = -1$. Therefore

$$\frac{-3x+2}{2x^2-3x+1} = \frac{-1}{2x-1} - \frac{-1}{x-1} = \frac{1}{1-2x} + \frac{1}{1-x} = \sum_{n=0}^{\infty} (2x)^n + \sum_{n=0}^{\infty} x^n = \sum_{n=0}^{\infty} (2^n+1)x^n$$

39. a. $f(0) = e^{(0^2)} \int_0^0 e^{(-t^2)}\,dt = 0$; $f'(x) = 2xe^{(x^2)} \int_0^x e^{-(t^2)}\,dt + e^{(x^2)}e^{-(x^2)} = 2xf(x) + 1$.

b. By part (a), we obtain $f''(x) = 2f(x) + 2xf'(x)$, $f^{(3)}(x) = 2f'(x) + 2f'(x) + 2xf''(x) = 2(2)f'(x) + 2xf''(x)$. In general, $f^{(n)}(x) = 2(n-1)f^{(n-2)}(x) + 2xf^{(n-1)}(x)$ for $n \ge 2$. Therefore $f^{(n)}(0) = 2(n-1)f^{(n-2)}(0)$ for $n \ge 2$. By part (a), $f(0) = 0$ and $f'(0) = 2(0)f(0) + 1 = 1$, so that for $k \ge 0$ we have $f^{(2k)}(0) = 2(2k-1)f^{(2k-2)}(0) = \cdots = 0$. In addition,

$$f^{(2n+1)}(0) = 2(2n)f^{(2n-1)}(0) = 2(2n)(2)(2n-2)f^{(2n-3)}(0) = 2^{2\cdot 2}n(n-1)f^{(2n-3)}(0)$$

$$= 2^{2\cdot 3}n(n-1)(n-2)f^{(2n-5)}(0) = \cdots = 2^{2n}n!\,f'(0) = 4^n n!$$

Consequently $f(x) = \sum_{n=0}^{\infty} [4^n n!/(2n+1)!]x^{2n+1}$.

41. a. $0 \le a_{n+1} = a_n + a_{n-1} \le 2a_n$. Thus $a_2 \le 2a_1 = 2$, $a_3 \le 2a_2 \le 4$, and, in general,

$$a_n \le 2a_{n-1} \le 4a_{n-2} \le \cdots \le 2^{n-1}a_2 \le 2^n a_1 = 2^n$$

Since the radius of convergence of $\sum_{n=1}^{\infty} 2^n x^n$ is $\frac{1}{2}$ (by the Ratio Test), it follows that $\sum_{n=1}^{\infty} a_n x^n$ converges for $|x| < \frac{1}{2}$. Thus the radius of convergence of $\sum_{n=1}^{\infty} a_n x^n$ is at least $\frac{1}{2}$.

b. $\sum_{n=1}^{\infty} a_n x^n - x\sum_{n=1}^{\infty} a_n x^n - x^2\sum_{n=1}^{\infty} a_n x^n = \left(a_1x + a_2x^2 + \sum_{n=3}^{\infty} a_n x^n\right)$

$$- \left(a_1x^2 + \sum_{n=3}^{\infty} a_{n-1}x^n\right) - \sum_{n=3}^{\infty} a_{n-2}x^n$$

$$= (x + x^2 - x^2) + \sum_{n=3}^{\infty} (a_n - a_{n-1} - a_{n-2})x^n$$

$$= x \quad \text{for } |x| < \tfrac{1}{2}$$

since $a_n = a_{n-1} + a_{n-2}$ for $n \geq 3$. Thus $\sum_{n=1}^{\infty} a_n x^n = x/(1 - x - x^2)$ for $|x| < \frac{1}{2}$.

SECTION 9.10

1. $\sqrt{1.05} = \sqrt{1 + \frac{1}{20}}$, so by (5) with $s = \frac{1}{2}$ and $x = \frac{1}{20}$, we have

$$\left|r_N\left(\frac{1}{20}\right)\right| \leq \frac{1}{2}\frac{1/20^{N+1}}{1 - 1/20} = \frac{1}{30 \cdot 20^N}$$

Then $|r_N(\frac{1}{20})| < 0.001$ if $N = 2$, so by taking the first 3 terms in (2) we obtain

$$\sqrt{1.05} \approx 1 + \tfrac{1}{2}(\tfrac{1}{20}) - \tfrac{1}{8}(\tfrac{1}{20})^2 \approx 1.02469$$

as the desired approximation.

3. $\sqrt[4]{83} = 3\sqrt[4]{1 + \frac{2}{81}}$, so by (5) with $s = \frac{1}{4}$ and $x = \frac{2}{81}$, we have

$$\left|r_N\left(\frac{2}{81}\right)\right| \leq \frac{1}{4}\frac{(\frac{2}{81})^{N+1}}{1 - \frac{2}{81}} = \frac{2^{N-1}}{79 \cdot 81^N}$$

Then $|r_N(\frac{2}{81})| < \frac{1}{3000}$ if $N = 1$, so by taking the first two terms in (2) we obtain

$$\sqrt[4]{83} \approx 3(1 + \tfrac{1}{4}(\tfrac{2}{81})) \approx 3.01852$$

as the desired approximation.

5. $\sqrt[6]{65} = 2\sqrt[6]{1 + \frac{1}{64}}$, so by (5) with $s = \frac{1}{6}$ and $x = \frac{1}{64}$, we have

$$\left|r_N\left(\frac{1}{64}\right)\right| \leq \frac{1}{6}\frac{(\frac{1}{64})^{N+1}}{1 - \frac{1}{64}} = \frac{1}{378.64^N}$$

Then $|r_N(\frac{1}{64})| < 0.0005$ if $N = 1$, so by taking the first two terms in (2) we obtain

$$\sqrt[6]{65} \approx 2(1 + \tfrac{1}{6}(\tfrac{1}{64})) \approx 2.00521$$

as the desired approximation.

7. $$\frac{1}{\sqrt{1+x}} = \sum_{n=0}^{\infty} \binom{-\frac{1}{2}}{n} x^n = 1 - \frac{1}{2}x + \frac{1}{2!}\left(\frac{-1}{2}\right)\left(\frac{-3}{2}\right)x^2 + \frac{1}{3!}\left(\frac{-1}{2}\right)\left(\frac{-3}{2}\right)\left(\frac{-5}{2}\right)x^3 + \cdots$$

9. $$(1 + x)^{-8/5} = \sum_{n=0}^{\infty} \binom{-\frac{8}{5}}{n} x^n = 1 - \frac{8}{5}x + \frac{1}{2!}\left(\frac{-8}{5}\right)\left(\frac{-13}{5}\right)x^2 + \frac{1}{3!}\left(\frac{-8}{5}\right)\left(\frac{-13}{5}\right)\left(\frac{-18}{5}\right)x^3 + \cdots$$

11. $$\frac{x}{\sqrt{1 - x^2}} = x\sum_{n=0}^{\infty} \binom{-\frac{1}{2}}{n}(-x^2)^n = \sum_{n=0}^{\infty} (-1)^n \binom{-\frac{1}{2}}{n} x^{2n+1}$$

$$= x + \frac{1}{2}x^3 + \frac{1}{2!}\left(\frac{-1}{2}\right)\left(\frac{-3}{2}\right)x^5 - \frac{1}{3!}\left(\frac{-1}{2}\right)\left(\frac{-3}{2}\right)\left(\frac{-5}{2}\right)x^7 + \cdots$$

13. $$\sqrt{1 - (x + 1)^2} = \sum_{n=0}^{\infty} \binom{\frac{1}{2}}{n}(-(x + 1)^2)^n = \sum_{n=0}^{\infty} (-1)^n \binom{\frac{1}{2}}{n}(x + 1)^{2n}$$

$$= 1 - \frac{1}{2}(x + 1)^2 + \frac{1}{2!}\left(\frac{1}{2}\right)\left(\frac{-1}{2}\right)(x + 1)^4 - \frac{1}{3!}\left(\frac{1}{2}\right)\left(\frac{-1}{2}\right)\left(\frac{-3}{2}\right)(x + 1)^6 + \cdots$$

15. $$\int_0^1 \sqrt{1 - x^2}\,dx = \int_0^1 \left[\sum_{n=0}^{\infty} \binom{\frac{1}{2}}{n}(-1)^n x^{2n}\right] dx = \sum_{n=0}^{\infty} \binom{\frac{1}{2}}{n}\frac{(-1)^n}{2n + 1}(1)^{2n} = \sum_{n=0}^{\infty} \binom{\frac{1}{2}}{n}\frac{(-1)^n}{2n + 1}$$

and by integration by trigonometric substitution with $x = \sin u$, we find that

$$\int_0^1 \sqrt{1 - x^2}\,dx = \int_0^{\pi/2} \sqrt{1 - \sin^2 u}\cos u\,du = \int_0^{\pi/2} \cos^2 u\,du$$

$$= \frac{1}{2}\int_0^{\pi/2} (1 + \cos 2u)\,du = \frac{1}{2}\left(u + \frac{1}{2}\sin 2u\right)\Big|_0^{\pi/2} = \frac{\pi}{4}$$

thus $$\sum_{n=0}^{\infty} \binom{\frac{1}{2}}{n}\frac{(-1)^n}{2n + 1} = \frac{\pi}{4}$$

17. Without loss of generality, assume in this problem that $a > 0$.

a. $$\int_0^x \sqrt{a^2 + t^2}\,dt = a\int_0^x \sqrt{1 + \left(\frac{t}{a}\right)^2}\,dt = a\int_0^x \left(\sum_{n=0}^{\infty} \binom{\frac{1}{2}}{n}\left(\frac{t}{a}\right)^{2n}\right) dt$$

$$= a\sum_{n=0}^{\infty} \binom{\frac{1}{2}}{n}\left(\int_0^x \left(\frac{t}{a}\right)^{2n} dt\right) = a\sum_{n=0}^{\infty} \binom{\frac{1}{2}}{n}\left(\frac{a}{2n + 1}\left(\frac{t}{a}\right)^{2n+1}\Bigg|_0^x\right)$$

$$= a^2\sum_{n=0}^{\infty} \binom{\frac{1}{2}}{n}\frac{1}{2n + 1}\left(\frac{x}{a}\right)^{2n+1} = \sum_{n=0}^{\infty} \binom{\frac{1}{2}}{n}\left(\frac{1}{a}\right)^{2n-1}\frac{1}{2n + 1}x^{2n+1}$$

Since $$\lim_{n\to\infty}\left|\frac{\binom{\frac{1}{2}}{n+1}\left(\frac{x}{a}\right)^{2n+3}\Big/(2n + 3)}{\binom{\frac{1}{2}}{n}\left(\frac{x}{a}\right)^{2n+1}\Big/(2n + 1)}\right| = \lim_{n\to\infty} \frac{|\frac{1}{2} - n|}{n + 1}\,\frac{2n + 1}{2n + 3}\left|\frac{x}{a}\right|^2 = \left|\frac{x}{a}\right|^2$$

the radius of convergence is a.

b. $$\int_0^x \sqrt{a^2 - t^2}\,dt = a\int_0^x \sqrt{1 - \left(\frac{t}{a}\right)^2}\,dt = a\int_0^x \left(\sum_{n=0}^{\infty} (-1)^n\binom{\frac{1}{2}}{n}\left(\frac{t}{a}\right)^{2n} dt\right)$$

$$= a\sum_{n=0}^{\infty} (-1)^n\binom{\frac{1}{2}}{n}\left(\int_0^x \left(\frac{t}{a}\right)^{2n} dt\right) = a^2\sum_{n=0}^{\infty} (-1)^n\binom{\frac{1}{2}}{n}\frac{1}{2n + 1}\left(\frac{x}{a}\right)^{2n+1}$$

$$= \sum_{n=0}^{\infty} (-1)^n\binom{\frac{1}{2}}{n}\left(\frac{1}{a}\right)^{2n-1}\frac{1}{2n + 1}x^{2n+1}$$

As in (a), the radius of convergence is a.

19. For $s = 0$ we have $\left|\binom{0}{n}\right| = 0 = \left|\binom{0}{n-1}\right|$. For $n = 1$ we have $\left|\binom{s}{1}\right| = |s| \le 1 = \left|\binom{s}{0}\right|$.

For $n > 1$ and $0 < |s| \le 1$, we have $|s - n + 1| \le n$, so that

$$\left|\frac{\binom{s}{n-1}}{\binom{s}{n}}\right| = \left|\frac{s(s-1)\cdots(s-n+2)/(n-1)!}{s(s-1)\cdots(s-n+1)/n!}\right| = \frac{n}{|s-n+1|} \ge 1$$

and thus $\left|\binom{s}{n}\right| \le \left|\binom{s}{n-1}\right|$.

21. From the figure below, maximum depth is $4000 - x$, where $x = \sqrt{4000^2 - 100^2} = 4000\sqrt{1 - (\frac{100}{4000})^2} = 4000\sqrt{1 - \frac{1}{1600}}$. By Exercise 20,

$$\sqrt{1 - \tfrac{1}{1600}} \approx 1 - 1/[2(1600)] = \tfrac{3199}{3200}$$

so that $x \approx 4000(\frac{3199}{3200}) = 3998.75$, and hence the desired approximation of the maximum depth is 1.25 (miles). (The maximum depth is actually 1.25020, accurate to 6 places.)

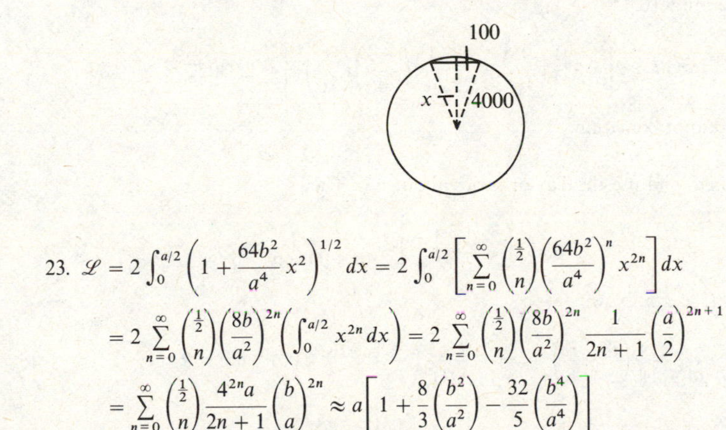

23. $$\mathscr{L} = 2\int_0^{a/2}\left(1 + \frac{64b^2}{a^4}x^2\right)^{1/2} dx = 2\int_0^{a/2}\left[\sum_{n=0}^{\infty}\binom{\frac{1}{2}}{n}\left(\frac{64b^2}{a^4}\right)^n x^{2n}\right]dx$$

$$= 2\sum_{n=0}^{\infty}\binom{\frac{1}{2}}{n}\left(\frac{8b}{a^2}\right)^{2n}\left(\int_0^{a/2} x^{2n}\,dx\right) = 2\sum_{n=0}^{\infty}\binom{\frac{1}{2}}{n}\left(\frac{8b}{a^2}\right)^{2n}\frac{1}{2n+1}\left(\frac{a}{2}\right)^{2n+1}$$

$$= \sum_{n=0}^{\infty}\binom{\frac{1}{2}}{n}\frac{4^{2n}a}{2n+1}\left(\frac{b}{a}\right)^{2n} \approx a\left[1 + \frac{8}{3}\left(\frac{b^2}{a^2}\right) - \frac{32}{5}\left(\frac{b^4}{a^4}\right)\right]$$

If $a = 500$ and $b = 40$, then

$$\mathscr{L} \approx 500[1 + \tfrac{8}{3}(\tfrac{40}{500})^2 - \tfrac{32}{5}(\tfrac{40}{500})^4] \approx 508.402 \text{ (feet)}$$

CHAPTER 9 REVIEW

1. By replacing 0.05 by e in the solution of Exercise 25 in Section 9.2, we obtain $\lim_{n\to\infty}(1 + e/n)^n = e^e$.

3. $$\lim_{n\to\infty}(\sqrt{n^2+n} - \sqrt{n^2-n}) = \lim_{n\to\infty}(\sqrt{n^2+n} - \sqrt{n^2-n})\left(\frac{\sqrt{n^2+n}+\sqrt{n^2-n}}{\sqrt{n^2+n}+\sqrt{n^2-n}}\right)$$

$$= \lim_{n\to\infty}\frac{2n}{\sqrt{n^2+n}+\sqrt{n^2-n}}$$

$$= \lim_{n\to\infty}\frac{2}{\sqrt{1+1/n}+\sqrt{1-1/n}}$$

$$= \frac{2}{2} = 1$$

5. Let $f(x) = 1/x$, and partition the interval $[1, 2]$ into n subintervals of equal length, so that $\Delta x_k = 1/n$. If t_k is the right-hand endpoints of the kth subinterval, then $t_k = 1 + k/n = (n+k)/n$, so that

$$f(t_k)\,\Delta x_k = \frac{n}{n+k}\cdot\frac{1}{n} = \frac{1}{n+k}$$

The corresponding Riemann sum is

$$\sum_{k=1}^{n} f(t_k)\,\Delta x_k = \sum_{k=1}^{n}\frac{1}{n+k} = \frac{1}{n+1} + \frac{1}{n+2} + \cdots + \frac{1}{2n}$$

It follows from (5) in Section 5.2 that

$$\lim_{n\to\infty}\left(\frac{1}{n+1} + \frac{1}{n+2} + \cdots + \frac{1}{2n}\right) = \lim_{n\to\infty}\sum_{k=1}^{n} f(t_k)\,\Delta x_k = \int_1^2 \frac{1}{x}\,dx = \ln 2$$

7. Since $1/(n + n\sqrt{n}) < 1/n^{3/2}$ for $n \ge 1$, and $\sum_{n=1}^{\infty} 1/n^{3/2}$ converges, it follows that $\sum_{n=1}^{\infty} 1/(n + n\sqrt{n})$ converges.

9. Since $\sqrt{n}/(n^2 + n + 1) \le \sqrt{n}/n^2 = 1/n^{3/2}$ for $n \ge 1$, and $\sum_{n=1}^{\infty} 1/n^{3/2}$ converges, it follows that $\sum_{n=1}^{\infty}\sqrt{n}/(n^2 + n + 1)$ converges.

11. Since $$\frac{6^n}{n^2(\ln n)^2} > \frac{6^n}{n^2(2n^{1/2})^2} = \frac{6^n}{4n^3}$$

and since $\lim_{n\to\infty}\sqrt[n]{6^n/4n^3} = \lim_{n\to\infty} 6/\sqrt[n]{4n^3} = 6$, it follows that $\sum_{n=2}^{\infty} 6^n/n^3$ diverges, and hence that $\sum_{n=2}^{\infty} 6^n/[n^2(\ln n)^2]$ diverges.

13. Since $$\lim_{n\to\infty}\frac{3^{n+1}/(n+1)^3}{3^n/n^3} = \lim_{n\to\infty} 3\left(\frac{n}{n+1}\right)^3 = 3$$

it follows that $\sum_{n=1}^{\infty} 3^n/n^3$ diverges.

15. Let $f(x) = \sqrt{x}/(x-3)$ for $x \ge 4$. Then

$$f'(x) = \frac{(x-3)/(2\sqrt{x}) - \sqrt{x}}{(x-3)^2} = \frac{-x-3}{2\sqrt{x}(x-3)^2} < 0$$

so f is decreasing on $[4, \infty]$. Thus $\{\sqrt{n}/(n-3)\}_{n=4}^{\infty}$ is a nonnegative, decreasing sequence, and

$$\lim_{n\to\infty}\frac{\sqrt{n}}{n-3} = \lim_{n\to\infty}\frac{1}{\sqrt{n} - 3/\sqrt{n}} = 0$$

Therefore $\sum_{n=4}^{\infty}(-1)^n\sqrt{n}/(n-3)$ converges.

17. Since $\sin^2(1/n) \le 1/n^2$ for $n \ge 1$, and $\sum_{n=1}^{\infty} 1/n^2$ converges, it follows that $\sum_{n=1}^{\infty} \sin^2(1/n)$ converges.

19. $$\ln 2 = \sum_{n=1}^{\infty} (-1)^{n+1}\frac{1}{n} = 1 - \frac{1}{2} + \frac{1}{3} - \frac{1}{4} + \frac{1}{5} - \frac{1}{6} + \cdots$$
$$= \left(1 - \frac{1}{2}\right) + \left(\frac{1}{3} - \frac{1}{4}\right) + \left(\frac{1}{5} - \frac{1}{6}\right) + \cdots = \frac{1}{1\cdot 2} + \frac{1}{3\cdot 4} + \frac{1}{5\cdot 6} + \cdots$$
$$= \sum_{n=0}^{\infty} \frac{1}{(2n+1)(2n+2)}$$

21. $$\sum_{n=0}^{\infty} (-1)^n \frac{2n+3}{(n+1)(n+2)} = \sum_{n=0}^{\infty} (-1)^n\left(\frac{1}{n+1} + \frac{1}{n+2}\right)$$
$$= \sum_{n=0}^{\infty} (-1)^n \frac{1}{n+1} + \sum_{n=0}^{\infty} (-1)^n \frac{1}{n+2}$$
$$= 1 + \sum_{n=2}^{\infty} (-1)^n \frac{1}{n} - \sum_{n=2}^{\infty} (-1)^n \frac{1}{n} = 1$$

23. Since $\sum_{n=1}^{\infty} x^n = x/(1-x)$ for $|x| < 1$, differentiation yields $\sum_{n=1}^{\infty} nx^{n-1} = 1/(1-x)^2$. Thus
$$(1-x)\sum_{n=1}^{\infty} nx^n = [(1-x)x]\sum_{n=1}^{\infty} nx^{n-1} = (1-x)\,x\frac{1}{(1-x)^2} = \frac{x}{1-x} \quad \text{for } |x| < 1$$

25. a. If $\lim_{n\to\infty} a_n = L$ and $\lim_{n\to\infty} b_n = M$, then $\lim_{n\to\infty}(b_n - a_n) = \lim_{n\to\infty} b_n - \lim_{n\to\infty} a_n = M - L$. Let $\varepsilon > 0$. Then there is a positive integer N such that if $n \ge N$, then $|(b_n - a_n) - (M - L)| < \varepsilon$, so that $(M - L) - \varepsilon < b_n - a_n < (M - L) + \varepsilon$. By hypothesis, $a_n \le b_n$, so that $b_n - a_n \ge 0$, and thus $0 < (M - L) + \varepsilon$, or $-\varepsilon < M - L$. Since ε is arbitrary and positive, $0 \le M - L$, so that $L \le M$, which is equivalent to $\lim_{n\to\infty} a_n \le \lim_{n\to\infty} b_n$.

b. Let $s_j = \sum_{n=1}^{j} a_n$ and $s_j' = \sum_{n=1}^{j} |a_n|$ for $j \ge 1$. Since $a_n \le |a_n|$ for $n \ge 1$, we have $s_j \le s_j'$ for $j \ge 1$. Moreover, $\{s_j\}_{j=1}^{\infty}$ and $\{s_j'\}_{j=1}^{\infty}$ converge by hypothesis, and thus by part (a), $\lim_{j\to\infty} s_j \le \lim_{j\to\infty} s_j'$, so that $\sum_{n=1}^{\infty} a_n \le \sum_{n=1}^{\infty} |a_n|$. Replacing a_n by $-a_n$, we find that $-\sum_{n=1}^{\infty} a_n = \sum_{n=1}^{\infty} (-a_n) \le \sum_{n=1}^{\infty} |a_n|$. Thus $|\sum_{n=1}^{\infty} a_n| \le \sum_{n=1}^{\infty} |a_n|$.

27. $$\lim_{n\to\infty} \sqrt[n]{\frac{3^n}{5^{2n}}|x|^{3n}} = \lim_{n\to\infty} \frac{3}{25}|x|^3 = \frac{3}{25}|x|^3$$

so $\sum_{n=0}^{\infty} (3^n/5^{2n})x^{3n}$ converges for $|x| < \sqrt[3]{\frac{25}{3}}$ and diverges for $|x| > \sqrt[3]{\frac{25}{3}}$. Since $\sum_{n=0}^{\infty} (-1)^n$ and $\sum_{n=0}^{\infty} 1$ diverge, the interval of convergence is $(-\sqrt[3]{\frac{25}{3}}, \sqrt[3]{\frac{25}{3}})$.

29. $$\lim_{n\to\infty} \frac{(n+1)^{2n+2}|x|^{n+1}/(2n+2)!}{n^{2n}|x|^n/(2n)!} = \lim_{n\to\infty} \left[\left(1 + \frac{1}{n}\right)^n\right]^2 \frac{n+1}{4n+2}|x| = \frac{e^2|x|}{4}$$

The radius of convergence is $4/e^2$

31. $f(x) = \sqrt{1 + x^4}$, $f'(x) = \dfrac{2x^3}{\sqrt{1+x^4}}$, $f''(x) = \dfrac{6x^2 + 2x^6}{(1+x^4)^{3/2}}$;

$f(0) = 1$, $f'(0) = 0$, $f''(0) = 0$; $p_2(x) = 1$.

33. $f^{(2k)}(x) = (-1)^k \sin x$ and $f^{(2k+1)}(x) = (-1)^k \cos x$, and $f^{(2k)}(\pi/4) = (-1)^k(\sqrt{2}/2)$ and $f^{(2k+1)}(\pi/4) = (-1)^k(\sqrt{2}/2)$ for $k \ge 0$; the Taylor series about $\pi/4$ is
$$\sum_{n=0}^{\infty} \frac{f^{(n)}(\pi/4)}{n!}\left(x - \frac{\pi}{4}\right)^n = \sum_{n=0}^{\infty} (-1)^n \frac{\sqrt{2}}{2}\left[\frac{(x - \pi/4)^{2n}}{(2n)!} + \frac{(x - \pi/4)^{2n+1}}{(2n+1)!}\right]$$

35. $$\frac{x-1}{x+1} = 1 - \frac{2}{x+1} = 1 - \sum_{n=0}^{\infty} 2(-1)^n x^n$$

37. $\sqrt[4]{17} = \sqrt[4]{16+1} = 2\sqrt[4]{1 + \frac{1}{16}}$; by (5) in Section 9.10 with $s = \frac{1}{4}$ and $x = \frac{1}{16}$,
$$\left|r_N\left(\frac{1}{16}\right)\right| \le \frac{1}{4}\frac{(1/16)^{N+1}}{1 - (1/16)} = \frac{1}{60\cdot 16^N} < 0.0005 \quad \text{if } N = 2$$

By (2) of Section 9.10,
$$\sqrt[4]{17} \approx 2\left[1 + \frac{1}{4}\left(\frac{1}{16}\right) + \frac{1}{2!}\left(\frac{1}{4}\right)\left(\frac{-3}{4}\right)\left(\frac{1}{16}\right)^2\right] \approx 2.03052$$

is the desired approximation.

39. Let $f(x) = \sin x$, and use the Taylor series about $\pi/6$. Then
$$|r_n(\pi/5)| = \left|\frac{f^{(n+1)}(t_x)}{(n+1)!}\left(\frac{\pi}{5} - \frac{\pi}{6}\right)^{n+1}\right| \le \frac{1}{(n+1)!}\left(\frac{\pi}{5} - \frac{\pi}{6}\right)^{n+1}$$

where $\pi/6 < t_x < \pi/5$. Now
$$\frac{1}{(n+1)!}\left(\frac{\pi}{5} - \frac{\pi}{6}\right)^{n+1} = \frac{1}{(n+1)!}\left(\frac{\pi}{30}\right)^{n+1} < 0.001 \quad \text{if } n = 2$$

Then $$\sin\frac{\pi}{5} \approx \frac{1}{2} + \frac{\sqrt{3}}{2}\left(\frac{\pi}{5} - \frac{\pi}{6}\right) + \frac{1}{2!}\left(\frac{-1}{2}\right)\left(\frac{\pi}{5} - \frac{\pi}{6}\right)^2 \approx 0.587948$$

is the desired approximation.

41. a. $\sum_{n=1}^{25} a_n = s_{25} = 2 - \dfrac{2}{25} = \dfrac{48}{25}$

b. $\lim_{j\to\infty} s_j = \lim_{j\to\infty}(2 - 2/j) = 2$, so $\sum_{n=1}^{\infty} a_n$ converges and its sum is 2.

c. Since $\sum_{n=1}^{\infty} a_n$ converges, $\lim_{n\to\infty} a_n = 0$.

d. $a_n = s_n - s_{n-1} = \left(2 - \dfrac{2}{n}\right) - \left(2 - \dfrac{2}{n-1}\right) = \dfrac{2}{n-1} - \dfrac{2}{n} = \dfrac{2}{n(n-1)}$

43. $\dfrac{1}{1-x^2} = \sum_{n=0}^{\infty} x^{2n}$, so that

$$\tanh^{-1} x = \int_0^x \frac{1}{1-t^2}\,dt = \int_0^x \left(\sum_{n=0}^{\infty} t^{2n}\right) dt = \sum_{n=0}^{\infty}\left(\int_0^x t^{2n}\,dt\right) = \sum_{n=0}^{\infty} \frac{x^{2n+1}}{2n+1}$$

45. Since $1/(1+x^{3/2}) = \sum_{n=0}^{\infty}(-1)^n(x^{3/2})^n$, which is an alternating series, we know by Theorem 9.17 that for any $N \ge 0$,

$$\left|\frac{1}{1+x^{3/2}} - \sum_{n=0}^{N}(-1)^n x^{3n/2}\right| = \left|\sum_{n=0}^{\infty}(-1)^n x^{3n/2} - \sum_{n=0}^{N}(-1)^n x^{3n/2}\right| \le |x|^{3(N+1)/2}$$

Thus

$$\left|\int_0^{1/4}\left(\frac{1}{1+x^{3/2}} - \sum_{n=0}^{N}(-1)^n x^{3n/2}\right)dx\right| \le \int_0^{1/4}\left|\frac{1}{1+x^{3/2}} - \sum_{n=0}^{N}(-1)^n x^{3n/2}\right| dx$$
$$\le \int_0^{1/4} x^{3(N+1)/2}\,dx = \frac{2x^{(3N+5)/2}}{3N+5}\bigg|_0^{1/4}$$
$$= \frac{2x}{3N+5}\left(\frac{1}{2}\right)^{3N+5}$$

Now $2/(3N+5)(\frac{1}{2})^{3N+5} < 0.001$ if $N = 1$. Thus

$$\int_0^{1/4} \frac{1}{1+x^{3/2}}\,dx \approx \int_0^{1/4}(1-x^{3/2})\,dx = (x - \tfrac{2}{5}x^{5/2})\Big|_0^{1/4} = \tfrac{1}{4} - \tfrac{1}{80} = \tfrac{19}{80} = 0.2375$$

CUMULATIVE REVIEW (CHAPTERS 1–8)

1. The conditions for applying l'Hôpital's Rule four times are met;

$$\lim_{x\to 0^-} \frac{\cos x + \frac{1}{2}x^2 - 1}{x^5} = \lim_{x\to 0^-} \frac{-\sin x + x}{5x^4} = \lim_{x\to 0^-} \frac{-\cos x + 1}{20x^3} = \lim_{x\to 0^-} \frac{\sin x}{60x^2}$$
$$= \lim_{x\to 0^-} \frac{\cos x}{120x} = -\infty$$

3. $\lim_{x\to 0^+} 3^x = \lim_{x\to 0^+} e^{x\ln 3} = e^0 = 1$ and $\lim_{x\to 0^+} \frac{1}{x^3} = \infty$, so $\lim_{x\to 0^+} \frac{3^x}{x^3} = \infty$.

5. a. The domain of f consists of all x for which $3x - 4 > 0$ (or $x > \frac{4}{3}$) and $\ln(3x-4) > 0$. But $\ln(3x-4) > 0$ if and only if $3x - 4 > 1$, or $x > \frac{5}{3}$. Thus the domain is $(\frac{5}{3}, \infty)$.

   b. $f'(x) = \dfrac{1}{\ln(3x-4)} \cdot \dfrac{1}{3x-4} \cdot (3) = \dfrac{3}{(3x-4)\ln(3x-4)}$

7. Differentiating the given equation implicitly, we obtain $e^x + 2e^{2x} - y^2(dy/dx) + (dy/dx) = 0$, so that $e^x + 2e^{2x} + (dy/dx)(1-y^2) = 0$. Thus if $(dy/dx)(1-y^2) = -6$ at (a, b), then $e^a + 2e^{2a} - 6 = 0$, or $2e^{2a} + e^a - 6 = 0$, or $(2e^a - 3)(e^a + 2) = 0$. Since $e^a + 2 > 0$, it follows that $e^a = \frac{3}{2}$, so that $a = \ln\frac{3}{2}$.

9. Since $A = \pi r^2$ and $dA/dt = \frac{1}{2}$, we have $\frac{1}{2} = dA/dt = 2\pi r(dr/dt)$, so that $dr/dt = 1/(4\pi r)$. Thus when $r = 1$, the radius is increasing at the rate of $1/(4\pi)$ foot per second.

11. By Exercise 53 in Section 7.1,

$$\int \frac{\cos x}{e^{3x}}\,dx = \int e^{-3x}\cos x\,dx = \frac{e^{-3x}}{10}(-3\cos x + \sin x) + C$$

13. Let $x = 4\sin u$, so that $dx = 4\cos u\,du$. If $x = 2\sqrt{2}$, then $u = \pi/4$, and if $x = 4$, then $u = \pi/2$. Thus

$$\int_{2\sqrt{2}}^{4} \frac{\sqrt{16-x^2}}{x^3}\,dx = \int_{\pi/4}^{\pi/2} \frac{4\cos u}{64\sin^3 u}\,4\cos u\,du = \frac{1}{4}\int_{\pi/4}^{\pi/2}\frac{\cos^2 u}{\sin^3 u}\,du = \frac{1}{4}\int_{\pi/4}^{\pi/2}\frac{1-\sin^2 u}{\sin^3 u}\,du$$
$$= \frac{1}{4}\int_{\pi/4}^{\pi/2}(\csc^3 u - \csc u)\,du.$$

By Exercise 28 of Section 7.2, $\int \csc^3 u\,du = -\frac{1}{2}\csc u\cot u - \frac{1}{2}\ln|\csc u + \cot u| + C$, and by Exercise 46 of Section 5.7, $\int \csc u\,du = -\ln|\csc u + \cot u| + C$. Therefore

$$\int_{2\sqrt{2}}^{4}\frac{\sqrt{16-x^2}}{x^3}\,dx = \tfrac{1}{4}\int_{\pi/4}^{\pi/2}(\csc^3 u - \csc u)\,du$$
$$= \tfrac{1}{4}(-\tfrac{1}{2}\csc u\cot u + \tfrac{1}{2}\ln|\csc u + \cot u|)\Big|_{\pi/4}^{\pi/2}$$
$$= \tfrac{1}{4}(\tfrac{1}{2}\sqrt{2} - \tfrac{1}{2}\ln(\sqrt{2}+1)) = \tfrac{1}{8}(\sqrt{2} - \ln(\sqrt{2}+1))$$

15. Let $u = 1 + e^x$, so that $du = e^x\,dx$. Then

$$\int \frac{e^x}{(1+e^x)^2}\,dx = \int \frac{1}{u^2}\,du = -\frac{1}{u} + C = \frac{-1}{1+e^x} + C$$

Thus

$$\int_0^{\infty}\frac{e^x}{(1+e^x)^2}\,dx = \lim_{b\to\infty}\int_0^b \frac{e^x}{(1+e^x)^2}\,dx = \lim_{b\to\infty}\frac{-1}{1+e^x}\bigg|_0^b = \lim_{b\to\infty}\left(\frac{-1}{1+e^b} + \frac{1}{2}\right) = \frac{1}{2}$$

and

$$\int_{-\infty}^{0}\frac{e^x}{(1+e^x)^2}\,dx = \lim_{a\to-\infty}\int_a^0 \frac{e^x}{(1+e^x)^2}\,dx = \lim_{a\to-\infty}\frac{-1}{1+e^x}\bigg|_a^0 = \lim_{a\to-\infty}\left(-\frac{1}{2} + \frac{1}{1+e^a}\right)$$
$$= -\tfrac{1}{2} + 1 = \tfrac{1}{2}$$

Thus the given integral converges, and

$$\int_{-\infty}^{\infty}\frac{e^x}{(1+e^x)^2}\,dx = \int_{-\infty}^{0}\frac{e^x}{(1+e^x)^2}\,dx + \int_0^{\infty}\frac{e^x}{(1+e^x)^2}\,dx = \frac{1}{2} + \frac{1}{2} = 1$$

17. $$\mathscr{L} = \int_0^{\pi/4}\sqrt{\left(\frac{dx}{dt}\right)^2 + \left(\frac{dy}{dt}\right)^2}\,dt = \int_0^{\pi/4}\sqrt{(\sec t\tan t)^2 + (\sec t)^2}\,dt$$
$$= \int_0^{\pi/4}\sqrt{\sec^2 t(\tan^2 t + 1)}\,dt = \int_0^{\pi/4}\sqrt{\sec^4 t}\,dt = \int_0^{\pi/4}\sec^2 t\,dt = \tan t\Big|_0^{\pi/4} = 1$$

19. $A = \frac{1}{2}\int_0^1[\theta^2 - (\theta^3)^2]\,d\theta = \frac{1}{2}\int_0^1(\theta^2 - \theta^6)\,d\theta = \frac{1}{2}(\frac{1}{3}\theta^3 - \frac{1}{7}\theta^7)\Big|_0^1 = \frac{2}{21}$

10

COMPLEX NUMBERS AND FUNCTIONS

SECTION 10.1

1. $3 + 9i$
3. $-1 + 10i$
5. $18 + 43i$
7. $\frac{10}{3} - 30i$
9. $10.2 + 9.8i$
11. $-i$
13. $\frac{1}{(1/2) + i} = \frac{1}{(1/2) + i} \cdot \frac{(1/2) - i}{(1/2) - i} = \frac{1}{5/4}\left(\frac{1}{2} - i\right) = \frac{2}{5} - \frac{4}{5}i$
15. $\frac{-3 + 5i}{3 + 4i} = \frac{-3 + 5i}{3 + 4i} \cdot \frac{3 - 4i}{3 - 4i} = \frac{1}{25}(11 + 27i) = \frac{11}{25} + \frac{27}{25}i$
17. $i^7 = i^4 i^3 = -i$
19. $i^{122} = i^{120} i^2 = -1$
21. $\sqrt{-36} = \sqrt{36}i = 6i$
23. $\sqrt{-12} = \sqrt{12}i = 2\sqrt{3}i$
25. $x = 9i$ or $-9i$
27. $x = \frac{-2 \pm \sqrt{4 - 8}}{2} = -1 + i$ or $-1 - i$
29. $x = \frac{4 \pm \sqrt{16 - 52}}{2} = 2 + 3i$ or $2 - 3i$
31. $x = \frac{-1 \pm \sqrt{1 - 4}}{2} = -\frac{1}{2} + \frac{1}{2}\sqrt{3}i$ or $-\frac{1}{2} - \frac{1}{2}\sqrt{3}i$
33. $x = \frac{-1 \pm \sqrt{1 - 8}}{2} = -\frac{1}{2} + \frac{1}{2}\sqrt{7}i$ or $-\frac{1}{2} - \frac{1}{2}\sqrt{7}i$
35. $x = \frac{2 \pm \sqrt{4 - 8}}{4} = \frac{1}{2} + \frac{1}{2}i$ or $\frac{1}{2} - \frac{1}{2}i$
37. $x = \frac{-6 \pm \sqrt{36 - 360}}{-18} = \frac{-6 \pm 18i}{-18} = \frac{1}{3} + i$ or $\frac{1}{3} - i$
39. $x = \frac{-2 \pm \sqrt{4 - 8}}{2/3} = \frac{-2 \pm 2i}{2/3} = -3 + 3i$ or $-3 - 3i$
41. $z^2 = \frac{-3 \pm \sqrt{9 + 16}}{2} = -\frac{3}{2} \pm \frac{5}{2} = -4$ or 1, so $z = 2i, -2i, 1$ or -1
43. If $z = x + iy$ and $w = u + iv$, then $\overline{z + w} = \overline{(x + u) + i(y + v)} = (x + u) - i(y + v) = (x - iy) + (u - iv) = \bar{z} + \bar{w}$.
45. $\frac{1}{2}(z + \bar{z}) = \frac{1}{2}[(a + bi) + (a - bi)] = \frac{1}{2}(2a) = a$; $(1/2)i(z - \bar{z}) = (1/2)i[(a + bi) - (a - bi)] = 1/2i(2bi) = b$
47. a. Since the solutions $(-b \pm \sqrt{b^2 - 4ac})/2a$ are pure imaginary, it follows that $b = 0$.
    b. Since the solutions $(-b \pm \sqrt{b^2 - 4ac})/2a$ are pure imaginary, it follows that $4ac > b^2$. But $b = 0$ from part (a), so that $ac > 0$.
49. Suppose that $(a + bi)^2 = -c$. Then $a^2 - b^2 + 2abi = -c$, so that $a^2 - b^2 = -c$ and $2ab = 0$. Thus $a = 0$ or $b = 0$. However, $c > 0$ by hypothesis, so that $0 > -c = a^2 - b^2$, which means that $b \neq 0$. Consequently $a = 0$, so that $-b^2 = -c$, and thus $b^2 = c$. Therefore $a + bi = bi = \sqrt{ci} = \sqrt{-c}$ or $a + bi = -bi = -\sqrt{ci} = -\sqrt{-c}$.

SECTION 10.2

1. a.

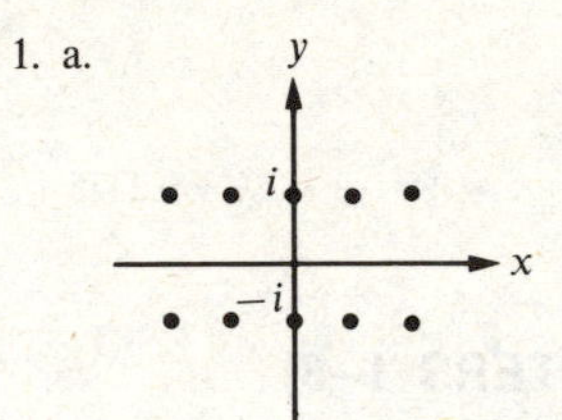

   b. They are symmetrically placed about the real line.

3.

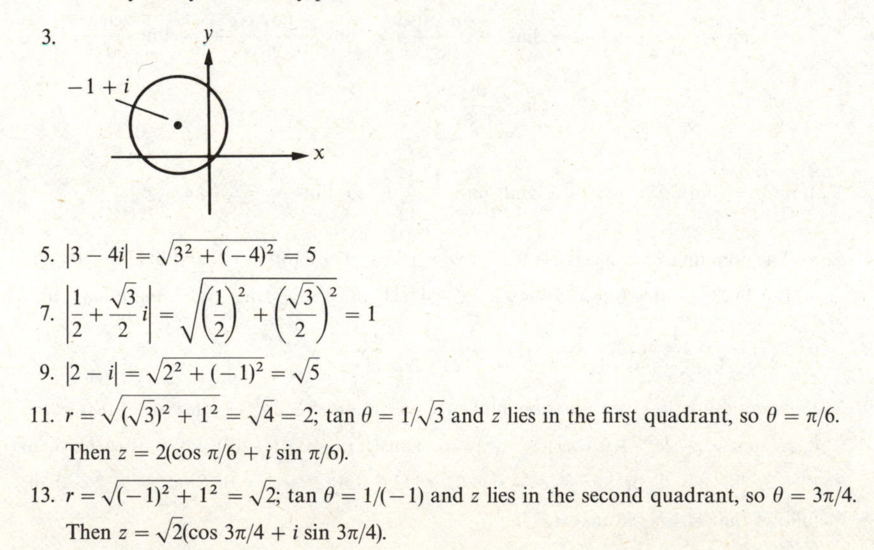

5. $|3 - 4i| = \sqrt{3^2 + (-4)^2} = 5$
7. $\left|\frac{1}{2} + \frac{\sqrt{3}}{2}i\right| = \sqrt{\left(\frac{1}{2}\right)^2 + \left(\frac{\sqrt{3}}{2}\right)^2} = 1$
9. $|2 - i| = \sqrt{2^2 + (-1)^2} = \sqrt{5}$
11. $r = \sqrt{(\sqrt{3})^2 + 1^2} = \sqrt{4} = 2$; $\tan\theta = 1/\sqrt{3}$ and z lies in the first quadrant, so $\theta = \pi/6$. Then $z = 2(\cos \pi/6 + i \sin \pi/6)$.
13. $r = \sqrt{(-1)^2 + 1^2} = \sqrt{2}$; $\tan\theta = 1/(-1)$ and z lies in the second quadrant, so $\theta = 3\pi/4$. Then $z = \sqrt{2}(\cos 3\pi/4 + i \sin 3\pi/4)$.
15. $r_1 = \sqrt{(\sqrt{3})^2 + (-1)^2} = 2$; $\tan\theta_1 = -1/\sqrt{3}$ and z_1 lies in the fourth quadrant, so $\theta_1 = 11\pi/6$. Thus $z_1 = 2(\cos 11\pi/6 + i \sin 11\pi/6)$. Next, $r_2 = \sqrt{2^2 + (2\sqrt{3})^2} = 4$; $\tan\theta_2 =$

$2\sqrt{3}/2 = \sqrt{3}$ and z_2 lies in the first quadrant, so $\theta_2 = \pi/3$. Thus $z_2 = 4(\cos \pi/3 + i \sin \pi/3)$. Then by (4), $z_1 z_2 = 8(\cos 13\pi/6 + i \sin 13\pi/6) = 8(\cos \pi/6 + i \sin \pi/6)$, and by (5), $z_1/z_2 = \frac{1}{2}(\cos 3\pi/2 + i \sin 3\pi/2)$.

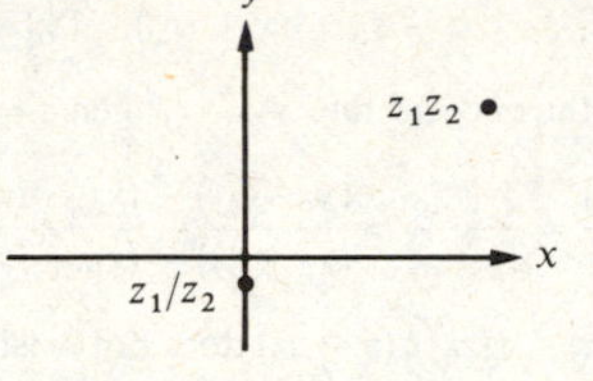

17. $r_1 = \sqrt{(-2)^2 + (-2)^2} = 2\sqrt{2}$; $\tan \theta_1 = -2/(-2) = 1$ and z_1 lies in the third quadrant, so $\theta_1 = 5\pi/4$. Thus $z_1 = 2\sqrt{2}(\cos 5\pi/4 + i \sin 5\pi/4)$. Next, $r_2 = \sqrt{(-3)^2 + 3^2} = 3\sqrt{2}$; $\tan \theta_2 = 3/(-3) = -1$ and z_2 lies in the second quadrant, so $\theta_2 = 3\pi/4$. Thus $z_2 = 3\sqrt{2}(\cos 3\pi/4 + i \sin 3\pi/4)$. Then by (4), $z_1 z_2 = 12(\cos 2\pi + i \sin 2\pi) = 12(\cos 0 + i \sin 0)$, and by (5), $z_1/z_2 = \frac{2}{3}(\cos \pi/2 + i \sin \pi/2)$.

19. If $z = -1 - i$, then by using the result for z_1 in Exercise 16, $z = \sqrt{2}(\cos 5\pi/4 + i \sin 5\pi/4)$. Therefore by De Moivre's Theorem, $(-1-i)^5 = z^5 = 2^{5/2}(\cos 25\pi/4 + i \sin 25\pi/4) = 2^{5/2}(\frac{1}{2}\sqrt{2} + \frac{1}{2}\sqrt{2}i) = 4 + 4i$.

21. If $z = -1 - \sqrt{3}i$, then $r = 2$ and $\theta = 4\pi/3$, so $z = 2(\cos 4\pi/3 + i \sin 4\pi/3)$. Thus the three 3rd roots are $2^{1/3}[\cos(4\pi/9 + 2\pi k/3) + i \sin(4\pi/9 + 2\pi k/3)]$ for $k = 0$, 1, and 2.

23. If $z = -1 + i$, then $r = \sqrt{2}$ and $\theta = 3\pi/4$, so $z = \sqrt{2}(\cos 3\pi/4 + i \sin 3\pi/4)$. Thus the three 3rd roots are $2^{1/3}[\cos(\pi/4 + 2\pi k/3) + i \sin(\pi/4 + 2\pi k/3)]$ for $k = 0$, 1, and 2.

25. If $z = -i$, then $z = 1(\cos 3\pi/2 + i \sin 3\pi/2)$. Thus the four 4th roots are

$$\cos\left(\frac{3\pi}{8} + \frac{2\pi k}{4}\right) + i \sin\left(\frac{3\pi}{8} + \frac{2\pi k}{4}\right)$$

for $k = 0$, 1, 2, and 3.

27. If $z = 1 - \sqrt{3}i$, then $r = 2$ and $\theta = 5\pi/3$, so $z = 2(\cos 5\pi/3 + i \sin 5\pi/3)$. Thus the five 5th roots are $2^{1/5}[\cos(\pi/3 + 2\pi k/5) + i \sin(\pi/3 + 2\pi k/5)]$ for $k = 0$, 1, 2, 3, and 4.

29. If $z = 1$, then $z = 1(\cos 0 + i \sin 0)$. Thus the six 6th roots are $\cos 2\pi k/6 + i \sin 2\pi k/6$ for $k = 0$, 1, 2, 3, 4, and 5. More simply, the roots are 1, $\frac{1}{2}\sqrt{3} + \frac{1}{2}$, $-\frac{1}{2}\sqrt{3} + \frac{1}{2}$, -1, $-\frac{1}{2}\sqrt{3} - \frac{1}{2}$, and $\frac{1}{2}\sqrt{3} - \frac{1}{2}$.

31. Let $z = x + yi$. Then $|z|^2 = x^2 + y^2$ and $z\bar{z} = (x + yi)(x - yi) = x^2 + y^2$. Therefore $|z|^2 = z\bar{z}$.

33. a. $-z = r[\cos(\theta + \pi) + i \sin(\theta + \pi)]$

    b. $\bar{z} = r[\cos(-\theta) + i \sin(-\theta)]$

35. a. Let $z = x + yi$. If $|1 - z| < 1$, then $1 > |1 - (x + yi)| = \sqrt{(1-x)^2 + y^2} \geq |1 - x|$, so that $x > 0$. Therefore $1 < |1 + x| \leq \sqrt{(1+x)^2 + y^2} = |1 + z|$.

    b. Replace z by $-z$ in part (a).

37. Since $z^n = 1$, we have $1 + z + z^2 + \cdots + z^{n-1} = (1 - z^n)/(1 - z) = (1 - 1)/(1 - z) = 0$.

SECTION 10.3

1. All complex numbers except 0

3. Since $z^2 + 1 = 0$ if $z = i$ or $-i$, the domain consists of all complex numbers except i and $-i$.

5. Since $1 - x^2 - y^2 + ixy = 0$ if $xy = 0$ and $1 = x^2 + y^2$, it follows that

$$\text{either } x = 0 \text{ and } y = 1 \text{ or } -1$$
$$\text{or} \quad y = 0 \text{ and } x = 1 \text{ or } -1$$

Thus the domain consists of all complex numbers except 1, -1, i, and $-i$.

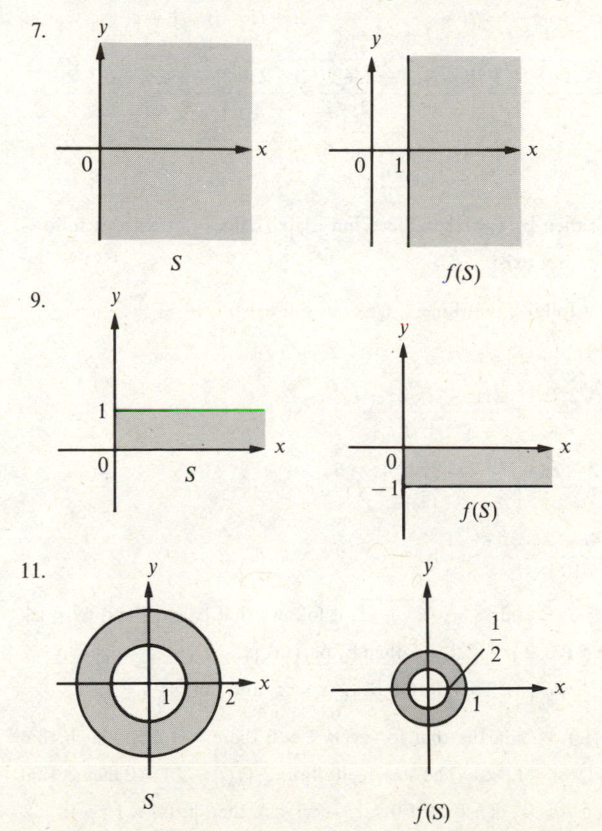

13. If $z = f(z) = z^2$, then $z = 0$ or 1. Therefore the fixed points of f are 0 and 1.

15. If $z = f(z) = 1/(1 - z)$, then $z(1 - z) = 1$. Therefore $z^2 - z + 1 = 0$, so that $z = (1 \pm \sqrt{1 - 4})/2 = \frac{1}{2} \pm \frac{1}{2}\sqrt{3}i$. Thus the fixed points of f are $\frac{1}{2} + \frac{1}{2}\sqrt{3}i$ and $\frac{1}{2} - \frac{1}{2}\sqrt{3}i$.

17. If $z = f(z) = z^3 + 3z$, then $z^3 + 2z = 0$, so that $z(z^2 + 2) = 0$. Therefore the fixed points of f are 0, $\sqrt{2}i$, and $-\sqrt{2}i$.

19. Let $f(z) = w$. Then $w = (z+1)/(z-1)$, so that $wz - w = z + 1$, and thus $wz - z = w + 1$, which means that $z = (w+1)/(w-1)$. Therefore $f^{-1}(z) = (z+1)/(z-1)$.

21. Since $f(f(z)) = \bar{\bar{z}} = z$, it follows that $f^{-1}(z) = f(z) = \bar{z}$.

23. If $z = x + yi$, then $f(z) = z^3 = (x+yi)^3 = (x^3 - 3xy^2) + (3x^2y - y^3)i$. Thus $u = x^3 - 3xy^2$ and $v = 3x^2y - y^3$.

25. If $z = x + yi$, then $f(z) = 1/z = 1/(x+iy) = (x-yi)/(x^2+y^2)$. Thus $u = x/(x^2+y^2)$ and $v = -y/(x^2+y^2)$.

27. $\lim_{z\to i}(2z^3 - 3z^2) = 2i^3 - 3i^2 = -2i + 3 = 3 - 2i$

29. $$\lim_{z\to 1+i}\left(z^2 + \frac{2}{z}\right) = (1+i)^2 + \frac{2}{1+i} = 2i + \frac{2}{1+i}\frac{1-i}{1-i} = 2i + (1-i) = 1 + i$$

31. $$\lim_{z\to -2}\frac{iz^2 + (1+i)z}{z+i} = \frac{i(-2)^2 + (1+i)(-2)}{(-2)+i} = \frac{-2+2i}{-2+i} = \frac{-2+2i}{-2+i}\frac{-2-i}{-2-i} = \frac{6}{5} - \frac{2}{5}i$$

33. $$\lim_{z\to 0}\frac{|z|^2}{z} = \lim_{z\to 0}\frac{z\bar{z}}{z} = \lim_{z\to 0}\bar{z} = 0$$

35. Let $z = x + yi$. If $y = 0$, then $|z|/z = |x|/x$. Since $\lim_{x\to 0}|x|/x$ does not exist, it follows that $\lim_{z\to 0}|z|/z$ also does not exist.

37. Since $\lim_{z\to 0}1/|z| = \infty$, it follows that $\lim_{z\to 0}1/z$ does not exist.

39. $f'(z) = -20z^4 + 6z - 2$

41. $$f'(z) = \frac{i(z^2+1) - (iz + \sqrt{2})2z}{(z^2+1)^2} = \frac{-iz^2 - 2\sqrt{2}z + i}{(z^2+i)^2}$$

43. $$f'(z) = \frac{(-1)(3-z^2) - (2-z)(-2z)}{(3-z^2)^2} = \frac{-z^2 + 4z - 3}{(3-z^2)^2};$$

$$f'(i) = \frac{-i^2 + 4i - 3}{(3-i^2)^2} = \frac{4i-2}{16} = -\frac{1}{8} + \frac{1}{4}i$$

45. a. Since $|x| = \sqrt{x^2}$, $|y| = \sqrt{y^2}$, and $|z| = \sqrt{x^2+y^2}$, it follows that $|x| \le |z|$ and $|y| \le |z|$.
b. Let $\varepsilon > 0$, and let $\delta = \varepsilon$. If $0 < |z - z_0| < \delta$, then by part (a), $|x - x_0| \le |z - z_0| < \delta = \varepsilon$ and $|y - y_0| \le |z - z_0| < \delta = \varepsilon$. Therefore $\lim_{z\to z_0} x = x_0$ and $\lim_{z\to z_0} y = y_0$.

47. The statement $\lim_{z\to z_0} f(z) = L$ means that for every $\varepsilon > 0$ there is a $\delta > 0$ such that if $0 < |z - z_0| < \delta$, then $|f(z) - L| < \varepsilon$. The statement $\lim_{z\to z_0}(f(z) - L) = 0$ means that for every $\varepsilon > 0$ there is a $\delta > 0$ such that if $0 < |z - z_0| < \delta$, then $|(f(z) - L) - 0| < \varepsilon$. Since $|(f(z) - L) - 0| = |f(z) - L|$, the two statements are equivalent.

49. Let $f(z) = z$. For any $\varepsilon > 0$, let $\delta = \varepsilon$. If $0 < |z - z_0| < \delta$, then $|f(z) - z_0| = |z - z_0| < \delta = \varepsilon$. Therefore $\lim_{z\to z_0} z = \lim_{z\to z_0} f(z) = z_0$.

51. Let $z_0 = x_0 + iy_0$. If $z \ne z_0$ and z is on the horizontal line $y = y_0$, then $z = x + iy_0$, so that

$$\frac{f(z) - f(z_0)}{z - z_0} = \frac{\bar{z} - \bar{z}_0}{z - z_0} = \frac{(x - iy_0) - (x_0 - iy_0)}{(x + iy_0) - (x_0 + iy_0)} = \frac{x - x_0}{x - x_0} = 1$$

If $z \ne z_0$ and z is on the vertical line $x = x_0$, then $z = x_0 + iy$, so that

$$\frac{f(z) - f(z_0)}{z - z_0} = \frac{\bar{z} - \bar{z}_0}{z - z_0} = \frac{(x_0 - iy) - (x_0 - iy_0)}{(x_0 + iy) - (x_0 + iy_0)} = \frac{-i(y - y_0)}{i(y - y_0)} = -1$$

Therefore $\lim_{z\to z_0}(f(z) - f(z_0))/(z - z_0)$ does not exist, and hence f is not differentiable at the point z_0.

SECTION 10.4

1. By Theorem 10.7,

$$\lim_{n\to\infty}\left(\frac{1}{n} + i\frac{\sin n}{n^2}\right) = \lim_{n\to\infty}\frac{1}{n} + i\lim_{n\to\infty}\frac{\sin n}{n^2} = 0 + 0i = 0$$

3. By Theorem 10.7,

$$\lim_{n\to\infty}\frac{2n + 3i}{n} = \lim_{n\to\infty}\left(2 + \frac{3i}{n}\right) = \lim_{n\to\infty} 2 + i\lim_{n\to\infty}\frac{3}{n} = 2 + 0i = 2$$

5. By the Geometric Series Theorem (Theorem 9.10),

$$\sum_{n=0}^{\infty}\left[\left(\frac{2}{3}\right)^n + i\left(\frac{1}{4}\right)^{n+1}\right] = \sum_{n=0}^{\infty}\left(\frac{2}{3}\right)^n + i\sum_{n=0}^{\infty}\left(\frac{1}{4}\right)^{n+1} = \sum_{n=0}^{\infty}\left(\frac{2}{3}\right)^n + \frac{1}{4}i\sum_{n=0}^{\infty}\left(\frac{1}{4}\right)^n$$

$$= \frac{1}{1 - 2/3} + \frac{1}{4}\cdot\frac{1}{1 - 1/4}i = 3 + \frac{1}{3}i$$

7. Since $\lim_{n\to\infty}\sqrt[n]{|2+i|^n} = \lim_{n\to\infty}|2+i| = |2+i| > 2$, the Root Test implies that the series diverges.

9. Since

$$\lim_{n\to\infty}\left|\frac{(n+2)i^{n+1}/(n+1)!}{(n+1)i^n/n!}\right| = \lim_{n\to\infty}\frac{n+2}{(n+1)^2} = 0$$

the Ratio Test implies that the series converges.

11. Since $|(ni)^n/n!| = n^n/n! \ge 1$ for all n, the nth Term Test implies that the series diverges.

13. Since

$$\lim_{n\to\infty}\left|\frac{2^{n+2}i^{n+1}z^{n+1}}{2^{n+1}i^n z^n}\right| = \lim_{n\to\infty} 2|z| = 2|z|$$

it follows that the series converges for $|z| < \frac{1}{2}$ and diverges for $|z| > \frac{1}{2}$. Thus $R = \frac{1}{2}$.

15. Since

$$\lim_{n\to\infty}\left|\frac{(n+1)!z^{n+2}/[2(n+1)+3]}{n!z^{n+1}/(2n+3)}\right| = \lim_{n\to\infty}\left|(n+1)\frac{2n+3}{2n+5}z\right| = \infty$$

for all $z \neq 0$, it follows that $R = 0$.

17. Since

$$\lim_{n\to\infty} \sqrt[n]{\left|\frac{n^2 z^n}{(3i)^n}\right|} = \lim_{n\to\infty} \frac{\sqrt[n]{n^2}\,|z|}{|3i|} = \frac{1}{3}|z|$$

it follows that the series converges for $|z| < 3$ and diverges for $|z| > 3$. Thus $R = 3$.

19. a. Since

$$\lim_{n\to\infty} \left|\frac{(-1)^{n+1} z^{n+1}/(n+1)}{(-1)^n z^n/n}\right| = \lim_{n\to\infty} \frac{n}{n+1}|z| = |z|$$

it follows that the series converges for $|z| < 1$ and diverges for $|z| > 1$. Thus $R = 1$.

b. Since

$$\lim_{n\to\infty} \left|\frac{(-1)^{n+1} z^{2n+3}/(2n+3)}{(-1)^n z^{2n+1}/(2n+1)}\right| = \lim_{n\to\infty} \frac{2n+1}{2n+3}|z|^2 = |z|^2$$

it follows that the series converges for $|z| < 1$ and diverges for $|z| > 1$. Therefore $R = 1$.

21. a. Since $\lim_{n\to\infty} |cz^{n+1}/cz^n| = |z|$, it follows that the series converges for $|z| < 1$ and diverges for $|z| > 1$. If $|z| = 1$, then $|cz^n| = |c|$ for all n, so by the nth Term Test the series diverges. Consequently the series converges only if $|z| < 1$.

b. If $|z| < 1$, then $\sum_{n=m}^{m+j-1} cz^n = cz^m \sum_{n=0}^{j-1} z^n = cz^m(1 - z^j)/(1 - z)$. Since $\lim_{j\to\infty} cz^m(1 - z^j)/(1 - z) = cz^m/(1 - z)$, it follows that $\sum_{n=m}^{\infty} cz^n = cz^m/(1 - z)$.

23. By (9) and (4),

$$\begin{aligned} e^{iz} - e^{-iz} &= (\cos z + i \sin z) - [\cos(-z) + i \sin(-z)] \\ &= (\cos z + i \sin z) - (\cos z - i \sin z) = 2i \sin z \end{aligned}$$

Thus $\sin z = (e^{iz} - e^{-iz})/2i$.

25. Letting $z = ix$ in Exercise 23, we obtain

$$\sin ix = \frac{1}{2i}(e^{-x} - e^{x}) = \frac{1}{i}\left(\frac{e^{-x} - e^{x}}{2}\right) = i\left(\frac{e^{x} - e^{-x}}{2}\right) = i \sinh x$$

for all real x.

27. Using (12) and (13), we find that

$$\begin{aligned} \sin z \cos w + \cos z \sin w &= \left[\frac{1}{2i}(e^{iz} - e^{-iz})\right]\left[\frac{1}{2}(e^{iw} + e^{-iw})\right] \\ &\quad + \left[\frac{1}{2}(e^{iz} + e^{-iz})\right]\left[\frac{1}{2i}(e^{iw} - e^{-iw})\right] \\ &= \frac{1}{4i}(2e^{iz}e^{iw} - 2e^{-iz}e^{-iw}) = \frac{1}{2i}(e^{i(z+w)} - e^{-i(z+w)}) = \sin(z + w) \end{aligned}$$

29. Let $z = x + iy$. Then by Exercise 28, $2 = \sin(x + iy)$ if and only if $2 = \sin x \cosh y + i \cos x \sinh y$. In that case, $\cos x \sinh y = 0$ and $2 = \sin x \cosh y$. Now if $\cos x \sinh y = 0$, then either $\cos x = 0$ or $\sinh y = 0$. Suppose that $\sinh y = 0$. Then $y = 0$, so $\cosh y = 1$, and thus $2 = \sin x \cosh y = \sin x$, which is impossible since $|\sin x| \leq 1$. Thus $\sinh y \neq 0$, and hence $\cos x = 0$. This means that $x = \frac{1}{2}\pi + n\pi$ for some integer n. Since $2 = \sin x \cosh y$ and $\cosh y \geq 0$, it follows that $\sin x \geq 0$, so that $x = \frac{1}{2}\pi + 2m\pi$ for an appropriate integer m. But $\sin(\frac{1}{2}\pi + 2m\pi) = 1$, so we need to solve $2 = \cosh y$ to find y. If $2 = \cosh y = (e^y + e^{-y})/2$, then $4 = e^y + e^{-y}$, which means that $0 = e^{2y} - 4e^y + 1$, and thus $e^y = (4 \pm \sqrt{16 - 4})/2 = 2 \pm \sqrt{3}$. Consequently $y = \ln(2 + \sqrt{3})$ or $y = \ln(2 - \sqrt{3})$. We conclude that $z_0 = \frac{1}{2}\pi + 2m\pi + (\ln(2 + \sqrt{3}))i$ or $z_0 = \frac{1}{2}\pi + 2m\pi + (\ln(2 - \sqrt{3}))i$.

31. $e^{z+2\pi i} \overset{(7)}{=} e^z e^{2\pi i} \overset{(10)}{=} e^z(\cos 2\pi + i \sin 2\pi) = e^z$

33. a. $\displaystyle \frac{d}{dz}\sin z = \frac{d}{dz}\left(\sum_{n=0}^{\infty} \frac{(-1)^n}{(2n+1)!} z^{2n+1}\right) = \sum_{n=0}^{\infty} \frac{(2n+1)(-1)^n}{(2n+1)!} z^{2n} = \sum_{n=0}^{\infty} \frac{(-1)^n}{(2n)!} z^{2n} = \cos z$

b. $\displaystyle \frac{d}{dz}e^z = \frac{d}{dz}\sum_{n=0}^{\infty} \frac{z^n}{n!} = \sum_{n=1}^{\infty} \frac{nz^{n-1}}{n!} = \sum_{n=1}^{\infty} \frac{z^{n-1}}{(n-1)!} = \sum_{n=0}^{\infty} \frac{z^n}{n!} = e^z$

SECTION 10.5

1. Since $f_{-i}(z) = z^2 - i$ and $a_0 = 0$, we find that $a_1 = 0^2 - i = -i$, $a_2 = (-i)^2 - i = -1 - i$, $a_3 = (-1 - i)^2 - i = i$, and $a_4 = i^2 - i = -1 - i$.

3. Since $c = -\frac{1}{2}$, we have $(\frac{1}{4} - c)^{1/2} = (\frac{3}{4})^{1/2} = \frac{1}{2}\sqrt{3}$, so that $|\frac{1}{2} - (\frac{1}{4} - c)^{1/2}| = |\frac{1}{2} - \frac{1}{2}\sqrt{3}| = \frac{1}{2}|1 - \sqrt{3}| < \frac{1}{2}$. Therefore (6) is satisfied, so that $-\frac{1}{2}$ is in $\mathcal{M}_0$.

5. Since $c = \frac{1}{4} - \frac{1}{2}i$, we have $(\frac{1}{4} - c)^{1/2} = (\frac{1}{2}i)^{1/2} = \frac{1}{2} + \frac{1}{2}i$, so that $|\frac{1}{2} + (\frac{1}{4} - c)^{1/2}| = |1 + \frac{1}{2}i| = \sqrt{1^2 + (\frac{1}{2})^2} \geq \frac{1}{2}$ and $|\frac{1}{2} - (\frac{1}{4} - c)^{1/2}| = |-\frac{1}{2}i| = \frac{1}{2}$. Therefore (6) is not satisfied, so that $\frac{1}{4} + \frac{1}{2}i$ is not in $\mathcal{M}_0$.

7. Let $a_0 = -1$. Since $f_{-1}(z) = z^2 - 1$, we have $a_1 = (-1)^2 - 1 = 0$ and $a_2 = 0^2 - 1 = -1 = a_0$. Therefore $\{-1, 0\}$ is a 2-cycle for f_{-1}. To show that the cycle is attracting, notice that $|f_{-1} \circ f_{-1}(z)| = (z^2 - 1)^2 - 1 = z^4 - 2z^2$, so that $(f_{-1} \circ f_{-1})'(z) = 4z^3 - 4z$. Thus $(f_{-1} \circ f_{-1})'(-1) = -4 + 4 = 0$. Since $|(f_{-1} \circ f_{-1})'(-1)| < 1$, we conclude that $\{-1, 0\}$ is an attracting 2-cycle.

9. By the Chain Rule, $(f \circ f \circ f)'(z) = f'(f(f(z)))f'(f(z))f'(z)$. Because $\{z_0, z_1, z_2\}$ is a 3-cycle, $f(f(z_0)) = z_2$ and $f(z_0) = z_1$. Thus $(f \circ f \circ f)'(z_0) = f'(z_2)f'(z_1)f'(z_0)$. By permuting the z_0, z_1, and z_2, we find that $(f \circ f \circ f)'(z_0) = f'(z_2)f'(z_1)f'(z_0) = (f \circ f \circ f)'(z_1) = (f \circ f \circ f)'(z_2)$. Therefore if z_0 is attracting for f, then z_1 and z_2 are also attracting for f.

11. Suppose that some iterate a_r of z_0 is periodic for f_0. Then there is an integer $s > r$ such that $a_s = a_r$. By (7), $(e^{2\pi i\theta})^{2^s} = (e^{2\pi i\theta})^{2^r}$, or equivalently, $e^{(2\pi i\theta)2^s} = e^{(2\pi i\theta)2^r}$. From (11) of Section 10.4, it follows that there is an integer m such that $(2\pi i\theta)2^s = (2\pi i\theta)2^r + 2\pi im$. Therefore $\theta \cdot 2^s = \theta \cdot 2^r + m$, so that $\theta = m/(2^s - 2^r)$. This means that θ is rational.

13. We have $f_c(c) = c^2 + c$. Notice that c is a fixed point of f_c only if $c^2 + c = c$, that is, $c^2 = 0$, so that $c = 0$. Thus if $c \neq 0$, then c is not a fixed point of f_c. We will show that

there is a nonzero number c such that $(f_c \circ f_c \circ f_c)(c) = c$, so that $\{c, f_c(c), (f_c \circ f_c)(c)\}$ is a 3-cycle that is not a fixed point. Notice that $(f_c \circ f_c)(c) = f_c(f_c(c)) = f_c(c^2 + c) = (c^2 + c)^2 + c$ and $(f_c \circ f_c \circ f_c)(c) = f_c((f_c \circ f_c)(c)) = f_c((c^2 + c)^2 + c) = [(c^2 + c)^2 + c]^2 + c$. Therefore the equation $(f_c \circ f_c \circ f_c)(c) = c$ becomes $[(c^2 + c)^2 + c]^2 + c = c$, or $[(c^2 + c)^2 + c]^2 = 0$, or $(c^2 + c)^2 + c = 0$. To find a nonzero value of c for which the last equation holds, let $g(x) = (x^2 + x)^2 + x$. Then $g(-2) = 2 > 0$ and $g(-1) = -1 < 0$. By the Intermediate Value Theorem, there is a number c in $(-2, -1)$ such that $g(c) = 0$, that is, $(c^2 + c)^2 + c = 0$. Therefore c is a nonzero fixed point of $f_c \circ f_c \circ f_c$, so that $\{c, f_c(c), (f_c \circ f_c)(c)\}$ is the required 3-cycle.

15. Since z_0 is an attracting fixed point, $f_c(z_0) = z_0$ and $|f_c'(z_0)| < 1$. Thus $|\lim_{z\to z_0} [f_c(z) - f_c(z_0)]/(z - z_0)| < 1$, which means that there is a number $r > 0$ such that if $0 < |z - z_0| < r$, then $|[f_c(z) - f_c(z_0)]/(z - z_0)| < 1$. This in turn means that for such z, $|f_c(z) - z_0| < |z - z_0|$.

17. If the nth iterates of z_0 and w_0 are the same, then (7) implies that $z_0^{2^n} = w_0^{2^n}$. Therefore (11) of Section 10.4 implies that $(2\pi i\theta)2^n = 2\pi i(\frac{1}{3})2^n + 2\pi i m$ for some integer m. Dividing by $(2\pi i)2^n$, we find that $\theta = \frac{1}{3} + m/2^n$. Conversely, if $\theta = \frac{1}{3} + m/2^n$, then $(2\pi i\theta)2^n = 2\pi i(\frac{1}{3})2^n + 2\pi i m$, so that $e^{(2\pi i\theta)2^n} = e^{(2\pi i(1/3))2^n}$, and hence $z_0^{2^n} = w_0^{2^n}$.

CHAPTER 10 REVIEW

1. $(2 - 3i)^2\overline{(-5 - 12i)} = (-5 - 12i)(-5 + 12i) = 25 + 144 = 169$

3. $\sqrt{-\frac{1}{9}} = \sqrt{\frac{1}{9}}\,i = \frac{1}{3}i$

5. $|\frac{3}{2} - 2i| = \sqrt{(\frac{3}{2})^2 + (-2)^2} = \sqrt{\frac{25}{4}} = \frac{5}{2}$

7. $x = \dfrac{-2 \pm \sqrt{4 - 40}}{4} = \dfrac{-2 \pm 6i}{4} = -\dfrac{1}{2} + \dfrac{3}{2}i$ or $-\dfrac{1}{2} - \dfrac{3}{2}i$

9.

y

x

3 − 4i

11. Let $z_1 = -2i$ and $z_2 = 1 - i$. First we will obtain polar forms for z_1 and z_2. For z_1 we have $z_1 = 2(\cos 3\pi/2 + i \sin 3\pi/2)$. For z_2 we notice that $r_2 = \sqrt{2}$; $\tan \theta_2 = -1/1 = -1$, and since z_2 lies in the fourth quadrant, $\theta_2 = 7\pi/4$. Thus $z_2 = \sqrt{2}(\cos 7\pi/4 + i \sin 7\pi/4)$, so that $z_2^3 = 2\sqrt{2}(\cos 21\pi/4 + i \sin 21\pi/4) = 2\sqrt{2}(\cos 5\pi/4 + i \sin 5\pi/4)$. Then by (5) of Section 10.2, $-2i/(1 - i)^3 = z_1/z_2^3 = (\cos \pi/4 + i \sin \pi/4)/\sqrt{2} = \frac{1}{2}\sqrt{2}(\cos \pi/4 + i \sin \pi/4)$.

13. Since $z = 1 = \cos 0 + i \sin 0$, it follows that the five 5th roots are $\cos 2\pi k/5 + i \sin 2\pi k/5$ for $k = 0, 1, 2, 3$, and 4.

15. Since $z = -16 = 16(\cos \pi + i \sin \pi)$, it follows that the four 4th roots are $2[\cos(\pi/4 + 2\pi k/4) + i \sin(\pi/4 + 2\pi k/4)]$ for $k = 0, 1, 2$, and 3. More simply, the roots are $\sqrt{2} + \sqrt{2}i$, $-\sqrt{2} + \sqrt{2}i$, $-\sqrt{2} - \sqrt{2}i$, and $\sqrt{2} - \sqrt{2}i$.

17. Since $z^4 + 5z^2 + 4 = 0$ if $z^2 = -4$ or $z^2 = -1$, which means that $z = -2i, 2i, -i$, or i, the domain consists of all complex numbers except $-2i, 2i, -i$, and i.

19. If $z = f(z) = 2z^2 + 3z + 5$, then $2z^2 + 2z + 5 = 0$, so that $z = \frac{1}{4}(-2 \pm \sqrt{4 - 40}) = -\frac{1}{2} + \frac{3}{2}i$ or $-\frac{1}{2} - \frac{3}{2}i$.

21. Since $|z/\bar{z}| = 1$ for all $z \neq 0$, it follows that $\lim_{z\to 0} |z^2/\bar{z}| = \lim_{z\to 0} (|z|\,|z/\bar{z}|) = \lim_{z\to 0} |z| \cdot \lim_{z\to 0} |z/\bar{z}| = 0 \cdot 1 = 0$. Therefore $\lim_{z\to 0} z^2/\bar{z} = 0$.

23. $f'(z) = -2e^{-2z} \cos z - e^{-2z} \sin z$

25. Since $|e^{in}| = |\cos n + i \sin n| = \sqrt{\cos^2 n + \sin^2 n} = 1$, it follows that $\lim_{n\to\infty} |e^{in}/n| = \lim_{n\to\infty} 1/n = 0$, so that $\lim_{n\to\infty} e^{in}/n = 0$.

27. Since

$$\lim_{n\to\infty} \left|\frac{(3 + i)^{n+1}z^{n+1}/(n + 1)^2}{(3 + i)^n z^n/n^2}\right| = \lim_{n\to\infty} \left[\left(\frac{n}{n + 1}\right)^2 \cdot |3 + i|\,|z|\right] = \sqrt{10}|z|$$

it follows that the series converges for $|z| < 1/\sqrt{10}$ and diverges for $|z| > 1/\sqrt{10}$. Thus the radius of convergence is $1/\sqrt{10}$.

29. By (10) in Section 10.4, $e^{-i\pi/6} = \cos(-\pi/6) + i \sin(-\pi/6) = \frac{1}{2}\sqrt{3} - \frac{1}{2}i$.

31. Let $z_0 = x_0 + iy_0$. We will show that if $z_0 \neq 0$, then $f'(z_0)$ does not exist. To that end, first let z approach z_0 along the horizontal line through (x_0, y_0), so that $z = x + iy_0$. For such values of z,

$$\frac{f(z) - f(z_0)}{z - z_0} = \frac{|z|^2 - |z_0|^2}{z - z_0} = \frac{(x^2 + y_0^2) - (x_0^2 + y_0^2)}{(x + iy_0) - (x_0 + iy_0)} = \frac{x^2 - x_0^2}{x - x_0} = x + x_0$$

which approaches the real number $2x_0$. Next, let z approach z_0 along the vertical line through (x_0, y_0), so that $z = x_0 + iy$. For such values of z,

$$\frac{f(z) - f(z_0)}{z - z_0} = \frac{|z|^2 - |z_0|^2}{z - z_0} = \frac{(x_0^2 + y^2) - (x_0^2 + y_0^2)}{(x_0 + iy) - (x_0 + iy_0)} = \frac{y^2 - y_0^2}{i(y - y_0)} = -i(y + y_0)$$

which approaches $-2iy_0$. Since $2x_0$ can equal $-2iy_0$ only if $x_0 = 0 = y_0$, it follows that if $z_0 \neq 0$, then $f'(z_0)$ does not exist. Finally, if $z_0 = 0$, then we obtain

$$f'(0) = \lim_{z\to 0} \frac{f(z) - f(0)}{z - 0} = \lim_{z\to 0} \frac{|z|^2}{z} = \lim_{z\to 0} |z| \frac{|z|}{z} = 0$$

Consequently $f'(z_0)$ exists if and only if $z_0 = 0$.

CUMULATIVE REVIEW (CHAPTERS 1–9)

1. $\lim_{x\to\infty} \dfrac{x^2+1}{x\sqrt{3x^2+1}} = \lim_{x\to\infty} \dfrac{1+1/x^2}{\sqrt{3+1/x^2}} = \dfrac{1}{\sqrt{3}}$

3. The conditions for applying l'Hôpital's Rule are met:

$$\lim_{x\to 0} \frac{1-\cos 4x}{3x^2} = \lim_{x\to 0}\frac{4\sin 4x}{6x} = \lim_{x\to 0}\frac{16\cos 4x}{6} = \frac{8}{3}$$

5. Notice that $f(x) = \int_{-x}^{x^2} \sin\sqrt{t^3+1}\,dt = \int_{-x}^{0}\sin\sqrt{t^3+1}\,dt + \int_0^{x^2}\sin\sqrt{t^3+1}\,dt$. Let $G(x) = \int_{-x}^{0}\sin\sqrt{t^3+1}\,dt$ and $H(x) = \int_0^{x^2}\sin\sqrt{t^3+1}\,dt$, so that $G(x) = -\int_0^{-x}\sin\sqrt{t^3+1}\,dt$. Then $G'(x) = [-\sin\sqrt{-x^3+1}](-1)$ and $H'(x) = (\sin\sqrt{x^6+1})(2x) = 2x\sin\sqrt{x^6+1}$. Therefore $f'(x) = G'(x) + H'(x) = \sin\sqrt{-x^3+1} + 2x\sin\sqrt{x^6+1}$.

7. a. Let x denote the horizontal distance between the person and the kite, and y the length of string let out. Then $x^2 + 30^2 = y^2$. We want to find dx/dt when $y = 50$. Now $(2x)(dx/dt) = (2y)(dy/dt)$. By hypothesis, $dy/dt = -4$, so that $dx/dt = (y/x)(dy/dt) = -4y/x$. Because the triangle is right-angled, when $y = 50$ we have $x = \sqrt{(50)^2-(30)^2} = 40$. Therefore at the moment when $y = 50$, we have $dx/dt = -4(50)/40 = -5$ (feet per second), so x is decreasing at 5 feet per second.

   b. Yes, because then $x = \sqrt{(45)^2-(30)^2} = 15\sqrt{5}$, so that when $y = 45$ we have $dx/dt = -4(45)/(15\sqrt{5}) = -12/\sqrt{5}$ (feet per second).

9. $f'(x) = \cos x\cos(x-c) - \sin x\sin(x-c) = \cos[x+(x-c)] = \cos(2x-c)$, so that $f'(x) = 0$ if $\cos(2x-c) = 0$. Therefore $2x - c = \frac{1}{2}\pi + n\pi$, so that $x = \frac{1}{2}c + \frac{1}{4}\pi + \frac{1}{2}n\pi$ for some integer n. Since $f''(x) = -2\sin(2x-c)$, it follows that $f''(\frac{1}{2}c + \frac{1}{4}\pi + \frac{1}{2}n\pi) = -2\sin(c + \frac{1}{2}\pi + n\pi - c) = -2\sin(\frac{1}{2}\pi + n\pi)$, so that $f''(\frac{1}{2}c + \frac{1}{4}\pi + \frac{1}{2}n\pi) < 0$ if n is even. In that case, $f(\frac{1}{2}c + \frac{1}{4}\pi + \frac{1}{2}n\pi)$ is a relative maximum value. Since f has period π, $f(\frac{1}{2}c + \frac{1}{4}\pi)$ is the maximum value of f.

11. a. Let $f(x) = a^x$. Then $f'(x) = (\ln a)a^x$, so that $f'(0) = \ln a$. By the definition of the derivative, $f'(0) = \lim_{h\to 0}(f(h)-f(0))/(h-0) = \lim_{h\to 0}(a^h-1)/h$. Thus $\lim_{h\to 0}(a^h-1)/h = \ln a$.

   b. Let $h = 1/n$. Then part (a) implies that $\ln a = \lim_{n\to\infty}\dfrac{a^{1/n}-1}{1/n} = \lim_{n\to\infty} n(\sqrt[n]{a}-1)$

13. Let $u = 1 + 4x$, so that $du = 4\,dx$ and $36x = 9(u-1)$. If $x = 0$, then $u = 1$; if $x = \frac{3}{4}$, then $u = 4$. Thus

$$\int_0^{3/4} 36x\sqrt{1+4x}\,dx = \int_1^4 9(u-1)u^{1/2}\frac{1}{4}\,du = \frac{9}{4}\int_1^4 (u^{3/2}-u^{1/2})\,du$$
$$= \frac{9}{4}\left(\frac{2}{5}u^{5/2} - \frac{2}{3}u^{3/2}\right)\Big|_1^4 = \frac{9}{4}\left[\left(\frac{64}{5}-\frac{16}{3}\right)-\left(\frac{2}{5}-\frac{2}{3}\right)\right] = \frac{87}{5}$$

15. $u = x^2$, $dv = e^{2x}\,dx$; $du = 2x\,dx$, $v = \frac{1}{2}e^{2x}$; $\int x^2/e^{-2x}\,dx = \int x^2e^{2x}\,dx = \frac{1}{2}x^2e^{2x} - \int xe^{2x}\,dx$. For $\int xe^{2x}\,dx$ we let $u = x$, $dv = e^{2x}\,dx$; $du = dx$, $v = \frac{1}{2}e^{2x}$; $\int xe^{2x}\,dx = \frac{1}{2}xe^{2x} - \int\frac{1}{2}e^{2x}\,dx = \frac{1}{2}xe^{2x} - \frac{1}{4}e^{2x} + C_1$. Thus $\int x^2/e^{-2x}\,dx = \frac{1}{2}x^2e^{2x} - \frac{1}{2}xe^{2x} + \frac{1}{4}e^{2x} + C$.

17. $\mathscr{L} = \int_0^{2\pi}\sqrt{(-3\sin t + 3\sin 3t)^2 + (3\cos t - 3\cos 3t)^2}\,dt$

$$= 3\int_0^{2\pi}\sqrt{2 - 2(\cos t\cos 3t + \sin t\sin 3t)}\,dt = 3\int_0^{2\pi}\sqrt{2-2\cos 2t}\,dt$$
$$= 6\int_0^{2\pi}\sqrt{\frac{1-\cos 2t}{2}}\,dt = 6\int_0^{2\pi}|\sin t|\,dt = 12\int_0^{\pi}\sin t\,dt = 12(-\cos t)\Big|_0^{\pi} = 24$$

19. Since

$$\lim_{n\to\infty}\frac{(n+1)!/(1.4)^{n+1}}{n!/(1.4)^n} = \lim_{n\to\infty}\frac{n+1}{1.4} = \infty$$

the series diverges by the Ratio Test.

21. Since

$$\lim_{n\to\infty}\left|\frac{5(n+1)^2x^{n+1}/[4^{(n+1)}((n+1)^3+2)]}{5n^2x^n/[4^n(n^3+2)]}\right| = \lim_{n\to\infty}\left(\frac{n+1}{n}\right)^2\frac{1}{4}\frac{n^3+2}{(n+1)^3+2}|x| = \frac{1}{4}|x|$$

the series converges for $|x| < 4$ and diverges for $|x| > 4$. For $x = 4$, the series $\sum_{n=1}^{\infty} 5n^2/(n^3+2)$ diverges by comparison with the series $\sum_{n=1}^{\infty} 1/n$. For $x = -4$, the series $\sum_{n=1}^{\infty} 5n^2(-1)^n/(n^3+2)$ converges by the Alternating Series Test. Therefore the interval of convergence is $[-4, 4)$.

11

CONIC SECTIONS

SECTION 11.1

1. $c = -2;\ y^2 = -8x$

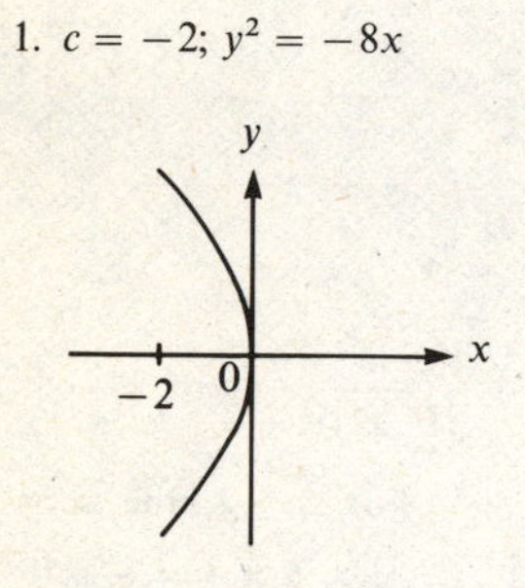

3. $c = -6;\ x^2 = -24y$

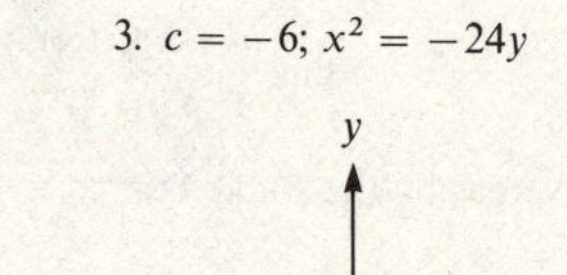

5. vertex is $(\frac{5}{2}, 0)$; $c = -\frac{5}{2}$;

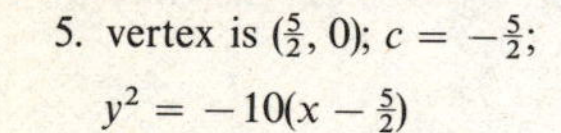

$y^2 = -10(x - \frac{5}{2})$

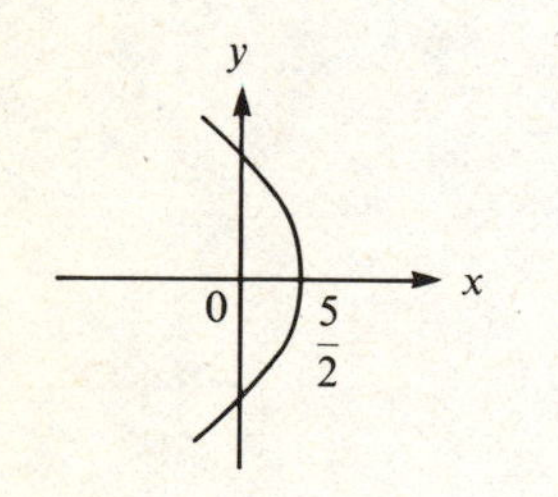

7. $c = 3;\ y^2 = 12(x - 1)$

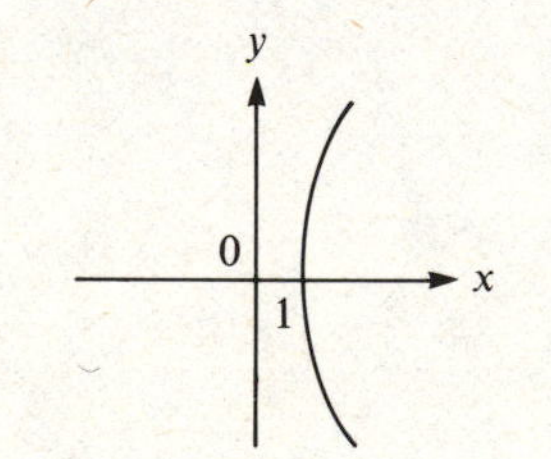

9. $c = 1;\ (x - 3)^2 = 4(y - 2)$

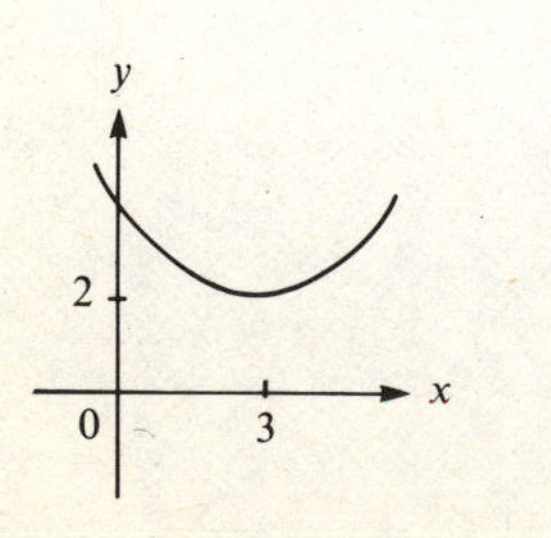

11. $y^2 = 4cx$ and $1^2 = 4c(-1)$, so $c = -\frac{1}{4}$;

$y^2 = -x$

13. $\sqrt{(x-3)^2 + (y-4)^2} = |y - 2|;\ (x-3)^2 + (y-4)^2 = (y-2)^2;$

$(x-3)^2 + y^2 - 8y + 16 = y^2 - 4y + 4;\ (x-3)^2 = 4(y-3)$

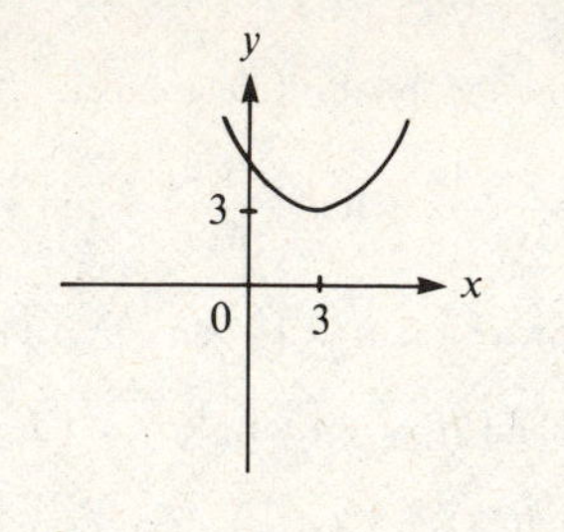

15. $c = \frac{3}{4}$, focus is $(\frac{3}{4}, 0)$; vertex is $(0, 0)$; directrix is $x = -\frac{3}{4}$; axis is $y = 0$

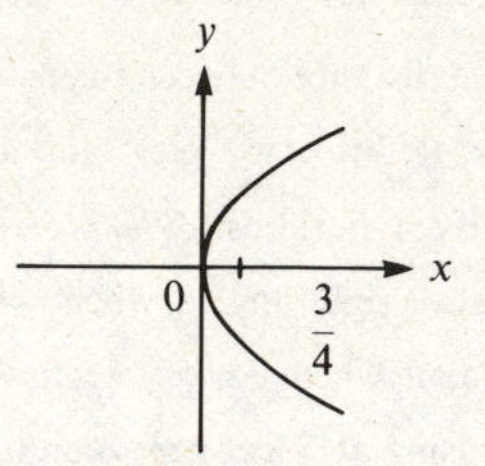

17. $c = \frac{1}{4}$, focus is $(0, \frac{1}{4})$; vertex is $(0, 0)$; directrix is $y = -\frac{1}{4}$; axis is $x = 0$

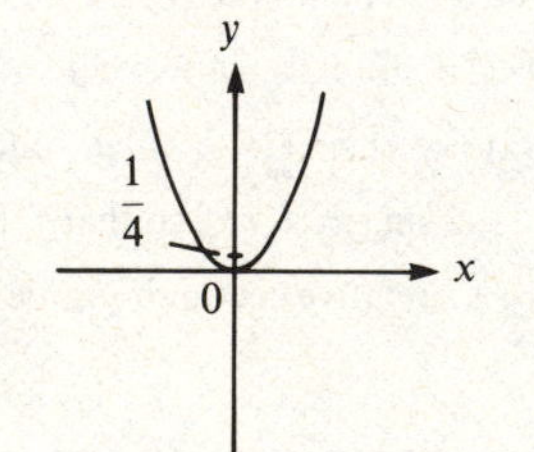

19. $c = \frac{1}{4}$, focus is $(1, -\frac{7}{4})$; vertex is $(1, -2)$; directrix is $y = -\frac{9}{4}$; axis is $x = 1$

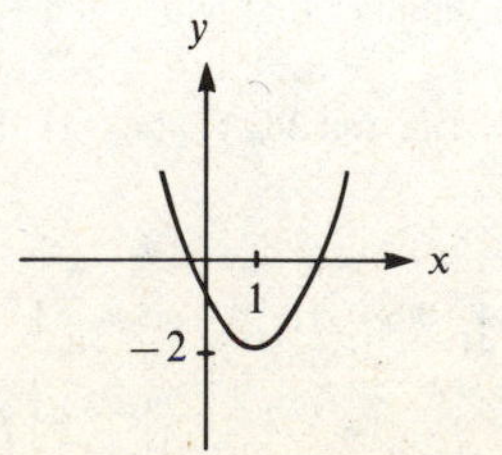

21. $c = \frac{3}{4}$, focus is $(2, \frac{15}{4})$; vertex is $(2, 3)$; directrix is $y = \frac{9}{4}$; axis is $x = 2$

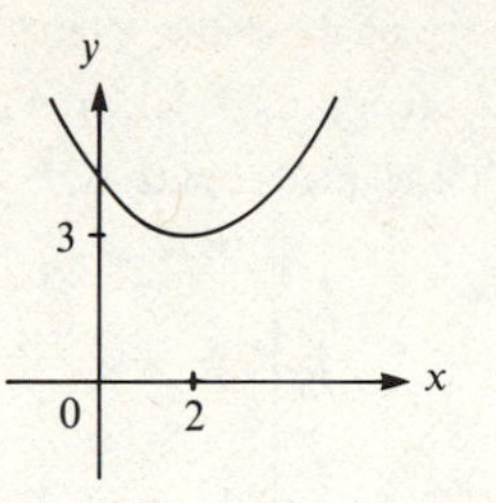

23. $0 = x^2 - 6x - 2y + 1 = (x^2 - 6x + 9) - 2y + 1 - 9 = (x - 3)^2 - 2(y + 4)$; $X = x - 3$, $Y = y + 4$; $X^2 = 2Y$

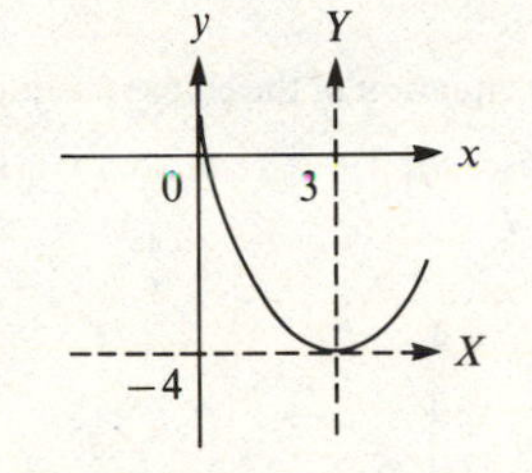

25. $0 = 3y^2 - 5x + 3y - \frac{17}{4} = (3y^2 + 3y + \frac{3}{4}) - 5x - \frac{17}{4} - \frac{3}{4} = 3(y + \frac{1}{2})^2 - 5(x + 1)$; $X = x + 1$, $Y = y + \frac{1}{2}$; $3Y^2 = 5X$, or $Y^2 = \frac{5}{3}X$

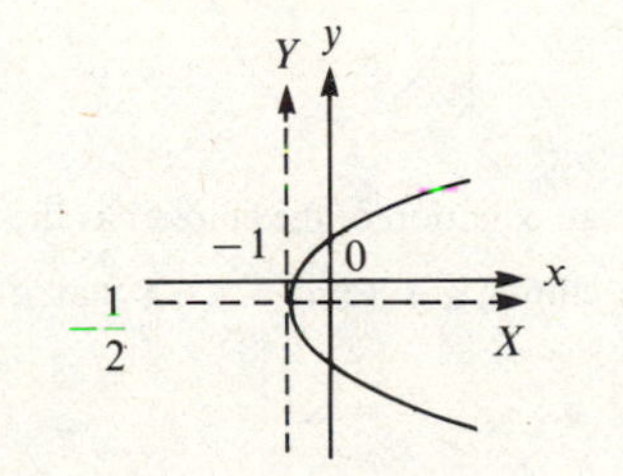

27. If $4y^2 - \sqrt{2}x + 2y = 1/\sqrt{2} - 1$, then $4(y^2 + \frac{1}{2}y + \frac{1}{16}) = \sqrt{2}x + 1/\sqrt{2} - 1 + \frac{1}{4}$, so that $4(y + \frac{1}{4})^2 = \sqrt{2}(x + \frac{1}{2}) - \frac{3}{4} = \sqrt{2}(x + \frac{1}{2} - 3\sqrt{2}/8)$; $X = x + \frac{1}{2} - 3\sqrt{2}/8$, $Y = y + \frac{1}{4}$; $4Y^2 = \sqrt{2}X$, or $Y^2 = (\sqrt{2}/4)X$

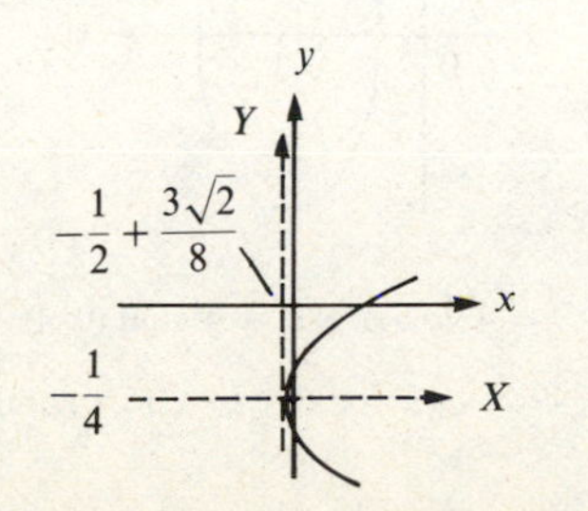

29. Since $y^2 = 4x$ for any point (x, y) on the parabola, the distance from (x, y) to $(1, 0)$ is $\sqrt{(x - 1)^2 + y^2} = \sqrt{(x - 1)^2 + 4x}$. The distance is minimized for the same nonnegative value of x that the square of the distance is, so let $f(x) = (x - 1)^2 + 4x$ for $x \geq 0$. Then $f'(x) = 2(x - 1) + 4 > 0$ for all $x \geq 0$. Thus $f(0)$ is the minimum value of f, so that $(0, 0)$ is the closest point to $(1, 0)$.

31. If the parabola has the form $x^2 = 4cy$ with $c > 0$, then the focus is $(0, c)$ and the endpoints of the latus rectum are $(2c, c)$ and $(-2c, c)$, and its length is $4c$. If $y^2 = 4cx$ with $c > 0$, then the endpoints of the latus rectum are $(c, 2c)$ and $(c, -2c)$, and its length is again $4c$.

33. The slope of the line $x + y = d$ is -1, and the slope of the parabola $x^2 = 2y$, or $y = x^2/2$, at $(x, x^2/2)$ is x. Thus if the line is tangential to the parabola, then $x = -1$, and $y = (-1)^2/2 = \frac{1}{2}$. Therefore $d = x + y = -1 + \frac{1}{2} = -\frac{1}{2}$, and the point of tangency is $(-1, \frac{1}{2})$.

35. The path is symmetric about the vertical line which passes through its highest point, which is located at $(300, 200)$. Thus an equation of the parabola is $(x - 300)^2 = 4c(y - 200)$, where c is such that $(0, 0)$ lies on the parabola (as does $(600, 0)$). Thus $(-300)^2 = 4c(-200)$, so that $c = 90{,}000/-800 = -225/2$. The equation of the parabola becomes $(x - 300)^2 = -450(y - 200)$.

37. Let an equation of the parabola be $x^2 = 4cy$, with c such that $(5, 5)$ lies on the parabola. Then $25 = 4 \cdot c \cdot 5$, so that $c = \frac{5}{4}$. Thus the distance from the focus to the center is $\frac{5}{4}$.

39. By Exercise 38 an equation of the parabola is $x^2 = (a^2/4b)y$. If $a = 3500$ and $b = 316$, then $x^2 = [(3500)^2/4(316)]y = (765{,}625/79)y$.

SECTION 11.2

1. $a = 3$, $c = 2$, $b = \sqrt{5}$; $\dfrac{x^2}{9} + \dfrac{y^2}{5} = 1$

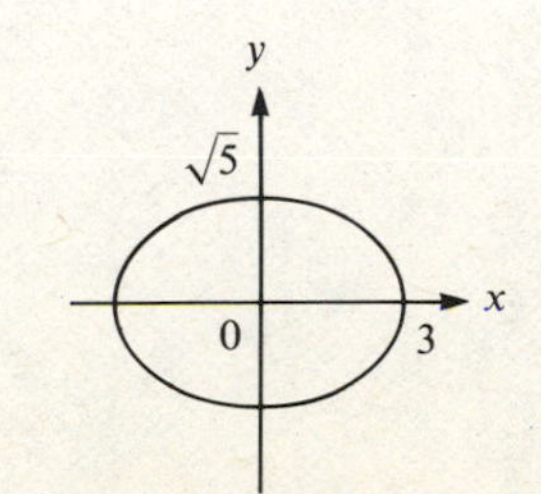

3. center is $(2, 0)$, $a = 2$, $c = 1$,

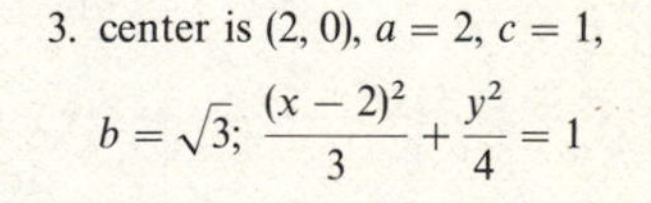

$b = \sqrt{3}$; $\dfrac{(x - 2)^2}{3} + \dfrac{y^2}{4} = 1$

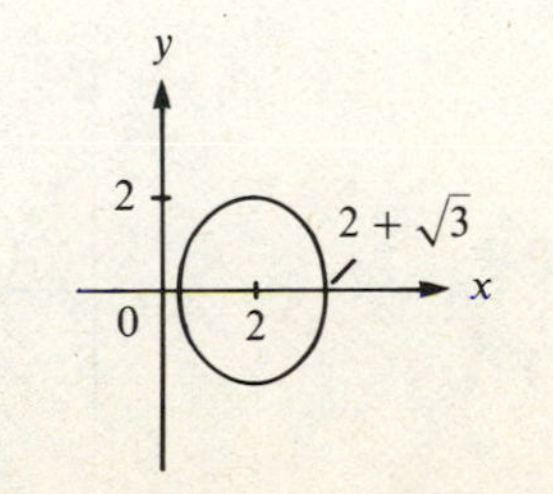

5. center is $(0, \pi)$, $b = \frac{9}{2}$, $c = 4$, $a = \sqrt{\frac{81}{4} + 16} = \frac{\sqrt{145}}{2}$; $\frac{4x^2}{145} + \frac{4(y-\pi)^2}{81} = 1$

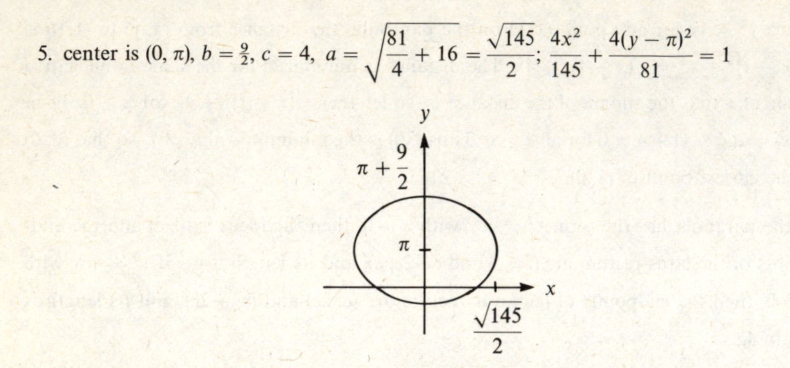

7. $a = 16$; $x^2/b^2 + y^2/256 = 1$. Since $(2\sqrt{3}, 4)$ lies on the ellipse, $12/b^2 + \frac{16}{256} = 1$, or $12/b^2 = \frac{15}{16}$. Thus $b^2 = 12(\frac{16}{15}) = \frac{64}{5}$, so an equation of the ellipse is $5x^2/64 + y^2/256 = 1$

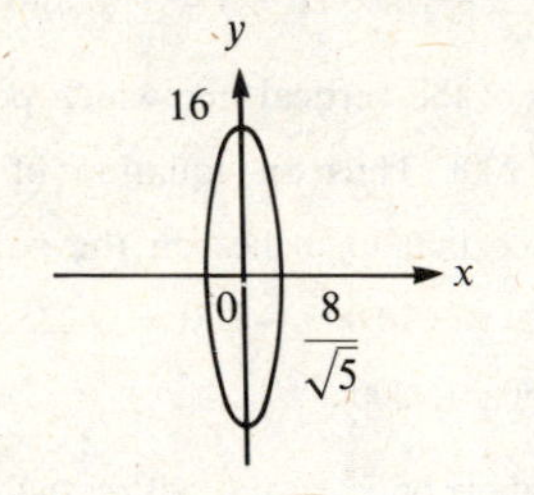

9. center is $(2, 0)$; $a = 3$, $c = 2$, $b = \sqrt{9-4} = \sqrt{5}$; $\frac{(x-2)^2}{9} + \frac{y^2}{5} = 1$

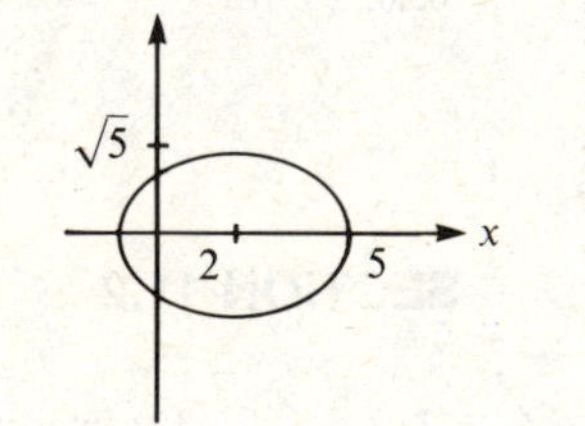

11. $a = 4$, $c = 3$, $b = \sqrt{16-9} = \sqrt{7}$; $\frac{(x-6)^2}{16} + \frac{(y-1)^2}{7} = 1$

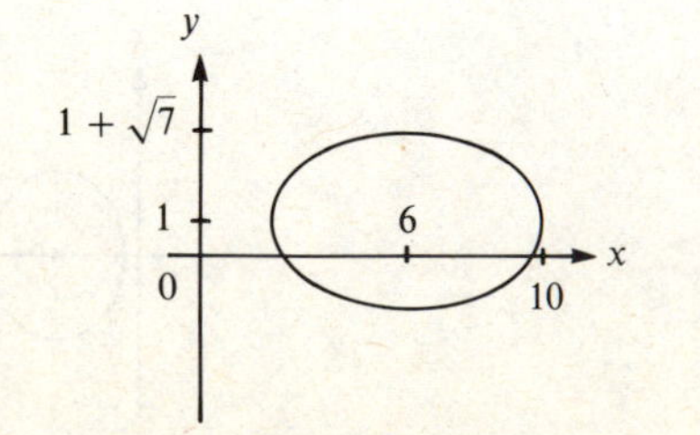

13. Let an equation of the ellipse be $x^2/p^2 + y^2/q^2 = 1$. Since $(-1, 1)$ is on the ellipse, $1/p^2 + 1/q^2 = 1$. Since $(\frac{1}{2}, -2)$ is on the ellipse, $1/4p^2 + 4/q^2 = 1$, so that $1/p^2 = 4 - 16/q^2$. Then $1 - 1/q^2 = 1/p^2 = 4 - 16/q^2$, so $3 = 15/q^2$. Thus $q = \sqrt{5}$, and consequently $p = \sqrt{5}/2$. The equation of the ellipse becomes $4x^2/5 + y^2/5 = 1$.

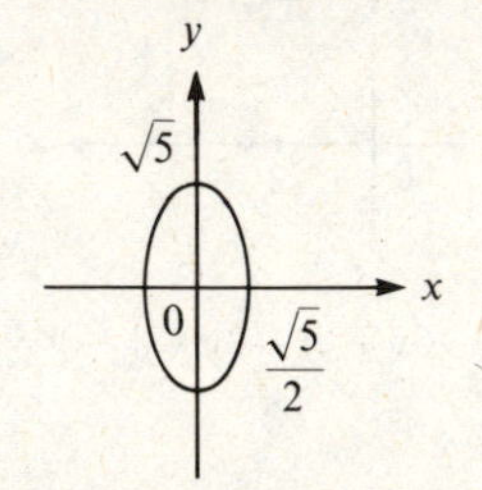

15. The center is $(3, 0)$, $a = 4$; an equation of the ellipse has the form $(x-3)^2/b^2 + y^2/16 = 1$. Since $(0, 0)$ is on the ellipse, $9/b^2 + \frac{0}{16} = 1$, so that $b = 3$. The equation becomes $(x-3)^2/9 + y^2/16 = 1$.

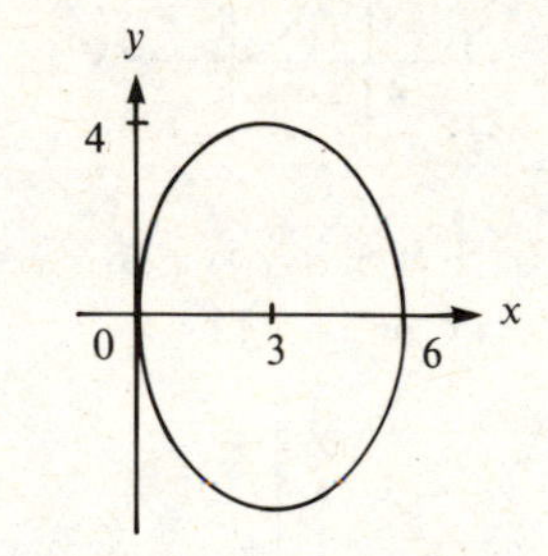

17. The center is $(3, 0)$, $b = 2$; an equation of the ellipse has the form $(x-3)^2/4 + y^2/a^2 = 1$. Since $(4, 2\sqrt{3})$ is on the ellipse, $\frac{1}{4} + 12/a^2 = 1$, so that $a = 4$. The equation becomes $(x-3)^2/4 + y^2/16 = 1$.

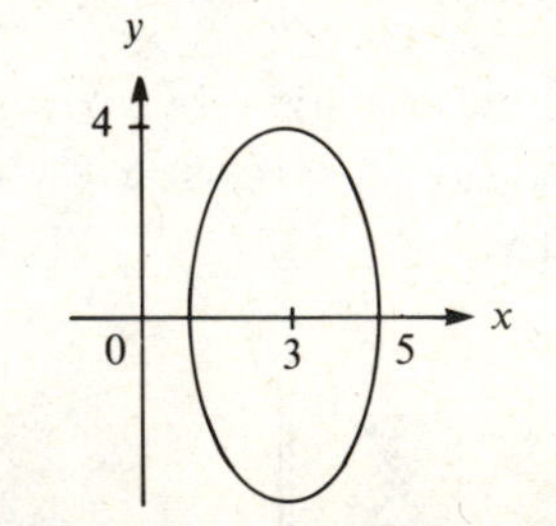

19. $a = 5$, $b = 3$, $c = \sqrt{25-9} = 4$; foci are $(0, -4)$ and $(0, 4)$; vertices are $(0, -5)$ and $(0, 5)$.

21. $a = 8$, $b = 7$, $c = \sqrt{64-49} = \sqrt{15}$; foci are $(-\sqrt{15}, 0)$ and $(\sqrt{15}, 0)$; vertices are $(-8, 0)$ and $(8, 0)$

23. $x^2/9 + y^2/3 = 1$; $a = 3$, $b = \sqrt{3}$, $c = \sqrt{9-3} = \sqrt{6}$; foci are $(-\sqrt{6}, 0)$ and $(\sqrt{6}, 0)$; vertices are $(-3, 0)$ and $(3, 0)$

25. $x^2/\frac{1}{4} + y^2/1 = 1$; $a = 1$, $b = \frac{1}{2}$, $c = \sqrt{1 - \frac{1}{4}} = \sqrt{3}/2$; foci are $(0, -\sqrt{3}/2)$ and $(0, \sqrt{3}/2)$; vertices are $(0, -1)$ and $(0, 1)$

27. $(x+1)^2/\frac{1}{25} + (y-3)^2/1 = 1$; $a = 1$, $b = \frac{1}{5}$, $c = \sqrt{1 - \frac{1}{25}} = 2\sqrt{6}/5$; center is $(-1, 3)$; foci are $(-1, 3 - 2\sqrt{6}/5)$ and $(-1, 3 + 2\sqrt{6}/5)$; vertices are $(-1, 2)$ and $(-1, 4)$

29. $x^2 - 2x + 1 + 2(y^2 - 2y + 1) = 4$, or $(x-1)^2 + 2(y-1)^2 = 4$, so that $(x-1)^2/4 + (y-1)^2/2 = 1$; $a = 2$, $b = \sqrt{2}$, $c = \sqrt{4-2} = \sqrt{2}$; center is $(1, 1)$; vertices are $(-1, 1)$ and $(3, 1)$; foci $(1 - \sqrt{2}, 1)$ and $(1 + \sqrt{2}, 1)$

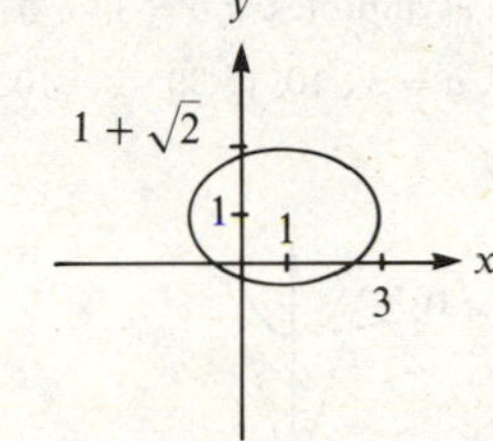

31. $(x^2 - 8x + 16) + 2y^2 = -12 + 16 = 4$, or $(x-4)^2 + 2y^2 = 4$, so that $(x-4)^2/4 + y^2/2 = 1$; $a = 2$, $b = \sqrt{2}$, $c = \sqrt{4-2} = \sqrt{2}$; center is $(4, 0)$; vertices are $(2, 0)$ and $(6, 0)$; foci are $(4 - \sqrt{2}, 0)$ and $(4 + \sqrt{2}, 0)$

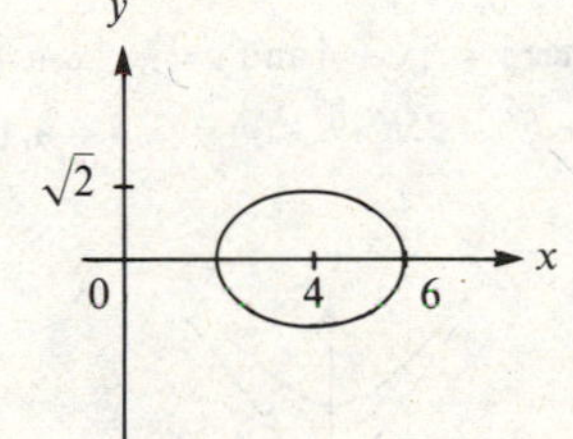

33. $(x^2 + 2x + 1) + 2(y^2 + 2y + 1) = 4$, or $(x+1)^2 + 2(y+1)^2 = 4$, so that $(x+1)^2/4 + (y+1)^2/2 = 1$; $a = 2$, $b = \sqrt{2}$, $c = \sqrt{4-2} = \sqrt{2}$; center is $(-1, -1)$; vertices are $(-3, -1)$ and $(1, -1)$; foci are $(-1 - \sqrt{2}, -1)$ and $(-1 + \sqrt{2}, -1)$

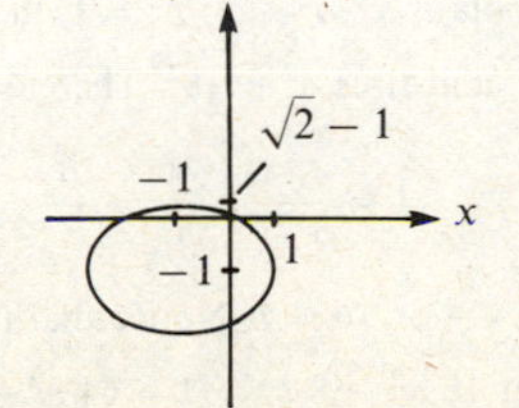

35. If the ellipse passes through the origin, then $(0-1)^2/4 + (0+2)^2/9 = c$, so that $c = \frac{1}{4} + \frac{4}{9} = \frac{25}{36}$.

37. Since the ellipse is in standard position, it has an equation of the form $x^2/a^2 + y^2/b^2 = 1$, for appropriate constants $a > 0$ and $b > 0$. Since the ellipse passes through $(0, 2)$, it follows that $2^2/b^2 = 1$, so $b = 2$. Thus $x^2/a^2 + y^2/4 = 1$. By implicit differentiation, we obtain

$$\frac{2x}{a^2} + \frac{2y}{4}\frac{dy}{dx} = 0, \quad \text{or} \quad \frac{y}{2}\frac{dy}{dx} = -\frac{2x}{a^2}$$

By hypothesis, if $x = -2$ and $y > 0$, then $dy/dx = 1/\sqrt{2}$, so that $y/2(1/\sqrt{2}) = -2(-2)/a^2$ and thus $y = 8\sqrt{2}/a^2$. Then $x^2/a^2 + y^2/b^2 = 1$ becomes

$$\frac{(-2)^2}{a^2} + \frac{(8\sqrt{2}/a^2)^2}{4} = 1, \quad \text{or} \quad \frac{4}{a^2} + \frac{32}{a^4} = 1$$

It follows that $4a^2 + 32 = a^4$, that is, $0 = a^4 - 4a^2 - 32 = (a^2 - 8)(a^2 + 4)$. Since $a > 0$, this means that $a = 2\sqrt{2}$. Consequently an equation of the ellipse is $x^2/8 + y^2/4 = 1$.

39. The ellipse is given by $x^2/2 + y^2/8 = 1$, so that by Exercise 38 with $a = \sqrt{2}$ and $b = 2\sqrt{2}$, an equation of the tangent line at (x_0, y_0) is $xx_0/2 + yy_0/8 = 1$. This will be the line $2x + y = d$ if $2/d = x_0/2$ and $1/d = y_0/8$, which means that $x_0/4 = 1/d = y_0/8$, or $y_0 = 2x_0$. Since (x_0, y_0) is on the ellipse, $4x_0^2 + (2x_0)^2 = 8$, so that $x_0 = \pm 1$ and $y_0 = \pm 2$. Then $d = 4$ or $d = -4$, and the points of tangency are $(1, 2)$ and $(-1, -2)$.

41. Let (c, y) be on the ellipse. If $y > 0$, then $y = b\sqrt{1 - c^2/a^2} = (b/a)\sqrt{a^2 - c^2} = b^2/a$, while if $y < 0$, then $y = -b\sqrt{1 - c^2/a^2} = -(b/a)\sqrt{a^2 - c^2} = -b^2/a$. Thus the length of the latus rectum is $b^2/a + b^2/a = 2b^2/a$.

43. By Exercise 42, we find that

$$r = \frac{D^2}{4A} + \frac{E^2}{4C} - F = \frac{64}{8} + \frac{256}{16} - F = 24 - F$$

a. If $F < 24$, then $r > 0$ and the graph of the equation is an ellipse.

b. If $F = 24$, then the graph is a point.

c. If $F > 24$, then the graph consists of no points.

45. Since $a = 228$, $b = 227$, the lengths of the major and minor axes are 456 and 454, respectively, so the ratio is $\frac{456}{454} = \frac{228}{227}$.

47. Let (x, y) be a point such that $x^2/a^2 + y^2/b^2 = 1$, and let $c^2 = a^2 - b^2$. Now $y^2 = b^2 - b^2x^2/a^2$, and the distances from (x, y) to $(-c, 0)$ and to $(c, 0)$ are

$$\sqrt{(x+c)^2 + y^2} = \sqrt{(x+c)^2 + b^2 - \frac{b^2x^2}{a^2}} = \sqrt{\left(x^2 - \frac{b^2x^2}{a^2}\right) + 2cx + (c^2 + b^2)}$$

$$= \sqrt{\frac{c^2x^2}{a^2} + 2cx + a^2}$$

and

$$\sqrt{(x-c)^2+y^2} = \sqrt{(x-c)^2+b^2-\frac{b^2x^2}{a^2}} = \sqrt{\left(x^2-\frac{b^2x^2}{a^2}\right)-2cx+(c^2+b^2)}$$
$$= \sqrt{\frac{c^2x^2}{a^2}-2cx+a^2}$$

respectively. As a result, the sum k of the distances from (x, y) to the foci is given by

$$k = \sqrt{\frac{c^2x^2}{a^2}+2cx+a^2} + \sqrt{\frac{c^2x^2}{a^2}-2cx+a^2}$$
$$= \left|\frac{cx}{a}+a\right| + \left|\frac{cx}{a}-a\right| = \frac{1}{a}(|a^2+cx|+|a^2-cx|)$$

Since $a^2 - cx$ and $a^2 + cx$ are both positive, $k = (1/a)[(a^2+cx)+(a^2-cx)] = 2a$. Thus (x, y) lies on the ellipse by Definition 11.2.

SECTION 11.3

1. $a = 4$, $c = 9$, $b = \sqrt{81-16} = \sqrt{65}$; $\dfrac{x^2}{16} - \dfrac{y^2}{65} = 1$

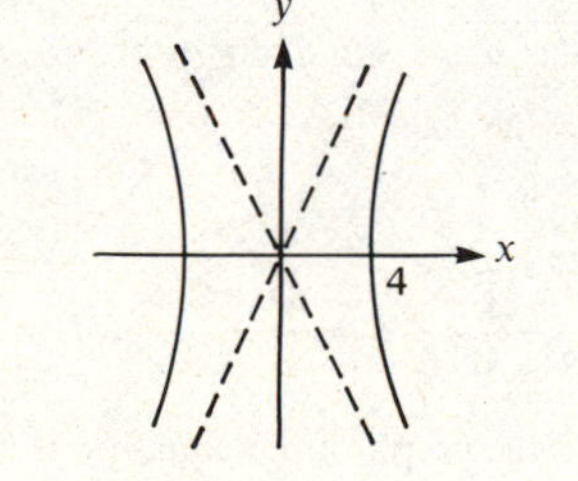

3. $a = \frac{1}{3}$, $c = 1$, $b = \sqrt{1-\frac{1}{9}} = \dfrac{2\sqrt{2}}{3}$, $\dfrac{9y^2}{1} - \dfrac{9x^2}{8} = 1$, or $9y^2 - \dfrac{9x^2}{8} = 1$

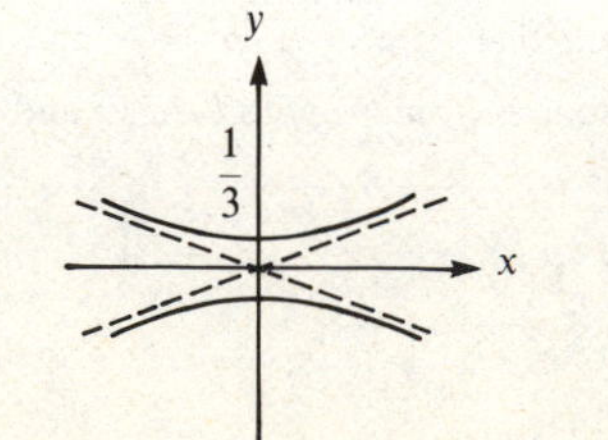

5. $a = 3$, $c = 5$, $b = \sqrt{25-9} = 4$; $\dfrac{y^2}{9} - \dfrac{x^2}{16} = 1$

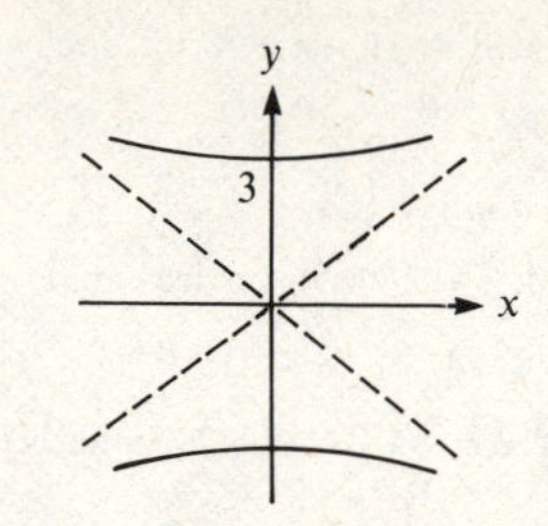

7. $c = 10$; since $y = \pm 3x$ are the asymptotes, $a/b = 3$, or $a = 3b$; thus $100 = c^2 = (3b)^2 + b^2 = 10b^2$, so $b = \sqrt{10}$, $a = 3\sqrt{10}$; $y^2/90 - x^2/10 = 1$

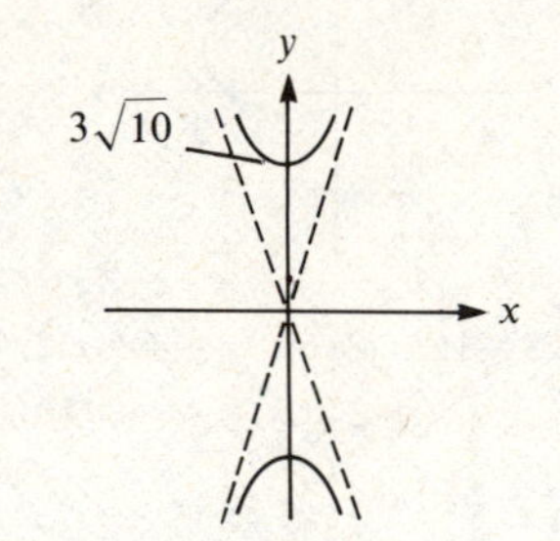

9. The center is $(0, 4)$, $c = 5$; since $y = \frac{4}{3}x + 4$ and $y = -\frac{4}{3}x + 4$ are the asymptotes, $a/b = \frac{4}{3}$, or $a = \frac{4}{3}b$; thus $25 = (\frac{4}{3}b)^2 + b^2 = 25b^2/9$, so $b = 3$, $a = 4$; $(y-4)^2/16 - x^2/9 = 1$.

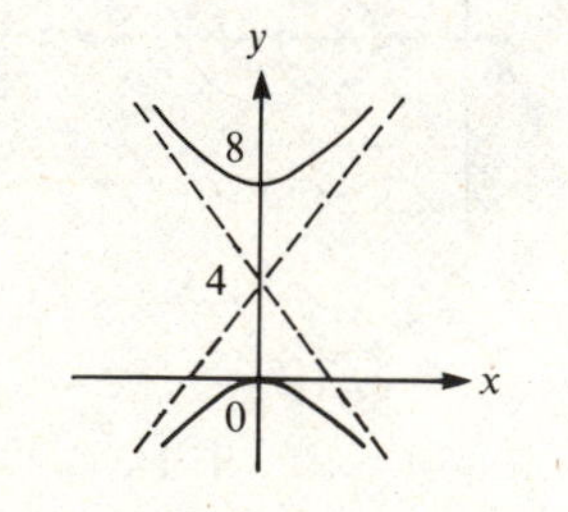

11. If an equation of the hyperbola is $x^2/a^2 - y^2/b^2 = 1$, then $20/a^2 - 64/b^2 = 1 = 8/a^2 - 16/b^2$, so that $12/a^2 = 48/b^2$, and thus $a^2 = \frac{1}{4}b^2$. Therefore

$$1 = \frac{8}{b^2/4} - \frac{16}{b^2} = \frac{16}{b^2}$$

so that $b = 4$, and $a = 2$; $x^2/4 - y^2/16 = 1$. Notice that if an equation of the hyperbola were $y^2/a^2 - x^2/b^2 = 1$, then $16/a^2 - 8/b^2 = 1 = 64/a^2 - 20/b^2$, so that $12/b^2 = 48/a^2$, and $a^2 = 4b^2$. But then $16/4b^2 - 8/b^2 = 1$, which is impossible. Thus there is no equation of the form $y^2/a^2 - x^2/b^2 = 1$ that satisfies the hypotheses.

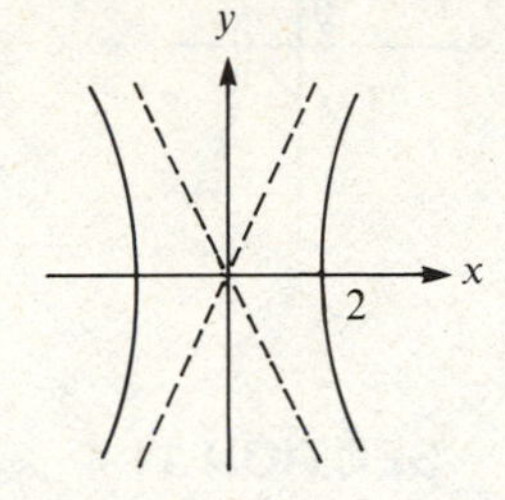

13. $a = 3, b = 4, c = 5$; vertices are $(-3, 0)$ and $(3, 0)$; foci are $(-5, 0)$ and $(5, 0)$; asymptotes are $y = \frac{4}{3}x$ and $y = -\frac{4}{3}x$

15. $a = 5, b = 7, c = \sqrt{25 + 49} = \sqrt{74}$; vertices are $(-5, 0)$ and $(5, 0)$; foci are $(-\sqrt{74}, 0)$ and $(\sqrt{74}, 0)$; asymptotes are $y = \frac{7}{5}x$ and $y = -\frac{7}{5}x$

17. $y^2/25 - x^2/\frac{25}{9} = 1$; $a = 5$, $b = \frac{5}{3}$, $c = \sqrt{25 + \frac{25}{9}} = 5\sqrt{10}/3$; vertices are $(0, -5)$ and $(0, 5)$; foci are $(0, -5\sqrt{10}/3)$ and $(0, 5\sqrt{10}/3)$; asymptotes are $y = 3x$ and $y = -3x$

19. $x^2/\frac{25}{4} - y^2/25 = 1$; $a = \frac{5}{2}$, $b = 5$, $c = \sqrt{\frac{25}{4} + 25} = 5\sqrt{5/2}$; vertices are $(-\frac{5}{2}, 0)$ and $(\frac{5}{2}, 0)$; foci are $(-5\sqrt{5/2}, 0)$ and $(5\sqrt{5/2}, 0)$; asymptotes are $y = 2x$ and $y = -2x$

21. center is $(-3, -1)$; $a = 5$, $b = 12$, $c = \sqrt{25 + 144} = 13$; vertices are $(-8, -1)$ and $(2, -1)$; foci are $(-16, -1)$ and $(10, -1)$; asymptotes are $y + 1 = \frac{12}{5}(x + 3)$ and $y + 1 = -\frac{12}{5}(x + 3)$

23. center is $(-\sqrt{2}, 3)$; $a = 1$, $b = 9$, $c = \sqrt{1 + 81} = \sqrt{82}$; vertices are $(-\sqrt{2} - 1, 3)$ and $(-\sqrt{2} + 1, 3)$; foci are $(-\sqrt{2} - \sqrt{82}, 3)$ and $(-\sqrt{2} + \sqrt{82}, 3)$; asymptotes are $y - 3 = 9(x + \sqrt{2})$ and $y - 3 = -9(x + \sqrt{2})$

25. $(x^2 + 6x + 9) - (y^2 - 12y + 36) = 9$, or $(x + 3)^2 - (y - 6)^2 = 9$, so that $(x + 3)^2/9 - (y - 6)^2/9 = 1$; $a = 3 = b$; center is $(-3, 6)$; vertices are $(-6, 6)$ and $(0, 6)$; asymptotes are $y - 6 = x + 3$ and $y - 6 = -(x + 3)$

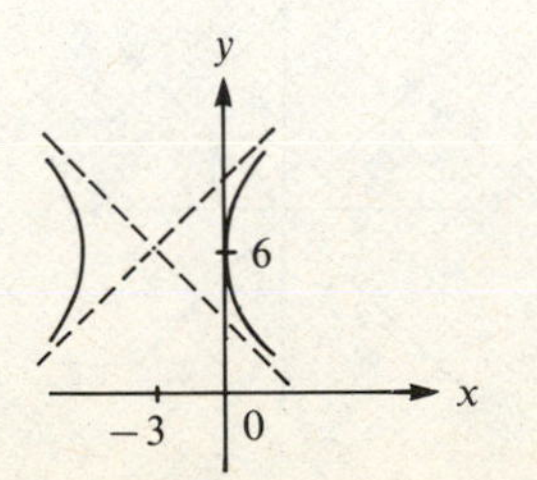

27. $4(x^2 - 2x + 1) - 9(y^2 + 4y + 4) = 36$, or $4(x - 1)^2 - 9(y + 2)^2 = 36$, so that $(x - 1)^2/9 - (y + 2)^2/4 = 1$; center is $(1, -2)$; $a = 3$, $b = 2$; vertices are $(-2, -2)$ and $(4, -2)$; asymptotes are $y + 2 = \frac{2}{3}(x - 1)$ and $y + 2 = -\frac{2}{3}(x - 1)$

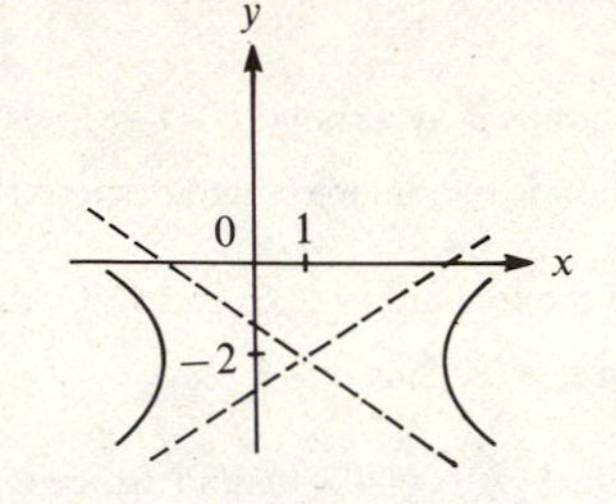

29. $(x^2 + 2\sqrt{2}x + 2) - (y^2 - 2\sqrt{2}y + 2) = 1$, or $(x + \sqrt{2})^2 - (y - \sqrt{2})^2 = 1$; center is $(-\sqrt{2}, \sqrt{2})$; $a = 1 = b$; vertices are $(-\sqrt{2} - 1, \sqrt{2})$ and $(-\sqrt{2} + 1, \sqrt{2})$; asymptotes are $y - \sqrt{2} = x + \sqrt{2}$ and $y - \sqrt{2} = -(x + \sqrt{2})$

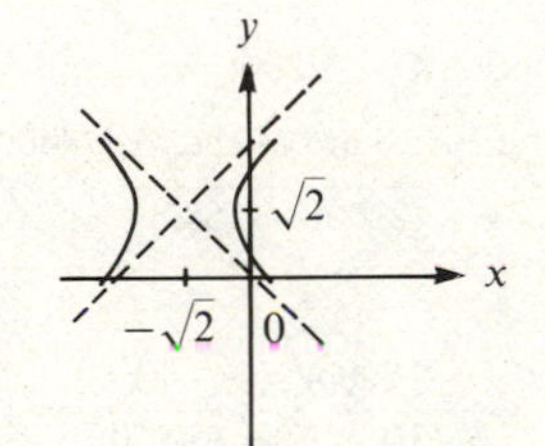

31. Differentiating the equation $x^2/a^2 - y^2/b^2 = 1$ implicitly, we obtain $2x/a^2 - (2y/b^2)(dy/dx) = 0$, so that $dy/dx = b^2x/(a^2y)$. At (x_0, y_0) we have $dy/dx = b^2x_0/(a^2y_0)$. An equation of the tangent line is $y - y_0 = [b^2x_0/(a^2y_0)](x - x_0)$. Multiplying both sides of this equation by y_0/b^2, we obtain

$$\frac{yy_0}{b^2} - \frac{y_0^2}{b^2} = \frac{xx_0}{a^2} - \frac{x_0^2}{a^2}, \quad \text{and thus} \quad \frac{xx_0}{a_2} - \frac{yy_0}{b^2} = \frac{x_0^2}{a^2} - \frac{y_0^2}{b^2} = 1$$

33. Assume that an equation of the hyperbola is $x^2/a^2 - y^2/b^2 = 1$. Then the asymptotes are $y = (b/a)x$ and $y = -(b/a)x$. Any line parallel to but distinct from the line $y = (b/a)x$ has an equation of the form $y = (b/a)x + d$, where $d \neq 0$. If this line and the hyperbola intersect at (x, y), then

$$\frac{x^2}{a^2} - \frac{((b/a)x + d)^2}{b^2} = 1$$

so that $-(2dx/(ab)) - d^2/b^2 = 1$, or $x = -[a(b^2 + d^2)/(2bd)]$. Thus

$$y = \frac{b}{a}\left[-\frac{a(b^2 + d^2)}{2bd}\right] + d = d - \frac{b^2 + d^2}{2d}$$

Thus the line intersects the hyperbola exactly once. Similar considerations apply to lines parallel to the other asymptote.

35. Let (c, y) be on the hyperbola. If $y > 0$, then $y = b\sqrt{c^2/a^2 - 1} = b^2/a$, while if $y < 0$, then $y = -b\sqrt{c^2/a^2 - 1} = -b^2/a$. Thus the length of the latus rectum is $b^2/a + b^2/a = 2b^2/a$.

37. In the notation of Exercise 36, $r = \dfrac{6^2}{4 \cdot 3} + \dfrac{8^2}{4(-4)} - F = -1 - F$.

   a. The graph of the equation is a hyperbola if $-1 - F \neq 0$, that is, if $F \neq -1$.

   b. The graph of the equation is two intersecting lines if $-1 - F = 0$, that is, if $F = -1$.

39. For a hyperbola, $c > a$. If the foci are $(-2, 0)$ and $(2, 0)$, then $c = 2$, and if the vertices are $(-3, 0)$ and $(3, 0)$, then $a = 3$. Thus $a > c$.

41. If Marian is at $(4400, 0)$, Jack at $(-4400, 0)$, and Bruce midway between Marian and Jack, we find that Bruce is at the origin. Since Bruce hears the thunder 1 second after Marian does, if (x, y) represents the point at which the lightning strikes, then

$$\sqrt{x^2 + y^2} - \sqrt{(x - 4400)^2 + y^2} = 1100$$

and this represents the equation $(x - 2200)^2/a^2 - y^2/b^2 = 1$, with $x > 2200$, and $a = \frac{1100}{2} = 550$. Since Marian and Bruce are at the foci, 4400 feet apart, $2c = 4400$, so that $c = 2200$. Thus $b = \sqrt{c^2 - a^2} = \sqrt{(2200)^2 - (550)^2} = 550\sqrt{15}$. Therefore the equation of the hyperbola becomes

$$\frac{(x - 2200)^2}{(550)^2} - \frac{y^2}{15(550)^2} = 1$$

From the result of Exercise 40, along with the above equation, Marian hears the thunder 4 seconds before Jack and 1 second before Bruce only if the lightning strikes a point (x, y) satisfying

$$\frac{x^2}{(2200)^2} - \frac{y^2}{3(2200)^2} = 1 \quad \text{and} \quad \frac{(x - 2200)^2}{(550)^2} - \frac{y^2}{15(550)^2} = 1$$

Solving for y^2 in the second equation yields $y^2 = 15(x - 2200)^2 - 15(550)^2$, and substituting in the first equation yields

$$\frac{x^2}{(2200)^2} - \frac{15(x - 2200)^2 - 15(550)^2}{3(2200)^2} = 1$$

This equation reduces to $4x^2 - 5(4400)x + 91(550)^2 = 0$, so by the quadratic formula,

$$x = \frac{5500 \pm \sqrt{25(16)(275)^2 - 4(91)(275)^2}}{2} = 3575 \text{ or } 1925$$

Since x must be greater than 2200, it follows that $x = 3575$, and thus $y^2 = 15(3575 - 2200)^2 - 15(550)^2$, so that $y = \pm 825\sqrt{35}$. Consequently lightning strikes at $(3575, 825\sqrt{35})$ or at $(3575, -825\sqrt{35})$.

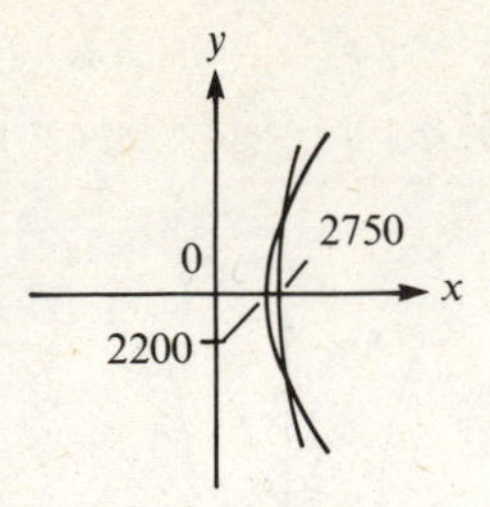

SECTION 11.4

1. $A = C$, so $\theta = \pi/4$; the equation becomes

$$\left(\frac{\sqrt{2}}{2}X - \frac{\sqrt{2}}{2}Y\right)\left(\frac{\sqrt{2}}{2}X + \frac{\sqrt{2}}{2}Y\right) = -4, \quad \text{or} \quad \frac{Y^2}{8} - \frac{X^2}{8} = 1$$

The conic section is a hyperbola.

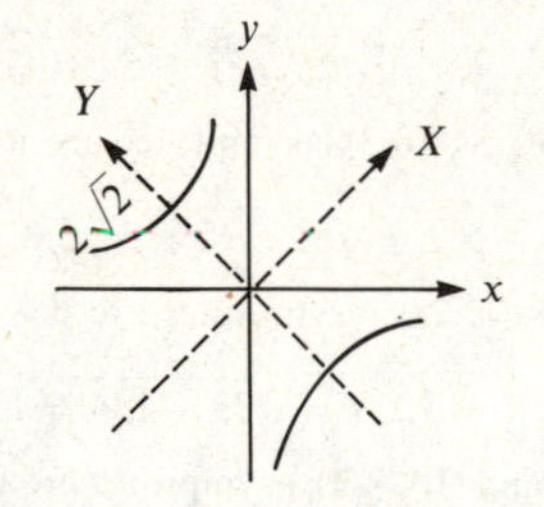

3. $A = C$, so $\theta = \pi/4$; the equation becomes

$$\left(\frac{\sqrt{2}}{2}X - \frac{\sqrt{2}}{2}Y\right)^2 - \left(\frac{\sqrt{2}}{2}X - \frac{\sqrt{2}}{2}Y\right)\left(\frac{\sqrt{2}}{2}X + \frac{\sqrt{2}}{2}Y\right) + \left(\frac{\sqrt{2}}{2}X + \frac{\sqrt{2}}{2}Y\right)^2 = 2$$

or

$$\frac{X^2}{4} + \frac{3Y^2}{4} = 1$$

The conic section is an ellipse.

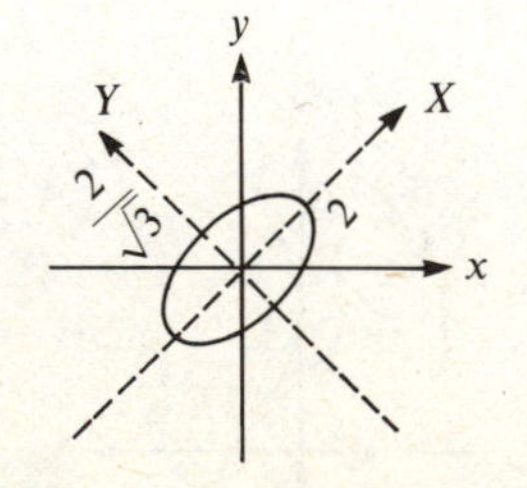

5. $A = 145$, $B = 120$, $C = 180$, $\tan 2\theta = 120/(145 - 180) = -\frac{24}{7}$; from Example 2, $\cos\theta = \frac{3}{5}$ and $\sin\theta = \frac{4}{5}$; the equation becomes $145(\frac{3}{5}X - \frac{4}{5}Y)^2 + 120(\frac{3}{5}X - \frac{4}{5}Y)(\frac{4}{5}X + \frac{3}{5}Y) + 180(\frac{4}{5}X + \frac{3}{5}Y)^2 = 900$, or $X^2/4 + Y^2/9 = 1$. The conic section is an ellipse.

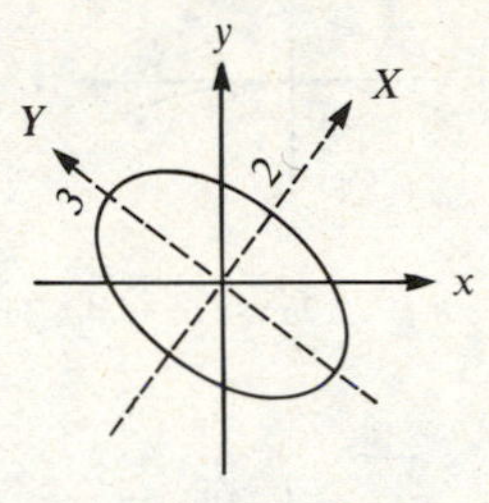

7. $A = 16$, $B = -24$, $C = 9$; $\tan 2\theta = -24/(16 - 9) = -\frac{24}{7}$; from Example 2, $\cos\theta = \frac{3}{5}$ and $\sin\theta = \frac{4}{5}$; the equation becomes $16(\frac{3}{5}X - \frac{4}{5}Y)^2 - 24(\frac{3}{5}X - \frac{4}{5}Y)(\frac{4}{5}X + \frac{3}{5}Y) + 9(\frac{4}{5}X + \frac{3}{5}Y)^2 - 5(\frac{3}{5}X - \frac{4}{5}Y) - 90(\frac{4}{5}X + \frac{3}{5}Y) + 25 = 0$, or $25Y^2 - 75X - 50Y + 25 = 0$, or $(Y - 1)^2 = 3X$. The conic section is a parabola.

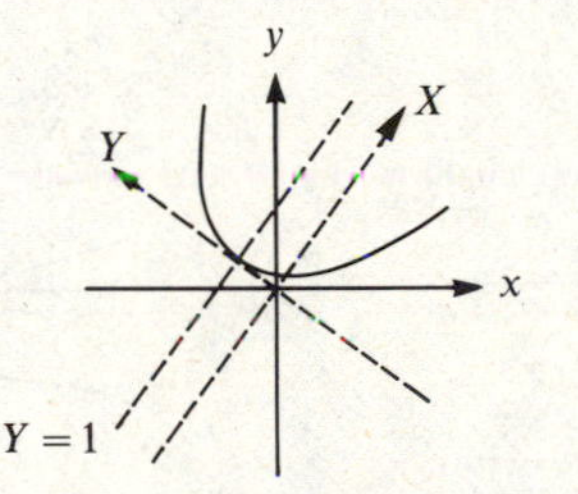

9. $A = 2$, $B = -72$, $C = 23$; $\tan 2\theta = -72/(2 - 23) = \frac{24}{7}$; by the method of Example 2, $\cos 2\theta = \frac{7}{25}$, $\cos\theta = \frac{4}{5}$ and $\sin\theta = \frac{3}{5}$; the equation becomes $2(\frac{4}{5}X - \frac{3}{5}Y)^2 - 72(\frac{4}{5}X - \frac{3}{5}Y)(\frac{3}{5}X + \frac{4}{5}Y) + 23(\frac{3}{5}X + \frac{4}{5}Y)^2 + 100(\frac{4}{5}X - \frac{3}{5}Y) - 50(\frac{3}{5}X + \frac{4}{5}Y) = 0$, or $-25X^2 + 50Y^2 + 50X - 100Y = 0$, or $2(Y - 1)^2 - (X - 1)^2 = 1$. The conic section is a hyperbola.

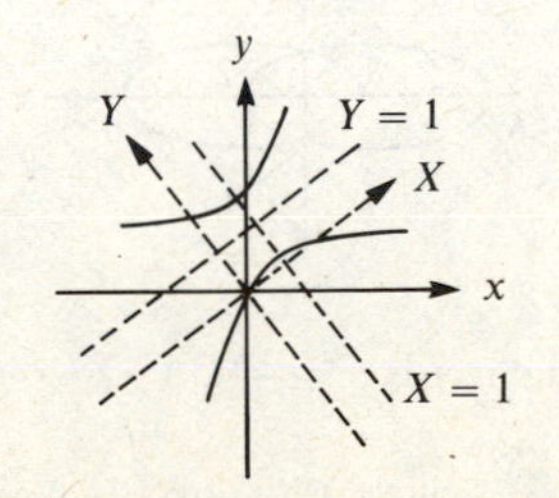

11. $A = 9$, $B = -24$, $C = 2$; $\tan 2\theta = -24/(9 - 2) = -\frac{24}{7}$; from Example 2, $\cos\theta = \frac{3}{5}$ and $\sin\theta = \frac{4}{5}$; the equation becomes $9(\frac{3}{5}X - \frac{4}{5}Y)^2 - 24(\frac{3}{5}X - \frac{4}{5}Y)(\frac{4}{5}X + \frac{3}{5}Y) + 2(\frac{4}{5}X + \frac{3}{5}Y)^2 = 0$, or $18Y^2 = 7X^2$; two intersecting lines $Y = \sqrt{\frac{7}{18}}X$ and $Y = -\sqrt{\frac{7}{18}}X$.

13. $A = 145$, $B = 120$, $C = 180$; $\tan 2\theta = 120/(145 - 180) = -\frac{24}{7}$; from Example 2, $\cos\theta = \frac{3}{5}$ and $\sin\theta = \frac{4}{5}$; the equation becomes

$$145(\tfrac{3}{5}X - \tfrac{4}{5}Y)^2 + 120(\tfrac{3}{5}X - \tfrac{4}{5}Y)(\tfrac{4}{5}X + \tfrac{3}{5}Y) + 180(\tfrac{4}{5}X + \tfrac{3}{5}Y)^2 = -900$$

or $225X^2 + 100Y^2 = -900$; no points on the graph.

15. Since $A = C$, we have $\theta = \pi/4$. The equation becomes

$$B\left(\frac{\sqrt{2}}{2}X - \frac{\sqrt{2}}{2}Y\right)\left(\frac{\sqrt{2}}{2}X + \frac{\sqrt{2}}{2}Y\right) + D\left(\frac{\sqrt{2}}{2}X - \frac{\sqrt{2}}{2}Y\right) + E\left(\frac{\sqrt{2}}{2}X + \frac{\sqrt{2}}{2}Y\right) + F = 0$$

or

$$\frac{1}{2}B(X^2 - Y^2) + \frac{\sqrt{2}}{2}(D + E)X + \frac{\sqrt{2}}{2}(E - D)Y + F = 0$$

When the squares are completed the equation will take the form $\frac{1}{2}B(X - a)^2 - \frac{1}{2}B(Y - b)^2 = c$. If $c \neq 0$, the graph is a hyperbola; if $c = 0$, the graph is two intersecting lines.

17. Let

$$\begin{aligned} B' &= -2A\cos\theta\sin\theta + B(\cos^2\theta - \sin^2\theta) + 2C\cos\theta\sin\theta \\ &= B(\cos^2\theta - \sin^2\theta) - 2(A - C)\sin\theta\cos\theta \end{aligned}$$

From (6) and the equation preceding (6), we find that

$$\begin{aligned} A' &= A\cos^2\theta + B\sin\theta\cos\theta + C\sin^2\theta \\ C' &= A\sin^2\theta - B\sin\theta\cos\theta + C\cos^2\theta \end{aligned}$$

Therefore

$$\begin{aligned} (B')^2 - 4A'C' &= [B(\cos^2\theta - \sin^2\theta) - 2(A - C)\sin\theta\cos\theta]^2 \\ &\quad - 4(A\cos^2\theta + B\sin\theta\cos\theta + C\sin^2\theta)(A\sin^2\theta \\ &\quad - B\sin\theta\cos\theta + C\cos^2\theta) \\ &= B^2(\cos^4\theta - 2\sin^2\theta\cos^2\theta + \sin^4\theta + 4\sin^2\theta\cos^2\theta) \\ &\quad - 4AB(\sin\theta\cos^3\theta - \sin^3\theta\cos\theta - \sin\theta\cos^3\theta + \sin^3\theta\cos\theta) \\ &\quad + 4BC(\sin\theta\cos^3\theta - \sin^3\theta\cos\theta - \sin\theta\cos^3\theta + \sin^3\theta\cos\theta) \\ &\quad + 4A^2(\sin^2\theta\cos^2\theta - \sin^2\theta\cos^2\theta) - 4AC(2\sin^2\theta\cos^2\theta \\ &\quad + \cos^4\theta + \sin^4\theta) + 4C^2(\sin^2\theta\cos^2\theta - \sin^2\theta\cos^2\theta) \\ &= B^2(\cos^2\theta + \sin^2\theta)^2 - 4AC(\cos^2\theta + \sin^2\theta)^2 \\ &= B^2 - 4AC \end{aligned}$$

But $B' = 0$ for the value of θ chosen in (8). Thus $B^2 - 4AC = -4A'C'$.

SECTION 11.5

1. $c = \sqrt{25 - 9} = 4$; $e = \dfrac{4}{5}$

3. $c = \sqrt{9 + 25} = \sqrt{34}$; $e = \dfrac{\sqrt{34}}{3}$

5. $\dfrac{x^2}{2} + \dfrac{y^2}{8} = 1$; $c = \sqrt{8 - 2} = \sqrt{6}$; $e = \dfrac{\sqrt{6}}{\sqrt{8}} = \dfrac{\sqrt{3}}{2}$

7. $\dfrac{(x-3)^2}{2} + \dfrac{(y+3)^2}{8} = 1$; $c = \sqrt{8 - 2} = \sqrt{6}$; $e = \dfrac{\sqrt{6}}{\sqrt{8}} = \dfrac{\sqrt{3}}{2}$

9. $\dfrac{(x-1)^2}{4} + \dfrac{(y-1)^2}{2} = 1$; $c = \sqrt{4 - 2} = \sqrt{2}$; $e = \dfrac{\sqrt{2}}{2}$

11. $\dfrac{(x+1)^2}{9} - \dfrac{(y+2)^2}{49} = 1$; $c = \sqrt{9 + 49} = \sqrt{58}$; $e = \dfrac{\sqrt{58}}{3}$

13. $c = 9$, $e = \dfrac{3}{5}$, $a = \dfrac{9}{3/5} = 15$, $b = \sqrt{225 - 81} = 12$; $\dfrac{x^2}{225} + \dfrac{y^2}{144} = 1$

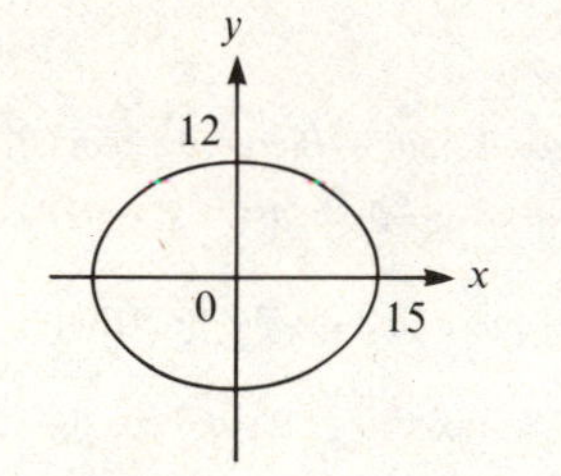

15. $c = 1$, $e = 2$, $a = \dfrac{1}{2}$, $b = \sqrt{1 - \dfrac{1}{4}} = \dfrac{\sqrt{3}}{2}$; $4y^2 - \dfrac{4x^2}{3} = 1$

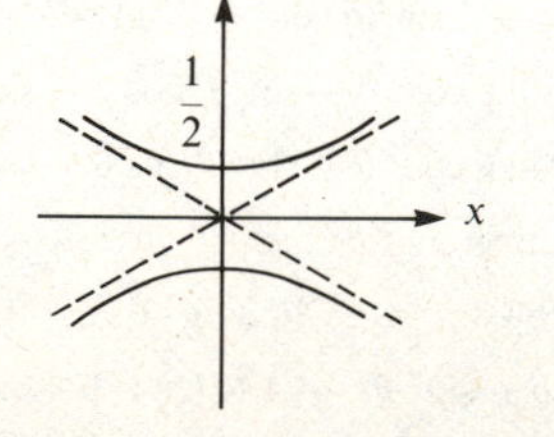

17. $a = 5$, $e = \dfrac{13}{5}$, $c = 5\left(\dfrac{13}{5}\right) = 13$, $b = \sqrt{169 - 25} = 12$; $\dfrac{(y-5)^2}{25} - \dfrac{x^2}{144} = 1$

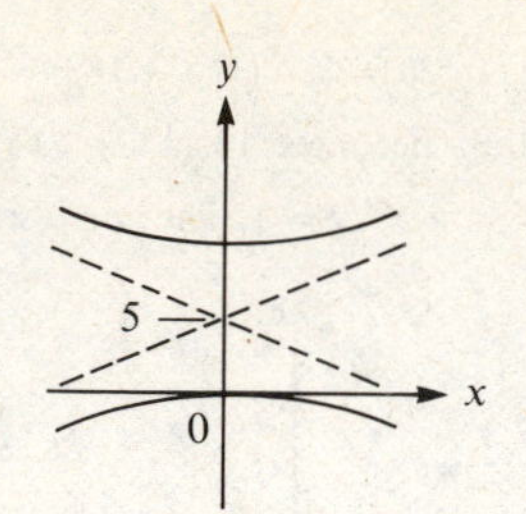

19. Since $c = ae = a\sqrt{17}$, we have $b = \sqrt{17a^2 - a^2} = 4a$. Thus an equation of the hyperbola is either

$$\frac{x^2}{a^2} - \frac{y^2}{16a^2} = 1 \quad \text{or} \quad \frac{y^2}{a^2} - \frac{x^2}{16a^2} = 1$$

Since the hyperbola passes through $(\sqrt{20}, 8)$, we find for the first of these equations that $20/a^2 - 64/16a^2 = 1$, or $a^2 = 16$, so that the equation becomes $x^2/16 - y^2/256 = 1$. For the second equation we find that $64/a^2 - 20/16a^2 = 1$, or $a^2 = \frac{251}{4}$, so that the equation becomes $4y^2/251 - x^2/1004 = 1$.

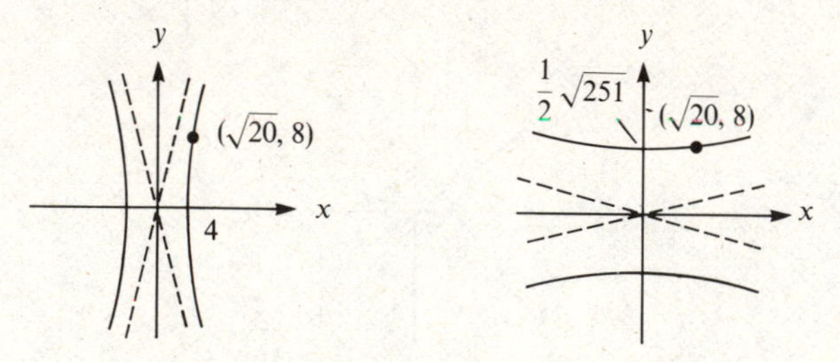

21. Since $a = 4$ and $e = \frac{1}{2}$, we have $c = 4(\frac{1}{2}) = 2$. Thus $b = \sqrt{16 - 4} = 2\sqrt{3}$, and also the center is located at either $(2, 0)$ or $(-2, 0)$. Therefore an equation of the ellipse is either $(x-2)^2/16 + y^2/12 = 1$ or $(x+2)^2/16 + y^2/12 = 1$.

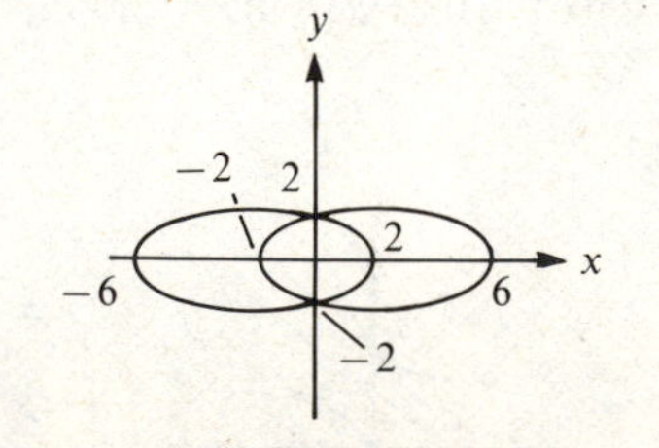

23. $e = 1$, $k = 1$; parabola with directrix $y = -1$

25. $r = \dfrac{5}{1 - \frac{3}{5}\sin\theta}$; $e = \dfrac{3}{5}$, $k = \dfrac{25}{3}$; ellipse with directrix $y = -\dfrac{25}{3}$

27. $r = \dfrac{\frac{1}{2}}{1 - \cos\theta}$; $e = 1$, $k = \dfrac{1}{2}$; parabola with directrix $x = -\dfrac{1}{2}$

29. $e = 1$, $k = 3$; parabola with directrix $y = 3$

31. Since $e = .017$ and $a = \frac{299}{2}$, we have $c = (\frac{299}{2})(.017) = 2.5415$, and thus

$$b = \sqrt{(\tfrac{299}{2})^2 - (2.5415)^2} \approx 149.48$$

An approximate equation of the orbit is $x^2/(149.5)^2 + y^2/(149.48)^2 = 1$, if we place the x axis along the major axis and the origin at the center of the orbit.

33. By the information in the solution of Exercise 32, the distance r between a planet and the sun placed at the origin can be given by $r = ek/(1 + e\cos\theta)$. From Exercise 32, the minimum distance is $ek/(1 + e)$. The maximum distance occurs when the denominator of $ek/(1 + e\cos\theta)$ is minimum, that is, when $\theta = \pi$, and thus the maximum distance is $ek/(1 - e)$. As in Exercise 32,

$$\frac{ek}{1+e} = a(1-e), \quad \text{and} \quad \frac{ek}{1-e} = \frac{a - ae^2}{1-e} = a(1+e)$$

For Neptune the minimum and maximum distances are given by $a(1 - e) = \frac{8{,}996}{2}(1 - .008) = 4462.016$ (million kilometers), and $a(1 + e) = \frac{8{,}996}{2}(1 + .008) = 4533.984$ (million kilometers), respectively. For Pluto the minimum and maximum distances are given by $a(1 - e) = \frac{11{,}800}{2}(1 - .249) = 4430.9$ (million kilometers), and $a(1 + e) = \frac{11{,}800}{2}(1 + .249) = 7369.1$ (million kilometers), respectively. Since the minimum distance from Pluto to the sun is less than the minimum distance from Neptune to the sun, and since the maximum distance from Pluto to the sun is greater than the maximum distance from Neptune to the sun, it is theoretically possible from the given information that the two planets could collide. But other factors seem to preclude such a collision.

CHAPTER 11 REVIEW

1. vertex is (1, 4); $c = -4$; $(y - 4)^2 = -16(x - 1)$

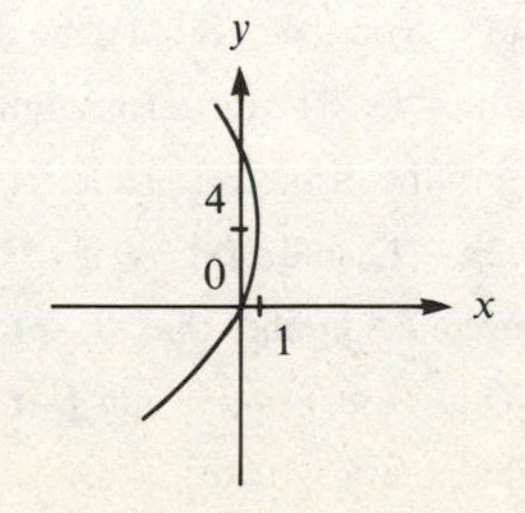

3. The focus is $(-2, b)$ for an appropriate value of b. Since the directrix is $y = -10$ and the point $(6, -2)$ lies on the parabola, the definition of a parabola implies that

$$\sqrt{(-2-6)^2 + (b+2)^2} = |-2 - (-10)| = 8$$

so that $b = -2$. Therefore the focus is $(-2, -2)$, so that the vertex is $(-2, -6)$ and $c = 4$. Thus an equation of the parabola is $(x + 2)^2 = 16(y + 6)$.

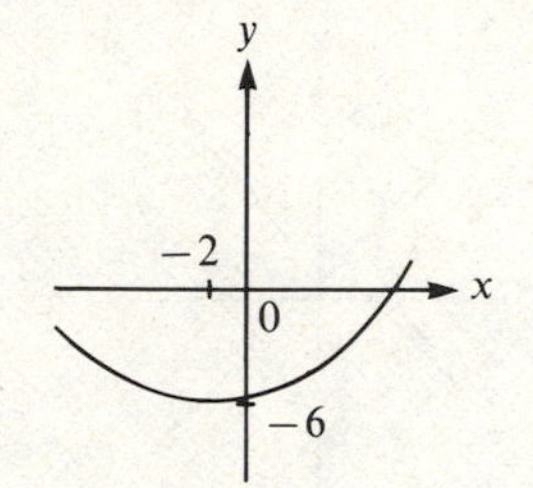

5. $a = 2\sqrt{2}$; an equation of the ellipse is $x^2/b^2 + y^2/8 = 1$. Since $(1, \sqrt{6})$ is on the ellipse, we have $1/b^2 + \frac{6}{8} = 1$, so that $b = 2$. Consequently an equation of the ellipse becomes $x^2/4 + y^2/8 = 1$.

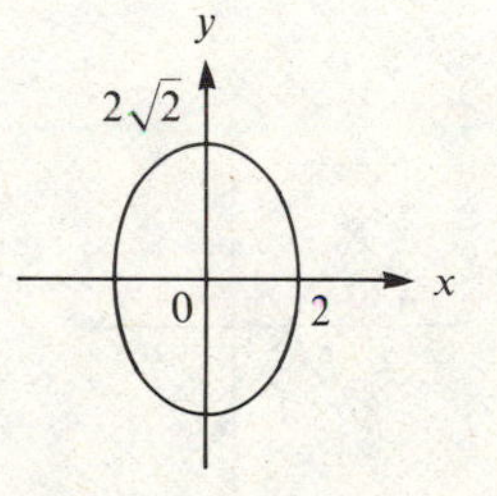

7. The center is $(1, -1)$; $a = 4$, $c = 2$, and $b = \sqrt{16 - 4} = 2\sqrt{3}$; $\dfrac{(x-1)^2}{12} + \dfrac{(y+1)^2}{16} = 1$

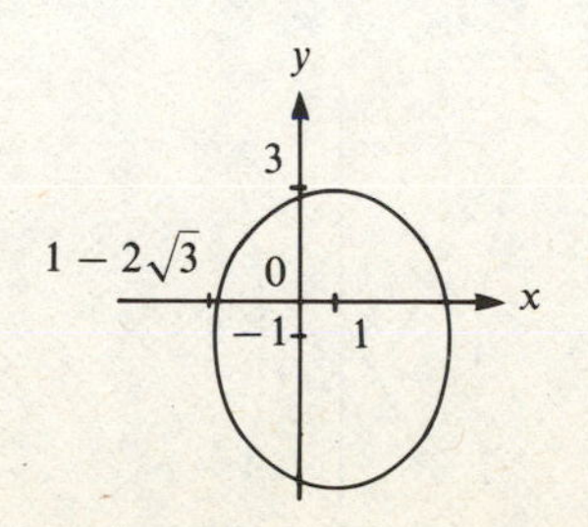

9. The center is $(-1, 2)$, $a = 2$; an equation of the hyperbola is $(x + 1)^2/4 - (y - 2)^2/b^2 = 1$. The asymptotes have equations $y - 2 = (b/2)(x + 1)$ and $y - 2 = -(b/2)(x + 1)$. Since the asymptotes are perpendicular,

$$\frac{b}{2} = \frac{-1}{-b/2} = \frac{2}{b}$$

so that $b = 2$. The equation of the hyperbola becomes $(x + 1)^2/4 - (y - 2)^2/4 = 1$.

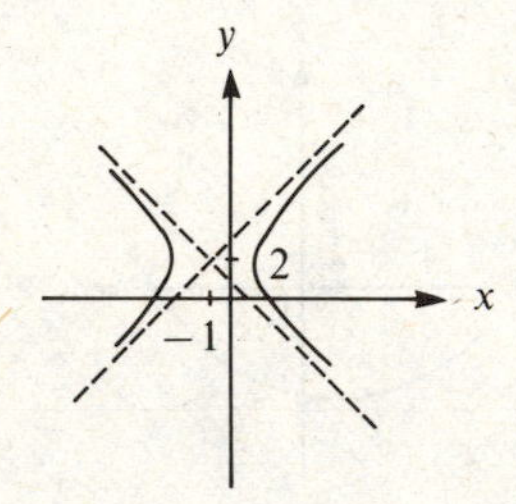

11. The distance from the vertex to the directrix is 2. Since $e = 2$, the distance from the vertex to the focus is $2 \cdot 2 = 4$. Thus the focus corresponding to the vertex $(-2, 0)$ is $(2, 0)$. The center is $(p, 0)$, for an appropriate value of p. Then $2 = c/a = (2 - p)/(-2 - p)$, so that $p = -6$, and the center is $(-6, 0)$. Thus $a = 4$, $c = 8$, and $b = \sqrt{64 - 16} = 4\sqrt{3}$. Therefore an equation of the hyperbola is $(x + 6)^2/16 - y^2/48 = 1$.

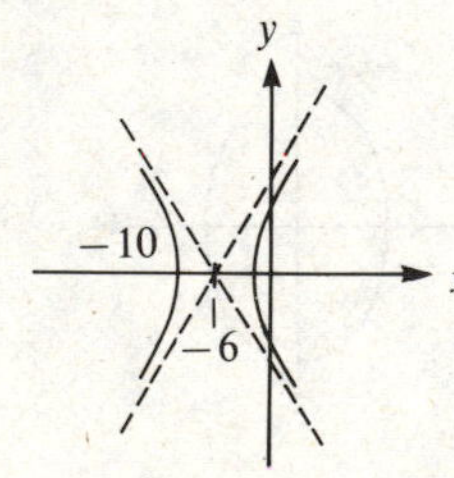

13. $49(x^2 + 2x + 1) - 9(y^2 + 4y + 4) = 441$, or $49(x + 1)^2 - 9(y + 2)^2 = 441$, so that

$$\frac{(x + 1)^2}{9} - \frac{(y + 2)^2}{49} = 1$$

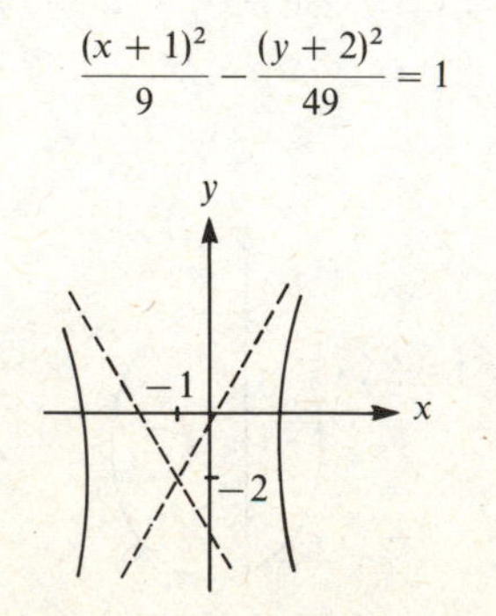

15. $9(x^2 - 4x + 4) + 5y - 15 = 0$, so that $9(x - 2)^2 = -5(y - 3)$

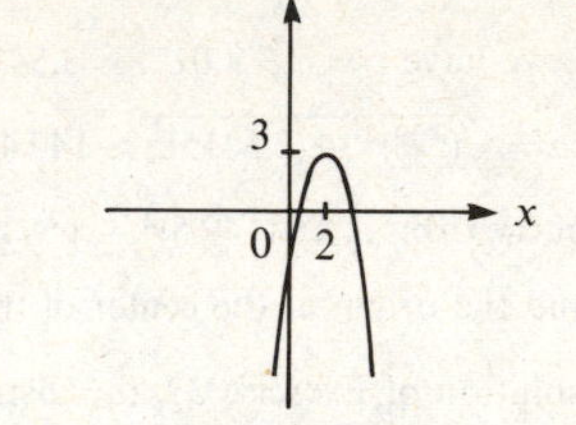

17. $A = 4$, $B = -24$, $C = 11$; $\tan 2\theta = -24/(4 - 11) = \frac{24}{7}$; by the method of Example 2 in Section 10.4, $\cos 2\theta = \frac{7}{25}$, $\cos\theta = \frac{4}{5}$, and $\sin\theta = \frac{3}{5}$; the equation becomes $4(\frac{4}{5}X - \frac{3}{5}Y)^2 - 24(\frac{4}{5}X - \frac{3}{5}y)(\frac{3}{5}X + \frac{4}{5}Y) + 11(\frac{3}{5}X + \frac{4}{5}Y)^2 + 40(\frac{4}{5}X - \frac{3}{5}Y) + 30(\frac{3}{5}X + \frac{4}{5}Y) - 45 = 0$, or $-5X^2 + 50X + 20Y^2 - 45 = 0$, or $(X - 5)^2/16 - Y^2/4 = 1$.

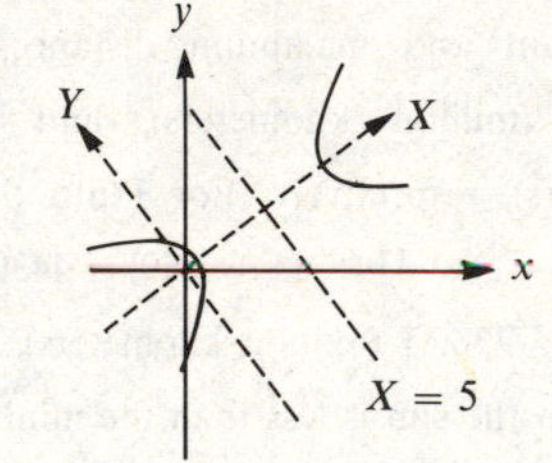

19. a. $\sqrt{(x - 2)^2 + (y - 4)^2} = 3|y|$, so that $(x - 2)^2 + (y - 4)^2 = 9y^2$, or $(x - 2)^2 - 8(y^2 + y + \frac{1}{4}) = -18$ or $4(y + \frac{1}{2})^2/9 - (x - 2)^2/18 = 1$

b. $\sqrt{(x - 2)^2 + (y - 4)^2} = |y|$, so that $(x - 2)^2 + (y - 4)^2 = y^2$, or $(x - 2)^2 - 8(y - 2) = 0$, or $(x - 2)^2 = 8(y - 2)$

c. $\sqrt{(x - 2)^2 + (y - 4)^2} = \frac{1}{2}|y|$, so that $(x - 2)^2 + (y - 4)^2 = \frac{1}{4}y^2$, or $(x - 2)^2 + \frac{3}{4}(y^2 - \frac{32}{3}y + \frac{256}{9}) = \frac{16}{3}$, or $3(x - 2)^2/16 + 9(y - 16/3)^2/64 = 1$

21. $r = \dfrac{\frac{3}{4}}{1 - \frac{1}{4}\cos\theta}$, so $e = \frac{1}{4}$, and hence the conic section is an ellipse.

23. Since $y^2 = -8x$ for any point (x, y) on the parabola, the distance from (x, y) to $(-10, 0)$ is $\sqrt{(x + 10)^2 + y^2} = \sqrt{(x + 10)^2 - 8x}$. The distance is minimized for the same nonpositive value of x that the square of the distance is, so let $f(x) = (x + 10)^2 - 8x$ for $x \le 0$. Then $f'(x) = 2(x + 10) - 8 = 2x + 12$, and $f'(x) = 0$ if $x = -6$. Since $f'(x) < 0$ if $x < -6$ and $f'(x) > 0$ if $x > -6$, Theorem 4.8 implies that $f(-6)$ is the minimum value of f on $(-\infty, 0]$, the domain of f. Thus $(-6, -4\sqrt{3})$ and $(-6, 4\sqrt{3})$ are the points on the parabola that are closest to the point $(-10, 0)$.

25. The focus divides the major axis into lines of length $a - c$ and $a + c$. Then

$$\sqrt{(a-c)(a+c)} = \sqrt{a^2 - c^2} = \sqrt{b^2} = b$$

Thus the length b of the minor axis is the required geometric mean.

CUMULATIVE REVIEW (CHAPTERS 1–10)

1. The conditions for applying l'Hôpital's Rule are met:

$$\lim_{x\to 0^+} \frac{\arcsin x^2}{\sqrt{x^3}} = \lim_{x\to 0^+} \frac{2x/\sqrt{1-x^4}}{\frac{3}{2}x^{1/2}} = \lim_{x\to 0^+} \frac{4x^{1/2}}{3\sqrt{1-x^4}} = 0$$

3. a. The domain consists of all t such that $t \neq 0$ and $3 + t \geq 0$. Thus the domain is the union of $[-3, 0)$ and $(0, \infty)$.

b. $$\lim_{t\to 0} f(t) = \lim_{t\to 0} \frac{\sqrt{3+t} - \sqrt{3}}{t} = \lim_{t\to 0} \frac{\sqrt{3+t} - \sqrt{3}}{t} \frac{\sqrt{3+t} + \sqrt{3}}{\sqrt{3+t} + \sqrt{3}} = \lim_{t\to 0} \frac{(3+t) - 3}{t(\sqrt{3+t} + \sqrt{3})}$$

$$= \lim_{t\to 0} \frac{1}{\sqrt{3+t} + \sqrt{3}} = \frac{1}{2\sqrt{3}} = \frac{1}{6}\sqrt{3}$$

5. Differentiating implicitly, we find that $2y(dy/dx) + \arcsin y + (x/\sqrt{1-y^2})(dy/dx) = 0$, so

$$\frac{dy}{dx} = \frac{-\arcsin y}{2y + x/\sqrt{1-y^2}} = -\frac{(\arcsin y)\sqrt{1-y^2}}{2y\sqrt{1-y^2} + x}$$

7. There is a positive constant c such that $S = cwd^2$. Since $d^2 + w^2 = 84^2$, we have $S = cw(84^2 - w^2)$. We need to find the values of w and d that yield the maximum value of S. Now $S'(w) = c(84)^2 - 3cw^2$, so that $S'(w) = 0$ if $w^2 = (84)^2/3$, or $w = 84/\sqrt{3} = 28\sqrt{3}$. Since $S''(w) = -6cw < 0$, the strength S is maximum for $w = 28\sqrt{3}$. Then $d^2 = (84)^2 - (84)^2/3 = \frac{2}{3}(84)^2$, so $d = 84\sqrt{\frac{2}{3}} = 28\sqrt{6}$. Consequently the width of the strongest beam is $28\sqrt{3}$ centimeters and the depth is $28\sqrt{6}$ centimeters.

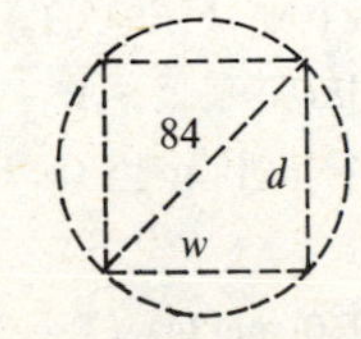

9. $f'(x) = e^{-x^2/2}$, so the slope of the tangent line at $(x, f(x))$ is $\frac{1}{4}$ if $e^{-x^2/2} = \frac{1}{4}$, or $-x^2/2 = \ln\frac{1}{4} = -\ln 4$, or $x^2 = 2\ln 4 = 4\ln 2$. Thus the tangent line at $(x, f(x))$ is $\frac{1}{4}$ if $x = -2\sqrt{\ln 2}$ or $x = 2\sqrt{\ln 2}$.

11. a. $f'(x) = e^{-x^2/2} > 0$ for all x, so f is strictly increasing on $(-\infty, \infty)$ and hence has an inverse.

b. Since $\int_0^1 e^{-t^2/2}\,dt = A$, we have $f(1) = A$. Thus

$$(f^{-1})'(A) = \frac{1}{f'(1)} = \frac{1}{e^{-1/2}} = e^{1/2}$$

13. Let $u = \sec x$, so that $du = \sec x \tan x\,dx$. Then

$$\int \frac{\sec x \tan x}{\sec^2 x + 1}\,dx = \int \frac{1}{u^2+1}\,du = \arctan u + C = \arctan(\sec x) + C$$

15. Let $t = x^3$, so that $dt = 3x^2\,dx$. Then

$$\int x^5 \sin x^3\,dx = \int \tfrac{1}{3}x^3(\sin x^3)(3x^2)\,dx = \int \tfrac{1}{3}t \sin t\,dt$$

We use integration by parts next with $u = \frac{1}{3}t$ and $dv = \sin t\,dt$, so that $du = \frac{1}{3}dt$ and $v = -\cos t$. Thus

$$\int x^5 \sin x^3\,dx = \int \tfrac{1}{3}t\sin t\,dt = -\tfrac{1}{3}t\cos t - \int(-\cos t)\tfrac{1}{3}\,dt$$
$$= -\tfrac{1}{3}t\cos t + \tfrac{1}{3}\sin t + C$$
$$= -\tfrac{1}{3}x^3\cos x^3 + \tfrac{1}{3}\sin x^3 + C$$

17. The graphs intersect at (x, y) if $x^2 = y = 2x$, that is, $x = 0$ or $x = 2$. Also, $x^2 \geq 2x$ on $[-1, 0]$ and $2x \geq x^2$ on $[0, 2]$. Thus

$$A = \int_{-1}^0 (x^2 - 2x)\,dx + \int_0^2 (2x - x^2)\,dx = (\tfrac{1}{3}x^3 - x^2)\Big|_{-1}^0 + (x^2 - \tfrac{1}{3}x^3)\Big|_0^2 = \tfrac{4}{3} + \tfrac{4}{3} = \tfrac{8}{3}$$

19. Let the vat be positioned with respect to the xy plane as in the figure. Since the water is to be pumped up to 4 feet above the top of the vat, $\ell = 8$. For $0 \leq y \leq 4$, the width of the horizontal cross section at y is $2\sqrt{y}$. Since the length of the cross section is 10, the cross-sectional area at y is $10(2\sqrt{y}) = 20\sqrt{y}$, so

$$W = \int_0^4 62.5(8-y)20\sqrt{y}\,dy = 1250\int_0^4 (8\sqrt{y} - y^{3/2})\,dy = 1250(\tfrac{16}{3}y^{3/2} - \tfrac{2}{5}y^{5/2})\Big|_0^4$$
$$= 1250\left(\frac{128}{3} - \frac{64}{5}\right) = \frac{112{,}000}{3} \text{ (foot-pounds)}$$

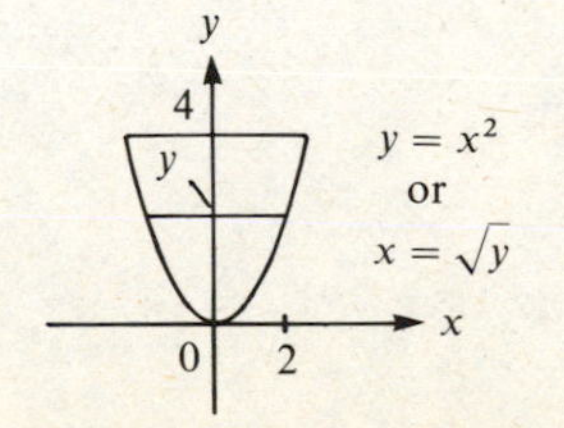

21. $\left|\dfrac{\sin(e^n)}{(n+1)^{3/2}}\right| \le \dfrac{1}{(n+1)^{3/2}} \le \dfrac{1}{n^{3/2}}$ and $\displaystyle\sum_{n=1}^{\infty} \frac{1}{n^{3/2}}$

is a convergent p series (with $p = \frac{3}{2}$). Thus by the Comparison Test, $\displaystyle\sum_{n=1}^{\infty} \frac{\sin(e^n)}{(n+1)^{3/2}}$ converges absolutely and hence converges.

23. $$\left|\frac{a_{n+1}}{a_n}\right| = \frac{(2^{n+1}|x|^{n+1})/\sqrt[3]{n+2}}{(2^n|x|^n)/\sqrt[3]{n+1}} = \frac{2|x|\sqrt[3]{n+1}}{\sqrt[3]{n+2}} = \frac{2|x|\sqrt[3]{1+1/n}}{\sqrt[3]{1+2/n}}$$

so $$\lim_{n\to\infty}\left|\frac{a_{n+1}}{a_n}\right| = \frac{2|x|\sqrt[3]{1+0}}{\sqrt[3]{1+0}} = 2|x|$$

Thus the given series converges for $|x| < \frac{1}{2}$ and diverges for $|x| > \frac{1}{2}$. For $x = -\frac{1}{2}$ the series becomes

$$\sum_{n=1}^{\infty} \frac{(-1)^n 2^n(-\frac{1}{2})^n}{\sqrt[3]{n+1}} = \sum_{n=1}^{\infty} \frac{1}{\sqrt[3]{n+1}}$$

which diverges by comparison with the p series $\sum_{n=1}^{\infty} 1/n^{1/3}$. For $x = \frac{1}{2}$ the series becomes

$$\sum_{n=1}^{\infty} \frac{(-1)^n 2^n(\frac{1}{2})^n}{\sqrt[3]{n+1}} = \sum_{n=1}^{\infty} \frac{(-1)^n}{\sqrt[3]{n+1}}$$

which converges by the Alternating Series Test. Therefore the interval of convergence of the given series is $(-\frac{1}{2}, \frac{1}{2}]$.

25. $f'(x) = e^x + xe^x = (x+1)e^x$; $f''(x) = e^x + (x+1)e^x = (x+2)e^x$. In general,

$$f^{(n)}(x) = (x+n)e^x$$

so that $f^{(n)}(1) = (n+1)e$. Consequently the Taylor series of f about 1 is

$$\sum_{n=0}^{\infty} \frac{(n+1)e}{n!}(x-1)^n$$

12

VECTORS, LINES, AND PLANES

SECTION 12.1

1. $|PQ| = \sqrt{(0-\sqrt{2})^2 + (1-0)^2 + (1-0)^2} = \sqrt{2+1+1} = 2$

3. $|PQ| = \sqrt{(0-(-3))^2 + (8-4)^2 + (7-(-5))^2} = \sqrt{9+16+144} = 13$

5. $|PQ| = \sqrt{(4-(-1))^2 + (2-3)^2 + (7-6)^2} = \sqrt{25+1+1} = 3\sqrt{3}$

7. $$|PQ| = \sqrt{(\sin x - 2\sin x)^2 + (2\cos x - \cos x)^2 + (0 - \tan x)^2}$$
$$= \sqrt{\sin^2 x + \cos^2 x + \tan^2 x} = \sqrt{1+\tan^2 x} = |\sec x|$$

9. Let $P = (3, 0, 2)$, $Q = (1, -1, 5)$, and $R = (5, 1, -1)$. Then

$$|PQ| = \sqrt{(1-3)^2 + (-1-0)^2 + (5-2)^2} = \sqrt{14}$$

and $|PR| = \sqrt{(5-3)^2 + (1-0)^2 + (-1-2)^2} = \sqrt{14}$. Thus P is equidistant from Q and R.

11. $(x-2)^2 + (y-1)^2 + (z+7)^2 = 25$

13. Completing the squares, we have $(x^2 - 2x + 1) + (y^2 - 4y + 4) + (z^2 + 6z + 9) = -10 + 1 + 4 + 9$, or $(x-1)^2 + (y-2)^2 + (z+3)^2 = 4$, which is an equation of a sphere with center $(1, 2, -3)$ and radius 2.

15. $x^2 + (y+2)^2 + (z+3)^2 \le 36$

17. $|PQ| = \sqrt{(2-1)^2 + (1-(-1))^2 + (-1-1)^2} = \sqrt{1+4+4} = 3$;
$|RP| = \sqrt{1^2 + (-1)^2 + 1^2} = \sqrt{3}$; $|RQ| = \sqrt{2^2 + 1^2 + (-1)^2} = \sqrt{6}$.
Thus $|RP|^2 + |RQ|^2 = 3 + 6 = 9 = |PQ|^2$, so by the Pythagorean Theorem, the triangle is a right triangle.

19. If (x, y, z) is equidistant from $(2, 1, 0)$ and $(4, -1, -3)$, then $(x-2)^2 + (y-1)^2 + z^2 = (x-4)^2 + (y+1)^2 + (z+3)^2$, so that $4x - 4y - 6z = 21$.

21. Let $P = (x_0, y_0, z_0)$, $Q = (x_1, y_1, z_1)$ and $R = (\frac{1}{2}(x_0 + x_1), \frac{1}{2}(y_0 + y_1), \frac{1}{2}(z_0 + z_1))$. Then

$$|PR|^2 = [\tfrac{1}{2}(x_0 + x_1) - x_0]^2 + [\tfrac{1}{2}(y_0 + y_1) - y_0]^2 + [\tfrac{1}{2}(z_0 + z_1) - z_0]^2$$
$$= \tfrac{1}{4}(x_1 - x_0)^2 + \tfrac{1}{4}(y_1 - y_0)^2 + \tfrac{1}{4}(z_1 - z_0)^2$$

and $$|QR|^2 = [\tfrac{1}{2}(x_0 + x_1) - x_1]^2 + [\tfrac{1}{2}(y_0 + y_1) - y_1]^2 + [\tfrac{1}{2}(z_0 + z_1) - z_1]^2 = \tfrac{1}{4}(x_0 - x_1)^2 + \tfrac{1}{4}(y_0 - y_1)^2 + \tfrac{1}{4}(z_0 - z_1)^2$$

Thus $|PR|^2 = |QR|^2$, so R is the midpoint of PQ.

23. By the result of Exercise 21, the midpoint of the line segment joining $(2, -1, 3)$ and $(4, 1, 7)$ is $(3, 0, 5)$, so the center of the sphere is $(3, 0, 5)$. The distance between $(2, -1, 3)$ and $(3, 0, 5)$ is $\sqrt{(3-2)^2 + (0-(-1))^2 + (5-3)^2} = \sqrt{1+1+4} = \sqrt{6}$, so an equation of the sphere is $(x-3)^2 + y^2 + (z-5)^2 = 6$.

SECTION 12.2

1. $\overrightarrow{PQ} = 3\mathbf{i} - 4\mathbf{j} + 10\mathbf{k}$

3. $\overrightarrow{PQ} = 3\mathbf{i} - 2\mathbf{j} + \sqrt{7}\mathbf{k}$

5. $\mathbf{a} + \mathbf{b} = \mathbf{i} - 3\mathbf{j} + \mathbf{k}$; $\mathbf{a} - \mathbf{b} = 3\mathbf{i} - 7\mathbf{j} + 19\mathbf{k}$; $c\mathbf{a} = 4\mathbf{i} - 10\mathbf{j} + 20\mathbf{k}$

7. $\mathbf{a} + \mathbf{b} = 2\mathbf{i} + \mathbf{j} + \mathbf{k}$; $\mathbf{a} - \mathbf{b} = 2\mathbf{i} - \mathbf{j} - \mathbf{k}$; $c\mathbf{a} = \frac{2}{3}\mathbf{i}$

9. $\|\mathbf{a}\| = \sqrt{1^2 + (-1)^2 + 1^2} = \sqrt{3}$

11. $\|\mathbf{b}\| = \sqrt{(-3)^2 + 4^2 + (-12)^2} = 13$

13. $\|\mathbf{c}\| = \sqrt{(\sqrt{2})^2 + (-1)^2 + 1^2} = 2$

15. $\dfrac{\mathbf{a}}{\|\mathbf{a}\|} = \dfrac{\mathbf{a}}{13} = -\dfrac{3}{13}\mathbf{i} + \dfrac{4}{13}\mathbf{j} - \dfrac{12}{13}\mathbf{k}$

17. $\dfrac{\mathbf{b}}{\|\mathbf{b}\|} = \dfrac{\mathbf{b}}{\sqrt{13}} = \dfrac{2}{\sqrt{13}}\mathbf{i} - \dfrac{3}{\sqrt{13}}\mathbf{j}$

19. a. If we place the vector $\mathbf{a}$ in the xy plane with its initial point at the origin and let its terminal point be (a_1, a_2), then $\mathbf{a} = a_1\mathbf{i} + a_2\mathbf{j}$. By the definition of $\cos\theta$ and $\sin\theta$, we have $\cos\theta = a_1/\|\mathbf{a}\|$ and $\sin\theta = a_2/\|\mathbf{a}\|$. Thus $\mathbf{a} = a_1\mathbf{i} + a_2\mathbf{j} = \|\mathbf{a}\|(\cos\theta\mathbf{i} + \sin\theta\mathbf{j})$.

 b. Let θ be the angle between the positive x axis and $\mathbf{u}$. Since $\|\mathbf{u}\| = 1$, it follows from part (a) that $\mathbf{u} = \cos\theta\mathbf{i} + \sin\theta\mathbf{j}$.

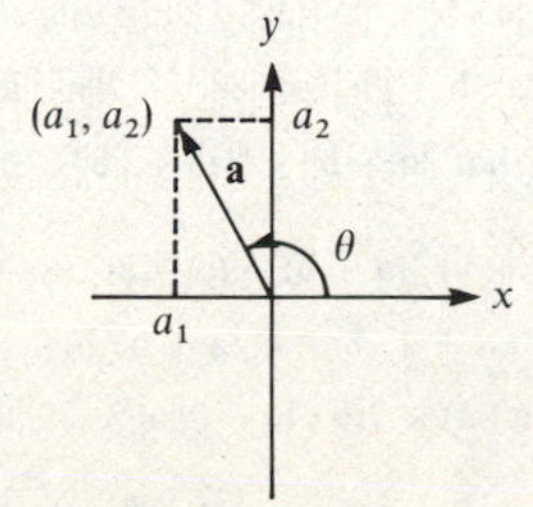

21. Assume that the quadrilateral lies in the xy plane with P at the origin as in the figure, and let $\mathbf{a}$ and $\mathbf{b}$ be vectors along the two sides that contain the origin. Then $\mathbf{a} + \mathbf{b}$ and $\mathbf{a} - \mathbf{b}$ lie along the diagonals. Since the diagonals bisect each other, we have $\overrightarrow{SR} = \frac{1}{2}(\mathbf{a} - \mathbf{b}) + \frac{1}{2}(\mathbf{a} + \mathbf{b}) = \mathbf{a} = \overrightarrow{PQ}$ and $\overrightarrow{QR} = -\frac{1}{2}(\mathbf{a} - \mathbf{b}) + \frac{1}{2}(\mathbf{a} + \mathbf{b}) = \mathbf{b} = \overrightarrow{PS}$. Thus the opposite sides of the quadrilateral are parallel, so that the quadrilateral is a parallelogram.

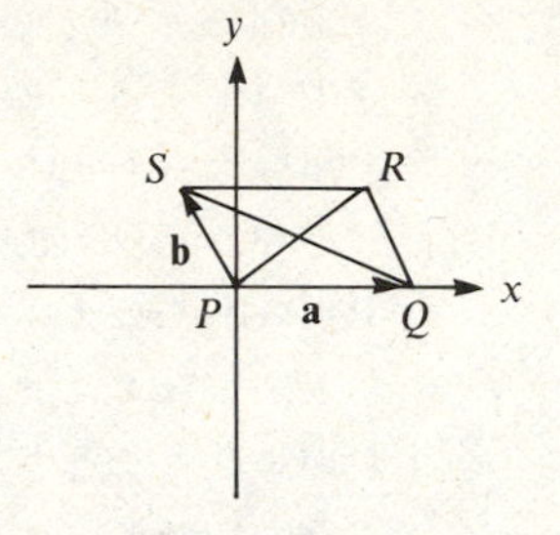

23. The force $\mathbf{F}_1$ exerted by the first child is given by $\mathbf{F}_1 = -20\mathbf{j}$, and the force $\mathbf{F}_2$ exerted by the second child is given by

$$\mathbf{F}_2 = 100\left(\cos\frac{\pi}{3}\mathbf{i} + \sin\frac{\pi}{3}\mathbf{j}\right) = 50(\mathbf{i} + \sqrt{3}\mathbf{j})$$

If $\mathbf{F}$ is the force exerted by the third child, and θ the angle $\mathbf{F}$ makes with the positive x axis, then $\mathbf{F} = \|\mathbf{F}\|(\cos\theta\mathbf{i} + \sin\theta\mathbf{j})$. If the total force exerted on the ball is to be $\mathbf{0}$, then $\mathbf{F}_1 + \mathbf{F}_2 + \mathbf{F} = \mathbf{0}$, so that $-20\mathbf{j} + 50(\mathbf{i} + \sqrt{3}\mathbf{j}) + \|\mathbf{F}\|(\cos\theta\mathbf{i} + \sin\theta\mathbf{j}) = \mathbf{0}$. Combining coefficients of $\mathbf{i}$ and $\mathbf{j}$, we find that

$$50 + \|\mathbf{F}\|\cos\theta = 0 \quad \text{and} \quad -20 + 50\sqrt{3} + \|\mathbf{F}\|\sin\theta = 0$$

so that $$\|\mathbf{F}\|\cos\theta = -50 \quad \text{and} \quad \|\mathbf{F}\|\sin\theta = 20 - 50\sqrt{3}$$

Therefore $$\tan\theta = \frac{\sin\theta}{\cos\theta} = \frac{20 - 50\sqrt{3}}{-50} = \frac{5\sqrt{3} - 2}{5}$$

and $$\mathbf{F} = \|\mathbf{F}\|(\cos\theta\mathbf{i} + \sin\theta\mathbf{j}) = \|\mathbf{F}\|\cos\theta\mathbf{i} + \|\mathbf{F}\|\sin\theta\mathbf{j} = -50\mathbf{i} + (20 - 50\sqrt{3})\mathbf{j}$$

25. Let $\mathbf{v}$ be the velocity of the airplane with respect to the ground, $\mathbf{v}_1$ the velocity of the airplane with respect to the air, and $\mathbf{v}_2$ the velocity of the air with respect to the ground. By the hint,

$$\mathbf{v} = \mathbf{v}_1 + \mathbf{v}_2 = 300\left(\cos\frac{\pi}{6}\mathbf{i} + \sin\frac{\pi}{6}\mathbf{j}\right) + 20\mathbf{j} = 150\sqrt{3}\mathbf{i} + 170\mathbf{j}$$

Thus $\|\mathbf{v}\| = \sqrt{(150\sqrt{3})^2 + (170)^2} = \sqrt{96{,}400} \approx 310.483$ (miles per hour).

27. If we let $q_1 = 1.6 \times 10^{-19}$, $\mathbf{u}_1 = (-\mathbf{j} - \mathbf{k})/\sqrt{2}$, $r_1 = 10^{-11}\sqrt{2}$, $q_2 = -1.6 \times 10^{-19}$, $\mathbf{u}_2 = (-\mathbf{i} - \mathbf{j})/\sqrt{2}$, and $r_2 = 10^{-12}\sqrt{2}$, then

$$\mathbf{F} = \frac{q_1(1)}{4\pi\varepsilon_0 r_1^2}\mathbf{u}_1 + \frac{q_2(1)}{4\pi\varepsilon_0 r_2^2}\mathbf{u}_2 = \frac{1.6 \times 10^{-19}}{4\pi\varepsilon_0(2 \times 10^{-22})}\left(\frac{-\mathbf{j} - \mathbf{k}}{\sqrt{2}}\right) + \frac{-1.6 \times 10^{-19}}{4\pi\varepsilon_0(2 \times 10^{-24})}\left(\frac{-\mathbf{i} - \mathbf{j}}{\sqrt{2}}\right)$$
$$= \frac{100\sqrt{2}}{\pi\varepsilon_0}(-\mathbf{j} - \mathbf{k} + 100\mathbf{i} + 100\mathbf{j}) = \frac{100\sqrt{2}}{\pi\varepsilon_0}(100\mathbf{i} + 99\mathbf{j} - \mathbf{k})$$

29. Let $m = m_1 + m_2 + \cdots + m_n$. By writing the vectors in terms of the corresponding components, we find that

$$\begin{aligned} m_1\overrightarrow{PP_1} + m_2\overrightarrow{PP_2} + \cdots + m_n\overrightarrow{PP_n} &= [m_1(x_1 - \bar{x})\mathbf{i} + m_1(y_1 - \bar{y})\mathbf{j}] \\ &\quad + [m_2(x_2 - \bar{x})\mathbf{i} + m_2(y_2 - \bar{y})\mathbf{j}] + \cdots \\ &\quad + [m_n(x_n - \bar{x})\mathbf{i} + m_n(y_n - \bar{y})\mathbf{j}] \\ &= [m_1(x_1 - \bar{x}) + m_2(x_2 - \bar{x}) + \cdots + m_n(x_n - \bar{x})]\mathbf{i} \\ &\quad + [m_1(y_1 - \bar{y}) + m_2(y_2 - \bar{y}) + \cdots + m_n(y_n - \bar{y})]\mathbf{j} \\ &= [(m_1x_1 + m_2x_2 + \cdots + m_nx_n) \\ &\quad - (m_1 + m_2 + \cdots + m_n)\bar{x}]\mathbf{i} \\ &\quad + [(m_1y_1 + m_2y_2 + \cdots + m_ny_n) \\ &\quad - (m_1 + m_2 + \cdots + m_n)\bar{y}]\mathbf{j} \\ &= [(m_1x_1 + m_2x_2 + \cdots + m_nx_n) - m\bar{x}]\mathbf{i} \\ &\quad + [(m_1y_1 + m_2y_2 + \cdots + m_ny_n) - m\bar{y}]\mathbf{j} \end{aligned}$$

By (3) of Section 8.7, the coefficients of **i** and **j** are 0, so that $m_1\overrightarrow{PP_1} + m_2\overrightarrow{PP_2} + \cdots + m_n\overrightarrow{PP_n} = \mathbf{0}$.

SECTION 12.3

1. $\mathbf{a} \cdot \mathbf{b} = (1)(2) + (1)(-3) + (-1)(4) = -5$; $\cos\theta = \dfrac{\mathbf{a} \cdot \mathbf{b}}{\|\mathbf{a}\|\,\|\mathbf{b}\|} = \dfrac{-5}{\sqrt{3}\sqrt{29}} = \dfrac{-5}{\sqrt{87}}$

3. $\mathbf{a} \cdot \mathbf{b} = (\sqrt{2})(-\sqrt{2}) + (4)(-\sqrt{3}) + (\sqrt{3})(2) = -2 - 2\sqrt{3}$;

$$\cos\theta = \frac{\mathbf{a} \cdot \mathbf{b}}{\|\mathbf{a}\|\,\|\mathbf{b}\|} = \frac{-2 - 2\sqrt{3}}{\sqrt{21} \cdot 3} = -\frac{2(1 + \sqrt{3})}{3\sqrt{21}}$$

5. a. Let $\mathbf{a} = -\sqrt{6}\mathbf{i} + \mathbf{j} - \mathbf{k}$ and $\mathbf{b} = \mathbf{i}$, and let θ be the angle between **a** and **b**. Then

$$\cos\theta = \frac{\mathbf{a} \cdot \mathbf{b}}{\|\mathbf{a}\|\,\|\mathbf{b}\|} = -\frac{\sqrt{6}}{\sqrt{8}\sqrt{1}} = -\sqrt{\frac{3}{4}} = -\frac{1}{2}\sqrt{3}, \quad \text{so} \quad \theta = \frac{5\pi}{6}$$

b. If $\mathbf{a} \cdot \mathbf{b} < 0$, then $\cos\theta < 0$ by (4), and hence $\pi/2 < \theta < \pi$.

7. $\mathbf{a} \cdot \mathbf{b} = (1)(1) + (-1)(1) = 0$; they are perpendicular.

9. $\mathbf{a} \cdot \mathbf{b} = (\sqrt{2})(-1) + (3)(\sqrt{2}) + (1)(5) = 2\sqrt{2} + 5 \neq 0$; they are not perpendicular.

11. $pr_{\mathbf{a}}\mathbf{b} = \dfrac{\mathbf{a} \cdot \mathbf{b}}{\|\mathbf{a}\|^2}\mathbf{a} = \dfrac{2}{9}(2\mathbf{i} - \mathbf{j} + 2\mathbf{k}) = \dfrac{4}{9}\mathbf{i} - \dfrac{2}{9}\mathbf{j} + \dfrac{4}{9}\mathbf{k}$

13. $pr_{\mathbf{a}}\mathbf{b} = \dfrac{\mathbf{a} \cdot \mathbf{b}}{\|\mathbf{a}\|^2}\mathbf{a} = \dfrac{1}{2}(\mathbf{i} + \mathbf{j})$

15. $\mathbf{a} \cdot \mathbf{a}' = 0 + 2 - 2 = 0$. Thus **a** and **a′** are perpendicular.

$$pr_{\mathbf{a}}\mathbf{b} = \frac{\mathbf{a} \cdot \mathbf{b}}{\|\mathbf{a}\|^2}\mathbf{a} = \frac{18}{6}\mathbf{a} = 3\mathbf{a} = 3(\mathbf{i} + 2\mathbf{j} - \mathbf{k})$$

$$pr_{\mathbf{a}'}\mathbf{b} = \mathbf{b} - pr_{\mathbf{a}}\mathbf{b} = 3\mathbf{i} + \mathbf{j} - 13\mathbf{k} - 3(\mathbf{i} + 2\mathbf{j} - \mathbf{k}) = -5\mathbf{j} - 10\mathbf{k} = -5\mathbf{a}'$$

Thus $\mathbf{b} = 3\mathbf{a} - 5\mathbf{a}'$.

17. $\mathbf{a} \cdot \mathbf{a}' = 4 - 24 + 20 = 0$. Thus **a** and **a′** are perpendicular.

$$pr_{\mathbf{a}}\mathbf{b} = \frac{\mathbf{a} \cdot \mathbf{b}}{\|\mathbf{a}\|^2}\mathbf{a} = \frac{-45}{45}\mathbf{a} = -\mathbf{a} = -(2\mathbf{i} - 4\mathbf{j} + 5\mathbf{k})$$

$$pr_{\mathbf{a}'}\mathbf{b} = \mathbf{b} - pr_{\mathbf{a}}\mathbf{b} = \mathbf{i} + 13\mathbf{j} + \mathbf{k} + (2\mathbf{i} - 4\mathbf{j} + 5\mathbf{k}) = 3\mathbf{i} + 9\mathbf{j} + 6\mathbf{k} = \tfrac{3}{2}\mathbf{a}'$$

Thus $\mathbf{b} = -\mathbf{a} + \frac{3}{2}\mathbf{a}'$.

19. $\mathbf{a} \cdot \mathbf{a} = (a_1)(a_1) + (a_2)(a_2) + (a_3)(a_3) = a_1^2 + a_2^2 + a_3^2 = \|\mathbf{a}\|^2$.

21. a.
$$\begin{aligned} \mathbf{a} \cdot (\mathbf{b} + \mathbf{c}) &= \mathbf{a} \cdot [(b_1 + c_1)\mathbf{i} + (b_2 + c_2)\mathbf{j} + (b_3 + c_3)\mathbf{k}] \\ &= a_1(b_1 + c_1) + a_2(b_2 + c_2) + a_3(b_3 + c_3) \\ &= (a_1b_1 + a_2b_2 + a_3b_3) + (a_1c_1 + a_2c_2 + a_3c_3) \\ &= \mathbf{a} \cdot \mathbf{b} + \mathbf{a} \cdot \mathbf{c} \end{aligned}$$

b. If **a** is perpendicular to **b** and **c**, then $\mathbf{a} \cdot \mathbf{b} = 0 = \mathbf{a} \cdot \mathbf{c}$, so that by (a), $\mathbf{a} \cdot (\mathbf{b} + \mathbf{c}) = \mathbf{a} \cdot \mathbf{b} + \mathbf{a} \cdot \mathbf{c} = 0 + 0 = 0$, which implies that **a** is perpendicular to $\mathbf{b} + \mathbf{c}$.

23. a. $\mathbf{a} \cdot \mathbf{b} = \|\mathbf{a}\|\,\|\mathbf{b}\| \cos\theta = \cos\theta$ is maximum if $\theta = 0$.

b. $\mathbf{a} \cdot \mathbf{b} = \|\mathbf{a}\|\,\|\mathbf{b}\| \cos\theta = \cos\theta$ is minimum if $\theta = \pi$.

c. $|\mathbf{a} \cdot \mathbf{b}| = \|\mathbf{a}\|\,\|\mathbf{b}\|\,|\cos\theta| = |\cos\theta|$ is minimum if $\theta = \pi/2$.

25. a.
$$\begin{aligned} \|\mathbf{a} + \mathbf{b}\|^2 &= (\mathbf{a} + \mathbf{b}) \cdot (\mathbf{a} + \mathbf{b}) = \mathbf{a} \cdot (\mathbf{a} + \mathbf{b}) + \mathbf{b} \cdot (\mathbf{a} + \mathbf{b}) \\ &= \mathbf{a} \cdot \mathbf{a} + \mathbf{a} \cdot \mathbf{b} + \mathbf{b} \cdot \mathbf{a} + \mathbf{b} \cdot \mathbf{b} = \|\mathbf{a}\|^2 + 2\mathbf{a} \cdot \mathbf{b} + \|\mathbf{b}\|^2 \end{aligned}$$

b. By part (a), $\|\mathbf{a} + \mathbf{b}\|^2 = \|\mathbf{a}\|^2 + \|\mathbf{b}\|^2$ if and only if $2\mathbf{a} \cdot \mathbf{b} = 0$, that is, **a** and **b** are perpendicular.

c. By Exercise 24(a), $\mathbf{a} \cdot \mathbf{b} \le |\mathbf{a} \cdot \mathbf{b}| \le \|\mathbf{a}\|\,\|\mathbf{b}\|$. Using this and part (a) of this exercise, we have $\|\mathbf{a} + \mathbf{b}\|^2 = \|\mathbf{a}\|^2 + 2\mathbf{a} \cdot \mathbf{b} + \|\mathbf{b}\|^2 \le \|\mathbf{a}\|^2 + 2\|\mathbf{a}\|\,\|\mathbf{b}\| + \|\mathbf{b}\|^2 = (\|\mathbf{a}\| + \|\mathbf{b}\|)^2$. Taking square roots, we find that $\|\mathbf{a} + \mathbf{b}\| \le \|\mathbf{a}\| + \|\mathbf{b}\|$.

27. a.
$$\begin{aligned} \|\mathbf{a} + \mathbf{b}\|^2 + \|\mathbf{a} - \mathbf{b}\|^2 &= (\mathbf{a} + \mathbf{b}) \cdot (\mathbf{a} + \mathbf{b}) + (\mathbf{a} - \mathbf{b}) \cdot (\mathbf{a} - \mathbf{b}) \\ &= (\mathbf{a} \cdot \mathbf{a} + \mathbf{a} \cdot \mathbf{b} + \mathbf{b} \cdot \mathbf{a} + \mathbf{b} \cdot \mathbf{b}) + (\mathbf{a} \cdot \mathbf{a} - \mathbf{a} \cdot \mathbf{b} - \mathbf{b} \cdot \mathbf{a} + \mathbf{b} \cdot \mathbf{b}) \\ &= 2(\mathbf{a} \cdot \mathbf{a}) + 2(\mathbf{b} \cdot \mathbf{b}) = 2\|\mathbf{a}\|^2 + 2\|\mathbf{b}\|^2 \end{aligned}$$

b.
$$\begin{aligned} \|\mathbf{a} + \mathbf{b}\|^2 - \|\mathbf{a} - \mathbf{b}\|^2 &= (\mathbf{a} + \mathbf{b}) \cdot (\mathbf{a} + \mathbf{b}) - (\mathbf{a} - \mathbf{b}) \cdot (\mathbf{a} - \mathbf{b}) \\ &= \mathbf{a} \cdot \mathbf{a} + \mathbf{a} \cdot \mathbf{b} + \mathbf{b} \cdot \mathbf{a} + \mathbf{b} \cdot \mathbf{b} - \mathbf{a} \cdot \mathbf{a} + \mathbf{a} \cdot \mathbf{b} + \mathbf{b} \cdot \mathbf{a} - \mathbf{b} \cdot \mathbf{b} \\ &= 2\mathbf{a} \cdot \mathbf{b} + 2\mathbf{b} \cdot \mathbf{a} = 4\mathbf{a} \cdot \mathbf{b} \end{aligned}$$

c. Let **a** and **b** be vectors along two adjacent sides of the parallelogram, as in Figure 12.34. If the diagonals have equal length, then $\|\mathbf{a}+\mathbf{b}\| = \|\mathbf{a}-\mathbf{b}\|$, so that by part (b), $\mathbf{a}\cdot\mathbf{b} = 0$. Thus the sides are perpendicular, which means that the parallelogram is a rectangle. Conversely, if the parallelogram is a rectangle, then **a** and **b** are perpendicular, so that $\mathbf{a}\cdot\mathbf{b} = 0$, and thus by part (b), $\|\mathbf{a}+\mathbf{b}\|^2 - \|\mathbf{a}-\mathbf{b}\|^2 = 0$, or $\|\mathbf{a}+\mathbf{b}\|^2 = \|\mathbf{a}-\mathbf{b}\|^2$. This means that the diagonals have equal length.

29. Let the cube be placed so its center is the origin, its faces are parallel to the coordinate planes, and its edges have length $2r$. Let **a** and **b** be as in the figure, so $\mathbf{a} = r\mathbf{i} + r\mathbf{j} + r\mathbf{k}$ and $\mathbf{b} = r\mathbf{i} + r\mathbf{j} - r\mathbf{k}$. If θ is the angle between **a** and **b**, then

$$\cos\theta = \frac{\mathbf{a}\cdot\mathbf{b}}{\|\mathbf{a}\|\,\|\mathbf{b}\|} = \frac{r^2 + r^2 - r^2}{(r\sqrt{3})(r\sqrt{3})} = \frac{1}{3}$$

so $\theta \approx 1.231$ (radians), or approximately $70.53°$.

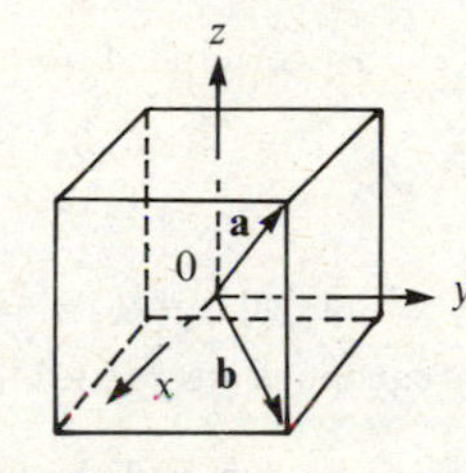

31. Let θ be the angle between **a** and **c**, and ϕ the angle between **c** and **b**. If $\mathbf{c} \neq \mathbf{0}$, then

$$\cos\theta = \frac{\mathbf{a}\cdot\mathbf{c}}{\|\mathbf{a}\|\,\|\mathbf{c}\|} = \frac{\mathbf{a}\cdot(\|\mathbf{b}\|\mathbf{a} + \|\mathbf{a}\|\mathbf{b})}{\|\mathbf{a}\|\,\|\mathbf{c}\|} = \frac{\|\mathbf{b}\|\mathbf{a}\cdot\mathbf{a} + \|\mathbf{a}\|\mathbf{a}\cdot\mathbf{b}}{\|\mathbf{a}\|\,\|\mathbf{c}\|}$$
$$= \frac{\|\mathbf{b}\|\,\|\mathbf{a}\|^2 + \|\mathbf{a}\|\mathbf{a}\cdot\mathbf{b}}{\|\mathbf{a}\|\,\|\mathbf{c}\|} = \frac{\|\mathbf{b}\|\,\|\mathbf{a}\| + \mathbf{a}\cdot\mathbf{b}}{\|\mathbf{c}\|}$$

and

$$\cos\phi = \frac{\mathbf{c}\cdot\mathbf{b}}{\|\mathbf{c}\|\,\|\mathbf{b}\|} = \frac{(\|\mathbf{b}\|\mathbf{a} + \|\mathbf{a}\|\mathbf{b})\cdot\mathbf{b}}{\|\mathbf{c}\|\,\|\mathbf{b}\|} = \frac{\|\mathbf{b}\|\mathbf{a}\cdot\mathbf{b} + \|\mathbf{a}\|\mathbf{b}\cdot\mathbf{b}}{\|\mathbf{c}\|\,\|\mathbf{b}\|}$$
$$= \frac{\|\mathbf{b}\|\mathbf{a}\cdot\mathbf{b} + \|\mathbf{a}\|\,\|\mathbf{b}\|^2}{\|\mathbf{c}\|\,\|\mathbf{b}\|} = \frac{\mathbf{a}\cdot\mathbf{b} + \|\mathbf{a}\|\,\|\mathbf{b}\|}{\|\mathbf{c}\|}$$

Thus $\cos\theta = \cos\phi$. Since $0 \le \theta \le \pi$ and $0 \le \phi \le \pi$, this means that $\theta = \phi$. Thus **c** bisects the angle formed by **a** and **b**.

33. Using the same terminology as that in Example 7, we find that

$$\overrightarrow{PQ} = 500\mathbf{i} \quad\text{and}\quad \mathbf{F} = 100\left(\cos\frac{\pi}{6}\mathbf{i} + \sin\frac{\pi}{6}\mathbf{j}\right) = 50\sqrt{3}\mathbf{i} + 50\mathbf{j}$$

Therefore $W = \mathbf{F}\cdot\overrightarrow{PQ} = (50\sqrt{3})500 = 25{,}000\sqrt{3}$ (foot-pounds).

SECTION 12.4

1. $\mathbf{a}\times\mathbf{b} = \begin{vmatrix} \mathbf{i} & \mathbf{j} & \mathbf{k} \\ 1 & 1 & 0 \\ 0 & 1 & 1 \end{vmatrix} = \mathbf{i} - \mathbf{j} + \mathbf{k}$

   Thus $\mathbf{c}\cdot(\mathbf{a}\times\mathbf{b}) = (-1)(1) + (-3)(-1) + (4)(1) = 6$.

3. $\mathbf{a}\times\mathbf{b} = \begin{vmatrix} \mathbf{i} & \mathbf{j} & \mathbf{k} \\ 2 & 3 & -1 \\ -1 & 4 & 5 \end{vmatrix} = 19\mathbf{i} - 9\mathbf{j} + 11\mathbf{k}$

   Thus $\mathbf{c}\cdot(\mathbf{a}\times\mathbf{b}) = (2)(19) + (3)(-9) + (4)(11) = 55$.

5. $\mathbf{a}\times\mathbf{b} = \begin{vmatrix} \mathbf{i} & \mathbf{j} & \mathbf{k} \\ 3 & 4 & 12 \\ 3 & 4 & 12 \end{vmatrix} = 0\mathbf{i} + 0\mathbf{j} + 0\mathbf{k} = \mathbf{0}$

   Thus $\mathbf{c}\cdot(\mathbf{a}\times\mathbf{b}) = (1)(0) + (1)(0) + (0)(0) = 0$

7. $\mathbf{a}\times\mathbf{b} = (a_2b_3 - a_3b_2)\mathbf{i} + (a_3b_1 - a_1b_3)\mathbf{j} + (a_1b_2 - a_2b_1)\mathbf{k}$
   $= -(b_2a_3 - b_3a_2)\mathbf{i} - (b_3a_1 - b_1a_3)\mathbf{j} - (b_1a_2 - b_2a_1)\mathbf{k} = -\mathbf{b}\times\mathbf{a}$

9. $\mathbf{0} = \mathbf{a}\times\mathbf{0} = \mathbf{a}\times(\mathbf{a}+\mathbf{b}+\mathbf{c}) = (\mathbf{a}\times\mathbf{a}) + (\mathbf{a}\times\mathbf{b}) + (\mathbf{a}\times\mathbf{c}) = (\mathbf{a}\times\mathbf{b}) + (\mathbf{a}\times\mathbf{c})$. Thus $\mathbf{a}\times\mathbf{b} = -(\mathbf{a}\times\mathbf{c}) = \mathbf{c}\times\mathbf{a}$ by Exercise 7. Similarly, $\mathbf{0} = \mathbf{b}\times\mathbf{0} = \mathbf{b}\times(\mathbf{a}+\mathbf{b}+\mathbf{c}) = (\mathbf{b}\times\mathbf{a}) + (\mathbf{b}\times\mathbf{b}) + (\mathbf{b}\times\mathbf{c}) = (\mathbf{b}\times\mathbf{a}) + (\mathbf{b}\times\mathbf{c})$. Thus $\mathbf{b}\times\mathbf{c} = -(\mathbf{b}\times\mathbf{a}) = \mathbf{a}\times\mathbf{b}$ by Exercise 7.

11. If $\mathbf{a} = 2\mathbf{i} - 3\mathbf{j} + 4\mathbf{k}$, $\mathbf{b} = \mathbf{i} + \mathbf{j} - \mathbf{k}$, and $\mathbf{c} = 4\mathbf{i} - \mathbf{j} - \mathbf{k}$, then the volume is $|\mathbf{a}\cdot(\mathbf{b}\times\mathbf{c})|$.

    Since $\mathbf{a}\cdot(\mathbf{b}\times\mathbf{c}) = \begin{vmatrix} 2 & -3 & 4 \\ 1 & 1 & -1 \\ 4 & -1 & -1 \end{vmatrix} = (-2 + 12 - 4) - (16 + 3 + 2) = -15$

    the volume is 15.

13. We have $\mathbf{a}\cdot(\mathbf{b}-\mathbf{c}) = \mathbf{a}\cdot\mathbf{b} - \mathbf{a}\cdot\mathbf{c} = 0$ and $\mathbf{a}\times(\mathbf{b}-\mathbf{c}) = \mathbf{a}\times\mathbf{b} - \mathbf{a}\times\mathbf{c} = \mathbf{0}$. If **b** and **c** were not equal, then $\mathbf{b}-\mathbf{c}$ would be nonzero and it would follow from the peeceding calculations and the fact that $\mathbf{a} \neq \mathbf{0}$ that **a** is both perpendicular to and parallel to $\mathbf{b}-\mathbf{c}$, which is impossible. Thus it does follow that $\mathbf{b} = \mathbf{c}$.

15. $\mathbf{b}\times\mathbf{c} = \begin{vmatrix} \mathbf{i} & \mathbf{j} & \mathbf{k} \\ \frac{1}{2} & 1 & -1 \\ 4 & -5 & 6 \end{vmatrix} = \mathbf{i} - 7\mathbf{j} - \frac{13}{2}\mathbf{k}$

    $\mathbf{a}\times(\mathbf{b}\times\mathbf{c}) = \begin{vmatrix} \mathbf{i} & \mathbf{j} & \mathbf{k} \\ 2 & -3 & 4 \\ 1 & -7 & -13/2 \end{vmatrix} = \frac{95}{2}\mathbf{i} + 17\mathbf{j} - 11\mathbf{k}$

    $\mathbf{a}\times(\mathbf{b}\times\mathbf{c}) = \mathbf{b}(\mathbf{a}\cdot\mathbf{c}) - \mathbf{c}(\mathbf{a}\cdot\mathbf{b}) = 47\mathbf{b} + 6\mathbf{c} = \frac{95}{2}\mathbf{i} + 17\mathbf{j} - 11\mathbf{k}$

17. If $\mathbf{u} \times \mathbf{a}$, $\mathbf{u}$ and $\mathbf{b}$ replace $\mathbf{a}$, $\mathbf{b}$, and $\mathbf{c}$ in (3), then $(\mathbf{u} \times \mathbf{a}) \cdot (\mathbf{u} \times \mathbf{b}) = [(\mathbf{u} \times \mathbf{a}) \times \mathbf{u}] \cdot \mathbf{b} = -[\mathbf{u} \times (\mathbf{u} \times \mathbf{a})] \cdot \mathbf{b}$. If $\mathbf{u}$ and $\mathbf{a}$ are perpendicular, then by (8), with $\mathbf{u}$ replacing $\mathbf{a}$ and $\mathbf{a}$ replacing $\mathbf{c}$, we find that $\mathbf{u} \times (\mathbf{u} \times \mathbf{a}) = -\|\mathbf{u}\|^2\mathbf{a}$, so $(\mathbf{u} \times \mathbf{a}) \cdot (\mathbf{u} \times \mathbf{b}) = -[-\|\mathbf{u}\|^2\mathbf{a}] \cdot \mathbf{b} = \|\mathbf{u}\|^2(\mathbf{a} \cdot \mathbf{b})$. Similarly, the formula holds if $\mathbf{u}$ and $\mathbf{b}$ are perpendicular, since $(\mathbf{u} \times \mathbf{a}) \cdot (\mathbf{u} \times \mathbf{b}) = (\mathbf{u} \times \mathbf{b}) \cdot (\mathbf{u} \times \mathbf{a})$ and $\mathbf{a} \cdot \mathbf{b} = \mathbf{b} \cdot \mathbf{a}$.

19. By the *bac* − *cab* rule,

$$\begin{aligned}\mathbf{a} \times (\mathbf{b} \times \mathbf{c}) + \mathbf{b} \times (\mathbf{c} \times \mathbf{a}) + \mathbf{c} \times (\mathbf{a} \times \mathbf{b}) &= [\mathbf{b}(\mathbf{a} \cdot \mathbf{c}) - \mathbf{c}(\mathbf{a} \cdot \mathbf{b})] + [\mathbf{c}(\mathbf{b} \cdot \mathbf{a}) - \mathbf{a}(\mathbf{b} \cdot \mathbf{c})] \\ &\quad + [\mathbf{a}(\mathbf{c} \cdot \mathbf{b}) - \mathbf{b}(\mathbf{c} \cdot \mathbf{a})] \\ &= \mathbf{b}(\mathbf{a} \cdot \mathbf{c} - \mathbf{c} \cdot \mathbf{a}) + \mathbf{c}(-\mathbf{a} \cdot \mathbf{b} + \mathbf{b} \cdot \mathbf{a}) \\ &\quad + \mathbf{a}(-\mathbf{b} \cdot \mathbf{c} + \mathbf{c} \cdot \mathbf{b}) = \mathbf{0}\end{aligned}$$

21. Using the coordinate system shown in Figure 12.37, with the stapler in the yz plane, we have

$$\overrightarrow{PQ} = \frac{3}{2}\left(\cos\frac{\pi}{6}\mathbf{j} + \sin\frac{\pi}{6}\mathbf{k}\right) = \frac{3}{2}\left(\frac{\sqrt{3}}{2}\mathbf{j} + \frac{1}{2}\mathbf{k}\right) = \frac{3}{4}(\sqrt{3}\mathbf{j} + \mathbf{k}) \quad \text{and} \quad \mathbf{F} = -32\mathbf{k}$$

Thus $\mathbf{M} = \overrightarrow{PQ} \times \mathbf{F} = -24\sqrt{3}\mathbf{i}$.

SECTION 12.5

1. $\mathbf{r}_0 = -2\mathbf{i} + \mathbf{j}$, so that a vector equation of the line is $\mathbf{r} = (-2 + 3t)\mathbf{i} + (1 - t)\mathbf{j} + 5t\mathbf{k}$, and thus parametric equations are $x = -2 + 3t$, $y = 1 - t$, $z = 5t$; $a = 3$, $b = -1$, $c = 5$, so that symmetric equations of the lines are $(x + 2)/3 = (y - 1)/-1 = z/5$.

3. $\mathbf{r}_0 = 3\mathbf{i} + 4\mathbf{j} + 5\mathbf{k}$, so that a vector equation of the line is $\mathbf{r} = (3 + \frac{1}{2}t)\mathbf{i} + (4 - \frac{1}{3}t)\mathbf{j} + (5 + \frac{1}{6}t)\mathbf{k}$, and thus parametric equations are $x = 3 + \frac{1}{2}t$, $y = 4 - \frac{1}{3}t$, $z = 5 + \frac{1}{6}t$; $a = \frac{1}{2}$, $b = -\frac{1}{3}$, $c = \frac{1}{6}$, so that symmetric equations of the line are $(x - 3)/\frac{1}{2} = (y - 4)/-\frac{1}{3} = (z - 5)/\frac{1}{6}$.

5. $\mathbf{r}_0 = 2\mathbf{i} + 5\mathbf{k}$, so that a vector equation of the line is $\mathbf{r} = 2\mathbf{i} + 2t\mathbf{j} + (5 + 3t)\mathbf{k}$, and thus parametric equations are $x = 2$, $y = 2t$, $z = 5 + 3t$; $a = 0$, $b = 2$, $c = 3$, so that symmetric equations of the line are $x = 2$ and $y/2 = (z - 5)/3$.

7. $\mathbf{r}_0 = 4\mathbf{i} + 2\mathbf{j} - \mathbf{k}$, so that a vector equation of the line is $\mathbf{r} = 4\mathbf{i} + (2 + t)\mathbf{j} - \mathbf{k}$, and thus parametric equations are $x = 4$, $y = 2 + t$, $z = -1$; $a = 0 = c$, $b = 1$, so that symmetric equations of the line are $x = 4$ and $z = -1$.

9. $x_0 = -1$, $y_0 = 1$, $z_0 = 0$, $a = -2 - (-1) = -1$, $b = 5 - 1 = 4$, $c = 7 - 0 = 7$, so that parametric equations for the line are $x = -1 - t$, $y = 1 + 4t$, $z = 7t$.

11. $x_0 = -1$, $y_0 = 1$, $z_0 = 0$, $a = 0$, $b = 0$, $c = 7$, so that symmetric equations for the line are $x = -1$ and $y = 1$.

13. The line through $(2, -1, 3)$ and $(0, 7, 9)$ is parallel to $\mathbf{L}_1$, where

$$\mathbf{L}_1 = (0 - 2)\mathbf{i} + (7 + 1)\mathbf{j} + (9 - 3)\mathbf{k} = -2\mathbf{i} + 8\mathbf{j} + 6\mathbf{k}$$

The line through $(-1, 0, 4)$ and $(2, 3, 1)$ is parallel to $\mathbf{L}_2$, where

$$\mathbf{L}_2 = (2 + 1)\mathbf{i} + (3 - 0)\mathbf{j} + (1 - 4)\mathbf{k} = 3\mathbf{i} + 3\mathbf{j} - 3\mathbf{k}$$

Since $\mathbf{L}_1 \cdot \mathbf{L}_2 = (-2)(3) + (8)(3) + (6)(-3) = 0$, $\mathbf{L}_1$ and $\mathbf{L}_2$ are perpendicular. Consequently the two lines are perpendicular.

15. The line through $(0, 0, 5)$ and $(1, -1, 4)$ is parallel to $\mathbf{L}_1$, where

$$\mathbf{L}_1 = (1 - 0)\mathbf{i} + (-1 - 0)\mathbf{j} + (4 - 5)\mathbf{k} = \mathbf{i} - \mathbf{j} - \mathbf{k}$$

The line with equation $x/7 = (y - 3)/4 = (z + 9)/3$ is parallel to $\mathbf{L}_2$, where $\mathbf{L}_2 = 7\mathbf{i} + 4\mathbf{j} + 3\mathbf{k}$. Since $\mathbf{L}_1 \cdot \mathbf{L}_2 = (1)(7) + (-1)(4) + (-1)(3) = 0$, $\mathbf{L}_1$ and $\mathbf{L}_2$ are perpendicular. Consequently the two lines are perpendicular.

17. By (3), the parametric equations have the form $x = x_0 + at$, $y = y_0 + bt$, $z = z_0 + ct$. Since P_1 corresponds to $t = 0$ and P_2 corresponds to $t = 2$, we have

$$\begin{array}{lll} -1 = x_0 + a \cdot 0, & -2 = y_0 + b \cdot 0, & -3 = z_0 + c \cdot 0 \\ 2 = x_0 + 2a, & -1 = y_0 + 2b, & 0 = z_0 + 2c \end{array}$$

Thus $x_0 = -1$, $y_0 = -2$, $z_0 = -3$, so that $a = \frac{1}{2}(2 - x_0) = \frac{3}{2}$, $b = \frac{1}{2}(-1 - y_0) = \frac{1}{2}$, $c = -\frac{1}{2}z_0 = \frac{3}{2}$. Thus the parametric equations are $x = -1 + \frac{3}{2}t$, $y = -2 + \frac{1}{2}t$, $z = -3 + \frac{3}{2}t$.

19. The point $P_0 = (1, -2, -1)$ is on the line, and the line is parallel to $\mathbf{i} - 2\mathbf{j} + 3\mathbf{k}$. If $P_1 = (5, 0, -4)$, then P_1 is not on the line, and $\overrightarrow{P_0P_1} = 4\mathbf{i} + 2\mathbf{j} - 3\mathbf{k}$, so by (5) the distance D is given by

$$D = \frac{\|(\mathbf{i} - 2\mathbf{j} + 3\mathbf{k}) \times (4\mathbf{i} + 2\mathbf{j} - 3\mathbf{k})\|}{\sqrt{1^2 + (-2)^2 + 3^2}} = \frac{\|15\mathbf{j} + 10\mathbf{k}\|}{\sqrt{14}} = 5\sqrt{\frac{13}{14}}$$

21. The line has a vector equation $\mathbf{r} = (-3 + 2t)\mathbf{i} + (-3 - 3t)\mathbf{j} + (3 + 5t)\mathbf{k}$, and if $P_1 = (0, 0, 0)$, then P_1 is not on the line. Let $P_0 = (-3, -3, 3)$, and let $\mathbf{L} = 2\mathbf{i} - 3\mathbf{j} + 5\mathbf{k}$, so that $\mathbf{r}$ and $\mathbf{L}$ are parallel. Then $\overrightarrow{P_0P_1} = 3\mathbf{i} + 3\mathbf{j} - 3\mathbf{k}$, and by (5) the distance D from P_1 to the given line is given by

$$D = \frac{\|(2\mathbf{i} - 3\mathbf{j} + 5\mathbf{k}) \times (3\mathbf{i} + 3\mathbf{j} - 3\mathbf{k})\|}{\sqrt{2^2 + (-3)^2 + 5^2}} = \frac{3\|-2\mathbf{i} + 7\mathbf{j} + 5\mathbf{k}\|}{\sqrt{38}} = 3\sqrt{\frac{39}{19}}$$

23. $P = (1, -1, 2)$ is on the first line; $Q = (0, 2, 3)$ is on the second line; $\mathbf{L} = 2\mathbf{i} - \mathbf{j} - 2\mathbf{k}$ is parallel to the first line. Thus

$$D = \frac{\|\mathbf{L} \times \overrightarrow{PQ}\|}{\|\mathbf{L}\|} = \frac{\|(2\mathbf{i} - \mathbf{j} - 2\mathbf{k}) \times (-\mathbf{i} + 3\mathbf{j} + \mathbf{k})\|}{\|2\mathbf{i} - \mathbf{j} - 2\mathbf{k}\|} = \frac{\|5\mathbf{i} + 5\mathbf{k}\|}{3} = \frac{5}{3}\sqrt{2}$$

25. $\sqrt{(x - 0)^2 + (y - y)^2 + (z - 0)^2} = \sqrt{2}$, or $x^2 + z^2 = 2$.

27. The line through (1, 4, 2) and (4, −3, −5) is parallel to $3\mathbf{i} - 7\mathbf{j} - 7\mathbf{k}$, and the line through (1, 4, 2) and (−5, −10, −8) is parallel to $-6\mathbf{i} - 14\mathbf{j} - 10\mathbf{k}$. Since these two vectors are not parallel, the three points do not lie on the same line.

29. Notice that $\mathbf{a} - \mathbf{b}$ lies on ℓ, so is parallel to $\mathbf{L}$. Thus $\mathbf{L} \times (\mathbf{a} - \mathbf{b}) = \mathbf{0}$. Since $\mathbf{0} = \mathbf{L} \times (\mathbf{a} - \mathbf{b}) = (\mathbf{L} \times \mathbf{a}) - (\mathbf{L} \times \mathbf{b})$, it follows that $\mathbf{L} \times \mathbf{a} = \mathbf{L} \times \mathbf{b}$.

SECTION 12.6

1. $x_0 = -1, y_0 = 2, z_0 = 3, a = -4, b = 15, c = -\frac{1}{2}$; $-4(x + 1) + 15(y - 2) - \frac{1}{2}(z - 3) = 0$, or $8x - 30y + z = -65$.

3. $x_0 = 9, y_0 = 17, z_0 = -7, a = 2, b = 0, c = -3$; $2(x - 9) + 0(y - 17) - 3(z + 7) = 0$, or $2x - 3z = 39$.

5. $x_0 = 2, y_0 = 3, z_0 = -5, a = 0, b = 1, c = 0, 0(x - 2) + 1(y - 3) + 0(z + 5) = 0$, or $y = 3$.

7. The point $P_0 = (-2, -1, -5)$ is on the line, and hence on the plane. Let $P_1 = (1, -1, 2)$, so that $\overrightarrow{P_0P_1} = 3\mathbf{i} + 7\mathbf{k}$. The vector $\mathbf{i} + \mathbf{j} + 2\mathbf{k}$ is parallel to the line but not parallel to $\overrightarrow{P_0P_1}$. For a normal to the plane we take

$$\mathbf{N} = \overrightarrow{P_0P_1} \times (\mathbf{i} + \mathbf{j} + 2\mathbf{k}) = (3\mathbf{i} + 7\mathbf{k}) \times (\mathbf{i} + \mathbf{j} + 2\mathbf{k}) = -7\mathbf{i} + \mathbf{j} + 3\mathbf{k}$$

An equation of the plane is $-7(x + 2) + 1(y + 1) + 3(z + 5) = 0$, or $7x - y - 3z = 2$.

9. A normal to the plane is given by $\mathbf{N} = 2\mathbf{i} + 5\mathbf{j} + 9\mathbf{k}$, so an equation of the plane is $2(x - 2) + 5(y - \frac{1}{2}) + 9(z - \frac{1}{3}) = 0$, or $4x + 10y + 18z = 19$.

11. a. The line ℓ is perpendicular to the vectors $2\mathbf{i} - 3\mathbf{j} + 4\mathbf{k}$ and $\mathbf{i} - \mathbf{k}$, which are normal to the two planes. Thus ℓ is parallel to $(2\mathbf{i} - 3\mathbf{j} + 4\mathbf{k}) \times (\mathbf{i} - \mathbf{k}) = 3\mathbf{i} + 6\mathbf{j} + 3\mathbf{k}$ and hence to $\mathbf{i} + 2\mathbf{j} + \mathbf{k}$. Since (1, 0, 0) is on the intersection of the two planes, a vector equation of ℓ is $\mathbf{r} = \mathbf{i} + t(\mathbf{i} + 2\mathbf{j} + \mathbf{k}) = (1 + t)\mathbf{i} + 2t\mathbf{j} + t\mathbf{k}$.

 b. A normal to the plane is given by $\mathbf{N} = \mathbf{i} + 2\mathbf{j} + \mathbf{k}$, so that an equation of the plane is $1(x + 9) + 2(y - 12) + 1(z - 14) = 0$, or $x + 2y + z = 29$.

13. The point $P_1 = (3, -1, 4)$ is not on the plane, whereas $P_0 = (0, 0, 5)$ is on the plane. Then $\overrightarrow{P_0P_1} = 3\mathbf{i} - \mathbf{j} - \mathbf{k}$. If $\mathbf{N} = 2\mathbf{i} - \mathbf{j} + \mathbf{k}$, then $\mathbf{N}$ is normal to the plane, and by Theorem 12.13, the distance D from P_1 to the plane is given by

$$D = \frac{|\mathbf{N} \cdot \overrightarrow{P_0P_1}|}{\|\mathbf{N}\|} = \frac{|(2\mathbf{i} - \mathbf{j} + \mathbf{k}) \cdot (3\mathbf{i} - \mathbf{j} - \mathbf{k})|}{\sqrt{2^2 + (-1)^2 + 1^2}} = \frac{6}{\sqrt{6}} = \sqrt{6}$$

15. If the plane passes through the origin, then $d = 0$, so that the distance from the origin to the plane and the number $|d|/\sqrt{a^2 + b^2 + c^2}$ are both 0. If the plane does not pass through the origin, let $P_1 = (0, 0, 0)$. Also assume that $c \neq 0$, and let $P_0 = (0, 0, d/c)$, so that P_0 is on the plane. By Theorem 12.13, the distance D from P_1 to the plane is given by

$$D = \frac{|(a\mathbf{i} + b\mathbf{j} + c\mathbf{k}) \cdot \overrightarrow{P_0P_1}|}{\sqrt{a^2 + b^2 + c^2}} = \frac{|(a\mathbf{i} + b\mathbf{j} + c\mathbf{k}) \cdot ((d/c)\mathbf{k})|}{\sqrt{a^2 + b^2 + c^2}} = \frac{|d|}{\sqrt{a^2 + b^2 + c^2}}$$

The same result follows if $a \neq 0$, or if $b \neq 0$.

17. Let (x, y, z) be on the plane. Then

$$\sqrt{(x - 3)^2 + (y - 1)^2 + (z - 5)^2} = \sqrt{(x - 5)^2 + (y + 1)^2 + (z - 3)^2}$$

Squaring both sides, we obtain $(x - 3)^2 + (y - 1)^2 + (z - 5)^2 = (x - 5)^2 + (y + 1)^2 + (z - 3)^2$. Simplifying, we obtain $x - y - z = 0$.

19. 21.

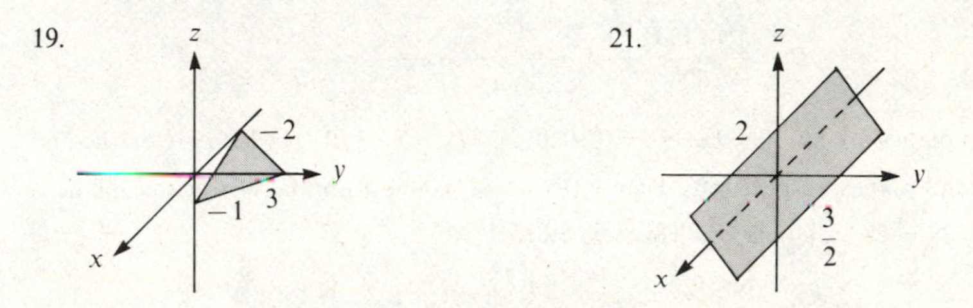

23. Let $P_0 = (2, 3, 2)$, $P_1 = (1, -1, -3)$, and $P_2 = (1, 0, -1)$. Then $\overrightarrow{P_0P_1} = -\mathbf{i} - 4\mathbf{j} - 5\mathbf{k}$ and $\overrightarrow{P_0P_2} = -\mathbf{i} - 3\mathbf{j} - 3\mathbf{k}$. Thus a normal $\mathbf{N}$ to the plane containing the three points is given by

$$\mathbf{N} = \overrightarrow{P_0P_1} \times \overrightarrow{P_0P_2} = (-\mathbf{i} - 4\mathbf{j} - 5\mathbf{k}) \times (-\mathbf{i} - 3\mathbf{j} - 3\mathbf{k}) = -3\mathbf{i} + 2\mathbf{j} - \mathbf{k}$$

Therefore an equation of the plane is $-3(x - 2) + 2(y - 3) - 1(z - 2) = 0$, or $3x - 2y + z = 2$. Since $3 \cdot 5 - 2 \cdot 9 + 5 = 2$, the fourth point (5, 9, 5) lies on the plane, so that all four given points lie on the same plane.

25. Since $(\mathbf{a} \times \mathbf{b}) \times (\mathbf{c} \times \mathbf{d})$ is perpendicular to $(\mathbf{a} \times \mathbf{b})$ and $(\mathbf{c} \times \mathbf{d})$, which are normal to P_1 and P_2 respectively, $(\mathbf{a} \times \mathbf{b}) \times (\mathbf{c} \times \mathbf{d})$ is parallel to all vectors that lie in both P_1 and P_2, and thus is parallel to the intersection of P_1 and P_2.

27. Normals of the planes in (a) – (d) are $\mathbf{i} + \mathbf{j} - \mathbf{k}$, $\mathbf{i} - \mathbf{j}$, $\mathbf{j} - \mathbf{k}$, and $\mathbf{i} + \mathbf{j}$. Since none of these is a multiple of another, no two planes are identical or parallel. Since $(\mathbf{i} + \mathbf{j} - \mathbf{k}) \cdot (\mathbf{i} - \mathbf{j}) = 0$ and $(\mathbf{i} - \mathbf{j}) \cdot (\mathbf{i} + \mathbf{j}) = 0$, the planes in (a) and (b) are perpendicular, as are the planes in (b) and (d).

29. a. The vector $\mathbf{i}$ is normal to the plane, so an equation of the plane is $1(x + 4) = 0$, or $x = -4$.

 b. The vector $\mathbf{j}$ is normal to the plane, so an equation of the plane is $1(y + 5) = 0$, or $y = -5$.

 c. The vector $\mathbf{k}$ is normal to the plane, so an equation of the plane is $1(z + 3) = 0$, or $z = -3$.

31. From the first and second equations, $x = 1 - y$ and $z = 2 - y$. Substituting in the third equation, we find that $3 = x + z = (1 - y) + (2 - y) = 3 - 2y$, so that $y = 0$. Thus $x = 1$ and $z = 2$. The point of intersection is $(1, 0, 2)$.

33. Adding the first and second equations, we obtain $3x = -1$, so that $x = -\frac{1}{3}$. The second equation becomes $3y + z = -\frac{5}{3}$. Subtracting the third equation from this equation, we obtain $y = -\frac{11}{6}$. Therefore $z = \frac{23}{6}$. The point is $(-\frac{1}{3}, -\frac{11}{6}, \frac{23}{6})$.

35. The two planes are parallel. Let $P_0 = (0, 0, 2)$ and $P_1 = (0, 0, \frac{1}{3})$, which are on the first and second planes, respectively. Then $\overrightarrow{P_0P_1} = -\frac{5}{3}\mathbf{k}$, and a normal $\mathbf{N}$ to either plane is given by $\mathbf{N} = \mathbf{i} - \mathbf{j} + \mathbf{k}$. By Theorem 12.13, the distance D is given by

$$D = \frac{|\mathbf{N} \cdot \overrightarrow{P_0P_1}|}{\|\mathbf{N}\|} = \frac{\frac{5}{3}}{\sqrt{1^2 + (-1)^2 + 1^2}} = \frac{5\sqrt{3}}{9}$$

37. The two planes are parallel. Let $P_0 = (\frac{5}{2}, 0, 0)$ and $P_1 = (-\frac{1}{4}, 0, 0)$, which are on the first and second planes, respectively. Then $\overrightarrow{P_0P_1} = -\frac{11}{4}\mathbf{i}$, and a normal $\mathbf{N}$ to either plane is given by $\mathbf{N} = 2\mathbf{i} - 3\mathbf{j} + 4\mathbf{k}$. By Theorem 12.13,

$$D = \frac{|\mathbf{N} \cdot \overrightarrow{P_0P_1}|}{\|\mathbf{N}\|} = \frac{\frac{11}{2}}{\sqrt{2^2 + (-3)^2 + 4^2}} = \frac{11}{58}\sqrt{29}$$

39. The vectors $\mathbf{N}_1 = \mathbf{j} - \mathbf{k}$ and $\mathbf{N}_2 = 4\mathbf{i} - \mathbf{j} - 2\mathbf{k}$ are normal to the first and second planes, respectively. Thus

$$\cos\theta = \frac{\mathbf{N}_1 \cdot \mathbf{N}_2}{\|\mathbf{N}_1\|\|\mathbf{N}_2\|} = \frac{1}{\sqrt{2}\sqrt{21}} = \frac{1}{42}\sqrt{42}$$

so $\theta \approx 1.416$ radians, or $\theta \approx 81.12°$.

CHAPTER 12 REVIEW

1. $2\mathbf{a} + \mathbf{b} - 3\mathbf{c} = 5\mathbf{i} - 10\mathbf{j} + 11\mathbf{k}$; $\mathbf{a} \times \mathbf{b} = \begin{vmatrix} \mathbf{i} & \mathbf{j} & \mathbf{k} \\ 2 & -3 & 1 \\ 1 & -1 & 0 \end{vmatrix} = \mathbf{i} + \mathbf{j} + \mathbf{k}$;

$\mathbf{c} \cdot (\mathbf{a} \times \mathbf{b}) = (\mathbf{j} - 3\mathbf{k}) \cdot (\mathbf{i} + \mathbf{j} + \mathbf{k}) = -2$;

$\mathbf{a} \times (\mathbf{b} \times \mathbf{c}) = \mathbf{b}(\mathbf{a} \cdot \mathbf{c}) - \mathbf{c}(\mathbf{a} \cdot \mathbf{b}) = -6\mathbf{b} - 5\mathbf{c} = -6\mathbf{i} + \mathbf{j} + 15\mathbf{k}$

3. $2\mathbf{a} + \mathbf{b} - 3\mathbf{c} = 11\mathbf{i} - 9\mathbf{j} + 6\mathbf{k}$; $\mathbf{a} \times \mathbf{b} = \begin{vmatrix} \mathbf{i} & \mathbf{j} & \mathbf{k} \\ 3 & -2 & 1 \\ 5 & -2 & 1 \end{vmatrix} = 2\mathbf{j} + 4\mathbf{k}$;

$\mathbf{c} \cdot (\mathbf{a} \times \mathbf{b}) = (\mathbf{j} - \mathbf{k}) \cdot (2\mathbf{j} + 4\mathbf{k}) = -2$;

$$\mathbf{a} \times (\mathbf{b} \times \mathbf{c}) = \mathbf{b}(\mathbf{a} \cdot \mathbf{c}) - \mathbf{c}(\mathbf{a} \cdot \mathbf{b}) = -3\mathbf{b} - 20\mathbf{c} = -15\mathbf{i} - 14\mathbf{j} + 17\mathbf{k}$$

5. Let $Q = (b_1, b_2, b_3)$. Then $\overrightarrow{PQ} = (b_1 - 1)\mathbf{i} + (b_2 + 2)\mathbf{j} + (b_3 - 3)\mathbf{k}$, so that $\overrightarrow{PQ} = \mathbf{a}$ if $b_1 - 1 = 2$, $b_2 + 2 = -2$, $b_3 - 3 = 1$, that is, if $b_1 = 3$, $b_2 = -4$, $b_3 = 4$. Thus $Q = (3, -4, 4)$.

7. Let $\mathbf{b} = 2\mathbf{i} - \mathbf{j} - \mathbf{k}$, $\mathbf{a} = 2\mathbf{j} + \mathbf{k}$, and $\mathbf{a}' = -20\mathbf{i} - 2\mathbf{j} + 4\mathbf{k}$. Note that $\mathbf{a} \cdot \mathbf{a} = -4 + 4 = 0$, so that $\mathbf{a}$ and $\mathbf{a}'$ are perpendicular.

$$pr_{\mathbf{a}}\mathbf{b} = \frac{\mathbf{a} \cdot \mathbf{b}}{\|\mathbf{a}\|^2}\mathbf{a} = -\frac{3}{5}\mathbf{a} = -\frac{3}{5}(2\mathbf{j} + \mathbf{k})$$

$$pr_{\mathbf{a}'}\mathbf{b} = \mathbf{b} - pr_{\mathbf{a}}\mathbf{b} = 2\mathbf{i} - \mathbf{j} - \mathbf{k} + \tfrac{3}{5}(2\mathbf{j} + \mathbf{k}) = 2\mathbf{i} + \tfrac{1}{5}\mathbf{j} - \tfrac{2}{5}\mathbf{k} = -\tfrac{1}{10}\mathbf{a}'$$

Thus $\mathbf{b} = -\frac{3}{5}\mathbf{a} - \frac{1}{10}\mathbf{a}'$.

9. If $P_0 = (\frac{1}{2}, \frac{1}{3}, 0)$, $P_1 = (1, 1, -1)$, and $P_2 = (-2, -3, 5)$, then $\overrightarrow{P_0P_1} = \frac{1}{2}\mathbf{i} + \frac{2}{3}\mathbf{j} - \mathbf{k}$, whereas $\overrightarrow{P_1P_2} = -3\mathbf{i} - 4\mathbf{j} + 6\mathbf{k}$. Thus $-6\overrightarrow{P_0P_1} = \overrightarrow{P_1P_2}$, so that the three points are collinear. Symmetric equations of the line are $(x - 1)/-3 = (y - 1)/-4 = (z + 1)/6$.

11. The line is parallel to $2\mathbf{i} - 3\mathbf{j} + 4\mathbf{k}$, which is normal to the plane. Since $\mathbf{r}_0 = -3\mathbf{i} - 3\mathbf{j} + \mathbf{k}$, a vector equation of the line is $\mathbf{r} = (-3 + 2t)\mathbf{i} + (-3 - 3t)\mathbf{j} + (1 + 4t)\mathbf{k}$.

13. Let $P_0 = (-1, 1, 1)$, $P_1 = (0, 2, 1)$, and $P_2 = (0, 0, \frac{3}{2})$. Then $\overrightarrow{P_0P_1} = \mathbf{i} + \mathbf{j}$ and $\overrightarrow{P_0P_2} = \mathbf{i} - \mathbf{j} + \frac{1}{2}\mathbf{k}$. Thus a normal $\mathbf{N}$ to the plane containing the three points is given by

$$\mathbf{N} = \overrightarrow{P_0P_1} \times \overrightarrow{P_0P_2} = \tfrac{1}{2}\mathbf{i} - \tfrac{1}{2}\mathbf{j} - 2\mathbf{k}$$

Therefore an equation of the plane is $\frac{1}{2}(x + 1) - \frac{1}{2}(y - 1) - 2(z - 1) = 0$, or $\frac{1}{2}x - \frac{1}{2}y - 2z = -3$. Since $\frac{1}{2}(13) - \frac{1}{2}(-1) - 2(5) = -3$, the fourth point $(13, -1, 5)$ lies on the plane, so that all four points lie on the same plane.

15. Since a normal $\mathbf{N}$ of the plane is perpendicular to the z axis, $\mathbf{N} = a\mathbf{i} + b\mathbf{j}$ for appropriate choices of a and b. Then an equation of the plane is $a(x - 3) + b(y + 1) + 0(z - 5) = 0$, or $ax + by = 3a - b$. Since $(7, 9, 4)$ is on the plane, $7a + 9b = 3a - b$, so that $4a = -10b$, or $a = -\frac{5}{2}b$. Thus an equation of the plane is $-\frac{5}{2}bx + by = -\frac{15}{2}b - b$, or $5x - 2y = 17$.

17. The point $P_1 = (1, -2, 5)$ is not on the plane, whereas $P_0 = (1, -2, 0)$ is on the plane. Then $\overrightarrow{P_0P_1} = 5\mathbf{k}$. If $\mathbf{N} = 3\mathbf{i} - 4\mathbf{j} + 12\mathbf{k}$, then $\mathbf{N}$ is normal to the plane, and the distance D from P_1 to the plane is given by

$$D = \frac{|\mathbf{N} \cdot \overrightarrow{P_0P_1}|}{\|\mathbf{N}\|} = \frac{|(3\mathbf{i} - 4\mathbf{j} + 12\mathbf{k}) \cdot (5\mathbf{k})|}{\sqrt{3^2 + (-4)^2 + (12)^2}} = \frac{60}{13}$$

19. a. Letting $y = 4$ and $z = 1$ in the first equation, we obtain $3x - 4 + 1 = 2$, or $x = \frac{5}{3}$. Thus the three planes have the point $(\frac{5}{3}, 4, 1)$ in common.

b. If (x, y, z) is on the first two planes, then $2x + y - 2z - 1 = 0 = 3x + y - z - 2$, so that $-z = x - 1$, or $x = 1 - z$. If (x, y, z) is on the first and third planes, then $2x + y - 2z - 1 = 0 = 2x - 2y + 2z$, so that $3y - 1 = 4z$, or $y = (4z + 1)/3$. Thus if (x, y, z) is on all three planes, then $0 = x - y + z = (1 - z) - ((4z + 1)/3) + z$, so that $z = \frac{1}{2}$. Thus $x = \frac{1}{2}$ and $y = 1$. Thus the planes have the point $(\frac{1}{2}, 1, \frac{1}{2})$ in common.

c. If (x, y, z) is on the first two planes, then $2x - 11y + 6z + 2 = 0 = 2x - 3y + 2z - 2$, or $z + 1 = 2y$. If (x, y, z) is on the first and third planes, then $2x - 11y + 6z + 2 = 0 = 2x - 9y + 5z + 1$, or $z + 1 = 2y$. Thus $z = 2y - 1$. Substituting $z = 2y - 1$ into the equation $2x - 3y + 2z = 2$, we obtain $2x - 3y + 2(2y - 1) = 2$, or $2x + y = 4$. Thus $x = -\frac{1}{2}y + 2$. If we let $y = t$, then $x = 2 - \frac{1}{2}t$, $y = t$, $z = -1 + 2t$ are parametric equations of the line common to the three planes.

21. An equation of the required plane is $a(x - a) + b(y - b) + c(z - c) = 0$, or

$$ax + by + cz = a^2 + b^2 + c^2$$

23. Using the notation in the figure, we have $\mathbf{a} = \mathbf{c} + \mathbf{d}$, $\mathbf{b} = -\mathbf{c} + \mathbf{d}$, and $\|\mathbf{c}\| = \|\mathbf{d}\| = r$, the radius of the circle. Thus

$$\mathbf{a} \cdot \mathbf{b} = (\mathbf{c} + \mathbf{d}) \cdot (-\mathbf{c} + \mathbf{d}) = -\mathbf{c} \cdot \mathbf{c} + \mathbf{c} \cdot \mathbf{d} - \mathbf{c} \cdot \mathbf{d} + \mathbf{d} \cdot \mathbf{d} = \|\mathbf{d}\|^2 - \|\mathbf{c}\|^2 = 0$$

so $\mathbf{a}$ and $\mathbf{b}$ are perpendicular. Therefore every angle inscribed in a semicircle is a right angle.

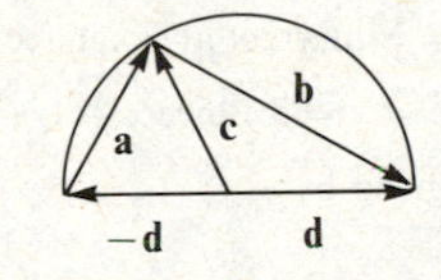

25. a. Since $\mathbf{a} \cdot \mathbf{c} = \mathbf{b} \cdot \mathbf{c}$ for all $\mathbf{c}$, we let $\mathbf{c} = \mathbf{a} - \mathbf{b}$ and find that $\|\mathbf{a} - \mathbf{b}\|^2 = (\mathbf{a} - \mathbf{b}) \cdot (\mathbf{a} - \mathbf{b}) = (\mathbf{a} - \mathbf{b}) \cdot \mathbf{c} = \mathbf{a} \cdot \mathbf{c} - \mathbf{b} \cdot \mathbf{c} = 0$. Thus $\mathbf{a} - \mathbf{b} = 0$, so that $\mathbf{a} = \mathbf{b}$.

b. If $\mathbf{a} \neq \mathbf{b}$, then $\mathbf{a} - \mathbf{b} \neq \mathbf{0}$, so we let $\mathbf{c}$ be a nonzero vector perpendicular to $\mathbf{a} - \mathbf{b}$. It follows that $\mathbf{0} = (\mathbf{a} \times \mathbf{c}) - (\mathbf{b} \times \mathbf{c}) = (\mathbf{a} - \mathbf{b}) \times \mathbf{c}$. Therefore $\|\mathbf{a} - \mathbf{b}\| \, \|\mathbf{c}\| = 0$, which is impossible since $\|\mathbf{a} - \mathbf{b}\| \neq 0$ and $\|\mathbf{c}\| \neq 0$. Thus $\mathbf{a} = \mathbf{b}$.

27. Letting $\mathbf{a} = \cos x\mathbf{i} + \sin x\mathbf{j}$ and $\mathbf{b} = \cos y\mathbf{i} - \sin y\mathbf{j}$, we find that $\mathbf{a} \cdot \mathbf{b} = \cos x \cos y - \sin x \cdot \sin y$. But by the definitions of $\mathbf{a}$ and $\mathbf{b}$, the angle between $\mathbf{a}$ and $\mathbf{b}$ is $x + y$, (see figure), so that $\mathbf{a} \cdot \mathbf{b} = \|\mathbf{a}\| \, \|\mathbf{b}\| \cos(x + y) = \cos(x + y)$. Thus

$$\cos(x + y) = \cos x \cos y - \sin x \sin y$$

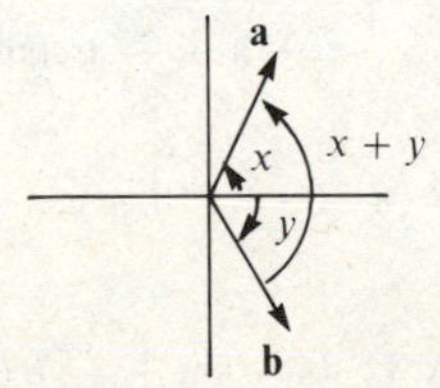

29. Using the fact that $\mathbf{a}$, $\mathbf{b}$, and $\mathbf{c}$ are pairwise perpendicular, we have

$$\|\mathbf{d}\|^2 = \mathbf{d} \cdot \mathbf{d} = (a\mathbf{a} + b\mathbf{b} + c\mathbf{c}) \cdot (a\mathbf{a} + b\mathbf{b} + c\mathbf{c}) = a^2(\mathbf{a} \cdot \mathbf{a}) + b^2(\mathbf{b} \cdot \mathbf{b}) + c^2(\mathbf{c} \cdot \mathbf{c})$$

Since $\mathbf{a}$, $\mathbf{b}$, and $\mathbf{c}$ are unit vectors, $\mathbf{a} \cdot \mathbf{a} = \mathbf{b} \cdot \mathbf{b} = \mathbf{c} \cdot \mathbf{c} = 1$, so that $a^2(\mathbf{a} \cdot \mathbf{a}) + b^2(\mathbf{b} \cdot \mathbf{b}) + c^2(\mathbf{c} \cdot \mathbf{c}) = a^2 + b^2 + c^2$, and thus $\|\mathbf{d}\| = \sqrt{a^2 + b^2 + c^2}$.

31. Consider a coordinate system with the forces applied at the origin, the 500 pound force $\mathbf{F}_1$ above the positive x axis and the 300 pound force along the line at an angle of $\pi/3$ with the positive x axis. Then $\mathbf{F}_1 = 500\mathbf{i}$ and $\mathbf{F}_2 = 300(\cos(\pi/3)\mathbf{i} + \sin(\pi/3)\mathbf{j}) = 150\mathbf{i} + 150\sqrt{3}\mathbf{j}$. Thus the resultant force $\mathbf{F} = \mathbf{F}_1 + \mathbf{F}_2 = 500\mathbf{i} + (150\mathbf{i} + 150\sqrt{3}\mathbf{j}) = 650\mathbf{i} + 150\sqrt{3}\mathbf{j}$. Therefore the magnitude of $\mathbf{F}$ is given by $\|\mathbf{F}\| = \sqrt{(650)^2 + (150)^2(3)} = 700$ (pounds). For the cosine of the angle θ between $\mathbf{F}$ and $\mathbf{F}_1$ we obtain

$$\cos\theta = \frac{\mathbf{F} \cdot \mathbf{F}_1}{\|\mathbf{F}\| \, \|\mathbf{F}_1\|} = \frac{650 \cdot 500}{700 \cdot 500} = \frac{13}{14}$$

33. Consider a coordinate system with the river flowing in the direction of the negative x axis and the motorboat traveling with increasing values of y. Let $\mathbf{v}_1$ be the velocity of the motorboat with respect to the water, and $\mathbf{v}_2$ the velocity of the river, so that $\mathbf{v}_1 + \mathbf{v}_2$ is the velocity of the motorboat with respect to the ground. Notice that $\mathbf{v}_2 = -5\mathbf{i}$ and $\mathbf{v}_1 = 10(\cos\theta\mathbf{i} + \sin\theta\mathbf{j})$, with θ to be chosen so that $\mathbf{v}_1 + \mathbf{v}_2 = c\mathbf{j}$ for an appropriate positive value of c. But $\mathbf{v}_1 + \mathbf{v}_2 = 10(\cos\theta\mathbf{i} + \sin\theta\mathbf{j}) - 5\mathbf{i} = (10\cos\theta - 5)\mathbf{i} + 10\sin\theta\mathbf{j}$, so that $10\cos\theta - 5 = 0$ and $10\sin\theta = c > 0$ if $\theta = \pi/3$. Thus the boat should be pointed at an angle of $\pi/3$ with respect to the shore. Also $\mathbf{v}_1 + \mathbf{v}_2 = 10\sin(\pi/3)\mathbf{j} = 5\sqrt{3}\mathbf{j}$. Therefore the $\frac{1}{2}$ mile width of the river is traveled at a speed of $5\sqrt{3}$ miles per hour in $1/(10\sqrt{3})$ hours, or approximately 3.46410 minutes.

35. If we let $q_1 = 3.2 \times 10^{-19}$, $\mathbf{u}_1 = -\mathbf{i}$, $r_1 = 10^{-12}$, $q_2 = -6.4 \times 10^{-19}$, $\mathbf{u}_2 = -\mathbf{j}$, $r_2 = 2 \times 10^{-12}$, $q_3 = 4.8 \times 10^{-19}$, $\mathbf{u}_3 = -\mathbf{k}$, and $r_3 = 3 \times 10^{-12}$, then

$$\begin{aligned}\mathbf{F} &= \frac{q_1(1)}{4\pi\varepsilon_0 r_1^2}\mathbf{u}_1 + \frac{q_2(1)}{4\pi\varepsilon_0 r_2^2}\mathbf{u}_2 + \frac{q_3(1)}{4\pi\varepsilon_0 r_3^2}\mathbf{u}_3 \\ &= \frac{3.2 \times 10^{-19}}{4\pi\varepsilon_0 10^{-24}}(-\mathbf{i}) + \frac{-6.4 \times 10^{-19}}{4\pi\varepsilon_0(4 \times 10^{-24})}(-\mathbf{j}) + \frac{4.8 \times 10^{-19}}{4\pi\varepsilon_0(9 \times 10^{-24})}(-\mathbf{k}) \\ &= \frac{10^4}{\pi\varepsilon_0}\left(-8\mathbf{i} + 4\mathbf{j} - \frac{4}{3}\mathbf{k}\right)\end{aligned}$$

CUMULATIVE REVIEW (CHAPTERS 1–11)

1. a. The domain consists of all x such that $(1 - 2x)/(1 - 3x) \geq 0$, or $\frac{2}{3}[(x - \frac{1}{2})/(x - \frac{1}{3})] \geq 0$. This occurs if $x \geq \frac{1}{2}$ and $x > \frac{1}{3}$, or if $x \leq \frac{1}{2}$ and $x < \frac{1}{3}$. Thus the domain is the union of $(-\infty, \frac{1}{3})$ and $[\frac{1}{2}, \infty)$.

b. $$\lim_{x \to -\infty} f(x) = \lim_{x \to -\infty} \sqrt{\frac{1 - 2x}{1 - 3x}} = \lim_{x \to -\infty} \sqrt{\frac{1/x - 2}{1/x - 3}} = \sqrt{\frac{0 - 2}{0 - 3}} = \sqrt{\frac{2}{3}}$$

3. $\lim_{x\to\infty} x^{\tan(1/x)} = \lim_{x\to\infty} e^{\tan(1/x)\ln x} = e^{\lim_{x\to\infty} \ln x/\cot(1/x)}$

By l'Hôpital's Rule,

$$\lim_{x\to\infty} \frac{\ln x}{\cot(1/x)} = \lim_{x\to\infty} \frac{1/x}{[-\csc^2(1/x)](-1/x^2)} = \lim_{x\to\infty} \sin\frac{1}{x}\frac{\sin(1/x)}{1/x} = 0\cdot 1 = 0$$

Thus $$\lim_{x\to\infty} x^{\tan(1/x)} = e^{\lim_{x\to\infty} \ln x/\cot(1/x)} = e^0 = 1$$

5. a. $$f'(x) = \frac{1}{(\pi/2) + \arctan x}\frac{1}{x^2+1} > 0 \quad \text{for all } x$$

Therefore f is strictly increasing, so f^{-1} exists. The domain of f^{-1} = the range of f. Since the range of arctan x is $(-\pi/2, \pi/2)$, it follows that the range of $\pi/2 + \arctan x$ is $(0, \pi)$ and the range of $\ln(\pi/2 + \arctan x)$ is $(-\infty, \ln \pi)$. Thus the domain of f^{-1} is $(-\infty, \ln \pi)$. The range of f^{-1} = the domain of $f = (-\infty, \infty)$.

b. $(f^{-1})'(\ln \pi/3) = 1/f'(a)$ for the value of a in $(-\infty, \infty)$ such that $f(a) = \ln \pi/3$, or by the definition of f, $\ln(\pi/2 + \arctan a) = \ln \pi/3$. This means that $\pi/2 + \arctan a = \pi/3$, so that $\arctan a = -\pi/6$ and hence $a = -\sqrt{3}/3$. Thus

$$(f^{-1})'\left(\ln\frac{\pi}{3}\right) = \frac{1}{f'(-\sqrt{3}/3)} = \frac{1}{\dfrac{1}{\pi/2 + \arctan(-\sqrt{3}/3)}\dfrac{1}{(-\sqrt{3}/3)^2+1}}$$
$$= \left(\frac{\pi}{2} - \frac{\pi}{6}\right)\left(\frac{1}{3}+1\right) = \frac{4\pi}{9}$$

c. $y = f^{-1}(x)$ if and only if $x = f(y) = \ln(\pi/2 + \arctan y)$ if and only if $e^x = \pi/2 + \arctan y$ if and only if $\arctan y = e^x - \pi/2$ if and only if $y = \tan(e^x - \pi/2)$. Therefore $f^{-1}(x) = \tan(e^x - \pi/2)$ for $x < \ln \pi$.

7. a. $\dfrac{dy}{dx} = \dfrac{dy}{dt}\dfrac{1}{dx/dt} = \dfrac{\sin t}{1-\cos t}$; $\cos t = 1 - y$ and $0 < t < \pi$,

so $\sin t = \sqrt{1-\cos^2 t} = \sqrt{1-(1-y)^2} = \sqrt{2y-y^2}$; thus

$$\frac{dy}{dx} = \frac{\sin t}{1-\cos t} = \frac{\sqrt{2y-y^2}}{y}$$

b. Since $\cos t = 1 - y$, we have $t = \arccos(1-y)$ for $0 < t < \pi$. Thus by part (a), $x = t - \sin t = \arccos(1-y) - \sqrt{2y-y^2}$.

c. By differentiating the equation for x in part (b) implicitly with respect to x, we obtain

$$1 = \frac{-1}{\sqrt{1-(1-y)^2}}\left(-\frac{dy}{dx}\right) - \frac{1}{2\sqrt{2y-y^2}}(2-2y)\frac{dy}{dx}$$

or $$1 = \frac{1}{\sqrt{2y-y^2}}\frac{dy}{dx} - \frac{1-y}{\sqrt{2y-y^2}}\frac{dy}{dx}, \quad \text{so} \quad \frac{dy}{dx} = \frac{\sqrt{2y-y^2}}{y}$$

9. $f'(x) = 5(x^2-5)(x^2-1)$; $f''(x) = 20x(x^2-3)$; relative maximum values are $f(-\sqrt{5}) = 0$ and $f(1) = 16$; relative minimum values are $f(-1) = -16$ and $f(\sqrt{5}) = 0$; increasing on $(-\infty, -\sqrt{5}]$, $[-1, 1]$, and $[\sqrt{5}, \infty]$; decreasing on $[-\sqrt{5}, -1]$ and $[1, \sqrt{5}]$; concave upward on $(-\sqrt{3}, 0)$ and $(\sqrt{3}, \infty)$, and concave downward on $(-\infty, -\sqrt{3})$ and $(0, \sqrt{3})$; inflection points are $(-\sqrt{3}, -4\sqrt{3})$, $(0, 0)$, and $(\sqrt{3}, 4\sqrt{3})$; symmetric with respect to the origin

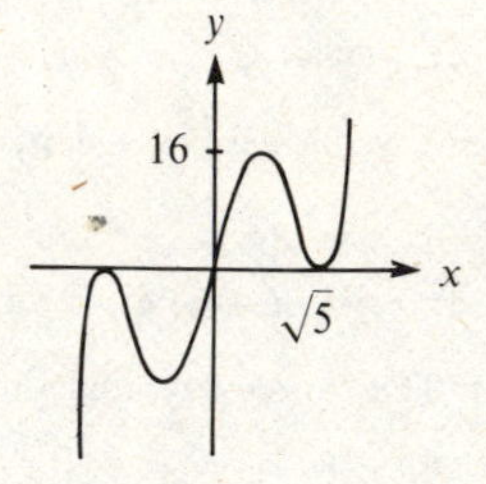

11. The graph of f and the line $y = \frac{5}{2}$ intersect at (x, y) if $x + 1/x = \frac{5}{2}$, or $2x^2 - 5x + 2 = 0$, or $(2x-1)(x-2) = 0$, or $x = \frac{1}{2}$ or $x = 2$. Since $x + 1/x \le \frac{5}{2}$ for $1 \le x \le 2$ and $x + 1/x \ge \frac{5}{2}$ for $2 \le x \le 3$, the area A is given by

$$A = \int_1^2 \left[\frac{5}{2} - \left(x + \frac{1}{x}\right)\right] dx + \int_2^3 \left[\left(x + \frac{1}{x}\right) - \frac{5}{2}\right] dx$$
$$= \left(\tfrac{5}{2}x - \tfrac{1}{2}x^2 - \ln x\right)\Big|_1^2 + \left(\tfrac{1}{2}x^2 + \ln x - \tfrac{5}{2}x\right)\Big|_2^3$$
$$= (1 - \ln 2) + (\ln 3 - \ln 2) = 1 + \ln \tfrac{3}{4}$$

13. $$\frac{x^3}{x^2-x+1} = x + 1 - \frac{1}{x^2-x+1} = x + 1 - \frac{1}{(x-\frac{1}{2})^2 + \frac{3}{4}}$$

Thus $$\int \frac{x^3}{x^2-x+1}\,dx = \int \left[x + 1 - \frac{1}{(x-\frac{1}{2})^2+\frac{3}{4}}\right] dx$$
$$= \frac{1}{2}x^2 + x - \frac{2}{\sqrt{3}}\arctan\left[\frac{2}{\sqrt{3}}\left(x - \frac{1}{2}\right)\right] + C$$
$$= \frac{1}{2}x^2 + x - \frac{2}{\sqrt{3}}\arctan\frac{2x-1}{\sqrt{3}} + C$$

15. Let $u = -3x^2$, so that $du = -6x\,dx$. If $x = 2$, then $u = -12$, and if $x = b$, then $u = -3b^2$. Thus

$$\int_2^\infty xe^{-x^2}\,dx = \lim_{b\to\infty}\int_2^b xe^{-x^2}\,dx = \lim_{b\to\infty}\int_{-12}^{-3b^2} e^u\left(-\tfrac{1}{6}\right)du = \lim_{b\to\infty}\left(-\tfrac{1}{6}e^u\right)\Big|_{-12}^{-3b^2}$$
$$= \lim_{b\to\infty} \tfrac{1}{6}\left(e^{-12} - e^{-3b^2}\right) = \tfrac{1}{6}e^{-12}$$

Therefore the integral converges and its value is $\frac{1}{6}e^{-12}$.

17. $\mathscr{L} = \int_{-1}^{1} \sqrt{(-e^{-t}\sin 2t + 2e^{-t}\cos 2t)^2 + (-e^{-t}\cos 2t - 2e^{-t}\sin 2t)^2}\, dt$

$$= \int_{-1}^{1} e^{-t}\sqrt{\sin^2 2t - 4\sin 2t\cos 2t + 4\cos^2 2t + \cos^2 2t + 4\cos 2t\sin 2t + 4\sin^2 2t}\, dt$$

$$= \int_{-1}^{1} e^{-t}\sqrt{5}\, dt = -\sqrt{5}\, e^{-t}\Big|_{-1}^{1} = \sqrt{5}(e - e^{-1})$$

19. For any c,

$$\lim_{n\to\infty} (\sqrt{n+c} - \sqrt{n}) = \lim_{n\to\infty} \frac{(\sqrt{n+c} - \sqrt{n})(\sqrt{n+c} + \sqrt{n})}{\sqrt{n+c} + \sqrt{n}} = \lim_{n\to\infty} \frac{n+c-n}{\sqrt{n+c} + \sqrt{n}}$$

$$= \lim_{n\to\infty} \frac{c}{\sqrt{n+c} + \sqrt{n}} = 0$$

Thus the limit exists for all c.

21. Since $$\lim_{n\to\infty} \frac{n^2}{\sqrt{n^4+2}} = \lim_{n\to\infty} \frac{1}{\sqrt{1 + 2/n^4}} = 1$$

$\lim_{n\to\infty} (-1)^n n^2/\sqrt{n^4+2}$ does not exist. By Theorem 9.9(b) the given series diverges.

23. By Theorem 9.28 and the uniqueness of Taylor series, the Taylor series for $\int_0^x f(t)\, dt$ is

$$\sum_{n=1}^{\infty} \int_0^x \frac{t^n}{n+2}\, dt = \sum_{n=1}^{\infty} \frac{x^{n+1}}{(n+1)(n+2)}$$

For $x \neq 0$, $$\lim_{n\to\infty} \frac{|x^{n+2}|/[(n+2)(n+3)]}{|x^{n+1}|/[(n+1)(n+2)]} = \lim_{n\to\infty} \frac{n+1}{n+3}|x| = |x|$$

By the Generalized Ratio Test, the series $\sum_{n=1}^{\infty} x^{n+1}/[(n+1)(n+2)]$ converges for $|x| < 1$ and diverges for $|x| > 1$. For $x = 1$ the series becomes $\sum_{n=1}^{\infty} 1/[(n+1)(n+2)]$, which converges by comparison with the p series $\sum_{n=1}^{\infty} 1/n^2$. For $x = -1$ the series becomes $\sum_{n=1}^{\infty} (-1)^n\, 1/[(n+1)(n+2)]$, which converges by the Alternating Series Test. Consequently the interval of convergence of $\sum_{n=1}^{\infty} x^{n+1}/[(n+1)(n+2)]$ is $[-1, 1]$.

13

VECTOR-VALUED FUNCTIONS

SECTION 13.1

1. domain: $(-\infty, \infty)$; $f_1(t) = t$, $f_2(t) = t^2$, $f_3(t) = t^3$

3. $\tanh t$ is defined for all t; domain: $(-\infty, -2)$, and $(-2, 2)$ and $(2, \infty)$; $f_1(t) = \tanh t$, $f_2(t) = 0$, $f_3(t) = -1/(t^2 - 4)$

5. $\mathbf{F}(t) = 2\sqrt{t}\,\mathbf{i} - 2t^{3/2}\mathbf{j} - (t^3 + 1)\mathbf{k}$. Thus the domain is $[0, \infty)$. Also $f_1(t) = 2\sqrt{t}$, $f_2(t) = -2t^{3/2}$, and $f_3(t) = -(t^3 + 1)$.

7. $(2\mathbf{F} - 3\mathbf{G})(t) = (2t - 3\cos t)\mathbf{i} + (2t^2 - 3\sin t)\mathbf{j} + (2t^3 - 3)\mathbf{k}$. Thus the domain is $(-\infty, \infty)$. Also $f_1(t) = 2t - 3\cos t$, $f_2(t) = 2t^2 - 3\sin t$, and $f_3(t) = 2t^3 - 3$.

9. $$(\mathbf{F} \times \mathbf{G})(t) = \frac{1}{\sqrt{t}}(1 - \cos t)\mathbf{i} - \frac{1}{\sqrt{t}}(t - \sin t)\mathbf{j} - t(t - \sin t)\mathbf{k}$$

Thus the domain is $(0, \infty)$. Also $f_1(t) = (1/\sqrt{t})(1 - \cos t)$, $f_2(t) = -(1/\sqrt{t})(t - \sin t)$, and $f_3(t) = -t(t - \sin t)$.

11. $(\mathbf{F} \circ g)(t) = \cos t^{1/3}\mathbf{i} + \sin t^{1/3}\mathbf{j} + \sqrt{t^{1/3} + 2}\,\mathbf{k}$. Thus the domain consists of all t for which $t^{1/3} + 2 \geq 0$, and hence is $[-8, \infty)$. Also $f_1(t) = \cos t^{1/3}$, $f_2(t) = \sin t^{1/3}$, and $f_3(t) = \sqrt{t^{1/3} + 2}$.

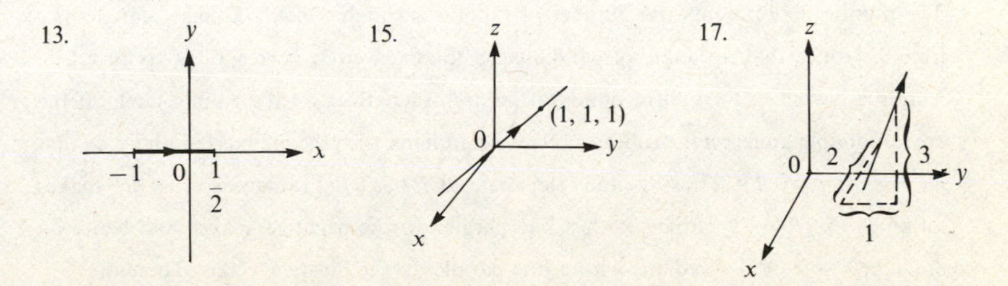

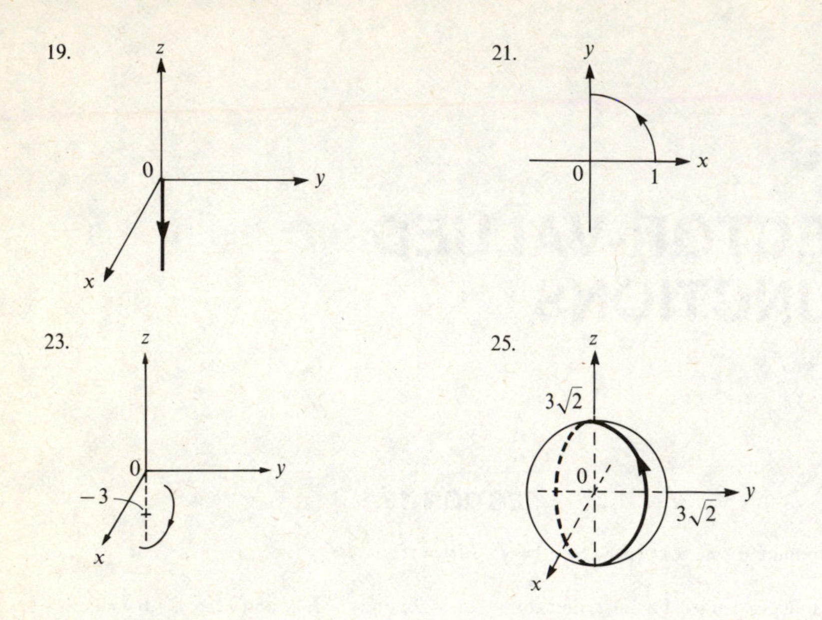

27. a. $(\mathbf{F}-\mathbf{G})(t)=(t-1)\mathbf{i}+(-2t+3)\mathbf{j}-4t\mathbf{k}$, so parametric equations are $x=t-1$, $y=-2t+3$, and $z=-4t$.

b. $(\mathbf{F}+3\mathbf{G})(t)=(5t+3)\mathbf{i}+(10t-5)\mathbf{j}$, so parametric equations are $x=5t+3$, $y=10t-5$, and $z=0$.

c. $(\mathbf{F}\circ g)(t)=2\cos t\mathbf{i}+(\cos t+1)\mathbf{j}-3\cos t\mathbf{k}$. Then $x=2\cos t$, $y=\cos t+1$, and $z=-3\cos t$ for all real t, which yields the line segment parameterized by $x=2t$, $y=t+1$, $z=-3t$ for $-1\le t\le 1$.

29. If (x, y, z) is on the curve, then $x=\cos \pi t$, $y=\sin \pi t$, and $z=t$. If in addition (x, y, z) is on the sphere $x^2+y^2+z^2=10$, then $(\cos \pi t)^2+(\sin \pi t)^2+t^2=10$, so that $t^2=9$, and thus $t=\pm 3$. If $t=3$, then $(x, y, z)=(\cos 3\pi, \sin 3\pi, 3)=(-1, 0, 3)$, and if $t=-3$, then $(x, y, z)=(\cos(-3\pi), \sin(-3\pi), -3)=(-1, 0, -3)$. The points of intersection are therefore $(-1, 0, 3)$ and $(-1, 0, -3)$.

31. $\mathbf{F}(t)$ can be written $\mathbf{a}+\mathbf{b}$, where $\mathbf{a}$ describes the motion of the center C of the circle and $\mathbf{b}$ describes the motion of P around the center. First of all, $\mathbf{a}=(r+b)\cos t\mathbf{i}+(r+b)\sin t\mathbf{j}$. The number t represents the number of radians through which OC has rotated since time 0. Notice that an angle of t radians on the fixed circle corresponds to an arc of length rt, which in turn corresponds to an arc of length rt on the rolling circle. If the corresponding angle on the rolling circle has α radians, then the arc has length $b\alpha$, so that $b\alpha=rt$, or $\alpha=(r/b)t$. Thus by time t the angle OCP has $(r/b)t$ radians, so that CP makes an angle of $(r/b)t+t$ radians with a line parallel to the negative x axis, and hence an angle of $\pi+(r/b)t+t$ radians with a line parallel to the positive x axis. Therefore

$$\mathbf{b}=CP=b\cos\left(\pi+\frac{r}{b}t+t\right)\mathbf{i}+b\sin\left(\pi+\frac{r}{b}t+t\right)\mathbf{j}$$
$$=-b\cos\left(\frac{r}{b}t+t\right)\mathbf{i}-b\sin\left(\frac{r}{b}t+t\right)\mathbf{j}$$

Consequently

$$\mathbf{F}(t)=\mathbf{a}+\mathbf{b}=[(r+b)\cos t\mathbf{i}+(r+b)\sin t\mathbf{j}]$$
$$+\left[-b\cos\left(\frac{r}{b}t+t\right)\mathbf{i}-b\sin\left(\frac{r}{b}t+t\right)\mathbf{j}\right]$$
$$=\left[(r+b)\cos t-b\cos\left(\frac{r+b}{b}t\right)\right]\mathbf{i}+\left[(r+b)\sin t-b\sin\left(\frac{r+b}{b}t\right)\right]\mathbf{j}$$

SECTION 13.2

1. $\lim_{t\to 4}(\mathbf{i}-\mathbf{j}+\mathbf{k})=\mathbf{i}-\mathbf{j}+\mathbf{k}$

3. $$\lim_{t\to\pi}(\tan t\mathbf{i}+3t\mathbf{j}-4\mathbf{k})=\left(\lim_{t\to\pi}\tan t\right)\mathbf{i}+\left(\lim_{t\to\pi}3t\right)\mathbf{j}+\left(\lim_{t\to\pi}(-4)\right)\mathbf{k}$$
$$=0\mathbf{i}+3\pi\mathbf{j}-4\mathbf{k}=3\pi\mathbf{j}-4\mathbf{k}$$

5. $$\lim_{t\to 2^-}\mathbf{F}(t)=\lim_{t\to 2^-}(5\mathbf{i}-\sqrt{2t^2+2t+4}\,\mathbf{j}+e^{-(t-2)}\mathbf{k})$$
$$=\left(\lim_{t\to 2^-}5\right)\mathbf{i}+\left(\lim_{t\to 2^-}-\sqrt{2t^2+2t+4}\right)\mathbf{j}+\left(\lim_{t\to 2^-}e^{-(t-2)}\right)\mathbf{k}$$
$$=5\mathbf{i}-4\mathbf{j}+\mathbf{k}$$

whereas

$$\lim_{t\to 2^+}\mathbf{F}(t)=\lim_{t\to 2^+}(t^2+1)\mathbf{i}+(4-t^3)\mathbf{j}+\mathbf{k}$$
$$=\left(\lim_{t\to 2^+}(t^2+1)\right)\mathbf{i}+\left(\lim_{t\to 2^+}(4-t^3)\right)\mathbf{j}+\left(\lim_{t\to 2^+}1\right)\mathbf{k}$$
$$=5\mathbf{i}-4\mathbf{j}+\mathbf{k}$$

Since $\lim_{t\to 2^-}\mathbf{F}(t)=\lim_{t\to 2^+}\mathbf{F}(t)=5\mathbf{i}-4\mathbf{j}+\mathbf{k}$, it follows that $\lim_{t\to 2}\mathbf{F}(t)$ exists and that $\lim_{t\to 2}\mathbf{F}(t)=5\mathbf{i}-4\mathbf{j}+\mathbf{k}$.

7. $$\lim_{t\to 0}(\mathbf{F}-\mathbf{G})(t)=\lim_{t\to 0}\left[(e^{-1/t^2}+\pi)\mathbf{i}+\left(\cos t-\frac{1+\cos t}{t}\right)\mathbf{j}+t^3\mathbf{k}\right]$$

Since $\lim_{t\to 0}(1+\cos t)/t$ and hence $\lim_{t\to 0}(\cos t-(1+\cos t)/t)$ does not exist, Theorem 13.4 implies that $\lim_{t\to 0}(\mathbf{F}-\mathbf{G})(t)$ does not exist.

9. $$\lim_{t\to 3}\left(\frac{t^2-5t+6}{t-3}\mathbf{i}+\frac{t^2-2t-3}{t-3}\mathbf{j}+\frac{t^2+4t-21}{t-3}\mathbf{k}\right)$$
$$=\lim_{t\to 3}[(t-2)\mathbf{i}+(t+1)\mathbf{j}+(t+7)\mathbf{k}]$$
$$=\left(\lim_{t\to 3}(t-2)\right)\mathbf{i}+\left(\lim_{t\to 3}(t+1)\right)\mathbf{j}+\left(\lim_{t\to 3}(t+7)\right)\mathbf{k}=\mathbf{i}+4\mathbf{j}+10\mathbf{k}$$

11. $(-\infty, \infty)$

13. Since $\sqrt{2t}$ is continuous on $[0, \infty)$, so is $\mathbf{F}$.

15. $\lim_{t\to 0^-} \mathbf{F}(t) = \lim_{t\to 0^-} \left(\frac{\sin t}{t}\mathbf{i} + \mathbf{j} + \mathbf{k}\right) = \mathbf{i} + \mathbf{j} + \mathbf{k}$

and $\lim_{t\to 0^+} \mathbf{F}(t) = \lim_{t\to 0^+} [(t^2+1)\mathbf{i} + \ln(t+3)\mathbf{j} + \mathbf{k}]$

$$= \left(\lim_{t\to 0^+}(t^2+1)\right)\mathbf{i} + \left(\lim_{t\to 0^+}\ln(t+3)\right)\mathbf{j} + \left(\lim_{t\to 0^+} 1\right)\mathbf{k}$$

$$= \mathbf{i} + \ln 3\mathbf{j} + \mathbf{k}$$

Thus $\mathbf{F}$ is not continuous at 0. However, $[(\sin t)/t]\mathbf{i} + \mathbf{j} + \mathbf{k}$ is continuous on $(-\infty, 0)$ and $(t^2+1)\mathbf{i} + \ln(t+3)\mathbf{j} + \mathbf{k}$ is continuous on $[0, \infty)$. Thus $\mathbf{F}$ is continuous on $(-\infty, 0)$ and $[0, \infty)$.

17. a. Since $\lim_{t\to 0^+} \frac{-\ln t}{t}$ does not exist, neither does the given limit.

b. $\lim_{t\to 1^+} [e^{1/(1-t)}\mathbf{i} + \sqrt{t-1}\,\mathbf{j} + \ln t\mathbf{k}] = \left(\lim_{t\to 1^+} e^{1/(1-t)}\right)\mathbf{i} + \left(\lim_{t\to 1^+}\sqrt{t-1}\right)\mathbf{j} + \left(\lim_{t\to 1^+}\ln t\right)\mathbf{k}$

$$= 0\mathbf{i} + 0\mathbf{j} + 0\mathbf{k} = \mathbf{0}$$

c. $\lim_{t\to 1^-} [\sqrt{1-t}\,\mathbf{i} - (1-t)\ln(1-t)\mathbf{j}] = \left(\lim_{t\to 1^-}\sqrt{1-t}\right)\mathbf{i} + \left(\lim_{t\to 1^-} -(1-t)\ln(1-t)\right)\mathbf{j}$

Since

$$\lim_{t\to 1^-}(1-t)\ln(1-t) = \lim_{u\to 0^+} u\ln u = \lim_{u\to 0^+}\frac{\ln u}{1/u} = \lim_{u\to 0^+}\frac{1/u}{-1/u^2} = \lim_{u\to 0^+}(-u) = 0$$

by l'Hôpital's Rule, we find that $(\lim_{t\to 1^-}\sqrt{1-t})\mathbf{i} + (\lim_{t\to 1^-} -(1-t)\ln(1-t))\mathbf{j} = 0\mathbf{i} + 0\mathbf{j} = \mathbf{0}$. Thus $(\lim_{t\to 1^-}(\sqrt{1-t}\,\mathbf{i} - (1-t)\ln(1-t)\mathbf{j})) = \mathbf{0}$.

19. a. By Theorem 13.5(a), $\lim_{t\to t_0}(\mathbf{F}+\mathbf{G})(t) = \lim_{t\to t_0}\mathbf{F}(t) + \lim_{t\to t_0}\mathbf{G}(t) = \mathbf{F}(t_0) + \mathbf{G}(t_0) = (\mathbf{F}+\mathbf{G})(t_0)$, so that $\mathbf{F}+\mathbf{G}$ is continuous at t_0.

b. By Theorem 13.5(c) with $f(t) = c$ for all t, $\lim_{t\to t_0}(c\mathbf{F})(t) = \lim_{t\to t_0} c \cdot \lim_{t\to t_0}\mathbf{F}(t) = c(\mathbf{F}(t_0)) = (c\mathbf{F})(t_0)$, so that $c\mathbf{F}$ is continuous at t_0.

c. Let f_1, f_2, and f_3 be the component functions of $\mathbf{F}$. Then

$$\|\mathbf{F}(t)\| = \sqrt{(f_1(t))^2 + (f_2(t))^2 + (f_3(t))^2}$$

Since f_1, f_2, and f_3 are continuous at t_0 by Theorem 13.7, and since the square root function is continuous on $(0, \infty)$, it follows that $\|\mathbf{F}\|$ is continuous at t_0.

d. By Theorem 13.5(d), $\lim_{t\to t_0}(\mathbf{F}\cdot\mathbf{G})(t) = \lim_{t\to t_0}\mathbf{F}(t)\cdot\lim_{t\to t_0}\mathbf{G}(t) = F(t_0)\cdot G(t_0) = (\mathbf{F}\cdot\mathbf{G})(t_0)$, so that $\mathbf{F}\cdot\mathbf{G}$ is continuous at t_0.

e. By Theorem 13.5(e), $\lim_{t\to t_0}(\mathbf{F}\times\mathbf{G})(t) = \lim_{t\to t_0}\mathbf{F}(t)\times\lim_{t\to t_0}\mathbf{G}(t) = \mathbf{F}(t_0)\times\mathbf{G}(t_0) = (\mathbf{F}\times\mathbf{G})(t_0)$, so that $\mathbf{F}\times\mathbf{G}$ is continuous at t_0.

SECTION 13.3

1. $\mathbf{F}'(t) = \mathbf{j} + 5t^4\mathbf{k}$

3. $\mathbf{F}'(t) = \frac{3}{2}(1+t)^{1/2}\mathbf{i} + \frac{3}{2}(1-t)^{1/2}\mathbf{j} + \frac{3}{2}\mathbf{k}$

5. $\mathbf{F}'(t) = \sec^2 t\mathbf{i} + \sec t\tan t\mathbf{k}$

7. $F'(t) = \sinh t\mathbf{i} + \cosh t\mathbf{j} - \frac{1}{2\sqrt{t}}\mathbf{k}$

9. $(4\mathbf{F} - 2\mathbf{G})(t) = (8\sec t - 6t)\mathbf{i} + (2t^2 - 12)\mathbf{j} + 12\csc t\mathbf{k}$, so that $(4\mathbf{F} - 2\mathbf{G})'(t) = (8\sec t\tan t - 6)\mathbf{i} + 4t\mathbf{j} - 12\csc t\cot t\mathbf{k}$

11. $(\mathbf{F}\times\mathbf{G})(t) = -3\ln t\mathbf{i} - 2\sec t\ln t\mathbf{j}$; $(\mathbf{F}\times\mathbf{G})'(t) = \frac{-3}{t}\mathbf{i} - \left(2\sec t\tan t\ln t + \frac{2}{t}\sec t\right)\mathbf{j}$

13. $(\mathbf{F}\times\mathbf{G})(t) = \left(t - \frac{3}{t}e^{-t}\right)\mathbf{k}$; $(\mathbf{F}\times\mathbf{G})'(t) = \left(1 + \frac{3}{t^2}e^{-t} + \frac{3}{t}e^{-t}\right)\mathbf{k}$

15. $(\mathbf{F}\circ g)'(t) = \mathbf{F}'(g(t))g'(t) = \left(\frac{1}{\sqrt{t}}\mathbf{i} - 8e^{2\sqrt{t}}\mathbf{j} + \frac{1}{(\sqrt{t})^2}\mathbf{k}\right)\frac{1}{2\sqrt{t}} = \frac{1}{2t}\mathbf{i} - \frac{4e^{2\sqrt{t}}}{\sqrt{t}}\mathbf{j} + \frac{1}{2}t^{-3/2}\mathbf{k}$

17. $\int\left(t^2\mathbf{i} - (3t-1)\mathbf{j} - \frac{1}{t^3}\mathbf{k}\right)dt = \frac{t^3}{3}\mathbf{i} - \left(\frac{3}{2}t^2 - t\right)\mathbf{j} + \frac{1}{2t^2}\mathbf{k} + \mathbf{C}$

19. $\int_0^1 (e^t\mathbf{i} + e^{-t}\mathbf{j} + 2t\mathbf{k})\,dt = e^t\Big|_0^1\mathbf{i} - e^{-t}\Big|_0^1\mathbf{j} + t^2\Big|_0^1\mathbf{k} = (e-1)\mathbf{i} + (1-e^{-1})\mathbf{j} + \mathbf{k}$

21. $\int_{-1}^1 [(1+t)^{3/2}\mathbf{i} + (1-t)^{3/2}\mathbf{j}]\,dt = \frac{2}{5}(1+t)^{5/2}\Big|_{-1}^1\mathbf{i} - \frac{2}{5}(1-t)^{5/2}\Big|_{-1}^1\mathbf{j} = \frac{8}{5}\sqrt{2}(\mathbf{i}+\mathbf{j})$

23. $\mathbf{v}(t) = -\sin t\mathbf{i} + \cos t\mathbf{j} - 32t\mathbf{k}$,
$\|\mathbf{v}(t)\| = \sqrt{(-\sin t)^2 + \cos^2 t + (-32t)^2} = \sqrt{1 + 1024t^2}$, $\mathbf{a}(t) = -\cos t\mathbf{i} - \sin t\mathbf{j} - 32\mathbf{k}$

25. $\mathbf{v}(t) = 2\mathbf{i} + 2t\mathbf{j} + \frac{1}{t}\mathbf{k}$, $\|\mathbf{v}(t)\| = \sqrt{4 + 4t^2 + \frac{1}{t^2}} = \frac{2t^2+1}{t}$, $\mathbf{a}(t) = 2\mathbf{j} - \frac{1}{t^2}\mathbf{k}$

27. $\mathbf{v}(t) = (e^t\sin t + e^t\cos t)\mathbf{i} + (e^t\cos t - e^t\sin t)\mathbf{j} + e^t\mathbf{k}$,
$\|\mathbf{v}(t)\| = \sqrt{(e^{2t}\sin^2 t + 2e^{2t}\sin t\cos t + e^{2t}\cos^2 t) + [e^{2t}\cos^2 t - 2e^{2t}\sin t\cos t + e^{2t}\sin^2 t] + e^{2t}} = e^t\sqrt{3}$,
$\mathbf{a}(t) = 2e^t\cos t\mathbf{i} - 2e^t\sin t\mathbf{j} + e^t\mathbf{k}$

29. $\mathbf{v}(t) = \int\mathbf{a}(t)\,dt = \int -32\mathbf{k}\,dt = -32t\mathbf{k} + \mathbf{C}$;
$\mathbf{r}(t) = \int\mathbf{v}(t)\,dt = \int(-32t\mathbf{k} + \mathbf{C})\,dt = -16t^2\mathbf{k} + \mathbf{C}t + \mathbf{C}_1$.
Since $\mathbf{v}(0) = \mathbf{v}_0 = \mathbf{i} + \mathbf{j}$, we have $\mathbf{C} = \mathbf{i} + \mathbf{j}$, so that $\mathbf{v}(t) = \mathbf{i} + \mathbf{j} - 32t\mathbf{k}$. Since $\mathbf{r}(0) = \mathbf{r}_0 = \mathbf{0}$, we have $\mathbf{C}_1 = \mathbf{0}$, so that $\mathbf{r}(t) = t\mathbf{i} + t\mathbf{j} - 16t^2\mathbf{k}$. Finally, $\|\mathbf{v}(t)\| = \sqrt{1 + 1 + (-32t)^2} = \sqrt{2 + 1024t^2}$.

31. $\mathbf{v}(t) = \int\mathbf{a}(t)\,dt = \int(-\cos t\mathbf{i} - \sin t\mathbf{j})\,dt = -\sin t\mathbf{i} + \cos t\mathbf{j} + \mathbf{C}$;
$\mathbf{r}(t) = \int\mathbf{v}(t)\,dt = \int(-\sin t\mathbf{i} + \cos t\mathbf{j} + \mathbf{C})\,dt = \cos t\mathbf{i} + \sin t\mathbf{j} + \mathbf{C}t + \mathbf{C}_1$.
Since $\mathbf{v}(0) = \mathbf{v}_0 = \mathbf{k}$, we have $\mathbf{k} = -\sin 0\mathbf{i} + \cos 0\mathbf{j} + \mathbf{C} = \mathbf{j} + \mathbf{C}$, so that $\mathbf{C} = \mathbf{k} - \mathbf{j}$, and thus $\mathbf{v}(t) = -\sin t\mathbf{i} + (\cos t - 1)\mathbf{j} + \mathbf{k}$. Since $\mathbf{r}(0) = \mathbf{r}_0 = \mathbf{i}$, we have $\mathbf{i} = \cos 0\mathbf{i} + \sin 0\mathbf{j} + (\mathbf{k} - \mathbf{j})0 + \mathbf{C}_1 = \mathbf{i} + \mathbf{C}_1$, so that $\mathbf{C}_1 = \mathbf{0}$, and thus $\mathbf{r}(t) = \cos t\mathbf{i} + \sin t\mathbf{j} + t(\mathbf{k} - \mathbf{j}) = \cos t\mathbf{i} + (\sin t - t)\mathbf{j} + t\mathbf{k}$. Finally, $\|\mathbf{v}(t)\| = \sqrt{\sin^2 t + (\cos t - 1)^2 + 1} = \sqrt{3 - 2\cos t}$.

33. $\mathbf{F}(t) = \left(\int_0^t u \tan u^3\, du\right)\mathbf{i} + \left(\int_0^t \cos e^u\, du\right)\mathbf{j} + \left(\int_0^t e^{(u^2)}\, du\right)\mathbf{k}$

and thus $\mathbf{F}'(t) = t \tan t^3\mathbf{i} + \cos e^t\mathbf{j} + e^{(t^2)}\mathbf{k}$.

35. $\|\mathbf{F}(t)\| = \sqrt{\dfrac{16t^2}{(1+4t^2)^2} + \dfrac{1-8t^2+16t^4}{(1+4t^2)^2}} = \sqrt{\dfrac{(1+4t^2)^2}{(1+4t^2)^2}} = 1$

so that by Corollary 13.11, $\mathbf{F}(t)\cdot\mathbf{F}'(t) = 0$ for all t.

37. $\mathbf{F}'(t) = \cos t\mathbf{i} + \sin t\mathbf{j}$ and $\mathbf{F}''(t) = -\sin t\mathbf{i} + \cos t\mathbf{j} = -\mathbf{F}(t)$. Since $\|\mathbf{F}(t)\| = \|\mathbf{F}''(t)\| = 1$, it follows that $\mathbf{F}(t)$ and $\mathbf{F}''(t)$ are parallel but have opposite, and hence dissimilar, directions, for all t.

39. By Theorem 13.10(e), $\dfrac{d}{dt}(\mathbf{F}\times\mathbf{F}')(t) = [\mathbf{F}'(t)\times\mathbf{F}'(t)] + [\mathbf{F}(t)\times\mathbf{F}''(t)] = \mathbf{F}(t)\times\mathbf{F}''(t)$.

41. By using Theorem 13.10(d) and then Theorem 13.10(e), we find that

$$\begin{aligned}\frac{d}{dt}[\mathbf{F}\cdot(\mathbf{G}\times\mathbf{H})] &= \frac{d\mathbf{F}}{dt}\cdot(\mathbf{G}\times\mathbf{H}) + \mathbf{F}\cdot\frac{d}{dt}(\mathbf{G}\times\mathbf{H})\\ &= \frac{d\mathbf{F}}{dt}\cdot(\mathbf{G}\times\mathbf{H}) + \mathbf{F}\cdot\left[\left(\frac{d\mathbf{G}}{dt}\times\mathbf{H}\right) + \left(\mathbf{G}\times\frac{d\mathbf{H}}{dt}\right)\right]\\ &= \frac{d\mathbf{F}}{dt}\cdot(\mathbf{G}\times\mathbf{H}) + \mathbf{F}\cdot\left(\frac{d\mathbf{G}}{dt}\times\mathbf{H}\right) + \mathbf{F}\cdot\left(\mathbf{G}\times\frac{d\mathbf{H}}{dt}\right)\end{aligned}$$

43. a. Initial position: $\mathbf{r}_0 = \mathbf{0}$; velocity: $\mathbf{v}(t) = 90\sqrt{2}\mathbf{i} + 90\sqrt{2}\mathbf{j} + (64 - 32t)\mathbf{k}$; initial velocity: $\mathbf{v}_0 = 90\sqrt{2}\mathbf{i} + 90\sqrt{2}\mathbf{j} + 64\mathbf{k}$

b. $\mathbf{r}(4) = 360\sqrt{2}\mathbf{i} + 360\sqrt{2}\mathbf{j} + 0\mathbf{k}$, so the height is 0 when $t = 4$; distance from initial position: $\|\mathbf{r}(4) - \mathbf{r}(0)\| = \|360\sqrt{2}\mathbf{i} + 360\sqrt{2}\mathbf{j}\| = 720$ (feet).

45. a. We choose a coordinate system so that the base of the container lies on the xy plane, with the origin as in the figure. By Example 10, the position of a water droplet t seconds after it leaves the container is given by $\mathbf{r}(t) = -\frac{1}{2}gt^2\mathbf{k} + t\mathbf{v}_0 + \mathbf{r}_0$. By assumption, $\mathbf{v}_0 = \sqrt{2gh}\mathbf{j}$ and $\mathbf{r}_0 = (H - h)\mathbf{k}$. Thus $\mathbf{r}(t) = -\frac{1}{2}gt^2\mathbf{k} + \sqrt{2gh}\,t\mathbf{j} + (H-h)\mathbf{k} = \sqrt{2gh}\,t\mathbf{j} + [(H-h) - \frac{1}{2}gt^2]\mathbf{k}$ until the droplet hits the floor.

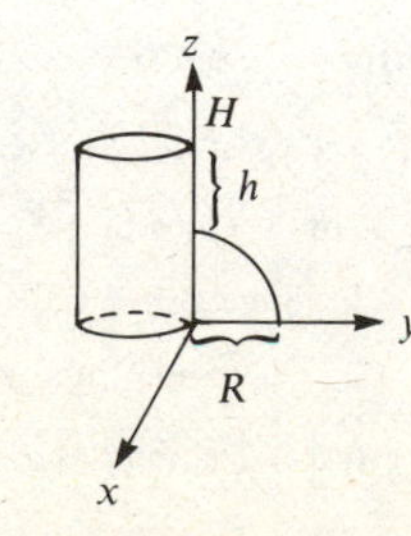

b. When a given droplet hits the floor, the $\mathbf{k}$ component of $\mathbf{r}$ is 0, so $[(H-h) - \frac{1}{2}gt^2] = 0$, and thus $t = \sqrt{2(H-h)/g}$. Then R, which is the $\mathbf{j}$ component of $\mathbf{r}$ at that instant, is given by $R = \sqrt{2gh}\sqrt{2(H-h)/g} = 2\sqrt{h(H-h)}$. Now R is maximized for the same value of h as R_1, where $R_1 = R^2$. Since $R_1(h) = 4h(H-h)$ and $R_1'(h) = 4H - 8h$, we find that $R_1'(h) = 0$ if $h = \frac{1}{2}H$. Since $R_1''(h) = -8 < 0$, it follows that R_1, and hence R, is maximized if $h = H/2$.

c. If $h = H/2$, then $R = 2\sqrt{(H/2)(H - H/2)} = 2(H/2) = H$.

47. Taking the radius of the earth to be 3960 miles, so the radius of the satellite's orbit is $3960 + 500 = 4460$ (miles), we use the comments following (6) to deduce the speed v_0 of the satellite:

$$v_0 = \sqrt{\frac{C}{r_0}} = \sqrt{\frac{32(3960)^2(5280)^2}{(4460)(5280)}} \approx 24{,}373.6 \text{ (feet per second)}$$

or approximately 16,618.4 miles per hour.

49. The bobsled moves $60(\frac{5280}{3600})$ feet per second, so by (4),

$$\|\mathbf{a}(t)\| = \frac{v_0^2}{r} = \frac{[60(\frac{5280}{3600})]^2}{100} = 77.44 \text{ (feet per second per second)}$$

SECTION 13.4

1. $\mathbf{r}'(t) = \mathbf{i} + 2t\mathbf{j} + 3t^2\mathbf{k}$; $\mathbf{r}'$ is continuous and is never $\mathbf{0}$, so $\mathbf{r}$ is smooth.

3. $\mathbf{r}'(t) = \begin{cases} -\mathbf{i}+\mathbf{j}+\mathbf{k} & \text{for } t < 0\\ \mathbf{i}+\mathbf{j}+\mathbf{k} & \text{for } t > 0\end{cases}$

so $\mathbf{r}'$ is continuous and nonzero on $(-\infty, 0)$ and on $(0, \infty)$. Since the one-sided derivatives at 0 are $-\mathbf{i}+\mathbf{j}+\mathbf{k}$ and $\mathbf{i}+\mathbf{j}+\mathbf{k}$ respectively, we conclude that $\mathbf{r}$ is piecewise smooth but not smooth.

5. $\mathbf{r}'(t) = \frac{3}{2}(1+t)^{1/2}\mathbf{i} - \frac{3}{2}(1-t)^{1/2}\mathbf{j} + \frac{3}{2}\mathbf{k}$ for $-1 < t < 1$, and the one-sided derivatives exist and make $\mathbf{r}'$ continuous and nonzero on $[-1, 1]$. Thus $\mathbf{r}$ is smooth.

7. $\mathbf{r}'(t) = -2\cos t\sin t\mathbf{i} + 2\sin t\cos t\mathbf{j} + 2t\mathbf{k}$; $\mathbf{r}$ is continuous and $\mathbf{r}'(t) = \mathbf{0}$ only if $t = 0$. Thus $\mathbf{r}$ is piecewise smooth.

9. $\mathbf{r}'(t) = (e^t - 1)\mathbf{i} + 2t\mathbf{j} + 3t^2\mathbf{k}$; $\mathbf{r}'$ is continuous and $\mathbf{r}'(t) = \mathbf{0}$ only if $t = 0$. Thus $\mathbf{r}$ is piecewise smooth.

11. The line segment is parallel to $7\mathbf{i} - 2\mathbf{j} + 4\mathbf{k}$ and starts at $(-3, 2, 1)$. Thus a parametrization is $\mathbf{r}(t) = (-3 + 7t)\mathbf{i} + (2 - 2t)\mathbf{j} + (1 + 4t)\mathbf{k}$ for $0 \le t \le 1$, and $\mathbf{r}$ is smooth.

13. $\mathbf{r}(t) = 6\cos t\mathbf{i} + 6\sin t\mathbf{j}$ for $0 \le t \le 2\pi$ is one such smooth parametrization.

15. $\mathbf{r}(t) = \cos t\mathbf{i} + \sin t\mathbf{j}$ for $0 \le t \le \pi$ is one such smooth parametrization.

17. Notice that the quarter circle in the xy plane that extends from $(\sqrt{2}/2, \sqrt{2}/2)$ to $(\sqrt{2}/2, -\sqrt{2}/2)$ corresponds to the quarter circle from $(1, \pi/4)$ to $(1, -\pi/4)$ in polar coordinates. This leads us to $\mathbf{r}(t) = \cos t\mathbf{i} - \sin t\mathbf{j} + 4\mathbf{k}$ for $-\pi/4 \le t \le \pi/4$, which is smooth.

19. $\mathbf{r}(t) = t\mathbf{i} + \tan t\mathbf{j}$ for $0 \le t \le \pi/4$ is one such smooth parametrization.

21. $\mathscr{L} = \int_0^{2\pi} \sqrt{[3\cos^2 t(-\sin t)]^2 + [3\sin^2 t\cos t]^2}\, dt = \int_0^{2\pi} 3|\sin t\cos t|\, dt$
$= 12\int_0^{\pi/2} \sin t\cos t\, dt = 6\sin^2 t\Big|_0^{\pi/2} = 6$

23. $\mathscr{L} = \int_{-1}^1 \sqrt{\tfrac{1}{4}(1+t) + \tfrac{1}{4}(1-t) + \tfrac{1}{4}}\, dt = \int_{-1}^1 \tfrac{1}{2}\sqrt{3}\, dt = \tfrac{1}{2}\sqrt{3}t\Big|_{-1}^1 = \sqrt{3}$

25. $\mathscr{L} = \int_0^1 \sqrt{e^{2t} + e^{-2t} + 2}\, dt = \int_0^1 \sqrt{(e^t + e^{-t})^2}\, dt = \int_0^1 (e^t + e^{-t})\, dt$
$= (e^t - e^{-t})\Big|_0^1 = e - e^{-1}$

27. $\mathscr{L} = \int_1^{\sqrt{8}} \sqrt{[36t^2(t^2-1)] + 36t^2 + 36t^2}\, dt = \int_1^{\sqrt{8}} 6\sqrt{t^4 + t^2}\, dt$
$= \int_1^{\sqrt{8}} 6t\sqrt{t^2+1}\, dt = 2(t^2+1)^{3/2}\Big|_1^{\sqrt{8}} = 54 - 4\sqrt{2}$

29. $\dfrac{ds}{dt} = \sqrt{4\cos^2 2t + 4\sin^2 2t + t} = \sqrt{4+t}$

31. $\dfrac{ds}{dt} = \sqrt{(\cos t - t\sin t)^2 + (\sin t + t\cos t)^2 + 1}$
$= \sqrt{\cos^2 t + \sin^2 t + t^2(\sin^2 t + \cos^2 t) + 1} = \sqrt{2 + t^2}$

33. $\dfrac{ds}{dt} = \sqrt{(1-\cos t)^2 + \sin^2 t + 1} = \sqrt{3 - 2\cos t}$

35. $\mathscr{L} = \int_0^\pi \sqrt{(a\sin\theta)^2 + (a\cos\theta)^2}\, d\theta = a\int_0^\pi 1\, d\theta = \pi a$

37. $\mathscr{L} = \int_{-1}^1 \sqrt{\cos^2 t + \sin^2 t + t^4}\, dt = \int_{-1}^1 \sqrt{1+t^4}\, dt$
$\approx \tfrac{2}{12}\left[\sqrt{2} + 4\sqrt{\tfrac{17}{16}} + 2\cdot 1 + 4\sqrt{\tfrac{17}{16}} + \sqrt{2}\right] \approx 2.17911$

39. $\mathscr{L} = \int_{-1}^1 \sqrt{1 + 4t^2 + 9t^4}\, dt \approx \tfrac{2}{12}\left[\sqrt{14} + 4\sqrt{\tfrac{41}{16}} + 2\cdot 1 + 4\sqrt{\tfrac{41}{16}} + \sqrt{14}\right] \approx 3.71493$

41. $\|\mathbf{r}(t)\| = \sqrt{\dfrac{(1-t^2)^2}{(1+t^2)^2} + \dfrac{4t^2}{(1+t^2)^2}} = 1$ for all t

so $\mathbf{r}$ parametrizes a portion of the circle $x^2 + y^2 = 1$. Let $f_1(t)$ denote the $\mathbf{i}$ component of $\mathbf{r}$. Since $f_1(0) = 1$ and $\lim_{t\to-\infty} f_1(t) = \lim_{t\to\infty} f_1(t) = -1$, we conclude from the Intermediate Value Theorem that f_1 takes all values in the interval $(-1, 1]$. Since the $\mathbf{j}$ component of $\mathbf{r}$ is positive for $t > 0$ and negative for $t < 0$, it follows that $\mathbf{r}$ parametrizes all the points on the circle between $(1, 0)$ and $(-1, 0)$. Finally, $(-1, 0)$ is not included in the parametrization since $(1-t^2)/(1+t^2) \ne -1$ for all t and hence $\mathbf{r}(t) \ne -\mathbf{i}$ for all t.

43. a. Assume that the radius of the circle is $r > 0$. Then $n = 8$ and $d = 2r$. Thus $\pi nd/8 = [\pi(8)(2r)]/8 = 2\pi r =$ circumference $\mathscr{L}$ of the circle.

b. $n = 18$, $d = \frac{1}{3}$, so $\mathscr{L} \approx [\pi(18)(\frac{1}{3})]/8 = 3\pi/4 \approx 2.35619$.

SECTION 13.5

1. $\mathbf{r}'(t) = 2t\mathbf{i} + 2\mathbf{j}$; $\|\mathbf{r}'(t)\| = \sqrt{(2t)^2 + 2^2} = 2\sqrt{t^2+1}$;

$$\mathbf{T}(t) = \frac{\mathbf{r}'(t)}{\|\mathbf{r}'(t)\|} = \frac{2t\mathbf{i} + 2\mathbf{j}}{2\sqrt{t^2+1}} = \frac{t}{\sqrt{t^2+1}}\mathbf{i} + \frac{1}{\sqrt{t^2+1}}\mathbf{j};$$

$$\mathbf{T}'(t) = \frac{\sqrt{t^2+1} - t^2/\sqrt{t^2+1}}{t^2+1}\mathbf{i} - \frac{t}{(t^2+1)^{3/2}}\mathbf{j} = \frac{1}{(t^2+1)^{3/2}}\mathbf{i} - \frac{t}{(t^2+1)^{3/2}}\mathbf{j};$$

$$\|\mathbf{T}'(t)\| = \sqrt{\frac{1}{(t^2+1)^3} + \frac{t^2}{(t^2+1)^3}} = \frac{1}{t^2+1};\quad \mathbf{N}(t) = \frac{\mathbf{T}'(t)}{\|\mathbf{T}'(t)\|} = \frac{1}{\sqrt{t^2+1}}\mathbf{i} - \frac{t}{\sqrt{t^2+1}}\mathbf{j}$$

3. $\mathbf{r}'(t) = -\sin t\mathbf{i} - \sin t\mathbf{j} + \sqrt{2}\cos t\mathbf{k}$;
$\|\mathbf{r}'(t)\| = \sqrt{(-\sin t)^2 + (-\sin t)^2 + (\sqrt{2}\cos t)^2} = \sqrt{2}$;

$$\mathbf{T}(t) = \frac{\mathbf{r}'(t)}{\|\mathbf{r}'(t)\|} = -\frac{\sin t}{\sqrt{2}}\mathbf{i} - \frac{\sin t}{\sqrt{2}}\mathbf{j} + \cos t\mathbf{k};\quad \mathbf{T}'(t) = -\frac{\cos t}{\sqrt{2}}\mathbf{i} - \frac{\cos t}{\sqrt{2}}\mathbf{j} - \sin t\mathbf{k};$$

$$\|\mathbf{T}'(t)\| = \sqrt{\left(\frac{-\cos t}{\sqrt{2}}\right)^2 + \left(\frac{-\cos t}{\sqrt{2}}\right)^2 + (-\sin t)^2} = 1;$$

$$\mathbf{N}(t) = \frac{\mathbf{T}'(t)}{\|\mathbf{T}'(t)\|} = -\frac{\cos t}{\sqrt{2}}\mathbf{i} - \frac{\cos t}{\sqrt{2}}\mathbf{j} - \sin t\mathbf{k}$$

5. $\mathbf{r}'(t) = 2\mathbf{i} + 2t\mathbf{j} + t^2\mathbf{k}$; $\|\mathbf{r}'(t)\| = \sqrt{2^2 + (2t)^2 + (t^2)^2} = \sqrt{4 + 4t^2 + t^4} = 2 + t^2$;

$$\mathbf{T}(t) = \frac{\mathbf{r}'(t)}{\|\mathbf{r}'(t)\|} = \frac{2}{2+t^2}\mathbf{i} + \frac{2t}{2+t^2}\mathbf{j} + \frac{t^2}{2+t^2}\mathbf{k};$$

$$\mathbf{T}'(t) = \frac{-4t}{(2+t^2)^2}\mathbf{i} + \frac{2(2+t^2) - 4t^2}{(2+t^2)^2}\mathbf{j} + \frac{2t(2+t^2) - 2t^3}{(2+t^2)^2}\mathbf{k}$$
$$= \frac{-4t}{(2+t^2)^2}\mathbf{i} + \frac{4 - 2t^2}{(2+t^2)^2}\mathbf{j} + \frac{4t}{(2+t^2)^2}\mathbf{k};$$

$$\|\mathbf{T}'(t)\| = \frac{1}{(2+t^2)^2}\sqrt{(-4t)^2 + (4-2t^2)^2 + (4t)^2} = \frac{1}{(2+t^2)^2}\sqrt{16 + 16t^2 + 4t^4} = \frac{2}{2+t^2};$$

$$\mathbf{N}(t) = \frac{\mathbf{T}'(t)}{\|\mathbf{T}'(t)\|} = \frac{-2t}{2+t^2}\mathbf{i} + \frac{2-t^2}{2+t^2}\mathbf{j} + \frac{2t}{2+t^2}\mathbf{k}$$

7. $\mathbf{r}'(t) = e^t\mathbf{i} - e^{-t}\mathbf{j} + \sqrt{2}\mathbf{k}$; $\|\mathbf{r}'(t)\| = \sqrt{e^{2t} + e^{-2t} + 2} = e^t + e^{-t}$;

$$\mathbf{T}(t) = \frac{\mathbf{r}'(t)}{\|\mathbf{r}'(t)\|} = \frac{e^t}{e^t + e^{-t}}\mathbf{i} - \frac{e^{-t}}{e^t + e^{-t}}\mathbf{j} + \frac{\sqrt{2}}{e^t + e^{-t}}\mathbf{k};$$

$$\mathbf{T}'(t) = \frac{e^t(e^t + e^{-t}) - e^t(e^t - e^{-t})}{(e^t + e^{-t})^2}\mathbf{i} - \frac{-e^{-t}(e^t + e^{-t}) - e^{-t}(e^t - e^{-t})}{(e^t + e^{-t})^2}\mathbf{j} - \frac{\sqrt{2}(e^t - e^{-t})}{(e^t + e^{-t})^2}\mathbf{k}$$
$$= \frac{2}{(e^t + e^{-t})^2}\mathbf{i} + \frac{2}{(e^t + e^{-t})^2}\mathbf{j} - \frac{\sqrt{2}(e^t - e^{-t})}{(e^t + e^{-t})^2}\mathbf{k};$$

$$\|\mathbf{T}'(t)\| = \frac{1}{(e^t + e^{-t})^2}\sqrt{2^2 + 2^2 + [\sqrt{2}(e^{-t} - e^t)]^2} = \frac{\sqrt{2e^{-2t} + 4 + 2e^{2t}}}{(e^t + e^{-t})^2} = \frac{\sqrt{2}}{e^t + e^{-t}};$$

$$\mathbf{N}(t) = \frac{\mathbf{T}'(t)}{\|\mathbf{T}'(t)\|} = \frac{\sqrt{2}}{e^t + e^{-t}}\mathbf{i} + \frac{\sqrt{2}}{e^t + e^{-t}}\mathbf{j} - \frac{e^t - e^{-t}}{e^t + e^{-t}}\mathbf{k}$$

9. $\mathbf{r}'(t) = 2\mathbf{i} + 2t\mathbf{j} + \frac{1}{t}\mathbf{k}$; $\|\mathbf{r}'(t)\| = \sqrt{2^2 + (2t)^2 + \left(\frac{1}{t}\right)^2} = \sqrt{4t^2 + 4 + \frac{1}{t^2}} = 2t + \frac{1}{t} = \frac{2t^2+1}{t}$;

$$\mathbf{T}(t) = \frac{\mathbf{r}'(t)}{\|\mathbf{r}'(t)\|} = \frac{2t}{2t^2+1}\mathbf{i} + \frac{2t^2}{2t^2+1}\mathbf{j} + \frac{1}{2t^2+1}\mathbf{k};$$

$$\mathbf{T}'(t) = \frac{2(2t^2+1) - 2t(4t)}{(2t^2+1)^2}\mathbf{i} + \frac{4t(2t^2+1) - 2t^2(4t)}{(2t^2+1)^2}\mathbf{j} - \frac{4t}{(2t^2+1)^2}\mathbf{k}$$

$$= \frac{2-4t^2}{(2t^2+1)^2}\mathbf{i} + \frac{4t}{(2t^2+1)^2}\mathbf{j} - \frac{4t}{(2t^2+1)^2}\mathbf{k};$$

$$\|\mathbf{T}'(t)\| = \frac{1}{(2t^2+1)^2}\sqrt{(2-4t^2)^2 + (4t)^2 + (-4t)^2} = \frac{\sqrt{16t^4 + 16t^2 + 4}}{(2t^2+1)^2} = \frac{2}{2t^2+1};$$

$$\mathbf{N}(t) = \frac{\mathbf{T}'(t)}{\|\mathbf{T}'(t)\|} = \frac{1-2t^2}{2t^2+1}\mathbf{i} + \frac{2t}{2t^2+1}\mathbf{j} - \frac{2t}{2t^2+1}\mathbf{k}$$

11. $\mathbf{v} = r(1 - \cos t)\mathbf{i} + r \sin t\mathbf{j}$; $\|\mathbf{v}\| = \sqrt{r^2(1-\cos t)^2 + r^2 \sin^2 t} = \sqrt{2r^2(1 - \cos t)}$;

$$a_{\mathbf{T}} = \frac{d\|\mathbf{v}\|}{dt} = \frac{r^2 \sin t}{\sqrt{2r^2(1-\cos t)}}; \quad \mathbf{a} = \frac{d\mathbf{v}}{dt} = r \sin t\mathbf{i} + r \cos t\mathbf{j};$$

$$\|\mathbf{a}\|^2 = r^2 \sin^2 t + r^2 \cos^2 t = r^2;$$

$$a_{\mathbf{N}} = \sqrt{\|\mathbf{a}\|^2 - a_{\mathbf{T}}^2} = \sqrt{r^2 - \frac{r^4 \sin^2 t}{2r^2(1-\cos t)}} = \sqrt{r^2 - \frac{r^2(1-\cos^2 t)}{2(1-\cos t)}} = \frac{|r|}{\sqrt{2}}\sqrt{1-\cos t}$$

13. $\mathbf{v} = 2\mathbf{i} + 2t\mathbf{j} + t^2\mathbf{k}$; $\|\mathbf{v}\| = \sqrt{4 + 4t^2 + t^4} = 2 + t^2$; $a_{\mathbf{T}} = \frac{d\|\mathbf{v}\|}{dt} = 2t$;

$\mathbf{a} = 2\mathbf{j} + 2t\mathbf{k}$; $\|\mathbf{a}\|^2 = 4(1+t^2)$; $a_{\mathbf{N}} = \sqrt{\|\mathbf{a}\|^2 - a_{\mathbf{T}}^2} = \sqrt{4(1+t^2) - 4t^2} = 2$

15. $\mathbf{v} = e^t\mathbf{i} - e^{-t}\mathbf{j} + \sqrt{2}\mathbf{k}$; $\|\mathbf{v}\| = \sqrt{e^{2t} + e^{-2t} + 2} = e^t + e^{-t}$; $a_{\mathbf{T}} = \frac{d\|\mathbf{v}\|}{dt} = e^t - e^{-t}$;

$\mathbf{a} = e^t\mathbf{i} + e^{-t}\mathbf{j}$; $\|\mathbf{a}\|^2 = e^{2t} + e^{-2t}$; $a_{\mathbf{N}} = \sqrt{\|\mathbf{a}\|^2 - a_{\mathbf{T}}^2} = \sqrt{(e^{2t} + e^{-2t}) - (e^t - e^{-t})^2} = \sqrt{2}$

17. $\mathbf{r}(t) = -4 \sin t\mathbf{j} + 4 \cos t\mathbf{k}$ for $0 \le t \le \pi$

19. $$\mathbf{r}(t) = \begin{cases} t\mathbf{i} + t^2\mathbf{j} & \text{for } 0 \le t \le 1 \\ (2-t)\mathbf{i} + (2-t)\mathbf{j} & \text{for } 1 \le t \le 2 \end{cases}$$

21. $$\mathbf{r}(t) = \begin{cases} (t+1)\mathbf{i} & \text{for } 0 \le t \le 1 \\ 2\cos\frac{\pi}{2}(t-1)\mathbf{i} + 2\sin\frac{\pi}{2}(t-1)\mathbf{j} & \text{for } 1 \le t \le 2 \\ (4-t)\mathbf{j} & \text{for } 2 \le t \le 3 \\ \sin\frac{\pi}{2}(t-3)\mathbf{i} + \cos\frac{\pi}{2}(t-3)\mathbf{j} & \text{for } 3 \le t \le 4 \end{cases}$$

23. Since $\mathbf{r}_1(1) = \mathbf{i} + 2\mathbf{j} + \mathbf{k} = \mathbf{r}_2(-1)$, the two curves intersect at $(1, 2, 1)$. Since $\mathbf{r}_1'(t) = \mathbf{i} + 2\mathbf{j} + 2t\mathbf{k}$ and $\mathbf{r}_2'(t) = 2t\mathbf{i} - \mathbf{j} - 2t\mathbf{k}$, we have $\mathbf{r}_1'(1) = \mathbf{i} + 2\mathbf{j} + 2\mathbf{k}$ and $\mathbf{r}_2'(-1) = -2\mathbf{i} - \mathbf{j} + 2\mathbf{k}$. Thus $\mathbf{T}_1(1) = \frac{1}{3}(\mathbf{i} + 2\mathbf{j} + 2\mathbf{k})$ and $\mathbf{T}_2(-1) = \frac{1}{3}(-2\mathbf{i} - \mathbf{j} + 2\mathbf{k})$, so that $\mathbf{T}_1(1) \cdot \mathbf{T}_2(-1) = \frac{1}{9}(-2 - 2 + 4) = 0$. Thus the two tangent vectors are perpendicular.

25. Since $a_{\mathbf{T}} = d\|\mathbf{v}\|/dt$, it follows from Theorem 4.7 that $a_{\mathbf{T}} = 0$ only if $\|\mathbf{v}\|$ is constant.

27. Let $\mathbf{r}(t) = t\mathbf{i} + \sin t\mathbf{j}$. Then $d\mathbf{r}/dt = \mathbf{i} + \cos t\mathbf{j}$ and $\|d\mathbf{r}/dt\| = \sqrt{1 + \cos^2 t}$. Therefore

$$\mathbf{T}(t) = \frac{1}{\sqrt{1+\cos^2 t}}\mathbf{i} + \frac{\cos t}{\sqrt{1+\cos^2 t}}\mathbf{j}$$

Thus $$\mathbf{T}'(t) = \frac{\sin t \cos t}{(1+\cos^2 t)^{3/2}}\mathbf{i} - \frac{\sin t}{(1+\cos^2 t)^{3/2}}\mathbf{j}$$

But $\mathbf{T}'(t) = \mathbf{0}$ if $t = n\pi$, so $\mathbf{N}(n\pi)$ fails to exist, for all integers n.

29. Using the results of Example 5, we have

$$\mathbf{B} = \mathbf{T} \times \mathbf{N} = \begin{vmatrix} \mathbf{i} & \mathbf{j} & \mathbf{k} \\ -\frac{2}{\sqrt{13}}\sin t & \frac{2}{\sqrt{13}}\cos t & \frac{3}{\sqrt{13}} \\ -\cos t & -\sin t & 0 \end{vmatrix} = \frac{3}{\sqrt{13}}\sin t\mathbf{i} - \frac{3}{\sqrt{13}}\cos t\mathbf{j} + \frac{2}{\sqrt{13}}\mathbf{k}$$

SECTION 13.6

1. From Exercise 1 in Section 13.5, $\kappa(t) = \frac{\|\mathbf{T}'(t)\|}{\|\mathbf{r}'(t)\|} = \frac{1/(t^2+1)}{2\sqrt{t^2+1}} = \frac{1}{2(t^2+1)^{3/2}}$.

3. From Exercise 3 in Section 13.5, $\kappa(t) = \frac{\|\mathbf{T}'(t)\|}{\|\mathbf{r}'(t)\|} = \frac{1}{\sqrt{2}}$.

5. From Exercise 5 in Section 13.5, $\kappa(t) = \frac{\|\mathbf{T}'(t)\|}{\|\mathbf{r}'(t)\|} = \frac{2/(2+t^2)}{2+t^2} = \frac{2}{(2+t^2)^2}$.

7. From Exercise 7 in Section 13.5, $\kappa(t) = \frac{\|\mathbf{T}'(t)\|}{\|\mathbf{r}'(t)\|} = \frac{\sqrt{2}/(e^t + e^{-t})}{e^t + e^{-t}} = \frac{\sqrt{2}}{(e^t + e^{-t})^2}$.

9. From Exercise 9 in Section 13.5, $\kappa(t) = \frac{\|\mathbf{T}'(t)\|}{\|\mathbf{r}'(t)\|} = \frac{2/(2t^2+1)}{(2t^2+1)/t} = \frac{2t}{(2t^2+1)^2}$.

11. $\mathbf{v} = \frac{d\mathbf{r}}{dt} = 2\mathbf{i} + 2t\mathbf{j}$; $\|\mathbf{v}\| = \sqrt{2^2 + (2t)^2} = 2\sqrt{1+t^2}$; $\mathbf{a} = \frac{d\mathbf{v}}{dt} = 2\mathbf{j}$;

$$\mathbf{v} \times \mathbf{a} = \begin{vmatrix} \mathbf{i} & \mathbf{j} & \mathbf{k} \\ 2 & 2t & 0 \\ 0 & 2 & 0 \end{vmatrix} = 4\mathbf{k}; \quad \|\mathbf{v} \times \mathbf{a}\| = 4; \quad \kappa = \frac{\|\mathbf{v} \times \mathbf{a}\|}{\|\mathbf{v}\|^3} = \frac{4}{(2\sqrt{1+t^2})^3} = \frac{1}{2(1+t^2)^{3/2}}$$

13. $\mathbf{v} = \frac{d\mathbf{r}}{dt} = e^t(\sin t + \cos t)\mathbf{i} + e^t(\cos t - \sin t)\mathbf{j} + \mathbf{k}$;

$$\|\mathbf{v}\| = \sqrt{e^{2t}(\sin t + \cos t)^2 + e^{2t}(\cos t - \sin t)^2 + 1}$$
$$= \sqrt{e^{2t}(\sin^2 t + 2\sin t\cos t + \cos^2 t + \cos^2 t - 2\cos t \sin t + \sin^2 t) + 1}$$
$$= \sqrt{2e^{2t} + 1};$$

$$\mathbf{a} = \frac{d\mathbf{v}}{dt} = [e^t(\sin t + \cos t) + e^t(\cos t - \sin t)]\mathbf{i} + [e^t(\cos t - \sin t) + e^t(-\sin t - \cos t)]\mathbf{j}$$
$$= 2e^t \cos t\mathbf{i} - 2e^t \sin t\mathbf{j};$$

$$\mathbf{v}\times\mathbf{a} = \begin{vmatrix} \mathbf{i} & \mathbf{j} & \mathbf{k} \\ e^t(\sin t + \cos t) & e^t(\cos t - \sin t) & 1 \\ 2e^t\cos t & -2e^t \sin t & 0 \end{vmatrix} = 2e^t \sin t\mathbf{i} + 2e^t\cos t\mathbf{j} - 2e^{2t}\mathbf{k};$$

$$\|\mathbf{v}\times\mathbf{a}\| = 2e^t\sqrt{\sin^2 t + \cos^2 t + (-e^t)^2} = 2e^t\sqrt{1 + e^{2t}};$$

$$\kappa = \frac{\|\mathbf{v}\times\mathbf{a}\|}{\|\mathbf{v}\|^3} = \frac{2e^t(1 + e^{2t})^{1/2}}{(2e^{2t} + 1)^{3/2}}$$

15. $\mathbf{v} = \dfrac{d\mathbf{r}}{dt} = \cos t\mathbf{i} - \sin t\mathbf{j} + t^{1/2}\mathbf{k}$; $\|\mathbf{v}\| = \sqrt{\cos^2 t + \sin^2 t + t} = \sqrt{1 + t}$;

$$\mathbf{a} = \frac{d\mathbf{v}}{dt} = -\sin t\mathbf{i} - \cos t\mathbf{j} + \frac{1}{2}t^{-1/2}\mathbf{k};$$

$$\mathbf{v}\times\mathbf{a} = \begin{vmatrix} \mathbf{i} & \mathbf{j} & \mathbf{k} \\ \cos t & -\sin t & t^{1/2} \\ -\sin t & -\cos t & \frac{1}{2}t^{-1/2} \end{vmatrix}$$

$$= (-\tfrac{1}{2}t^{-1/2}\sin t + t^{1/2}\cos t)\mathbf{i} + (-t^{1/2}\sin t - \tfrac{1}{2}t^{-1/2}\cos t)\mathbf{j} - \mathbf{k};$$

$$\|\mathbf{v}\times\mathbf{a}\| = \sqrt{(-\tfrac{1}{2}t^{-1/2}\sin t + t^{1/2}\cos t)^2 + (-t^{1/2}\sin t - \tfrac{1}{2}t^{-1/2}\cos t)^2 + 1}$$

$$= \sqrt{\tfrac{1}{4}t^{-1}\sin^2 t - \sin t\cos t + t\cos^2 t + t\sin^2 t + \sin t \cos t + \tfrac{1}{4}t^{-1}\cos^2 t + 1}$$

$$= \sqrt{\tfrac{1}{4}t^{-1} + 1 + t} = \sqrt{\frac{1 + 4t + 4t^2}{4t}} = \frac{2t + 1}{2\sqrt{t}};$$

$$\kappa = \frac{\|\mathbf{v}\times\mathbf{a}\|}{\|\mathbf{v}\|^3} = \frac{(2t + 1)/(2\sqrt{t})}{(\sqrt{1 + t})^3} = \frac{2t + 1}{2\sqrt{t}(1 + t)^{3/2}}$$

17. By the solution of Exercise 16, $\kappa = 6/(4 + 5\cos^2 t)^{3/2}$; for $t_0 = \pi/2$, $\kappa = 6/4^{3/2} = \frac{3}{4}$, so that $\rho = 1/\kappa = \frac{4}{3}$.

19. $x = t$, $\dfrac{dx}{dt} = 1$, $\dfrac{d^2x}{dt^2} = 0$; $y = \dfrac{1}{3}t^3$, $\dfrac{dy}{dt} = t^2$, $\dfrac{d^2y}{dt^2} = 2t$;

$$\kappa = \frac{\left|\dfrac{dx}{dt}\dfrac{d^2y}{dt^2} - \dfrac{d^2x}{dt^2}\dfrac{dy}{dt}\right|}{\left[\left(\dfrac{dx}{dt}\right)^2 + \left(\dfrac{dy}{dt}\right)^2\right]^{3/2}} = \frac{2|t|}{(1 + t^4)^{3/2}};\ \text{for } t_0 = 1,\ \kappa = \frac{2}{2^{3/2}} = \frac{1}{\sqrt{2}},\ \text{so that } \rho = \frac{1}{\kappa} = \sqrt{2}.$$

21. $\dfrac{dy}{dx} = \dfrac{1}{x}$, $\dfrac{d^2y}{dx^2} = -\dfrac{1}{x^2}$; $\kappa = \dfrac{|d^2y/dx^2|}{[1 + (dy/dx)^2]^{3/2}} = \dfrac{1/x^2}{(1 + 1/x^2)^{3/2}} = \dfrac{x}{(1 + x^2)^{3/2}}$

23. $\dfrac{dy}{dx} = -\dfrac{1}{x^2}$, $\dfrac{d^2y}{dx^2} = \dfrac{2}{x^3}$; $\kappa = \dfrac{|d^2y/dx^2|}{[1 + (dy/dx)^2]^{3/2}} = \dfrac{-2/x^3}{(1 + 1/x^4)^{3/2}} = \dfrac{-2x^3}{(1 + x^4)^{3/2}}$

25. $\dfrac{dy}{dx} = e^x = \dfrac{d^2y}{dx^2}$; $\kappa = \dfrac{|d^2y/dx^2|}{[1 + (dy/dx)^2]^{3/2}} = \dfrac{e^x}{(1 + e^{2x})^{3/2}}$;

$$\frac{d\kappa}{dx} = \frac{e^x(1 + e^{2x})^{3/2} - e^x(\frac{3}{2})(1 + e^{2x})^{1/2}(2e^{2x})}{(1 + e^{2x})^3} = \frac{e^x - 2e^{3x}}{(1 + e^{2x})^{5/2}} = \frac{e^x(1 - 2e^{2x})}{(1 + e^{2x})^{5/2}}$$

$d\kappa/dx = 0$ if $1 - 2e^{2x} = 0$, or $x = -\frac{1}{2}\ln 2$. Since $d\kappa/dx > 0$ for $x < -\frac{1}{2}\ln 2$, and $d\kappa/dx < 0$ for $x > -\frac{1}{2}\ln 2$, κ is maximum at $(-\frac{1}{2}\ln 2, \sqrt{2}/2)$.

27. $\mathbf{v} = -\sin t\mathbf{i} + \cos t\mathbf{j} + \mathbf{k}$; $\|\mathbf{v}\| = \sqrt{\sin^2 t + \cos^2 t + 1} = \sqrt{2}$;

$$\mathbf{a} = -\cos t\mathbf{i} - \sin t\mathbf{j};\ \mathbf{v}\times\mathbf{a} = \begin{vmatrix} \mathbf{i} & \mathbf{j} & \mathbf{k} \\ -\sin t & \cos t & 1 \\ -\cos t & -\sin t & 0 \end{vmatrix} = \sin t\mathbf{i} - \cos t\mathbf{j} + \mathbf{k};$$

$$\|\mathbf{v}\times\mathbf{a}\| = \sqrt{\sin^2 t + \cos^2 t + 1} = \sqrt{2};\ \kappa = \frac{\|\mathbf{v}\times\mathbf{a}\|}{\|\mathbf{v}\|^3} = \frac{\sqrt{2}}{(\sqrt{2})^3} = \frac{1}{2}$$

29. a. By Definition 13.19, $\|d\mathbf{T}/dt\| = \kappa\|d\mathbf{r}/dt\| = \kappa\|\mathbf{v}\|$, so that by (5) of Section 13.5, $a_{\mathbf{N}} = \|\mathbf{v}\|\,\|d\mathbf{T}/dt\| = \kappa\|\mathbf{v}\|^2$.

b. Since the graph of the sine function has an inflection point at $(\pi, 0)$, Exercise 28 implies that $\kappa = 0$ at $(\pi, 0)$. Thus by part (a), $a_{\mathbf{N}} = 0$ at $(\pi, 0)$.

31. By Exercise 30,

$$\kappa(\theta) = \frac{|2(3\cos 3\theta)^2 - \sin 3\theta(-9\sin 3\theta) + \sin^2 3\theta|}{[(3\cos 3\theta)^2 + \sin^2 3\theta]^{3/2}}$$

$$= \frac{18\cos^2 3\theta + 10\sin^2 3\theta}{(9\cos^2 3\theta + \sin^2 3\theta)^{3/2}} = \frac{8\cos^2 3\theta + 10}{(8\cos^2 3\theta + 1)^{3/2}}$$

33. a. $\theta'(t) = \kappa(t)$; $\mathbf{v} = \cos\theta(t)\mathbf{i} + \sin\theta(t)\mathbf{j}$; $\|\mathbf{v}\| = \sqrt{\cos^2\theta(t) + \sin^2\theta(t)} = 1$;

$\mathbf{a} = (-\sin\theta(t))\theta'(t)\mathbf{i} + (\cos\theta(t))\theta'(t)\mathbf{j} = -\kappa(t)\sin\theta(t)\mathbf{i} + \kappa(t)\cos\theta(t)\mathbf{j}$;

$$\mathbf{v}\times\mathbf{a} = \begin{vmatrix} \mathbf{i} & \mathbf{j} & \mathbf{k} \\ \cos\theta(t) & \sin\theta(t) & 0 \\ -\kappa(t)\sin\theta(t) & \kappa(t)\cos\theta(t) & 0 \end{vmatrix} = \kappa(t)\mathbf{k};$$

$\|\mathbf{v}\times\mathbf{a}\| = \kappa(t)$. Thus the curvature is $\|\mathbf{v}\times\mathbf{a}\|/\|\mathbf{v}\|^3 = \kappa(t)$.

b. Taking $a = 0$ in part (a), we find that $\theta(t) = \int_0^t (1/\sqrt{1 - u^2})\,du = \arcsin t$, so that

$$\int_0^t \cos\theta(u)\,du = \int_0^t \cos(\arcsin u)\,du = \int_0^t \sqrt{1 - u^2}\,du$$

$$\overset{u = \sin w}{=} \int_0^{\arcsin t} \sqrt{1 - \sin^2 w}\cos w\,dw = \int_0^{\arcsin t} \cos^2 w\,dw$$

$$= \int_0^{\arcsin t} (\tfrac{1}{2} + \tfrac{1}{2}\cos 2w)\,dw = (\tfrac{1}{2}w + \tfrac{1}{4}\sin 2w)\Big|_0^{\arcsin t}$$

$$= \tfrac{1}{2}\arcsin t + \tfrac{1}{2}\sin(\arcsin t)\cos(\arcsin t)$$

$$= \frac{1}{2}\arcsin t + \frac{t}{2}\sqrt{1 - t^2}$$

$$\int_0^t \sin\theta(u)\,du = \int_0^t \sin(\arcsin u)\,du = \int_0^t u\,du = \tfrac{1}{2}t^2$$

Thus the desired parameterization is

$$\mathbf{r}(t) = \left(\frac{1}{2}\arcsin t + \frac{t}{2}\sqrt{1 - t^2}\right)\mathbf{i} + \frac{1}{2}t^2\mathbf{j} \quad \text{for } -1 < t < 1$$

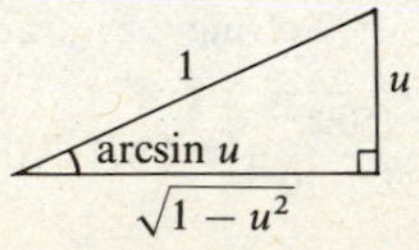

1
u
$\arcsin u$
$\sqrt{1 - u^2}$

c. Taking $a = 0$ in part (a), we find that $\theta(t) = \int_0^t (1/(1+u^2))\,du = \arctan t$, so that

$$\int_0^t \cos\theta(u)\,du = \int_0^t \cos(\arctan u)\,du = \int_0^t \frac{1}{\sqrt{1+u^2}}\,du$$

$$\overset{u=\tan w}{=} \int_0^{\arctan t} \frac{1}{\sqrt{1+\tan^2 w}}\sec^2 w\,dw = \int_0^{\arctan t} \sec w\,dw$$

$$= \ln|\sec w + \tan w|\Big|_0^{\arctan t} = \ln|\sec(\arctan t) + \tan(\arctan t)|$$

$$= \ln(t + \sqrt{1+t^2});$$

$$\int_0^t \sin\theta(u)\,du = \int_0^t \sin(\arctan u)\,du = \int_0^t \frac{u}{\sqrt{1+u^2}}\,du = \sqrt{1+u^2}\Big|_0^t = \sqrt{1+t^2} - 1$$

Thus the desired parameterization is $\mathbf{r}(t) = \ln(t + \sqrt{1+t^2})\mathbf{i} + (\sqrt{1+t^2} - 1)\mathbf{j}$

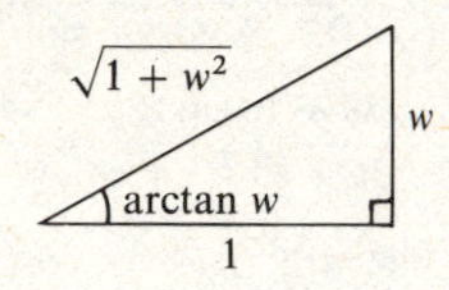

35. $a_{\mathbf{T}} = d\|\mathbf{v}\|/dt = (0-81)/(9-0) = -9$. Since $\|\mathbf{v}(0)\| = 81$ and $d\|\mathbf{v}\|/dt = -9$, we have $\|\mathbf{v}\| = 81 - 9t$. Since the radius of the circle is 729, it follows that $\kappa = 1/\rho = \frac{1}{729}$, so that by Exercise 29(a), $a_{\mathbf{N}} = \kappa\|\mathbf{v}\|^2 = \frac{1}{729}(81-9t)^2 = (9-t)^2/9$. Thus $\|\mathbf{a}\| = \sqrt{a_{\mathbf{T}}^2 + a_{\mathbf{N}}^2} = \sqrt{81 + (9-t)^4/81}$.

SECTION 13.7

1. a. Since $\mathbf{u}$ is a unit vector, $\mathbf{u}$ and $d\mathbf{u}/dt$ are perpendicular by Corollary 12.11. Thus by (6),

$$p = \|p\mathbf{k}\| = \left\|r^2\left(\mathbf{u} \times \frac{d\mathbf{u}}{dt}\right)\right\| = r^2\|\mathbf{u}\|\left\|\frac{d\mathbf{u}}{dt}\right\|\sin\frac{\pi}{2} = r^2\left\|\frac{d\mathbf{u}}{dt}\right\|$$

b. Since $dr/dt = 0$ when r is minimum, (5) implies that $d\mathbf{r}/dt = r(d\mathbf{u}/dt)$ when r is minimum.

c. Since r_0 is the minimum value of r, and v_0 is the corresponding speed, it follows from (a) and (b) that

$$p = r_0^2\left\|\frac{d\mathbf{u}}{dt}\right\| = r_0\left\|r_0\frac{d\mathbf{u}}{dt}\right\| = r_0\left\|\frac{d\mathbf{r}}{dt}\right\| = r_0 v_0$$

3. If the orbit is circular, then r is constant, so that $r = r_0$. From (13) and Exercise 1(c) we have $r_0^2 = p^4/G^2M^2 = r_0^4 v_0^4/G^2M^2$. Solving for v_0, we find that $v_0 = \sqrt{GM/r_0}$.

5. By (23), $T = \sqrt{\dfrac{4\pi^2a^3}{GM}} = \sqrt{\dfrac{4\pi^2(5000)^3}{1.237 \times 10^{12}}} \approx 1.99733$ (hours).

7. By (22),

$$c = \sqrt{a^2 - b^2} = \sqrt{\frac{P^4G^2M^2}{(G^2M^2 - w^2)^2} - \frac{p^4}{G^2M^2 - w^2}} = \sqrt{\frac{G^2M^2}{G^2M^2 - w^2} - 1}\,\frac{p^2}{\sqrt{G^2M^2 - w^2}}$$

$$= \frac{w}{\sqrt{G^2M^2 - w^2}}\,\frac{p^2}{\sqrt{G^2M^2 - w^2}} = \frac{wp^2}{G^2M^2 - w^2}$$

Since c is the distance from the center of the ellipse to either focus and since the center is $(-wp^2/(G^2M^2 - w^2), 0)$ by (21), it follows that one focus is the origin, where the sun is located.

9. a. Since $d\mathbf{r}/dt$ is perpendicular to $\mathbf{k}$, (9) implies that

$$\|GM\mathbf{u} + \mathbf{w}_1\| = \left\|\frac{d\mathbf{r}}{dt} \times p\mathbf{k}\right\| = p\left\|\frac{d\mathbf{r}}{dt}\right\|\|\mathbf{k}\|\sin\frac{\pi}{2} = p\left\|\frac{d\mathbf{r}}{dt}\right\|$$

Thus $\|d\mathbf{r}/dt\| = (1/p)\|GM\mathbf{u} + \mathbf{w}_1\|$.

b. Since $\mathbf{w}_1 = w\mathbf{i}$ and $\mathbf{u} = \cos\theta\mathbf{i} + \sin\theta\mathbf{j}$, we have

$$\|GM\mathbf{u} + \mathbf{w}_1\| = \|(GM\cos\theta + w)\mathbf{i} + GM\sin\theta\mathbf{j}\|$$

$$= \sqrt{(GM\cos\theta + w)^2 + G^2M^2\sin^2\theta}$$

$$= \sqrt{G^2M^2 + w^2 + 2GMw\cos\theta}$$

Thus $\|GM\mathbf{u} + \mathbf{w}_1\|$ is maximum when $\cos\theta = 1$, or $\theta = 0$, and the maximum value is $GM + w$. By Exercises 9(a), 8, and 1(c) we find that the maximum speed is

$$\frac{GM + w}{p} = \frac{r_0v_0^2}{p} = \frac{r_0v_0^2}{r_0v_0} = v_0$$

11. a. We have $a - c = r_0 = 100 + 3960 = 4060$ and $2a = 3100 + 100 + 2(3960) = 11{,}120$, so that $a = 5560$ and $c = 5560 - 4060 = 1500$. Thus

$$b = \sqrt{a^2 - c^2} = \sqrt{5560^2 - 1500^2} \approx 5353.84$$

By equations (20) and (23) we have

$$p = \frac{2\pi ab}{T} = 2\pi ab\sqrt{\frac{GM}{4\pi^2a^3}} = b\sqrt{\frac{GM}{a}} \approx 5353.84\sqrt{\frac{1.237 \times 10^{12}}{5560}}$$

$$\approx 7.98570 \times 10^7$$

By Exercise 1(c),

$$v_0 = \frac{p}{r_0} \approx \frac{7.98570 \times 10^7}{4060} \approx 19{,}669.2 \text{ (miles per hour)}$$

b. By Exercise 8, $w = r_0v_0^2 - GM \approx 4060\,(19{,}669.2)^2 - 1.237 \times 10^{12} \approx 3.33722 \times 10^{11}$. At aphelion $\mathbf{u} = -\mathbf{i}$ so that $\|GM\mathbf{u} + \mathbf{w}_1\| = \|-GM\mathbf{i} + w\mathbf{i}\| = GM - w$. Thus, by Exercise 9(a) the minimum velocity is given by

$$\left\|\frac{d\mathbf{r}}{dt}\right\| = \frac{1}{p}\|GM\mathbf{u} + \mathbf{w}_1\| = \frac{GM - w}{p} \approx \frac{1.237 \times 10^{12} - 3.33722 \times 10^{11}}{7.98570 \times 10^7}$$

$$\approx 11{,}311.2 \text{ (miles per hour)}$$

13. Since $GM = w$ for a parabolic orbit, it follows from Exercise 8 that $GM = w = r_0 v_0^2 - GM$, so that $v_0 = \sqrt{2GM/r_0}$.

15. Let F_e be the magnitude of the gravitational force exerted by the earth on the spacecraft, F_m the magnitude of the gravitational force exerted by the moon on the spacecraft, m the mass of the spacecraft, and M_m the mass of the moon. By Newton's Law of Gravitation,

$$\frac{F_e}{F_m} = \frac{GM_e m/(240{,}000 - 4080)^2}{GM_m m/(4080)^2} = \frac{(4080)^2}{(M_m/M_e)(235{,}920)^2} \approx \frac{(4080)^2}{(.0123)(235{,}920)^2} \approx .024316$$

17. $\mathbf{L}'(t) = \dfrac{d}{dt}(\mathbf{r} \times m\mathbf{v}) = \dfrac{d}{dt}\left(\mathbf{r} \times m\dfrac{d\mathbf{r}}{dt}\right) = m\dfrac{d}{dt}\left(\mathbf{r} \times \dfrac{d\mathbf{r}}{dt}\right)$. Thus by (3), $\mathbf{L}'(t) = \mathbf{0}$.

CHAPTER 13 REVIEW

1.

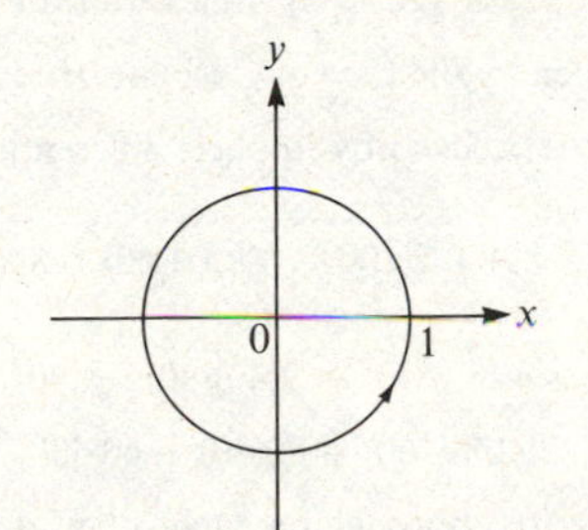

3. $(\mathbf{F} \times \mathbf{G})(t) = \begin{vmatrix} \mathbf{i} & \mathbf{j} & \mathbf{k} \\ t & 1 & 0 \\ 0 & 1 & t \end{vmatrix} = t\mathbf{i} - t^2\mathbf{j} + t\mathbf{k}$

$$[(\mathbf{F} \times \mathbf{G}) \times \mathbf{H}](t) = \begin{vmatrix} \mathbf{i} & \mathbf{j} & \mathbf{k} \\ t & -t^2 & t \\ 0 & t & 0 \end{vmatrix} = -t^2\mathbf{i} + t^2\mathbf{k}$$

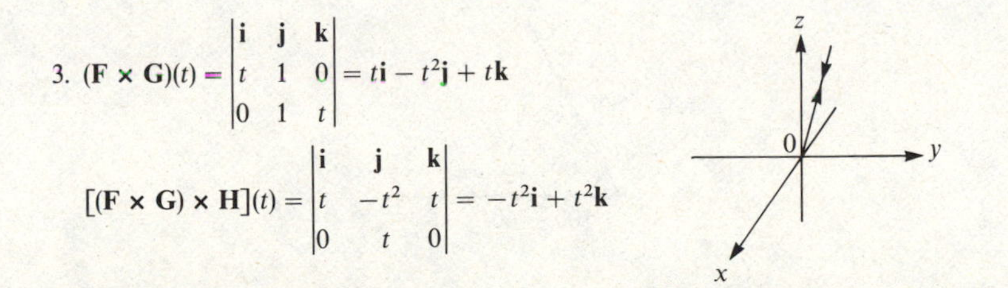

5. a. $(\mathbf{F} \cdot \mathbf{G})(t) = t^3$; $(\mathbf{F} \cdot \mathbf{G})'(t) = 3t^2$

b. $(\mathbf{F} \times \mathbf{G})(t) = \begin{vmatrix} \mathbf{i} & \mathbf{j} & \mathbf{k} \\ \frac{1}{t} & t & 0 \\ 0 & t^2 & \frac{-1}{t^2} \end{vmatrix} = \dfrac{-1}{t}\mathbf{i} + \dfrac{1}{t^3}\mathbf{j} + t\mathbf{k}$; $(\mathbf{F} \times \mathbf{G})'(t) = \dfrac{1}{t^2}\mathbf{i} - \dfrac{3}{t^4}\mathbf{j} + \mathbf{k}$

7. $$\int \left(\tan 2\pi t\mathbf{i} + \sec^2 2\pi t\mathbf{j} + \frac{4}{1+t^2}\mathbf{k}\right) dt = \left(\int \tan 2\pi t\, dt\right)\mathbf{i} + \left(\int \sec^2 2\pi t\, dt\right)\mathbf{j} + \left(\int \frac{4}{1+t^2}\, dt\right)\mathbf{k}$$
$$= -\frac{1}{2\pi}\ln|\cos 2\pi t|\mathbf{i} + \frac{1}{2\pi}\tan 2\pi t\mathbf{j} + 4\arctan t\mathbf{k} + \mathbf{C}$$

9. a. $\mathscr{L} = \int_0^{3\pi} \sqrt{(e^t\cos t - e^t\sin t)^2 + (e^t\sin t + e^t\cos t)^2}\, dt = \int_0^{3\pi} e^t\sqrt{2}\, dt$
$= e^t\sqrt{2}\Big|_0^{3\pi} = \sqrt{2}(e^{3\pi} - 1)$

b. As in part (a) but with the limits of integration altered,

$$\mathscr{L} = \int_{-2\pi}^{1} e^t\sqrt{2}\, dt = e^t\sqrt{2}\Big|_{-2\pi}^{1} = \sqrt{2}(e - e^{-2\pi})$$

11. $\mathbf{r} = [(\frac{3}{2}t)^{2/3} - 1]\mathbf{i} + \frac{2}{3}[(\frac{3}{2}t)^{2/3} - 1]^{3/2}\mathbf{j}$;
$\mathbf{v} = \frac{2}{3}(\frac{3}{2}t)^{-1/3}(\frac{3}{2})\mathbf{i} + [(\frac{3}{2}t)^{2/3} - 1]^{1/2}(\frac{2}{3})(\frac{3}{2}t)^{-1/3}(\frac{3}{2})\mathbf{j} = (\frac{3}{2}t)^{-1/3}\mathbf{i} + [(\frac{3}{2}t)^{2/3} - 1]^{1/2}(\frac{3}{2}t)^{-1/3}\mathbf{j}$;
$\|\mathbf{v}\| = (\frac{3}{2}t)^{-1/3}\sqrt{1 + [(\frac{3}{2}t)^{2/3} - 1]} = (\frac{3}{2}t)^{-1/3}(\frac{3}{2}t)^{1/3} = 1$

13. $\mathbf{v} = (3 - 3t^2)\mathbf{i} + 6t\mathbf{j} + (3 + 3t^2)\mathbf{k}$;
$\|\mathbf{v}\| = \sqrt{(3 - 3t^2)^2 + (6t)^2 + (3 + 3t^2)^2} = \sqrt{18 + 36t^2 + 18t^4} = 3\sqrt{2}(1 + t^2)$;
$\mathbf{a} = -6t\mathbf{i} + 6\mathbf{j} + 6t\mathbf{k}$;

$$\mathbf{v} \times \mathbf{a} = \begin{vmatrix} \mathbf{i} & \mathbf{j} & \mathbf{k} \\ 3 - 3t^2 & 6t & 3 + 3t^2 \\ -6t & 6 & 6t \end{vmatrix} = 18(t^2 - 1)\mathbf{i} - 36t\mathbf{j} + 18(t^2 + 1)\mathbf{k}$$

$\|\mathbf{v} \times \mathbf{a}\| = 18\sqrt{(t^2 - 1)^2 + (-2t)^2 + (t^2 + 1)^2} = 18\sqrt{2t^4 + 4t^2 + 2} = 18\sqrt{2}(t^2 + 1)$;

$$\kappa = \frac{\|\mathbf{v} \times \mathbf{a}\|}{\|\mathbf{v}\|^3} = \frac{18\sqrt{2}(t^2 + 1)}{[3\sqrt{2}(1 + t^2)]^3} = \frac{1}{3(1 + t^2)^2}$$

15. a. $y = \dfrac{1}{x}$, $\dfrac{dy}{dx} = -\dfrac{1}{x^2}$, $\dfrac{d^2y}{dx^2} = \dfrac{2}{x^3}$; $\kappa = \dfrac{|d^2y/dx^2|}{[1 + (dy/dx)^2]^{3/2}} = \dfrac{2/x^3}{(1 + 1/x^4)^{1/2}} = \dfrac{2x^3}{(x^4 + 1)^{3/2}}$

b. $\dfrac{d\kappa}{dx} = \dfrac{6x^2(x^4 + 1)^{3/2} - 2x^3(\frac{3}{2})(x^4 + 1)^{1/2}(4x^3)}{(x^4 + 1)^3} = \dfrac{6x^2(1 - x^4)}{(x^4 + 1)^{5/2}}$

Since $d\kappa/dx > 0$ for $x < 1$ and $d\kappa/dx < 0$ for $x > 1$, it follows from Theorem 4.11 that κ is maximum for $x = 1$. Thus the maximum value of κ is $\kappa(1) = 2/2^{3/2} = \sqrt{2}/2$.

c. Using the value of κ at (1, 1) from part (b), we have $\rho(1) = 1/\kappa(1) = \sqrt{2}$. Thus the radius of curvature at (1, 1) is $\sqrt{2}$.

17. $\mathbf{v} = e^t(\cos t - \sin t)\mathbf{i} + e^t(\sin t + \cos t)\mathbf{j} + e^t\mathbf{k}$;

$\mathbf{a} = e^t(\cos t - \sin t - \sin t - \cos t)\mathbf{i} + e^t(\sin t + \cos t + \cos t - \sin t)\mathbf{j} + e^t\mathbf{k}$

$= -2e^t \sin t\mathbf{i} + 2e^t \cos t\mathbf{j} + e^t\mathbf{k}$;

$\|\mathbf{v}\| = \sqrt{e^{2t}(\cos t - \sin t)^2 + e^{2t}(\sin t + \cos t)^2 + e^{2t}} = e^t\sqrt{3}$

$$\mathbf{v} \times \mathbf{a} = \begin{vmatrix} \mathbf{i} & \mathbf{j} & \mathbf{k} \\ e^t(\cos t - \sin t) & e^t(\sin t + \cos t) & e^t \\ -2e^t \sin t & 2e^t \cos t & e^t \end{vmatrix}$$

$= e^{2t}(\sin t - \cos t)\mathbf{i} - e^{2t}(\sin t + \cos t)\mathbf{j} + 2e^{2t}\mathbf{k}$;

$\|\mathbf{v} \times \mathbf{a}\| = e^{2t}\sqrt{(\sin t - \cos t)^2 + (\sin t + \cos t)^2 + 2^2} = e^{2t}\sqrt{6}$

$$\kappa = \frac{\|\mathbf{v} \times \mathbf{a}\|}{\|\mathbf{v}\|^3} = \frac{e^{2t}\sqrt{6}}{(e^t\sqrt{3})^3} = \frac{\sqrt{2}}{3e^t};\ \rho = \frac{1}{\kappa} = \frac{3\sqrt{2}e^t}{2}$$

19. $\mathbf{r}'(t) = (1 - \cos t)\mathbf{i} + \sin t\mathbf{j} + 2\cos\frac{t}{2}\mathbf{k}$;

$$\|\mathbf{r}'(t)\| = \sqrt{(1 - \cos t)^2 + \sin^2 t + 4\cos^2\frac{t}{2}} = \sqrt{1 - 2\cos t + \cos^2 t + \sin^2 t + 4\cos^2\frac{t}{2}}$$

$= \sqrt{2 - 2\cos t + 2 + 2\cos t} = 2$;

$$\mathbf{T}(t) = \frac{\mathbf{r}'(t)}{\|\mathbf{r}'(t)\|} = \frac{1}{2}(1 - \cos t)\mathbf{i} + \frac{1}{2}\sin t\mathbf{j} + \cos\frac{t}{2}\mathbf{k};$$

$$\mathbf{T}'(t) = \frac{1}{2}\sin t\mathbf{i} + \frac{1}{2}\cos t\mathbf{j} - \frac{1}{2}\sin\frac{t}{2}\mathbf{k};$$

$$\|\mathbf{T}'(t)\| = \frac{1}{2}\sqrt{\sin^2 t + \cos^2 t + \sin^2\frac{t}{2}} = \frac{1}{2}\sqrt{1 + \sin^2\frac{t}{2}} = \frac{1}{2}\sqrt{\frac{3}{2} - \frac{1}{2}\cos t}$$

$$= \frac{\sqrt{2}}{4}\sqrt{3 - \cos t};$$

$$\mathbf{N}(t) = \frac{\mathbf{T}'(t)}{\|\mathbf{T}'(t)\|} = \frac{\sqrt{2}\sin t}{\sqrt{3 - \cos t}}\mathbf{i} + \frac{\sqrt{2}\cos t}{\sqrt{3 - \cos t}}\mathbf{j} - \frac{\sqrt{2}\sin(t/2)}{\sqrt{3 - \cos t}}\mathbf{k};$$

$$\kappa(t) = \frac{\|\mathbf{T}'(t)\|}{\|\mathbf{r}'(t)\|} = \frac{\sqrt{2}}{8}\sqrt{3 - \cos t}$$

21. Since $\|\mathbf{v}\| = 1$ and $\mathbf{a} = d\mathbf{v}/dt$, it follows from Corollary 13.11 that $\mathbf{v} \cdot \mathbf{a} = \mathbf{v} \cdot d\mathbf{v}/dt = 0$. Since $\|\mathbf{v}\| = \|\mathbf{a}\| = 1$, we conclude that $\mathbf{v}$ and $\mathbf{a}$ are perpendicular. Thus $\|\mathbf{v} \times \mathbf{a}\| = \|\mathbf{v}\|\,\|\mathbf{a}\| \sin(\pi/2) = 1$, so that $\kappa = \|\mathbf{v} \times \mathbf{a}\|/\|\mathbf{v}\|^3 = 1$.

CUMULATIVE REVIEW (CHAPTERS 1–12)

1. $\lim_{x\to -3^+} \frac{1}{|x - 3|} = \frac{1}{6}$ and $\lim_{x\to -3^+} \frac{1}{x + 3} = \infty$; $\lim_{x\to -3^+} \left(\frac{1}{x + 3} - \frac{1}{|x - 3|}\right) = \infty$.

3. For $x = 0$, $\lim_{y\to 0} \frac{x^2 - y^3}{x^2 + y^2} = \lim_{y\to 0} \left(-\frac{y^3}{y^2}\right) = \lim_{y\to 0}(-y) = 0$.

For $x \neq 0$, $\lim_{y\to 0} \frac{x^2 - y^3}{x^2 + y^2} = \frac{x^2 - 0^3}{x^2 + 0^2} = 1$.

5. $f'(x) = \frac{1}{\sqrt{1 - ((\sin x)/5)^2}}\left(\frac{\cos x}{5}\right) = \frac{\cos x}{\sqrt{25 - \sin^2 x}}$

7. Rewriting the given equation, we obtain $x^3 - 2y^3 = 6x^2 + 6y^2$. By implicit differentiation, $3x^2 - 6y^2(dy/dx) = 12x + 12y(dy/dx)$. The tangent is horizontal at (x, y) provided that $dy/dx = 0$, which means that $3x^2 = 12x$, and thus $x = 0$ or $x = 4$. If $x = 0$, then $y \neq 0$ since $(x^3 - 2y^3)/(x^2 + y^2) = 6$, and therefore the equation $x^3 - 2y^3 = 6x^2 + 6y^2$ becomes $-2y = 6$, so that $y = -3$. If $x = 4$, then the equation becomes $64 - 2y^3 = 96 + 6y^2$, or $y^3 + 3y^2 + 16 = 0$, or $(y + 4)(y^2 - y + 4) = 0$, so $y = -4$. Therefore the tangent is horizontal at $(0, -3)$ and $(4, -4)$.

9. At t hours after noon, the train traveling 60 miles per hour is $100 - 60t$ miles from the junction, and the other train is $120 - 80t$ miles from the junction. If D is the distance between the two trains, then by the Law of Cosines, $D^2 = (120 - 80t)^2 + (100 - 60t)^2 - 2(120 - 80t)(100 - 60t)\cos(\pi/3)$, so by implicit differentiation, we find that

$$2D\frac{dD}{dt} = 2(120 - 80t)(-80) + 2(100 - 60t)(-60) + 80(100 - 60t) + 60(120 - 80t)$$

At 1 p.m., $t = 1$, so $D^2 = 40^2 + 40^2 - 2(40)(40)\frac{1}{2} = 40^2$, and thus $D = 40$. Therefore $80(dD/dt) = 2(40)(-80) + 2(40)(-60) + 80(40) + 60(40)$, which yields $dD/dt = -70$. Thus at 1 p.m. the trains are approaching each other at the rate of 70 miles per hour.

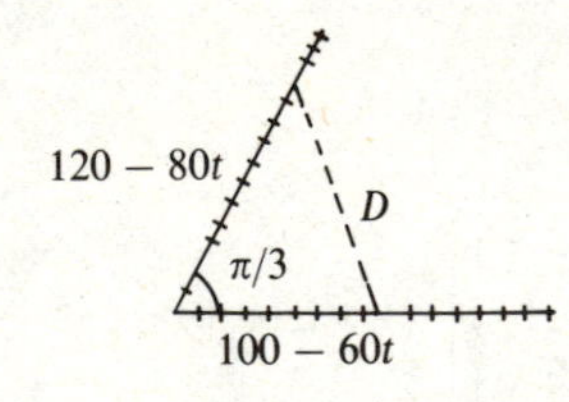

11. Using the notation in the diagram we have $2x + y - 3 = 77$, or $y = 80 - 2x$. Thus the area A is given by $A = xy = x(80 - 2x) = 80x - 2x^2$. Therefore $A'(x) = 80 - 4x$, and $A'(x) = 0$ for $x = 20$. Since $A''(x) = -4 < 0$, A is maximum for $x = 20$. Then $y = 80 - 2(20) = 40$. Thus the fence should be 20 feet long perpendicular to the house, and 40 feet long parallel to the house. If the gate is to be placed on one of the sides perpendicular to the house, as in the second diagram, we find that $x + (x - 3) + y = 77$, so that once again $y = 80 - 2x$, which yields the same dimensions as before.

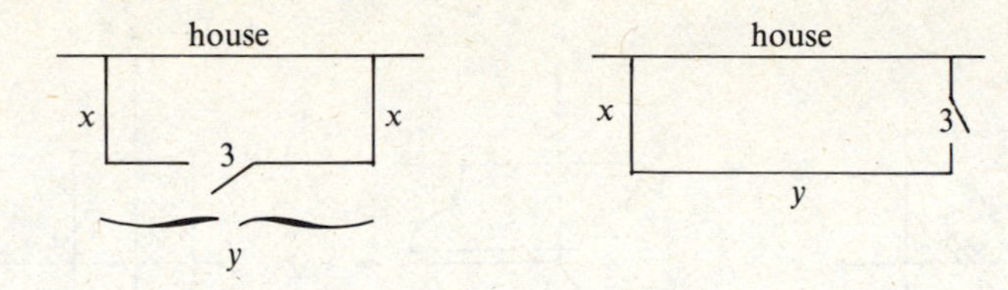

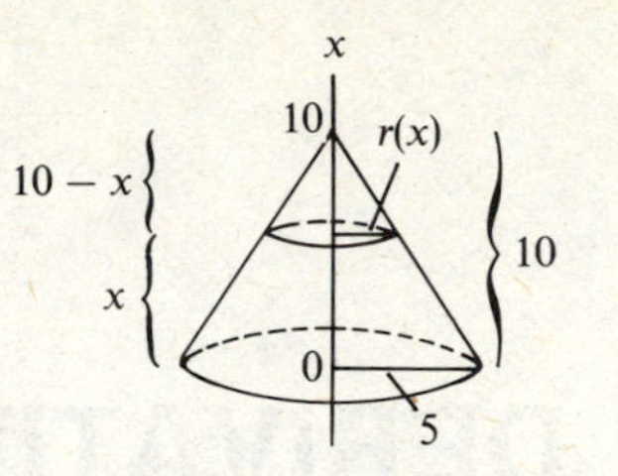

13. Let $f(t)$ be the amount of the substance remaining after t years. Then $f(t) = f(0)e^{kt}$ for some k, and we need to find the half-life h, which is $-(1/k)\ln 2$. Now by hypothesis, $f(5) = .9f(0)$. Thus $.9f(0) = f(5) = f(0)e^{k5}$, so $e^{5k} = .9$, or $5k = \ln .9$, or $k = \frac{1}{5}\ln .9$. Then $\frac{1}{2}f(0) = f(h) = f(0)e^{kh}$, so $\frac{1}{2} = e^{kh}$, or $kh = \ln\frac{1}{2} = -\ln 2$, or $h = -(1/k)\ln 2 = -5(\ln 2/\ln .9) \approx 32.8941$. Thus the half-life is approximately 32.8941 years.

15. $\dfrac{x}{(x+2)(x^2+6)} = \dfrac{A}{x+2} + \dfrac{Bx+C}{x^2+6}$; $x = A(x^2+6) + (Bx+C)(x+2)$;

$A + B = 0,\ 2B + C = 1,\ 6A + 2C = 0;\ A = -\frac{1}{5},\ B = \frac{1}{5},\ C = \frac{3}{5}$;

$$\int \frac{x}{(x+2)(x^2+6)}\,dx = \int\left(\frac{-1}{5(x+2)} + \frac{x+3}{5(x^2+6)}\right)dx = -\frac{1}{5}\int\frac{1}{x+2}\,dx + \frac{1}{5}\int\frac{x}{x^2+6}\,dx + \frac{3}{5}\int\frac{1}{x^2+6}\,dx$$
$$= -\frac{1}{5}\ln|x+2| + \frac{1}{10}\ln(x^2+6) + \frac{3}{5\sqrt{6}}\arctan\frac{x}{\sqrt{6}} + C$$

17. Let $u = \sqrt{x}$, so that $du = \dfrac{1}{2\sqrt{x}}\,dx$. If $x = b$, then $u = \sqrt{b}$, and if $x = \pi^2/4$, then $u = \pi/2$.

Thus

$$\int_0^{\pi^2/4}\frac{\cos\sqrt{x}}{\sqrt{x}}\,dx = \lim_{b\to 0^+}\int_b^{\pi^2/4}\frac{\cos\sqrt{x}}{\sqrt{x}}\,dx = \lim_{b\to 0^+}\int_{\sqrt{b}}^{\pi/2}(\cos u)2\,du$$
$$= \lim_{b\to 0^+} 2\int_{\sqrt{b}}^{\pi/2}\cos u\,du = \lim_{b\to 0^+} 2\sin u\Big|_{\sqrt{b}}^{\pi/2} = \lim_{b\to 0^+}(2 - \sin\sqrt{b}) = 2$$

Thus the integral converges and its value is 2.

19. From the diagram, $r(x)/(10-x) = \frac{5}{10} = \frac{1}{2}$, so that $r(x) = (10-x)/2$ and therefore $A(x) = \pi[r(x)]^2 = (\pi/4)(10-x)^2$. Since a particle of water x feet from the bottom is to be raised $13 - x$ feet, we find that

$$W = 62.5\int_0^4 (13-x)\left[\frac{\pi}{4}(10-x)^2\right]dx = \frac{62.5\pi}{4}\int_0^4(-x^3 + 33x^2 - 360x + 1300)\,dx$$
$$= \frac{62.5\pi}{4}\left(-\frac{1}{4}x^4 + 11x^3 - 180x^2 + 1300x\right)\Big|_0^4 = 46{,}250\pi \text{ (foot-pounds)}$$

21. Since $(-1)^n n/(n+1) > 0$ for n even and $(-1)^n n/(n+1) < 0$ for n odd, the only possible limit is 0. But $\lim_{n\to\infty}|(-1)^n n/(n+1)| = \lim_{n\to\infty} n/(n+1) = 1$, so the sequence cannot approach 0. Thus $\lim_{n\to\infty}(-1)^n n/(n+1)$ does not exist, so the given sequence diverges.

23. $$\lim_{n\to\infty}\frac{\dfrac{(n+1)^{n+1}}{3^{n+1}(n+1)!}}{n^n/(3^n n!)} = \lim_{n\to\infty}\frac{(n+1)^{n+1}}{n^n}\frac{3^n n!}{3^{n+1}(n+1)!} = \lim_{n\to\infty}\frac{(n+1)^{n+1}}{3n^n(n+1)} = \lim_{n\to\infty}\frac{(n+1)^n}{3n^n}$$
$$= \lim_{n\to\infty}\frac{1}{3}\left(1+\frac{1}{n}\right)^n = \frac{e}{3} < 1$$

By the Ratio Test the series converges. Since the terms are positive, the series converges absolutely.

25. $$f(x) = -\frac{2x}{(1+x^2)^2} = \frac{d}{dx}\left(\frac{1}{1+x^2}\right) = \frac{d}{dx}\left(\sum_{n=0}^{\infty}(-1)^n x^{2n}\right) = \sum_{n=1}^{\infty}(-1)^n(2n)x^{2n-1}$$

Since $$\lim_{n\to\infty}\left|\frac{(-1)^{n+1}(2n+2)x^{2n+1}}{(-1)^n(2n)x^{2n-1}}\right| = \lim_{n\to\infty}\frac{(2n+2)}{(2n)}|x|^2 = |x|^2$$

the radius of convergence of the series is 1. For $x = 1$ the series becomes $\sum_{n=1}^{\infty}(-1)^n 2n$, which diverges; for $x = -1$ the series becomes $-\sum_{n=1}^{\infty}(-1)^n(2n)$, which also diverges. Thus the interval of convergence of $\sum_{n=1}^{\infty}(-1)^n(2n)x^{2n-1}$ is $(-1, 1)$.

27. Let P_1, P_2, and P_3 be the points $(1, -1, 2)$, $(2, 3, -1)$, and $(0, 2, 0)$, respectively. Then $\overrightarrow{P_1P_2} = \mathbf{i} + 4\mathbf{j} - 3\mathbf{k}$ and $\overrightarrow{P_1P_3} = -\mathbf{i} + 3\mathbf{j} - 2\mathbf{k}$, so that

$$\overrightarrow{P_1P_2}\times\overrightarrow{P_1P_3} = \begin{vmatrix}\mathbf{i} & \mathbf{j} & \mathbf{k}\\ 1 & 4 & -3\\ -1 & 3 & -2\end{vmatrix} = [-8-(-9)]\mathbf{i} + [3-(-2)]\mathbf{j} + [3-(-4)]\mathbf{k}$$
$$= \mathbf{i} + 5\mathbf{j} + 7\mathbf{k}$$

Since $\overrightarrow{P_1P_2}\times\overrightarrow{P_1P_3}$ is perpendicular to the plane, and $(1, -1, 2)$ lies on the plane, an equation of the plane is $1(x-1) + 5(y+1) + 7(z-2) = 0$, or $x + 5y + 7z = 10$.

14
PARTIAL DERIVATIVES

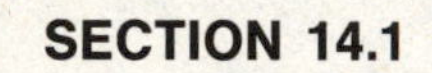

SECTION 14.1

1. all (x, y) such that $x \geq 0$ and $y \geq 0$

3. all (x, y) such that $x \neq 0$ and $y \neq 0$

5. all (x, y) such that $x^2 + y^2 \geq 25$

7. all (x, y) such that $x + y \neq 0$

9. all (x, y, z) such that $x^2 + y^2 + z^2 \leq 1$

11. all (x, y, z) such that $x \neq 0$, $y \neq 0$, and $z \neq 0$

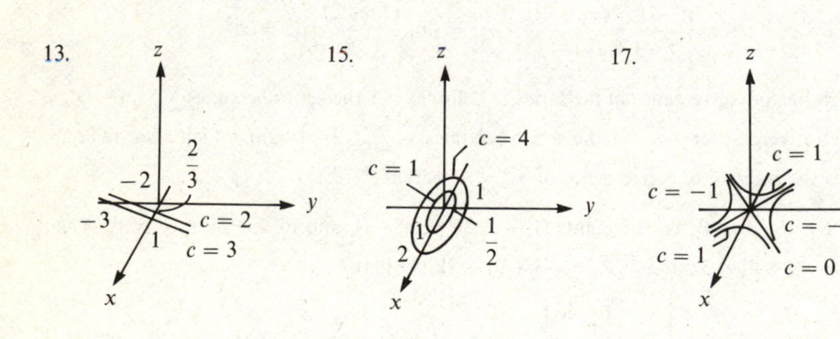

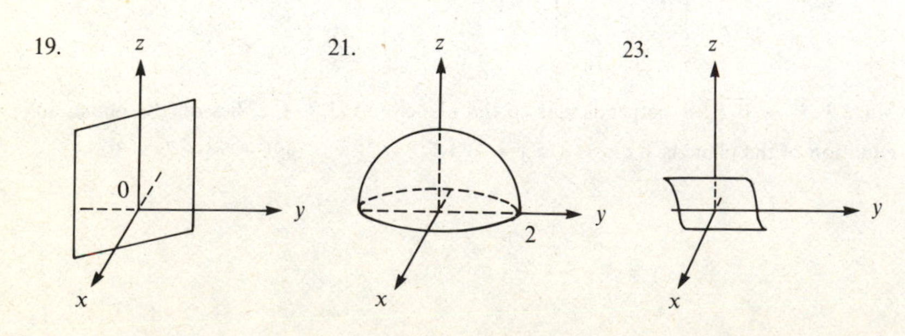

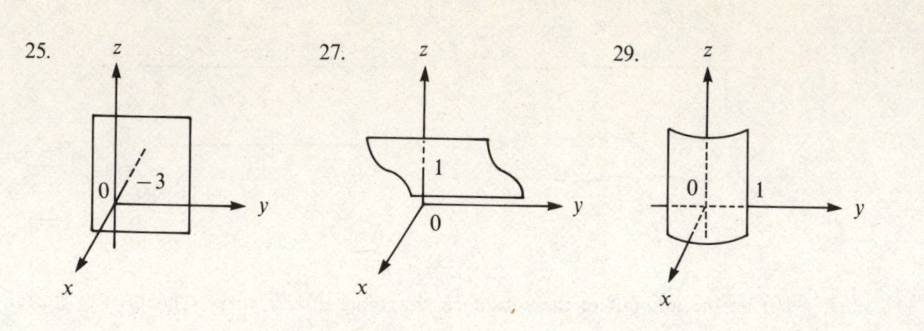

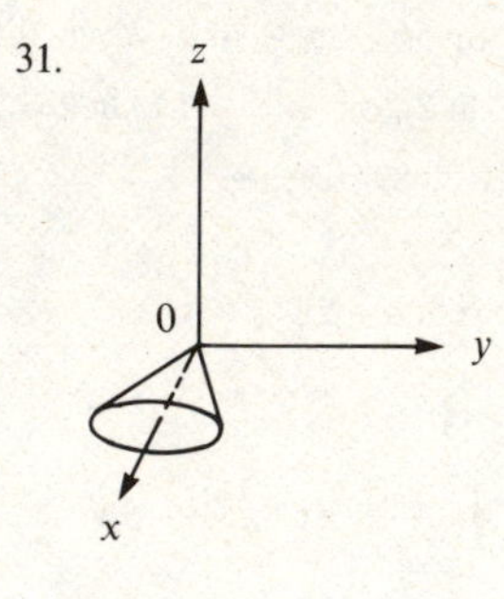

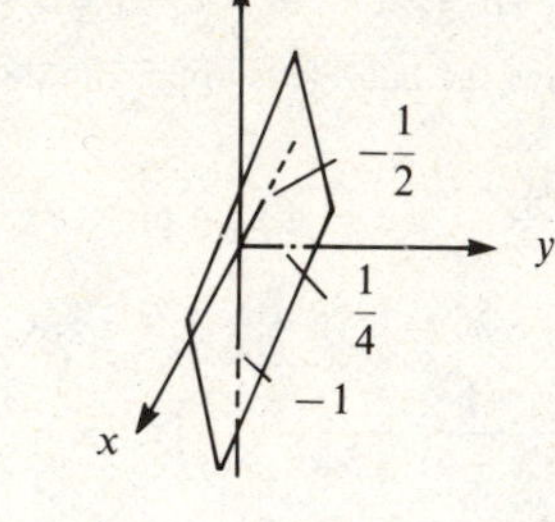

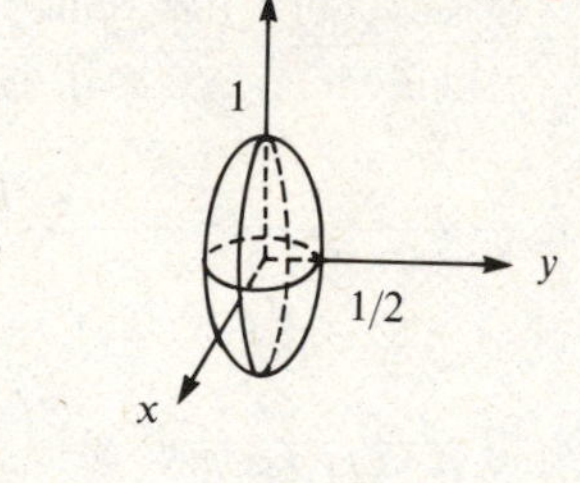

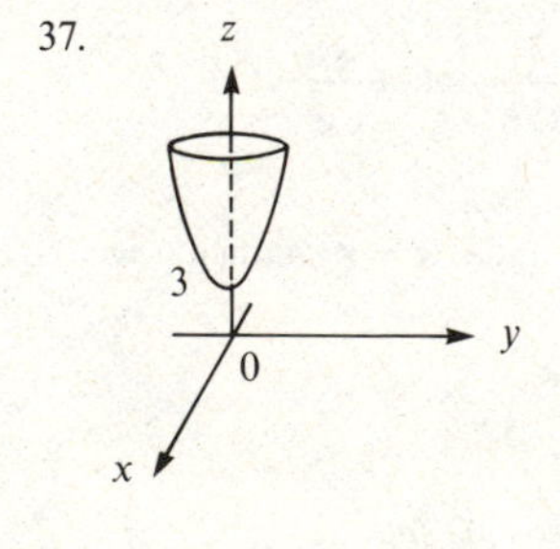

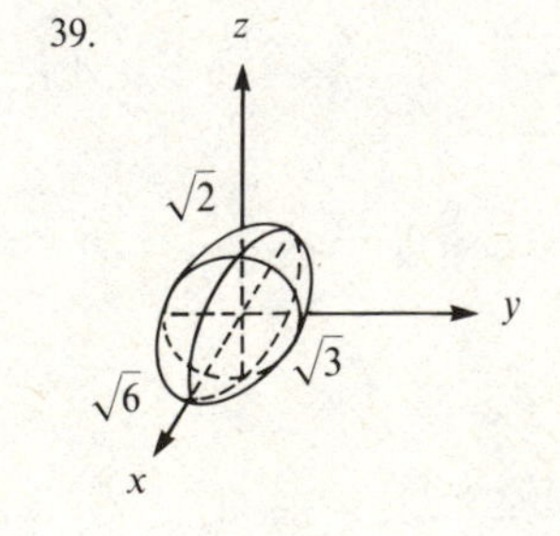

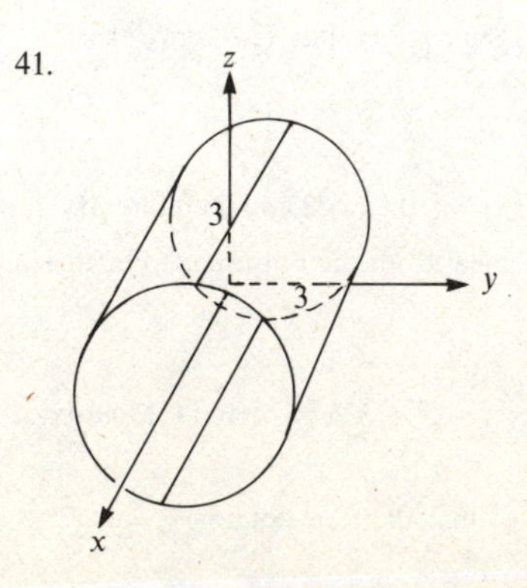

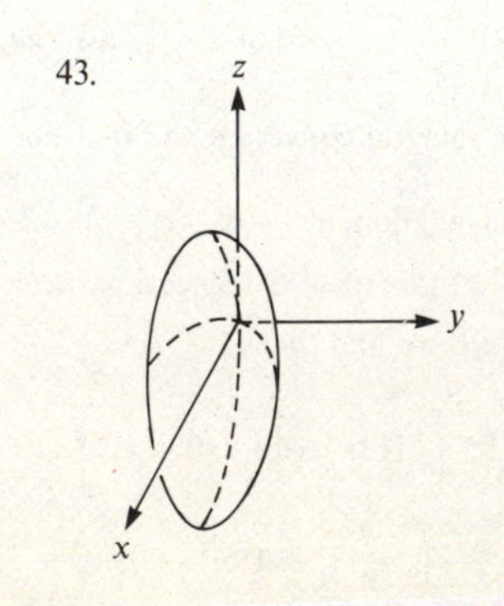

45.

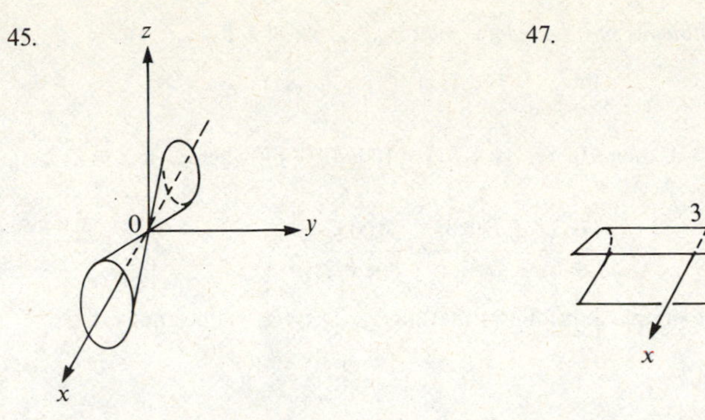

47.

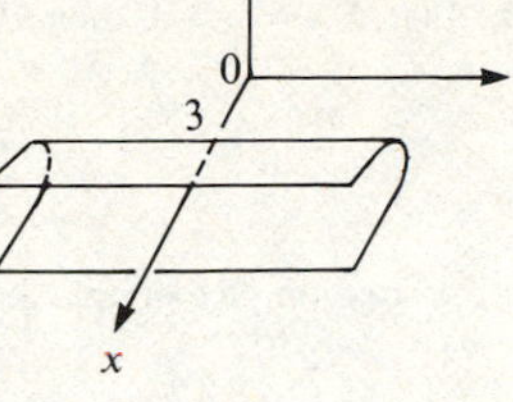

49.

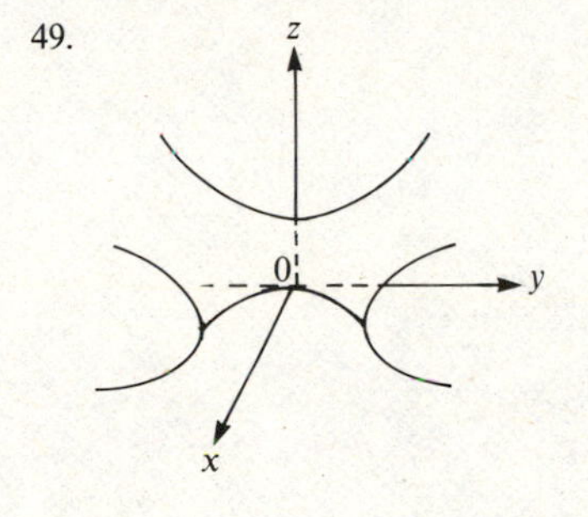

51.

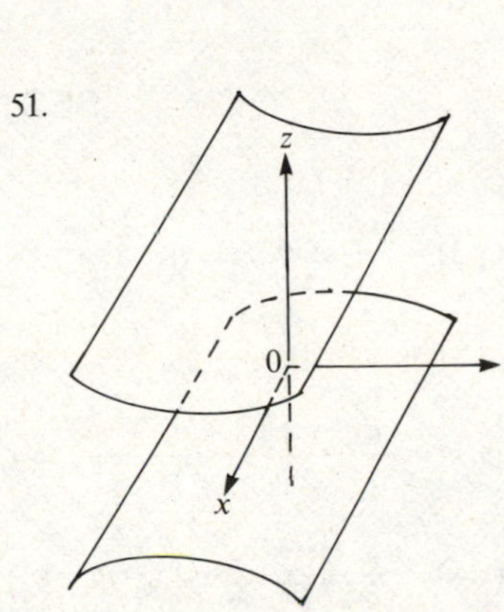

53.

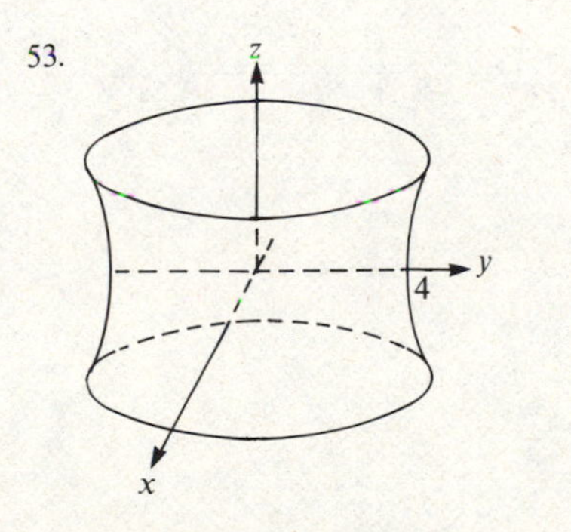

55.

57. Let $f(x, y) = x^2 + 6y$ and $g(t) = \ln t$.

59. Let $f(x, y) = xy$ and $g(t) = t^2 + 5t + 10$.

61. $h = \dfrac{v}{\pi r^2}$ for $r > 0$ and $v > 0$

63. Let x and y denote the dimensions of the base and z the height. Then the surface area is given by $S = xy + 2xz + 2yz$ for $x > 0$, $y > 0$, and $z > 0$.

65. The cost in dollars is given by $C = \frac{3}{10}xy$ for $x > 0$ and $y > 0$.

67. For any positive constant b, the level surface $E(x, y, z) = b$ consists of all (x, y, z) such that $c/\sqrt{x^2 + y^2} = b$, or $x^2 + y^2 = c^2/b^2$. Thus the level surfaces are circular cylinders.

69. For any constant b, the level curve $T(x, y) = b$ consists of all (x, y) such that $x \geq 0$, $y \geq 0$, and $xy = b$. Thus the level curves are either hyperbolas or two intersecting lines.

SECTION 14.2

1. $\lim_{(x,y)\to(2,4)} (x + \frac{1}{2}) = 2 + \frac{1}{2} = \frac{5}{2}$

3. $$\lim_{(x,y)\to(1,0)} \frac{x^2 - xy + 1}{x^2 + y^2} = \frac{1^2 - (1)(0) + 1}{1^2 + 0^2} = 2$$

5. $$\lim_{(x,y,z)\to(2,1,-1)} \frac{2x^2y - xz^2}{y^2 - xz} = \frac{2(2)^2(1) - (2)(-1)^2}{1^2 - (2)(-1)} = 2$$

7. $$\lim_{(x,y)\to(2,1)} \frac{x^3 + 2x^2y - xy - 2y^2}{x + 2y} = \lim_{(x,y)\to(2,1)} \frac{(x + 2y)(x^2 - y)}{x + 2y} = \lim_{(x,y)\to(2,1)} (x^2 - y)$$
$$= 2^2 - 1 = 3$$

9. Since $\lim_{(x,y,z)\to(\pi/2,-\pi/2,0)} (x + y + z) = \pi/2 - \pi/2 + 0 = 0$ and the cosine function is continuous, it follows from the substitution formula that
$$\lim_{(x,y,z)\to(\pi/2,-\pi/2,0)} \cos(x + y + z) = \cos 0 = 1$$

11. Since $\lim_{(x,y)\to(0,0)} (x^2 + y^2) = 0 + 0 = 0$ and $\lim_{t\to 0} (\sin t)/t = 1$, it follows from the substitution formula that
$$\lim_{(x,y)\to(0,0)} \frac{\sin(x^2 + y^2)}{x^2 + y^2} = \lim_{t\to 0} \frac{\sin t}{t} = 1$$

13. From Example 4 we have
$$\lim_{(x,y)\to(0,0)} \frac{x^3}{x^2 + y^2} = 0 = \lim_{(x,y)\to(0,0)} \frac{y^3}{x^2 + y^2}$$
Thus
$$\lim_{(x,y)\to(0,0)} xy\frac{x^2 - y^2}{x^2 + y^2} = \lim_{(x,y)\to(0,0)} \left[y\frac{x^3}{x^2 + y^2} - x\frac{y^3}{x^2 + y^2}\right] = (0)(0) - (0)(0) = 0$$

15. For any $\varepsilon > 0$ let $\delta = \varepsilon$. If $0 < \sqrt{x^2 + y^2 + z^2} < \delta$, then $x = \sqrt{x^2} \leq \sqrt{x^2 + y^2 + z^2} < \delta = \varepsilon$, so that
$$\left|\frac{x^3}{x^2 + y^2 + z^2}\right| = \left|x\frac{x^2}{x^2 + y^2 + z^2}\right| \leq |x|1 < \varepsilon$$
Thus
$$\lim_{(x,y,z)\to(0,0,0)} \frac{x^3}{x^2 + y^2 + z^2} = 0$$
Similarly,
$$\lim_{(x,y,z)\to(0,0,0)} \frac{y^3}{x^2 + y^2 + z^2} = 0 = \lim_{(x,y,z)\to(0,0,0)} \frac{z^3}{x^2 + y^2 + z^2}$$
It follows that
$$\lim_{(x,y,z)\to(0,0,0)} \frac{x^3 + y^3 + z^3}{x^2 + y^2 + z^2} = \lim_{(x,y,z)\to(0,0,0)} \left[\frac{x^3}{x^2 + y^2 + z^2} + \frac{y^3}{x^2 + y^2 + z^2} + \frac{z^3}{x^2 + y^2 + z^2}\right]$$
$$= 0 + 0 + 0 = 0$$

17. Notice that if $y = 0$ and $x \neq 0$, then $xy/(x^2+y^2) = 0$, whereas if $x = y \neq 0$, then

$$\frac{xy}{x^2+y^2} = \frac{x^2}{2x^2} = \frac{1}{2}$$

By the remark preceding Example 3, it follows that $\lim_{(x,y)\to(0,0)} xy/(x^2+y^2)$ does not exist.

19. Since $\lim_{(x,y)\to(1,1)} (x-y) = 1-1 = 0$ and the arcsine function is continuous, it follows from the substitution formula that $\lim_{(x,y)\to(1,1)} \arcsin(x-y) = \arcsin 0 = 0$. Since g is continuous, we conclude that $\lim_{(x,y)\to(1,1)} g(f(x,y)) = g(0) = -1$.

21. f is continuous because it is a polynomial.

23. f is continuous because it is a rational function.

25. Let $h(x,y,z) = e^x + e^{yz}$ and $g(t) = \ln t$. Then $f = g \circ h$. Since h and g are continuous, f is also continuous.

27. a. Since $f(x,0) = 0$ for all x and $f(0,y) = 0$ for all y, f is continuous in each variable separately at $(0, 0)$.

b. Notice that if $y = 0$, then $f(x,y) = 0$, whereas if $x = y \neq 0$, then $f(x,y) = \frac{1}{2}$. By the remark preceding Example 3, it follows that $\lim_{(x,y)\to(0,0)} f(x,y)$ does not exist. Thus f is not continuous at $(0, 0)$.

29. Notice that if $y = 0$ and $x \neq 0$, then $f(x,y) = 0$, whereas if $y = x^3 \neq 0$, then $f(x,y) = \frac{1}{2}$. By the remark preceding Example 3, it follows that $\lim_{(x,y)\to(0,0)} f(x,y)$ does not exist. Therefore f is not continuous at $(0, 0)$.

31. Notice that if $y = 0$ and $x \neq 0$, then $f(x,y) = 1$, whereas if $x = y \neq 0$, then $f(x,y) = \frac{1}{2}$. By the remark preceding Example 3, it follows that $\lim_{(x,y)\to(0,0)} f(x,y)$ does not exist.

33. If $|y| < |x^3|$, then $|f(x,y)| = |y|/|x| < |x^3|/|x| = x^2$. Thus for any $\varepsilon > 0$ we choose $\delta = \sqrt{\varepsilon}$. If (x, y) is in R and $0 < \sqrt{x^2+y^2} < \delta$, we have $|f(x,y)| < x^2 < \varepsilon$. It follows that $\lim_{(x,y)\to(0,0)} f(x,y) = 0$.

35. Since $f(x,y) = 0$ for $x > 0$, the only possible value for $\lim_{(x,y)\to(0,0)} f(x,y)$ is 0. But $f(0,0) = 2$. Thus either $\lim_{(x,y)\to(0,0)} f(x,y)$ does not exist or it is different from $f(0,0)$. Either way f is not continuous at $(0, 0)$. Thus f is not continuous on R.

37. Since f is constant on R, f is continuous on R.

39. a. A point P is called a *boundary point* of a set D if every ball centered at P contains points inside R and points outside R. The *boundary* of D is the collection of its boundary points.

b. A number L is the *limit* of a function f at a boundary point (x_0, y_0, z_0) of D if for every $\varepsilon > 0$ there is a number $\delta > 0$ such that

if (x, y, z) is in D and $0 < \sqrt{(x-x_0)^2+(y-y_0)^2+(z-z_0)^2} < \delta$

then $|f(x,y,z) - L| < \varepsilon$

c. A function f is *continuous at a boundary point* (x_0, y_0, z_0) of D if

$$\lim_{(x,y,z)\to(x_0,y_0,z_0)} f(x,y,z) = f(x_0,y_0,z_0)$$

41. Notice that if $x = y > 0$, then $100ax/(ax+by) = 100a/(a+b)$, whereas if $x = 2y > 0$, then

$$\frac{100ax}{ax+by} = \frac{200a}{2a+b} = \frac{100a}{a+b/2}$$

As in the solution of Example 3, it follows that $\lim_{(x,y)\to(0,0)} f(x,y)$ does not exist.

SECTION 14.3

1. $f_x(x,y) = x^{1/2}$, $f_y(x,y) = 0$

3. $f_x(x,y) = 2 + 6xy^4$, $f_y(x,y) = 12x^2y^3$

5. $$g_u(u,v) = \frac{(u^2+v^2)3u^2 - (u^3+v^3)2u}{(u^2+v^2)^2} = \frac{u^4+3u^2v^2-2uv^3}{(u^2+v^2)^2},$$

$$g_v(u,v) = \frac{(u^2+v^2)3v^2 - (u^3+v^3)2v}{(u^2+v^2)^2} = \frac{v^4+3u^2v^2-2u^3v}{(u^2+v^2)^2}$$

7. $$f_x(x,y) = \frac{-x}{\sqrt{4-x^2-9y^2}}, \quad f_y(x,y) = \frac{-9y}{\sqrt{4-x^2-9y^2}}$$

9. $$\frac{\partial z}{\partial x} = \frac{1}{2\sqrt{(1-x^{2/3})^3-y^2}}[3(1-x^{2/3})^2]\left(\frac{-2}{3}x^{-1/3}\right) = \frac{-(1-x^{2/3})^2}{x^{1/3}\sqrt{(1-x^{2/3})^3-y^2}},$$

$$\frac{\partial z}{\partial y} = \frac{-y}{\sqrt{(1-x^{2/3})^3-y^2}}$$

11. $$\frac{\partial z}{\partial x} = 3(\sin x^2y)^2(\cos x^2y)(2xy) = 6xy\sin^2 x^2y\cos x^2y,$$

$$\frac{\partial z}{\partial y} = 3(\sin x^2y)^2(\cos x^2y)(x^2) = 3x^2\sin^2 x^2y\cos x^2y$$

13. $f_x(x,y,z) = 2xy^5 + z^2$, $f_y(x,y,z) = 5x^2y^4$, $f_z(x,y,z) = 2xz$

15. $$f_x(x,y,z) = \frac{xy+yz+zx-(x+y+z)(y+z)}{(xy+yz+zx)^2} = \frac{-(y^2+yz+z^2)}{(xy+yz+zx)^2},$$

$$f_y(x,y,z) = \frac{xy+yz+zx-(x+y+z)(x+z)}{(xy+yz+zx)^2} = \frac{-(z^2+zx+x^2)}{(xy+yz+zx)^2},$$

$$f_z(x,y,z) = \frac{xy+yz+zx-(x+y+z)(y+x)}{(xy+yz+zx)^2} = \frac{-(x^2+xy+y^2)}{(xy+yz+zx)^2}$$

17. $$\frac{\partial w}{\partial x} = e^x(\cos y + \sin z), \quad \frac{\partial w}{\partial y} = e^x(-\sin y) = -e^x\sin y, \quad \frac{\partial w}{\partial z} = e^x\cos z$$

19. $\dfrac{\partial w}{\partial x} = \dfrac{1}{\sqrt{1 - 1/(1 + xyz^2)^2}}\left(\dfrac{-1}{(1 + xyz^2)^2}\right)yz^2 = \dfrac{-yz^2}{(1 + xyz^2)^2\sqrt{1 - (1/(1 + xyz^2))^2}}$,

$\dfrac{\partial w}{\partial y} = \dfrac{1}{\sqrt{1 - 1/(1 + xyz^2)^2}}\left(\dfrac{-1}{(1 + xyz^2)^2}\right)xz^2 = \dfrac{-xz^2}{(1 + xyz^2)^2\sqrt{1 - (1/(1 + xyz^2))^2}}$,

$\dfrac{\partial w}{\partial z} = \dfrac{1}{\sqrt{1 - 1/(1 + xyz^2)^2}}\left(\dfrac{-1}{(1 + xyz^2)^2}\right)2xyz = \dfrac{-2xyz}{(1 + xyz^2)^2\sqrt{1 - (1/(1 + xyz^2))^2}}$

21. $f_x(x, y) = \dfrac{4x}{\sqrt{4x^2 + y^2}}$, $f_x(2, -3) = \dfrac{4(2)}{\sqrt{4(2)^2 + (-3)^2}} = \dfrac{8}{5}$;

$f_y(x, y) = \dfrac{y}{\sqrt{4x^2 + y^2}}$, $f_y(2, -3) = \dfrac{(-3)}{\sqrt{4(2)^2 + (-3)^2}} = -\dfrac{3}{5}$

23. $f_x(x, y, z) = 2e^{2x-4y-z}$, $f_x(0, -1, 1) = 2e^{2(0)-4(-1)-1} = 2e^3$;

$f_y(x, y, z) = -4e^{2x-4y-z}$, $f_y(0, -1, 1) = -4e^{2(0)-4(-1)-1} = -4e^3$;

$f_z(x, y, z) = -e^{2x-4y-z}$, $f_z(0, -1, 1) = -e^{2(0)-4(-1)-1} = -e^3$

25. $f_x(0, 0) = \lim_{h\to 0} \dfrac{f(h, 0) - f(0, 0)}{h - 0} = \lim_{h\to 0} \dfrac{0 - 0}{h - 0} = \lim_{h\to 0} 0 = 0$;

$f_y(0, 0) = \lim_{h\to 0} \dfrac{f(0, h) - f(0, 0)}{h - 0} = \lim_{h\to 0} \dfrac{0 - 0}{h - 0} = \lim_{h\to 0} 0 = 0$

27. Let y be fixed, and let $g(x) = f(x, y)$, $h(u) = \int_\pi^u \sin t^2\, dt$, and $k(x) = x^2 + y^2$. Then $g(x) = f(x, y) = \int_\pi^{x^2+y^2} \sin t^2\, dt = h(k(x))$ and $h'(u) = \sin u^2$, so that $f_x(x, y) = g'(x) = h'(k(x))k'(x) = [\sin (k(x))^2]2x = 2x \sin (x^2 + y^2)^2$. Interchanging the roles of x and y in the preceding discussion, we find that $f_y(x, y) = 2y \sin (x^2 + y^2)^2$.

29. $f_x(x, y) = 6x - \sqrt{2}y^2$, $f_{xy}(x, y) = -2\sqrt{2}y$; $f_y(x, y) = -2\sqrt{2}xy + 5y^4$, $f_{yx}(x, y) = -2\sqrt{2}y$

31. $f_x(x, y) = \dfrac{x}{\sqrt{x^2 + y^2}}$, $f_{xy}(x, y) = \dfrac{-xy}{(x^2 + y^2)^{3/2}}$;

$f_y(x, y) = \dfrac{y}{\sqrt{x^2 + y^2}}$, $f_{yx}(x, y) = \dfrac{-xy}{(x^2 + y^2)^{3/2}}$

33. $f_x(x, y, z) = -yz \sin xy$, $f_{xy}(x, y, z) = -z \sin xy - xyz \cos xy$;

$f_y(x, y, z) = -xz \sin xy$, $f_{yx}(x, y, z) = -z \sin xy - xyz \cos xy$

35. $f_x(x, y) = e^{x-2y}$, $f_{xx}(x, y) = e^{x-2y}$; $f_y(x, y) = -2e^{x-2y}$, $f_{yy}(x, y) = 4e^{x-2y}$

37. $f_x(x, y, z) = \dfrac{x}{\sqrt{x^2 + y^2 + z^2}}$,

$f_{xx}(x, y, z) = \dfrac{\sqrt{x^2 + y^2 + z^2} - \dfrac{x^2}{\sqrt{x^2 + y^2 + z^2}}}{x^2 + y^2 + z^2} = \dfrac{y^2 + z^2}{(x^2 + y^2 + z^2)^{3/2}}$,

$f_y(x, y, z) = \dfrac{y}{\sqrt{x^2 + y^2 + z^2}}$,

$f_{yy}(x, y, z) = \dfrac{\sqrt{x^2 + y^2 + z^2} - \dfrac{y^2}{\sqrt{x^2 + y^2 + z^2}}}{x^2 + y^2 + z^2} = \dfrac{x^2 + z^2}{(x^2 + y^2 + z^2)^{3/2}}$;

$f_z(x, y, z) = \dfrac{z}{\sqrt{x^2 + y^2 + z^2}}$,

$f_{zz}(x, y, z) = \dfrac{\sqrt{x^2 + y^2 + z^2} - \dfrac{z^2}{\sqrt{x^2 + y^2 + z^2}}}{x^2 + y^2 + z^2} = \dfrac{x^2 + y^2}{(x^2 + y^2 + z^2)^{3/2}}$

39. Let $f(x, y) = x^2 + 16y^2$. Then $f_x(x, y) = 2x$, so that $f_x(-3, 1) = 2(-3) = -6$, so the line in question has equations $y = 1$ and $z - 25 = -6(x + 3)$.

41. $f_x(x, y) = -1$ and $f_y(x, y) = 0$, so $\sqrt{f_x^2(x, y) + f_y^2(x, y) + 1} = \sqrt{(-1)^2 + 0^2 + 1} = \sqrt{2}$.

43. $f_x(x, y) = \dfrac{x}{\sqrt{x^2 + y^2}}$ and $f_y(x, y) = \dfrac{y}{\sqrt{x^2 + y^2}}$,

so $\sqrt{f_x^2(x, y) + f_y^2(x, y) + 1} = \sqrt{\left(\dfrac{x}{\sqrt{x^2 + y^2}}\right)^2 + \left(\dfrac{y}{\sqrt{x^2 + y^2}}\right)^2 + 1} = \sqrt{2}$

45. $f_x(0, 0) = \lim_{h\to 0} \dfrac{f(h, 0) - f(0, 0)}{h - 0} = \lim_{h\to 0} \dfrac{0 - 0}{h} = \lim_{h\to 0} 0 = 0$,

and $f_y(0, 0) = \lim_{h\to 0} \dfrac{f(0, h) - f(0, 0)}{h - 0} = \lim_{h\to 0} \dfrac{0 - 0}{h} = \lim_{h\to 0} 0 = 0$

By Exercise 27 of Section 14.2, f is not continuous at (0, 0).

47. $\dfrac{\partial z}{\partial x} = -ae^{-ay} \sin ax$, $\dfrac{\partial^2 z}{\partial x^2} = -a^2e^{-ay} \cos ax$, $\dfrac{\partial z}{\partial y} = -ae^{-ay} \cos ax$;

thus $\dfrac{\partial^2 z}{\partial x^2} = -a^2e^{-ay} \cos ax = a\dfrac{\partial z}{\partial y}$

49. $u_x = 2x$, $v_y = 2x$, so $u_x = v_y$; $u_y = -2y$, $v_x = 2y$, so $u_y = -v_x$.

51. $u_x = e^x \cos y$, $v_y = e^x \cos y$, so $u_x = v_y$; $u_y = -e^x \sin y$, $v_x = e^x \sin y$, so $u_y = -v_x$.

53. $\dfrac{\partial z}{\partial x} = \dfrac{2x}{x^2 + y^2}$; $\dfrac{\partial^2 z}{\partial x^2} = \dfrac{2y^2 - 2x^2}{(x^2 + y^2)^2}$, $\dfrac{\partial z}{\partial y} = \dfrac{2y}{x^2 + y^2}$, $\dfrac{\partial^2 z}{\partial y^2} = \dfrac{2x^2 - 2y^2}{(x^2 + y^2)^2}$,

so $\dfrac{\partial^2 z}{\partial x^2} + \dfrac{\partial^2 z}{\partial y^2} = \dfrac{2y^2 - 2x^2}{(x^2 + y^2)^2} + \dfrac{2x^2 - 2y^2}{(x^2 + y^2)^2} = 0$.

55. By Example 3, $f_x(x, y) = (x^4y + 4x^2y^3 - y^5)/(x^2 + y^2)^2$, so

$$f_{xy}(x, 0) = \lim_{h\to 0} \frac{f_x(x, h) - f_x(x, 0)}{h - 0} = \lim_{h\to 0} \frac{x^4h + 4x^2h^3 - h^5}{h(x^2 + h^2)^2} = \lim_{h\to 0} \frac{x^4 + 4x^2h^2 - h^4}{(x^2 + h^2)^2} = 1 \quad \text{if } x \neq 0$$

Thus $\lim_{x\to 0} f_{xy}(x, 0) = 1$. Since $f_{xy}(0, 0) = -1$ by Example 7, it follows that f_{xy} is not continuous at (0, 0).

57. Let y be fixed in $[c, d]$, and let $f(x) = M(x, y)$. Then $f'(x) = (\partial M/\partial x)(x, y)$ for $a \le x \le b$, so that by the Fundamental Theorem of Calculus,

$$\int_a^b \frac{\partial M}{\partial x}(x, y)\,dx = \int_a^b f'(x)\,dx = f(x)\Big|_a^b = f(b) - f(a) = M(b, y) - M(a, y)$$

The other formula follows in a completely analogous fashion.

59. a. $\dfrac{\partial u}{\partial x} = ae^{ax+bt}$, $\dfrac{\partial^2 u}{\partial x^2} = a^2e^{ax+bt}$, and $\dfrac{\partial u}{\partial t} = be^{ax+bt}$, so that $\dfrac{b}{a^2}\dfrac{\partial^2 u}{\partial x^2} = be^{ax+bt} = \dfrac{\partial u}{\partial t}$.

b. $\dfrac{\partial u}{\partial x} = \dfrac{1}{\sqrt{t}}e^{-x^2/at}\left(\dfrac{-2x}{at}\right) = \dfrac{-2x}{at^{3/2}}e^{-x^2/at}$,

$$\frac{\partial^2 u}{\partial x^2} = \frac{-2}{at^{3/2}}e^{-x^2/at} - \frac{2x}{at^{3/2}}e^{-x^2/at}\left(\frac{-2x}{at}\right) = \left(\frac{4x^2}{a^2t^{5/2}} - \frac{2}{at^{3/2}}\right)e^{-x^2/at},$$

and $\dfrac{\partial u}{\partial t} = \dfrac{-1}{2t^{3/2}}e^{-x^2/at} + \dfrac{1}{t^{1/2}}e^{-x^2/at}\left(\dfrac{x^2}{at^2}\right) = \left(\dfrac{x^2}{at^{3/2}} - \dfrac{1}{2t^{3/2}}\right)e^{-x^2/at}$,

so that $\dfrac{\partial u}{\partial t} = \dfrac{a}{4}\dfrac{\partial^2 u}{\partial x^2}$.

61. $\dfrac{\partial a}{\partial m_1} = \dfrac{(m_1 + m_2) - (m_1 - m_2)}{(m_1 + m_2)^2}g = \dfrac{2m_2}{(m_1 + m_2)^2}g$

and $\dfrac{\partial a}{\partial m_2} = \dfrac{-(m_1 + m_2) - (m_1 - m_2)}{(m_1 + m_2)^2}g = \dfrac{-2m_1}{(m_1 + m_2)^2}g$

so that $m_1\dfrac{\partial a}{\partial m_1} + m_2\dfrac{\partial a}{\partial m_2} = \dfrac{2m_1m_2}{(m_1 + m_2)^2}g - \dfrac{2m_1m_2}{(m_1 + m_2)^2}g = 0$.

63. $\dfrac{\partial K}{\partial m} = \dfrac{1}{2}v^2$, $\dfrac{\partial K}{\partial v} = mv$, and $\dfrac{\partial^2 K}{\partial v^2} = m$, so that $\dfrac{\partial K}{\partial m}\dfrac{\partial^2 K}{\partial v^2} = (\tfrac{1}{2}v^2)(m) = \tfrac{1}{2}mv^2 = K$.

65. If $\mu = 1.333$, then $i_\mu = \arcsin\sqrt{\dfrac{4 - \mu^2}{3}} = \arcsin\sqrt{\dfrac{4 - (1.333)^2}{3}} \approx 1.03691$. By (3),

$$\theta(\mu, i_\mu) = 4\arcsin\left(\frac{\sin i_\mu}{\mu}\right) - 2i_\mu \approx .734402$$

or approximately 42.1°.

67. $R = c\dfrac{2\pi r(h + 2r)}{\pi r^2(h + \frac{4}{3}r)} = \dfrac{6c(h + 2r)}{3hr + 4r^2}$;

$$\frac{\partial R}{\partial r} = \frac{12c(3hr + 4r^2) - 6c(h + 2r)(3h + 8r)}{(3hr + 4r^2)^2} = \frac{-48chr - 18ch^2 - 48cr^2}{(3hr + 4r^2)^2} < 0;$$

$$\frac{\partial R}{\partial h} = \frac{6c(3hr + 4r^2) - 6c(h + 2r)3r}{(3hr + 4r^2)^2} = \frac{-12cr^2}{(3hr + 4r^2)^2} < 0.$$

69. a. $f(tx, ty) = (tx)^\alpha(ty)^\beta = t^{\alpha+\beta}x^\alpha y^\beta = t^{\alpha+\beta}f(x, y)$

b. $\dfrac{\partial z}{\partial x} = \alpha x^{\alpha-1}y^\beta$, so $\dfrac{1}{z}\dfrac{\partial z}{\partial x} = \dfrac{1}{x^\alpha y^\beta}(\alpha x^{\alpha-1}y^\beta) = \dfrac{\alpha}{x}$; $\dfrac{\partial z}{\partial y} = \beta x^\alpha y^{\beta-1}$, so

$\dfrac{1}{z}\dfrac{\partial z}{\partial y} = \dfrac{1}{x^\alpha y^\beta}(\beta x^\alpha y^{\beta-1}) = \dfrac{\beta}{y}$;

$$x\frac{\partial z}{\partial x} + y\frac{\partial z}{\partial y} = x(\alpha x^{\alpha-1}y^\beta) + y(\beta x^\alpha y^{\beta-1}) = (\alpha + \beta)x^\alpha y^\beta = (\alpha + \beta)z$$

71. a. Since the tax on each unit is t, the total tax on x units is tx, so the profit is given by $P(x, t) = P_0(x) - tx$.

b. By (a), $\dfrac{\partial P}{\partial x}(x, t) = P_0'(x) - t$ and thus $\dfrac{\partial^2 P}{\partial x^2}(x, t) = P_0''(x)$.

c. By (6), $0 = \dfrac{\partial P}{\partial x}(f(t), t)$ and by (b), $\dfrac{\partial P}{\partial x}(f(t), t) = P_0'(f(t)) - t$, so $P_0'(f(t)) - t = 0$.

d. From (c), $P_0''(f(t))f'(t) - 1 = 0$, so with the help of (b) and (7) we find that

$$f'(t) = \frac{1}{P_0''(f(t))} = \frac{1}{\dfrac{\partial^2 P}{\partial x^2}(f(t), t)} < 0$$

SECTION 14.4

1. $\dfrac{dz}{dt} = \dfrac{\partial z}{\partial x}\dfrac{dx}{dt} + \dfrac{\partial z}{\partial y}\dfrac{dy}{dt} = 4x\left(\dfrac{1}{2\sqrt{t}}\right) - 9y^2(2e^{2t}) = 4\sqrt{t}\left(\dfrac{1}{2\sqrt{t}}\right) - 9e^{4t}(2e^{2t}) = 2 - 18e^{6t}$

3. $\dfrac{dz}{dt} = \dfrac{\partial z}{\partial x}\dfrac{dx}{dt} + \dfrac{\partial z}{\partial y}\dfrac{dy}{dt} = (\cos x - y\sin xy)(2t) + (-x\sin xy)(0) = 2t(\cos t^2 - \sin t^2)$

5. $\dfrac{dz}{dt} = \dfrac{\partial z}{\partial x}\dfrac{dx}{dt} + \dfrac{\partial z}{\partial y}\dfrac{dy}{dt} = \dfrac{1}{\sqrt{2x - 4y}}\left(\dfrac{1}{t}\right) + \dfrac{-2}{\sqrt{2x - 4y}}(-9t^2) = \dfrac{1 + 18t^3}{t\sqrt{2\ln t - 4(1 - 3t^3)}}$

7. $\dfrac{\partial z}{\partial u} = \dfrac{\partial z}{\partial x}\dfrac{\partial x}{\partial u} + \dfrac{\partial z}{\partial y}\dfrac{\partial y}{\partial u} = \left(\dfrac{-4}{x^2y} - \dfrac{1}{y}\right)(2u) + \left(\dfrac{-4}{xy^2} + \dfrac{x}{y^2}\right)(v)$

$= \left(\dfrac{-4}{u^5v} - \dfrac{1}{uv}\right)(2u) + \left(\dfrac{-4}{u^4v^2} + \dfrac{1}{v^2}\right)(v) = \dfrac{-12 - u^4}{u^4v}$;

$\dfrac{\partial z}{\partial v} = \dfrac{\partial z}{\partial x}\dfrac{\partial x}{\partial v} + \dfrac{\partial z}{\partial y}\dfrac{\partial y}{\partial v} = \left(\dfrac{-4}{x^2y} - \dfrac{1}{y}\right)(0) + \left(\dfrac{-4}{xy^2} + \dfrac{x}{y^2}\right)(u) = \left(\dfrac{-4}{u^4v^2} + \dfrac{1}{v^2}\right)(u) = \dfrac{-4 + u^4}{u^3v^2}$

9. $\dfrac{\partial z}{\partial u} = \dfrac{\partial z}{\partial x}\dfrac{\partial x}{\partial u} + \dfrac{\partial z}{\partial y}\dfrac{\partial y}{\partial u} = \dfrac{2x}{x^2 - y^2}(1) - \dfrac{2y}{x^2 - y^2}(2u) = \dfrac{2(u - v) - 4u(u^2 + v^2)}{(u - v)^2 - (u^2 + v^2)^2}$

$\dfrac{\partial z}{\partial v} = \dfrac{\partial z}{\partial x}\dfrac{\partial x}{\partial v} + \dfrac{\partial z}{\partial y}\dfrac{\partial y}{\partial v} = \dfrac{2x}{x^2 - y^2}(-1) - \dfrac{2y}{x^2 - y^2}(2v) = \dfrac{-2(u - v) - 4v(u^2 + v^2)}{(u - v)^2 - (u^2 + v^2)^2}$

11. $\dfrac{\partial z}{\partial r}=\dfrac{\partial z}{\partial u}\dfrac{\partial u}{\partial r}+\dfrac{\partial z}{\partial v}\dfrac{\partial v}{\partial r}=(2\cos 2u\cos 3v)(2(r+s))+(-3\sin 2u\sin 3v)(2(r-s))$

$=4(r+s)\cos[2(r+s)^2]\cos[3(r-s)^2]-6(r-s)\sin[2(r+s)^2]\sin[3(r-s)^2]$

$\dfrac{\partial z}{\partial s}=\dfrac{\partial z}{\partial u}\dfrac{\partial u}{\partial s}+\dfrac{\partial z}{\partial v}\dfrac{\partial v}{\partial s}=(2\cos 2u\cos 3v)(2(r+s))+(-3\sin 2u\sin 3v)(-2(r-s))$

$=4(r+s)\cos[2(r+s)^2]\cos[3(r-s)^2]+6(r-s)\sin[2(r+s)^2]\sin[3(r-s)^2]$

13. $\dfrac{\partial z}{\partial r}=\dfrac{\partial z}{\partial u}\dfrac{\partial u}{\partial r}+\dfrac{\partial z}{\partial v}\dfrac{\partial v}{\partial r}=(e^v-ve^{-u})\left(\dfrac{1}{r}\right)+(ue^v+e^{-u})\left(\dfrac{s}{r}\right)$

$=\dfrac{1}{r}\left(r^s-\dfrac{s}{r}\ln r\right)+\dfrac{s}{r}\left(r^s\ln r+\dfrac{1}{r}\right)$

$\dfrac{\partial z}{\partial s}=\dfrac{\partial z}{\partial u}\dfrac{\partial u}{\partial s}+\dfrac{\partial z}{\partial v}\dfrac{\partial v}{\partial s}=(e^v-ve^{-u})(0)+(ue^v+e^{-u})(\ln r)=\left(r^s\ln r+\dfrac{1}{r}\right)\ln r$

15. $\dfrac{dw}{dt}=\dfrac{\partial w}{\partial x}\dfrac{dx}{dt}+\dfrac{\partial w}{\partial y}\dfrac{dy}{dt}+\dfrac{\partial w}{\partial z}\dfrac{dz}{dt}=\left(\dfrac{1}{y}+\dfrac{z}{x^2}\right)(\cos t)-\dfrac{x}{y^2}(-\sin t)-\dfrac{1}{x}(\sec^2 t)$

$=1+\csc t+\tan^2 t-\csc t\sec^2 t$

17. $\dfrac{dw}{dt}=\dfrac{\partial w}{\partial x}\dfrac{dx}{dt}+\dfrac{\partial w}{\partial y}\dfrac{dy}{dt}+\dfrac{\partial w}{\partial z}\dfrac{dz}{dt}$

$=\dfrac{x}{\sqrt{x^2+y^2+z^2}}(e^t)+\dfrac{y}{\sqrt{x^2+y^2+z^2}}(-e^{-t})+\dfrac{z}{\sqrt{x^2+y^2+x^2}}(2)$

$=\dfrac{e^{2t}-e^{-2t}+4t}{\sqrt{e^{2t}+e^{-2t}+4t^2}}$

19. $\dfrac{dw}{dt}=\dfrac{\partial w}{\partial x}\dfrac{dx}{dt}+\dfrac{\partial w}{\partial y}\dfrac{dy}{dt}+\dfrac{\partial w}{\partial z}\dfrac{dz}{dt}$

$=(y^2z^3\cos xy^2z^3)(3)+(2xyz^3\cos xy^2z^3)(\tfrac{1}{2}t^{-1/2})+(3xy^2z^2\cos xy^2z^3)(\tfrac{1}{3}t^{-2/3})$

$=9t^2\cos 3t^3$

21. $\dfrac{\partial w}{\partial u}=\dfrac{\partial w}{\partial x}\dfrac{\partial x}{\partial u}+\dfrac{\partial w}{\partial y}\dfrac{\partial y}{\partial u}+\dfrac{\partial w}{\partial z}\dfrac{\partial z}{\partial u}=\dfrac{-yz(2x+y)}{x^2(x+y)^2}(2u)+\dfrac{\partial w}{\partial y}(0)+\dfrac{y}{x(x+y)}(2u)$

$=\dfrac{-v^2(u^2-v^2)(2u^2+v^2)2u+v^2(2u)[u^2(u^2+v^2)]}{u^4(u^2+v^2)^2}=\dfrac{2v^2(-u^4+2u^2v^2+v^4)}{u^3(u^2+v^2)^2}$

$\dfrac{\partial w}{\partial v}=\dfrac{\partial w}{\partial x}\dfrac{\partial x}{\partial v}+\dfrac{\partial w}{\partial y}\dfrac{\partial y}{\partial v}+\dfrac{\partial w}{\partial z}\dfrac{\partial z}{\partial v}=\dfrac{\partial w}{\partial x}(0)+\dfrac{(x^2+xy)z-yzx}{x^2(x+y)^2}(2v)+\dfrac{y}{x(x+y)}(-2v)$

$=\dfrac{u^4(u^2-v^2)(2v)-v^2(2v)[u^2(u^2+v^2)]}{u^4(u^2+v^2)^2}=\dfrac{2v(u^4-2u^2v^2-v^4)}{u^2(u^2+v^2)^2}$

23. $\dfrac{\partial w}{\partial u}=\dfrac{\partial w}{\partial x}\dfrac{\partial x}{\partial u}+\dfrac{\partial w}{\partial y}\dfrac{\partial y}{\partial u}+\dfrac{\partial w}{\partial z}\dfrac{\partial z}{\partial u}=\dfrac{y}{x}(ve^u)+(\ln xz)(2uv^4)+\dfrac{y}{z}e^v$

$=\dfrac{u^2v^4}{ve^u}(ve^u)+2uv^4\ln(ve^uue^v)+\dfrac{u^2v^4}{ue^v}e^v=uv^4[1+u+2\ln(uve^ue^v)]$

$\dfrac{\partial w}{\partial v}=\dfrac{\partial w}{\partial x}\dfrac{\partial x}{\partial v}+\dfrac{\partial w}{\partial y}\dfrac{\partial y}{\partial v}+\dfrac{\partial w}{\partial z}\dfrac{\partial z}{\partial v}=\dfrac{y}{x}(e^u)+(\ln xz)(4u^2v^3)+\dfrac{y}{z}(ue^v)$

$=\dfrac{u^2v^4}{ve^u}(e^u)+4u^2v^3\ln(ve^uue^v)+\dfrac{u^2v^4}{ue^v}(ue^v)=u^2v^3[1+v+4\ln(uve^ue^v)]$

25. Let $z=x^3+4x^2y-3xy^2+2y^3+5$. Then $\dfrac{dy}{dx}=\dfrac{-\partial z/\partial x}{\partial z/\partial y}=\dfrac{-3x^2-8xy+3y^2}{4x^2-6xy+6y^2}$.

27. Let $z=x^2+y^2+\sin xy^2$. Then $\dfrac{dy}{dx}=\dfrac{-\partial z/\partial x}{\partial z/\partial y}=\dfrac{-2x-y^2\cos xy^2}{2y+2xy\cos xy^2}$.

29. Let $z=x^2-\dfrac{y^2}{y^2-1}$. Then $\dfrac{dy}{dx}=\dfrac{-\partial z/\partial x}{\partial z/\partial y}=\dfrac{-2x}{2y/(y^2-1)^2}=\dfrac{-x(y^2-1)^2}{y}$.

31. Let $w=x-yz+\cos xyz-2$. Then

$$\frac{\partial z}{\partial x}=\frac{-\partial w/\partial x}{\partial w/\partial z}=\frac{-1+yz\sin xyz}{-y-xy\sin xyz}\quad\text{and}\quad\frac{\partial z}{\partial y}=\frac{-\partial w/\partial y}{\partial w/\partial z}=\frac{z+xz\sin xyz}{-y-xy\sin xyz}$$

33. If we let $u=x-y$, then $z=f(u)$, so that

$$\frac{\partial z}{\partial x}=\frac{dz}{du}\frac{\partial u}{\partial x}=\frac{dz}{du}\quad\text{and}\quad\frac{\partial z}{\partial y}=\frac{dz}{du}\frac{\partial u}{\partial y}=-\frac{dz}{du}$$

Thus $dz/dx=-\partial z/\partial y$.

35. If $u=y+ax$ and $v=y-ax$, then $z=f(u)+g(v)$, so that

$$\frac{\partial z}{\partial x}=\frac{\partial z}{\partial u}\frac{\partial u}{\partial x}+\frac{\partial z}{\partial v}\frac{\partial v}{\partial x}=a\frac{df}{du}-a\frac{dg}{dv}\quad\text{and}\quad\frac{\partial z}{\partial y}=\frac{\partial z}{\partial u}\frac{\partial u}{\partial y}+\frac{\partial z}{\partial v}\frac{\partial v}{\partial y}=\frac{df}{du}+\frac{dg}{dv}$$

It follows that

$$\frac{\partial^2 z}{\partial x^2}=\frac{\partial}{\partial x}\left(\frac{\partial z}{\partial x}\right)=\frac{\partial}{\partial x}\left(a\frac{df}{du}-a\frac{dg}{dv}\right)=\left[\frac{\partial}{\partial u}\left(a\frac{df}{du}-a\frac{dg}{dv}\right)\right]\frac{\partial u}{\partial x}$$
$$+\left[\frac{\partial}{\partial v}\left(a\frac{df}{du}-a\frac{dg}{dv}\right)\right]\frac{\partial v}{\partial x}=\left(a\frac{d^2f}{du^2}-0\right)(a)+\left(0-a\frac{d^2g}{dv^2}\right)(-a)$$
$$=a^2\left(\frac{d^2f}{du^2}+\frac{d^2g}{dv^2}\right)$$

Similarly,

$$\frac{\partial^2 z}{\partial y^2}=\frac{\partial}{\partial y}\left(\frac{\partial z}{\partial y}\right)=\frac{\partial}{\partial y}\left(\frac{df}{du}+\frac{dg}{dv}\right)=\frac{\partial}{\partial u}\left(\frac{df}{du}+\frac{dg}{dv}\right)\frac{\partial u}{\partial y}+\frac{\partial}{\partial v}\left(\frac{df}{du}+\frac{dg}{dv}\right)\frac{\partial v}{\partial y}$$
$$=\left(\frac{d^2f}{du^2}+0\right)(1)+\left(0+\frac{d^2g}{dv^2}\right)(1)=\frac{d^2f}{du^2}+\frac{d^2g}{dv^2}$$

Thus $\partial^2z/\partial x^2=a^2(\partial^2z/\partial y^2)$.

37. We have $$\frac{\partial w}{\partial s} = \frac{\partial w}{\partial x}\frac{\partial x}{\partial s} + \frac{\partial w}{\partial y}\frac{\partial y}{\partial s} = \frac{\partial w}{\partial x} e^s \cos t + \frac{\partial w}{\partial y} e^s \sin t$$

and $$\frac{\partial w}{\partial t} = \frac{\partial w}{\partial x}\frac{\partial x}{\partial t} + \frac{\partial w}{\partial y}\frac{\partial y}{\partial t} = -\frac{\partial w}{\partial x} e^s \sin t + \frac{\partial w}{\partial y} e^s \cos t$$

Thus

$$\frac{\partial^2 w}{\partial s^2} = \frac{\partial}{\partial s}\left(\frac{\partial w}{\partial s}\right) = \frac{\partial}{\partial s}\left(\frac{\partial w}{\partial x} e^s \cos t + \frac{\partial w}{\partial y} e^s \sin t\right)$$
$$= \left[\frac{\partial}{\partial s}\left(\frac{\partial w}{\partial x}\right)\right] e^s \cos t + \frac{\partial w}{\partial x} e^s \cos t + \left[\frac{\partial}{\partial s}\left(\frac{\partial w}{\partial y}\right)\right] e^s \sin t + \frac{\partial w}{\partial y} e^s \sin t$$

and

$$\frac{\partial^2 w}{\partial t^2} = \frac{\partial}{\partial t}\left(\frac{\partial w}{\partial t}\right) = \frac{\partial}{\partial t}\left(-\frac{\partial w}{\partial x} e^s \sin t + \frac{\partial w}{\partial y} e^s \cos t\right)$$
$$= -\left[\frac{\partial}{\partial t}\left(\frac{\partial w}{\partial x}\right)\right] e^s \sin t - \frac{\partial w}{\partial x} e^s \cos t + \left[\frac{\partial}{\partial t}\left(\frac{\partial w}{\partial y}\right)\right] e^s \cos t - \frac{\partial w}{\partial y} e^s \sin t$$

But

$$\frac{\partial}{\partial s}\left(\frac{\partial w}{\partial x}\right) = \left[\frac{\partial}{\partial x}\left(\frac{\partial w}{\partial x}\right)\right]\frac{\partial x}{\partial s} + \left[\frac{\partial}{\partial y}\left(\frac{\partial w}{\partial x}\right)\right]\frac{\partial y}{\partial s} = \frac{\partial^2 w}{\partial x^2} e^s \cos t + \frac{\partial^2 w}{\partial y \partial x} e^s \sin t$$
$$\frac{\partial}{\partial s}\left(\frac{\partial w}{\partial y}\right) = \left[\frac{\partial}{\partial x}\left(\frac{\partial w}{\partial y}\right)\right]\frac{\partial x}{\partial s} + \left[\frac{\partial}{\partial y}\left(\frac{\partial w}{\partial y}\right)\right]\frac{\partial y}{\partial s} = \frac{\partial^2 w}{\partial x \partial y} e^s \cos t + \frac{\partial^2 w}{\partial y^2} e^s \sin t$$
$$\frac{\partial}{\partial t}\left(\frac{\partial w}{\partial x}\right) = \left[\frac{\partial}{\partial x}\left(\frac{\partial w}{\partial x}\right)\right]\frac{\partial x}{\partial t} + \left[\frac{\partial}{\partial y}\left(\frac{\partial w}{\partial x}\right)\right]\frac{\partial y}{\partial t} = -\frac{\partial^2 w}{\partial x^2} e^s \sin t + \frac{\partial^2 w}{\partial y \partial x} e^s \cos t$$

and

$$\frac{\partial}{\partial t}\left(\frac{\partial w}{\partial y}\right) = \left[\frac{\partial}{\partial x}\left(\frac{\partial w}{\partial y}\right)\right]\frac{\partial x}{\partial t} + \left[\frac{\partial}{\partial y}\left(\frac{\partial w}{\partial y}\right)\right]\frac{\partial y}{\partial t} = -\frac{\partial^2 w}{\partial x \partial y} e^s \sin t + \frac{\partial^2 w}{\partial y^2} e^s \cos t$$

Thus

$$\frac{\partial^2 w}{\partial s^2} + \frac{\partial^2 w}{\partial t^2} = \left[\left(\frac{\partial^2 w}{\partial x^2} e^s \cos t + \frac{\partial^2 w}{\partial y \partial x} e^s \sin t\right) e^s \cos t + \frac{\partial w}{\partial x} e^s \cos t\right.$$
$$\left. + \left(\frac{\partial^2 w}{\partial x \partial y} e^s \cos t + \frac{\partial^2 w}{\partial y^2} e^s \sin t\right) e^s \sin t + \frac{\partial w}{\partial y} e^s \sin t\right]$$
$$+ \left[-\left(-\frac{\partial^2 w}{\partial x^2} e^s \sin t + \frac{\partial^2 w}{\partial y \partial x} e^s \cos t\right) e^s \sin t - \frac{\partial w}{\partial x} e^s \cos t\right.$$
$$\left. + \left(-\frac{\partial^2 w}{\partial x \partial y} e^s \sin t + \frac{\partial^2 w}{\partial y^2} e^s \cos t\right) e^s \cos t - \frac{\partial w}{\partial y} e^s \sin t\right]$$
$$= e^{2s}\left(\frac{\partial^2 w}{\partial x^2} + \frac{\partial^2 w}{\partial y^2}\right)$$

Thus $$\frac{\partial^2 w}{\partial x^2} + \frac{\partial^2 w}{\partial y^2} = e^{-2s}\left(\frac{\partial^2 w}{\partial s^2} + \frac{\partial^2 w}{\partial t^2}\right)$$

39. Since $f(tx, ty) = \tan\,[(tx)^2 + (ty)^2]/(txty) = f(x, y) = t^0 f(x, y)$, we find from Exercise 38 with $n = 0$ that $xf_x(x, y) + yf_y(x, y) = 0$.

41. By Exercise 17 of Section 14.2, f is not continuous at (0, 0). From Exercise 40, it follows that f is not differentiable at (0, 0).

43. $$\frac{dQ}{dt} = \frac{\partial Q}{\partial r}\frac{dr}{dt} + \frac{\partial Q}{\partial p}\frac{dp}{dt} = \left(\frac{4\pi p r^3}{8\ell\eta}\right)\left(\frac{1}{10}\right) + \left(\frac{\pi r^4}{8\ell\eta}\right)\left(\frac{-1}{5}\right) = \frac{\pi r^3}{40\ell\eta}(2p - r)$$

45. $$\frac{dF}{dt} = \frac{\partial F}{\partial m}\frac{dm}{dt} + \frac{\partial F}{\partial r}\frac{dr}{dt} = \frac{GM}{r^2}\frac{dm}{dt} - \frac{2GMm}{r^3}\frac{dr}{dt}$$

Thus if $dm/dt = -40$, $r = 6400$, and $dr/dt = 100$, we have

$$\frac{dF}{dt} = \frac{GM}{(6400)^2}(-40) - \frac{2GMm}{(6400)^3}(100) = \frac{-GM}{(6400)^2}\left(40 + \frac{m}{32}\right)$$

SECTION 14.5

1. $f_x(x, y) = 4x - 3y$, $f_y(x, y) = -3x + 2y$, and $\|\mathbf{a}\| = 1$;

$$D_{\mathbf{a}} f(1, 1) = f_x(1, 1)\left(\frac{1}{\sqrt{2}}\right) + f_y(1, 1)\left(\frac{1}{\sqrt{2}}\right) = \frac{1}{\sqrt{2}} - \frac{1}{\sqrt{2}} = 0$$

3. $$f_x(x, y) = \frac{2x(x^2 + y^2) - (x^2 - y^2)2x}{(x^2 + y^2)^2} = \frac{4xy^2}{(x^2 + y^2)^2},$$
$$f_y(x, y) = \frac{-2y(x^2 + y^2) - (x^2 - y^2)2y}{(x^2 + y^2)^2} = \frac{-4x^2 y}{(x^2 + y^2)^2},$$
and $\|\mathbf{a}\| = 1$; $D_{\mathbf{a}} f(3, 4) = f_x(3, 4)\left(\frac{1}{2}\right) + f_y(3, 4)\left(\frac{-\sqrt{3}}{2}\right) = \frac{96 + 72\sqrt{3}}{625}$

5. $f_x(x, y) = 0$, $f_y(x, y) = 4e^{4y}$, and $\|\mathbf{a}\| = 4$, so that $\mathbf{u} = \mathbf{a}/\|\mathbf{a}\| = \mathbf{i}$;

$D_{\mathbf{u}} f(\frac{1}{2}, \frac{1}{4}) = f_x(\frac{1}{2}, \frac{1}{4})(1) + f_y(\frac{1}{2}, \frac{1}{4})(0) = 0$

7. $f_x(x, y) = \sec^2(x + 2y)$, $f_y(x, y) = 2\sec^2(x + 2y)$, and $\|\mathbf{a}\| = \sqrt{41}$,

so that $\mathbf{u} = \dfrac{\mathbf{a}}{\|\mathbf{a}\|} = \dfrac{-4}{\sqrt{41}}\mathbf{i} + \dfrac{5}{\sqrt{41}}\mathbf{j}$;

$$D_{\mathbf{u}} f\left(0, \frac{\pi}{6}\right) = f_x\left(0, \frac{\pi}{6}\right)\left(\frac{-4}{\sqrt{41}}\right) + f_y\left(0, \frac{\pi}{6}\right)\left(\frac{5}{\sqrt{41}}\right) = \frac{-16}{\sqrt{41}} + \frac{40}{\sqrt{41}} = \frac{24}{41}\sqrt{41}$$

9. $f_x(x, y, z) = 3x^2y^2z$, $f_y(x, y, z) = 2x^3yz$, $f_z(x, y, z) = x^3y^2$, and $\|\mathbf{a}\| = 3$,

so that $\mathbf{u} = \dfrac{\mathbf{a}}{\|\mathbf{a}\|} = \dfrac{2}{3}\mathbf{i} - \dfrac{1}{3}\mathbf{j} - \dfrac{2}{3}\mathbf{k}$;

$$D_{\mathbf{u}} f(2, -1, 2) = f_x(2, -1, 2)\left(\frac{2}{3}\right) + f_y(2, -1, 2)\left(\frac{-1}{3}\right) + f_z(2, -1, 2)\left(\frac{-2}{3}\right)$$
$$= 16 + \tfrac{32}{3} - \tfrac{16}{3} = \tfrac{64}{3}$$

11. $f_x(x, y, z) = \dfrac{(x+y+z)-(x-y-z)}{(x+y+z)^2} = \dfrac{2y+2z}{(x+y+z)^2}$,

$f_y(x, y, z) = \dfrac{-(x+y+z)-(x-y-z)}{(x+y+z)^2} = \dfrac{-2x}{(x+y+z)^2}$,

$f_z(x, y, z) = \dfrac{-(x+y+z)-(x-y-z)}{(x+y+z)^2} = \dfrac{-2x}{(x+y+z)^2}$, and $\|\mathbf{a}\| = \sqrt{6}$,

so that $\mathbf{u} = \dfrac{\mathbf{a}}{\|\mathbf{a}\|} = \dfrac{-2}{\sqrt{6}}\mathbf{i} - \dfrac{1}{\sqrt{6}}\mathbf{j} - \dfrac{1}{\sqrt{6}}\mathbf{k}$;

$$D_{\mathbf{u}}f(2, 1, -1) = f_x(2, 1, -1)\left(\frac{-2}{\sqrt{6}}\right) + f_y(2, 1, -1)\left(\frac{-1}{\sqrt{6}}\right) + f_z(2, 1, -1)\left(\frac{-1}{\sqrt{6}}\right)$$

$$= 0 + \frac{1}{\sqrt{6}} + \frac{1}{\sqrt{6}} = \frac{\sqrt{6}}{3}$$

13. $f(x, y, z) = yz2^x = yze^{x \ln 2}$; $f_x(x, y, z) = (\ln 2)\, yze^{x \ln 2} = yz2^x \ln 2$,

$f_y(x, y, z) = z2^x$, $f_z(x, y, z) = y2^x$, $\|\mathbf{a}\| = \sqrt{5}$, so that $\mathbf{u} = \dfrac{\mathbf{a}}{\|\mathbf{a}\|} = \dfrac{2}{\sqrt{5}}\mathbf{j} - \dfrac{1}{\sqrt{5}}\mathbf{k}$;

$$D_{\mathbf{u}}f(1, -1, 1) = f_x(1, -1, 1)(0) + f_y(1, -1, 1)\left(\frac{2}{\sqrt{5}}\right) + f_z(1, -1, 1)\left(\frac{-1}{\sqrt{5}}\right)$$

$$= 0 + \frac{4}{\sqrt{5}} + \frac{2}{\sqrt{5}} = \frac{6}{5}\sqrt{5}$$

15. Let $r = f_x(1, 2)$ and $s = f_y(1, 2)$. Since $\|\mathbf{a}\| = \sqrt{2}$ and $\|\mathbf{b}\| = 3\sqrt{2}$, let

$$\mathbf{u_a} = \frac{\mathbf{a}}{\|\mathbf{a}\|} = \frac{1}{\sqrt{2}}\mathbf{i} - \frac{1}{\sqrt{2}}\mathbf{j} \quad \text{and} \quad \mathbf{u_b} = \frac{\mathbf{b}}{\|\mathbf{b}\|} = \frac{1}{\sqrt{2}}\mathbf{i} + \frac{1}{\sqrt{2}}\mathbf{j}$$

Then $\quad 6\sqrt{2} = D_{\mathbf{u_a}}f(1, 2) = f_x(1, 2)\left(\dfrac{1}{\sqrt{2}}\right) + f_y(1, 2)\left(\dfrac{-1}{\sqrt{2}}\right) = \dfrac{r}{\sqrt{2}} - \dfrac{s}{\sqrt{2}}$

and $\quad -2\sqrt{2} = D_{\mathbf{u_b}}f(1, 2) = f_x(1, 2)\left(\dfrac{1}{\sqrt{2}}\right) + f_y(1, 2)\left(\dfrac{1}{\sqrt{2}}\right) = \dfrac{r}{\sqrt{2}} + \dfrac{s}{\sqrt{2}}$

Adding, we obtain $4\sqrt{2} = 2r/\sqrt{2}$ so that $r = 4$, and thus $s = -8$. Consequently $f_x(1, 2) = 4$ and $f_y(1, 2) = -8$.

17. Let $\mathbf{a} = 2\mathbf{i} - \mathbf{j}$. Then $\|\mathbf{a}\| = \sqrt{4+1} = \sqrt{5}$, so we will find $D_{\mathbf{u}}f(2, -1)$, where

$$\mathbf{u} = \frac{1}{\|\mathbf{a}\|}\mathbf{a} = \frac{2}{\sqrt{5}}\mathbf{i} - \frac{1}{\sqrt{5}}\mathbf{j}$$

Since $f_x(x, y) = (y/x)\cosh(y \ln x)$ and $f_y(x, y) = (\ln x)\cosh(y \ln x)$, it follows that

$$D_{\mathbf{u}}f(2, -1) = f_x(2, -1)\left(\frac{2}{\sqrt{5}}\right) + f_y(2, -1)\left(-\frac{1}{\sqrt{5}}\right)$$

$$= \left[-\frac{1}{2}\cosh(-\ln 2)\right]\left(\frac{2}{\sqrt{5}}\right) + [(\ln 2)\cosh(-\ln 2)]\left(-\frac{1}{\sqrt{5}}\right)$$

$$= -\frac{\sqrt{5}}{4}(1 + \ln 2)$$

SECTION 14.6

1. $\operatorname{grad} f(x, y) = 3\mathbf{i} - 5\mathbf{j}$

3. $\operatorname{grad} g(x, y) = -2e^{-2x}\ln(y-4)\mathbf{i} + \dfrac{e^{-2x}}{y-4}\mathbf{j}$

5. $\operatorname{grad} f(x, y, z) = 4x\mathbf{i} - 2y\mathbf{j} - 8z\mathbf{k}$

7. $\operatorname{grad} g(x, y, z) = \dfrac{-(-x+z)-(-x+y)(-1)}{(-x+z)^2}\mathbf{i} + \dfrac{1}{-x+z}\mathbf{j} - \dfrac{-x+y}{(-x+z)^2}\mathbf{k}$

$$= \frac{y-z}{(-x+z)^2}\mathbf{i} + \frac{1}{-x+z}\mathbf{j} + \frac{x-y}{(-x+z)^2}\mathbf{k}$$

9. $\operatorname{grad} f(x, y) = \dfrac{5x+2y-(x+3y)(5)}{(5x+2y)^2}\mathbf{i} + \dfrac{3(5x+2y)-(x+3y)(2)}{(5x+2y)^2}\mathbf{j}$

$$= \frac{-13y}{(5x+2y)^2}\mathbf{i} + \frac{13x}{(5x+2y)^2}\mathbf{j}$$

$\operatorname{grad} f(-1, \tfrac{3}{2}) = -\tfrac{39}{8}\mathbf{i} - \tfrac{13}{4}\mathbf{j}$

11. $\operatorname{grad} g(x, y) = \left[\ln(x+y) + \dfrac{x}{x+y}\right]\mathbf{i} + \dfrac{x}{x+y}\mathbf{j}$

$\operatorname{grad} g(-2, 3) = -2\mathbf{i} - 2\mathbf{j}$

13. $\operatorname{grad} f(x, y, z) = -ze^{-x}\tan y\mathbf{i} + ze^{-x}\sec^2 y\mathbf{j} + e^{-x}\tan y\mathbf{k}$

$\operatorname{grad} f(0, \pi, -2) = -2\mathbf{j}$

15. $\operatorname{grad} f(x, y) = e^x(\cos y + \sin y)\mathbf{i} + e^x(-\sin y + \cos y)\mathbf{j}$, so $\operatorname{grad} f(0, 0) = \mathbf{i} + \mathbf{j}$. Consequently the direction in which f increases most rapidly at $(0, 0)$ is $\mathbf{i} + \mathbf{j}$.

17. $\operatorname{grad} f(x, y) = 6x\mathbf{i} + 8y\mathbf{j}$, so that $\operatorname{grad} f(-1, 1) = -6\mathbf{i} + 8\mathbf{j} = 2(-3\mathbf{i} + 4\mathbf{j})$. Thus the direction in which f increases most rapidly at $(-1, 1)$ is $-3\mathbf{i} + 4\mathbf{j}$.

19. $\operatorname{grad} f(x, y, z) = e^x\mathbf{i} + e^y\mathbf{j} + 2e^{2z}\mathbf{k}$,

so that $\operatorname{grad} f(1, 1, -1) = e\mathbf{i} + e\mathbf{j} + 2e^{-2}\mathbf{k} = e(\mathbf{i} + \mathbf{j} + 2e^{-3}\mathbf{k})$. Thus the direction in which f increases most rapidly at $(1, 1, -1)$ is $\mathbf{i} + \mathbf{j} + 2e^{-3}\mathbf{k}$.

21. $\operatorname{grad} f(x, y) = \pi y\cos \pi xy\mathbf{i} + \pi x\cos \pi xy\mathbf{j}$,

so that $\operatorname{grad} f(\tfrac{1}{2}, \tfrac{2}{3}) = (\pi/3)\mathbf{i} + (\pi/4)\mathbf{j} = \pi/12\,(4\mathbf{i} + 3\mathbf{j})$. Thus the direction in which f decreases most rapidly at $(\tfrac{1}{2}, \tfrac{2}{3})$ is $-4\mathbf{i} - 3\mathbf{j}$.

23. $\operatorname{grad} f(x, y, z) = \dfrac{1}{y+z}\mathbf{i} - \dfrac{x-z}{(y+z)^2}\mathbf{j} + \dfrac{-(y+z)-(x-z)}{(y+z)^2}\mathbf{k}$

$$= \frac{1}{y+z}\mathbf{i} + \frac{z-x}{(y+z)^2}\mathbf{j} - \frac{y+x}{(y+z)^2}\mathbf{k}$$

so that $\operatorname{grad} f(-1, 1, 3) = \tfrac{1}{4}\mathbf{i} + \tfrac{1}{4}\mathbf{j} = \tfrac{1}{4}(\mathbf{i} + \mathbf{j})$. Thus the direction in which f decreases most rapidly at $(-1, 1, 3)$ is $-\mathbf{i} - \mathbf{j}$.

25. Let $f(x, y) = \sin \pi xy$. Since the graph of the given equation is a level curve of f, Theorem 14.16 implies that grad $f(\frac{1}{6}, 2)$ is normal to the graph at $(\frac{1}{6}, 2)$. Since grad $f(x, y) = \pi y \cos \pi xy\mathbf{i} + \pi x \cos \pi xy\mathbf{j}$, so that grad $f(\frac{1}{6}, 2) = \pi\mathbf{i} + (\pi/12)\mathbf{j}$, we find that $\pi\mathbf{i} + (\pi/12)\mathbf{j}$ is normal to the graph at $(\frac{1}{6}, 2)$.

27. By (2), $f_x(-2, 1)\mathbf{i} + f_y(-2, 1)\mathbf{j} - \mathbf{k}$ is normal to the graph of f at $(-2, 1, 16)$. Since $f_x(x, y) = 6x$ and $f_y(x, y) = 8y$, so $f_x(-2, 1) = -12$ and $f_y(-2, 1) = 8$, it follows that $-12\mathbf{i} + 8\mathbf{j} - \mathbf{k}$ is normal to the graph at $(-2, 1, 16)$.

29. By (2), $f_x(0, 2)\mathbf{i} + f_y(0, 2)\mathbf{j} - \mathbf{k}$ is normal to the graph of f at $(0, 2, 1)$. Since $f_x(x, y) = -2x$ and $f_y(x, y) = 0$, so $f_x(0, 2) = 0 = f_y(0, 2)$, it follows that $-\mathbf{k}$ is normal to the graph at $(0, 2, 1)$.

31. Since $f_x(x, y) = y - 1$ and $f_y(x, y) = x + 1$, so $f_x(0, 2) = 1$ and $f_y(0, 2) = 1$, it follows from (3) that an equation of the plane tangent at $(0, 2, 7)$ is $(x - 0) + (y - 2) - (z - 7) = 0$, or $x + y - z = -5$.

33. Since $g_x(x, y) = \pi y \cos \pi xy$ and $g_y(x, y) = \pi x \cos \pi xy$, so

$$g_x(-\sqrt{2}, \sqrt{2}) = \pi\sqrt{2},\ g_y(-\sqrt{2}, \sqrt{2}) = -\pi\sqrt{2},\ g_x\left(-\frac{1}{2}, \frac{1}{3}\right) = \frac{\pi\sqrt{3}}{6}$$

and

$$g_y\left(-\frac{1}{2}, \frac{1}{3}\right) = \frac{-\pi\sqrt{3}}{4}$$

it follows from (3) that an equation of the plane tangent at $(-\sqrt{2}, \sqrt{2}, 0)$ is

$$\pi\sqrt{2}(x + \sqrt{2}) - \pi\sqrt{2}(y - \sqrt{2}) - (z - 0) = 0, \quad \text{or} \quad \pi\sqrt{2}x - \pi\sqrt{2}y - z = -4\pi$$

and that an equation of the plane tangent at $(-\frac{1}{2}, \frac{1}{3}, -1/2)$ is

$$\frac{\pi\sqrt{3}}{6}\left(x + \frac{1}{2}\right) - \frac{\pi\sqrt{3}}{4}\left(y - \frac{1}{3}\right) - \left(z + \frac{1}{2}\right) = 0$$

or

$$\frac{\pi\sqrt{3}}{6}x - \frac{\pi\sqrt{3}}{4}y - z = \frac{-\pi\sqrt{3}}{6} + \frac{1}{2}$$

35. Since $f_x(x, y) = 2(2 + x - y)$ and $f_y(x, y) = -2(2 + x - y)$, so $f_x(3, -1) = 12$ and $f_y(3, -1) = -12$, it follows from (3) that an equation of the plane tangent at $(3, -1, 36)$ is

$$12(x - 3) - 12(y + 1) - (z - 36) = 0, \quad \text{or} \quad 12x - 12y - z = 12$$

37. Since $f_x(x, y) = 2x/(x^2 + y^2)$ and $f_y(x, y) = 2y/(x^2 + y^2)$, so $f_x(-1, 0) = -2$, $f_y(-1, 0) = 0$, $f_x(-1, 1) = -1$, and $f_y(-1, 1) = 1$, it follows from (3) that an equation of the plane tangent at $(-1, 0, 0)$ is $-2(x + 1) + 0(y - 0) - (z - 0) = 0$, or $2x + z = -2$, and that an equation of the plane tangent at $(-1, 1, \ln 2)$ is $-(x + 1) + (y - 1) - (z - \ln 2) = 0$, or $x - y + z = \ln 2 - 2$.

39. If $f(x, y, z) = x^2 + y^2 + z^2$, then $x^2 + y^2 + z^2 = 1$ is a level surface of f. Since $f_x(x, y, z) = 2x$, $f_y(x, y, z) = 2y$, and $f_z(x, y, z) = 2z$, so

$$f_x\left(\frac{1}{2}, -\frac{1}{2}, -\frac{1}{\sqrt{2}}\right) = 1, \quad f_y\left(\frac{1}{2}, -\frac{1}{2}, -\frac{1}{\sqrt{2}}\right) = -1, \quad \text{and} \quad f_z\left(\frac{1}{2}, -\frac{1}{2}, -\frac{1}{\sqrt{2}}\right) = -\sqrt{2}$$

it follows that an equation of the plane tangent at $(\frac{1}{2}, -\frac{1}{2}, -1/\sqrt{2})$ is

$$\left(x - \frac{1}{2}\right) - \left(y + \frac{1}{2}\right) - \sqrt{2}\left(z + \frac{1}{\sqrt{2}}\right) = 0, \quad \text{or} \quad x - y - \sqrt{2}z = 2$$

41. If $f(x, y, z) = xyz$, then the given level surface is a level surface of f. Since $f_x(x, y, z) = yz$, $f_y(x, y, z) = xz$, and $f_z(x, y, z) = xy$, so $f_x(\frac{1}{2}, -2, -1) = 2$, $f_y(\frac{1}{2}, -2, -1) = -\frac{1}{2}$, and $f_z(\frac{1}{2}, -2, 1) = -1$, it follows that an equation of the plane tangent at $(\frac{1}{2}, -2, -1)$ is $2(x - \frac{1}{2}) - \frac{1}{2}(y + 2) - (z + 1) = 0$, or $2x - \frac{1}{2}y - z = 3$.

43. If $f(x, y, z) = ye^{xy} + z^2$, then the given level surface is a level surface of f. Since $f_x(x, y, z) = y^2e^{xy}$, $f_y(x, y, z) = e^{xy} + xye^{xy}$, and $f_z(x, y, z) = 2z$, so $f_x(0, -1, 1) = 1$, $f_y(0, -1, 1) = 1$, and $f_z(0, -1, 1) = 2$, it follows that an equation of the plane tangent at $(0, -1, 1)$ is $(x - 0) + (y + 1) + 2(z - 1) = 0$, or $x + y + 2z = 1$.

45. If $f(x, y, z) = \ln\sqrt{x^2 + y^2 + z^2} = \frac{1}{2}\ln(x^2 + y^2 + z^2)$, then the given level surface is a level surface of f. Since

$$f_x(x, y, z) = \frac{x}{x^2 + y^2 + z^2}$$

$$f_y(x, y, z) = \frac{y}{x^2 + y^2 + z^2}, \quad \text{and} \quad f_z(x, y, z) = \frac{z}{x^2 + y^2 + z^2}$$

so $f_x(0, -1, 0) = 0$, $f_y(0, -1, 0) = -1$ and $f_z(0, -1, 0) = 0$, it follows that an equation of the plane tangent at $(0, -1, 0)$ is $0(x - 0) - (y + 1) + 0(z - 0) = 0$, or $y = -1$.

47. Let $f(x, y) = 9 - 4x^2 - y^2$. Then $f_x(x, y) = -8x$ and $f_y(x, y) = -2y$. By (2), a vector normal to the tangent plane at any point (x_0, y_0, z_0) on the graph of f is $-8x_0\mathbf{i} - 2y_0\mathbf{j} - \mathbf{k}$. If the tangent plane is parallel to the plane $z = 4y$, whose normal is $-4\mathbf{j} + \mathbf{k}$, then $-8x_0\mathbf{i} - 2y_0\mathbf{j} - \mathbf{k} = c(-4\mathbf{j} + \mathbf{k})$ for some constant c. It follows that $c = -1$, so that $x_0 = 0$ and $y_0 = -2$. The corresponding point on the paraboloid is $(0, -2, 5)$.

49. Let $f(x, y) = xy - 2$. Then $f_x(x, y) = y$ and $f_y(x, y) = x$. By (2), a vector normal to the plane tangent to the graph of f at $(1, 1, -1)$ is $\mathbf{i} + \mathbf{j} - \mathbf{k}$. Let $g(x, y, z) = x^2 + y^2 + z^2$. Then grad $g(x, y, z) = 2x\mathbf{i} + 2y\mathbf{j} + 2z\mathbf{k}$. By Definition 14.17, a vector normal to the plane tangent to the level surface $g(x, y, z) = 3$ at $(1, 1, -1)$ is $2\mathbf{i} + 2\mathbf{j} - 2\mathbf{k}$. Since the vectors normal to the two tangent planes are parallel, and since both surfaces pass through $(1, 1, -1)$, the two tangent planes at $(1, 1, -1)$ are identical.

51. Let $f(x, y) = x^2 + 4y^2 - 12$. Then $f_x(x, y) = 2x$ and $f_y(x, y) = 8y$, so that by (2) a vector normal to the plane tangent to the graph of f at $(-3, -1, 1)$ is $-6\mathbf{i} - 8\mathbf{j} - \mathbf{k}$. Let $g(x, y) = \frac{1}{8}(4x + y^2 + 19)$. Then $g_x(x, y) = \frac{1}{2}$ and $g_y(x, y) = \frac{1}{4}y$, so that a vector normal to the plane tangent to the graph of g at $(-3, -1, 1)$ is $\frac{1}{2}\mathbf{i} - \frac{1}{4}\mathbf{j} - \mathbf{k}$. Since $(-6\mathbf{i} - 8\mathbf{j} - \mathbf{k}) \cdot (\frac{1}{2}\mathbf{i} - \frac{1}{4}\mathbf{j} - \mathbf{k}) = -3 + 2 + 1 = 0$, the two normal vectors and hence the two planes tangent at $(-3, -1, 1)$ are perpendicular. Thus the two surfaces are normal at $(-3, -1, 1)$.

53. Let $f(x, y, z) = x^2/a^2 + y^2/b^2 + z^2/c^2$. Then $f_x(x, y, z) = 2x/a^2$, $f_y(x, y, z) = 2y/b^2$, and $f_z(x, y, z) = 2z/c^2$, so that by Definition 14.17 an equation of the plane tangent to the level surface $f(x, y, z) = 1$ at (x_0, y_0, z_0) is

$$\frac{2x_0}{a^2}(x - x_0) + \frac{2y_0}{b^2}(y - y_0) + \frac{2z_0}{c^2}(z - z_0) = 0$$

or

$$\frac{xx_0}{a^2} + \frac{yy_0}{b^2} + \frac{zz_0}{c^2} = \frac{x_0^2}{a^2} + \frac{y_0^2}{b^2} + \frac{z_0^2}{c^2} = 1$$

55. Since $f_x(x, y) = g(y/x) - (y/x)g'(y/x)$ and $f_y(x, y) = g'(y/x)$, it follows from (3) that an equation of the plane tangent to the graph of f at $(x_0, y_0, x_0 g(y_0/x_0))$ is

$$\left[g\left(\frac{y_0}{x_0}\right) - \frac{y_0}{x_0}g'\left(\frac{y_0}{x_0}\right)\right](x - x_0) + g'\left(\frac{y_0}{x_0}\right)(y - y_0) - \left(z - x_0 g\left(\frac{y_0}{x_0}\right)\right) = 0$$

or

$$\left[g\left(\frac{y_0}{x_0}\right) - \frac{y_0}{x_0}g'\left(\frac{y_0}{x_0}\right)\right]x + g'\left(\frac{y_0}{x_0}\right)y - z = x_0 g\left(\frac{y_0}{x_0}\right) - y_0 g'\left(\frac{y_0}{x_0}\right) + y_0 g'\left(\frac{y_0}{x_0}\right) - x_0 g\left(\frac{y_0}{x_0}\right) = 0$$

Thus the origin lies on the tangent plane.

57. Let $f(x, y, z) = z^2 - x^2 - y^2$. Then grad $f(x, y, z) = -2x\mathbf{i} - 2y\mathbf{j} + 2z\mathbf{k}$. By Definition 14.17 the vector $-2x_0\mathbf{i} - 2y_0\mathbf{j} + 2z_0\mathbf{k}$ is normal to the plane tangent to the level surface $f(x, y, z) = 0$ at any point (x_0, y_0, z_0) on the level surface. Thus if x_0, y_0, and z_0 are nonzero, then equations of the normal line are

$$\frac{x - x_0}{-2x_0} = \frac{y - y_0}{-2y_0} = \frac{z - z_0}{2z_0}$$

from which it follows that $(0, 0, 2z_0)$ lies on the normal line and hence that the normal line intersects the z axis. If $x_0 = 0$, and y_0 and z_0 are nonzero, then equations of the normal line are $x_0 = 0$ and $(y - y_0)/(-2y_0) = (z - z_0)/(2z_0)$, from which it again follows that the normal line intersects the z axis at $(0, 0, 2z_0)$. All other cases are handled analogously.

59. $2x - 2z\dfrac{\partial z}{\partial x} = 0$, so that $\dfrac{\partial z}{\partial x} = \dfrac{x}{z}$; $-2y - 2z\dfrac{\partial z}{\partial y} = 0$, so that $\dfrac{\partial z}{\partial y} = -\dfrac{y}{z}$. Thus

$$\left.\frac{\partial z}{\partial x}\right|_{(\sqrt{2},0,1)} = \sqrt{2}, \quad\text{and}\quad \left.\frac{\partial z}{\partial y}\right|_{(\sqrt{2},0,1)} = 0$$

so that an equation of the plane tangent to the given level surface at $(\sqrt{2}, 0, 1)$ is $\sqrt{2}(x - \sqrt{2}) + 0(y - 0) - (z - 1) = 0$, or $\sqrt{2}x - z = 1$.

61. $\dfrac{1}{x} + \dfrac{1}{z}\dfrac{\partial z}{\partial x} = 0$, so that $\dfrac{\partial z}{\partial x} = -\dfrac{z}{x}$; $\dfrac{1}{y} + \dfrac{1}{z}\dfrac{\partial z}{\partial y} = 0$, so that $\dfrac{\partial z}{\partial y} = -\dfrac{z}{y}$. Thus

$$\left.\frac{\partial z}{\partial x}\right|_{(1,1,e)} = -e \quad\text{and}\quad \left.\frac{\partial z}{\partial y}\right|_{(1,1,e)} = -e$$

so that an equation of the plane tangent to the given level surface at $(1, 1, e)$ is $-e(x - 1) - e(y - 1) - (z - e) = 0$, or $ex + ey + z = 3e$.

63. The temperature $T(x, y)$ at any point (x, y) on the plate is given by $T(x, y) = c/\sqrt{x^2 + y^2}$, where c is a positive constant. Then

$$\text{grad } T(x, y) = \frac{-cx}{(x^2 + y^2)^{3/2}}\mathbf{i} - \frac{cy}{(x^2 + y^2)^{3/2}}\mathbf{j}, \quad\text{so that}\quad \text{grad } T(3, 2) = \frac{-3c}{(13)^{3/2}}\mathbf{i} - \frac{2c}{(13)^{3/2}}\mathbf{j}$$

Thus T is decreasing most rapidly in the direction of

$$\frac{-1}{\left\|\dfrac{-3c}{(13)^{3/2}}\mathbf{i} - \dfrac{2c}{(13)^{3/2}}\mathbf{j}\right\|}\left(\frac{-3c}{(13)^{3/2}}\mathbf{i} - \frac{2c}{(13)^{3/2}}\mathbf{j}\right) = \frac{3}{\sqrt{13}}\mathbf{i} + \frac{2}{\sqrt{13}}\mathbf{j}$$

This is the direction in which the ant should crawl.

65. grad $T(x, y, z) = (3x^2y + 6xy^2z)\mathbf{i} + (x^3 + 6x^2yz)\mathbf{j} + 3x^2y^2\mathbf{k}$, so that grad $T(1, 1, -1) = -3\mathbf{i} - 5\mathbf{j} + 3\mathbf{k}$. Let

$$\mathbf{u} = \frac{\text{grad } T(1, 1, -1)}{\|\text{grad } T(1, 1, -1)\|} = -\frac{3}{\sqrt{43}}\mathbf{i} - \frac{5}{\sqrt{43}}\mathbf{j} + \frac{3}{\sqrt{43}}\mathbf{k}$$

Then

$$D_{\mathbf{u}}T(1, 1, -1) = T_x(1, 1, -1)\left(-\frac{3}{\sqrt{43}}\right) + T_y(1, 1, -1)\left(-\frac{5}{\sqrt{43}}\right) + T_z(1, 1, -1)\left(\frac{3}{\sqrt{43}}\right)$$

$$= \frac{9}{\sqrt{43}} + \frac{25}{\sqrt{43}} + \frac{9}{\sqrt{43}} = \sqrt{43}$$

SECTION 14.7

1. Let $x_0 = 3$, $y_0 = 4$, $h = .01$, and $k = .03$. Since

$$f_x(x, y) = \frac{x}{\sqrt{x^2 + y^2}} \quad\text{and}\quad f_y(x, y) = \frac{y}{\sqrt{x^2 + y^2}}$$

we have $f_x(3, 4) = \frac{3}{5}$ and $f_y(3, 4) = \frac{4}{5}$. Also $f(3, 4) = 5$. By (6),

$$f(3.01, 4.03) \approx 5 + \tfrac{3}{5}(.01) + \tfrac{4}{5}(.03) = 5.03$$

3. Let $x_0 = 0$, $y_0 = 1$, $h = -.03$, and $k = -.02$. Since $f_x(x, y) = 2x/(x^2 + y^2)$ and $f_y(x, y) = 2y/(x^2 + y^2)$, we have $f_x(0, 1) = 0$ and $f_y(0, 1) = 2$. Also $f(0, 1) = 0$. By (6),

$$f(-.03, .98) \approx 0 + 0(-.03) + 2(-.02) = -.04$$

5. Let $x_0 = \pi$, $y_0 = .25$, $h = -.01\pi$, and $k = -.01$. Since $f_x(x, y) = y \sec^2 xy$ and $f_y(x, y) = x \sec^2 xy$, we have $f_x(\pi, .25) = \frac{1}{4}\sec^2(\pi/4) = \frac{1}{2}$ and $f_y(\pi, .25) = 2\pi$. Also $f(\pi, .25) = \tan(\pi/4) = 1$. By (6),

$$f(.99\pi, .24) \approx 1 + \tfrac{1}{2}(-.01\pi) + 2\pi(-.01) = 1 - .025\pi \approx .921460$$

7. Let $x_0 = 3$, $y_0 = 4$, $z_0 = 12$, $h = .01$, $k = .02$, and $\ell = -.02$. Since

$$f_x(x, y, z) = \frac{x}{\sqrt{x^2 + y^2 + z^2}}, \qquad f_y(x, y, z) = \frac{y}{\sqrt{x^2 + y^2 + z^2}},$$

and

$$f_z(x, y, z) = \frac{z}{\sqrt{x^2 + y^2 + z^2}}$$

we have $f_x(3, 4, 12) = \frac{3}{13}$, $f_y(3, 4, 12) = \frac{4}{13}$, and $f_z(3, 4, 12) = \frac{12}{13}$. Also $f(3, 4, 12) = 13$. By (7), $f(3.01, 4.02, 11.98) \approx 13 + \frac{3}{13}(.01) + \frac{4}{13}(.02) + \frac{12}{13}(-.02) = 12.99$

9. Let $f(x, y) = \sqrt[4]{x^3 + y^3}$, $x_0 = 2$, $y_0 = 2$, $h = -.1$, and $k = .1$. Since

$$f_x(x, y) = \frac{3x^2}{4(x^3 + y^3)^{3/4}} \quad \text{and} \quad f_y(x, y) = \frac{3y^2}{4(x^3 + y^3)^{3/4}}$$

we have $f_x(2, 2) = \frac{3}{8}$ and $f_y(2, 2) = \frac{3}{8}$. Also $f(2, 2) = 2$. By (6),

$$\sqrt[4]{(1.9)^3 + (2.1)^3} = f(1.9, 2.1) \approx 2 + \tfrac{3}{8}(-.1) + \tfrac{3}{8}(.1) = 2$$

11. Let $f(x, y) = e^x \ln y$, $x_0 = 0$, $y_0 = 1$, $h = .1$, and $k = -.1$. Since $f_x(x, y) = e^x \ln y$ and $f_y(x, y) = e^x/y$, we have $f_x(0, 1) = 0$ and $f_y(0, 1) = 1$. Also $f(0, 1) = 0$. By (6), $e^{\cdot 1} \ln .9 = f(.1, .9) \approx 0 + 0(.1) + 1(-.1) = -.1$

13. Since $\partial f/\partial x = 9x^2 - 2xy$ and $\partial f/\partial y = -x^2 + 1$, we have $df = (9x^2 - 2xy)\,dx + (-x^2 + 1)\,dy$.

15. Since $\partial f/\partial x = 2x$ and $\partial f/\partial y = 2y$, we have $df = 2x\,dx + 2y\,dy$.

17. Since $\partial f/\partial x = \tan y - y \csc^2 x$ and $\partial f/\partial y = x \sec^2 y + \cot x$, we have $df = (\tan y - y \csc^2 x)\,dx + (x \sec^2 y + \cot x)\,dy$

19. Since

$$\frac{\partial f}{\partial x} = \frac{xz^2}{\sqrt{1 + x^2 + y^2}}, \quad \frac{\partial f}{\partial y} = \frac{yz^2}{\sqrt{1 + x^2 + y^2}}, \quad \text{and} \quad \frac{\partial f}{\partial z} = 2z\sqrt{1 + x^2 + y^2}$$

we have
$$df = \frac{xz^2}{\sqrt{1 + x^2 + y^2}}\,dx + \frac{yz^2}{\sqrt{1 + x^2 + y^2}}\,dy + 2z\sqrt{1 + x^2 + y^2}\,dz$$

21. Since $\partial f/\partial x = e^{y^2 - z^2}$, $\partial f/\partial y = 2yxe^{y^2 - z^2}$, and $\dfrac{\partial f}{\partial z} = -2zxe^{y^2 - z^2}$, we have

$df = e^{y^2 - z^2}\,dx + 2yxe^{y^2 - z^2}\,dy - 2zxe^{y^2 - z^2}\,dz$

23. Since
$$\frac{\partial R}{\partial R_1} = \frac{R_2(R_1 + R_2) - R_1 R_2}{(R_1 + R_2)^2} = \frac{R_2^2}{(R_1 + R_2)^2}$$

and
$$\frac{\partial R}{\partial R_2} = \frac{R_1(R_1 + R_2) - R_1 R_2}{(R_1 + R_2)^2} = \frac{R_1^2}{(R_1 + R_2)^2}$$

we have $\dfrac{\partial R}{\partial R_1}(2, 6) = \frac{9}{16}$ and $\dfrac{\partial R}{\partial R_2}(2, 6) = \frac{1}{16}$. Since $R(2, 6) = \frac{3}{2}$, if we let $h = .013$ and $k = -.028$, then by (6),

$$R(2.013, 5.972) \approx \frac{3}{2} + \frac{9}{16}(.013) + \frac{1}{16}(-.028) = \frac{3}{2} + \frac{.089}{16} \approx 1.50556 \text{ (ohms)}$$

25. Let the dimensions of the box be x, y, and z, respectively, and let the surface area be $f(x, y, z)$. Furthermore, let $x_0 = 3$, $y_0 = 4$, $z_0 = 12$, $h = .019$, $k = -.021$, and $\ell = -.027$. Because the box has a top, $f(x, y, z) = 2xy + 2xz + 2yz$, so that $f_x(x, y, z) = 2y + 2z$, $f_y(x, y, z) = 2x + 2z$, and $f_z(x, y, z) = 2x + 2y$, and hence $f_x(3, 4, 12) = 32$, $f_y(3, 4, 12) = 30$, and $f_z(3, 4, 12) = 14$. Since $f(3, 4, 12) = 192$, we find from (7) that

$$f(3.019, 3.979, 11.973) \approx 192 + 32(.019) + 30(-.021) + 14(-.027) = 191.6$$

SECTION 14.8

1. $f_x(x, y) = 2x - 6$ and $f_y(x, y) = 4y + 8$, so $f_x(x, y) = 0 = f_y(x, y)$ if $x = 3$ and $y = -2$. Thus $(3, -2)$ is a critical point. Next, $f_{xx}(x, y) = 2$, $f_{yy}(x, y) = 4$, and $f_{xy}(x, y) = 0$. Thus $D(3, -2) = f_{xx}(3, -2)f_{yy}(3, -2) - [f_{xy}(3, -2)]^2 = (2)(4) - 0^2 = 8$. Since $D(3, -2) > 0$ and $f_{xx}(3, -2) > 0$, f has a relative minimum value at $(3, -2)$.

3. $f_x(x, y) = 2x + 6y - 6$ and $f_y(x, y) = 6x + 4y + 10$, so $f_x(x, y) = 0 = f_y(x, y)$ if $2x + 6y - 6 = 0$ and $6x + 4y + 10 = 0$. Solving for x in the first equation and substituting for it in the second equation yields $6(3 - 3y) + 4y + 10 = 0$, or $y = 2$. Then $x = -3$, so $(-3, 2)$ is a critical point. Next, $f_{xx}(x, y) = 2$, $f_{yy}(x, y) = 4$, and $f_{xy}(x, y) = 6$. Thus $D(-3, 2) = f_{xx}(-3, 2)f_{yy}(-3, 2) - [f_{xy}(-3, 2)]^2 = (2)(4) - 6^2 = -28$. Since $D(-3, 2) < 0$, f has a saddle point at $(-3, 2)$.

5. $k_x(x, y) = -2x - 2y + 6$ and $k_y(x, y) = -2x - 4y - 10$, so $k_x(x, y) = 0 = k_y(x, y)$ if $-2x - 2y + 6 = 0$ and $-2x - 4y - 10 = 0$. Solving for y in the first equation and substituting for it in the second equation yields $-2x - 4(-x + 3) - 10 = 0$, or $x = 11$. Then $y = -8$, so $(11, -8)$ is a critical point. Next, $k_{xx}(x, y) = -2$, $k_{yy}(x, y) = -4$, and $k_{xy}(x, y) = -2$. Thus $D(11, -8) = k_{xx}(11, -8)k_{yy}(11, -8) - [k_{xy}(11, -8)]^2 = (-2)(-4) - (-2)^2 = 4$. Since $D(11, -8) > 0$ and $k_{xx}(11, -8) < 0$, k has a relative maximum value at $(11, -8)$.

7. $f_x(x, y) = 2xy - 2y = 2y(x - 1)$ and $f_y(x, y) = x^2 - 2x + 4y - 15$, so $f_x(x, y) = 0 = f_y(x, y)$ if $2y(x - 1) = 0$ and $x^2 - 2x + 4y - 15 = 0$. From the first equation we see that $y = 0$ or $x = 1$. If $y = 0$, it follows from the second equation that $x^2 - 2x - 15 = 0$, so that $x = -3$ or $x = 5$. Thus $(-3, 0)$ and $(5, 0)$ are critical points. If $x = 1$, it follows from the second equation that $1 - 2 + 4y - 15 = 0$, or $y = 4$. Thus $(1, 4)$ is also a critical point. Next, $f_{xx}(x, y) = 2y$, $f_{yy}(x, y) = 4$, and $f_{xy}(x, y) = 2x - 2$. Thus

$$D(-3, 0) = f_{xx}(-3, 0)f_{yy}(-3, 0) - [f_{xy}(-3, 0)]^2 = (0)(4) - (-8)^2 = -64,$$
$$D(5, 0) = f_{xx}(5, 0)f_{yy}(5, 0) - [f_{xy}(5, 0)]^2 = (0)(4) - 8^2 = -64$$
and
$$D(1, 4) = f_{xx}(1, 4)f_{yy}(1, 4) - [f_{xy}(1, 4)]^2 = (8)(4) - 0^2 = 32$$

Since $D(1, 4) > 0$ and $f_{xx}(1, 4) > 0$, f has a relative minimum value at $(1, 4)$. Since $D(-3, 0) < 0$ and $D(5, 0) < 0$, f has saddle points at $(-3, 0)$ and at $(5, 0)$.

9. $f_x(x, y) = 6x - 3y^2$ and $f_y(x, y) = -6xy + 3y^2 + 6y = 3y(-2x + y + 2)$, so $f_x(x, y) = 0 = f_y(x, y)$ if $6x - 3y^2 = 0$ and $3y(-2x + y + 2) = 0$. From the first equation we see that $x = \frac{1}{2}y^2$, so from the second equation, $y = 0$ or $0 = -2x + y + 2 = -y^2 + y + 2$, so that $y = 0$, $y = -1$, or $y = 2$. The corresponding values of x are 0, $\frac{1}{2}$, and 2. Thus $(0, 0)$, $(\frac{1}{2}, -1)$, and $(2, 2)$ are critical points. Next, $f_{xx}(x, y) = 6$, $f_{yy}(x, y) = -6x + 6y + 6$, and $f_{xy}(x, y) = -6y$. Thus $D(0, 0) = f_{xx}(0, 0)f_{yy}(0, 0) - [f_{xy}(0, 0)]^2 = (6)(6) - 0^2 = 36$, $D(\frac{1}{2}, -1) = f_{xx}(\frac{1}{2}, -1)f_{yy}(\frac{1}{2}, -1) - [f_{xy}(\frac{1}{2}, -1)]^2 = (6)(-3) - 6^2 = -54$, and $D(2, 2) = f_{xx}(2, 2)f_{yy}(2, 2) - [f_{xy}(2, 2)]^2 = (6)(6) - (-12)^2 = -108$. Since $D(0, 0) > 0$ and $f_{xx}(0, 0) > 0$, f has a relative minimum value at $(0, 0)$. Since $D(\frac{1}{2}, -1) < 0$ and $D(2, 2) < 0$, f has saddle points at $(\frac{1}{2}, -1)$ and $(2, 2)$.

11. $f_x(x, y) = 4y + 4xy - y^2 = y(4 + 4x - y)$ and $f_y(x, y) = 4x + 2x^2 - 2xy = 2x(2 + x - y)$, so $f_x(x, y) = 0 = f_y(x, y)$ if $y(4 + 4x - y) = 0$ and $2x(2 + x - y) = 0$. From the first equation we see that either $y = 0$ or $y = 4 + 4x$. If $y = 0$, the second equation implies that $x = 0$ or $x = -2$. Thus $(0, 0,)$ and $(-2, 0)$ are critical points. If $y = 4 + 4x$, the second equation implies that either $x = 0$ or $0 = 2 + x - y = 2 + x - (4 + 4x)$, so that $x = 0$ or $x = -\frac{2}{3}$. The corresponding values of y are 4 and $\frac{4}{3}$. Thus $(0, 4)$ and $(-\frac{2}{3}, \frac{4}{3})$ are critical points. Next, $f_{xx}(x, y) = 4y$, $f_{yy}(x, y) = -2x$, and $f_{xy}(x, y) = 4 + 4x - 2y$. Thus

$$D(0, 0) = f_{xx}(0, 0)f_{yy}(0, 0) - [f_{xy}(0, 0)]^2 = (0)(0) - 4^2 = -16$$
$$D(-2, 0) = f_{xx}(-2, 0)f_{yy}(-2, 0) - [f_{xy}(-2, 0)]^2 = (0)(4) - (-4)^2 = -16$$
$$D(0, 4) = f_{xx}(0, 4)f_{yy}(0, 4) - [f_{xy}(0, 4)]^2 = (16)(0) - (-4)^2 = -16$$

and $D(-\frac{2}{3}, \frac{4}{3}) = f_{xx}(-\frac{2}{3}, \frac{4}{3})f_{yy}(-\frac{2}{3}, \frac{4}{3}) - [f_{xy}(-\frac{2}{3}, \frac{4}{3})]^2 = (\frac{16}{3})(\frac{4}{3}) - (-\frac{4}{3})^2 = \frac{16}{3}$. Since $D(-\frac{2}{3}, \frac{4}{3}) > 0$ and $f_{xx}(-\frac{2}{3}, \frac{4}{3}) > 0$, f has a relative minimum value at $(-\frac{2}{3}, \frac{4}{3})$. Since $D(0, 0) < 0$, $D(0, 4) < 0$, and $D(-2, 0) < 0$, f has saddle points at $(0, 0)$, $(0, 4)$, and $(-2, 0)$.

13. $f_x(x, y) = 2x$ and $f_y(x, y) = -2ye^{y^2}$, so $f_x(x, y) = 0 = f_y(x, y)$ if $x = 0$ and $y = 0$. Thus $(0, 0)$ is a critical point. Next, $f_{xx}(x, y) = 2$, $f_{yy}(x, y) = -2e^{y^2} - 4y^2e^{y^2}$, and $f_{xy}(x, y) = 0$. Thus $D(0, 0) = f_{xx}(0, 0)f_{yy}(0, 0) - [f_{xy}(0, 0)]^2 = (2)(-2) - (0)^2 = -4$. Since $D(0, 0) < 0$, it follows that f has a saddle point at $(0, 0)$.

15. $k_x(x, y) = e^x \sin y$ and $k_y(x, y) = e^x \cos y$, so $k_x(x, y) = 0 = k_y(x, y)$ if $e^x \sin y = 0$ and $e^x \cos y = 0$, or $\sin y = 0$ and $\cos y = 0$, which is impossible. Thus k has no critical points.

17. $f_u(u, v) = 1$ if $u > 0$, $f_u(u, v) = -1$ if $u < 0$, and $f_u(u, v)$ does not exist if $u = 0$. Similarly, $f_v(u, v) = 1$ if $v > 0$, $f_v(u, v) = -1$ if $v < 0$, and $f_v(u\ v)$ does not exist if $v = 0$. Thus any point of the form $(0, v)$ or $(u, 0)$ is a critical point of f. Since either f_{uu} or f_{vv} does not exist at each such point, the Second Partials Test does not apply. Notice that $f(0, 0) = 0$ and $f(u, v) \geq 0$ for all u and v, so that f has a (relative) minimum value at $(0, 0)$. However, at any other critical point (u, v), either $f_u(u, v) \neq 0$ or $f_v(u, v) \neq 0$, so that f does not have a relative extreme value or a saddle point at any critical point except $(0, 0)$.

19. $f_x(x, y) = ye^{xy}$ and $f_y(x, y) = xe^{xy}$, so $f_x(x, y) = 0 = f_y(x, y)$ if $x = 0$ and $y = 0$. Thus $(0, 0)$ is a critical point. Next, $f_{xx}(x, y) = y^2e^x$, $f_{yy}(x, y) = x^2e^{xy}$, and $f_{xy}(x, y) = e^{xy} + xye^{xy}$. Thus $D(0, 0) = f_{xx}(0, 0)f_{yy}(0, 0) - [f_{xy}(0, 0)]^2 = (0)(0) - 1^2 = -1$. Since $D(0, 0) < 0$, f has a saddle point at $(0, 0)$.

21. $f_x(x, y) = \cos x$ and $f_y(x, y) = \cos y$, so $f_x(x, y) = 0 = f_y(x, y)$ if $\cos x = 0$ and $\cos y = 0$. Thus f has a critical point at any point of the form $(\pi/2 + m\pi, \pi/2 + n\pi)$, where m and n are integers. Next, $f_{xx}(x, y) = -\sin x$, $f_{yy}(x, y) = -\sin y$, and $f_{xy}(x, y) = 0$. Thus

$$\begin{aligned} D(\pi/2 + m\pi, \pi/2 + n\pi) &= f_{xx}(\pi/2 + m\pi, \pi/2 + n\pi)f_{yy}(\pi/2 + m\pi, \pi/2 + n\pi) \\ &\quad - [f_{xy}(\pi/2 + m\pi, \pi/2 + n\pi)]^2 \\ &= (-\sin(\pi/2 + m\pi))(-\sin(\pi/2 + n\pi)) - 0^2 \\ &= \sin(\pi/2 + m\pi)\sin(\pi/2 + n\pi) \end{aligned}$$

Since $D(\pi/2 + m\pi, \pi/2 + n\pi) = 1 > 0$ and $f_{xx}(\pi/2 + m\pi, \pi/2 + n\pi) = -1 < 0$ if m and n are both even, f has a relative maximum value at the corresponding point $(\pi/2 + m\pi, \pi/2 + n\pi)$. Since $D(\pi/2 + m\pi, \pi/2 + n\pi) = 1$ and $f_{xx}(\pi/2 + m\pi, \pi/2 + n\pi) = 1 > 0$ if m and n are both odd, f has a relative minimum value at the corresponding point $(\pi/2 + m\pi, \pi/2 + n\pi)$. Since $D(\pi/2 + m\pi, \pi/2 + n\pi) = -1 < 0$ if either m or n is odd and the other is even, f has a saddle point at the corresponding point $(\pi/2 + m\pi, \pi/2 + n\pi)$.

23. $f_x(x, y) = 2a(y + ax + b)$ and $f_y(x, y) = 2(y + ax + b)$, so $f_x(x, y) = 0 = f_y(x, y)$ if $y + ax + b = 0$, or $y = -(ax + b)$. Thus any point of the form $(x, -(ax + b))$ is a critical point. Next, $f_{xx}(x, y) = 2a^2$, $f_{yy}(x, y) = 2$, and $f_{xy}(x, y) = 2a$. Thus $D(x, -(ax + b)) = f_{xx}(x, -(ax + b))f_{yy}(x, -(ax + b)) - [f_{xy}(x, -(ax + b))]^2 = (2a^2)(2) - (2a)^2 = 0$, so the Second Partials Test yields no conclusion. However, $f(x, -(ax + b)) = 0$ and $f(x, y) \geq 0$ for all x and y. Thus f has a (relative) minimum value at any point of the form $(x, -(ax + b))$, that is, at any point on the line $y = -(ax + b) = -ax - b$.

25. $f_x(x, y) = 2x$ and $f_y(x, y) = -2y$, so $f_x(x, y) = 0 = f_y(x, y)$ if $x = 0$ and $y = 0$. Thus $(0, 0)$ is a critical point in R. On the boundary $x^2 + y^2 = 1$ of R we have $y^2 = 1 - x^2$, so that $f(x, y) = x^2 - (1 - x^2) = 2x^2 - 1$ for $-1 \leq x \leq 1$. The maximum value of $2x^2 - 1$ on $[-1, 1]$ is 1 and occurs for $x = 1$ and $x = -1$, and the minimum value is -1 and occurs for $x = 0$. Since $f(0, 0) = 0$, it follows that the maximum value of f on R is $f(1, 0) = f(-1, 0) = 1$, and the minimum value is $f(0, 1) = f(0, -1) = -1$.

27. $f_x(x, y) = 2\cos x$ and $f_y(x, y) = -3\sin y$, so $f_x(x, y) = 0 = f_y(x, y)$ for (x, y) in R if $x = \pi/2$ and $y = 0$. Thus $(\pi/2, 0)$ is a critical point of f in R. From the figure we observe that for (x, y) on ℓ_1 we have $y = -\pi/2$ and $\cos(-\pi/2) = 0$, so that $f(x, y) = 2\sin x$. Thus

on ℓ_1 the maximum value of f is $f(\pi/2, -\pi/2) = 2 \sin \pi/2 = 2$ and the minimum value is $f(0, -\pi/2) = 2 \sin 0 = 0$. The same extreme values are obtained for ℓ_3. For (x, y) on ℓ_2 we have $x = \pi$, so that $f(x, y) = 3 \cos y$. Thus on ℓ_2 the maximum value of f is $f(\pi, 0) = 3 \cos 0 = 3$ and the minimum value is $f(\pi, -\pi/2) = 3 \cos(-\pi/2) = 0$. The same extreme values are obtained for ℓ_4. Since $f(\pi/2, 0) = 2 \sin \pi/2 + 3 \cos 0 = 2 + 3 = 5$, the maximum value of f on R is $f(\pi/2, 0) = 5$ and the minimum value is $f(\pi, -\pi/2) = 0$.

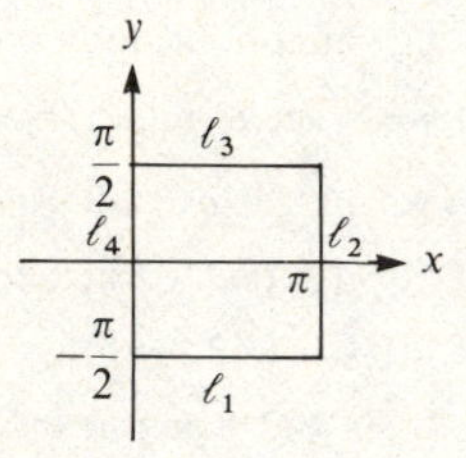

29. Let $P = xyz$ for any nonnegative numbers x, y, and z satisfying $x + y + z = 48$. Then $z = 48 - x - y$, so $P = xy(48 - x - y)$. We seek the maximum value of P on the triangular region R consisting of all (x, y) for which $x \geq 0$, $y \geq 0$, and $x + y \leq 48$. Such a maximum value exists by the Maximum-Minimum Theorem. Since $P = 0$ if $x = 0$, $y = 0$, or $x + y = 48$, the maximum value of P on R does not occur on the boundary of R and hence must occur at a critical point in the interior of R. $\partial P/\partial x = y(48 - x - y) - xy = y(48 - 2x - y)$ and $\partial P/\partial y = x(48 - x - y) - xy = x(48 - x - 2y)$, so $\partial P/\partial x = 0 = \partial P/\partial y$ if $y(48 - 2x - y) = 0$ and $x(48 - x - 2y) = 0$. From the first equation we see that $y = 0$ or $48 - 2x - y = 0$, and from the second equation we see that $x = 0$ or $48 - x - 2y = 0$. Since $x \neq 0$ and $y \neq 0$ at each interior point of R, it follows that a point (x, y) in the interior of R is a critical point of P only if $48 - 2x - y = 0$ and $48 - x - 2y = 0$. Solving for y in the first of these equations, we obtain $y = 48 - 2x$. Substituting for y in the second equation gives us $48 - x - 2(48 - 2x) = 0$, so that $x = 16$. Then $y = 16$, so $(16, 16)$ is the only critical point of P in R, and the corresponding value of z is 16 also. The maximum product must occur at a critical point, so it is $P(16, 16) = 4096$.

31. Set up a coordinate system with the origin at the center of the sphere and with the coordinate planes parallel to the sides of the box. If r is the radius of the sphere, then an equation of the sphere is $x^2 + y^2 + z^2 = r^2$. If (x, y, z) is the vertex of the box with $x \geq 0$, $y \geq 0$, and $z \geq 0$, then $z = \sqrt{r^2 - x^2 - y^2}$, and the volume V of the box is given by $V = xyz = xy\sqrt{r^2 - x^2 - y^2}$. We seek the maximum value of V on the set R of points (x, y) satisfying $x \geq 0$, $y \geq 0$, and $x^2 + y^2 \leq r^2$. Such a maximum value exists by the Maximum-Minimum Theorem. Since $V = 0$ if $x = 0$, $y = 0$, or $x^2 + y^2 = r^2$, V does not have its maximum value on R at any boundary point of R. Thus the maximum value of V on R occurs at a critical point in the interior of R.

$$\frac{\partial V}{\partial x} = y\sqrt{r^2 - x^2 - y^2} - \frac{x^2 y}{\sqrt{r^2 - x^2 - y^2}} \quad \text{and} \quad \frac{\partial V}{\partial y} = x\sqrt{r^2 - x^2 - y^2} - \frac{xy^2}{\sqrt{r^2 - x^2 - y^2}}$$

so $\partial V/\partial x = 0 = \partial V/\partial y$ if

$$y\sqrt{r^2 - x^2 - y^2} - \frac{x^2 y}{\sqrt{r^2 - x^2 - y^2}} = 0 \quad \text{and} \quad x\sqrt{r^2 - x^2 - y^2} - \frac{xy^2}{\sqrt{r^2 - x^2 - y^2}} = 0$$

Since $x \neq 0$ and $y \neq 0$ at any interior point of R, we conclude from these equations that

$$0 = \sqrt{r^2 - x^2 - y^2} - \frac{x^2}{\sqrt{r^2 - x^2 - y^2}} = \frac{r^2 - x^2 - y^2 - z^2}{\sqrt{r^2 - x^2 - y^2}} \quad \text{and}$$

$$0 = \sqrt{r^2 - x^2 - y^2} - \frac{y^2}{\sqrt{r^2 - x^2 - y^2}} = \frac{r^2 - x^2 - y^2 - y^2}{\sqrt{r^2 - x^2 - y^2}}$$

Thus $r^2 - x^2 - y^2 - x^2 = 0 = r^2 - x^2 - y^2 - y^2$, or $x^2 = y^2$, or $x = y$. Therefore $0 = r^2 - x^2 - y^2 - x^2 = r^2 - 3x^2$, so that $x = r/\sqrt{3}$. Then $y = r/\sqrt{3}$, so the only critical point of S in R is $(r/\sqrt{3}, r/\sqrt{3})$. The maximum value must occur at a critical point, so it occurs for $x = r/\sqrt{3} = y$. This means that $z = \sqrt{r^2 - x^2 - y^2} = r/\sqrt{3}$ and thus the box is a cube.

33. The distance from a point to the origin will be minimum if and only if the square of the distance from the point to the origin is minimum. To simply the calculations, we will minimize the square of the distance. Thus for points (x, y, z) satisfying $x + y + z = 48$, we seek the minimum value of $x^2 + y^2 + z^2 = x^2 + y^2 + (48 - x - y)^2$. Let $f(x, y) = x^2 + y^2 + (48 - x - y)^2$. Since $f(0, 0) = (48)^2$ and $f(x, y) > (48)^2$ if $x^2 + y^2 > (48)^2$, and since the Maximum-Minimum Theorem implies that f has a minimum value on the disk $x^2 + y^2 \leq (48)^2$, it follows that f has a minimum value. Since the domain of f is the entire xy plane, the minimum value of f occurs at a critical point. $f_x(x, y) = 2x - 2(48 - x - y)$ and $f_y(x, y) = 2y - 2(48 - x - y)$. Thus $f_x(x, y) = 0 = f_y(x, y)$ if $2x - 2(48 - x - y) = 0$ and $2y - 2(48 - x - y) = 0$, so $x = y$. Then $0 = 2x - 2(48 - x - y) = 2x - 2(48 - 2x)$, so that $x = 16$. Then $y = 16$ so the only critical point of f is $(16, 16)$. The minimum value of f must occur at a critical point, so it occurs at $(16, 16)$. Since $z = 16$ if $x = 16 = y$, $(16, 16, 16)$ is the point the sum of whose coordinates is 48 and whose distance to the origin is minimum.

We seek the maximum value of A on the rectangular region R consisting of all (x, θ) for which $0 \leq x \leq \ell/2$ and $0 \leq \theta \leq \pi/2$. Such a maximum value exists by the Maximum-

35. a. $f_x(x, y) = e^{-y^2}(6x^2 - 6x) + e^{-y}(6x^2 - 6x) = (e^{-y^2} + e^{-y})(6x^2 - 6x)$ and $f_y(x, y) = -2ye^{-y^2}(2x^3 - 3x^2 + 1) - e^{-y}(2x^3 - 3x^2)$. To find the critical points of f, suppose that $f_x(x, y) = 0$ and $f_y(x, y) = 0$. Since $f_x(x, y) = 0$ and $e^{-y^2} + e^{-y} > 0$, it follows that $6x^2 - 6x = 0$, that is, $x = 0$ or $x = 1$. If $x = 0$, then $0 = f_y(x, y) = -2ye^{-y^2}$, so that $y = 0$. If $x = 1$, then $0 = f_y(x, y) = e^{-y}$, which is impossible. Therefore the only critical point is $(0, 0)$.

b. $f_{xx}(x, y) = (e^{-y^2} + e^{-y})(12x - 6)$, $f_{yy}(x, y) = (-2e^{-y^2} + 4y^2e^{-y^2})(2x^3 - 3x^2 + 1) + e^{-y}(2x^3 - 3x^2)$, and $f_{xy}(x, y) = (-2ye^{-y^2} - e^{-y})(6x^2 - 6x)$. Thus $D(0, 0) = f_{xx}(0, 0)f_{yy}(0, 0) - [f_{xy}(0, 0)]^2 = (-12)(-2) - 0^2 = 24 > 0$. Since $f_{xx}(0, 0) = -12 < 0$, f has a relative maximum value at $(0, 0)$.

c. Since $f(2, y) = 5e^{-y^2} + 4e^{-y} > 4e^{-y}$, and since $\lim_{y \to -\infty} 4e^{-y} = \infty$, it follows that $\lim_{y \to -\infty} f(2, y) = \infty$. Thus f has no maximum value.

37. $f(m, b) = [0 - (0 + b)]^2 + [-1 - (m + b)]^2 + [1 - (-2m + b)]^2 = b^2 + 1 + 2(m + b) + (m + b)^2 + 1 - 2(-2m + b) + (-2m + b)^2 = 2 + 6m + 5m^2 - 2mb + 3b^2$. $f_m(m, b) = 6 + 10m - 2b$ and $f_b(m, b) = -2m + 6b$. Thus $f_m(m, b) = 0 = f_b(m, b)$ if $6 + 10m - 2b = 0$ and $-2m + 6b = 0$. Solving for m in the second equation and substituting in the first equation, we obtain $0 = 6 + 10m - 2b = 6 + 10(3b) - 2b$, so that $b = -\frac{3}{14}$. Then $m = -\frac{9}{14}$. Thus an equation of the line of best fit is $y = -\frac{9}{14}x - \frac{3}{14}$.

39. Let x be the length in feet of the front and back walls, y the length of the side walls, and z the height of the gymnasium. Then $xyz = 960{,}000$, so that $z = 960{,}000/xy$. If a is the cost per square foot of the side and back walls and the floor, then the total cost C is given by $C = a(2xz + xz + 2yz + xy + \frac{3}{2}xy) = a(3xz + 2yz + \frac{5}{2}xy) = a(2{,}880{,}000/y + 1{,}920{,}000/x + \frac{5}{2}xy)$. Although the Maximum-Minimum Theorem does not apply, C has a minimum value, which occurs at a critical point. Now $\partial C/\partial x = a(-1{,}920{,}000/x^2 + \frac{5}{2}y)$ and $\partial C/\partial y = a(-2{,}880{,}000/y^2 + \frac{5}{2}x)$. Thus $\partial C/\partial x = 0 = (\partial C/\partial y)(x, y)$ if $-1{,}920{,}000/x^2 + \frac{5}{2}y = 0$ and $-2{,}880{,}000/y^2 + \frac{5}{2}x = 0$, so that $y = 768{,}000/x^2$ and $x = 1{,}152{,}000/y^2$. Thus $x = 1{,}152{,}000/y^2 = (1{,}152{,}000)x^4/(768{,}000)^2 = x^4/512{,}000$. Since $x \neq 0$, we have $x^3 = 512{,}000$, or $x = 80$. Then $y = 768{,}000/(80)^2 = 120$, so the only critical point of C is $(80, 120)$. Thus the maximum value of C occurs for $x = 80$ and $y = 120$. The corresponding value of z is given by $z = 960{,}000/(80)(120) = 100$. Consequently the front and back walls should be 80 feet long, the side walls should be 120 feet wide, and the gymnasium should be 100 feet tall.

41. Since the cross section is a trapezoid, its area is given by

$$A = \tfrac{1}{2}[(\ell - 2x) + (\ell - 2x + 2x\cos\theta)](x \sin\theta) = \ell x \sin\theta - 2x^2 \sin\theta + x^2 \sin\theta\cos\theta$$

Minimum Theorem. First we find the critical points in the interior of R. Now $\partial A/\partial x = \ell \sin\theta - 4x\sin\theta + 2x\sin\theta\cos\theta$ and $\partial A/\partial\theta = \ell x\cos\theta - 2x^2\cos\theta + x^2(\cos^2\theta - \sin^2\theta)$. Thus $\partial A/\partial x = 0 = \partial A/\partial\theta$ if $\ell\sin\theta - 4x\sin\theta + 2x\sin\theta\cos\theta = 0$ and $\ell x\cos\theta - 2x^2\cos\theta + x^2(\cos^2\theta - \sin^2\theta) = 0$. Since $\sin\theta \neq 0$ if (x, θ) is in the interior of R, we can cancel $\sin\theta$ in the first equation and solve for ℓ, which means that $\ell = 4x - 2x\cos\theta$. Substituting for ℓ in the second equation, we obtain $(4x - 2x\cos\theta)x\cos\theta - 2x^2\cos\theta + x^2(\cos^2\theta - \sin^2\theta) = 0$, or $2x^2\cos\theta - x^2\cos^2\theta - x^2\sin^2\theta = 0$, or $2x^2\cos\theta - x^2 = 0$. Since $x \neq 0$ if (x, θ) is in the interior of R, we may cancel x^2 in the last equation and obtain $2\cos\theta - 1 = 0$, or $\cos\theta = \frac{1}{2}$, or $\theta = \pi/3$. Then $\ell = 4x - 2x\cos(\pi/3) = 3x$, so $x = \ell/3$. Thus the only critical point in the interior of R is $(\ell/3, \pi/3)$. The corresponding area is given by

$$A\left(\frac{\ell}{3}, \frac{\pi}{3}\right) = (\ell)\left(\frac{\ell}{3}\right)\left(\frac{\sqrt{3}}{2}\right) - 2\left(\frac{\ell}{3}\right)^2\left(\frac{\sqrt{3}}{2}\right) + \left(\frac{\ell}{3}\right)^2\left(\frac{\sqrt{3}}{2}\right)\left(\frac{1}{2}\right) = \frac{\ell^2\sqrt{3}}{12}$$

The boundary of R consists of the four line segments ℓ_1, ℓ_2, ℓ_3, and ℓ_4 shown in the figure. On ℓ_1, $x = 0$, so $A = 0$. On ℓ_2, $\theta = 0$, so $A = 0$. On ℓ_3, $x = \ell/2$, so $A = (\ell^2/4)\sin\theta\cos\theta = \ell^2/8 \sin 2\theta$ for $0 \leq \theta \leq \pi/2$. Since $dA/d\theta = \frac{1}{4}\ell^2\cos 2\theta$, we find that $dA/d\theta = 0$ for $\theta = \pi/4$. It follows that the maximum value of A on ℓ_3 is $A(\ell/2, 0) = 0$, $A(\ell/2, \pi/4) = \ell^2/8$, or $A(\ell/2, \pi/2) = 0$. Thus the maximum value of A on ℓ_3 is $\ell^2/8$. On ℓ_4, $\theta = \pi/2$, so $A = \ell x - 2x^2$ for $0 \leq x \leq \ell/2$. Since $dA/dx = \ell - 4x$, we find that $dA/dx = 0$ if $x = \ell/4$. It follows that the maximum value of A on ℓ_4 is $A(0, \pi/2) = 0$, $A(\ell/4, \pi/2) = \ell^2/8$, or $A(\ell/2, \pi/2) = 0$. Comparing the value of A at the critical point $(\ell/3, \pi/3)$ with the maximum values of A on ℓ_1, ℓ_2, ℓ_3, and ℓ_4, and noting that $\sqrt{3}/12 > \frac{1}{8}$, we conclude that the maximum value of A is obtained if $x = \ell/3$ and $\theta = \pi/3$.

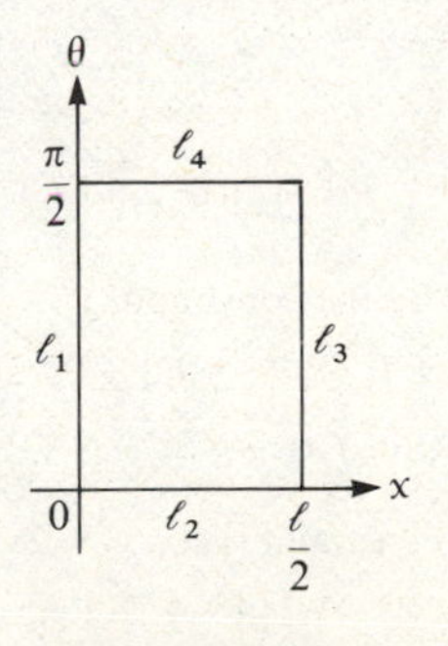

SECTION 14.9

1. Let $g(x, y) = x^2 + y^2$, so the constraint is $g(x, y) = x^2 + y^2 = 4$. Next, grad $f(x, y) = \mathbf{i} + 2y\mathbf{j}$ and grad $g(x, y) = 2x\mathbf{i} + 2y\mathbf{j}$, so that grad $f(x, y) = \lambda$ grad $g(x, y)$ if

$$1 = 2x\lambda \quad \text{and} \quad 2y = 2y\lambda$$

By the second equation, $y = 0$ or $\lambda = 1$. By the constraint,

$$\text{if} \quad y = 0, \quad \text{then } x^2 + 0^2 = 4, \quad \text{so } x = \pm 2$$

$$\text{if} \quad \lambda = 1, \quad \text{then } 1 = 2x, \quad \text{so } x = \frac{1}{2} \quad \text{and} \quad y = \pm\sqrt{4 - \frac{1}{4}} = \pm\frac{\sqrt{15}}{2}$$

The possible extreme values of f are $f(2, 0) = 2$, $f(-2, 0) = -2$, $f(\frac{1}{2}, \sqrt{15}/2) = \frac{17}{4}$, and $f(\frac{1}{2}, -\sqrt{15}/2) = \frac{17}{4}$. The maximum value is $\frac{17}{4}$ and the minimum value is -2.

3. Let $g(x, y) = x^2 + y^2$, so the constraint is $g(x, y) = x^2 + y^2 = 1$. Next, grad $f(x, y) = 3x^2\mathbf{i} + 6y^2\mathbf{j}$ and grad $g(x, y) = 2x\mathbf{i} + 2y\mathbf{j}$, so that grad $f(x, y) = \lambda$ grad $g(x, y)$ if

$$3x^2 = 2x\lambda \quad \text{and} \quad 6y^2 = 2y\lambda$$

By the first equation, either $x = 0$ or $x = \frac{2}{3}\lambda$, and by the second equation, either $y = 0$ or $y = \frac{1}{3}\lambda$. Now if $x = 0$, then by the constraint, $y = \pm 1$, and if $y = 0$, then by the constraint, $x = \pm 1$. Otherwise $x = \frac{2}{3}\lambda$ and $y = \frac{1}{3}\lambda$, so the constraint becomes $(\frac{2}{3}\lambda)^2 + (\frac{1}{3}\lambda)^2 = 1$. Then $\lambda = \pm 3/\sqrt{5}$, so that $x = \pm 2/\sqrt{5}$ and $y = \pm 1/\sqrt{5}$. The possible extreme values are $f(0, 1) = 2$, $f(0, -1) = -2$, $f(1, 0) = 1$, $f(-1, 0) = -1$,

$$f\left(\frac{2}{\sqrt{5}}, \frac{1}{\sqrt{5}}\right) = \frac{2}{\sqrt{5}}, \; f\left(\frac{-2}{\sqrt{5}}, \frac{-1}{\sqrt{5}}\right) = \frac{-2}{\sqrt{5}}, \; f\left(\frac{2}{\sqrt{5}}, \frac{-1}{\sqrt{5}}\right) = \frac{6}{5\sqrt{5}}, \; \text{and } f\left(\frac{-2}{\sqrt{5}}, \frac{1}{\sqrt{5}}\right) = \frac{-6}{5\sqrt{5}}$$

The maximum value is 2 and the minimum value is -2.

5. Let $g(x, y, z) = x^2 + y^2 + 4z^2$, so the constraint is $g(x, y, z) = x^2 + y^2 + 4z^2 = 6$. Next, grad $f(x, y, z) = yz\mathbf{i} + xz\mathbf{j} + xy\mathbf{k}$ and grad $g(x, y, z) = 2x\mathbf{i} + 2y\mathbf{j} + 8z\mathbf{k}$, so that grad $f(x, y, z) = \lambda$ grad $g(x, y, z)$ if

$$yz = 2x\lambda, \quad xz = 2y\lambda, \quad \text{and} \quad xy = 8z\lambda$$

If $x = 0$ or $y = 0$ or $z = 0$ or $\lambda = 0$, then $f(x, y, z) = 0$. Assume henceforth that x, y, z, and λ are nonzero. Then

$$\lambda = \frac{yz}{2x} = \frac{xz}{2y} = \frac{xy}{8z}$$

so that $x^2 = y^2$ and $y^2 = 4z^2$. By the constraint, $y^2 + y^2 + y^2 = 6$, so $y = \pm\sqrt{2}$. Thus $x = \pm\sqrt{2}$ and $z = \pm\sqrt{2}/2$. Since $f(\pm\sqrt{2}, \pm\sqrt{2}, \pm\sqrt{2}/2) = \pm\sqrt{2}$, the maximum value of f is $\sqrt{2}$ and the minimum value of f is $-\sqrt{2}$.

7. Let $g(x, y) = 2x^2 + \frac{3}{2}y^2$, so the constraint is $g(x, y) = 2x^2 + \frac{3}{2}y^2 = \frac{3}{2}$. Next grad $f(x, y) = 8x\mathbf{i} + (3y^2 + 3)\mathbf{j}$, and grad $g(x, y) = 4x\mathbf{i} + 3y\mathbf{j}$, so that grad $f(x, y) = \lambda$ grad $g(x, y)$ if

$$8x = 4x\lambda \quad \text{and} \quad 3y^2 + 3 = 3y\lambda$$

From the first equation, either $x = 0$ or $\lambda = 2$. If $x = 0$, then from the constraint, $2 \cdot 0^2 + \frac{3}{2}y^2 = \frac{3}{2}$, so that $y = 1$ or $y = -1$. If $\lambda = 2$, then from the second equation, $3y^2 + 3 = 6y$, so $3y^2 - 6y + 3 = 0$, and thus $y = 1$. But by the constraint, if $y = 1$, then $2x^2 + \frac{3}{2} \cdot 1^2 = \frac{3}{2}$, so $x = 0$. Since $f(0, 1) = 11$ and $f(0, -1) = 3$, the minimum value of f is 3.

9. Let $g(x, y, z) = x + y + z$, so the constraint becomes $g(x, y, z) = x + y + z = \frac{11}{12}$. Next, grad $f(x, y, z) = 4x^3\mathbf{i} + 32y^3\mathbf{j} + 108z^3\mathbf{k}$ and grad $g(x, y, z) = \mathbf{i} + \mathbf{j} + \mathbf{k}$, so that grad $f(x, y, z) = \lambda$ grad $g(x, y, z)$ if

$$4x^3 = \lambda, \quad 32y^3 = \lambda, \quad \text{and} \quad 108z^3 = \lambda$$

Thus $4x^3 = 32y^3$, so that $y = \frac{1}{2}x$, and $4x^3 = 108z^3$, so that $z = \frac{1}{3}x$. The constraint becomes $x + \frac{1}{2}x + \frac{1}{3}x = \frac{11}{12}$, which means that $x = \frac{1}{2}$, and thus $y = \frac{1}{4}$ and $z = \frac{1}{6}$. Then $f(\frac{1}{2}, \frac{1}{4}, \frac{1}{6}) = (\frac{1}{2})^4 + 8(\frac{1}{4})^4 + 27(\frac{1}{6})^4 = \frac{11}{96}$ is the minimum value of f.

11. First we use Lagrange multipliers to find the possible extreme values of f on the circle $x^2 + y^2 = 4$. Let $g(x, y) = x^2 + y^2$, so the constraint becomes $g(x, y) = x^2 + y^2 = 4$. Since grad $f(x, y) = 4x\mathbf{i} + (2y + 2)\mathbf{j}$ and grad $g(x, y) = 2x\mathbf{i} + 2y\mathbf{j}$, it follows that grad $f(x, y) = \lambda$ grad $g(x, y)$ if

$$4x = 2x\lambda \quad \text{and} \quad 2y + 2 = 2y\lambda$$

From the first equation, $x = 0$ or $\lambda = 2$. If $x = 0$, then from the constraint, $0^2 + y^2 = 4$, so that $y = \pm 2$. If $\lambda = 2$, then from the second equation, $2y + 2 = 4y$, so that $y = 1$, and then the constraint implies that $x^2 + 1 = 4$, or $x = \pm\sqrt{3}$. The possible extreme values of f on the circle $x^2 + y^2 = 4$ are $f(0, 2) = 5$, $f(0, -2) = -3$, and $f(\pm\sqrt{3}, 1) = 6$. For the interior $x^2 + y^2 < 4$ of the disk, we find that

$$f_x(x, y) = 4x \quad \text{and} \quad f_y(x, y) = 2y + 2$$

so that $f_x(x, y) = 0 = f_y(x, y)$ only if $x = 0$ and $y = -1$. But $f(0, -1) = -4$. Consequently the maximum value and minimum value of f on the disk $x^2 + y^2 \le 4$ are 6 and -4, respectively.

13. First we use Lagrange multipliers to find the possible extreme values of f on the ellipse $2x^2 + y^2 = 4$. Let $g(x, y) = 2x^2 + y^2$, so the constraint is $g(x, y) = 2x^2 + y^2 = 4$. Since grad $f(x, y) = y\mathbf{i} + x\mathbf{j}$ and grad $g(x, y) = 4x\mathbf{i} + 2y\mathbf{j}$, it follows that grad $f(x, y) = \lambda$ grad $g(x, y)$ if

$$y = 4x\lambda \quad \text{and} \quad x = 2y\lambda$$

Because of the constraint, $y \ne 0$, since if $y = 0$ then the second equation would then imply that $x = 0$ also. Thus $\lambda = x/2y$, so that the first equation becomes $y = 4x(x/2y)$, and therefore $y^2 = 2x^2$. The constraint then becomes $2x^2 + 2x^2 = 4$, so $x = \pm 1$ and hence $y = \pm\sqrt{2}$. The possible extreme values of f on the ellipse are $f(1, \sqrt{2}) = \sqrt{2}$, $f(1, -\sqrt{2}) = -\sqrt{2}$, $f(-1, \sqrt{2}) = -\sqrt{2}$, and $f(-1, -\sqrt{2}) = \sqrt{2}$. For the interior $2x^2 + y^2 < 4$ of the ellipse, we find that

$$f_x(x, y) = y \quad \text{and} \quad f_y(x, y) = x$$

so that $f_x(x, y) = 0 = f_y(x, y)$ only if $x = 0$ and $y = 0$. But $f(0, 0) = 0$. Consequently the maximum and minimum values of f on $2x^2 + y^2 \le 4$ are $\sqrt{2}$ and $-\sqrt{2}$, respectively.

15. The distance from (x, y, z) to the origin is minimum if and only if the square of the distance is minimum, so let $f(x, y, z) = x^2 + y^2 + z^2$. Also let $g(x, y, z) = x^2 - yz$, so the constraint becomes $g(x, y, z) = x^2 - yz = 1$. Since grad $f(x, y, z) = 2x\mathbf{i} + 2y\mathbf{j} + 2z\mathbf{k}$ and grad $g(x, y, z) = 2x\mathbf{i} - z\mathbf{j} - y\mathbf{k}$, it follows that grad $f(x, y, z) = \lambda$ grad $g(x, y, z)$ if

$$2x = 2x\lambda, \quad 2y = -z\lambda, \quad \text{and} \quad 2z = -y\lambda$$

From the first equation, $x = 0$ or $\lambda = 1$. If $x = 0$, the constraint becomes $-yz = 1$, or $z = -1/y$. Then the second equation becomes $2y = \lambda/y$, so $\lambda = 2y^2$. This means that the third equation becomes $2(-1/y) = -y(2y^2)$, so $1 = y^4$, and thus $y = \pm 1$, and hence $z = \mp 1$. If $x \ne 0$, then $\lambda = 1$, so the third equation becomes $2z = -y$, and the second equation then becomes $2(-2z) = -z$, which means that $z = 0$, and thus $y = 0$ and $x = \pm 1$. The possible minimum values of f are $f(0, 1, -1) = 2$, $f(0, -1, 1) = 2$, $f(1, 0, 0) = 1$, and $f(-1, 0, 0) = 1$. Consequently the points on $x^2 - yz = 1$ closest to the origin are $(1, 0, 0)$ and $(-1, 0, 0)$.

17. Let x and y be the acute angles, so that $x + y = \pi/2$. Let $f(x, y) = \sin x \sin y$ and $g(x, y) = x + y$, so the constraint is $g(x, y) = x + y = \pi/2$. Since grad $f(x, y) =$ $\cos x \sin y\mathbf{i} + \sin x \cos y\mathbf{j}$ and grad $g(x, y) = \mathbf{i} + \mathbf{j}$, it follows that grad $f(x, y) =$ λ grad $g(x, y)$ if

$$\cos x \sin y = \lambda \quad \text{and} \quad \sin x \cos y = \lambda$$

Then $\sin x \cos y - \cos x \sin y = 0$, so $\sin (x - y) = 0$, or $x = y$. Then the constraint becomes $x + x = \pi/2$, or $x = \pi/4$ and $y = \pi/4$. The maximum value of f is $f(\pi/4, \pi/4) = \frac{1}{2}$.

19. Let (x, y, z) be on the sphere in the first octant, and let $f(x, y, z) = 1/6xyz$ denote the volume of the corresponding tetrahedron with vertex at (x, y, z). Let $g(x, y, z) = x^2 + y^2 + z^2$, so that the constraint is $g(x, y, z) = x^2 + y^2 + z^2 = 1$. Since grad $f(x, y, z) =$ $-1/(6x^2yz)\mathbf{i} - 1/(6xy^2z)\mathbf{j} - 1/(6xyz^2)\mathbf{k}$ and grad $g(x, y, z) = 2x\mathbf{i} + 2y\mathbf{j} + 2z\mathbf{k}$, it follows that grad $f(x, y, z) = \lambda$ grad $g(x, y, z)$ if

$$\frac{-1}{6x^2yz} = 2x\lambda, \quad \frac{-1}{6xy^2z} = 2y\lambda, \quad \text{and} \quad \frac{-1}{6xyz^2} = 2z\lambda$$

Then $\lambda \ne 0$, and the three equations in turn yield $-1/(6xyz\lambda) = x^2 = y^2 = z^2$. Since $x > 0$, $y > 0$, and $z > 0$, this means that $x = y = z$. From the constraint, $x^2 + x^2 + x^2 =$ 1, so that $x = 1/\sqrt{3}$, and thus $y = 1/\sqrt{3}$ and $z = 1/\sqrt{3}$. The minimum volume is given by $f(1/\sqrt{3}, 1/\sqrt{3}, 1/\sqrt{3}) = \frac{1}{2}\sqrt{3}$.

21. Let x, y, and z be as in the figure, with $x > 0$, $y > 0$, and $z > 0$. Then the surface area is given by $S(x, y, z) = 3xz + 7yz + xy$. Let V denote the volume, so that $V(x, y, z) =$ xyz. The constraint becomes $V(x, y, z) = xyz = 12$. Since grad $S(x, y, z) = (3z + y)\mathbf{i} +$ $(7z + x)\mathbf{j} + (3x + 7y)\mathbf{k}$ and grad $V(x, y, z) = yz\mathbf{i} + xz\mathbf{j} + xy\mathbf{k}$, it follows that grad $S(x, y, z) = \lambda$ grad $V(x, y, z)$ if

$$3z + y = yz\lambda, \quad 7z + x = xz\lambda, \quad \text{and} \quad 3x + 7y = xy\lambda$$

Solving for λ in the three equations, and noting that $x > 0$, $y > 0$, and $z > 0$, we obtain

$$\lambda = \frac{3z + y}{yz} = \frac{3}{y} + \frac{1}{z}, \quad \lambda = \frac{7z + x}{xz} = \frac{7}{x} + \frac{1}{z}, \quad \text{and} \quad \lambda = \frac{3x + 7y}{xy} = \frac{3}{y} + \frac{7}{x}$$

so that
$$\frac{3}{y} + \frac{1}{z} = \frac{7}{x} + \frac{1}{z} \quad \text{and} \quad \frac{7}{x} + \frac{1}{z} = \frac{3}{y} + \frac{7}{x}$$

Thus $x = \frac{7}{3}y$ and $z = \frac{1}{3}y$, so the constraint becomes $(\frac{7}{3}y)(y)(\frac{1}{3}y) = 12$, so that $y = (\frac{108}{7})^{1/3}$, and hence $x = \frac{7}{3}(\frac{108}{7})^{1/3} = (196)^{1/3}$, and $z = \frac{1}{3}(\frac{108}{7})^{1/3} = (\frac{4}{7})^{1/3}$. These dimensions yield the minimum surface area.

23. The volume of the pipe is given by $V(r, \ell) = \pi r^2 \ell$, with $r > 0$ and $\ell > 0$. Since $2r + \frac{1}{2}\ell =$ $3 \sec (\pi/4) = 3\sqrt{2}$, let $g(r, \ell) = 2r + \frac{1}{2}\ell$, so that the constraint becomes $g(r, \ell) = 2r + \frac{1}{2}\ell =$ $3\sqrt{2}$. Since grad $V(r, \ell) = 2\pi r\ell\mathbf{i} + \pi r^2\mathbf{j}$ and grad $g(r, \ell) = 2\mathbf{i} + \frac{1}{2}\mathbf{j}$, it follows that grad $V(r, \ell) = \lambda$ grad $g(r, \ell)$ if

$$2\pi r\ell = 2\lambda \quad \text{and} \quad \pi r^2 = \tfrac{1}{2}\lambda$$

Solving for λ in the two equations, we obtain $\pi r\ell = \lambda = 2\pi r^2$. Since $r \ne 0$, this means that $\ell = 2r$. From the constraint, $2r + \frac{1}{2}(2r) = 3\sqrt{2}$, so that $r = \sqrt{2}$, and thus $\ell = 2\sqrt{2}$. The pipe has maximum volume if it has a radius of $\sqrt{2}$ meters and is $2\sqrt{2}$ meters long.

25. The volume of the box is given by $V(x, y, z) = xyz$, and the constraint is $V(x, y, z) =$ $xyz = V$ for a constant V. Since grad $E(x, y, z) = -h^2/(4mx^3)\mathbf{i} - h^2/(4my^3)\mathbf{j} - h^2/(4mz^3)\mathbf{k}$ and grad $V(x, y, z) = yz\mathbf{i} + xz\mathbf{j} + xy\mathbf{k}$, it follows that grad $E(x, y, z) = \lambda$ grad $V(x, y, z)$ if

$$\frac{-h^2}{4mx^3} = yz\lambda, \quad \frac{-h^2}{4my^3} = xz\lambda, \quad \text{and} \quad \frac{-h^2}{4mz^3} = xy\lambda$$

Thus $-h^2/(4mxyz\lambda) = x^2 = y^2 = z^2$, so since x, y, and z are positive, $x = y = z$. The constraint becomes $x^3 = V$, so $x = y = z = V^{1/3}$, and the box is a cube whose sides are $V^{1/3}$ units long.

27. Let A be a units from the horizontal line and let B be b units from the horizontal line. Then the distance light travels in the medium containing A is $\sqrt{x^2 + a^2}$ and the distance light travels in the medium containing B is $\sqrt{y^2 + b^2}$. Thus the total time $t(x, y)$ required for light to travel from A to B is given by $t(x, y) = \sqrt{x^2 + a^2}/v + \sqrt{y^2 + b^2}/u$, and we wish to minimize t. If $D(x, y) = x + y$, then the constraint is $D(x, y) = x + y = \ell$. Since grad $t(x, y) = x/(v\sqrt{x^2 + a^2})\mathbf{i} + y/(u\sqrt{y^2 + b^2})\mathbf{j}$ and grad $D(x, y) = \mathbf{i} + \mathbf{j}$, it follows that grad $t(x, y) = \lambda$ grad $D(x, y)$ if

$$\frac{x}{v\sqrt{x^2+a^2}} = \lambda \quad \text{and} \quad \frac{y}{u\sqrt{y^2+b^2}} = \lambda$$

which implies that
$$\frac{x}{v\sqrt{x^2+a^2}} = \frac{y}{u\sqrt{y^2+b^2}}$$

Since $x/\sqrt{x^2+a^2} = \sin\theta$ and $y/\sqrt{y^2+b^2} = \sin\phi$, it follows that

$$\frac{\sin\theta}{v} = \frac{\sin\phi}{u}$$

29. a. Let $g(x, y) = ax + by$, so the constraint is $g(x, y) = ax + by = c$. Since grad $f(x_0, y_0) = f_x(x_0, y_0)\mathbf{i} + f_y(x_0, y_0)\mathbf{j}$ and grad $g(x_0, y_0) = a\mathbf{i} + b\mathbf{j}$, it follows that grad $f(x_0, y_0) = \lambda$ grad $g(x_0, y_0)$ if

$$f_x(x_0, y_0) = a\lambda \quad \text{and} \quad f_y(x_0, y_0) = b\lambda$$

If a and b are nonzero, then

$$\frac{f_x(x_0, y_0)}{a} = \frac{f_y(x_0, y_0)}{b} = \lambda$$

b. If $f(x, y) = x^\alpha y^\beta$, then $f_x(x_0, y_0) = \alpha x_0^{\alpha-1} y_0^\beta$ and $f_y(x_0, y_0) = \beta x_0^\alpha y_0^{\beta-1}$. By part (a) we have

$$\frac{\alpha x_0^{\alpha-1} y_0^\beta}{a} = \frac{\beta x_0^\alpha y_0^{\beta-1}}{b} = \lambda$$

Canceling terms, we conclude that $y_0/x_0 = \beta a/(\alpha b)$.

31. Let (x, y, z) be on the paraboloid and the plane. The distance from (x, y, z) to the origin is minimum if and only if the square of the distance is minimum, so let $f(x, y, z) = x^2 + y^2 + z^2$. Let $g_1(x, y, z) = x^2 + y^2 + z$ and $g_2(x, y, z) = x + 2y$. The constraints are $g_1(x, y, z) = x^2 + y^2 + z = \frac{3}{2}$ and $g_2(x, y, z) = x + 2y = 1$. Since grad $f(x, y, z) = 2x\mathbf{i} + 2y\mathbf{j} + 2z\mathbf{k}$, grad $g_1(x, y, z) = 2x\mathbf{i} + 2y\mathbf{j} + \mathbf{k}$, and grad $g_2(x, y, z) = \mathbf{i} + 2\mathbf{j}$, it follows that grad $f(x, y, z) = \lambda$ grad $g_1(x, y, z) + \mu$ grad $g_2(x, y, z)$ if

$$2x = 2x\lambda + \mu, \quad 2y = 2y\lambda + 2\mu, \quad \text{and} \quad 2z = \lambda \qquad (*)$$

Solving the first two equations in (*) for μ, we obtain $2x - 2x\lambda = \mu = y - y\lambda$, and thus $2x(1 - \lambda) = y(1 - \lambda)$. If $\lambda = 1$ then the third equation in (*) yields $z = \frac{1}{2}$. With $z = \frac{1}{2}$ and the fact that $x = 1 - 2y$ by the second constraint, the first constraint becomes $(1 - 2y)^2 + y^2 + \frac{1}{2} = \frac{3}{2}$, or $1 - 4y + 5y^2 = 1$. Thus $5y^2 = 4y$, so $y = 0$ or $y = \frac{4}{5}$. If $y = 0$ then $x = 1 - 2(0) = 1$, and if $y = \frac{4}{5}$ then $x = 1 - 2(\frac{4}{5}) = -\frac{3}{5}$. For $(1, 0, \frac{1}{2})$ the distance is $\sqrt{f(1, 0, \frac{1}{2})} = \sqrt{1^2 + (\frac{1}{2})^2} = \frac{1}{2}\sqrt{5}$. For $(-\frac{3}{5}, \frac{4}{5}, \frac{1}{2})$ the distance is

$$\sqrt{f(-\tfrac{3}{5}, \tfrac{4}{5}, \tfrac{1}{2})} = \sqrt{(-\tfrac{3}{5})^2 + (\tfrac{4}{5})^2 + (\tfrac{1}{2})^2} = \sqrt{\tfrac{9}{25} + \tfrac{16}{25} + \tfrac{1}{4}} = \sqrt{\tfrac{125}{100}} = \tfrac{1}{2}\sqrt{5}$$

Next, if $\lambda \neq 1$, then the equation $2x(1 - \lambda) = y(1 - \lambda)$ becomes $2x = y$, so since $x + 2y = 1$ by the second constraint, we obtain $x + 2(2x) = 1$; thus $x = \frac{1}{5}$, so $y = \frac{2}{5}$, and by the first constraint, $z = \frac{3}{2} - (\frac{1}{5})^2 - (\frac{2}{5})^2 = \frac{13}{10}$. For $(\frac{1}{5}, \frac{2}{5}, \frac{13}{10})$ the distance is

$$\sqrt{f(\tfrac{1}{5}, \tfrac{2}{5}, \tfrac{13}{10})} = \sqrt{(\tfrac{1}{5})^2 + (\tfrac{2}{5})^2 + (\tfrac{13}{10})^2} = \sqrt{\tfrac{189}{100}} = \tfrac{1}{10}\sqrt{189}$$

Since $\frac{1}{2}\sqrt{5} < \frac{1}{10}\sqrt{189}$, the minimum distance is $\frac{1}{2}\sqrt{5}$.

CHAPTER 14 REVIEW

1. all (x, y) such that $\frac{1}{4}x^2 + \frac{1}{25}y^2 \le 1$

3. all (x, y, z) such that $x - y + z > 0$

5. The level curves are the hyperbolas $xy = 1$ and $xy = -1$, respectively.

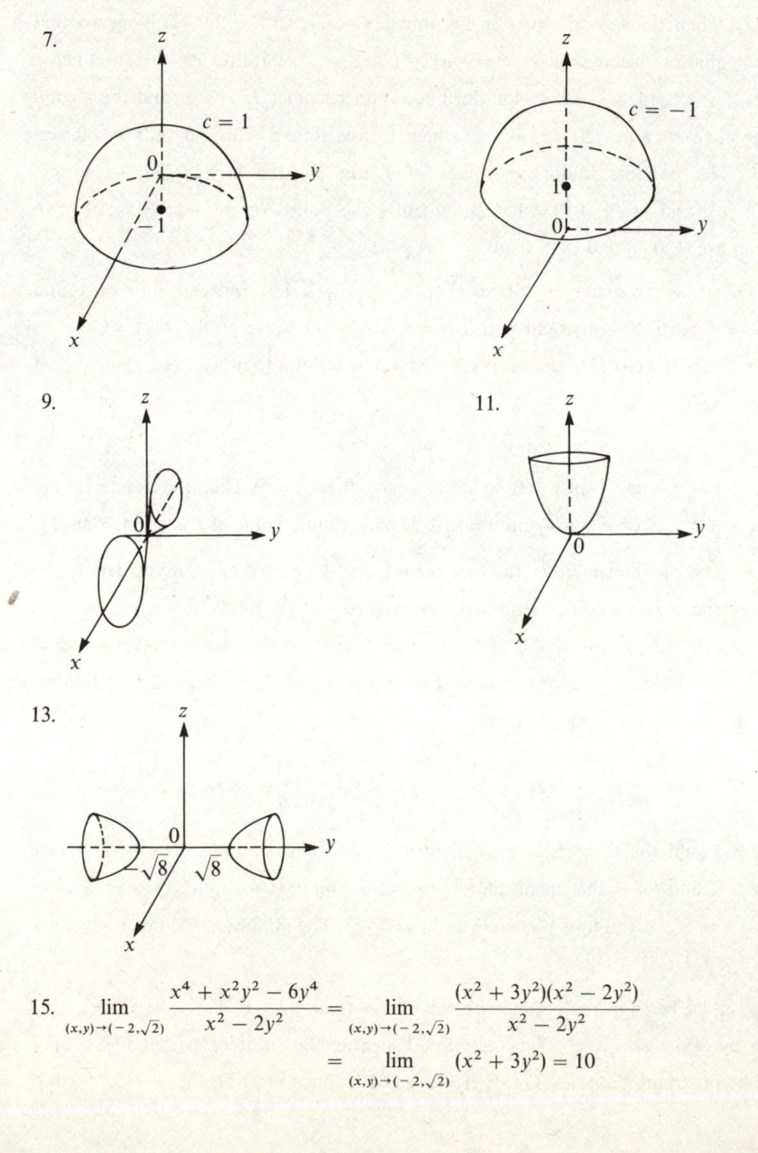

15. $$\lim_{(x,y)\to(-2,\sqrt{2})} \frac{x^4 + x^2y^2 - 6y^4}{x^2 - 2y^2} = \lim_{(x,y)\to(-2,\sqrt{2})} \frac{(x^2 + 3y^2)(x^2 - 2y^2)}{x^2 - 2y^2} = \lim_{(x,y)\to(-2,\sqrt{2})} (x^2 + 3y^2) = 10$$

17. $f_x(x, y) = 12x^2$, $f_y(x, y) = -6y$

19. $f_x(x, y, z) = 2xe^{x^2} \ln (y^2 - 3z)$, $f_y(x, y, z) = \dfrac{2ye^{x^2}}{y^2 - 3z}$, $f_z(x, y, z) = \dfrac{-3e^{x^2}}{y^2 - 3z}$

21. $k_x(x, y, z) = \frac{5}{2}[\sqrt{z} \tan (x^2 + y)]^{3/2}(2x\sqrt{z} \sec^2 (x^2 + y))$
$= 5xz^{5/4}[\tan^{3/2} (x^2 + y)] \sec^2 (x^2 + y)$,
$k_y(x, y, z) = \frac{5}{2}z^{5/4} [\tan^{3/2} (x^2 + y)] \sec^2 (x^2 + y)$,
$k_z(x, y, z) = \dfrac{5}{2}[\sqrt{z} \tan (x^2 + y)]^{3/2}\left[\dfrac{1}{2}\dfrac{1}{\sqrt{z}} \tan (x^2 + y)\right] = \dfrac{5}{4}\sqrt[4]{z} \tan^{5/2} (x^2 + y)$

23. $f_y(x, y) = 1 + 2xe^y$ and $g_x(x, y) = 1 + 2xe^y$, so $f_y = g_x$.

25. $g(u, v) = \ln \dfrac{u^2}{e^v} = \ln u^2 - \ln e^v = 2 \ln u - v$; $g_u(u, v) = \dfrac{2}{u}$, $g_{uu}(u, v) = \dfrac{-2}{u^2}$, $g_{uv}(u, v) = 0$;
$g_v(u, v) = -1$, $g_{vv}(u, v) = g_{vu}(u, v) = 0$.

27. $\dfrac{\partial z}{\partial x} = -be^{-ay} \sin bx$, $\dfrac{\partial^2 z}{\partial x^2} = -b^2e^{-ay} \cos bx$, $\dfrac{\partial z}{\partial y} = -ae^{-ay} \cos bx$, $\dfrac{\partial^2 z}{\partial y^2} = a^2e^{-ay} \cos bx$;
thus $a^2 \dfrac{\partial^2 z}{\partial x^2} + b^2 \dfrac{\partial^2 z}{\partial y^2} = -a^2b^2e^{-ay} \cos bx + a^2b^2e^{-ay} \cos bx = 0$.

29. $\dfrac{dz}{dt} = \dfrac{\partial z}{\partial x}\dfrac{dx}{dt} + \dfrac{\partial z}{\partial y}\dfrac{dy}{dt} = (2xe^{x^2})(2t) + (-\frac{1}{2}e^{y/2})(3t^2 - 1) = 4t^3e^{t^4} - \frac{1}{2}(3t^2 - 1)e^{(t^3 - t)/2}$

31. $\dfrac{dw}{dt} = \dfrac{\partial w}{\partial x}\dfrac{dx}{dt} + \dfrac{\partial w}{\partial y}\dfrac{dy}{dt} + \dfrac{\partial w}{\partial z}\dfrac{dz}{dt}$

$$= \frac{x}{\sqrt{x^2 + y^2z^4}}(2) + \frac{yz^4}{\sqrt{x^2 + y^2z^4}}(3t^2) + \frac{2y^2z^3}{\sqrt{x^2 + y^2z^4}}\left(\frac{-1}{t^2}\right) = \frac{\sqrt{5}t}{|t|}$$

33. $\partial z/\partial x = 2x\, f'(x^2 + y^2)$ and $\partial z/\partial y = 2y\, f'(x^2 + y^2)$. Thus
$y(\partial z/\partial x) - x(\partial z/\partial y) = 2xy\, f'(x^2 + y^2) - 2xy\, f'(x^2 + y^2) = 0$.

35. Treating z as a function of x and y and taking partial derivatives with respect to x, we find that

$$yz + xy\frac{\partial z}{\partial x} - \frac{1}{x^2yz} - \frac{1}{xyz^2}\frac{\partial z}{\partial x} = 3z^2\frac{\partial z}{\partial x}$$

Solving for $\partial z/\partial x$, we obtain

$$\frac{\partial z}{\partial x} = \frac{yz - 1/x^2yz}{3z^2 + (1/xyz^2) - xy} = \frac{z(x^2y^2z^2 - 1)}{x(3xyz^4 + 1 - x^2y^2z^2)}$$

Interchanging the roles of x and y, we obtain

$$\frac{\partial z}{\partial y} = \frac{z(x^2y^2z^2 - 1)}{y(3xyz^4 + 1 - x^2y^2z^2)}$$

37. $f_x(x, y, z) = \dfrac{-2x}{(x^2 + y^2 + z^2)^2}$, $f_y(x, y, z) = \dfrac{-2y}{(x^2 + y^2 + z^2)^2}$, $f_z(x, y, z) = \dfrac{-2z}{(x^2 + y^2 + z^2)^2}$,
and $\|\mathbf{a}\| = \sqrt{3}$, so that $\mathbf{u} = \dfrac{1}{\sqrt{3}}\mathbf{i} - \dfrac{1}{\sqrt{3}}\mathbf{j} - \dfrac{1}{\sqrt{3}}\mathbf{k}$;

$$D_{\mathbf{u}}f(-1, 0, 2) = f_x(-1, 0, 2)\left(\frac{1}{\sqrt{3}}\right) + f_y(-1, 0, 2)\left(\frac{-1}{\sqrt{3}}\right) + f_z(-1, 0, 2)\left(\frac{-1}{\sqrt{3}}\right)$$
$$= \frac{2}{25\sqrt{3}} + 0 + \frac{4}{25\sqrt{3}} = \frac{2}{25}\sqrt{3}$$

39. $\operatorname{grad} f(x, y) = 2e^{2x} \ln y\mathbf{i} + e^{2x}\dfrac{1}{y}\mathbf{j}$, so that $\operatorname{grad} f(0, 1) = \mathbf{j}$.

41. $\operatorname{grad} f(x, y) = -y \sin xy\mathbf{i} - x \sin xy\mathbf{j}$, so that $\operatorname{grad} f(\frac{1}{2}, \pi) = -\pi\mathbf{i} - \frac{1}{2}\mathbf{j}$. Thus the direction in which f increases most rapidly at $(\frac{1}{2}, \pi)$ is $-\pi\mathbf{i} - \frac{1}{2}\mathbf{j}$.

43. Let $f(x, y) = \arctan (x^2 + y)$. Since the graph of the given equation is a level curve of f, Theorem 14.15 implies that $\operatorname{grad} f(\frac{1}{2}, \frac{3}{4})$ is normal to the graph at $(\frac{1}{2}, \frac{3}{4})$. Since

$$\operatorname{grad} f(x, y) = \frac{2x}{1 + (x^2 + y)^2}\mathbf{i} + \frac{1}{1 + (x^2 + y)^2}\mathbf{j}$$

so that $\operatorname{grad} f(\frac{1}{2}, \frac{3}{4}) = \frac{1}{2}\mathbf{i} + \frac{1}{2}\mathbf{j}$, we find that $\frac{1}{2}\mathbf{i} + \frac{1}{2}\mathbf{j}$ is normal to the graph at $(\frac{1}{2}, \frac{3}{4})$.

45. Since $f_x(x, y) = -1/x^2$ and $f_y(x, y) = 1/y^2$, so that $f_x(-\frac{1}{2}, \frac{1}{3}) = -4$ and $f_y(-\frac{1}{2}, \frac{1}{3}) = 9$, the vector $-4\mathbf{i} + 9\mathbf{j} - \mathbf{k}$ is normal to the graph of f at $(-\frac{1}{2}, \frac{1}{3}, -5)$, and an equation of the plane tangent at $(-\frac{1}{2}, \frac{1}{3}, -5)$ is $-4(x + \frac{1}{2}) + 9(y - \frac{1}{3}) - (z + 5) = 0$, or $-4x + 9y - z = 10$.

47. Since $f_x(x, y) = 3x/\sqrt{3x^2 + 2y^2 + 2}$ and $f_y(x, y) = 2y/\sqrt{3x^2 + 2y^2 + 2}$, so that $f_x(4, -5) = \frac{6}{5}$ and $f_y(4, -5) = -1$, the vector $\frac{6}{5}\mathbf{i} - \mathbf{j} - \mathbf{k}$ is normal to the graph of f at $(4, -5, 10)$, and an equation of the plane tangent at $(4, -5, 10)$ is $\frac{6}{5}(x - 4) - (y + 5) - (z - 10) = 0$, or $6x - 5y - 5z = -1$.

49. If $f(x, y, z) = xe^{yz} - 2y$, then $xe^{yz} - 2y = -1$ is a level surface of f. Since $f_x(x, y, z) = e^{yz}$, $f_y(x, y, z) = xze^{yz} - 2$, and $f_z(x, y, z) = xye^{yz}$, so $f_x(1, 1, 0) = 1$, $f_y(1, 1, 0) = -2$, and $f_z(1, 1, 0) = 1$, it follows that an equation of the plane tangent at $(1, 1, 0)$ is $(x - 1) - 2(y - 1) + (z - 0) = 0$, or $x - 2y + z = -1$.

51. Let $f(x, y, z) = 2x^3 + y - z^2$. Then $f_x(x, y, z) = 6x^2$, $f_y(x, y, z) = 1$, and $f_z(x, y, z) = -2z$, so a vector normal to the plane tangent at any point (x_0, y_0, z_0) on the level surface $2x^3 + y - z^2 = 5$ is $6x_0^2\mathbf{i} + \mathbf{j} - 2z_0\mathbf{k}$. If the tangent plane is parallel to the plane $24x + y - 6z = 3$, whose normal is $24\mathbf{i} + \mathbf{j} - 6\mathbf{k}$, then $6x_0^2\mathbf{i} + \mathbf{j} - 2z_0\mathbf{k} = c(24\mathbf{i} + \mathbf{j} - 6\mathbf{k})$ for some c. It follows that $c = 1$, so that $x_0 = \pm 2$ and $z_0 = 3$. Because (x_0, y_0, z_0) is on the level surface $2x^3 + y - z^2 = 5$ we find that if $x_0 = 2$ and $z_0 = 3$, then $y_0 = -2$, and

if $x_0 = -2$ and $z_0 = 3$, then $y_0 = 30$. The required points on the level surface are $(2, -2, 3)$ and $(-2, 30, 3)$.

53. Let $f(x, y) = x^2 + 4y^2$. Then $f_x(x, y) = 2x$ and $f_y(x, y) = 8y$, so that $f_x(2, 0) = 4$ and $f_y(2, 0) = 0$. Thus a vector normal to the plane tangent to the graph of f at $(2, 0, 4)$ is $4\mathbf{i} - \mathbf{k}$. Next, let $g(x, y) = 4x + y^2 - 4$. Then $g_x(x, y) = 4$ and $g_y(x, y) = 2y$, so that $g_x(2, 0) = 4$ and $g_y(2, 0) = 0$. Thus a vector normal to the plane tangent to the graph of g at $(2, 0, 4)$ is $4\mathbf{i} - \mathbf{k}$. Since the normals are identical and the planes both pass through $(2, 0, 4)$, the two planes are identical.

55. Let $x_0 = 1$, $y_0 = 0$, $h = -.03$, and $k = .05$. Since

$$f_x(x, y) = \frac{1}{1 + \left(\frac{x}{1+y}\right)^2}\left(\frac{1}{1+y}\right) = \frac{1+y}{(1+y)^2 + x^2}$$

and

$$f_y(x, y) = \frac{1}{1 + \left(\frac{x}{1+y}\right)^2}\left(\frac{-x}{(1+y)^2}\right) = \frac{-x}{(1+y)^2 + x^2}$$

we have $f_x(1, 0) = \frac{1}{2}$ and $f_y(1, 0) = -1/2$. Also $f(1, 0) = \pi/4$. Thus

$$f(.97, .05) \approx \frac{\pi}{4} + \frac{1}{2}(-.03) - \frac{1}{2}(.05) = \frac{\pi}{4} - .04 \approx .745398$$

57. Since $f(x, y) = \ln(x/y) = \ln x - \ln y$, we have $\partial f/\partial z = 1/x$ and $\partial f/\partial y = -1/y$, so that $df = (1/x)\,dx - (1/y)\,dy$.

59. $f_x(x, y) = 2x - 2$ and $f_y(x, y) = 2y - 4$, so $f_x(x, y) = 0 = f_y(x, y)$ if $x = 1$ and $y = 2$. Thus $(1, 2)$ is a critical point. Next, $f_{xx}(x, y) = 2$, $f_{yy}(x, y) = 2$, and $f_{xy}(x, y) = 0$. Thus $D(1, 2) = f_{xx}(1, 2)f_{yy}(1, 2) - [f_{xy}(1, 2)]^2 = (2)(2) - 0^2 = 4$. Since $D(1, 2) > 0$ and $f_{xx}(1, 2) > 0$, f has a relative minimum value at $(1, 2)$.

61. $f_x(x, y) = y - 16/x^3$ and $f_y(x, y) = x - 16/y^3$, so $f_x(x, y) = 0 = f_y(x, y)$ if $y - 16/x^3 = 0$ and $x - 16/y^3 = 0$. Solving for y in the first equation and substituting for it in the second equation yields $x - 16(x^3/16)^3 = 0$, so that $x^8 = 256$, and thus $x = \pm 2$. Then $y = \pm 2$, so that $(2, 2)$ and $(-2, -2)$ are critical points. Next, $f_{xx}(x, y) = 48/x^4$, $f_{yy}(x, y) = 48/x^4$, and $f_{xy}(x, y) = 1$. Thus $D(2, 2) = f_{xx}(2, 2)f_{yy}(2, 2) - [f_{xy}(2, 2)]^2 = (3)(3) - 1^2 = 8$. Since $D(2, 2) > 0$ and $f_{xx}(2, 2) > 0$, f has a relative minimum value at $(2, 2)$. Finally, $D(-2, -2) = f_{xx}(-2, -2)f_{yy}(-2, -2) - [f_{xy}(-2, -2)]^2 = (3)(3) - 1^2 = 8$, and since $D(-2, -2) > 0$ and $f_{xx}(-2, -2) > 0$, f has a relative minimum value at $(-2, -2)$.

63. Let $g(x, y) = 3x^2 + y^2$, so the constraint is $g(x, y) = 3x^2 + y^2 = 3$. Since $\operatorname{grad} f(x, y) = (6x - y)\mathbf{i} + (-x + 2y)\mathbf{j}$ and $\operatorname{grad} g(x, y) = 6x\mathbf{i} + 2y\mathbf{j}$, it follows that $\operatorname{grad} f(x, y) = \lambda \operatorname{grad} g(x, y)$ if

$$6x - y = 6x\lambda \quad \text{and} \quad -x + 2y = 2y\lambda$$

If $x = 0$, then by the first equation, $y = 0$, and the constraint is contradicted. Similarly, if $y = 0$, then $x = 0$ by the second equation, and the constraint is contradicted. Thus $x \neq 0$ and $y \neq 0$, so the two equations can be solved for λ:

$$\lambda = \frac{6x - y}{6x} = 1 - \frac{y}{6x} \quad \text{and} \quad \lambda = \frac{-x + 2y}{2y} = \frac{-x}{2y} + 1$$

so that $y/6x = x/2y$, and hence $y^2 = 3x^2$. The constraint becomes $3x^2 + 3x^2 = 3$, so $x = \pm 1/\sqrt{2}$, and therefore $y = \pm\sqrt{\frac{3}{2}}$. The possible extreme values of f are

$$f\left(\frac{1}{\sqrt{2}}, \sqrt{\frac{3}{2}}\right) = f\left(\frac{-1}{\sqrt{2}}, -\sqrt{\frac{3}{2}}\right) = 3 - \frac{\sqrt{3}}{2}$$

and

$$f\left(\frac{-1}{\sqrt{2}}, \sqrt{\frac{3}{2}}\right) = f\left(\frac{1}{\sqrt{2}}, -\sqrt{\frac{3}{2}}\right) = 3 + \frac{\sqrt{3}}{2}$$

The maximum value is $3 + \sqrt{3}/2$ and the minimum value is $3 - \sqrt{3}/2$.

65. Let $g(x, y, z) = x^2 + y^2 - z$, so the constraint is $g(x, y, z) = x^2 + y^2 - z = 2$. Since

$$\operatorname{grad} f(x, y, z) = \frac{2x}{z^2 + 5}\mathbf{i} + \frac{2y}{z^2 + 5}\mathbf{j} - \frac{2z(x^2 + y^2)}{(z^2 + 5)^2}\mathbf{k} \quad \text{and} \quad \operatorname{grad} g(x, y, z) = 2x\mathbf{i} + 2y\mathbf{j} - \mathbf{k}$$

it follows that $\operatorname{grad} f(x, y, z) = \lambda \operatorname{grad} g(x, y, z)$ if

$$\frac{2x}{z^2 + 5} = 2x\lambda, \quad \frac{2y}{z^2 + 5} = 2y\lambda, \quad \text{and} \quad \frac{-2z(x^2 + y^2)}{(z^2 + 5)^2} = -\lambda \qquad (*)$$

From the first equation, $x = 0$ or $\lambda = 1/(z^2 + 5)$. Assume that $\lambda = 1/(z^2 + 5)$. Then the third equation is transformed into $2z(x^2 + y^2) = z^2 + 5$. With the help of the constraint this means that $2z(2 + z) = z^2 + 5$, so that $z^2 + 4z - 5 = 0$, and hence $z = 1$ or $z = -5$. If $z = 1$, then from the constraint, $x^2 + y^2 = 2 + 1 = 3$, so $f(x, y, 1) = (x^2 + y^2)/(1^2 + 5) = \frac{3}{6} = \frac{1}{2}$. If $z = -5$, then from the constraint, $x^2 + y^2 = 2 - 5 = -3$, which can't be true. Finally we assume that $\lambda \neq 1/(z^2 + 5)$. Then by the first two equations in (*) we have $x = 0$ and $y = 0$, respectively. The constraint becomes $0^2 + 0^2 - z = 2$, so $z = -2$. But $f(0, 0, -2) = 0$. Consequently the maximum value of f is $\frac{1}{2}$ and the minimum value is 0.

67. Let (x, y, z) be a vertex of the parallelepiped that is on the ellipsoid and in the first octant. Let $V(x, y, z) = 8xyz$ denote the volume of the parallelepiped. Let $g(x, y, z) = x^2 + 4y^2 + 9z^2$, so the constraint is $g(x, y, z) = x^2 + 4y^2 + 9z^2 = 36$. Since

$$\operatorname{grad} V(x, y, z) = 8yz\mathbf{i} + 8xz\mathbf{j} + 8xy\mathbf{k} \quad \text{and} \quad \operatorname{grad} g(x, y, z) = 2x\mathbf{i} + 8y\mathbf{j} + 18z\mathbf{k}$$

it follows that $\operatorname{grad} V(x, y, z) = \lambda \operatorname{grad} g(x, y, z)$ if

$$8yz = 2x\lambda, \quad 8xz = 8y\lambda, \quad \text{and} \quad 8xy = 18z\lambda$$

If x, y, or z is zero, then the volume is 0. Otherwise, $x > 0$, $y > 0$, and $z > 0$, and we can solve for λ in the three equations:

$$\lambda = \frac{4yz}{x}, \quad \lambda = \frac{xz}{y}, \quad \text{and} \quad \lambda = \frac{4xy}{9z} \qquad (*)$$

From the first two equations in (*) we obtain $4yz/x = xz/y$, so that $y = x/2$; from the first and third equations in (*) we obtain $4yz/x = 4xy/9z$, so that $z = x/3$. The constraint becomes $x^2 + 4(x/2)^2 + 9(x/3)^2 = 36$, so that $x = 2\sqrt{3}$. Then $y = \sqrt{3}$ and $z = 2\sqrt{3}/3$. Consequently the dimensions of the parallelepiped with largest volume are $4\sqrt{3}$, $2\sqrt{3}$, and $\frac{4}{3}\sqrt{3}$.

69. $\dfrac{\partial T}{\partial \dot{\theta}} = m\ell^2\dot{\theta}$, so that $\dfrac{d}{dt}\left(\dfrac{\partial T}{\partial \dot{\theta}}\right) = \left[\dfrac{\partial}{\partial \dot{\theta}}\left(\dfrac{\partial T}{\partial \dot{\theta}}\right)\right]\dfrac{d\dot{\theta}}{dt} = \left[\dfrac{\partial}{\partial \dot{\theta}}(m\ell^2\dot{\theta})\right]\dfrac{d\dot{\theta}}{dt} = m\ell^2\ddot{\theta}$

Also $\partial V/\partial\theta = mg\ell\theta$. Thus Lagrange's equation becomes $m\ell^2\ddot{\theta} + m\ell\theta = 0$, or rather, $\ddot{\theta} + (g/\ell)\theta = 0$.

CUMULATIVE REVIEW (CHAPTERS 1–13)

1. The conditions for applying l'Hôpital's Rule are satisfied;

$$\lim_{h\to 0} \frac{\sqrt{x-h} - \sqrt{x}}{h} = \lim_{h\to 0} \frac{\dfrac{1}{2\sqrt{x-h}}(-1) - 0}{1} = -\frac{1}{2\sqrt{x}} \quad \text{for } x > 0$$

3. Since $dy/dx = 8x$, an equation of the line tangent to the parabola at $(a, 4a^2)$ is $y - 4a^2 = 8a(x - a)$. If $(0, -2)$ lies on the tangent line, then $-2 - 4a^2 = 8a(0 - a)$, so that $4a^2 = 2$, and thus $a = \sqrt{2}/2$ or $a = -\sqrt{2}/2$. Therefore equations of the tangent lines that pass through $(0, -2)$ are $y - 2 = 4\sqrt{2}(x - \sqrt{2}/2)$ (or equivalently $y = 4\sqrt{2}x - 2$) and $y - 2 = -4\sqrt{2}(x + \sqrt{2}/2)$ (or equivalently $y = -4\sqrt{2}x - 2$).

5. $f'(x) = n \cos nx \sin^n x + n \sin nx \sin^{n-1} x \cos x$

$= n \sin^{n-1} x (\cos nx \sin x + \sin nx \cos x) = n(\sin^{n-1} x)(\sin (n+1)x)$

7. Differentiating implicitly, we have $1/(2\sqrt{x}) + (1/(2\sqrt{y}))(dy/dx) = 0$, so that $dy/dx = -\sqrt{y}/\sqrt{x}$. Differentiating implicitly again, and using the formula for dy/dx and the given equation, we find that

$$\frac{d^2y}{dx^2} = \frac{-(1/(2\sqrt{y}))(dy/dx)\sqrt{x} + \sqrt{y}(1/(2\sqrt{x}))}{x} = \frac{\frac{1}{2} + \sqrt{y}/(2\sqrt{x})}{x} = \frac{\sqrt{x} + \sqrt{y}}{2x^{3/2}} = \frac{9}{2}x^{-3/2}$$

9. $f'(x) = \dfrac{2x}{(1-x^2)^2}$; $f''(x) = \dfrac{2 + 6x^2}{(1-x^2)^3}$; relative minimum value is $f(0) = 2$; increasing on $[0, 1)$ and $(1, \infty)$, and decreasing on $(-\infty, -1)$ and $(-1, 0]$; concave upward on $(-1, 1)$, and concave downward on $(-\infty, -1)$ and $(1, \infty)$; vertical asymptotes are $x = -1$ and $x = 1$; horizontal asymptote is $y = 1$; symmetric with respect to the y axis.

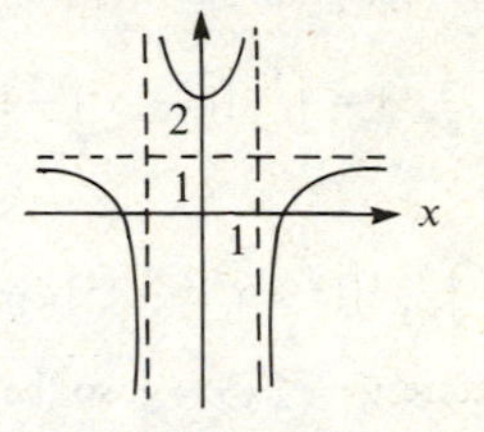

11. The curves intersect at (x, y) if $2x^3 + 2x^2 + 10x = y = x^3 - 3x^2 + 4x$, that is,

$$x^3 + 5x^2 + 6x = 0, \quad \text{or} \quad x(x+2)(x+3) = 0, \quad \text{or} \quad x = -3, x = -2, \quad \text{or} \quad x = 0$$

Moreover, $2x^3 + 2x^2 + 10x \geq x^3 - 3x^2 + 4x$ for $-3 \leq x \leq -2$ and $x^3 - 3x^2 + 4x \geq 2x^3 + 2x^2 + 10x$ for $-2 \leq x \leq 0$. Thus

$$\begin{aligned} A &= \int_{-3}^{-2} [(2x^3 + 2x^2 + 10x) - (x^3 - 3x^2 + 4x)]\,dx \\ &\quad + \int_{-2}^{0} [(x^3 - 3x^2 + 4) - (2x^3 + 2x^2 + 10x)]\,dx \\ &= \int_{-3}^{-2} (x^3 + 5x^2 + 6x)\,dx + \int_{-2}^{0} (-x^3 - 5x^2 - 6x)\,dx \\ &= \left(\tfrac{1}{4}x^4 + \tfrac{5}{3}x^3 + 3x^2\right)\Big|_{-3}^{-2} + \left(-\tfrac{1}{4}x^4 - \tfrac{5}{3}x^3 - 3x^2\right)\Big|_{-2}^{0} \\ &= \left(4 - \tfrac{40}{3} + 12\right) - \left(\tfrac{81}{4} - 45 + 27\right) - \left(-4 + \tfrac{40}{3} - 12\right) = \tfrac{37}{12} \end{aligned}$$

13. $\displaystyle\int (\cos^2\theta - \cos^3\theta)\,d\theta = \int \left[\tfrac{1}{2} + \tfrac{1}{2}\cos 2\theta - (1 - \sin^2\theta)\cos\theta\right] d\theta = \tfrac{1}{2}\theta + \tfrac{1}{4}\sin 2\theta$

$\displaystyle - \int (1 - \sin^2\theta)\cos\theta\,d\theta$

Let $u = \sin\theta$, so that $du = \cos\theta\,d\theta$. Then

$$-\int (1 - \sin^2\theta)\cos\theta\,d\theta = -\int (1 - u^2)\,du = -u + \tfrac{1}{3}u^3 + C = -\sin\theta + \tfrac{1}{3}\sin^3\theta + C$$

Thus $\displaystyle\int (\cos^2\theta - \cos^3\theta)\,d\theta = \tfrac{1}{2}\theta + \tfrac{1}{4}\sin 2\theta - \sin\theta + \tfrac{1}{3}\sin^3\theta + C$.

15. Let $u = 3 + x^4$, so that $du = 4x^3\,dx$. If $x = 1$, then $u = 4$, and if $x = b$, then $u = 3 + b^4$. Thus

$$\begin{aligned} \int_1^\infty \frac{x^3}{(3+x^4)^{3/2}}\,dx &= \lim_{b\to\infty} \int_1^b \frac{x^3}{(3+x^4)^{3/2}}\,dx \overset{u=3+x^4}{=} \lim_{b\to\infty} \int_4^{3+b^4} \frac{1}{u^{3/2}}\frac{1}{4}\,du \\ &= \frac{1}{4}\lim_{b\to\infty} \int_4^{3+b^4} \frac{1}{u^{3/2}}\,du = -\frac{1}{2}\lim_{b\to\infty} \frac{1}{u^{1/2}}\Big|_4^{3+b^4} \\ &= -\frac{1}{2}\lim_{b\to\infty}\left[\frac{1}{(3+b^4)^{1/2}} - \frac{1}{2}\right] = \frac{1}{4} \end{aligned}$$

It follows that the given improper integral converges, and its value is $\frac{1}{4}$.

17. The region is symmetric with respect to the y axis, so $\bar{x} = 0$ by symmetry. The graphs intersect at (x, y) if $2 - x^2 = y = 1$, or $x^2 = 1$, so $x = -1$ or $x = 1$. Thus R is the region between the graphs of $y = 1$ and $y = 2 - x^2$ on $[-1, 1]$. Thus the area A of R is given

by $$A = \int_{-1}^{1} [(2 - x^2) - 1]\, dx = \int_{-1}^{1} (1 - x^2)\, dx = \left(x - \tfrac{1}{3}x^3\right)\Big|_{-1}^{1} = \tfrac{4}{3}$$

and

$$\mathcal{M}_x = \int_{-1}^{1} \tfrac{1}{2}[(2 - x^2)^2 - 1^2]\, dx = \tfrac{1}{2}\int_{-1}^{1} (3 - 4x^2 + x^4)\, dx = \tfrac{1}{2}\left(3x - \tfrac{4}{3}x^3 + \tfrac{1}{5}x^5\right)\Big|_{-1}^{1} = \tfrac{28}{15}$$

Therefore $\bar{y} = \frac{28/15}{4/3} = \frac{7}{5}$, so the center of gravity of R is $(0, \frac{7}{5})$.

19. The circles intersect at (r, θ) if $r = 0$ or if $\sin\theta = r = \sqrt{3}\cos\theta$, so that $\tan\theta = \sqrt{3}$, that is, $\theta = \pi/3$. Thus

$$A = A_1 + A_2 = \frac{1}{2}\pi\left(\frac{1}{2}\right)^2 + \int_{\pi/3}^{\pi/2} \frac{1}{2}[(\sin\theta)^2 - (\sqrt{3}\cos\theta)^2]\, d\theta$$
$$= \frac{\pi}{8} + \frac{1}{2}\int_{\pi/3}^{\pi/2} (\sin^2\theta - 3\cos^2\theta)\, d\theta = \frac{\pi}{8} + \frac{1}{2}\int_{\pi/3}^{\pi/2}\left(\frac{1}{2} - \frac{1}{2}\cos 2\theta - \frac{3}{2} - \frac{3}{2}\cos 2\theta\right) d\theta$$
$$= \frac{\pi}{8} + \frac{1}{2}(-\theta - \sin 2\theta)\Big|_{\pi/3}^{\pi/2} = \frac{\pi}{8} - \frac{\pi}{12} + \frac{\sqrt{3}}{4} = \frac{\pi}{24} + \frac{\sqrt{3}}{4}$$

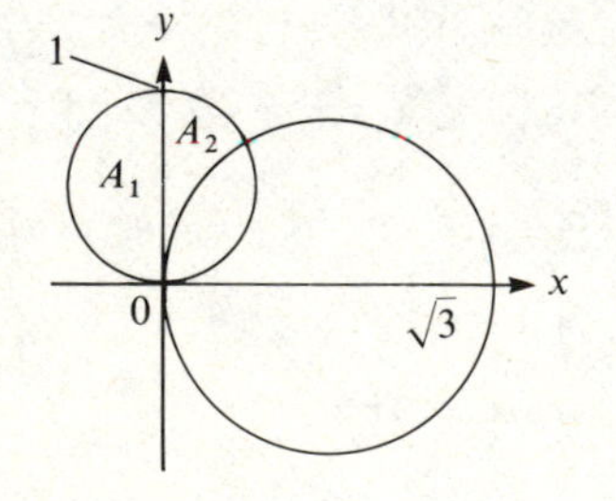

21. The jth partial sum s_j is given by

$$s_j = \left(\frac{1}{\sqrt{1}} - \frac{1}{\sqrt{2}}\right) + \left(\frac{1}{\sqrt{2}} - \frac{1}{\sqrt{3}}\right) + \cdots + \left(\frac{1}{\sqrt{j}} - \frac{1}{\sqrt{j+1}}\right) = 1 - \frac{1}{\sqrt{j+1}}$$

Therefore $$\sum_{n=1}^{\infty}\left(\frac{1}{\sqrt{n}} - \frac{1}{\sqrt{n+1}}\right) = \lim_{j\to\infty} s_j = \lim_{j\to\infty}\left(1 - \frac{1}{\sqrt{j+1}}\right) = 1$$

23. By (3) of Section 9.8, $e^x = \sum_{n=0}^{\infty} (1/n!)x^n$ for all x, so

$$f(x) = 2x^3 e^{-x^3} = 2x^3 \sum_{n=0}^{\infty} \frac{1}{n!}(-x^3)^n = \sum_{n=0}^{\infty} \frac{2}{n!}(-1)^n x^{3n+3}$$

for all x. By the Differentiation Theorem for power series,

$$f'(x) = \frac{d}{dx}\sum_{n=0}^{\infty} \frac{2}{n!}(-1)^n x^{3n+3} = \sum_{n=0}^{\infty} \frac{2(-1)^n(3n+3)}{n!} x^{3n+2} \quad \text{for all } x$$

In particular, the radius of convergence of each series is ∞.

25. a. Since $\mathbf{T}(t)$ is parallel to $\mathbf{r}'(t) = \mathbf{i} + 2t\mathbf{j} + 2t^2\mathbf{k}$, $\mathbf{T}'(t)$ is parallel to $\mathbf{i} - \mathbf{j} + \frac{1}{2}\mathbf{k}$ only if $2t = -1$ and $2t^2 = \frac{1}{2}$, that is, $t = -\frac{1}{2}$.

b. By (3) in Section 13.6, $K(t) = \|\mathbf{v}(t) \times \mathbf{a}(t)\|/\|\mathbf{v}(t)\|^3$. Since $\mathbf{v}(t) = \mathbf{r}'(t) = \mathbf{i} + 2t\mathbf{j} + 2t^2\mathbf{k}$ and $\mathbf{a}(t) = \mathbf{v}'(t) = 2\mathbf{j} + 4t\mathbf{k}$, we have $\|\mathbf{v}(t)\| = \sqrt{1 + (2t)^2 + (2t^2)^2} = 1 + 2t^2$ and

$$\mathbf{v}(t) \times \mathbf{a}(t) = \begin{vmatrix} \mathbf{i} & \mathbf{j} & \mathbf{k} \\ 1 & 2t & 2t^2 \\ 0 & 2 & 4t \end{vmatrix} = 4t^2\mathbf{i} - 4t\mathbf{j} + 2\mathbf{k}$$

Therefore $\|\mathbf{v}(t) \times \mathbf{a}(t)\| = \sqrt{(4t^2)^2 + (-4t)^2 + 2^2} = 2(2t^2 + 1)$. Thus

$$K(t) = \frac{\|\mathbf{v}(t) \times \mathbf{a}(t)\|}{\|\mathbf{v}(t)\|^3} = \frac{2}{(2t^2 + 1)^2}$$

Since $K(0) = 2 \geq 2/(2t^2 + 1)^2$ for all t, it follows that K is maximum for $t = 0$.

15 MULTIPLE INTEGRALS

SECTION 15.1

1. a. $\sum_{k=1}^{3} f(x_k, y_k)\Delta A_k = f(-2, 0)\Delta A_1 + f(-2, 1)\Delta A_2 + f(-1, 0)\Delta A_3$

$= 4(1) + 3(1) + 4(1) = 11$

b. $\sum_{k=1}^{3} f(x_k, y_k)\Delta A_k = f(-1, 1)\Delta A_1 + f(-1, 2)\Delta A_2 + f(0, 1)\Delta A_3$

$= 3(1) + 2(1) + 3(1) = 8$

3. $\int_0^1 \int_{-1}^1 x\,dy\,dx = \int_0^1 xy\Big|_{-1}^1 dx = \int_0^1 2x\,dx = x^2\Big|_0^1 = 1$

5. $\int_0^1 \int_0^1 e^{x+y}\,dy\,dx = \int_0^1 e^{x+y}\Big|_0^1 dx = \int_0^1 (e^{x+1} - e^x)\,dx = (e^{x+1} - e^x)\Big|_0^1 = e^2 - 2e + 1$

7. $\int_0^1 \int_x^{x^2} 1\,dy\,dx = \int_0^1 y\Big|_x^{x^2} dx = \int_0^1 (x^2 - x)\,dx = (\frac{1}{3}x^3 - \frac{1}{2}x^2)\Big|_0^1 = -\frac{1}{6}$

9. $\int_0^1 \int_0^y x\sqrt{y^2 - x^2}\,dx\,dy = \int_0^1 -\frac{1}{3}(y^2 - x^2)^{3/2}\Big|_0^y dy = \int_0^1 \frac{1}{3}y^3\,dy = \frac{1}{12}y^4\Big|_0^1 = \frac{1}{12}$

11. $\int_0^2 \int_0^{\sqrt{4-y^2}} x\,dx\,dy = \int_0^2 \frac{1}{2}x^2\Big|_0^{\sqrt{4-y^2}} dy = \int_0^2 \frac{1}{2}(4 - y^2)\,dy = (2y - \frac{1}{6}y^3)\Big|_0^2 = \frac{8}{3}$

13. $\int_0^{2\pi} \int_0^1 r\sqrt{1 - r^2}\,dr\,d\theta = \int_0^{2\pi} -\frac{1}{3}(1 - r^2)^{3/2}\Big|_0^1 d\theta = \int_0^{2\pi} \frac{1}{3}\,d\theta = \frac{1}{3}\theta\Big|_0^{2\pi} = \frac{2}{3}\pi$

15. $\int_1^3 \int_0^x \frac{2}{x^2 + y^2}\,dy\,dx = \int_1^3 \frac{2}{x} \arctan \frac{y}{x}\Big|_0^x dx = \int_1^3 \frac{\pi}{2x}\,dx = \frac{\pi}{2} \ln x\Big|_1^3 = \frac{\pi}{2} \ln 3$

17. $\int_0^1 \int_0^x e^{(x^2)}\,dy\,dx = \int_0^1 ye^{(x^2)}\Big|_0^x dx = \int_0^1 xe^{(x^2)}\,dx = \frac{1}{2}e^{(x^2)}\Big|_0^1 = \frac{1}{2}(e - 1)$

19. $\int_0^{\pi/4} \int_0^{\cos y} e^x \sin y\,dx\,dy = \int_0^{\pi/4} e^x \sin y\Big|_0^{\cos y} dy = \int_0^{\pi/4} (e^{\cos y} \sin y - \sin y)\,dy$

$= (-e^{\cos y} + \cos y)\Big|_0^{\pi/4} = e - e^{\sqrt{2}/2} + \frac{\sqrt{2}}{2} - 1$

21. $\iint_R (x + y)\,dA = \int_0^2 \int_{2x}^4 (x + y)\,dy\,dx = \int_0^2 (xy + \frac{1}{2}y^2)\Big|_{2x}^4 dx = \int_0^2 (4x + 8 - 4x^2)\,dx$

$= (2x^2 + 8x - \frac{4}{3}x^3)\Big|_0^2 = \frac{40}{3}$

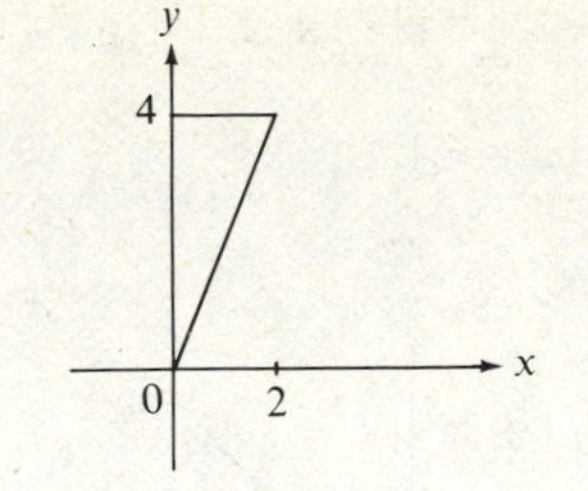

23. $\iint_R x\,dA = \int_3^5 \int_1^x x\,dy\,dx = \int_3^5 xy\Big|_1^x dx = \int_3^5 (x^2 - x)\,dx = (\frac{1}{3}x^3 - \frac{1}{2}x^2)\Big|_3^5 = \frac{74}{3}$

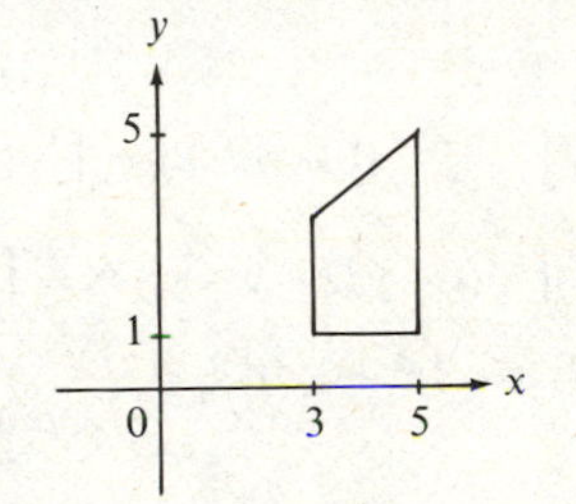

25. The lines intersect at (x, y) such that $5 + x = y = -x + 7$, so that $2x = 2$, or $x = 1$. Since $5 + x \geq 7 - x$ for x in $[1, 10]$, we have

$$\iint_R (3x - 5)\,dA = \int_1^{10} \int_{7-x}^{5+x} (3x - 5)\,dy\,dx = \int_1^{10} (3x - 5)y\Big|_{7-x}^{5+x} dx$$

$$= \int_1^{10} (3x - 5)(2x - 2)\,dx = \int_1^{10} (6x^2 - 16x + 10)\,dx$$

$$= (2x^3 - 8x^2 + 10x)\Big|_1^{10} = 1296$$

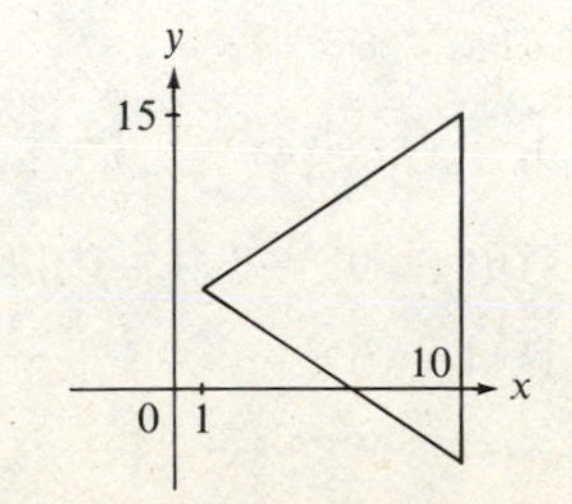

27. $\iint_R 1\,dA = \int_\pi^{2\pi}\int_{\sin x}^{1+x} 1\,dy\,dx = \int_\pi^{2\pi} y\Big|_{\sin x}^{1+x} dx = \int_\pi^{2\pi} (1 + x - \sin x)\,dx$

$= (x + \frac{1}{2}x^2 + \cos x)\Big|_\pi^{2\pi} = \frac{3}{2}\pi^2 + \pi + 2$

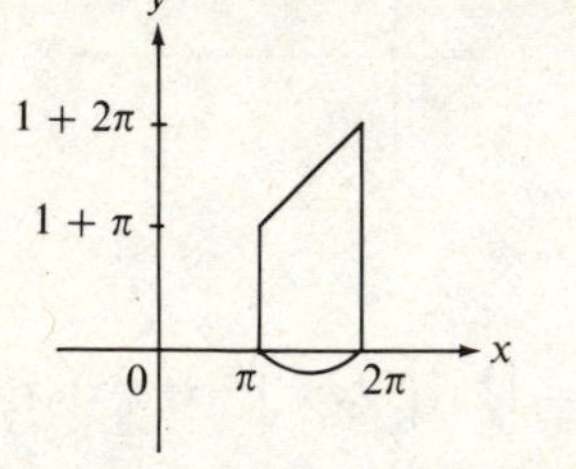

29. The graphs intersect at (x, y) such that $y^2 = x = 2 - y$, so that $y^2 + y - 2 = 0$, or $y = -2$ or $y = 1$. Thus

$$\iint_R (1 - y)\,dA = \int_{-2}^1 \int_{y^2}^{2-y} (1 - y)\,dx\,dy = \int_{-2}^1 (1 - y)x\Big|_{y^2}^{2-y} dy$$

$$= \int_{-2}^1 (1 - y)(2 - y - y^2)\,dy = \int_{-2}^1 (2 - 3y + y^3)\,dy$$

$$= (2y - \frac{3}{2}y^2 + \frac{1}{4}y^4)\Big|_{-2}^1 = \frac{27}{4}$$

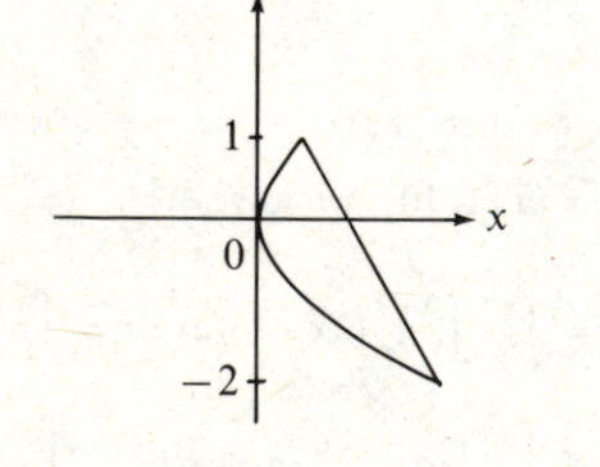

31. The line and the circle intersect at (x, y) such that $(1 - x)^2 = y^2 = 1 - x^2$, or $2x^2 - 2x = 0$, or $x = 0$ or $x = 1$. Since R lies above the line $y = 1 - x$, it follows that R lies above the x axis. Thus the part of the circle $x^2 + y^2 = 1$ that bounds R is the graph of $y = \sqrt{1 - x^2}$ for $0 \le x \le 1$. Thus

$$\iint_R xy^2\,dA = \int_0^1 \int_{1-x}^{\sqrt{1-x^2}} xy^2\,dy\,dx = \int_0^1 \frac{1}{3}xy^3\Big|_{1-x}^{\sqrt{1-x^2}} dx$$

$$= \int_0^1 \left[\frac{1}{3}x(1 - x^2)^{3/2} - \frac{1}{3}x(1 - x^3)\right] dx$$

$$= \int_0^1 \left[\frac{1}{3}x(1 - x^2)^{3/2} - \frac{1}{3}x + x^2 - x^3 + \frac{1}{3}x^4\right] dx$$

$$= \left[-\frac{1}{15}(1 - x^2)^{5/2} - \frac{1}{6}x^2 + \frac{1}{3}x^3 - \frac{1}{4}x^4 + \frac{1}{15}x^5\right]\Big|_0^1 = \frac{1}{20}$$

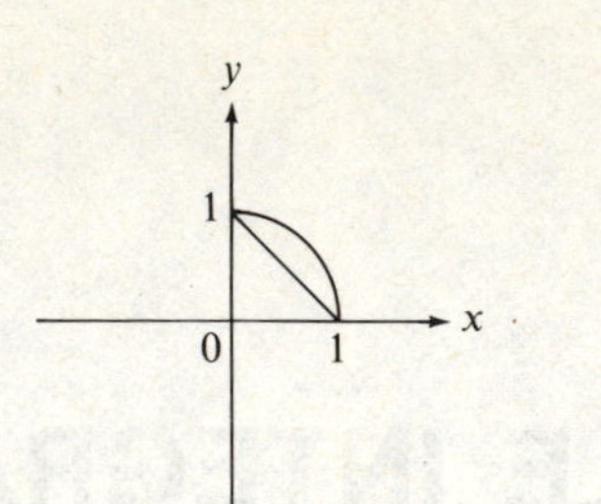

33. The graphs intersect at (x, y) such that $x^3 + x^2 + 1 = y = x^3 + x + 1$, so that $x^2 = x$, or $x = 0$ or $x = 1$. Since $x^3 + x^2 + 1 \ge x^3 + x + 1$ for $-1 \le x \le 0$ and $x^3 + x + 1 \ge x^3 + x^2 + 1$ for $0 \le x \le 1$, we have

$$\iint_R x^2\,dA = \int_{-1}^0 \int_{x^3+x+1}^{x^3+x^2+1} x^2\,dy\,dx + \int_0^1 \int_{x^3+x^2+1}^{x^3+x+1} x^2\,dy\,dx$$

$$= \int_{-1}^0 x^2 y\Big|_{x^3+x+1}^{x^3+x^2+1} dx + \int_0^1 x^2 y\Big|_{x^3+x^2+1}^{x^3+x+1} dx$$

$$= \int_{-1}^0 (x^4 - x^3)\,dx + \int_0^1 (x^3 - x^4)\,dx = (\frac{1}{5}x^5 - \frac{1}{4}x^4)\Big|_{-1}^0 + (\frac{1}{4}x^4 - \frac{1}{5}x^5)\Big|_0^1$$

$$= \frac{9}{20} + \frac{1}{20} = \frac{1}{2}$$

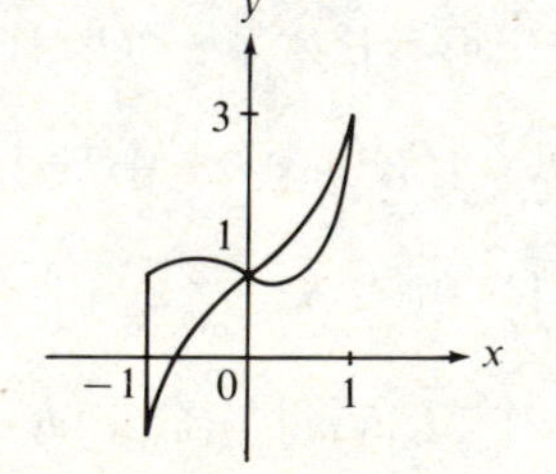

35. The intersection of the solid region with the xy plane is the region R bounded by the lines $x + 2y = 6$ (or $y = \frac{1}{2}(6 - x)$), $x = 0$, and $y = 0$. Thus

$$V = \iint_R \frac{1}{3}(6 - x - 2y)\,dA = \int_0^6 \int_0^{(1/2)(6-x)} \frac{1}{3}(6 - x - 2y)\,dy\,dx$$

$$= \int_0^6 \frac{1}{3}(6y - xy - y^2)\Big|_0^{(1/2)(6-x)} dx = \int_0^6 \frac{1}{12}(6 - x)^2\,dx = -\frac{1}{36}(6 - x)^3\Big|_0^6 = 6$$

37. The intersection of the solid region with the xy plane is the region R bounded by the lines $x + y = 1$ (or $y = 1 - x$), $x = 0$, and $y = 0$. Thus

$$V = \iint_R (x^2 + y^2)\,dA = \int_0^1 \int_0^{1-x} (x^2 + y^2)\,dy\,dx = \int_0^1 (x^2 y + \frac{1}{3}y^3)\Big|_0^{1-x} dx$$

$$= \int_0^1 \left[x^2 - x^3 + \frac{1}{3}(1 - x)^3\right] dx = \left[\frac{1}{3}x^3 - \frac{1}{4}x^4 - \frac{1}{12}(1 - x)^4\right]\Big|_0^1 = \frac{1}{6}$$

39. The intersection of the solid region with the xy plane is the vertically simple region R between the graphs of $y = 0$ and $y = \sqrt{4 - x^2}$ on $[0, 2]$. Thus

$$V = \iint_R y\,dA = \int_0^2 \int_0^{\sqrt{4-x^2}} y\,dy\,dx = \int_0^2 \tfrac{1}{2}y^2\Big|_0^{\sqrt{4-x^2}} dx = \int_0^2 (2 - \tfrac{1}{2}x^2)\,dx$$
$$= (2x - \tfrac{1}{6}x^3)\Big|_0^2 = \tfrac{8}{3}$$

41. The intersection of the solid region with the xy plane is the region R in the first quadrant that lies inside the circle $x^2 + y^2 = 1$. Thus R is the vertically simple region between the graphs of $y = 0$ and $y = \sqrt{1 - x^2}$ on $[0, 1]$, so that

$$V = \iint_R \sqrt{1 - x^2}\,dA = \int_0^1 \int_0^{\sqrt{1-x^2}} \sqrt{1 - x^2}\,dy\,dx = \int_0^1 y\sqrt{1 - x^2}\Big|_0^{\sqrt{1-x^2}} dx$$
$$= \int_0^1 (1 - x^2)\,dx = (x - \tfrac{1}{3}x^3)\Big|_0^1 = \tfrac{2}{3}$$

43. The intersection of the solid region with the xy plane is the region R between the graphs of $y = x$ and $y = x^3$. These intersect at (x, y) such that $x = y = x^3$, so that $x = -1$, $x = 0$, or $x = 1$. Since $x^3 \ge x$ for $-1 \le x \le 0$ and $x \ge x^3$ for $0 \le x \le 1$, we have

$$V = \iint_R xy\,dA = \int_{-1}^0 \int_x^{x^3} xy\,dy\,dx + \int_0^1 \int_{x^3}^x xy\,dy\,dx$$
$$= \int_{-1}^0 \tfrac{1}{2}xy^2\Big|_x^{x^3} dx + \int_0^1 \tfrac{1}{2}xy^2\Big|_{x^3}^x dx = \int_{-1}^0 \tfrac{1}{2}(x^7 - x^3)\,dx + \int_0^1 \tfrac{1}{2}(x^3 - x^7)\,dx$$
$$= \tfrac{1}{2}(\tfrac{1}{8}x^8 - \tfrac{1}{4}x^4)\Big|_{-1}^0 + \tfrac{1}{2}(\tfrac{1}{4}x^4 - \tfrac{1}{8}x^8)\Big|_0^1 = \tfrac{1}{8}$$

45. 47. 49.

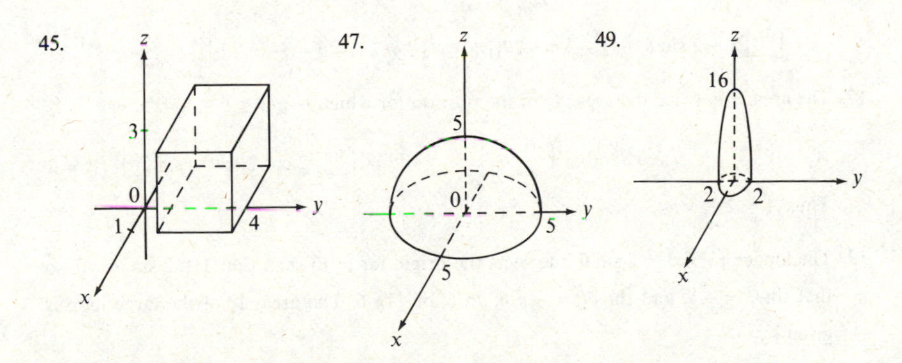

51. Since $x^2 \ge \sqrt{x}$ for $1 \le x \le 4$, we have

$$A = \int_1^4 \int_{\sqrt{x}}^{x^2} 1\,dy\,dx = \int_1^4 y\Big|_{\sqrt{x}}^{x^2} dx = \int_1^4 (x^2 - \sqrt{x})\,dx = (\tfrac{1}{3}x^3 - \tfrac{2}{3}x^{3/2})\Big|_1^4 = \tfrac{49}{3}$$

53. The parabolas intersect at (x, y) such that $y^2 = x = 32 - y^2$, or $y^2 = 16$, or $y = -4$ or $y = 4$. Since $32 - y^2 \ge y^2$ for $-4 \le y \le 4$, we have

$$A = \int_{-4}^4 \int_{y^2}^{32-y^2} 1\,dx\,dy = \int_{-4}^4 x\Big|_{y^2}^{32-y^2} dy = \int_{-4}^4 (32 - 2y^2)\,dy = (32y - \tfrac{2}{3}y^3)\Big|_{-4}^4 = \tfrac{512}{3}$$

55. The line $y = x$ and the parabola $x = 2 - y^2$ intersect in the first quadrant at (x, y) such that $y \ge 0$ and $y = x = 2 - y^2$, or $y^2 + y - 2 = 0$, which means that $y = 1$. Since $x = 2 - y^2$ and $y \ge 0$ imply that $y = \sqrt{2 - x}$, and since $\sqrt{2 - x} \ge x$ for $-2 \le x \le 1$, we have

$$A = \int_0^1 \int_x^{\sqrt{2-x}} 1\,dy\,dx = \int_0^1 y\Big|_x^{\sqrt{2-x}} dx = \int_0^1 (\sqrt{2 - x} - x)\,dx$$
$$= [-\tfrac{2}{3}(2 - x)^{3/2} - \tfrac{1}{2}x^2]\Big|_0^1 = \tfrac{4}{3}\sqrt{2} - \tfrac{7}{6}$$

57. $\int_0^1 \int_y^1 e^{(x^2)}\,dx\,dy = \int_0^1 \int_0^x e^{(x^2)}\,dy\,dx = \int_0^1 ye^{(x^2)}\Big|_0^x dx = \int_0^1 xe^{(x^2)}\,dx = \tfrac{1}{2}e^{(x^2)}\Big|_0^1 = \tfrac{1}{2}(e - 1)$

59. $\int_0^2 \int_{1+y^2}^5 ye^{(x-1)^2}\,dx\,dy = \int_1^5 \int_0^{\sqrt{x-1}} ye^{(x-1)^2}\,dy\,dx = \int_1^5 \tfrac{1}{2}y^2e^{(x-1)^2}\Big|_0^{\sqrt{x-1}} dx$
$= \int_1^5 \tfrac{1}{2}(x - 1)e^{(x-1)^2}\,dx = \tfrac{1}{4}e^{(x-1)^2}\Big|_1^5 = \tfrac{1}{4}(e^{16} - 1)$

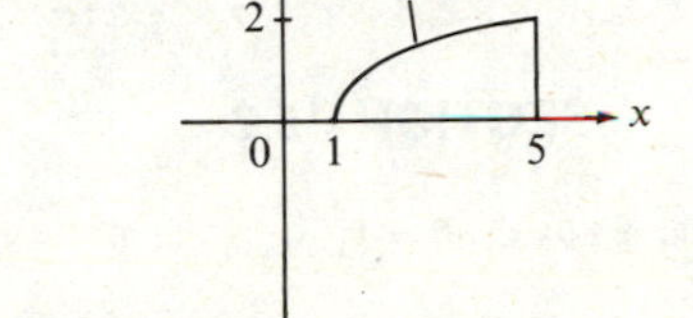

61. $\int_1^e \int_0^{\ln x} y\,dy\,dx = \int_0^1 \int_{e^y}^e y\,dx\,dy = \int_0^1 xy\Big|_{e^y}^e dy = \int_0^1 (ey - ye^y)\,dy$
$= (\tfrac{1}{2}ey^2 - ye^y + e^y)\Big|_0^1 = \tfrac{1}{2}e - 1$

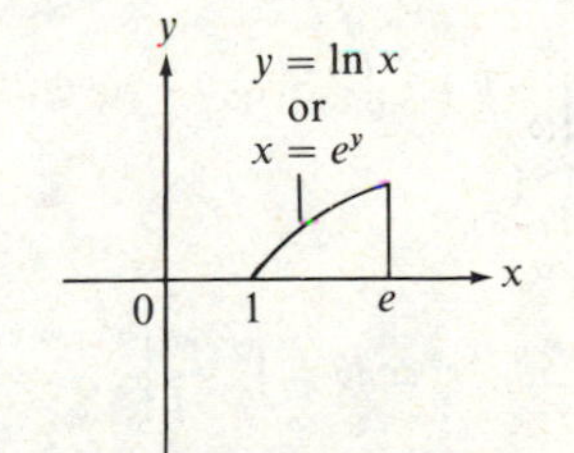

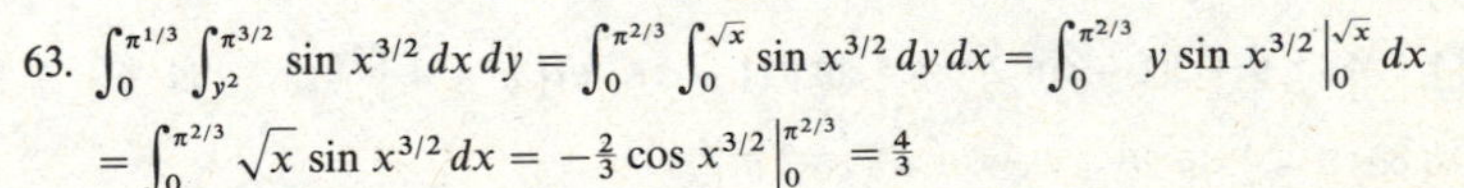

63. $\int_0^{\pi^{1/3}} \int_{y^2}^{\pi^{2/3}} \sin x^{3/2}\,dx\,dy = \int_0^{\pi^{2/3}} \int_0^{\sqrt{x}} \sin x^{3/2}\,dy\,dx = \int_0^{\pi^{2/3}} y \sin x^{3/2}\Big|_0^{\sqrt{x}} dx$
$= \int_0^{\pi^{2/3}} \sqrt{x} \sin x^{3/2}\,dx = -\tfrac{2}{3}\cos x^{3/2}\Big|_0^{\pi^{2/3}} = \tfrac{4}{3}$

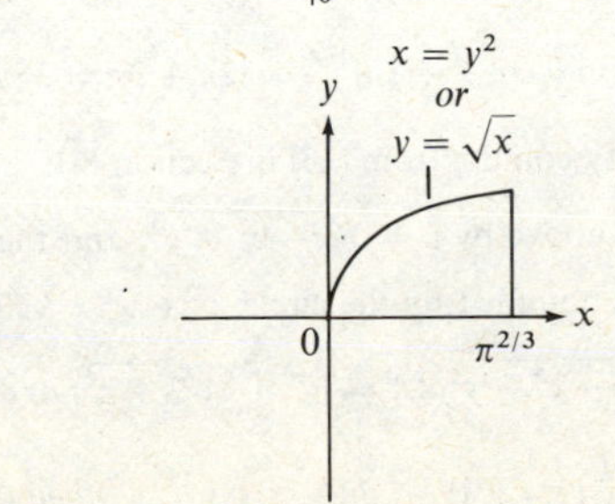

65. a. If $\mathscr{P} = \{R_1, R_2, \ldots, R_n\}$ and $L_f(\mathscr{P}) = U_f(\mathscr{P})$, then it follows from the definitions of $L_f(\mathscr{P})$ and $U_f(\mathscr{P})$ that $m_1 = M_1, m_2 = M_2, \ldots, m_n = M_n$. From these equations it follows that f is constant on $R_1, R_2, \ldots, R_n$, and hence on R. In that case, if a is any point in R, then for $1 \le k \le n$ we have $m_k = f(a) = M_k$, so that $L_f(\mathscr{P}) = m_1\,\Delta A_1 + m_2\,\Delta A_2 + \cdots + m_n\,\Delta A_n = f(a)(\Delta A_1 + \Delta A_2 + \cdots + \Delta A_n) = f(a)(\text{area of } R)$. Similarly, $U_f(\mathscr{P}) = f(a)(\text{area of } R)$. Thus

$$\iint_R f(x, y)\,dA = L_f(\mathscr{P}) = U_f(\mathscr{P}) = f(a)(\text{area of } R)$$

where a is any point in R.

b. In the same way, f is constant on each R_k, so is constant on R. Also

$$\iint_R f(x, y)\,dA = L_f(\mathscr{P}) = U_f(\mathscr{P}) = f(a)(\text{area of } R)$$

where a is any point in R.

SECTION 15.2

1. $$\iint_R xy\,dA = \int_0^{2\pi}\int_0^5 (r\cos\theta)(r\sin\theta)r\,dr\,d\theta = \int_0^{2\pi}\int_0^5 r^3\sin\theta\cos\theta\,dr\,d\theta$$
$$= \int_0^{2\pi}\left(\frac{r^4}{4}\sin\theta\cos\theta\right)\bigg|_0^5 d\theta = \int_0^{2\pi}\frac{625}{4}\sin\theta\cos\theta\,d\theta = \frac{625}{8}\sin^2\theta\bigg|_0^{2\pi} = 0$$

3. $$\iint_R (x+y)\,dA = \int_0^{\pi/3}\int_0^2 (r\cos\theta + r\sin\theta)r\,dr\,d\theta = \int_0^{\pi/3}\int_0^2 r^2(\cos\theta + \sin\theta)\,d\theta$$
$$= \int_0^{\pi/3}\left[\frac{r^3}{3}(\cos\theta + \sin\theta)\right]\bigg|_0^2 d\theta = \frac{8}{3}\int_0^{\pi/3}(\cos\theta + \sin\theta)\,d\theta = \frac{8}{3}(\sin\theta - \cos\theta)\bigg|_0^{\pi/3}$$
$$= \frac{8}{3}\left(\frac{\sqrt{3}}{2} + \frac{1}{2}\right) = \frac{4}{3}(\sqrt{3} + 1)$$

5. $$\iint_R (x^2 + y^2)\,dA = \int_0^{2\pi}\int_0^{2(1+\sin\theta)} r^3\,dr\,d\theta = \int_0^{2\pi}\frac{r^4}{4}\bigg|_0^{2(1+\sin\theta)} d\theta = 4\int_0^{2\pi}(1+\sin\theta)^4\,d\theta$$
$$= 4\int_0^{2\pi}(1 + 4\sin\theta + 6\sin^2\theta + 4\sin^3\theta + \sin^4\theta)\,d\theta$$
$$= 4\int_0^{2\pi}[1 + 4\sin\theta + 3(1 - \cos 2\theta) + 4\sin\theta\,(1 - \cos^2\theta)]\,d\theta + 4\int_0^{2\pi}\sin^4\theta\,d\theta$$
$$= 4(4\theta - 4\cos\theta - \tfrac{3}{2}\sin 2\theta - 4\cos\theta + \tfrac{4}{3}\cos^3\theta)\Big|_0^{2\pi}$$
$$+ 4(-\tfrac{1}{4}\sin^3\theta\cos\theta - \tfrac{3}{8}\sin\theta\cos\theta + \tfrac{3}{8}\theta)\Big|_0^{2\pi} = 32\pi + 3\pi = 35\pi$$

with the third to last equality coming from (13) in Section 8.1.

7. The hemisphere is bounded above by $z = \sqrt{9 - x^2 - y^2}$, and the region R over which the integral is to be taken is bounded by the circle $x^2 + y^2 = 9$. Thus

$$V = \iint_R \sqrt{9 - x^2 - y^2}\,dA = \int_0^{2\pi}\int_0^3 \sqrt{9 - r^2}\,r\,dr\,d\theta$$
$$= -\tfrac{1}{3}\int_0^{2\pi}(9 - r^2)^{3/2}\Big|_0^3 d\theta = -\tfrac{1}{3}\int_0^{2\pi} -27\,d\theta = 18\pi$$

9. The region R over which the integral is to be taken is bounded by $x^2 + y^2 = 1$. Thus

$$V = \iint_R (4 + x + 2y)\,dA = \int_0^{2\pi}\int_0^1 (4 + r\cos\theta + 2r\sin\theta)r\,dr\,d\theta$$
$$= \int_0^{2\pi}\int_0^1 (4r + r^2\cos\theta + 2r^2\sin\theta)\,dr\,d\theta = \int_0^{2\pi}\left(2r^2 + \frac{r^3}{3}\cos\theta + \frac{2}{3}r^3\sin\theta\right)\bigg|_0^1 d\theta$$
$$= \int_0^{2\pi}(2 + \tfrac{1}{3}\cos\theta + \tfrac{2}{3}\sin\theta)\,d\theta = (2\theta + \tfrac{1}{3}\sin\theta - \tfrac{2}{3}\cos\theta)\Big|_0^{2\pi} = 4\pi$$

11. The solid region is bounded above by $z = \sqrt{4 - x^2 - y^2}$, and the region R over which the integral is to be taken is bounded by the circles $x^2 + y^2 = 1$ and $x^2 + y^2 = 4$. Thus

$$V = \iint_R \sqrt{4 - x^2 - y^2}\,dA = \int_0^{2\pi}\int_1^2 \sqrt{4 - r^2}\,r\,dr\,d\theta$$
$$= \int_0^{2\pi} -\tfrac{1}{3}(4 - r^2)^{3/2}\Big|_1^2 d\theta = \sqrt{3}\int_0^{2\pi} 1\,d\theta = 2\sqrt{3}\pi$$

13. The solid region is bounded above by $z^2 = x^2 + y^2$, whose equation can be rewritten as $z = r$. The region R over which the integral is to be taken is bounded by $x^2 + y^2 - 4x = 0$, whose equation in polar coordinates is $r = 4\cos\theta$. Thus

$$V = \iint_R \sqrt{x^2 + y^2}\,dA = \int_{-\pi/2}^{\pi/2}\int_0^{4\cos\theta} r^2\,dr\,d\theta = \int_{-\pi/2}^{\pi/2}\frac{r^3}{3}\bigg|_0^{4\cos\theta} d\theta = \int_{-\pi/2}^{\pi/2}\frac{64}{3}\cos^3\theta\,d\theta$$
$$= \tfrac{64}{3}\int_{-\pi/2}^{\pi/2}\cos\theta\,(1 - \sin^2\theta)\,d\theta = \tfrac{64}{3}(\sin\theta - \tfrac{1}{3}\sin^3\theta)\Big|_{-\pi/2}^{\pi/2} = \tfrac{256}{9}$$

15. $$A = \int_0^{2\pi}\int_0^{2+\sin\theta} r\,dr\,d\theta = \int_0^{2\pi}\frac{r^2}{2}\bigg|_0^{2+\sin\theta} d\theta = \int_0^{2\pi}\left(2 + 2\sin\theta + \frac{1}{2}\sin^2\theta\right)d\theta$$
$$= \int_0^{2\pi}[2 + 2\sin\theta + (\tfrac{1}{4} - \tfrac{1}{4}\cos 2\theta)]\,d\theta = (\tfrac{9}{4}\theta - 2\cos\theta - \tfrac{1}{8}\sin 2\theta)\Big|_0^{2\pi} = \tfrac{9}{2}\pi$$

17. The area A is twice the area A_1 of the portion for which $-\pi/4 \le \theta \le \pi/4$. Now

$$A_1 = \int_{-\pi/4}^{\pi/4}\int_0^{\sqrt{4\cos 2\theta}} r\,dr\,d\theta = \int_{-\pi/4}^{\pi/4}\frac{r^2}{2}\bigg|_0^{\sqrt{4\cos 2\theta}} d\theta = \int_{-\pi/4}^{\pi/4} 2\cos 2\theta\,d\theta = \sin 2\theta\Big|_{-\pi/4}^{\pi/4} = 2$$

Thus $A = 2A_1 = 4$.

19. The limaçon $r = 1 + 2\sin\theta$ intersects the origin for (r, θ) such that $1 + 2\sin\theta = 0$, so that $\sin\theta = -\frac{1}{2}$, and thus $\theta = -\pi/6, 7\pi/6$, or $11\pi/6$. The area A_1 of the large loop is given by

$$A_1 = \int_{-\pi/6}^{7\pi/6}\int_0^{1+2\sin\theta} r\,dr\,d\theta = \int_{-\pi/6}^{7\pi/6}\frac{r^2}{2}\bigg|_0^{1+2\sin\theta} d\theta = \int_{-\pi/6}^{7\pi/6}\left(\frac{1}{2} + 2\sin\theta + 2\sin^2\theta\right)d\theta$$
$$= \int_{-\pi/6}^{7\pi/6}[\tfrac{1}{2} + 2\sin\theta + (1 - \cos 2\theta)]\,d\theta = (\tfrac{3}{2}\theta - 2\cos\theta - \tfrac{1}{2}\sin 2\theta)\Big|_{-\pi/6}^{7\pi/6} = 2\pi + \tfrac{3}{2}\sqrt{3}$$

For $7\pi/6 \le \theta \le 11\pi/6$, $1 + 2\sin\theta \le 0$, so the area A_2 of the small loop is given by

$$A_2 = \int_{7\pi/6}^{11\pi/6}\int_0^{-(1+2\sin\theta)} r\,dr\,d\theta = \int_{7\pi/6}^{11\pi/6}\frac{r^2}{2}\bigg|_0^{-(1+2\sin\theta)} d\theta$$
$$= \int_{7\pi/6}^{11\pi/6}(\tfrac{1}{2} + 2\sin\theta + 2\sin^2\theta)\,d\theta = (\tfrac{3}{2}\theta - 2\cos\theta - \tfrac{1}{2}\sin 2\theta)\Big|_{7\pi/6}^{11\pi/6} = \pi - \tfrac{3}{2}\sqrt{3}$$

Thus the area A of the region inside the large loop and outside the small loop is given by $A = A_1 - A_2 = \pi + 3\sqrt{3}$.

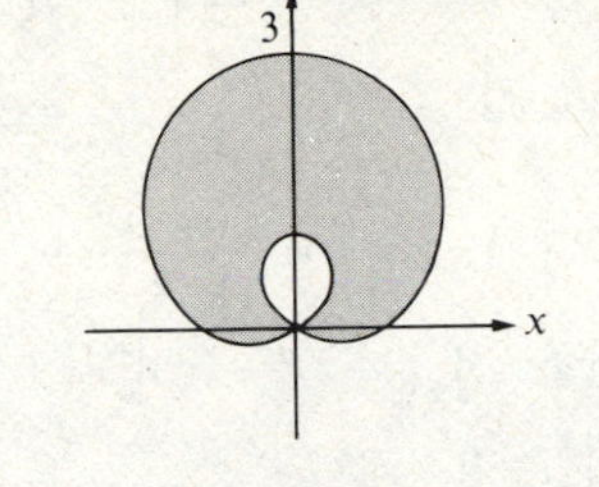

21. The spirals intersect on $[0, 3\pi]$ only for $\theta = 0$. Thus the area is given by

$$A = \int_0^{3\pi}\int_{e^\theta}^{e^{2\theta}} r\,dr\,d\theta = \int_0^{3\pi} \frac{r^2}{2}\Big|_{e^\theta}^{e^{2\theta}} d\theta = \int_0^{3\pi} \tfrac{1}{2}(e^{4\theta} - e^{2\theta})\,d\theta$$

$$= (\tfrac{1}{8}e^{4\theta} - \tfrac{1}{4}e^{2\theta})\Big|_0^{3\pi} = \tfrac{1}{8}(e^{12\pi} - 2e^{6\pi} + 1)$$

23. $\displaystyle\int_0^1\int_y^{\sqrt{2-y^2}} 1\,dx\,dy = \int_0^{\pi/4}\int_0^{\sqrt{2}} r\,dr\,d\theta = \int_0^{\pi/4}\frac{r^2}{2}\Big|_0^{\sqrt{2}} d\theta = \int_0^{\pi/4} 1\,d\theta = \frac{1}{4}\pi$

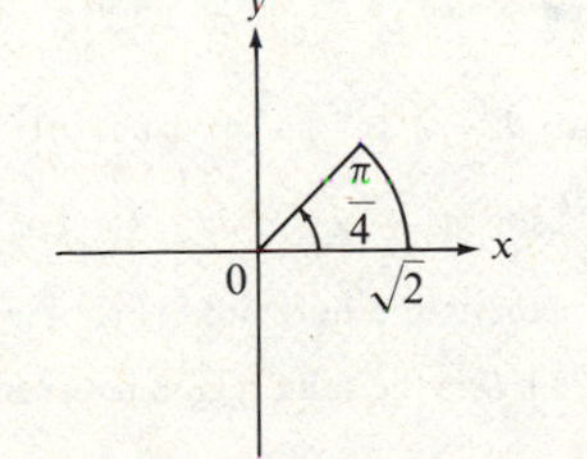

25. $\displaystyle\int_0^1\int_0^{\sqrt{1-y^2}} \sin(x^2 + y^2)\,dx\,dy = \int_0^{\pi/2}\int_0^1 (\sin r^2) r\,dr\,d\theta = \int_0^{\pi/2} -\tfrac{1}{2}\cos r^2\Big|_0^1 d\theta$

$$= \int_0^{\pi/2}\left(\frac{1}{2} - \frac{1}{2}\cos 1\right)d\theta = \frac{\pi}{4}(1 - \cos 1)$$

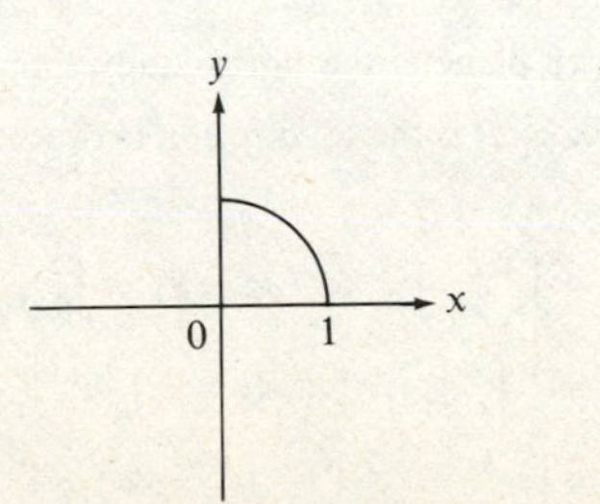

27. $\displaystyle\int_0^1\int_0^{\sqrt{1-x^2}} e^{-(x^2+y^2)}\,dy\,dx = \int_0^{\pi/2}\int_0^1 e^{-r^2} r\,dr\,d\theta = \int_0^{\pi/2} -\tfrac{1}{2}e^{-r^2}\Big|_0^1 d\theta$

$$= \int_0^{\pi/2}\frac{1}{2}(1 - e^{-1})\,d\theta = \frac{\pi}{4}(1 - e^{-1})$$

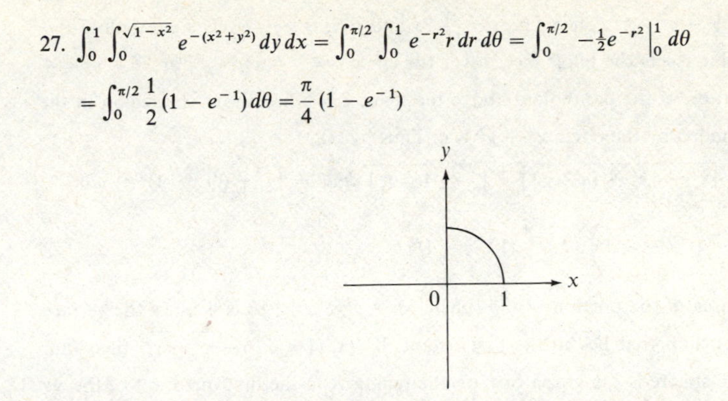

29. If $y = \sqrt{x - x^2}$, then $y^2 = x - x^2$, so $x^2 + y^2 = x$, which becomes $r = \cos\theta$ in polar coordinates. Thus

$$\int_0^1\int_{-\sqrt{x-x^2}}^{\sqrt{x-x^2}} (x^2 + y^2)\,dy\,dx = \int_{-\pi/2}^{\pi/2}\int_0^{\cos\theta} r^3\,dr\,d\theta = \int_{-\pi/2}^{\pi/2}\frac{r^4}{4}\Big|_0^{\cos\theta} d\theta = \frac{1}{4}\int_{-\pi/2}^{\pi/2}\cos^4\theta\,d\theta$$

$$= \tfrac{1}{4}(\tfrac{1}{4}\cos^3\theta\sin\theta + \tfrac{3}{8}\cos\theta\sin\theta + \tfrac{3}{8}\theta)\Big|_{-\pi/2}^{\pi/2} = \tfrac{3}{32}\pi$$

with the next to last equality coming from (13) in Section 7.1.

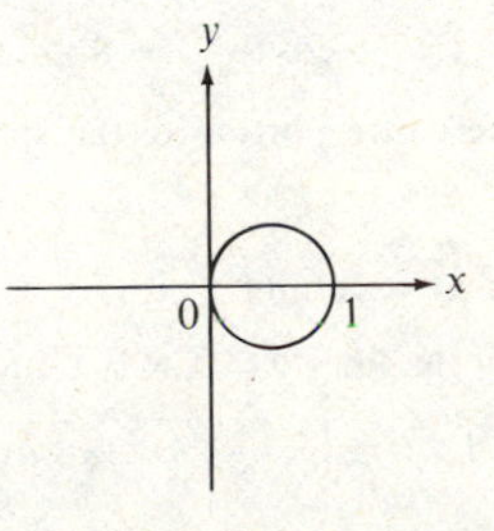

31. Make the substitution $u = x/\sqrt{2}$ and use the result of Exercise 30(c):

$$\int_{-\infty}^{\infty} e^{-x^2/2}\,dx = \int_{-\infty}^{\infty} e^{-u^2}\sqrt{2}\,du = \sqrt{2}\sqrt{\pi} = \sqrt{2\pi}$$

SECTION 15.3

1. Let $f(x, y) = \frac{1}{3}(6 - x - 2y)$. Then $f_x(x, y) = -\frac{1}{3}$ and $f_y(x, y) = -\frac{2}{3}$. The portion of the plane $z = \frac{1}{3}(6 - x - 2y)$ in the first octant lies over the region R in the first quadrant of the xy plane bounded by the lines $x = 0$, $y = 0$, and $x + 2y = 6$. Thus by (1),

$$S = \iint_R \sqrt{\left(-\frac{1}{3}\right)^2 + \left(-\frac{2}{3}\right)^2 + 1}\,dA = \int_0^6\int_0^{(1/2)(6-x)} \frac{\sqrt{14}}{3}\,dy\,dx = \int_0^6 \frac{\sqrt{14}}{3}\,y\Big|_0^{(1/2)(6-x)} dx$$

$$= \int_0^6 \frac{\sqrt{14}}{6}(6 - x)\,dx = \frac{\sqrt{14}}{6}\left(6x - \frac{1}{2}x^2\right)\Big|_0^6 = 3\sqrt{14}$$

3. Let $f(x, y) = 9 - x^2 - y^2$. Then $f_x(x, y) = -2x$ and $f_y(x, y) = -2y$. The paraboloid $z = 9 - x^2 - y^2$ intersects the plane $z = 5$ over the circle $5 = 9 - x^2 - y^2$, or $x^2 + y^2 = 4$. Thus the portion of the paraboloid above the plane $z = 5$ lies above the region in the xy plane bounded by the circle $x^2 + y^2 = 4$. Thus by (1),

$$S = \iint_R \sqrt{4x^2 + 4y^2 + 1}\, dA = \int_0^{2\pi} \int_0^2 r\sqrt{4r^2 + 1}\, dr\, d\theta = \int_0^{2\pi} \tfrac{1}{12}(4r^2 + 1)^{3/2}\Big|_0^2 d\theta$$

$$= \int_0^{2\pi} \frac{1}{12}\left(17^{3/2} - 1\right) d\theta = \frac{\pi}{6}(17^{3/2} - 1)$$

5. The surface area of the portion of the sphere $x^2 + y^2 + z^2 = 16$ is 4 times the surface area of the portion that lies in the first octant. If $f(x, y) = \sqrt{16 - x^2 - y^2}$, then that portion of the sphere is the graph of f on the region R in the first quadrant of the xy plane bounded by the x axis and the circle $x^2 - 4x + y^2 = 0$ (whose equation in polar coordinates is $r = 4 \cos \theta$). By (1), the surface area S of the graph of f on R is given as an improper integral by

$$S = \iint_R \sqrt{\frac{x^2}{16 - x^2 - y^2} + \frac{y^2}{16 - x^2 - y^2} + 1}\, dA = \iint_R \sqrt{\frac{x^2 + y^2 + 16 - x^2 - y^2}{16 - x^2 - y^2}}\, dA$$

$$= \iint_R \frac{4}{\sqrt{16 - x^2 - y^2}}\, dA = \int_0^{\pi/2} \int_0^{4\cos\theta} \frac{4r}{\sqrt{16 - r^2}}\, dr\, d\theta = \int_0^{\pi/2} -4\sqrt{16 - r^2}\Big|_0^{4\cos\theta} d\theta$$

$$= \int_0^{\pi/2} 16(1 - \sin\theta)\, d\theta = 16(\theta + \cos\theta)\Big|_0^{\pi/2} = 8\pi - 16$$

Thus the surface area of the entire portion of the sphere that is inside the cylinder is $4(8\pi - 16) = 32(\pi - 2)$.

7. Let $f(x, y) = x^2$. Then $f_x(x, y) = 2x$ and $f_y(x, y) = 0$. The region R bounded by the given triangle is bounded by the lines $y = 0$, $x = 1$, and $y = x$. Thus by (1),

$$S = \iint_R \sqrt{4x^2 + 0^2 + 1}\, dA = \int_0^1 \int_0^x \sqrt{4x^2 + 1}\, dy\, dx = \int_0^1 y\sqrt{4x^2 + 1}\Big|_0^x dx$$

$$= \int_0^1 x\sqrt{4x^2 + 1}\, dx = \tfrac{1}{12}(4x^2 + 1)^{3/2}\Big|_0^1 = \tfrac{1}{12}(5^{3/2} - 1)$$

9. If $x^2 + y^2 + z^2 = 14z$, then $x^2 + y^2 + (z - 7)^2 = 49$, or $z = 7 \pm \sqrt{49 - x^2 - y^2}$. The sphere and the paraboloid intersect at (x, y, z) such that $5z = x^2 + y^2 = 14z - z^2$, so that $z^2 - 9z = 0$, or $z = 0$ or $z = 9$, and consequently $x^2 + y^2 = 0$ or $x^2 + y^2 = 45$. Thus the portion of the sphere that is inside the paraboloid lies over the region R in the xy plane bounded by the circle $x^2 + y^2 = 45$. Since $z \geq 7$ if (x, y, z) is a point on the sphere inside the given paraboloid, we let $f(x, y) = 7 + \sqrt{49 - x^2 - y^2}$. Then $f_x(x, y) = -x/\sqrt{49 - x^2 - y^2}$ and $f_y(x, y) = -y/\sqrt{49 - x^2 - y^2}$. Thus by (1),

$$S = \iint_R \sqrt{\frac{x^2}{49 - x^2 - y^2} + \frac{y^2}{49 - x^2 - y^2} + 1}\, dA = \iint_R \frac{7}{\sqrt{49 - x^2 - y^2}}\, dA$$

$$= \int_0^{2\pi} \int_0^{\sqrt{45}} \frac{7r}{\sqrt{49 - r^2}}\, dr\, d\theta = \int_0^{2\pi} -7\sqrt{49 - r^2}\Big|_0^{\sqrt{45}} d\theta = \int_0^{2\pi} 35\, d\theta = 70\pi$$

SECTION 15.4

1. $$\int_0^3 \int_{-1}^1 \int_2^4 (y - xz)\, dz\, dy\, dx = \int_0^3 \int_{-1}^1 \left(yz - \frac{xz^2}{2}\right)\Big|_2^4 dy\, dx = \int_0^3 \int_{-1}^1 (2y - 6x)\, dy\, dx$$

$$= \int_0^3 (y^2 - 6xy)\Big|_{-1}^1 dx = \int_0^3 (-12x)\, dx = -6x^2\Big|_0^3 = -54$$

3. $$\int_{-1}^1 \int_0^x \int_{x-y}^{x+y} (z - 2x - y)\, dz\, dy\, dx = \int_{-1}^1 \int_0^x \left(\frac{z^2}{2} - 2xz - yz\right)\Big|_{x-y}^{x+y} dy\, dx$$

$$= \int_{-1}^1 \int_0^x \left[\frac{(x + y)^2}{2} - 2x(x + y) - y(x + y) - \frac{(x - y)^2}{2} + 2x(x - y) + y(x - y)\right] dy\, dx$$

$$= \int_{-1}^1 \int_0^x (-2xy - 2y^2)\, dy\, dx = \int_{-1}^1 (-xy^2 - \tfrac{2}{3}y^3)\Big|_0^x dx$$

$$= \int_{-1}^1 -\tfrac{5}{3}x^3\, dx = -\tfrac{5}{12}x^4\Big|_{-1}^1 = 0$$

5. $$\int_0^{\ln 3} \int_0^1 \int_0^y (z^2 + 1)\, e^{(y^2)}\, dx\, dz\, dy = \int_0^{\ln 3} \int_0^1 (z^2 + 1)\, e^{(y^2)}\, x\Big|_0^y dz\, dy$$

$$= \int_0^{\ln 3} \int_0^1 (z^2 + 1)\, ye^{(y^2)}\, dz\, dy = \int_0^{\ln 3} \left[\left(\frac{z^3}{3} + z\right) ye^{(y^2)}\right]\Big|_0^1 dy$$

$$= \tfrac{4}{3} \int_0^{\ln 3} ye^{(y^2)}\, dy = \tfrac{2}{3}e^{(y^2)}\Big|_0^{\ln 3} = \tfrac{2}{3}e^{(\ln 3)^2} - \tfrac{2}{3}$$

7. $$\int_{-15}^{13} \int_1^e \int_0^{1/\sqrt{x}} z(\ln x)^2\, dz\, dx\, dy = \int_{-15}^{13} \int_1^e \frac{z^2}{2}(\ln x)^2\Big|_0^{1/\sqrt{x}} dx\, dy = \int_{-15}^{13} \int_1^e \frac{1}{2x}(\ln x)^2\, dx\, dy$$

$$= \int_{-15}^{13} \tfrac{1}{6}(\ln x)^3\Big|_1^e dy = \int_{-15}^{13} \tfrac{1}{6}\, dy = \tfrac{14}{3}$$

9. $$\int_0^{\pi/2} \int_0^{\pi/2} \int_0^{\sin z} x^2 \sin y\, dx\, dy\, dz = \int_0^{\pi/2} \int_0^{\pi/2} \frac{x^3}{3} \sin y\Big|_0^{\sin z} dy\, dz$$

$$= \int_0^{\pi/2} \int_0^{\pi/2} \tfrac{1}{3}\sin^3 z \sin y\, dy\, dz = \tfrac{1}{3} \int_0^{\pi/2} (-\sin^3 z \cos y)\Big|_0^{\pi/2} dz$$

$$= \tfrac{1}{3} \int_0^{\pi/2} \sin^3 z\, dz = \tfrac{1}{3} \int_0^{\pi/2} \sin z\,(1 - \cos^2 z)\, dz = \tfrac{1}{3}(-\cos z + \tfrac{1}{3}\cos^3 z)\Big|_0^{\pi/2} = \tfrac{2}{9}$$

11. The region R in the xy plane is the horizontally simple region between the graphs of $x = -y$ and $x = y$ on $[0, 1]$; D is the solid region between the graphs of $z = 0$ and $z = y$ on R. Thus

$$\iiint_D e^y\, dV = \int_0^1 \int_{-y}^y \int_0^y e^y\, dz\, dx\, dy = \int_0^1 \int_{-y}^y ze^y\Big|_0^y dx\, dy = \int_0^1 \int_{-y}^y ye^y\, dx\, dy$$

$$= \int_0^1 xye^y\Big|_{-y}^y dy = \int_0^1 2y^2e^y\, dy \overset{\text{parts}}{=} 2y^2e^y\Big|_0^1 - \int_0^1 4ye^y\, dy$$

$$\overset{\text{parts}}{=} 2e - \left(4ye^y\Big|_0^1\right) + 4\int_0^1 e^y\, dy = 2e - 4e + \left(4e^y\Big|_0^1\right) = 2e - 4$$

13. The region R in the xy plane is the horizontally simple region between the graph of $x = 1$ and $x = 3$ on $[0, 2]$; D is the solid region between the graphs of $z = -2$ and $z = 0$ on R. Thus

$$\iiint_D ye^{xy}\, dV = \int_0^2 \int_1^3 \int_{-2}^0 ye^{xy}\, dz\, dx\, dy = \int_0^2 \int_1^3 zye^{xy}\Big|_{-2}^0 dx\, dy$$

$$= 2\int_0^2 \int_1^3 ye^{xy}\, dx\, dy = 2\int_0^2 e^{xy}\Big|_1^3 dy = 2\int_0^2 (e^{3y} - e^y)\, dy$$

$$= (\tfrac{2}{3}e^{3y} - 2e^y)\Big|_0^2 = \tfrac{2}{3}e^6 - 2e^2 + \tfrac{4}{3}$$

15. The region R in the xy plane is the portion of the region $x^2 + y^2 \le 1$ that is in the first quadrant, so is the vertically simple region between the graph of $y = \sqrt{1-x^2}$ and the x axis on $[0, 1]$; D is the solid region between the graphs of $z = \sqrt{x^2+y^2}$ and $z = 1$ on R. Thus

$$\iiint_D zy\,dV = \int_0^1 \int_0^{\sqrt{1-x^2}} \int_{\sqrt{x^2+y^2}}^1 zy\,dz\,dy\,dx = \int_0^1 \int_0^{\sqrt{1-x^2}} \frac{z^2}{2}y\bigg|_{\sqrt{x^2+y^2}}^1 dy\,dx$$
$$= \frac{1}{2}\int_0^1 \int_0^{\sqrt{1-x^2}} [y - y(x^2+y^2)]\,dy\,dx = \frac{1}{2}\int_0^1 \int_0^{\sqrt{1-x^2}} (y - x^2y - y^3)\,dy\,dx$$
$$= \frac{1}{2}\int_0^1 \left(\frac{y^2}{2} - \frac{x^2y^2}{2} - \frac{y^4}{4}\right)\bigg|_0^{\sqrt{1-x^2}} dx$$
$$= \frac{1}{2}\int_0^1 \left(\frac{1-x^2}{2} - \frac{x^2(1-x^2)}{2} - \frac{(1-x^2)^2}{4}\right) dx$$
$$= \frac{1}{8}\int_0^1 (1 - 2x^2 + x^4)\,dx = \frac{1}{8}\left(x - \frac{2}{3}x^3 + \frac{1}{5}x^5\right)\bigg|_0^1 = \frac{1}{15}$$

17. The region R in the xy plane is a square, and is the horizontally simple region between the graphs of $y = -1$ and $y = 1$ on $[-1, 1]$; D is the solid region between the graphs of $z = 0$ and $z = \sqrt{9 - x^2 - y^2}$ on R. Thus

$$\iiint_D z\,dV = \int_{-1}^1 \int_{-1}^1 \int_0^{\sqrt{9-x^2-y^2}} z\,dz\,dy\,dx = \int_{-1}^1 \int_{-1}^1 \frac{z^2}{2}\bigg|_0^{\sqrt{9-x^2-y^2}} dy\,dx$$
$$= \frac{1}{2}\int_{-1}^1 \int_{-1}^1 (9 - x^2 - y^2)\,dy\,dx = \frac{1}{2}\int_{-1}^1 \left(9y - x^2y - \frac{y^3}{3}\right)\bigg|_{-1}^1 dx$$
$$= \int_{-1}^1 \left(\frac{26}{3} - x^2\right) dx = \left(\frac{26}{3}x - \frac{x^3}{3}\right)\bigg|_{-1}^1 = \frac{50}{3}$$

19. At any point (x, y, z) of intersection of the cone and cylinder, x and y must satisfy

$$x^2 + y^2 = z^2 = 1 - x^2$$

so that $y^2 = 1 - 2x^2$, and thus $y = -\sqrt{1-2x^2}$ or $y = \sqrt{1-2x^2}$ for $-1/\sqrt{2} \le x \le 1/\sqrt{2}$. Therefore the region R in the xy plane is the vertically simple region between the graphs of $y = -\sqrt{1-2x^2}$ and $y = \sqrt{1-2x^2}$ on $[-1/\sqrt{2}, 1/\sqrt{2}]$. Then D is the solid region between the graphs of $z = \sqrt{x^2+y^2}$ and $z = \sqrt{1-x^2}$ on R. Thus

$$\iiint_D 3xy\,dV = \int_{-1/\sqrt{2}}^{1/\sqrt{2}} \int_{-\sqrt{1-2x^2}}^{\sqrt{1-2x^2}} \int_{\sqrt{x^2+y^2}}^{\sqrt{1-x^2}} 3xy\,dz\,dy\,dx$$
$$= \int_{-1/\sqrt{2}}^{1/\sqrt{2}} \int_{-\sqrt{1-2x^2}}^{\sqrt{1-2x^2}} 3xyz\bigg|_{\sqrt{x^2+y^2}}^{\sqrt{1-x^2}} dy\,dx$$
$$= \int_{-1/\sqrt{2}}^{1/\sqrt{2}} \int_{-\sqrt{1-2x^2}}^{\sqrt{1-2x^2}} 3xy\left(\sqrt{1-x^2} - \sqrt{x^2+y^2}\right) dy\,dx$$
$$= \int_{-1/\sqrt{2}}^{1/\sqrt{2}} \left[\frac{3}{2}xy^2\sqrt{1-x^2} - x(x^2+y^2)^{3/2}\right]\bigg|_{-\sqrt{1-2x^2}}^{\sqrt{1-2x^2}} dx = \int_{-1/\sqrt{2}}^{1/\sqrt{2}} 0\,dx = 0$$

21. Since the graphs of $y = x$ and $y = 2 - x$ intersect for (x, y) such that $x = y = 2 - x$, so that $x = 1$, the region R in the xy plane is the vertically simple region between the graphs of $y = x$ and $y = 2 - x$ on $[0, 1]$. Then D is the solid region between the graphs of $z = 0$ and $z = 10 + x + y$ on R. Thus

$$V = \iiint_D 1\,dV = \int_0^1 \int_x^{2-x} \int_0^{10+x+y} 1\,dz\,dy\,dx = \int_0^1 \int_x^{2-x} z\bigg|_0^{10+x+y} dy\,dx$$
$$= \int_0^1 \int_x^{2-x} (10 + x + y)\,dy\,dx = \int_0^1 \left(10y + xy + \frac{y^2}{2}\right)\bigg|_x^{2-x} dx$$
$$= \int_0^1 \left[10(2-x) + x(2-x) + \frac{(2-x)^2}{2} - 10x - x^2 - \frac{x^2}{2}\right] dx$$
$$= \int_0^1 (22 - 20x - 2x^2)\,dx = \left(22x - 10x^2 - \frac{2}{3}x^3\right)\bigg|_0^1 = \frac{34}{3}$$

23. Since the graphs of $y = x^2$ and $y = x$ intersect for (x, y) such that $x^2 = y = x$, so that $x = 0$ or $x = 1$, the region R in the xy plane is the vertically simple region between the graphs of $y = x^2$ and $y = x$ on $[0, 1]$. Then D is the solid region between the graphs of $z = -2$ and $z = 4(x^2 + y^2)$ on R. Thus

$$V = \iiint_D 1\,dV = \int_0^1 \int_{x^2}^x \int_{-2}^{4(x^2+y^2)} 1\,dz\,dy\,dx = \int_0^1 \int_{x^2}^x z\bigg|_{-2}^{4(x^2+y^2)} dy\,dx$$
$$= \int_0^1 \int_{x^2}^x (4x^2 + 4y^2 + 2)\,dy\,dx = \int_0^1 \left(4x^2y + \frac{4}{3}y^3 + 2y\right)\bigg|_{x^2}^x dx$$
$$= \int_0^1 \left[4x^2(x - x^2) + \frac{4}{3}(x^3 - x^6) + 2(x - x^2)\right] dx$$
$$= \int_0^1 \left(-\frac{4}{3}x^6 - 4x^4 + \frac{16}{3}x^3 - 2x^2 + 2x\right) dx$$
$$= \left(-\frac{4}{21}x^7 - \frac{4}{5}x^5 + \frac{4}{3}x^4 - \frac{2}{3}x^3 + x^2\right)\bigg|_0^1 = \frac{71}{105}$$

25. The cone and plane intersect for (x, y, z) such that $h^2(x^2 + y^2) = z^2 = h^2$, so that $x^2 + y^2 = 1$. Thus the region R in the xy plane is circular, and is the vertically simple region between the graphs of $y = -\sqrt{1-x^2}$ and $y = \sqrt{1-x^2}$ on $[-1, 1]$. Then D is the solid region between the graphs of $z = h\sqrt{x^2+y^2}$ and $z = h$ on R. Thus

$$V = \iiint_D 1\,dV = \int_{-1}^1 \int_{-\sqrt{1-x^2}}^{\sqrt{1-x^2}} \int_{h\sqrt{x^2+y^2}}^h 1\,dz\,dy\,dx$$
$$= \int_{-1}^1 \int_{-\sqrt{1-x^2}}^{\sqrt{1-x^2}} z\bigg|_{h\sqrt{x^2+y^2}}^h dy\,dx = \int_{-1}^1 \int_{-\sqrt{1-x^2}}^{\sqrt{1-x^2}} \left(h - h\sqrt{x^2+y^2}\right) dy\,dx$$
$$= \int_{-1}^1 \int_{-\sqrt{1-x^2}}^{\sqrt{1-x^2}} h\,dy\,dx - \int_{-1}^1 \int_{-\sqrt{1-x^2}}^{\sqrt{1-x^2}} h\sqrt{x^2+y^2}\,dy\,dx$$

Now

$$\int_{-1}^1 \int_{-\sqrt{1-x^2}}^{\sqrt{1-x^2}} h\,dy\,dx = \int_{-1}^1 hy\bigg|_{-\sqrt{1-x^2}}^{\sqrt{1-x^2}} dx = \int_{-1}^1 2h\sqrt{1-x^2}\,dx$$
$$\overset{x=\sin u}{=} 2h\int_{-\pi/2}^{\pi/2} \sqrt{1 - \sin^2 u}\cos u\,du = 2h\int_{-\pi/2}^{\pi/2} \cos^2 u\,du$$
$$= 2h\int_{-\pi/2}^{\pi/2} \left(\frac{1}{2} + \frac{1}{2}\cos 2u\right) du = h\left(u + \frac{1}{2}\sin 2u\right)\bigg|_{-\pi/2}^{\pi/2} = \pi h$$

Next, by changing to polar coordinates we find that

$$\int_{-1}^1 \int_{-\sqrt{1-x^2}}^{\sqrt{1-x^2}} h\sqrt{x^2+y^2}\,dy\,dx = \int_0^{2\pi} \int_0^1 hr^2\,dr\,d\theta = \int_0^{2\pi} \frac{hr^3}{3}\bigg|_0^1 d\theta = \int_0^{2\pi} \frac{h}{3}\,d\theta = \frac{2\pi}{3}h$$

Adding our results, we conclude that $V = \pi h - (2\pi/3)h = \frac{1}{3}\pi h$.

27. The plane $2x + y + z = 1$ intersects the xy plane for (x, y, z) such that $0 = z = 1 - 2x - y$, so that $y = 1 - 2x$. Thus the region R in the xy plane is the vertically simple region between the graphs of $y = 0$ and $y = 1 - 2x$ on $[0, \frac{1}{2}]$. Then D is the solid region between the graphs of $z = 0$ and $z = 1 - 2x - y$ on R.

a. Since $\delta(x, y, z) = z$, the total mass m is given by

$$m = \int_0^{1/2}\int_0^{1-2x}\int_0^{1-2x-y} z\,dz\,dy\,dx = \int_0^{1/2}\int_0^{1-2x} \left.\frac{z^2}{2}\right|_0^{1-2x-y} dy\,dx$$
$$= \tfrac{1}{2}\int_0^{1/2}\int_0^{1-2x} (1-2x-y)^2\,dy\,dx = \tfrac{1}{2}\int_0^{1/2} \left.-\tfrac{1}{3}(1-2x-y)^3\right|_0^{1-2x} dx$$
$$= \tfrac{1}{6}\int_0^{1/2} (1-2x)^3\,dx = \left.-\tfrac{1}{48}(1-2x)^4\right|_0^{1/2} = \tfrac{1}{48}$$

b. Since $\delta(x, y, z) = 2y$, the total mass m is given by

$$m = \int_0^{1/2}\int_0^{1-2x}\int_0^{1-2x-y} 2y\,dz\,dy\,dx = \int_0^{1/2}\int_0^{1-2x} \left.2yz\right|_0^{1-2x-y} dy\,dx$$
$$= \int_0^{1/2}\int_0^{1-2x} 2y(1-2x-y)\,dy\,dx = \int_0^{1/2}\int_0^{1-2x} [2y(1-2x) - 2y^2]\,dy\,dx$$
$$= \int_0^{1/2} \left[y^2(1-2x) - \tfrac{2}{3}y^3\right]\Big|_0^{1-2x} dx = \int_0^{1/2} [(1-2x)^3 - \tfrac{2}{3}(1-2x)^3]\,dx$$
$$= \int_0^{1/2} \tfrac{1}{3}(1-2x)^3\,dx = \left.-\tfrac{1}{24}(1-2x)^4\right|_0^{1/2} = \tfrac{1}{24}$$

c. Since $\delta(x, y, z) = x$, the total mass m is given by

$$m = \int_0^{1/2}\int_0^{1-2x}\int_0^{1-2x-y} x\,dz\,dy\,dx = \int_0^{1/2}\int_0^{1-2x} \left.xz\right|_0^{1-2x-y} dy\,dx$$
$$= \int_0^{1/2}\int_0^{1-2x} x(1-2x-y)\,dy\,dx = \int_0^{1/2}\int_0^{1-2x} [x(1-2x) - xy]\,dy\,dx$$
$$= \int_0^{1/2} \left[x(1-2x)y - \frac{xy^2}{2}\right]\Bigg|_0^{1-2x} dx = \int_0^{1/2} \left[x(1-2x)^2 - \frac{x}{2}(1-2x)^2\right] dx$$
$$= \int_0^{1/2} \left(\frac{x}{2} - 2x^2 + 2x^3\right) dx = \left(\frac{x^2}{4} - \frac{2}{3}x^3 + \frac{1}{2}x^4\right)\Bigg|_0^{1/2} = \frac{1}{96}$$

29. First $\rho(x, y, z) = z$. Next, R is circular, and is the vertically simple region between the graphs of $y = -\sqrt{9 - x^2}$ and $y = \sqrt{9 - x^2}$ on $[-3, 3]$. Finally, D is the solid region between the graphs of $z = 0$ and $z = \sqrt{9 - x^2 - y^2}$ on R. Thus the total charge is given by

$$q = \iiint_D z\,dV = \int_{-3}^{3}\int_{-\sqrt{9-x^2}}^{\sqrt{9-x^2}}\int_0^{\sqrt{9-x^2-y^2}} z\,dz\,dy\,dx = \int_{-3}^{3}\int_{-\sqrt{9-x^2}}^{\sqrt{9-x^2}} \left.\frac{z^2}{2}\right|_0^{\sqrt{9-x^2-y^2}} dy\,dx$$
$$= \frac{1}{2}\int_{-3}^{3}\int_{-\sqrt{9-x^2}}^{\sqrt{9-x^2}} (9 - x^2 - y^2)\,dy\,dx = \frac{1}{2}\int_{-3}^{3} \left[(9-x^2)y - \frac{y^3}{3}\right]\Bigg|_{-\sqrt{9-x^2}}^{\sqrt{9-x^2}} dx$$
$$= \tfrac{1}{2}\int_{-3}^{3} \tfrac{4}{3}(9-x^2)^{3/2}\,dx \overset{x = 3\sin u}{=} \tfrac{2}{3}\int_{-\pi/2}^{\pi/2} 27\cos^3 u(3\cos u)\,du = 54\int_{-\pi/2}^{\pi/2} \cos^4 u\,du$$
$$= 54(\tfrac{1}{4}\cos^3 u \sin u + \tfrac{3}{8}\cos u \sin u + \tfrac{3}{8}u)\Big|_{-\pi/2}^{\pi/2} = \tfrac{81}{4}\pi$$

with the next to last equality coming from (13) in Section 7.1.

31. a. The sheet and the plane $z = x$ intersect for (x, y, z) such that $x^2 = z = x$, so that $x = 0$ or $x = 1$. Thus the region R in the xy plane is the square bounded by the lines $y = 0$ and $y = 1$ on $[0, 1]$. Finally, D is the solid region between the graphs of $z = x^2$ and $z = x$ on R. Therefore the volume is given by

$$V = \iiint_D 1\,dV = \int_0^1\int_0^1\int_{x^2}^{x} 1\,dz\,dy\,dx = \int_0^1\int_0^1 z\Big|_{x^2}^{x} dy\,dx = \int_0^1\int_0^1 (x - x^2)\,dy\,dx$$
$$= \int_0^1 (x - x^2)y\Big|_0^1 dx = \int_0^1 (x - x^2)\,dx = \left(\frac{x^2}{2} - \frac{x^3}{3}\right)\Bigg|_0^1 = \frac{1}{6}$$

Consequently the average value of f on D is given by

$$\frac{1}{V}\iiint_D (x + y + z)\,dV = 6\int_0^1\int_0^1\int_{x^2}^{x} (x + y + z)\,dz\,dy\,dx$$
$$= 6\int_0^1\int_0^1 [(x + y)z + \tfrac{1}{2}z^2]\Big|_{x^2}^{x} dy\,dx$$
$$= 6\int_0^1\int_0^1 [-\tfrac{1}{2}x^4 - x^3 + \tfrac{3}{2}x^2 + (x - x^2)y]\,dy\,dx$$
$$= 6\int_0^1 [(-\tfrac{1}{2}x^4 - x^3 + \tfrac{3}{2}x^2)y + (x - x^2)(\tfrac{1}{2}y^2)]\Big|_0^1 dx$$
$$= 6\int_0^1 (-\tfrac{1}{2}x^4 - x^3 + x^2 + \tfrac{1}{2}x)\,dx$$
$$= 6(-\tfrac{1}{10}x^5 - \tfrac{1}{4}x^4 + \tfrac{1}{3}x^3 + \tfrac{1}{4}x^2)\Big|_0^1 = \tfrac{7}{5}$$

b. The paraboloids intersect for (x, y, z) such that $2 - x^2 - y^2 = z = x^2 + y^2$, that is, $x^2 + y^2 = 1$. Thus R is circular, and is the vertically simple region between the graph of $y = \sqrt{1 - x^2}$ and the x axis on $[0, 1]$. Finally, D is the solid region between the graphs of $z = x^2 + y^2$ and $z = 2 - x^2 - y^2$ on R. Therefore the volume is given by

$$V = \iiint_D 1\,dV = \int_0^1\int_0^{\sqrt{1-x^2}}\int_{x^2+y^2}^{2-x^2-y^2} 1\,dz\,dy\,dx = \int_0^1\int_0^{\sqrt{1-x^2}} (2 - 2x^2 - 2y^2)\,dy\,dx$$

Converting to polar coordinates, we find that

$$V = \int_0^1\int_0^{\sqrt{1-x^2}} (2 - 2x^2 - 2y^2)\,dy\,dx = \int_0^{\pi/2}\int_0^1 (2 - 2r^2)r\,dr\,d\theta$$
$$= \int_0^{\pi/2} \left(r^2 - \frac{1}{2}r^4\right)\Bigg|_0^1 d\theta = \int_0^{\pi/2} \frac{1}{2}\,d\theta = \frac{1}{2}\cdot\frac{\pi}{2} = \frac{1}{4}\pi$$

Consequently the average value of f on D is given by

$$\frac{1}{V}\iiint_D xy\,dV = \frac{4}{\pi}\int_0^1\int_0^{\sqrt{1-x^2}}\int_{x^2+y^2}^{2-x^2-y^2} xy\,dz\,dy\,dx = \frac{4}{\pi}\int_0^1\int_0^{\sqrt{1-x^2}} xyz\Big|_{x^2+y^2}^{2-x^2-y^2} dy\,dx$$
$$= \frac{4}{\pi}\int_0^1\int_0^{\sqrt{1-x^2}} (2xy - 2x^3y - 2xy^3)\,dy\,dx$$
$$= \frac{4}{\pi}\int_0^1 \left(xy^2 - x^3y^2 - \frac{1}{2}xy^4\right)\Bigg|_0^{\sqrt{1-x^2}} dx$$
$$= \frac{4}{\pi}\int_0^1 \left[x(1 - x^2) - x^3(1 - x^2) - \frac{1}{2}x(1 - x^2)^2\right] dx$$
$$= \frac{4}{\pi}\int_0^1 \left(\frac{1}{2}x - x^3 + \frac{1}{2}x^5\right) dx = \frac{4}{\pi}\left(\frac{1}{4}x^2 - \frac{1}{4}x^4 + \frac{1}{12}x^6\right)\Bigg|_0^1 = \frac{1}{3\pi}$$

SECTION 15.5

1. $r \sin\theta = -4$

3. $r(\cos\theta + \sin\theta) + z = 3$

5. $r^2 + z = 1$

7. $4r^2 = z^2$, or $z = 2r$

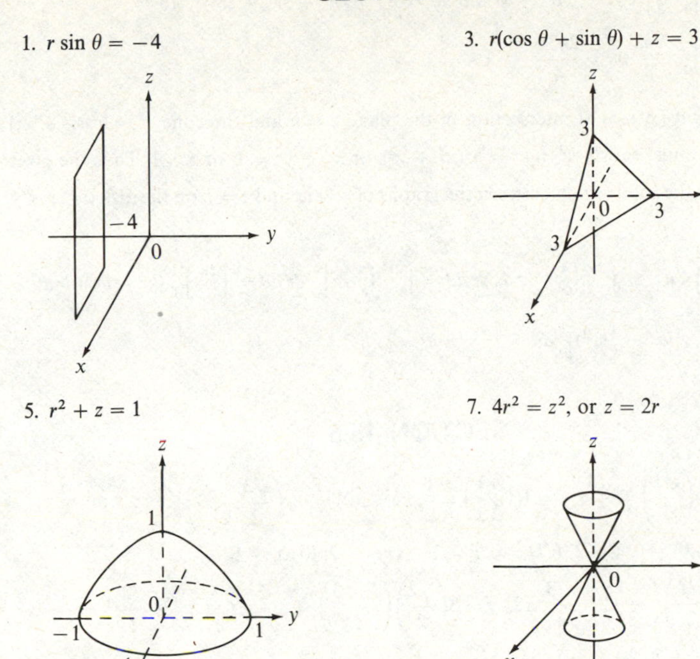

9. $\int_0^{2\pi}\int_1^2\int_0^5 e^z r\,dz\,dr\,d\theta = \int_0^{2\pi}\int_1^2 e^z r\Big|_0^5 dr\,d\theta = \int_0^{2\pi}\int_1^2 (e^5-1)r\,dr\,d\theta$
$= \int_0^{2\pi} \frac{1}{2}(e^5-1)r^2\Big|_1^2 d\theta = \int_0^{2\pi} \frac{3}{2}(e^5-1)\,d\theta = 3\pi(e^5-1)$

11. $\int_{-\pi/2}^0\int_0^{2\sin\theta}\int_0^{r^2} r^2\cos\theta\,dz\,dr\,d\theta = \int_{-\pi/2}^0\int_0^{2\sin\theta} zr^2\cos\theta\Big|_0^{r^2} dr\,d\theta$
$= \int_{-\pi/2}^0\int_0^{2\sin\theta} r^4\cos\theta\,dr\,d\theta = \int_{-\pi/2}^0 \frac{1}{5}r^5\cos\theta\Big|_0^{2\sin\theta} d\theta = \int_{-\pi/2}^0 \frac{32}{5}\sin^5\theta\cos\theta\,d\theta$
$= \frac{16}{15}\sin^6\theta\Big|_{-\pi/2}^0 = -\frac{16}{15}$

13. $\int_{-\pi/4}^{\pi/4}\int_0^{1-2\cos^2\theta}\int_0^1 r\sin\theta\,dz\,dr\,d\theta = \int_{-\pi/4}^{\pi/4}\int_0^{1-2\cos^2\theta} zr\sin\theta\Big|_0^1 dr\,d\theta$
$= \int_{-\pi/4}^{\pi/4}\int_0^{1-2\cos^2\theta} r\sin\theta\,dr\,d\theta = \int_{-\pi/4}^{\pi/4} \frac{1}{2}r^2\sin\theta\Big|_0^{1-2\cos^2\theta} d\theta$
$= \int_{-\pi/4}^{\pi/4} (\frac{1}{2} - 2\cos^2\theta + 2\cos^4\theta)\sin\theta\,d\theta = (-\frac{1}{2}\cos\theta + \frac{2}{3}\cos^3\theta - \frac{2}{5}\cos^5\theta)\Big|_{-\pi/4}^{\pi/4} = 0$

15. D is the solid region between the graphs of $z = 0$ and $z = \sqrt{1-r^2}$ on the region bounded by the polar graphs of $r = 0$ and $r = 1$ for $0 \le \theta \le \pi/2$. Therefore

$$\iiint_D z\,dV = \int_0^{\pi/2}\int_0^1\int_0^{\sqrt{1-r^2}} zr\,dz\,dr\,d\theta = \int_0^{\pi/2}\int_0^1 \frac{1}{2}z^2r\Big|_0^{\sqrt{1-r^2}} dr\,d\theta$$
$$= \int_0^{\pi/2}\int_0^1 \left(\frac{1}{2}r - \frac{1}{2}r^3\right) dr\,d\theta = \int_0^{\pi/2}\left(\frac{1}{4}r^2 - \frac{1}{8}r^4\right)\Big|_0^1 d\theta = \int_0^{\pi/2}\frac{1}{8}\,d\theta = \frac{\pi}{16}$$

17. D is the solid region between the graphs of $z = 0$ and $z = \sqrt{4-r^2}$ on the quarter disk $0 \le r \le 2$ with $0 \le \theta \le \pi/2$. Therefore

$$\iiint_D xz\,dV = \int_0^{\pi/2}\int_0^2\int_0^{\sqrt{4-r^2}} r^2z\cos\theta\,dz\,dr\,d\theta = \int_0^{\pi/2}\int_0^2 \frac{1}{2}r^2z^2\cos\theta\Big|_0^{\sqrt{4-r^2}} dr\,d\theta$$
$$= \int_0^{\pi/2}\int_0^2 (2r^2 - \tfrac{1}{2}r^4)\cos\theta\,dr\,d\theta = \int_0^{\pi/2}(\tfrac{2}{3}r^3 - \tfrac{1}{10}r^5)\cos\theta\Big|_0^2 d\theta$$
$$= \int_0^{\pi/2}\tfrac{32}{15}\cos\theta\,d\theta = \tfrac{32}{15}\sin\theta\Big|_0^{\pi/2} = \tfrac{32}{15}$$

19. The surface $z = \sqrt{r}$ and the plane $z = 1$ intersect over the circle $r = 1$ in the xy plane. Thus the given region D is the solid region between the graphs of $z = \sqrt{r}$ and $z = 1$ on the disk $0 \le r \le 1$. Therefore

$$V = \iiint_D 1\,dV = \int_0^{2\pi}\int_0^1\int_{\sqrt{r}}^1 r\,dz\,dr\,d\theta = \int_0^{2\pi}\int_0^1 rz\Big|_{\sqrt{r}}^1 dr\,d\theta = \int_0^{2\pi}\int_0^1 (r - r^{3/2})\,dr\,d\theta$$
$$= \int_0^{2\pi}(\tfrac{1}{2}r^2 - \tfrac{2}{5}r^{5/2})\Big|_0^1 d\theta = \int_0^{2\pi}\tfrac{1}{10}\,d\theta = \tfrac{1}{5}\pi$$

21. The given region D is the solid region between the graphs of $z = 0$ and $z = e^{-r^2}$ on the disk $0 \le r \le 1$. Therefore

$$V = \iiint_D 1\,dV = \int_0^{2\pi}\int_0^1\int_0^{e^{-r^2}} r\,dz\,dr\,d\theta = \int_0^{2\pi}\int_0^1 rz\Big|_0^{e^{-r^2}} dr\,d\theta$$
$$= \int_0^{2\pi}\int_0^1 re^{-r^2}\,dr\,d\theta = \int_0^{2\pi} -\tfrac{1}{2}e^{-r^2}\Big|_0^1 d\theta = \int_0^{2\pi}\tfrac{1}{2}(1 - e^{-1})\,d\theta = \pi(1 - e^{-1})$$

23. At any point (x, y, z) of intersection of the sphere $x^2 + y^2 + z^2 = 4$ and the cone $z^2 = 3x^2 + 3y^2$, x and y must satisfy $x^2 + y^2 + 3x^2 + 3y^2 = 4$, or $r = 1$. Thus the given region D is the solid region between the graphs of $z = \sqrt{3}r$ and $z = \sqrt{4-r^2}$ on the disk $0 \le r \le 1$. Therefore

$$V = \iiint_D 1\,dV = \int_0^{2\pi}\int_0^1\int_{\sqrt{3}r}^{\sqrt{4-r^2}} r\,dz\,dr\,d\theta = \int_0^{2\pi}\int_0^1 rz\Big|_{\sqrt{3}r}^{\sqrt{4-r^2}} dr\,d\theta$$
$$= \int_0^{2\pi}\int_0^1 (r\sqrt{4-r^2} - \sqrt{3}r^2)\,dr\,d\theta = \int_0^{2\pi}\left[\frac{-1}{3}(4-r^2)^{3/2} - \frac{\sqrt{3}}{3}r^3\right]\Bigg|_0^1 d\theta$$
$$= \int_0^{2\pi}\left(\frac{8}{3} - \frac{4\sqrt{3}}{3}\right)d\theta = \frac{8\pi}{3}(2 - \sqrt{3})$$

25. The cone $z = r$ intersects the planes $z = 1$ and $z = 2$ over the circles $r = 1$ and $r = 2$, respectively, in the xy plane. Thus the given solid region D consists of the regions D_1 and D_2, where D_1 is the solid region between the graphs of $z = 1$ and $z = 2$ on the disk $0 \le r \le 1$, and D_2 is the solid region between the graphs of $z = r$ and $z = 2$ on the ring $1 \le r \le 2$. Therefore

$$V = \iiint_D 1\,dV = \iiint_{D_1} 1\,dV + \iiint_{D_2} 1\,dV = \int_0^{2\pi}\int_0^1\int_1^2 r\,dz\,dr\,d\theta + \int_0^{2\pi}\int_1^2\int_r^2 r\,dz\,dr\,d\theta$$
$$= \int_0^{2\pi}\int_0^1 rz\Big|_1^2 dr\,d\theta + \int_0^{2\pi}\int_1^2 rz\Big|_r^2 dr\,d\theta = \int_0^{2\pi}\int_0^1 r\,dr\,d\theta + \int_0^{2\pi}\int_1^2 (2r - r^2)\,dr\,d\theta$$
$$= \int_0^{2\pi}\tfrac{1}{2}r^2\Big|_0^1 d\theta + \int_0^{2\pi}(r^2 - \tfrac{1}{3}r^3)\Big|_1^2 d\theta = \int_0^{2\pi}\tfrac{1}{2}\,d\theta + \int_0^{2\pi}\tfrac{2}{3}\,d\theta = \pi + \tfrac{4}{3}\pi = \tfrac{7}{3}\pi$$

27. At any point (x, y, z) of intersection of the plane $z = y$ and the paraboloid $z = x^2 + y^2$, x, y, and z must satisfy $y = z = x^2 + y^2$, or $r \sin\theta = r^2$, or $r = \sin\theta$. Thus the given region D is the solid region between the graphs of $z = r^2$ and $z = r\sin\theta$ on the region bounded by the circle $r = \sin\theta$. Therefore

$$V = \iiint_D 1\,dV = \int_0^\pi \int_0^{\sin\theta} \int_{r^2}^{r\sin\theta} r\,dz\,dr\,d\theta = \int_0^\pi \int_0^{\sin\theta} rz\Big|_{r^2}^{r\sin\theta} dr\,d\theta$$
$$= \int_0^\pi \int_0^{\sin\theta} (r^2\sin\theta - r^3)\,dr\,d\theta = \int_0^\pi \left(\tfrac{1}{3}r^3\sin\theta - \tfrac{1}{4}r^4\right)\Big|_0^{\sin\theta} d\theta$$
$$= \int_0^\pi \frac{1}{12}\sin^4\theta\,d\theta = \frac{1}{12}\left(-\frac{1}{4}\sin^3\theta\cos\theta - \frac{3}{8}\sin\theta\cos\theta + \frac{3}{8}\theta\right)\Big|_0^\pi = \frac{\pi}{32}$$

with the next to last equality coming from (12) in Section 7.1.

29. The given region D is the solid region between the graphs of $z = -\sqrt{a^2 - r^2}$ and $z = \sqrt{a^2 - r^2}$ on the region bounded by the circle $r = a\sin\theta$. Therefore

$$V = \iiint_D 1\,dV = \int_0^\pi \int_0^{a\sin\theta} \int_{-\sqrt{a^2-r^2}}^{\sqrt{a^2-r^2}} r\,dz\,dr\,d\theta = \int_0^\pi \int_0^{a\sin\theta} rz\Big|_{-\sqrt{a^2-r^2}}^{\sqrt{a^2-r^2}} dr\,d\theta$$
$$= \int_0^\pi \int_0^{a\sin\theta} 2r\sqrt{a^2 - r^2}\,dr\,d\theta = \int_0^\pi -\tfrac{2}{3}(a^2 - r^2)^{3/2}\Big|_0^{a\sin\theta} d\theta = \int_0^\pi \tfrac{2}{3}a^3(1 - |\cos^3\theta|)\,d\theta$$
$$= \int_0^{\pi/2} \tfrac{2}{3}a^3(1 - \cos^3\theta)\,d\theta + \int_{\pi/2}^\pi \tfrac{2}{3}a^3(1 + \cos^3\theta)\,d\theta$$
$$= \tfrac{2}{3}a^3 \int_0^{\pi/2} [1 - (1 - \sin^2\theta)\cos\theta]\,d\theta + \tfrac{2}{3}a^3 \int_{\pi/2}^\pi [1 + (1 - \sin^2\theta)\cos\theta]\,d\theta$$
$$= \tfrac{2}{3}a^3\left(\theta - \sin\theta + \tfrac{1}{3}\sin^3\theta\right)\Big|_0^{\pi/2} + \tfrac{2}{3}a^3\left(\theta + \sin\theta - \tfrac{1}{3}\sin^3\theta\right)\Big|_{\pi/2}^\pi$$
$$= \frac{2}{3}a^3\left(\frac{\pi}{2} - \frac{2}{3}\right) + \frac{2}{3}a^3\left(\frac{\pi}{2} - \frac{2}{3}\right) = \frac{4}{3}a^3\left(\frac{\pi}{2} - \frac{2}{3}\right)$$

31. The given region D is the solid region between the graphs of $z = 0$ and $z = r^2$ on the region in the first quadrant bounded by the x axis and the polar graph of $r = 2\sqrt{\cos\theta}$. Therefore

$$V = \iiint_D 1\,dV = \int_0^{\pi/2} \int_0^{2\sqrt{\cos\theta}} \int_0^{r^2} r\,dz\,dr\,d\theta = \int_0^{\pi/2} \int_0^{2\sqrt{\cos\theta}} rz\Big|_0^{r^2} dr\,d\theta$$
$$= \int_0^{\pi/2} \int_0^{2\sqrt{\cos\theta}} r^3\,dr\,d\theta = \int_0^{\pi/2} \tfrac{1}{4}r^4\Big|_0^{2\sqrt{\cos\theta}} d\theta = \int_0^{\pi/2} 4\cos^2\theta\,d\theta$$
$$= \int_0^{\pi/2} (2 + 2\cos 2\theta)\,d\theta = (2\theta + \sin 2\theta)\Big|_0^{\pi/2} = \pi$$

33. At any point (x, y, z) of intersection of the cone $z^2 = 9x^2 + 9y^2$ and the plane $z = 9$, x, y, and z must satisfy $81 = z^2 = 9x^2 + 9y^2$, or $r = 3$. Thus the object occupies the solid region D between the graphs of $z = 3r$ and $z = 9$ on the disk $0 \le r \le 3$. Since the mass density is given by $\delta(x, y, z) = 9 - z$, the total mass m is given by

$$m = \iiint_D (9 - z)\,dV = \int_0^{2\pi} \int_0^3 \int_{3r}^9 (9 - z)r\,dz\,dr\,d\theta = \int_0^{2\pi} \int_0^3 \left(9z - \tfrac{1}{2}z^2\right)r\Big|_{3r}^9 dr\,d\theta$$
$$= \int_0^{2\pi} \int_0^3 \left(\tfrac{81}{2}r - 27r^2 + \tfrac{9}{2}r^3\right)dr\,d\theta = \int_0^{2\pi} \left(\tfrac{81}{4}r^2 - 9r^3 + \tfrac{9}{8}r^4\right)\Big|_0^3 d\theta$$
$$= \int_0^{2\pi} \frac{243}{8}\,d\theta = \frac{243\pi}{4}$$

35. At any point (x, y, z) of intersection of the plane $z = h$ and the cone $z^2 = h^2(x^2 + y^2)$, x, y, and z must satisfy $h^2 = z^2 = h^2(x^2 + y^2)$, or $x^2 + y^2 = 1$, or $r = 1$. Thus the given region D is the solid region between the graphs of $z = hr$ and $z = h$ on the disk $0 \le r \le 1$. Therefore

$$V = \iiint_D 1\,dV = \int_0^{2\pi} \int_0^1 \int_{hr}^h r\,dz\,dr\,d\theta = \int_0^{2\pi} \int_0^1 rz\Big|_{hr}^h dr\,d\theta = \int_0^{2\pi} \int_0^1 (hr - hr^2)\,dr\,d\theta$$
$$= \int_0^{2\pi} \left(\tfrac{1}{2}hr^2 - \tfrac{1}{3}hr^3\right)\Big|_0^1 d\theta = \int_0^{2\pi} \tfrac{1}{6}h\,d\theta = \tfrac{1}{3}\pi h$$

SECTION 15.6

1. a. $x = 1(1)\left(\frac{\sqrt{3}}{2}\right) = \frac{\sqrt{3}}{2}$, $y = 1(1)\left(\frac{1}{2}\right) = \frac{1}{2}$, $z = 1(0) = 0$; $\left(\frac{\sqrt{3}}{2}, \frac{1}{2}, 0\right)$

b. $x = 2(0)(0) = 0$, $y = 2(0)(1) = 0$, $z = 2(-1) = -2$; $(0, 0, -2)$

c. $x = 3\left(\frac{\sqrt{2}}{2}\right)\left(\frac{-1}{2}\right) = \frac{-3}{4}\sqrt{2}$, $y = 3\left(\frac{\sqrt{2}}{2}\right)\left(\frac{-\sqrt{3}}{2}\right) = \frac{-3}{4}\sqrt{6}$, $z = 3\left(\frac{\sqrt{2}}{2}\right) = \frac{3}{2}\sqrt{2}$; $\left(\frac{-3}{4}\sqrt{2}, \frac{-3}{4}\sqrt{6}, \frac{3}{2}\sqrt{2}\right)$

d. $x = \frac{1}{2}\left(\frac{\sqrt{3}}{2}\right)\left(\frac{-\sqrt{2}}{2}\right) = \frac{-\sqrt{6}}{8}$, $y = \frac{1}{2}\left(\frac{\sqrt{3}}{2}\right)\left(\frac{-\sqrt{2}}{2}\right) = \frac{-\sqrt{6}}{8}$, $z = \frac{1}{2}\left(\frac{1}{2}\right) = \frac{1}{4}$; $\left(\frac{-\sqrt{6}}{8}, \frac{-\sqrt{6}}{8}, \frac{1}{4}\right)$

e. $x = 1(0)\left(\frac{-\sqrt{3}}{2}\right) = 0$, $y = 1(0)\left(\frac{-1}{2}\right) = 0$, $z = 1(1) = 1$; $(0, 0, 1)$

f. $x = 5(1)(1) = 5$, $y = 5(1)(0) = 0$, $z = 5(0) = 0$; $(5, 0, 0)$

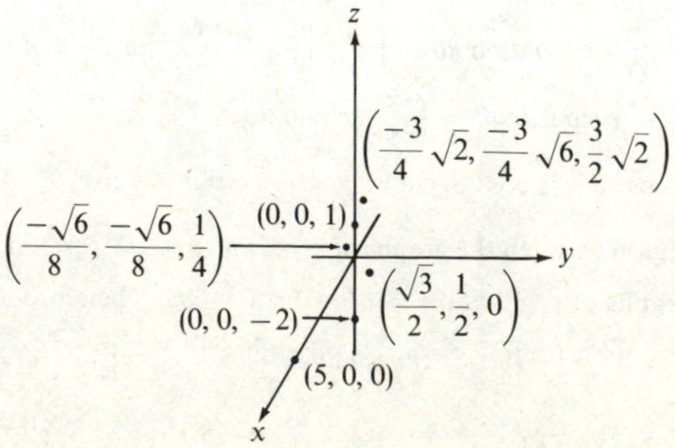

z
$\left(\frac{-3}{4}\sqrt{2}, \frac{-3}{4}\sqrt{6}, \frac{3}{2}\sqrt{2}\right)$
$(0, 0, 1)$
$\left(\frac{-\sqrt{6}}{8}, \frac{-\sqrt{6}}{8}, \frac{1}{4}\right)$
y
$\left(\frac{\sqrt{3}}{2}, \frac{1}{2}, 0\right)$
$(0, 0, -2)$
$(5, 0, 0)$
x

3. $\int_0^{2\pi}\int_0^{\pi/4}\int_0^1 \rho^2\sin\phi\,d\rho\,d\phi\,d\theta = \int_0^{2\pi}\int_0^{\pi/4}\left(\frac{\rho^3}{3}\sin\phi\right)\Big|_0^1 d\phi\,d\theta = \int_0^{2\pi}\int_0^{\pi/4}\frac{1}{3}\sin\phi\,d\phi\,d\theta$

$= \int_0^{2\pi}\left(\frac{-1}{3}\cos\phi\right)\Big|_0^{\pi/4} d\theta = \int_0^{2\pi}\frac{1}{6}(2-\sqrt{2})\,d\theta = \frac{\pi}{3}(2-\sqrt{2})$

5. $\int_0^{\pi}\int_{\pi/2}^{\pi}\int_1^2 \rho^4\sin^2\phi\cos^2\theta\,d\rho\,d\phi\,d\theta = \int_0^{\pi}\int_{\pi/2}^{\pi}\left(\frac{\rho^5}{5}\sin^2\phi\cos^2\theta\right)\Big|_1^2 d\phi\,d\theta$

$= \frac{31}{5}\int_0^{\pi}\int_{\pi/2}^{\pi}\sin^2\phi\cos^2\theta\,d\phi\,d\theta = \frac{31}{5}\int_0^{\pi}\int_{\pi/2}^{\pi}\left(\frac{1}{2}-\frac{1}{2}\cos 2\phi\right)\cos^2\theta\,d\phi\,d\theta$

$= \frac{31}{5}\int_0^{\pi}\left[\left(\frac{1}{2}\phi-\frac{1}{4}\sin 2\phi\right)\cos^2\theta\right]\Big|_{\pi/2}^{\pi} d\theta = \frac{31}{20}\pi\int_0^{\pi}\cos^2\theta\,d\theta = \frac{31}{20}\pi\int_0^{\pi}\left(\frac{1}{2}+\frac{1}{2}\cos 2\theta\right)d\theta$

$= \frac{31}{20}\pi\left(\frac{1}{2}\theta+\frac{1}{4}\sin 2\theta\right)\Big|_0^{\pi} = \frac{31}{40}\pi^2$

7. $\int_{\pi/4}^{\pi/3}\int_0^{\theta}\int_0^{9\sec\phi} \rho\cos^2\phi\cos\theta\,d\rho\,d\phi\,d\theta = \int_{\pi/4}^{\pi/3}\int_0^{\theta}\left(\frac{\rho^2}{2}\cos^2\phi\cos\theta\right)\Big|_0^{9\sec\phi} d\phi\,d\theta$

$= \int_{\pi/4}^{\pi/3}\int_0^{\theta}\frac{81}{2}\cos\theta\,d\phi\,d\theta = \int_{\pi/4}^{\pi/3}\left(\phi\,\frac{81}{2}\cos\theta\right)\Big|_0^{\theta} d\theta = \frac{81}{2}\int_{\pi/4}^{\pi/3}\theta\cos\theta\,d\theta$

$\overset{\text{parts}}{=} \frac{81}{2}(\theta\sin\theta)\Big|_{\pi/4}^{\pi/3} - \frac{81}{2}\int_{\pi/4}^{\pi/3}\sin\theta\,d\theta = \frac{81}{2}(\theta\sin\theta)\Big|_{\pi/4}^{\pi/3} + \frac{81}{2}\cos\theta\Big|_{\pi/4}^{\pi/3}$

$= \frac{27}{4}\pi\left(\sqrt{3}-\frac{3}{4}\sqrt{2}\right) + \frac{81}{4}(1-\sqrt{2})$

9. D is the collection of all points with spherical coordinates (ρ, ϕ, θ) such that $0 \le \theta \le 2\pi$, $0 \le \phi \le \pi$, and $2 \le \rho \le 3$. Since $x^2 = \rho^2\sin^2\phi\cos^2\theta$, we find that

$$\iiint_D x^2\,dV = \int_0^{2\pi}\int_0^{\pi}\int_2^3 (\rho^2\sin^2\phi\cos^2\theta)\rho^2\sin\phi\,d\rho\,d\phi\,d\theta$$

$$= \int_0^{2\pi}\int_0^{\pi}\left(\frac{\rho^5}{5}\sin^3\phi\cos^2\theta\right)\Big|_2^3 d\phi\,d\theta = \frac{211}{5}\int_0^{2\pi}\int_0^{\pi}\sin^3\phi\cos^2\theta\,d\phi\,d\theta$$

$$= \frac{211}{5}\int_0^{2\pi}\int_0^{\pi}\sin\phi\,(1-\cos^2\phi)\cos^2\theta\,d\phi\,d\theta$$

$$= \frac{211}{5}\int_0^{2\pi}\left(-\cos\phi+\frac{1}{3}\cos^3\phi\right)\cos^2\theta\Big|_0^{\pi} d\theta$$

$$= \frac{844}{15}\int_0^{2\pi}\cos^2\theta\,d\theta = \frac{844}{15}\int_0^{2\pi}\left(\frac{1}{2}+\frac{1}{2}\cos 2\theta\right)d\theta$$

$$= \frac{844}{15}\left(\frac{1}{2}\theta+\frac{1}{4}\sin 2\theta\right)\Big|_0^{2\pi} = \frac{844}{15}\pi$$

11. In spherical coordinates the cone has equation $\rho\cos\phi = \sqrt{3}\rho\sin\phi$, so that $\cot\phi = \sqrt{3}$, and hence $\phi = \pi/6$. Then D is the collection of all points with spherical coordinates (ρ, ϕ, θ) such that $0 \le \theta \le 2\pi$, $0 \le \phi \le \pi/6$, and $3 \le \rho \le 9$. Since $1/(x^2+y^2+z^2) = 1/\rho^2$, we find that

$$\iiint_D \frac{1}{x^2+y^2+z^2}\,dV = \int_0^{2\pi}\int_0^{\pi/6}\int_3^9 \frac{1}{\rho^2}(\rho^2\sin\phi)\,d\rho\,d\phi\,d\theta = \int_0^{2\pi}\int_0^{\pi/6}(\rho\sin\phi)\Big|_3^9 d\phi\,d\theta$$

$$= \int_0^{2\pi}\int_0^{\pi/6} 6\sin\phi\,d\phi\,d\theta = 6\int_0^{2\pi}(-\cos\phi)\Big|_0^{\pi/6} d\theta$$

$$= 6\int_0^{2\pi}\left(1-\frac{\sqrt{3}}{2}\right)d\theta = 6\pi(2-\sqrt{3})$$

13. In spherical coordinates the planes $x = \sqrt{3}y$ and $x = y$ have equations $\theta = \pi/6$ and $\theta = \pi/4$, respectively. Thus D is the collection of all points with spherical coordinates (ρ, ϕ, θ) such that $\pi/6 \le \theta \le \pi/4$, $0 \le \phi \le \pi/2$, and $0 \le \rho \le 4$. Since $\sqrt{z} = \sqrt{\rho\cos\phi}$, we find that

$$\iiint_D \sqrt{z}\,dV = \int_{\pi/6}^{\pi/4}\int_0^{\pi/2}\int_0^4 \sqrt{\rho\cos\phi}\,(\rho^2\sin\phi)\,d\rho\,d\phi\,d\theta$$

$$= \int_{\pi/6}^{\pi/4}\int_0^{\pi/2}\left(\frac{2}{7}\rho^{7/2}\cos^{1/2}\phi\sin\phi\right)\Big|_0^4 d\phi\,d\theta$$

$$= \int_{\pi/6}^{\pi/4}\int_0^{\pi/2}\frac{256}{7}\cos^{1/2}\phi\sin\phi\,d\phi\,d\theta = \frac{256}{7}\int_{\pi/6}^{\pi/4}\frac{-2}{3}\cos^{3/2}\phi\Big|_0^{\pi/2} d\theta$$

$$= \frac{512}{21}\int_{\pi/6}^{\pi/4} 1\,d\theta = \left(\frac{512}{21}\right)\left(\frac{\pi}{12}\right) = \frac{128}{63}\pi$$

15. In spherical coordinates the cone has equation $\rho^2\cos^2\phi = \rho^2\sin^2\phi$, so that $\phi = \pi/4$. D is the collection of all points with spherical coordinates (ρ, ϕ, θ) such that $0 \le \theta \le 2\pi$, $0 \le \phi \le \pi/4$, and $0 \le \rho \le 2$. Thus

$$V = \iiint_D 1\,dV = \int_0^{2\pi}\int_0^{\pi/4}\int_0^2 \rho^2\sin\phi\,d\rho\,d\phi\,d\theta = \int_0^{2\pi}\int_0^{\pi/4}\left(\frac{\rho^3}{3}\sin\phi\right)\Big|_0^2 d\phi\,d\theta$$

$$= \int_0^{2\pi}\int_0^{\pi/4}\frac{8}{3}\sin\phi\,d\phi\,d\theta = \int_0^{2\pi}\left(-\frac{8}{3}\cos\phi\right)\Big|_0^{\pi/4} d\theta = \int_0^{2\pi}\left(\frac{8}{3}-\frac{4}{3}\sqrt{2}\right)d\theta = \frac{8}{3}\pi(2-\sqrt{2})$$

17. In spherical coordinates the cone has equation $3\rho^2\cos^2\phi = \rho^2\sin^2\phi$, so that $\cot\phi = 1/\sqrt{3}$, and thus $\phi = \pi/3$. Therefore D is the collection of all points with spherical coordinates (ρ, ϕ, θ) such that $0 \le \theta \le 2\pi$, $\pi/3 \le \phi \le \pi$, and $1 \le \rho \le 2$. Thus

$$V = \iiint_D 1\,dV = \int_0^{2\pi}\int_{\pi/3}^{\pi}\int_1^2 \rho^2\sin\phi\,d\rho\,d\phi\,d\theta = \int_0^{2\pi}\int_{\pi/3}^{\pi}\left(\frac{\rho^3}{3}\sin\phi\right)\Big|_1^2 d\phi\,d\theta$$

$$= \int_0^{2\pi}\int_{\pi/3}^{\pi}\frac{7}{3}\sin\phi\,d\phi\,d\theta = \int_0^{2\pi}\left(-\frac{7}{3}\cos\phi\right)\Big|_{\pi/3}^{\pi} d\theta = \int_0^{2\pi}\frac{7}{2}\,d\theta = 7\pi$$

19. In spherical coordinates the cone has equation $\rho^2\sin^2\phi = \rho^2\cos^2\phi$, so that $\phi = \pi/4$, and the cylinder has equation $\rho^2\sin^2\phi = 4$, so that $\rho\sin\phi = 2$, and thus $\rho = 2\csc\phi$. Therefore D is the collection of all points with spherical coordinates (ρ, ϕ, θ) such that $0 \le \theta \le 2\pi$, $\pi/4 \le \phi \le \pi/2$, and $0 \le \rho \le 2\csc\phi$. Thus

$$V = \iiint_D 1\,dV = \int_0^{2\pi}\int_{\pi/4}^{\pi/2}\int_0^{2\csc\phi} \rho^2\sin\phi\,d\rho\,d\phi\,d\theta = \int_0^{2\pi}\int_{\pi/4}^{\pi/2}\left(\frac{\rho^3}{3}\sin\phi\right)\Big|_0^{2\csc\phi} d\phi\,d\theta$$

$$= \int_0^{2\pi}\int_{\pi/4}^{\pi/2}\frac{8}{3}\csc^2\phi\,d\phi\,d\theta = \int_0^{2\pi}\left(-\frac{8}{3}\cot\phi\right)\Big|_{\pi/4}^{\pi/2} d\theta = \int_0^{2\pi}\frac{8}{3}\,d\theta = \frac{16}{3}\pi$$

21. In spherical coordinates the plane $z = -4\sqrt{3}$ has equation $\rho\cos\phi = -4\sqrt{3}$, or $\rho = -4\sqrt{3}\sec\phi$. Thus the plane $z = -4\sqrt{3}$ and the sphere $x^2+y^2+z^2 = 64$ intersect at

points having spherical coordinates (ρ, ϕ, θ) satisfying $8 = \rho = -4\sqrt{3} \sec \phi$, so that $\sec \phi = -2/\sqrt{3}$, or $\phi = 5\pi/6$. Thus the points that are inside the sphere $x^2 + y^2 + z^2 = 8$ and lie *below* the plane $z = -4\sqrt{3}$ have spherical coordinates (ρ, ϕ, θ) such that $0 \le \theta \le 2\pi$, $5\pi/6 \le \phi \le \pi$, and $-4\sqrt{3} \sec \phi \le \rho \le 8$. The volume V_1 of the set of such points is given by

$$V_1 = \int_0^{2\pi} \int_{5\pi/6}^{\pi} \int_{-4\sqrt{3}\sec\phi}^{8} \rho^2 \sin \phi \, d\rho \, d\phi \, d\theta = \int_0^{2\pi} \int_{5\pi/6}^{\pi} \frac{\rho^3}{3} \sin \phi \Big|_{-4\sqrt{3}\sec\phi}^{8} d\phi \, d\theta$$
$$= \int_0^{2\pi} \int_{5\pi/6}^{\pi} \left(\frac{512}{3} \sin \phi + 64\sqrt{3} \sec^3 \phi \sin \phi \right) d\phi \, d\theta$$
$$= \int_0^{2\pi} \left(-\frac{512}{3} \cos \phi + \frac{32\sqrt{3}}{\cos^2 \phi} \right)\Big|_{5\pi/6}^{\pi} d\theta = \int_0^{2\pi} \left(\frac{512}{3} - \frac{288}{3}\sqrt{3} \right) d\theta$$
$$= \frac{64\pi}{3}(16 - 9\sqrt{3})$$

Thus $V = \frac{4}{3}\pi 8^3 - V_1 = (64\pi/3)(32 - (16 - 9\sqrt{3})) = (64\pi/3)(16 + 9\sqrt{3})$.

23. Since $\rho \ge 0$ and $0 \le \phi \le \pi$, it follows that if $\rho = \cos \phi$, then $0 \le \phi \le \pi/2$. Thus D is the collection of all points with spherical coordinates (ρ, ϕ, θ) such that $0 \le \theta \le 2\pi$, $0 \le \phi \le \pi/2$, and $0 \le \rho \le \cos \phi$. Thus

$$V = \iiint_D 1 \, dV = \int_0^{2\pi} \int_0^{\pi/2} \int_0^{\cos\phi} \rho^2 \sin \phi \, d\rho \, d\phi \, d\theta = \int_0^{2\pi} \int_0^{\pi/2} \left(\frac{\rho^3}{3} \sin \phi \right)\Big|_0^{\cos\phi} d\phi \, d\theta$$
$$= \int_0^{2\pi} \int_0^{\pi/2} \tfrac{1}{3} \cos^3 \phi \sin \phi \, d\phi \, d\theta = \int_0^{2\pi} (-\tfrac{1}{12} \cos^4 \phi)\Big|_0^{\pi/2} d\theta = \int_0^{2\pi} \tfrac{1}{12} d\theta = \tfrac{1}{6}\pi$$

Notice that D is a ball with radius $\frac{1}{2}$, centered at $(0, 0, \frac{1}{2})$.

25. D is the collection of all points with spherical coordinates (ρ, ϕ, θ) such that $0 \le \theta \le 2\pi$, $0 \le \phi \le \pi$, and $2 \le \rho \le 4$. Since $\delta(x, y, z) = 1/\sqrt{x^2 + y^2 + z^2}$, the total mass m is given by

$$m = \int_0^{2\pi} \int_0^{\pi} \int_2^4 \frac{1}{\rho} (\rho^2 \sin \phi) \, d\rho \, d\phi \, d\theta = \int_0^{2\pi} \int_0^{\pi} \left(\frac{\rho^2}{2} \sin \phi \right)\Big|_2^4 d\phi \, d\theta$$
$$= \int_0^{2\pi} \int_0^{\pi} 6 \sin \phi \, d\phi \, d\theta = \int_0^{2\pi} (-6 \cos \phi)\Big|_0^{\pi} d\theta = \int_0^{2\pi} 12 \, d\theta = 24\pi$$

SECTION 15.7

1. By symmetry, $\bar{x} = 0$. The graphs of $y = 5$ and $y = 1 + x^2$ intersect at (x, y) such that $5 = y = 1 + x^2$, so that $x = -2$ or $x = 2$. Consequently

$$\mathcal{M}_x = \iint_R y \, dA = \int_{-2}^{2} \int_{1+x^2}^{5} y \, dy \, dx = \int_{-2}^{2} \tfrac{1}{2} y^2 \Big|_{1+x^2}^{5} dx = \int_{-2}^{2} (12 - x^2 - \tfrac{1}{2}x^4) \, dx$$
$$= (12x - \tfrac{1}{3}x^3 - \tfrac{1}{10}x^5)\Big|_{-2}^{2} = \tfrac{544}{15}$$

Since

$$A = \int_{-2}^{2} \int_{1+x^2}^{5} 1 \, dy \, dx = \int_{-2}^{2} y \Big|_{1+x^2}^{5} dx = \int_{-2}^{2} (4 - x^2) \, dx = (4x - \tfrac{1}{3}x^3)\Big|_{-2}^{2} = \tfrac{32}{3}$$

it follows that $\bar{y} = \mathcal{M}_x/A = \frac{544/32}{15/3} = \frac{17}{5}$. Thus $(\bar{x}, \bar{y}) = (0, \frac{17}{5})$.

3. The given region is symmetric with respect to the lines $y = x$ and $x + y = 1$. Thus $\bar{x} = \bar{y}$ and $\bar{x} + \bar{y} = 1$, so that $(\bar{x}, \bar{y}) = (\frac{1}{2}, \frac{1}{2})$.

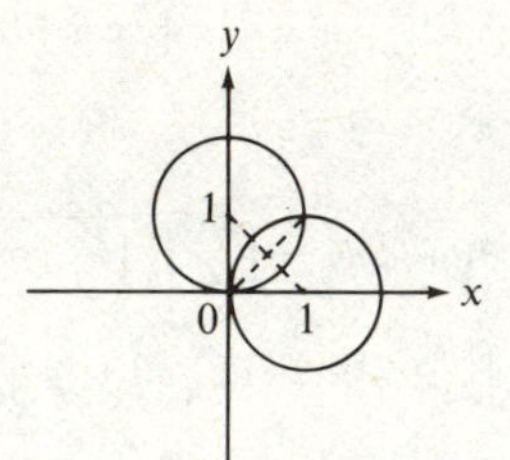

5. By symmetry, $\bar{y} = 0$, Now

$$\mathcal{M}_y = \iint_R x \, dA = \int_0^{2\pi} \int_0^{1+\cos\theta} r^2 \cos \theta \, dr \, d\theta = \int_0^{2\pi} \tfrac{1}{3} r^3 \cos \theta \Big|_0^{1+\cos\theta} d\theta$$
$$= \int_0^{2\pi} (\tfrac{1}{3} \cos \theta + \cos^2 \theta + \cos^3 \theta + \tfrac{1}{3} \cos^4 \theta) \, d\theta$$
$$= \int_0^{2\pi} [\tfrac{1}{3} \cos \theta + \tfrac{1}{2} + \tfrac{1}{2} \cos 2\theta + (1 - \sin^2 \theta) \cos \theta] \, d\theta + \int_0^{2\pi} \tfrac{1}{3} \cos^4 \theta \, d\theta$$
$$= (\tfrac{1}{3} \sin \theta + \tfrac{1}{2}\theta + \tfrac{1}{4} \sin 2\theta + \sin \theta - \tfrac{1}{3} \sin^3 \theta)\Big|_0^{2\pi}$$
$$+ (\tfrac{1}{12} \cos^3 \theta \sin \theta + \tfrac{1}{8} \cos \theta \sin \theta + \tfrac{1}{8}\theta)\Big|_0^{2\pi} = \tfrac{5}{4}\pi$$

with the next to last equality coming from (13) in Section 7.1. Since

$$A = \int_0^{2\pi} \int_0^{1+\cos\theta} r \, dr \, d\theta = \int_0^{2\pi} \tfrac{1}{2} r^2 \Big|_0^{1+\cos\theta} d\theta = \int_0^{2\pi} (\tfrac{1}{2} + \cos \theta + \tfrac{1}{2} \cos^2 \theta) \, d\theta$$
$$= \int_0^{2\pi} (\tfrac{1}{2} + \cos \theta + \tfrac{1}{4} + \tfrac{1}{4} \cos 2\theta) \, d\theta = (\tfrac{3}{4}\theta + \sin \theta + \tfrac{1}{8} \sin 2\theta)\Big|_0^{2\pi} = \tfrac{3}{2}\pi$$

it follows that $\bar{x} = \mathcal{M}_y/A = (5\pi/4)/(3\pi/2) = \frac{5}{6}$. Thus $(\bar{x}, \bar{y}) = (\frac{5}{6}, 0)$.

7. By symmetry, $\bar{x} = \bar{y} = 0$. Taking $\delta = 1$, we have

$$\mathcal{M}_{xy} = \iiint_D z \, dV = \int_0^{2\pi} \int_0^a \int_0^{\sqrt{a^2 - r^2}} zr \, dz \, dr \, d\theta = \int_0^{2\pi} \int_0^a \tfrac{1}{2} z^2 r \Big|_0^{\sqrt{a^2-r^2}} dr \, d\theta$$
$$= \int_0^{2\pi} \int_0^a (\tfrac{1}{2}a^2 r - \tfrac{1}{2}r^3) \, dr \, d\theta = \int_0^{2\pi} (\tfrac{1}{4}a^2 r^2 - \tfrac{1}{8}r^4)\Big|_0^a d\theta = \int_0^{2\pi} \tfrac{1}{8}a^4 \, d\theta = \tfrac{1}{4}\pi a^4$$

Since $m = \frac{2}{3}\pi a^3$, it follows that $\bar{z} = \mathcal{M}_{xy}/m = (\pi a^4/4)/(2\pi a^3/3) = \frac{3}{8}a$. Thus $(\bar{x}, \bar{y}, \bar{z}) = (0, 0, \frac{3}{8}a)$.

9. By symmetry, $\bar{x} = \bar{y} = 0$. The plane $z = 1$ and the cone $z^2 = 9x^2 + 9y^2$ intersect over the circle $x^2 + y^2 = \frac{1}{9}$. Taking $\delta = 1$, we have

$$\mathcal{M}_{xy} = \iiint_D z\,dV = \int_0^{2\pi}\int_0^{1/3}\int_{3r}^1 zr\,dz\,dr\,d\theta = \int_0^{2\pi}\int_0^{1/3} \tfrac{1}{2}z^2r\Big|_{3r}^1 dr\,d\theta$$

$$= \int_0^{2\pi}\int_0^{1/3} (\tfrac{1}{2}r - \tfrac{9}{2}r^3)\,dr\,d\theta = \int_0^{2\pi} (\tfrac{1}{4}r^2 - \tfrac{9}{8}r^4)\Big|_0^{1/3} d\theta = \int_0^{2\pi} \tfrac{1}{72}\,d\theta = \tfrac{1}{36}\pi$$

Since

$$m = \iiint_D 1\,dV = \int_0^{2\pi}\int_0^{1/3}\int_{3r}^1 r\,dz\,dr\,d\theta = \int_0^{2\pi}\int_0^{1/3} rz\Big|_{3r}^1 dr\,d\theta$$

$$= \int_0^{2\pi}\int_0^{1/3} (r - 3r^2)\,dr\,d\theta = \int_0^{2\pi} (\tfrac{1}{2}r^2 - r^3)\Big|_0^{1/3} d\theta = \int_0^{2\pi} \tfrac{1}{54}\,d\theta = \tfrac{1}{27}\pi$$

it follows that $\bar{z} = \mathcal{M}_{xy}/m = (\pi/36)/(\pi/27) = \frac{3}{4}$. Thus $(\bar{x}, \bar{y}, \bar{z}) = (0, 0, \frac{3}{4})$.

11. The paraboloids $z = 1 - x^2 - y^2$ and $z = x^2 + y^2$ intersect at (x, y, z) such that $1 - x^2 - y^2 = z = x^2 + y^2$, so that $x^2 + y^2 = \frac{1}{2}$. Taking $\delta = 1$, we have

$$\mathcal{M}_{xy} = \iiint_D z\,dV = \int_0^{\pi/2}\int_0^{1/\sqrt{2}}\int_{r^2}^{1-r^2} zr\,dz\,dr\,d\theta = \int_0^{\pi/2}\int_0^{1/\sqrt{2}} \tfrac{1}{2}z^2r\Big|_{r^2}^{1-r^2} dr\,d\theta$$

$$= \int_0^{\pi/2}\int_0^{1/\sqrt{2}} (\tfrac{1}{2}r - r^3)\,dr\,d\theta = \int_0^{\pi/2} (\tfrac{1}{4}r^2 - \tfrac{1}{4}r^4)\Big|_0^{1/\sqrt{2}} d\theta = \int_0^{\pi/2} \tfrac{1}{16}\,d\theta = \tfrac{1}{32}\pi$$

$$\mathcal{M}_{xz} = \iiint_D y\,dV = \int_0^{\pi/2}\int_0^{1/\sqrt{2}}\int_{r^2}^{1-r^2} r^2\sin\theta\,dz\,dr\,d\theta = \int_0^{\pi/2}\int_0^{1/\sqrt{2}} r^2z\sin\theta\Big|_{r^2}^{1-r^2} dr\,d\theta$$

$$= \int_0^{\pi/2}\int_0^{1/\sqrt{2}} (r^2 - 2r^4)\sin\theta\,dr\,d\theta = \int_0^{\pi/2} (\tfrac{1}{3}r^3 - \tfrac{2}{5}r^5)\sin\theta\Big|_0^{1/\sqrt{2}} d\theta$$

$$= \int_0^{\pi/2} \frac{\sqrt{2}}{30}\sin\theta\,d\theta = \frac{-\sqrt{2}}{30}\cos\theta\Big|_0^{\pi/2} = \frac{\sqrt{2}}{30}$$

and

$$\mathcal{M}_{yz} = \iiint_D x\,dV = \int_0^{\pi/2}\int_0^{1/\sqrt{2}}\int_{r^2}^{1-r^2} r^2\cos\theta\,dz\,dr\,d\theta = \int_0^{\pi/2}\int_0^{1/\sqrt{2}} r^2z\cos\theta\Big|_{r^2}^{1-r^2} dr\,d\theta$$

$$= \int_0^{\pi/2}\int_0^{1/\sqrt{2}} (r^2 - 2r^4)\cos\theta\,dr\,d\theta = \int_0^{\pi/2} (\tfrac{1}{3}r^3 - \tfrac{2}{5}r^5)\cos\theta\Big|_0^{1/\sqrt{2}} d\theta$$

$$= \int_0^{\pi/2} \frac{\sqrt{2}}{30}\cos\theta\,d\theta = \frac{\sqrt{2}}{30}\sin\theta\Big|_0^{\pi/2} = \frac{\sqrt{2}}{30}$$

Since

$$m = \iiint_D 1\,dV = \int_0^{\pi/2}\int_0^{1/\sqrt{2}}\int_{r^2}^{1-r^2} r\,dz\,dr\,d\theta = \int_0^{\pi/2}\int_0^{1/\sqrt{2}} rz\Big|_{r^2}^{1-r^2} dr\,d\theta$$

$$= \int_0^{\pi/2}\int_0^{1/\sqrt{2}} (r - 2r^3)\,dr\,d\theta = \int_0^{\pi/2} (\tfrac{1}{2}r^2 - \tfrac{1}{2}r^4)\Big|_0^{1/\sqrt{2}} d\theta = \int_0^{\pi/2} \tfrac{1}{8}\,d\theta = \tfrac{1}{16}\pi$$

it follows that

$$\bar{x} = \frac{\mathcal{M}_{yz}}{m} = \frac{\sqrt{2}/30}{\pi/16} = \frac{8\sqrt{2}}{15\pi}, \quad \bar{y} = \frac{\mathcal{M}_{xz}}{m} = \frac{\sqrt{2}/30}{\pi/16} = \frac{8\sqrt{2}}{15\pi}, \quad \text{and} \quad \bar{z} = \frac{\mathcal{M}_{xy}}{m} = \frac{\pi/32}{\pi/16} = \frac{1}{2}$$

Thus $(\bar{x}, \bar{y}, \bar{z}) = (8\sqrt{2}/15\pi, 8\sqrt{2}/15\pi, \frac{1}{2})$.

13. By symmetry, $\bar{x} = \bar{y} = 0$. An equation of the plane that contains the face of the pyramid in the first octant is $x + y + z/2 = 1$, or $z = 2(1 - x - y)$. Thus, taking $\delta = 1$, we have

$$\mathcal{M}_{xy} = \iiint_D z\,dV = 4\int_0^1\int_0^{1-x}\int_0^{2(1-x-y)} z\,dz\,dy\,dx = \int_0^1\int_0^{1-x} 2z^2\Big|_0^{2(1-x-y)} dy\,dx$$

$$= \int_0^1\int_0^{1-x} 8(1-x-y)^2\,dy\,dx = \int_0^1 -\tfrac{8}{3}(1-x-y)^3\Big|_0^{1-x} dx$$

$$= \int_0^1 \tfrac{8}{3}(1-x)^3\,dx = -\tfrac{2}{3}(1-x)^4\Big|_0^1 = \tfrac{2}{3}$$

Since

$$m = \iiint_D 1\,dV = 4\int_0^1\int_0^{1-x}\int_0^{2(1-x-y)} 1\,dz\,dy\,dx = \int_0^1\int_0^{1-x} 4z\Big|_0^{2(1-x-y)} dy\,dx$$

$$= \int_0^1\int_0^{1-x} 8(1-x-y)\,dy\,dx = \int_0^1 -4(1-x-y)^2\Big|_0^{1-x} dx$$

$$= \int_0^1 4(1-x)^2\,dx = -\tfrac{4}{3}(1-x)^3\Big|_0^1 = \tfrac{4}{3}$$

it follows that $\bar{z} = \mathcal{M}_{xy}/m = \frac{2}{3}/\frac{4}{3} = \frac{1}{2}$. Thus $(\bar{x}, \bar{y}, \bar{z}) = (0, 0, \frac{1}{2})$.

15. Since $\delta(x, y, z)$ does not depend on x or y and since the given region is symmetric with respect to the yz plane and the xz plane, we have $\bar{x} = 0 = \bar{y}$. Next

$$\mathcal{M}_{xy} = \iiint_D z\delta(x, y, z)\,dV = \iiint_D z(z^2 + 1)\,dV = 0$$

so that $\bar{z} = \mathcal{M}_{xy}/m = 0/m = 0$. Thus $(\bar{x}, \bar{y}, \bar{z}) = (0, 0, 0)$.

17. $$\mathcal{M}_{xy} = \iiint_D z\delta(x, y, z)\,dV = \int_0^2\int_0^2\int_0^2 z(1+x)\,dz\,dy\,dx = \int_0^2\int_0^2 \tfrac{1}{2}z^2(1+x)\Big|_0^2 dy\,dx$$

$$= \int_0^2\int_0^2 2(1+x)\,dy\,dx = \int_0^2 2(1+x)y\Big|_0^2 dx = \int_0^2 4(1+x)\,dx = (4x + 2x^2)\Big|_0^2 = 16$$

$$\mathcal{M}_{xz} = \iiint_D y\delta(x, y, z)\,dV = \int_0^2\int_0^2\int_0^2 y(1+x)\,dz\,dy\,dx = \int_0^2\int_0^2 y(1+x)z\Big|_0^2 dy\,dx$$

$$= \int_0^2\int_0^2 2y(1+x)\,dy\,dx = \int_0^2 y^2(1+x)\Big|_0^2 dx = \int_0^2 4(1+x)\,dx = (4x + 2x^2)\Big|_0^2 = 16$$

and

$$\mathcal{M}_{yz} = \iiint_D x\delta(x, y, z)\,dV = \int_0^2\int_0^2\int_0^2 x(1+x)\,dz\,dy\,dx = \int_0^2\int_0^2 x(1+x)z\Big|_0^2 dy\,dx$$

$$= \int_0^2\int_0^2 2x(1+x)\,dy\,dx = \int_0^2 2x(1+x)y\Big|_0^2 dx = \int_0^2 4x(1+x)\,dx$$

$$= (2x^2 + \tfrac{4}{3}x^3)\Big|_0^2 = \tfrac{56}{3}$$

Since

$$m = \iiint_D \delta(x, y, z)\,dV = \int_0^2\int_0^2\int_0^2 (1+x)\,dz\,dy\,dx = \int_0^2\int_0^2 (1+x)z\Big|_0^2 dy\,dx$$

$$= \int_0^2\int_0^2 2(1+x)\,dy\,dx = \int_0^2 2(1+x)y\Big|_0^2 dx = \int_0^2 4(1+x)\,dx = (4x + 2x^2)\Big|_0^2 = 16$$

it follows that $\bar{x} = \mathcal{M}_{yz}/m = \frac{56}{3}/16 = \frac{7}{6}$, $\bar{y} = \mathcal{M}_{xz}/m = \frac{16}{16} = 1$, and $\bar{z} = \mathcal{M}_{xy}/m = \frac{16}{16} = 1$. Thus $(\bar{x}, \bar{y}, \bar{z}) = (\frac{7}{6}, 1, 1)$.

19. Since $\delta(x, y, z) = \sqrt{x^2 + y^2}$, we have

$$\mathcal{M}_{xy} = \iiint_D z\delta(x, y, z)\,dV = \iiint_D z\sqrt{x^2 + y^2}\,dV = \int_0^{2\pi}\int_0^3\int_0^{\sqrt{9-r^2}} zr^2\,dz\,dr\,d\theta$$
$$= \int_0^{2\pi}\int_0^3 \tfrac{1}{2}z^2r^2\Big|_0^{\sqrt{9-r^2}}\,dr\,d\theta = \int_0^{2\pi}\int_0^3 (\tfrac{9}{2}r^2 - \tfrac{1}{2}r^4)\,dr\,d\theta$$
$$= \int_0^{2\pi} (\tfrac{3}{2}r^3 - \tfrac{1}{10}r^5)\Big|_0^3\,d\theta = \int_0^{2\pi} \tfrac{81}{5}\,d\theta = \tfrac{162}{5}\pi$$

By symmetry $\quad \mathcal{M}_{xz} = \iiint_D y\delta(x, y, z)\,dV = \iiint_D y\sqrt{x^2 + y^2}\,dV = 0$

and $\quad \mathcal{M}_{yz} = \iiint_D x\delta(x, y, z)\,dV = \iiint_D x\sqrt{x^2 + y^2}\,dV = 0$

Since

$$m = \iiint_D \delta(x, y, z)\,dV = \iiint_D \sqrt{x^2 + y^2}\,dV = \int_0^{2\pi}\int_0^3\int_0^{\sqrt{9-r^2}} r^2\,dz\,dr\,d\theta$$
$$= \int_0^{2\pi}\int_0^3 r^2z\Big|_0^{\sqrt{9-r^2}}\,dr\,d\theta = \int_0^{2\pi}\int_0^3 r^2\sqrt{9 - r^2}\,dr\,d\theta$$
$$\overset{r=3\sin u}{=} \int_0^{2\pi}\int_0^{\pi/2} (9\sin^2 u)(3\cos u)(3\cos u)\,du\,d\theta = \int_0^{2\pi}\int_0^{\pi/2} \tfrac{81}{4}\sin^2 2u\,du\,d\theta$$
$$= \int_0^{2\pi}\int_0^{\pi/2} \tfrac{81}{4}(\tfrac{1}{2} - \tfrac{1}{2}\cos 4u)\,du\,d\theta = \int_0^{2\pi} \tfrac{81}{4}(\tfrac{1}{2}u - \tfrac{1}{8}\sin 4u)\Big|_0^{\pi/2}\,d\theta = \int_0^{2\pi} \tfrac{81}{16}\pi\,d\theta = \tfrac{81}{8}\pi^2$$

it follows that $\bar{x} = \mathcal{M}_{yz}/m = 0/m = 0$, $\bar{y} = \mathcal{M}_{xz}/m = 0/m = 0$, and $\bar{z} = \mathcal{M}_{xy}/m = (162\pi/5)/(81\pi^2/8) = 16/5\pi$. Thus $(\bar{x}, \bar{y}, \bar{z}) = (0, 0, 16/5\pi)$.

21. Since $\delta(x, y, z)$ does not depend on x or y and since the region occupied by the juice is symmetric with respect to the yz plane and the xz plane, we have $\bar{x} = 0 = \bar{y}$. Next,

$$\mathcal{M}_{xy} = \iiint_D z\delta(x, y, z)\,dV = \iiint_D az(40 - z)\,dV = \int_0^{2\pi}\int_0^4\int_0^{20} az(40 - z)r\,dz\,dr\,d\theta$$
$$= a\int_0^{2\pi}\int_0^4 \left(20z^2 - \frac{1}{3}z^3\right)r\Big|_0^{20}\,dr\,d\theta = a\int_0^{2\pi}\int_0^4 \frac{16{,}000}{3}r\,dr\,d\theta$$
$$= a\int_0^{2\pi} \frac{8{,}000}{3}r^2\Big|_0^4\,d\theta = a\int_0^{2\pi} \frac{128{,}000}{3}\,d\theta = \frac{256{,}000}{3}\pi a$$

Since

$$m = \iiint_D \delta(x, y, z)\,dV = \iiint_D a(40 - z)\,dV = \int_0^{2\pi}\int_0^4\int_0^{20} a(40 - z)r\,dz\,dr\,d\theta$$
$$= a\int_0^{2\pi}\int_0^4 (40z - \tfrac{1}{2}z^2)r\Big|_0^{20}\,dr\,d\theta = a\int_0^{2\pi}\int_0^4 600r\,dr\,d\theta$$
$$= a\int_0^{2\pi} 300r^2\Big|_0^4\,d\theta = a\int_0^{2\pi} 4800\,d\theta = 9600\pi a$$

it follows that $\bar{z} = \mathcal{M}_{xy}/m = (256{,}000\pi a/3)/(9600\pi a) = \frac{80}{9}$. Thus $(\bar{x}, \bar{y}, \bar{z}) = (0, 0, \frac{80}{9})$.

23. $I_x = \iiint_D (y^2 + z^2)5\,dV = \int_0^{2\pi}\int_0^\pi\int_0^5 (\rho^2\sin^2\phi\sin^2\theta + \rho^2\cos^2\phi)5\rho^2\sin\phi\,d\rho\,d\phi\,d\theta$

$$= \int_0^{2\pi}\int_0^\pi \rho^5(\sin^2\phi\sin^2\theta + \cos^2\phi)\sin\phi\Big|_0^5\,d\phi\,d\theta$$
$$= \int_0^{2\pi}\int_0^\pi 3125[(1 - \cos^2\phi)\sin^2\theta + \cos^2\phi]\sin\phi\,d\phi\,d\theta$$
$$= 3125\int_0^{2\pi} [(-\cos\phi + \tfrac{1}{3}\cos^3\phi)\sin^2\theta - \tfrac{1}{3}\cos^3\phi]\Big|_0^\pi\,d\theta$$
$$= 3125\int_0^{2\pi} (\tfrac{4}{3}\sin^2\theta + \tfrac{2}{3})\,d\theta = 3125\int_0^{2\pi} (\tfrac{2}{3} - \tfrac{2}{3}\cos 2\theta + \tfrac{2}{3})\,d\theta$$
$$= 3125\left(\frac{4}{3}\theta - \frac{1}{3}\sin 2\theta\right)\Big|_0^{2\pi} = \frac{25{,}000}{3}\pi$$

By symmetry of the region and the mass density, $I_y = (25{,}000/3)\pi = I_z$.

25. $I_x = \iiint_D (y^2 + z^2)2\,dV = \int_0^{2\pi}\int_0^2\int_0^6 2(r^2\sin^2\theta + z^2)r\,dz\,dr\,d\theta$

$$= \int_0^{2\pi}\int_0^2 (2r^3z\sin^2\theta + \tfrac{2}{3}z^3r)\Big|_0^6\,dr\,d\theta = \int_0^{2\pi}\int_0^2 (12r^3\sin^2\theta + 144r)\,dr\,d\theta$$
$$= \int_0^{2\pi} (3r^4\sin^2\theta + 72r^2)\Big|_0^2\,d\theta = \int_0^{2\pi} (48\sin^2\theta + 288)\,d\theta$$
$$= \int_0^{2\pi} (24 - 24\cos 2\theta + 288)\,d\theta = (312\theta - 12\sin 2\theta)\Big|_0^{2\pi} = 624\pi$$

By the symmetry of the region and the mass density, $I_y = 624\pi$. Finally,

$$I_z = \iiint_D (x^2 + y^2)2\,dV = \int_0^{2\pi}\int_0^2\int_0^6 2r^3\,dz\,dr\,d\theta = \int_0^{2\pi}\int_0^2 2r^3z\Big|_0^6\,dr\,d\theta$$
$$= \int_0^{2\pi}\int_0^2 12r^3\,dr\,d\theta = \int_0^{2\pi} 3r^4\Big|_0^2\,d\theta = \int_0^{2\pi} 48\,d\theta = 96\pi$$

SECTION 15.8

1. $$\frac{\partial(x, y)}{\partial(u, v)} = \begin{vmatrix} \frac{\partial x}{\partial u} & \frac{\partial x}{\partial v} \\ \frac{\partial y}{\partial u} & \frac{\partial y}{\partial v} \end{vmatrix} = \begin{vmatrix} 3 & -4 \\ \frac{1}{2} & \frac{1}{6} \end{vmatrix} = (3)\left(\frac{1}{6}\right) - (-4)\left(\frac{1}{2}\right) = \frac{5}{2}$$

3. $$\frac{\partial(x, y)}{\partial(u, v)} = \begin{vmatrix} \frac{\partial x}{\partial u} & \frac{\partial x}{\partial v} \\ \frac{\partial y}{\partial u} & \frac{\partial y}{\partial v} \end{vmatrix} = \begin{vmatrix} v & u \\ 2u & 2v \end{vmatrix} = (v)(2v) - (u)(2u) = 2v^2 - 2u^2$$

5. $$\frac{\partial(x, y)}{\partial(u, v)} = \begin{vmatrix} \frac{\partial x}{\partial u} & \frac{\partial x}{\partial v} \\ \frac{\partial y}{\partial u} & \frac{\partial y}{\partial v} \end{vmatrix} = \begin{vmatrix} 0 & e^v \\ e^v & ue^v \end{vmatrix} = (0)(ue^v) - (e^v)(e^v) = -e^{2v}$$

7. $$\frac{\partial(x, y, z)}{\partial(u, v, w)} = \begin{vmatrix} \frac{\partial x}{\partial u} & \frac{\partial x}{\partial v} & \frac{\partial x}{\partial w} \\ \frac{\partial y}{\partial u} & \frac{\partial y}{\partial v} & \frac{\partial y}{\partial w} \\ \frac{\partial z}{\partial u} & \frac{\partial z}{\partial v} & \frac{\partial z}{\partial w} \end{vmatrix} = \begin{vmatrix} a & 0 & 0 \\ 0 & b & 0 \\ 0 & 0 & 1 \end{vmatrix}$$
$$= a[(b)(1) - (0)(0)] + 0[(0)(0) - (0)(1)] + 0[(0)(0) - (b)(0)] = ab$$

9. First, $\dfrac{\partial(x,y)}{\partial(u,v)} = \begin{vmatrix} \dfrac{\partial x}{\partial u} & \dfrac{\partial x}{\partial v} \\ \dfrac{\partial y}{\partial u} & \dfrac{\partial y}{\partial v} \end{vmatrix} = \begin{vmatrix} 3 & 1 \\ 1 & 0 \end{vmatrix} = (3)(0) - (1)(1) = -1$

To find S we observe that for $y = 1$ we have $1 = y = u$; for $y = \frac{1}{4}x$ we have $u = \frac{1}{4}(3u + v)$, so that $\frac{1}{4}u = \frac{1}{4}v$, or $u = v$; for $x - 3y = e$ we have $(3u + v) - 3u = e$, or $v = e$. Consequently S is the region in the uv plane bounded by the lines $u = 1$, $u = v$, and $v = e$ (see the figure). By (9),

$$\iint_R \frac{y}{x-3y}\,dA = \iint_S \frac{u}{(3u+v)-3u}\left|\frac{\partial(x,y)}{\partial(u,v)}\right| dA = \int_1^e \int_1^v \frac{u}{v}|-1|\,du\,dv = \int_1^e \frac{u^2}{2}\frac{1}{v}\bigg|_1^v dv$$

$$= \int_1^e \left(\frac{v}{2} - \frac{1}{2v}\right) dv = \left(\frac{v^2}{4} - \frac{1}{2}\ln v\right)\bigg|_1^e = \left(\frac{e^2}{4} - \frac{1}{2}\right) - \left(\frac{1}{4} - 0\right) = \frac{1}{4}(e^2 - 3)$$

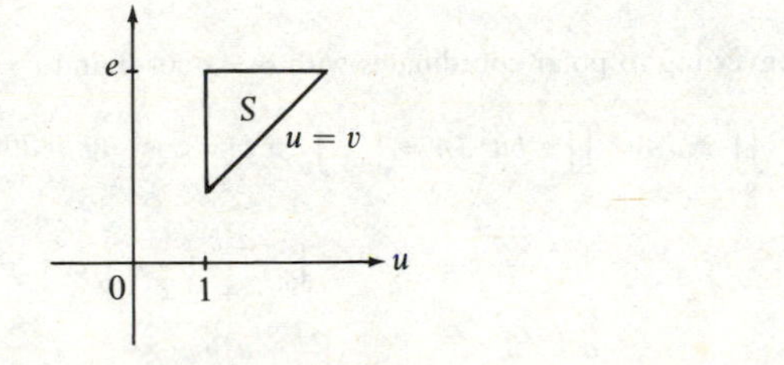

11. First, $\dfrac{\partial(x,y)}{\partial(u,v)} = \begin{vmatrix} \dfrac{\partial x}{\partial u} & \dfrac{\partial x}{\partial v} \\ \dfrac{\partial y}{\partial u} & \dfrac{\partial y}{\partial v} \end{vmatrix} = \begin{vmatrix} \frac{3}{5} & \frac{1}{5} \\ -\frac{2}{5} & \frac{1}{5} \end{vmatrix} = \left(\frac{3}{5}\right)\left(\frac{1}{5}\right) - \left(\frac{1}{5}\right)\left(-\frac{2}{5}\right) = \frac{1}{5}$

To find S we observe that for $x - y = 2$ we have $\frac{1}{5}(3u + v) - \frac{1}{5}(v - 2u) = 2$, so that $u = 2$; for $x - y = -1$ we have $\frac{1}{5}(3u + v) - \frac{1}{5}(v - 2u) = -1$, so that $u = -1$; for $2x + 3y = 1$ we have $\frac{2}{5}(3u + v) + \frac{3}{5}(v - 2u) = 1$, so that $v = 1$; for $2x + 3y = 0$ we have $\frac{2}{5}(3u + v) + \frac{3}{5}(v - 2u) = 0$, so that $v = 0$. Thus S is the square region bounded by the lines $u = -1$, $u = 2$, $v = 1$ and $v = 0$. By (9),

$$\iint_R xy^2\,dA = \iint_S [\tfrac{1}{5}(3u+v)][\tfrac{1}{5}(v-2u)]^2\left|\frac{\partial(x,y)}{\partial(u,v)}\right| dA$$

$$= \iint_S \tfrac{1}{125}(3u+v)(v^2 - 4uv + 4u^2)(\tfrac{1}{5})\,dA$$

$$= \tfrac{1}{625}\int_0^1\int_{-1}^2 (12u^3 - 8u^2v - uv^2 + v^3)\,du\,dv$$

$$= \tfrac{1}{625}\int_0^1 (3u^4 - \tfrac{8}{3}u^3v - \tfrac{1}{2}u^2v^2 + uv^3)\Big|_{-1}^2 dv$$

$$= \tfrac{1}{625}\int_0^1 (45 - 24v - \tfrac{3}{2}v^2 + 3v^3)\,dv = \tfrac{1}{625}(45v - 12v^2 - \tfrac{1}{2}v^3 + \tfrac{3}{4}v^4)\Big|_0^1 = \tfrac{133}{2500}$$

13. First,

$$\frac{\partial(x,y)}{\partial(u,v)} = \begin{vmatrix} \dfrac{\partial x}{\partial u} & \dfrac{\partial x}{\partial v} \\ \dfrac{\partial y}{\partial u} & \dfrac{\partial y}{\partial v} \end{vmatrix} = \begin{vmatrix} \sec v & u\sec v\tan v \\ \tan v & u\sec^2 v \end{vmatrix} = (\sec v)(u\sec^2 v) - (u\sec v\tan v)(\tan v)$$

$$= u\sec v(\sec^2 v - \tan^2 v) = u\sec v$$

To find S we observe that for $x^2 - y^2 = 1$ we have $1 = x^2 - y^2 = u^2\sec^2 v - u^2\tan^2 v = u^2$, and since $u > 0$ by hypothesis, $u = 1$; similarly, for $x^2 - y^2 = 4$ we have $u = 2$; for $x = 2y$ we have $u\sec v = 2u\tan v$, or $u/\cos v = (2u\sin v)/\cos v$; since $u > 0$ by hypothesis, we have $\sin v = \frac{1}{2}$, and since $0 < v < \pi/2$, $v = \pi/6$; similarly, for $x = \sqrt{2}y$ we have $\sin v = 1/\sqrt{2} = \frac{1}{2}\sqrt{2}$, and thus $v = \pi/4$. Consequently S is the rectangular region in the uv plane bounded by the lines $u = 1$, $u = 2$, $v = \pi/6$ and $v = \pi/4$. By (9),

$$\iint_R \frac{y}{x}e^{x^2-y^2}\,dA = \iint_S \frac{u\tan v}{u\sec v}e^{u^2\sec^2 v - u^2\tan^2 v}\left|\frac{\partial(x,y)}{\partial(u,v)}\right| dA = \iint_S \frac{\tan v}{\sec v}e^{u^2}u\sec v\,dA$$

$$= \iint_S ue^{u^2}\tan v\,dA = \int_{\pi/6}^{\pi/4}\int_1^2 ue^{u^2}\tan v\,du\,dv = \int_{\pi/6}^{\pi/4} \tfrac{1}{2}e^{u^2}\tan v\Big|_1^2 dv$$

$$= \int_{\pi/6}^{\pi/4} \tfrac{1}{2}(e^4 - e)\tan v\,dv = -\tfrac{1}{2}(e^4 - e)\ln\cos v\Big|_{\pi/6}^{\pi/4}$$

$$= -\tfrac{1}{2}(e^4 - e)\left(\ln\frac{\sqrt{2}}{2} - \ln\frac{\sqrt{3}}{2}\right) = \tfrac{1}{2}(e^4 - e)\ln\sqrt{\tfrac{3}{2}} = \tfrac{1}{4}(e^4 - e)\ln\tfrac{3}{2}$$

15. First,

$$\frac{\partial(x,y)}{\partial(u,v)} = \begin{vmatrix} \dfrac{\partial x}{\partial u} & \dfrac{\partial x}{\partial v} \\ \dfrac{\partial y}{\partial u} & \dfrac{\partial y}{\partial v} \end{vmatrix} = \begin{vmatrix} \cosh v & u\sinh v \\ \sinh v & u\cosh v \end{vmatrix} = (\cosh v)(u\cosh v) - (u\sinh v)(\sinh v)$$

$$= u(\cosh^2 v - \sinh^2 v) = u$$

To find S we observe that for $x^2 - y^2 = 1$ we have $1 = u^2\cosh^2 v - u^2\sinh^2 v = u^2$, and since $u > 0$ by hypothesis, $u = 1$; similarly, for $x^2 - y^2 = 4$ we have $u = 2$; for $y = 0$ we have $0 = y = u\sinh v$, and since $u > 0$ by hypothesis, $v = 0$; for $y = \frac{3}{5}x$ we have $u\sinh v = \frac{3}{5}u\cosh v$, and since $u > 0$ by hypothesis, $5\sinh v = 3\cosh v$, or $5(e^v - e^{-v}) = 3(e^v + e^{-v})$, or $2e^v = 8e^{-v}$, or $e^{2v} = 4$, so $v = \ln 2$. Consequently S is the rectangular region in the uv plane bounded by the lines $u = 1$, $u = 2$, $v = 0$ and $v = \ln 2$. By (9),

$$\iint_R e^{x^2-y^2}\,dA = \iint_S e^{u^2\cosh^2 v - u^2\sinh^2 v}\left|\frac{\partial(x,y)}{\partial(u,v)}\right| dA = \iint_S e^{u^2}(u)\,dA = \int_0^{\ln 2}\int_1^2 ue^{u^2}\,du\,dv$$

$$= \int_0^{\ln 2}\frac{1}{2}e^{u^2}\Big|_1^2 dv = \int_0^{\ln 2}\frac{1}{2}(e^4 - e)\,dv = \frac{v}{2}(e^4 - e)\bigg|_0^{\ln 2} = \frac{1}{2}(e^4 - e)\ln 2$$

17. First,

$$\frac{\partial(x, y, z)}{\partial(u, v, w)} = \begin{vmatrix} \frac{\partial x}{\partial u} & \frac{\partial x}{\partial v} & \frac{\partial x}{\partial w} \\ \frac{\partial y}{\partial u} & \frac{\partial y}{\partial v} & \frac{\partial y}{\partial w} \\ \frac{\partial z}{\partial u} & \frac{\partial z}{\partial v} & \frac{\partial z}{\partial w} \end{vmatrix} = \begin{vmatrix} -\frac{v}{u^2}\cos w & \frac{1}{u}\cos w & -\frac{v}{u}\sin w \\ -\frac{v}{u^2}\sin w & \frac{1}{u}\sin w & \frac{v}{u}\cos w \\ 0 & 2v & 0 \end{vmatrix}$$

$$= -\frac{v}{u^2}\cos w\left[\left(\frac{1}{u}\sin w\right)(0) - \left(\frac{v}{u}\cos w\right)(2v)\right]$$

$$+\frac{1}{u}\cos w\left[\left(\frac{v}{u}\cos w\right)(0) - \left(-\frac{v}{u^2}\sin w\right)(0)\right]$$

$$-\frac{v}{u}\sin w\left[\left(-\frac{v}{u^2}\sin w\right)(2v) - \left(\frac{1}{u}\sin w\right)(0)\right]$$

$$= \frac{2v^3}{u^3}\cos^2 w + \frac{2v^3}{u^3}\sin^2 w = \frac{2v^3}{u^3}$$

To find E we observe that for $z = x^2 + y^2$ we have $v^2 = ((v/u)\cos w)^2 + ((v/u)\sin w)^2 = v^2/u^2$, so since $u > 0$ and $1 \le z = v^2 \le 4$ by hypothesis, $u = 1$; similarly, for $z = 4(x^2 + y^2)$ we have $u = 2$; for $z = 1$ we have $v^2 = 1$, so $v = 1$; similarly, for $z = 4$ we have $v = 2$. Consequently E is the solid region in space bounded by the planes $u = 1$, $u = 2$, $v = 1$, and $v = 2$, and such that $0 \le w \le \pi/2$. By (15),

$$\iiint_D (x^2 + y^2)\, dV = \iiint_E \left[\left(\frac{v}{u}\cos w\right)^2 + \left(\frac{v}{u}\sin w\right)^2\right]\left|\frac{\partial(x, y, z)}{\partial(u, v, w)}\right| dV = \iiint_E \frac{v^2}{u^2}\frac{2v^3}{u^3}\, dV$$

$$= \int_0^{\pi/2}\int_1^2\int_1^2 2\frac{v^5}{u^5}\, du\, dv\, dw = \int_0^{\pi/2}\int_1^2 2v^5\left(-\frac{1}{4u^4}\right)\bigg|_1^2\, dv\, dw$$

$$= \int_0^{\pi/2}\int_1^2 \tfrac{15}{32}v^5\, dv\, dw = \int_0^{\pi/2} \tfrac{5}{64}v^6\Big|_1^2\, dw = \int_0^{\pi/2} \tfrac{315}{64}\, dw = \tfrac{315}{128}\pi$$

19. Let T_R be defined by $u = x - 2y$ and $v = x + 2y$. To find S we observe that for $x - 2y = 1$ we have $u = 1$; for $x - 2y = 2$ we have $u = 2$; for $x + 2y = 1$ we have $v = 1$; for $x + 2y = 3$ we have $v = 3$. Consequently S is the rectangular region in the uv plane bounded by the lines $u = 1$, $u = 2$, $v = 1$ and $v = 3$. Next,

$$\frac{\partial(u, v)}{\partial(x, y)} = \begin{vmatrix} \frac{\partial u}{\partial x} & \frac{\partial u}{\partial y} \\ \frac{\partial v}{\partial x} & \frac{\partial v}{\partial y} \end{vmatrix} = \begin{vmatrix} 1 & -2 \\ 1 & 2 \end{vmatrix} = (1)(2) - (-2)(1) = 4$$

By (9) and (14),

$$\iint_R \left(\frac{x - 2y}{x + 2y}\right)^3 dA = \iint_S \left(\frac{u}{v}\right)^3 \left|\frac{\partial(x, y)}{\partial(u, v)}\right| dA = \int_1^3\int_1^2 \frac{u^3}{v^3}\frac{1}{4}\, du\, dv = \int_1^3 \frac{1}{v^3}\frac{u^4}{16}\bigg|_1^2\, dv$$

$$= \int_1^3 \frac{15}{16}\frac{1}{v^3}\, dv = -\frac{15}{32}\frac{1}{v^2}\bigg|_1^3 = -\frac{15}{32}\left(\frac{1}{9} - 1\right) = \frac{5}{12}$$

21. Let T_R be defined by $u = x/a$ and $v = y/b$. To find S we observe that for $x^2/a^2 + y^2/b^2 = 1$ we have $u^2 + v^2 = 1$, so that S is the unit disk $u^2 + v^2 \le 1$ in the uv plane. Next,

$$\frac{\partial(u, v)}{\partial(x, y)} = \begin{vmatrix} \frac{\partial u}{\partial x} & \frac{\partial u}{\partial y} \\ \frac{\partial v}{\partial x} & \frac{\partial v}{\partial y} \end{vmatrix} = \begin{vmatrix} \frac{1}{a} & 0 \\ 0 & \frac{1}{b} \end{vmatrix} = \left(\frac{1}{a}\right)\left(\frac{1}{b}\right) - (0)(0) = \frac{1}{ab}$$

Solving for x in terms of a, we have $x = au$. Then by (9) and (14),

$$\iint_R x^2\, dA - \iint_S (au)^2 \left|\frac{\partial(x, y)}{\partial(u, v)}\right| dA = \iint_S a^2u^2(ab)\, dA = \iint_S a^3bu^2\, dA$$

Converting to polar coordinates with $u = r\cos\theta$ and $v = r\sin\theta$, we find that

$$\iint_R x^2\, dA = \iint_S a^3bu^2\, dA = \int_0^{2\pi}\int_0^1 a^3b(r^2\cos^2\theta)r\, dr\, d\theta = \int_0^{2\pi} a^3b(\cos^2\theta)\frac{r^4}{4}\bigg|_0^1\, d\theta$$

$$= \int_0^{2\pi} \frac{a^3b}{4}\cos^2\theta\, d\theta = \int_0^{2\pi} \frac{a^3b}{4}\left(\frac{1}{2} + \frac{1}{2}\cos 2\theta\right) d\theta$$

$$= \frac{a^3b}{4}\left(\frac{\theta}{2} + \frac{1}{4}\sin 2\theta\right)\bigg|_0^{2\pi} = \frac{a^3b\pi}{4}$$

23. Let T_R be defined by $u = y - x$ and $v = y + x$. To find S we observe that for $x + y = 1$ we have $v = 1$; for $x + y = 2$ we have $v = 2$; for $x = 0$ we have $u = y$ and $v = y$, so that $u = v$; for $y = 0$ we have $u = -x$ and $v = x$, so that $u = -v$. Consequently S is the trapezoidal region in the uv plane bounded by the lines $v = 1$, $v = 2$, $u = v$ and $u = -v$ (see the figure). Next,

$$\frac{\partial(u, v)}{\partial(x, y)} = \begin{vmatrix} \frac{\partial u}{\partial x} & \frac{\partial u}{\partial y} \\ \frac{\partial v}{\partial x} & \frac{\partial v}{\partial y} \end{vmatrix} = \begin{vmatrix} -1 & 1 \\ 1 & 1 \end{vmatrix} = (-1)(1) - (1)(1) = -2$$

By (9) and (14),

$$\iint_R \sin\left[\pi\left(\frac{y - x}{y + x}\right)\right] dA = \iint_S \sin\left(\pi\frac{u}{v}\right)\left|\frac{\partial(x, y)}{\partial(u, v)}\right| dA = \int_1^2\int_{-v}^{v} \frac{1}{2}\sin\left(\pi\frac{u}{v}\right) du\, dv$$

$$= \int_1^2 -\frac{v}{2\pi}\cos\left(\pi\frac{u}{v}\right)\bigg|_{-v}^{v}\, dv = \int_1^2 -\frac{v}{2\pi}(\cos\pi - \cos(-\pi))\, dv$$

$$= \int_1^2 0\, dv = 0$$

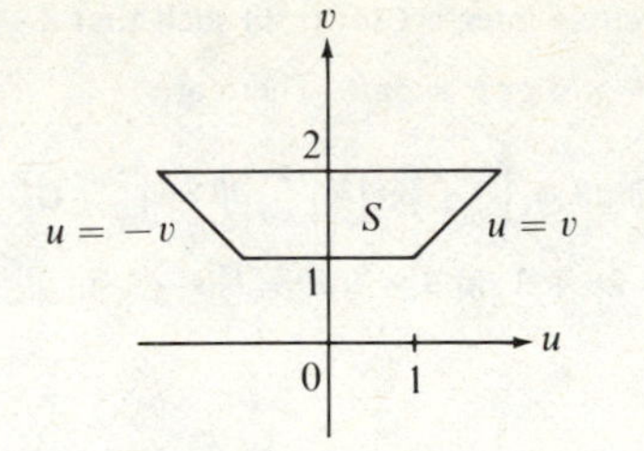

25. Let T_R be defined by $u = 2x - y$ and $v = x + y$. To find S, we observe that for $y = 2x$ we have $2x - y = 0$, so that $u = 0$; for $x + y = 1$ we have $v = 1$; for $x + y = 2$ we have $v = 2$. For $x = 2y$ we solve for x and y in terms of u and v to obtain $u + v = (2x - y) + (x + y) = 3x$, so that $x = \frac{1}{3}(u + v)$, and $2v - u = 2(x + y) - (2x - y) = 3y$, so that $y = \frac{1}{3}(2v - u)$. Thus for $x = 2y$ we have $\frac{1}{3}(u + v) = \frac{2}{3}(2v - u)$, so that $u = v$. Consequently S is the trapezoidal region in the uv plane bounded by the lines $u = 0$, $v = 1$, $v = 2$ and $u = v$ (see the figure). Next,

$$\frac{\partial(x, y)}{\partial(u, v)} = \begin{vmatrix} \dfrac{\partial x}{\partial u} & \dfrac{\partial x}{\partial v} \\ \dfrac{\partial y}{\partial u} & \dfrac{\partial y}{\partial v} \end{vmatrix} = \begin{vmatrix} \frac{1}{3} & \frac{1}{3} \\ -\frac{1}{3} & \frac{2}{3} \end{vmatrix} = \left(\frac{1}{3}\right)\left(\frac{2}{3}\right) - \left(\frac{1}{3}\right)\left(-\frac{1}{3}\right) = \frac{1}{3}$$

By (9),

$$\iint_R e^{(2x-y)/(x+y)}\,dA = \iint_S e^{u/v}\left|\frac{\partial(x, y)}{\partial(u, v)}\right| dA = \int_1^2 \int_0^v e^{u/v}\,\frac{1}{3}\,du\,dv = \int_1^2 \frac{v}{3}\,e^{u/v}\Big|_0^v$$
$$= \int_1^2 \frac{e-1}{3}\,v\,dv = \frac{1}{6}(e-1)v^2\Big|_1^2 = \frac{1}{2}(e-1)$$

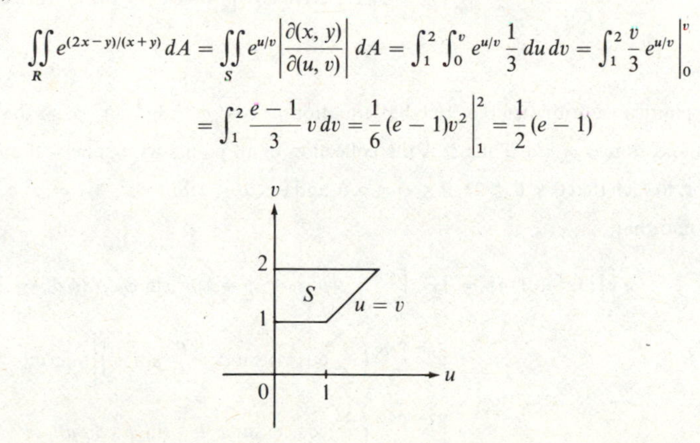

27. First, $A = \iint_R 1\,dA$. Next,

$$\frac{\partial(x, y)}{\partial(u, v)} = \begin{vmatrix} \dfrac{\partial x}{\partial u} & \dfrac{\partial x}{\partial v} \\ \dfrac{\partial y}{\partial u} & \dfrac{\partial y}{\partial v} \end{vmatrix} = \begin{vmatrix} \cosh v & u \sinh v \\ \sinh v & u \cosh v \end{vmatrix} = (\cosh v)(u \cosh v) - (u \sinh v)(\sinh v) = u$$

Since $a > 0$, we have $x^2 \ge x^2 - y^2 \ge a^2 > 0$, so that $u > 0$. To find S we observe that for $x^2 - y^2 = a^2$ we have $a^2 = u^2 \cosh^2 v - u^2 \sinh^2 v = u^2$, so since $a \ge 0$ and $u \ge 0$ we have $u = a$; similarly, if $x^2 - y^2 = b^2$ we have $u = b$; for $y = 0$ we have $0 = u \sinh v$, so since $u \ge 0$ we have $v = 0$; for $y = \frac{1}{2}x$ we have $u \sinh v = \frac{1}{2}u \cosh v$, so since $u > 0$ we have $\sinh v = \frac{1}{2}\cosh v$, so that $e^v - e^{-v} = \frac{1}{2}(e^v + e^{-v})$, or $\frac{1}{2}e^v = \frac{3}{2}e^{-v}$, or $e^{2v} = 3$, or $v = \frac{1}{2}\ln 3$. Consequently S is the rectangular region in the uv plane bounded by the lines $u = a$, $u = b$, $v = 0$ and $v = \frac{1}{2}\ln 3$. By (9) and (14),

$$A = \iint_R 1\,dA = \iint_S \left|\frac{\partial(x, y)}{\partial(u, v)}\right| dA = \int_a^b \int_0^{(1/2)\ln 3} u\,dv\,du = \int_a^b \frac{1}{2}(\ln 3)u\,du$$
$$= \frac{1}{2}(\ln 3)\frac{u^2}{2}\Big|_a^b = \frac{1}{4}(b^2 - a^2)\ln 3$$

CHAPTER 15 REVIEW

1. $\int_0^1 \int_x^{3x} y e^{(x^3)}\,dy\,dx = \int_0^1 \frac{y^2}{2} e^{(x^3)}\Big|_x^{3x} dx = \int_0^1 4x^2 e^{(x^3)}\,dx = \frac{4}{3}e^{(x^3)}\Big|_0^1 = \frac{4}{3}(e-1)$

3. $\int_{-1}^1 \int_0^2 \int_{2x}^{5x} e^{xy}\,dz\,dy\,dx = \int_{-1}^1 \int_0^2 z e^{xy}\Big|_{2x}^{5x} dy\,dx = \int_{-1}^1 \int_0^2 3x e^{xy}\,dy\,dx = \int_{-1}^1 3e^{xy}\Big|_0^2 dx$
$= \int_{-1}^1 3(e^{2x} - 1)\,dx = (\frac{3}{2}e^{2x} - 3x)\Big|_{-1}^1 = \frac{3}{2}(e^2 - e^{-2}) - 6$

5. The region R of integration is the vertically simple region between the graphs of $y = \sqrt{x}$ and $y = 1$ on $[0, 1]$. It is also the horizontally simple region between the graphs of $x = 0$ and $x = y^2$ on $[0, 1]$. Thus

$$\int_0^1 \int_{\sqrt{x}}^1 e^{(y^3)}\,dy\,dx = \int_0^1 \int_0^{y^2} e^{(y^3)}\,dx\,dy = \int_0^1 x e^{(y^3)}\Big|_0^{y^2} dy = \int_0^1 y^2 e^{(y^3)}\,dy = \tfrac{1}{3}e^{(y^3)}\Big|_0^1$$
$$= \tfrac{1}{3}(e-1)$$

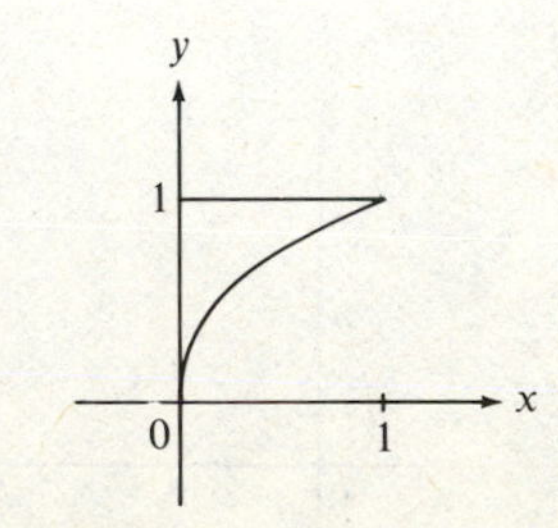

7. The region R of integration is the vertically simple region between the graphs of $y = x$ and $y = \sqrt{3}$ on $[1, \sqrt{3}]$. It is also the horizontally simple region between the graphs of $x = 1$ and $x = y$ on $[1, \sqrt{3}]$. Thus

$$\int_1^{\sqrt{3}} \int_x^{\sqrt{3}} \frac{x}{(x^2+y^2)^{3/2}}\,dy\,dx = \int_1^{\sqrt{3}} \int_1^y \frac{x}{(x^2+y^2)^{3/2}}\,dx\,dy = \int_1^{\sqrt{3}} \frac{-1}{(x^2+y^2)^{1/2}}\bigg|_1^y\,dy$$

$$= \int_1^{\sqrt{3}} \left[\frac{1}{(1+y^2)^{1/2}} - \frac{1}{\sqrt{2}y}\right] dy$$

$$= \int_1^{\sqrt{3}} \frac{1}{(1+y^2)^{1/2}}\,dy - \frac{\sqrt{2}}{2}\int_1^{\sqrt{3}} \frac{1}{y}\,dy$$

$$\overset{y=\tan u}{=} \int_{\pi/4}^{\pi/3} \sec u\,du - \frac{\sqrt{2}}{2}\ln y\bigg|_1^{\sqrt{3}}$$

$$= \ln|\sec u + \tan u|\bigg|_{\pi/4}^{\pi/3} - \frac{\sqrt{2}}{2}\ln\sqrt{3} = \ln\frac{2+\sqrt{3}}{1+\sqrt{2}} - \frac{\sqrt{2}}{4}\ln 3$$

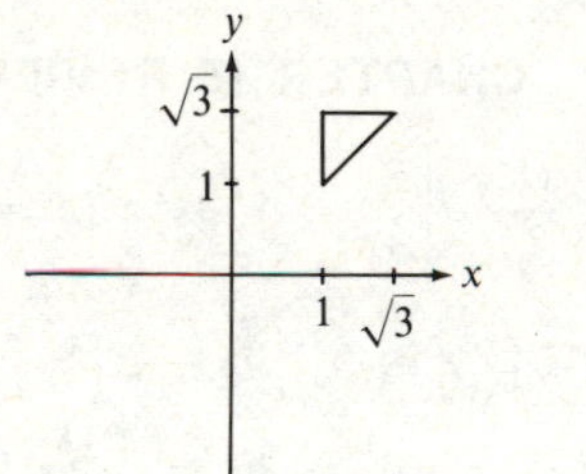

9. The graphs of $\sqrt{x} + \sqrt{y} = \sqrt{a}$ and $x + y = a$ intersect for (x, y) such that $a - x = y = (\sqrt{a} - \sqrt{x})^2 = a + x - 2\sqrt{a}\sqrt{x}$, so that $x = \sqrt{a}\sqrt{x}$, and thus $x = 0$ or $x = a$. Since $a - x \geq (\sqrt{a} - \sqrt{x})^2$ for $0 \leq x \leq a$, we have

$$A = \int_0^a \int_{(\sqrt{a}-\sqrt{x})^2}^{a-x} 1\,dy\,dx = \int_0^a y\bigg|_{(\sqrt{a}-\sqrt{x})^2}^{a-x} dx = \int_0^a [(a-x) - (a + x - 2\sqrt{a}\sqrt{x})]\,dx$$

$$= \int_0^a (-2x + 2\sqrt{a}\sqrt{x})\,dx = \left(-x^2 + \tfrac{4}{3}\sqrt{a}x^{3/2}\right)\bigg|_0^a = \tfrac{1}{3}a^2$$

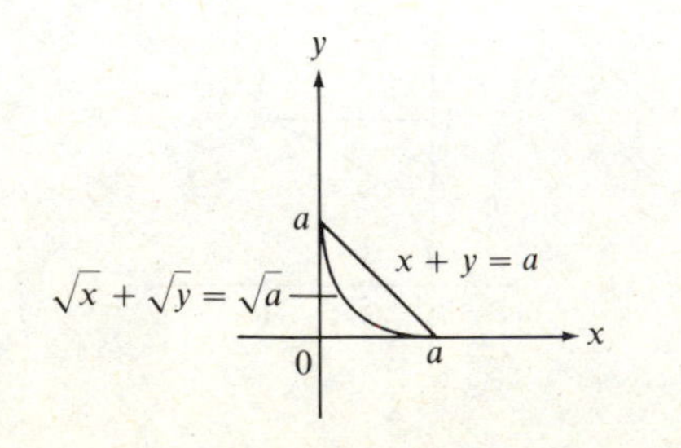

11. The limaçon and the circle intersect for (r, θ) such that $3 - \sin\theta = r = 5\sin\theta$, so that $\sin\theta = \frac{1}{2}$, and thus $\theta = \pi/6$ or $\theta = 5\pi/6$. Therefore

$$A = \int_{\pi/6}^{5\pi/6} \int_{3-\sin\theta}^{5\sin\theta} r\,dr\,d\theta = \int_{\pi/6}^{5\pi/6} \tfrac{1}{2}r^2\bigg|_{3-\sin\theta}^{5\sin\theta} d\theta = \int_{\pi/6}^{5\pi/6} (12\sin^2\theta + 3\sin\theta - \tfrac{9}{2})\,d\theta$$

$$= \int_{\pi/6}^{5\pi/6} (6 - 6\cos 2\theta + 3\sin\theta - \tfrac{9}{2})\,d\theta = (\tfrac{3}{2}\theta - 3\sin 2\theta - 3\cos\theta)\bigg|_{\pi/6}^{5\pi/6} = \pi + 6\sqrt{3}$$

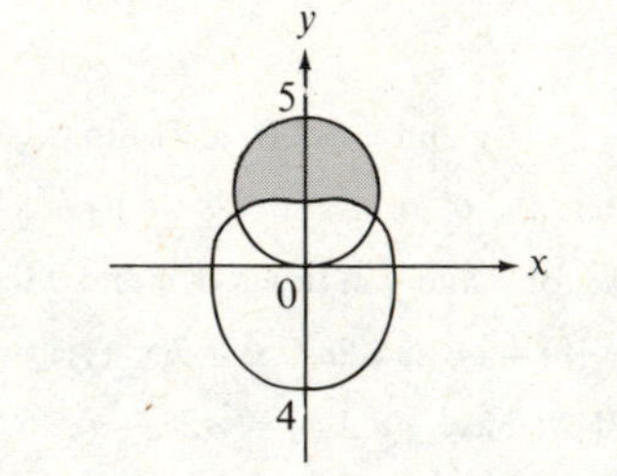

13. The lines $y = 5 + x$ and $y = -x + 7$ intersect for (x, y) such that $5 + x = y = -x + 7$, so that $x = 1$. Thus R is the vertically simple region between the graphs of $y = 5 + x$ and $y = -x + 7$ on $[0, 1]$. Therefore

$$\iint_R (3x - 5)\,dA = \int_0^1 \int_{5+x}^{-x+7} (3x-5)\,dy\,dx = \int_0^1 (3x-5)y\bigg|_{5+x}^{-x+7} dx$$

$$= \int_0^1 (3x-5)(2-2x)\,dx = \int_0^1 (-6x^2 + 16x - 10)\,dx$$

$$= (-2x^3 + 8x^2 - 10x)\bigg|_0^1 = -4$$

15. In spherical coordinates the cone has equation $\rho^2\cos^2\phi = 3\rho^2\sin^2\phi$, so that $\cot\phi = \sqrt{3}$, and thus $\phi = \pi/6$. Thus D is the collection of all points with spherical coordinates (ρ, ϕ, θ) such that $0 \leq \theta \leq 2\pi$, $0 \leq \phi \leq \pi/6$, and $0 \leq \rho \leq 2$. Since $z^2 + 1 = \rho^2\cos^2\phi + 1$, we find that

$$\iiint_D (z^2+1)\,dV = \int_0^{2\pi} \int_0^{\pi/6} \int_0^2 (\rho^2\cos^2\phi + 1)\rho^2\sin\phi\,d\rho\,d\phi\,d\theta$$

$$= \int_0^{2\pi} \int_0^{\pi/6} \left(\frac{\rho^5}{5}\cos^2\phi\sin\phi + \frac{\rho^3}{3}\sin\phi\right)\bigg|_0^2 d\phi\,d\theta$$

$$= \int_0^{2\pi} \int_0^{\pi/6} \left(\frac{32}{5}\cos^2\phi\sin\phi + \frac{8}{3}\sin\phi\right) d\phi\,d\theta$$

$$= \int_0^{2\pi} \left(\frac{-32}{15}\cos^3\phi - \frac{8}{3}\cos\phi\right)\bigg|_0^{\pi/6} d\theta$$

$$= \int_0^{2\pi} \left[\frac{32}{15}\left(1 - \frac{3\sqrt{3}}{8}\right) + \frac{8}{3}\left(1 - \frac{\sqrt{3}}{2}\right)\right] d\theta$$

$$= \left(\frac{72 - 32\sqrt{3}}{15}\right)2\pi = \frac{16\pi}{15}(9 - 4\sqrt{3})$$

17. The region R in the xy plane is the vertically simple circular region in the xy plane between the graphs of $y = -\sqrt{9-x^2}$ and $y = \sqrt{9-x^2}$ on $[-3, 3]$; D is the solid region between the graphs of $z = -\sqrt{9-x^2-y^2}$ and $z = 0$ on R. Thus

$$\iiint_D xyz\,dV = \int_{-3}^{3}\int_{-\sqrt{9-x^2}}^{\sqrt{9-x^2}}\int_{-\sqrt{9-x^2-y^2}}^{0} xyz\,dz\,dy\,dx$$
$$= \int_{-3}^{3}\int_{-\sqrt{9-x^2}}^{\sqrt{9-x^2}} \tfrac{1}{2}xyz^2\Big|_{-\sqrt{9-x^2-y^2}}^{0} dy\,dx$$
$$= \int_{-3}^{3}\int_{-\sqrt{9-x^2}}^{\sqrt{9-x^2}} -\tfrac{1}{2}xy(9-x^2-y^2)\,dy\,dx$$
$$= \int_{-3}^{3} \tfrac{1}{8}x(9-x^2-y^2)^2\Big|_{-\sqrt{9-x^2}}^{\sqrt{9-x^2}} dx = \int_{-3}^{3} 0\,dx = 0$$

19. Let $f(x, y) = xy$. Then $f_x(x, y) = y$ and $f_x(x, y) = x$. The portion of the surface $z = xy$ that is inside the cylinder $x^2 + y^2 = 1$ lies over the disk R in the xy plane bounded by the circle $r = 1$. Thus

$$S = \iint_R \sqrt{y^2+x^2+1}\,dA = \int_0^{2\pi}\int_0^1 \sqrt{r^2+1}\,r\,dr\,d\theta = \int_0^{2\pi} \tfrac{1}{3}(r^2+1)^{3/2}\Big|_0^1 d\theta$$
$$= \int_0^{2\pi} \tfrac{1}{3}(2\sqrt{2}-1)\,d\theta = \frac{2\pi}{3}(2\sqrt{2}-1)$$

21. The paraboloid and cone intersect for (x, y, z) such that $z = x^2 + y^2 = z^2$, so that $z = 0$ or $z = 1$, and thus $x^2 + y^2 = 0$ or $x^2 + y^2 = 1$. In polar coordinates R is the region in the xy plane between the graphs of $r = 0$ and $r = 1$ on $[0, 2\pi]$. Then D is the solid region between the graphs of $z = x^2 + y^2 = r^2$ and $z = \sqrt{x^2+y^2} = r$ on R. Therefore

$$V = \iiint_D 1\,dV = \int_0^{2\pi}\int_0^1\int_{r^2}^{r} r\,dz\,dr\,d\theta = \int_0^{2\pi}\int_0^1 rz\Big|_{r^2}^{r} dr\,d\theta = \int_0^{2\pi}\int_0^1 (r^2-r^3)\,dr\,d\theta$$
$$= \int_0^{2\pi}\left(\frac{r^3}{3}-\frac{r^4}{4}\right)\Big|_0^1 d\theta = \int_0^{2\pi}\frac{1}{12}\,d\theta = \frac{1}{6}\pi$$

23. In polar coordinates R is the region in the xy plane between the graphs of $r = 0$ and $r = 4\sin\theta$ on $[0, \pi]$; D is the solid region between the graphs of $z = 0$ and $z = r$ on R. Thus

$$V = \iiint_D 1\,dV = \int_0^{\pi}\int_0^{4\sin\theta}\int_0^{r} r\,dz\,dr\,d\theta = \int_0^{\pi}\int_0^{4\sin\theta} rz\Big|_0^{r} dr\,d\theta = \int_0^{\pi}\int_0^{4\sin\theta} r^2\,dr\,d\theta$$
$$= \int_0^{\pi}\frac{r^3}{3}\Big|_0^{4\sin\theta} d\theta = \int_0^{\pi}\tfrac{64}{3}\sin^3\theta\,d\theta = \tfrac{64}{3}\int_0^{\pi}(1-\cos^2\theta)\sin\theta\,d\theta$$
$$= \tfrac{64}{3}\left(-\cos\theta + \tfrac{1}{3}\cos^3\theta\right)\Big|_0^{\pi} = \tfrac{256}{9}$$

25. The paraboloid and sheet intersect for (x, y, z) such that $4x^2 + y^2 = z = 16 - 3y^2$, so that $x^2 + y^2 = 4$. In polar coordinates R is the circular region in the xy plane between the graphs of $r = 0$ and $r = 2$ on $[0, 2\pi]$. Then D is the solid region between the graphs of $z = 4x^2 + y^2$ and $z = 16 - 3y^2$ on R. Thus

$$V = \iiint_D 1\,dV = \int_0^{2\pi}\int_0^2\int_{r^2+3r^2\cos^2\theta}^{16-3r^2\sin^2\theta} r\,dz\,dr\,d\theta = \int_0^{2\pi}\int_0^2 rz\Big|_{r^2+3r^2\cos^2\theta}^{16-3r^2\sin^2\theta} dr\,d\theta$$
$$= \int_0^{2\pi}\int_0^2 r(16-4r^2)\,dr\,d\theta = \int_0^{2\pi}(8r^2-r^4)\Big|_0^2 d\theta = \int_0^{2\pi} 16\,d\theta = 32\pi$$

27. The region R in the xy plane is the horizontally simple region between the lines $x = 0$ and $x = 6 - 3y$ on $[0, 2]$; D is the solid region between the graphs of $z = 0$ and $z = x + 2y$ on R. Thus

$$V = \iiint_D 1\,dV = \int_0^2\int_0^{6-3y}\int_0^{x+2y} 1\,dz\,dx\,dy = \int_0^2\int_0^{6-3y} z\Big|_0^{x+2y} dx\,dy$$
$$= \int_0^2\int_0^{6-3y}(x+2y)\,dx\,dy = \int_0^2\left(\frac{x^2}{2}+2yx\right)\Big|_0^{6-3y} dy$$
$$= \int_0^2\left[\tfrac{1}{2}(6-3y)^2 + 12y - 6y^2\right]dy = \left[\tfrac{-1}{18}(6-3y)^3 + 6y^2 - 2y^3\right]\Big|_0^2 = 20$$

29. The sphere and paraboloid intersect for (x, y, z) such that $3z + 21 = 49 - z^2$, so that $z^2 + 3z - 28 = 0$, and thus $z = -7$ or $z = 4$. If $z = -7$, then $x^2 + y^2 = 0$, so that $r = 0$, and if $z = 4$, then $x^2 + y^2 = 33$, so that $r = \sqrt{33}$. In polar coordinates R is the circular region in the xy plane between the graphs of $r = 0$ and $r = \sqrt{33}$ on $[0, 2\pi]$. Then D is the solid region between the graphs of $z = \frac{1}{3}r^2 - 7$ and $z = \sqrt{49-r^2}$ on R. Thus

$$V = \iiint_D 1\,dV = \int_0^{2\pi}\int_0^{\sqrt{33}}\int_{(r^2/3)-7}^{\sqrt{49-r^2}} r\,dz\,dr\,d\theta = \int_0^{2\pi}\int_0^{\sqrt{33}} rz\Big|_{(r^2/3)-7}^{\sqrt{49-r^2}} dr\,d\theta$$
$$= \int_0^{2\pi}\int_0^{\sqrt{33}}\left(r\sqrt{49-r^2} - \tfrac{1}{3}r^3 + 7r\right)dr\,d\theta$$
$$= \int_0^{2\pi}\left[-\tfrac{1}{3}(49-r^2)^{3/2} - \tfrac{1}{12}r^4 + \tfrac{7}{2}r^2\right]\Big|_0^{\sqrt{33}} d\theta = \int_0^{2\pi}\tfrac{471}{4}\,d\theta = \tfrac{471}{2}\pi$$

31. Since the region D occupied by the solid is symmetric with respect to the yz plane and the xz plane, and since $\delta(x, y, z) = z$, so that $\delta(x, y, z)$ is independent of x and y, we have $\bar{x} = \bar{y} = 0$. The cone $z = \sqrt{x^2+y^2}$ and the plane $z = 3$ intersect for (x, y, z) such that $\sqrt{x^2+y^2} = z = 3$, and thus $r = 3$. Consequently D is the solid region between the graphs of $z = r$ and $z = 3$ on the disk $0 \le r \le 3$. Therefore

$$\mathscr{M}_{xy} = \iiint_D z\delta(x, y, z)\,dV = \int_0^{2\pi}\int_0^3\int_r^3 z^2 r\,dz\,dr\,d\theta = \int_0^{2\pi}\int_0^3 \tfrac{1}{3}z^3 r\Big|_r^3 dr\,d\theta$$
$$= \int_0^{2\pi}\int_0^3\left(9r - \tfrac{1}{3}r^4\right)dr\,d\theta = \int_0^{2\pi}\left(\tfrac{9}{2}r^2 - \tfrac{1}{15}r^5\right)\Big|_0^3 d\theta = \int_0^{2\pi}\tfrac{243}{10}\,d\theta = \tfrac{243}{5}\pi$$

Since
$$m = \iiint_D \delta(x, y, z)\,dV = \int_0^{2\pi}\int_0^3\int_r^3 zr\,dz\,dr\,d\theta = \int_0^{2\pi}\int_0^3 \tfrac{1}{2}z^2 r\Big|_r^3 dr\,d\theta$$
$$= \int_0^{2\pi}\int_0^3\left(\tfrac{9}{2}r - \tfrac{1}{2}r^3\right)dr\,d\theta = \int_0^{2\pi}\left(\tfrac{9}{4}r^2 - \tfrac{1}{8}r^4\right)\Big|_0^3 d\theta = \int_0^{2\pi}\tfrac{81}{8}\,d\theta = \tfrac{81}{4}\pi$$

we have $\bar{z} = \mathscr{M}_{xy}/m = (243\pi/5)/(81\pi/4) = \frac{12}{5}$. Thus $(\bar{x}, \bar{y}, \bar{z}) = (0, 0, \frac{12}{5})$.

33. By symmetry, $\bar{x} = \bar{y} = 0$. The paraboloid $z = x^2 + y^2$ and the upper nappe of the cone $z^2 = x^2 + y^2$ intersect for (r, θ, z) such that $z = r^2 = z^2$, so that $z = 0$ or $z = 1$, and thus $r = 0$ or $r = 1$. Therefore the given region D is the solid region between the graphs of $z = r^2$ and $z = r$ on the disk $0 \le r \le 1$. Taking $\delta = 1$, we have

$$\mathcal{M}_{xy} = \iiint_D z\,dV = \int_0^{2\pi}\int_0^1\int_{r^2}^r zr\,dz\,dr\,d\theta = \int_0^{2\pi}\int_0^1 \tfrac{1}{2}z^2r\Big|_{r^2}^r\,dr\,d\theta$$
$$= \int_0^{2\pi}\int_0^1 \tfrac{1}{2}(r^3 - r^5)\,dr\,d\theta = \int_0^{2\pi} \tfrac{1}{2}\left(\tfrac{1}{4}r^4 - \tfrac{1}{6}r^6\right)\Big|_0^1\,d\theta = \int_0^{2\pi}\tfrac{1}{24}\,d\theta = \tfrac{1}{12}\pi$$

Since $\delta = 1$, we have

$$m = \iiint_D 1\,dV = \int_0^{2\pi}\int_0^1\int_{r^2}^r r\,dz\,dr\,d\theta = \int_0^{2\pi}\int_0^1 rz\Big|_{r^2}^r\,dr\,d\theta = \int_0^{2\pi}\int_0^1 (r^2 - r^3)\,dr\,d\theta$$
$$= \int_0^{2\pi}\left(\tfrac{1}{3}r^3 - \tfrac{1}{4}r^4\right)\Big|_0^1\,d\theta = \int_0^{2\pi}\tfrac{1}{12}\,d\theta = \tfrac{1}{6}\pi$$

and thus $\bar{z} = \mathcal{M}_{xy}/m = (\pi/12)/(\pi/6) = \frac{1}{2}$. Therefore $(\bar{x}, \bar{y}, \bar{z}) = (0, 0, \frac{1}{2})$.

35. First,

$$\frac{\partial(x,y)}{\partial(u,v)} = \begin{vmatrix} \dfrac{\partial x}{\partial u} & \dfrac{\partial x}{\partial v} \\[2mm] \dfrac{\partial y}{\partial u} & \dfrac{\partial y}{\partial v}\end{vmatrix} = \begin{vmatrix} 1 & 1 \\[1mm] \dfrac{-v}{(u+v)^2} & \dfrac{u}{(u+v)^2}\end{vmatrix} = (1)\left(\frac{u}{(u+v)^2}\right) - (1)\left(\frac{-v}{(u+v)^2}\right) = \frac{1}{u+v}$$

To find S we observe that for $xy = 1$ we have $(u + v)(v/(u + v)) = 1$, or $v = 1$; similarly, for $xy = 2$ we have $v = 2$. For $x(1 - y) = 1$ we have $(u + v)(1 - v/(u + v)) = 1$, or $u + v - v = 1$, or $u = 1$; similarly, for $x(1 - y) = 2$ we have $u = 2$. Thus S is the square region in the uv plane bounded by the lines $u = 1$, $u = 2$, $v = 1$, and $v = 2$. By (9) in Section 15.8,

$$\iint_R x\,dA = \iint_S (u+v)\left|\frac{\partial(x,y)}{\partial(u,v)}\right| dA = \int_1^2\int_1^2 (u+v)\frac{1}{u+v}\,du\,dv = \int_1^2\int_1^2 1\,du\,dv = 1$$

37. Let T_R be defined by $u = x - y$ and $v = x + y$. To find S we observe that for $x - y = 1$ we have $u = 1$; for $x - y = 3$ we have $u = 3$; for $x + y = 2$ we have $v = 2$; for $x + y = 4$ we have $v = 4$. Consequently S is the square region in the uv plane bounded by the lines $u = 1$, $u = 3$, $v = 2$, and $v = 4$. Solving for x and y in terms of u and v, we find that $u + v = (x - y) + (x + y) = 2x$, so that $x = \frac{1}{2}(u + v)$, and $v - u = (x + y) - (x - y) = 2y$, so that $y = \frac{1}{2}(v - u)$. Then T is defined by $x = \frac{1}{2}(u + v)$ and $y = \frac{1}{2}(v - u)$. Thus

$$\frac{\partial(x,y)}{\partial(u,v)} = \begin{vmatrix} \dfrac{\partial x}{\partial u} & \dfrac{\partial x}{\partial v} \\[2mm] \dfrac{\partial y}{\partial u} & \dfrac{\partial y}{\partial v}\end{vmatrix} = \begin{vmatrix} \frac{1}{2} & \frac{1}{2} \\ -\frac{1}{2} & \frac{1}{2}\end{vmatrix} = \left(\tfrac{1}{2}\right)\left(\tfrac{1}{2}\right) - \left(\tfrac{1}{2}\right)\left(-\tfrac{1}{2}\right) = \tfrac{1}{2}$$

By (9) in Section 15.8,

$$\iint_R (2x - y^2)\,dA = \iint_S \left[(u+v) - \tfrac{1}{4}(v-u)^2\right]\left|\frac{\partial(x,y)}{\partial(u,v)}\right| dA$$
$$= \int_2^4\int_1^3 \left(u + v - \tfrac{1}{4}v^2 + \tfrac{1}{2}uv - \tfrac{1}{4}u^2\right)\tfrac{1}{2}\,du\,dv$$
$$= \int_2^4 \left(\tfrac{1}{4}u^2 + \tfrac{1}{2}uv - \tfrac{1}{8}uv^2 + \tfrac{1}{8}u^2v - \tfrac{1}{24}u^3\right)\Big|_1^3\,dv$$
$$= \int_2^4 \left(\tfrac{11}{12} + 2v - \tfrac{1}{4}v^2\right)dv = \left(\tfrac{11}{12}v + v^2 - \tfrac{1}{12}v^3\right)\Big|_2^4 = \tfrac{55}{6}$$

CUMULATIVE REVIEW (CHAPTERS 1–14)

1. If $x \ne 0$, $\displaystyle\lim_{h\to 0}\frac{2x^2h - 5xh^2 - 6h^3}{3xh + 2h^2} = \lim_{h\to 0}\frac{2x^2 - 5xh - 6h^2}{3x + 2h} = \frac{2x^2}{3x} = \frac{2}{3}x$

   If $x = 0$, $\displaystyle\lim_{h\to 0}\frac{2x^2h - 5xh^2 - 6h^3}{3xh + 2h^2} = \lim_{h\to 0}\frac{-6h^3}{2h^2} = \lim_{h\to 0}(-3h) = 0$

   Thus $\displaystyle\lim_{h\to 0}\frac{2x^2h - 5xh^2 - 6h^3}{3xh + 2h^2} = \frac{2}{3}x$ for all x

3. a. The domain of f consists of all x for which $x^2 - 4 \ge 0$ (that is, $|x| \ge 2$) and $x + \sqrt{x^2 - 4} \ge 0$. Since $x + \sqrt{x^2 - 4} < 0$ for $x \le -2$, and since $x + \sqrt{x^2 - 4} > 0$ for $x \ge 2$, the domain of f is $[2, \infty)$.

   b. $f'(x) = \dfrac{1}{2\sqrt{x + \sqrt{x^2 - 4}}}\left(1 + \dfrac{x}{\sqrt{x^2 - 4}}\right)$

5. Differentiating the given equation implicitly, we obtain $2x - y - x(dy/dx) + 2y(dy/dx) = 0$, so if $dy/dx = 1$, then $2x - y - x + 2y = 0$, or $y = -x$. Then the original equation becomes $x^2 - x(-x) + (-x)^2 = 4$, or $3x^2 = 4$, so that $x = -\frac{2}{3}\sqrt{3}$ or $x = \frac{2}{3}\sqrt{3}$. Thus the two points at which the slope of the tangent line is 1 are $(-\frac{2}{3}\sqrt{3}, \frac{2}{3}\sqrt{3})$ and $(\frac{2}{3}\sqrt{3}, -\frac{2}{3}\sqrt{3})$.

7. Let x and y be as in the figure. Since the volume of the toolshed is to be 1512 cubic feet, we have $7xy = 1512$, so that $y = 216/x$. Let k be the cost per square foot of the sides. Then the cost C of the toolshed is given by

$$C = \underset{\text{sides}}{14xk} + \underset{\text{front and back}}{14y\left(\tfrac{3}{2}k\right)} + \underset{\text{top}}{xy(2k)} = k\left[14x + 21\left(\frac{216}{x}\right) + 432\right]$$

   We are to minimize C. Now $C'(x) = k[14 - (21)(216)/x^2]$, so that $C'(x) = 0$ if $14x^2 = (21)(216)$, or $x = 18$. Since $C''(x) = k(42)(216)/x^3 > 0$ for $x > 0$, C is minimized for $x = 18$. Then $y = \frac{216}{18} = 12$. Thus for minimal cost the front should be 12 feet wide and the sides should be 18 feet wide.

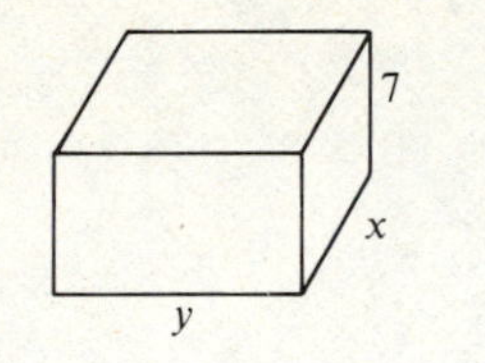

9. $f'(x) = \dfrac{\ln x - 1}{(\ln x)^2}$; $f''(x) = \dfrac{2 - \ln x}{x(\ln x)^3}$

relative minimum value is $f(e) = e$; increasing on $[e, \infty)$, and decreasing on $(0, 1)$ and $(1, e]$; concave upward on $(1, e^2)$, and concave downward on $(0, 1)$ and (e^2, ∞); inflection point is $(e^2, \frac{1}{2}e^2)$; vertical asymptote is $x = 1$; $\lim_{x \to 0^+} (x/\ln x) = 0$.

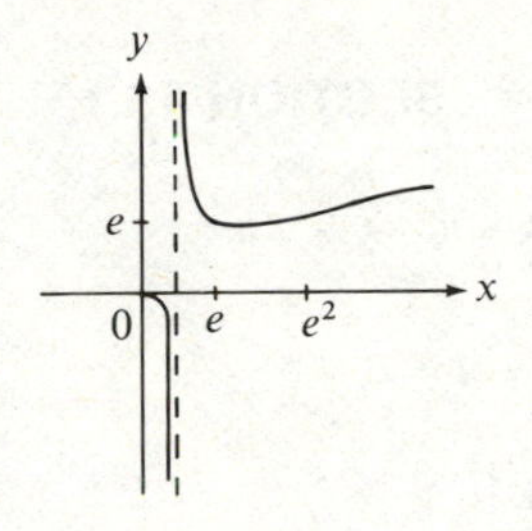

11. $$\int \tan x \sin^2 x \cos^5 x \, dx = \int \frac{\sin x}{\cos x} \sin^2 x \cos^5 x \, dx = \int \sin^3 x \cos^4 x \, dx$$
$$= \int (1 - \cos^2 x) \cos^4 x \sin x \, dx$$

Let $u = \cos x$, so that $du = -\sin x \, dx$. Then

$$\int \tan x \sin^2 x \cos^5 x \, dx = \int (1 - \cos^2 x) \cos^4 x \sin x \, dx = \int (1 - u^2) u^4 (-1) \, du$$
$$= -\tfrac{1}{5}u^5 + \tfrac{1}{7}u^7 + C = -\tfrac{1}{5}\cos^5 x + \tfrac{1}{7}\cos^7 x + C$$

13. $$\int_{1/2}^{1} \frac{1}{\sqrt{1 - x^2}} \, dx = \lim_{b \to 1^-} \int_{1/2}^{b} \frac{1}{\sqrt{1 - x^2}} \, dx = \lim_{b \to 1^-} \left(\arcsin x \Big|_{1/2}^{b} \right)$$
$$= \lim_{b \to 1^-} (\arcsin b - \arcsin \tfrac{1}{2}) = \tfrac{1}{2}\pi - \tfrac{1}{6}\pi = \tfrac{1}{3}\pi$$

Thus the region has finite area, and the area of the region is $\pi/3$.

15. $$\mathscr{L} = \int_0^{\pi/2} \sqrt{\left(\frac{dx}{dt}\right)^2 + \left(\frac{dy}{dt}\right)^2} \, dt = \int_0^{\pi/2} \sqrt{(\sin^{1/2} t \cos t)^2 + (\cos t)^2} \, dt$$
$$= \int_0^{\pi/2} \sqrt{\sin t \cos^2 t + \cos^2 t} \, dt = \int_0^{\pi/2} \sqrt{\sin t + 1} \cos t \, dt$$

Let $u = \sin t + 1$, so that $du = \cos t \, dt$. If $t = 0$, then $u = 1$, and if $t = \pi/2$, then $u = 2$. Thus $\mathscr{L} = \int_0^{\pi/2} \sqrt{\sin t + 1} \cos t \, dt = \int_1^2 \sqrt{u} \, du = \frac{2}{3}u^{3/2}\Big|_1^2 = \frac{2}{3}(2^{3/2} - 1)$.

17. $$\frac{a_{k+1}}{a_k} = \frac{\dfrac{[(k+1)!]^3(27)^{k+1}}{(3k+3)!}}{\dfrac{(k!)^3(27)^k}{(3k)!}} = 27\left(\frac{(k+1)!}{k!}\right)^3 \frac{(3k)!}{(3k+3)!} = \frac{27(k+1)^3}{(3k+1)(3k+2)(3k+3)}$$
$$= \frac{9(k+1)^2}{(3k+1)(3k+2)} = \frac{3k+3}{3k+1} \cdot \frac{3k+3}{3k+2} > 1 \cdot 1 = 1$$

Thus $a_{k+1} > a_k$, so the given sequence is increasing.

19. $$\lim_{n \to \infty} \left| \frac{[(-1)^{n+1}/(n+2)]x^{2n+2}}{[(-1)^n/(n+1)]x^{2n}} \right| = \lim_{n \to \infty} \frac{n+1}{n+2} |x|^2 = |x|^2$$

By the Generalized Ratio Test, the power series converges for $|x| < 1$ and diverges for $|x| > 1$. For $x = -1$ and for $x = 1$ the series becomes $\sum_{n=1}^{\infty} (-1)^n/(n+1)$, which converges by the Alternating Series Test. Thus the interval of convergence of the given power series is $[-1, 1]$.

21. We set up a coordinate system so that the sack moves from the origin 0 to the point $Q = (0, 6)$. Then the force $\mathbf{F}$ is given by

$$\mathbf{F} = 100\left(\cos \frac{\pi}{4} \mathbf{i} + \sin \frac{\pi}{4} \mathbf{j}\right) = 50\sqrt{2}(\mathbf{i} + \mathbf{j})$$

The work is given by

$$W = \mathbf{F} \cdot \overrightarrow{OQ} = 50\sqrt{2}(\mathbf{i} + \mathbf{j}) \cdot 6\mathbf{j} = (50\sqrt{2})(6) = 300\sqrt{2} \text{ (foot-pounds)}$$

23. a. $\mathbf{r}'(t) = 3t^2\mathbf{i} + 6\mathbf{j} + 6t\mathbf{k}$; $\|\mathbf{r}'(t)\| = \sqrt{(3t^2)^2 + 6^2 + (6t)^2} = \sqrt{9t^4 + 36t^2 + 36} = 3(t^2 + 2)$;

$$\mathbf{T}(t) = \frac{\mathbf{r}'(t)}{\|\mathbf{r}'(t)\|} = \frac{3t^2\mathbf{i} + 6\mathbf{j} + 6t\mathbf{k}}{3(t^2 + 2)} = \frac{t^2}{t^2 + 2}\mathbf{i} + \frac{2}{t^2 + 2}\mathbf{j} + \frac{2t}{t^2 + 2}\mathbf{k}.$$

Therefore $\mathbf{T}(t)$ is parallel to the y axis if $t = 0$.

b. $$\mathbf{T}'(t) = \frac{2t(t^2 + 2) - 2t(t^2)}{(t^2 + 2)^2}\mathbf{i} - \frac{2(2t)}{(t^2 + 2)^2}\mathbf{j} + \frac{2(t^2 + 2) - 2t(2t)}{(t^2 + 2)^2}\mathbf{k}$$
$$= \frac{4t}{(t^2 + 2)^2}\mathbf{i} - \frac{4t}{(t^2 + 2)^2}\mathbf{j} + \frac{4 - 2t^2}{(t^2 + 2)^2}\mathbf{k};$$
$$\|\mathbf{T}'(t)\| = \frac{1}{(t^2 + 2)^2}\sqrt{(4t)^2 + (-4t)^2 + (4 - 2t^2)^2} = \frac{\sqrt{4t^4 + 16t^2 + 16}}{(t^2 + 2)^2} = \frac{2}{t^2 + 2};$$
$$\mathbf{N}(t) = \frac{\mathbf{T}'(t)}{\|\mathbf{T}'(t)\|} = \frac{2t}{t^2 + 2}\mathbf{i} - \frac{2t}{t^2 + 2}\mathbf{j} + \frac{2 - t^2}{t^2 + 2}\mathbf{k}.$$

c. $\mathbf{T}(1) = \frac{1}{3}\mathbf{i} + \frac{2}{3}\mathbf{j} + \frac{2}{3}\mathbf{k}$; $\mathbf{N}(1) = \frac{2}{3}\mathbf{i} - \frac{2}{3}\mathbf{j} + \frac{1}{3}\mathbf{k}$; $\mathbf{T}(1) \times \mathbf{N}(1) = \begin{vmatrix} \mathbf{i} & \mathbf{j} & \mathbf{k} \\ \frac{1}{3} & \frac{2}{3} & \frac{2}{3} \\ \frac{2}{3} & -\frac{2}{3} & \frac{1}{3} \end{vmatrix} = \frac{2}{3}\mathbf{i} + \frac{1}{3}\mathbf{j} - \frac{2}{3}\mathbf{k}$.

$\mathbf{T}(1) \times \mathbf{N}(1)$ is perpendicular to both $\mathbf{T}(1)$ and $\mathbf{N}(1)$, but has positive $\mathbf{j}$ component. Also, $\|\mathbf{T}(1) \times \mathbf{N}(1)\| = \sqrt{(\frac{2}{3})^2 + (\frac{1}{3})^2 + (-\frac{2}{3})^2} = 1$. Therefore $-\frac{2}{3}\mathbf{i} - \frac{1}{3}\mathbf{j} + \frac{2}{3}\mathbf{k}$ is the unit vector with negative $\mathbf{j}$ component that is perpendicular to both $\mathbf{T}(1)$ and $\mathbf{N}(1)$.

25. a.

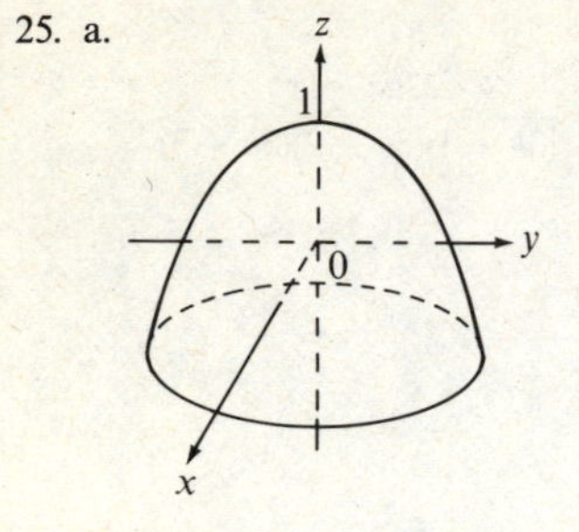

b. Let $f(x, y) = 1 - x^2 - y^2$. Notice that $f_x(x, y) = -2x$, so that $f_x(-1, 1) = 2$; also $f_y(x, y) = -2y$, so that $f_y(-1, 1) = -2$. Therefore by (2) in Section 13.6, the vector $f_x(-1, 1)\mathbf{i} + f_y(-1, 1)\mathbf{j} - \mathbf{k} = 2\mathbf{i} - 2\mathbf{j} - \mathbf{k}$ is normal to the paraboloid at $(-1, 1, -1)$. Consequently symmetric equations for the normal line ℓ through $(-1, 1, -1)$ are

$$\frac{x+1}{2} = \frac{y-1}{-2} = \frac{z+1}{-1}$$

16 CALCULUS OF VECTOR FIELDS

SECTION 16.1

1. $\operatorname{curl} \mathbf{F}(x, y) = \left(\frac{\partial N}{\partial x} - \frac{\partial M}{\partial y}\right)\mathbf{k} = (0 - 0)\mathbf{k} = \mathbf{0}$

$\operatorname{div} \mathbf{F}(x, y) = \frac{\partial M}{\partial x} + \frac{\partial N}{\partial y} = 1 + 1 = 2$

3. $\operatorname{curl} \mathbf{F}(x, y, z) = \begin{vmatrix} \mathbf{i} & \mathbf{j} & \mathbf{k} \\ \frac{\partial}{\partial x} & \frac{\partial}{\partial y} & \frac{\partial}{\partial z} \\ y & z & x \end{vmatrix} = -\mathbf{i} - \mathbf{j} - \mathbf{k}$

$\operatorname{div} \mathbf{F}(x, y, z) = \frac{\partial M}{\partial x} + \frac{\partial N}{\partial y} + \frac{\partial P}{\partial z} = 0 + 0 + 0 = 0$

5. $\operatorname{curl} \mathbf{F}(x, y, z) = \begin{vmatrix} \mathbf{i} & \mathbf{j} & \mathbf{k} \\ \frac{\partial}{\partial x} & \frac{\partial}{\partial y} & \frac{\partial}{\partial z} \\ x^2 & y^2 & z^2 \end{vmatrix} = 0\mathbf{i} + 0\mathbf{j} + 0\mathbf{k} = \mathbf{0}$

$\operatorname{div} \mathbf{F}(x, y, z) = \frac{\partial M}{\partial x} + \frac{\partial N}{\partial y} + \frac{\partial P}{\partial z} = 2x + 2y + 2z = 2(x + y + z)$

7. $\operatorname{curl} \mathbf{F}(x, y, z) = \begin{vmatrix} \mathbf{i} & \mathbf{j} & \mathbf{k} \\ \frac{\partial}{\partial x} & \frac{\partial}{\partial y} & \frac{\partial}{\partial z} \\ \frac{-x}{z} & \frac{-y}{z} & \frac{1}{z} \end{vmatrix} = \frac{-y}{z^2}\mathbf{i} + \frac{x}{z^2}\mathbf{j} + 0\mathbf{k} = \frac{-y}{z^2}\mathbf{i} + \frac{x}{z^2}\mathbf{j}$

$\operatorname{div} \mathbf{F}(x, y, z) = \frac{\partial M}{\partial x} + \frac{\partial N}{\partial y} + \frac{\partial P}{\partial z} = -\frac{1}{z} - \frac{1}{z} - \frac{1}{z^2} = -\frac{2}{z} - \frac{1}{z^2}$

9. $\text{curl}\,\mathbf{F}(x, y, z) = \begin{vmatrix} \mathbf{i} & \mathbf{j} & \mathbf{k} \\ \dfrac{\partial}{\partial x} & \dfrac{\partial}{\partial y} & \dfrac{\partial}{\partial z} \\ e^x \cos y & e^x \sin y & z \end{vmatrix} = 0\mathbf{i} + 0\mathbf{j} + (e^x \sin y + e^x \sin y)\mathbf{k} = 2e^x \sin y\mathbf{k}$

$$\text{div}\,\mathbf{F}(x, y, z) = \frac{\partial M}{\partial x} + \frac{\partial N}{\partial y} + \frac{\partial P}{\partial z} = e^x \cos y + e^x \cos y + 1 = 2e^x \cos y + 1$$

11. $\dfrac{\partial^2 f}{\partial x^2} + \dfrac{\partial^2 f}{\partial y^2} = 2 - 2 = 0$

13. $\dfrac{\partial^2 f}{\partial x^2} + \dfrac{\partial^2 f}{\partial y^2} + \dfrac{\partial^2 f}{\partial z^2} = 2 + 2 - 4 = 0$

15. $\partial N/\partial x = e^y = \partial M/\partial y$, so $\mathbf{F}$ is the gradient of some function f, and

$$\frac{\partial f}{\partial x} = e^y \quad \text{and} \quad \frac{\partial f}{\partial y} = xe^y + y \qquad (*)$$

Integrating both sides of the first equation in (*) with respect to x, we obtain $f(x, y) = xe^y + g(y)$. Taking partial derivatives with respect to y, we find that $\partial f/\partial y = xe^y + dg/dy$. Comparing this with the second equation in (*), we conclude that $dg/dy = y$, so that $g(y) = y^2/2 + C$. Therefore $f(x, y) = xe^y + y^2/2 + C$.

17. Since $\partial N/\partial x = -y \sin xy$ and $\partial M/\partial y = x \cos xy$, $\mathbf{F}$ is not a gradient.

19. By Example 6, $\mathbf{F}$ is the gradient of some function f, and

$$\frac{\partial f}{\partial x} = 2xyz, \qquad \frac{\partial f}{\partial y} = x^2 z, \quad \text{and} \quad \frac{\partial f}{\partial z} = x^2 y + 1 \qquad (*)$$

Integrating both sides of the first equation in (*) with respect to x, we obtain $f(x, y, z) = x^2 yz + g(y, z)$. Taking partial derivatives with respect to y, we find that $\partial f/\partial y = x^2 z + \partial g/\partial y$. Comparing this with the second equation in (*), we deduce that $\partial g/\partial y = 0$, so that g is constant with respect to y. Thus $f(x, y, z) = x^2 yz + h(z)$. Taking partial derivatives with respect to z, we obtain $\partial f/\partial z = x^2 y + dh/dz$. Comparing this with the third equation in (*), we conclude that $dh/dz = 1$, so that $h(z) = z + C$. Therefore $f(x, y, z) = x^2 yz + z + C$.

21. Since $\partial P/\partial y = 0$ and $\partial N/\partial z = y$, $\mathbf{F}$ is not a gradient.

23. Since $\partial P/\partial y = 0$ and $\partial N/\partial z = 2z$, $\mathbf{F}$ is not a gradient.

25. a. grad (fg) is a vector field.
    b. grad $\mathbf{F}$ is meaningless.
    c. curl (grad f) is a vector field.
    d. grad (div $\mathbf{F}$) is a vector field.
    e. curl (curl $\mathbf{F}$) is a vector field.
    f. div (grad f) is a function of several variables.
    g. (grad f) $\times$ (curl $\mathbf{F}$) is a vector field.
    h. div (curl (grad f)) is a function of several variables.
    i. curl (div (grad f)) is meaningless.

27. Let $\mathbf{F} = M\mathbf{i} + N\mathbf{j} + P\mathbf{k}$. Then

$$\text{curl}\,\mathbf{F} = \left(\frac{\partial P}{\partial y} - \frac{\partial N}{\partial z}\right)\mathbf{i} + \left(\frac{\partial M}{\partial z} - \frac{\partial P}{\partial x}\right)\mathbf{j} + \left(\frac{\partial N}{\partial x} - \frac{\partial M}{\partial y}\right)\mathbf{k}$$

so that

$$\text{div}\,(\text{curl}\,\mathbf{F}) = \left(\frac{\partial^2 P}{\partial x \partial y} - \frac{\partial^2 N}{\partial x \partial z}\right) + \left(\frac{\partial^2 M}{\partial y \partial z} - \frac{\partial^2 P}{\partial y \partial x}\right) + \left(\frac{\partial^2 N}{\partial z \partial x} - \frac{\partial^2 M}{\partial z \partial y}\right)$$

Since all second partials are continuous by assumption,

$$\frac{\partial^2 M}{\partial y \partial z} = \frac{\partial^2 M}{\partial z \partial y}, \quad \frac{\partial^2 N}{\partial x \partial z} = \frac{\partial^2 N}{\partial z \partial x}, \quad \text{and} \quad \frac{\partial^2 P}{\partial x \partial y} = \frac{\partial^2 P}{\partial y \partial x}$$

Thus div (curl $\mathbf{F}$) = 0.

29. Let $\mathbf{F} = M\mathbf{i} + N\mathbf{j} + P\mathbf{k}$. Then $f\mathbf{F} = fM\mathbf{i} + fN\mathbf{j} + fP\mathbf{k}$, so that

$$\begin{aligned} \text{div}\,(f\mathbf{F}) &= \left(f\frac{\partial M}{\partial x} + \frac{\partial f}{\partial x}M\right) + \left(f\frac{\partial N}{\partial y} + \frac{\partial f}{\partial y}N\right) + \left(f\frac{\partial P}{\partial z} + \frac{\partial f}{\partial z}P\right) \\ &= f\left(\frac{\partial M}{\partial x} + \frac{\partial N}{\partial y} + \frac{\partial P}{\partial z}\right) + \left(\frac{\partial f}{\partial x}M + \frac{\partial f}{\partial y}N + \frac{\partial f}{\partial z}P\right) = f\,\text{div}\,\mathbf{F} + (\text{grad}\,f)\cdot\mathbf{F} \end{aligned}$$

31. Let $\mathbf{F} = M\mathbf{i} + N\mathbf{j} + P\mathbf{k}$. Then $f\mathbf{F} = fM\mathbf{i} + fN\mathbf{j} + fP\mathbf{k}$ so that

$$\begin{aligned} \text{curl}\,f\mathbf{F} &= \begin{vmatrix} \mathbf{i} & \mathbf{j} & \mathbf{k} \\ \dfrac{\partial}{\partial x} & \dfrac{\partial}{\partial y} & \dfrac{\partial}{\partial z} \\ fM & fN & fP \end{vmatrix} \\ &= \left(\frac{\partial f}{\partial y}P + f\frac{\partial P}{\partial y} - \frac{\partial f}{\partial z}N - f\frac{\partial N}{\partial z}\right)\mathbf{i} + \left(\frac{\partial f}{\partial z}M + f\frac{\partial M}{\partial z} - \frac{\partial f}{\partial x}P - f\frac{\partial P}{\partial x}\right)\mathbf{j} \\ &\quad + \left(\frac{\partial f}{\partial x}N + f\frac{\partial N}{\partial x} - \frac{\partial f}{\partial y}M - f\frac{\partial M}{\partial y}\right)\mathbf{k} \\ &= f\left[\left(\frac{\partial P}{\partial y} - \frac{\partial N}{\partial z}\right)\mathbf{i} + \left(\frac{\partial M}{\partial z} - \frac{\partial P}{\partial x}\right)\mathbf{j} + \left(\frac{\partial N}{\partial x} - \frac{\partial M}{\partial y}\right)\mathbf{k}\right] \\ &\quad + \left[\left(\frac{\partial f}{\partial y}P - \frac{\partial f}{\partial z}N\right)\mathbf{i} + \left(\frac{\partial f}{\partial z}M - \frac{\partial f}{\partial x}P\right)\mathbf{j} + \left(\frac{\partial f}{\partial x}N - \frac{\partial f}{\partial y}M\right)\mathbf{k}\right] \\ &= f\,(\text{curl}\,\mathbf{F}) + (\text{grad}\,f) \times \mathbf{F} \end{aligned}$$

33. Let $\mathbf{F} = a\mathbf{i} + b\mathbf{j} + c\mathbf{k}$, where a, b, and c are constants. Then

$$(\mathbf{F} \times \mathbf{G})(x, y, z) = \begin{vmatrix} \mathbf{i} & \mathbf{j} & \mathbf{k} \\ a & b & c \\ x & y & z \end{vmatrix} = (bz - cy)\mathbf{i} + (cx - az)\mathbf{j} + (ay - bx)\mathbf{k}$$

and thus

$$[\text{curl}\,(\mathbf{F} \times \mathbf{G})](x, y, z) = \begin{vmatrix} \mathbf{i} & \mathbf{j} & \mathbf{k} \\ \dfrac{\partial}{\partial x} & \dfrac{\partial}{\partial y} & \dfrac{\partial}{\partial z} \\ bz - cy & cx - az & ay - bx \end{vmatrix}$$
$$= [a - (-a)]\mathbf{i} + [b - (-b)]\mathbf{j} + [c - (-c)]\mathbf{k} = 2\mathbf{F}(x, y, z)$$

35. Let $\mathbf{F} = \text{grad}\, f$ and $\mathbf{G} = \text{grad}\, g$. Then

$$\mathbf{F} + \mathbf{G} = \left(\frac{\partial f}{\partial x}\mathbf{i} + \frac{\partial f}{\partial y}\mathbf{j} + \frac{\partial f}{\partial z}\mathbf{k}\right) + \left(\frac{\partial g}{\partial x}\mathbf{i} + \frac{\partial g}{\partial y}\mathbf{j} + \frac{\partial g}{\partial z}\mathbf{k}\right)$$
$$= \left(\frac{\partial f}{\partial x} + \frac{\partial g}{\partial x}\right)\mathbf{i} + \left(\frac{\partial f}{\partial y} + \frac{\partial g}{\partial y}\right)\mathbf{j} + \left(\frac{\partial f}{\partial z} + \frac{\partial g}{\partial z}\right)\mathbf{k}$$
$$= \frac{\partial(f + g)}{\partial x}\mathbf{i} + \frac{\partial(f + g)}{\partial y}\mathbf{j} + \frac{\partial(f + g)}{\partial z}\mathbf{k} = \text{grad}\,(f + g)$$

Thus $\mathbf{F} + \mathbf{G}$ is conservative.

37. $\text{grad}\, f(x, y, z) = (m\omega^2/2)(2x\mathbf{i} + 2y\mathbf{j} + 2z\mathbf{k}) = m\omega^2(x\mathbf{i} + y\mathbf{j} + z\mathbf{k}) = \mathbf{F}(x, y, z)$. Therefore f is a potential function for $\mathbf{F}$.

39. If $\mathbf{F} = \text{curl}\,\mathbf{G}$, then by Exercise 27, $\text{div}\,\mathbf{F} = \text{div}\,(\text{curl}\,\mathbf{G}) = 0$, so that $\mathbf{F}$ is solenoidal.

41. Let $\mathbf{G} = M\mathbf{i} + N\mathbf{j} + P\mathbf{k}$, and assume that $\mathbf{F}(x, y, z) = 8\mathbf{i} = \text{curl}\,\mathbf{G}(x, y, z)$. Then

$$\frac{\partial P}{\partial y} - \frac{\partial N}{\partial z} = 8, \quad \frac{\partial M}{\partial z} - \frac{\partial P}{\partial x} = 0, \quad \text{and} \quad \frac{\partial N}{\partial x} - \frac{\partial M}{\partial y} = 0$$

Notice that if $P(x, y, z) = 8y$ and $M(x, y, z) = 0 = N(x, y, z)$, then the three equations are satisfied. Consequently if we let $\mathbf{G}(x, y, z) = 8y\mathbf{k}$, then $\text{curl}\,\mathbf{G}(x, y, z) = 8\mathbf{i} = \mathbf{F}(x, y, z)$. Thus $\mathbf{G}$ is a vector potential of $\mathbf{F}$. (Other solutions are also possible.)

43. a. $\text{grad}\,\mathbf{T}(x, y, z) = -4x\mathbf{i} - 2y\mathbf{j} - 8z\mathbf{k}$. Since each component is continuous, so is grad $\mathbf{T}$.

b. $\|\text{grad}\,\mathbf{T}(x, y, z)\| = \sqrt{16x^2 + 4y^2 + 64z^2}$. Then $\|\text{grad}\,\mathbf{T}(1, 0, 0)\| = 4$ and $\|\text{grad}\,\mathbf{T}(0, 1, 0)\| = 2$. Since $(1, 0, 0)$ and $(0, 1, 0)$ are the same distance from the origin, and since the magnitudes of grad $\mathbf{T}$ at these points are distinct, grad $\mathbf{T}$ is not a central force field.

SECTION 16.2

1. Since $x(t) = 2t^{3/2}$, $y(t) = t^2$, $z(t) = 0$, and $\|d\mathbf{r}/dt\| = \sqrt{(3t^{1/2})^2 + (2t)^2 + 0^2} = \sqrt{9t + 4t^2}$, (4) implies that

$$\int_C (9 + 8y^{1/2})\,ds = \int_0^1 (9 + 8t)\sqrt{9t + 4t^2}\,dt = \tfrac{2}{3}(9t + 4t^2)^{3/2}\Big|_0^1 = \tfrac{26}{3}\sqrt{13}$$

3. Since $x(t) = t$, $y(t) = t^3$, $z(t) = 0$, and $\|d\mathbf{r}/dt\| = \sqrt{1^2 + (3t^2)^2 + 0^2} = \sqrt{1 + 9t^4}$, (4) implies that

$$\int_C y\,ds = \int_{-1}^0 t^3\sqrt{1 + 9t^4}\,dt = \tfrac{1}{54}(1 + 9t^4)^{3/2}\Big|_{-1}^0 = \tfrac{1}{54}(1 - 10^{3/2})$$

5. Since $x(t) = e^t$, $y(t) = e^{-t}$, $z(t) = \sqrt{2}t$, and $\|d\mathbf{r}/dt\| = \sqrt{(e^t)^2 + (-e^{-t})^2 + (\sqrt{2})^2} = \sqrt{e^{2t} + 2 + e^{-2t}} = \sqrt{(e^t + e^{-t})^2} = e^t + e^{-t}$, (4) implies that

$$\int_C 2xyz\,ds = \int_0^1 2(e^t)(e^{-t})(\sqrt{2}t)(e^t + e^{-t})\,dt = \int_0^1 2\sqrt{2}t(e^t + e^{-t})\,dt$$
$$\overset{\text{parts}}{=} 2\sqrt{2}t(e^t - e^{-t})\Big|_0^1 - \int_0^1 2\sqrt{2}(e^t - e^{-t})\,dt$$
$$= 2\sqrt{2}(e - e^{-1}) - [2\sqrt{2}(e^t + e^{-t})]\Big|_0^1 = 4\sqrt{2}(1 - e^{-1})$$

7. Since $x(t) = \cos t$, $y(t) = \sin t$, $z(t) = t^{3/2}$, and $\|d\mathbf{r}/dt\| = \sqrt{(-\sin t)^2 + (\cos t)^2 + (\frac{3}{2}t^{1/2})^2} = \sqrt{1 + \frac{9}{4}t}$, (4) implies that

$$\int_C (1 + \tfrac{9}{4}z^{2/3})^{1/4}\,ds = \int_0^{20/3} (1 + \tfrac{9}{4}t)^{1/4}\sqrt{1 + \tfrac{9}{4}t}\,dt$$
$$= \int_0^{20/3} (1 + \tfrac{9}{4}t)^{3/4}\,dt = \tfrac{16}{63}(1 + \tfrac{9}{4}t)^{7/4}\Big|_0^{20/3} = \tfrac{2032}{63}$$

9. The triangular path is composed of the three line segments C_1, C_2, and C_3 parametrized respectively by $\mathbf{r}_1(t) = (-1 + t)\mathbf{i} + t\mathbf{j}$ for $0 \le t \le 1$, $\mathbf{r}_2(t) = (1 - t)\mathbf{j} + t\mathbf{k}$ for $0 \le t \le 1$, and $\mathbf{r}_3(t) = -t\mathbf{i} + (1 - t)\mathbf{k}$ for $0 \le t \le 1$. Since $x(t) = -1 + t$, $y(t) = t$, $z(t) = 0$, and $\|d\mathbf{r}_1/dt\| = \sqrt{1^2 + 1^2 + 0^2} = \sqrt{2}$ for C_1, (4) implies that

$$\int_{C_1} (y + 2z)\,ds = \int_0^1 [t + 2(0)]\sqrt{2}\,dt = \int_0^1 \sqrt{2}t\,dt = \frac{\sqrt{2}}{2}t^2\Big|_0^1 = \frac{\sqrt{2}}{2}$$

Since $x(t) = 0$, $y(t) = 1 - t$, $z(t) = t$, and $\|d\mathbf{r}_2/dt\| = \sqrt{0^2 + (-1)^2 + 1^2} = \sqrt{2}$ for C_2, (4) implies that

$$\int_{C_2} (y + 2z)\,ds = \int_0^1 (1 - t + 2t)\sqrt{2}\,dt = \int_0^1 \sqrt{2}(1 + t)\,dt = \sqrt{2}(t + \tfrac{1}{2}t^2)\Big|_0^1 = \tfrac{3}{2}\sqrt{2}$$

Since $x(t) = -t$, $y(t) = 0$, $z(t) = 1 - t$, and $\|d\mathbf{r}_3/dt\| = \sqrt{(-1)^2 + 0^2 + (-1)^2} = \sqrt{2}$ for C_3, (4) implies that

$$\int_{C_3} (y + 2z)\,ds = \int_0^1 [0 + 2(1 - t)]\sqrt{2}\,dt = \int_0^1 2\sqrt{2}(1 - t)\,dt = 2\sqrt{2}(t - \tfrac{1}{2}t^2)\Big|_0^1 = \sqrt{2}$$

It follows that

$$\int_C (y + 2z)\,ds = \int_{C_1} (y + 2z)\,ds + \int_{C_2} (y + 2z)\,ds + \int_{C_3} (y + 2z)\,ds$$
$$= \frac{\sqrt{2}}{2} + \frac{3}{2}\sqrt{2} + \sqrt{2} = 3\sqrt{2}$$

11. Since $x(t) = 5$, $y(t) = -\sin t$, $z(t) = -\cos t$, and $d\mathbf{r}/dt = -\cos t\,\mathbf{j} + \sin t\,\mathbf{k}$, (7) implies that

$$\int_C \mathbf{F} \cdot d\mathbf{r} = \int_0^{\pi/4} (-\cos t\,\mathbf{i} + \sin t\,\mathbf{j} - 5\mathbf{k}) \cdot (-\cos t\,\mathbf{j} + \sin t\,\mathbf{k})\,dt$$
$$= \int_0^{\pi/4} (-\sin t \cos t - 5 \sin t)\,dt = (-\tfrac{1}{2}\sin^2 t + 5\cos t)\Big|_0^{\pi/4} = \tfrac{5}{2}\sqrt{2} - \tfrac{21}{4}$$

13. Since $x(t) = \cos t$, $y(t) = \sin t$, $z(t) = 2t$, and $d\mathbf{r}/dt = -\sin t\mathbf{i} + \cos t\mathbf{j} + 2\mathbf{k}$, (7) implies that

$$\int_C \mathbf{F} \cdot d\mathbf{r} = \int_0^{\pi/2} (\sin t\mathbf{i} + \cos t \sin t\mathbf{j} + 8t^3\mathbf{k}) \cdot (-\sin t\mathbf{i} + \cos t\mathbf{j} + 2\mathbf{k})\, dt$$
$$= \int_0^{\pi/2} (-\sin^2 t + \cos^2 t \sin t + 16t^3)\, dt = \int_0^{\pi/2} (-\tfrac{1}{2} + \tfrac{1}{2}\cos 2t + \cos^2 t \sin t + 16t^3)\, dt$$
$$= (-\tfrac{1}{2}t + \tfrac{1}{4}\sin 2t - \tfrac{1}{3}\cos^3 t + 4t^4)\Big|_0^{\pi/2} = -\tfrac{1}{4}\pi + \tfrac{1}{4}\pi^4 + \tfrac{1}{3}$$

15. $\int_{-C} \mathbf{F} \cdot d\mathbf{r} = -\int_C \mathbf{F} \cdot d\mathbf{r} = -\pi$, by the solution of Exercise 14.

17. Since $x(t) = \frac{1}{2}$, $y(t) = 2$, $z(t) = -\ln(\cosh t)$, and $d\mathbf{r}/dt = -(\sinh t/\cosh t)\mathbf{k}$, we have $\mathbf{F}(x(t), y(t), z(t)) \cdot d\mathbf{r}/dt = 0$, so that by (7), $\int_C \mathbf{F} \cdot d\mathbf{r} = \int_0^{\pi/6} 0\, dt = 0$.

19. Since $x(t) = e^{-t}$, $y(t) = e^t$, $z(t) = t$, $dx/dt = -e^{-t}$, $dy/dt = e^t$, and $dz/dt = 1$, (9) implies that

$$\int_C y\, dx - x\, dy + xyz^2\, dz = \int_0^1 [(e^t)(-e^{-t}) - (e^{-t})(e^t) + (e^{-t}e^t t^2)(1)]\, dt$$
$$= \int_0^1 (-2 + t^2)\, dt = (-2t + \tfrac{1}{3}t^3)\Big|_0^1 = -\tfrac{5}{3}$$

21. $\int_C e^x\, dx + xy\, dy + xyz\, dz = -\int_{-C} e^x\, dx + xy\, dy + xyz\, dz = e^{-1} - e - \frac{2}{3}$ by the solution of Exercise 20.

23. Since $x(t) = t + 1$, $y(t) = t - 1$, $z(t) = t^2$, $dx/dt = 1$, $dy/dt = 1$, and $dz/dt = 2t$, (9) implies that

$$\int_C xy\, dx + (x + z)\, dy + z^2\, dz = \int_{-1}^2 [(t + 1)(t - 1)(1) + (t + 1 + t^2)(1) + (t^4)(2t)]\, dt$$
$$= \int_{-1}^2 (2t^5 + 2t^2 + t)\, dt = (\tfrac{1}{3}t^6 + \tfrac{2}{3}t^3 + \tfrac{1}{2}t^2)\Big|_{-1}^2 = \tfrac{57}{2}$$

25. Since C is parametrized by $\mathbf{r}(t) = \cos t\mathbf{i} + \sin t\mathbf{j}$ for $0 \le t \le \pi/2$, and since $x(t) = \cos t$, $y(t) = \sin t$, $dx/dt = -\sin t$, and $dy/dt = \cos t$, the two-dimensional version of (9) implies that

$$\int_C \frac{1}{1 + x^2}\, dx + \frac{2}{1 + y^2}\, dy = \int_0^{\pi/2} \left[\frac{1}{1 + \cos^2 t}(-\sin t) + \frac{2}{1 + \sin^2 t}(\cos t)\right] dt$$
$$= [\arctan(\cos t) + 2\arctan(\sin t)]\Big|_0^{\pi/2} = \tfrac{1}{4}\pi$$

27. Since $x(t) = t$, $y(t) = t^2$, $z(t) = t^3$, $dx/dt = 1$, $dy/dt = 2t$, and $dz/dt = 3t^2$, (9) implies that

$$\int_C x \ln\left(\frac{xz}{y}\right) dx + \cos\left(\frac{\pi xy}{z}\right) dy = \int_1^2 [(t \ln t^2)(1) + (\cos \pi)(2t)]\, dt = \int_1^2 (2t \ln t - 2t)\, dt$$
$$\overset{\text{parts}}{=} t^2 \ln t\Big|_1^2 - \int_1^2 t\, dt - t^2\Big|_1^2 = 4 \ln 2 - 3 - \tfrac{1}{2}t^2\Big|_1^2 = 4 \ln 2 - \tfrac{9}{2}$$

29. a. The curve C is composed of the line segments C_1 and C_2 parametrized respectively by $\mathbf{r}_1(t) = -5t\mathbf{j}$ for $0 \le t \le 1$, and $\mathbf{r}_2(t) = (-5 + 6t)\mathbf{j} + t\mathbf{k}$ for $0 \le t \le 1$. By (9),

$$\int_C y\, dx + z\, dy + x\, dz = \int_{C_1} y\, dx + z\, dy + x\, dz + \int_{C_2} y\, dx + z\, dy + x\, dz$$
$$= \int_0^1 [(-5t)(0) + (0)(-5) + (0)(0)]\, dt + \int_0^1 [(-5 + 6t)(0) + (t)(6) + (0)(1)]\, dt$$
$$= \int_0^1 0\, dt + \int_0^1 6t\, dt = 3t^2\Big|_0^1 = 3$$

b. The curve C is composed of the line segments C_1 and C_2 parametrized respectively by $\mathbf{r}_1(t) = t\mathbf{i}$ for $0 \le t \le 1$, and $\mathbf{r}_2(t) = (1 - t)\mathbf{i} + t\mathbf{j} + t\mathbf{k}$ for $0 \le t \le 1$. By (9),

$$\int_C y\, dx + z\, dy + x\, dz = \int_{C_1} y\, dx + z\, dy + x\, dz + \int_{C_2} y\, dx + z\, dy + x\, dz$$
$$= \int_0^1 [(0)(1) + (0)(0) + (t)(0)]\, dt + \int_0^1 [(t)(-1) + (t)(1) + (1 - t)(1)]\, dt$$
$$= \int_0^1 0\, dt + \int_0^1 (1 - t)\, dt = (t - \tfrac{1}{2}t^2)\Big|_0^1 = \tfrac{1}{2}$$

31. If the first component of a parametrization is constant, then $dx/dt = 0$, so that by the first formula in (10),

$$\int_C M(x, y, z)\, dx = \int_a^b M(x(t), y(t), z(t)) \frac{dx}{dt}\, dt = \int_a^b 0\, dt = 0$$

33. Since the density at (x, y, z) is given by $f(x, y, z) = x^2 + y^2 + z^2$,

$$m = \int_C (x^2 + y^2 + z^2)\, ds = \int_\pi^{2\pi} (\sin^2 t + \cos^2 t + 16t^2)\sqrt{17}\, dt$$
$$= \sqrt{17} \int_\pi^{2\pi} (1 + 16t^2)\, dt = \sqrt{17}(t + \tfrac{16}{3}t^3)\Big|_\pi^{2\pi} = \sqrt{17}(\pi + \tfrac{112}{3}\pi^3)$$

35. a. The line segment C from $(0, 0, 0)$ to $(1, 1, 1)$ is parametrized by $\mathbf{r}(t) = t\mathbf{i} + t\mathbf{j} + t\mathbf{k}$ for $0 \le t \le 1$. By (6), $W = \int_C \mathbf{F} \cdot d\mathbf{r}$. Since $x(t) = t$, $y(t) = t$, $z(t) = t$, and $d\mathbf{r}/dt = \mathbf{i} + \mathbf{j} + \mathbf{k}$, (7) implies that

$$W = \int_C \mathbf{F} \cdot d\mathbf{r} = \int_0^1 (t\mathbf{i} + 2t\mathbf{j} + 0\mathbf{k}) \cdot (\mathbf{i} + \mathbf{j} + \mathbf{k})\, dt = \int_0^1 (t + 2t)\, dt = \tfrac{3}{2}t^2\Big|_0^1 = \tfrac{3}{2}$$

b. Let C be the curve parametrized by $\mathbf{r}$. By (6), $W = \int_C \mathbf{F} \cdot d\mathbf{r}$. Since $x(t) = \sin(\pi t/2)$, $y(t) = \sin(\pi t/2)$, $z(t) = t$, and $d\mathbf{r}/dt = (\pi/2)\cos(\pi t/2)\mathbf{i} + (\pi/2)\cos(\pi t/2)\mathbf{j} + \mathbf{k}$, (7) implies that

$$W = \int_C \mathbf{F} \cdot d\mathbf{r} = \int_0^1 \left[\sin\frac{\pi t}{2}\mathbf{i} + 2t\mathbf{j} + \left(\sin\frac{\pi t}{2} - t\right)\mathbf{k}\right] \cdot \left[\frac{\pi}{2}\cos\frac{\pi t}{2}\mathbf{i} + \frac{\pi}{2}\cos\frac{\pi t}{2}\mathbf{j} + \mathbf{k}\right] dt$$
$$= \int_0^1 \left(\frac{\pi}{2}\sin\frac{\pi t}{2}\cos\frac{\pi t}{2} + \pi t\cos\frac{\pi t}{2} + \sin\frac{\pi t}{2} - t\right) dt$$
$$= \left[\frac{1}{2}\sin^2\left(\frac{\pi t}{2}\right) - \frac{2}{\pi}\cos\frac{\pi t}{2} - \frac{1}{2}t^2\right]\Bigg|_0^1 + \int_0^1 \pi t\cos\frac{\pi t}{2}\, dt$$
$$\overset{\text{parts}}{=} \frac{2}{\pi} + 2t\sin\frac{\pi t}{2}\Big|_0^1 - \int_0^1 2\sin\frac{\pi t}{2}\, dt = \frac{2}{\pi} + 2 + \frac{4}{\pi}\cos\frac{\pi t}{2}\Big|_0^1 = 2 - \frac{2}{\pi}$$

37. When the height of the painter is z, the weight of the painter and pail is $120 + (30 - \frac{1}{10}z) = 150 - \frac{1}{10}z$. Thus $\mathbf{F}(x, y, z) = -(150 - \frac{1}{10}z)\mathbf{k} = (\frac{1}{10}z - 150)\mathbf{k}$. By (6) and (7),

$$W = \int_C \mathbf{F} \cdot d\mathbf{r} = \int_0^{8\pi} \left[\frac{1}{10}\left(\frac{25t}{\pi}\right) - 150\right]\mathbf{k} \cdot \left[-50\sin t\mathbf{i} + 50\cos t\mathbf{j} + \frac{25}{\pi}\mathbf{k}\right] dt$$
$$= \int_0^{8\pi} \frac{25}{\pi}\left[\frac{1}{10}\left(\frac{25t}{\pi}\right) - 150\right] dt = \frac{25}{\pi}\left(\frac{5}{4\pi}t^2 - 150t\right)\Bigg|_0^{8\pi} = -28{,}000 \text{ (foot-pounds)}$$

39. a. If the force acts in a direction normal to the path, then $\mathbf{F}(x, y, z) = f(x, y, z)\mathbf{N}(x, y, z)$, where $\mathbf{N}$ is the normal vector at (x, y, z) and f is a function of three variables. Then by (5),

$$W = \int_C \mathbf{F}(x, y, z) \cdot \mathbf{T}(x, y, z)\, ds = \int_C f(x, y, z)\mathbf{N}(x, y, z) \cdot \mathbf{T}(x, y, z)\, ds = \int_C 0\, ds = 0$$

b. Since the gravitational field of the earth is a central force field and since the normal vector at any point on the circular orbit also points toward the origin, it follows that the gravitational force acts in a direction normal to the path of the object. Consequently by part (a), the work done by the force is 0.

SECTION 16.3

Note: In Exercises 1–7 of this section, the curve C lies in a region without holes, so conditions 1–4 are equivalent. In particular, if curl $\mathbf{F} = \mathbf{0}$, then $\int_C \mathbf{F} \cdot d\mathbf{r}$ is independent of path.

1. Let $\mathbf{F}(x, y) = (e^x + y)\mathbf{i} + (x + 2y)\mathbf{j} = M(x, y)\mathbf{i} + N(x, y)\mathbf{j}$. Since $\partial N/\partial x = 1 = \partial M/\partial y$, so curl $\mathbf{F} = \mathbf{0}$, the line integral is independent of path. Let grad $f(x, y) = \mathbf{F}(x, y)$. Then

$$\frac{\partial f}{\partial x} = e^x + y \quad \text{and} \quad \frac{\partial f}{\partial y} = x + 2y \qquad (*)$$

Integrating both sides of the first equation in (*) with respect to x, we obtain $f(x, y) = e^x + xy + g(y)$. Taking partial derivatives with respect to y, we find that $\partial f/\partial y = x + dg/dy$. Comparing this with the second equation in (*), we conclude that $dg/dy = 2y$, so that $g(y) = y^2 + C$. If $C = 0$, then $f(x, y) = e^x + xy + y^2$. Thus

$$\int_C (e^x + y)\, dx + (x + 2y)\, dy = f(2, 3) - f(0, 1) = (e^2 + 15) - 2 = e^2 + 13$$

3. Let $\mathbf{F}(x, y, z) = y\mathbf{i} + (x + z)\mathbf{j} + y\mathbf{k} = M(x, y, z)\mathbf{i} + N(x, y, z)\mathbf{j} + P(x, y, z)\mathbf{k}$. Since $\partial P/\partial y = 1 = \partial N/\partial z$, $\partial M/\partial z = 0 = \partial P/\partial x$, and $\partial N/\partial x = 1 = \partial M/\partial y$, so curl $\mathbf{F} = \mathbf{0}$, the line integral is independent of path. Let grad $f(x, y, z) = \mathbf{F}(x, y, z)$. Then

$$\frac{\partial f}{\partial x} = y, \quad \frac{\partial f}{\partial y} = x + z, \quad \text{and} \quad \frac{\partial f}{\partial z} = y \qquad (*)$$

Integrating both sides of the first equation in (*) with respect to x, we obtain $f(x, y, z) = xy + g(y, z)$. Taking partial derivatives with respect to y, we find that $\partial f/\partial y = x + \partial g/\partial y$. Comparing this with the second equation in (*), we deduce that $\partial g/\partial y = z$, so that by integrating with respect to y we obtain $g(y, z) = yz + h(z)$. Thus $f(x, y, z) = xy + yz + h(z)$. Taking partial derivatives with respect to z, we find that $\partial f/\partial z = y + dh/dz$. Comparing this with the third equation in (*), we conclude that $dh/dz = 0$, so that $h(z) = C$. If $C = 0$, then $f(x, y, z) = xy + yz$. Since $\mathbf{r}(0) = -\mathbf{i} + \mathbf{j}$ and $\mathbf{r}(\frac{1}{2}) = -\frac{5}{3}\mathbf{i} + \mathbf{k}$, it follows that the initial point of C is $(-1, 1, 0)$ and the terminal point is $(-\frac{5}{3}, 0, 1)$. Thus

$$\int_C y\, dx + (x + z)\, dy + y\, dz = f(-\tfrac{5}{3}, 0, 1) - f(-1, 1, 0) = 0 - (-1) = 1$$

5. Let

$$\mathbf{F}(x, y, z) = \frac{x}{1 + x^2 + y^2 + z^2}\mathbf{i} + \frac{y}{1 + x^2 + y^2 + z^2}\mathbf{j} + \frac{z}{1 + x^2 + y^2 + z^2}\mathbf{k}$$
$$= M(x, y, z)\mathbf{i} + N(x, y, z)\mathbf{j} + P(x, y, z)\mathbf{k}$$

Since

$$\frac{\partial P}{\partial y} = \frac{-2yz}{(1 + x^2 + y^2 + z^2)^2} = \frac{\partial N}{\partial z}, \qquad \frac{\partial M}{\partial z} = \frac{-2xz}{(1 + x^2 + y^2 + z^2)^2} = \frac{\partial P}{\partial x}$$

and

$$\frac{\partial N}{\partial x} = \frac{-2xy}{(1 + x^2 + y^2 + z^2)^2} = \frac{\partial M}{\partial y}$$

so curl $\mathbf{F} = \mathbf{0}$, the line integral is independent of path. Let grad $f(x, y, z) = \mathbf{F}(x, y, z)$. Then

$$\frac{\partial f}{\partial x} = \frac{x}{1 + x^2 + y^2 + z^2}, \quad \frac{\partial f}{\partial y} = \frac{y}{1 + x^2 + y^2 + z^2}, \quad \text{and} \quad \frac{\partial f}{\partial z} = \frac{z}{1 + x^2 + y^2 + z^2}$$

Integrating both sides of the first equation in (*) with respect to x, we obtain $f(x, y, z) = \frac{1}{2}\ln(1 + x^2 + y^2 + z^2) + g(y, z)$. Taking partial derivatives with respect to y, we find that $\partial f/\partial y = y/(1 + x^2 + y^2 + z^2) + \partial g/\partial y$. Comparing this with the second equation in (*), we deduce that $\partial g/\partial y = 0$, so that integrating with respect to y yields $g(y, z) = h(z)$. Thus $f(x, y, z) = \frac{1}{2}\ln(1 + x^2 + y^2 + z^2) + h(z)$. Taking partial derivatives with respect to z, we find that $\partial f/\partial z = z/(1 + x^2 + y^2 + z^2) + dh/dz$. Comparing this with the third equation in (*), we conclude that $dh/dz = 0$, so that integration yields $h(z) = C$. If $C = 0$, then $f(x, y, z) = \frac{1}{2}\ln(1 + x^2 + y^2 + z^2)$. Since $\mathbf{r}(0) = \mathbf{0}$ and $\mathbf{r}(1) = \mathbf{i} + \mathbf{j} + \mathbf{k}$, it follows that the initial and terminal points of C are $(0, 0, 0)$ and $(1, 1, 1)$, respectively. Then

$$\int_C \frac{x}{1 + x^2 + y^2 + z^2}\, dx + \frac{y}{1 + x^2 + y^2 + z^2}\, dy + \frac{z}{1 + x^2 + y^2 + z^2}\, dz$$
$$= f(1, 1, 1) - f(0, 0, 0) = \tfrac{1}{2}\ln 4 - 0 = \ln 2$$

7. Let $\mathbf{F}(x, y, z) = e^{-x}\ln y\,\mathbf{i} - (e^{-x}/y)\mathbf{j} + z\mathbf{k} = M(x, y, z)\mathbf{i} + N(x, y, z)\mathbf{j} + P(x, y, z)\mathbf{k}$. Since $\partial P/\partial y = 0 = \partial N/\partial z$, $\partial M/\partial z = 0 = \partial P/\partial x$, and $\partial N/\partial x = e^{-x}/y = \partial M/\partial y$, so curl $\mathbf{F} = \mathbf{0}$, the line integral is independent of path. Let grad $f(x, y, z) = \mathbf{F}(x, y, z)$. Then

$$\frac{\partial f}{\partial x} = e^{-x}\ln y, \quad \frac{\partial f}{\partial y} = \frac{-e^{-x}}{y}, \quad \text{and} \quad \frac{\partial f}{\partial z} = z \qquad (*)$$

Integrating both sides of the first equation in (*) with respect to x, we obtain $f(x, y, z) = -e^{-x}\ln y + g(y, z)$. Taking partial derivatives with respect to y, we find that $\partial f/\partial y = -e^{-x}/y + \partial g/\partial y$. Comparing this with the second equation in (*), we deduce that $\partial g/\partial y = 0$, so integration yields $g(y, z) = h(z)$. Then $f(x, y, z) = -e^{-x}\ln y + h(z)$. Taking partial derivatives with respect to z, we find that $\partial f/\partial z = dh/dz$. Comparing this with the third equation in (*), we conclude that $dh/dz = z$, so integration yields $h(z) = \frac{1}{2}z^2 + C$.

If $C = 0$, then we find that $f(x, y, z) = -e^{-x}\ln y + \frac{1}{2}z^2$. Since $\mathbf{r}(0) = -\mathbf{i} + \mathbf{j} + \mathbf{k}$ and $\mathbf{r}(1) = e\mathbf{j} + 2\mathbf{k}$, it follows that the initial and terminal points of C and $(-1, 1, 1)$ and $(0, e, 2)$, respectively. Then

$$\int_C e^{-x}\ln y\,dx - \frac{e^{-x}}{y}\,dy + z\,dz = f(0, e, 2) - f(-1, 1, 1) = 1 - \tfrac{1}{2} = \tfrac{1}{2}$$

9. Let $\mathbf{F}(x, y, z) = f(x)\mathbf{i} + g(y)\mathbf{j} + h(z)\mathbf{k} = M(x, y, z)\mathbf{i} + N(x, y, z)\mathbf{j} + P(x, y, z)\mathbf{k}$. Since $\partial P/\partial y = 0 = \partial N/\partial z$, $\partial M/\partial z = 0 = \partial P/\partial x$, and $\partial N/\partial x = 0 = \partial M/\partial y$, it follows that curl $\mathbf{F} = \mathbf{0}$, so that $\int_C \mathbf{F}\cdot d\mathbf{r} = \int_C f(x)\,dx + g(y)\,dy + h(z)\,dz$ is independent of path.

11. a. Let $h(u) = \int g(u)\,du$ and $f(x, y, z) = \frac{1}{2}h(x^2 + y^2 + z^2)$. If $u = x^2 + y^2 + z^2$, then

$$\begin{aligned}\text{grad } f &= \frac{\partial f}{\partial x}\mathbf{i} + \frac{\partial f}{\partial y}\mathbf{j} + \frac{\partial f}{\partial z}\mathbf{k} = \frac{1}{2}\frac{dh}{du}\frac{\partial u}{\partial x}\mathbf{i} + \frac{1}{2}\frac{dh}{du}\frac{\partial u}{\partial y}\mathbf{j} + \frac{1}{2}\frac{dh}{du}\frac{\partial u}{\partial z}\mathbf{k}\\ &= \tfrac{1}{2}g(x^2 + y^2 + z^2)(2x)\mathbf{i} + \tfrac{1}{2}g(x^2 + y^2 + z^2)(2y)\mathbf{j} + \tfrac{1}{2}g(x^2 + y^2 + z^2)(2z)\mathbf{k}\\ &= g(x^2 + y^2 + z^2)(x\mathbf{i} + y\mathbf{j} + z\mathbf{k}) = \mathbf{F}(x, y, z)\end{aligned}$$

Thus $\mathbf{F}$ is conservative.

b. Since curl $\mathbf{F}$ = curl (grad f) from (a), and since curl (grad f) $= \mathbf{0}$ from (5) in Section 16.1, it follows that curl $\mathbf{F} = \mathbf{0}$, so $\mathbf{F}$ is irrotational.

13. Since the electric field $\mathbf{E}$ is conservative, the work is independent of path. Thus we may assume that the charge moves in a straight line. We set up a coordinate system so that the electron is at the origin and the charge moves from $(0, 0, 10^{-11})$ to $(0, 0, 10^{-12})$. At this time we recall from Section 16.1 that

$$\mathbf{E}(x, y, z) = \frac{(-1.6)\times 10^{-19}}{4\pi\varepsilon_0(x^2 + y^2 + z^2)^{3/2}}(x\mathbf{i} + y\mathbf{j} + z\mathbf{k})$$

Now if

$$f(x, y, z) = \frac{(1.6)\times 10^{-19}}{(x^2 + y^2 + z^2)^{1/2}}$$

then grad $f(x, y, z) = \mathbf{E}(x, y, z)$, so

$$\begin{aligned}W &= \int_C \mathbf{E}\cdot d\mathbf{r} = f(0, 0, 10^{-12}) - f(0, 0, 10^{-11}) = \frac{(1.6)\times 10^{-19}}{4\pi\varepsilon_0}(10^{12} - 10^{11})\\ &\approx 1293.8 \text{ (joules)}\end{aligned}$$

15. Let $t = a$ when the speed of the object is 50 meters per second, and $t = b$ when the speed of the object is 10 meters per second. The change in kinetic energy (in joules) is

$$\frac{m}{2}\|\mathbf{r}'(b)\|^2 - \frac{m}{2}\|\mathbf{r}'(a)\|^2 = \frac{5}{2}(10)^2 - \frac{5}{2}(50)^2 = -6000$$

It follows from the Law of Conservation of Energy that the change in kinetic energy is the negative of the change in potential energy, so this latter change is 6000 joules.

SECTION 16.4

In Exercises 1–18, R denotes the region enclosed by C.

1. By Green's Theorem,

$$\int_C M(x, y)\,dx + N(x, y)\,dy = \iint_R \left(\frac{\partial N}{\partial x} - \frac{\partial M}{\partial y}\right)dA = \iint_R (0 - 1)\,dA = -(\text{area of } R) = -\pi$$

3. By Green's Theorem,

$$\begin{aligned}\int_C M(x, y)\,dx + N(x, y)\,dy &= \iint_R \left(\frac{\partial N}{\partial x} - \frac{\partial M}{\partial y}\right)dA = \iint_R (\tfrac{3}{2}x^{1/2} - x)\,dA\\ &= \int_0^1\int_0^1 (\tfrac{3}{2}x^{1/2} - x)\,dx\,dy = \int_0^1 (x^{3/2} - \tfrac{1}{2}x^2)\Big|_0^1\,dy = \int_0^1 \tfrac{1}{2}\,dy = \tfrac{1}{2}\end{aligned}$$

5. By Green's Theorem,

$$\begin{aligned}\int_C M(x, y)\,dx + N(x, y)\,dy &= \iint_R \left(\frac{\partial N}{\partial x} - \frac{\partial M}{\partial y}\right)dA\\ &= \iint_R [3x(x^2 + y^2)^{1/2} - 3y(x^2 + y^2)^{1/2}]\,dA = \int_0^{2\pi}\int_0^1 3r^3(\cos\theta - \sin\theta)\,dr\,d\theta\\ &= \int_0^{2\pi} \tfrac{3}{4}r^4(\cos\theta - \sin\theta)\Big|_0^1\,d\theta = \int_0^{2\pi} \tfrac{3}{4}(\cos\theta - \sin\theta)\,d\theta = \tfrac{3}{4}(\sin\theta + \cos\theta)\Big|_0^{2\pi} = 0\end{aligned}$$

7. By Green's Theorem,

$$\begin{aligned}\int_C y\,dx - x\,dy &= \iint_R \left[\frac{\partial}{\partial x}(-x) - \frac{\partial y}{\partial y}\right]dA = \iint_R -2\,dA = \int_0^{2\pi}\int_0^{1-\cos\theta} -2r\,dr\,d\theta\\ &= \int_0^{2\pi} -r^2\Big|_0^{1-\cos\theta}\,d\theta = \int_0^{2\pi}(-1 + 2\cos\theta - \cos^2\theta)\,d\theta\\ &= \int_0^{2\pi}(-1 + 2\cos\theta - \tfrac{1}{2} - \cos 2\theta)\,d\theta = (-\tfrac{3}{2}\theta + 2\sin\theta - \tfrac{1}{2}\sin 2\theta)\Big|_0^{2\pi} = -3\pi\end{aligned}$$

9. By Green's Theorem,

$$\begin{aligned}\int_C e^x\sin y\,dx + e^x\cos y\,dy &= \iint_R \left[\frac{\partial}{\partial x}(e^x\cos y) - \frac{\partial}{\partial y}(e^x\sin y)\right]dA\\ &= \iint_R (e^x\cos y - e^x\cos y)\,dA = \iint_R 0\,dA = 0\end{aligned}$$

11. By Green's Theorem,

$$\begin{aligned}\int_C xy\,dx + (\tfrac{1}{2}x^2 + xy)\,dy &= \iint_R \left[\frac{\partial}{\partial x}(\tfrac{1}{2}x^2 + xy) - \frac{\partial}{\partial y}(xy)\right]dA = \iint_R (x + y - x)\,dA\\ &= \iint_R y\,dA = \int_{-1}^1\int_0^{(1/2)\sqrt{1-x^2}} y\,dy\,dx = \int_{-1}^1 \tfrac{1}{2}y^2\Big|_0^{(1/2)\sqrt{1-x^2}}\,dx = \int_{-1}^1 \tfrac{1}{8}(1 - x^2)\,dx\\ &= \tfrac{1}{8}(x - \tfrac{1}{3}x^3)\Big|_{-1}^1 = \tfrac{1}{6}\end{aligned}$$

13. By Green's Theorem,

$$\int_C (\cos^3 x + e^x)\,dx + e^y\,dy = \iint_R \left[\frac{\partial}{\partial x}(e^y) - \frac{\partial}{\partial y}(\cos^3 x + e^x)\right]dA = \iint_R 0\,dA = 0$$

15. By Green's Theorem,

$$\int_C \mathbf{F}\cdot d\mathbf{r} = \int_C y\,dx + 3x\,dy = \iint_R \left[\frac{\partial}{\partial x}(3x) - \frac{\partial y}{\partial y}\right] dA = \iint_R (3-1)\,dA = 2(\text{area of } R) = 8\pi$$

17. By Green's Theorem,

$$\int_C \mathbf{F}\cdot d\mathbf{r} = \int_C y\sin x\,dx - \cos x\,dy = \iint_R \left[\frac{\partial}{\partial x}(-\cos x) - \frac{\partial}{\partial y}(y\sin x)\right] dA$$

$$= \iint_R (\sin x - \sin x)\,dA = \iint_R 0\,dA = 0$$

19. Let C be the boundary of the given region R, with C oriented counterclockwise. Then C is composed of C_1 and $-C_2$, where C_1 is the interval $[0, 2\pi]$, parametrized by $\mathbf{r}_1(t) = t\mathbf{i}$ for $0 \le t \le 2\pi$, and C_2 is the cycloid parametrized by the given vector-valued function $\mathbf{r}$. Thus by (4),

$$A = \tfrac{1}{2}\int_C x\,dy - y\,dx = \tfrac{1}{2}\int_{C_1} x\,dy - y\,dx - \tfrac{1}{2}\int_{C_2} x\,dy - y\,dx$$

$$= \tfrac{1}{2}\int_0^{2\pi} [(t)(0) - (0)(1)]\,dt - \tfrac{1}{2}\int_0^{2\pi} [(t - \sin t)(\sin t) - (1 - \cos t)(1 - \cos t)]\,dt$$

$$= -\tfrac{1}{2}\int_0^{2\pi} (t\sin t - \sin^2 t - 1 + 2\cos t - \cos^2 t)\,dt$$

$$= -\tfrac{1}{2}\int_0^{2\pi} (-2 + 2\cos t)\,dt - \tfrac{1}{2}\int_0^{2\pi} t\sin t\,dt$$

$$\overset{\text{parts}}{=} -\tfrac{1}{2}(-2t + 2\sin t)\Big|_0^{2\pi} + \tfrac{1}{2}t\cos t\Big|_0^{2\pi} - \tfrac{1}{2}\int_0^{2\pi} \cos t\,dt = 2\pi + \pi - \tfrac{1}{2}\sin t\Big|_0^{2\pi} = 3\pi$$

21. Let C be the boundary of the given region, with C oriented in the counterclockwise direction. Then C consists of C_1, C_2, and C_3, where C_1 is the curve parametrized by the given vector-valued function $\mathbf{r}$, C_2 is the curve parametrized by $\mathbf{r}_2(t) = (1 - t)\mathbf{i} + \frac{1}{4}\mathbf{j}$ for $0 \le t \le 1$, and C_3 is the curve parametrized by $\mathbf{r}_3(t) = (\frac{1}{4} - t)\mathbf{j}$ for $0 \le t \le \frac{1}{4}$. Thus by (4),

$$A = \int_C x\,dy = \int_{C_1} x\,dy + \int_{C_2} x\,dy + \int_{C_3} x\,dy$$

$$= \int_0^{1/2} (\sin \pi t)(1 - 2t)\,dt + \int_0^1 (1 - t)(0)\,dt + \int_0^{1/4} (0)(-1)\,dt$$

$$= \int_0^{1/2} \sin \pi t\,dt - 2\int_0^{1/2} t\sin \pi t\,dt$$

$$\overset{\text{parts}}{=} -\frac{1}{\pi}\cos \pi t\Big|_0^{1/2} + \frac{2}{\pi}t\cos \pi t\Big|_0^{1/2} - 2\int_0^{1/2} \frac{1}{\pi}\cos \pi t\,dt = \frac{1}{\pi} - \frac{2}{\pi^2}\sin \pi t\Big|_0^{1/2} = \frac{1}{\pi} - \frac{2}{\pi^2}$$

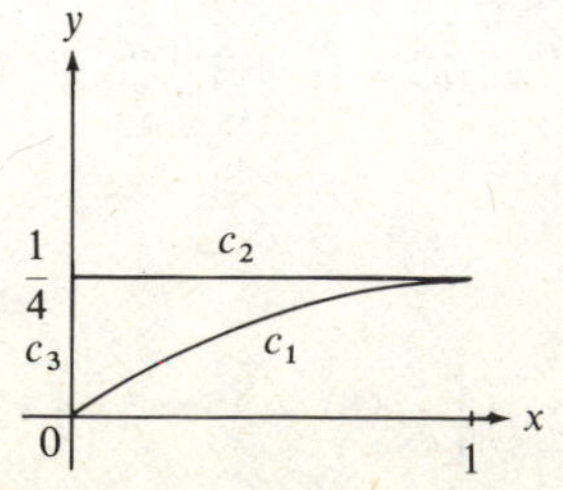

23. Let R_1 and R_2 be the regions enclosed by C_1 and C_2, respectively. Then by Green's Theorem,

$$\int_{C_1} M(x, y)\,dx + N(x, y)\,dy = \iint_{R_1} \left(\frac{\partial N}{\partial x} - \frac{\partial M}{\partial y}\right) dA = \iint_{P_1} 0\,dA = 0$$

and

$$\int_{C_2} M(x, y)\,dx + N(x, y)\,dy = \iint_{R_2} \left(\frac{\partial N}{\partial x} - \frac{\partial M}{\partial y}\right) dA = \iint_{R_2} 0\,dA = 0$$

Thus

$$\int_{C_1} M(x, y)\,dx + N(x, y)\,dy = \int_{C_2} M(x, y)\,dx + N(x, y)\,dy$$

25. a. Applying (6) with $\mathbf{F}$ replaced by $g\mathbf{F}$, we obtain

$$\int_C g\mathbf{F}\cdot\mathbf{n}\,ds = \iint_R \text{div}\,(g\mathbf{F})\,dA = \iint_R (g\,\text{div}\,\mathbf{F} + (\text{grad}\,g)\cdot\mathbf{F})\,dA$$

b. Applying part (a) with $\mathbf{F}$ replaced by grad f, and using the fact that div (grad f) $= \nabla^2 f$, we obtain

$$\int_C g(\text{grad}\,f)\cdot\mathbf{n}\,ds = \iint_R [g\,\text{div}\,(\text{grad}\,f) + (\text{grad}\,g)\cdot(\text{grad}\,f)]\,dA$$

$$= \iint_R [g\nabla^2 f + (\text{grad}\,g)\cdot(\text{grad}\,f)]\,dA$$

c. From (8) we obtain

$$\int_C (g\,\text{grad}\,f - f\,\text{grad}\,g)\cdot\mathbf{n}\,ds = \int_C g(\text{grad}\,f)\cdot\mathbf{n}\,ds - \int_C f(\text{grad}\,g)\cdot\mathbf{n}\,ds$$

$$\iint_R [g\nabla^2 f + (\text{grad}\,g)\cdot(\text{grad}\,f)]\,dA - \iint_R [f\nabla^2 g + (\text{grad}\,f)\cdot(\text{grad}\,g)]\,dA$$

$$= \iint_R (g\nabla^2 f - f\nabla^2 g)\,dA$$

27. a. The line segment C is parametrized by $\mathbf{r}(t) = [x_1 + (x_2 - x_1)t]\mathbf{i} + [y_1 + (y_2 - y_1)t]\mathbf{j}$ for $0 \le t \le 1$. Thus

$$\tfrac{1}{2}\int_C x\,dy - y\,dx = \tfrac{1}{2}\int_0^1 \{[x_1 + (x_2 - x_1)t](y_2 - y_1) - [y_1 + (y_2 - y_1)t](x_2 - x_1)\}\,dt$$

$$= \tfrac{1}{2}\int_0^1 [x_1(y_2 - y_1) - y_1(x_2 - x_1)]\,dt = \tfrac{1}{2}(x_1y_2 - x_2y_1)$$

b. Let C be the polygon oriented counterclockwise. Then we have $C = C_1 + C_2 + \cdots + C_n$, where for $1 \le k \le n - 1$, C_k is the line segment joining (x_k, y_k) to (x_{k+1}, y_{k+1}) and C_n is the line segment joining (x_n, y_n) to (x_1, y_1). Then by (4) and part (a),

$$A = \tfrac{1}{2}\int_C x\,dy - y\,dx = \tfrac{1}{2}\int_{C_1} x\,dy - y\,dx + \tfrac{1}{2}\int_{C_2} x\,dy - y\,dx + \cdots + \int_{C_n} x\,dy - y\,dx$$

$$= \tfrac{1}{2}(x_1y_2 - x_2y_1) + \tfrac{1}{2}(x_2y_3 - x_3y_2) + \cdots + \tfrac{1}{2}(x_{n-1}y_n - x_ny_{n-1}) + \tfrac{1}{2}(x_ny_1 - x_1y_n)$$

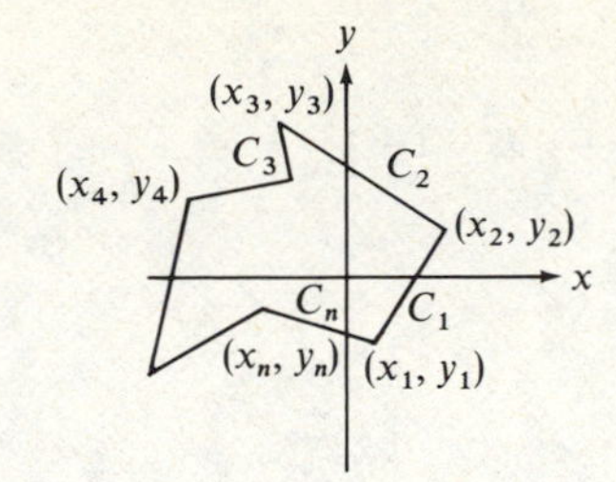

c. Taking $(x_1, y_1) = (0, 0)$, $(x_2, y_2) = (1, 0)$, $(x_3, y_3) = (2, 3)$, and $(x_4, y_4) = (-1, 1)$ in part (b), we obtain $A = \frac{1}{2}(0) + \frac{1}{2}(3) + \frac{1}{5}(5) + \frac{1}{2}(0) = 4$.

29. a. Since $\mathbf{r} = x\mathbf{i} + y\mathbf{j}$, we have $d\mathbf{r}/d\tau = (dx/d\tau)\mathbf{i} + (dy/d\tau)\mathbf{j}$, so that $\mathbf{r} \times d\mathbf{r}/d\tau = (x\,dy/d\tau - y\,dx/d\tau)\mathbf{k}$. Thus (4) of Section 13.7, with τ replacing t, implies that $x\,dy/d\tau - y\,dx/d\tau = p$.

b. Since C_1 is the line segment joining (x, y) to $(0, 0)$, and C_2 is the line segment joining $(0, 0)$ to (x_0, y_0), Exercise 27(a) implies that

$$\frac{1}{2}\int_{C_1} x\,dy - y\,dx = \frac{1}{2}[(x)(0) - (0)(y)] = 0 \quad \text{and}$$

$$\frac{1}{2}\int_{C_2} x\,dy - y\,dx = \frac{1}{2}[(0)(y_0) - (x_0)(0)] = 0$$

Since C_3 is parametrized by $\mathbf{r}(\tau) = x(\tau)\mathbf{i} + y(\tau)\mathbf{j}$ for $t_0 \le \tau \le t$, we have

$$\frac{1}{2}\int_{C_3} x\,dy - y\,dx = \frac{1}{2}\int_{t_0}^{t}\left(x\frac{dy}{d\tau} - y\frac{dx}{d\tau}\right)d\tau$$

Therefore (4) (in the present section) implies that

$$A(t) = \frac{1}{2}\int_C x\,dy - y\,dx = \frac{1}{2}\int_{C_1} x\,dy - y\,dx + \frac{1}{2}\int_{C_2} x\,dy - y\,dx + \frac{1}{2}\int_{C_3} x\,dy - y\,dx$$

$$= \frac{1}{2}\int_{t_0}^{t}\left(x\frac{dy}{d\tau} - y\frac{dx}{d\tau}\right)d\tau$$

c. From (a) and (b) we have

$$A(t) = \frac{1}{2}\int_{t_0}^{t}\left(x\frac{dy}{d\tau} - y\frac{dx}{d\tau}\right)d\tau = \frac{1}{2}\int_{t_0}^{t} p\,d\tau = \frac{p}{2}\tau\Big|_{t_0}^{t} = \frac{p}{2}(t - t_0)$$

Thus $dA/dt = p/2$.

SECTION 16.5

1. Let R be the region in the xy plane between the graphs of $y = 0$ and $y = (6 - 2x)/3$ on $[0, 3]$. If $f(x, y) = 6 - 2x - 3y$ for (x, y) in R, then Σ is the graph of f on R. Since

$$\sqrt{f_x^2(x, y) + f_y^2(x, y) + 1} = \sqrt{(-2)^2 + (-3)^2 + 1} = \sqrt{14}$$

it follows from (3) that

$$\iint_\Sigma x\,dS = \iint_R x\sqrt{14}\,dA = \sqrt{14}\int_0^3\int_0^{(6-2x)/3} x\,dy\,dx$$

$$= \sqrt{14}\int_0^3 xy\Big|_0^{(6-2x)/3} dx = \sqrt{14}\int_0^3 (2x - \tfrac{2}{3}x^2)\,dx = \sqrt{14}(x^2 - \tfrac{2}{9}x^3)\Big|_0^3 = 3\sqrt{14}$$

3. Let R be the circular region in the xy plane bounded by the circle $r = 2$ in polar coordinates. If $f(x, y) = 3x - 2$ for (x, y) in R, then Σ is the graph of f on R. Since

$$\sqrt{f_x^2(x, y) + f_y^2(x, y) + 1} = \sqrt{3^2 + 1} = \sqrt{10}$$

it follows from (3) that

$$\iint_\Sigma (2x^2 + 1)\,dS = \iint_R (2x^2 + 1)\sqrt{10}\,dA = \int_0^{2\pi}\int_0^2 (2r^2\cos^2\theta + 1)\sqrt{10}\,r\,dr\,d\theta$$

$$= \sqrt{10}\int_0^{2\pi} (\tfrac{1}{2}r^4\cos^2\theta + \tfrac{1}{2}r^2)\Big|_0^2 d\theta = \sqrt{10}\int_0^{2\pi} (8\cos^2\theta + 2)\,d\theta$$

$$= \sqrt{10}\int_0^{2\pi} (4 + 4\cos 2\theta + 2)\,d\theta = \sqrt{10}(6\theta + 2\sin 2\theta)\Big|_0^{2\pi} = 12\pi\sqrt{10}$$

5. The paraboloid and plane intersect for (x, y, z) such that $x^2 + y^2 = z = y$, which in polar coordinates means that $r^2 = r\sin\theta$, so that $r = \sin\theta$ for $0 \le \theta \le \pi$. Let R be the region in the xy plane bounded by $r = \sin\theta$ on $[0, \pi]$ in polar coordinates. If $f(x, y) = x^2 + y^2$ for (x, y) in R, then Σ is the graph of f on R. Since

$$\sqrt{f_x^2(x, y) + f_y^2(x, y) + 1} = \sqrt{(2x)^2 + (2y)^2 + 1} = \sqrt{4x^2 + 4y^2 + 1}$$

it follows from (3) that

$$\iint_\Sigma \sqrt{4x^2 + 4y^2 + 1}\,dS = \iint_R \sqrt{4x^2 + 4y^2 + 1}\,\sqrt{4x^2 + 4y^2 + 1}\,dA$$

$$= \int_0^\pi\int_0^{\sin\theta} (4r^2 + 1)r\,dr\,d\theta = \int_0^\pi (r^4 + \tfrac{1}{2}r^2)\Big|_0^{\sin\theta} d\theta = \int_0^\pi (\sin^4\theta + \tfrac{1}{2}\sin^2\theta)\,d\theta$$

$$= \int_0^\pi (\sin^4\theta + \tfrac{1}{4} - \tfrac{1}{4}\cos 2\theta)\,d\theta$$

$$= (-\tfrac{1}{4}\sin^3\theta\cos\theta - \tfrac{3}{8}\sin\theta\cos\theta + \tfrac{3}{8}\theta + \tfrac{1}{4}\theta - \tfrac{1}{8}\sin 2\theta)\Big|_0^\pi = \tfrac{5}{8}\pi$$

with the next to last equality coming from (12) in Section 7.1.

7. Let R be the region in the xy plane between the graphs of $y = 0$ and $y = 2$ on $[0, 3]$. If $f(x, y) = 4 - y^2$ for (x, y) in R, then Σ is the graph of f on R. Since

$$\sqrt{f_x^2(x, y) + f_y^2(x, y) + 1} = \sqrt{(-2y)^2 + 1} = \sqrt{4y^2 + 1}$$

it follows from (3) that

$$\iint_\Sigma y\,dS = \iint_R y\sqrt{4y^2 + 1}\,dA = \int_0^3\int_0^2 y\sqrt{4y^2 + 1}\,dy\,dx = \int_0^3 \tfrac{1}{12}(4y^2 + 1)^{3/2}\Big|_0^2 dx$$

$$= \int_0^3 \tfrac{1}{12}(17^{3/2} - 1)\,dx = \tfrac{1}{4}(17^{3/2} - 1)$$

9. Let R be the region in the xy plane bounded by the circle $r = 2$ in polar coordinates, and for $0 < b < 2$, let R_b be the region bounded by $r = b$. If $f(x, y) = \sqrt{4 - x^2 - y^2}$ for (x, y) in R, then Σ is the graph of f on R. Since

$$\sqrt{f_x^2(x, y) + f_y^2(x, y) + 1} = \sqrt{\left(\frac{-x}{\sqrt{4 - x^2 - y^2}}\right)^2 + \left(\frac{-y}{\sqrt{4 - x^2 - y^2}}\right)^2 + 1}$$
$$= \frac{2}{\sqrt{4 - x^2 - y^2}} \quad \text{for } (x, y) \text{ in } R_b$$

it follows from (3) and the solution of Example 3 that

$$\iint_\Sigma z(x^2 + y^2)\,dS = \lim_{b \to 2^-} \iint_{R_b} [(x^2 + y^2)\sqrt{4 - x^2 - y^2}]\frac{2}{\sqrt{4 - x^2 - y^2}}\,dA$$
$$= \lim_{b \to 2^-} \int_0^{2\pi}\int_0^b 2r^3\,dr\,d\theta = \lim_{b \to 2^-} \int_0^{2\pi} \tfrac{1}{2}r^4\Big|_0^b\,d\theta = \lim_{b \to 2^-} \int_0^{2\pi} \tfrac{1}{2}b^4\,d\theta = \lim_{b \to 2^-} \pi b^4 = 16\pi$$

11. Let Σ be composed of the six surfaces Σ_1, Σ_2, Σ_3, Σ_4, Σ_5, and Σ_6, as in the figure. For Σ_1 and Σ_2 we use the region R_1 in the xy plane bounded by $y = 0$ and $y = 1$ on $[0, 1]$. If $f(x, y) = 0$, then $\sqrt{f_x^2(x, y) + f_y^2(x, y) + 1} = 1$, and

$$\iint_{\Sigma_1} (x + y)\,dS = \iint_{R_1} (x + y)\,dA = \int_0^1\int_0^1 (x + y)\,dy\,dx = \int_0^1 \left(xy + \frac{y^2}{2}\right)\Big|_0^1\,dx$$
$$= \int_0^1 (x + \tfrac{1}{2})\,dx = (\tfrac{1}{2}x^2 + \tfrac{1}{2}x)\Big|_0^1 = 1$$

If $f(x, y) = 1$, then $\sqrt{f_x^2(x, y) + f_y^2(x, y) + 1} = 1$, and as above, $\iint_{\Sigma_2} (x + y)\,dS = \iint_{R_1} (x + y)\,dA = 1$. For Σ_3 and Σ_4 we use the region R_3 in the xz plane bounded by $z = 0$ and $z = 1$ on $[0, 1]$. If $f(x, z) = 0$, then $\sqrt{f_x^2(x, z) + f_z^2(x, z) + 1} = 1$, and

$$\iint_{\Sigma_3} (x + y)\,dS = \iint_{R_3} x\,dA = \int_0^1\int_0^1 x\,dz\,dx = \int_0^1 xz\Big|_0^1\,dx = \int_0^1 x\,dx = \tfrac{1}{2}x^2\Big|_0^1 = \tfrac{1}{2}$$

If $f(x, z) = 1$, then $\sqrt{f_x^2(x, z) + f_z^2(x, z) + 1} = 1$, and

$$\iint_{\Sigma_4} (x + y)\,dS = \iint_{R_3} (x + 1)\,dA = \int_0^1\int_0^1 (x + 1)\,dz\,dx = \int_0^1 (xz + z)\Big|_0^1\,dx = \int_0^1 (x + 1)\,dx$$
$$= (\tfrac{1}{2}x^2 + x)\Big|_0^1 = \tfrac{3}{2}$$

For Σ_5 and Σ_6 we use the region R_5 in the yz plane bounded by $z = 0$ and $z = 1$ on $[0, 1]$. If $f(y, z) = 0$, then $\sqrt{f_y^2(y, z) + f_z^2(y, z) + 1} = 1$, and

$$\iint_{\Sigma_5} (x + y)\,dS = \iint_{R_5} y\,dA = \int_0^1\int_0^1 y\,dz\,dy = \int_0^1 yz\Big|_0^1\,dy = \int_0^1 y\,dy = \tfrac{1}{2}y^2\Big|_0^1 = \tfrac{1}{2}$$

Finally, if $f(y, z) = 1$, then $\sqrt{f_y^2(y, z) + f_z^2(y, z) + 1} = 1$, and

$$\iint_{\Sigma_6} (x + y)\,dS = \iint_{R_5} (1 + y)\,dA = \int_0^1\int_0^1 (1 + y)\,dz\,dy = \int_0^1 (z + yz)\Big|_0^1\,dy = \int_0^1 (1 + y)\,dy$$
$$= (y + \tfrac{1}{2}y^2)\Big|_0^1 = \tfrac{3}{2}$$

Consequently

$$\iint_\Sigma (x + y)\,dS = \sum_{k=1}^{6} \iint_{\Sigma_k} (x + y)\,dS = 1 + 1 + \tfrac{1}{2} + \tfrac{3}{2} + \tfrac{1}{2} + \tfrac{3}{2} = 6$$

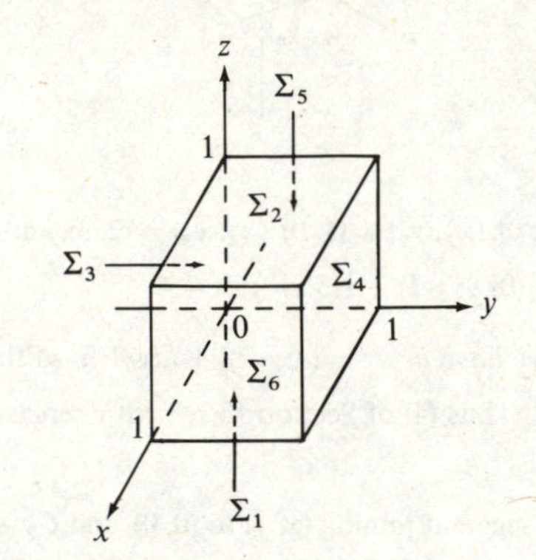

13. Let R be the region in the xy plane bounded by $r = \frac{1}{4}$ and $r = 2$ in polar coordinates, and let $f(x, y) = 2\sqrt{x^2 + y^2}$ for (x, y) in R. Then the funnel occupies the region Σ which is the graph of f on R. The mass of the funnel is given by

$$m = \iint_\Sigma \delta(x, y, z)\,dS = \iint_\Sigma (6 - z)\,dS$$
$$= \iint_R (6 - 2\sqrt{x^2 + y^2})\sqrt{\left(\frac{2x}{\sqrt{x^2 + y^2}}\right)^2 + \left(\frac{2y}{\sqrt{x^2 + y^2}}\right)^2 + 1}\,dA$$
$$= \iint_R (6 - 2\sqrt{x^2 + y^2})\sqrt{5}\,dA = \int_0^{2\pi}\int_{1/4}^2 (6 - 2r)\sqrt{5}\,r\,dr\,d\theta$$
$$= \sqrt{5}\int_0^{2\pi} (3r^2 - \tfrac{2}{3}r^3)\Big|_{1/4}^2\,d\theta = \sqrt{5}\int_0^{2\pi} \tfrac{623}{96}\,d\theta = \tfrac{623}{48}\sqrt{5}\pi$$

15. The surface Σ of the tank consists of the top part, Σ_1, and the bottom part, Σ_2, as in the figure. Let R be the region in the xy plane bounded by the circle $r = 10$, and for $0 < b < 10$ let R_b be the region in the xy plane bounded by the circle $r = b$. If $f(x, y) = \sqrt{100 - x^2 - y^2}$, then Σ_1 is the graph of f on R, and $\Sigma_2 = R$. Moreover, $z_0 = 10$. By the formula for the force F on Σ and by (3),

$$F = \iint_\Sigma (62.5)(z_0 - z)\,dS = \iint_{\Sigma_1} (62.5)(10 - z)\,dS + \iint_{\Sigma_2} (62.5)(10 - z)\,dS$$
$$= (62.5)\lim_{b \to 10^-} \iint_{R_b} (10 - \sqrt{100 - x^2 - y^2})$$
$$\times \sqrt{\left(\frac{-x}{\sqrt{100 - x^2 - y^2}}\right)^2 + \left(\frac{-y}{\sqrt{100 - x^2 - y^2}}\right)^2 + 1}\,dA + (62.5)\iint_R (10 - 0)\,dA$$

$$= (62.5) \lim_{b\to 10^-} \left[\iint_{R_b} \frac{100}{\sqrt{100 - x^2 - y^2}}\, dA - \iint_{R_b} 10\, dA \right] + (62.5) \iint_R 10\, dA$$

$$= (62.5) \lim_{b\to 10^-} \iint_{R_b} \frac{100}{\sqrt{100 - x^2 - y^2}}\, dA = (62.5) \lim_{b\to 10^-} \int_0^{2\pi}\int_0^b \frac{100}{\sqrt{100 - r^2}}\, r\, dr\, d\theta$$

$$= (62.5) \lim_{b\to 10^-} \int_0^{2\pi} -100\sqrt{100 - r^2}\Big|_0^b\, d\theta$$

$$= (62.5) \lim_{b\to 10^-} \int_0^{2\pi} (-100\sqrt{100 - b^2} + 1000)\, d\theta$$

$$= (62.5)(2\pi) \lim_{b\to 10^-} (-100\sqrt{100 - b^2} + 1000) = 125{,}000\pi \text{ (pounds)}$$

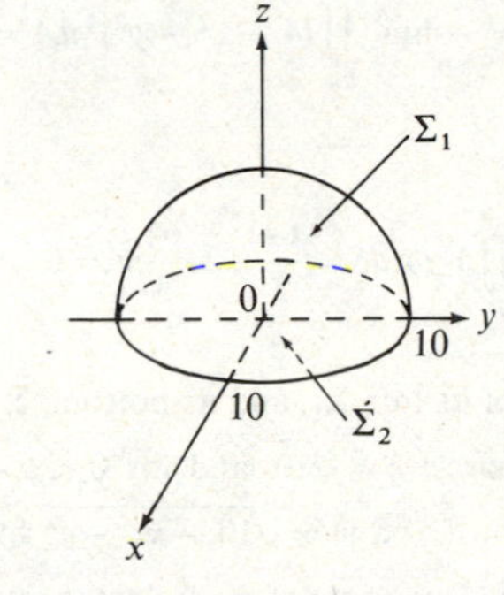

17. The surface Σ of the tank consists of the top hemisphere, Σ_1 and the bottom hemisphere, Σ_2. Let R be the region in the xy plane bounded by the circle $r = 10$, and for $0 < b < 10$ let R_b be the region in the xy plane bounded by the circle $r = b$. If $f(x, y) = \sqrt{100 - x^2 - y^2}$, then Σ_1 is the graph of f on R, and Σ_2 is the graph of $-f$ on R. Moreover, $z_0 = 10$. Thus by (3),

$$F = \iint_\Sigma (62.5)(z_0 - z)\, dS = \iint_{\Sigma_1} (62.5)(10 - z)\, dS + \iint_{\Sigma_2} (62.5)(10 - z)\, dS$$

$$= 62.5 \iint_R (10 - \sqrt{100 - x^2 - y^2}) \sqrt{\left(\frac{-x}{\sqrt{100 - x^2 - y^2}}\right)^2 + \left(\frac{-y}{\sqrt{100 - x^2 - y^2}}\right)^2 + 1}\, dA$$

$$+ 62.5 \iint_R (10 + \sqrt{100 - x^2 - y^2}) \sqrt{\left(\frac{x}{\sqrt{100 - x^2 - y^2}}\right)^2 + \left(\frac{y}{\sqrt{100 - x^2 - y^2}}\right)^2 + 1}\, dA$$

$$= 62.5 \iint_R 20 \sqrt{\frac{x^2}{100 - x^2 - y^2} + \frac{y^2}{100 - x^2 - y^2} + 1}\, dA = 12{,}500 \iint_R \frac{1}{\sqrt{100 - x^2 - y^2}}\, dA$$

$$= 12{,}500 \lim_{b\to 10^-} \int_0^{2\pi}\int_0^b \frac{r}{\sqrt{100 - r^2}}\, dr\, d\theta = 12{,}500 \lim_{b\to 10^-} \int_0^{2\pi} -\sqrt{100 - r^2}\Big|_0^b\, d\theta$$

$$= 12{,}500 \lim_{b\to 10^-} \int_0^{2\pi} (-\sqrt{100 - b^2} + 10)\, d\theta = 25{,}000\pi \lim_{b\to 10^-} (-\sqrt{100 - b^2} + 10)$$

$$= 250{,}000\pi \text{ (pounds)}$$

SECTION 16.6

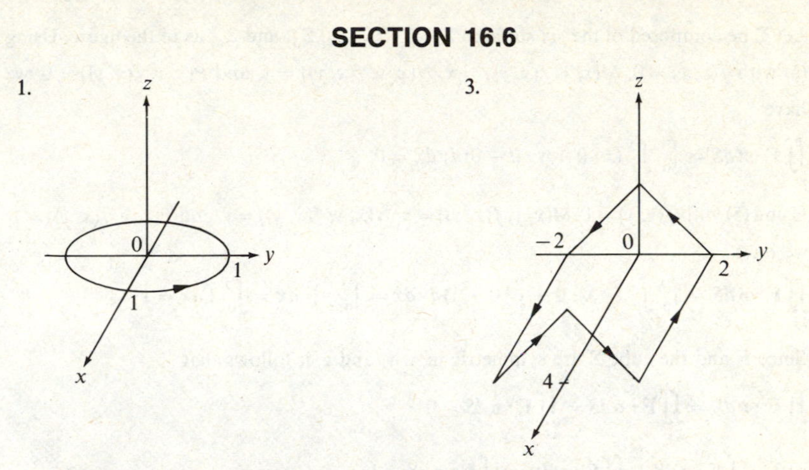

5. Let R be the region in the xy plane bounded by the circle $r = 3$ in polar coordinates, and let $f(x, y) = 9 - x^2 - y^2$ for (x, y) in R. We use (5) with $f_x(x, y) = -2x$, $f_y(x, y) = -2y$, $M(x, y, f(x, y)) = y$, $N(x, y, f(x, y)) = -x$, and $P(x, y, f(x, y)) = 8$, and find that

$$\iint_R \mathbf{F}\cdot\mathbf{n}\, dS = \iint_R [-y(-2x) + x(-2y) + 8]\, dA = \iint_R 8\, dA = 8 \text{ (area of } R) = 72\pi$$

7. Let R be the region in the xy plane between the graphs of $y = -\sqrt{1 - x^2}$ and $y = \sqrt{1 - x^2}$ on $[-1, 1]$, and R_b the region between the graphs of $y = -\sqrt{b^2 - x^2}$ and $y = \sqrt{b^2 - x^2}$ on $[-b, b]$ for $0 < b < 1$. Let $f(x, y) = -\sqrt{1 - x^2 - y^2}$ for (x, y) in R. We use (5) with $f_x(x, y) = x/\sqrt{1 - x^2 - y^2}$, $f_y(x, y) = y/\sqrt{1 - x^2 - y^2}$, $M(x, y, f(x, y)) = 1$, $N(x, y, f(x, y)) = 1$, and $P(x, y, f(x, y)) = 2$, and find that

$$\iint_\Sigma \mathbf{F}\cdot\mathbf{n}\, dS = \lim_{b\to 1^-} \iint_{R_b} \left[\frac{-x}{\sqrt{1 - x^2 - y^2}} - \frac{y}{\sqrt{1 - x^2 - y^2}} + 2\right] dA$$

Notice that

$$\lim_{b\to 1^-} \iint_{R_b} \frac{-y}{\sqrt{1 - x^2 - y^2}}\, dA = \lim_{b\to 1^-} \int_{-b}^b \int_{-\sqrt{b^2 - x^2}}^{\sqrt{b^2 - x^2}} \frac{-y}{\sqrt{1 - x^2 - y^2}}\, dy\, dx$$

$$= \lim_{b\to 1^-} \int_{-b}^b \sqrt{1 - x^2 - y^2}\Big|_{-\sqrt{b^2 - x^2}}^{\sqrt{b^2 - x^2}}\, dx = 0$$

Similarly,

$$\lim_{b\to 1^-} \iint_{R_b} \frac{-x}{\sqrt{1 - x^2 - y^2}}\, dA = 0$$

Finally,

$$\lim_{b\to 1^-} \iint_{R_b} 2\, dA = \lim_{b\to 1^-} \int_0^{2\pi}\int_0^b 2r\, dr\, d\theta = \lim_{b\to 1^-} \int_0^{2\pi} r^2\Big|_0^b\, d\theta$$

$$= \lim_{b\to 1^-} \int_0^{2\pi} b^2\, d\theta = \lim_{b\to 1^-} 2\pi b^2 = 2\pi$$

Thus $\iint_\Sigma \mathbf{F}\cdot\mathbf{n}\, ds = 0 + 0 + 2\pi = 2\pi$.

9. Let Σ be composed of the six surfaces $\Sigma_1, \Sigma_2, \Sigma_3, \Sigma_4, \Sigma_5$, and Σ_6, as in the figure. Using (6) with $f(x, y) = 0$, $M(x, y, f(x, y)) = x$, $N(x, y, f(x, y)) = y$, and $P(x, y, f(x, y)) = 0$, we have

$$\iint_{\Sigma} \mathbf{F} \cdot \mathbf{n}\, dS = \int_0^1 \int_0^1 (x \cdot 0 + y \cdot 0 - 0)\, dy\, dx = 0$$

Using (5) with $f(x, y) = 1$, $M(x, y, f(x, y)) = x$, $N(x, y, f(x, y)) = y$, and $P(x, y, f(x, y)) = 1$, we have

$$\iint_{\Sigma_2} \mathbf{F} \cdot \mathbf{n}\, dS = \int_0^1 \int_0^1 (-x \cdot 0 - y \cdot 0 + 1)\, dy\, dx = \int_0^1 y\Big|_0^1 dx = \int_0^1 1\, dx = 1$$

Since $\mathbf{F}$ and the cube Σ are symmetric in x, y, and z, it follows that

$$\iint_{\Sigma} \mathbf{F} \cdot \mathbf{n}\, dS = \iint_{\Sigma_3} \mathbf{F} \cdot \mathbf{n}\, dS = \iint_{\Sigma_5} \mathbf{F} \cdot \mathbf{n}\, dS = 0$$

and $\displaystyle\iint_{\Sigma_2} \mathbf{F} \cdot \mathbf{n}\, dS = \iint_{\Sigma_4} \mathbf{F} \cdot \mathbf{n}\, dS = \iint_{\Sigma_6} \mathbf{F} \cdot \mathbf{n}\, dS = 1$

Thus

$$\iint_{\Sigma} \mathbf{F} \cdot \mathbf{n}\, dS = \sum_{k=1}^{6} \iint_{\Sigma_k} \mathbf{F} \cdot \mathbf{n}\, dS = 0 + 1 + 0 + 1 + 0 + 1 = 3$$

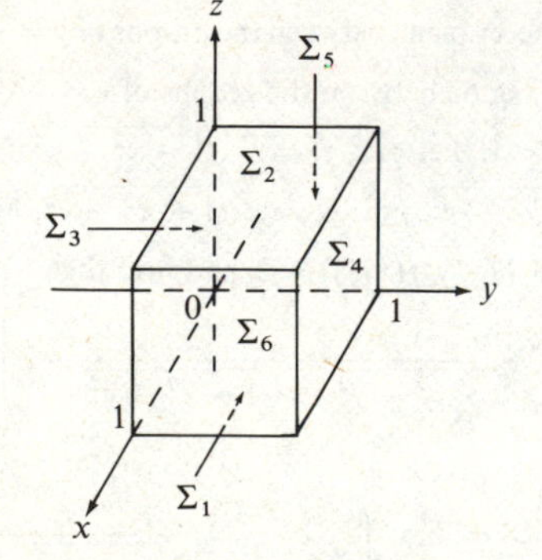

11. Let Σ be composed of the top half, Σ_1, and the bottom half, Σ_2. Let R be the region in the xy plane bounded by the circle $r = 2$ in polar coordinates, and let R_b be the region bounded by the circle $r = b$ for $0 < b < 2$. For Σ_1, let $f(x, y) = \sqrt{4 - x^2 - y^2}$ for (x, y) in R. Using (5) with $f_x(x, y) = -x/\sqrt{4 - x^2 - y^2}$, $f_y(x, y) = -y/\sqrt{4 - x^2 - y^2}$, $M(x, y, f(x, y)) = -y$, $N(x, y, f(x, y)) = x$, and $P(x, y, f(x, y)) = (\sqrt{4 - x^2 - y^2})^4$, we have

$$\iint_{\Sigma_1} \mathbf{F} \cdot \mathbf{n}\, dS = \lim_{b \to 2^-} \iint_{R_b} \left[y\left(\frac{-x}{\sqrt{4 - x^2 - y^2}}\right) - x\left(\frac{-y}{\sqrt{4 - x^2 - y^2}}\right) + (\sqrt{4 - x^2 - y^2})^4 \right] dA$$

$$= \lim_{b \to 2^-} \iint_{R_b} (4 - x^2 - y^2)^2\, dA = \lim_{b \to 2^-} \int_0^{2\pi} \int_0^b (4 - r^2)^2 r\, dr\, d\theta$$

$$= \lim_{b \to 2^-} \int_0^{2\pi} -\tfrac{1}{6}(4 - r^2)^3 \Big|_0^b d\theta = \lim_{b \to 2^-} \int_0^{2\pi} -\tfrac{1}{6}[(4 - b^2)^3 - 64]\, d\theta$$

$$= \lim_{b \to 2^-} \frac{\pi}{3}[64 - (4 - b^2)^3] = \frac{64}{3}\pi$$

For Σ_2 let $f(x, y) = -\sqrt{4 - x^2 - y^2}$ for (x, y) in R. Using (6) with $f_x(x, y) = x/\sqrt{4 - x^2 - y^2}$, $f_y(x, y) = y/\sqrt{4 - x^2 - y^2}$, M and N as above, and $P(x, y, f(x, y)) = (-\sqrt{4 - x^2 - y^2})^4$, we have

$$\iint_{\Sigma_2} \mathbf{F} \cdot \mathbf{n}\, dS = \lim_{b \to 2^-} \iint_{R_b} \left[-y\left(\frac{x}{\sqrt{4 - x^2 - y^2}}\right) + x\left(\frac{y}{\sqrt{4 - x^2 - y^2}}\right) - (-\sqrt{4 - x^2 - y^2})^4 \right] dA = -\lim_{b \to 2^-} \iint_{R_b} (4 - x^2 - y^2)^2\, dA = -\frac{64}{3}\pi$$

calculated as above. Thus

$$\iint_{\Sigma} \mathbf{F} \cdot \mathbf{n}\, dS = \iint_{\Sigma_1} \mathbf{F} \cdot \mathbf{n}\, dS + \iint_{\Sigma_2} \mathbf{F} \cdot \mathbf{n}\, dS = \frac{64}{3}\pi - \frac{64}{3}\pi = 0$$

13. The sphere Σ is composed of its top, Σ_1, and its bottom, Σ_2. Let R be the region in the xy plane bounded by the circle $r = \sqrt{10}$, and for $0 < b < \sqrt{10}$ let R_b be the region bounded by the circle $r = b$. If $f(x, y) = \sqrt{10 - x^2 - y^2}$ for (x, y) in R, then Σ_1 is the graph of f on R. By (1) and (5) the rate of mass flow through Σ_1 is given by

$$\iint_{\Sigma_1} \delta\mathbf{v} \cdot \mathbf{n}\, dS = \lim_{b \to \sqrt{10}^-} \iint_{R_b} 50\left[-x\left(\frac{-x}{\sqrt{10 - x^2 - y^2}}\right) - y\left(\frac{-y}{\sqrt{10 - x^2 - y^2}}\right) + \sqrt{10 - x^2 - y^2} \right] dA = \lim_{b \to \sqrt{10}^-} \int_0^{2\pi} \int_0^b \frac{500}{\sqrt{10 - r^2}} r\, dr\, d\theta$$

$$= \lim_{b \to \sqrt{10}^-} \int_0^{2\pi} -500\sqrt{10 - r^2}\Big|_0^b d\theta = \lim_{b \to \sqrt{10}^-} \int_0^{2\pi} (-500\sqrt{10 - b^2} + 500\sqrt{10})\, d\theta$$

$$= \lim_{b \to \sqrt{10}^-} 2\pi(-500\sqrt{10 - b^2} + 500\sqrt{10}) = 1000\pi\sqrt{10}$$

By (1) and (6) with $f(x, y) = -\sqrt{10 - x^2 - y^2}$ for (x, y) in R we find that the rate of mass flow through Σ_2 is given by

$$\iint_{\Sigma_2} \delta\mathbf{v} \cdot \mathbf{n}\, dS = \iint_{\Sigma_1} \delta\mathbf{v} \cdot \mathbf{n}\, dS = 1000\pi\sqrt{10}$$

Thus the total mass flow through the sphere Σ is

$$\iint_{\Sigma_1} \delta\mathbf{v} \cdot \mathbf{n}\, dS + \iint_{\Sigma_2} \delta\mathbf{v} \cdot \mathbf{n}\, dS = 2000\pi\sqrt{10}$$

15. Let R be the region in the xy plane between the graphs of $x = -y$ and $x = y$ on $[0, 1]$, and let R_b be the region between the graphs of $x = -y$ and $x = y$ on $[b, 1]$ for $0 < b < 1$. If $f(x, y) = \sqrt{y^2 - x^2}$, then Σ is the graph of f on R. By (5),

$$\iint_\Sigma \mathbf{E}\cdot\mathbf{n}\,dS = \lim_{b\to 0^+}\iint_{R_b}\left[-x\left(\frac{-x}{\sqrt{y^2-x^2}}\right) - y\left(\frac{y}{\sqrt{y^2-x^2}}\right) + 0\right]dA$$

$$= \lim_{b\to 0^+}\int_b^1\int_{-y}^{y} -\sqrt{y^2-x^2}\,dx\,dy \overset{x=y\sin u}{=} \lim_{b\to 0^+}\int_b^1\int_{-\pi/2}^{\pi/2} -(y\cos u)(y\cos u)\,du\,dy$$

$$= \lim_{b\to 0^+}\int_b^1\int_{-\pi/2}^{\pi/2} -y^2(\tfrac{1}{2}+\tfrac{1}{2}\cos 2u)\,du\,dy = \lim_{b\to 0^+}\int_b^1 -y^2(\tfrac{1}{2}u+\tfrac{1}{4}\sin 2u)\Big|_{-\pi/2}^{\pi/2}dy$$

$$= \lim_{b\to 0^+}\int_b^1 -\tfrac{1}{2}\pi y^2\,dy = \lim_{b\to 0^+}(-\tfrac{1}{6}\pi y^3)\Big|_b^1 = \lim_{b\to 0^+}\tfrac{1}{6}\pi(b^3-1) = -\tfrac{1}{6}\pi$$

17. Let (x_0, y_0, z_0) be a point in space with $z_0 > 0$, and consider a rectangular parallelepiped Σ centered at $(x_0, y_0, 0)$, with height $2z_0$ and base area 1, with top Σ_t, bottom Σ_b, and sides Σ_s. Let R be the region in the xy plane inside the parallelepiped. By symmetry there is a function f of a single variable such that $\mathbf{E}(x, y, z) = f(z)\mathbf{k}$ and such that $f(-z) = -f(z)$ for $z \neq 0$. Since the area of R is 1, the total charge inside the parallelepiped is σ, so Gauss's Law implies that

$$\frac{\sigma}{\varepsilon_0} = \frac{q}{\varepsilon_0} = \iint_\Sigma \mathbf{E}\cdot\mathbf{n}\,dS = \iint_{\Sigma_t}\mathbf{E}\cdot\mathbf{n}\,dS + \iint_{\Sigma_b}\mathbf{E}\cdot\mathbf{n}\,dS + \iint_{\Sigma_s}\mathbf{E}\cdot\mathbf{n}\,dS$$

Now

$$\iint_{\Sigma_t}\mathbf{E}\cdot\mathbf{n}\,dS = \iint_{\Sigma_t}\{[f(z_0)\mathbf{k}]\cdot\mathbf{k}\}\,dS = \iint_{\Sigma_t} f(z_0)\,dS = f(z_0)\iint_R 1\,dA = f(z_0)$$

and

$$\iint_{\Sigma_b}\mathbf{E}\cdot\mathbf{n}\,dS = \iint_{\Sigma_b}\{[-f(z_0)\mathbf{k}]\cdot(-\mathbf{k})\}\,dS = f(z_0)\iint_R 1\,dA = f(z_0)$$

On Σ_s the normal $\mathbf{n}$ is $\pm\mathbf{i}$ or $\pm\mathbf{j}$, and thus $\iint_{\Sigma_s}\mathbf{E}\cdot\mathbf{n}\,dS = 0$. Consequently $\sigma/\varepsilon_0 = f(z_0) + f(z_0) + 0$, so $f(z_0) = \sigma/2\varepsilon_0$ if $z_0 > 0$. By symmetry, $f(z_0) = -\sigma/2\varepsilon_0$ if $z_0 < 0$. Therefore the electric field is given by

$$\mathbf{E}(x, y, z) = \begin{cases} \dfrac{\sigma}{2\varepsilon_0}\mathbf{k} & \text{for } z > 0 \\[2mm] \dfrac{-\sigma}{2\varepsilon_0}\mathbf{k} & \text{for } z < 0 \end{cases}$$

19. Assume that the center of the sphere is the origin. By symmetry $\mathbf{E}$ is constant on spheres centered at the origin, and is normal to the spheres, so there is a function f of a single variable such that

$$\mathbf{E}(x, y, z) = f(\sqrt{x^2+y^2+z^2})\,\frac{x\mathbf{i}+y\mathbf{j}+z\mathbf{k}}{\sqrt{x^2+y^2+z^2}} \quad \text{for } x^2+y^2+z^2 > 0$$

If (x_0, y_0, z_0) is not the origin, let Σ be the sphere centered at the origin with radius $\sqrt{x_0^2+y_0^2+z_0^2}$. Then

$$\iint_\Sigma \mathbf{E}\cdot\mathbf{n}\,dS = \iint_\Sigma f(\sqrt{x_0^2+y_0^2+z_0^2})\,dS = f(\sqrt{x_0^2+y_0^2+z_0^2})\iint_\Sigma 1\,dS$$
$$= f(\sqrt{x_0^2+y_0^2+z_0^2})4\pi(x_0^2+y_0^2+z_0^2)$$

Let $x_0^2+y_0^2+z_0^2 \le a^2$, and let D be the interior of Σ. By hypothesis the total charge inside Σ is given by

$$q = \iiint_D \rho_0\,dV = \rho_0[\tfrac{4}{3}\pi(x_0^2+y_0^2+z_0^2)^{3/2}]$$

By Gauss's Law this means that

$$\frac{\rho_0}{\varepsilon_0}\left[\frac{4}{3}\pi(x_0^2+y_0^2+z_0^2)^{3/2}\right] = \frac{q}{\varepsilon_0} = \iint_\Sigma \mathbf{E}\cdot\mathbf{n}\,dS = f(\sqrt{x_0^2+y_0^2+z_0^2})4\pi(x_0^2+y_0^2+z_0^2)$$

so that

$$f(\sqrt{x_0^2+y_0^2+z_0^2}) = \frac{\rho_0}{3\varepsilon_0}(x_0^2+y_0^2+z_0^2)^{1/2}$$

Now let $x_0^2+y_0^2+z_0^2 > a^2$, and again let D be the interior of Σ. By hypothesis the total charge inside Σ is given by $q = \frac{4}{3}\pi a^3\rho_0$. By Gauss's Law this means that

$$\frac{4}{3}\pi a^3\frac{\rho_0}{\varepsilon_0} = \frac{q}{\varepsilon_0} = \iint_\Sigma \mathbf{E}\cdot\mathbf{n}\,dS = f(\sqrt{x_0^2+y_0^2+z_0^2})4\pi(x_0^2+y_0^2+z_0^2)$$

so that

$$f(\sqrt{x_0^2+y_0^2+z_0^2}) = \frac{\rho_0}{3\varepsilon_0}\,\frac{a^3}{x_0^2+y_0^2+z_0^2}$$

Consequently the electric field is given by

$$\mathbf{E}(x, y, z) = \begin{cases} \dfrac{\rho_0}{3\varepsilon_0}(x\mathbf{i}+y\mathbf{j}+z\mathbf{k}) & \text{for } 0 \le x^2+y^2+z^2 \le a^2 \\[2mm] \dfrac{\rho_0}{3\varepsilon_0}\dfrac{a^3}{(x^2+y^2+z^2)^{3/2}}(x\mathbf{i}+y\mathbf{j}+z\mathbf{k}) & \text{for } x^2+y^2+z^2 > a^2 \end{cases}$$

SECTION 16.7

1. $$\text{curl}\,\mathbf{F}(x, y, z) = \begin{vmatrix} \mathbf{i} & \mathbf{j} & \mathbf{k} \\ \dfrac{\partial}{\partial x} & \dfrac{\partial}{\partial y} & \dfrac{\partial}{\partial z} \\ z & x & y \end{vmatrix} = \mathbf{i}+\mathbf{j}+\mathbf{k}$$

Let $f(x, y) = 1 - x^2 - y^2$, and let R be the region in the first quadrant of the xy plane bounded by the lines $x = 0$, $y = 0$, and the circle $x^2 + y^2 = 1$. Then Σ is the graph of f on R. Thus by Stokes's Theorem and (6) of Section 16.6,

$$\int_C \mathbf{F}\cdot d\mathbf{r} = \iint_\Sigma(\text{curl}\,\mathbf{F})\cdot\mathbf{n}\,dS = \iint_R[(1)(-2x)+(1)(-2y)-1]\,dA$$

$$= \int_0^{\pi/2}\int_0^1(-2r\sin\theta - 2r\cos\theta - 1)r\,dr\,d\theta = \int_0^{\pi/2}(-\tfrac{2}{3}r^3\sin\theta - \tfrac{2}{3}r^3\cos\theta - \tfrac{1}{2}r^2)\Big|_0^1 d\theta$$

$$= \int_0^{\pi/2}\left(-\frac{2}{3}\sin\theta - \frac{2}{3}\cos\theta - \frac{1}{2}\right)d\theta = \left(\frac{2}{3}\cos\theta - \frac{2}{3}\sin\theta - \frac{1}{2}\theta\right)\Big|_0^{\pi/2} = -\frac{4}{3} - \frac{\pi}{4}$$

3. $\operatorname{curl} \mathbf{F}(x, y, z) = \begin{vmatrix} \mathbf{i} & \mathbf{j} & \mathbf{k} \\ \dfrac{\partial}{\partial x} & \dfrac{\partial}{\partial y} & \dfrac{\partial}{\partial z} \\ 2y & 3z & -2x \end{vmatrix} = -3\mathbf{i} + 2\mathbf{j} - 2\mathbf{k}$

Let $f(x, y) = \sqrt{1 - x^2 - y^2}$ and let R be the region in the first quadrant of the xy plane bounded by the lines $x = 0$, $y = 0$, and the circle $x^2 + y^2 = 1$. Then Σ is the graph of f on R. For $0 \le b < 1$, let R_b be the part of R inside the circle $x^2 + y^2 = b^2$. Then by Stokes's Theorem and (5) of Section 16.6.

$$\int_C \mathbf{F} \cdot d\mathbf{r} = \iint_\Sigma (\operatorname{curl} \mathbf{F}) \cdot \mathbf{n}\, dS$$

$$= \lim_{b \to 1^-} \iint_{R_b} \left[-(-3)\frac{-x}{\sqrt{1 - x^2 - y^2}} - (2)\frac{-y}{\sqrt{1 - x^2 - y^2}} - 2 \right] dA$$

$$= \lim_{b \to 1^-} \left[-3 \iint_{R_b} \frac{x}{\sqrt{1 - x^2 - y^2}}\, dA + 2 \iint_{R_b} \frac{y}{\sqrt{1 - x^2 - y^2}}\, dA - 2(\text{area of } R_b) \right]$$

Since the two double integrals have the same form but with x and y interchanged, and since R_b is symmetric in x and y, the two double integrals are equal. Therefore

$$\int_C \mathbf{F} \cdot d\mathbf{r} = \lim_{b \to 1^-} \left[-\iint_{R_b} \frac{x}{\sqrt{1 - x^2 - y^2}}\, dA - 2(\text{area of } R_b) \right]$$

Now

$$\iint_{R_b} \frac{x}{\sqrt{1 - x^2 - y^2}}\, dA = \int_0^b \int_0^{\pi/2} \frac{r \cos\theta}{\sqrt{1 - r^2}}\, r\, d\theta\, dr = \int_0^b \frac{r^2}{\sqrt{1 - r^2}} \sin\theta \Big|_0^{\pi/2} dr$$

$$= \int_0^b \frac{r^2}{\sqrt{1 - r^2}}\, dr \overset{r = \sin u}{=} \int_0^{\arcsin b} \frac{\sin^2 u}{\cos u} \cos u\, du = \int_0^{\arcsin b} \sin^2 u\, du$$

$$= \left(\tfrac{1}{2}u + \tfrac{1}{4}\sin 2u\right)\Big|_0^{\arcsin b} = \left(\tfrac{1}{2}u + \tfrac{1}{2}\sin u \cos u\right)\Big|_0^{\arcsin b} = \tfrac{1}{2}\arcsin b + \tfrac{1}{2}b\cos(\arcsin b)$$

Thus

$$\int_C \mathbf{F} \cdot d\mathbf{r} = \lim_{b \to 1^-} \left[-\iint_{R_b} \frac{x}{\sqrt{1 - x^2 - y^2}}\, dA - 2(\text{area of } R_b) \right]$$

$$= \lim_{b \to 1^-} \left[-\frac{1}{2}\arcsin b - \frac{1}{2} b\cos(\arcsin b) - 2\left(\frac{\pi}{4}\right) \right] = -\frac{1}{2}\left(\frac{\pi}{2}\right) - \frac{1}{2} b \cos\frac{\pi}{2} - 2\left(\frac{\pi}{4}\right)$$

$$= -\frac{\pi}{4} - \frac{\pi}{2} = -\frac{3}{4}\pi$$

5. $\operatorname{curl} \mathbf{F}(x, y, z) = \begin{vmatrix} \mathbf{i} & \mathbf{j} & \mathbf{k} \\ \dfrac{\partial}{\partial x} & \dfrac{\partial}{\partial y} & \dfrac{\partial}{\partial z} \\ y & -x & z \end{vmatrix} = -2\mathbf{k}$

Let Σ_1 be the part of Σ on the plane $z = 1$, and Σ_2 the part of Σ on the cylinder $x^2 + y^2 = 1$. On Σ_1, $\mathbf{n} = \mathbf{k}$, so that $(\operatorname{curl} \mathbf{F}) \cdot \mathbf{n} = (-2\mathbf{k}) \cdot \mathbf{k} = -2$. On Σ_2, $\mathbf{n}$ is perpendicular to the z axis, so that $(\operatorname{curl} \mathbf{F}) \cdot \mathbf{n} = 0$. Thus by Stokes's Theorem,

$$\int_C \mathbf{F} \cdot d\mathbf{r} = \iint_\Sigma (\operatorname{curl} \mathbf{F}) \cdot \mathbf{n}\, dS = \iint_{\Sigma_1} (\operatorname{curl} \mathbf{F}) \cdot \mathbf{n}\, dS + \iint_{\Sigma_2} (\operatorname{curl} \mathbf{F}) \cdot \mathbf{n}\, dS = \iint_{\Sigma_1} -2\, dS + 0$$

$$= -2(\text{area of } \Sigma_1) = -2\pi$$

7. $\operatorname{curl} \mathbf{F}(x, y, z) = \begin{vmatrix} \mathbf{i} & \mathbf{j} & \mathbf{k} \\ \dfrac{\partial}{\partial x} & \dfrac{\partial}{\partial y} & \dfrac{\partial}{\partial z} \\ xz & y^2 & x^2 \end{vmatrix} = -x\mathbf{j}$

Let Σ be the part of the plane $x + y + z = 5$ that lies inside the cylinder $x^2 + \frac{1}{4}y^2 = 1$, and let Σ be oriented with normal $\mathbf{n}$ directed upward. If $f(x, y) = 5 - x - y$, then Σ is the graph of f on the region R in the xy plane bounded by the ellipse $x^2 + \frac{1}{4}y^2 = 1$. Thus by Stokes's Theorem and (5) of Section 16.6.

$$\int_C \mathbf{F} \cdot d\mathbf{r} = \iint_\Sigma (\operatorname{curl} \mathbf{F}) \cdot \mathbf{n}\, dS = \iint_R [-(0)(-1) - (-x)(-1) + 0]\, dA$$

$$= \int_{-2}^2 \int_{-\sqrt{1 - y^2/4}}^{\sqrt{1 - y^2/4}} -x\, dx\, dy = \int_{-2}^2 -\tfrac{1}{2}x^2 \Big|_{-\sqrt{1 - y^2/4}}^{\sqrt{1 - y^2/4}} dy = \int_{-2}^2 0\, dy = 0$$

9. $\operatorname{curl} \mathbf{F}(x, y, z) = \begin{vmatrix} \mathbf{i} & \mathbf{j} & \mathbf{k} \\ \dfrac{\partial}{\partial x} & \dfrac{\partial}{\partial y} & \dfrac{\partial}{\partial z} \\ y(x^2 + y^2) & -x(x^2 + y^2) & 0 \end{vmatrix} = -4(x^2 + y^2)\mathbf{k}$

Let Σ be the part of the plane $z = y$ that is bounded by C, and let Σ be oriented with normal $\mathbf{n}$ directed upward. If $f(x, y) = y$, then Σ is the graph of f on the region R in the xy plane bounded by the rectangle with vertices $(0, 0)$, $(1, 0)$, $(1, 1)$, and $(0, 1)$. Thus by Stokes's Theorem and (5) in Section 16.6,

$$\int_C \mathbf{F} \cdot d\mathbf{r} = \iint_\Sigma (\operatorname{curl} \mathbf{F}) \cdot \mathbf{n}\, dS = \iint_R [-(0)(0) - (0)(1) - 4(x^2 + y^2)]\, dA$$

$$= \int_0^1 \int_0^1 -4(x^2 + y^2)\, dy\, dx = \int_0^1 -4(x^2 y + \tfrac{1}{3}y^3)\Big|_0^1 dx = \int_0^1 -4(x^2 + \tfrac{1}{3})\, dx$$

$$= -4(\tfrac{1}{3}x^3 + \tfrac{1}{3}x)\Big|_0^1 = -\tfrac{8}{3}$$

11. $\operatorname{curl} \mathbf{F}(x, y, z) = \begin{vmatrix} \mathbf{i} & \mathbf{j} & \mathbf{k} \\ \dfrac{\partial}{\partial x} & \dfrac{\partial}{\partial y} & \dfrac{\partial}{\partial z} \\ z - y & y & x \end{vmatrix} = \mathbf{k}$

Let Σ be the part of the sphere $x^2 + y^2 + z^2 = 1$ that is contained in the cylinder $r = \cos\theta$ and lies above the xy plane, and let Σ be oriented with normal $\mathbf{n}$ directed upward. If $f(x, y) = \sqrt{1 - x^2 - y^2}$, then Σ is the graph of f on the region in the xy plane bounded by the circle $r = \cos\theta$. Thus by Stokes's Theorem and (5) of Section 16.6,

$$\int_C \mathbf{F} \cdot d\mathbf{r} = \iint_\Sigma (\text{curl } \mathbf{F}) \cdot \mathbf{n}\, dS = \iint_R \left[-(0)\frac{-x}{\sqrt{1 - x^2 - y^2}} - (0)\frac{-y}{\sqrt{1 - x^2 - y^2}} + 1 \right] dA$$

$$= \iint_R 1\, dA = \text{area of } R = \frac{\pi}{4}$$

13. $$\text{curl } \mathbf{F}(x, y, z) = \begin{vmatrix} \mathbf{i} & \mathbf{j} & \mathbf{k} \\ \dfrac{\partial}{\partial x} & \dfrac{\partial}{\partial y} & \dfrac{\partial}{\partial z} \\ x^2 + z & y^2 + x & z^2 + y \end{vmatrix} = \mathbf{i} + \mathbf{j} + \mathbf{k}$$

The sphere $x^2 + y^2 + z^2 = 1$ and the cone $z = \sqrt{x^2 + y^2}$ intersect at (x, y, z) such that $z \geq 0$ and $1 - x^2 - y^2 = z^2 = x^2 + y^2$, so that $x^2 + y^2 = \frac{1}{2}$ and thus $z = \sqrt{2}/2$. Making use of (1), we let Σ be the part of the plane $z = \sqrt{2}/2$ that lies inside the cone $z = \sqrt{x^2 + y^2}$, and let Σ be oriented with normal $\mathbf{n}$ directed upward. If $f(x, y) = \sqrt{2}/2$, then Σ is the graph of f on the region R in the xy plane bounded by the circle $x^2 + y^2 = \frac{1}{2}$. Thus by Stokes's Theorem and (5) of Section 16.6,

$$\int_C \mathbf{F} \cdot d\mathbf{r} = \iint_\Sigma (\text{curl } \mathbf{F}) \cdot \mathbf{n}\, dS = \iint_R [-(1)(0) - (1)(0) + 1]\, dA = \text{area of } R = \tfrac{1}{2}\pi$$

15. $$\text{curl } \mathbf{F}(x, y, z) = \begin{vmatrix} \mathbf{i} & \mathbf{j} & \mathbf{k} \\ \dfrac{\partial}{\partial x} & \dfrac{\partial}{\partial y} & \dfrac{\partial}{\partial z} \\ x & x^2 + y^2 + z^2 & z(y^4 - 1) \end{vmatrix} = (4zy^3 - 2z)\mathbf{i} + 2x\mathbf{k}$$

Let Σ_1 be the rectangular region in the xy plane bounded by the lines $x = -1$, $x = 1$, $y = -1$, and $y = 1$, and orient Σ_1 with normal $\mathbf{n} = \mathbf{k}$. Then Σ and Σ_1 induce the same orientation on their common boundary. Thus by (9),

$$\iint_\Sigma (\text{curl } \mathbf{F}) \cdot \mathbf{n}\, dS = \iint_{\Sigma_1} (\text{curl } \mathbf{F}) \cdot \mathbf{n}\, dS = \iint_{\Sigma_1} [(4zy^3 - 2z)\mathbf{j} + 2x\mathbf{k}] \cdot \mathbf{k}\, dS = \iint_{\Sigma_1} 2x\, dS$$

$$= \int_{-1}^1 \int_{-1}^1 2x\, dy\, dx = \int_{-1}^1 2xy\Big|_{-1}^1 dx = \int_{-1}^1 4x\, dx = 2x^2\Big|_{-1}^1 = 0$$

17. $$\text{curl } \mathbf{F}(x, y, z) = \begin{vmatrix} \mathbf{i} & \mathbf{j} & \mathbf{k} \\ \dfrac{\partial}{\partial x} & \dfrac{\partial}{\partial y} & \dfrac{\partial}{\partial z} \\ x \sin z & xy & yz \end{vmatrix} = z\mathbf{i} + x\cos z\,\mathbf{j} + y\mathbf{k}$$

Let Σ_1 be the face of the cube in the plane $z = 1$, and let Σ_1 be oriented with normal $\mathbf{n} = -\mathbf{k}$. Then Σ and Σ_1 induce the same orientation on their common boundary. Thus by (9),

$$\iint_\Sigma (\text{curl } \mathbf{F}) \cdot \mathbf{n}\, dS = \iint_{\Sigma_1} (\text{curl } \mathbf{F}) \cdot \mathbf{n}\, dS = \iint_{\Sigma_1} [(z\mathbf{i} + x\cos z\,\mathbf{j} + y\mathbf{k}) \cdot (-\mathbf{k})]\, dS = \iint_{\Sigma_1} -y\, dS$$

$$= \int_0^1 \int_0^1 -y\, dy\, dx = \int_0^1 -\tfrac{1}{2}y^2\Big|_0^1 dx = \int_0^1 -\tfrac{1}{2}\, dx = -\tfrac{1}{2}$$

19. $$\text{curl } \mathbf{F}(x, y, z) = \begin{vmatrix} \mathbf{i} & \mathbf{j} & \mathbf{k} \\ \dfrac{\partial}{\partial x} & \dfrac{\partial}{\partial y} & \dfrac{\partial}{\partial z} \\ xz^2 & x^3 & \cos xz \end{vmatrix} = (2xz + z\sin xz)\mathbf{j} + 3x^2\mathbf{k}$$

Let Σ_1 be the disk in the xy plane bounded by the circle $x^2 + y^2 = 1$, and let Σ_1 be oriented with normal $\mathbf{n} = -\mathbf{k}$. Then Σ and Σ_1 induce the same orientation on their common boundary. Thus by (9),

$$\iint_\Sigma (\text{curl } \mathbf{F}) \cdot \mathbf{n}\, dS = \iint_{\Sigma_1} (\text{curl } \mathbf{F}) \cdot \mathbf{n}\, dS = \iint_{\Sigma_1} [(2xz + z\sin xz)\mathbf{j} + 3x^2\mathbf{k}] \cdot (-\mathbf{k})\, dS$$

$$= \iint_{\Sigma_1} -3x^2\, dS = \int_0^{2\pi} \int_0^1 -3r^3\cos^2\theta\, dr\, d\theta = \int_0^{2\pi} -\tfrac{3}{4}r^4\cos^2\theta\Big|_0^1 d\theta$$

$$= \int_0^{2\pi} -\tfrac{3}{4}\cos^2\theta\, d\theta = \int_0^{2\pi} -(\tfrac{3}{8} + \tfrac{3}{8}\cos 2\theta)\, d\theta = -(\tfrac{3}{8}\theta + \tfrac{3}{16}\sin 2\theta)\Big|_0^{2\pi} = -\tfrac{3}{4}\pi$$

21. Since F is a constant vector field, curl $\mathbf{F}(x, y, z) = \mathbf{0}$. Thus by Stokes's Theorem,

$$\int_C \mathbf{F} \cdot d\mathbf{r} = \iint_\Sigma (\text{curl } \mathbf{F}) \cdot \mathbf{n}\, dS = \iint_\Sigma \mathbf{0} \cdot \mathbf{n}\, dS = \iint_\Sigma 0\, dS = 0$$

23. $$\text{curl } \mathbf{v}(x, y, z) = \begin{vmatrix} \mathbf{i} & \mathbf{j} & \mathbf{k} \\ \dfrac{\partial}{\partial x} & \dfrac{\partial}{\partial y} & \dfrac{\partial}{\partial z} \\ x^3 & -zy & x \end{vmatrix} = y\mathbf{i} - \mathbf{j}$$

The sphere $x^2 + y^2 + z^2 = 1$ and the plane $z = y$ intersect at (x, y, z) such that $1 = x^2 + y^2 + z^2 = x^2 + 2y^2$. Let Σ_1 be the part of the plane $z = y$ that is enclosed by the boundary C of Σ, and orient Σ_1 with normal $\mathbf{n}$ directed upward. Then Σ and Σ_1 induce the same orientation on C. Let $f(x, y) = y$, and let R be the region in the xy plane bounded by the ellipse $x^2 + 2y^2 = 1$. Then Σ_1 is the graph of f on R. By the definition of circulation, by Stokes's Theorem, and by (5) in Section 16.6, the circulation of the fluid is given by

$$\int_C \mathbf{v} \cdot d\mathbf{r} = \iint_{\Sigma_1} (\text{curl } \mathbf{v}) \cdot \mathbf{n}\, dS = \iint_R [-(y)(0) - (-1)(1) + 0]\, dA = \iint_R 1\, dA$$

$$= \text{area of } R = \frac{\pi}{\sqrt{2}}$$

SECTION 16.8

1. Yes

3. Yes

5. No

7. No

9. div $\mathbf{F}(x, y, z) = 2x + x - 2x = x$

$$\iint_\Sigma \mathbf{F} \cdot \mathbf{n}\, dS = \iiint_D \text{div}\, \mathbf{F}(x, y, z)\, dV = \iiint_D x\, dV = \int_0^1 \int_0^{1-x} \int_0^{1-x-y} x\, dz\, dy\, dx$$

$$= \int_0^1 \int_0^{1-x} xz \Big|_0^{1-x-y} dy\, dx = \int_0^1 \int_0^{1-x} (x - x^2 - xy)\, dy\, dx$$

$$= \int_0^1 [(x - x^2)y - \tfrac{1}{2}xy^2] \Big|_0^{1-x} dx = \int_0^1 [x(1-x)^2 - \tfrac{1}{2}x(1-x)^2]\, dx$$

$$= \tfrac{1}{2} \int_0^1 (x - 2x^2 + x^3)\, dx = \tfrac{1}{2}(\tfrac{1}{2}x^2 - \tfrac{2}{3}x^3 + \tfrac{1}{4}x^4) \Big|_0^1 = \tfrac{1}{24}$$

11. div $\mathbf{F}(x, y, z) = 1 + 1 + 1 = 3$

$$\iint_\Sigma \mathbf{F} \cdot \mathbf{n}\, dS = \iiint_D \text{div}\, \mathbf{F}(x, y, z)\, dV = \iiint_D 3\, dV = 3(\text{volume of } D) = \frac{3\pi}{4}$$

13. div $\mathbf{F}(x, y, z) = 1 + 1 + 1 = 3$

$$\iint_\Sigma \mathbf{F} \cdot \mathbf{n}\, dS = \iiint_D \text{div}\, \mathbf{F}(x, y, z)\, dV = \iiint_D 3\, dV = 3(\text{volume of } D) = 2\pi$$

15. div $\mathbf{F}(x, y, z) = 2x + 2y + 2z$

$$\iint_\Sigma \mathbf{F} \cdot \mathbf{n}\, dS = \iiint_D \text{div}\, \mathbf{F}(x, y, z)\, dV = \iiint_D (2x + 2y + 2z)\, dV$$

$$= 2 \int_0^{2\pi} \int_0^2 \int_0^2 [r^2(\cos\theta + \sin\theta) + zr]\, dz\, dr\, d\theta$$

$$= 2 \int_0^{2\pi} \int_0^2 [r^2 z(\cos\theta + \sin\theta) + \tfrac{1}{2}z^2 r] \Big|_0^2 dr\, d\theta = 4 \int_0^{2\pi} \int_0^2 [r^2(\cos\theta + \sin\theta) + r]\, dr\, d\theta$$

$$= 4 \int_0^{2\pi} [\tfrac{1}{3}r^3(\cos\theta + \sin\theta) + \tfrac{1}{2}r^2] \Big|_0^2 d\theta = 8 \int_0^{2\pi} [\tfrac{4}{3}(\cos\theta + \sin\theta) + 1]\, d\theta$$

$$= [\tfrac{32}{3}(\sin\theta - \cos\theta) + 8\theta] \Big|_0^{2\pi} = 16\pi$$

17. At any point (x, y, z) of intersection of the plane $z = 2x$ and the paraboloid $z = x^2 + y^2$, x, y, and z must satisfy $2x = z = x^2 + y^2$, or in polar coordinates, $2r\cos\theta = r^2$, or $r = 2\cos\theta$. Thus the given region D is the solid region between the graphs of $z = r^2$ and $z = 2r\cos\theta$ on the region in the xy plane bounded by $r = 2\cos\theta$ for $-\pi/2 \le \theta \le \pi/2$. Since div $\mathbf{F}(x, y, z) = 3xy(x^2 + y^2)^{1/2} - 3xy(x^2 + y^2)^{1/2} + 1 = 1$, we have

$$\iint_\Sigma \mathbf{F} \cdot \mathbf{n}\, dS = \iiint_D \text{div}\, \mathbf{F}(x, y, z)\, dV = \iiint_D 1\, dV = \int_{-\pi/2}^{\pi/2} \int_0^{2\cos\theta} \int_{r^2}^{2r\cos\theta} r\, dz\, dr\, d\theta$$

$$= \int_{-\pi/2}^{\pi/2} \int_0^{2\cos\theta} rz \Big|_{r^2}^{2r\cos\theta} dr\, d\theta = \int_{-\pi/2}^{\pi/2} \int_0^{2\cos\theta} (2r^2\cos\theta - r^3)\, dr\, d\theta$$

$$= \int_{-\pi/2}^{\pi/2} (\tfrac{2}{3}r^3\cos\theta - \tfrac{1}{4}r^4) \Big|_0^{2\cos\theta} d\theta = \int_{-\pi/2}^{\pi/2} (\tfrac{16}{3}\cos^4\theta - 4\cos^4\theta)\, d\theta = \int_{-\pi/2}^{\pi/2} \tfrac{4}{3}\cos^4\theta\, d\theta$$

$$= \tfrac{4}{3}(\tfrac{1}{4}\cos^3\theta\sin\theta + \tfrac{3}{8}\cos\theta\sin\theta + \tfrac{3}{8}\theta) \Big|_{-\pi/2}^{\pi/2} = \tfrac{1}{2}\pi$$

with the next to last equality coming from (13) in Section 7.1.

19. div $\mathbf{F}(x, y, z) = -2 + 4 - 7 = -5$

$$\iint_\Sigma \mathbf{F} \cdot \mathbf{n}\, dS = \iiint_D \text{div}\, \mathbf{F}(x, y, z)\, dV = \iiint_D -5\, dV = \int_0^{2\pi} \int_1^2 \int_{-\sqrt{4-r^2}}^{\sqrt{4-r^2}} -5r\, dz\, dr\, d\theta$$

$$= \int_0^{2\pi} \int_1^2 -5rz \Big|_{-\sqrt{4-r^2}}^{\sqrt{4-r^2}} dr\, d\theta = \int_0^{2\pi} \int_1^2 -10r\sqrt{4 - r^2}\, dr\, d\theta = \int_0^{2\pi} \tfrac{10}{3}r(4 - r^2)^{3/2} \Big|_1^2 d\theta$$

$$= \int_0^{2\pi} -\tfrac{10}{3}3^{3/2}\, d\theta = -20\pi\sqrt{3}$$

21. div $\mathbf{F}(x, y, z) = 2x + 1 - 4z$

$$\iint_\Sigma \mathbf{F} \cdot \mathbf{n}\, dS = \iiint_D \text{div}\, \mathbf{F}(x, y, z)\, dV = \iiint_D (2x + 1 - 4z)\, dV$$

$$= \int_{-\sqrt{2}}^{\sqrt{2}} \int_0^{2-y^2} \int_0^x (2x + 1 - 4z)\, dz\, dx\, dy = \int_{-\sqrt{2}}^{\sqrt{2}} \int_0^{2-y^2} [(2x + 1)z - 2z^2] \Big|_0^x dx\, dy$$

$$= \int_{-\sqrt{2}}^{\sqrt{2}} \int_0^{2-y^2} [(2x + 1)x - 2x^2]\, dx\, dy = \int_{-\sqrt{2}}^{\sqrt{2}} \int_0^{2-y^2} x\, dx\, dy = \int_{-\sqrt{2}}^{\sqrt{2}} \tfrac{1}{2}x^2 \Big|_0^{2-y^2} dy$$

$$= \int_{-\sqrt{2}}^{\sqrt{2}} \tfrac{1}{2}(4 - 4y^2 + y^4)\, dy = \tfrac{1}{2}(4y - \tfrac{4}{3}y^3 + \tfrac{1}{5}y^5) \Big|_{-\sqrt{2}}^{\sqrt{2}} = \tfrac{32}{15}\sqrt{2}$$

23. div $\mathbf{F}(x, y, z) = (3x^2 + y^2 + z^2) + (x^2 + 3y^2 + z^2) + 0 = 4x^2 + 4y^2 + 2z^2$

$$\iint_\Sigma \mathbf{F} \cdot \mathbf{n}\, dS = \iiint_D \text{div}\, \mathbf{F}(x, y, z)\, dV = \iiint_D (4x^2 + 4y^2 + 2z^2)\, dV$$

$$= \int_0^{2\pi} \int_0^\pi \int_0^3 (2\rho^2 + 2\rho^2\sin^2\phi)\rho^2\sin\phi\, d\rho\, d\phi\, d\theta = \int_0^{2\pi} \int_0^\pi \tfrac{2}{5}\rho^5 (\sin\phi + \sin^3\phi) \Big|_0^3 d\phi\, d\theta$$

$$= \tfrac{486}{5} \int_0^{2\pi} \int_0^\pi (\sin\phi + \sin^3\phi)\, d\phi\, d\theta = \tfrac{486}{5} \int_0^{2\pi} \int_0^\pi [\sin\phi + \sin\phi\,(1 - \cos^2\phi)]\, d\phi\, d\theta$$

$$= \tfrac{486}{5} \int_0^{2\pi} (-2\cos\phi + \tfrac{1}{3}\cos^3\phi) \Big|_0^\pi d\theta = \tfrac{486}{5} \int_0^{2\pi} \tfrac{10}{3}\, d\theta = 648\pi$$

25. $$\iint_\Sigma \mathbf{F} \cdot \mathbf{n}\, dS = \iiint_D \text{div}\, \mathbf{F}(x, y, z)\, dV = \iiint_D \text{div}\,(\text{curl}\, \mathbf{G})(x, y, z)\, dV = \iiint_D 0\, dV = 0$$

by (4) in Section 16.1.

27. Let R be the region in the xy plane bounded by the circle $r = a$. Let Σ_1 denote the graph of $z = (h/a)\sqrt{x^2 + y^2}$ for (x, y) in R, and Σ_2 the graph of $z = h$ on R. Let Σ consist of Σ_1 and Σ_2. By Exercise 26, the volume of the conical region is given by

$$V = \tfrac{1}{3} \iint_\Sigma \mathbf{F} \cdot \mathbf{n}\, dS = \tfrac{1}{3} \iint_{\Sigma_1} \mathbf{F} \cdot \mathbf{n}\, dS + \tfrac{1}{3} \iint_{\Sigma_2} \mathbf{F} \cdot \mathbf{n}\, dS$$

with $\mathbf{F}(x, y, z) = x\mathbf{i} + y\mathbf{j} + z\mathbf{k}$. To evaluate $\iint_{\Sigma_1} \mathbf{F} \cdot \mathbf{n}\, dS$ we let $f(x, y) = (h/a)\sqrt{x^2 + y^2}$, and for $0 < b < a$, we let R_b be the region in the xy plane bounded by the circle $r = b$. We use (6) in Section 16.6 with $f_x(x, y) = (h/a)(x/\sqrt{x^2 + y^2})$, $f_y(x, y) = (h/a)(y/\sqrt{x^2 + y^2})$, $M(x, y, f(x, y)) = x$, $N(x, y, f(x, y)) = y$, and $P(x, y, f(x, y)) = (h/a)\sqrt{x^2 + y^2}$, and conclude that

$$\iint_{\Sigma_1} \mathbf{F} \cdot \mathbf{n}\, dS = \lim_{b \to a^-} \iint_{R_b} \left[x\left(\frac{h}{a} \frac{x}{\sqrt{x^2 + y^2}}\right) + y\left(\frac{h}{a} \frac{y}{\sqrt{x^2 + y^2}}\right) - \frac{h}{a}\sqrt{x^2 + y^2} \right] dA$$

$$= \lim_{b \to a^-} \iint_{R_b} 0\, dA = 0$$

Next,

$$\iint_{\Sigma_2} \mathbf{F}\cdot\mathbf{n}\,dS = \iint_{\Sigma_2} (x\mathbf{i} + y\mathbf{j} + h\mathbf{k})\cdot\mathbf{k}\,dS = \iint_R h\,dA = \pi a^2 h$$

Therefore

$$V = \tfrac{1}{3}\iint_{\Sigma} \mathbf{F}\cdot\mathbf{n}\,dS = \tfrac{1}{3}\iint_{\Sigma_1} \mathbf{F}\cdot\mathbf{n}\,dS + \tfrac{1}{3}\iint_{\Sigma_2} \mathbf{F}\cdot\mathbf{n}\,dS = 0 + \tfrac{1}{3}\pi a^2 h = \tfrac{1}{3}\pi a^2 h$$

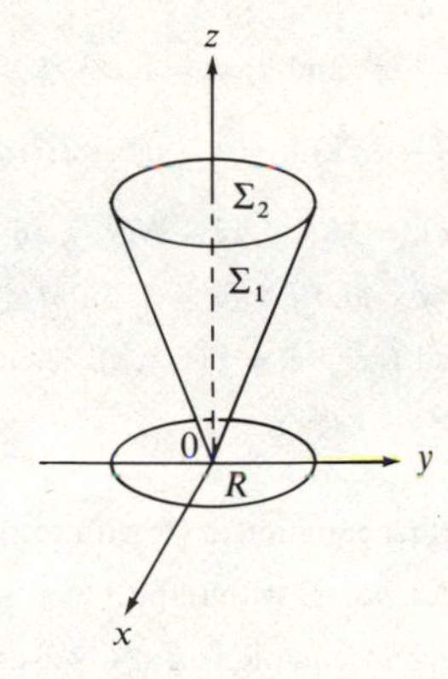

29. If $\mathbf{F}$ is constant, then div $\mathbf{F} = 0$, so that by the Divergence Theorem,

$$\iint_{\Sigma} \mathbf{F}\cdot\mathbf{n}\,dS = \iiint_D \text{div}\,\mathbf{F}(x, y, z)\,dV = \iiint_D 0\,dV = 0$$

CHAPTER 16 REVIEW

1. Let $\mathbf{a} = a_1\mathbf{i} + a_2\mathbf{j} + a_3\mathbf{k}$ and $\mathbf{b} = b_1\mathbf{i} + b_2\mathbf{j} + b_3\mathbf{k}$. Then

$$\mathbf{b}\times\mathbf{r} = \begin{vmatrix} \mathbf{i} & \mathbf{j} & \mathbf{k} \\ b_1 & b_2 & b_3 \\ x & y & z \end{vmatrix} = (b_2 z - b_3 y)\mathbf{i} + (b_3 x - b_1 z)\mathbf{j} + (b_1 y - b_2 x)\mathbf{k}$$

so that $\mathbf{a}\cdot(\mathbf{b}\times\mathbf{r}) = a_1(b_2 z - b_3 y) + a_2(b_3 x - b_1 z) + a_3(b_1 y - b_2 x)$. Thus
grad $[\mathbf{a}\cdot(\mathbf{b}\times\mathbf{r})] = (a_2 b_3 - a_3 b_2)\mathbf{i} + (a_3 b_1 - a_1 b_3)\mathbf{j} + (a_1 b_2 - a_2 b_1)\mathbf{k} = \mathbf{a}\times\mathbf{b}$.

3. To begin with,

$$\frac{\partial f}{\partial x} = y^2 - y\sin xy \quad\text{and}\quad \frac{\partial f}{\partial y} = 2xy - x\sin xy \qquad (*)$$

Integrating both sides of the first equation in (*) with respect to x, we obtain $f(x, y) = xy^2 + \cos xy + g(y)$. Taking partial derivatives with respect to y, we find that $\partial f/\partial y = 2xy - x\sin xy + dg/dy$. Comparing this with the second equation in (*), we conclude that $dg/dy = 0$, so that $g(y) = C$. Therefore $f(x, y) = xy^2 + \cos xy + C$.

5. Since $x(t) = \cos t$, $y(t) = \sin t$, $z(t) = t$, and $\|d\mathbf{r}/dt\| = \sqrt{(-\sin t)^2 + \cos^2 t + 1^2} = \sqrt{2}$, (4) in Section 16.2 implies that

$$\int_C (xy + z^2)\,ds = \int_{\pi/4}^{3\pi/4} (\cos t\sin t + t^2)\sqrt{2}\,dt = \left(\tfrac{1}{2}\sin^2 t + \tfrac{1}{3}t^3\right)\sqrt{2}\Big|_{\pi/4}^{3\pi/4} = \frac{13\sqrt{2}}{96}\pi^3$$

7. Since $x(t) = t$, $y(t) = \cos t$, $z(t) = \sin t$, $dx/dt = 1$, $dy/dt = -\sin t$, and $dz/dt = \cos t$, (9) of Section 16.2 implies that

$$\int_C xy\,dx + z\cos x\,dy + z\,dz = \int_0^{\pi/2} [(t\cos t)(1) + (\sin t\cos t)(-\sin t) + (\sin t)(\cos t)]\,dt$$
$$= \int_0^{\pi/2} t\cos t\,dt + \int_0^{\pi/2} (-\sin^2 t\cos t + \sin t\cos t)\,dt$$
$$\overset{\text{parts}}{=} t\sin t\Big|_0^{\pi/2} - \int_0^{\pi/2} \sin t\,dt + \left(-\tfrac{1}{3}\sin^3 t + \tfrac{1}{2}\sin^2 t\right)\Big|_0^{\pi/2}$$
$$= \frac{\pi}{2} + \cos t\Big|_0^{\pi/2} + \frac{1}{6} = \frac{\pi}{2} - \frac{5}{6}$$

9. The curve C is parametrized by $\mathbf{r}(t) = t\mathbf{i} + t^2\mathbf{j}$ for $0 \le t \le 1$. Since $x(t) = t$, $y(t) = 2t$, and $\|d\mathbf{r}/dt\| = \sqrt{1^2 + (2t)^2} = \sqrt{1 + 4t^2}$, it follows from (4) in Section 16.2 that

$$\int_C x\,ds = \int_0^1 t\sqrt{1 + 4t^2}\,dt = \tfrac{1}{12}(1 + 4t^2)^{3/2}\Big|_0^1 = \tfrac{1}{12}(5^{3/2} - 1)$$

11. Let $\mathbf{F}(x, y, z) = e^x\cos z\,\mathbf{i} + y\mathbf{j} - e^x\sin z\,\mathbf{k} = M(x, y, z)\mathbf{i} + N(x, y, z)\mathbf{j} + P(x, y, z)\mathbf{k}$. Since $\partial P/\partial y = 0 = \partial N/\partial z$, $\partial M/\partial z = -e^x\sin z = \partial P/\partial x$, and $\partial N/\partial x = 0 = \partial M/\partial y$, the line integral is independent of path. Let grad $f(x, y, z) = \mathbf{F}(x, y, z)$. Then

$$\frac{\partial f}{\partial x} = e^x\cos z, \quad \frac{\partial f}{\partial y} = y, \quad\text{and}\quad \frac{\partial f}{\partial z} = -e^x\sin z \qquad (*)$$

Integrating both sides of the first equation in (*) with respect to x, we obtain $f(x, y, z) = e^x\cos z + g(y, z)$. Taking partial derivatives with respect to y, we find that $\partial f/\partial y = \partial g/\partial y$. Comparing this with the second equation in (*), we deduce that $\partial g/\partial y = y$, so that $g(y, z) = \frac{1}{2}y^2 + h(z)$. Thus $f(x, y, z) = e^x\cos z + \frac{1}{2}y^2 + h(z)$. Taking partial derivatives with respect to z, we find that $\partial f/\partial z = -e^x\sin z + dh/dz$. Comparing this with the third equation in (*), we conclude that $dh/dz = 0$, so that $h(z) = C$. If $C = 0$, then $f(x, y, z) = e^x\cos z + \frac{1}{2}y^2$. Since $\mathbf{r}(0) = \mathbf{i}$ and $\mathbf{r}(1) = \mathbf{i} + \mathbf{j} + \mathbf{k}$, it follows that the initial and terminal points of C are (1, 0, 0) and (1, 1, 1), respectively. Thus

$$\int_C e^x\cos z\,dx + y\,dy - e^x\sin z\,dz = f(1, 1, 1) - f(1, 0, 0) = \left(e\cos 1 + \tfrac{1}{2}\right) - e$$
$$= \tfrac{1}{2} + e(\cos 1 - 1)$$

13. Since div $\mathbf{F}(x, y, z) = 1 + x + 1 = x + 2$, the Divergence Theorem implies that

$$\iint_{\Sigma} \mathbf{F}\cdot\mathbf{n}\,dS = \iiint_D \text{div}\,\mathbf{F}(x, y, z)\,dV = \iiint_D (x + 2)\,dV = \int_0^1\int_0^{1-y^2}\int_0^{1+x} (x + 2)\,dz\,dx\,dy$$
$$= \int_0^1\int_0^{1-y^2} (x + 2)z\Big|_0^{1+x} dx\,dy = \int_0^1\int_0^{1-y^2} (x^2 + 3x + 2)\,dx\,dy$$
$$= \int_0^1 \left(\tfrac{1}{3}x^3 + \tfrac{3}{2}x^2 + 2x\right)\Big|_0^{1-y^2} dy = \int_0^1 \left[\tfrac{1}{3}(1 - y^2)^3 + \tfrac{3}{2}(1 - y^2)^2 + 2(1 - y^2)\right]dy$$
$$= \int_0^1 \left(-\tfrac{1}{3}y^6 + \tfrac{5}{2}y^4 - 6y^2 + \tfrac{23}{6}\right)dy = \left(-\tfrac{1}{21}y^7 + \tfrac{1}{2}y^5 - 2y^3 + \tfrac{23}{6}y\right)\Big|_0^1 = \tfrac{16}{7}$$

15. Since div $\mathbf{F}(x, y, z) = yz + 2z$, the Divergence Theorem implies that

$$\iint_\Sigma \mathbf{F}\cdot\mathbf{n}\,dS = \iiint_D \operatorname{div}\mathbf{F}(x, y, z)\,dV = \iiint_D (yz + 2z)\,dV$$

$$= \int_0^{2\pi}\int_0^2\int_{-2}^3 (r\sin\theta + 2)zr\,dz\,dr\,d\theta = \int_0^{2\pi}\int_0^2 (r^2\sin\theta + 2r)\frac{z^2}{2}\bigg|_{-2}^3\,dr\,d\theta$$

$$= \int_0^{2\pi}\int_0^2 \tfrac{5}{2}(r^2\sin\theta + 2r)\,dr\,d\theta = \int_0^{2\pi}\left(\tfrac{5}{6}r^3\sin\theta + \tfrac{5}{2}r^2\right)\bigg|_0^2 d\theta = \int_0^{2\pi}\left(\tfrac{20}{3}\sin\theta + 10\right)d\theta$$

$$= \left(-\tfrac{20}{3}\cos\theta + 10\theta\right)\bigg|_0^{2\pi} = 20\pi$$

17. The paraboloid $z = -1 + x^2 + y^2$ and the plane $z = 1$ intersect at (x, y, z) such that $-1 + x^2 + y^2 = z = 1$, which means that $x^2 + y^2 = 2$, or in polar coordinates, $r = \sqrt{2}$. Let R be the region in the xy plane bounded by $r = \sqrt{2}$, and let $f(x, y) = -1 + x^2 + y^2$ for (x, y) in R. We use (6) in Section 16.6, with $f_x(x, y) = 2x$, $f_y(x, y) = 2y$, $M(x, y, f(x, y)) = y$, $N(x, y, f(x, y)) = -x$, and $P(x, y, f(x, y)) = -1 + x^2 + y^2$, and find that

$$\iint_\Sigma \mathbf{F}\cdot\mathbf{n}\,dS = \iint_R [(y)(2x) + (-x)(2y) - (-1 + x^2 + y^2)]\,dA = \iint_R (1 - x^2 - y^2)\,dA$$

$$= \int_0^{2\pi}\int_0^{\sqrt{2}} (1 - r^2)r\,dr\,d\theta = \int_0^{2\pi}\left(\tfrac{1}{2}r^2 - \tfrac{1}{4}r^4\right)\bigg|_0^{\sqrt{2}} d\theta = \int_0^{2\pi} 0\,d\theta = 0$$

19. Let Σ_1 be the part of the plane $z = 1$ inside the sphere $x^2 + y^2 + z^2 = 2$, and let Σ_1 be oriented with normal $\mathbf{n} = -\mathbf{k}$. Let R be the region in the xy plane bounded by the circle $x^2 + y^2 = 1$, or in polar coordinates $r = 1$. Since Σ and Σ_1 induce the same orientation on their common boundary, (9) in Section 16.7 implies that

$$\iint_\Sigma (\operatorname{curl}\mathbf{F})\cdot\mathbf{n}\,dS = \iint_{\Sigma_1} (\operatorname{curl}\mathbf{F})\cdot\mathbf{n}\,ds = \iint_{\Sigma_1}\left\{\left[\frac{\partial}{\partial x}(-y^3) - \frac{\partial}{\partial y}(x^3y)\right]\mathbf{k}\right\}\cdot -\mathbf{k}\,dS$$

$$= \iint_R x^3\,dS = \iint_R x^3\,dA = \int_0^{2\pi}\int_0^1 r^4\cos^3\theta\,dr\,d\theta = \int_0^{2\pi}\tfrac{1}{5}r^5\cos^3\theta\bigg|_0^1 d\theta$$

$$= \int_0^{2\pi}\tfrac{1}{5}\cos^3\theta\,d\theta = \int_0^{2\pi}\tfrac{1}{5}(1 - \sin^2\theta)\cos\theta\,d\theta = \tfrac{1}{5}\left(\sin\theta - \tfrac{1}{3}\sin^3\theta\right)\bigg|_0^{2\pi} = 0$$

21. Since div $\mathbf{F}(x, y, z) = \sec^2 x - (1 + \tan^2 x) - 6 = -6$, the Divergence Theorem implies that

$$\iint_\Sigma \mathbf{F}\cdot\mathbf{n}\,dS = \iiint_D \operatorname{div}\mathbf{F}(x, y, z)\,dV = \iiint_D -6\,dV$$

$$= -6(\text{volume of the hemisphere}) = -6\left(\tfrac{2}{3}\pi 2^3\right) = -32\pi$$

23. Let $\mathbf{F}(x, y, z) = y\mathbf{i} + y\mathbf{j} + x^2\mathbf{k}$. Then

$$\operatorname{curl}\mathbf{F}(x, y, z) = \begin{vmatrix} \mathbf{i} & \mathbf{j} & \mathbf{k} \\ \dfrac{\partial}{\partial x} & \dfrac{\partial}{\partial y} & \dfrac{\partial}{\partial z} \\ y & y & x^2 \end{vmatrix} = -2x\mathbf{j} - \mathbf{k}$$

The surfaces $z = x^2 + y^2$ and $z = 1 - y^2$ intersect at (x, y, z) such that $x^2 + y^2 = z = 1 - y^2$, so that $x^2 + 2y^2 = 1$. Let $f(x, y) = 1 - y^2$, let Σ be the graph of f on the region R in the xy plane bounded by the ellipse $x^2 + 2y^2 = 1$, and let Σ be oriented with normal $\mathbf{n}$ directed upward. Then by Stokes's Theorem and (5) of Section 16.6,

$$\int_C y\,dx + y\,dy + x^2\,dz = \int_C \mathbf{F}\cdot d\mathbf{r} = \iint_\Sigma (\operatorname{curl}\mathbf{F})\cdot\mathbf{n}\,dS = \iint_R [-(-2x)(-2y) - 1]\,dA$$

$$= \int_{-1}^1\int_{-\sqrt{(1-x^2)/2}}^{\sqrt{(1-x^2)/2}} (-4xy - 1)\,dy\,dx = \int_{-1}^1 (-2xy^2 - y)\bigg|_{-\sqrt{(1-x^2)/2}}^{\sqrt{(1-x^2)/2}} dx$$

$$= -\sqrt{2}\int_{-1}^1 \sqrt{1 - x^2}\,dx = -\sqrt{2}(\text{area of semicircle of radius } 1) = -\frac{\sqrt{2}}{2}\pi$$

25. Since $x(t) = t$, $y(t) = 0$, $z(t) = -t^3$, and $d\mathbf{r}/dt = \mathbf{i} - 3t^2\mathbf{k}$, (7) in Section 16.2 implies that

$$\int_C \mathbf{F}\cdot d\mathbf{r} = \int_{-1}^1 (t^4\mathbf{i} - t^3\mathbf{k})\cdot(\mathbf{i} - 3t^2\mathbf{k})\,dt = \int_{-1}^1 (t^4 - 3t^5)\,dt = \left(\tfrac{1}{5}t^5 - \tfrac{1}{2}t^6\right)\bigg|_{-1}^1 = \tfrac{2}{5}$$

27. Let $\mathbf{F}(x, y, z) = y\mathbf{i} + x\mathbf{j} + z^3\mathbf{k} = M(x, y, z)\mathbf{i} + N(x, y, z)\mathbf{j} + P(x, y, z)\mathbf{k}$. Since $\partial P/\partial y = 0 = \partial N/\partial z$, $\partial M/\partial z = 0 = \partial P/\partial x$, and $\partial N/\partial x = 1 = \partial M/\partial y$, the line integral $\int_C \mathbf{F}\cdot d\mathbf{r}$ is independent of path. Let grad $f(x, y, z) = \mathbf{F}(x, y, z)$. Then

$$\frac{\partial f}{\partial x} = y, \quad \frac{\partial f}{\partial y} = x, \quad \text{and} \quad \frac{\partial f}{\partial z} = z^3 \qquad (*)$$

Integrating both sides of the first equation in (*) with respect to x, we obtain $f(x, y, z) = xy + g(y, z)$. Taking partial derivatives with respect to y, we find that $\partial f/\partial y = x + \partial g/\partial y$. Comparing this with the second equation in (*), we deduce that $\partial g/\partial y = 0$, so that $g(y, z) = h(z)$. Thus $f(x, y, z) = xy + h(z)$. Taking partial derivatives with respect to z, we find that $\partial f/\partial z = dh/dz$. Comparing this with the third equation in (*), we conclude that $dh/dz = z^3$, so that $h(z) = \frac{1}{4}z^4 + C$. If $C = 0$, then $f(x, y, z) = xy + \frac{1}{4}z^4$.

a. Let C denote the line segment from $(1, 0, 0)$ to $(0, 1, \pi)$. By the definition of work and the Fundamental Theorem of Line Integrals,

$$W = \int_C \mathbf{F}\cdot d\mathbf{r} = f(0, 1, \pi) - f(1, 0, 0) = \tfrac{1}{4}\pi^4 - 0 = \tfrac{1}{4}\pi^4$$

b. Let C_1 denote the given curve. Since the integral is independent of path, it follows from part (a) and the definition of work that

$$W = \int_{C_1} \mathbf{F}\cdot d\mathbf{r} = \int_C \mathbf{F}\cdot d\mathbf{r} = \tfrac{1}{4}\pi^4$$

CUMULATIVE REVIEW (CHAPTERS 1–15)

1. $$\lim_{x\to-\infty}\frac{1}{x^2 + 2}\sqrt{\frac{x^5 - 1}{2x + 1}} = \lim_{x\to-\infty}\frac{1}{x^2 + 2}\sqrt{\frac{x^5(1 - 1/x^5)}{x(2 + 1/x)}} = \lim_{x\to-\infty}\frac{x^2}{x^2 + 2}\sqrt{\frac{1 - 1/x^5}{2 + 1/x}}$$

$$= \lim_{x\to-\infty}\frac{1}{1 + 2/x^2}\sqrt{\frac{1 - 1/x^5}{2 + 1/x}} = \sqrt{\frac{1}{2}} = \frac{\sqrt{2}}{2}$$

3. $f'(x) = 3x^2\cos(x^3)$, so $f'(0) = 0$; $f''(x) = 6x\cos(x^3) - [3x^2\sin(x^3)](3x^2) = 6x\cos(x^3) - 9x^4\sin(x^3)$, so $f''(0) = 0$; $f^{(3)}(x) = 6\cos(x^3) - [6x\sin(x^3)](3x^2) - 36x^3\sin(x^3) - [9x^4\cos(x^3)](3x^2)$, so $f^{(3)}(0) = 6$. Thus 3 is the smallest positive integer n such that $f^{(n)}(0) \neq 0$.

5. Using the notation in the figure, we find by similar triangles that $r(x)/x = (1/2)/1 = \frac{1}{2}$, that is, $r(x) = \frac{1}{2}x$. Therefore the volume V of the part of the drill that has penetrated the plank is given by $V = \frac{1}{3}\pi(r(x))^2 x = (\pi/3)(\frac{1}{2}x)^2 x = \pi x^3/12$. By implicit differentiation, $dV/dt = (\pi x^2/4)\,dx/dt$. Since $dx/dt = \frac{1}{12}$, we have

$$\left.\frac{dV}{dt}\right|_{x=2/3} = \frac{\pi(2/3)^2}{4}\cdot\frac{1}{12} = \frac{\pi}{108}$$

Thus the volume of the drilled hole increases at the rate of $\pi/108$ cubic inches per second when the drill tip is $\frac{2}{3}$ inch in the wood.

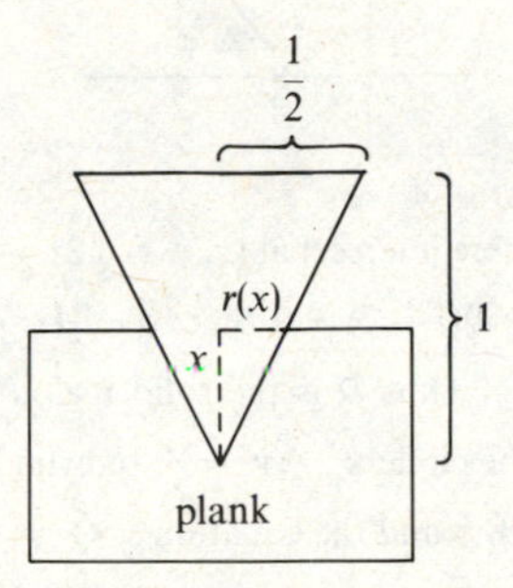

7. $f'(x) = \dfrac{2 - x^2}{(x^2 + 2x + 2)^2}$; $f''(x) = \dfrac{2(x^3 - 6x - 4)}{(x^2 + 2x + 2)^3} = \dfrac{2(x + 2)(x - 1 - \sqrt{3})(x - 1 + \sqrt{3})}{(x^2 + 2x + 2)^3}$;

relative maximum value is $f(\sqrt{2}) = \sqrt{2}/(4 + 2\sqrt{2}) = \frac{1}{2}(\sqrt{2} - 1)$; relative minimum value is $f(-\sqrt{2}) = -\sqrt{2}/(4 - 2\sqrt{2}) = -\frac{1}{2}(\sqrt{2} + 1)$; strictly increasing on $[-\sqrt{2}, \sqrt{2}]$, and strictly decreasing on $(-\infty, -\sqrt{2}]$ and $[\sqrt{2}, \infty)$; concave upward on $(-2, 1 - \sqrt{3})$ and $(1 + \sqrt{3}, \infty)$, and concave downward on $(-\infty, -2)$ and $(1 - \sqrt{3}, 1 + \sqrt{3})$; inflection points are $(-2, -1)$, $(1 - \sqrt{3}, (-1 - \sqrt{3})/4)$ and $(1 + \sqrt{3}, (-1 + \sqrt{3})/4)$; horizontal asymptote is $y = 0$.

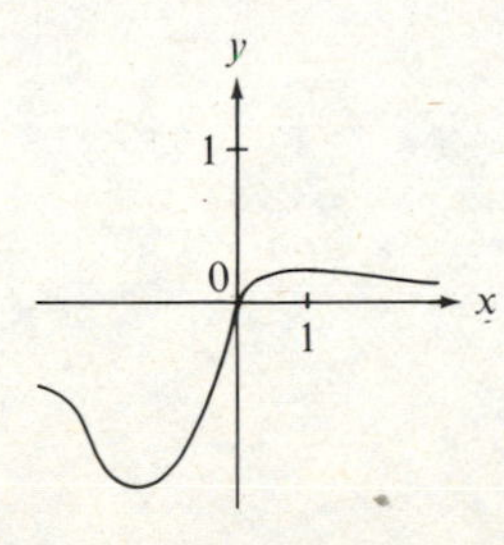

9. To integrate by parts, let $u = x^2$ and $dv = \sec^2 x \tan x\,dx$, so that $du = 2x\,dx$ and $v = \frac{1}{2}\tan^2 x$. Then

$\int x^2 \sec^2 x \tan x\,dx = \frac{1}{2}x^2\tan^2 x - \int 2x(\frac{1}{2}\tan^2 x)\,dx = \frac{1}{2}x^2\tan^2 x - \int x(\sec^2 x - 1)\,dx$

Integrating by parts a second time, with $u = x$ and $dv = (\sec^2 x - 1)\,dx$, so that $du = dx$ and $v = \tan x - x$, we find that $\int x(\sec^2 x - 1)\,dx = x(\tan x - x) - \int(\tan x - x)\,dx = x\tan x - x^2 + \ln|\cos x| + \frac{1}{2}x^2 + C_1$. Thus $\int x^2\sec^2 x\tan x\,dx = \frac{1}{2}x^2\tan^2 x - \int x(\sec^2 x - 1)\,dx = \frac{1}{2}x^2\tan^2 x - x\tan x + x^2 - \ln|\cos x| - \frac{1}{2}x^2 + C = \frac{1}{2}x^2\tan^2 x - x\tan x - \ln|\cos x| + \frac{1}{2}x^2 + C$.

11. First we will find $\int (x^2/\sqrt{4 - x^2})\,dx$. Let $x = 2\sin u$, so that $dx = 2\cos u\,du$;

$$\int \frac{x^2}{\sqrt{4 - x^2}}\,dx = \int \frac{4\sin^2 u}{\sqrt{4 - 4\sin^2 u}}(2\cos u)\,du = \int \frac{8\sin^2 u\cos u}{2\cos u}\,du = 4\int \sin^2 u\,du$$

$$= 4\int\left(\frac{1}{2} - \frac{1}{2}\cos 2u\right)du = 2u - \sin 2u + C = 2u - 2\sin u\cos u + C$$

$$= 2\arcsin\frac{x}{2} - \frac{x}{2}\sqrt{4 - x^2} + C$$

Therefore

$$\int_0^2 \frac{x^2}{\sqrt{4 - x^2}}\,dx = \lim_{b\to 2^-}\int_0^b \frac{x^2}{\sqrt{4 - x^2}}\,dx = \lim_{b\to 2^-}\left(2\arcsin\frac{x}{2} - \frac{x}{2}\sqrt{4 - x^2}\right)\Bigg|_0^b$$

$$= \lim_{b\to 2^-}\left(2\arcsin\frac{b}{2} - \frac{b}{2}\sqrt{4 - b^2}\right) = 2\arcsin 1 = 2\left(\frac{\pi}{2}\right) = \pi$$

Thus the given integral converges, and its value is π.

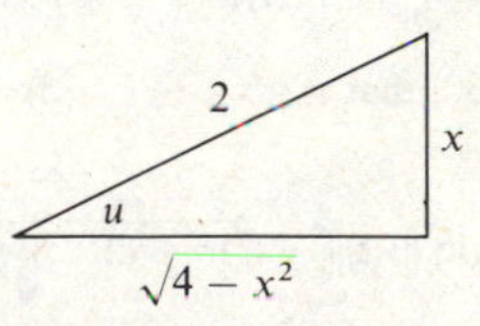

13. a. Since $x = r\cos\theta$, $y = r\sin\theta$, and $x^2 + y^2 = r^2$, the given equation is equivalent to $(r^2)^3 = 3(r\cos\theta)^2(r\sin\theta)^2 = 3r^4\cos^2\theta\sin^2\theta$, or $r^2 = 3\cos^2\theta\sin^2\theta = \frac{3}{4}\sin^2 2\theta$, or $r = (\sqrt{3}/2)\sin 2\theta$. The graph is a four-leaved rose, sketched in the figure.

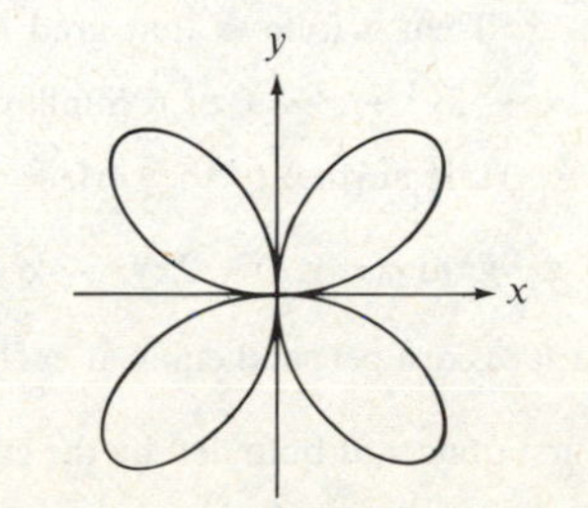

b. From part (a), $|r| \le \sqrt{3}/2$ and $r = \sqrt{3}/2$ for $\theta = \pi/4$. Therefore the maximum distance between points on the graph of the equation and the origin is $\sqrt{3}/2$.

15. Since

$$\lim_{n\to\infty} \sqrt[n]{\left|(-1)^n\left(1+\frac{1}{n}\right)^{-n^2}\right|} = \lim_{n\to\infty}\left(1+\frac{1}{n}\right)^{-n} = \lim_{n\to\infty}\frac{1}{(1+1/n)^n} = \frac{1}{e} < 1$$

the given series converges absolutely.

17. Since $e^x = \sum_{n=0}^{\infty}(1/n!)x^n$ for all x, we have

$$e^{-t^2/2} = \sum_{n=0}^{\infty}\frac{1}{n!}\left(-\frac{t^2}{2}\right)^n = \sum_{n=0}^{\infty}\frac{(-1)^n}{n!2^n}t^{2n} \quad \text{for all } t$$

By the Integration Theorem for Power Series,

$$\int_0^1 e^{-t^2/2}\,dt = \int_0^1 \sum_{n=0}^{\infty}\frac{(-1)^n}{n!2^n}t^{2n}\,dt = \sum_{n=0}^{\infty}\int_0^1 \frac{(-1)^n}{n!2^n}t^{2n}\,dt = \sum_{n=0}^{\infty}\left(\frac{(-1)^n}{n!2^n(2n+1)}\right)t^{2n+1}\bigg|_0^1$$

$$= \sum_{n=0}^{\infty}\frac{(-1)^n}{n!2^n(2n+1)}$$

Since this series is alternating, it follows from the Alternating Series Test Theorem (Theorem 9.17) that

$$\frac{5}{6} = s_2 \le \sum_{n=0}^{\infty}\frac{(-1)^n}{n!2^n(2n+1)} \le s_3 = \frac{103}{120}$$

19. a. $$\mathscr{L} = \int_0^b \sqrt{\left(\frac{dx}{dt}\right)^2 + \left(\frac{dy}{dt}\right)^2 + \left(\frac{dz}{dt}\right)^2}\,dt = \int_0^b \sqrt{1^2 + (2t^{1/2})^2 + (2t)^2}\,dt$$

$$= \int_0^b \sqrt{1 + 4t + 4t^2}\,dt = \int_0^b (1+2t)\,dt = (t+t^2)\Big|_0^b = b + b^2$$

Thus $\mathscr{L} = 30$ if $b + b^2 = 30$, that is, $(b+6)(b-5) = 0$, so 5 is the positive value of b that makes $\mathscr{L} = 30$.

b. $\mathbf{v}(t) = \mathbf{r}'(t) = \mathbf{i} + 2t^{1/2}\mathbf{j} + 2t\mathbf{k}$; $\|\mathbf{v}(t)\| = \sqrt{1^2 + (2t^{1/2})^2 + (2t)^2} = 1 + 2t$;

$$a_{\mathbf{T}} = \frac{d}{dt}\|\mathbf{v}(t)\| = 2.$$

c. $\mathbf{a}(t) = \mathbf{v}'(t) = t^{-1/2}\mathbf{j} + 2\mathbf{k}$; $\|\mathbf{a}(t)\| = \sqrt{(t^{-1/2})^2 + 2^2} = \sqrt{t^{-1} + 4}$; $a_{\mathbf{N}} = \sqrt{\|\mathbf{a}\|^2 - a_{\mathbf{T}}^2}$

$= \sqrt{(t^{-1} + 4) - 2^2} = t^{-1/2}$, so that $a_{\mathbf{N}}$ is decreasing on $(0, \infty)$.

21. Let $f(x, y, z) = x^2 - 2y^2 + z^2$. Then it follows that grad $f(x, y, z) = 2x\mathbf{i} - 4y\mathbf{j} + 2z\mathbf{k}$ is normal to the level surface $x^2 - 2y^2 + z^2 = 0$ of f. Similarly, let $g(x, y, z) = xyz$, so that grad $g(x, y, z) = yz\mathbf{i} + xz\mathbf{j} + xy\mathbf{k}$ is normal to the surface $xyz = 1$. Since

$$\text{grad } f(x, y, z) \cdot \text{grad } g(x, y, z) = 2xyz - 4xyz + 2xyz = 0$$

it follows that the given surfaces are perpendicular at each point of intersection.

23. Let R be the region in the first quadrant bounded by the curve $x = y^{1/3}$ (or equivalently, $y = x^3$) and the lines $y = 1$ and $x = 2$. Thus

$$\int_1^8 \int_{y^{1/3}}^2 \cos\left(\frac{x^4}{4} - x\right)dx\,dy = \iint_R \cos\left(\frac{x^4}{4} - x\right)dA = \int_1^2 \int_1^{x^3} \cos\left(\frac{x^4}{4} - x\right)dy\,dx$$

$$= \int_1^2 y\cos\left(\frac{x^4}{4} - x\right)\bigg|_1^{x^3}dx = \int_1^2 (x^3 - 1)\cos\left(\frac{x^4}{4} - x\right)dx = \sin\left(\frac{x^4}{4} - x\right)\bigg|_1^2$$

$$= \sin 2 - \sin\left(-\frac{3}{4}\right) = \sin 2 + \sin\frac{3}{4}$$

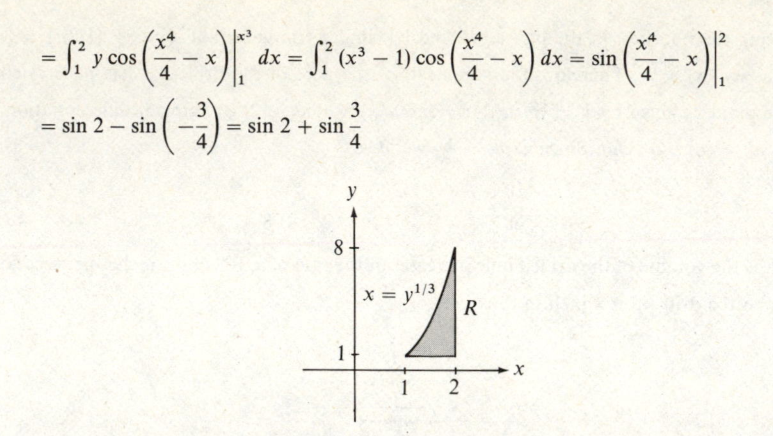

25. The paraboloid and the sphere intersect at (x, y, z) if $2z = x^2 + y^2$ and $x^2 + y^2 + z^2 = 8$, so that $2z + z^2 = 8$, or $(z+4)(z-2) = 0$. Since $z = \frac{1}{2}(x^2 + y^2) \ge 0$, it follows that $z = 2$ and hence that $x^2 + y^2 = 4$. Thus D is the solid region that lies over the region R in the xy plane bounded by the circle $x^2 + y^2 = 4$. In cylindrical coordinates the equation $2z = x^2 + y^2$ becomes $z = \frac{1}{2}r^2$, and the equation $x^2 + y^2 + z^2 = 8$ becomes $r^2 + z^2 = 8$, or $z = \sqrt{8 - r^2}$. Therefore

$$\iiint_D z\,dV = \int_0^{2\pi}\int_0^2\int_{r^2/2}^{\sqrt{8-r^2}} zr\,dz\,dr\,d\theta = \int_0^{2\pi}\int_0^2 \tfrac{1}{2}z^2r\Big|_{r^2/2}^{\sqrt{8-r^2}}dr\,d\theta$$

$$= \int_0^{2\pi}\int_0^2 \tfrac{1}{2}r(8 - r^2 - \tfrac{1}{4}r^4)\,dr\,d\theta = \int_0^{2\pi}\int_0^2 (4r - \tfrac{1}{2}r^3 - \tfrac{1}{8}r^5)\,dr\,d\theta$$

$$= \int_0^{2\pi}\left(2r^2 - \frac{1}{8}r^4 - \frac{1}{48}r^6\right)\bigg|_0^2 d\theta = \int_0^{2\pi}\frac{14}{3}\,d\theta = \frac{28\pi}{3}$$

17 DIFFERENTIAL EQUATIONS

SECTION 17.1

1. If $y = \frac{1}{3}e^{3x}$, then $dy/dx = 3(\frac{1}{3}e^{3x}) = e^{3x}$, so that $dy/dx = e^{3x}$.

3. If $y = \tan x + \sec x$ for $0 < x < \pi/2$, then $dy/dx = \sec^2 x + \sec x \tan x$, so that $2(dy/dx) - y^2 = 2(\sec^2 x + \sec x \tan x) - (\tan x + \sec x)^2 = 2\sec^2 x + 2\sec x \tan x - (\tan^2 x + 2\sec x \tan x + \sec^2 x) = \sec^2 x - \tan^2 x = 1$. Thus $2(dy/dx) - y^2 = 1$.

5. If $y = \sinh x$, then $dy/dx = \cosh x$, and $d^2y/dx^2 = \sinh x$. Thus $d^2y/dx^2 = y$.

7. If $y = e^{-x} + \sin x$, then $dy/dx = -e^{-x} + \cos x$ and $d^2y/dx^2 = e^{-x} - \sin x$. Thus $d^2y/dx^2 + y = (e^{-x} - \sin x) + (e^{-x} + \sin x) = 2e^{-x}$, so that $d^2y/dx^2 + y = 2e^{-x}$.

9. If $y = e^{ax}\sin bx$, then $dy/dx = ae^{ax}\sin bx + be^{ax}\cos bx$, and $d^2y/dx^2 = a^2e^{ax}\sin bx + 2abe^{ax}\cos bx - b^2e^{ax}\sin bx$. Thus $d^2y/dx^2 - 2a(dy/dx) + (a^2+b^2)y = (a^2e^{ax}\sin bx + 2abe^{ax}\cos bx - b^2e^{ax}\sin bx) - 2a(ae^{ax}\sin bx + be^{ax}\cos bx) + (a^2+b^2)e^{ax}\sin bx = 0$, so that $d^2y/dx^2 - 2a(dy/dx) + (a^2+b^2)y = 0$.

11. If $y = e^{-4t}$, then $dy/dt = -4e^{-4t}$, $d^2y/dt^2 = 16e^{-4t}$, and $d^3y/dt^3 = -64e^{-4t}$. Thus $d^3y/dt^3 + 64y = -64e^{-4t} + 64e^{-4t} = 0$, so that $d^3y/dt^3 + 64y = 0$.

13. If $y = -2e^{-3x}$, then $dy/dx = 6e^{-3x}$, so that $dy/dx + 5y = 6e^{-3x} + 5(-2e^{-3x}) = -4e^{-3x}$. Thus y satisfies the differential equation. Since $y(0) = -2e^{-3(0)} = -2$, y also satisfies the initial condition.

15. If $y = 1/x - 1$, then $dy/dx = -1/x^2$, and $d^2y/dx^2 = 2/x^3$. Thus
$$x^3\frac{d^2y}{dx^2} + x^2\frac{dy}{dx} - xy = x^3\left(\frac{2}{x^3}\right) + x^2\left(-\frac{1}{x^2}\right) - x\left(\frac{1}{x} - 1\right) = x$$
so that y satisfies the differential equation. Since $y(1) = \frac{1}{1} - 1 = 0$ and $y'(1) = -\frac{1}{1} = -1$, y also satisfies the initial conditions.

17. If $y = x\int_0^x \sqrt{1+t^4}\,dt$, then $dy/dx = \int_0^x \sqrt{1+t^4}\,dt + x\sqrt{1+x^4}$. Thus
$$x\frac{dy}{dx} - y = x\left(\int_0^x \sqrt{1+t^4}\,dt + x\sqrt{1+x^4}\right) - x\int_0^x \sqrt{1+t^4}\,dt = x^2\sqrt{1+x^4}$$
so that y satisfies the differential equation. Since $y(0) = 0\int_0^0 \sqrt{1+t^4}\,dt = 0$ and $y'(0) = \int_0^0 \sqrt{1+t^4}\,dt + 0\sqrt{1+0^4} = 0$, y also satisfies the initial conditions.

19. If $y^2 = 8x + 16$, then $2y(dy/dx) = 8$, so that $dy/dx = 8/(2y) = 4/y$, and thus
$$2x\frac{dy}{dx} + y\left(\frac{dy}{dx}\right)^2 = 2x\left(\frac{4}{y}\right) + y\left(\frac{4}{y}\right)^2 = \frac{8x}{y} + \frac{16}{y} = \frac{8x+16}{y} = \frac{y^2}{y} = y$$
Consequently y satisfies the differential equation.

21. If $N = N_0e^{kt} + (c/k)(e^{kt} - 1)$, then $dN/dt = kN_0e^{kt} + ce^{kt}$. Thus
$$kN + c = (kN_0e^{kt} + ce^{kt} - c) + c = kN_0e^{kt} + ce^{kt} = dN/dt$$
so that N is a solution of the differential equation.

SECTION 17.2

1. $\dfrac{dy}{dx} = \dfrac{x}{y}$, so $y\,dy = x\,dx$; thus $\frac{1}{2}y^2 = \frac{1}{2}x^2 + C_1$, or $y^2 - x^2 = c$.

3. $\dfrac{2y}{y^2+1}\dfrac{dy}{dx} = \dfrac{1}{x^2}$, so $\dfrac{2y}{y^2+1}\,dy = \dfrac{1}{x^2}\,dx$;
thus $\ln(y^2+1) = -1/x + C_1$, so that $y^2 + 1 = e^{-(1/x)+C_1}$, which means that $y^2 = -1 + Ce^{-1/x}$, with $C > 0$.

5. $(y^2 - 3)\dfrac{dy}{dt} = 1$, so $(y^2-3)\,dy = dt$; thus $\frac{1}{3}y^3 - 3y = t + C$.

7. $(1+x^2)\,dy = (1+y^2)\,dx$, so $\dfrac{1}{1+y^2}\,dy = \dfrac{1}{1+x^2}\,dx$; thus $\arctan y = \arctan x + C$.

9. $\dfrac{1+e^x}{1-e^{-y}}\,dy + e^{x+y}\,dx = 0$, so $\dfrac{e^{-y}}{1-e^{-y}}\,dy + \dfrac{e^x}{1+e^x}\,dx = 0$;
thus $\ln|1-e^{-y}| + \ln(1+e^x) = C_1$, so that if $C_1 = \ln C$, then $\ln|1-e^{-y}| + \ln(1+e^x) = \ln C$, or $\ln[|1-e^{-y}|(1+e^x)] = \ln C$. Therefore $|1-e^{-y}|(1+e^x) = C$, or $|1-e^{-y}| = C/(1+e^x)$.

11. $y^2x\dfrac{dy}{dx} - x + 1 = 0$, so $y^2\,dy = \dfrac{1}{x}(x-1)\,dx = \left(1 - \dfrac{1}{x}\right)dx$;
thus $\frac{1}{3}y^3 = x - \ln|x| + C$. If $y(1) = 3$, then $\frac{1}{3}(3)^3 = 1 - \ln 1 + C$, so that $C = 8$. Therefore the particular solution with $y(1) = 3$ is $\frac{1}{3}y^3 = x - \ln|x| + 8$.

13. $\sqrt{x^2+1}\,\dfrac{dy}{dx} = \dfrac{x}{y}$, so $y\,dy = \dfrac{x}{\sqrt{x^2+1}}\,dx$; thus $\frac{1}{2}y^2 = \sqrt{x^2+1} + C$.
If $y(\sqrt{3}) = 2$, then $\frac{1}{2}(2)^2 = \sqrt{(\sqrt{3})^2 + 1} + C$, so that $C = 0$. Therefore the particular solution with $y(\sqrt{3}) = 2$ is $\frac{1}{2}y^2 = \sqrt{x^2+1}$, or $y = \sqrt{2\sqrt{x^2+1}}$.

15. $e^{-2y}\,dy = (x-2)\,dx$, and thus $-\frac{1}{2}e^{-2y} = \frac{1}{2}x^2 - 2x + C$. If $y(0) = 0$, then $-\frac{1}{2}e^0 = 0 - 0 + C$, so that $C = -\frac{1}{2}$. Therefore the particular solution with $y(0) = 0$ is $-\frac{1}{2}e^{-2y} = \frac{1}{2}x^2 - 2x - \frac{1}{2}$, or $y = -\frac{1}{2}\ln(-x^2 + 4x + 1)$.

17. a. Since $v = y/x$, we have $y = xv$, so that $dy/dx = x\,(dv/dx) + v$.

    b. Using the definition of g and part (a), we have $g(v) = g(y/x) = f(x, y) = dy/dx = x(dv/dx) + v$, so that $g(v) - v = x(dv/dx)$, and thus $1/(g(v) - v)\,dv = (1/x)\,dx$.

19. $f(x, y) = \dfrac{x+y}{x} = 1 + \dfrac{y}{x} = g(y/x)$, where $g(v) = 1 + v$.

By (8), $1/((1+v) - v)\,dv = (1/x)\,dx$, so that $dv = (1/x)\,dx$, and thus $v = \ln|x| + C_1$. If $\ln C = C_1$, then since $v = y/x$, we conclude that $y/x = \ln|x| + \ln C = \ln C|x|$, so that $y = x\ln C|x|$, with $C > 0$.

SECTION 17.3

1. Since $\dfrac{\partial}{\partial x}(-3xy^2) = -3y^2 = \dfrac{\partial}{\partial y}(2x - y^3)$, the differential equation is exact. Thus there is a function f such that $\partial f/\partial x = 2x - y^3$ and $\partial f/\partial y = -3xy^2$. Then $f(x, y) = x^2 - xy^3 + g(y)$ for a suitable function g, so that $\partial f/\partial y = -3xy^2 + dg/dy$. Since $\partial f/\partial y = -3xy^2$, it follows that $dg/dy = 0$, and thus $g(y) = C_1$. Consequently the general solution is $x^2 - xy^3 = C$.

3. Since $\dfrac{\partial}{\partial x}(3x^2y^2 + e^{-x} - 4) = 6xy^2 - e^{-x} = \dfrac{\partial}{\partial y}(2xy^3 - ye^{-x})$, the differential equation is exact. Thus there is a function f such that $\partial f/\partial x = 2xy^3 - ye^{-x}$ and $\partial f/\partial y = 3x^2y^2 + e^{-x} - 4$. Then $f(x, y) = x^2y^3 + ye^{-x} + g(y)$, for a suitable function g, so that $\partial f/\partial y = 3x^2y^2 + e^{-x} + dg/dy$. Since $\partial f/\partial y = 3x^2y^2 + e^{-x} - 4$, it follows that $dg/dy = -4$, so that $g(y) = -4y + C_1$. Consequently the general solution is

$$x^2y^3 + ye^{-x} - 4y = C$$

5. The differential equation is equivalent to $2x - y/x^2 + (1/x + \cos 2y)(dy/dx) = 0$. Since $\dfrac{\partial}{\partial x}(1/x + \cos 2y) = -1/x^2 = \dfrac{\partial}{\partial y}(2x - y/x^2)$, this differential equation is exact. Thus there is a function f such that $\partial f/\partial x = 2x - y/x^2$ and $\partial f/\partial y = 1/x + \cos 2y$. Then $f(x, y) = x^2 + y/x + g(y)$, for a suitable function g, so that $\partial f/\partial y = 1/x + dg/dy$. Since $\partial f/\partial y = 1/x + \cos 2y$, it follows that $dg/dy = \cos 2y$, so that $g(y) = \frac{1}{2}\sin 2y + C_1$. Consequently the general solution is $x^2 + y/x + \frac{1}{2}\sin 2y = C$.

7. The differential equation is equivalent to $(\pi y - \sin x) + (\pi x + \arcsin y)(dy/dx) = 0$. Since $\dfrac{\partial}{\partial x}(\pi x + \arcsin y) = \pi = \dfrac{\partial}{\partial y}(\pi y - \sin x)$; this differential equation is exact. Thus there is a function f such that $\partial f/\partial x = \pi y - \sin x$ and $\partial f/\partial y = \pi x + \arcsin y$. Then $f(x, y) = \pi xy + \cos x + g(y)$, for a suitable function g, so that $\partial f/\partial y = \pi x + dg/dy$. Since $\partial f/\partial y = \pi x + \arcsin y$, it follows that $dg/dy = \arcsin y$, so that by integration by parts,

$$g(y) = \int \arcsin y\,dy = y\arcsin y - \int \frac{y}{\sqrt{1-y^2}}\,dy = y\arcsin y + \sqrt{1-y^2} + C_1$$

Consequently the general solution is $\pi xy + \cos x + y\arcsin y + \sqrt{1-y^2} = C$.

9. Since $\dfrac{\partial}{\partial x}((\ln x)/y + \sin y) = 1/(xy) = \dfrac{\partial}{\partial y}((\ln y)/x)$, the differential equation is exact. Thus there is a function f such that $\partial f/\partial x = (\ln y)/x$ and $\partial f/\partial y = (\ln x)/y + \sin y$. Then $f(x, y) = \ln x \ln y + g(y)$, for a suitable function g, so that $\partial f/\partial y = (\ln x)/y + dg/dy$. Since $\partial f/\partial y = (\ln x)/y + \sin y$, it follows that $dg/dy = \sin y$, and thus $g(y) = -\cos y + C_1$. Consequently the general solution is $\ln x \ln y - \cos y = C$.

11. Since $\dfrac{\partial}{\partial x}(1 + \tan x \sec y \tan y) = \sec^2 x \sec y \tan y = \dfrac{\partial}{\partial y}(\sec^2 x \sec y)$, the differential equation is exact. Thus there is a function f such that $\partial f/\partial x = \sec^2 x \sec y$ and $\partial f/\partial y = 1 + \tan x \sec y \tan y$. Then $f(x, y) = \tan x \sec y + g(y)$, for a suitable function g, so that $\partial f/\partial y = \tan x \sec y \tan y + dg/dy$. Since $\partial f/\partial y = 1 + \tan x \sec y \tan y$, it follows that $dg/dy = 1$, and thus $g(y) = y + C_1$. Consequently the general solution is

$$\tan x \sec y + y = C$$

13. Since $\dfrac{\partial}{\partial x}(e^x \sin y + \frac{1}{4}x^4 + \sec^2 y) = e^x \sin y + x^3 = \dfrac{\partial}{\partial y}(x^3 y - e^x \cos y)$, the differential equation is exact. Thus there is a function f such that $\partial f/\partial x = x^3 y - e^x \cos y$ and $\partial f/\partial y = e^x \sin y + \frac{1}{4}x^4 + \sec^2 y$. Then $f(x, y) = \frac{1}{4}x^4 y - e^x \cos y + g(y)$, for a suitable function g, so that $\partial f/\partial y = \frac{1}{4}x^4 + e^x \sin y + dg/dy$. Since $\partial f/\partial y = e^x \sin y + \frac{1}{4}x^4 + \sec^2 y$, it follows that $dg/dy = \sec^2 y$, so that $g(y) = \tan y + C_1$. Consequently the general solution is $\frac{1}{4}x^4 y - e^x \cos y + \tan y = C$.

15. Since $\dfrac{\partial}{\partial x}(8xy) = 8y = \dfrac{\partial}{\partial y}(4y^2)$, the differential equation is exact. Thus there is a function f such that $\partial f/\partial x = 4y^2$ and $\partial f/\partial y = 8xy$. Then $f(x, y) = 4xy^2 + g(y)$, for a suitable function g, so that $\partial f/\partial y = 8xy + dg/dy$. Since $\partial f/\partial y = 8xy$, it follows that $dg/dy = 0$, and thus $g(y) = C_1$. Consequently the general solution is $4xy^2 = C$. If $y(3) = \sqrt{2}/2$, then $4(3)(\sqrt{2}/2)^2 = C$, so that $C = 6$. Therefore the particular solution with $y(3) = \sqrt{2}/2$ is $4xy^2 = 6$, or $xy^2 = \frac{3}{2}$, so $y = \sqrt{3/(2x)}$.

17. Since

$$\frac{\partial}{\partial x}\left(\frac{2xy}{1+y^2}\right) = \frac{2y}{1+y^2} = \frac{\partial}{\partial y}[\ln(1+y^2)]$$

the differential equation is exact. Thus there is a function f such that $\partial f/\partial x = \ln(1+y^2)$ and $\partial f/\partial y = 2xy/(1+y^2)$. Then $f(x, y) = x\ln(1+y^2) + g(y)$, for a suitable function g, so that $\partial f/\partial y = 2xy/(1+y^2) + dg/dy$. Since $\partial f/\partial y = 2xy/(1+y^2)$, it follows that $dg/dy = 0$, and thus $g(y) = C_1$. Consequently the general solution is $x\ln(1+y^2) = C$. If $y(2) = \sqrt{e-1}$, then $2\ln[1+(\sqrt{e-1})^2] = C$, so that $C = 2$. Therefore the particular solution with $y(2) = \sqrt{e-1}$ is $x\ln(1+y^2) = 2$, so that $1+y^2 = e^{2/x}$, and thus $y = \sqrt{e^{2/x}-1}$.

19. Since $\dfrac{\partial}{\partial x}(\frac{1}{4}e^{x^2}\cos 3y) = \frac{1}{2}xe^{x^2}\cos 3y$ and $\dfrac{\partial}{\partial y}(axe^{x^2}\sin 3y) = 3axe^{x^2}\cos 3y$, it follows that $\dfrac{\partial}{\partial x}(\frac{1}{4}e^{x^2}\cos 3y) = \dfrac{\partial}{\partial y}(axe^{x^2}\sin 3y)$ if $\frac{1}{2}xe^{x^2}\cos 3y = 3axe^{x^2}\cos 3y$, that is, if $a = \frac{1}{6}$. The exact differential equation is $\frac{1}{6}xe^{x^2}\sin 3y + \frac{1}{4}e^{x^2}\cos 3y(dy/dx) = 0$. Thus there is a function f such that $\partial f/\partial x = \frac{1}{6}xe^{x^2}\sin 3y$ and $\partial f/\partial y = \frac{1}{4}e^{x^2}\cos 3y$. Then $f(x, y) = \frac{1}{12}e^{x^2}\sin 3y + g(y)$, for a suitable function g, so that $\partial f/\partial y = \frac{1}{4}e^{x^2}\cos 3y + dg/dy$. Since $\partial f/\partial y = \frac{1}{4}e^{x^2}\cos 3y$, it follows that $dg/dy = 0$, so that $g(y) = C_1$. Consequently the general solution is $\frac{1}{12}e^{x^2}\sin 3y = C$.

21. If we multiply both sides of the differential equation by x^2+y^2, then it becomes $y^2 + 2xy(dy/dx) = 0$, or $y^2\,dx + 2xy\,dy = 0$. Since $\dfrac{\partial}{\partial x}(2xy) = 2y = \dfrac{\partial}{\partial y}(y^2)$, the new differential equation is exact. Thus there is a function f such that $\partial f/\partial x = y^2$ and $\partial f/\partial y = 2xy$. Then $f(x, y) = xy^2 + g(y)$, for a suitable function g, so that $\partial f/\partial y = 2xy + dg/dy$. Since $\partial f/\partial y = 2xy$, it follows that $dg/dy = 0$, and thus $g(y) = C_1$. Consequently the general solution of the differential equation $y^2\,dx + 2xy\,dy = 0$, and hence of $y^2/(x^2+y^2) + 2xy/(x^2+y^2)(dy/dx) = 0$, is $xy^2 = C$.

SECTION 17.4

1. $P(x) = 1/x^2$ and $Q(x) = 0$. Since $-1/x$ is an antiderivative of P, $S(x) = -1/x$, so (3) implies that $y = e^{1/x}\int e^{-1/x}0\,dx = e^{1/x}(C) = Ce^{1/x}$.

3. $P(x) = 2$ and $Q(x) = 4$. Since $2x$ is an antiderivative of P, $S(x) = 2x$, so (3) implies that $y = e^{-2x}\int e^{2x}4\,dx = 4e^{-2x}\int e^{2x}\,dx = 4e^{-2x}(\frac{1}{2}e^{2x} + C_1) = 2 + Ce^{-2x}$.

5. $P(x) = -a$ and $Q(x) = f(x)$. Since $-ax$ is an antiderivative of P, $S(x) = -ax$, so (3) implies that $y = e^{ax}\int e^{-ax}f(x)\,dx$.

7. $P(x) = 6x^5$ and $Q(x) = x^5$. Since x^6 is an antiderivative of P, $S(x) = x^6$, so (3) implies that $y = e^{-x^6}\int e^{x^6}(x^5)\,dx = \frac{1}{6}e^{-x^6}\int(6x^5)e^{x^6}\,dx = \frac{1}{6}e^{-x^6}(e^{x^6} + C_1) = \frac{1}{6} + Ce^{-x^6}$.

9. $P(x) = -1$ and $Q(x) = 1/(1-e^{-x})$. Since $-x$ is an antiderivative of P, $S(x) = -x$, so (3) implies that

$$y = e^x\int e^{-x}\frac{1}{1-e^{-x}}\,dx = e^x(\ln|1-e^{-x}| + C)$$

11. $P(x) = \tan x$ and $Q(x) = \tan x$. Since $-\ln\cos x$ is an antiderivative of P for $-\pi/2 < x < \pi/2$ (since $\cos x > 0$ for such x), we have $S(x) = -\ln\cos x$. Then (3) implies that $y = e^{\ln\cos x}\int e^{-\ln\cos x}\tan x\,dx = \cos x\int\sec x\tan x\,dx = \cos x(\sec x + C) = 1 + C\cos x$.

13. The equation is equivalent to $dy/dt + (1/t)y = \sin t^2$, for which $P(t) = 1/t$ and $Q(t) = \sin t^2$. Since $\ln t$ is an antiderivative of P for $t > 0$, $S(t) = \ln t$, so (3) implies that

$$y = e^{-\ln t}\int e^{\ln t}\sin t^2\,dt = \frac{1}{t}\int t\sin t^2\,dt = \frac{1}{t}\left(-\frac{1}{2}\cos t^2 + C\right) = -\frac{1}{2t}\cos t^2 + \frac{C}{t}$$

15. $P(x) = 5$ and $Q(x) = -4e^{-3x}$. Since $5x$ is an antiderivative of P, $S(x) = 5x$, so (3) implies that $y = e^{-5x}\int e^{5x}(-4e^{-3x})\,dx = -4e^{-5x}\int e^{2x}\,dx = -4e^{-5x}(\frac{1}{2}e^{2x} + C)$. If $y(0) = -4$, then $-4 = -4e^{-5(0)}(\frac{1}{2}e^{2(0)} + C)$, so that $C = \frac{1}{2}$. Therefore the particular solution with $y(0) = -4$ is $y = -4e^{-5x}(\frac{1}{2}e^{2x} + \frac{1}{2}) = -2e^{-3x} - 2e^{-5x}$.

17. The equation is equivalent to $dy/dx + (1/\cos x)y = 1/\cos x$, so that $P(x) = 1/\cos x = \sec x = Q(x)$. Since $\ln(\sec x + \tan x)$ is an antiderivative of P for $0 < x < \pi/2$, $S(x) = \ln(\sec x + \tan x)$, so (3) implies that

$$y = e^{-\ln(\sec x+\tan x)}\int e^{\ln(\sec x+\tan x)}\sec x\,dx = \frac{1}{\sec x + \tan x}\int(\sec x + \tan x)\sec x\,dx$$

$$= \frac{1}{\sec x + \tan x}\int(\sec^2 x + \tan x\sec x)\,dx = \frac{1}{\sec x + \tan x}(\tan x + \sec x + C)$$

$$= 1 + \frac{C}{\sec x + \tan x}$$

If $y(\pi/4) = 2$, then $2 = 1 + C/(\sqrt{2}+1)$, so that $C = \sqrt{2}+1$. Therefore the particular solution with $y(\pi/4) = 2$ is $y = 1 + (\sqrt{2}+1)/(\sec x + \tan x)$.

19. The equation is equivalent to $dI/dt + (R/L)I = (1/L)e^t$, so $P(t) = R/L$ and $Q(t) = (1/L)e^t$. Since $(R/L)t$ is an antiderivative of P, $S(t) = (R/L)t$, so (3) implies that

$$I = e^{-Rt/L}\int e^{Rt/L}\left(\frac{1}{L}e^t\right)dt = \frac{1}{L}e^{-Rt/L}\int e^{(R/L+1)t}\,dt$$

$$= \frac{1}{L}e^{-Rt/L}\left(\frac{1}{R/L+1}e^{(R/L+1)t} + C_1\right) = \frac{1}{R+L}e^t + Ce^{-Rt/L}$$

21. The equation is equivalent to $dI/dt + (R/L)I = (1/L)\cos t$, so $P(t) = R/L$ and $Q(t) = (1/L)\cos t$. Since $(R/L)t$ is an antiderivative of P, $S(t) = (R/L)t$, so (3) implies that $I = (1/L)e^{-(R/L)t}\int e^{(R/L)t}\cos t\,dt$. Integrating twice by parts, we find that

$$\int e^{(R/L)t}\cos t\,dt = e^{(R/L)t}\sin t - \frac{R}{L}\int e^{(R/L)t}\sin t\,dt$$

$$= e^{(R/L)t}\sin t + \frac{R}{L}e^{(R/L)t}\cos t - \frac{R^2}{L^2}\int e^{(R/L)t}\cos t\,dt$$

Combining integrals involving $\cos t$, we obtain

$$\int e^{(R/L)t}\cos t\,dt = \frac{L^2}{L^2+R^2}\left(e^{(R/L)t}\sin t + \frac{R}{L}e^{(R/L)t}\cos t\right) + C_1$$

$$= \frac{L}{L^2+R^2}e^{(R/L)t}(L\sin t + R\cos t) + C_1$$

Therefore the general solution is

$$I = \frac{1}{L}e^{-(R/L)t}\left[\frac{L}{L^2+R^2}e^{(R/L)t}(L\sin t + R\cos t) + C_1\right]$$

$$= \frac{1}{L^2+R^2}(L\sin t + R\cos t) + Ce^{-(R/L)t}$$

If $I(0) = R/(L^2+R^2)$ for the solution of the differential equation, then

$$\frac{R}{L^2+R^2} = \frac{1}{L^2+R^2}(L\sin 0 + R\cos 0) + Ce^{-(R/L)0} = \frac{R}{L^2+R^2} + C$$

so that $C = 0$. Therefore the particular solution is

$$I = \frac{1}{L^2+R^2}(L\sin t + R\cos t)$$

SECTION 17.5

1. The characteristic equation is $s^2 - 5s - 14 = 0$, with roots $s = 7$ and $s = -2$, so Case 1 applies. By (6) the general solution is $y = C_1e^{7x} + C_2e^{-2x}$.

3. The characteristic equation is $s^2 + 2s - 24 = 0$, with roots $s = -6$ and $s = 4$, so Case 1 applies. By (6) the general solution is $y = C_1e^{-6x} + C_2e^{4x}$.

5. The characteristic equation is $s^2 + 10s + 25 = 0$, with root $s = -5$, so Case 2 applies. By (8) the general solution is $y = C_1e^{-5x} + C_2xe^{-5x}$.

7. The characteristic equation is $s^2 + 9 = 0$, so $b = 0$ and $c = 9$, and thus $b^2 - 4c = -36 < 0$. Therefore Case 3 applies. If $u = -b/2 = 0$ and $v = \frac{1}{2}\sqrt{4c - b^2} = \frac{1}{2}\sqrt{4\cdot 9} = 3$, then by (11) the general solution is $y = C_1\sin 3x + C_2\cos 3x$.

9. The characteristic equation is $s^2 + 3s + 3 = 0$, so $b = 3 = c$, and thus $b^2 - 4c = 3^2 - 4\cdot 3 = -3 < 0$. Therefore Case 3 applies. If $u = -b/2 = -3/2$ and $v = \frac{1}{2}\sqrt{4c - b^2} = \frac{1}{2}\sqrt{4\cdot 3 - 3^2} = \frac{1}{2}\sqrt{3}$, then by (11) the general solution is

$$y = C_1e^{-3x/2}\sin\tfrac{1}{2}\sqrt{3}x + C_2e^{-3x/2}\cos\tfrac{1}{2}\sqrt{3}x$$

11. Dividing by 6, we obtain $d^2y/dt^2 - \frac{4}{6}y = 0$, or $d^2y/dt^2 - \frac{2}{3}y = 0$, whose characteristic equation is $s^2 - \frac{2}{3} = 0$. Its roots are $s = \sqrt{2/3}$ and $s = -\sqrt{2/3}$, so Case 1 applies. By (6) the general solution is $y = C_1e^{\sqrt{2/3}t} + C_2e^{-\sqrt{2/3}t}$.

13. Dividing by 4, we obtain $d^2y/dx^2 + 3(dy/dx) + \frac{9}{4}y = 0$, whose characteristic equation is $s^2 + 3s + \frac{9}{4} = 0$, or $4s^2 + 12s + 9 = 0$, or $(2s+3)^2 = 0$. The root is $s = -\frac{3}{2}$, and thus Case 2 applies. By (8) the general solution is $y = C_1e^{-3x/2} + C_2xe^{-3x/2}$.

15. The characteristic equation is $s^2 - 2s - 15 = 0$, with roots $s = 5$ and $s = -3$, so Case 1 applies. By (6) the general solution is $y = C_1e^{5x} + C_2e^{-3x}$. If $y(0) = 1$ and $y'(0) = -1$, then $1 = y(0) = C_1 + C_2$. Since $dy/dx = 5C_1e^{5x} - 3C_2e^{-3x}$, we have $-1 = y'(0) = 5C_1 - 3C_2$. Therefore $C_1 = 1 - C_2$, so $-1 = 5(1 - C_2) - 3C_2 = 5 - 8C_2$, and thus $C_2 = \frac{3}{4}$, so that $C_1 = 1 - \frac{3}{4} = \frac{1}{4}$. Consequently the particular solution is $y = \frac{1}{4}e^{5x} + \frac{3}{4}e^{-3x}$.

17. The characteristic equation is $s^2 - 10s + 25 = 0$, with root $s = 5$, so Case 2 applies. By (8) the general solution is $y = C_1e^{5x} + C_2xe^{5x}$. If $y(1) = 0$ and $y'(1) = e^5$, then $0 = y(1) = C_1e^5 + C_2e^5$, so $C_1 + C_2 = 0$. Since $dy/dx = 5C_1e^{5x} + C_2e^{5x} + 5C_2xe^{5x}$, we have $e^5 = y'(1) = 5C_1e^5 + C_2e^5 + 5C_2e^5$, so $1 = 5C_1 + 6C_2$. Therefore $C_1 = -C_2$, so $1 = 5(-C_2) + 6C_2 = C_2$, and thus $C_1 = -1$. Consequently the particular solution is $y = -e^{5x} + xe^{5x}$.

19. Dividing by 3, we obtain $d^2y/dx^2 + \frac{8}{3}(dy/dx) - y = 0$, whose characteristic equation is $s^2 + \frac{8}{3}s - 1 = 0$, or $3s^2 + 8s - 3 = 0$. By the quadratic formula, $s = (-8 \pm \sqrt{64 + 36})/6 = (-8 \pm 10)/6$, so the roots are $s = -3$ and $s = \frac{1}{3}$, and thus Case 1 applies. By (6) the general solution is $y = C_1e^{-3x} + C_2e^{x/3}$. If $y(0) = 2$ and $y'(0) = -2$, then $2 = y(0) = C_1 + C_2$. Since $dy/dx = -3C_1e^{-3x} + \frac{1}{3}C_2e^{x/3}$, we have $-2 = y'(0) = -3C_1 + \frac{1}{3}C_2$. Therefore $C_1 = 2 - C_2$, so that $-2 = -3(2 - C_2) + \frac{1}{3}C_2 = -6 + \frac{10}{3}C_2$, and thus $C_2 = \frac{6}{5}$, and $C_1 = 2 - C_2 = 2 - \frac{6}{5} = \frac{4}{5}$. Consequently the particular solution is $y = \frac{4}{5}e^{-3x} + \frac{6}{5}e^{x/3}$.

21. If $y = C_1y_1 + C_2y_2$ with y_1 and y_2 solutions, then

$$\frac{d^2y}{dx^2} + b\frac{dy}{dx} + cy = \left(C_1\frac{d^2y_1}{dx^2} + C_2\frac{d^2y_2}{dx^2}\right) + b\left(C_1\frac{dy_1}{dx} + C_2\frac{dy_2}{dx}\right) + c(C_1y_1 + C_2y_2)$$

$$= C_1\left(\frac{d^2y_1}{dx^2} + b\frac{dy_1}{dx} + cy_1\right) + C_2\left(\frac{d^2y_2}{dx^2} + b\frac{dy_2}{dx} + cy_2\right) = 0 + 0 = 0$$

Thus $C_1y_1 + C_2y_2$ is also a solution.

23. By the hypotheses and (12) we have $0.01(d^2q/dt^2) + 2(dq/dt) + (1/0.005)q = 0$, or equivalently, $d^2q/dt^2 + 200(dq/dt) + 20{,}000q = 0$. The characteristic equation is $s^2 + 200s + 20{,}000 = 0$, so $b = 200$ and $c = 20{,}000$, and $b^2 - 4c = (200)^2 - 80{,}000 < 0$. Therefore Case 3 applies. If $u = -b/2 = -100$ and $v = \frac{1}{2}\sqrt{4c - b^2} = \frac{1}{2}\sqrt{4(20{,}000) - (200)^2} = 100$, then by (11) the general solution is $q = C_1e^{-100t}\sin 100t + C_2e^{-100t}\cos 100t$. By hypothesis, $q(0) = 1$ and $q'(1) = 100$. It follows that $1 = q(0) = C_2$, so $q = e^{-100t}(C_1 \sin 100t + \cos 100t)$. Since $dq/dt = -100e^{-100t}(C_1 \sin 100t + \cos 100t) + e^{-100t}(100C_1 \cos 100t - 100 \sin 100t)$, we have $100 = q'(0) = -100 + 100C_1$, so that $C_1 = 2$. Consequently the charge is given by $q = 2e^{-100t}\sin 100t + e^{-100t}\cos 100t$ for $t \geq 0$.

SECTION 17.6

1. The homogeneous equation is $d^2y/dx^2 + 5(dy/dx) + 4y = 0$, with characteristic equation $s^2 + 5s + 4 = 0$, whose roots are $s = -1$ and $s = -4$. Thus $y_1 = e^{-x}$ and $y_2 = e^{-4x}$. Next $g(x) = 3$, so by (9) and (10),

$$u_1'(x) = \frac{-e^{-4x}(3)}{e^{-x}(-4e^{-4x}) - (-e^{-x})(e^{-4x})} = e^x$$

$$u_2'(x) = \frac{e^{-x}(3)}{e^{-x}(-4e^{-4x}) - (-e^{-x})(e^{-4x})} = -e^{4x}$$

Therefore $u_1(x) = e^x$ and $u_2(x) = -\frac{1}{4}e^{4x}$, so that by (5), $y_p = e^xe^{-x} + (-\frac{1}{4}e^{4x})e^{-4x} = 1 - \frac{1}{4} = \frac{3}{4}$. By (4), the general solution is $y = \frac{3}{4} + C_1e^{-x} + C_2e^{-4x}$.

3. The homogeneous equation is $d^2y/dx^2 - 2(dy/dx) - 3y = 0$, with characteristic equation $s^2 - 2s - 3 = 0$, whose roots are $s = 3$ and $s = -1$. Thus $y_1 = e^{3x}$ and $y_2 = e^{-x}$. Next $g(x) = e^x$, so by (9) and (10),

$$u_1'(x) = \frac{-e^{-x}e^x}{e^{3x}(-e^{-x}) - 3e^{3x}e^{-x}} = \frac{1}{4}e^{-2x}$$

$$u_2'(x) = \frac{e^{3x}e^x}{e^{3x}(-e^{-x}) - 3e^{3x}e^{-x}} = -\frac{1}{4}e^{2x}$$

Therefore $u_1(x) = -\frac{1}{8}e^{-2x}$ and $u_2(x) = -\frac{1}{8}e^{2x}$, so that by (5), $y_p(x) = -\frac{1}{8}e^{-2x}e^{3x} - \frac{1}{8}e^{2x}e^{-x} = -\frac{1}{4}e^x$. By (4), the general solution is $y = -\frac{1}{4}e^x + C_1e^{3x} + C_2e^{-x}$.

5. The homogeneous equation is $d^2y/dx^2 - dy/dx = 0$, with characteristic equation $s^2 - s = 0$, whose roots are $s = 0$ and $s = 1$. Thus $y_1 = 1$ and $y_2 = e^x$. Next, $g(x) = 2x - 3$, so by (9) and (10),

$$u_1'(x) = \frac{-e^x(2x - 3)}{1 \cdot e^x - 0 \cdot e^x} = -2x + 3$$

$$u_2'(x) = \frac{1(2x - 3)}{1 \cdot e^x - 0 \cdot e^x} = (2x - 3)e^{-x}$$

Therefore $u_1(x) = -x^2 + 3x$, and since

$$\int (2x - 3)e^{-x}\,dx = 2\int xe^{-x}\,dx - 3\int e^{-x}\,dx \overset{\text{parts}}{=} 2\left(-xe^{-x} + \int e^{-x}\,dx\right) + 3e^{-x}$$
$$= -2xe^{-x} - 2e^{-x} + 3e^{-x} + C = -2xe^{-x} + e^{-x} + C$$

we have $u_2(x) = -2xe^{-x} + e^{-x}$. By (5), $y_p = (-x^2 + 3x)1 + (-2xe^{-x} + e^{-x})e^x = -x^2 + 3x - 2x + 1 = -x^2 + x + 1$. By (4), the general solution is $y = -x^2 + x + 1 + C_1 + C_2e^x$.

7. The homogeneous equation is $d^2y/dx^2 + y = 0$, with characteristic equation $s^2 + 1 = 0$. This falls under Case 3, with $b = 0$ and $c = 1$. Since $u = -b/2 = 0$ and $v = \frac{1}{2}\sqrt{4c - b^2} = \frac{1}{2}\sqrt{4 \cdot 1 - 0^2} = 1$, it follows that the general solution of the homogeneous equation is $y = C_1 \sin x + C_2 \cos x$. Thus $y_1 = \sin x$ and $y_2 = \cos x$. Next, $g(x) = \csc x \cot x$, so by (9) and (10),

$$u_1'(x) = \frac{-\cos x \csc x \cot x}{\sin x(-\sin x) - \cos x \cos x} = \cot^2 x = \csc^2 x - 1$$

$$u_2'(x) = \frac{\sin x \csc x \cot x}{\sin x(-\sin x) - \cos x \cos x} = -\cot x$$

Therefore $u_1(x) = -\cot x - x$ and $u_2 = -\ln \sin x$, so that by (5), $y_p = (-\cot x - x)\sin x - (\ln \sin x)\cos x = -\cos x - x \sin x - \cos x \ln \sin x$. By (4) the general solution is

$$y = -\cos x - x \sin x - \cos x \ln \sin x + C_1 \sin x + C_2' \cos x$$
$$= -x \sin x - \cos x \ln \sin x + C_1 \sin x + C_2 \cos x.$$

9. The homogeneous equation is $d^2y/dt^2 + 9y = 0$, with characteristic equation $s^2 + 9 = 0$. This falls under Case 3, with $b = 0$ and $c = 9$. Since $u = -b/2 = 0$ and $v = \frac{1}{2}\sqrt{4c - b^2} = \frac{1}{2}\sqrt{4 \cdot 9 - 0^2} = 3$, it follows that the general solution of the homogeneous equation is $y = C_1 \sin 3t + C_2 \cos 3t$. Thus $y_1 = \sin 3t$ and $y_2 = \cos 3t$. Next $g(t) = 3t$, so by (9) and (10),

$$u_1'(t) = \frac{-(\cos 3t)(3t)}{\sin 3t(-3 \sin 3t) - 3 \cos 3t \cos 3t} = t \cos 3t$$

$$u_2'(t) = \frac{(\sin 3t)(3t)}{\sin 3t(-3 \sin 3t) - 3 \cos 3t \cos 3t} = -t \sin 3t$$

By integration by parts we have $\int t \cos 3t\,dt = \frac{1}{3}t \sin 3t - \int \frac{1}{3}\sin 3t\,dt = \frac{1}{3}t \sin 3t + \frac{1}{9}\cos 3t + C'$ and $\int -t \sin 3t\,dt = \frac{1}{3}t \cos 3t - \int \frac{1}{3}\cos 3t\,dt = \frac{1}{3}t \cos 3t - \frac{1}{9}\sin 3t + C''$. Thus we take $u_1(t) = \frac{1}{3}t \sin 3t + \frac{1}{9}\cos 3t$ and $u_2(t) = \frac{1}{3}t \cos 3t - \frac{1}{9}\sin 3t$, so that by (5), $y_p = (\frac{1}{3}t \sin 3t + \frac{1}{9}\cos 3t)\sin 3t + (\frac{1}{3}t \cos 3t - \frac{1}{9}\sin 3t)\cos 3t = \frac{1}{3}t \sin^2 3t + \frac{1}{3}t \cos^2 3t = \frac{1}{3}t$. By (4), the general solution is $y = \frac{1}{3}t + C_1 \sin 3t + C_2 \cos 3t$.

11. The homogeneous equation is $d^2y/dt^2 + 9y = 0$, with characteristic equation $s^2 + 9 = 0$. This falls under Case 3, with $b = 0$ and $c = 9$. Since $u = -b/2 = 0$ and $v = \frac{1}{2}\sqrt{4c - b^2} = \frac{1}{2}\sqrt{4 \cdot 9 - 0^2} = 3$, it follows that the general solution of the homogeneous equation is $y = C_1 \sin 3t + C_2 \cos 3t$. Thus $y_1 = \sin 3t$ and $y_2 = \cos 3t$. Next, $g(t) = \sin 3t$, so by (9) and (10),

$$u_1'(t) = \frac{-\cos 3t \sin 3t}{\sin 3t\,(-3 \sin 3t) - (3 \cos 3t)(\cos 3t)} = \frac{1}{3} \sin 3t \cos 3t = \frac{1}{6} \sin 6t$$

$$u_2'(t) = \frac{\sin 3t \sin 3t}{\sin 3t\,(-3 \sin 3t) - (3 \cos 3t)(\cos 3t)} = -\frac{1}{3} \sin^2 3t = -\frac{1}{3}\left(\frac{1}{2} - \frac{1}{2} \cos 6t\right)$$

Thus $u_1(t) = -\frac{1}{36} \cos 6t$ and $u_2(t) = -\frac{1}{6}t + \frac{1}{36} \sin 6t$, so that by (5), $y_p = (-\frac{1}{36} \cos 6t) \sin 3t + (-\frac{1}{6}t + \frac{1}{36} \sin 6t) \cos 3t = -\frac{1}{6}t \cos 3t + \frac{1}{36} \sin 3t$. By (4), the general solution is

$$\begin{aligned} y &= -\tfrac{1}{6}t \cos 3t + \tfrac{1}{36} \sin 3t + C_1' \sin 3t + C_2 \cos 3t \\ &= -\tfrac{1}{6}t \cos 3t + C_1 \sin 3t + C_2 \cos 3t \end{aligned}$$

13. The homogeneous equation is $d^2y/dx^2 - 2(dy/dx) + y = 0$, with characteristic equation $s^2 - 2s + 1 = 0$, whose root is $s = 1$. Thus $y_1 = e^x$ and $y_2 = xe^x$. Next, $g(x) = (1/x)e^x$, so by (9) and (10),

$$u_1'(x) = \frac{-(xe^x)((1/x)e^x)}{(e^x)(e^x + xe^x) - (e^x)(xe^x)} = \frac{-e^{2x}}{e^{2x}} = -1$$

$$u_2'(x) = \frac{(e^x)((1/x)e^x)}{(e^x)(e^x + xe^x) - (e^x)(xe^x)} = \frac{(1/x)e^{2x}}{e^{2x}} = \frac{1}{x}$$

Therefore $u_1(x) = -x$ and $u_2(x) = \ln |x|$, so that by (5), $y_p = (-x)(e^x) + (\ln |x|)(xe^x) = -xe^x + xe^x \ln |x|$. By (4), the general solution is $y = -xe^x + xe^x \ln |x| + C_1e^x + C_2'xe^x = xe^x \ln |x| + C_1e^x + C_2xe^x$.

15. The homogeneous equation is $d^2y/dx^2 - 5(dy/dx) + 6y = 0$, with characteristic equation $s^2 - 5s + 6 = 0$, whose roots are $s = 2$ and $s = 3$. Therefore $y_1 = e^{2x}$ and $y_2 = e^{3x}$. Next, $g(x) = x^3e^{2x}$, so by (9) and (10),

$$u_1'(x) = \frac{-e^{3x}(x^3e^{2x})}{e^{2x}(3e^{3x}) - (2e^{2x})e^{3x}} = \frac{-x^3e^{5x}}{e^{5x}} = -x^3$$

$$u_2'(x) = \frac{e^{2x}(x^3e^{2x})}{e^{2x}(3e^{3x}) - (2e^{2x})e^{3x}} = \frac{x^3e^{4x}}{e^{5x}} = x^3e^{-x}$$

Thus $u_1(x) = -\frac{1}{4}x^4$. By successive integrations by parts, we obtain

$$\begin{aligned} \int x^3e^{-x}\,dx &= -x^3e^{-x} + \int 3x^2e^{-x}\,dx = -x^3e^{-x} - 3x^2e^{-x} + \int 6xe^{-x}\,dx \\ &= -x^3e^{-x} - 3x^2e^{-x} - 6xe^{-x} + \int 6e^{-x}\,dx \\ &= -x^3e^{-x} - 3x^2e^{-x} - 6xe^{-x} - 6e^{-x} + C \end{aligned}$$

so it follows that $u_2(x) = -(x^3 + 3x^2 + 6x + 6)e^{-x}$. Therefore by (5), $y_p = -\frac{1}{4}x^4e^{2x} - (x^3 + 3x^2 + 6x + 6)e^{-x}(e^{3x}) = -\frac{1}{4}x^4e^{2x} - (x^3 + 3x^2 + 6x + 6)e^{2x}$. By (4) the general solution is

$$\begin{aligned} y &= -(\tfrac{1}{4}x^4 + x^3 + 3x^2 + 6x + 6)e^{2x} + C_1'e^{2x} + C_2e^{3x} \\ &= -(\tfrac{1}{4}x^4 + x^3 + 3x^2 + 6x)e^{2x} + C_1e^{2x} + C_2e^{3x} \end{aligned}$$

17. We have $(0.04)(d^2q/dt^2) + 5(dq/dt) + (1/0.01)q = E(t)$, so that $d^2q/dt^2 + 125(dq/dt) + 2500q = E(t)$. The characteristic equation is $s^2 + 125s + 2500 = 0$, so $(s + 100)(s + 25) = 0$. Thus $q_1 = e^{-100t}$ and $q_2 = e^{-25t}$.

a. Here $g(t) = E(t) = e^{-50t}$, so that by (9) and (10),

$$u_1'(t) = \frac{(-e^{-25t})(e^{-50t})}{(e^{-100t})(-25e^{-25t}) - (-100e^{-100t})(e^{-25t})} = \frac{-e^{-75t}}{75e^{-125t}} = -\frac{1}{75}e^{50t}$$

$$u_2'(t) = \frac{(e^{-100t})(e^{-50t})}{(e^{-100t})(-25e^{-25t}) - (-100e^{-100t})(e^{-25t})} = \frac{e^{-150t}}{75e^{-125t}} = \frac{1}{75}e^{-25t}$$

Thus $u_1(t) = -\frac{1}{3750}e^{50t}$ and $u_2(t) = -\frac{1}{1875}e^{-25t}$, so by (5), $q_p = (-\frac{1}{3750}e^{50t})(e^{-100t}) + (-\frac{1}{1875}e^{-25t})(e^{-25t}) = -\frac{1}{1250}e^{-50t}$. By (4), the general solution is $q = -\frac{1}{1250}e^{-50t} + C_1e^{-100t} + C_2e^{-25t}$ for $t \geq 0$.

b. Here $g(t) = E(t) = \sin 50t$, so that by (9) and (10),

$$\begin{aligned} u_1'(t) &= \frac{(-e^{-25t}) \sin 50t}{(e^{-100t})(-25e^{-25t}) - (-100e^{-100t})(e^{-25t})} \\ &= \frac{-e^{-25t} \sin 50t}{75e^{-125t}} = -\frac{1}{75}e^{100t} \sin 50t \end{aligned}$$

$$\begin{aligned} u_2'(t) &= \frac{(e^{-100t}) \sin 50t}{(e^{-100t})(-25e^{-25t}) - (-100e^{-100t})(e^{-25t})} \\ &= \frac{e^{-100t} \sin 50t}{75e^{-125t}} = \frac{1}{75}e^{25t} \sin 50t \end{aligned}$$

By Exercise 53(a) in Section 7.1,

$$\begin{aligned} u_1(t) &= -\frac{1}{75}\int e^{100t} \sin 50t\,dt = -\frac{1}{75}\left[\frac{e^{100t}}{100^2 + 50^2}(100 \sin 50t - 50 \cos 50t)\right] \\ &= \frac{1}{18{,}750}e^{100t}(\cos 50t - 2 \sin 50t) \end{aligned}$$

and

$$\begin{aligned} u_2(t) &= \frac{1}{75}\int e^{25t} \sin 50t\,dt = \frac{1}{75}\left[\frac{e^{25t}}{25^2 + 50^2}(25 \sin 50t - 50 \cos 50t)\right] \\ &= \frac{1}{9375}e^{25t}(\sin 50t - 2 \cos 50t) \end{aligned}$$

Consequently by (5),

$$q_p = \frac{1}{18{,}750}(\cos 50t - 2\sin 50t) + \frac{1}{9375}(\sin 50t - 2\cos 50t) = -\frac{1}{6250}\cos 50t$$

By (4), the general solution is $q = -\frac{1}{6250}\cos 50t + C_1e^{-100t} + C_2e^{-25t}$ for $t \geq 0$.

SECTION 17.7

1. As in Example 1, we obtain the differential equation $d^2x/dt^2 + 64x = 0$, which has the solution $x = C_1 \sin 8t + C_2 \cos 8t$ for appropriate C_1 and C_2. By hypothesis, $x(0) = \frac{1}{6}$, so that $\frac{1}{6} = C_1 \sin 0 + C_2 \cos 0 = C_2$. Also $dx/dt = 8C_1 \cos 8t - 8C_2 \sin 8t$, so that since $x'(0) = 0$ by hypothesis, we have $0 = 8C_1 \cos 0 - 8C_2 \sin 0 = 8C_1$, so $C_1 = 0$. Thus $x = \frac{1}{6}\cos 8t$.

3. From Exercise 1 and the hypothesis that $x(0) = \frac{1}{6}$, we obtain $x = C_1 \sin 8t + C_2 \cos 8t = C_1 \sin 8t + \frac{1}{6}\cos 8t$. Since $dx/dt = 8C_1 \cos 8t - \frac{4}{3}\sin 8t$ and since $x'(0) = -\frac{1}{6}$ by hypothesis, we have $-\frac{1}{6} = 8C_1 - \frac{4}{3}\sin 0 = 8C_1$, so that $C_1 = -\frac{1}{48}$. Therefore the solution is $x = -\frac{1}{48}\sin 8t + \frac{1}{6}\cos 8t$.

5. We have $m = \text{weight}/g = 1.6/32 = \frac{1}{20}$, $k = \frac{1}{4}$, and $p = 0.2 = \frac{1}{5}$. Therefore (8) becomes $d^2x/dt^2 + (\frac{1}{5}/\frac{1}{20})(dx/dt) + (\frac{1}{4}/\frac{1}{20})x = 0$, that is, $d^2x/dt^2 + 4(dx/dt) + 5x = 0$. The characteristic equation is $s^2 + 4s + 5 = 0$, which falls under Case 3 of Section 17.5 with $b = 4$ and $c = 5$. Since $u = -b/2 = -2$ and $v = \frac{1}{2}\sqrt{4c - b^2} = \frac{1}{2}\sqrt{4 \cdot 5 - 4^2} = \frac{1}{2}\sqrt{4} = 1$, we conclude from (11) of Section 17.5 that $x = C_1e^{-2t}\sin t + C_2e^{-2t}\cos t$ for appropriate C_1 and C_2. By hypothesis, $x(0) = -\frac{1}{12}$, so that $-\frac{1}{12} = C_1e^0 \sin 0 + C_2e^0 \cos 0 = C_2$. Also $dx/dt = -(2C_1 + C_2)e^{-2t}\sin t + (C_1 - 2C_2)e^{-2t}\cos t$. Since $x'(0) = -\frac{1}{6}$ by hypothesis, we have $-\frac{1}{6} = -(2C_1 + C_2)e^0 \sin 0 + (C_1 - 2C_2)e^0 \cos 0 = C_1 - 2C_2$. Thus $C_1 = -\frac{1}{6} + 2C_2 = -\frac{1}{6} + 2(-\frac{1}{12}) = -\frac{1}{3}$. Consequently $x = -\frac{1}{3}e^{-2t}\sin t - \frac{1}{12}e^{-2t}\cos t$.

7. We first calculate the mass m, which must satisfy $p^2 = 4km$. By hypothesis we have $p = \frac{1}{5}$ and $k = \frac{1}{4}$, so that $\frac{1}{25} = 4(\frac{1}{4})m = m$. Since weight $= g(\text{mass}) = 32(\frac{1}{25}) = 1.28$, a weight of 1.28 pounds must be attached to achieve critical damping.

9. a. Since $e^{r_1t} > 0$ and $e^{r_2t} > 0$ for all t, it follows that if $C_1 > 0$ and $C_2 > 0$, then $f(t) = C_1t^{r_1t} + C_2e^{r_2t} > 0$, whereas if $C_1 < 0$ and $C_2 < 0$, then $f(t) = C_1e^{r_1t} + C_2e^{r_2t} < 0$. Either way, $f(t) \neq 0$ for all t.

   b. If C_1 and C_2 have opposite signs, then $-C_2/C_1 > 0$. Now $f(t) = 0$ provided that $0 = C_1e^{r_1t} + C_2e^{r_2t}$, so that $C_1e^{r_1t} = -C_2e^{r_2t}$, and thus $e^{(r_1-r_2)t} = -C_2/C_1$, which means that $(r_1 - r_2)t = \ln(-C_2/C_1)$, and finally, $t = \ln(-C_2/C_1)/(r_1 - r_2)$.

SECTION 17.8

1. Let $y = \sum_{n=0}^{\infty} c_nx^n$. Then $dy/dx = \sum_{n=1}^{\infty} nc_nx^{n-1} = \sum_{n=0}^{\infty}(n+1)c_{n+1}x^n$. Since $dy/dx - 5y = 0$, this means that

$$\sum_{n=0}^{\infty}(n+1)c_{n+1}x^n - 5\sum_{n=0}^{\infty}c_nx^n = 0$$

so that $\sum_{n=0}^{\infty}[(n+1)c_{n+1} - 5c_n]x^n = 0$. By Corollary 9.27, $(n+1)c_{n+1} - 5c_n = 0$, that is, $c_{n+1} = [5/(n+1)]c_n$. This yields

$$c_1 = \frac{5}{1}c_0 \qquad c_3 = \frac{5}{3}c_2 = \frac{5^3}{3!}c_0$$

$$c_2 = \frac{5}{2}c_1 = \frac{5^2}{1 \cdot 2}c_0 \qquad c_4 = \frac{5}{4}c_3 = \frac{5^4}{4!}c_0$$

In general for any positive integer n we have $c_n = (5^n/n!)c_0$. Thus a series solution is $y = c_0\sum_{n=0}^{\infty}(5^n/n!)x^n$.

3. Let $y = \sum_{n=0}^{\infty} c_nx^n$. Then $dy/dx = \sum_{n=1}^{\infty} nc_nx^{n-1}$ and $d^2y/dx^2 = \sum_{n=2}^{\infty} n(n-1)c_nx^{n-2}$. Since $d^2y/dx^2 + y = 0$, this means that

$$\sum_{n=2}^{\infty} n(n-1)c_nx^{n-2} + \sum_{n=0}^{\infty}c_nx^n = 0$$

or equivalently, $$\sum_{n=0}^{\infty}(n+2)(n+1)c_{n+2}x^n + \sum_{n=0}^{\infty}c_nx^n = 0$$

so that $$\sum_{n=0}^{\infty}[(n+2)(n+1)c_{n+2} + c_n]x^n = 0$$

By Corollary 9.27, $(n+2)(n+1)c_{n+2} + c_n = 0$, that is, $c_{n+2} = [1/[(n+2)(n+1)]]\,c_n$ for $n \geq 0$. Thus

$$c_2 = -\frac{1}{2 \cdot 1}c_0 \qquad c_4 = -\frac{1}{4 \cdot 3}c_2 = (-1)^2\frac{1}{4 \cdot 3 \cdot 2}c_0$$

$$c_3 = -\frac{1}{3 \cdot 2}c_1 \qquad c_5 = -\frac{1}{5 \cdot 4}c_3 = (-1)^2\frac{1}{5 \cdot 4 \cdot 3 \cdot 2}c_1$$

In general for any positive integer n we have

$$c_{2n} = (-1)^n\frac{1}{(2n)!}c_0 \quad \text{and} \quad c_{2n+1} = (-1)^n\frac{1}{(2n+1)!}c_1$$

Consequently a series solution is

$$y = c_0\sum_{n=0}^{\infty}\frac{(-1)^n}{(2n)!}x^{2n} + c_1\sum_{n=0}^{\infty}\frac{(-1)^n}{(2n+1)!}x^{2n+1}$$

5. Let $y = \sum_{n=0}^{\infty} c_n x^n$. Then $dy/dx = \sum_{n=1}^{\infty} nc_n x^{n-1}$ and $d^2y/dx^2 = \sum_{n=2}^{\infty} n(n-1)c_n x^{n-2}$. Since $d^2y/dx^2 + x(dy/dx) + y = 0$, this means that

$$\sum_{n=2}^{\infty} n(n-1)c_n x^{n-2} + x\sum_{n=1}^{\infty} nc_n x^{n-1} + \sum_{n=0}^{\infty} c_n x^n = 0$$

or equivalently,

$$\sum_{n=0}^{\infty} (n+2)(n+1)c_{n+2}x^n + \sum_{n=1}^{\infty} nc_n x^n + \sum_{n=0}^{\infty} c_n x^n = 0$$

so that $$2c_2 + c_0 + \sum_{n=1}^{\infty} [(n+2)(n+1)c_{n+2} + (n+1)c_n]x^n = 0$$

By Corollary 9.27, $2c_2 + c_0 = 0$, that is, $c_2 = -\frac{1}{2}c_0$

and

$$(n+2)(n+1)c_{n+2} + (n+1)c_n = 0, \quad \text{that is} \quad c_{n+2} = -\frac{1}{n+2}c_n \quad \text{for } n \geq 1$$

Thus $$c_2 = -\frac{1}{2}c_0 \qquad c_4 = -\frac{1}{4}c_2 = (-1)^2\frac{1}{4\cdot 2}c_0$$

$$c_3 = -\frac{1}{3}c_1 \qquad c_5 = -\frac{1}{5}c_3 = (-1)^2\frac{1}{5\cdot 3}c_1$$

In general for any positive integer n we have

$$c_{2n} = (-1)^n\frac{1}{(2n)(2n-2)\cdots 4\cdot 2}c_0 = (-1)^n\frac{1}{2^n n!}c_0$$

and

$$c_{2n+1} = (-1)^n\frac{1}{(2n+1)(2n-1)\cdots 5\cdot 3}c_1 = (-1)^n\frac{(2n)(2n-2)\cdots 4\cdot 2}{(2n+1)!}c_1$$
$$= \frac{(-1)^n 2^n n!}{(2n+1)!}c_1$$

Therefore a series solution is

$$y = c_0\sum_{n=0}^{\infty}(-1)^n\frac{1}{2^n n!}x^{2n} + c_1\sum_{n=0}^{\infty}(-1)^n\frac{2^n n!}{(2n+1)!}x^{2n+1}$$

7. Let $y = \sum_{n=0}^{\infty} c_n x^n$. Then $dy/dx = \sum_{n=1}^{\infty} nc_n x^{n-1}$. Since $dy/dx = xy$, this means that

$$\sum_{n=1}^{\infty} nc_n x^{n-1} = x\sum_{n=0}^{\infty} c_n x^n$$

or equivalently,

$$\sum_{n=0}^{\infty}(n+1)c_{n+1}x^n - \sum_{n=1}^{\infty} c_{n-1}x^n = 0, \quad \text{so that} \quad c_1 + \sum_{n=1}^{\infty}[(n+1)c_{n+1} - c_{n-1}]x^n = 0$$

By Corollary 9.27, $c_1 = 0$, and $(n+1)c_{n+1} - c_{n-1} = 0$ for $n \geq 1$, which means that $c_{n+2} = [1/(n+2)]c_n$ for $n \geq 0$. Since $c_1 = 0$, this implies that $c_n = 0$ for any odd positive integer n. Also

$$c_2 = \frac{1}{2}c_0 \qquad c_6 = \frac{1}{6}c_4 = \frac{1}{6\cdot 4\cdot 2}c_0$$

$$c_4 = \frac{1}{4}c_2 = \frac{1}{4\cdot 2}c_0 \qquad c_8 = \frac{1}{8}c_6 = \frac{1}{8\cdot 6\cdot 4\cdot 2}c_0$$

In general for any positive integer n we have

$$c_{2n} = \frac{1}{(2n)(2n-2)\cdots 4\cdot 2}c_0 = \frac{1}{2^n n!}c_0$$

Therefore a series solution is $y = c_0\sum_{n=0}^{\infty}[1/(2^n n!)]x^{2n}$. If $y(0) = 1$, then $c_0 = 1$, so the particular solution is $y = \sum_{n=0}^{\infty}[1/(2^n n!)]x^{2n}$.

9. Let $y = \sum_{n=0}^{\infty} c_n x^n$. Then $dy/dx = \sum_{n=1}^{\infty} nc_n x^{n-1}$ and $d^2y/dx^2 = \sum_{n=2}^{\infty} n(n-1)c_n x^{n-2}$. Since $x^2(d^2y/dx^2) - 6y = 0$, this means that

$$x^2\sum_{n=2}^{\infty} n(n-1)c_n x^{n-2} - 6\sum_{n=0}^{\infty} c_n x^n = 0$$

or equivalently,

$$\sum_{n=2}^{\infty} n(n-1)c_n x^n - 6\sum_{n=0}^{\infty} c_n x^n = 0, \quad \text{so that} \quad -6c_0 - 6c_1 x + \sum_{n=2}^{\infty}[n(n-1)c_n - 6c_n]x^n = 0$$

By Corollary 9.27, $c_0 = 0$, $c_1 = 0$, and $[n(n-1) - 6]c_n = 0$ for $n \geq 2$. Since $n(n-1) - 6 = 0$ only if $n = 3$, it follows that $c_n = 0$ for $n = 2$ and $n \geq 4$. Thus a series solution is $y = c_3 x^3$. If $y(1) = 5$, then $5 = y(1) = c_3(1)^3 = c_3$, so that the particular solution is $y = 5x^3$.

11. Let $y = \sum_{n=0}^{\infty} c_n x^n$. Then $dy/dx = \sum_{n=1}^{\infty} nc_n x^{n-1}$ and $d^2y/dx^2 = \sum_{n=2}^{\infty} n(n-1)c_n x^{n-2}$. Since $d^2y/dx^2 - x(dy/dx) - y = 0$, this means that

$$\sum_{n=2}^{\infty} n(n-1)c_n x^{n-2} - x\sum_{n=1}^{\infty} nc_n x^{n-1} - \sum_{n=0}^{\infty} c_n x^n = 0$$

or equivalently,

$$\sum_{n=0}^{\infty}(n+2)(n+1)c_{n+2}x^n - \sum_{n=1}^{\infty} nc_n x^n - \sum_{n=0}^{\infty} c_n x^n = 0$$

that is, $$2c_2 - c_0 + \sum_{n=1}^{\infty}[(n+2)(n+1)c_{n+2} - (n+1)c_n]x^n = 0$$

By Corollary 9.27, $c_2 = \frac{1}{2}c_0$ and $(n+2)(n+1)c_{n+2} - (n+1)c_n = 0$, so that $c_{n+2} = [1/(n+2)]c_n$ for $n \geq 0$. Thus

$$c_3 = \frac{1}{3}c_1 \qquad c_5 = \frac{1}{5}c_3 = \frac{1}{5\cdot 3}c_1$$

$$c_4 = \frac{1}{4}c_2 = \frac{1}{4\cdot 2}c_0 \qquad c_6 = \frac{1}{6}c_4 = \frac{1}{6\cdot 4\cdot 2}c_0$$

In general for any positive integer n we have

$$c_{2n} = \frac{1}{(2n)(2n-2)\cdots 4\cdot 2}c_0 = \frac{1}{2^n n!}c_0$$

and

$$c_{2n+1} = \frac{1}{(2n+1)(2n-1)\cdots 5\cdot 3}c_1 = \frac{(2n)(2n-2)\cdots 4\cdot 2}{(2n+1)!}c_1 = \frac{2^n n!}{(2n+1)!}c_1$$

Thus a series solution is

$$y = \sum_{n=0}^{\infty} c_{2n}x^{2n} + \sum_{n=0}^{\infty} c_{2n+1}x^{2n+1} = c_0 \sum_{n=0}^{\infty} \frac{1}{2^n n!}x^{2n} + c_1 \sum_{n=0}^{\infty} \frac{2^n n!}{(2n+1)!}x^{2n+1}$$

If $y(0) = 1$, then $c_0 = 1$. Next,

$$\frac{dy}{dx} = c_0 \sum_{n=1}^{\infty} \frac{2n}{2^n n!}x^{2n-1} + c_1 \sum_{n=0}^{\infty} \frac{2^n n!(2n+1)}{(2n+1)!}x^{2n}$$

Thus if $y'(0) = 0$, then $c_1 = 0$. Consequently the particular solution is

$$y = \sum_{n=0}^{\infty} [1/(2^n n!)]x^{2n}$$

SECTION 17.9

1. $2y + 5 = 0$ if $y = -\frac{5}{2}$. Thus $y = -\frac{5}{2}$ is the constant solution.

3. $y^2 - 2y - 15 = 0$ if $(y-5)(y+3) = 0$, so that $y = 5$ or $y = -3$. Thus $y = 5$ and $y = -3$ are the constant solutions.

5. $y^2 e^{-y} = 0$ if $y = 0$. Thus $y = 0$ is the constant solution.

7. $f(y) = 1 - y$; $f'(y) = -1$; $f'(y(t))\frac{dy}{dt} = (-1)\frac{dy}{dt}$

   Constant solution: $y = 1$. If $y(0) = 2$ then $f(y(0)) = f(2) = -1$, so y is strictly decreasing. By (5), the graph is concave upward. If $y(0) = -2$ then $f(y(0)) = f(-2) = 3$, so y is strictly increasing. By (5), the graph is concave downward.

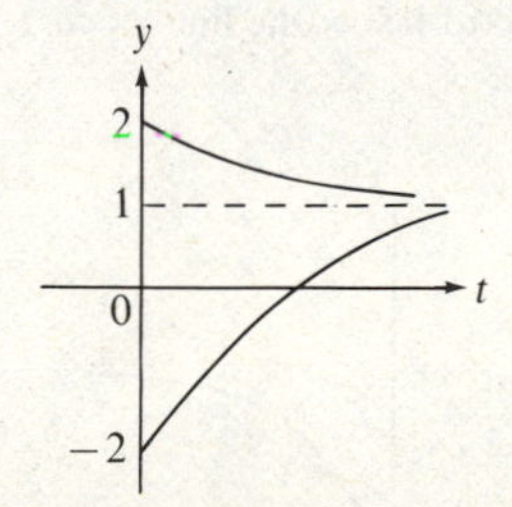

9. $f(y) = y^2 - 2y = y(y-2)$; $f'(y) = 2y - 2 = 2(y-1)$; $f'(y(t))\frac{dy}{dt} = 2(y(t)-1)\frac{dy}{dt}$

   Constant solutions: $y = 0$ and $y = 2$. If $y(0) = \frac{5}{2}$ then $f(y(0)) = f(\frac{5}{2}) = \frac{5}{4}$, so y is strictly increasing. By (5), the graph is concave upward. If $y(0) = \frac{3}{2}$ then $f(y(0)) = f(\frac{3}{2}) = -\frac{3}{4}$, so y is strictly decreasing. By (5), the graph is concave downward above the line $y = 1$ and concave upward below the line $y = 1$. If $y(0) = \frac{1}{2}$ then $f(y(0)) = f(\frac{1}{2}) = -\frac{3}{4}$, so y is strictly decreasing. By (5), the graph is concave upward. If $y(0) = -\frac{1}{2}$ then $f(y(0)) = f(-\frac{1}{2}) = \frac{5}{4}$, so y is strictly increasing. By (5) the graph is concave downward.

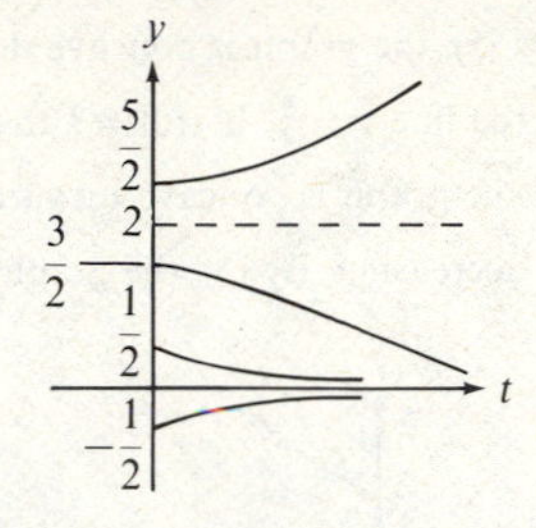

11. $f(y) = y^2 + 2y + 4 = (y+1)^2 + 3$; $f'(y) = 2y + 2 = 2(y+1)$; $f'(y(t))\frac{dy}{dt} = 2(y(t)+1)\frac{dy}{dt}$

    No constant solutions. If $y(0) = -\frac{2}{3}$ then $f(y(0)) = f(-\frac{2}{3}) = \frac{28}{9}$, so y is strictly increasing. By (5), the graph is concave upward. If $y(0) = -3$ then $f(y(0)) = f(-3) = 7$, so y is strictly increasing. By (5), the graph is concave downward below the line $y = -1$ and concave upward above the line $y = -1$.

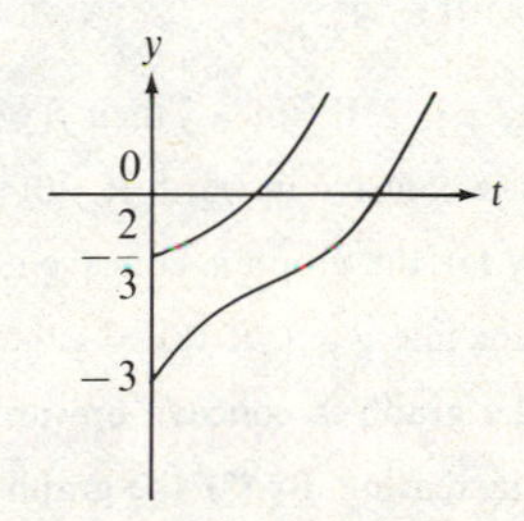

13. $f(y) = y^3$; $f'(y) = 3y^2$; $f'(y(t))\frac{dy}{dt} = 3[y(t)]^2\frac{dy}{dt}$

    Constant solution: $y = 0$. If $y(0) = 1$ then $f(y(0)) = f(1) = 1$, so y is strictly increasing. By (5), the graph is concave upward. If $y(0) = -1$ then $f(y(0)) = f(-1) = -1$, so y is strictly decreasing. By (5), the graph is concave downward.

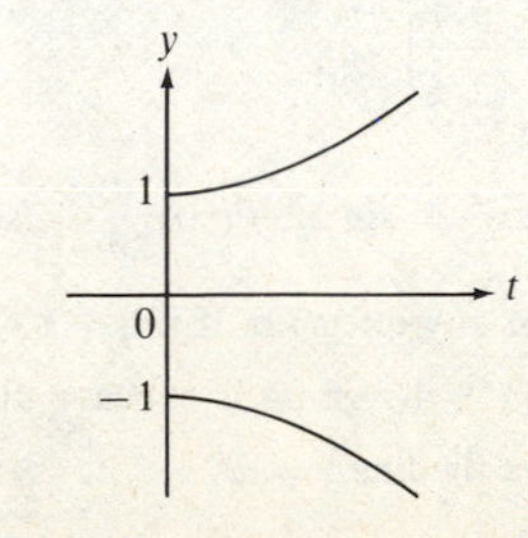

15. $f(y) = y^2(y-1)$; $f'(y) = 3y^2 - 2y = 3y(y - \frac{2}{3})$; $f'(y(t))\dfrac{dy}{dt} = 3y(t)(y(t) - \frac{2}{3})\dfrac{dy}{dt}$

Constant solutions: $y = 0$ and $y = 1$. If $y(0) = \frac{3}{2}$ then $f(y(0)) = f(\frac{3}{2}) = \frac{9}{8}$, so y is strictly increasing. By (5), the graph is concave upward. If $y(0) = \frac{3}{4}$ then $f(y(0)) = f(\frac{3}{4}) = -\frac{9}{64}$, so y is strictly decreasing. By (5), the graph is concave downward above the line $y = \frac{2}{3}$ and concave upward below the line $y = \frac{2}{3}$. If $y(0) = \frac{1}{6}$ then $f(y(0)) = f(\frac{1}{6}) = -\frac{5}{216}$, so y is strictly decreasing. By (5), the graph is concave upward. If $y(0) = -\frac{1}{2}$ then $f(y(0)) = f(-\frac{1}{2}) = -\frac{3}{8}$, so y is strictly decreasing. By (5), the graph is concave downward.

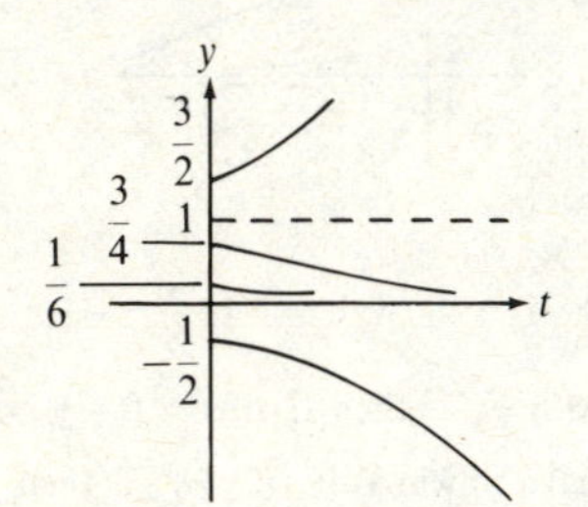

17. $f(y) = y^3(y-2)^3$; $f'(y) = 3y^2(y-2)^3 + 3y^3(y-2)^2 = 6y^2(y-1)(y-2)^2$;

$$f'(y(t))\frac{dy}{dt} = 6(y(t))^2(y(t)-1)(y(t)-2)^2\frac{dy}{dt}$$

Constant solutions: $y = 0$ and $y = 2$. If $y(0) = \frac{5}{2}$ then $f(y(0)) = f(\frac{5}{2}) = \frac{125}{64}$, so y is strictly increasing. By (5), the graph is concave upward. If $y(0) = \frac{3}{2}$ then $f(y(0)) = f(\frac{3}{2}) = -\frac{27}{64}$, so y is strictly decreasing. By (5), the graph is concave downward above the line $y = 1$ and concave upward below the line $y = 1$. If $y(0) = \frac{1}{2}$ then $f(y(0)) = f(\frac{1}{2}) = -\frac{27}{64}$, so y is strictly decreasing. By (5), the graph is concave upward. If $y(0) = -\frac{1}{2}$ then $f(y(0)) = f(-\frac{1}{2}) = \frac{125}{64}$, so y is strictly increasing. By (5), the graph is concave downward.

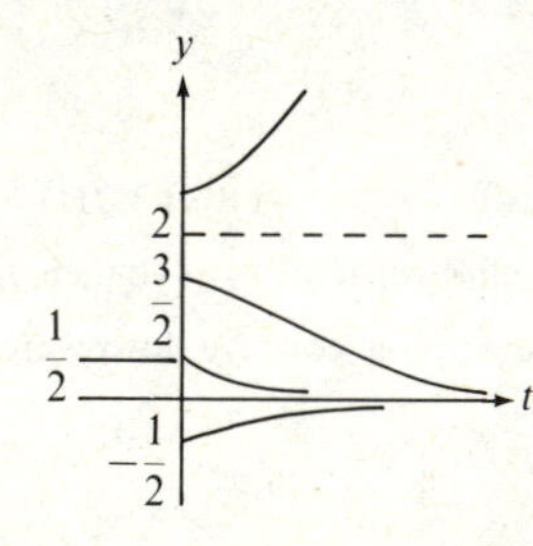

19. $f(y) = \sin^2 y$; $f'(y) = 2\sin y\cos y = \sin 2y$; $f'(y(t))\dfrac{dy}{dt} = (\sin 2y(t))\dfrac{dy}{dt}$

Constant solutions: $y = n\pi$ for any integer n. If $y(0) = \pi/4$ then $f(y(0)) = \sin^2(\pi/4) = \frac{1}{2}$, so y is strictly increasing. By (5), the graph is concave upward below the line $y = \pi/2$ and concave downward above the line $y = \pi/2$.

If $y(0) = 3\pi/4$ then $f(y(0)) = f(3\pi/4) = \sin^2(3\pi/4) = \frac{1}{2}$, so y is strictly increasing. By (5), the graph is concave downward. If $y(0) = 5\pi/4$ then $f(y(0)) = f(5\pi/4) = \sin^2(5\pi/4) = \frac{1}{2}$, so y is strictly increasing. By (5), the graph is concave upward below the line $y = 3\pi/2$ and concave downward above the line $y = 3\pi/2$. If $y(0) = 7\pi/4$ then $f(y(0)) = f(7\pi/4) = \sin^2(7\pi/4) = \frac{1}{2}$, so y is strictly increasing. By (5), the graph is concave downward.

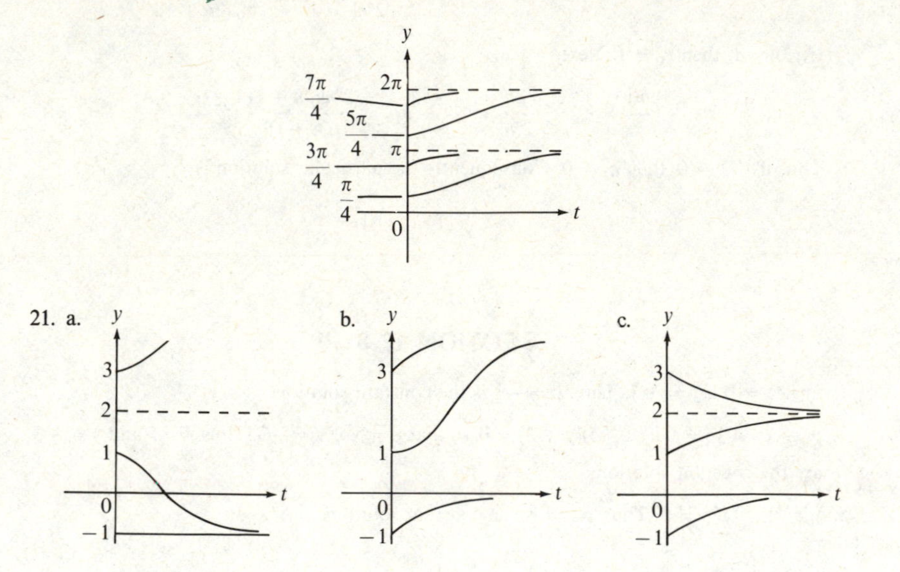

23. $f(y) = ky(a-y)$; $f'(y) = k(a-y) - ky = k(a-2y)$; $f'(y(t))\dfrac{dy}{dt} = k(a - 2y(t))\dfrac{dy}{dt}$

Constant solutions: $y = 0$ and $y = a$.

a. If $0 < y(0) < a/2$ then $f(y(0)) = ky(0)(a - y(0)) > 0$, so y is strictly increasing. By (5), the graph is concave upward below the line $y = a/2$ and concave downward above the line $y = a/2$.

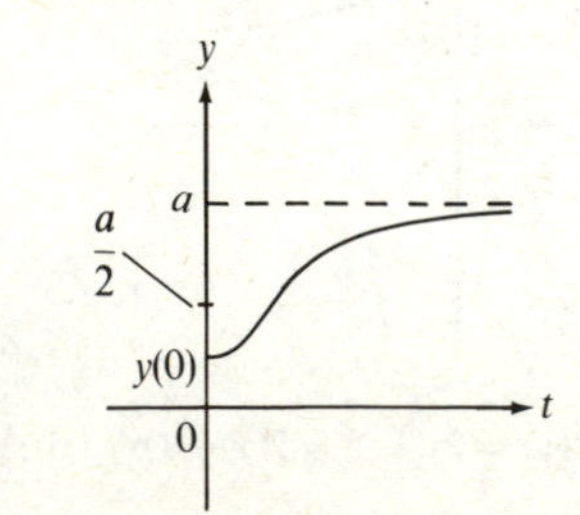

b. If $a/2 < y(0) < a$ then $f(y(0)) = ky(0)(a - y(0)) > 0$, so y is strictly increasing. By (5), the graph is concave downward.

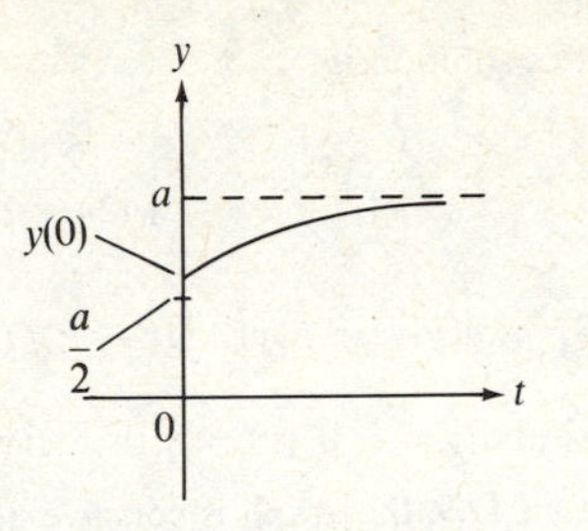

CHAPTER 17 REVIEW

1. $P(x) = 2/x$ and $Q(x) = x^2 + 6$. Since $2 \ln x$ is an antiderivative of P for $x > 0$, $S(x) = 2 \ln x$, so by (3) in Section 17.4, the general solution is

$$y = e^{-2 \ln x} \int e^{2 \ln x}(x^2 + 6)\, dx = \frac{1}{x^2} \int x^2(x^2 + 6)\, dx = \frac{1}{x^2} \int (x^4 + 6x^2)\, dx$$
$$= \frac{1}{x^2}\left(\frac{1}{5}x^5 + 2x^3 + C\right) = \frac{1}{5}x^3 + 2x + \frac{C}{x^2}$$

3. Since
$$\frac{\partial}{\partial x} \frac{y}{\sqrt{x^2 + y^2}} = \frac{-xy}{(x^2 + y^2)^{3/2}} = \frac{\partial}{\partial y} \frac{x}{\sqrt{x^2 + y^2}}$$
the differential equation is exact. Thus there is a function f such that $\partial f/\partial x = x/\sqrt{x^2 + y^2}$ and $\partial f/\partial y = y/\sqrt{x^2 + y^2}$. Then $f(x, y) = \sqrt{x^2 + y^2} + g(y)$, for a suitable function g, so that $\partial f/\partial y = y/\sqrt{x^2 + y^2} + dg/dy$. Since $\partial f/\partial y = y/\sqrt{x^2 + y^2}$, it follows that $dg/dy = 0$, so that $g(y) = C_1$. Consequently the general solution is $\sqrt{x^2 + y^2} = C_2$, or $x^2 + y^2 = C$. (This solution could also be obtained by first multiplying both sides of the given equation by $\sqrt{x^2 + y^2}$ and then finding the function f.)

5. The characteristic equation is $s^2 - 4s + 8 = 0$, so $b = -4$ and $c = 8$, and thus $b^2 - 4c = 16 - 32 < 0$. Therefore Case 3 of Section 17.5 applies. If $u = -b/2 = 2$ and $v = \frac{1}{2}\sqrt{4c - b^2} = \frac{1}{2}\sqrt{32 - 16} = 2$, then by (11) of Section 17.5 the general solution is $y = C_1 e^{2x} \sin 2x + C_2 e^{2x} \cos 2x$.

7. $P(x) = \cot x$ and $Q(x) = \csc x$. Since $\ln \sin x$ is an antiderivative of P on $(0, \pi)$, $S(x) = \ln \sin x$, so by (3) in Section 17.4, the general solution is

$$y = e^{-\ln \sin x} \int e^{\ln \sin x} \csc x\, dx = \frac{1}{\sin x} \int \sin x \csc x\, dx = \frac{1}{\sin x} \int 1\, dx = \frac{1}{\sin x}(x + C)$$

9. Since $\dfrac{\partial}{\partial x}(\sinh 2x \sinh 2y) = 2 \cosh 2x \sinh 2y = \dfrac{\partial}{\partial y}(\cosh 2x \cosh 2y)$, the differential equation is exact. Thus there is a function f such that $\partial f/\partial x = \cosh 2x \cosh 2y$ and $\partial f/\partial y = \sinh 2x \sinh 2y$. Then $f(x, y) = \frac{1}{2} \sinh 2x \cosh 2y + g(y)$ for a suitable function g, so that $\partial f/\partial y = \sinh 2y \sinh 2y + dg/dy$. Since $\partial f/\partial y = \sinh 2x \sinh 2y$, it follows that $dg/dy = 0$, so that $g(y) = C_1$. Consequently the general solution is $\frac{1}{2} \sinh 2x \cosh 2y = C_2$, and thus $\sinh 2x \cosh 2y = C$.

11. The differential equation is equivalent to the separable differential equation $(1/x^2)\, dx + ye^{-y^2}\, dy = 0$. By integration we obtain the general solution $-1/x - \frac{1}{2}e^{-y^2} = C_1$, or $1/x + \frac{1}{2}e^{-y^2} = C$.

13. The homogeneous equation is $d^2y/dx^2 + 4y = 0$, with characteristic equation $s^2 + 4 = 0$. This falls under Case 3 of Section 17.5 with $b = 0$ and $c = 4$. Since $u = -b/2 = 0$ and $v = \frac{1}{2}\sqrt{4c - b^2} = \frac{1}{2}\sqrt{16} = 2$, it follows from (11) in Section 17.5 that the general solution of the homogeneous equation is $y = C_1 \sin 2x + C_2 \cos 2x$. Thus $y_1 = \sin 2x$ and $y_2 = \cos 2x$. Next, $g(x) = \cos 2x$, so by (9) and (10) of Section 17.6,

$$u_1'(x) = \frac{-\cos 2x \cos 2x}{\sin 2x\,(-2 \sin 2x) - 2 \cos 2x\,(\cos 2x)} = \frac{1}{2}\cos^2 2x = \frac{1}{4} + \frac{1}{4}\cos 4x$$
$$u_2'(x) = \frac{\sin 2x \cos 2x}{\sin 2x\,(-2 \sin 2x) - 2 \cos 2x\,(\cos 2x)} = -\frac{1}{2}\sin 2x \cos 2x = -\frac{1}{4}\sin 4x$$

Therefore $u_1(x) = \frac{1}{4}x + \frac{1}{16}\sin 4x$ and $u_2(x) = \frac{1}{16}\cos 4x$, so that by (5) in Section 17.6, a particular solution y_p is given by

$$y_p = (\tfrac{1}{4}x + \tfrac{1}{16}\sin 4x)\sin 2x + \tfrac{1}{16}\cos 4x \cos 2x$$
$$= \tfrac{1}{4}x \sin 2x + \tfrac{1}{16}(\cos 4x \cos 2x + \sin 4x \sin 2x)$$
$$= \tfrac{1}{4}x \sin 2x + \tfrac{1}{16}\cos(4x - 2x) = \tfrac{1}{4}x \sin 2x + \tfrac{1}{16}\cos 2x$$

Consequently by (4) in Section 17.6, the general solution is given by $y = \frac{1}{4}x \sin 2x + \frac{1}{16}\cos 2x + C_1 \sin 2x + C_2' \cos 2x = \frac{1}{4}x \sin 2x + C_1 \sin 2x + C_2 \cos 2x$.

15. The differential equation is equivalent to the separable differential equation $2x\, dx = [y/(y + 1)]\, dy$. Since $y/(y + 1) = (y + 1 - 1)/(y + 1) = 1 - 1/(y + 1)$, we obtain the general solution by integration: $x^2 = y - \ln|y + 1| + C$. If $y(0) = -2$, then $0 = -2 - \ln|-2 + 1| + C$, so that $C = 2$, and thus the particular solution is given by $y - \ln|y + 1| = x^2 - 2$.

17. The characteristic equation is $s^2 - 2s + 3 = 0$, so $b = -2$ and $c = 3$, and thus $b^2 - 4c = 4 - 12 < 0$. Therefore Case 3 of Section 17.5 applies. If $u = -b/2 = 1$ and $v = \frac{1}{2}\sqrt{4c - b^2} = \frac{1}{2}\sqrt{12 - 4} = \sqrt{2}$, then by (11) of Section 17.5 the general solution is $y = C_1 e^x \sin \sqrt{2}x + C_2 e^x \cos \sqrt{2}x$. If $y(0) = 1$, then $1 = y(0) = C_2$. Next, $dy/dx = C_1 e^x \sin \sqrt{2}x + \sqrt{2}C_1 e^x \cos \sqrt{2}x + C_2 e^x \cos \sqrt{2}x - \sqrt{2}C_2 e^x \sin \sqrt{2}x$, so if $y'(0) = 3$ then $3 = y'(0) = \sqrt{2}C_1 + C_2 = \sqrt{2}C_1 + 1$, and thus $C_1 = 2/\sqrt{2} = \sqrt{2}$. Consequently the particular solution is $y = \sqrt{2}e^x \sin \sqrt{2}x + e^x \cos \sqrt{2}x$.

19. $y(1 + x^2)\, dy + (y^2 + 1)\, dx = 0$, so $[y/(y^2 + 1)]\, dy = -[1/(1 + x^2)]\, dx$, a separable differential equation. By integration we obtain the general solution $\frac{1}{2}\ln(y^2 + 1) = -\arctan x + C$. If $y(0) = \sqrt{3}$ then $\frac{1}{2}\ln(3 + 1) = 0 + C = C$, so $C = \frac{1}{2}\ln 4 = \frac{1}{2}\ln 2^2 = \ln 2$, and thus the particular solution is $\frac{1}{2}\ln(y^2 + 1) + \arctan x = \ln 2$.

21. $P(x) = \tan x$, and $Q(x) = \sec x$. Since $-\ln \cos x$ is an antiderivative of P on $(-\pi/2, \pi/2)$, $S(x) = -\ln \cos x$, so by (3) in Section 17.4, the general solution is

$$y = e^{\ln \cos x} \int e^{-\ln \cos x} \sec x \, dx = \cos x \int \frac{1}{\cos x} \sec x \, dx = \cos x \int \sec^2 x \, dx$$

$$= \cos x \, (\tan x + C) = \sin x + C \cos x$$

If $y(0) = \pi/4$, then $\pi/4 = \sin 0 + C \cos 0 = C$, so the particular solution is $y = \sin x + (\pi/4) \cos x$.

23. $(\frac{3}{2}x^2 - 3y)\,dy = x(x^4 - 3y)\,dx$, so $x(x^4 - 3y)\,dx - (\frac{3}{2}x^2 - 3y)\,dy = 0$. Since $\frac{\partial}{\partial x}[-(\frac{3}{2}x^2 - 3y)] = -3x = \frac{\partial}{\partial y}[x(x^4 - 3y)]$, this differential equation is exact. Thus there is a function f such that $\partial f/\partial x = x(x^4 - 3y) = x^5 - 3xy$ and $\partial f/\partial y = -\frac{3}{2}x^2 + 3y$. Then $f(x, y) = \frac{1}{6}x^6 - \frac{3}{2}x^2 y + g(y)$, for a suitable function g, so that $\partial f/\partial y = -\frac{3}{2}x^2 + dg/dy$. Since $\partial f/\partial y = -\frac{3}{2}x^2 + 3y$, it follows that $dg/dy = 3y$, so that $g(y) = \frac{3}{2}y^2 + C_1$. Consequently the general solution is $\frac{1}{6}x^6 - \frac{3}{2}x^2 y + \frac{3}{2}y^2 = C$. If $y(-1) = 2$, then $C = \frac{1}{6}(-1)^6 - \frac{3}{2}(-1)^2(2) + \frac{3}{2}(2^2) = \frac{19}{6}$.

25. Let $y = \sum_{n=0}^{\infty} c_n x^n$. Then $dy/dx = \sum_{n=1}^{\infty} n c_n x^{n-1}$ and $d^2y/dx^2 = \sum_{n=2}^{\infty} n(n-1)c_n x^{n-2}$. Since $d^2y/dx^2 - 3x(dy/dx) - 3y = 0$, this means that

$$\sum_{n=2}^{\infty} n(n-1)c_n x^{n-2} - 3x \sum_{n=1}^{\infty} n c_n x^{n-1} - 3 \sum_{n=0}^{\infty} c_n x^n = 0$$

or equivalently,

$$\sum_{n=0}^{\infty} (n+2)(n+1)c_{n+2} x^n - 3 \sum_{n=1}^{\infty} n c_n x^n - 3 \sum_{n=0}^{\infty} c_n x^n = 0$$

so that

$$2c_2 - 3c_0 + \sum_{n=1}^{\infty} [(n+2)(n+1)c_{n+2} - 3(n+1)c_n] x^n = 0$$

By Corollary 9.27, $2c_2 - 3c_0 = 0$, that is, $c_2 = \frac{3}{2}c_0$, and for $n \geq 1$ we have $(n+2)(n+1)c_{n+2} - 3(n+1)c_n = 0$, that is, $c_{n+2} = [3/(n+2)]c_n$. Thus

$$c_3 = \frac{3}{3}c_1 = c_1 \qquad c_5 = \frac{3}{5}c_3 = \frac{3^2}{5 \cdot 3}c_1$$

$$c_4 = \frac{3}{4}c_2 = \frac{3}{4}\frac{3}{2}c_0 = \frac{3^2}{4 \cdot 2}c_0 \qquad c_6 = \frac{3}{6}c_4 = \frac{3^3}{6 \cdot 4 \cdot 2}c_0$$

In general for any positive integer n we have

$$c_{2n} = \frac{3^n}{(2n)(2n-2)\cdots 4 \cdot 2}c_0 = \frac{3^n}{2^n n!}c_0$$

and

$$c_{2n+1} = \frac{3^n}{(2n+1)(2n-1)\cdots 5 \cdot 3}c_1 = \frac{3^n(2n)(2n-2)\cdots 4 \cdot 2}{(2n+1)!}c_1$$

$$= \frac{3^n 2^n n!}{(2n+1)!}c_1 = \frac{6^n n!}{(2n+1)!}c_1$$

Therefore the general series solution is

$$y = c_0 \sum_{n=0}^{\infty} \frac{3^n}{2^n n!} x^{2n} + c_1 \sum_{n=0}^{\infty} \frac{6^n n!}{(2n+1)!} x^{2n+1}$$

27. $f(y) = 2y^2 - 7y + 5 = (2y-5)(y-1)$; $f'(y) = 4y - 7$; $f'(y(t))\frac{dy}{dt} = (4y(t) - 7)\frac{dy}{dt}$

Constant solutions: $y = \frac{5}{2}$ and $y = 1$. If $y(0) = \frac{1}{2}$ then $f(y(0)) = f(\frac{1}{2}) = 2$, so y is strictly increasing. By (5) in Section 17.9, the graph is concave downward.

If $y(0) = \frac{3}{2}$ then $f(y(0)) = f(\frac{3}{2}) = -1$, so y is strictly decreasing. By (5) in Section 17.9, the graph is concave upward. If $y(0) = 2$ then $f(y(0)) = f(2) = -1$, so y is strictly decreasing. By (5) in Section 17.9, the graph is concave downward above the line $y = \frac{7}{4}$ and concave upward below the line $y = \frac{7}{4}$. If $y(0) = 3$ then $f(y(0)) = f(3) = 2$, so y is strictly increasing. By (5) in Section 17.9, the graph is concave upward.

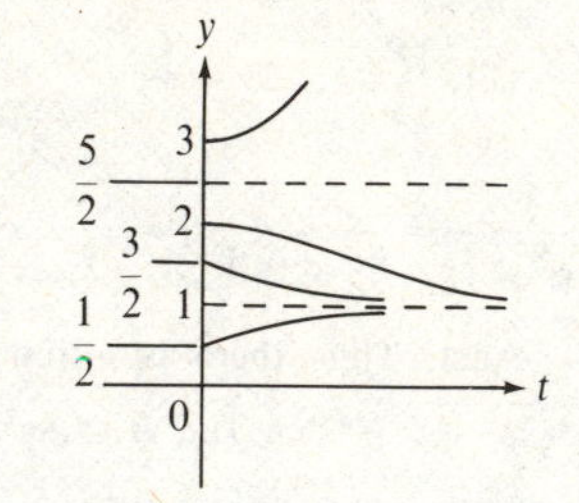

29. $f(y) = ye^y$; $f'(y) = e^y + ye^y = (1+y)e^y$; $f'(y(t))\frac{dy}{dt} = (1 + y(t))e^{y(t)}\frac{dy}{dt}$

Constant solution: $y = 0$. If $y(0) = -2$ then $f(y(0)) = f(-2) = -2e^{-2}$, so y is strictly decreasing. By (5) in Section 17.9, the graph is concave upward. If $y(0) = -\frac{1}{2}$ then $f(y(0)) = f(-\frac{1}{2}) = -\frac{1}{2}e^{-1/2}$, so y is strictly decreasing. By (5) in Section 17.9, the graph is concave downward above the line $y = -1$ and concave upward below the line $y = -1$. If $y(0) = \frac{1}{2}$ then $f(y(0)) = f(\frac{1}{2}) = \frac{1}{2}e^{1/2}$, so y is strictly increasing. By (5) in Section 17.9, the graph is concave upward.

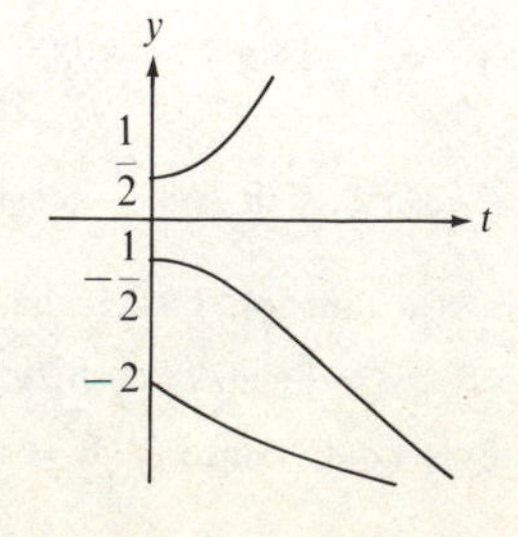

31. By (2) in Section 17.7, we have $k/m = g/e$, and by hypothesis $e = \frac{1}{2}$. Thus $k/m = 32/\frac{1}{2} = 64$. Therefore by (4) in Section 17.7, $d^2x/dt^2 + 64x = 0$, which falls under Case 3 of Section 17.5 with $b = 0$ and $c = 64$.
If $u = -(b/2) = 0$ and $v = \frac{1}{2}\sqrt{4c - b^2} = \frac{1}{2}\sqrt{256 - 0} = 8$, it follows from (11) in Section 17.5 that the general solution is $x = C_1 \sin 8t + C_2 \cos 8t$. By hypothesis $x(0) = \frac{1}{12}$, so that $\frac{1}{12} = C_2$. Next, $dx/dt = 8C_1 \cos 8t - 8C_2 \sin 8t$. By hypothesis, $x'(0) = \frac{1}{24}$, so that $\frac{1}{24} = x'(0) = 8C_1$, and thus $C_1 = \frac{1}{192}$. Consequently $x = \frac{1}{192} \sin 8t + \frac{1}{12} \cos 8t$.

33. Let $f(y) = ay^2 - by = y(ay - b)$, so that the zeros of f are 0 and b/a. If $0 < y(0) < b/a$, then $f(y(0)) = y(0)[ay(0) - b] < y(0)[b - b] = 0$. By the discussion preceding Example 2 in Section 17.9, this means that if $0 < y(0) < b/a$, then y must be decreasing, and therefore $(t, y(t))$ must approach the horizontal line $y = 0$ as t approaches ∞. This means that $\lim_{t \to \infty} y(t) = 0$, so that the species becomes extinct.

APPENDIX

1. The least upper bound is 1; the greatest lower bound is -1.

3. The least upper bound is π; the greatest lower bound is 0.

5. The least upper bound is 5; the greatest lower bound is 0.

7. The least upper bound is 1; the greatest lower bound is 0.

9. Assume that the set S of positive integers had an upper bound. By the Least Upper Bound Axiom S would then have a least upper bound M. For any number n in S, $n + 1$ is in S, so that by the definition of M, we have $n + 1 \le M$. Thus $n \le M - 1$, and consequently $M - 1$ is an upper bound of S. This contradicts the property of M that M is the least upper bound. Therefore S has no upper bound.

11. Let S be a set that is bounded below, and let T be the set of all numbers of the form $-s$, for s in S. Since S is bounded below, T is bounded above, so by the Least Upper Bound Axiom, there is a least upper bound M for T. If s is in S, then $-s$ is in T, so $-s \le M$, and thus $s \ge -M$. Consequently $-M$ is a lower bound of S. If N is any lower bound of S, then $-N$ is an upper bound of T, so $-N \ge M$ since M is the least upper bound of T. Therefore $N \le -M$. Thus $-M$ is the greatest lower bound of S.

13. Let $\varepsilon = 1$, and let δ be any positive number less than 1. If $x = \delta$ and $y = \frac{1}{2}\delta$, then

$$|x - y| = \tfrac{1}{2}\delta < \delta, \quad \text{and} \quad \left|\frac{1}{x} - \frac{1}{y}\right| = \left|\frac{1}{\delta} - \frac{2}{\delta}\right| = \frac{1}{\delta} > 1 = \varepsilon$$

Thus $1/x$ is not uniformly continuous on $(0, 1)$.

B C 1 D 2 E 3 F 4 G 5 H 6 I 7 J 8